INTERNATIONAL HANDBOOK OF RESEARCH ON ENVIRONMENTAL EDUCATION

The environment and contested notions of sustainability are increasingly topics of public interest, political debate, and legislation across the world. Environmental education journals now publish research from a wide variety of methodological traditions that show linkages between the environment, health, development, and education. The growth in scholarship makes this an opportune time to review and synthesize the knowledge base of the environmental education (EE) field.

The purpose of this 51-chapter handbook is not only to illuminate the most important concepts, findings, and theories that have been developed by EE research, but also to examine critically the historical progression of the field, its current debates and controversies, what is still missing from the EE research agenda, and where that agenda might be headed.

Published for the American Educational Research Association (AERA) by Routledge Publishers.

Robert B. Stevenson, James Cook University, Australia

Michael Brody, Montana State University, USA

Justin Dillon, King's College London, UK

Arjen E.J. Wals, Wageningen University, The Netherlands

International Handbook of Research on Environmental Education

Edited by

Robert B. Stevenson

Michael Brody

Justin Dillon

Arjen E.J. Wals

NEW YORK AND LONDON

First published 2013
by Routledge
711 Third Avenue, New York, NY 10017

Simultaneously published in the UK
by Routledge
2 Park Square, Milton Park, Abingdon, Oxon OX14 4RN

Routledge is an imprint of the Taylor & Francis Group, an informa business

Library of Congress Cataloging-in-Publication Data
International handbook of research on environmental education/edited by Robert Stevenson ... [et al.].
p. cm.
Includes bibliographical references and index.
1. Environmental education—Handbooks, manuals, etc. 2. Environmental sciences—Study and teaching—Handbooks, manuals, etc.
I. Stevenson, Robert.
GE70.I584 2012
363.70071--dc23
2012029041

ISBN: 978-0-415-89238-4 (hbk)
ISBN: 978-0-415-89239-1 (pbk)
ISBN: 978-0-203-81333-1 (ebk)

Typeset in Times LT Std
Project Managed and Typeset by diacriTech

Printed and bound in the United States of America by Edwards Brothers, Inc.

This handbook is dedicated to the pioneers in the field of environmental education who had the foresight that education and learning are crucial in finding pathways that allow humanity to live on this planet in an equitable and just way without compromising its carrying capacity while maintaining the integrity of all species.

Contents

Foreword xi

Acknowledgments xiii

1 Introduction: An Orientation to Environmental Education and the Handbook 1
Robert B. Stevenson, Arjen E.J. Wals, Justin Dillon, and Michael Brody

Part A. Conceptualizing Environmental Education as a Field of Inquiry

Section I. Historical, Contextual, and Theoretical Orientations That Have Shaped Environmental Education Research
Annette Gough

2 The Emergence of Environmental Education Research: A "History" of the Field 13
Annette Gough

3 Socioecological Approaches to Environmental Education and Research: A Paradigmatic Response to Behavioral Change Orientations 23
Regula Kyburz-Graber

4 Thinking Globally in Environmental Education: A Critical History 33
Noel Gough

5 Selected Trends in Thirty Years of Doctoral Research in Environmental Education in *Dissertation Abstracts International* From Collections Prepared in the United States of America 45
Thomas Marcinkowski, Jennifer Bucheit, Vanessa Spero-Swingle, Christine Linsenbardt, Jennifer Engelhardt, Marianne Stadel, Richard Santangelo, and Katherine Guzmon

6 Transformation, Empowerment, and the Governing of Environmental Conduct: Insights to be Gained From a "History of the Present" Approach 63
Jo-Anne Ferreira

Section II. Normative Dimensions of Environmental Education Research: Conceptions of Education and Environmental Ethics
Bob Jickling and Arjen E.J. Wals

7 Probing Normative Research in Environmental Education: Ideas About Education and Ethics 74
Bob Jickling and Arjen E.J. Wals

8 Self, Environment, and Education: Normative Arisings 87
Michael Bonnett

9 A Critical Theory of Place-Conscious Education 93
David A. Greenwood

10 Learning From Hermit Crabs, Mycelia, and Banyan: Schools as Centers of Critical Inquiry and Renormatization 101
Heesoon Bai and Serenna Romanycia

11 Why We Need a Language of (Environmental) Education 108
Lesley Le Grange

12 Environmental Ethics as Processes of Open-Ended, Pluralistic, Deliberative Enquiry 115
Lausanne Olvitt

Section III. Analyses of Environmental Education Discourses and Policies
Ian Robottom and Robert B. Stevenson

13 The Politics of Needs and Sustainability Education 126
Lesley Le Grange

14 Languages and Discourses of Education, Environment, and Sustainable Development 133
Tom Berryman and Lucie Sauvé

15 Researching Tensions and Pretensions in Environmental/Sustainability Education Policies: From Critical to Civically Engaged Policy Scholarship 147
Robert B. Stevenson

16 Changing Discourses in EE/ESD: A Role for Professional Self-Development 156
Ian Robottom

17 Connecting Vocational and Technical Education With Sustainability 163
Alberto Arenas and Fernando Londoño

18 Trends, Junctures, and Disjunctures in Latin American Environmental Education Research 171
Edgar González Gaudiano and Leonir Lorenzetti

19 EE Policies in Three Chinese Communities: Challenges and Prospects for Future Development 178
Lee Chi Kin John, Wang Shun Mei, and Yang Guang

Part B. Research on Environmental Education Curriculum, Learning, and Assessment: Processes and Outcomes

Section IV. Curriculum Research in Environmental Education
Heila Lotz-Sisitka

20 Traditions and New Niches: An Overview of Environmental Education Curriculum and Learning Research 194
Heila Lotz-Sisitka, John Fien, and Mphemelang Ketlhoilwe

21 Environmental Education in a Cultural Context 206
Albert Zeyer and Elin Kelsey

22 Place-Based Education: Practice and Impacts 213
Gregory A. Smith

23 Getting the Picture: From the Old Reflection—Hearing Pictures and Telling Tales, to the New Reflection—Seeing Voices and Painting Scenes 221
Tony Shallcross and John Robinson

24 Moinho D'Água: Environmental Education, Participation, and Autonomy in Rural Areas 231
João Luiz de Moraes Hoeffel, Almerinda B. Fadini, M.K. Machado, J.C. Reis, and F.B. Lima

Section V. Research on Learning Processes in Environmental Education
Justin Dillon, Joe E. Heimlich, and Elin Kelsey

25 Environmental Learning: Insights From Research Into the Student Experience 243
Cecilia Lundholm, Nick Hopwood, and Mark Rickinson

26 Conventional and Emerging Learning Theories: Implications and Choices for Educational Researchers With a Planetary Consciousness 253
Arjen E.J. Wals and Justin Dillon

27 Belief to Behavior: A Vital Link 262
Joe E. Heimlich, Preethi Mony, and Victor Yocco

28 Landscapes as Contexts for Learning 275
Carol B. Brandt

Section VI. Evaluation and Analysis of Environmental Education Programs, Materials, and Technologies and the Assessment of Learners and Learning
Michael Brody and Martin Storksdieck

29 Research on the Long-Term Impacts of Environmental Education 289
Kendra Liddicoat and Marianne E. Krasny

30 Advancing Environmental Education Program Evaluation: Insights From a Review of Behavioral Outcome Evaluations 298
Michaela Zint

31 National Assessments of Environmental Literacy: A Review, Comparison, and Analysis 310
Thomas Marcinkowski, Donghee Shin, Kyung-Im Noh, Maya Negev, Gonen Sagy, Yaakov Garb, Bill McBeth, Harold Hungerford, Trudi Volk, Ron Meyers, and Mehmet Erdogan

32 Geospatial Technologies: The Present and Future Roles of Emerging Technologies in Environmental Education 331
Michael Barnett, James G. MaKinster, Nancy M. Trautmann, Meredith Houle Vaughn, and Sheron Mark

33 Sustainability Education: Theory and Practice 349
Sarah Holdsworth, Ian Thomas, and Kathryn Hegarty

34 Learning From Neighboring Fields: Conceptualizing Outcomes of Environmental Education Within the Framework of Free-Choice Learning Experiences 359
Lynn D. Dierking, John H. Falk, and Martin Storksdieck

Part C. Issues of Framing, Doing, and Assessing in Environmental Education Research

Section VII. Moving Margins in Environmental Education Research
Constance L. Russell and Leesa Fawcett

35 Researching Differently: Generating a Gender Agenda for Research in Environmental Education 375
Annette Gough

36 The Representation of Indigenous Knowledges 384
Soul Shava

37 Educating for Environmental Justice 394
Randolph Haluza-DeLay

38 Indigenous Environmental Education Research in North America: A Brief Review 404
Greg Lowan-Trudeau

39 Three Degrees of Separation: Accounting for Naturecultures in Environmental Education Research 409
Leesa Fawcett

Section VIII. Philosophical and Methodological Perspectives
Paul Hart

40 (Un)timely Ecophenomenological Framings of Environmental Education Research 424
Phillip G. Payne

41 Children as Active Researchers: The Potential of Environmental Education Research Involving Children 438
Elisabeth Barratt Hacking, Amy Cutter-Mackenzie, and Robert Barratt

42 Collaborative Ecological Inquiry: Where Action Research Meets Sustainable Development 459
Hilary Bradbury-Huang and Ken Long

43 Critical Action Research and Environmental Education: Conceptual Congruencies and Imperatives in Practice 469
Robert B. Stevenson and Ian Robottom

44 A Feminist Poststructural Approach to Environmental Education Research 480
Bronwyn Davies

45 Suited: Relational Learning and Socioecological Pedagogies 487
Marcia McKenzie, Kim Butcher, Dustin Fruson, Michelle Knorr, Joshua Stone, Scott Allen, Teresa Hill, Jeremy Murphy, Sheelah McLean, Jean Kayira, and Vince Anderson

46 Greening the Knowledge Economy: Ecosophy, Ecology, and Economy 498
Michael A. Peters

47 Preconceptions and Positionings: Can We See Ourselves Within Our Own Terrain? 507
Paul Hart

Section IX. Insights, Gaps, and Future Directions in Environmental Education Research

48 The Evolving Characteristics of Environmental Education Research 512
Robert B. Stevenson, Justin Dillon, Arjen E.J. Wals, and Michael Brody

49 Identifying Needs in Environmental Education Research 518
Alan Reid and William Scott

50 Handbooks of Environmental Education Research: For Further Reading and Writing 529
Alan Reid and Phillip G. Payne

51 Tentative Directions for Environmental Education Research in Uncertain Times 542
Arjen E.J. Wals, Robert B. Stevenson, Michael Brody, and Justin Dillon

Author Index 549

Subject Index 563

The Editors 571

The Contributors 573

Foreword

This significant, informative, and engaging book on research in environmental education is being published at a critical time in U.S. history. It is a time when competing ideas and perspectives about how to respond to questions and concerns about the environment are hotly debated and often ridiculed. These competing and contentious ideas make it difficult to identify the research, innovations, and programs that can be used to create a sustainable environment and prevent further damage to it. The field of environmental education is an appropriate focal point for many of the concerns that are being raised about the environment. This *Handbook* will be instrumental in identifying what is known in environmental education research. It will also help researchers identify questions that need to be answered as issues ranging from climate change and its effect on biodiversity to nature conservation are widely discussed.

The editors of the *International Handbook of Research on Environmental Education* have collected in a single volume a wealth of information including statistics, analyses, and studies that describe the role that research can play in helping educators, researchers, policy makers, and students better understand the environmental issues that people around the world are confronting. This *Handbook* will be a particularly important resource for the research community and those trying to comprehend and to advance research on environmental education. Information that will help graduate students craft thoughtful and insightful research questions as well as information about research methods and techniques are interspersed throughout this *Handbook.* In addition, the *Handbook* provides a framework and a knowledge base for decision makers who need to make informed and thoughtful decisions about environmental education issues, and educators who need to make curricular decisions about what students need to know and understand about the environment.

The 51 chapters in this *Handbook* include a mix of established and new voices in environmental education research. They include scholars from six continents and 15 countries. These myriad voices created a *Handbook* that is, as the editors point out, "attentive to the diverse populations served by contemporary educational systems as well as to opportunities to engage individuals and communities in nonformal and informal learning contexts." The *Handbook*'s chapters are grouped into nine sections. Each section addresses important dimensions of environmental education research, ranging from its historical and theoretical foundations to issues related to gender, race, and colonialism. In addition to chapters that establish the knowledge base in environmental education, the *Handbook* also includes chapters that discuss ethics, the politics of sustainability, environmental justice, and other topics that link research in environmental education to the work being done by scholars in contiguous fields. The result is a volume that encompasses new and divergent perspectives and insights on environmental education issues and research.

The AERA Handbook Series in Education Research was designed and implemented by the AERA Books Editorial Board. When the proposal for the *International Handbook of Research on Environmental Education* was approved for inclusion in the series, the members of the board were Robert E. Floden, Patrick B. Forsyth, Felice J. Levine, Gary J. Natriello, Carol Camp Yeakey, myself, and Robert J. Sternberg, who chaired the publications committee. The board now consists of D. Jean Clandinin, Gilberto Q. Conchas, Robert E. Floden, Mary M. Juzwik, Felice J. Levine, Simon W. Marginson, Mariana Souto-Manning, Olga M. Welch, and myself.

The Handbook Series is part of a comprehensive AERA books publication program that aims to publish works that advance knowledge, expand access to significant research and research analyses and syntheses, and promote knowledge utilization. The series specifically seeks to publish volumes of excellence that are conceptually and substantively distinct. The volumes in the series "offer state-of-the-art knowledge and the foundation to advance research to scholars and students in education research and related social science fields." When the Books Editorial Board issued its call for proposals for handbooks in education research, the editors of the *International Handbook of Research on Environmental Education* were among the first to respond. Their proposal was accepted after a substantive review and a revision process directed by the Books Editorial Board.

The following criteria outlined in guidelines for preparing handbook proposals were used to review the proposal for the *International Handbook of Research on Environmental Education*. First and foremost, the board examined the proposed *Handbook* in terms of whether it would provide an opportunity for readers to take stock of and advance their thinking about current and future directions of environmental education research. Second, the board focused on the extent to which the proposed *Handbook* would draw on the strongest research—including research both within and outside the United States. Third, the board was interested in the ability of the editors of the proposed *Handbook* to bring together a team of authors who could assess the knowledge base of environmental education research and do so with respect to the diverse populations served by contemporary educational systems. Finally, the board reviewed the proposed content of the *Handbook* to get a sense of the book's scope and the extent to which it would include a "critical analysis of the strengths and limitations of extant studies as well as address the essential tools and elements for research progress."

At the end of the review process, the board enthusiastically approved and moved the *International Handbook of Research on Environmental Education* into development. Support was given to the editors during the manuscript development process, and when the manuscript was complete, it was reviewed and approved for publication by the board. We are very pleased to make this comprehensive, well-conceptualized, and theoretically strong *Handbook* available to readers interested in research on environmental education. It advances both theory and practice in the field of environmental education and will help strengthen the quality of environmental education research as well as help educators better understand, identify, and design and implement curriculum on environmental issues.

On behalf of the AERA Books Editorial Board, I want to thank editors Robert B. Stevenson, Michael Brody, Justin Dillon, and Arjen E.J. Wals for their substantial investment in this research *Handbook* and for producing a timely and significant volume. Special thanks is also due to Todd Reitzel who worked closely with the board and the editors in bringing this *Handbook* to fruition. Finally I wish to thank the many authors and reviewers for their important contributions to this important publication.

Cherry A. McGee Banks
Chair, AERA Books Editorial Board
University of Washington Bothell

Acknowledgments

This first AERA *International Handbook of Research on Environmental Education* represents the collective endeavors of countless individuals. The credit for what you hold in your hands or see on your screen is shared by authors, section editors, critical friends, the broader (and wonderfully diverse) community of scholars within the field of environmental education, and colleagues at the American Educational Research Association and Routledge.

The book has its roots in what is now called the Environmental Education (EE) Special Interest Group of the American Educational Research Association (AERA) but which began life in the early 1990s as the Ecological and Environmental Education SIG. This group has grown to be one of the strongest SIGs in AERA and continues to provide a social and intellectual home for researchers in environmental education. We acknowledge the leadership of that group over the past two and a half decades.

The strength of the EE SIG encouraged Felice Levine, AERA's Executive Director, and the AERA Books Editorial Board to invite a contribution to their education research handbook series. It has to be said that this was some years ago and the book has taken a long time to emerge. Without their initial invitation and their unstinting support, this book would be nothing but an idea in a few people's heads.

The length of time from invitation to delivery partly represents the fact that there was an extensive process of consultation and collaboration with many environmental education scholars across the world. As we spell out in the Introduction, we held a series of meetings to discuss the structure and the content of the *Handbook* at two AERA annual meetings, at the research symposium of the North American Association for Environmental Education, at the World Environmental Education Congress in South Africa, and at the Invitational Seminar on Research and Development in Environmental and Health Education in Switzerland. These meetings were characterized by thoughtful dialogue, critical insight, and a depth of unbridled collegiality which has supported us through the development of the *Handbook*.

From the dialogue and discussion emerged an open call for authors. Gradually a sense of order and cohesion emerged and, together with a devoted and determined group of section editors, the long process of writing, critique, editing, and pulling together led to this, the first ever AERA *International Handbook of Research on Environmental Education*.

To all those who helped us with the planning, who submitted ideas and abstracts, who wrote the chapters, who reviewed submissions and edited the sections, and who took the final drafts and turned them into something beyond words, we offer our unconditional thanks. To those who supported us through this process emotionally, we offer our unconditional love.

As coeditors we take responsibility for the errors and biases that you might find within the nine sections and more than 50 chapters. We have done our best and trust that future editors, authors, and publishers will build on the foundations that this volume represents.

Robert B. Stevenson
Michael Brody
Justin Dillon
Arjen E.J. Wals

1

Introduction

An Orientation to Environmental Education and the Handbook

Robert B. Stevenson
James Cook University, Australia

Arjen E.J. Wals
Wageningen University, The Netherlands; Cornell University, USA

Justin Dillon
King's College London, UK

Michael Brody
Montana State University, USA

Although the field of environmental education (EE) has a history of over forty years—and much longer if forerunners such as nature study, outdoor and conservation education are included—it has received considerably more attention in recent years as contested notions of environment and sustainability have become common topics of conversation among the public, the subject of media interest, and the focus of much political debate and legislation. Systemic linkages between environment, health, climate, poverty, development, and education have become more widely accepted as the years have passed. Therefore, this handbook was developed at an opportune time to take stock of and consolidate what we know and don't know as a field, and to demarcate the limits of our (un)certainties. More specifically, the purpose of the handbook is not only to illuminate the most important understandings that have been developed by environmental education research, but also to critically examine the ways in which the field has changed over the decades, the current debates and controversies, what is still missing from the environmental education research agenda, and where that agenda might and could be headed in the future. Environmental education as a field of inquiry is conceptualized from a range of vantage points, including historical, theoretical, and ethical perspectives; discourse, policy, curriculum, learning, and assessment are examined from an EE-perspective; and key issues are raised of framing, doing, and assessing the missing voices in environmental education research.

Characteristics of Environmental Education

Before discussing the structure, processes of development, and ways of engaging with the handbook, for those new to the field, we first offer a brief background on some conceptions and characteristics of environmental education.

An early (hence the sexist language) and often quoted (particularly in Europe and Australia) definition of environmental education states that:

> Environmental education is a process of recognizing values and clarifying concepts in order to develop skills and attitudes necessary to understand and appreciate the interrelatedness among man, his culture and his biophysical surroundings. Environmental education also entails practice in decision-making and self-formulating of a code of behaviour about issues concerning environmental quality. (Martin, 1975, p. 21)

Following the establishment of the United Nations Environment Program, a workshop of environmental educators from UNESCO countries produced the Belgrade Charter which identified the goal of EE as:

> To develop a world population that is aware of, and concerned about, the environment and its associated problems, and which has the knowledge, skills, attitudes, motivations and commitment to work individually and collectively toward solutions of current problems and prevention of new ones.

The notable addition to earlier conceptions of EE was an emphasis on action, which was reinforced in a later intergovernmental conference in Tbilisi. Besides developing critical thinking, problem solving, and decision-making skills in relation to quality-of-life issues, the Tbilisi Declaration emphasized that students should be "actively involved at all levels in working toward the resolution of environmental problems" (Tbilisi Declaration, 1978, p. 18).

In the late 1980s new discourses or slogans of sustainable development and sustainability which emerged in international policy circles gave rise to education for sustainable development (ESD) and education for sustainability (EfS), which have come to replace environmental education as the dominant discourse in the most national policy arenas. However, the ambiguities and multiple interpretations of these terms have been the subject of much analysis and debate among scholars in the field. For a more detailed history of the field and some of the contested terrain of its formulation, see Gough (1997). In summary, the conceptions emerging from these conferences and promulgated in policy documents have been influential in shaping the discourse, conceptualizing the field and providing a source of debate on a range of issues for theory and practice—many of which are examined in the following chapters in this handbook.

Among many characteristics that have been identified as associated with environmental (or sustainability) education, we emphasize five characteristics on which there seems to be broad consensus—albeit with multiple interpretations—and which have important implications for conceptualizing and constructing research in the field. First, environmental education embraces normative questions as environmental issues are fundamentally normative or value-laden by nature, as Jickling and Wals point out in their introduction to Section 2 of this handbook. Second, the interdisciplinary nature of people-society-environment relationships compels environmental education to be interdisciplinary: a position which is explicitly articulated in the "triple bottom line" discourse of sustainable development and sustainability referencing environmental, sociocultural, and economic dimensions. Third, environmental education is concerned not only with knowledge and understanding, and attitudes and values, but also includes developing the agency of learners in participating and taking action on environmental and sustainability issues. Education traditionally equips students to understand, and at best conceptualize, problems, but rarely to enact solutions (Shawcross & Robinson, Chapter 24). Fourth, the field encompasses education and learning that not only takes place within formal institutional settings but also within nonformal or informal public domain settings (Hart & Nolan, 1999). The boundaries between these settings, however, are beginning to break down, an issue we discuss in the final concluding chapter. Finally, environmental education has both a global and local orientation given that the scale of environmental issues ranges from the local to the global.

The History and Purpose of the Handbook

Once dominated by a small group of empirical-analytical researchers publishing in the *Journal of Environmental Education,* a relatively small circulation North American journal founded in 1969, there are now a number of journals across the world publishing research from a wide variety of traditions and drawing on multiple methodologies. Although still an emerging field, we believe—and the AERA Books Editorial Board clearly agreed—that the scope, sophistication, and richness of the scholarship of environmental education warranted the production of a first *International Handbook of Research on Environmental Education.* This richness is reflected in the scholarship presented in the following chapters by a diverse group of international researchers who are both mindful and critical of various histories of environmental education research and the work of those who have contributed to their creation.

This *International Handbook of Research on Environmental Education*, following AERA's educational research handbook guidelines, offers a current "state-of-the-art" assessment of the substance and robustness of the knowledge base derived from relevant areas of inquiry in this field and is intended to provide a foundation for advancing the thinking of scholars and students about future directions for environmental education research. It attempts to provide a comprehensive treatment of major current lines of research on environmental education and its close relatives (education for sustainability, sustainability education, and education for sustainable development) and to examine the relationship of environmental education research to educational research in general and educational research in particular that overlaps, intersects, or borders with environmental education (i.e. educational research focusing on science, social studies, health, development, social justice, citizenship, peace, and conflict). However, as several section editors in their introduction make clear, any history of a field represents only a snapshot of the many histories that could be created and reflect the perspectives of those whose voices are included.

Contributions include philosophically and empirically grounded research (of all genres) that critically examines the conceptualization, discourses, policies, programs, processes, structures, and research approaches to environmental education in the broadest sense. The handbook attempts to be comprehensive by addressing histories, contexts, methodologies, ontologies, epistemologies, and literacies that help establish the knowledge base of research on environmental education at a metalevel, and identifies possible futures and future directions of environmental education research.

Over the years environmental education has been researched by scholars who bring a variety of disciplinary (and interdisciplinary) perspectives to the field, among them education and its subfields (e.g., educational psychology, sociology of education, curriculum studies), environmental and natural sciences (e.g., biology, ecology), and environmental social sciences (e.g., environmental psychology, sociology, and philosophy). One of our missions in this volume was to bring together divergent perspectives, methodologies, methods, and findings of this broad community of scholars. Too often their work fails to cross borders and expand the horizons of those working in contiguous fields.

Contributors were encouraged, as much as possible, to make explicit and critically examine the assumptions underlying their perspective or particular vantage point, and to address how these assumptions might shape their contribution. This reflexivity hopefully makes the text more transformative in that contributions have been written in such a way that we hope you, the reader, can engage with the text meaningfully, allowing you to distill your own lessons learned and to mirror them with your own assumptions, knowledge base, and experience. Inevitably, of course, section editors have brought their own ideas about what it means to operate at a metalevel or to be a "reflexive writer" or to create a text that is "transformative," but we hope that this emphasis has produced a handbook that is neither a "show-and-tell" text, nor a collection of "best-practices" (nor viewed as the definitive work on environmental education research). Rather, we intend for the book to do justice to this still evolving field with, as Annette Gough states in her introduction to the first section, its somewhat "fuzzy boundaries" and multiple "interpretations of its foundations and documents." We also hope that the handbook is engaging, and that it invites and inspires readers to participate in reflecting on their own interpretations of these newly created, as well as past, textual documentations of the field and in determining future directions of the field.

The Process of Development of the Handbook

A handbook of this size and scope obviously entails a long and elaborated process of gestation and production. An outline of this process can help reveal the strengths and limitations of the final product. First, an extensive process of consultation and collaboration with numerous experienced and new environmental education scholars around the world preceded the development of a proposal for the handbook. A series of meetings were arranged and held in 2006 and 2007 at international environmental education research conferences on three different continents (2006 and 2007 meetings in North America of the Ecological & Environmental Education SIG at the AERA annual conference and the research symposium of the North American Association of Environmental Education conference, the World Environmental Education Congress in South Africa, and the Invitational Seminar on Research in Environmental and Health Education in Switzerland). These meetings led to an agreed need for and vision of a handbook on environmental education research and helped identify the editorial team, a supporting advisory group, and potential contributors. The Ecological & Environmental Education (now the Environmental Education) SIG was particularly central during these times to the development and support of the handbook as it was envisaged from early on to be an AERA publication. Meetings were then held with the AERA Books Editorial Board to discuss the idea of the proposed handbook before a formal prospectus was developed. Discussions of potential inclusions and issues being faced were held and reports on progress were presented each year at the business meeting of the Ecological & Environmental Education at the AERA annual meeting throughout the development process.

A call for contributors was widely distributed in September 2007, including to participants at the above conferences and members of major environmental education professional associations around the world. Proposals were invited for either: (1) a section that focuses on a broad research topic or issue and included a number of chapters addressing that topic or issue; or (2) individual chapters on a specific research topic or issue. Contributions were sought of conceptually and/or empirically grounded research (of all genres) that critically examined the conceptualization, discourses, policies, programs, processes, structures, and research approaches to environmental education (and education for sustainable development or sustainability). Eleven potential areas were identified by the editors, in consultation with a group of editorial advisors, to provide some guidelines for contributors as to possible topics and issues for both section and chapter proposals (and provide an approximation of the envisaged size and scope of the handbook), but other topics or issues identified by contributors also were invited. These eleven were not envisaged as necessarily representing the final structure or sections of the handbook and in fact the final structure was reduced to nine sections. These eleven (now nine) areas were organized within three broad themes of: conceptualizing EE as a field of inquiry; EE curriculum, teaching, assessment, and learning: processes and outcomes; and issues of framing, doing, and assessing EE research.

Interested contributors were invited to forward detailed abstracts (500 words) of their proposals. Over eighty individual chapter proposals and three proposals for whole sections were received. Proposals were submitted from authors on six continents with approximately equal numbers from Australasia (all but one from Australia), Europe, and North America. Six proposals each were received from Africa (all from South Africa) and Asia, but only two from South America. Unfortunately, many of the submitted proposals were very specific and focused on individual studies that were more suited to journal articles than to a broader chapter within a handbook of research. A first review of proposals by the editors identified thirty-nine individual chapter proposals that were considered possibilities for expansion of the abstract into full papers. Additional contributors, who had not submitted proposals, were then identified who either were known and respected for their scholarship in particular areas within the tentative handbook framework or could address evident gaps. Invited proposals focused in particular on the two sections in which there was a lack of proposals, namely *Analyses of EE Discourses and Policies* and *Research on Learning Processes in EE*, which were not surprising given that these are areas that researchers have noted as receiving limited attention in the field (Rickinson, 2001; Stevenson, 2006). In two other sections there was also a lack of individual proposals but we were fortunate in receiving section proposals for which the section editors invited contributors. These were *environmental conceptions, philosophies and ethics that situate the environmental in EE research* (subsequently renamed *Normative dimensions of environmental education research*) and *methodological issues in doing research* (subsequently renamed *philosophies and methodologies of environmental education research*). These two sections focused on conceptions/philosophies of education and environmental ethics in the first case, and philosophical and methodological conceptualizations of approaches to environmental education research in the second. The absence of individual proposals in these two areas of inquiry was perhaps more surprising, although the number of scholars working in them is relatively small.

After considering these invited contributors and addressing overlap with and among the submitted proposals, the initial list of thirty-nine individual submissions was reduced to twenty-six and complemented with sixteen invited chapters, including those invited by the three sets of authors who submitted section proposals, in addition to chapters authored by one or more of the editors. All three section proposals were accepted after the editors' discussion and negotiation with the submitting section editors. Section editors were invited for the remaining five major sections (with individual handbook editors coediting several of these sections), with the four handbook editors serving as editors for the final section. Contributors were selected on the basis of the quality and appropriateness of their submissions for a handbook on research, as well as, particularly in the case of invited contributors, their past research contributions to the field; the result is a mix of well known senior and a number of both midcareer and promising emerging scholars, as well as a range of diverse cultural contexts in which these scholars are working.

Having completed the identification of chapter authors and section editors, a set of guidelines for both groups were developed. All chapters were peer reviewed with each section editor(s) serving a similar function to a guest editor of a special issue of a journal, by seeking two reviewers of each chapter and then providing feedback and suggestions to authors. One of the handbook editors served as liaison with the various section editors during this process. A final draft of the whole handbook was reviewed by two anonymous reviewers selected by AERA.

The result of this process is that contributing first authors come from six continents (Africa, Asia, Australasia, Europe, North and South America) and fifteen countries and their work is situated in formal (early childhood, primary/elementary, secondary, and postsecondary), nonformal and informal education settings. Thus, the research agenda of contributors should be attentive to the diverse populations served by contemporary educational systems as well as to opportunities to engage individuals and communities in nonformal and informal learning contexts.

In addition to drawing on contributors from a diverse range of contexts, our initial intent was also to attempt to be deliberately expansive by including research from outside the field of education research as well as within. Unfortunately, we were unsuccessful in this endeavor for several reasons, and perhaps we could be viewed as being somewhat naïve in our ambitions. It speaks to the challenges of boundary crossing that scholars without a background in the literature of environmental education were uncomfortable in offering a contribution, although the accountability pressures in universities where chapters in edited or only publications within one's disciplinary count may also have played a role. We also arguably could be criticized for being insufficiently persistent in our efforts to identify and elicit appropriate authors from our outside our own networks and professional communities. Other limitations of this handbook, as with any other handbook of research, will be surfaced by readers and critics.

Structure of the Handbook

Given the extensive number and diversity of the handbook chapters, readers will probably appreciate a brief overview of its organization in order to determine entry points for accessing writings connected to their own interests. The handbook is organized in three layers: three parts, nine sections, and fifty-one chapters. The parts are described below, while each section (except the last or concluding section) is introduced by its editors, offering the reader a way of identifying those chapters that may be of particular interest. The three parts are broadly intended to speak respectively to research on: (a) historical, theoretical, and ethical foundations of environmental education; (b) the major dimensions of educational policy and practice (discourse, curriculum, learning, and assessment); and (c) issues related to methodologies, marginalized groups, and the strengths, limitations, and gaps of extant studies. Stated another way, the sections have been organized to try to capture scholarship focused on the important place of historical and theoretical perspectives, normative considerations, and language and discourse in environmental education before examining research in the areas of curriculum, learning, and assessment. Voices that have been marginalized from this research and methodological (and ontological and epistemological) orientations to this research are then examined before concluding in the final section with an effort to distil from the previous chapters the meanings of the current characteristics of environmental education research, the strengths and limitations of the field, and possible courses for charting a research agenda for the future.

Part A: *Conceptualizing Environmental Education as a Field of Inquiry,* comprises three sections that examine some of the historical, contextual, and theoretical orientations which have shaped environmental education and environmental education research; the normative dimensions of environmental education research, focusing on conceptions and philosophies of education and environmental ethics that situate the educational and the environmental in environmental education research; and analyzes of environmental education discourses and policies. Section 1, as the editor points out, provides a starting point for readers to consider "some of the theories, contexts and histories that have shaped the field," recognizing that the authors' theoretical orientations "inform, and have been informed by, contextual and historical perspectives that each author brings to their work" (Gough). Appropriately given the intellectual traditions currently shaping the field, these theoretical orientations span behaviorist, critical, and poststructuralist approaches.

The editors of Section 2 argue that many, if not most, environmental issues are fundamentally normative (or value laden) in nature and that in considering normative issues, the task "is no longer a question of attacking false universalisms but of overcoming relativism and the fragmentation of the social" (Delanty, 1999, p. 3). They add that attention to normative questions is underrepresented in the research literature and tackling such questions implicitly involves uncertainty and risk. The contributors to this section examine normative dimensions of the key concepts of ethics and education (and schooling) and the relationships between them as the editors envisage the task is to reimagine places for environmental and educational philosophies and ethics within environmental education research.

The chapters in Section 3 focus on critical analyses of language/discourses and policies in environmental education as well as the intersection of these discourses and policies. These analyses include: deconstructing components of the discourse of sustainable development and education for sustainable development (ESD), the shifting and competing discourses of international and national policy contexts, the intersection of ESD discourse with other educational discourses (such as professional development, vocational and technical education) and discourses as policy and slogans.

Part B: *Research on Environmental Education Curriculum, Learning, and Assessment* consists of three sections: research on curriculum issues in environmental education; research on learning processes in environmental education; and evaluation of environmental education programs, materials, and technologies and the assessment of learners and learning.

Section 4 discusses environmental education curriculum research historically and methodologically and deals with such knowledge interests as learner participation in the curriculum process, cultural change through curriculum, and place-based environment-related education. In addition, some emerging new niches for environmental education research within the wider curriculum research landscape are mapped out.

Section 5 highlights the shift in research on environmental learning from a focus on knowledge, attitudes, and behavior, particularly as outcomes of educational interventions, to processes of learning, especially those that recognize learners as active agents who respond cognitively and emotionally to their environmental education experiences. Chapters examine how learners make sense of their environmental education in formal settings, the power of landscape to link learners' place and identity, and calls for more transformative, more social approaches to learning. Collectively the section, in the words of its editors, offers "a set of models and theories of learning with examples of their implications for teachers and learners; ... etc and, more importantly a sense that the field is moving toward broader and deeper views of the role of learning in empowering individuals to be reflexive and socially critical."

In the final section in Part B, the evaluation and assessment of goals and outcomes of environmental education are situated within the larger socio cultural, economic, political, and environmental context in which environmental education takes place. The authors discuss a number of challenges facing evaluation and assessment, from taking a long term perspective of significant life experiences on learning outcomes, to the potential of geospatial technologies to play a transformative role in environmental education, to addressing the difficulties of determining attribution of outcomes across time, space and experiences.

Finally, the three sections in Part C: *Issues of Framing, Doing, and Assessing Environmental Education Research,* address: Issues of marginalization; philosophical and methodological perspectives; and trends, gaps, and future directions in environmental education research. Section 7 is concerned with the particular voices that have been at the margins of environmental education research. Authors examine how such issues as gender, race, and colonialism (in contributing to the relative dearth of indigenous perspectives) fail to be addressed in most environmental education research and why they are important. The section editors convincingly argue that social class and disability have received much less attention in environmental education research than in the wider education field and that sexuality "remains firmly at the margins" (Russell & Fawcett, Section 8 introduction).

In Section 8, a wide (but obviously not comprehensive) range of philosophical perspectives and methodological approaches (e.g., interpretive, critical, and postcritical) that characterize the environmental education research field are presented with the intent of challenging "readers to look beyond their own perspectives" (Hart, introduction). As Hart argues in the section introduction, this focus signifies not only that EE research has a history of embracing diverse methodologies, but also that "the complex task of grounding environmental education inquiry in philosophical and methodological work must precede more practical tasks of fieldwork" and methods given that "the problems of educational research are philosophical in origin and substance."

The final Section 9 of the handbook, whose contributors were the last to be identified (as explained in the section introduction), includes a critical analysis of the strengths and limitations of extant studies, identifies gaps in the field's present knowledge base and research approaches, and offers an analysis of issues confronting the field and possibilities for future directions.

Engaging With the Handbook

There were, of course, many other possible structures and conceptualizations of the field that might have been used for framing this handbook. No doubt some alternative organizational structures—both those which we considered and many which we did not even think of—will emerge as readers engage with the text. Others will engage less with the categorization of different dimensions of the field and more relationally with specific authors with whose work they are familiar, or wish to become familiar, or through responses to the language in titles. Some of these readings will lead to new conceptualizations and new contributions to the production of knowledge of environmental education from and about environmental education research. Primarily, however, our hope is that this example of global knowledge synthesis and production might serve as a catalyst for what we have already indicated was a guideline to contributors and therefore should resonate throughout many chapters in this handbook, and that is the notion of reflexivity. In other words, we invite and encourage readers to make explicit the assumptions underlying their own perspectives on environmental education research and to critically examine how these assumptions might shape their own current or emerging perspectives.

References

Delanty, G. (1999). *Social theory in a changing world: Conceptions of modernity*. Cambridge, UK: Polity Press.

Gough, A. (1997). *Education and the environment: Policy, trends and the problems of marginalisation*. Melbourne, VIC: Australian Council for Educational Research.

Hart, P., & Nolan, K. (1999) A critical analysis of research in environmental education. *Studies in Science Education, 34*, 1–69.

Martin, G. C. (1975). A review of objectives for environmental education. In G. Martin & K. Wheeler (Eds.), *Insights into environmental education*. Edinburgh, UK: Oliver & Boyd.

Rickinson, M. (2001). Learners and learning in environmental education: A critical review of the evidence. *Environmental Education Research, 7*(3), 207–320.

Stevenson, R. B. (2006). Tensions and transitions in policy discourse: Re-contextualizing a de-contextualized EE/ESD debate. *Environmental Education Research, 12*(3), 277–290.

Tbilisi Declaration (1978). *Toward an action plan: A report on the Tbilisi intergovernmental conference on environmental education*. Washington, DC: US Government Printing Office.

Part A

Conceptualizing Environmental Education as a Field of Inquiry

Section I

Historical, Contextual, and Theoretical Orientations That Have Shaped Environmental Education Research

Introduction

ANNETTE GOUGH
RMIT University, Australia

Where have we come from, what is shaping research in the field and where are we going? Environmental education is an emerging field, albeit one that has been around for over forty years. Yet it is not an easy field to tie down (if one would want to): its boundaries are fuzzy and interpretations of its documents, foundations, and directions are multiple.

The chapters selected for inclusion in this section provide some varying views on the historical, contextual, and theoretical orientations that have shaped environmental education research. These have been written by some of the leading researchers in environmental education from across three continents, each writing from the perspective of their own methodological and epistemological orientations and geographic locations, and their chapters are generally reflexive in that they endeavor to make explicit and address how their assumptions and experiences might affect the content of their contributions. Indeed, in many ways this section initiates the performance of the suggestion from Noel Gough, in the conclusion to his chapter in this section—that "thinking globally" in environmental education research might best be understood as a process of constructing transcultural "spaces" in which scholars from different localities collaborate in reframing and decentering their own knowledge traditions and negotiate trust in each other's contributions to their collective work—but there is still much to be done as the field and research about it continues to evolve and transform.

The historical orientations discussed in this section trace the field of environmental education to Rachel Carson's (1962) *Silent Spring* and similar works of that period. Tom Marcinkowski and colleagues also remind us of the educational movements that preceded and created favorable conditions for environmental education to arise in the late 1960s: the Nature Study Movement (ca.1900); Outdoor Education Movement (ca.1920); and Conservation Education Movement (ca.1930). There was also the popularization of Ecology Education in the 1950s and the Education for Sustainable Development Movement of the 1990s.

Several of the chapters argue that we need to be aware of different notions of our history and how it influences us, a perspective that is developed by Jo-Anne Ferreira. In particular, she highlights the importance of having "an understanding of how we have come to be what we are where we are, and therefore an understanding of how we might become something other than what we are, and do something other than what we do, now and into the future." Consistent with this view, Noel Gough's critical history troubles several of the "taken-for-granteds" in the field, such as the slogan "Think globally: Act locally," and asserts that *we have never thought globally*.

Annette Gough draws attention to her "history" chapter as being her story of her understanding of where the environmental education movement has come from. Her chapter provides a "history of the present" (Foucault, 1980) in the form of a genealogy which traces the emergence of the field that has become environmental education through a critical analysis of the definitions of the 1960s and 1970s and the reorientation toward education for sustainable development, particularly in the past decade. She also traces the changes in the orientation of environmental education research from behavioral change discourses to socioecological critical and poststructuralist approaches, and to a position where all research methodologies can now be considered valuable and appropriate, depending on the questions, issues, and problems being researched. Regula Kyburz-Graber expands on the history of this middle period through her tracing of the emergence of socioecological approaches to environmental education in the 1990s, as it was increasingly recognized that "the individual change concept of environmental education and a predominant natural science perspective on environmental problems are simplifications and do not form adequate approaches to the multilayered challenge of environmental problems."

The chapter by Tom Marcinkowski and colleagues describes historical trends in dissertation research, which is differently informed after having read the preceding chapters. It presents a review of selected trends in a segment of the international (USA and Canadian) doctoral literature in and related to environmental education, between 1971 and 2000. It builds on his earlier work with Mrazek (1996) and other reviews of dissertations such as Iozzi (1981). Marcinkowski et al. argue that as a dissertation is a graduation requirement, a broad review of dissertations can offer insights into the thinking and interest of doctoral students as emerging researchers. The review is generally based on collections of dissertations prepared by others and nearly all appear in *Dissertation Abstracts International*. However, this is a limited source which tends not to sample outside of North America; even for the period 1991–2000 only twenty-two dissertations are listed outside of Canada and the United States, and so Marcinkowski et al. conclude that "great caution should be taken in any attempt to generalize the results of this review beyond these two nations."

Nevertheless, as a snapshot of the field as reflected in the analysis of doctoral dissertations over a thirty year period, this chapter does reflect the changing nature of the field over time, particular when read in conjunction with other reviews of the research literature of the field developed by Hart and Nolan (1999) and Rickinson (2001, 2003). Marcinkowski et al. are able to trace a number of trends that are important historically, contextually, and theoretically as the field evolved over this period. Consistent with the chapters by Regula Kyburz-Graber and Noel Gough, and Annette Gough (1997), they note the long association of environmental education with science education—and the rhetoric/reality gap within the US literature between environmental education rhetoric, which calls for interdisciplinary approaches, and environmental education practice, where science is the dominant subject into which environmental education has been infused over this thirty-year period. They associate environmental education particularly with scientific knowledge, through the definition of environmental education they adopted and the search terms they used in *Dissertation Abstracts International*. Importantly, they did expand their terms for the 1990s to reflect a broader concept of environmental education, consistent with a growing sustainable development perspective and to take account of the increasing number of dissertations in these areas. Their analysis also traces the increasing numbers of dissertations in the field over the decades and the changing research methodologies of the dissertations from dominantly quantitative (69 percent in the first decade, 18 percent in the most recent) to a wide variety of methodological approaches (e.g., nearly 25 percent used only qualitative and critical methodologies in the 1990s compared with less than 10 percent in other decades). There was also a change in focus from the selection and organization of content to a focus on conditions that affect learning, the relationship of conceptual learning to other cognitive and affective outcomes, and the influence of constructivism.

The context for environmental education is concern for the quality of the total environment: "natural and built, technological and social (economic, political, technological, cultural-historical, moral, aesthetic)" and "to create new patterns of behaviour of individuals, groups and society as a whole toward the environment" (UNESCO, 1978, p. 27). This is a broader definition of environment than that encapsulated in one of the earliest definitions of the field: "*Environmental education* is aimed at producing a citizenry that is *knowledgeable* concerning the biophysical environment and its associated problems, aware of *how* to help solve these problems and *motivated* to work toward their solution" (Stapp et al., 1969, pp. 30–31, emphasis in the original). And definitions have continued to evolve as education for sustainable development emerges.

The contextual orientations that are discussed by the authors in this section are generally grounded in Western literature and perspective, and perhaps even a Northern American perspective; however several of the authors draw attention to this limitation of their work. For example, Annette Gough draws attention to how the foundational discourses of environmental education are "man-made" discourses due to the absence of women in their formulation and because of the anthropocentric and androcentric gaze of modernist science that separates "man" and "nature" and associates "woman" with "nature," an argument that she has further developed elsewhere in this handbook. Noel Gough offers a critical history of approaches to "thinking globally" in environmental education research. In particular, he argues that environmental education research could be enhanced by understanding Western science as one among many local knowledge traditions, and, if this was to occur, "global knowledge production in/for environmental education might then be understood as creating conditions under which local knowledge traditions can be performed together, rather than as creating a global 'common market' in which representations of local knowledge must be translated into (or exchanged for) the terms of a universal discourse."

The theoretical orientations discussed in these chapters span behaviorist, critical, and poststructuralist writings, each of which can, and should, inform current and future research in the field of environmental education. In this regard, the ordering of the chapters in this section is significant—Annette Gough's historical perspectives on the field provide a framing of the socioecological approaches to environmental education which informs a reading of Regula Kyburz-Graber's chapter. Kyburz-Graber builds on the history provided by Annette Gough and discusses the development of socioecological approaches to environmental education research, particularly from a "being critical" (Carr & Kemmis, 1986) perspective. According to Kyburz-Graber, socioecological approaches "build on the assumption that there is no determined set of goals that describes how an environmentally sound and sustainable world should be." Rather, such goals emerge from ongoing social processes of critical reflection on world views and on mainstream attempts at problem solving, with learners and teachers as part of these processes.

Jo-Anne Ferreira's chapter complements the earlier chapters in this section by drawing attention to the influence of Western societies' ways of thinking and acting on environmental problems, particularly the need to investigate why environmental education desires to empower individuals to transform themselves into informed and active environmental citizens as a more appropriate means to achieve a sustainable society, rather than other more traditional and intentional means of behavior change. She poses questions that are designed to make the familiar in environmental education visible and illuminate the micro-practices in which environmental educators engage, similar to the critical history of "thinking globally" provided by Noel Gough.

There is much more that could be said about the historical, contextual, and theoretical orientations that have shaped environmental education research, and fortunately other sections and chapters in this handbook also provide guidance in this direction, particularly those that focus on philosophical and methodological perspectives (Section 8) and moving the margins in environmental education (Section 7) which should be seen as complements to this section.

Implications for Future Research in Environmental Education

The future directions for research in relation to historical and theoretical orientations to environmental education suggested by the chapters in this section are consistent with the suggestions proposed by Hart and Nolan (1999, p. 41), that environmental education research needs to:

- begin to address critical and feminist and postmodern challenges
- strive for more in-depth qualitative analyses, and
- move outside the academy and develop partnerships with schools and communities.

The trends in thirty years of doctoral dissertations discussed by Marcinkowski et al. have indicated that environmental education research is definitely beginning to address critical and feminist challenges, and some chapters included in this section have posed postmodernist challenges (although Noel Gough, Jo-Anne Ferreira, and Annette Gough would probably see them more as poststructuralist). The trends also indicated that there is an increase in in-depth qualitative analysis occurring in dissertations, and this is being reflected in the literature of the field as well as the directions being argued for in this section. A silence in this section has been the school-community partnership context for environmental education research, but this is taken up in other sections of this Handbook.

Hart and Nolan's suggestions complement, but perhaps do not go as far as, the guiding principles for research in environmental education that I have proposed as a way forward (Gough 1999, p. 153) and which are reflected in most of the chapters in this section, with the exception of the trends chapter:

- to recognize that knowledge is partial, multiple, and contradictory;
- to draw attention to racism and gender blindness in environmental education;
- to develop a willingness to listen to silenced voices and to provide opportunities for them to be heard; and
- to develop understandings of the stories of which we are a part and our abilities to deconstruct them.

Related to these principles, there are a number of key themes for future research directions in the field that emerge from these chapters, with a dominant one being the importance of interrogating the blind spots and blank spots (Wagner, 1993) that characterize our research in the field to date—to problematize the orthodoxies of our beliefs and practices (Ferreira), to think globally *without* enacting some form of epistemological imperialism (Noel Gough), and to profoundly question real-life situations in view of socially constructed human-nature relationships (Kyburz-Graber).

Rather than championing empowerment as the alternative to more traditional approaches to behavior change (Ferreira), it might be more fruitful to explore in what learning settings and how learners come to critically investigate and reflect upon environmental questions and what kinds of contextual knowledges they gain through their inquiries (Kyburz-Graber), while being aware that we should not be privileging Western knowledge systems but rather engaging in global knowledge production in/for environmental education that creates conditions under which local knowledge traditions can be performed together (Noel Gough).

What has emerged from this section is the importance of reflexivity in our practices as environmental education researchers, and that we need to continue to "open up new avenues for recognizing the workings of power in the ways we construct our world and its possibilities . . . [and toward] developing more effective social change practices" (Lather, 1991, p. 100).

Conclusion

As Ferreira challenges us in this regard, "undertaking a history of rationalities of rule in environmental education can help us to understand the ways in which we as environmental educators work to govern our own and others conduct." Each of the chapters here has, in its own way, discussed the ways we work to govern our own and others conduct. What is clear is that there has been a move away from the "longing for 'one true story' that has been the psychic motor for Western science" (Harding, 1986, p. 193), and a recognition of the need to be aware of the historical, contextual, and theoretical orientations to environmental education that inform and influence our work as researchers.

The theoretical orientations of these chapters inform, and are informed by, the contextual and historical perspectives that each of the authors bring to their work. As a result, the chapters in this section provide a starting point for reflection on the theories, contexts, and history that have shaped the field that we now recognize as environmental education research and indicate possibilities for its future directions.

References

Carr, W., & Kemmis, S. (1986). *Becoming critical: Knowing through action research*. Revised edition. Geelong, VIC: Deakin University. (First published 1983).

Carson, R. (1962). *Silent Spring*. Greenwich, Conn.: Fawcett.

Foucault, M. (1980). In C. Gordon (Ed.), *Power/knowledge*. London, UK: Harvester.

Gough, A. (1997). *Education and the environment: Policy, trends and the problems of marginalisation*. Australian Education Review Series No. 39. Melbourne, VIC: Australian Council for Educational Research.

Gough, A. (1999). Recognizing women in environmental education pedagogy and research: Towards an ecofeminist poststructuralist perspective. *Environmental Education Research, 5*(2), 143–161.

Harding, S. (1986). *The science question in feminism*. Ithaca, NY: Cornell University Press.

Hart, P., & Nolan, K. (1999). A critical analysis of research in environmental education. *Studies in Science Education, 34*, 1–69.

Iozzi, L. A. (Ed.). (1981). *Research in environmental education 1971–1980*. Columbus, OH: ERIC Clearinghouse for Science, Mathematics and Environmental Education.

Lather, P. (1991). *Getting smart: Feminist research and pedagogy with/in the postmodern*. New York, NY, and London, UK: Routledge.

Marcinkowski, T., & Mrazek, R. (Eds.). (1996). *Research in environmental education, 1981–1990*. Troy, OH: North American Association for Environmental Education.

Rickinson, M. (2001). Learners and learning in environmental education: A critical review of the evidence. *Environmental Education Research, 7*(3), 207–320.

Rickinson, M. (2003). Reviewing research evidence in environmental education: some methodological reflections and challenges. *Environmental Education Research, 9*(2), 257–271.

Stapp, W. B., et al. (1969). The concept of environmental education. *Journal of Environmental Education, 1*(1), 30–31.

UNESCO. (1978). *Intergovernmental Conference on Environmental Education: Tbilisi (USSR), October 14–26 1977. Final Report*. Paris: Author.

Wagner, J. (1993). Ignorance in educational research: Or, how can you *not* know that? *Educational Researcher, 22*(5), 15–23.

2

The Emergence of Environmental Education Research

A "History" of the Field

Annette Gough
RMIT University, Australia

This "history" of environmental education traces the emergence of the field in formal education and educational research. The word "a" is intentionally employed in the chapter title because this is my story of my understanding of where the movement has come from and what has informed it. This chapter is also historical research: a curriculum history (Hamilton, 1990) in the form of a genealogy following Foucault (1980).

This chapter documents the emergence of environmental education research and includes a discussion of the archive of the movement, particularly the early statements that were made to describe and proscribe the movement as they helped to frame the emergence of research in the field through international meetings on the environment and environmental education convened by the United Nations and its agencies since 1972. This chapter also acknowledges and endeavors not to duplicate previous reviews of research in the field (such as Hart & Nolan, 1999; Iozzi, 1981, 1984; Rickinson, 2001, 2003; and the authors included in Mrazek, 1993; Stevenson & Dillon, 2010; Zandvleit, 2009 as well as this handbook) but rather provides a "history of the present" (Foucault, 1980) through tracing the changes in the orientation of environmental research from behavioral change discourses to socioecological and poststructuralist approaches.

The Declaration and Recommendations from the 1977 Tbilisi UNESCO-UNEP Intergovernmental Conference on Environmental Education (UNESCO, 1978) in many ways formalized the field of environmental education. They also provided the fundamental principles for the statements and proposals on "promoting education, public awareness and training" in *Agenda 21*, the global action plan from the United Nations Conference on Environment and Development (UNCED), which was held at Rio de Janeiro, Brazil in June 1992. However, at UNCED there was also a significant shift in terminology away from environmental education to refer to education as "critical for promoting sustainable development and improving the capacity of the people to address environment and development issues" (1992, para. 36.3), which came to be known as education for sustainable development. The most recent in this succession of meetings was the UNESCO World Conference on Education for Sustainable Development—Moving into the Second Half of the UN Decade, part of the United Nations Decade of Education for Sustainable Development (2005–2014), which was held in Bonn, Germany in 2009. The resultant Bonn Declaration (UNESCO, 2009) describes education for sustainable development over ten paragraphs and specifies action in formal, nonformal, informal, vocational, and teacher education.

This change in name from environmental education to education for sustainable development education was seen by some as "not in the best interest of the stability of environmental education" (Knapp, 1995, p. 9), but it has proceeded, reinforced at the World Summit on Sustainable Development, held in Johannesburg, South Africa in 2002 and enshrined in the United Nations Decade on Education for Sustainable Development 2005–2014 and its associated activities. The emergence of Education for Sustainable Development has recently been documented by Wals (2009) and key issues in sustainable development and learning are discussed by Scott and Gough (2003, 2004) and Jones, Selby, and Sterling. (2010) to name just a few. This is not the place to expand on the debates around the changing terminology: it is sufficient to say that most who researched environmental education now undertake research which is often categorized as education for sustainable development, and many see Education for Sustainable Development as a contributor to an enhanced relevance of environmental education (e.g., Fien, 2002; Wals, 2009; and the Bonn Declaration [UNESCO, 2009]).

Environmental Education Emerges

The field that has become environmental education arose out of the growing awareness of the threat of environmental degradation in the 1960s. Throughout the decade of the 1960s scientists increasingly drew attention to the growing scientific and ecological problems of the environment and the need for greater public awareness of these problems (see, e.g., Carson, 1962; Ehrlich, 1968; Goldsmith, Allen, Allaby, Davoll, & Lawrence, 1972; Hardin, 1968). The problems were seen as the increasing contamination of land, air, and water, the growth in world population, and the continuing depletion of natural resources. These problems were formally recognized in the 1972 United Nations Declaration on the Human Environment (as cited in Greenall & Womersley, 1977, p. 15):

> We see around us growing evidence of man-made [sic][1] harm in many regions of the earth; dangerous levels of pollution in water, air, earth and living things; major and undesirable disturbances to the ecological balance of the biosphere; destruction and depletion of irreplaceable resources; and gross deficiencies in the man-made [sic] environment of human settlement.

The scientists' calls were for more information about the state of the environment, and for education, albeit from a Western (and male) perspective (see A. Gough, 1994, 1999; N. Gough, 2003 and this handbook). For example, Rachel Carson (1962, p. 30) argued that, "The public must decide whether it wishes to continue on the present road, and it can do so only when in full possession of the facts." At the 1972 United Nations Conference on the Human Environment the importance of education was asserted. In the prelude to recommendations for international action it stated that, "Education and training on environmental problems are vital to the long-term success of environmental policies because they are the only means of mobilizing an enlightened and responsible population, and of securing the manpower [sic] needed for practical action programmes" (Greenall & Womersley, 1977, p. 16).

Environmental problems were often seen as scientific problems which science and technology could solve, but increasingly even the scientists themselves were arguing that science and technology were not enough. For example, urban biologist Stephen Boyden (1970) argued that:

> The suggestion that all our problems will be solved through further scientific research is not only foolish, but in fact dangerous … the environmental changes of our time have arisen out of the tremendous intensification of the interaction between cultural and natural processes. They can neither be considered as problems to be left to the natural scientists, nor as problems to be left to those concerned professionally with the phenomena of culture … all sections of the community have a role to play, certain key groups have, at the present time, a special responsibility. (p. 18)

He saw educational institutions as being at the top of the list of key groups, and charged them with providing students with an awareness of the threats to the human species and stimulating thinking and discussion on the social and biological problems facing humanity while avoiding "the implication in teaching that all the answers to any problems that man [sic] may have lie simply in further intensification of scientific and technological effort" (Boyden, 1970, p. 19).

Scientists were not the only ones putting pressure "towards using education to help restore and maintain a viable life-support system … The pressures come from government and from advocates of a variety of disparate positions concerning environmental needs" (Lucas, 1979, p. 3). The role of the mass media in drawing the public's attention to the environmental situation is highlighted by Schoenfeld who founded the *Journal of Environmental Education*, a journal "devoted to research and development in conservation communication," in 1969. However, the scientists were strong in their calls for education as a necessary component of any solution to the environmental crisis. Schoenfeld (1975, p. 45) states the position succinctly: "it is a cadre of scientific leaders that sets the environmental agenda in this country [USA]," and elsewhere, and, as previously mentioned, Western scientists such as Carson, Ehrlich, Goldsmith, and Hardin were putting education on the environmental agenda.

Their statements about education supported the concern expressed by Stenhouse that there is a danger of an educational lobby in environmental education. He defined a lobby "as a pressure group seeking to influence the curriculum of schools in the light of a social rather than an educational concern," and was concerned that "a lobby does not consider the wider educational issues adequately; it overstresses a particular social concern and tries to influence the curriculum toward that concern" (1977, p. 36). However, his concern went unheeded for many years.

According to Wheeler (1975, p. 15), the term "environmental education" was first used in the United States. The first usage of the term in the United Kingdom was in March 1965 at a conference at the University of Keele. Here it was agreed that environmental education "should become an essential part of the education of *all* citizens, not only because of the importance of their understanding something of their environment but because of its immense educational potential in assisting the emergence of a scientifically literate nation" (Wheeler, 1975, p. 8). The relationship between science education and environmental education was implicit. Also included in the conference recommendations was that "fundamental and operational education research, with participation by teachers, should be intensified to determine more exactly the content of environmental education and methods of teaching best suited to modern needs" (as quoted in Goodson, 1983, p. 118).

Framing the Field

The descriptions of the requirements of environmental education which emerged in the late 1960s and early 1970s were concerned with introducing ecological (environmental) content into educational curricula at all levels, promoting technical training and stimulating general awareness of environmental problems. The recommendations were similar whether they came from a 1968 UNESCO Biosphere Conference (Goodson, 1983) or from the 1970 Australian Academy of Science conference (Evans & Boyden, 1970). However, the statements were more exhortations than specifications.

Around this time there were many individuals, groups, and organizations proposing definitions of environmental education in attempts to clarify their intents. Bill Stapp and a group of colleagues in the School of Natural Resources at the University of Michigan developed a definition for a new educational approach "that effectively educates man regarding his relationship to the total environment" which they called "environmental education": "*Environmental education* is aimed at producing a citizenry that is *knowledgeable* concerning the biophysical environment and its associated problems, aware of *how* to help solve these problems and *motivated* to work toward their solution" (Stapp et al., 1969, pp. 30–31, emphasis in the original). This definition, together with four objectives of environmental education, was published in the first issue of the *Journal of Environmental Education*. The four objectives were to help individuals acquire (Stapp et al., 1969, p. 31).[2]

1. A clear understanding that man [sic] is an inseparable part of a system, consisting of man [sic], culture, and the biophysical environment, and that man [sic] has the ability to alter the interrelationships of this system.
2. A broad understanding of the biophysical environment both natural and man-made [sic] and its role in contemporary society.
3. A fundamental understanding of the biophysical environmental problems confronting man [sic], how these problems can be solved, and the responsibility of citizens and government to work toward their solution.
4. Attitudes of concern for the quality of the biophysical environment which will motivate citizens to participate in biophysical environmental problem-solving.

Stapp et al. argued that this educational approach was different from conservation education which was seen as being oriented primarily to basic resources, not focused on the community environment and its associated problems, and not emphasizing "the role of the citizen in working, both individually and collectively, toward the solution of problems that affect our well being" (1969, p. 30). He proposed a curriculum development model which he brought to Australia in 1970 when he spoke at the Australian Academy of Science conference (Evans & Boyden, 1970). This model focuses on curriculum development procedures, "with a consequent emphasis on administrative strategies rather than philosophical analysis" (Linke, 1980, pp. 34–35), an orientation which has dominated much of the environmental education discourse.

The Stapp et al. (1969) definition and objectives for environmental education formed the basis for a number of other conceptions of the field. For example, in September 1970 the International Union for the Conservation of Nature and Natural Resources (IUCN) convened an International Working Meeting on Environmental Education in the School Curriculum, in Nevada, United States which accepted a definition of environmental education which was to become widely used in subsequent years (cited, e.g., in Linke, 1980, pp. 26–27):

> Environmental education is the process of recognizing values and clarifying concepts in order to develop skills and attitudes necessary to understand and appreciate the interrelatedness among man [sic], his [sic] culture and his [sic] biophysical surroundings. Environmental education also entails practice in decision-making and self-formulating of a code of behavior about issues concerning environmental quality.

Given the types of definitions of environmental education that were emerging, using terms such as "man," "biophysical," "ecosystems," and "ecological principles," it is perhaps not surprising that science education was frequently seen as the place for environmental education, generally in the form of ecological concepts, to be incorporated in the school curriculum. However environmental education was not seen as an educational priority by education departments in the way that it was seen as a scientific or social priority by scientists, environmentalists, and academics. Rather, it was treated as yet another lobby group wanting space in an already overcrowded curriculum (Gough, 1997).

The use of sexist language in these early statements about environmental education was problematic. Hamilton (1991) argued that the use of "man" and "he" is exclusionary of women as well as being ambiguous and Gilligan's research indicates that, while women identify themselves in terms of relationships, "individual achievement rivets the male imagination, and great ideas or distinctive activity defines the standard of self-assessment and success" (1982, p. 163). While some women would probably argue that activities by the males of the human species have been a major factor in the deterioration of the environment, it is important that all humans are encompassed by environmental education statements. Thus it is important that one significant difference between the *Belgrade Charter* and previous formulations of environmental education was its use of nonsexist language. For the first time neither "man" nor "he" was in the statements, although there was a "man-made" in the guiding principles of the *Belgrade Charter* which became "built" in the Tbilisi guiding principles. This change could be due to 1975 being International Women's Year and the United Nations' guidelines for nonsexist writing taking effect (assuming there was some), but is unlikely because "man" still occurs in the papers from Belgrade, *Trends in Environmental Education* (UNESCO, 1977). However the terms "man" and "he" re-emerged in the Declaration from the Tbilisi Conference (UNESCO, 1978, p. 24) where the opening sentence stated, "In the last few decades, man has, through his power to transform his environment, wrought accelerated changes in the balance of nature."

However, the foundational discourses of environmental education are "man-made" discourses at least two levels—because of the absence of women in their formulation and because of the modernist science that separates "man" and "nature" and associates "woman" with "nature." The genderedness of the discourses also permeates their epistemology—not only are nonmale perspectives not valued, but the epistemology, being consistent with modernist science, views knowledge as universal, consistent, and coherent and the subject of knowledge as culturally and historically disembodied or invisible and homogeneous and unitary (Gough, 1994).

An Emerging Research Agenda

Although debate about the various educational research methodologies had been ongoing in the wider educational community for twenty years or more, and a range of research methodologies previously been applied to research in environmental education, this had been without necessarily "engaging seriously" the ideologies of research methodology until around 1990. As Robottom and Hart (1993, p. 18) argue, "the issue of the relationship between the respective ideologies of research methodology on the one hand and the substantive field of environmental education on the other had not been engaged seriously within the field of environmental education" until a symposium on "Contesting Paradigms in Environmental Education Research" was held at the 1990 annual conference of the North American Association for Environmental Education (the papers from which were published with others in Mrazek [1993] under the less controversial title of "alternative paradigms").

This notwithstanding, echoing Schwab (1969), Robottom and Hart (1993, p. 3) observed "that environmental education research is somewhat moribund" for two main reasons. Firstly, they noted "that in some areas at least it is firmly in the grip of an insular research paradigm" which is "characterized by the ideologies of determinism and individualism," that is, positivist inquiries. Secondly, they argued that "the research paradigm adopted in much environmental education research, however well intentioned may actually counter the achievement of some of [the] purposes" of environmental education "espoused in the founding of modern environmental education some two decades ago" (p. 3) and there is a large body of evidence to support their argument about positivist inquiries being the dominant form of research in environmental education in North America up until this time.

Behavior Change as the Focus for Environmental Education Research

Since the earliest definitions of environmental education, there has been a concern for evaluating the effectiveness of the programs that have been developed to achieve the perceived "goals" of environmental education. For example, Arthur Lucas, who developed the descriptive model of environmental education as education *in*, *about*, and *for* the environment as an attempt to understand the range of meanings that were being given to the concept of environmental education as part of his 1972 doctoral thesis (Lucas, 1979), later argued that (1980, pp. 20–21):

> There is a clear need for empirical science education research that
>
> i. tests the effectiveness of programs by examining their effects *on behaviour*, and not on attitude change alone;
> ii. establishes causal relationships, not merely correlations between knowledge, behavioural intentions, and behaviour, in particular examining the causal links between behaviour/attitudes and ... factual knowledge and conceptual knowledge;
> iii. investigates the effectiveness of techniques of changing behaviour in producing stable, environmentally sound behaviour, given that this behaviour can be determined; and
> iv. attempts longitudinal studies of formal and informal environmental education on adult behaviour.

Another early environmental education researcher, Linke (as reported in Collins, Gray, and Johnston, c.1984, p. 318), was also concerned with finding a successful system for environmental education programs:

> What of evaluation in EE? There has to be a methodology for EE; a set of strategies. Such a set is predicated on a set of values. Thence there was a need for matching action with values. There has to be a system, in which there might be different approaches, such as a teacher-centered approach.

The belief that there is *a* methodology, *a* system, *a* conceptual framework, *a* set of goals for environmental education was a shared goal of many early environmental educators (Gough, 1994). For example, environmental education research in the United States has for some time been concerned with ways of identifying predictors of responsible environmental behavior, as Howe and Disinger (1991, p. 5) assert: "the bottom-line purpose of environmental education in the view of most of its supporters and many of its practitioners, is the development of responsible individual and societal environmental behavior." They trace their assertion to "the standard (and still most often cited) definition of environmental education (Stapp et al., 1969) [which] makes a clear statement to this effect" (p. 5) and note that "much of the research-and-development work focused on fostering responsible environmental behavior ... in the United States has been conducted by Hungerford and his co-workers" (p. 6). It is therefore not unexpected that Hungerford and his colleagues made statements such as "the solutions to environmental problems do not lie in traditional technological approaches but rather in the alteration of human behavior" (Culen et al., 1986, p. 24), "the ultimate aim of education is to shape human behavior" (Hungerford & Volk, 1990, p. 8), and "responsible environmental behavior has been cited as the ultimate goal of environmental education" (Ramsey & Hungerford, 1989, p. 29). For Hungerford and Volk (1990, p. 9), an environmentally responsible citizen is defined as:

> One who has (1) an awareness and sensitivity to the total environment and its allied problems [and/or issues], (2) a basic understanding of the environment and its allied problems [and/or issues], (3) feelings of concern for the environment and motivation for actively participating in environmental improvement and protection, (4) skills for identifying and solving environmental problems [and/or issues], and (5) active involvement at all levels in working toward resolution of environmental problems [and/or issues].

Robottom and Hart (1993, p. 41) comment that, in the research by Hungerford and his colleagues, the variables "are all characteristics of individual human beings, focusing the research squarely and almost exclusively upon the individual ... the research rarely takes into account the historical, social, and political context within which the environmental acts of individuals and groups have meaning." Such statements are consistent with Huckle's (1983, p. 61) comment that "values education is rooted in liberal philosophy which focuses on the perceived social and political needs of the individual ... [which] encourages personal decision making on social issues but is often ambivalent towards the political system." Huckle (1986, p. 13) has also likened the focus on changing behaviors to "an evangelical mission. People are to be converted; their hearts and minds, their values changed ... [but] it gives values a prominence they do not deserve and overlooks issues of power. Values are primarily shaped by the material circumstance within which people live."

More recent research around proenvironmental behaviors and actions has sought to more carefully define the terms and to take them beyond an individualistic focus. Kollmuss and Agyeman (2002, p. 240) define "proenvironmental behavior" as the sort of behavior "that consciously seeks to minimize the negative impact of one's actions on the natural and built world," where "behavior" only refers to those personal actions that are directly related to environmental improvement, and, for Jensen (2002, p. 326), action in environmental education embraces indirect as well as direct actions; "an action is targeted at a change: a change in one's own lifestyle, in the school, in the local or in global society." Kollmuss and Agyeman (2002, p. 239) have argued that "the question of what shapes proenvironmental behavior is such a complex one that it cannot be visualized through one single framework or diagram." Such a view is consistent with the expanding notions of research in environmental education and the realization that changing behaviors is not a simple process and that behavior change reflects nonlinear rather than linear theories of knowledge.

In contrast with this perspective, much of the earliest research in environmental education in the United States took the form of positivist inquiries which worked from the premise that the key goal of environmental education is the acquisition of responsible environmental behavior (Howe & Disinger, 1991; Hungerford & Volk, 1990; Ramsey, Hungerford, & Volk, 1992), and the researchers referred to the fields of behavioral and social psychology for their authority in terms of pedagogical organization and practice (Marcinkowski, 1993a, 1993b; Robottom & Hart, 1993).

In the first report of the National Commission on Environmental Education Research (Iozzi, 1981, p. xiii), research was defined as "investigations employing systematic methods to study or interpret phenomena. It is data-based and employs valid observations with an intent to generalize results or build new models," and environmental education research "includes components of efforts concerned with developing or analyzing environmental awareness, valuing, or problem-solving behavior," over 90 percent of which employed quantitative methods. Thus it is to be expected that this is the type of research which was encouraged in the North American context, and this was confirmed in *A Summary of Research in Environmental Education, 1971–1982* (Iozzi, 1984), where 70 percent of research was classified as descriptive, which Iozzi (1984, p. 9) saw as "most reasonable and logical (as) ... EE was really an emerging area of inquiry. Now that the ground work has been done, EE researchers need to begin to emphasize a more directed type of research employing more vigorous research designs and methods." He saw the 18 percent of studies classified as "true experimental" as "most encouraging": "This could be a sign that the field is, in fact, maturing" (1984, p. 10). Others, such as Robottom and Hart (1993), saw the application of such methodologies to environmental education as most inappropriate. However, Connell (1997) has critiqued the naive antipositivism of Robottom and Hart (1993) and the way they failed to distinguish between quasi-experimental and experimental research and between postpositivisms within positivist research. The argument advanced by Connell (1997, p. 130) opens up space for legitimizing all types of research methodologies where researchers "do what they do well and where methodologies are selected to meet clearly identified research needs, balanced with a clear understanding of the social, political and philosophical contexts in which they are located."

Expanding Notions of Research in Environmental Education

While environmental education research of the 1970s and 1980s was dominated by applied science (quantitative) methods, popularized particularly by American researchers, there was a slow and steady evolution away from these methods toward postpositivist methodologies (Palmer, 1998). In their review of environmental education research at the end of the twentieth century, Hart and Nolan (1999, pp. 1–2) similarly concluded that, "What is clear after careful and extensive reading of the literature … is that environmental education research is a more complex and controversial field than it was a decade ago." They argue that in the 1990s conceptions of research in environmental education broadened beyond applied science designs to include interpretive, critical, and postmodern inquiries, and note that, in this more recent research, questions are being asked about the fundamental intents and purposes of the research, and about methods and methodologies as well as epistemologies and ontologies. This perspective was supported by Fien (2002) who, in reference to research on sustainability in higher education, illustrated how all research methodologies can be valuable, depending on the questions, issues, and problems. It was also further expanded by Gough and Gough (2004) in their review of environmental education research in southern Africa. In particular, they drew attention to "the colonial/postcolonial dilemmas of interpretation that arise when writing is also expected to perform … 'cultural translation'" as the South African researchers try to "explore a world in all its depth" while at the same time "having to explain it to outsiders" (p. 410).

These are very different visions of research in environmental education from the simple focus on changing behaviors. They recognize that behavior change reflects nonlinear rather than linear theories of knowledge. Linear epistemologies strive for one coherent knowledge system, are underpinned by assumptions of universality, foundation, homogeneity, monotony and clarity, and consider unknowns, or conflicting knowings, as "not-yet-resolved-but-in-principle-resolvable" imperfections (Bauman, 1993, p. 8). In contrast, achieving behavior change relates more with nonlinear theories of knowledge which "accept unknowns as well as plurality, dissent and conflicting knowledge claims as central and inevitable components to understanding knowledge construction, deconstruction and reconstruction processes" (Ward, 2002, p. 29).

Examples of interpretivist research in environmental education exist from the 1980s onward. For example, Greenall (1981, 1987) adopted a hermeneutic or interpretive inquiry approach and looked "for assumptions and meanings beneath the texture of everyday life" and viewed "reality as intersubjectively constituted and shared within a historical, political and social context" (Schubert, 1986, p. 181) in case studies charting the political history and development of environmental education in Australia. In North America, interpretive research in environmental education was not so readily accepted. For example, Cantrell (1993) believed that she still had to argue a case for research other than of the positivist kind: "For years, inquirers have hammered with one research paradigm, positivism, as if all topics of inquiry were nails. Currently, the tools for realizing the full potential for environmental education remain locked in the tool box" (p. 81). However, Hart and Nolan (1999) found a significant increase in these types of studies in the 1990s, including ethnographic and phenomenological research in environmental education.

While interpretivist inquiries in environmental education continued to grow in number, some researchers argued that interpretive inquiries "do not sufficiently incorporate political action" (Schubert, 1986, p. 82). They moved on to critical enquiries as a more appropriate research methodology for environmental education and to argue for a socially critical approach that takes a more holistic perspective as being a desirable direction for environmental education—see, for example, Fien (1991, 1993), Greenall Gough (1990), Greenall Gough and Robottom (1993), Huckle (1983, 1991), Robottom (1990, 1991, 1992, 1993), and Stevenson (1987, 1993). These researchers supported the assertion that environmental education "should adopt a critical approach to encourage careful awareness of the various factors involved in the situation," involve learners in planning their learning experiences, "utilise diverse learning environments and a broad array of educational approaches to teaching/learning," and "focus on current and potential environmental situations" (UNESCO, 1980, p. 27). For example, Robottom and Hart argued that research in environmental education should be compatible with the educational worldview it seeks to promote and support, and thus argue in favor of critical inquiries: "Just as environmental education programs ought to adopt an environmentally and socially critical approach, so provision ought to be made to allow researchers and practitioners, as educational inquirers, to adopt a similarly critical form of educational inquiry" (1993, p. 51). However, as argued by Connell (1997) and Fien (2002), critical research is not the only legitimate form of research in environmental education, and it is noteworthy that Hart (1996) also engages in other types of research, including interpretive research on teacher thinking in environmental education.

At this time the level of development of critical research in environmental education varied between researchers. Huckle (1991) and Greenall Gough (1991) mainly listed the characteristics of emancipatory or critical pedagogy and argued for these as being the appropriate model for the development of environmental education, "seeking to empower pupils so that they can democratically transform society" (Huckle, 1991, p. 54). Stevenson (1987, p. 79) also argued at the theoretical level, suggesting "a new definition of the role of the teacher and … changes in the organisational conditions" of schools if environmental education is to become a reality in schools. Robottom and Hart (1993) provided a different perspective in that they call on actual examples of participatory action research in environmental education (such as Greenall Gough & Robottom, 1993; Robottom, 1991) to support their argument for a socially critical approach to educational inquiry. They saw such an approach as fostering "the development of independent critical and creative thinking in relation to environmental issues as the aspiration of environmental education, [and] would have all research participants involved in critical and creative thinking in relation to research action" (Robottom & Hart, 1993, p. 52). Since the early 1990s examples of action research and participatory action research have grown exponentially (see, e.g., Posch, 1994; Wals & Albas, 1997, and as documented by Hart & Nolan, 1999) although such approaches are not without their critics (see, e.g., Walker, 1997).

Reflecting changes in educational research in general and changes in society, other developments in environmental education research which were at the opposite extreme to the search for a single method or approach are those which are categorized as postmodern or poststructuralist research studies: "Poststructuralism holds that there is no final knowledge; 'the contingency and historical moment of all readings' means that, whatever the object of our gaze, it 'is contested, temporal and emergent'" (Lather, 1991, p. 111). According to Lather (1991, p. 112):

> Poststructuralism views research as an enactment of power relations; the focus is on the development of a mutual, dialogic production of a multi-voice, multi-centered discourse. Research practices are viewed as much more inscriptions of legitimation than procedures that help us get closer to some "truth" capturable via language.

There are a number of research studies which have taken up the production of multivoice, multicentered discourses which reflect the foregrounding of indigenous knowledge (see, e.g., O'Donoghue & Neluvhalani, 2002), postcolonial perspectives (see, e.g., N. Gough, 2000, 2003), feminist perspectives (see, e.g., A. Gough, 1999; Gough & Whitehouse, 2003) postcolonial studies on language, place, and being (see, e.g., Cloete, 2010; Whitehouse, 2003), and the more recent work on socioecological resilience which examines interactions between society and nature, and between society and science (see, e.g., McKenzie, 2004; Colucci-Gray Camino, Barbiero, & Gray 2006; Morehouse et al., 2008; Krasny Lundholm, & Plummer, 2011). These newer directions also reflect the broader conceptions of education for sustainable development and the need to recognize other perspectives in related educational research.

From a "History of the Present" to the Future

This chapter has focused on some of the historical, contextual, and theoretical orientations that have helped shape the fields known as environmental education and environmental education research in the form of a "history of the present." I have discussed how environmental education had its origins in the concerns of scientists about the state of the world's environment but it now takes ten paragraphs to explain education for sustainable development in the Bonn Declaration the 2009 UNESCO World Conference on Education for Sustainable Development. I also discussed how environmental education research initially was dominated by quantitative applied science type research with a significant focus on knowledge-attitudes-behavior change, but now a thousand flowers bloom and we find a wide range of research methodologies being engaged in environmental education research—ranging from positivist to interpretive to critical to poststructuralist and including indigenous, postcolonial and feminist, among others.

What is likely to influence environmental education research in the future? Interestingly, different reviewers have congruent views.

The implementation scheme for the UN Decade of Education for Sustainable Development 2005–2014 (UNESCO, 2004) acknowledged that "aspects such as the adoption of values and changes in behavior cannot be adequately captured by numbers alone" (p. 41) and noted that a broader range of approaches to research are needed:

> Ethnographic approaches will enable a close look to be taken at specific communities in terms of changed behaviors, awareness of the values of sustainable development, and adoption of new practices. Longitudinal studies as well as community-wide ethnographic studies and analyses will provide data and will show the multiple connections in people's lives between the changes, values, practices, behaviors and relationships which sustainable development implies. (p. 42)

Hart and Nolan (1999, p. 42) similarly concluded that:

> Future research in environmental education must work to systematically examine the myths that underpin our thought and practice in school systems—on the structures and ethos that must be created to support teachers in their quest to examine their beliefs and to understand how to support new practices that are consistent with new beliefs.

In particular, they argue that environmental education research needs to (p. 41):

- begin to address critical and feminist and postmodern challenges,
- strive for more in-depth qualitative analyses, and
- move outside the academy and develop partnerships with schools and communities.

Their argument is consistent with Rickinson's (2003, p. 267) realizations that "research evidence will rarely translate easily into simple ingredients for developing environmental education practice or policy, particularly as 'factual information cannot, in itself, tell us what should be done' (Foster & Hammersley, 1998, p. 621)" and that there is "the danger that 'practical recommendations [can] effectively close down discussion of those issues' with possible negative consequences for the development of educational provision and reflective practice" (Foster & Hammersley, 1998, p. 624).

In closing this chapter I am reminded of Hazlett's (1979, p. 133) caution that "(t)he nation tends to reduce political, social, and economic problems to educational ones and claims to expect schools to cure present ills and provide for a brighter tomorrow for individuals and the collectivity." This reflects the origins of environmental education and provides a challenge for the future of environmental education research.

Acknowledgments

While the ultimate responsibility for the content of this chapter is mine, I would like to acknowledge the invaluable advice provided by the readers of the chapter, my much appreciated fellow academics, John Fien, Noel Gough, and Hilary Whitehouse. The chapter is better for their contributions.

Notes

1. A persistent practical problem in studying the "history" of the field is the frequent use of the term "man" to refer to persons of both sexes in the literature of science, science education, and environmental education. This practice is a focus of feminist concern: "Feminists agree that to use 'man' (or the generic masculine) to refer to people is iniquitous, ambiguous and exclusive … For example Simone de Beauvoir in *The Second Sex* (1953) argues that the use of man in language shows us how woman is defined as Other in our culture" (Humm, 1989, p. 125). Recent research on masculine bias in the attribution of personhood supports the argument that the use of the terms "he" and "man" to refer to people of both sexes is "ambiguous, exclusionary, and even detrimental" (Hamilton, 1991, p. 393). While I have been known to use the term "man" in an all-encompassing sense in the past (see, e.g., Greenall, 1978, 1980), I have for some time been offended by its usage and have avoided doing so. Unfortunately this is not the case with some other writers I have included in this chapter, which presents a problem in quoting from their texts. The frequent use of "[sic]" is tedious but I believe it is important that the use of the universal "man" in these statements be acknowledged.
2. The language in this statement reflects its science groundings through the use of terms such as "man" and "man-made" in the supposed scientific sense of being inclusive of both genders, and "biophysical environment," a very scientific term. Similar phrasing was used in the 1970 IUCN definition of environmental education, also cited in this chapter, where reference was made to "man, his culture and his biophysical surroundings" (in Linke, 1980, pp. 26–27).

References

Bauman, Z. (1993). *Postmodern ethics*. Oxford, UK: Blackwell.

Boyden, S. V. (1970). Environmental change: Perspectives and responsibilities. In J. Evans & S. Boyden (Eds.), *Education and the environmental crisis* (pp. 9–22). Canberra, ACT: Australian Academy of Science.

Cantrell, D. C. (1993). Alternative paradigms in environmental education: The interpretive perspective. In R. Mrazek (Ed.), *Alternative paradigms in environmental education research* (pp. 81–104). Troy, OH: North American Association for Environmental Education.

Carson, R. (1962). *Silent spring*. Greenwich, CT: Fawcett.

Cloete, E. (in press). Going to the bush: Language, power and the conserved environment in southern Africa. *Environmental Education Research*.

Collins, D., Gray, C., & Johnston, J. (Eds.). (undated, c.1984). Urban environmental education. In *Proceedings of the Third National Conference of the Australian Association for Environmental Education, 27–31 August 1984, Sydney*. Sydney, NSW: Australian Association for Environmental Education.

Colucci-Gray, L., Camino, E., Barbiero, G., & Gray, D. (2006). From scientific literacy to sustainability literacy: An ecological framework for education. *Science Education, 90*(2), 227–252.

Connell, S. (1997). Empirical-analytical methodological research in environmental education: response to a negative trend in methodological and ideological discussions. *Environmental Education Research, 3*(2), 117–132.

Culen, G. R., Hungerford, H. R., Tomera, A. N., Sivek, D. J., Harrington, M., & Squillo, M. (1986). A comparison of environmental perceptions and behaviors of five discrete populations. *Journal of Environmental Education, 17*(3), 24–32.

de Beauvoir, S. (1972). *The second sex*. Harmondsworth: Penguin. (First published in French in 1949 and in English in 1953).

Ehrlich, P. R. (1968). *The population bomb*. New York, NY: Ballantyne.

Evans, J., & Boyden, S. (Eds.). (1970). *Education and the environmental crisis*. Canberra, ACT: Australian Academy of Science.

Fien, J. (1991). Towards school-level curriculum enquiry in environmental education. *Australian Journal of Environmental Education, 7*, 17–29.

Fien, J. (1993). *Education for the environment: Critical curriculum theorising and environmental education*. Geelong, VIC: Deakin University.

Fien, J. (2002). Advancing sustainability in higher education: issues and opportunities for research. *Higher Education Policy, 15*, 143–152.

Foster, P., & Hammersley, M. (1998). A Review of Reviews: Structure and function in reviews of educational research. *British Educational Research Journal, 5*(4), 609–628.

Foucault, M. (1980). In C. Gordan (Ed.), *Power/knowledge*. London, UK: Harvester.

Gilligan, C. (1982). *In a different voice: Psychological theory and women's development*. Cambridge, MA/London: Harvard University Press.

Goldsmith, E. Allen, R., Allaby, M., Davoll, J., & Lawrence, S. (1972). Blueprint for survival. *The Ecologist, 2*, 1–50.

Goodson, I. F. (1983). *School subjects and curriculum change*. London/Sydney: Croom Helm.

Gough, A. (1994). *Fathoming the fathers in environmental education: A feminist poststructuralist analysis* (Unpublished doctoral dissertation). Deakin University, Geelong, VIC.

Gough, A. (1997). *Education and the environment: Policy, trends and the problems of marginalisation*. Australian Education Review Series No. 39. Melbourne, VIC: Australian Council for Educational Research.

Gough, A. (1999). Recognizing women in environmental education pedagogy and research: Toward an ecofeminist poststructuralist perspective. *Environmental Education Research, 5*(2), 143–161.

Gough, A., & Gough, N. (2004). Environmental education research in southern Africa: Dilemmas of interpretation. *Environmental Education Research, 10*(3), 409–424.

Gough, A., & Whitehouse, H. (2003). The "nature" of environmental education research from a feminist poststructuralist standpoint. *Canadian Journal of Environmental Education, 8*, 31–43.

Gough, N. (2000). Interrogating silence: Environmental education research as postcolonialist textwork. *Australian Journal of Environmental Education, 15/16*, 113–120.

Gough, N. (2003). Thinking globally in environmental education: Implications for internationalizing curriculum inquiry. In W. F. Pinar (Ed.), *International handbook of curriculum research* (pp. 53–72). Mahwah, NJ: Lawrence Erlbaum.

Greenall, A. (1978). *Environmental education teachers' handbook*. Melbourne, VIC: Longman-Cheshire.

Greenall, A. (1980). *Environmental education for schools or how to catch environmental education*. Canberra, ACT: Curriculum Development Centre.

Greenall, A. (1981). *Environmental education in Australia: Phenomenon of the seventies (a case study in national curriculum development)*. Occasional Paper No. 7. Canberra, ACT: Curriculum Development Centre.

Greenall, A. (1987). A political history of environmental education in Australia: Snakes and ladders. In I. Robottom (Ed.), *Environmental education: Practice and possibility* (pp. 3–21). Geelong, VIC: Deakin University.

Greenall, A., & Womersley, J. (Eds.). (1977). *Development of environmental education in Australia—Key issues*. Canberra, ACT: Curriculum Development Centre.

Greenall Gough, A. (1990). Red and green: Two case studies in learning through ecopolitical action. *Curriculum Perspectives, 10*(2), 60–65.

Greenall Gough, A. (1991). Recycling the flavour of the month: Environmental education on its third time around. *Australian Journal of Environmental Education, 7*, 92–98.

Greenall Gough, A., & Robottom, I. (1993). Towards a socially critical environmental education: Water quality studies in a coastal school. *Journal of Curriculum Studies, 25*(4), 301–316.

Hamilton, D. (1990). *Curriculum history*. Geelong, VIC: Deakin University.

Hamilton, M. C. (1991). Masculine bias in the attribution of personhood. *Psychology of Women Quarterly, 15*, 393–402.

Hardin, G. (1968). The tragedy of the commons. *Science, 162*, 1243–1248.

Hart, P. (1996). Problematizing enquiry in environmental education: Issues of method in a study of teacher thinking and practice. *Canadian Journal of Environmental Education, 1*(1), 56–89.

Hart, P., & Nolan, K. (1999). A critical analysis of research in environmental education. *Studies in Science Education, 34*, 1–69.

Hazlett, J. S. (1979). Conceptions of curriculum history. *Journal of Curriculum Studies, 9*(2), 129–135.

Howe, R. W., & Disinger, J. F. (1991). Environmental education research news. *The Environmentalist, 11*(1), 5–8.

Huckle, J. (1983). Environmental education. In J. Huckle (Ed.), *Geographical education: Reflection and action* (pp. 99–111). Oxford, UK: Oxford University Press.

Huckle, J. (1986). Ten red questions to ask green teachers. *Green Teacher, 1*(2), 11–15.

Huckle, J. (1991). Education for sustainability: Assessing pathways to the future. *Australian Journal of Environmental Education, 7*, 43–62.

Humm, M. (1989). *The dictionary of feminist theory*. Hemel Hempstead, Herts: Harvester Wheatsheaf.

Hungerford, H. R., & Volk, T. L. (1990). Changing learner behavior through environmental education. *Journal of Environmental Education, 21*(3), 8–21.

Iozzi, L. A. (Ed.). (1981). *Research in environmental education 1971–1980*. Columbus, OH: ERIC Clearinghouse for Science, Mathematics and Environmental Education.

Iozzi, L. A. (Ed.). (1984). *A summary of research in environmental education, 1971–1982*. Columbus, OH: ERIC Clearinghouse for Science, Mathematics and Environmental Education. Monographs in Environmental Education and Environmental Studies, #2.

Jensen, B. B. (2002). Knowledge, action and pro-environmental behaviour. *Environmental Education Research, 8*(3), 325–334.

Jones, P., Selby, D., & Sterling, S. (Eds.). (2010). *Sustainability education: Perspectives and practice across higher education*. London, UK: Earthscan.

Knapp, D. (1995). Twenty years after Tbilisi: UNESCO inter-regional workshop on re-orienting environmental education for sustainable development. *Environmental Communicator, 25*(6), 9.

Kollmuss, A., & Agyeman, J. (2002). Mind the gap: Why do people act environmentally and what are the barriers to pro-environmental behavior? *Environmental Education Research, 8*(3), 239–260.

Krasny, M. E., Lundholm, C., & Plummer, R. (Eds.). (2011). *Resilience in social-ecological systems: The role of learning and education*. London: Routledge.

Lather, P. (1991). *Getting smart: Feminist research and pedagogy with/in the postmodern*. New York/London: Routledge.

Linke, R. D. (1980). *Environmental education in Australia*. Sydney, NSW: Allen and Unwin.

Lucas, A. M. (1979). *Environment and environmental education: Conceptual issues and curriculum implications*. Melbourne, VIC: Australian International Press and Publications.

Lucas, A. M. (1980). Science and environmental education: Pious hopes, self praise and disciplinary chauvinism. *Studies in Science Education, 7*, 1–26.

Marcinkowski, T. (1993a). A contextual review of the "quantitative paradigm" in EE research. In R. Mrazek (Ed.), *Alternative paradigms in environmental education research* (pp. 29–78). Troy, OH: North American Association for Environmental Education.

Marcinkowski, T. (1993b). A response to "Beyond Behaviorism." In R. Mrazek (Ed.), *Alternative paradigms in environmental education research* (pp. 313–322). Troy, OH: North American Association for Environmental Education.

McKenzie, M. (2004). The "willful contradiction" of poststructural socio-ecological education. *Canadian Journal of Environmental Education, 9*, 177–190.

Morehouse, B. J., et al. (2008). Science and socioecological resilience: Examples from the Arizona-Sonora Border. *Environmental Science & Policy, 11*(3), 272–284.

Mrazek, R. (Ed.). (1993). *Alternative paradigms in environmental education research*. Troy, OH: North American Association for Environmental Education.

O'Donoghue, R., & Neluvhalani, E. (2002). Indigenous knowledge and the school curriculum: A review of developing methods and methodological perspectives. In E. Janse van Rensburg, J. Hattingh, H. Lotz-Sisitka, & R. O'Donoghue (Eds.), *Environmental education, ethics and action in southern Africa* (pp. 121–134). Pretoria, ZA: Human Sciences Research Council.

Palmer, J. (1998). *Environmental education in the 21st century: Theory, practice, progress and promise*. London, UK: Routledge.

Posch, P. (1994). Action research in environmental education. *Educational Action Research, 1*(3), 447–455.

Ramsey, J. M., & Hungerford, H. (1989). The effects of issue investigation and action training on environmental behavior in seventh grade students. *Journal of Environmental Education, 20*(4), 29–35.

Ramsey, J. M., Hungerford, H. R., & Volk, T. L. (1992). Environmental education in the K-12 curriculum: Finding a niche. *Journal of Environmental Education, 23*(2), 35–45.

Rickinson, M. (2001). Learners and learning in environmental education: A critical review of the evidence. *Environmental Education Research, 7*(3), 207–320.

Rickinson, M. (2003). Reviewing research evidence in environmental education: Some methodological reflections and challenges. *Environmental Education Research, 9*(2), 257–271.

Robottom, I. (1990). Environmental education: Reconstructing the curriculum for social responsibility. *New Education, XII*(1), 61–77.

Robottom, I. (1991). Technocratic environmental education: A critique and some alternatives. *Journal of Experiential Education, 14*(1), 20–26.

Robottom, I. (1992). Matching the purposes of environmental education with consistent approaches to research and professional development. *Australian Journal of Environmental Education, 8*, 133–146.

Robottom, I. (1993). Beyond behaviorism: Making EE research educational. In R. Mrazek (Ed.), *Alternative paradigms in environmental education research* (pp. 133–143). Troy, OH: North American Association for Environmental Education.

Robottom, I., & Hart, P. (1993). *Research in environmental education: Engaging the debate*. Geelong, VIC: Deakin University.

Schoenfeld, C. A. (1975). National environmental education perspectives. In R. J. H. Schafer & J. F. Disinger (Eds.), *Environmental education: Perspectives and prospectives* (pp. 43–45). Columbus, OH: ERIC/SMEAC.

Schubert, W. H. (1986). *Curriculum: Perspective, paradigm, and possibility*. New York, NY: Macmillan.

Schwab, J. J. (1969). The practical: A language for curriculum. *School Review, 78*, 1–24.

Scott, W., & Gough, S. (2003). *Sustainable development and learning: Framing the issues*. London/New York: RoutledgeFalmer.

Scott, W., & Gough, S. (Eds.). (2004). *Key issues in sustainable development and learning: A critical review*. London/New York: RoutledgeFalmer.

Stapp, W. B., et al. (1969). The concept of environmental education. *Journal of Environmental Education, 1*(1), 30–31.

Stenhouse, L. (1977). Aspirations and realities in environmental education. In R. D. Linke (Ed.), *Education and the human environment* (pp. 35–49). Canberra, ACT: Curriculum Development Centre.

Stevenson, R. B. (1987). Schooling and environmental education: Contradictions in purpose and practice. In I. Robottom (Ed.), *Environmental education: Practice and possibility* (pp. 69–82). Geelong, VIC: Deakin University.

Stevenson, R. B. (1993). Becoming compatible: Curriculum and environmental thought. *Journal of Environmental Education, 24*(2), 4–9.

Stevenson, R. B., & Dillon, J. (Eds.). (2010). *Engaging environmental education: Learning, culture and agency*. Rotterdam: Sense Publishers.

United Nations Conference on Environment and Development. (1992). *Agenda 21*. Rio de Janeiro, Brazil: Author. Final, advanced version as adopted by the Plenary on 14 June 1992.

UNESCO. (1977). *Trends in environmental education*. Paris: Author.

UNESCO. (1978). *Intergovernmental conference on environmental education: Tbilisi (USSR), 14–26 October 1977*. Final Report. Paris: Author.

UNESCO. (1980). *Environmental education in the light of the Tbilisi conference*. Paris: Author.

UNESCO. (2004). United Nations Decade of Education for Sustainable Development 2005–2014. Draft Implementation Scheme. October 2004. Retrieved from portal.unesco.org/education/en/file_download.php/03f375b07798a2a55dcdc39db7aa8211Final+IIS.pdf

UNESCO World Conference on Education for Sustainable Development. (2009). Bonn Declaration. Retrieved from www.esd-world-conference-2009.org/en/whats-new/news-detail/item/bonn-declaration-now-available-in-8-languages.html

Walker, K. (1997). Challenging critical theory in environmental education, *Environmental Education Research, 3*(2), 155–162.

Wals, A. (2009). *Review of contexts and structures for education for sustainable development 2009*. Paris: UNESCO.

Wals, A. E. J., & Albas, A. H. (1997). School-based research and development of environmental education: A case study. *Environmental Education Research, 3*(3), 253–267.

Ward, M. (2002). Environmental management: Expertise, uncertainty, responsibility. In E. Janse van Rensburg, J. Hattingh, H. Lotz-Sisitka, & R. O'Donoghue (Eds.), *Environmental education, ethics and action in Southern Africa* (pp. 28–35). Pretoria: Human Sciences Research Council.

Wheeler, K. (1975). The genesis of environmental education. In G. C. Martin & K. Wheeler (Eds.), *Insights into environmental education* (pp. 2–19). Edinburgh, UK: Oliver and Boyd.

Whitehouse, H. (2002). Landshaping: A concept for exploring the constructions of environmental meanings within tropical Australia. *Australian Journal of Environmental Education, 18*, 57–62.

Zandvleit, D. B. (Ed.). (2009). *Diversity in environmental education research*. Rotterdam: Sense Publishers.

3

Socioecological Approaches to Environmental Education and Research

A Paradigmatic Response to Behavioral Change Orientations

Regula Kyburz-Graber
University of Zurich, Switzerland

Introduction

Environmental education as a new demand for educational systems has been launched in many countries toward the end of the sixties in the twentieth century as a response to growing fears about the degradation of the environment. What was in a first attempt more or less a private initiative of engaged biology and geography teachers became a more political initiative in the context of the political strategies concerning environmental protection framed at the United Nations Conference on the Human Environment in Stockholm 1972. Certainly, there have been nature education movements before that time, assumedly emerging in the first half of the twentieth century. However, the evolving educational debate on *environmental* education may be seen as closely linked to the evolving political debate on environmental protection (Fensham, 1978). The Stockholm declaration comprehending twenty-six principles which calls "upon Governments and peoples to exert common efforts for the preservation and improvement of the human environment, for the benefit of all the people and for their posterity" entails in principle 19 the role of education within the process of environmental protection as "essential in order to broaden the basis for an enlightened opinion and responsible conduct by individuals, enterprises and communities in protecting and improving the environment in its full human dimension" (United Nations Environmental Programme [UNEP], 1972).

Five years later environmental education was declared as an indispensable instrument for improving the environment at the United Nations Education, Scientific, & Cultural Organization—United Nations Environmental Programme (UNESCO-UNEP) Conference on Environmental Education in Tbilisi 1977. The demand in principle 19 of the Stockholm Declaration that "education should broaden the basis for a responsible conduct by individuals, enterprises and communities" was taken up in the goals formulated in recommendation 2:

> The goals of environmental education are:
> a. to foster clear awareness of, and concern about, economic, social, political and ecological interdependence in urban and rural areas;
> b. to provide every person with opportunities to acquire the knowledge, values, attitudes, commitment, and skills needed to protect and improve the environment;
> c. to create new patterns of behavior of individuals, groups and society as a whole towards the environment. (UNESCO-UNEP, 1978, p. 24)

With this declaration, the ultimate goal for environmental education of behavioral change for the benefit of the society was addressed to individuals, groups, and society. Education was seen as a main instrument to solve environmental problems. Additional to the political systems, within the field there are numerous stakeholders concerned with environmental problems—e.g., nature protection organizations, water supply businesses, power plant businesses, consumer organizations, gene technology businesses, environmental researchers, and others—and they are all addressing educational institutions to respond to the societal demands in order to help solve environmental problems.

There have been conceptualized and explored various ways as reactions to societal requests concerning environmental problems, raised by stakeholders, environmentalists, educators themselves or environmental education researchers. Some important strands are (compare the seven traditional currents, Sauvé, 2005):

- promoting individual behavioral change through improving educational strategies and research (Hines,

Hungerford, & Tomera, 1986/1987; Hungerford & Volk, 1990; Sia, Hungerford, & Tomera, 1985/1986; Zelezny, 1999); for an overview of German research publications see Lehmann (1999); a "retrospect and prospect" in Reid and Scott (2006); and recent review by Heimlich and Ardoin (2008), attempted to provide a foundation for behavior related discussions in environmental education
- enhancing environmental awareness through environmental literacy usually closely linked to natural science literacy (e.g., Bybee & Mau, 1986; Giordan & Souchon, 1991; Volk & McBeth, 1997)
- enhancing nature-human-relationship and ecological awareness through nature experience (e.g., Bögeholz, 2006; Chawla, 1998; Lehmann, 1999)
- promoting action competence through action oriented learning (e.g., Bolscho & Michelsen, 1999; Eulefeld, Bolscho, Rode, Rost, & Seybold, 1993; Eulefeld, Bolscho, Rost & Seybold, 1988; Jensen & Schnack, 1997; an overview over the German tradition give Nickel & Reid, 2006)
- promoting ethical reflection and awareness of cultural contexts and diversity (e.g., Agyeman, 2002; Greenwood & McKenzie, 2009; Hargrove, 2000; Jickling, 2004a, 2004b)
- educating for becoming critical through transformative and critical education (e.g., Heid, 1992; O'Donoghue & McNaught, 1991; Posch, 2003; Robottom, 1987; Robottom & Hart, 1993; Stapp, 1974; Wals, 1993; contributions in Mrazek, 1993, such as Del Pilar Jimenes Silva, 1993; Gough, 1993; Hart, 1993; Robottom, 1993).

Socioecological approaches to environmental education have been developed within the "becoming critical" strand, outlined in this chapter.

The Nature of Socioecological Approaches

Socioecological approaches build on the assumption that the individual change concept of environmental education and a predominant natural science perspective on environmental problems are simplifications and do not form adequate approaches to the multilayered challenge of environmental problems. The notion "socioecological" has been founded on findings from case studies in secondary schools, which gave a strong argument for environmental education focused on critical reflection of real life situations regarding the interface of social and natural sciences (Kyburz-Graber, Rigendinger, Hirsch Hadorn, & Werner Zentner, 1997a, 1997b). In the context of science education, a related concept was developed, the socio*scientific* issues approach, which shows similarities albeit not explicitly linked to environmental education questions (Cross & Price, 1996; Laugksch, 2000; Ratcliffe, 2007; Zeidler, Sadler, & Simmons, 2005). The notion "socioecological" seems to be implicitly existent in various concepts of critical environmental education. Thus, the whole range of related concepts may be summarized as "socioecological approaches." During the last years, the notion "socioecological" seems to have been evolving particularly as an approach to research on the global level. It is used to describe the challenge of global resilience management of human-environment systems for which the notion of socioecological systems is used (Walker et al., 2002; Young et al., 2006).

The socioecological approaches to environmental education emerged as ways for dealing with conflicting multilayered demands concerning the environment in the nineties of the last century when interdisciplinary environmental research programs were conducted (e.g., Swiss priority program environment 1992–2000, the German research program "Man and Global Environmental Change" 1995–2001). Out of those research findings and similar initiatives, these key issues for socioecological perspectives on environmental education were described:

- questioning the nature of natural and social sciences and its interfaces;
- questioning the ways how knowledge is produced, reflected, disseminated, and continuously interpreted; and
- inquiring the interfaces between sciences and social questions.

The characteristics will be outlined providing arguments for the foundation of socioecological approaches illustrated by research studies as examples for ways of knowing in environmental education. It is argued in the following sections that socioecological approaches are

- *constructive*: learners and researchers construct and reconstruct knowledge on the basis of their own inquiries and case studies, and adopt environmental problems as socially constructed
- *reflective*: learners and researchers approach learning processes as reflection on ways of knowing and mediating knowledge
- *critical*: learners and researchers approach phenomenon and notions of environmental problems in critical and relational dimensions, including questioning historical and future perspectives
- *participatory*: learners, teachers, and researchers cooperatively interact while being aware of different interests and needs

Socioecological Approaches Are Constructive in the Way Environmental Problems Are Explored

A key aspect of socioecological approaches is the way in which environmental problems are addressed, inquired, analyzed, interpreted, and what conclusions are drawn regarding solutions.

It has been and still is a widely agreed approach to environmental problems to analyze what causes those problems, to analyze how they could be avoided and how their effects can be mitigated (Haeberli, Gessler, Grossenbacher-Mansuy, & Lehmann Pollheimer, 2002). The preferred measures are consistently seen in technological progress and changing people's behaviors. With increased interdisciplinary inquiries researchers are recognizing that the concept of environmental problems has to be extended into ontological, epistemological, and methodology questions embracing not only nature science but as well social, economic, ethical, philosophical, and other aspects. This broadened perspective was also described in the concept of "environmental issues" (Ramsey, Hungerford, & Volk, 1989).

A most important part of environmental problems is to point out the difference between observable and measurable facts and their interpretations. Thus, what is the meaning of the term "environmental problem" is not merely the observed facts, but as well what people and experts infer and interpret from the facts, which causes and effects they attribute to them, and what kind of solutions they are searching for (Hirsch, 1995).

In a wider sense, the notion "environmental problem" has to include the ontological question of which "reality" is assumed to be researched, that is, the world of the people concerned, or their view of the world; the world of nature science which sees its strengths in restrictions to objectively observable facts, or the world of culture documented in literature, historical sources, arts and music, or other worlds. (See the typology of conceptions of the environment and its relations to educational paradigm given by Sauvé, 1996.)

Environmental problems also include epistemological questions of how knowledge on environmental problems is generated. With reference to environmental problems, it is generally thought that the most relevant knowledge is produced by nature scientific approaches following a positivist research strategy: observing, measuring, setting hypotheses, conducting experiments (Robottom, 2003). As Robottom states:

> Environmental issues do not fundamentally consist of objectively existing facts that are more usually the concern of science and science education. If it is recognised that environmental issues actually consist of differences of opinion among human beings about the appropriateness of certain environmental actions, then it can be seen that environmental issues are best understood within a social discourse rather than a scientific discourse. (Robottom, 2003, p. 34)

From an interpretive or critical perspective, environmental problems as they are perceived and mediated in society cannot be taken for granted but they have to be explored as concepts of individuals and social groups—are they well educated or not, scientifically experienced or not—who construct and reconstruct their views and beliefs against their biographical and contextual background (see the three images of environmental education described by Robottom & Hart, 1993). Individuals are constantly interpreting and evaluating their observations, experiences and, in particular, what they hear, read, and see. In the strict sense, every oral or written expression of human activities has to be seen as a social construction. Thus, approaching environmental problems requires not only scientific knowledge but also inquiries on how people interpret and value what they came to know about a specific environmental problem. Extended ways of gaining knowledge about environmental problems are biographies, life stories, literature, all kinds of historical and current documents, oral history stories, individual experiences, and so on. Thus, environmental problems cannot be adequately approached if it is not taken into account that they are shaped by interests, needs, values, interpretations, conditions, and social contexts of the people concerned, as well as by arguments and views suggesting how environmental problems might be solved (Haeberli et al., 2002; Wals, 1993).

An Example for a Socially Constructed Environmental Problem: The Loss of Biodiversity

There is probably no doubt that the loss of biodiversity is a worldwide and severe environmental problem touching economic, health, global inequality, and long-term ecological aspects. It entails questions regarding how food can be produced in a sustainable way in the long run for an increasing world population without damaging existing species as gene reservoirs; additional aspects are the need of pharmaceutics on the basis of plants, sustaining the dynamics of food chains, economic changes due to loss of certain species (e.g., in fish grounds), the loss of attractiveness to tourists due to substantial changes in features of the landscape and its affective and aesthetic meanings for people, and others. Probably almost everyone asked would reinforce the claim that sustaining biodiversity is an important environmental problem in general. Still, in the details, there are many differing views about which species are in danger, if it matters at all, about spatial and temporal aspects of loss and the effects of loss of biodiversity, about what should and could be done, and not least, about what the role(s) and responsibility(s) of individuals and social groups in this process might be. If the loss of biodiversity is taken as a fact backed merely by natural science research findings the conclusion might be close that the problem can be solved through better education of people in terms of knowledge transmission and raising awareness (to follow these arguments see, e.g., Lindemann-Matthies, 2002).

If the loss of biodiversity is looked at as a socially constructed environmental problem, many more aspects related to human behavior have to be deliberated:

Even if individuals recognized how their behavior might have an impact on the decline of biodiversity—which is not directly obvious, everyone is generally confessing that he or she loves nature and is caring for nature—they will probably not be able to draw conclusions and to change anything in their life in favor of saving biodiversity, perhaps blaming the collective responsibility which is difficult to capture (for the problem of individual and collective responsibility and behavior a number of research projects have been conducted and synthesized in Haeberli et al., 2002). In a wider perspective, an environmental impact may be interpreted not only as the effect of inappropriate behavior, but also as the unexpected side effect of human activities. It may be argued that what people do or do not do is usually not decided according to environmental attitudes, but according to events in their private, professional, and public lives that primarily determine the actions people take. They usually make decisions in their daily lives based on their need, or desire, to be successful in their work, to drive to work, to produce goods, to go shopping or to enjoy their leisure time. They may not be blamed because they usually do not harm the environment on purpose. Whether the environment is attacked by those activities—and if so and how—is then dependent on subjective judgments and is supported or opposed by social groups, political strategies, and media reports (Haeberli et al., 2002; Kyburz-Graber, Hofer, & Wolfensberger, 2006).

It may be concluded that:

- exclusively blaming individuals for their thinking, feeling, and acting in view of how environmental problems develop and evolve is an overly simplifying strategy which can contribute to helplessness or point out the lacking political will of politicians to initiate democratic processes and to take adequate measures;
- people who are getting insight in the interdependence of society and ecology and who are improving their understanding of the ways how social and political processes develop are likely to be more critical participants in such processes; and
- generating knowledge about and within those interrelations in the context of real cases including real life complexities can provide excellent learning offers for students and teachers (see Kyburz-Graber, Hart, Posch, & Robottom, 2006; OECD-CERI, 1991, 1995).

Those conclusions constitute a key element of socioecological approaches to environmental education.

Socioecological Approaches Are Reflective in View of Learning Processes and Ways of Knowing

Education has traditionally been assigned the role of transmitting existing knowledge and social values that are widely approved and accepted by society (Posch, 2003). In the last years it has not only grown the awareness that commonly agreed and approved knowledge is becoming rare not least because of widespread pluralism and increasing knowledge production; but parallel to this, the knowledge transmission metaphor of learning has been replaced by a constructivist learning theory: learning is seen as an active knowledge construction process by the learner. The learning individual interacts in this process with knowledge provided from outside and contextual knowledge generated by him- or herself. The constructivist position emphasizes the fact that the individual is integrating new knowledge and experiences in existing structures in his or her mind, which have been implied through previous experiences, and learning processes. Bringing those aspects together, we claim that learning processes neither can be seen as uncritically accepting knowledge produced in unknown contexts, nor can it be seen as a passive knowledge reproducing process (Duffy & Jonassen, 1992; Gerstenmaier & Mandl, 1995; Knuth & Cunningham, 1993; Resnick, 1991).

Not astonishingly, there can be traced a coincidence of how the notion of learning has been developed in the last years into an individual and social reflectiveness, and the way how the notion of knowledge and knowledge production has developed into an understanding as a more critical and reflective interaction with information, not least in the context of environmental problems. This has also implications on the way that teachers and learners interact and how they both learn by sharing and exchanging knowledge.

Claiming that learning is an open and constructive, active, critical, and reflective process asks for a skeptical position against every attempt which tries to prove evidence for "best" educational approaches. Rather, it has to be assumed that environmental education pedagogy is highly contextual, depending on teachers' and students' previous experiences, on their local environments, school culture, and current societal trends. Thus, it might be more adequate to be interested in getting in-depth views of how environmental learning situations develop in this context, how teachers and learners get actively involved in social situations related to environmental problems, how they explore those issues and critically reflect about them.

Those are ways how socioecological approaches to environmental education set reflexivity on ways of knowing and processes of learning into educational practice.

Socioecological Approaches Are Critical in Ways of Looking at Environmental Problems and Environmental Education

So far, it has been shown that socioecological approaches to environmental education allow participants in educational settings to explore environmental problems as social constructions and subjective, contextual interpretations of the world. Searching beyond the surface about reasons, conditions, interests, and power relationships of people concerned evoke critical stances on how and why environmental situations are like they are. "Criticality" is thus another key element in environmental education related to socioecological approaches. It is a crucial part within the debate on environmental education as behavioral change or socially critical education (see, e.g., contributions in Mrazek, 1993; Robottom & Hart, 1993; Robottom & Sauvé, 2003; Sauvé, 1997; Walker, 1997).

"Critical" in the context of education may be related to discourses on "critical thinking" as well as to discourses on "critical pedagogy" (Wolfensberger, 2008). Wolfensberger traces the discourse on "critical thinking" back to historical roots of philosophy. In this tradition, the nature of "critical thinking" is seen as an intellectual process in which statements, arguments, or whole theories are approved in terms of coherence with logical deductions. In this tradition, John Dewey's work can be seen which is referred to in the evolving discourse on "critical thinking." In his book *How We Think* published in first edition 1910, John Dewey describes the key notion of "reflective thinking" as:

> that operation in which present facts suggest other facts (or truths) in such a way as to induce belief in what is suggested on the ground of real relation in the things themselves. (Dewey, 1933)

Thinking is seen as an ongoing process of perceiving, relating, and approving facts, beliefs, and values with the aim of developing comprehensive concepts of the world. John Dewey postulates a critical attitude as a precondition for reflective thinking in the sense that doubts and skeptical attitudes will induce questioning situations and reflecting problems.

Wolfensberger outlines how the "critical thinking" discourse has been taken up by Ennis (1962) and is reflected in numerous and ongoing endeavors to define concepts of "critical thinking" and develop strategies to promote them in educational settings (Brookfield, 1987; McPeck, 1981; Paul, 1990; Siegel, 1988). Critical thinking is defined as "reationale reflection" (Ennis, 1987), a combination of rationale and critical attitude (Siegel, 1988), questioning assumptions behind values, ideas, and actions (Brookfield, 1987), "reflective skepticism" linked with specialized knowledge (Brookfield, 1987; McPeck, 1981). In a similar way, Dubs claims that the main characteristic of "critical thinking in a wider sense" is reflection on basic assumptions and values (Dubs, 1992). "Critical thinking" according to Dubs is the highest and most comprehensive form of rationale thinking. In line with a number of environmental education researchers (Barnes & Todd, 1995; Lemke, 1990; Mortimer & Scott, 2000; Roth & Lucas, 1997; Wals, 1997) and from a discourse theoretical perspective, we claim that discourses in education can provide those environmental learning situations where critical reflection can take place.

While the discourses on "critical thinking" are related to philosophical traditions of "Critical Rationalism" and "Pragmatism," the "critical pedagogy" discourse on the other hand is linked to postulates of "Critical Theory" (Burbules & Berk, 1999). The common basis of the two conceptions of "critical" is the assumption that reality exists, but most people have more or less subjectively distorted views of it. A critical approach is aimed to engage people in questioning situations and analyzing multilayered perspectives on the world. According to the two positions of critique such inquiries may be conducted in different ways: "Critical thinking" may help people to rationally analyze existing knowledge and examine underpinning assumptions and beliefs. On the contrary, "critical pedagogy" claims that analyzing reality means to explore and expose structures in society that privilege certain social groups while other groups are excluded or oppressed. Social inequality, power relationships, interests guiding research and knowledge production, and emancipation are notions in the context of "critical pedagogy." While "critical thinking" is needed for critical analysis, "critical pedagogy" goes beyond it claiming that societal institutions producing and implementing knowledge, and related ideologies have to be questioned *and* transformed. "Critical pedagogy" links reflection and action, interpretation and social change (Burbules & Berk, 1999, p. 52; see also Karr & Kemmis, 1994).

Socioecological approaches to environmental education share with the two outlined positions of "criticality" the ontological assumption that there is a reality beyond human interpretation. But in contrast to the "critical thinking" and the "critical pedagogy" position the socioecological approaches follow the constructivist epistemological position that reality cannot be recognized itself. This is based on the constructivist assumption that knowledge on reality is an ongoing constructing and reconstructing process of all kind of actors according to their subjective experiences and their social, political, and cultural contexts. The aim of a socioecological approach to environmental education is to allow

> Young people to explore social issues in the real world by questioning values, perceptions, conditions and opinions. While doing this, they look behind and beyond the norms of social boundaries and critically question what is generally understood as objective scientific facts. (Kyburz-Graber, 1999, p. 416f)

Talking about "critically question what is generally understood as objective scientific fact" (Kyburz-Graber, 1999, p. 416) socioecological approaches adopt a position of the "critical thinking" philosophy. On the other hand, socioecological approaches adopt more the position of "critical pedagogy," criticizing the behaviorist individualistic change position in environmental education and arguing for an emancipative education aimed at self determination and empowerment for responsible judgment and action. Furthermore, the adoption of a "critical pedagogy" position in socioecological approaches is mirrored in the critique of social and scientific discourses on scientific and technological progress excluding inquiries on underpinning economic interests, ethical arguments for restrictions, and social inequalities. However, in contrast to the "critical pedagogy" position socioecological approaches do not ultimately claim that schools should initiate and participate in social change activities. They hold to a more pragmatic position regarding the role of education in society: environmental education should provide learning situations for critical inquiry and reflection and thus distinctively contribute to students' capacity to critically question, explore, and interpret social developments (for a newly launched debate on "schooling and environmental education in view of contradictions in purpose and practice," see Stevenson, 2007a, 2007b).

Following Wolfensberger (2008) in his reflections, it may be concluded that socioecological approaches neither are restricted to a "critical thinking" position excluding social change discourses, nor do they entirely adopt a "critical pedagogy" position radically demanding an active role of education in social transformation processes. The notion of "criticality" in socioecological approaches may be characterized as reflective, deliberative, pondering, discourse oriented in view of critically discussing observed as well as medially transmitted facts, generally agreed knowledge, ways of knowledge production; individual and social values, beliefs, interests and interpretations, and their implications for social developments. This notion of "criticality" is close to what Burbules and Berk (1999) describe as an "alternative criticality" in terms of thinking "outside a framework of conventional understandings" (p. 59) and "willingness, to engage in such conversations" (p. 61). As necessary conditions they see the motivation of the individuals but as well "challenging and supportive relations" and "contexts of difference that present us with the possibility of thinking otherwise" (Burbules & Berk, 1999, p. 62).

Socioecological Approaches Are Participatory in Learning and Research

Socioecological approaches are based on the assumption that learners are able to generate knowledge on complex environmental issues if they have the opportunity to explore, analyze, and interpret conditions, reasons, and expressions of human actions in real-life situations, getting in personal contact with people concerned and discussing possible developments.

Meaningful starting points for socioecological environmental education are real-life situations in which people are involved in their daily lives: family households, communities, political institutions, businesses, schools, supermarkets, recreation areas, etc. (Kyburz-Graber et al., 1997a; Kyburz-Graber, Hofer, & Wolfensberger, 2006). It has been shown in case studies on socioecological approaches in senior high schools that students and teachers as well as teachers and researchers come to manifold insights and are able to generate knowledge on environmental problems when they interact with real-life situations, reflect on values and value systems, explore conditions of action, work on possibilities for individual and structural change, and share and reflect their experiences and emerging knowledge (Kyburz-Graber, 2004; Kyburz-Graber et al., 2006).

Given the ontological and epistemological basis of socioecological approaches as constructive, reflective and critical, all kind of methodologies can be appropriate for inquiries on teaching and learning processes on environmental problems, depending on the focus of interest and research questions: narrative inquiry on experiences in real-life situations; phenomenological studies on the way how students and/or teachers experience a new role in teaching-learning settings on real-life experiences with environmental problems; ethnographic studies on a real-life situation where people are dealing with an environmental problem; action research concerning the way how people act in a specific situation.

A Case Study as an Example for a Research Study on Socioecological Approaches to Environmental Education

An example of a research study is a case analysis of five tandem-teams each formed of a biology and a history or philosophy teacher on secondary level who embarked on a socioecological curriculum development project taking three years in all. The research team invited the teachers to initiate a socioecological approach to a topic situated at the interface between natural science and society that was meaningful for the students in their actual context and could be linked to real-life situations regarding current private and public debates. Topics chosen were e.g., "Potential Benefits and Risks of New Biotechnologies"; "Consumption and Sustainable Development"; "Images of Nature."

The socioecological approach allowed the teachers to explore alternative ways of environmental education in the way that the students should be offered the opportunity to reflect on concurring perspectives regarding environmental problems, consider value judgments and interests, ways of knowing and uncertainties in knowledge production. The students' experiences with their focused inquiries on such questions and their reflections were shared in classroom discussions, which were video recorded, transcribed, and analyzed to get an in-depth view on contents and ways of discussion. The transcripts were analyzed in terms of descriptive statistics on interactions, arguments and contents, and by interpretative reconstruction as rich text (Wolfensberger, 2008; Kyburz-Graber et al., 2006).

Three topics were profoundly analyzed as case studies (Wolfensberger, 2008).

The classroom discussions evolved not surprisingly in highly different ways in terms of objectives, content, and interaction style. However, cross case analysis revealed a few similarities that allowed conclusions to be drawn regarding critical reflection on socioecological issues: The participants of classroom discussions were mostly concerned with raising facts related to the discussed topic and less with values, norms, feelings, and experiences.

Attempts to metacognitive reflection were visible, albeit only in few phases.

The interpretative reconstruction of the classroom discussions on "potential benefits and risks of new biotechnologies" revealed among participating students a concept of "division of labor" between natural science and social science/philosophy (Wolfensberger, 2008). Further, it produced an image of research that sees the researcher exclusively representing natural science and males, with an untamed urge to generate knowledge. Discussants expected that progress in natural science would contribute to determine ethical norms. Natural science research was attested the right of progress in favor of solving problems in the world. On the other hand, there were voices expressing the fear of an uncontrollable drive in research. Images of natural science and social science emerged also in classroom discussions of the project "images of nature." The reconstructed science images seem to attribute subjective, value, and culture based knowledge to social science, and objective value-free knowledge to natural science. Comparing the three case studies there was evidence that the students were engaged in lively discussions but did not go as much into depth in terms of a critical reflection that is associated with socioecological approaches to environmental education (Wolfensberger, 2008).

One might conclude that the claim of critical reflection on environmental issues is too sophisticated and not practicable except with high experienced students. Seen from a different angle, in the light of learning to become critical, it may be argued that socioecological approaches provide learning environments where students and teachers are challenged to interact with controversial real-life topics and reflect on causes and reasons.

The Potential of Socioecological Approaches to Environmental Education as Paradigmatic Response to Behavioral Change Orientations

A main feature of socioecological approaches is the critical reflection on environmental problems in a wide sense: questioning the nature of science and social sciences and its interfaces; questioning the ways that knowledge is produced, reflected, disseminated, and continuously interpreted; and questioning the interfaces between sciences and social questions.

Those questions seem to open up meaningful topical areas for exploring socioecological issues providing those "contexts of difference" and "communicative opportunities" (Burbules & Berk, 1999, p. 62) that can be seen as a prerequisite for critical approaches. This is a paradigmatically different account to environmental education compared to changing attitudes and behaviors (Robottom & Hart, 1993). Robottom and Hart (1993) critically point out that the changing behavior paradigm is built on the assumption that there is a common "view of what the world should be like and how such a state can be attained" (Robottom & Hart, 1993, p. 35). Socioecological approaches, in contrary, build on the assumption that there is no determined set of goals that describe how an environmentally sound and sustainable world should be. It is rather an ongoing social process of critical reflection on worldviews and on mainstream attempts to problem solving. Learners and teachers are part of this social process.

Critical reflection in the sense of socioecological approaches has to be engaged and supported. It may be assumed that students develop their ability to reflect critically on epistemological questions of knowledge production and the role of sciences embedded in social contexts if the ability is explicitly promoted and linked to experiences with real-life situations. Stating this, socioecological approaches are coming close to science education concepts that assign critical alternatives to traditional science education: the "nature of science" and "socioscientific issues" movements (Aikenhead, 1997; Geddis, 1991; Lederman, 1992; Lemke, 1990; Osborne, Erduran & Simon, 2004; Sadler & Zeidler, 2004). Both movements emphasize the reflection on ways of knowing and the interpretation of knowledge embedded in social and cultural contexts, and the distinction between observations and inferences.

However, critical reflection in socioecological approaches is considerably more than learning how to use arguments in discussions, and it goes beyond the reflection on ways of scientific knowledge production. It profoundly questions real-life situations in view of socially constructed human-nature interrelationships.

Socioecological approaches do not provide strategies of 'what works' but they show that teachers and students adopt a critical perspective if they work on local environmental problems (Kyburz-Graber et al., 2006), raising questions about the human-nature relation and the meaning of and the knowledge on "nature" in various scientific and cultural contexts, and in historical and future perspectives. Instead of learning isolated facts, students are encouraged to start socioecological inquiries in real-life situations in which people are involved in their daily lives, and where students will discover inconsistencies, controversial issues, contextually interpreted knowledge. They will start to reflect on interests, beliefs, values, and the basis of knowledge and power. And not least, they start thinking on "nature," they raise more or less unreflected metaphors linked with "nature," they start reflecting on interdependencies between culture, nature, and economy. In short: it seems that the nucleus of socioecological can be seen in critical reflection on interrelationships between how "nature" is conceived and valued on one hand, and political interests and knowledge production and dissemination in society, on the other hand. This includes not least the critical analysis of the role of natural science and the interface between natural science and society. Critical views on the ways societies come to know about environmental problems may reveal values, beliefs, and varieties of "truths" underpinning how problems are identified and approached.

It seems to be a promising perspective for future research to explore in what learning settings and how learners come to critically investigate and reflect environmental questions and what kind of contextual knowledge they gain through their inquiries. Linked to such learning situations there will emerge another range of challenging questions concerning the way in which teachers and whole schools develop conditions for socioecological learning processes and what supports or impedes them in their endeavors.

References

Agyeman, J. (Ed.), (2002). Culturing environmental education: From first nation to frustration. *Canadian Journal of Environmental Education, 7*(1), 5–12.

Aikenhead, G. S. (1997). Student views on the influence of culture on science. *International Journal of Science Education, 19*(4), 419–428.

Barnes, D., & Todd, F. (1995). *Communication and learning revisited: Making meaning through talk.* Portsmouth, NH: Boynton/Cook Publishers.

Bögeholz, S. (2006). Nature experience and its importance for environmental knowledge, values and action: Recent German empirical contributions. *Environmental Education Research, 12*(1), 65–84.

Bolscho, D., & Michelsen, G. (Eds.), (1999) Methods in environmental education research. *Methoden der Umweltbildungsforschung*. [Methods in environmental education research]. Opladen, Germany: Leske + Budrich.

Brookfield, S. (1987). *Developing critical thinkers.* San Francisco, CA/ London, UK: Joessey-Bass.

Burbules, N. C., & Berk, R. (1999). Critical thinking and critical pedagogy: Relations, differences, limits. In T. S. Popkewitz & L. Fendler (Eds.), *Critical theories in education. Changing terrains of knowledge and politics* (pp. 45–65). New York, NY, and London, UK: Routledge.

Bybee, R., & Mau, T. (1986). Science and technology related global problems: An international survey of science educators. *Journal of Research in Science Teaching, 23*(7), 599–618.

Chawla, L. (1998). Significant life experiences: A review of research on sources of environmental sensitivity. *The Journal of Environmental Education, 29*(3), 11–21.

Cross, R. T., & Price, R. F. (1996) Science teachers' social conscience and the role of controversial issues in the teaching of science. *Journal of Research in Science Teaching, 33*(3), 319–333.

Del Pilar Jimenes Silva, M. (1993). A psychological and ethical approach to environmental education. In R. Mrazek (Ed.), *Alternative paradigms in environmental education research* (pp. 239–248). Troy, OH: North American Association for Environmental Education.

Dewey, J. (1933). *How we think.* Boston, MA: D. C. Heath.

Dubs, R. (1992). Die Förderung des kritischen Denkens im Unterricht [Promoting critical thinking in the classroom]. *Bildungsforschung und Bildungspraxis, 14*(1), 28–56.

Duffy, T. M., & Jonassen, D. H. (1992). *Constructivism and the technology of instruction: A conversation*. Hillsdale, NJ: Lawrence Erlbaum.

Ennis, R. H. (1962). A concept of critical thinking. *Harvard Educational Review, 32*(1), 81–111.

Ennis, R. H. (1987). A taxonomy of critical thinking dispositions and abilities. In J. B. Baron & R. J. Sternberg (Eds.), *Teaching thinking skills: Theory and practice* (pp. 9–26). New York, NY: W. H. Freeman.

Eulefeld, G., Bolscho, D., Rode, H., Rost, J., & Seybold, H. (1993). *Entwicklung der Praxis schulischer Umwelterziehung in Deutschland* [Developments in environmental education in schools in Germany]. Kiel: IPN.

Eulefeld, G., Bolscho, D., Rost, J., & Seybold, H. (1988). *Praxis der Umwelterziehung in der Bundesrepublik Deutschland* [Practice in environmental education in Germany]. Kiel: IPN.

Fensham, P. (1978). Stockholm to Tbilisi—the evolution of environmental education. *Prospects, 8*(4), 446–455.

Geddis, A. N. (1991). Improving the quality of science classroom discourse on controversial issues. *Science Education, 75*(2), 169–183.

Gerstenmaier, J., & Mandl, H. (1995). Wissenserwerb unter konstruktivistischer Perspektive [Acquiring knowledge in a constructivist perspective] Zeitschrift für Pädagogik, 41, 867–888.

Giordan, A., & Souchon, C. (1991). *Une éducation pour l'environnement* [Education for the environment]. Nice: Z'éditions.

Gough, N. (1993). Narrative inquiry and critical pragmatism: Liberating research in environmental education. In R. Mrazek (Ed.), *Alternative paradigms in environmental education research* (pp. 175–197). Troy, OH: North American Association for Environmental Education.

Greenwood, D. A., & McKenzie, M. (Eds.), (2009). Context, experience and the socioecological: Inquiries into practice. *Canadian Journal of Environmental Education, 14*(1), 5–14.

Haeberli, R., Gessler, R., Grossenbacher-Mansuy, W., & Lehmann Pollheimer, D. (Eds.), (2002). *Vision Lebensqualität. Nachhaltige Entwicklung—ökologisch notwendig, ökonomisch klug, gesellschaftlich möglich* [The vision of life quality. Sustainable development-ecologically necessary, economically prudent, socially possible]. Final report of the Swiss priority program environment. Zürich, HB: vdf Hochschulverlag.

Hargrove, E. (2000). Toward teaching environmental ethics: Exploring problems in the language of evolving social values. *Canadian Journal of Environmental Education, 5*(1), 114–133.

Hart, P. (1993). Alternative paradigms in environmental education research: Paradigm of critically reflective inquiry. In R. Mrazek (Ed.), *Alternative paradigms in environmental education research* (pp. 107–130). Troy, OH: North American Association for Environmental Education.

Heid, H. (1992). *Ökologie als Bildungsfrage?* [Ecology as an educational question?]. *Zeitschrift für Pädagogik, 38*(1), 113–138.

Heimlich, J. E., & Ardoin, N. M. (2008). Understanding behavior to understand behavior change: A literature review. *Environmental Education Research, 14*(3), 215–237.

Hines, J. M., Hungerford, H. R., & Tomera A. N. (1986/1987). Analysis and synthesis of research on responsible environmental responsible environmental behavior: A meta-analysis. *The Journal of Environmental Education, 18*(2), 1–8.

Hirsch, G. (1995). *Beziehungen zwischen Umweltforschung und disziplinärer Forschung* [Relations between environmental research and discipline-related research]. *GAIA, 4*(5–6), 302–314.

Hungerford, H. R., & Volk, T. L. (1990). Changing learner behavior through envirnomental education. *The Journal of Environmental Education, 21*(3), 8–22.

Jensen, B. B., & Schnack, K. (1997). The action competence approach in environmental education. *Environmental Education Research, 3*(2), 163–178.

Jickling, B. (2004a). Making ethics an everyday activity: How can we reduce the barriers? *Canadian Journal of Environmental Education, 9*(1), 11–28.

Jickling, B. (Ed.), (2004b). Making ethics an everyday activity: How can we reduce the barriers? *Canadian Journal of Environmental Education, 9*(1), 11–26.

Karr, W., & Kemmis, S. (1994). *Becoming critical. Education, knowledge and action research.* Geelong, Victoria: Deakin University Press.

Knuth, R. A., & Cunningham, D. J. (1993). Tools for constructivism. In T. M. Duffy, J. Lowyk, & D. H. Jonassen, (Eds.), *Designing environments for constructive learning.* Berlin: Springer. Kapitel 13.

Kyburz-Graber, R. (1999). Environmental education as critical education: How teachers and students handle the challenge. *Cambridge Journal of Education, 29*(3), 415–432.

Kyburz-Graber, R. (2004). Does case-study methodology lack rigour? The need for quality criteria for sound case-study research, as illustrated by a recent case in secondary and higher education. *Environmental Education Research, 10*(1), 53–65.

Kyburz-Graber, R., Hart, P., Posch, P., & Robottom, I. (Eds.). (2006). *Reflective practice in teacher education. Learning from case studies of environmental education.* Bern: Peter Lang.

Kyburz-Graber, R., Hofer, K., & Wolfensberger, B. (2006). Studies on a socioecological approach to environmental education: A contribution to a critical position in the education for sustainable development discourse. *Environmental Education Research, 12*(1), 101–114.

Kyburz-Graber, R., Rigendinger, L., Hirsch Hadorn, G., & Werner Zentner, K. (1997a). A socioecological approach to interdisciplinary environmental education in senior high schools. *Environmental Education Research, 3*(1), 17–28.

Kyburz-Graber, R., Rigendinger, L., Hirsch Hadorn, G., & Werner Zentner, K. (1997b). *Sozio-ökologische Umweltbildung* [Socioecological environmental education]. Hamburg: Krämer.

Laugksch, R. C. (2000) Scientific literacy: A conceptual overview. *Science Education, 84*(1), 71–94.

Lederman, N. G. (1992). Students' and teachers' conceptions of the nature of science: A review of the research. *Journal of Research in Science Teaching, 29*(4), 331–359.

Lehmann, J. (1999) *Befunde empirischer Forschung zu Umweltbildung und Umweltbewusstsein* [Findings from empirical research on environmental education and environmental awareness]. Opladen: Leske + Budrich.

Lemke, J. (1990) *Talking science: Language, learning, and values.* Norwood, US: Ablex.

Lindemann-Matthies, P. (2002). The influence of an educational program on children's perception of biodiversity. *The Journal of Environmental Education, 33*, 22–31.

McPeck, J. (1981). *Critical thinking and education.* New York, NY: St. Martin's Press.

Mortimer, E., & Scott, P. (2000). Analysing discourse in the science classroom. In R. Millar, J. Leach, & J. Osborne (Eds.), *Improving science education. The contribution of research* (pp. 126–142). Buckingham, UK/Philadelphia, PA: Open University Press.

Mrazek, R. (Ed.). (1993). *Alternative paradigms in environmental education research.* Troy, OH: North American Association for Environmental Education.

Nickel. J., & Reid, A. (2006). Environmental education in three German countries: Tensions and challenges for research and development. *Environmental Education Research, 12*(1), 129–148.

O'Donoghue, R., & McNaught, C. (1991). Environmental education: The development of a curriculum through "grass-roots" reconstructive action, *International Journal of Science Education, 13*(4), 391–404.

OECD-CERI. (1995). *Environmental learning for the 21st century.* Paris: OECD.

OECD-CERI. (1991). *Environment, schools and active learning* (Vol. 17(2), pp. 31–40). Paris: OECD.

Osborne, J., Erduran, S., & Simon, S. (2004). Enhancing the quality of argumentation in school science. *Journal of Research in Scince Teaching, 41*(10), 994–1020.

Paul, R. (1990). *Critical thinking: What every person needs in a rapidly changing world.* Sonoma, US: Sonoma State University.

Posch, P. (2003). Reflective rationality and action research. In R. Kyburz-Graber, P. Posch, & U. Peter (Eds.), *Challenges in teacher education. Interdisciplinarity and environmental education* (pp. 63–80). Innsbruck/Wien: Studienverlag.

Ratcliffe, M. (2009). The place of socioscientific issues in citizenship education. In A. Ross (Ed.), *Human rights and citizenship education (pp. 12–16). London: CiCe.*

Ramsey, J., Hungerford, H., & Volk, T. (1989). The effects of issue investigation and action training on environmental issues. *Journal of Environmental Education, 21*(1), 26–30.

Reid, A., & Scott, W. (Eds.). (2006). Special issue: Researching education and the environment: Retrospect and prospect. *Environmental Education Research, 12*, 3–4.

Resnick, L. B. (1991). Shared cognition: Thinking as social practice. In L. B. Resnick, J. M. Levione, & S. D. Teasley (Eds.), *Perspectives on socially shared cognition* (pp. 1–20). Washington, DC: American Psychological Association.

Robottom, I. (Ed.)., (1987). *Environmental education: Practice and possibility.* Geelong, Victoria: Deakin University Press.

Robottom, I. (1993). Beyond ISM: Making environmental education research educational. In R. Mrazek (Ed.), *Alternative paradigms in environmental education research* (pp. 133–143). Troy, OH: North American Association for Environmental Education.

Robottom, I. (2003). Shifts in understanding environmental education. In R. Kyburz-Graber, P. Posch, & U. Peter (Eds.), *Challenges in teacher education. Interdisciplinarity and environmental education* (pp. 34–40). Innsbruck/Wien: Studienverlag.

Robottom, I., & Hart, P. (1993). *Research in environmental education. Engaging the debate.* Geelong, Victoria: Deakin University Press.

Robottom, I., & Sauvé, L. (2003). Reflecting on participatory research in environmental education. *Canadian Journal of Environmental Education, 8*(1), 111–128.

Roth, W. M., & Lucas, K. B. (1997). From "truth" to "invented reality": A discourse analysis of high school physics students' talk about scientific knowledge. *Journal of Research in Science Teaching, 34*(2), 145–179.

Sadler, T. D., & Zeidler, D. L. (2004). The morality of socioscientific issues: Construal and resolution of genetic engineering dilemmas. *Science Education, 88*(1), 4–27.

Sauvé, L. (1996). Environmental education and sustainable development: A further appraisal. *Canadian Journal of Environmental Education, 1*, 7–34.

Sauvé, L. (1997). L'approche critique en éducation relative à l'environnement [The critical approach to environmental education]. *Revue des sciences de l'éducation, 23*(1), 169–187.

Sauvé, L. (2005). Currents in environmental education: Mapping a complex and evolving. *Canadian Journal of Environmental Education, 10*(1), 11–37.

Sia, A. P., Hungerford, H. R., & Tomera, A. N. (1985/1986). Selected predictors of responsible environmental. *The Journal of Environmental Education, 17*(2), 31–40.

Siegel, H. (1988). *Educating reason, rationality, critical thinking, and education.* New York, NY/London, UK: Routledge.

Stapp, W. (1974). Historical setting of environmental education. In J. Stapp, W. B., Wals, A. E. J., & S. L. Stankorb (Eds.), (1996). *Environmental education for empowerment. Action research and community problem solving.* Dubuque, Iowa: Kendall/Hunt Publishing Company.

Stevenson, R. B. (2007a). Schooling and environmental education: Contradictions in purpose and practice. *Environmental Education Research, 13*(2), 139–153. (Original paper published 1987).

Stevenson, R. B. (Ed.). (2007b). Revisiting schooling and environmental education: Contradictions in purpose and practice. Editorial. *Environmental Education Research, 13*(2), 129–138.

United Nations Environmental Programme. (1972). *Report on the United Nations Conference on the Human Environment.* Geneva, Switzerland: Author.

United Nations Education, Scientific, & Cultural Organization—United Nations Environmental Programme. (1978). *Final Report Intergovernmental Conference on Environmental Education, Tbilisi 1977.* Paris: Author.

Volk, T., & McBeth, W. (1997). *Environmental literacy in the United States.* Washington, DC: North American Association for Environmental Education.

Walker, B., Carpenter, S., Anderies, J., Abel, N., Cumming, G. S., Janssen, M., . . . Pritchard, R. (2002). Resilience management in social-ecological systems: a working hypothesis for a participatory approach. *Conservation Ecology 6*(1), 14. Retrieved from http://www.consecol.org/vol6/iss1/art14/

Walker, K. (1997). Challenging critical theory in environmental education. *Environmental Education Research, 3*(2), 155–162.

Wals, A. E. J. (1993). Critical phenomenology and environmental education research. In R. Mrazek (Ed.), *Alternative paradigms in environmental education research* (pp. 153–174). Troy, OH: North American Association for Environmental Education.

Wals, A. E. J. (1997). Alternatives to national standards for environmental education: Process-based quality assessment. *Canadian Journal of Environmental Education, 2*(1), 7–27.

Wolfensberger, B. (2008). *Über Natur, Wissenschaft und Gesellschaft reden. Eine empirisch-qualitative Untersuchung von Klassengesprächen über Themen im Schnittbereich von Naturwissenschaften, Umwelt und Gesellschaft* [Classroom discussions on socioecological issues: three case studies from secondary school]. (Dissertation Thesis, Universität Zürich).

Young, O. R., Berkhout, F., Gallopin, G. C., Janssen, M. A., Ostrom, E., & Leeuw, S. V. D. (2006). The globalization of socioecological systems: An agenda for scientific research. *Global Environmental Change, 16*, 304–316.

Zeidler, D., Sadler, T. D., & Simmons, M. L. (2005). Beyond STS: A research-based framework for socioscientific issues education. *Science Education, 89*, 357–377.

Zelezny, L. (1999). Educational interventions that improve environmental behaviors: A meta-analysis. *The Journal of Environmental Education, 31*(1), 5–14.

4

Thinking Globally in Environmental Education

A Critical History

Noel Gough
La Trobe University, Australia

This essay offers a critical history of approaches to "thinking globally" in environmental education research,[1] with particular reference to the production, reproduction, and widespread circulation of assumptions about the universal applicability of modern Western science.[2] Although the transnational character of many environmental problems and issues demands that we "think globally," I argue that environmental education research could be enhanced by understanding Western science as one among many local knowledge traditions. Global knowledge production in/for environmental education might then be understood as creating conditions under which local knowledge traditions can be performed together, rather than as creating a global "common market" in which representations of local knowledge must be translated into (or exchanged for) the terms of a universal discourse.

Thinking Globally in Environmental Education: First Approximations

Think globally. Act locally. These familiar exhortations have circulated within the slogan system of environmental education for nearly four decades. They almost always appear as a pair, but environmental educators have not necessarily translated them into practice in comparable or commensurate ways. Many educational programs incorporate local action on environmental issues, but evidence of "thinking globally" is more elusive, equivocal, and problematic. Researchers can readily observe learners performing a school energy audit, participating in a recycling project, propagating locally indigenous plants to revegetate a degraded site, and so on. But what constitutes compelling evidence of learners, teachers, curriculum developers, and researchers "thinking globally?" I take a pragmatic approach, which is to clarify meanings by reference to consequences.[3] In this I follow Charles Sanders Peirce's (1931–58) explanation of how a pragmatic logic can be used to analyze the meaning of a concept:

> The word pragmatism was invented to express a certain maxim of logic . . . The maxim is intended to furnish a method for the analysis of concepts . . . The method prescribed in the maxim is to trace out in the imagination the conceivable practical consequences—that is, the consequences for deliberate, self-controlled conduct—of the affirmation or denial of the concept. (¶ 191)[4]

Pragmatically, the task of clarifying what environmental educators *mean* when they aspire to "thinking globally" requires an answer to the following question: in practical and performative terms, what imaginable, conceivable consequences follow from environmental educators affirming that they are "thinking globally"? However, as Cleo Cherryholmes (1993) notes, consequences "are of interest and can be assessed only in terms of purposes" (p. 3). Thus, we must also ask: what purposes does "thinking globally" serve in environmental education? How defensible are these purposes?

According to Ruth and William Eblen (1994), molecular biologist (and Nobel Laureate) René Dubos coined the phrase "think globally, act locally" in 1972, when he chaired the group of scientific experts advising the United Nations Conference on the Human Environment held in Stockholm.[5] We might thus interpret the launch of the UNESCO-UNEP[6] International Environmental Education Programme (IEEP) in 1974 as an early (post-Stockholm) manifestation of "thinking globally" in environmental education. This intergovernmental program sponsored many projects that promoted and supported local and regional educational action in response to concerns about the quality of the global environment. However, the consequences of "thinking globally" in such projects are subject to the differential power relations that accompany intergovernmental cooperation (or the appearance thereof) and some critics argue that the IEEP produced a neo-colonialist discourse in environmental education by systematically privileging Western (and especially the United States) interests and perspectives (see, e.g., Annette Gough, 1999).

By the mid-1980s, "think globally, act locally" had become an uncontested axiom of environmental education.[7] For example, *Earthrights: Education as if the Planet Really Mattered* (Greig, Pike, & Selby, 1987) invokes "think globally, act locally" as a taken-for-granted principle, without any citation of its author(ity). As their titles suggest, many other texts published in the late 1980s valorize variations on this principle, including *Global Teacher, Global Learner* (Pike & Selby, 1987), *Living in a Global Environment* (Fien, 1989), *Making Global Connections* (Hicks & Steiner, 1989), and the World Wide Fund for Nature's (WWF) Global Environmental Education Programme (e.g., Huckle, 1988).[8] These texts equate "thinking globally" with knowing and caring about the global dimensions and significance of environmental problems and issues; for example, John Huckle (1988) writes:

> Starting with products such as a tin of corned beef, a packet of potato crisps or a unit of electricity, teachers and pupils are encouraged to trace commodity chains and recognise their connections to such environmental issues as deforestation in Amazonia, the draining of wetlands in Britain and the debate over acid rain in Europe. (p. 2)

All of these texts infer that the purpose of "thinking globally" is to encourage learners and teachers to "recognize their connections" between their (local) experiences and conditions elsewhere in the world, and all draw attention to the far-reaching—and environmentally harmful—effects of Western imperialism, colonialism, and industrialization. However, they do not, for the most part, question the privileged status of the Western knowledge systems and ways of thinking within which their truth claims are produced. Some scholars in disciplines relevant to environmental education raise such questions, such as Lynn White (1967), whose influential appraisal of the historical roots of the twentieth century ecological crisis in the Christian Middle Ages questions the merits of Judeo-Christian attitudes to nature relative to other traditions. Although interest in non-Western worldviews, such as Buddhism, increased within the English-speaking world throughout the 1970s and '80s, there was little substantial scholarship on such matters until the publication of J. Baird Callicott and Roger Ames' (1989) edited collection, *Nature in Asian Traditions of Thought: Essays in Environmental Philosophy*, which compares Chinese, Japanese, Buddhist and Indian worldviews with those that predominate in the West.

Callicott and Ames's (1989) conclusions concerning the relative merits and reciprocal effects of Eastern[9] and Western environmental thought are, at least superficially, sympathetic to the former. For example, they assert that, "Eastern traditions of thought represent nature and the relationship of people to nature, in ways that cognitively resonate with contemporary ecological and environmental ideals" and that "the brute fact that environmental degradation is rampant in much of Asia" is best explained by the "intellectual colonization" of the East by the West (p. 279). Some of the assumptions underlying these conclusions are contestable (there is ample evidence of serious environmental destruction in parts of premodern Asia, and a number of Asian cultures have attempted to "conquer" nature in much the same ways that Westerners have done), but I am more concerned that Callicott and Ames (1989) ignore their own complicity in intellectually colonizing textual practices. One sign of intellectual colonization is what Susan Hawthorne (1999) calls the "unmarked category" (p. 121). For example, in the informational domains of the Internet, US addresses are unmarked but all other places are identified by the final term: au for Australia, sg for Singapore, za for South Africa, and so on. Unmarked cultural categories, such as whiteness in most Western societies, are especially troublesome for those of us who reside in them because they designate power and privilege. In Callicott and Ames's (1989) comparison of "Eastern traditions of thought" with "contemporary ecological and environmental ideals" (p. 279), the unmarked category is "Western": that is, they tacitly privilege contemporary *Western* "ecological and environmental ideals" as criteria that in some way validate Eastern philosophies (they also diminish "Eastern traditions of thought" by inferring that they are not "contemporary").

Thinking Globally in Environmental Education: Recent Positions

To bring my discussion of "thinking globally" into more recent times, I will focus on some of the ways in which this concept is deployed in a specific research publication, namely, *Environment, Education and Society in the Asia-Pacific: Local Traditions and Global Discourses*, edited by David Yencken, John Fien, and Helen Sykes (2000b). This book brings together some of the significant findings of a substantial comparative study of attitudes to nature and ecological sustainability, particularly among young people, in twelve sites in the Asia-Pacific region.[10] Some of the key questions explored in this research concern the relative influence of, and relationships between, local traditions and practices and global environmental discourses. Indeed, Yencken (2000) begins Chapter 1 by restating—and then inverting—Dubos's maxim:

> To protect the planet, we have long been told to think globally and act locally. But we can readily see that there are as many reasons to think locally and act globally. If we do not think locally, we may ignore rich sources of environmental knowledge and devalue local understanding and experience of environmental problems. If we do not act globally, we will never solve the big issues of the global commons: atmospheric and ocean pollution and the impacts of environmental degradation across national boundaries. Sustainability has many local and global dimensions. (p. 4)

Yencken (2000) provides a thoughtful and culturally sensitive overview of the various attitudes toward nature found in both the Eastern and Western nations of the Asia-Pacific region. He focuses not only on contemporary ecopolitical positions in the sites studied but also reviews the history of Western engagement with the environmental philosophies of Eastern cultures. He concludes by expressing his hopes for "the emergence of a global ideology of nature that transcends individual cultures" (p. 23):

> The environmental problems now facing the world are global problems stemming from the process of industrialization and capitalist development that has been taking place in every country, albeit at different speeds and intensities. We therefore need contemporary concepts to help frame both the nature of the problems and their likely solution . . . These concepts (sustainability, ecology, biodiversity, natural capital, intergenerational equity, precautionary principle and the like) and working models and techniques (metabolism, ecological footprint, natural step, environmental space, industrial ecology, etc.) need to gain widespread international acceptance. They should be developed cooperatively by scientists, environmental thinkers, local communities and others working hand in hand, with contributions from all cultures. (pp. 24–25)

Although Yencken clearly respects "contributions from all cultures," he nevertheless privileges (albeit implicitly) Western science as the prime source of the "contemporary concepts" that "need to gain widespread international acceptance." Many of the concepts that Yencken lists—ecology, biodiversity, metabolism—are already foreclosed by their production within Western scientific discourses, and so I find it difficult to imagine how they could be "developed cooperatively . . . with contributions from all cultures." Three assumptions underlying Yencken's position deserve critical scrutiny.

First, Yencken's use of the term "contemporary" is problematic, in part because some of the concepts to which he refers already have long histories in some cultures, such as the emphasis on intergenerational equity in the oral traditions of a number of Native American peoples. A more serious difficulty is that Yencken accompanies his conflation of "contemporary concepts" and Western science with the suggestion that they have displaced "traditional" ideas only in Western cultures. For example, Yencken (2000) describes the "great environmental awakening" and "new consciousness of Spaceship Earth" (p. 13) in the 1960s that led many people in Western industrialized nations to recognize that some of their environmentally damaging behaviors were rooted in Judeo-Christian traditions. Yencken (2000) cites research suggesting that these traditions of environmental thought have been superseded by a form of "contemporary environmentalism" that constitutes a "single cultural consensus about the environment" in the United States and Europe (p. 14). Although Yencken (2000) rejects attempts "to project Western priorities onto Eastern countries or Eastern traditions into Western cultures," he also asserts that "Western cultures undoubtedly have . . . much to learn from Asian traditional attitudes to nature in the same way that Eastern cultures have much to learn from Western environmentalism" (p. 25). In this formulation, *Western* environmentalism is tacitly "contemporary" but "traditional" attitudes to nature persist into the present in *Eastern* cultures. The unstated corollaries of these assertions are that Western cultures have little to learn from *contemporary* Asian attitudes to nature and that Eastern cultures have little to learn from *traditional* Western environmental thought, (which is not restricted to Judeo-Christian traditions, but also includes more ancient mythologies and archetypes such as those associated with the Green Man (see, e.g., Anderson, 1990).

A second difficulty with Yencken's (2000) formulation of "contemporary concepts" is the assumption that they can meaningfully be shared across cultures in ways that might be helpful in framing and resolving global environmental problems. For example, the term "ecology" does not command shared meaning even *within* the Western scientific communities that have shaped its conceptual development. To which and to whose "ecology" is Yencken referring? Many environmental education programs continue to privilege the particular variant of systems ecology pioneered by Eugene Odum (1953), such as the current (2005–2011) study design for the subject Environmental Science in the Victorian Certificate of Education (Victorian Curriculum and Assessment Authority, 2004), which presents an atomistic and reductionist view of large-scale ecosystem structure and function. For example, in Unit 1: The Environment, the first area of study is "Ecological components and interaction" and begins with this summary:

> The Earth's structure may be classified into four major categories: hydrosphere, lithosphere, atmosphere and biosphere. This area of study examines the processes occurring within the spheres of the Earth and the interactions that occur in and between the ecological components of each major category. (p. 12)

The second and third areas of study in Unit 1, "Environmental Change" and "Ecosystems" respectively, focus on the ecosystem as the primary unit for analysis. Neither the arbitrary categorical separation of the "spheres" nor the emphasis on ecosystems is consistent with many contemporary approaches to environmental analysis. For example, Donald Worster (1993) describes how many ecologists, over the previous two decades and more, have repudiated Odum's portrayal of orderly and predictable processes of ecological succession, yet this is an explicit item of curriculum content in environmental science. Studies such as those collected by Steward Pickett and P.S. White (1985) clearly demonstrate that the very concept of the ecosystem has receded in usefulness and, where the word "ecosystem" remains in use, that it has lost its former implications of order, equilibrium, and predictable succession.[11] Similarly, Andrew Jamison

(1993) documents "the failure of systems ecology to contribute very much to the actual solution of environmental problems" and concludes that "systems ecology today is only one (and not even the most significant one at that) of a number of competing ecological paradigms" (p. 202). Why does a school environmental science course in the year 2009 privilege an approach to ecology that many environmentalists have long regarded as a "failure?" And if there are "a number of competing ecological paradigms" within contemporary Western environmental science, how does Yencken see "ecology" functioning as a concept that might help to "frame both the nature of the problems and their likely solution" when it is at the same time a site of conceptual contestation?

These questions bring me to a third troubling aspect of Yencken's (2000) position, namely, his apparent belief in the *possibility*—and perhaps even the *necessity*—of a unitary and universal understanding of nature that "transcends individual cultures" (p. 23) and his apparent acceptance of Western science as the best approximation to such an understanding that humans have imagined to date. Yencken, Fien, and Sykes (2000a) elaborate this position in a subsequent chapter in which they are at pains both to recognize and respect feminist, postcolonialist, and multiculturalist critiques of modern Western science. Nevertheless, they maintain the view that a culturally transcendent environmental science is possible—that what they name as "science" provides the key to both thinking and acting globally. For example, Yencken et al. (2000a) assert: "It is generally accepted that most scientific research takes place within global theoretical assumptions" (p. 30). This is a seriously misleading statement, because many of the feminist, postcolonialist, and multiculturalist critiques that these authors claim to respect do *not* accept that the "theoretical assumptions" within which "most scientific research takes place" are "global." Indeed, one extreme way to characterize these critiques is to paraphrase Bruno Latour (1993) and assert that *we have never thought globally.*[12]

Western Science: Thinking Locally, Acting Imperially

Until relatively recently in human history, the social activities that produce distinctive forms of knowledge have for the most part been localized. The knowledges generated by these activities have thus carried what Sandra Harding (1994) calls the idiosyncratic "cultural fingerprints" (p. 304) of the times and places in which they were constructed. The knowledge signified by the English word "science" is no exception, but the global reach of European imperialism has given Western science the *appearance* of universal truth and rationality. Thus, many people (regardless of their locations) assume that it lacks the cultural fingerprints that are more conspicuous in knowledge systems that retain their ties to specific localities, such as the place-specific understandings of nature produced in many indigenous societies.[13] This occlusion of the cultural determinants of Western science contributes to what Harding (1993) calls an increasingly visible form of "scientific illiteracy," namely, "the Eurocentrism or androcentrism of many scientists, policymakers, and other highly educated citizens that severely limits public understanding of science as a fully social process":

> In particular, there are few aspects of the "best" science educations that enable anyone to grasp how nature-as-an-object-of-knowledge is always cultural . . . These elite science educations rarely expose students to systematic analyses of the social origins, traditions, meanings, practices, institutions, technologies, uses, and consequences of the natural sciences that ensure the fully historical character of the results of scientific research. (p. 1)

Over the past several decades, various processes of political, economic, and cultural globalization, such as the increasing traffic in trade, travel, and telecommunications crisscrossing the world, have made some multicultural perspectives on "nature-as-an-object-of-knowledge" more visible, including the indigenous knowledge systems popularized by terms such as the "wisdom of the elders" (Knudtson & Suzuki, 1992). The publication in English of studies in Islamic science (e.g., Sardar, 1989), and other postcolonial perspectives on the antecedents and effects of Western science (e.g., Petitjean, Jami, & Moulin, 1992; Sardar, 1988), raises further questions about the interrelationships of science and culture. However, globalization simultaneously (and contradictorily) encourages both cultural homogenization *and* the commodification of cultural difference within a transnational common market of knowledge and information that remains dominated by Western science, technology, and capital.

I suspect that many environmental education researchers are unaware of the subtle and insidious ways in which their textual practices sustain assumptions about the universality of Western science. That is why I focus much of my critical attention here on Yencken et al. (2000b), colleagues for whom I have nothing but respect, and whose respect for non-Western cultures is unequivocally sincere. Nevertheless, I argue that for all of their undeniably good intentions, these researchers maintain a culturally imperialistic view of science through the use of rhetorical strategies that privilege Western scientists' representations of "reality" and reproduce the conceit that the knowledge Western science produces is (or can be) universal.

For example, one way in which Yencken et al. (2000a) privilege Western science is to stipulate its uniqueness—"we depend on science for the formal analysis of the physical world and the monitoring of environmental change" (p. 32)—and to infer that its unique object ("the physical world") somehow renders it acultural: "*While* science is culturally shaped . . . environmental science is *nevertheless* dealing with physical reality" (p. 32, my emphasis).[14] Yencken et al. (2000a) clearly intend the word "formal" to signify something special about Western science, since they repeat and amplify this claim: "we rely on science for the formal analysis of environmental conditions and change. We have no more informed source to depend upon" (p. 33). Yencken et al. (2000a) imply a universal "we" but their assertions are culture-bound. Are they suggesting that non-Western knowledge traditions *ignore* "the formal analysis of the physical world" and are *not* concerned with "monitoring environmental change"? Or are they merely saying that non-Western analyses of the physical world and environmental change are "informal"? How do they distinguish between what is "formal" and what is not? What makes Western science an "informed source"? "Informed" by what (and/or by whom)?

Yencken et al. (2000a) overstate the uniqueness of Western science. David Peat's (1997) discussion of Blackfoot knowledge traditions demonstrates that Western cultures have no monopoly on forms of knowledge production that have the qualities that Yencken et al. attribute to "science." For example, Peat (1997) describes "the nature of Blackfoot reality" as "far wider than our own, yet firmly based within the natural world of vibrant, living things . . . a reality of rocks, trees, animals and energies":

> Once our European world saw nature in a similar way . . . [but] our consciousness has narrowed to the extent that matter is separated from spirit and we seek our reality in an imagined elsewhere of abstractions, Platonic realms, mathematical elegance, and physical laws.
>
> The Blackfoot know of no such fragmentation. Not only do they speak with rocks and trees, they are also able to converse with that which remains invisible to us, a world of what could be variously called spirits, or powers, or simply energies. However, these forces are not the occupants of a mystical or abstract domain, they remain an essential aspect of the natural, material world. (pp. 566–567)[15]

I am not suggesting that the Blackfoot view of reality is in any way superior (or inferior) to Western environmental science. My point is that Blackfoot people analyze the physical world (and more) and monitor environmental change in ways that are no less "formal" than ours. They, like us, are interested in "dealing with physical reality." They too rely on their knowledge traditions "for the formal analysis of environmental conditions and change."

Other cultures have developed ways of "dealing with physical reality" and "monitoring environmental change" that may be different, but no less effective, than those privileged in modern industrialized nations, and insinuating that they are neither "formal" nor "informed" serves no useful purpose. For example, David Turnbull (1991, 2000) points out that people from southeast Asia began to systematically colonize and transform the islands of the southwest Pacific some ten thousand years before the alleged "birth of civilization" (as Eurocentric historians describe it) in the Mediterranean basin. The Micronesian navigators combined knowledge of sea currents, marine life, weather, winds, and star patterns to produce a sophisticated and complex body of natural knowledge which, combined with their proficiency in constructing large sea-going canoes, enabled them to transport substantial numbers of people and materials over great distances in hazardous conditions. They were thus able to seek out new islands across vast expanses of open ocean and to establish enduring cultures throughout the Pacific by rendering the islands habitable through the introduction of new plants and animals. Although the knowledge system constructed by these people did not involve the use of either writing or mathematics, it is patronizing and indefensible to suggest that it was any less concerned with "physical reality" than Western science, or that it lacked a "formal analysis of environmental conditions."

Indeed, some anthropologists are convinced that indigenous people decipher "physical reality" using homologous assumptions to Western scientists, including a disposition to use systematic empirical inquiry as a means of revealing the inherent orderliness of nature. For example, Brent Berlin's (1990) field research suggests that the biological classification systems developed by many indigenous groups are "intellectualist"—that is, driven by curiosity about natural order and structure—rather than motivated only by a need to know which organisms are useful for practical purposes. Berlin therefore sees the difference between, say, Linnaean taxonomy and an indigenous classification system as chiefly one of degree: assisted by European imperialism, Linnaeus had access to a much larger sample of organisms than taxonomists who sampled relatively small sites and classified fewer organisms. But, given the vast numbers of organisms populating the earth, no system of classification—including contemporary Western phylogenies—can claim universality. Reviewing a number of similar anthropological studies, Susantha Goonatilake (1998) concludes:

> The world, it appears, is thus littered with indigenous starting points for potential trajectories of knowledge—trajectories which, if they were developed, would have led to different explorations of physical reality. The existence of all this anthropological evidence does not solve the problem of Western ethnocentricity or of the distinctive rise of Western science, but it does help to further problematize them. (pp. 70–71)

If the knowledge produced by Western scientists is applied only in cultural sites dominated by Western science, then their claim to its universality might be a relatively harmless conceit. However, attempts to generate global knowledge of environmental problems, such as climate change, draw increasing attention to the cultural biases and limits of Western science.

For example, Brian Wynne (1994, pp. 172–173) reports that up to the early 1990s the Intergovernmental Panel on Climate Change (IPCC) used models that equated global warming mainly with carbon emissions and largely ignored other factors such as cloud behavior, marine algal fixing of atmospheric carbon, and natural methane production. Western scientists and policy makers represented the IPCC models as a means for producing universally warranted conclusions, whereas many non-Western observers saw them as reflecting the interests of industrialized nations in obscuring the exploitation, domination, and social inequities underlying global environmental degradation. But if global warming is understood as a problem for *all* of the world's peoples, then we need to find ways in which *all* of the world's knowledge systems—Western, Blackfoot, Islamic, whatever—can jointly produce appropriate understandings and responses. I will not presume to suggest (indeed, I cannot imagine) what a Blackfoot or Islamic contribution to such jointly produced knowledge might be, but I am willing to assert that a coexistence of knowledge systems is unlikely to be enabled by the adherents of any one local knowledge tradition claiming that we *must* "rely" and "depend" on theirs.

The successive failures of the Kyoto Climate Change Summit in December 1997, The Hague World Conference on Climate Change in November 2000 and, most recently, the fifteenth United Nations Climate Change Conference in Copenhagen in December 2009, to reach effective transnational agreements on limiting greenhouse gas emissions demonstrate the difficulty of turning the rhetoric of "thinking globally" into tangible environmental action. Press reports from these conferences demonstrate how deeply the putative "global science" of climate change is enmeshed in local contexts, even among Western nations. This is not just because the conclusions Western scientists draw about aspects of global warming—such as how the vegetation in forests and farm crops function as "carbon sinks"—are contradictory or controversial, but also because the same "scientific facts" produce different meanings for different people. Thus, for example, Simon Mann (2000) reports that "the definition of a forest" was among the areas of disagreement that preoccupied negotiators at The Hague conference for several days, and at the time of writing this essay more than 1,400 documents addressing the definition of a forest can be found on the United Nations Framework Convention on Climate Change (UNFCCC) website.[16] I suspect that the impulse to attempt such a definition results from the false hope that some useful scientific truth claims can be made about all forests in the world, and their effects on atmospheric warming, regardless of their location. But each forest's local history and contingencies will uniquely determine the quantities of atmospheric carbon it fixes and solar heat it absorbs and radiates.

However, as an environmental educator I am less concerned about the warrant for Western scientific knowledge of the relationship between global warming and, say, atmospheric carbon fixing by vegetation, than with the conflation of Western science and "global science." Press reports and educational texts alike give the impression that the concept of "carbon sink" is now a legitimate component of "thinking globally" (and scientifically) about climate change. For example, one of the required outcomes of Unit 3: Ecological Issues: Energy and Biodiversity, in Victoria's Year 12 Environmental Science subject (Victorian Curriculum and Assessment Authority, 2004), is that students "should be able to describe the principles of energy and relate them to the contribution of one fossil and one nonfossil energy source to the enhanced greenhouse effect" (p. 19). To achieve this outcome, students must demonstrate knowledge of, among other things, "options for reducing the enhanced greenhouse effect, including National Greenhouse Strategy, Kyoto protocol, increasing energy efficiency, emission trading and vegetation sinks" (p. 20). This list implies that "emission trading and vegetation sinks" have some equivalence or comparability with "increasing energy efficiency" as "options for reducing the enhanced greenhouse effect," which gives them a global legitimacy they do not deserve. The "scientific facts" of carbon fixing by plants do not justify the metaphorical representation of forests as carbon "sinks." The "sink" metaphor is a rhetorical device for recruiting "scientific facts" to assist the political efforts of industrialized nations to discount their greenhouse gas emissions.

The authors of Victoria's Year 12 Environmental Science study design insinuate that terms such as "emission trading" and "carbon sinks" have global currency—that they are part of the semiotic apparatus that supports "thinking globally." But emission trading and carbon sinks are terms for thinking locally—terms that allow Western politicians and bureaucrats to represent mysterious[17] physical realities in the familiar language of economic rationalism. Examples such as these lead me to dispute Yencken et al.'s (2000a) claims, quoted previously, that "we depend on science for the formal analysis of the physical world and the monitoring of environmental change" and that "while science is culturally shaped . . . environmental science is nevertheless dealing with physical reality" (p. 32). We cannot depend on Western science alone because environmental science deals not only with physical reality but also with "culturally shaped" representations of this reality. Pretending that these representations are acultural is an imperialist act—an act of attempted intellectual colonization.

How Can We Think Globally?

My story so far is a cautionary tale. In Jon Wagner's (1993) terms, I have tried to identify some of the "blind spots and blank spots" that configure the "collective ignorance" of Western-enculturated environmental education researchers as we struggle to enact defensible ways of thinking globally.[18] In Wagner's schema, what we "know enough to question but not answer" are blank spots; what we "don't know well enough to even ask about or care about" are blind spots—"areas in which existing theories, methods, and perceptions actually keep us from seeing phenomena as clearly as we might." Much of the research reported by Yencken et al. (2000b) and their coresearchers clearly responds to blank spots in our emerging understandings of the complexities that arise from the interreferencing of local traditions and global discourses of environmental education. My principal concern here is with the blind spots that might still remain in the vision of even the most culturally sensitive scholars. The detectable traces of Western scientific imperialism in Yencken et al.'s (2000b) work underscore the difficulties we face when we attempt, as Patti Lather puts it, "to decolonize the space of academic discourse that is accessed by our privilege" (quoted in Pinar & Reynolds, 1992, p. 254). How can we think globally *without* enacting some form of epistemological imperialism?

As Lorraine Code (2000) observes, "addressing epistemological questions along a local-global spectrum raises timeworn questions about relativism versus absolutism" (p. 68). For example, David Hess (1995) argues that understanding science and technology in a multicultural world demands that we think in terms of "social constructivism" and "cultural relativism," but he explicitly rejects the need to invoke epistemological, metaphysical, or moral relativism. However, Code (2000) argues that "responsible global thinking *requires* not just cultural relativism but a *mitigated epistemological relativism* conjoined with a 'healthy skepticism' " (emphases in original):

> I am working with a deflated conception of relativism remote from the "anything goes" refrain which antirelativists inveigh against it. It is "mitigated" in its recognition that knowledge-construction is always constrained by the resistance of material and human-social realities to just any old fashioning or making. Yet . . . it is relativist in acknowledging "the plurality of criteria of knowledge . . . and deny[ing] the possibility of knowing absolute, objective, universal truth."[19] Its "healthy skepticism" in this context manifests itself in response to excessive and irresponsible global pretensions, whose excesses have to be communally debated and negotiated with due regard to local specificities and global implications. (p. 69)

Code's "mitigated epistemological relativism" resembles the "constrained constructivism" advocated by Katherine Hayles (1993): "within the representations we construct, some are ruled out by constraints, others are not" (p. 33); by ruling out some possibilities, "constraints enable scientific inquiry to tell us something about reality and not only about ourselves" (p. 32). Hayles emphasizes that constraints do not (and cannot) tell us what reality *is*, but they enable us to distinguish representations that are consistent with reality from those that are not. For example, Newtonian mechanics represents gravity as a mutual attraction between masses whereas Einstein's general theory of relativity represents it as an effect of the curvature of space. Hayles (1993) refers also to a Native American belief that objects fall because the spirit of Mother Earth calls out to kindred spirits in other bodies. But no representation of gravity that, as Code puts it, is "constrained by . . . material and human-social realities," would predict that when someone steps off a cliff s/he will remain suspended in midair. Different cultures interpret these constraints in different ways, but they operate multiculturally—and globally—to eliminate some constructions. Hayles (1993) notes that for any given phenomenon there will always be other representations, unknown or unimaginable, that are consistent with reality:

> The representations we present for falsification are limited by what we can imagine, which is to say, by the prevailing modes of representation within our culture, history, and species . . . Neither cut free from reality nor existing independent of human perception, the world as constrained constructivism sees it is the result of active and complex engagements between reality and human beings. Constrained constructivism invites—indeed cries out for—cultural readings of science, since the representations presented for disconfirmation have everything to do with prevailing cultural and disciplinary assumptions. (pp. 33–34)

As I argue elsewhere in greater detail (Gough, 2007), constrained constructivism offers a philosophical position that problematizes the nondiscursive "reality" of nature without collapsing into antirealist language games. Constrained constructivism is not "anything goes" but neither does it disallow representations that fail to meet criteria that disguise their Eurocentric and androcentric biases behind claims for universality. But, as my discussion of systems ecology demonstrates, many environmental educators often seem to do the precise opposite of what Hayles suggests by requiring learners to *confirm* representations that conform to "cultural and disciplinary assumptions" that no longer prevail even in the West.[20]

The literatures that I find most useful for thinking critically about "thinking globally"—and about the articulations between global and local knowledge production—are broadly speaking those that Harding (1998) calls post-Kuhnian and postcolonial science and technology studies,[21] especially Turnbull's (1994, 1997, 2000) work. From the postcolonialist and anti-imperialist standpoints that Harding and Turnbull share, all knowledge systems (including Western science) are always situated and constituted initially within specific sets of local practices, conditions and cultural values.[22] Turnbull's (1997) approach focuses particular attention on the *activities* involved in producing knowledge in particular social spaces, that is, on the "contingent processes of making assemblages and linkages, of creating spaces in which knowledge is possible" (p. 553).

Turnbull uses numerous examples to demonstrate how particular knowledge spaces are constructed from heterogeneous assemblages of people, skills, local knowledge and equipment linked by various social strategies and "technical devices which may include maps, templates, diagrams and drawings, but are typically techniques for spatial visualisation" (p. 553).[23] A major analytic advantage of Turnbull's spatialized perspective is that, because all knowledge systems have localness in common, many of the small but significant differences between them can be explained in terms of the different types of work—of *performance*—involved in constructing assemblages of people, practices, theories, and instruments in a given space. Some knowledge traditions move and assemble their products through art, ceremony, and ritual,[24] whereas Western science's accomplishments result from forming disciplinary societies, building instruments, standardizing techniques, and writing articles. Turnbull (1997) concludes that each form of knowledge production entails processes of knowledge assembly through "making connections and negotiating equivalences between the heterogeneous components while simultaneously establishing a social order of trust and authority resulting in a knowledge space" (p. 553). This spatialized analytic perspective provides a basis for comparing and framing knowledge traditions.

Turnbull (2000) analyzes knowledge construction among diverse groups of people in different locations and times, including medieval masons, Polynesian navigators, cartographers, malariologists, and turbulence engineers. He demonstrates that, in each case, their achievements are better understood performatively—as diverse, messy, contingent, unplanned, and arational combinations of social and technical practices—rather than as the result of logical, orderly, rational planning or a dependence on internal epistemological features to which "universal" validity can be ascribed. The purpose of Turnbull's emphasis on analyzing knowledge systems comparatively in terms of spatiality and performance is to find ways in which diverse knowledge traditions can coexist rather than one displacing others or being absorbed into an imperialist archive. Two examples of Western scientists attempting to displace knowledge spaces constructed by Indonesian rice farmers demonstrate the significance of Turnbull's (1997) analysis for "thinking globally" in environmental education research.

The development of high-yielding rice varieties in the 1960s shifted Indonesia from being a net importer of rice to being one of the world's largest rice exporters. In Java this necessitated the use of massive quantities of fertilizers and pesticides and abandoning indigenous rice strains. However, insect pests began to reach plague proportions in the monocrop environment and increasing the amounts of pesticides made the problem worse. The solution was to ban fertilizer and pesticide imports and to introduce "integrated pest management," which Turnbull (1997) describes as an:

> approach to pest control which recognises there will always be pests and the best way to manage them is to ensure that the populations of competing insects remain in balance. For this system to work, the local farmers had to become local experts, they had to monitor the insect populations on their own farms and to use locally appropriate rice strains (p. 599).

A somewhat similar reversal occurred in Bali where rice had traditionally been grown under an irrigation system controlled by the temples. The Indonesian government believed this to be old fashioned and superstitious and introduced Western scientific methods of water control and distribution as Turnbull (1997) notes:

> The result was the same as in Java: initial success followed by a crash in production. So they brought in more Western experts, but this time they included a rather unusual anthropologist and a computer expert. Between them they were able to show on the computer screen how the old system of temple control worked and why it was the most efficient. This resulted in the knowledge and power being given back to the local people while satisfying the central government's yen for high-tech solutions. (pp. 559–560)

These examples suggest that the globalization of knowledge production depends on creating spaces in which local knowledge traditions can be "reframed, decentred and the social organization of trust can be negotiated" (Turnbull 1997, pp. 560–561)—spaces created through "negotiation between spaces, where contrasting rationalities can work together but without the notion of a single transcendent reality" (Turnbull 2000, p. 228).[25] The production of such spaces is, in Turnbull's (1997) view, "crucially dependent" on "the reinclusion of the performative side of knowledge":

> Knowledge, in so far as it is portrayed as essentially a form of representation, will tend towards universal homogenous information at the expense of local knowledge traditions. If knowledge is recognised as both representational and performative it will be possible to create a space in which knowledge traditions can be performed together. (pp. 560–561)

Turnbull invites us to be suspicious of importing and exporting representations that are disconnected from the performative work that generated them. For example, representing forests as "carbon sinks" arises in Western industrialized nations because their emissions of greenhouse gases are of sufficient magnitude to motivate and make meaningful the work of producing "sinks" to which excessive atmospheric carbon can be removed. The resistance of some developing nations to accepting carbon sinks as a way for Western nations to discount their greenhouse gas emissions is only to be expected, because the "sink" metaphor has no cultural purchase in their localities. Global knowledge must therefore be coproduced—or as Yencken (2000) puts it, "developed cooperatively"—but its legitimacy cannot be tied to any one culture's social and political traditions for conferring legitimacy on knowledge construction.

If we think about coproducing knowledge in transcultural spaces, it becomes clear that some of the most revered processes of Western knowledge production will not necessarily appear to be trustworthy. For example, many of the truth claims that constitute Western scientific knowledge of nature are produced under laboratory conditions.[26] But, as Code (2000) argues, developing "methodological strategies for ecologicallyframed global thinking" requires a more "naturalized" epistemology than laboratory work assumes:

> I maintain that the laboratory is neither the only nor the best place for epistemologists to study "natural" human knowing in order to elaborate epistemologies that maintain clearer continuity with cognitive experiences—"natural knowings"—than orthodox *a priori*-normative epistemologies do. I advocate turning attention to how knowledge is made and circulated in situations with a greater claim to the elusive label "natural." My interests are in ways of gathering empirical evidence and in assumptions about the scope of evidence as it plays into regulative theories. My contention, briefly, is that evidence gathered from more mundane sites of knowledge production can afford better, if messier, starting points for naturalistic inquiry than much of laboratory evidence, for it translates more readily into settings where knowing matters in people's lives and the politics of knowledge are enacted. (p. 71)[27]

For example, despite claims for the "objectivity" of experimental methods, the methodological principle of controlling variables produces knowledge that can be incomprehensible in locations where this principle is not taken for granted. Again, as Code (2000) notes: "Descriptions, mappings, and judgments that separate evidence from extraneous "noise" are always value-saturated, products of some one's or some group's location and choice; hence always contestable" (p. 71).

In light of the above considerations, I suggest that "thinking globally" in environmental education research might best be understood as a process of constructing transcultural "spaces" in which scholars from different localities collaborate in reframing and decentering their own knowledge traditions and negotiate trust in each other's contributions to their collective work. For those of us who work in Western knowledge traditions, a first step must be to represent and perform our distinctive approaches to knowledge production in ways that authentically demonstrate their localness. We might not be able to speak—or think—from outside our own Eurocentrism, but we can continue to ask questions about how our specifically Western ways of "acting locally" (in the production of knowledge) might be performed *with* other local knowledge traditions. By coproducing global knowledge in transcultural spaces, we can, I believe, help to make both the limits *and strengths* of the local knowledge tradition we call Western science increasingly visible.

Acknowledgments

I thank Terry Carson, Annette Gough, and Bob Jickling for constructive comments on earlier drafts of this essay.

Notes

1. Quoting William Reid (1981), I understand research to be "any means by which a discipline or art develops, tests, and renews itself" (p. 1).
2. I realize that this formulation— "modern *Western* science" rather than just "science" or "modern science"—introduces a problematic "West versus the rest" dualism that might appear to overlook the historical influences of other cultures (e.g., Islamic, Indian, Chinese, etc.) on its development. However, I also want to emphasize that I am referring to the "science" that was uniquely coproduced with industrial capitalism in a particular time/place (seventeenth century northwestern Europe) and to the cultural characteristics of that enterprise that have endured to dominate Western (and many non-Western) understandings of science as a result of Euro-American imperialism.
3. Although my methodology is broadly situated within the discipline of analytic philosophy, I distance myself from what Michael Peters (2004) calls "the conservatism, apoliticism and ahistoricism of analytic philosophy that has denied its own history until very recently" (p. 218). Rather, I am disposed toward Gilles Deleuze and Félix Guattari's (1994) conceptualization of philosophy as *geophilosophy*, which complicates philosophical questions by tying them to their historical and spatial specificities.
4. This passage is quoted from the draft of a book review that Peirce wrote circa 1904. H. Standish Thayer (1981) regards the long paragraph from which this passage is taken as "the clearest and most complete single statement of what pragmatism is that Peirce ever wrote" (p. 493).
5. Many other sources identify Dubos as the author of this phrase, but the Eblens add weight to their claim by including his 1972 essay, "Think Globally, Act Locally," in their *Encyclopedia of the Environment*.
6. UNEP: United Nations Environment Programme.
7. Environmental educators were not alone in consolidating this aphorism. In *Harvard Business Review,* Theodore Levitt (1983) used "Think global. Act local" in arguing that "the globalization of markets is at hand" (p. 92). Of course, the imperative to think globally has a longer history. For example, in 1967 Marshall McLuhan noted that with the advent of an electronic information environment, "all the territorial aims and objectives of business and politics [tend] to become illusory" (McLuhan & Fiore, 1967, p. 5). A Google search reveals the extent to which "Think global. Act local" continues to be a popular trope in environmental, economic, and other discourses.
8. For convenience, I cite only one volume in the WWF Global Environmental Education Programme, which consists of four multi-volume modules.
9. "Eastern" appears to be Callicott and Ames's shorthand for a number of "Asian traditions of thought." As an "Other" of Western it is as problematic as "Oriental" (see, e.g., Said, 1978).
10. The sites were: Australia, Brunei, Fiji, India, Indonesia, Japan, New Zealand, Papua New Guinea, the Philippines, Singapore, South China and Thailand.
11. See also Robert Ulanowicz (1997, 2009), who emphasizes that chance, disarray, and randomness are necessary conditions for creative advance, emergence and autonomy in the natural world.
12. This deliberately provocative formulation is inspired by the title of Latour's (1993) book, *We Have Never Been Modern*. I am usually reluctant to use terms like "we" (which implies that I can speak for others) and "never" (which suggests an absolutism that I cannot defend), but I believe that the provocation is defensible in the circumstances I describe.
13. See, for example, Tom Jay's (1986) eloquent account of the place of salmon in the lives of Northwest Coast Native Americans, in which he shows how salmon stories incorporate moral understandings of self, community, earth, and the interrelationships among them.
14. I suggest that the rhetorical effect of the words I emphasize here (*"While . . . nevertheless"*) is to invite readers to accept (without further explanation) that "dealing with physical reality" somehow sets limits on the extent to which knowledge is "culturally shaped."

15. Although Peat (1997) refers to "Blackfoot physics," he clearly understands that Blackfoot people do not fragment their understandings into specialized categories such as "physics"—a term that might have no equivalent in Blackfoot vocabulary. Carol Geddes, a Yukon First Nation woman, asserts a similar perspective in relation to school curricula: "We would never have a subject called environmental ethics; it is simply part of the story" (quoted in Wren et al., 1995, p. 32).
16. http://unfccc.int (accessed January 5, 2010)
17. I use the term "mysterious" because I suspect that very few of the people involved in negotiating political positions on emission trading and on discounting emissions by counting carbon sinks have even a rudimentary understanding of the molecular biology and cellular physiology of atmospheric carbon fixing by plants.
18. I realize that the term "collective ignorance" is provocative and might even be offensive to some of my colleagues, which is why I have deliberately used "we" in this sentence so as to include myself in this accusation.
19. The quoted words are Peter Novick's (1988, p. 167).
20. I am aware that expressions such as "mitigated epistemological relativism," "constrained constructivism," and "the representations we present for falsification" might only make sense within certain modes of Western scholarship to which I have access and which my own education has privileged. The very awkwardness of some of these locutions exemplifies the difficulties we (Western scholars) face in representing the complexities toward which they gesture. For example, although I agree with the spirit (as I interpret it) of Hayles' assertion that "constrained constructivism invites—indeed cries out for—cultural readings of science," it only "works" for cultures in which the distinctive conceptual category of "science" exists.
21. From some standpoints, "postcolonial science and technology studies" might seem to be an oxymoron, but I suggest that this can be avoided by adopting a nontotalizing view of postcolonialism. As Helen Verran (2001) writes: "Postcolonialism is not a break with colonialism, a history begun when a particular 'us,' who are not 'them,' suddenly coalesces as opposition to colonizer . . . Postcolonialism is the ambiguous struggling through and with colonial pasts in making different futures" (p. 38).
22. There are some subtle and thought-provoking differences between Harding's and Turnbull's positions. Harding emphasizes the universalizing tendencies that accompany the "travel" of knowledges beyond the localities in which they were initially produced, whereas Turnbull is more concerned with how trust is established between heterogeneous knowledges that "arrive" (or are produced) in the same space. For an extended discussion of these differences see Gough (2003).
23. Turnbull's linking of social strategies and technical devices is consistent with Latour's (1992) contention that there are no purely "social" relations; rather, there are "sociotechnical" relations, embedded in and performed by a range of different materials—human, technical, "natural," and textual.
24. See also Jim Cheney's (1999) interpretations of the role of stories, ceremonies and rituals in the intergenerational passing down of "modes of action" (p. 149) in Native American communities.
25. Stanley Jeyaraja Tambiah (1990) and Edward Soja (1996) name the type of space that Turnbull envisages as "a third space," whereas Homi Bhabha (1994) calls it "an interstitial space" (p. 312).
26. I write "under laboratory conditions" rather than "in laboratories" because Western scientists typically try to create (or assume) laboratory conditions wherever they work. Indeed, Latour (1983) notes that a large proportion of national budgets for scientific activity is contributed to supporting international agencies that maintain standard weights and measures so that, in effect, the world at large can be treated as a giant laboratory.
27. On the idea of "messy" starting points for inquiry see also John Law (2004).

References

Anderson, W. (1990). *Green man: The archetype of our oneness with the earth*. San Francisco, CA: Harper Collins.

Berlin, B. (1990). The chicken and the egg-head revisited: Further evidence for the intellectualist bases of ethnobiological classification. In D. A. Posey & W. L. Overal (Eds.), *Ethnobiology: Implications and applications. Proceedings of the first international congress of ethnobiology* (Vol. 1, pp. 19–35). Belem, Brazil: Museo Emilio Goeldi.

Bhabha, H. K. (1994). *The location of culture*. New York, NY: Routledge.

Callicott, J. B., & Ames, R. T. (Eds.). (1989). *Nature in Asian traditions of thought: Essays in environmental philosophy*. Albany, NY: State University of New York Press.

Cheney, J. (1999). The journey home. In A. Weston (Ed.), *An invitation to environmental philosophy* (pp. 141–167). New York, NY: Oxford University Press.

Cherryholmes, C. (1993). Reading research. *Journal of Curriculum Studies, 25*(1), 1–32.

Code, L. (2000). How to think globally: Stretching the limits of imagination. In U. Narayan & S. Harding (Eds.), *Decentering the center: Philosophy for a multicultural, postcolonial, and feminist world* (pp. 67–79). Bloomington, IN and Indianapolis, IN: Indiana University Press.

Deleuze, Gs, & Guattari, F. (1994). *What is philosophy?* (G. Burchell & H. Tomlinson, Trans.). London, UK: Verso.

Eblen, R. A., & Eblen, W. R. (Eds.), (1994). *The encyclopedia of the environment*. Boston, MA and New York, NY: Houghton Mifflin.

Fien, J. (Ed.) (1989). *Living in a global environment: Classroom activities in development education*. Brisbane: Australian Geography Teachers Association.

Goonatilake, S. (1998). *Toward a global science: Mining civilizational knowledge*. Bloomington, IN and Indianapolis, IN: Indiana University Press.

Gough, A. (1999). Recognising women in environmental education pedagogy and research: Toward an ecofeminist poststructuralist perspective. *Environmental Education Research, 5*(2), 143–161.

Gough, N. (2003). Thinking globally in environmental education: implications for internationalizing curriculum inquiry. In W. F. Pinar (Ed.), *International handbook of curriculum research* (pp. 53–72). Mahwah, NJ: Lawrence Erlbaum Associates.

Gough, N. (2007). All around the world: science education, constructivism, and globalisation. In B. Atweh, M. Borba, A. C. Barton, N. Gough, C. Keitel, C. Vistro-Yu, & R. Vithal (Eds.), *Internationalisation and globalisation in mathematics and science education* (pp. 39–5). Dordrecht: Springer.

Greig, S., Pike, G., & Selby, D. (1987). *Earthrights: Education as if the planet really mattered*. London, UK: World Wildlife Fund and Kogan Page.

Harding, S. (1993). Introduction: Eurocentric scientific illiteracy—a challenge for the world community. In S. Harding (Ed.), *The "racial" economy of science: Toward a democratic future* (pp. 1–22). Bloomington and Indianapolis: Indiana University Press.

Harding, S. (1994). Is science multicultural? Challenges, resources, opportunities, uncertainties. *Configurations: A Journal of Literature, Science, and Technology, 2*(2), 301–330.

Harding, S. (1998). *Is science multicultural? Postcolonialisms, feminisms, and epistemologies*. Bloomington, IN and Indianapolis, IN: Indiana University Press.

Hawthorne, S. (1999). Connectivity: cultural practices of the powerful or subversion from the margins? In S. Hawthorne & R. Klein (Eds.), *CyberFeminism: Connectivity, critique and creativity* (pp. 119–133). North Melbourne: Spinifex Press.

Hayles, N. K. (1993). Constrained constructivism: locating scientific inquiry in the theater of representation. In G. Levine (Ed.), *Realism and representation: Essays on the problem of realism in relation to* science, literature and culture (pp. 27–43). Madison, WI: University of Wisconsin Press.

Hess, D. J. (1995). *Science and technology in a multicultural world: The cultural politics of facts and artifacts*. New York, NY: Columbia University Press.

Hicks, D., & Steiner, M. (Eds.). (1989). *Making global connections: A world studies workbook*. Edinburgh, UK: Oliver and Boyd.

Huckle, J. (1988). *What we consume: The teachers' handbook*. Richmond, Surrey: Richmond Publishing Company.

Jamison, A. (1993). National political cultures and the exchange of knowledge: the case of systems ecology. In E. Crawford, T. Shinn, & S. Sörlin (Eds.), *Denationalizing science: The contexts of international scientific practice* (pp. 187–208). Dordrecht: Kluwer.

Jay, T. (1986). The salmon of the heart. In F. Wilcox & J. Gorsline (Eds.), *Working the woods, working the sea* (pp. 101–124). Port Townsend, Washington: Empty Bowl.

Knudtson, P., & Suzuki, D. (1992). *Wisdom of the elders*. St Leonards, NSW: Allen & Unwin.

Latour, B. (1983). Give me a laboratory and I will raise the world. In K. D. Knorr-Cetina & M. Mulkay (Eds.), *Science observed: Perspectives on the social study of science* (pp. 141–170). London, UK: Sage Publications.

Latour, B. (1992). Where are the missing masses? Sociology of a few mundane artefacts. In W. E. Bijker & J. Law (Eds.), *Shaping technology, building society: Studies in sociotechnical change* (pp. 225–258). Cambridge, MA: The MIT Press.

Latour, B. (1993). *We have never been modern* (Catherine Porter, Trans.). Cambridge, MA: Harvard University Press.

Law, J. (2004). *After method: Mess in social science research*. London, UK: Routledge.

Levitt, T. (1983). The globalization of markets. *Harvard Business Review, 83*(3), 92–102.

Mann, S. (2000, November 22). Hill puts case on greenhouse gas. *The Age*, p. 15.

McLuhan, M., & Fiore, Q. (1967). *The medium is the massage*. New York, NY: Bantam Books.

Novick, P. (1988). *That noble dream: The "objectivity" question and the american historical profession*. Cambridge, MA: Cambridge University Press.

Odum, E. P., & Odum, H. T. (1953). *Fundamentals of ecology*. Philadelphia, PA: W.B. Saunders.

Peat, F. D. (1997). Blackfoot physics and European minds. *Futures, 29*(6), 563–573.

Peirce, C. S. (1931–1958). *Collected papers of Charles Sanders Pierce* (Vol. 8). Cambridge, MA: Harvard University Press.

Peters, M. (2004). Geophilosophy, education and the pedagogy of the concept. *Educational Philosophy and Theory, 36*(3), 217–231.

Petitjean, P., Jami, C., & Moulin, A. M. (Eds.). (1992). *Science and empires: Historical studies about scientific development and european expansion*. Dordrecht: Kluwer.

Pickett, S. T. A., & White, P. S. (Eds.). (1985). *The ecology of natural disturbance and patch dynamics*. Orlando, FL: Academic Press.

Pike, G., & Selby, D. (1987). *Global teacher, global learner*. London, UK: Hodder and Stoughton.

Pinar, W. F., & Reynolds, William M. (1992). Appendix: Genealogical notes—the history of phenomenology and poststructuralism in curriculum studies. In W. F. Pinar & W. M. Reynolds (Eds.), *Understanding curriculum as phenomenological and deconstructed text* (pp. 237–261). New York, NY: Teachers College Press.

Reid, W. A. (1981). *The practical, the theoretic, and the conduct of curriculum research*. Paper presented at the Annual Meeting of the American Educational Research Association, Los Angeles US.

Said, E. W. (1978). *Orientalism*. London, UK: Routledge and Kegan Paul.

Sardar, Z. (1989). *Explorations in Islamic Science*. London, UK and New York, NY: Mansell.

Sardar, Z. (Ed.). (1988). *The revenge of athena: Science, exploitation and the third world*. London, UK and New York, NY: Mansell.

Soja, E. (1996). *Thirdspace: Journeys to Los Angeles and other real-and-imagined places*. Cambridge, MA: Blackwell.

Tambiah, S. J. (1990). *Magic, science, religion, and the scope of rationality*. Cambridge, MA: Cambridge University Press.

Thayer, H. S. (1981). *Meaning and action: A critical history of pragmatism* (2nd ed.). Indianapolis, IN: Hackett Publishing Co.

Turnbull, D. (1991). *Mapping the world in the mind: An investigation of the unwritten knowledge of the micronesian navigators*. Geelong, VIC: Deakin University.

Turnbull, D. (1994). Local knowledge and comparative scientific traditions. *Knowledge and Policy, 6*(3/4), 29–54.

Turnbull, D. (1997). Reframing science and other local knowledge traditions. *Futures, 29*(6), 551–562.

Turnbull, D. (2000). *Masons, tricksters and cartographers: Comparative studies in the sociology of scientific and indigenous knowledge*. Amsterdam: Harwood Academic Publishers.

Ulanowicz, R. E. (1997). *Ecology: The ascendant perspective*. New York, NY: Columbia University Press.

Ulanowicz, R. E. (2009). *A third window: Natural life beyond Newton and Darwin*. West Conshohocken, PA: Templeton Foundation Press.

Verran, H. (2001). *Science and an African logic*. Chicago, IL and London, UK: University of Chicago Press.

Victorian Curriculum and Assessment Authority. (2004). *Environmental science: Study design*. Melbourne, VIC: Victorian Curriculum and Assessment Authority.

Wagner, J. (1993). Ignorance in educational research: or, how can you *not* know that? *Educational Researcher, 22*(5), 15–23.

White, L. (1967). The historical roots of our ecological crisis. *Science*, (155), 1204–1207.

Worster, D. (1993). *The wealth of nature: Environmental history and the ecological imagination*. New York, NY: Oxford University Press.

Wren, L., Jackson, M., Morris, H., Geddes, C., Tlen, D., & Kassi, N. (1995). What is a good way to teach children and young adults to respect the land? A panel discussion. In B. Jickling (Ed.), *A colloquium on environment, ethics, and education* (pp. 32–48). Whitehorse, Yukon, Canada: Yukon College.

Wynne, B. (1994). Scientific knowledge and the global environment. In M. Redclift & T. Benton (Eds.), *Social theory and the global environment* (pp. 169–189). London, UK: Routledge.

Yencken, D. (2000). Attitudes to nature in the East and West. In D. Yencken, J. Fien, & H. Sykes (Eds.), *Environment, education and society in the Asia-Pacific: Local traditions and global discourses* (pp. 4–27). London, UK and New York, NY: Routledge.

Yencken, D., Fien, J., & Sykes, H. (2000a). The research. In D. Yencken, J. Fien, & H. Sykes (Eds.), *Environment, education and society in the Asia-Pacific: Local traditions and global discourses* (pp. 28–50). London, UK and New York, NY: Routledge.

Yencken, D., Fien, J., & Sykes, H. (Eds.). (2000b). *Environment, education and society in the Asia-Pacific: Local traditions and global discourses*. London, UK and New York, NY: Routledge.

Author

Noel Gough is Foundation Professor of Outdoor and Environmental Education at La Trobe University, Victoria, Australia. Previously, he held senior academic appointments at the University of Canberra and Deakin University, and visiting fellowships at universities in Canada, South Africa, and the UK. His scholarship focuses on research methodology and curriculum studies, with particular reference to environmental education, science education, internationalization and globalization. In 1997 he received the inaugural Australian Museum

Eureka Prize for Environmental Education Research. He is the author or editor of five books, including *Curriculum Visions* (2002) and *Internationalisation and Globalisation in Mathematics and Science Education* (2007), and has published more than 100 book chapters and journal articles. He is the founding editor of *Transnational Curriculum Inquiry*, and a member of the editorial boards of eleven other international research journals. He is also a past President (2008) of the Australian Association for Research in Education.

5

Selected Trends in Thirty Years of Doctoral Research in Environmental Education in *Dissertation Abstracts International* From Collections Prepared in the United States of America

Thomas Marcinkowski
Florida Institute of Technology, USA

Jennifer Bucheit
Florida Institute of Technology, USA

Vanessa Spero-Swingle
University of Florida, USA

Christine Linsenbardt
Florida Institute of Technology, USA

Jennifer Engelhardt
Florida Institute of Technology, USA

Marianne Stadel
Florida Institute of Technology, USA

Richard Santangelo
Florida Institute of Technology, USA

Katherine Guzmon
Florida Institute of Technology, USA

Introduction

In many universities, a dissertation is a graduation requirement, evidence of scholarship (Patton, 2002), and a rite of passage into each doctoral student's chosen field. Thus, a review of dissertations can offer insights into the background and interests of doctoral students as emerging researchers, as well as into university programs in which dissertations were completed. Further, reviews may help identify scholarly communities within the field into which doctoral students have been socialized and inducted. Beyond this, reviews such as this may provide doctoral students with insights into topics, questions, and methods that have received substantial and limited attention over time.

Problem, Purpose, Scope, and Methods

On an international scale, there have been limited attempts to review doctoral dissertation research in and related to environmental education (EE) since the early 1980s, almost thirty years ago. A variety of factors appear to contribute to this, including the ever-increasing number of dissertations in and related to EE (hereafter "in EE"), the ever-expanding complexity of substantive and methodological features of these studies (Hart & Nolan, 1999), the variety of languages in which they are published, the absence of a single comprehensive index or online database for all dissertations, the fiscal and material resources required to maintain collections and conduct periodic reviews, and the prolonged commitment that reviews require from a team of researchers with diverse backgrounds (e.g., nationalities, languages, ethnic and cultural backgrounds, paradigmatic and methodological expertise, genders). As important as such an international effort would be, the necessary resources are not yet in place. Consequently, a review of that magnitude is well beyond the scope of this chapter.

The purpose of this chapter is to address limited attention to dissertations in past reviews of EE research by reviewing selected trends in a segment of the doctoral dissertation literature (hereafter "dissertations") in EE. As described below, the dissertations included in this review were found in collections of research in EE for the 1971–2000 period, nearly all of which were drawn from *Dissertation Abstracts International (DAI)*. Even though these collections were reasonably comprehensive and included more than 2,000 dissertations, more than 98 percent were completed at universities in the United States and, to a lesser extent, Canada. As a result, this review reflects trends in North American and, particularly, US dissertations.

In the absence of recent reviews of this literature, or reviews covering this time period and number of studies, there were few sources in EE to guide such a review. Thus, several methodological decisions were made. First, a decision was made to use content analysis, a method for studying the characteristics of written or visual communications such as journal articles, textbooks, newspapers, and TV programs (Ary, Jacobs, Razevieh, & Sorenson, 2006; Frankel & Wallen, 2000; Patton, 2002). Content analyses "may be done in a quantitative research framework with variables that are specified a priori and numbers that are generated to enable the researcher to draw conclusions about these specified variables" or "may be done in an emergent framework" (Ary et al., 2006, p. 464). The former was termed "classical content analysis" (Ryan & Bernard, 2000, p. 785), and consists of a flexible series of steps: (a) specifying the phenomena to be investigated; (b) selecting the sources to be reviewed; (c) developing a framework of variables or themes for coding sources (i.e., mutually exclusive coding categories); (d) developing a sampling plan for analyzing those sources; (e) developing a coding form for collecting observations from those samples using those coding categories, and procedures for using that coding form; (f) coding each sample using that form and those procedures; (g) organizing coding data in charts; and (h) describing and/or further analyzing of charted data (Ary et al., 2006, p. 465; Ryan & Bernard, 2000, p. 785).

Second, for steps (a) and (b), this review was delimited to collections of dissertations from 1971 to 2000 that appeared in Iozzi (1981), Marcinkowski and Mrazek (1996), Bucheit, Dorn, and Matheny (2006), and Spero (2006), with one exception. For step (d), these collections were used in their entirety, and the placement of studies into categories was based on available dissertation titles, abstracts, and descriptors in each collection and available through *DAI.*

Third, for step (c), a framework for each of seven content analyses was developed from prior analyses of trends, expanded and refined during the review, and finalized using iterative procedures similar to those employed by Hart and Nolan (1999). The first content analysis focused on the academic field and type of research in each dissertation so as to depict the variety of research in these collections, and to segment out dissertations in EE for further analysis. The next six content analyses were used to determine which, if any, of the following were apparent in each dissertation in EE: (2) the type of formal, nonformal, and/or informal program featured; (3) historical educational movements that influenced EE (Disinger, 1981, 1983; Stapp, 1974; Swan, 1984); (4) educational movements deemed contemporaries or subfields of EE (Disinger, 1983); (5) schools subjects featured in K-12 studies (Disinger, 1981, 1989); (6) EE goals emphasized and/or learning outcomes assessed (Childress, 1978; Rickinson, 2001; Simmons, 1991; Volk & McBeth, 1997); and (7) the research methodologies utilized (Hart & Nolan, 1999).

Fourth, for steps (e), (f), and (g), Marcinkowski, Engelhardt, and Stadel developed a coding form to guide each content analysis and a spreadsheet with separate columns for information from each dissertation pertinent to each analysis. The use of spreadsheets allowed team members to color code dissertations about which there were coding questions, and then meet weekly to discuss those questions, refine coding categories as needed, and record final coding decisions. To further ensure consistency, only one content analysis was undertaken at a time.

Lastly, for step (h), a table was prepared to present the results of each content analysis in the form of frequency counts of dissertations in each framework category and subcategory for each ten-year collection.

Why Dissertation Research?

The primary response to "Why?" was mentioned in the Introduction; i.e., the limited attention to dissertation research in reviews of EE research since the early 1980s. Table 5.1 was prepared to inspect this assertion. It summarizes key features of twenty collections and reviews of research in EE using a framework drawn from Bassey (2000), Rickinson (2001), and Marcinkowski (2003). *Collections* contain only descriptive information about individual studies (e.g., bibliographies with key words, annotations, full abstracts), while *reviews* identify patterns, present critiques, and/or summarize findings within and across studies. As a whole, these twenty collections and reviews encompass a major segment of research in EE published from the 1960s through 2007.

In Table 5.1, it is evident that the level of inclusion of dissertations varied over time. Up to 1985, reviews included a sizable number of dissertations. However, after 1985, only three publications did so; two were collections (Lee, 1998; Marcinkowski & Mrazek, 1996) and the third was a review of trends (Hart & Nolan, 1999). None of the seven syntheses of research findings since 1984 included more than six dissertations. One may speculate that the difficulties researchers face in identifying, accessing, and reading large numbers of potentially relevant dissertations have become too great. Regardless, few dissertations have been included in reviews of EE research over the last twenty-five to thirty years, so the potential or actual value of this body of work remains largely uninspected.

Dissertation Abstracts International

Dissertation Abstracts International (*DAI*) was the primary source of dissertations used to prepare collections by Iozzi (1981) and Marcinkowski and Mrazek (1996), and the only source used to prepare collections by Bucheit, et al. (2006) and Spero (2006). Thus, *DAI* was the source

TABLE 5.1
Analysis of Selected Collections and Reviews of Research in and Related to Environmental Education

General Features of the Collection and/or Review						
Author (Date)	**Time Period**[a]	**Full Refs.**	**Study Descriptions**	**Study Critiques**	**Synthesis**	**Dissertations**[b]
Roth and Helgeson (1972)	1950–1972	Yes	narrative		narrative	~70
Voelker, Heal, and Horvat (1972)	1969–1972	Yes	annotation			~40
Roth (1976)	1973–1976	Yes	narrative		narrative	~60
Winzler and Cherem (1978)	1976–1978	Yes	annotation			~10
Swan and Stilson (1980)	1937–1979	Yes				~600
Iozzi (1981)	1971–1980	Yes	abstract + descriptors			~90
Iozzi (1984)[c]	1971–1982	Yes	narrative		narrative	~55
Moore and Gross (1984)	1978–1984	Yes	subject index			~50
Hines, Hungerford, and Tomera (1986/87)	1971–1983	Yes			meta-analysis	two
Leeming, Dwyer, Porter, and Cobern (1993)	1974–1992	Yes	narrative + charts	each study		none
Marcinkowski and Mrazek (1996)	1971–1980	Yes	descriptors			~310
	1981–1990	Yes	narrative + descriptors			310
Volk and McBeth (1997)	1977–1995	Yes	narrative + charts		vote-count	three
Chawla (1998)	1980–1997	Yes	narrative + charts		narrative	four
Lee (1998)[d]	1949–1997	Yes				~30
Hart and Nolan (1999)	1990–1999	Yes	narrative	study trends		~110
Zelezny (1999)	1974–1999	Yes	charts	each study	meta-analysis	none
Rickinson (2001)	1993–1999	Yes	narrative	each study	narrative	two
Osbaldiston (2004)	1976–2000	Yes	charts		meta-analysis	none
Sward and Marcinkowski (2001)	1980–1998	Yes	narrative			six
Bamberg and Moser (2007)	1995–2007	Yes			meta-analysis	one

[a]"Time Period" refers to the years in which studies included in each collection/review were published.
[b]"Dissertations" refers to the level of inclusion of doctoral dissertations in each collection/review. These estimates are based on counts of dissertations referenced, and do not include references for journal articles based on dissertation studies.
[c]This review lists only studies from 1981 to 1982. All cited dissertations studies are included in Iozzi (1981).
[d]In Lee (1998), this count reflects only those dissertation listed in the section titled "Visitor Studies, Evaluation, and Research."

of the vast majority of dissertations in, and was consulted extensively throughout, this review.

The origins of *DAI* can be traced to 1938 when Power started University Microfilms International (UMI). Soon, UMI was microfilming dissertations and publishing abstracts in *Microfilm Abstracts*, a catalog of dissertations. In 1951, the Association of Research Libraries endorsed UMI's efforts. In 1985, UMI was sold to Bell & Howell, which created the ProQuest Company in 2001. In 2006, ProQuest, which now included UMI as a subsidiary, was sold to the Cambridge Information Group (Waltz, 2008).

Over time, changes in product, organization, and ownership broadened the temporal and geographic scope of UMI interests and *DAI*. Consequently, UMI has become "the largest repository of theses and dissertations produced at U.S. and Canadian universities . . . [and], in effect, the official repository of dissertations and theses for the national libraries of Canada and the United States" (Waltz, 2008, p. 1). Further, *DAI* now functions as ". . . a definitive subject, title, and author guide to virtually every [North] American dissertation accepted at an accredited institution since 1861 . . . In addition, since 1988, the database has included citations for dissertations from 50 British universities . . . collected and filmed at *The British Document Supply Centre*" (Dialog, 2008, p. 1).

Two sections of *DAI* were used to prepare these collections and support this review: Section A, which contains dissertations in the social sciences and humanities, including education; and Section B, which contains dissertations in the sciences and engineering. Between 1971 and 2000, dissertations in EE were included in both sections. In 1976, Section C, European dissertations, was started, although it was not available to those who prepared these collections or conducted this review. In 1988, Section C was renamed *Worldwide Dissertations* and was expanded to include citations and abstracts for dissertations from around the world in all disciplines, organized by subject, and indexed using English keywords. Since then, Section C has been included in Dissertation Abstracts Online, Dissertation Abstracts Ondisc, and the annual *Comprehensive Dissertation Index* (Corey, 2008; Dialog, 2008).

Today, *DAI* is included in a set of products and services offered by ProQuest's UMI Dissertation Publishing service that include: monthly print and microfiche collections of citations and abstracts for dissertations, indexed by author and keywords; *ProQuest Theses and Dissertations*, an online searchable database containing nearly 24,000 citations that predate 1900, and abstracts since 1980; as well as a reprint service through which copies of dissertations may be purchased using UMI index numbers.

Collections of Dissertation Research in EE That Served as Sources for This Review

This review is based on collections prepared by Iozzi (1981), Marcinkowski and Mrazek (1996), Bucheit et al. (2006), and Spero (2006). However, these could not be used in their original form; they were reorganized and integrated to support these content analyses, as described below. In addition, each collection was prepared with a somewhat different conception of EE and using slightly different methods (e.g., search terms and coding descriptors were adjusted to reflect changes in the field and in this literature). These conceptions and methods influenced the dissertations included in each collection and in this review, and therefore the results of these content analyses. For these reason, these collections are described below.

Dissertations from 1971 to 1980 Two collections of research in EE included dissertations from the 1971–1980 period. The first (Iozzi, 1981) was prepared by the Research Commission of the North American Association for Environmental Education (NAAEE), with support from the Educational Resources Information Center (ERIC) Center at The Ohio State University. In general, studies selected for inclusion reflected the following definition of EE research: "Includes components of efforts concerned with developing or analyzing environmental awareness, valuing, or problem-solving behavior" (1981, p. xiii). Section II included citations and abstracts for eighty-eight dissertations contained in *DAI*, Volumes *36*(7) through *42(*6) (1976–1980), or in the database at this ERIC Center. Abstracts were reprinted with permission from UMI. Each study was assigned an index number and coded with descriptors in each category: EE goals, EE facilities, education level, learning domain, kind of study, research methods, EE methods, EE curriculum mode, learner traits, and EE content area. Finally, descriptors and assigned numbers were used to create an index of studies.

This research commission also prepared the second collection (Marcinkowski & Mrazek, 1996). The definition of EE research presented in this collection was broadened to: "Includes, but is not limited to, components of efforts aimed at developing and/or measuring environmental awareness, ecological and issue-related scientific knowledge, issue investigation and decision-making skills, the empowerment of learners as change agents, citizenship behavior (responsible environmental behavior), and allied variables" (1996, p. iv). Section II contained citations for 308 dissertations from the 1971–1980 period located in *DAI* Volumes 31(7) through 41(6), other dissertation indexes (e.g., *Canadian Theses*, *American Doctoral Dissertations*), and other published sources (Moore & Gross, 1985; Swan & Stilson, 1980; Winzler & Cherem, 1978), but excluded those in Iozzi (1981). In *DAI,* searches were run using multiple search terms: conservation, ecology, energy, environment, marine, nature/natural, pollution, population, and resource.

This allowed relevant dissertations with and without "education" in the title to be found.

Each study was assigned a unique number, coded using an expanded set of descriptors, and indexed as in Iozzi (1981).

Over 2006–2007, Linsenbardt merged dissertations from the 1971–1980 period included in Iozzi (1981) and in Marcinkowski and Mrazek (1996) into one alphabetized set. This merged collection consisted of a total of 394 dissertations in EE.

Dissertations from 1981 to 1990 Section III in Marcinkowski and Mrazek (1996) contained citations for 307 dissertations for the 1981–1990 period located in *DAI* Volumes 41(7) through 51(5) (1981–1990) or in other sources consulted in the 1971–1980 search. These dissertations were identified, assigned numbers, coded, and indexed following procedures used to prepare Section II in Marcinkowski and Mrazek (1996).

Dissertations from 1991 to 2000 For the 1991–2000 period, three collections of dissertations in EE were used in this review. First, Malikova worked with Marcinkowski to search for and select, prepare bibliographic references for, and code with descriptors 147 dissertations from 1991 to 1992 (*DAI*,

Volumes 51(5) through 52(12)). Prior to this, Marcinkowski had received feedback on the 1971–1980 and 1981–1990 collections of dissertations in Marcinkowski and Mrazek (1996), some indicting that these collections included studies beyond the scope of EE. To address this, Marcinkowski and Malikova developed a framework for categorizing dissertations. Once dissertations in these volumes were identified and categorized, they were assigned numbers, coded using an expanded set of descriptors, and indexed following procedures used in Marcinkowski and Mrazek (1996).

Bucheit et al. (2006) prepared the second collection, which included 801 dissertations from 1992 to 1998 (*DAI* Volumes 53–58). Each searched two volumes of *DAI,* but discussed procedures and questions as a team. After analyzing Malikova's work using *DAI* microfiche and the ProQuest database, they modified procedures used in Marcinkowski and Mzarek (1996) and by Malikova. First, they used an expanded set of search terms that reflected a slightly broader conception of EE, including biodiversity, global/global change, outdoor, and sustainable/ity. Second, Malikova's categories of studies were expanded to include *Related Resource and Environmental Policy Research* due to the apparent increase in the number of dissertations pertaining to environmental policy and planning. Third, based on analyses of studies in Malikova's collection, new coding descriptors were added (e.g., qualitative research, content/document analysis, case study research, field research, thinking skills, learning styles, adoption of innovations, global change education, education for sustainability, and philosophy). Finally, they checked these categorizing and coding decisions using inter-rater reliability procedures (Gay & Airasian, 2003).

The third collection included 394 dissertations from 1999 to 2000 (*DAI* Volumes 59–60), and was prepared by Spero (2006). She worked with Bucheit et al. (2006), and used nearly identical procedures. Thus, the entire 1991–2000 collection contained 1,342 dissertations.

The Geographic Scope of Dissertation Research, 1971–1980

Given recent efforts to expand the geographic scope of *DAI*, one might infer that these collections included dissertations completed outside the United States. To inspect this, Santangelo and Guzmon conducted a detailed analysis. The results are summarized in Table 5.2. For the merged 1971–1980 and the 1981–1990 collection, nearly all dissertations were completed at US universities; only six were completed at Canadian universities, two at European universities, and one each at African and Australian universities. However, the expansion of *DAI* is apparent in the combined 1991–2000 collection, which included 107 dissertations completed at Canadian universities (8%) and eighteen completed at European, Australian, and African universities.

However, such counts reveal nothing about international doctoral students or dissertation studies conducted in other countries. While the former is difficult to determine, the latter is not. To inspect the latter, Santangelo and Guzmon used information from dissertation titles and abstracts to identify dissertations conducted inside and outside the nation in which it was defended. From the 1970s through the 1990s, the number of studies conducted abroad but defended in US universities rose from 17 to 296 and in Canadian universities rose from none to 21. Thus, while the awarding institution indicates these collections draw primarily from US universities, it is misleading to think these collections did not include studies conducted outside of the United States.

Limitations of This Review

Several limitations are apparent in this review. First, these content analyses relied upon dissertation titles, abstracts, and descriptors. Full dissertations were rarely available or read. Therefore, when questions about study characteristics arose, abstracts were consulted. However, abstracts for dissertations prior to 1980 are not available through ProQuest, only through *DAI* in hardcopy and microfiche, and Iozzi (1981). Further, even when abstracts were available, the clarity and quality of information in each varied widely. Thus, limited information for dissertations, particularly in the merged 1971–1980 collection, may have contributed to some errors in coding and categorizing dissertations. Second, as described below, it was difficult to categorize some dissertations due to the ways in which they were framed and conducted. Third, due to the scope of this review and to practical constraints, the depth of each content analysis is limited, so the results were used to describe general trends in these dissertations as they related to trends in the field. Further, no attempts were made to critique or summarize the findings of these studies (Table 5.1). While further analyses would be informative, these were beyond the resources available for this review. Finally, as noted above, more than 98 percent of dissertations in these collections were from US and Canadian universities. Thus, caution should be taken not to generalize the results of this review beyond dissertations completed in these two nations.

Content Analysis 1: Dissertation Studies by Field and Type

The first content analysis required the establishment of parameters for each category to ensure reasonably fair and consistent categorization. Iterative procedures similar to those employed by Hart and Nolan (1999) were used to expand and refine the categories apparent in 1991–2000 collections.

The category of primary interest was *Category A, Dissertations in and Closely Related to EE.* Dissertations in this category included those that: (a) had *environmental education* in their title; (b) reflected historical forerunners and contemporaries of EE within the United States (Disinger, 1981, 1983; Stapp, 1974; Swan, 1985); (c) reflected ecological or environmental content; and/or (d) reflected recent developments (e.g., sustainability, global warming/

TABLE 5.2
Selected General and Geographic Features of Collections of Dissertation Studies

A. General Features			
Time Period	1971–1980	1981–1990	1991–2000
Dissertation Abstracts International	Vols. 31–41	Vols. 41–51	Vols. 51–60
Number of Dissertations Referenced*	394	307	1,342
B. Geographic Features			
1. Total Studies in US Universities	391	300	1,216
a. Conducted in the US	376	268	920
b. Conducted in Canada and Mexico	0	2	19
c. Conducted in Latin Am. and Caribbean	0	5	55
d. Conducted in Europe	3	1	36
e. Conducted in the Middle East	1	4	8
f. Conducted in Africa	2	10	46
g. Conducted in Asia	6	9	107
h. Conducted in Australia and Oceania	1	0	7
i. Conducted in Multiple Regions	4	1	18
2. Total Studies in Canadian Universities	3	3	107
a. Conducted in Canada	3	3	86
b. Conducted in Latin Am. and Caribbean	0	0	2
c. Conducted in Africa	0	0	5
d. Conducted in Asia	0	0	10
e. Conducted in Multiple Regions	0	0	4
3. Studies in Universities Outside North Am.	0	4	19

**Note*. The number for 1971–1980 includes dissertations listed in Iozzi (1981) and Marcinkowski and Mrazek (1996). The number for 1981–1990 includes dissertations listed in Marcinkowski and Mrazek (1996). The number for 1991–2000 includes dissertations listed in Bucheit et al. (2006) and Spero (2006).

climate change). Further, these studies had to: (e) involve subjects in or associated with formal or nonformal education programs; and (f) address educational programs or features of them (e.g., curricula, materials, teaching, learning, assessment, evaluation).

Each of the remaining categories differed from *Category A* on at least two of these characteristics. *Category B, Other Related Educational Research*, included educational studies that reflected characteristics (e) and (f), but not environmental topics in (a–d) (e.g., Blankenship, 1990; Blevins, 1994; Moorefeld, 1994). Conversely, *Categories C* through *F* included studies that did not reflect characteristics (e) and (f), but did reflect characteristics (b, c, or d). More specifically, *Category C, Related Social and Behavioral Research*, included studies whose subjects were not associated with educational programs (e.g., Chatelain, 1981; Peters-Grant, 1987). *Category D, Related Historical, Literary, and Media Research*, included studies that made primary use of documents (e.g., Hirt, 1992; Ortiz, 1987; Sauvage, 1994). *Category E, Related Conservation/Environmental Policy and Planning Research* (e.g., Balakrishnan, 1995; Buhrs, 1992; Iacofano, 1986), and *Category F, Research Related to Conservation/Environmental Philosophy, Ethics, Religious Traditions, and Spirituality* (e.g., Booth, 1993; Mante, 1995; Steverson, 1992), each included studies that placed primary emphasis on topics, methods, and materials associated with that category.

Dissertations in *Categories B–F* were included for two reasons. First, dissertations in these areas were apparent in each collection, so this analysis helped determine the extent to which these collections included dissertations in these categories. Second, the field of EE has dynamic relationships with sectors, academic fields, and educational practices reflected in *Categories B–F*. For example, dissertations in *Categories C* and *E* often reflected the status of and changes in environmental/sustainability views, policies, and practices within governments and societies that can influence and be influenced by EE. Further, dissertations in *Categories D* and *H* were relevant to practices in environmental studies and environmental communications, two higher education fields closely allied with EE in the United States (e.g., within NAAEE as sections, conference strands, and proceedings, and monographs).

This framework was used to code each dissertation. As with many content analyses and associated frameworks, a relatively small percent of dissertations reflected two or more categories (below). These studies complicated the categorization process, but also offered insights into how doctoral researchers have explored new and complex questions that cross-disciplinary and -methodological boundaries. Examples included:

- historical studies of education (Morgan, 1998), of social and environmental conditions (Brown, 1995; Gelobter, 1995), of policy (Duke, 2000; Soffar, 1975; Thompson, 1995), and of philosophical or religious contributions (Leunes, 1978; Williams, 1992);
- studies of media influences on social perceptions (Hester, 1998); and
- studies of social perceptions of or influences on policy and planning (Onyenekuw, 1993; Wyatt, 1977).

Nonetheless, for the vast majority of the more than 2,000 dissertations in these collections, these categories were sufficient to permit this analysis of broad trends in these collections (Table 5.3).

Of the dissertations in the 1971–1980 collection, 270 were included in *Category A*, *Research in and Closely Related to EE*, and 76 in *Category C*, *Related Social and Behavioral Research*. By comparison, relatively few studies were included in *Categories B, D, E*, or *F*.

A comparison of the 1971–1980 and 1981–1990 collections indicated the number and percent of dissertations in *Category A* declined from 271 (69%) to 179 (58%). A further comparison of these collections indicated that while the number of dissertations in all other categories remained reasonably stable (Table 5.3), the variety of fields (e.g., topics) and types (e.g., methodological approaches) apparent in dissertations in *Categories B, C, D, E, and F* appeared to expand.

Finally, the number of dissertations in the 1991–2000 collection was more than four times that in the 1981–1990 collection. Thus, even though the number of studies in *Category A* approximated the number in earlier collections, the percent of studies in this category declined to about 18 percent. In this period, the number of dissertations in all other categories appeared to increase substantially (Table 5.3), and the variety of fields and types apparent in *Categories B, C, D, E*, and *F* continued to diversify.

In summary, the frequency counts in Table 5.3 reflect the varied definitions of EE, search terms, and coding procedures used to prepare each original collection. In light

TABLE 5.3
Contents of Collections of Dissertation Studies, by Study Category

I. General Features			
Time Period	1971–1980	1981–1990	1991–2000
Dissertation Abstracts International	Vols. 31–41	Vols. 41–51	Vols. 51–60
Total Number of Dissertations Referenced	394	307	1,342
II. Study Category*			
A. Research in and Closely Related to Environmental Education	271 (69%)	179 (58%)	248 (18%)
B. Other Related Educational Research	22 (6%)	23 (7%)	116 (9%)
C. Related Social and Behavioral Research	76 (19%)	89 (29%)	421 (31%)
D. Related Historical, Literary, and Media Research	16 (4%)	5 (2%)	182 (14%)
E. Research Related to Conservation/Environmental Policy and Planning	6 (2%)	7 (2%)	302 (22%)
F. Research Related to Conservation/Environment in Philosophy, Ethics, Religious Traditions, and Spirituality	3 (1%)	3 (1%)	74 (5%)

**Note*. Due to the use of rounding, percentages for each collection may not total 100%.

of the search terms and procedures used, one may infer that these counts approximate the number of dissertations in *Category A* through *Category D* for collections from each period. However, this does not apply to the number of dissertations in *Category E* and *Category F* due to the absence of specific search terms for these in the 1971–1980 and 1981–1990 collections, and the use of search terms for these in the 1991–2000 collection.

Content Analyses of Substantive Features of Dissertations in Category A, Dissertations in and Closely Related to Environmental Education

Content Analysis 2: Trends in Dissertations, by Sector This analysis of the sectors reflected in dissertations relied on information about study subjects and settings. The *Formal Education* sector included studies of learners in Pre-K-12 schools, pre- and in-service teachers, higher education environmental programs, and other higher education courses and students, as well as adult, continuing, and extension programs. The *Nonformal* and *Informal* sectors were treated separately in light of traditions within the field of EE (NAAEE, 2004c; Norland & Sommers, 2005). The *Nonformal Education* sector included studies of specific educational and/or environmental settings other than schools, including nature/environmental centers, camps and resident outdoor programs, museums, zoos and aquaria, gardens and herbaria, and parks and other natural areas protected or managed by government agencies. The *Informal Education* sector included studies not tied to any specific setting, and involved electronic and print media, the Internet, materials distributed by nongovernmental organizations (NGOs), and community events. Finally, while *State and Provincial Agencies* is rarely considered a separate sector, it was included to identify studies of EE planning and programming by state/provincial agencies. These results are summarized in Table 5.4.

The results presented in Table 5.4 reveal several reasonably stable trends across these three collections. Of the studies in *Category A* in each collection, approximately 43–49% focused on PreK-12 students and programs. Across collections, a smaller number of studies focused on pre- and in-service teachers (16–27%) and other higher education programs and students (6–11%). This reflects ongoing interest within the EE and environmental science/studies communities in formal education (NAAEE, 2004a, 2004b).

In the 1971–1980 collection, relatively few dissertation studies focused on nonformal education, although studies of nonformal programs and settings increased over 1981–1990 and into 1991–2000 (Table 5.4). The growing number of studies appears to coincide with increasing attention to program evaluation in the nonformal sector (Bennett, 1977; Chenery & Hammerman, 1984/85; Norland & Somers, 2005; Pandion Systems, 2006; Wiltz, 2000).

In each collection, the number of dissertation studies of informal education never exceeded five. This small number of studies may be due, in part, to the inclusion of some media studies in *Category C, Related Social and Behavioral Research* (i.e., media studies that did not meet all criteria for Category A) and *Category D, Related Historical, Literary, and Media Research* (e.g., content analyses and historical studies of media coverage of environmental topics).

Lastly, in the 1971–1980 collection, there were seven dissertation studies of state/provincial agencies. These studies coincided with (the United States) federal funding for *State Master Planning* in EE in the early 1970s (Rocchio & Lee, 1974). In the 1981–1990 and 1991–2000 collections, the number of dissertation studies of state/provincial agencies remained about the same, despite cuts in federal funding after 1973 (Young, 1992). It is noteworthy that studies in later collections focused less on planning and more on monitoring the status of EE implementation in K-12 schools.

Content Analysis 3: Trends Related to Historical Educational Movements In the United States, a number of educational movements preceded and created favorable conditions for EE to arise in the late 1960s. These are known as the Nature Study Movement (ca. 1900; see Bailey, 1909; Jackman, 1903), Outdoor Education Movement (ca. 1920s; see Hammerman & Hammerman, 1973; Smith, 1968), and Conservation Education Movement (ca. 1930s; see Funderburk, 1948; Lively & Priess, 1956). The history of guiding in national parks and other scenic areas (ca. 1910s) influenced and was influenced by each of these movements, and has since grown into the modern-day field of environmental interpretation (Ham, 1992; Sharpe, 1982; Tilden, 1957). In addition, shortly before the period in which EE arose within the United States, there was a popularization of ecology education at the university level (ca. 1950s) and the K-12 level (ca. 1960s). The contributions of these and other educational movements to the purposes and goals of, infrastructure for, contents of, and practices in EE have been substantial, and continue to influence EE programming in the United States (Disinger, 1983; Kirk, 1977; Stapp, 1974; Swan, 1984).

Since publication of the *Brundtland Commission Report* (World Commission on Environment and Development, 1987), and the report from the 1992 World Conference on Environment and Development (United Nations, 1992), another educational movement has arisen on a global scale; i.e., Education For Sustainable Development (ESD). Just as the educational movements noted above contributed to EE, so EE is commonly viewed as part of and contributing to ESD (Sato, 2006; United Nations Education, Scientific, and Cultural Organization [UNESCO], 1997).

This analysis focused on dissertations in EE that pertained to these historical educational movements, and the emergent ESD movement. It remains difficult to determine the degree to which dissertations in these collections reflected both EE and these other movements due to loose, overlapping, and evolving sets of defining characteristics for each (Disinger, 1983; Hart, 1981; Hesselink, van

TABLE 5.4
Contents of Collections of Dissertation Studies in Category A, by Sector

I. General Features			
Time Period	1971–1980	1981–1990	1991–2000
Dissertation Abstracts International	Vols. 31–41	Vols. 41–51	Vols. 51–60
Total Number of Dissertations Referenced	394	307	1,342
Number of Dissertation in *Category A*	271	179	248
II. Study Category*			
A. Formal Education			
1. K-12	121 (45%)	87 (49%)	107 (43%)
2. Teacher Education	43 (16%)	48 (27%)	51 (21%)
3. Environmental Studies, Science, and Professions	16 (6%)	19 (11%)	22 (9%)
4. Other Higher Education	22 (8%)	5 (3%)	11 (4%)
5. Adult, Continuing, and Extension Ed.	4 (1%)	8 (4%)	16 (6%)
B. Nonformal Education			
1. Nature/Environmental Centers	2 (<1%)	4 (2%)	5 (2%)
2. Camp and Resident Outdoor Programs	8 (3%)	8 (4%)	11 (4%)
3. Museums	1 (<1%)	4 (2%)	4 (2%)
4. Zoos, Aquaria, Botanical Gardens, and Herbaria	0	2 (1%)	2 (<1%)
5. State/Provincial Sites	1 (<1%)	2 (1%)	0
6. National and International Sites	3 (1%)	3 (2%)	0
7. Other (e.g., school grounds, local community sites, wilderness)	3 (1%)	4 (2%)	10 (4%)
C. Informal Education:			
1. TV and Radio	2 (<1%)	1 (1%)	1 (<1%)
2. Newspapers and Magazines	0	1 (1%)	1 (<1%)
3. Internet	0	0	3 (1%)
4. NGO Magazines and Newsletters	0	0	0
5. Community Events	0	0	0
D. State and Provincial Agencies	7 (3%)	7 **(4%)**	5 (2%)

**Notes.* (1) Studies in which nearly equivalent attention was given to multiple settings and/or subjects were included in counts twice (e.g., studies of K-12 students in resident outdoor programs: A.1 and B.2; studies of teachers and their students: A.1 and A.2). (2) Percentages are based on the number of dissertations in Category A. (3) Because this analysis included only studies related to sectors, percentages will sum to less than 100%.

Kampen, & Wals, 2000; Schoenfeld, 1969; Tanner, 1974). The results of this analysis are summarized in Table 5.5.

There was limited attention to influences of the Nature Study Movement in these collections. Explicit attention was more apparent in studies in the 1971–1980 collection (e.g., Kraus, 1973; Minton, 1980), than in the 1981–1990 and 1991–2000 collections (e.g., studies of natural history museums, nature centers, and the teaching of natural history).

In all three collections, substantial attention was given to outdoor education. Across collections, several studies focused on the effects of outdoor programs on cognitive and affective outcomes, and several compared the effects of indoor versus outdoor programming. Studies explored the use of school grounds and nonschool sites for field trips, residential programs, and extended field studies. In the 1981–1990 and 1991–2000 collections, this range of outdoor settings expanded (i.e., wilderness to urban), as did the populations served.

In each collection, substantial attention was given to conservation education. Through the mid-1970s, explicit attention was given to traditions within the Conservation education movement, although relatively few studies appeared after that (e.g., Stephen, 1984). Implicit attention was apparent in studies of energy conservation education in the later 1970s, and remained prominent in the 1981–1990 collection. In these collections, studies also focused on other specific conservation issues such as wildlife (e.g., loss of biodiversity), and on natural resource education in K-12, higher education, and interpretation programs (i.e., including those for minority populations in the U.S.).

Across collections, there also was a steady level of interest in Ecology Education. In the 1971–1980 collection, studies focused on the selection and organization of content, materials

TABLE 5.5
Contents of Collections of Dissertation Studies in Category A, by Historical Trend

I. General Features			
Time Period	1971–1980	1981–1990	1991–2000
Dissertation Abstracts International	Vols. 31–41	Vols. 41–51	Vols. 51–60
Total Number of Dissertations Referenced	394	307	1,342
Number of Dissertation in *Category A*	271	179	248
II. Historical Trends*			
A. Nature Study and Natural History	5 (2%)	8 (4%)	8 (3%)
B. Outdoor Education	28 (10%)	18 (10%)	23 (9%)
C. Conservation and Resource Management Education	27 (10%)	24 (13%)	25 (10%)
D. Ecology or Ecological Education	28 (10%)	28 (16%)	24 (10%)
E. Environmental Interpretation	10 (4%)	6 (3%)	8 (3%)
F. Education for Sustainable Development	0	1 (<1%)	14 (6%)

**Notes*. (1) Numerous professionals in EE have written about educational movements that set the stage for EE in the U.S. (A–E) or about the educational movement that has grown out of EE, ESD (F). (2) Percentages are based on the number of dissertations in Category A. (3) Because this analysis included only studies related to prior and subsequent educational movements, percentages will sum to less than 100%.

for class and field use, and assessment and evaluation associated with conceptual learning. In the 1981–1990 collection, this expanded to include studies of theories of learning pertinent to instruction and assessment, conditions that influenced learning, and the relationship of conceptual learning to other cognitive and affective outcomes. Dissertations in the 1991–2000 collection reflect broader tensions within the field of ecology over "applied ecology" as well as the influence of constructivism (e.g., active learning, meaningful learning, misconceptions).

The number of dissertations in environmental interpretation remained small but stable across these collections. As with research in EE, studies focused on definitional features and goals of the field, higher education programs for interpreters, studies of interpreters, methods for preparing trails/exhibits and programs, contributions of interpretation to resources conservation and management, program evaluation, and studies of visitor interests, activities, and learning.

The rise of ESD is apparent in Table 5.5. Only one study of ESD was found in these collections prior to 1990, and few studies appeared prior to 1993. The studies in the 1991–2000 collection focus on what constitutes sustainability, definitional features of ESD and education for sustainable agriculture, and the preparation of teachers and materials for ESD. It is noteworthy that several dissertations investigated the infusion of sustainability themes into EE, rather than on ESD per se.

Content Analysis 4: Trends Related to Concurrent Educational Movements The fourth content analysis focused on what Disinger (1981) and others referred to as concurrent educational movements; i.e., contemporaries or subfields of EE in the United States during the 1970s and 1980s. These included population, marine, energy, pollution, environmental health, and global education. In general, there is a considerable overlap between EE and each of these, although the degree of overlap has fluctuated over time, for many studies in these areas were included in *Category B* (i.e., educational studies involving nonenvironmental topics). These fluctuations are apparent in Table 5.6.

There was a steady decline in research population education; all four studies in the 1991–2000 collection were conducted at one university under the same advisor. There was a similar decrease in dissertations in pollution education, although studies of education on global warming/climate change appeared in the 1991–2000 collection. Interest in a wider range of global concerns was reflected in a noticeable increase of global education studies in the 1991–2000 collection. Few studies in environmental health education were apparent, although this could be due, in part, to the absence of health as a search term in the preparation of these collections. In marine education, there was a modest increase in the number of studies in the 1981–1990 collection, and in energy education, a substantial increase in this same collection, although there were decreases in each in the 1991–2000 collection.

Content Analysis 5: Trends in the K-12 Sector, by School Subject At least within the United States, there has been a gap between EE rhetoric, which calls for interdisciplinary approaches (UNESCO, 1978; Hart, 1981), and EE practice, where science is the dominant subject into

TABLE 5.6
Contents of Collections of Dissertation Studies in Category A, by Topical Emphasis

I. General Features			
Time Period	1971–1980	1981–1990	1991–2000
Dissertation Abstracts International	Vols. 31–41	Vols. 41–51	Vols. 51–60
Total Number of Dissertations Referenced	394	307	1,342
Number of Dissertation in *Category A*	271	179	248
II. Topical Emphasis*			
A. Population Education	18 (7%)	9 (5%)	4 (2%)
B. Marine Education	5 (2%)	10 (6%)	4 (2%)
C. Energy Education	14 (5%)	46 (26%)	5 (2%)
D. Pollution Education	5 (2%)	6 (3%)	3 (1%)
E. Environmental Health Education	2 (<1%)	3 (2%)	1 (<1%)
F. Global Education	2 (<1%)	0	10 (4%)
G. Other	1 (<1%)	1 (<1%)	6 (2%)

**Notes*. (1) Beginning in the 1970s, a variety of educational approaches related to EE were developed and advanced, but were often called something other than EE (Disinger, 1981, 1983). Typically, these reflected a topic area related to specific aspects of nature (e.g., marine environments) and/or of environmental problems and issues (e.g., population, pollution, energy, health). (2) Percentages are based on the number of dissertations in Category A. (3) Because this analysis included only studies related to concurrent educational movements, percentages will sum to less than 100%.

which EE has been infused over this thirty-year period (Childress, 1976, 1978; Disinger, 1981, 1989; Disinger & Bosquet, 1982; Lucas, 1981). In other nations, other school subjects have served as this dominant subject, for example geography in the United Kingdom (W. Scott, personal communication, January 24, 2009). This analysis was undertaken to determine if this dominance was reflected in the dissertation literature, as well as the extent to which dissertations featured the infusion of EE into single and multiple school subjects (Table 5.7).

The results indicate that across these collections, dissertations in K-12 education focused on science in greater numbers than any other school subject. While social studies has been identified as the second most common subject into which EE is infused (Disinger, 1981), relatively few dissertations focused on this subject in any of these collections. Similarly, very few studies were found in language arts, mathematics, PE/health, or the arts. However, the number of dissertations in "other subjects," primarily related to vocational education (e.g., agriculture, industrial, technology), was larger than for any subject other than science. Lastly, while there appeared to be an increase in the number of studies that focused on multiple subjects after 1980, these counts were relatively small, and some studies should be considered multidisciplinary (i.e., treatment in separate subjects) rather than interdisciplinary (i.e., planned integration across subjects).

Content Analysis 6: Trends in Goal Orientations and Learning Outcomes The framework used in the analysis of goals and outcomes was adapted from the Tbilisi categories of objectives (UNESCO, 1978). In addition, efforts were made to identify dissertations that focused on more than one goal and/or outcome (Table 5.8).

Consistent with prior reviews (Iozzi, 1984; Rickinson, 2001; Volk & McBeth, 1997), there was substantial attention to attitudes as a goal/outcome of EE across collections. For example, while the greatest number of dissertations focused on multiple goals/outcomes, a majority of those included an assessment of attitudes (1971–1980: 56 of 86; 1981–1990: 35 of 65). Further, the next greatest number of dissertations focused on the development and assessment of affective characteristics of learners. Again, more than half of those dissertations focused on attitudes, although there were studies of environmental sensitivity (Chawla, 1998), values and the valuing process, moral reasoning, assumption of personal responsibility, locus of control, and verbal commitment or intention.

In comparison, moderate attention was paid to the development of learners' cognitive awareness or knowledge, either of ecology or environmental problems/issues. Further, as reflected in recent reviews (Rickinson, 2001; Volk & McBeth, 1997), limited attention was given to skill development and application, either alone or as one of multiple goals/outcomes (i.e., fewer than twelve per collection). Lastly, while it appears as if very limited attention was given to participation, service, and action (Table 5.8), it was more common for this to be one of several goals/outcomes in a given study, such as studies of knowledge and behavior or of attitudes and behavior (Simmons, 1991). More careful analysis revealed numerous examples of faulty the knowledgeawareness/attitudebehavior (K-A-B) model as the basis for EE programming and assessment (Kollmuss & Agyeman, 2002). Since the later 1980s, dissertations reflect a wider range of goals/outcomes, including dimensions of

TABLE 5.7
Contents of Collections of Dissertation Studies in Category A, by School Subject (K-12)

I. General Features			
Time Period	1971–1980	1981–1990	1991–2000
Dissertation Abstracts International	Vols. 31–41	Vols. 41–51	Vols. 51–60
Total Number of Dissertations Referenced	394	307	1,342
Number of Dissertation in *Category A*	271	179	248
II. School Subject*			
A. Science	40 (18%)	33 (18%)	32 (13%)
B. Social Studies	6 (2%)	0	2 (<1%)
C. Language Arts	1 (<1%)	1 (<1%)	1 (<1%)
D. Mathematics	0	0	1 (<1%)
E. Physical Education/Health	2 (1%)	1 (<1%)	0
F. Fine and Performing Arts	0	0	1 (<1%)
G. Other School Subjects	9 (3%)	4 (2%)	10 (4%)
H. Multiple Subjects	4 (1%)	11 (6%)	11 (4%)

**Notes*. (1) "Other" includes studies conducted in subject areas such as vocational education, agriculture education, and technology education. "Multiple" refers to studies that emphasized two or more school subjects, as is common in interdisciplinary and multidisciplinary programming. (2) Percentages are based on the number of dissertations in Category A. (3) Because this analysis included only studies involving K-12 education, percentages will sum to less than 100%.

TABLE 5.8
Contents of Collections of Dissertation Studies in Category A, by Goal Orientation and Learning Outcome

I. General Features			
Time Period	1971–1980	1981–1990	1991–2000
Dissertation Abstracts International	Vols. 31–41	Vols. 41–51	Vols. 51–60
Total Number of Dissertations Referenced	394	307	1,342
Number of Dissertation in *Category A*	271	179	248
II. Goal Orientations and Learning Outcomes*			
A. EE Focused on Ecological Awareness or Knowledge	14 (5%)	8 (4%)	6 (2%)
B. EE Focused on Awareness or Knowledge of Environmental Problems and Issues	28 (10%)	15 (8%)	11 (4%)
C. EE Focused on Affective Development	57 (21%)	47 (26%)	64 (26%)
D. EE Focused on Problem- and IssueRelated Skills	3 (1%)	2 (1%)	2 (1%)
E. EE Focused on Participation, Service, Action, and Behavior Change	3 (1%)	2 (1%)	4 (2%)
F. EE Focused on Multiple Goals/Outcomes	86 (32%)	65 (36%)	86 (35%)

**Notes*. (1) These "Goal Orientations and Learning Outcomes" reflect the categories of objectives agreed upon at the UNESCO Intergovernmental Conference held in Tbilisi, USSR in 1977, and reaffirmed in subsequent UNESCO meetings on EE: Awareness and Knowledge (A and B), Attitudes (C), Skills (D), and Participation (E). (2) As called for in the literature (UNESCO, 1978; Hart, 1981), the last cluster (F) includes studies that addressed multiple goals and/or assessed participants on multiple outcomes (e.g., surveys of environmental knowledge and attitude: B and C; environmental awareness programs in which attitudes were assessed: B and C). (3) Percentages are based on the number of dissertations in Category A. (4) Because this analysis included only studies involving these goals/outcomes, percentages sum to less than 100%.

knowledge, affect, and skill beyond those commonly associated with the K-A-B model.

Content Analysis 7: Methodological Features of Dissertations in Category A, Dissertation in and Closely Related to Environmental Education This final content analysis focused on how studies in these collections were undertaken. Primary attention was given to research paradigms and associated methodologies and then, as deemed necessary by Hart and Nolan (1999), to research purpose. The former was a simple analysis of studies that employed quantitative, qualitative, and "mixed" methods methodologies. *Quantitative* methodologies included those that used predetermined designs, procedures, instruments, and statistical analyses in descriptive, correlation, causal-comparative, experimental, and related studies. The primary reference for what was counted as *qualitative* was Denzin and Lincoln (2000), which presents a wide range of philosophical, methodological, political, and ethical perspectives on and approaches to social and socially critical research. In the 1971–1980 and 1981–1990 collections, *mixed methods* studies were those that made prominent use of quantitative and qualitative methods for data collection and analysis. In the 1991–2000 collection, this referred to studies that contained "quantitative and qualitative" in the title or abstract. This difference in methods reflects the shift from limited awareness and acceptance of qualitative research within the EE community in the United States before the late 1980s, to a wider awareness and acceptance of it after 1990 (Mrazek, 1993). Despite ongoing discussions and debates about how to frame paradigmatic and methodological differences (e.g., Connell, 2006; Dillon & Wals, 2006; Hart, 1993b; Jickling 1993; Marcinkowski, 1993; Mrazek, 1993; Robottom & Hart, 1993), no attempt was made to conduct in-depth analyses of quantitative, qualitative, or mixed methods studies (e.g., to determine if doctoral researchers understood assumptions underlying their choice of methodology and methods).

The second part of this analysis was a review of studies that reflected each of six different purposes (Table 5.9). Several of these were used by Hart and Nolan (1999) (e.g., focused inquiries; large descriptive studies), while several emerged during this analysis (e.g., educational development studies; analyses of materials).

The results of this analysis indicated that quantitative methodologies (e.g., survey, Delphi and Q-sort, document analysis, correlation and prediction, ex post facto, one-group evaluation, experimental, and aptitude-treatment interaction studies) were used in about two-thirds of the dissertations within the 1971–1980 and 1981–1990 collections, but declined to 49 percent in the 1991–2000 collection. Of these, experimental studies were the most numerous in each collection (1971–1980: 67 of 175; 1981–1990: 40 of 122; 1991–2000: 38 of 122).

In the 1971–1980 and 1981–1990 collections, the number of studies that used qualitative and/or critical methodologies was less than 10 percent. However, with the growing use and acceptance of qualitative and critical methodologies in education since the 1980s, possibly aided by early advocacy efforts within EE (Mrazek, 1993; Robottom & Hart, 1993), the number of dissertations using only these methodologies increased to nearly 25 percent in the 1991–2000 collection. Over these three collections, there also was a noticeable increase in the variety of qualitative and critical methodologies, including those used in historical, ethnographic, case, hermeneutic, grounded theory, participatory action research, critical theory, and feminist studies.

The rise in the number of mixed methods studies in the 1981–1990 and 1991–2000 collections (Table 5.9) seems to reflect a growing willingness to use mixed methods (e.g., Johnson & Onwuegbuzie, 2004), despite cautions against this (e.g., Dillon & Wals, 2006).

As found by Hart and Nolan (1999), while some researchers employed methodologies and methods that reflected paradigms, others did not. These tended to cluster around several reasonably distinct research purposes. For example, the greatest number of studies in each collection included *educational development studies* (i.e., studies involving the development of models, plans, programs, curricula, units, modules, and instruments). While these studies often involve some form of "front-end" evaluation, design and development process, peer/professional review, pilot and field tests, and refinement, what was done and how each was undertaken varied from one product and context to another. The large number of development studies may be due to several factors, including: (a) the limited availability of suitable educational resources when EE was a young field; and (b) because EE was rarely part of the formal K-12 and teacher education systems in the United States over these periods, evidence from studies such as these was commonly viewed as a means to overcome struggles for legitimacy and inclusion.

The second most frequently addressed purpose included *large descriptive studies* of programs, teachers, and learners (Hart & Nolan, 1999). The third most frequently addressed purpose pertained to *definitional features and trends in EE*. Studies in the 1971–1980 collection were attempting to define and provide guidance to the field, while several of the studies in the 1991–2000 collection began to focus on the EESD/ESD relationship. *Focused inquiries* (Hart & Nolan, 1999) increased noticeably in the 1991–2000 collection, reflecting the growth of qualitative methods within the field. *Analyses of instructional resources* such as textbooks and kits were more common within the 1971–1980 and 1981–1990 collections than in the 1991–2000 collection. In the latter, attention appeared to shift toward studies of electronic resources (York, 1999), studies of materials in use (Klein, 1994), and tools to evaluate curricula (Minner, 1998). Lastly, while most dissertations contained a review of relevant theory and research, only one dissertation in these collections focused exclusively on the *review and synthesis of research* (Hines, 1985), which was replicated recently (Bamberg & Moser, 2007). Due to the increasing volume and diversity of research in and closely related to EE, review-and-synthesis studies are both more daunting

TABLE 5.9

Contents of Collections of Dissertation Studies in Category A, by Research Methodology

I. General Features			
Time Period	1971–1980	1981–1990	1991–2000
Dissertation Abstracts International	Vols. 31–41	Vols. 41–51	Vols. 51–60
Total Number of Dissertations Referenced	394	307	1,342
Number of Dissertation in *Category A*	271	179	248
II. Research Methodologies*			
A. Methodologies Allied with Paradigms			
1. Studies Using Quantitative Methods	175 (65%)	122 (68%)	122 (49%)
2. Studies Using Qualitative Methods	12 (4%)	16 (9%)	57 (23%)
3. Studies Using Mixed Methods	1 (<1%)	24 (13%)	35 (14%)
B. Methodologies Allied with Purpose			
1. Studies of Definitional Features of and Trends in EE	8 (3%)	8 (4%)	19 (8%)
2. Focused Inquiries	3 (1%)	3 (2%)	16 (6%)
3. Large-Scale Descriptive Studies	27 (10%)	24 (13%)	24 (10%)
4. Educational Development Studies	82 (30%)	41 (23%)	51 (21%)
5. Analyses of Instructional Resources	8 (3%)	7 (4%)	2 (1%)
6. Reviews and Summaries of Research	0	1 (<1%)	0

**Notes.* (1) Methodologies are split into two major categories to reflect (A) paradigms, as well as (B) selected clusters of methods allied with the study's purpose. Under (A), Denzin and Lincoln (2000) was used as the primary reference for A.2 and A.3. For (B), these clusters were adapted from Hart and Nolan (1999). (2) Percentages are based on the number of dissertations in Category A. (3) Because this analysis included only studies with abstracts presented sufficient information about these methodologies, percentages sum to less than 100%.

and more necessary. It is noteworthy that since 2000, several dissertations have presented a review and synthesis of research (Osbaldiston, 2004).

Implications and Recommendations

Those who have reviewed large segments of research in EE (e.g., Hart & Nolan, 1999; Rickinson, 2001; Volk & McBeth, 1997) have faced many of the questions, challenges, and concerns apparent in this chapter. Those arose due to the changing nature of the field of EE over the period of this review (1971–2000); i.e., it became more complex and dynamic. Content analyses were used to identify trends in dissertations that reflect some of the complexities and dynamics apparent within the United States and, to a lesser extent, Canada. The extent to which these (and other) conditions and trends may be apparent in other parts of the world remains an open question; one best answered by professionals familiar with conditions and trends in EE in their country or region.

In the context of Table 5.1, a broad review of this kind may be viewed as a *next step* that follows the preparation of reasonably comprehensive collections of research such as those used here. *Additional steps* lie beyond this, including in-depth analyses, critiques, and syntheses of research. In this review, limited attempts were made to analyze dissertations and trends in any depth, and none to critique or synthesize this body of research (i.e., to take these next steps). As a result, reviews of this kind usually raise more questions than they answer; much remains to be learned from reviews of other trends, in-depth analyses, critiques, and syntheses of this and other large segments of the EE research literature.

The following recommendations address delimitations and limitations apparent in this review.

- The source material, methods, frameworks, and results described in this review should be subjected to peer review, and refined as needed. To the extent possible, such a review should make use of full dissertations.

- The three collections of dissertations in EE used in this review cover the period from 1971 to 2000, primarily in the United States and Canada. A comparable collection for 2001–2010 should be prepared, and then subjected to comparable content analyses so as to extend the currency and relevance of these results.
- Whether for the 1971–2000 or the 1971–2010 period, these collections of research, should be subjected to more in-depth analyses and critiques of substantive and methodological features of studies, as well as syntheses of findings.
- Broad and reasonably comprehensive collections of research similar to those used in this review are needed for other countries and regions, particularly those that have lengthy and well-established research traditions in EE. Further, once available, those collections should serve as source material for broad analyses of trends, in-depth analyses, critiques, and syntheses of the research in those collections.

In addition, recommendations for further analysis and research may be offered in light of the results of these content analyses.

- Content Analysis 1: Within *Category A*, more in-depth analysis and further study is needed into the influences of "forerunners" of EE on contemporary EE research and practices (e.g., the extent to which these appear to support or hinder advances in the field), as well as into the evolving dynamics of the EEESD relationship in theory and in practice.
- Content Analysis 1: Dissertations in *Categories B–F* were included and reviewed due to their relationship or relevance to EE, at least within the United States. More detailed analyses and studies of these relationships are needed (e.g., environmental/ sustainability views, policies, and practices within agencies and sectors of society that influence and are influenced by EE; uses of studies by/in college/university environmental studies and environmental communication programs).
- Content Analysis 2: As of 2000, few dissertations were found for nonformal EE. There is a need to determine whether this number expanded during the 2001–2010 period (e.g., given apparent increases in such programs, and calls for increased accountability). Beyond this, various kinds of studies appear needed (e.g., stakeholder involvement, new target audiences, effects of instructional approaches such as team teaching and service-learning, longer-term impacts on participants, wider impacts on community factors such as resilience and on environmental factors).
- Content Analysis 2: In light of developments around the world, studies of recent National EE initiatives are needed, as are studies focused on recent State/ Provincial EE initiatives (e.g., development of *State Environmental Literacy Plans* in the United States in anticipation of *No Child Left Inside* legislation).
- Content Analysis 4: Building upon experience in countries such as Denmark, there is a need to analyze research in health education, particularly environmental health education, for its relationship and relevance to EE in countries where such research is limited.
- Content Analyses 5: In light of the apparent disparity between calls for interdisciplinary approaches to EE and the number of studies that reflect a single-subject emphasis (e.g., science education in the United States), there is a need for more detailed analyses and studies of this in K-12 education (e.g., conditions that hinder and approaches that foster interdisciplinarity).
- Content Analysis 6: In light of the ongoing emphasis on goals/objectives and outcomes related to knowledge and attitudes (Iozzi, 1981, 1984), there is a need for more detailed analysis and studies that focus on other dimensions of affect such as environmental sensitivity (Chawla, 1998), assumption of personal responsibility, locus of control, and intention (Simmons, 1991, 1995), as well as on the development, application, and transfer of skills such as critical thinking, inquiry/investigation, decision-making, and problem-solving/action skills (NAAEE, 2004a; Simmons 1995).
- Content Analysis 7: In light of continued use of quantitative, and increased use of qualitative and mixed methods, more detailed analyses are needed into doctoral student understanding of and choices related to research paradigms, methodologies, and methods (e.g., explicit or implicit attention to paradigmatic assumptions, appropriateness of selected methodologies and methods to he various kinds of research questions posed, other influences on methodological decisions such as constraints, practicalities, or convenience).

As noted in the introduction to this chapter, if the field is to take advantage of the intellectual capital apparent in journal, graduate, and fugitive studies, large-scale, ongoing efforts are needed to collect and review the EE research literature. However, as found by those who have undertaken large-scale reviews, the field is well beyond the point of relying on one or two professionals to do a credible job of this magnitude and complexity. While the barriers to do so are considerable, the educational and environmental needs we face beg for this type of diligent and sustained commitment on the part of national, regional, and global EE research communities.

References*

Ary, D., Jacobs, L., Razevieh, A., & Sorensen, C. (2006). *Introduction to research in education* (7th ed.). Belmont, CA: Wadsworth.

Bailey, L. (1909). *The nature study idea* (3rd ed.). New York, NY: MacMillan.

Note: Asterisks refer to dissertations from these collections. A full list of references for the dissertations reviewed is available, upon request, from the first author.

*Balakrishnan, U. (1995). International regimes on global warming. (Doctoral dissertation, University of Notre Dame, 1994). *Dissertation Abstracts International, 55*(7), 2133A. (UMI No. DA9434997).

Bamberg, S., & Moser, G. (2007). Twenty years after Hines, Hungerford, and Tomera: A new meta-analysis of psycho-social determinants of pro-environmental behavior. *Journal of Environmental Psychology, 27*(1), 14–25.

Bassey, M. (2000). Reviews of educational research. *Research Intelligence, 71*, 22–29.

Bennett, D. (1977). 12: The evaluation of environmental education learning. In J. Aldrich & A. Blackburn (Eds.), *Trends in Environmental Education* (pp. 193–211). Paris, France: UNESCO.

*Blankenship, G. (1990). Classroom climate, global knowledge, global attitudes, and civic attitudes. (Doctoral dissertation, Emory University, 1990). *Dissertation Abstracts International, 51*(5), 1569A. (UMI No. DA9027896).

*Blevins, J. E. K. (1994). Interdependence: A facet of global education as presented in elementary school social studies textbooks. (Doctoral dissertation, Auburn University, 1993). *Dissertation Abstracts International, 54*(8), 2884A. (UMI No. DA9402069).

*Booth, A. L. (1993). Learning to walk in beauty: Critical comparisons in ecophilosophy focusing on bioregionalism, deep ecology, ecological feminism, and Native American ecological consciousness. (Doctoral dissertation, The University of Wisconsin—Madison, 1992). *Dissertation Abstracts International, 53*(9), 4575B. (UMI No. DA9221899).

*Brown, M. L. (1995). Smoky Mountains story: Human values and environmental transformation in a southern bioregion, 1900–1950. (Doctoral dissertation, University of Kentucky, 1995). *Dissertation Abstracts International, 56*(5), 1943A. (UMI No. DA9530171).

Bucheit, J., Dorn, T., & Matheny, K. (2006). *Collection and coding of dissertation studies in environmental education and related fields, 1991–2000*. Melbourne, Florida, US: Unpublished master's research project, Florida Institute of Technology.

*Buhrs, T. (1992). Working within limits: The role of the commission for the environment in environmental policy development in New Zealand. (Doctoral dissertation, University of Auckland, 1991). *Dissertation Abstracts International, 52*(8), 3067A. (UMI No. DA9201572).

*Chatelain, L. (1981). Residential energy conservation practices, weatherization features in dwellings and electrical energy consumption in rural and urban households in Utah. (Doctoral dissertation, Florida State University, 1981). *Dissertation Abstracts International, 42*(1), 159B. (UMI No. DEN81-13258).

Chawla, L. (1998). Significant life experiences: A review of research on sources of environmental sensitivity. *The Journal of Environmental Education, 29*(3), 11–21.

Chenery, M., & Hammerman, W. (1984/85). Current practice in the evaluation of resident outdoor education programs: Report of a national survey. *The Journal of Environmental Education, 16*(2), 35–42.

Childress, R. (1976). Evaluation strategies and methodologies utilized in public school environmental education programs and projects—A report from a national study. In R. Marlett (Ed.), *Current issues in environmental education - II; selected papers from the fifth annual conference of the national association for environmental education* (pp. 23–34). Columbus, OH: ERIC Science, Mathematics, and Environmental Education Clearinghouse.

Childress, R. (1978). Public school environmental education curricula: A national profile. *The Journal of Environmental Education, 9*(3), 2–10.

Connell, S. (2006). Empirical-analytical methodological research in environmental education: Response to a negative trend in methodological and ideological discussions. *Environmental Education Research, 12*(3–4), 523–538.

Corey, D. (2008). *Dissertation Abstracts International*. Retrieved from http://www.netmlmarticles.com/Atricle/Dissertation-Abstracts-International/10603.

Denzin, N., & Lincoln, Y. (2000). *Handbook of qualitative research* (2nd ed.). Thousand Oaks, CA: Sage Publications, Inc.

Dialog, a ProQuest Company. (2008). *Dissertation abstracts online*. Retrieved from http://library.dialog.com/bluesheets/html/bl0035.html.

Dillon, J., & Wals, A. (2006). On the danger of blurring methods, methodologies and ideologies in environmental education research. *Environmental Education Research, 12* (3–4), 549–558.

Disinger, J. (1981). Environmental education in the K-12 schools: A national survey. In A. Sacks, et al. (Eds.), *Current issues vii, the yearbook of environmental education and environmental studies* (pp. 141–156). Columbus, OH: ERIC Science, Mathematics, and Environmental Education Clearinghouse.

Disinger, J. (1983). Environmental education's definitional problem. (ERIC Information Bulletin #2). Columbus, OH: ERIC Science, Mathematics, and Environmental Education Clearinghouse.

Disinger, J. (1989). The current status of environmental education in US school curricula. *Contemporary Education, 60*(3), 126–136.

Disinger, J., & Bousquet, W. (1982). Environmental education and the state education agencies: A report of a survey. *The Journal of Environmental Education, 13*(2), 13–22.

*Duke, D. F. (2000). Unnatural union: Soviet environmental policies, 1950–1991. (Doctoral dissertation, University of Alberta (Canada), 1999). *Dissertation Abstracts International, 60*(8), 3089A. (UMI No. DANQ39522).

Frankel, J., & Wallen, N. (2000). *How to design and evaluate research in education* (4th ed.). Boston, MA: McGraw-Hill Higher Education.

Funderburk, R. (1948). *The history of conservation education in the US* Nashville, TN: George Peabody College for Teachers.

Gay, L., & Airasian, P. (2003). *Educational research: Competencies for analysis and applications* (7th ed.). Upper Saddle River, NJ: Pearson Education, Inc.

*Gelobter, M. (1995). Race, class, and outdoor air pollution: The dynamics of environmental discrimination from 1970 to 1990. (Doctoral dissertation, University of California, Berkeley, 1993). *Dissertation Abstracts International, 55*(7), 2623B. (UMI No. DA9430490).

Gross, M., & Moore, D. (1985). *An interpretive research bibliography, 1978–1984*. College of Natural Resources, University of Wisconsin-Stevens Point.

Ham, S. (1992). *Environmental interpretation*. Golden, CO: North American Press.

Hammerman, D., & Hammerman, W. (1973). *Outdoor education: A book of readings*. Minneapolis, MN: Burgess Pub. Co.

Hart, E. P. (1981). Identification of key characteristics of environmental education. *The Journal of Environmental Education, 13*(1), 12–16.

Hart, E. P. (1993a). Alternative perspectives in environmental education research: Paradigm of critically reflective inquiry. In R. Mrazek (Ed.), *Alternative paradigms in environmental education research* (pp. 107–130). Troy, OH: North American Association for Environmental Education.

Hart, E. P. (1993b). Thoughts on research paradigms in EE. In R. Mrazek (Ed.), *Alternative paradigms in environmental education research* (pp. 301–302). Troy, OH: North American Association for Environmental Education.

Hart, E. P., & Nolan, K. (1999). A critical analysis of research in environmental education. *Studies in Science Education, 34*, 1–69.

Hesselink, F., van Kampen, P., & Wals, A. (2000). *ESDebate: International debate on education for sustainable development*. Gland, Switzerland: International Union for Conservation of Nature.

*Hester, J. B. (1998). The environment issue, 1987–1991: A time series analysis of TV news, real-world cues, and public opinion. (Doctoral dissertation, University of Alabama). *Dissertation Abstracts International, 59*(4), 995A. (UMI No. DA9831327).

*Hines, J. M. (1985). An analysis and synthesis of research on responsible environmental behavior. (Doctoral dissertation, Southern Illinois University at Carbondale, 1984). *Dissertation Abstracts International, 46*(3), 665A. (UMI No. DER85-10027).

Hines, J. M., Hungerford, H. R., & Tomera, A. N. (1986/87). An analysis and synthesis of research on responsible environmental behavior. *The Journal of Environmental Education*, *18*(2), 1–8.

*Hirt, P. W. (1992). A conspiracy of optimism: Sustained yield, multiple use, and intensive management on the national forests, 1945–1991. (Doctoral dissertation, University of Arizona, 1991). *Dissertation Abstracts International, 52*(10), 3702A. (UMI No. DA 9210288).

*Iacofano, D. S. (1986). Public involvement in environmental planning: A proactive theory for program design and management. (Doctoral dissertation, University of California—Berkeley, 1986). *Dissertation Abstracts International, 48*(5), 1333A. (UMI No. DET87-18024).

Inkowski, T. (1993). A contextual review of the "quantitative paradigm" in EE research. In R. Mrazek (Ed.), *Alternative paradigms in environmental education research* (pp. 29–79). Troy, OH: North American Association for Environmental Education.

Iozzi, L. (Ed.). (1981). *Research in environmental education,* 1971–1980. Columbus, OH: ERIC Science, Mathematics, and Environmental Education Clearinghouse. (ERIC Document Reproduction Service No. ED214762).

Iozzi, L. (Ed.). (1984). A summary of research in environmental education, 1971–1982. The second report of the National Commission on Environmental Education Research. *Monographs in Environmental Education and Environmental Studies,* (Vol. 2). Columbus, OH: ERIC Science, Mathematics, and Environmental Education Clearinghouse. (ERIC Document Reproduction Service No. ED259879).

Jackman, W. (1903). *Nature study for common schools* (3rd ed.). Bloomington, IL: National Society for the Study of Education.

Jickling, R. (1993). Thinking beyond paradigms in EE research. In R. Mrazek (Ed.), *Alternative paradigms in environmental education research* (pp. 307–309). Troy, OH: North American Association for Environmental Education.

Johnson, R., & Onwuegbuzie, A. (2004). Mixed methods research: A paradigm whose time has come. *Educational Researcher, 33*(7), 14–26.

Kirk, J. (1977). The quantum theory of environmental education. In R. McCabe (Ed.), *Current issues in environmental education - III* (pp. 29–36). Columbus, OH: ERIC Science, Mathematics, and Environmental Education Clearinghouse. (ERIC Document Reproduction Service No. ED 150 018).

Kollmuss, A., & Agyeman, J. (2002). Mind the gap: Why do people act environmentally and what are the barriers to pro-environmental behavior? *Environmental Education Research*, *8*(3), 239–260.

*Klein, P. A. (1994). Promoting geographic inquiry into environmental issues: An assessment of instructional materials for secondary geography education. (Doctoral dissertation, University of Colorado, 1993). *Dissertation Abstracts International, 55*(4), 1061A. (UMI No. DA9423510).

*Kraus, M. (1973). Science education in the National Parks of the United States: A descriptive study of the development of science education programs and facilities by the National Park Service and the relationship of these to the advent of nature study and conservation education in America. (Doctoral dissertation, New York University, 1973). *Dissertation Abstracts International, 34*(6), 3178A. (UMI No. 73-30086).

Lee, C. (Ed.). (1998). *Bibliography of interpretive research*. Ft. Collins, CO: National Association for Interpretation.

Leeming, F., Dwyer, W., Porter, B., & Cobern, M. (1993). Outcome research in environmental education: A critical review. *The Journal of Environmental Education*, *24*(4), 8–21.

*Leunes, B. L. (1978). The conservation philosophy of Stewart L. Udall, 1961–1968. (Doctoral dissertation, Texas A&M University, 1977). *Dissertation Abstracts International*, *38*(8), 5006A. (UMI No. DDK77-32170).

Lively, C., & Priess, J. (1956). *Conservation education in the U.S.* New York, NY: Ronald Press.

Lucas, A. (1981). The role of science education in educating for the environment. *The Journal of Environmental Education, 12*(2), 31–37.

*Mante, J. O. Y. (1995). Towards an ecological, Christian theology of creation in an African context. (Doctoral dissertation, The Claremont Graduate School, 1994). *Dissertation Abstracts International, 55*(8), 2442A. (UMI No. DA9502336).

Marcinkowski, T. (2003). Commentary on Rickinson's "Learners and learning in environmental education: A critical review of the evidence" (EER, 7(3)). *Environmental Education Research, 9*(2), 181–213.

Marcinkowski, T., & Mrazek, R. (Eds.). (1996). *Research in environmental education,* 1981–1990. Troy, OH: North American Association for Environmental Education.

*Minner, D. D. (1998). Environmental education curriculum evaluation questionnaire: A reliability and validity study. (Doctoral dissertation, The Pennsylvania State University, 1997). *Dissertation Abstracts International, 58*(12), 4543A. (UMI No. DA9817540).

*Minton, T. G. (1980). The history of the nature-study movement and its role in the development of environmental education. (Doctoral dissertation, University of Massachusetts Amherst, 1980). *Dissertation Abstracts International, 41*(3), 967A. (UMI No. DEM80-19480).

*Moorefield, D. L. (1994). A comparative study of experiential learning utilizing indoor-centered training and outdoor-centered training. (Doctoral dissertation, Texas Women's University, 1994). *Dissertation Abstracts International, 55*(6), 1479A. (UMI No. DA9428317).

*Morgan, P. A. (1998). Liberty Hyde Bailey: Pioneer and prophet of an ecological philosophy of education. (Doctoral dissertation, Columbia University). *Dissertation Abstracts International*, *59*(5), 1506A. (UMI No. DA9834358).

Mrazek, R. (Ed.). (1993). Alternative paradigms in environmental education research. *Monographs in Environmental Education and Environmental Studies,* (Vol. 3). Troy, OH: North American Association for Environmental Education.

Norland, E., & Somers, C. (Eds.). (2005). Evaluating nonformal education programs and settings. *New Directors for Evaluation*, *108,* 1–83.

North American Association for Environmental Education. (2004a). *Excellence in environmental education: Guidelines for learning (K-12)*. Washington, DC: Author.

North American Association for Environmental Education. (2004b). *Guidelines for the preparation and professional development of environmental educators*. Washington, DC: Author.

North American Association for Environmental Education. (2004c). *Nonformal environmental education programs: Guidelines for excellence*. Washington, DC: Author.

*Onyenekwu, C. M. (1993). Evidence and argumentation styles of scientists/engineers, business/economists, and political scientists associated with environmental protection public policy decisions. (Doctoral dissertation, The University of Oklahoma, 1993). *Dissertation Abstracts International, 54*(1), 151B. (UMI No. DA9315793).

*Ortiz, J. A. (1987). Energy education in women's magazines: A seventy year history, 1914–1984. (Doctoral dissertation, State University of New York at Buffalo, 1987). *Dissertation Abstracts International, 48*(10), 2747A. (UMI No. DEV87-27729).

Osbaldiston, R. (2004). Meta-analysis of the responsible environmental behavior literature. (Doctoral dissertation, University of Missouri—Columbia, 2004). *Dissertation Abstracts International*, *65*(8), 4340B. (UMI No. AAT 3144447).

Pandion Systems, Inc. (2006). *Best practices guide to program evaluation*. Alexandria, VA: Recreational Boating and Fishing Foundation.

Patton, M. Q. (2002). *Qualitative research and evaluation* (3rd ed.). Thousand Oaks, CA: Sage Publications, Inc.

*Peters-Grant, V. M. (1987). The influence of life experiences in the vocational interests of volunteer environmental workers. (Doctoral dissertation, University of Maine, 1986). *Dissertation Abstracts International, 47*(10), 3744A. (UMI No. DET87-03323).

Rickinson, M. (2001). Special Issue: Learners and learning in environmental education: A critical review of the evidence. *Environmental Education Research*, *7*(3), 208–320.

Robottom, I., & Hart, P. (1993). *Research in environmental education: Engaging the debate*. Geelong, Victoria, Australia: Deakin University Press.

Rocchio, R., & Lee, E. (1974). *On being a master planner . . . A step by step guide*. Columbus, OH: ERIC Science, Mathematics, and Environmental Education Clearinghouse.

Roth, R. (1976). *A review of research related to environmental education, 1973–1976*. Columbus, OH: ERIC Science, Mathematics, and Environmental Education Clearinghouse. (ERIC Document Reproduction Service No. ED135647).

Roth, R., & Helgeson, S. (1972). *A review of research related to environmental education*. Columbus, OH: ERIC Science, Mathematics, and Environmental Education Clearinghouse. (ERIC Document Reproduction Service No. ED068359).

Ryan, G., & Bernard, H. (2000). 29: Data management and analysis methods. In N. Denzin & Y. Lincoln (Eds.). *Handbook of qualitative research* (2nd ed., pp. 679–802). Thousand oaks, CA: Sage Publications.

Sato, M. (2006). *Evolving environmental education and its relation to EPD and ESD*. Paper presented at the UNESCO Expert Meeting on Education for Sustainable Development (ESD): Reorienting Education to Address Sustainability, Kanchanaburi, Thailand.

*Sauvage, J. L. (1994). Popular environmental theater in America, 1977–1993. (Doctoral dissertation, New York University, 1994). *Dissertation Abstracts International, 55*(4), 804A. (UMI No. DA9422944).

Schoenfeld, C. (1969). What's new about environmental education? *Environmental Education, 1*(1), 1–4.

Sharpe, G. (1982). *Interpreting the environment*. New York, NY: J. Wiley & Sons.

Simmons, D. (1991). Are we meeting the goal of responsible environmental behavior? An examination of nature and environmental center goals. *The Journal of Environmental Education, 22*(3), 16–21.

Simmons, D. (1995). Working Paper #2: Developing a framework for National Environmental Education Standards. In *Papers on the Development of Environmental Education Standards* (pp. 10–58). Troy, OH: North American Association for Environmental Education.

Smith, J. (1968). Where we have been—What we are—What we will become. *Journal of Outdoor Education, 2* (Winter), 3–6.

*Soffar, A. J. (1975). Differing views on the Gospel of Efficiency: Conservation controversies between agriculture and interior, 1898–1938. (Doctoral dissertation Texas Tech University, 1974). *Dissertation Abstracts International, 35*(11), 7237A. (UMI No. 75-07430).

Spero, V. (2006). *Collection and coding of dissertation studies in environmental education and related fields,* 1991–2000*; A continuation*. Melbourne, Florida, US: Unpublished master's research project, Florida Institute of Technology.

Stapp, W. (1974). Historical setting of environmental education. In J. Swan & W. Stapp (Eds.), *Environmental education* (pp. 42–49). New York, NY: J. Wiley and Sons.

*Stephen, C. R. (1984). Changes in the status of conservation education in selected institutions of higher education in the southeastern United States. (Doctoral dissertation, The Ohio State University, 1984). *Dissertation Abstracts International, 45*(6), 1713A. (UMI No. DEQ84-19016).

*Steverson, B. K. (1992). A critique of ecocentric environmental ethics. (Doctoral dissertation, Tulane University, 1991). *Dissertation Abstracts International, 52*(10), 3628A. (UMI No. DA9209668).

Swan, M. (1984). Forerunners of environmental education. In N. McInniss & D. Albrecht (Eds.), *What makes education environmental?* (pp. 4–20). Medford, NJ: Plexus Pub. Co.

Swan, M., & Stilson, J. (1980). *Dissertations in ECO-Education*. (Taft Campus Occasional Paper XV.) Oregon, IL: Lorado Taft Field Campus, Northern Illinois University.

Sward, L., & Marcinkowski, T. (2001). Environmental sensitivity: A review of the research, 1980–1998. In H. Hungerford, W. Bluhm, T. Volk, & J. Ramsey (Eds.). *Essential Readings in Environmental Education* (pp. 277–288). Champaign, IL: Stipes Publishing, L.L.C.

Tanner, R. (1974). *Ecology, environment, and education*. Lincoln, NE: Professional Educator's Publications, Inc.

*Thompson, S. J. (1995). Water resource development policy in agriculture: A comparative-historical analysis of state intervention in irrigate food-crop production in Malaysia, Sri Lanka and the western United States, 1850–1980. (Doctoral dissertation, Cornell University, 1992). *Dissertation Abstracts International, 55*(7), 2174A. (UMI No. DA9501359).

Tilden. F. (1957). *Interpreting our heritage*. Chapel Hill, NC: University of North Carolina Press.

United Nations. (1992). Chapter 36: Promoting education, public awareness and training. In *Earth Summit: Agenda 21, The United Nations programme of action from Rio*. New York, NY: Author.

United Nations Education, Scientific, and Cultural Organization. (1978). *Final Report: Intergovernmental Conference on Environmental Education*. Paris, France: Author.

United Nations Education, Scientific, and Cultural Organization. (1997). *Educating for a sustainable future: A transdisciplinary vision for concerted action*. Paris, France.

Voelker, A. M., Heal, F. A., & Horvat, R. E. (1973). *Environmental education-related research,* 1969–1972*: An annotated bibliography*. Madison, WI: Center for Environmental Communications and Education Studies, University of Wisconsin-Madison. (ERIC Document Reproduction Service No. ED 103280).

Volk, T., & McBeth, W. (1997). *Environmental literacy in the United States*. (Report Funded by the U.S. Environmental Protection Agency, and Submitted to the Environmental Education and Training Partnership, North American Association for Environmental Education). Washington, DC: North American Association for Environmental Education.

Waltz, M. E. (2008). *Repository profile: ProQuest UMI dissertation publishing*. Retrieved from http://www.crl.edu/PDF/umi_dissertations.pdf.

*Williams, D. C. (1992). The range of light: John Muir, Christianity, and nature in the post-Darwinian world. (Doctoral dissertation, Texas Tech University, 1992). *Dissertation Abstracts International, 53*(4), 1258A. (UMI No. DA9226297).

Wiltz, K. (Ed.). (2000). *Proceedings of the teton summit for program evaluation in nonformal environmental education*. Jackson Hole, WY: Teton Science School, and Columbus, OH: The Ohio State University.

Winzler, E., & Cherem, G. (1978). *An interpretive research bibliography*. Derwood, MD: Association of Interpretive Naturalists.

World Commission on Environment and Development. (1987). *Our common future*. Oxford, UK: Oxford University Press.

*Wyatt, R. C. (1977). Exploring organizational attempts to affect environmental policy in an urban area: An analysis of the basis for success or failure. (Doctoral dissertation, University of Texas at Austin, 1976). *Dissertation Abstracts International, 37*(8), 5374A. (UMI No. DBJ77-4004).

*York, K. J. (1999). Efficacy of the World Wide Web in K-12 environmental education. (Doctoral dissertation, Michigan State University, 1998). *Dissertation Abstracts International, 60*(3), 713A. (UMI No. DA9922395).

*Young, J. S. (1992). The Environmental Education Act of 1970: A study of federal efforts as curriculum intervention. (Doctoral dissertation, Syracuse University, 1992). *Dissertation Abstracts International, 53*(6), 1824A. (UMI No. DA9229708).

Zelezny, L. (1999). Educational interventions that improve environmental behaviors: A meta-analysis. *The Journal of Environmental Education, 31*(1), 5–14.

6

Transformation, Empowerment, and the Governing of Environmental Conduct

Insights to be Gained From a "History of the Present" Approach

Jo-Anne Ferreira
Griffith University, Australia

Introduction

Transforming individuals and their environmental behavior is generally regarded as the ultimate goal of environmental education (see, e.g., Beringer, 2006; Brody & Ryu, 2006; Gough, 1997; Hungerford & Volk, 1990; Jickling, 2004). However, the notion of "behavior change" is problematic for the field of environmental education, regarded as it is by many as too technical or technicist an approach to change. This has led to calls to instead empower individuals to transform themselves into informed and active environmental citizens, thereby leading, it is assumed, to the transformation of the behavior of whole communities and entire societies. As I have argued elsewhere (Ferreira, 2009), a belief in the need to empower and transform has become so well established within environmental education that it seems "wrong" to question either the goal of transformation or empowerment as the means through which this is to be achieved. Why is this the case? Why does the field of environmental education see empowerment as a more appropriate (even the most appropriate) means to achieve a sustainable society than other—more traditional and intentional—means of behavior change? Prevailing answers to such questions include that empowerment allows for individuals to freely choose a new path—to be empowered to transform themselves—rather than be manipulated to behave in particular ways through, for example, advertising campaigns or economic incentives; and that behavior change strategies that focus on individuals (as opposed to individuals as members of a community) and on specific behaviors do not provide a way for individuals to transform their whole self into an environmental being or environmental citizen.

Alternative insights on this issue are offered through the use of a "history of the present" approach. Historians of the present argue that we are better able to understand our current circumstances once we begin to "interrogate the 'rationality' of the present" (Gordon, 1980, p. 242), that is, to understand how our ways of thinking about and doing things—what Michel Foucault (1991) termed our "governmentalities"—have come to be. What such an approach offers environmental education is the possibility of understanding both *how* we have come to believe certain things, and what the *effects* of these beliefs (our rationalities or mentalities) are on our thinking and practice. By viewing empowerment and transformation through the lens that a history of the present approach provides, we are able to illuminate how such rationalities work to shape, rule or govern the ways in which we think and act. The focus of the questioning here is different: not a concern with *why* empowerment and transformation are the chosen rationalities, nor a questioning of the merit or value of empowerment and transformation, but rather a concern with *how*—how have we in environmental education come to see empowerment as the most appropriate means to achieve behavior change and transformation of the individual and society as the most appropriate goal.

Undertaking such a history stands in contrast to traditional approaches to undertaking histories, especially those that see our history as a glorious march of progress toward a sustainable future. A history of the present, for example, does not see the present as the apogee of human development, nor does it reduce the present to particular causal principles, such as ideologies. Nor do such histories seek to decode, interpret, or explain. Histories of the present seek instead to *diagnose* the present by using history to question how our established ways of knowing and ways of doing have come to be. It is an approach that allows for the minute, the down-to-earth, the mundane, indeed the normal, to be described in such a way that the "relations of power" (Foucault, 1980, p. 142) through which we are governed, and through which we govern others and ourselves, are made visible.

Governing and government here does not refer solely to "the government"—that is, to the social organizations that administer populations—but also to the government of the self—that is, to the techniques we as environmental educators use for governing our own and others conduct. As Mitchell Dean (1999, p. 11) noted, "[Government is] any more or less calculated and rational activity, undertaken by a multiplicity of authorities and agencies, employing a variety of techniques and forms of knowledge, that seeks to shape conduct by working through our desires, aspirations, interests, and beliefs, for definite but shifting ends and with a diverse set of relatively unpredictable consequences, effects, and outcomes." In this sense, our work as environmental educators involves efforts to both govern the conduct of others and to educate others about how to govern their own conduct. *How* we do this becomes clear when we investigate and describe our rationalities of rule and their effects. Without such an understanding it is difficult for us to bring our ways of thinking and ways of doing into question. It is my contention that without such questioning of our rationalities, we will be unable to understand the effects of our thinking on our practices.

This chapter thus investigates the ways in which undertaking histories of the present studies may be useful to the field of environmental education. This is achieved through a discussion of "how to" undertake a history of the present, illustrated with the sorts of questions we would need to ask in order to make visible how our ways of thinking in environmental education have become "normal"—that is, become habits and beliefs that are difficult to challenge and change. Such a redescription also demonstrates how histories of the present work to "lower" the level of analysis from, for example, debates over ideology, by illuminating the technical conditions and mundane circumstances out of which our current beliefs in environmental education—our accepted ways of knowing and ways of doing—have emerged. As noted previously, an understanding of the "how" allows us to also understand how we might bring about changes to our ways of knowing and ways of doing.

A cautionary note: histories of the present provide us with a tool and framework for making visible a different perspective from orthodox explanations. Such histories do not claim, as some research approaches do, to provide us with an essential truth or universal answer. What is illuminated instead by such an approach is that our rationalities—our regimes of truth—are not predestined but are instead contingent and variable—and therefore amenable to change. This is because histories of the present take history and context into account in a way that makes it difficult for universal claims to be made. While histories of the present may be critcized for not "providing the answer," I consider they offer something more: an understanding of how we have come to be what we are where we are, and therefore an understanding of how we might become something other than what we are, and do something other than what we do, now and into the future. For environmental education, histories of the present offer, for example, a way for us to problematize the orthodoxy of our beliefs and practices—even those that may appear at first glance to be unorthodox, radical challenges to the status quo, as I have argued elsewhere (Ferreira, 2009).

I begin this chapter by explaining what is meant by a "rationality of rule," and then describe characteristic concerns of a history of the present. This description is illustrated with examples from environmental education to demonstrate how undertaking a history of the present is able to illuminate not only the field's rationalities of rule but also, importantly, their effects on our practices.

The Ordinary and the Everyday

Rationalities of rule are so much a part of the fabric of our collective thought that they are difficult for us grasp. For example, one could argue that it is part of our collective thought that our purpose as environmental educators is to empower individuals to become informed and active environmental citizens. While we might for the most part agree on this purpose, we seldom if ever question it. This is because we do not tend to think about our own unquestioned assumptions, and if we do, then we struggle to accept that these are operating as rationalities that rule or govern our ways of thinking and our ways of acting, both as a field, and as individual practitioners. As Dean (1999, p. 16) notes, this is because "the thought involved in practices of government is collective and relatively taken for granted, i.e., not usually open to questioning by its practitioners." It is this that makes such rationalities so powerful, and why we need to always question and problematize them.

Undertaking a history of the present allows us to begin to see that our desire as environmental educators to shape individuals as environmental citizens is part of the governmentality of modern liberal forms of governance; forms of governance that prefer not to govern through top-down repressive means, but instead in a pastoral and productive fashion seek to "conduct the conduct" of particular populations to achieve particular ends (Foucault, 1982, pp. 214–215). Modern liberal modes of rule can be seen as pastoral and productive because they seek to direct conscience and fashion from the population—in a caring rather than repressive manner—citizens who are able to manage their own conduct; citizens who are civilized, self-reflective, self-governing persons. Such governing is undertaken not only by "the government" but by all of us—either through governing our own conduct (our efforts to live in a more environmental manner, e.g.) or through governing the conduct of others (through our efforts as educators to shape our students as informed and active environmental citizens, e.g.). It is only through an understanding of how this governance of one's own and others conduct occurs, that that we are able to make visible, for example, how empowerment has come to be a strategy used by environmental educators for governing the conduct of particular populations for particular purposes.

A Quasi-Roadmap

Studies seeking to understand our present through our past are indebted to Foucault whose work allows us to see that "that-which-is" may not be "that-which-is," that is, to unsettle the ordinary and taken-for-granted. However, this does not mean that histories of the present slavishly follow a series of methodological moves or rely on particular methods as promulgated by "the master." This is because, as Clare O'Farrell (2005, p. 50) notes, "Foucault continually changed and refined his concepts, not only on a major scale, but in very minute and subtle ways, something which makes his work extremely difficult to systematise for the purposes of a methodical and wholesale application." Rather, as Gordon (1980, p. 258) states, "[what] Foucault may have to offer is a set of possible tools, tools for the identification of the conditions of possibility which operate through the obviousness and enigmas of our present, tools perhaps also for the eventual modification of those conditions."

Nonetheless, four characteristic concerns can be identified in Foucault's investigations. The first of these is a concern with genealogy, that is, with the ways in which understanding the past can help us to understand the peculiar ways our seemingly objective truths, rationales and assumptions have come to *order* the present. It is such "enquiries into our past [that make] intelligible the 'objective conditions' of our social present" (Gordon, 1980, p. 233). So, we might ask ourselves, how has it come to be that environmental education seeks to shape the whole person as an active and environmental citizen, not only their environmentally problematic conduct? What are the historical shifts that have occurred? Annette Gough's chapter in this collection presents an example of such a study, with her genealogy of the emergence of environmental education research.

The second concern we see in Foucault's oeuvre is with archaeology, that is, with the ways in which quite particular material and historical conditions have legitimized—ordered, appropriated, excluded—some discourses and not others (Gordon, 1980). How is it, for example, that the field has come to prefer constructivist approaches to behavior change, such as empowerment, to behaviorist approaches such as behavior modification (as we seen in social marketing approaches, e.g.)? In what ways has empowerment come to be legitimized and behaviorist approaches excluded? What is the history of the concepts of empowerment and transformation: in what historical conditions do they emerge as key concepts for environmental educators? In a previous study (Ferreira, 2009), I have described, for example, the ways in which education *for* the environment has become legitimized in environmental education.

The third concern is with an "ethical" dimension: with the ways in which we are constituted and constitute ourselves as ethical subjects (Gordon, 1980). The concern here is with how we, as individualized subjects, have become both objects and subjects in need of governance. Of interest in a history of the present study, for example, would be the techniques used (by ourselves to work on ourselves and by others to work on us) to shape ourselves as informed and active environmental citizens, and how the informed and active environmental citizen has become a form of subject for us to aspire to (Ferreira, 2000).

The final concern is one that seeks to understand "the proper use to be made of the concept of power, and of the mutual enwrapping, interaction and interdependence of power and knowledge" (Gordon, 1980, p. 233). Here we might ask: how is power enacted when we work to shape ourselves and/or our students as informed and active environmental citizens? Indeed, we could also ask how power is operationalized when we, as environmental educators, seek to empower others to become such subjects.

Foucault also outlined five "methodological precautions" for his studies. The first of these is that power should be understood through its *effects*, that is, "at its extremities, in its ultimate destinations, with those points where it becomes capillary" (Foucault, 1980, p. 96). What, we might ask, is the effect on the subject of techniques that seek to empower individuals to transform themselves into informed and active citizens? How do such techniques enable the operation of power?

Secondly, power should be examined "at the point where it is in direct and immediate relationship with that which we can provisionally call its object, its target, its field of application, there—that is to say—where it installs itself and produces real effects" (Foucault, 1980, p. 97). Here we would seek to identify the precise points at and through which power is able to have an effect in shaping the subject as an informed and active environmental citizen.

Thirdly, the self needs to be understood as "an effect of [relations of] power, and at the same time, or precisely to the extent to which it is that effect, it is the element of its articulation. The individual which power has constituted is at the same time its vehicle" (Foucault, 1980, p. 98). This is one of the most challenging insights Foucault's studies offer as it requires us to think differently about our own subjectivity—not as necessarily natural and given but as something constituted by and through power.

Fourthly, in order to understand how mechanisms of power come into being, and continue to take hold, power must be examined not from the "top down" but from "its infinitesimal mechanisms" (Foucault, 1980, p. 99). Attention must, Nikolas Rose argues, therefore be paid to "the humble, the mundane, the little shifts in our ways of thinking and understanding, the small and contingent struggles, tensions and negotiations" (1999, p. 11). For Foucault, both our past and our present are made more visible to us through an examination at the "molecular" level rather than at the level of "historically significant events. Here we might seek to understand what the little shifts in our ways of thinking" are that have enabled empowerment to become the means du jour for the field of environmental education.

Foucault's final methodological precaution is that we need to keep in mind that power is not ideological. He argues that power is "both much more and much less than ideology. It is the production of effective instruments for the formation and accumulation of knowledge—methods of observation, techniques of registration, procedures for investigation and research, apparatuses of control. All this means that power, when it is exercised through these subtle mechanisms, cannot but evolve, organise and put into circulation a knowledge, or rather apparatuses of knowledge, which are not ideological constructs" (Foucault, 1980, p. 102). What is important here then, is not so much ideological discussions and debates but rather coming to understand how knowledge in environmental education is "strategically shaped and organised by exercises of power" (O'Farrell, 2005, p. 87).

An Analytics of Government

While histories of the present do not fall neatly together into a neo-discipline, it is nonetheless possible to identify a common approach and set of concerns to such studies, as outlined above. From this emerges a quasi-methodology referred to as "an analytics of government" (Dean, 1994, 1999; Kendall & Wickham, 1999; Rose, 1999). Foucault's initial ideas about how to undertake studies of our governmentalities have underpinned efforts by Foucault scholars—in particular the Anglo-Foucauldians—to develop a methodological framework for examining governmentality "in action." An analytics of government, though, remains quasi-methodological. In Dean's (1999, p. 3) words, "[t]here is no one governmentality paradigm. There is no one common way of using the intellectual tools being produced by workers in the area. There are no prescribed limits to the intellectual formations of which studies of governmentality can be a part or to the empirical areas in which they can be developed . . . In short, there is no single volume that surveys this literature, presenting its major concepts, providing an overview of its historical perspectives, or making intelligible its contribution to the analysis of present styles of government."

Nonetheless, an analytics of government is justified in so far as it provides 'a [quite specific] purchase for critical thought upon particular problems in the present' (Rose, 1999, p. 9). It is a means for investigating the practices that direct conduct, and for investigating the forms of thought or mentalities that guide such practices (Dean, 1999, pp. 36–40). To this end, an analytics of government seeks to ask certain sorts of questions—troubling questions–of those aspects of our present that seem most self-evident and natural to us, such as the desire to empower individuals to become environmental citizens.

An analytics of government thus encourages a certain disposition toward the field of inquiry, a disposition that seeks "to connect questions of government, politics and administration to the space of bodies, lives, selves and persons" (Dean, 1999, p. 12). To understand government, according to Dean (1999, p. 12), we must "analyse those practices that try to shape, sculpt, mobilize and work through the choices, desires, aspirations, needs, wants and lifestyles of individuals and groups." In this way, an analytics of government diagnoses the present, in order to make the familiar and self-evident both visible and historical.

The benefit to a field such as environmental education of making the familiar visible and historical is that it works to "lower" the debate in a field. For example, understanding government as a process through which definite but limited improvements rather than utopian transformations are possible, may cause us to redirect our efforts away from a search for ultimate (normative) goals or transcendent principles that should or should not direct the ways in which we govern and are governed. This is because an analytics of government is concerned with understanding how we govern and how we are governed. An analytics of government aims to gain a purchase on these governmental regimes by clarifying their forms of thought—their mentalities—and by examining the effects of these at their point of application: "Foucault thus implied that, rather than framing investigations in terms of state or politics, it might be more productive to investigate the formation and transformation of theories, proposals, strategies and technologies for the "conduct of conduct." Such studies of government would address the dimension of our history composed by the invention, contestation, operationalization and transformation of more or less rationalized schemes, programmes, techniques and devices which seek to shape conduct so as to achieve certain ends" (Rose, 1999, p. 3). What, in environmental education, are our schemes, programs, techniques, and devices—which seek to shape conduct toward certain ends—and what are their effects?

Analytics of government are, in this sense, empirical. They engage in an "empiricism of the surface," allowing for a description of "the differences in what is said, how it is said, and what it allows to be said" (Rose, 1999, p. 57). What are we—as a field of endeavor seeking to "reorient education" and "reorient society"—to make, for example, of government objectives such as: "Communities around Australia are empowered to work effectively toward sustainability by having the information and resources to enable them to act" (Australian Government, 2009, p. 26)? What is being said here? How is it being said? What does it allow to be said or not said?

What is called for here is a shift from *why* questions to *how* questions. How, for example, have we come to believe that the only defensible way, educationally, to address the environmental crisis is through the empowerment of individuals to become informed and active environmental citizens? How is it that even "the government" has come to be calling for the same thing? Studies undertaking an analytics of government therefore begin by identifying and examining the specific situations in which particular activities come to be understood as problematic. They ask *how* questions such as "How do we think about how we govern?" and "How are we governed?" Such studies do not see the practices of government as utopian expressions of pure principles or a priori but rather as complex regimes of practices. That is, the practices of government are not seen as simply the expression of a government's values. Rather, values are examined for the part they play in the rationality and rhetoric of government. Such studies are distinctive, therefore, in that they turn away from globalizing positions that see government as either good or bad (Dean, 1999, pp. 27–36).

The *how* questions of an analytics of government therefore display a quite particular perspective on relations of power, knowledge, and the self (Dean, 1999, pp. 27–38; Rose, 1999, pp. 4–60). They ask:

- How does a particular state of affairs become problematized?
- How are particular governmental strategies or modes of governing developed and deployed to address problems?
- How are these modes of governing able to appear in certain times and places, that is, how are they influenced by their setting?
- How do these modes of governing constitute us as subjects who are both governable and self-governing?

The characteristic concerns and *how* questions of an analytics of government thus call for a focus on the *effects* of ways of our thinking, in particular, how our ways of thinking shape what can be thought and what can be done in environmental education. What might it mean for us as environmental educators if our efforts to empower individuals to become environmentally informed and active citizens were to be recognized as governmental techniques for equipping individuals with an ensemble of culturally acceptable practices that enable them to live their lives in an environmentally sustainable manner? How, for example, would this challenge our thinking about what can be thought and what can be done in environmental education?

Illuminating the Ordinary

In order to make the familiar in environmental education visible, a range of questions that seek to illuminate the actual techniques—the micro-practices in which environmental educators engage, be this at the level of constructing discourse or at the level of empowering their students to take on new "orders of living" (Weber, 1958)—would need to be answered. Specific questions would need to be asked of the histories, theories, and practices of contemporary environmental education. If, for example, we want to understand how we in environmental education have come to see the development of informed and active environmental citizens as our principal goal, we would need to ask:

- *How did environmental conduct come to be understood as a problem?* That is, how was it that at a particular point in time the environmental conduct of individuals came to be of concern? What historical events, social movements, documents, and so on make environmental conduct a social/cultural/governmental concern? Where does this concern emerge and how does it spread, that is, what is the role that the setting and context plays in the emergence of environmental conduct as an issue of concern? Why in the 1960s in the West, for example, did the environmental conduct of individuals come to be seen as a major problem?
- *How does the setting in which environmental education emerges frame the responses to the problem of environmental conduct?* How is it, for example, that the empowerment of individuals, rather than legislation, comes to be seen as a solution to this newly identified problem of individuals' environmental conduct? How does education come to be seen as a suitable response to this problem? What historical events, social movements, documents, and so on are evident in particular settings and how might they influence the means used to address this newly identified problem of individuals' environmental conduct?
- *How do these responses govern the field, both in their development and their deployment?* In identifying the empowerment of individuals as a solution to the problem of environmental conduct, what is left out? How might empowerment, seen as a solution, work to limit other possible ways of addressing the problem of environmental conduct?
- *How do these responses work to fashion new forms of conduct, indeed, new personas?* How can empowerment be understood as a productive exercise of governmental power? That is, how does empowerment work to fashion and shape individuals into new beings called environmental citizens? How are environmental citizens both the object and the subject of power?

Such questions are designed to illuminate what Dean (1999, p. 40) refers to as "inconvenient facts." Could it be, for example, that a technique such as empowerment—traditionally thought of as a means for liberating repressed subjects—may be more accurately defined as a governmental technique for conducting the conduct of citizens? If this is the case, what are the implications for environmental education theory and practice?

Conclusion

Examining the historical in our present ways of thinking and ways of doing in environmental education allows us to ask new questions of our work. To answer these questions, however, involves us approaching environmental education not as a quasi-religious cause but rather as a body of knowledge in and through which governmental relations of power operate. At stake are some firmly and fondly held beliefs about the world and about our role as educators.

References

Australian Government. (2009). *Living sustainably: The Australian government's national action plan for education for sustainability.* Canberra, ACT: Department of Environment, Water, Heritage and the Arts.

Beringer, A. (2006). Reclaiming a sacred cosmology: Seyyed Hossein Nasr, the perennial philosophy, and sustainability education. *Canadian Journal of Environmental Education, 11*, 26–42.

Brody, S. D., & Ryu, H.-C. (2006). Measuring the educational impacts of a graduate course on sustainable development. *Environmental Education Research, 12*(2), 179–199.

Dean, M. (1994). *Critical and effective histories: Foucault's methods and historical sociology.* London, UK: Routledge.

Dean, M. (1999). *Governmentality: Power and rule in modern society.* London, UK: Sage Publications.

Ferreira, J. (2000). Learning to govern oneself: Environmental education pedagogy and the formation of an ethical subject. *Australian Journal of Environmental Education, 16*, 31–36.

Ferreira, J. (2009). Unsettling orthodoxies: Education for the environment/ for sustainability. *Environmental Education Research, 15*(5), 607–620.

Foucault, M. (1980). "Power and strategies" and "Two lectures." In C. Gordon (Ed.), *Michel Foucault—Power/knowledge: Selected interviews and other writings, 1972–1977* (pp. 78–108). New York, Brighton: Pantheon Books, Harvester Press.

Foucault, M. (1982). The subject and power. In H. Dreyfus & P. Rabinow (Eds.), *Michel Foucault: Beyond structuralism and hermeneutics* (pp. 208–226). Brighton, UK: Harvester.

Foucault, M. (1991). Governmentality. In G. Burchill, C. Gordon, & P. Miller (Eds.), *The Foucault effect: Studies in governmentality with two lectures by and an interview with Michel Foucault* (pp. 87–104). Chicago, IL: The University of Chicago Press.

Gordon, C. (1980). Preface and afterword. In C. Gordon (Ed.), *Michel Foucault—Power/Knowledge: Selected interviews and other writings, 1972–1977* (pp. vii–x, 229–260). New York, Brighton: Pantheon Books, Harvester Press.

Gough, A. (1997). *Education and the environment: Policy, trends and the problems of marginalisation.* Camberwell, VIC: The Australian Council for Educational Research.

Hungerford, H. R., & Volk, T. L. (1990). Changing learner behavior through environmental education. *Journal of Environmental Education, 21*(3), 8–21.

Jickling, B. (2004). Making ethics an everyday activity: How can we reduce the barriers? *Canadian Journal of Environmental Education, 9*, 11–26.

Kendall, G., & Wickham, G. (1999). *Using Foucault's methods.* London, Thousand Oaks, CA: Sage Publications.

O'Farrell, C. (2005). *Michel Foucault.* London, UK: Sage Publications.

Rose, N. (1999). *Powers of freedom: Reframing political thought.* Cambridge, UK: Cambridge University Press.

Weber, M. (1958). *The protestant ethic and the spirit of capitalism.* New York, NY: Charles Scribner's Sons.

Section II

Normative Dimensions of Environmental Education Research

Conceptions of Education and Environmental Ethics

Bob Jickling
Lakehead University, Canada

Arjen E.J. Wals
Wageningen University, The Netherlands; Cornell University, USA

Introduction

First, and perhaps most obviously, education and environmental education are normative ideas. By normative, we mean ideas, and associated achievements, that are valued in particular context. That is, they are about value as opposed to fact. Education is something that is widely valued, individually and collectively. Environmental education is also valued in varying degrees by growing numbers educators and political bodies as it is increasingly referenced in policy documents (educators are likely to be familiar with examples from their own regions). However, it can also be said, that environmental education is valuable in that it draws attention to overlooked educational priorities (Jickling, 1992, 2003, 2005a; Jickling & Spork, 1998). After all, education itself is at the heart of this endeavor. It has long been acknowledged that these are essentially contested concepts that develop and change over time suggesting a fluidity of meaning that shifts across a range of contexts (Barrow & Milburn, 1986; Barrow & Woods, 1982; Hart, Jickling, & Kool, 1999; Jickling, 1997; Peters, 1966, 1973; Walsh, 1993; Williams, 1976). These concepts are, and ought to be, continually recreated (Jickling & Wals, 2008). This section seeks to reivigorate that process.

Second, environmental issues are also fundamentally normative in nature. Considering ethical issues, that is considering the nature of "good" and how people should carry themselves in the world, is essential to the examination of critical standpoints and implicit value laden stances.[1] Hence, the task in this section is to reimagine places for environmental philosophies and ethics within environmental education research. The emergence environmental ethics as a formal field of inquiry suggests that the nature of ethics itself is also in a process of continual recreation (i.e., Leopold, 1966/1949; Weston, 1992).

Questions about the nature of education and ethics are, however, problematic. In a research era infused with a postmodern and poststructural mood there is skepticism about concepts that are sometimes associated with objective truths, broad generalizations, and meta-narratives. Despite these concerns there are no political vacuums. A relative absence of normative dialog leaves the conceptual terrain open to misunderstanding, mischief, and abuse. This section aims to generate new openings for researchers who find normative questions in environmental education inescapable.

Reopening Normative Questions

Critical theory, in various manifestations, is a powerful force in environmental education—perhaps even providing a dominant research agenda in some parts of the world, and in some research circles (Ferreira, 2007, 2009). So, it is refreshing to find concerns about normative issues being raised from *within* this body of literature. For example, in commenting on poststructuralism, Delanty (1999, p. 3) is of the view that "this movement is now at an end, having largely accomplished its objective—the relativization of identity and knowledge and the demonstration of the limits of the intellectual categories of the nineteenth century." He adds that for the twenty-first century, "it is no longer a question of attacking false universalisms but of overcoming relativism and the fragmentation of the social" (p. 3).

Sayer (2000) similarly argues that, "The massive imbalance between sophistication of positive social science and the poverty, at least outside political theory, of normative thinking is intolerable . . ." (p. 186). A key problem for Sayer comes from Nietzschean poststructuralism that reduces problems of justice and morality to simple problems of interest. Sayer argues:

> Such a radical reduction is not sustainable: those who sneer at values and morality get as upset as anyone else when someone treats them improperly. Moral discourse is indeed sometimes little more than a camouflage or legitimation of power, often hypocritical; but again, a bad use of such discourse need not drive out a good use. (p. 177)

For Sayer, considering ethical issues—concerning the nature of good and how people should treat one another—is essential to the examination of social sciences' critical standpoints and implicit normative stances. Politics and ethics may be similar in some contexts, but this might mark a point of departure in much contemporary discourse.

For the now seething poststructuralists among the readership of this introduction, we offer an interpretation of Delanty's provocative claims. In a sense, we do not think what he says is complete. There are certainly abundant instances of unequal distribution of power and authority, exploitation, marginalization, colonialism, and imposition of false universals and other readings of social contexts to maintain a viable body of research.

Yet, in another more theoretical way, Delanty opens, or reopens, lines of normative inquiry. As he suggests, many contemporary researchers are rightly suspicious of grand meta-narratives, but less sure where to go next. Delanty and Sayer both point toward normative questions as terrain for more theorizing and practice.

At the Intersection of Education and Ethics

As a starting point, we suggest that environmental education exists at the intersection of two normative ideas—education and ethics—and suggest that this intersection describes an area of research that deserves more attention.

To begin, imagine posing the question, "What is education?" In our experience, questions like this frequently evoke responses like: "Whose version of education should count?" "Education isn't value free," and "Education doesn't really have any meaning [read relevance] any more." Note, that these responses, anecdotally at least, seem to support Delanty's claim that poststructuralists have done their work! In fact these kinds of replies aren't new, but rather have become slogans, clichés, and conversation stoppers. They roll off the tongue but don't lead anywhere. If we cannot do better than this then we ***are*** left with relativism and fragmentation.

When talking about ethics in environmental education there can be similarly unhelpful responses. These range from, "Nobody's going to tell my kid what to believe," to a limited framing as "the ethics of research," to the relative merits of various positions extant in the literature ranging across topics such as animal rights, deep ecology, ecofeminism, and the new environmental paradigm. There are, of course, other ways to conceive of ethics, especially in an educational setting.

As a minimal response, it seems reasonable to expect researchers to disclose their assumptions—that is their operational conceptions of education and ethics. Whether consciously revealed and examined, or not, we all respond in our teaching and research in ways that reflect some kind of philosophical assumptions about what these concepts are. Given that qualitative researchers, in particular, have become accustomed to declaring their sociological assumptions and contextualizing their own positioning within the research context, it is a logical extension to also declare philosophical positioning on key issues like education. Doing so would provide three advantages. First, as in present qualitative research, clarity about one's own positioning would enable researchers and readers to better interpret issues that might be otherwise thought of as bias. Second, such an ongoing practice would assist in opening up an area of undervalued research. And, third, this practice could energize discussion about normative questions in our field.

So, What About Education? In addition to the clichéd responses described above, there appear to be two additional reactions. The first is to avoid the question, "What is education?" and second is to provide a substitute. The 2008 meeting of AERA's Environmental and Ecological Education Special Interest Group provided an opportunity to witness both. A question for panelists discussing environmental learning asked how they would link their work to the concept of education. It was initially rejected by the session chair. As he rightly pointed out, it was a large a question. Nevertheless, this struck me, and a few others, as an odd turn of events. Most good symposia do deal with large questions. If this had been a large question, but one that reflected more mainstream interests, would it have been spurned? While this was not done in bad faith, the effect still was to put aside an inconvenient question.

Fortunately, with a little persistence, the question did reach the table, whereupon one prominent educator embraced the substitution option. He seemed incredulous that such a question would be asked, replying that he "finds 'learning' a much richer concept." The critical point in this case is that learning is *not* a richer concept than education. They are different in nature, and it is a category mistake to compare them this way, or to substitute one for the other. The speaker found research about learning to be personally more fulfilling, but his own disenchantment over questions about education do not make them inherently less important.

We mention this example to underscore how deeply conditioned the avoidance of normative questions can be. A relative absence of dialog about "education" leaves the terrain open to advocates of other concepts like "learning." This is not to say that study of learning is not important, it is. But, it does not answer questions like, "How do you conceive of education?" "In theory? "In practice?" Or, alternatively, "What are your assumptions about education?" This concern is developed by Le Grange (2004) who argues that the rise of the language of learning now enables the redescription of the process of education in terms more like that of an economic transaction—with the learner as consumer and the educational institution as provider. Alternatively, he suggests, the language of education opens more space for complex understandings of the nature of environmental issues and for recognizing that "environmental knowledge is produced in interdependent and interactive relationships between teachers and learners who engage critically with information, issues, and problems often resulting in unintended outcomes" (p. 139). This section mirrors the kinds of concerns expressed by La Grange and appeals to environmental educators to reengage with the language of education.

When we do avoid questions about education the concept loses meaning; it becomes anything to anybody, and this isn't good for environmental education. Not only do we see a proliferation of well intended, but questionable curricula, but there is also pushback when the status quo feels pressure.

There were, for example, concerted assaults on environmental education by Michael Sanera and colleagues (Sanera, 1998; Sanera & Shaw, 1996; see also Cushman, 1997). And, more recent mainstream critiques on education for sustainable development and sustainability (Butcher, 2007). Although we see examples of half-truths, misleading information, and rhetorical excess in Sanera's work, we believe it did gain traction, at least in part, because those he was criticizing fuzzy ideas about education as reflected in some environmental education literature and, in some cases, an arguably dogma-driven visions of environmental education. Ideas and visions about environmental education were vulnerable. Even in debate with Sanera (1998, *Canadian Journal of Environmental Education, 3*), many environmental educators responding to Sanera's criticisms remained open to further critique about their own visions of the nature and aims of education. While this may seem to be history to some, the point is that environmental education remains vulnerable when its educational aims are poorly conceived and articulated.

So, What About Ethics? This is another big question raised in a rather limited space, and we can only point to some key features about how education and ethics might intersect. First, the word "ethics," like other words, is used in a variety of ways. This point not withstanding, ethics is centrally about questions like: What is a good life? What is a good way to live? What should I do? How should I live? Or, how should I live in the context of the larger good?

Questions about ethics often evoke performative elements as illustrated by the kinds of questions First Nations colleagues ask when considering ethics: What can we do to ennoble ourselves? What can we do so that people will tell good stories about us when we are gone? Or, how can we carry on our lives so that in the end, we will have accomplished what the Creator wanted for us? (e.g., Profeit-LeBlanc, 1996).

More recently, and in keeping with the present postmodern mood (Noddings, 2007), emphasis has shifted from truths to useful narratives, and questions look more like: "How can we tell (or enact) good action guiding stories, or useful fictions?" (e.g., Hickory, 2004; Jickling 2004, 2005b; Lotz-Sisitka, 2008).

Given the richness of these questions, and present skepticism about meta-narratives, ethics should not be limited to doctrines, lists of rules, or codes of professional conduct, as legitimate as these usages might be in some contexts. If one of our premises is accepted, that "education" is not about indoctrination or propaganda (and associated ideas), then ethics in the educational context would be less dogmatic and more concerned with using the questions posed as prompts for exploring controversy, dissonance, and unconventional ideas, and imagining new possibilities. Seen this way, it becomes evermore clear how important questions open up at this intersection of education and ethics. How education is conceived, in turn shapes interpretations of ethics, and vice versa.

It is at this imaginative and interdisciplinary nexus of education and ethics that theoretical and practical space is opened that can allow transcendence of traditional ideas and contemporary clichés (Egan, 1997; Judson, 2008). Fortunately, as fields of inquiry, ethics, and environmental ethics are represented by rich bodies of literature and practice. And, attention is given to opening new directions for inquiry and experimentation (e.g., Cheney & Weston, 1999; Jickling, 2007; Weston, 1992, 2004, 2007).

This Section

We understand that it is disturbingly difficult to respond to difficulties the world faces and to understand how environmental education research can provide helpful insights. Authors in this section cannot answer this challenge themselves. However, we would like to briefly make four general points that might help to orient some aspects of future research. First, for environmental educators research always involves normative questions, explicitly or implicitly. Second, attention to normative questions—that is questions about value laden ideas—is underrepresented in our research literature. Third, tackling normative questions implicitly involves uncertainty and risk, and they can be inconvenient—at least in the present context! And, fourth, key normative questions for environmental education researchers concern "ethics" and "education." Not only do these questions open (or reopen) important areas for inquiry; but, we feel that we avoid them at our peril. For, research insights at the intersection of education and ethics are potential tools that can enable us to teach, inquire, and ultimately live as if the world mattered.

We also understand that advancing normative research will require more than conducting evermore research within traditions as framed by historical methodologies extant in environmental education. More likely it will involve scholarship that is more generous toward research traditions that stretch us beyond our familiar territories and comfort zones; it will demand serious "border crossing" between research traditions (Russell, 2006). With this in mind, chapter authors in this section have been chosen for their diversity in research backgrounds and their predilections for broad research interests. They have been invited to explore a complex intersection of normative questions, and in doing so, to explore bridges and resonances and bridges between research traditions. In turn we expect that their work will create many openings for future research.

The chapters in this section, in various ways, all examine normative dimensions of key concepts in environmental education: ethics, education, schooling, and/or challenging and nuanced relationships between them. The first of these is our own which connects with some of our earlier joint work (Jickling & Wals, 2008). It takes up the possibility, building on Russell (2006), that effective engagement with "ideas about education and ethics" will likely require methodological border crossing. To this end, we develop an analytical heuristic that invites researchers, from across research traditions, to consider their own normative understanding of these key concepts. We then apply our own heuristic to an analysis of the educational merit of "education for sustainable development" and the ethics merit of the Earth Charter. As a heuristic, its intent is generative—to engage people with tensions that are implicit in the concepts education and ethics, and relationships between the two. It also intends to challenge researchers to frame and reframe their own perspectives about these concepts and to generate new lines of inquiry.

Like ourselves the, authors of the chapters 10 and 11 are concerned with understanding a need to examine everyday concepts. Lesley Le Grange develops and probes the concept of education while Heesoon Bai and Serenna Romanycia challenge researchers to conceive more carefully of the role of schools. First, Le Grange argues that language is important. He claims that we need a language of education, and more particularly environmental education, especially in light of a rise in a language of learning. He argues that language matters, that it matters whether we use the term "environmental learning" instead of "environmental education," because they are different ideas that, in practice, produce different effects. He then illustrates why he believes that a language of learning is inadequate in responding to the complex problems and risks that characterize the global environmental crisis. He then makes his case for a language of environmental education that shifts the focus from the kind of human produced to a focus of the "being" of the human being. He proposes a language of environmental education that entails coming into presence in a world of difference and plurality, where difference and plurality extend beyond the realm of human beings.

Heesoon Bai and Serenna Romanycia are concerned with schools as institutions of cultural transmission and, hence, agents for normatizing and normalizing norms of this society. At present, they argue, these norms are an indictment of the dominant global culture and civilization. As an antidote, they suggest that schools can be transformed into awakened institutions that systematically examine cultural norms and deliberate on means to change them. To do this, they draw on the ethics of care to create conditions in which new social norms can emerge through everyday activities, and schools can be renormatized in the key of ecological ethics. In the end they argue that we can collectively and deliberately choose to make schools institutions of renormatization and shape them to be more radically critical and critically aware, questioning, and responsive to a changing world and a challenging reality.

Chapters 8 and 9 are concerned with relationships; for David Greenwood this is presented as between people and place, and Michael Bonnett probes relationships between people and a greater whole. Greenwood's chapter sets out to provide environmental education with a conceptual framework for place-conscious education that bridges cultural and ecological analysis, that is responsive to diverse and changing cultural contexts, and that challenges educators to rethink the assumptions of schooling in the context of the places we inhabit and leave behind. He argues that the conceptual power for this framework is derived from twin educational aims, decolonization and reinhabitation. For Greenwood, decolonization enables the critiques of cultural practices and their underlying ontological and epistemological assumptions while reinhabitation suggests a need to reimagine and recover ecologically conscious relationships between people and place. Educationally, the emphasis is on enabling a sociological ethic to emerge from a learning relationship between cultural and natural landscapes. He observes that this place-conscious learning can be blocked through education, schooling, or other cultural experiences. However, he argues, that educational theory and practice that fails to be inclusive of person-place relationships is miseducative—with consequences for people, culture, and environment.

Michael Bonnett argues that, in a sense, education is necessarily concerned with the development of individuals and that these individuals are both locally emplaced and part of a greater whole. Ethical concern arises from antecedent relationships with individuals' environments and their already being in the world; that is, individuality-in-relationships. There is nothing purely objective or passive about a place. From these assumptions he poses the question: What would it be, then, for education to respect the emplaced individual? While acknowledging that individuals constitute themselves within a cultural context, he argues that the self is never exclusively a creature of its cultural constitution. The normative character of environmental education would become salient through examining what it means for education to respect and engage the embodied person—ontologically as an individual, yet in indissoluble relationship with a greater whole.

This section fittingly ends with a chapter by Lausanne Olvitt, who reflects on the challenges of doing research that exists at the intersection of environmental ethics and education. When currently facing with unprecedented socioecological crises, she observes that general responses have defied the application of a singular ethical system or value theory. Rather, she argues that ethical practices are negotiated daily in the diverse sociocultural, economic, and political spaces that locate people's authentic *in situ* struggles to live well in the world. She acknowledges that this view of ethics as a process of negotiation and responsiveness is a radical departure to modernist conceptions that reduce ethics to lists of duties and obligations. Rather, she argues that what is needed are more deliberative and transformative ways of engaging with the diversity and complexities of people-environment relationships. Seen this way, environmental ethics is an emergent field of inquiry that is dynamic and exploratory, which embraces uncertainty, is never satisfied, and is restless to learn more.

We hope that this section will open up avenues for scholars, both established and emerging, with an interest in environmental education, to engage in normative research. For some, perhaps many, a prerequisite for such engagement might be overcoming "normativity anxiety" or their fear to openly ask normative questions and to make explicit their own assumptions and biases.

Note

1. Ethics and politics can be closely linked. This section focuses on ethics to more consciously bring attention to these questions, which are infrequently taken up in more politically oriented discourses.

References

Butcher, J. (2007). Are you sustainability literate? *Spiked*, September 13. Retrieved January 13, 2009, from http://www.spiked-online.com/index.php?/site/article/3821/

Barrow, R., & Milburn, G. (1986). *A critical dictionary of educational concepts*. Brighton, UK: Wheatsheaf Books.

Barrow, R., & Woods, R. (1982). *An introduction to philosophy of education* (2nd ed.). London, UK: Methuen.

Canadian Journal of Environmental Education (1998). *3*.

Cheney, J., & Weston, A. (1999). Environmental ethics as environmental etiquette: Toward an ethics-based epistemology. *Environmental Ethics, 21*(2), 115–134.

Cushman, J. (1997). Critics rise up against environmental education. *New York Times*, April 22, 1997. Retrieved January 15, 2009, from http://query.nytimes.com/gst/fullpage.html?res=9E05EFD7143EF931A15757C0A961958260#

Delanty, G. (1999). *Social theory in a changing world: Conceptions of modernity*. Cambridge, UK: Polity Press.

Egan, K. (1997). *The educated mind: How cognitive tools shape our understanding*. Chicago, IL: University of Chicago Press.

Ferreira. J., (2007). *An unorthodox account of failure and success in environmental education*. Unpublished Phd dissertation. Griffith University, Brisbane, Australia.

Ferreira, J. (2009) Unsettling orthodoxies: Education for the environment/for sustainability. *Environmental Education Research, 15*(5), 607–620.

Hart, P., Jickling, B., & Kool, R. (1999). Starting points: Questions of quality in environmental education. *Canadian Journal of Environmental Education*, *4*, 104–124.

Hickory, S. (2004). Everyday ethics as comedy and story: A collage. *Canadian Journal of Environmental Education, 9*, 71–81.

Jickling, B. (1992). Why I don't want my children to be educated for sustainable development. *Journal of Environmental Education, 23*(4), 5–8.

Jickling, B. (1997). If environmental education is to make sense for teachers, we had better rethink how we define it! *Canadian Journal of Environmental Education, 2*, 86–103.

Jickling, B. (2003). Environmental education and environmental advocacy: revisited. *Journal of Environmental Education, 34*(2), 20–27.

Jickling, B. (2004). Making ethics an everyday activity: How can we reduce the barriers? *Canadian Journal of Environmental Education, 9*, 11–30.

Jickling, B. (2005a). Education and advocacy: A troubling relationship. In E. A. Johnson & M. Mappin (Eds.), *Environmental education and advocacy: Changing perspectives of ecology and education* (pp. 91–113). Cambridge, UK: Cambridge University Press.

Jickling, B. (2005b). Ethics research in environmental education. *Southern African Journal of Environmental Education, 22*, 20–34.

Jickling, B. (2007). Research in a changing world: Normative questions and questions that matter. *Southern African Journal of Environmental Education, 24*, 108–118.

Jickling, B., & Spork, H. (1998). Environmental education for the environment: A critique. *Environmental Education Research, 4*(3), 309–327.

Jickling, B., & Wals, A. E. J. (2008). Globalization and environmental education: Looking beyond sustainability and sustainable development. *Journal of Curriculum Studies, 40*(1), 1–21.

Judson, G. (2008). *Imaginative ecological education*. Unpublished doctoral thesis, Simon Fraser University.

Le Grange, L. (2004). Against environmental learning: Why we need a language of environmental education. *Southern African Journal of Environmental Education, 21*, 134–140.

Leopold, A. (1966). *A Sand County almanac*. Oxford, UK: Oxford University Press. (Original work published in 1949.)

Lotz-Sisitka, H. (2008). Utopianism and educational processes in the UN Decade of Education for Sustainable Development. *Canadian Journal of Environmental Education, 13*(1), 134–152.

Noddings, N. (2007). *Philosophy of education* (2nd ed.). Boulder, CO: Westview Press.

Peters, R. S. (1966). *Ethics and education*. London, UK: George Allen & Unwin.

Peters, R. S. (1973). Aims of education: A conceptual inquiry. In R. S. Peters (Ed.), *The philosophy of education* (pp. 11–57). Oxford, UK: Oxford University Press.

Profeit-LeBlanc, L. (1996). Transferring wisdom through storytelling. In B. Jickling (ed.), *A colloquium on environment, ethics, and education* (pp. 14–19). Whitehorse, YT: Yukon College.

Russell, C. (2006). Working across and with methodological difference in environmental education research. *Environmental Education Research, 12*(3–4), 403–412.

Sanera, M. (1998). Environmental education: Promise and performance. *Canadian Journal of Environmental Education, 3*, 9–26.

Sanera, M., & Shaw, J. (1996). *Facts not fear: A parent's guide to teaching children about the environment*. Washington, DC: Regnery.

Sayer, A. 2000. *Realism and social science*. London, UK: Sage Publications.

Walsh, P. (1993). *Education and meaning: Philosophy in practice*. London, UK: Cassell.

Weston, A. (1992). Before environmental ethics. *Environmental Ethics, 14*(4), 321–338.

Weston, A. (2004). What if teaching went wild? *Canadian Journal of Environmental Education, 9*, 31–46.

Weston, A. (2007). *A twenty-first century ethical toolbox* (2nd ed.). New York, NY: Oxford University Press.

Williams, R. (1976). *Keywords: A vocabulary of culture and society*. London, UK: Fontana.

7

Probing Normative Research in Environmental Education

Ideas About Education and Ethics

Bob Jickling
Lakehead University, Canada

Arjen E.J. Wals
Wageningen University, The Netherlands; Cornell University, USA

Education in Environmental Education Research

Environmental education research has, historically, been punctuated by periods of contestation. One notable episode occurred during the early 1990s when the field's research output was dominated by positivist methodologies (Mrazek, 1993). This was reflected in the editorial direction of the *Journal of Environmental Education,* which published research papers and viewpoints. The former category was reserved for work that met approved expectations while work from other research traditions was reduced to "viewpoint" status (e.g., Jickling, 1992; Robottom & Hart, 1995). A debate ensued, largely at annual conferences of the North American Association for Environmental Education, particularly in 1990, and in the publication, *Alternative Paradigms in Environmental Education Research* (Mrazek, 1993). Alternative conceptions of research were most often framed by the Habermasian typology of positivist, interpretivist, and critical research (e.g., Hart, 1993; Robottom, 1993).

Jo-Anne Ferreira's (2009) paper, "Unsettling orthodoxies: education for the environment/for sustainability," reflects on another contestation initiated by Jickling and Spork's (1998) paper, "Environmental education for the environment: A critique." Jickling and Spork became concerned when they noticed traces of the Australian-rooted conception of education "in," "about," and "for" the environment (Lucas, 1979, 1980, 1995) slipping into a broader international discourse with little reference to its antecedent discussions. In some instances this manifested itself in constructions that employed education to advance ideologies such as "red-green environmentalism" and "education for sustainable development." The Jickling and Spork (1998) analysis of education and its application in environmental education were meant to disrupt emerging orthodoxies in the field. However, they were also pointing to a methodological blank spot—something left out of the critical theory tradition arising in environmental education (also largely omitted in the positivist tradition). They were also raising questions that did not find a comfortable home in the Habermasian typology.

From the perspective of the analytic tradition of educational philosophy, the whole paper (Jickling & Spork, 1998) is a reflexive exercise that considers the nature of education and its practice. The questions at the core of the analysis are: "What is education?" and, in light of that analysis, "Is this educational?" However, these questions tend to fall outside of the methodological strategies in the critical tradition. This analysis should certainly have responded more vigorously to critical theorist demands for more ideological clarity—or at least a clearer statement of the value laden nature of their positions (Dillon & Wals, 2006). On one hand, a failure to acknowledge the importance of the analytical questions while, on the other hand, a failure to explicitly address ideological perspectives led to contentions and impasse. However, as these two research traditions are asking different questions it is not surprising that there are such discrepancies in understanding. For many, critical theory was "seen as inherently political and progressive" and researchers regarded "their intellectual efforts as an integral part of a social and political struggle for emancipation" (Glock, 2008, p. 183). As attractive as this is, it was worrisome that the end result might just be replacing one form of hegemony with another. The struggle for emancipation seemed to displace critical questioning about what constituted good education. We hope this chapter can lead to a better understanding of a more generous and productive kind of scholarship that can reach across these methodological divides (i.e., Russell, 2006). This does not mean banishing scholarly argumentation, but is, rather, an attempt to better understand differences.

Without the same overt focus on issues of power and control, analytical philosophy is often criticized as attempting to be politically neutral and value free (i.e., Glock, 2008; Noddings, 2007). It is a mistake to paint such a broad picture of analytical philosophy. R. S. Peters, for example, grappled with values in his landmark book *Ethics and Education* (1966; see also Peters, 1973). Further, Hans-Johann Glock (2008) is flabbergasted by *prima facie* the idea that analytic philosophy is apolitical or conservative. He cites Bertrand Russell's political engagement on the side of the oppressed as evidence to the contrary. Still there was, and is, failure in understanding, and we suggest this is due to differing orthodoxies within the respective traditions.

John Fien (1993, 2000), working in the critical theory tradition could not see what Jickling and Spork (1998) were *for*. Similarly, Ferreira (2007, 2009) rightly claims that education is an achievement word that reflects some kind of end. She is concerned that Jickling and Spork argue, impossibly, "that education should be free of specified ends" (Ferreira, 2007, p. 325). Again what may be self-evident in the analytic traditions of philosophy of education, fails communicate across the bounds of research traditions.

What Jickling and Spork (1998) were unabashedly *for* was education. Education as a concept and an achievement can be circumscribed, and/or defined, by ideas like the educated person, or by a capacity for critical thinking. Seen this way we don't have education for critical thinking—critical thinking is part of what education is, it is intrinsic to the concept. In hindsight, they should have said (among other things) (a) "that education should be free of specified ends that are external to the idea of education itself"; (b) "if education is inherently about achievement of critical thinking then it would be a contradiction to expect the educated person to comply with a prespecified end in that this would require the suppression of critical thinking." Commentators from critical theory tradition, have sometimes expressed an expectation that there should be an ideology that would signify a substitute for educational ends; if Jickling and Spork did not like red-green environmentalism or sustainable development, what would they prefer? (e.g., Fien, 1993, 2000; Huckle, 2010). However, from the analytic tradition, this just doesn't make sense as suggested by point (b) above.

If the language of ideology must be used then Jickling and Spork's (1998) ideology is education, and all that educators expect of this concept. It is true that in practice educational choices, in content and pedagogy, also carry another layer of values that shape learning, but that does not render a goal like self-directed critical thinking unworthy. It does mean that it will be imperfectly attained, and that constant vigilance will be required to make the most of the aim (Jickling, 2003, 2005a).

Given a predilection for questions arising from the analytic philosophy in this example, it should not be interpreted as an apology for the analytic tradition. While this tradition does provide important questions, and a penchant for clarity of thought and writing ("What do you mean?" being a favorite probe), this can be too much of a good thing (Zwicky, 1992, 2003). Put another way, Jan Zwicky (1992) describes thinking analytically as like going on a diet:

> Its aim is a healthy austerity of thought, a certain trimness of mind. But carried too far, we're only left with a skeleton. Carried too far, too often, we lose the sense that something is amiss when the patient exhibits no life. We come to take pride in our cases of polished bone. (p. 154)

Zwicky's lyric philosophy suggests that there are important research paths that fall outside of both the analytic and critical theory traditions.

One approach that is difficult to categorize, yet shows promise to reach across analytic and critical boundaries, and beyond, is Deleuze and Guattari's (1994) idea that concepts need to be constructed. Their somewhat enigmatic presentation can be read as a need for altogether new concepts, but also as a constructivist orientation to existing ideas. There is not space to explore these possibilities in this chapter, but their work promises to be generative.

In the meantime, we, as two authors from different research traditions but with overlapping interests, have been experimenting with ways to bring normative questions to light in ways that encourages participants to reach across research boundaries, and to construct meaning for themselves. While this work is not based on Deleuze and Guattari (1994) it does, we feel, resonate with their work in useful ways. In this chapter we draw on a heuristic that evolved over time (Jickling & Wals, 2008; Wals & Jickling, 2002) to explore the concepts of "education" and "ethics" and how they might be understood in the context of environmental education. We then engage the heuristic to examine "education for sustainable development" and the "Earth Charter" and their relationships to education and ethics.

While we are interested in the larger concepts of education and ethics, and their possible relationships within environmental education, boundaries are not clear and language usage shifts between environmental education and education, and between environmental ethics and ethics. When we talk about environmental education and environmental ethics we bring ourselves home with more narrow conceptions that reflect interests within our field. When we shift to talk about education and ethics we are probing nuances derived from the broader concepts.

Environmental Ethics and Education

Environmental ethics, as a formal field of study, has been around since the late 1970s—that is, as marked by the inauguration of its first journal (Hargrove, 2004). The word "field" here is used in a provisional way. We recognize that boundaries are permeable and that environmental ethics, thought of this way, rests on preceding histories and oral traditions. Still, much interesting thinking has accrued and insights gleaned since environmental ethics began to coalesce as an academic and research endeavor. However, the field has been criticized, too. Some critics, suspicious of metanarratives and universalizing theory, have argued that environmental ethics as a field is highly moralistic and un-reflexive about the full extent of its normalizing effects (Darier, 1999). They argue that it is about the application of moral codes. There is much to consider here.

Interestingly, Aldo Leopold (1949/1966), one of the field's predecessors anticipated these very discussions. In this *Sand County Almanac*, first published in 1949, Leopold is aware that ethics is importantly also a process. Similarly, Anthony Weston (1992) argues that environmental ethics is in an "originary" stage in the process of continual development. Ecofeminist scholars Such as Karen Warren (1990) and Val Plumwood (1991, 1993, 2002), have also argued against abstract and objective codes of ethics.

It is clear that there are divergent perspectives about the nature of environmental ethics. On one hand there are those who see it as a field consumed with moralizing and experiments concerned with universalizing discourse, on the other, there are strong voices promoting environmental ethics as a process that produces, at best, tentative results and involves experimentation and uncertainty. These contesting perspectives will be discussed more fully in later sections of this chapter.

An Emergent Heuristic

From the preceding discussion, several points become apparent. First, education and ethics are not singular or precise ideas. Indeed, they appear to be flexible, permissive, uncertain, and sometimes vision dependent to the extent that they are incommensurate in their variants, changing, and even an idea for each of us to make up our own minds about, or to construct on our own (cf. Deleuze and Guattari 1994; Jickling 1997; Walsh, 1993). Second, we (authors) see emergent parallels between ways these two concepts are used. For example, environmental education and environmental ethics are both used in ways that suggests guidance toward some predetermined vision, yet they can also signify processes whereby the learners' achievement will lean toward self-directed construction of meaning. One way to frame these contesting views is to see these concepts as existing along a continuum characterized as essentially transmissive at one end of the spectrum and as socioconstructivist /transformative at the other (Shepard, 2000).

By transmissive, we mean that education and/or ethics is fundamentally about transmission of facts, skills, and values to students. Here, content and learning outcomes are predetermined and prescribed by experts. Working within this transmissive perspective, much contemporary rhetoric rests on the assumption that education and ethics can be instruments getting one's "message" into impressionable young minds—for implanting a particular agenda. In this case, education leads to an authoritatively created destination.

Socioconstructivist and transformative education and/or ethics, on the other hand, reflect a belief that education is an emergent achievement. Here, knowledge and understanding are seen as coconstructed within a social context—new learning is shaped by prior knowledge and diverging cultural perspectives and is, as such, transformative. This characterization is more open and provides space for some autonomy and self-determination on the part of the learner (Leopold, 1949/1966; O'Sullivan, 1949/1999, 2002; Shepard 2000; Weston, 1992).

Viewed another way, and along another continuum, we can ask whether education and or ethics is mostly about social reproduction at one end, or more about enabling socially active citizenry at the other. Such distinctions can reflect the way practitioners imagine citizens interacting within society, and its democratic practices.

If social reproduction were expected, then "educated" or "ethically considerate" citizens would be expected to work efficiently within existing social frameworks. They would be obedient, deferential, and compliant in taking their place within hierarchical and authoritative social structures and power relationships. From this vantage, individuals are content to participate in democratic processes at electoral intervals, while decision makers and supporting bureaucracies make daily choices.

If social engagement was expected, then "educated and/or ethically considerate citizens" would actively participate in ongoing decision-making processes within their communities. They would be democratic practitioners in the sense that democracy is more than selecting a government, but rather:

> . . . a form of associated living, of conjoint communicated experience. The extension in space of the number of individuals who participate in an interest so that each has to refer his own action to that of others, and to consider the actions of others to give point and direction to his own, is equivalent to the breaking down of those barriers of class race and national territory which kept [people] from perceiving the full import of their activity. (Dewey, 1916/1966, p. 87)

Dewey suggested that more numerous and more varied points of contact denote a greater diversity of stimuli to which an individual can respond. According to his view, democracy in education, and presumably ethics, is crucial in realizing a sense of self, a sense of other, and a sense of community; it creates space for self-determination, as individuals and/or members of groups, and greater degrees of autonomous thinking in a social context.

Using these two parallel conceptions about the "acquisition" of education and ethics, and the two differing views of citizens, we have constructed a heuristic as

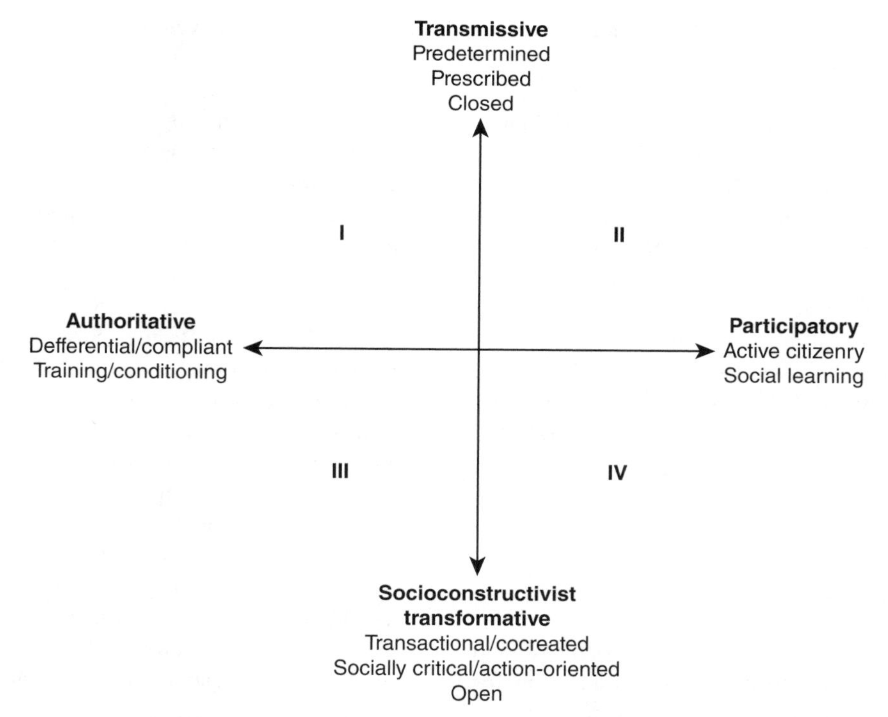

Figure 7.1 Positioning of ideas about "acquisition of ideas about education and/or ethics" alongside the social role of the "citizen." (Based on Jickling & Wals, 2008.)

pictured in Figure 7.1. For analytical purposes, we suggest that the dynamics framed by ideas represented along these two intersecting axes create interpretive possibilities within each of the four quadrants delineated.

Exploring the nature of concepts like education and ethics is a complex and messy business and a two-axes-heuristic is not sufficient to capture the shape and scope of the entire enterprise. That we have called it a heuristic, rather than a framework, is important. As a heuristic, its intent is generative—to engage people with tensions that can be found within environmental education and to challenge them to frame and reframe their own perspectives and questions. Its intent is also to engage people with the tensions inherent in the heuristic itself—to redesign it, to use alternative axes, or to use it as a stepping stone to construct an altogether alternative analytical tool.

The construction of this heuristic also represents an attempt by the authors to reach across research traditions by drawing on each other's differing research backgrounds and interests. In doing so, we are attempting to provide accessible and productive access to important normative questions in environmental education practice and research. In the next section we employ specific examples to explore some dynamics associated with these normative ideas.

The Heuristic at Work

Education for sustainable development and the Earth Charter are two initiatives often associated with environmental education that can be examined using the heuristic presented here. The first probes the broader concept of education. The second enables a parallel exploration of the concept of ethics. A brief description of the emergence of these concepts is presented together with an analysis of implications of various conceptions.

Sustainable Development: Some Background Related to Education Interest in Sustainable Development has, from the outset, been coupled with educational prescriptions. For example, the World Commission on Environment and Development argues in its report (WCED), *Our Common Future* (1987), that:

> Sustainable development has been described here in general terms. How are individuals in the real world to be persuaded or made to act in the common interest? The answer lies partly in education, institutional development, and law enforcement. (p. 46)

For these authors, education is a tool to "persuade or make" people act in a particular way—in this case, some notion of the "common interest."

Similarly, the action plan from the 1992 World Conference on Environment and Development in Rio de Janeiro, *Agenda 21* (UNCED, 1992), asserts that education should be reoriented toward sustainable development, and that:

> Both formal and non-formal education are indispensable to changing people's attitudes so that they have the capacity to assess and address their sustainable development concerns. (p. 36.3)

The plan of implementation of the World Summit on Sustainable Development at Johannesburg in 2002 (United

Nations, 2002a, p. 116) states that, "Education is critical for promoting sustainable development." Then, in December 2002, the United Nations passed *Resolution 57/254* (United Nations, 2002b) that declared a Decade of Education for Sustainable Development (DESD) beginning in 2005. Interestingly, aside from the preamble that recalls the 1992 United Nations Conference on Environment and Development, the resolution makes no reference to "environment," "environmental," "ecology," or "ecological."

Throughout this period there have been concerted efforts to transform environmental education into education for sustainable development (Jickling, 1992; Jickling & Spork, 1998; Jickling & Wals, 2008). We are reminded by Winter (2007), however, that education for sustainable development,

> . . . is not a neutral term: it embraces a particular policy commitment, however uncontroversial this may on the face of it seem, and, like any policy commitment, this inevitably has ethical and political dimensions. (p. 337)

It can be said, however, that within UN-circles and UNESCO documents the rhetoric around education for sustainable development has softened over time and with that spaces are opening up yet again for "environment," "environmental," "ecology," or "ecological, and indeed for environmental education" (Wals, 2009). With this necessarily brief background, we are convinced that education is affected by globalizing ideologies. Whether or not "sustainable development" is an appropriate aim for education depends on how education is conceptualized. And, there is considerable evidence to suggest that educators are divided on this point (see, e.g., Butcher, 2007; Hesselink, van Kempen, & Wals, 2000; Scott & Gough, 2007; Winter, 2007).

Out of the preceding discussion, three realms of possibility have emerged that can serve to focus analysis and discussion around relationships between sustainable development and education (See Fig. 7.2). We describe Quadrant I as "Big Brother Sustainable Development," reminiscent of George Orwell's (1949/1989) metaphor for extreme state control where even language used by citizens was controlled by the "thought police."

Quadrants II and III, differ in some characteristics, but, in the end share important qualities. In Quadrant II participatory approaches to learning are taken up, yet the approach delineated by this conceptual space also tilts toward transmissive goals. In Quadrant III socio-constructivist and/or transformative goals are moderated by authoritative approaches to teaching. It can be argued that while participatory learning and socioconstructivist goals promise possibilities for transcending education for sustainable development, the transmissive and authoritative tendencies still constrain possibilities. As such, both Quadrants II and III suggest a kind of "feel good sustainable development" in the sense that citizens are given a limited, possibly false,

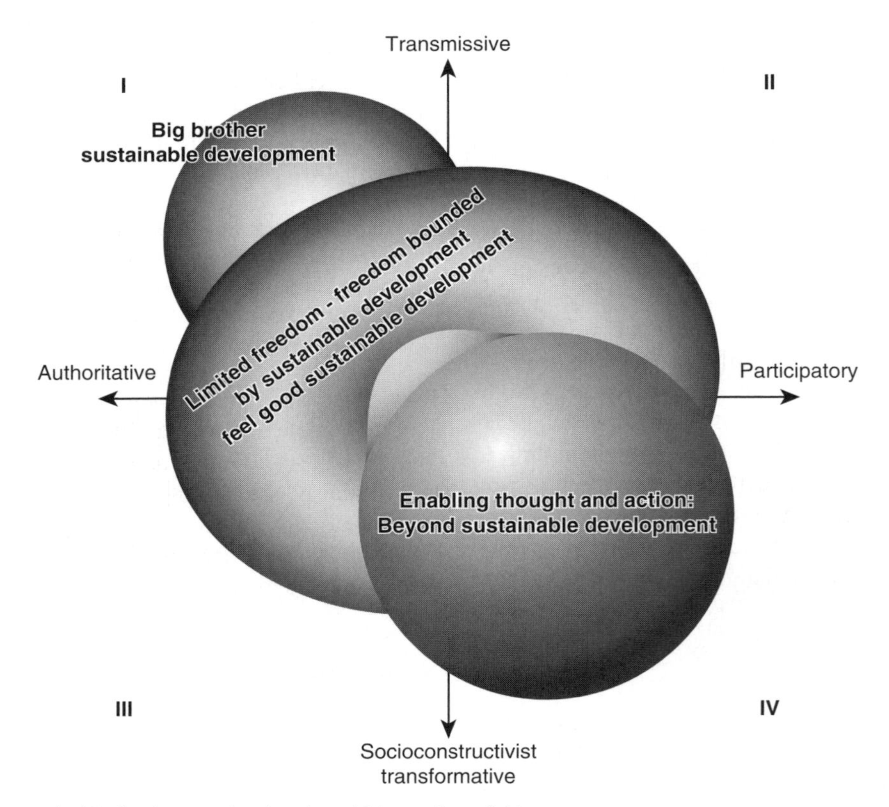

Figure 7.2 Positioning sustainable development in education within two force fields.

sense of control over their ability to shape the future while, in fact, authorities of all kinds remain in control. We discuss these two quadrants together under the heading, "Limited freedom—Freedom bounded by sustainable development."

Finally, we discuss Quadrant IV possibilities under the heading, "Enabling thought and action—Beyond sustainable development." Relationships such as these do not occur in such flat, one-dimensional, depictions. With this in mind, we have redrawn the heuristic in a three dimensional fashion to convey more multidimensional and dynamic relationships.

Big Brother Sustainable Development In Quadrant I, education is characterized as a tool or an instrument that can help realize the sustainable development agenda; "education" inculcates the preferred message, agenda, ideology, or consumer preference. This view of education is both instrumental and deterministic in that some segment of society decides what is best and it uses education as a tool to disseminate its conception of "best."

Examples of this thinking have been revealed in the statements above that describe educational missions that aim to change peoples' attitudes so that they have capacity to address their sustainable development concerns, or to persuade or make them act in particular ways. In a similar vein one advocate has asserted that education:

> [S]hould be able to cope with determining and implanting these broad guiding principles [of sustainability] at the heart of ESD [education for sustainable development]. (Hopkins, 1998, p. 172)

Many educators find these sentiments a misrepresentation of their task, because of the implication that some segment of society decides what is best—a preferred ideology—and then uses education as a tool to implant this "best ideology" into the minds of students. Sustainable development, they argue, is a normative term and placing it as a desired outcome can be seen as reducing education to indoctrination (see, e.g., Hattingh, 2002; Jickling, 1992, 2001).

Other proponents of sustainable development assert that internationally, "there will be no single 'right' definition of ESD, but there will be overall agreement on the concept of sustainable development that education addresses" (Rgozzi, 2003, p. 4). From a critical stance, Tom Berryman (1999, p. 59) warns of the totalizing nature of statements that assert there "will be overall agreement." He argues that such statements "despise other perspectives that don't share this single vision." He then goes on to say that "[s]ustainable development is but one story that cannot claim to embrace all the varied environmental visions." To allow for contextual differences yet insist that these exist within an overall understanding, or particular conceptual framework, just pays "lip service" to difference; it is patronizing.

Following Berryman's critique, Aminata Traoré (2002a, 2002b), a writer from Mali, provides a particularly vivid, interpretation of this "big brother" thinking. In discussing the impact of globalizing trends and ideas, and the effects they have upon her culture, she denounces the imposition of totalizing ideas as "viol de l'imaginaire"—rape of the imagination. With this in mind, we wonder about the impact of sustainable development talk on the cultures and capacities of citizens in many countries. Why is it that we are all supposed to use the same language to talk about the same ideas? And, what is the impact of such global hegemony?

Limited Freedom—Freedom Bounded by Sustainable Development Educators might choose to work in territory demarcated by Quadrants II and III for pragmatic reasons. Government support and funding often is more readily available for projects that are aligned with the sustainable development agenda. Such pragmatism does not necessarily preclude the emergence of transformative learning activities. However, it is also possible that an a priori commitment to sustainable development can, in the end, stifle education.

The United Kingdom has, for example, has been particularly active in developing policy initiatives designed to embed education for sustainable development in school curricula and in Higher Education (Butcher, 2007; Scott & Gough, 2007; Winter, 2007). For the purposes of constructing and illustrating our heuristic it is instructive to consider Winter's (2007) assessment of the *National Curriculum Handbook for Secondary Teachers in England Key Stages 3 and 4* (Department for Education and Skills [DfES] and Qualifications and Curriculum Authority [QCA], 1999). She directs us first to consider the reference to sustainable development identified in the second aim:

> The school curriculum should . . . develop their [pupils'] awareness and understanding of, and respect for, the environments in which they live, and secure their commitment to sustainable development at a personal, local, national and global level. (DfES & QCA, 1999, p. 11; cited in Winter, 2007, p. 341)

She then describes the purpose of education for sustainable development as described in the same *National Curriculum Handbook*:

> Education for sustainable development enables pupils to develop the knowledge, skills, understanding and values to participate in decisions about the way we do things individually and collectively, both locally and globally, that will improve the quality of life now without damaging the planet for the future. (DfES & QCA, 1999, p. 25; cited in Winter, 2007, p. 341)

In the first quotation, the irrefutable commitment to sustainable development is enshrined, while the second quotation speaks about developing knowledge, skill, understanding, and values that can enable emerging citizens to participate in local and global decision making.

At first glance, the second quotation appears to appeal to laudatory educational aims; however, a closer look reveals hidden tensions among these ideas. According to Winter's analysis, these policy statements mask underlying political and ethical issues implicit in the idea of sustainable development and they fail to bring critical attention to the concept itself. In other words, they propose to enable students to become thoughtful participants in political decision making—but within a given political framework. We think this is a good example of "feel good sustainable development" in that it creates an illusion that future citizens will be active participants in shaping their future when, in fact, the future has been prescribed by ideologically driven ideas of sustainable development. Freedom to think outside of the sustainable development box is limited.

Winter (2007) echoes our concerns through her analysis of the *National Curriculum Handbook* and its apparent failure to integrate opportunities for critical and deconstructive questioning of sustainable development and education for sustainable development. In turn this concern points to the fourth quadrant of our heuristic that describes educational territory lying beyond sustainable development.

Enabling Thought and Action—Beyond Sustainable Development Quadrant IV concerns itself with enabling environmental thought and emerging ideas. Inspired by socioconstructivist and transformative views about education and actively engaged citizens, it points to possibilities beyond sustainable development. Here, sustainable development is seen as just one stepping-stone in the continuing emergence of environmental thought. These emergent concepts will be useful to discuss, critique, and employ as devices to stimulate effective and creative dissonance across disciplinary boundaries. They will not be seen as preeminent ideas—or organizational frameworks—but just as more or less useful conceptual tools.

It is worth noting that when Rachel Carson (1962) wrote *Silent Spring*, no one had heard of deep ecology. When Arne Næss coined the term deep-ecology, nobody had heard of the term sustainable development. When sustainable development became popular (WCED, 1987) ecofeminism, with origins in earlier feminist writing, was beginning to make inroads into fields like environmental ethics and education. And, more recently, the idea of ethics-based epistemology (Cheney & Weston, 1999) has been a generative in both ethics and epistemology.

An explicit example of moving beyond education for sustainable development is found in Chet Bowers (2002) paper "Toward an Eco-Justice Pedagogy." Here he writes about "eco-justice" and explores questions about philosophy, ethics, and justice. Like others working in this heuristic location, he appears to feel no need to place these issues within the sustainable development agenda. He is interested in cultural perspectives, active engagement with contemporary issues, and action, without feeling the need to convince anyone that they ought to do so within a sustainable development framework. Recently Gruenewald and Smith (2007) have placed emphasis on place-based education. We believe that environmental thought, and education, is dynamic and evolving. With this in mind, educators aiming to operate in Quadrant IV would find it counterproductive to build a sustainable development fence around their work.

The Earth Charter: Some Background Related to Ethics and Environmental Education The Earth Charter is, as its website states, "a declaration of fundamental ethical principles for building a just, sustainable and peaceful global society in the twenty-first century" (Earth Charter Initiative, n.d., The Earth Charter Initiative, p. 1). As such, it is a bold initiative that promises to challenge "our" values and to choose a better way. The charter was conceived, as a civil society initiative, by Maurice Strong (Secretary-General of the Rio Summit in 1992) and Mikhail Gorbachev (former leader of the USSR). As such it is the product of an extensive, decade-long, consultation that involved worldwide and cross-cultural dialogs about common goals and shared values. The final text was approved at a meeting of the Earth Charter Commission at the UNESCO headquarters in March 2000. It is currently available on the worldwide web in fifty-two languages (Earth Charter Initiative, n.d., Read the Charter). The commission was cochaired by Strong and Gorbachev and comprised of a diverse group of twenty-three eminent persons from all the major regions of the world. Steven Rockefeller chaired the international drafting committee. The dissemination, adoption, use and implementation of the Earth Charter is managed by the Earth Charter International Secretariat located at the University for Peace in San José, Costa Rica.

The Earth Charter holds appeal for many educators (i.e., Gruenewald, 2004) because it explicitly links the caring for the earth and caring for people as two dimensions of the same task, as stated in the preamble:

> We must join together to bring forth a sustainable global society founded on respect for nature, universal human rights, economic justice, and a culture of peace. Towards this end, it is imperative that we, the peoples of Earth, declare our responsibility to one another, to the greater community of life, and to future generations. (Earth Charter Initiative, n.d., Read the Charter, p. 1)

For some, this ambitious work of socioecological synthesis gives this document a vision that invites environmental educators to re-examine the significance of their work in new, and fundamentally new ways Blenkinsop & Beeman, 2008; Gruenewald, 2004). It is interesting to note, however, that the Earth Charter International Secretariat also states on its website that: "At a time when education for sustainable development has become essential, the Earth Charter provides a very valuable educational instrument" (Earth Charter Initiative, n.d., What is the

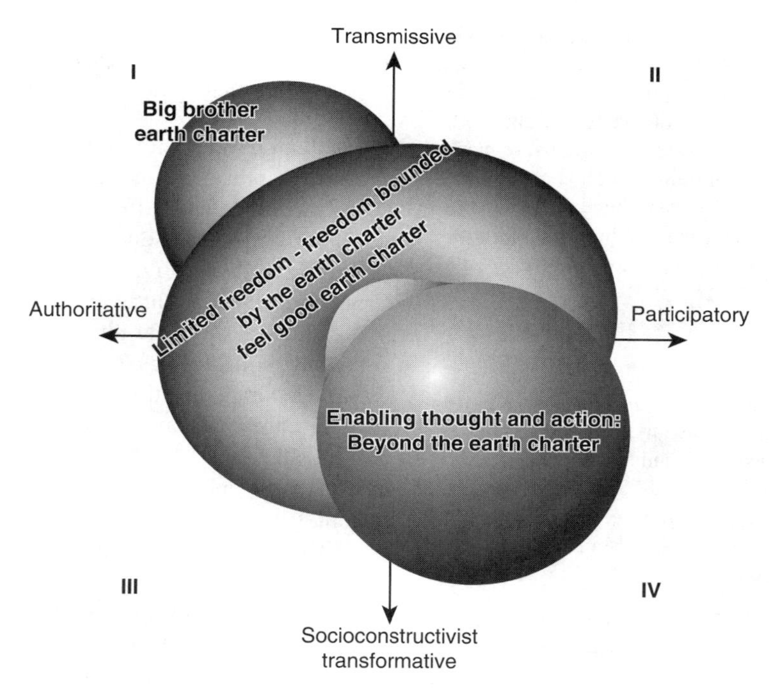

Figure 7.3 Positioning the Earth Charter in ethics within two force fields.

Earth Charter? p. 6). It seems, that the Earth Charter is an educational instrument serving the "truth" of education for sustainable development that was problematized in the last section.

Whether or not the Earth Charter could be seen as an appropriate aim for ethics depends on how ethics is conceptualized. Drawing again from the heuristic dynamic described in Figure 7.1, we suggest three emergent realms of possibility that can serve to focus analysis and discussion around relationships between the Earth Charter and ethics (See Fig. 7.3). We describe Quadrant 1 as Big Brother Earth Charter to reflect a prescriptive interpretation and implementation.

In Quadrant II participatory approaches to ethics are taken up, yet the approach delineated by this conceptual space also tilts toward transmissive goals. In Quadrant III socioconstructivist and/or transformative goals are moderated by authoritative approaches to ethics. It can be argued that while these goals promise possibilities for transcending the Earth Charter prescriptions, the transmissive and authoritative tendencies still constrain possibilities. As such, both Quadrants II and III suggest a kind of "feel good Earth Charter" in the sense that citizens are given a limited, or false, sense of control over their ability to shape the future while, in fact, authorities of all kinds remain in control. We discuss these two Quadrants together under the heading "Limited freedom—freedom bounded by the Earth Charter." Quadrant IV examines broad and expanding possibilities under the heading, "Enabling thought and action beyond the Earth Charter."

Big Brother Earth Charter In Quadrant I, ethics lean toward a transmitted message and a compliant citizenry. In this sense ethics is most often expressed as a code that describes what are thought to be appropriate values and expected conduct. Ethics is a product to be implemented. Such products are often presented as professional codes of ethics—with the expectation that they be followed within their communities. The Earth Charter vision and goal statements nod in this direction. For example, the vision states:

> We envision individuals, organizations, businesses, governments, and multilateral institutions throughout the world, including the United Nations General Assembly and UN agencies, acknowledging the Earth Charter, embracing its values and principles, and working collaboratively to build just, sustainable, and peaceful societies. (Earth Charter Initiative, n.d., Mission, vision and goals, p. 2)

And one of the goals is:

> To promote the use of the Earth Charter as an ethical guide and the implementation of its principles by civil society, business, and government. (Earth Charter Initiative, n.d., Mission, vision and goals, p. 4)

While it can be argued that such codes can be useful—many like to know that doctors, for example, are expected to comply with ethical standards of practice. This is certainly not the only legitimate conception of ethics.

Educators can be sensitive to this limited sense of ethics. We recall one African scholar who, when presented a "pitch" for the Earth Charter, commented that "it sounded like another salvation narrative." Ursula van Harmelen (2003) elaborates on this concern,

> To impose any ethical framework, no matter how "good" it may be, without subjecting it to constant critical scrutiny and challenge is a denial of human freedom to make informed choices. The imposition of an ethical framework as a given is to reify not only that framework, but to mask possible interpretations of that framework that may in fact be corruptions of the original ideals. Thus, if we were to accept the Earth Charter as a code of ethics without critical and informed analysis this would seem to me to be a form of indoctrination. (p. 125)

She later adds,

> By all means let us use the ethical framework of the Earth Charter in our educational endeavours. But let it be one set of ethics that should be explored and examined alongside other set, so that if individuals do subscribe to this particular ethical code they have done so based on an informed and reasoned choice, so that they understand their choice, its implications and its consequences. (pp. 125–126)

It could, for example, be said that educational programs that only use the Earth Charter as a critical referent, fall into this quadrant (i.e., Vilela de Araujo, Ramirez Ramirez, Hernandez Rojas, & Briceno Lobo, 2005).

Limited Freedom—Freedom Bounded by the Earth Charter Educators Joseph Weakland and Peter Blaze Corcoran (2009) have responded to the concerns outlined by van Harmelen (2003). They say that educators "must also encourage students to engage actively and critically with the text, rather than to proselytize its values" (p. 154). They further claim that:

> The Earth Charter is a normative statement of shared values. However, we see the text not as programmatic or totalizing structure for teaching about sustainability, but as a plurality of ideas about sustainability broadly conceived. (p. 155)

Still, they also quote Brendan Mackey, current Cochair of the Earth Charter International Council who claims that:

> Values education is an often-contested theme in education due to legitimate concern about "which values" and "whose values" are being promoted. These concerns can be accommodated so long as the values represent core values that are life-affirming, promote human dignity, advance environmental protection and social and economic justice, and respect cultural and ecological diversity and integrity. The Earth Charter can validly lay claim to represent such a core set of values, particularly given the participatory and multicultural processes that underpinned the drafting of the document. (Mackey, 2001, p. 9)

Weakland and Corcoran (2009) then add that:

> For university faculty members and students alike, the Earth Charter provides: a comprehensive and validated description of the necessary and sufficient conditions for sustainable development; a statement of specific principles to guide sustainable and ethical actions; and a call to collaborate on behalf of ecological integrity, social and economic justice and a culture of peace. (p. 155)

The bottom line here is that the Earth Charter, as an ethics code, is presented as necessary and sufficient. In spite of encouragements to engage critically with the Earth Charter, not to proselytize, and to apply it in many ways through a diverse range of settings, the program is still bounded by the Earth Charter. The freedom to explore the charter's inherent plurality and teach with it in many ways is still limited. It may feel good, but educators are not encouraged to think outside of the "Charter box." To be fair, they did say in one passage that the Earth Charter "can aid educators alongside other texts . . ." (p. 155). But, what these other texts were, and whether they included contesting visions of ethics, was never made clear.

There are also Earth Charter advocates that point in directions that seem to lead educators beyond. For example Gruenewald (2004) believes that it is a radical document representing a vision that can serve to challenge all educators, and particularly environmental educators, to re-examine the purpose context, and scope of their work. Further, it "is a call to action for educators to examine their assumptions and practices within a socioecological ethical framework" (p. 98). Similarly, Blenkinsop and Beeman (2008) declare that the Earth Charter and the process of its creation, represents intent to do things differently. And, they hold that and educational responses to the charter should be as radical as the document itself. For them, the charter "proposes a paradigmatic shift in the way we understand our environment and, if we are to take this message seriously, a similar shift is required of our classes, schools and educational systems" (p. 72). While it isn't clear how closely the authors of these two papers wish to tie themselves to the Earth Charter itself, they do point to its heuristic value in enabling educators to look beyond present expectations, standards, and ethical codes. This inclination leads to the next section that explores ethics possibilities beyond the Earth Charter.

Enabling Thought and Action—Beyond the Earth Charter There is ample literature to suggest that environmental ethics can meaningfully and fruitfully be conceived as something, beyond the Earth Charter, something other than a code. For example, Aldo Leopold (1949/1966) once wrote, "nothing so important as an ethic is ever 'written'" (p. 263). He spoke of it as an ongoing social evolution, and one that never stops. Implicit in this descriptions are participants who are constantly engaged in the reworking of relationships between themselves and, as he termed it then, the land. Implicit also, is an eschewing of any presumptions of a "true" or "final" ethic. Then he adds:

> Only the most superficial student of history supposes that Moses 'wrote' the Decalogue; it evolved in the in the

> minds of a thinking community, and Moses wrote a tentative summary of it for a 'seminar.' I say tentative because evolution never stops. (p. 263)

Anthony Weston picks up this theme in a series of papers that begin in the mid-1980s (1985, 1991) and come into focus in a paper titled, "Before environmental ethics" (1992). Following Leopold, he argues that we have no idea where this field will take us (1992). For Weston, environmental ethics is in need of a great deal of exploration. We should expect, at best, a long period of experimentation and uncertainty. So, it is clear, that from within the field of environmental ethics (and before) there are strong voices promoting ethics as a process that produces, at best, tentative results and involves experimentation and uncertainty. Perhaps, following authors like Blenkinsop and Beeman (2008) and Gruenewald (2004), researchers will find that the Earth Charter can be a useful tool, alongside other narratives in environmental ethics, to enable this exploration—without circumscribing it.

Elaborating, Weston (1992) argues that our challenge is not to systematize environmental values, but rather to create the "space" for environmental values to evolve. By space he is speaking about the social, psychological, and phenomenological preconditions that are needed to enable this evolutionary process. He is also speaking about the conceptual, experiential, and physical freedom to move and think. Here Weston is concerned that individuals and groups *can* actually begin to create, or coevolve, new values through everyday practices. And our job might thus be seen as "enabling environmental practice."

Feminist scholars have also troubled the notion of universal codes of ethics. Classic works by writers such as Karen Warren (1990) and Val Plumwood (1991, 1993, 2002) argue that environmental ethics is not an objective or disinterested, but rather, emerges as theory in process. According to these accounts, and other landmark writing such as Jim Cheney's (1989) paper on postmodern environmental ethics, ethics need to be contextualized, narrated, and hold a place for feelings and emotional understanding. A rich and growing body of feminist literature promises to lead researchers into promising new territory.

For now, we add one more line critique that problematizes ethics as codes. Arne Næss (1988, 2002) observes that human capacity for loving, based on moral duties or compliance with codes, is very limited. Often in such circumstances we act against our inclination, but comply out of respect for the moral law. His alternative is to seek ways to do what is right because of positive inclinations—out of joy, for example. And, when we do so, we perform a beautiful act. John Livingston (1994) and Sigmund Bauman (1993, 2008) are similarly concerned about moral codes. For Livingston, they are unknown in nature and, as human creations, are more like prosthetic devices. For him it is important to develop an extended consciousness—beyond the mere self. Bauman argues that complying with moral codes reduces responsibilities for one's own moral actions; codes actively erode our moral impulses. For Næss and Bauman, it is more important to find ways to develop people's inclinations, or moral impulses, than to develop their morals.

The examples in this section, reflecting Quadrant IV of our environmental ethics heuristic (Fig. 7.3), point to conceptions of ethics more concerned with processes of engagement, experimentation, and transformation. They also encourage continued exploration of what ethics, and ethics research in environmental education, can become (Jickling, 2004, 2005b). With these directions in mind, it would be inconsistent to remained fixed on the Earth Charter as an overarching organizing concept or ultimate aim.

A Few Cautions

According to the analysis presented here, educational work or environmental ethics located in Quadrant I and/or Quadrants II and III of our heuristic could reflect soft pretensions to participatory democracy and socioconstructed understanding and transformation. With expectations for overall agreement about sustainable development and/or the Earth Charter, and no meaningful opportunity to dissent, power slips from citizens (in spite of civil society initiatives) to those that promote these agendas. And, ideologues are unshakable in the belief that they are on the trail to truth—and to the solution to our problems. They have no doubt. Consider the following quotation:

> It [sustainable development] will even need to accommodate a disillusioned (and one would hope one day small) minority who think sustainable development is a plot, a trick or a bore . . . (Scott & Gough, 2003, p. 77)

This, of course, would come as no surprise to John Ralston Saul (1995) who claims that authoritarians, and their courtiers, are fond of order and contemptuous of legitimate doubters. Put another way, this contempt for legitimate doubters can be seen as a way to normalize a particular ideological orientation and undermine democracy.

While our analysis offers critique of two major ideas, education for sustainable development and the Earth Charter as a code of ethics, researchers need to be vigilant. If environmental education is conceived as an invitation to consider what is currently missing in education and ethics, then there are many generative research possibilities. However, environmental education is also susceptible to co-option by other ideological aims as potentially limiting as education for sustainable development and the Earth Charter. The heuristic presented here can similarly be used to analyse varied and disparate visions of environmental education.

We are most concerned about tendencies toward obedience—acquiescence in the face of hegemonic discourses. As Saul says, "Equilibrium [characterized by a balance between ideological stances and capacity for resistance to these ideologies], in the Western Experience, is dependent not just on criticism, but on nonconformism in the public place" (1995, p. 190). Accordingly, we tend to look toward socioconstructive and transformative opportunities projected in Quadrant IV of our heuristic for more coherence between our educational and ethics aims. Thus, we are nodding in the direction of our own assumptions about education and ethics. Here we anticipate more opportunity for critical reflections, active engagement with emerging ideas, and space for nonconformity—and hence opportunities for more authentic democratic engagement. To be clear, we are not intent, in our analysis, on holding back either environmental education or environmental ethics—or inspired action for that matter. This is not about foot dragging. Rather, we wonder why environmental educators should be satisfied with such meagre aims as sustainable development and the Earth Charter when education can offer so much more, and potentially more radical engagement with socioenvironmental issues.

However, we offer another caution. The heuristic presented here is designed to help researchers clarify their conceptions of education and ethics, especially in the context of environmental education and, from there clarify their aims and research programs. It is not meant to set standards for day-to-day practice. Many practitioners and researchers may ultimately aspire to the aims represented by one of these four quadrants, yet at times draw from the others in practice. Again we can look to Arne Næss (Næss & Jickling, 2000) who said, "to those who would like to be consistent: 'It's a high ideal to be consistent. And, you will achieve it when you die—not before'" (p. 58). Like Næss, we think the work we suggest is an ongoing process, sometimes requiring small steps, reflexivity, and practical experiments.

Although we work from the position that sustainable development and the Earth Charter can be seen as just one of many stepping-stones in environmental thought, we believe that this does not negate the generative potential of the heuristic to be used by others to conduct their own analyses. We see it as a tool that can be used to critique current discourses, evaluate new initiatives, and find one's own place within present debates. We also see it as a tool that can illuminate tensions between explicit aims—in this case educational and participatory—and implicit messages in policies and practices. Finally, we encourage readers to adapt, develop, or reinvent this heuristic to suit their own needs and to aid in their own reflection on education and environmental ethics. Perhaps in this way, a small bridge can be built between sometimes-divergent research traditions.

References

Bauman, Z. (1993). *Postmodern ethics*. Oxford, UK: Blackwell Publishers.

Bauman, Z. (2008). *Does ethics have a chance in a world of consumers*. Cambridge, MA: Harvard University Press.

Berryman, T. (1999). Relieving modern day atlas of an illusory burden: Abandoning the hypermodern fantasy of education to manage the globe. *Canadian Journal of Environmental Education, 4*, 50–68.

Blenkinsop, S., & Beeman, C. (2008). The Earth Charter, a radical document: A pedagogical response. *In Factis Pax, 2*(1), 69–87.

Bowers, C. A. (2002). Toward an eco-justice pedagogy. *Environmental Education Research, 8*(1), 21–34.

Butcher, J. (2007). *Are you sustainability literate? Spiked*. Retrieved from www.spiked-online.com/index.php?/site/article/3821/

Carson, R. (1962). *Silent spring*. Boston, MA: Houghton Mifflin.

Cheney, J. (1989). Postmodern environmental ethics: Ethics as bioregional narrative. *Environmental Ethics, 11*(2), 117–134.

Cheney, J., & Weston, A. (1999). Environmental ethics as environmental etiquette: Toward an ethics-based epistemology. *Environmental Ethics, 21*(2), 115–134.

Darier, É. (Ed.). (1999). *Discourses of the environment*. Oxford, UK: Blackwell.

Deleuze, G., & Guattari, F. (1994). *What is philosophy?* (G. Burchell & H. Tomlinson, Trans.). London, UK: Verso.

Department for Education and Skills and Qualifications and Curriculum Authority. (1999). *The national curriculum handbook for secondary teachers in England key stages 3 and 4*. London, UK: DfES.

Dewey, J. (1966). *Democracy and education: An introduction to the philosophy of education*. New York, NY: The Free Press. (Original work published 1916).

Dillon, J., & Wals, A. E. J. (2006). On the dangers of blurring methods, methodologies and ideologies in environmental education research. *Environmental Education Research, 12*(3/4), 549–558.

Earth Charter Initiative. (n.d.). *The Earth charter initiative*. Retrieved from www.earthcharterinaction.org/content/

Earth Charter Initiative. (n.d.). Read the charter. Retrieved from www.earthcharterinaction.org/content/pages/Read-the-Charter.html (Accessed 4.11.10).

Earth Charter Initiative. (n.d.). *What is the Earth Charter?* Retrieved from www.earthcharterinaction.org/content/pages/What-is-the-Earth-Charter%3F.html

Earth Charter Initiative. (n.d.). *Mission, vision, and goals*. Retrieved from www.earthcharterinaction.org/content/pages/Mission%2C-Vision-and-Goals.html

Ferreira, J. (2007). *An unorthodox account of failure and success in environmental education* (Unpublished PHD Dissertation). Griffith University, Brisbane, Australia.

Ferreira, J. (2009) Unsettling orthodoxies: Education for the environment/for sustainability. *Environmental Education Research, 15*(5), 607–620.

Fien, J. (1993). *Education for the environment: Critical curriculum theorising and environmental education*. Geelong, VIC: Deakin University Press.

Fien, J. (2000). "Education for the environment: A critique"—an analysis. *Environmental Education Research, 6*(2), 179–327.

Glock, H-J. (2008). *What is analytic philosophy?* Cambridge, UK: Cambridge University Press.

Gruenewald, D. (2004). A Foucauldian analysis of environmental education: Toward the socioecological challenge of the Earth Charter. *Curriculum Inquiry, 34*(1), 71–107.

Gruenewald, D., & Smith, G. (2007). *Place-based education in the global age*. New York, NY: Lawrence Erlbaum Associates.

Hargrove, E. (2004). After twenty-five years. *Environmental Ethics, 26*(1), 3–4.

Hart, P. (1993). Alternative perspectives in environmental education research. In R. Mrazek (Ed.), *Alternative paradigms in environmental education research* (pp. 107–130). Troy, OH: The North American Association for Environmental Education.

Hattingh, J. (2002). On the imperative of sustainable development: A philosophical and ethical appraisal. In J. Hattingh, H. Lotz-Sisitka, & R. O'Donoghue (Eds.), *Environmental education, ethics and action in Southern Africa* (pp. 5–16). Pretoria, SA: Human Sciences Research Council Publishers.

Hesselink, F., van Kempen, P.P., & Wals, A.E.J. (2000). *ESDebate: International online debate on education for sustainable development.* Gland, Switzerland: International Union for the Conservation of Nature.

Hopkins, C. (1998). Environment and society: Education and public awareness for sustainability. In M. Scoullos (Ed.), *Environment and society: Education and public awareness for sustainability* (pp. 169–172). *Proceedings of the Thessaloniki International conference organized by UNESCO and the Government of Greece* (8–12 December 1997). Athens: University of Athens, Mediterranean Information Office for Environment, Culture and Sustainable Development (MIO-ECSDE), and Ministry for the Environment, Ministry of Education.

Huckle, J. (2010). ESD and the current crisis of capitalism: Teaching beyond green new deals. *Journal of Education for Sustainable Development, 4*(1), 135–142.

Jickling, B. (1992). Why I don't want my children to be educated for sustainable development. *Journal of Environmental Education, 23*(4), 5–8.

Jickling, B. (1997). If environmental education is to make sense for teachers, we had better rethink how we define it! *Canadian Journal of Environmental Education, 2*, 86–103.

Jickling, B. (2001). Environmental thought, the language of sustainability, and digital watches. *Environmental Education Research, 7*(2), 167–180.

Jickling, B. (2003). Environmental education and environmental advocacy: Revisited. *Journal of Environmental Education, 34*(2), 20–27.

Jickling, B. (2004). Making ethics an everyday activity: How can we reduce the barriers? *Canadian Journal of Environmental Education, 9*, 11–30.

Jickling, B. (2005a). Education and advocacy: A troubling relationship. In E. A. Johnson & M. Mappin (Eds.), *Environmental education and advocacy: Changing perspectives of ecology and education* (pp. 91–113). Cambridge, UK: Cambridge University Press.

Jickling, B. (2005b). Ethics research in environmental education. *Southern African Journal of Environmental Education, 22*, 20–34.

Jickling, B., & Spork, H. (1998). Environmental education for the environment: A critique. *Environmental Education Research, 4*(3), 309–327.

Jickling, B., & Wals, A. E. J. (2008). Globalization and environmental education: Looking beyond sustainability and sustainable development. *Journal of Curriculum Studies, 40*(1), 1–21.

Leopold, A. (1966). *A Sand County almanac: With essays on conservation from Round River*. New York, NY: Sierra Club/Ballantine. (Original work published 1949).

Livingston, J. (1994). *Rogue primate: An exploration of human domestication*. Toronto, ON: Key Porter Books.

Lucas, A. M. (1979). *Environment and environmental education: Conceptual issues and curriculum implications*. Melbourne: Australian International Press and Publications.

Lucas, A. M. (1980). Science and environmental education: Pious hopes, self praise and disciplinary chauvinism. *Studies in Science Education, 7*(1), 1–26.

Lucas, A. M. (1995). *Beware of slogans!* Unpublished paper presented to a British Council Seminar, Environmental education: From policy to practice, King's College London, UK.

Mackey, B. (2001). *Update on the Earth Charter education programme. Earth Charter Education Bulletin*, (pp. 8–9). Retrieved from www.earthcharterinaction.org/invent/images/uploads/2001%2012%20EC%20Bulletin.pdf

Mrazek, R. (1993). *Alternative paradigms in environmental education research*. Troy, OH: The North American Association for Environmental Education.

Næss, A. (1988). Self realization: An ecological approach to being in the world. In J. Seed, J. Macy, P. Fleming, & A. Næss (Eds.), *Thinking like a mountain: Towards a council of all beings* (pp. 19–30). Gabriola Island, BC: New Society Publishers.

Næss, A. (2002). *Life's philosophy: Reason and feeling in a deeper world*. Athens, GA: University of Georgia Press.

Næss, A., & Jickling, B. (2000). Deep ecology and education: A conversation with Arne Næss. *Canadian Journal of Environmental Education, 5*, 48–62.

Noddings, N. (2007). *Philosophy of education* (2nd ed.). Boulder, CO: Westview Press.

Orwell, G. (1989). *Nineteen eighty-four*. London, UK: Penguin Books. (Original work published in 1949).

O'Sullivan, E. (1999). *Transformative learning: Educational vision for the 21st century*. Toronto, ON: University of Toronto. (Original work published 1949).

O'Sullivan, E. (2002). What kind of education should you experience at a university. *Canadian Journal of Environmental Education, 7*(2), 54–72.

Peters, R. S. (1966). *Ethics and education*. London, UK: George Allen & Unwin.

Peters, R. S. (1973). Aims of education: A conceptual inquiry. In R. S. Peters (Ed.), *The philosophy of education* (pp. 11–57). Oxford, UK: Oxford University Press.

Plumwood, V. (1991). Nature, self, and gender: Feminism, environmental philosophy, and the critique of reason. *Hypatia, 6*(1), 3–27.

Plumwood, V. (1993). *Feminism and the mastery of nature*. New York, NY: Routledge.

Plumwood, V. (2002). *Environmental culture: The ecological crisis of reason*. London, UK: Routledge.

Rgozzi, M. J. (2003) UNESCO and the international decade of education for sustainable development (2005–2015). *Connect, 28* (1–2), 1–7.

Robottom, I. (1993). Beyond behaviourism: Making EE research educational. In R. Mrazek (Ed.), *Alternative paradigms in environmental education research* (pp. 133–143). Troy, OH: The North American Association for Environmental Education.

Robottom, I., & Hart, P. (1995). Behaviourist EE research: Environmentalism as individualism. *The Journal of Environmental Education, 26*(2), 5–9.

Russell, C. (2006). Working across and with methodological difference in environmental education research. *Environmental Education Research, 12*(3 & 4), 403–412.

Saul, J. R. (1995). *The unconscious civilization*. Concord, ON: Anasi.

Scott, W., & Gough, S. (2003). *Sustainable development and learning: Framing the issues*. London, UK: RoutledgeFalmer.

Scott, W., & Gough, S. (2007). Universities and sustainable development: The necessity for barriers to change. *Perspectives: Policy & Practice in Higher Education, 11*(4), 109–118.

Shepard, L. A. (2000). The role of assessment in a learning culture. *Educational Researcher, 29*(7), 4–14.

Traoré, A. (2002a). Contestation au sommet de Johannesburg: l'oppression du développement. *Le Monde Diplomatique* [Challenging the Johannesburg Summit: The oppression of development]. Retrieved from www.monde-diplomatique.fr/2002/09/TRAORE/16825

Traoré, A. (2002b). *Le viol de l'imaginaire* [The rape of the imagination]. Paris: Fayard/Actes Sud.

United Nations. (2002a). *Johannesburg Plan of Implementation*. Retrieved from http://www.un.org/esa/sustdev/documents/WSSD_POI_PD/English/POIToc.htm

United Nations. (2002b). Resolution 57/254. *United Nations decade of education for sustainable development*. New York, NY: United Nations. Retrieved from http://portal.unesco.org/education/en/file_download.php/299680c3c67a50454833fb27fa42d95bUNresolutionen.pdf

United Nations Conference on Environment and Development. (1992). *Agenda 21, the United Nations programme of action from Rio*. New York, NY: UN Department of Public Information. Retrieved from www.un.org/esa/sustdev/documents/agenda21/index.htm

Van Harmelen, U. (2003). Education, ethics, and values: A response to Peter Blaze Corcoran's keynote address, EEASA 2003. *Southern African Journal of environmental education, 20*, 124–128.

Vilela de Araujo, M., Ramirez Ramirez, E., Hernandez Rojas, L., & Briceno Lobo, C. (2005). *Teaching a sustainable lifestyle with the Earth Charter*. San Jose, Costa Rica: The Earth Charter Initiative. Retrieved from http://www.earthcharter.nl/upload/cms/230_Teachers_guidebook.pdf

Wals, A. E. J. (2009). *Learning for a sustainable world: Review of contexts and structures for ESD*. Paris, France: UNESCO.

Wals, A. E. J., & Jickling, B. (2002). "Sustainability" in higher education from doublethink and newspeak to critical thinking and meaningful learning. *Higher Education Policy, 15*(2), 121–131.

Walsh, P. (1993). *Education and meaning: Philosophy in practice*. London, UK: Cassell.

Warren, K. (1990). The power and promise of ecological feminism. *Environmental Ethics, 12*(2), 125–146.

Weakland, J. P., & Corcoran, P. B. (2009). The Earth Charter in higher education for sustainability. *Journal of Education for Sustainable Development, 3*(2), 151–158.

Weston, A. (1985). Beyond intrinsic value: Pragmatism in environmental ethics. *Environmental Ethics, 7*(4), 321–339.

Weston, A. (1991). Nonanthropocentrism in a thoroughly anthropocentrized world. *The Trumpeter, 8*(3), 108–112.

Weston, A. (1992). Before environmental ethics. *Environmental Ethics, 14*(4), 321–338.

Winter, C. (2007). Education for sustainable development and the secondary curriculum in English schools: rhetoric or reality? *Cambridge Journal of Education, 37*(3), 337–354.

World Commission on Environment and Development. (1987). *Our common future*. Oxford, UK: Oxford University Press.

Zwicky, J. (1992). *Lyric philosophy*. Toronto, ON: University of Toronto Press.

Zwicky, J. (2003). *Wisdom and metaphor*. Kentville, NS: Gaspereau Press.

8

Self, Environment, and Education

Normative Arisings

Michael Bonnett
Cambridge University and University of Bath, UK

Introduction

I begin by recapping some key senses in which education and environmental education can be considered to be normative.

First, there is an important sense of education that expresses evaluative ideas concerning both what it is fundamentally worthwhile to learn or engage in, and how individuals (and students in particular) should be conceived and treated. This normative conception of education, made explicit by, for example, Richard Peters (1966), is certainly not the only way in which education can be conceived, and as stated here it leaves open the substantive resolution of the normative issues that it raises. Historically there have been differences and debate over how this should be achieved, but as a general approach it expresses a longstanding evaluative notion of education that goes back to the ancient Greeks and that recurs in the thought of many subsequent noteworthy educational thinkers such as Rousseau (1762/1911), Dewey (1902), and Whitehead (1932/1962). Also, it is presupposed by any who speak of "good," "bad," "indifferent," or "miseducation" and it is the conception that informs this paper. Second, environmental education, too, involves normative criteria, for example, for identifying (a) what count as the pressing environmental issues to be addressed, and (b) how one's relationship with the environment—and in particular with nature—is to be understood, and what is to count as a "right" relationship in this respect.

It is argued that while in some senses the above separation of general education from environmental education presents an artificial dichotomy, it can be productive to explore how giving precedence to the second set of issues can illuminate the first set. More broadly speaking, and for the moment taking a traditional approach to such matters, it will be held that there are two seminal underlying questions that are in reciprocal relationship:

- What is the ethical basis of environmental concerns?
- What is the environmental basis of ethical concerns?

The first of these questions makes the point that there are few, if any, actions that do not have an ethical dimension deriving from their intentions and consequences. It has been held that all behavior can be evaluated ethically and, insofar as environmental concerns are precursors to action, they always express this ethical dimension to some degree through either raising or presupposing ideas about how we ought to behave toward the things that we encounter in the environment. Hence we are invited to address the question of the nature and weight of the ethical significance of different aspects of the nonhuman world, and what our underlying stance toward them should be. A key issue here would be in what senses, if any, aspects of the nonhuman world should be considered to be bearers of intrinsic moral value as against being merely of instrumental value.

The second question draws attention to the fact that ethical concern does not arise in some pure form—say, as articulated in a set of pristine universal rational principles—but rather that our antecedent relationship with our environment—our always already being in a world—conditions all understanding, including the ethical. Our concern is with such and such an issue in such and such a context, all of which derives its intelligibility from an underlying form of sensibility (to our environment) in which we are thoroughly embedded. Hence there is an intimate reciprocity between ethical and environmental concern that fundamentally gives rise to the character of our caring.

In the context of this paper the two questions outlined above lead to a third:

> How should the interplay of the ethical and the environmental condition our understanding of education?

Harking back to the previously identified normative sense of education, for the purposes of this paper I interpret education as involving a holistic transformation of individuals. I take it that a central aspiration of education should be to contribute to a process of personal growth where this latter implies developing engagements with the world of an increasingly perceptive and therefore receptive character. So conceived, education is intimately related to, and reflective of, the ways in which the above interplay is understood in its many aspects and nuances. To delineate some important elements of this interplay is the business of the next section.

Self and Environment: Dimensions of the Holistic Self

Relational notions of the self are a strong feature of the literature of environmental concern, appearing in a particularly powerful (some would say overblown) form in deep green thinkers such as Arne Naess (1989) and Freya Mathews (1994). In the light of, and by way of extending, the above discussion, it is helpful to explore two important senses in which the self is rightly understood in the kind of relational terms foregrounded by certain strands of environmental debate. The first of these is the idea of the self as locally emplaced; the second is the idea of the self as part of a "greater whole." The considerations that arise are set alongside, and explored in terms of elucidating, important senses in which the self is experienced as possessing an internal integrity. This in turn raises a key issue: that of the character of the self's essential inherence in the world. If, as previously mentioned, we focus on a sense of education that is at some level necessarily concerned with the development of individuals, it is important to have an adequate understanding of the individuality involved, i.e., the nature of that individuality that is individuality-in-relationship.

The Self as Locally Emplaced

Elsewhere (Bonnett, 2009c), I have explored the argument highlighted by some strands of environmental discourse that among the variety of understandings of the self, those that recognize the salience of place in its constitution provide important insights. Phenomenologically, it seems clear that it is not simply a contingent matter that our sense of our own existence is conditioned by our sense of where we are, rather it is that place is an ontological condition of experience. In experience, nothing occurs unplaced. We experience ourselves as always already beings in a world and that world is not one of some even spatio-temporal space governed by universal abstract laws, but one of distinct neighborhoods characterized by changing local ambiences, events, and significances. As such, we are always (if often largely implicitly) to some degree claimed by the neighborhoods in which we find ourselves. That is to say feelings, perceptions, attitudes, moods, arise in relationship to (and often in direct response to) a neighborhood and some neighborhoods can be of such personal significance as to determine incisively aspects of our way of being when we *are* present there. One's home, the place of a certain seminal encounter, the place where a mother's child is buried can lay claim to us in this way. In an important sense we *are* the person that is so claimed, for us to be otherwise would require an act of severance. And such claims are achieved through relationships of mutual anticipation between self and its environment that is both bodily and cerebral.

Take the (rather extreme) negative example of someone who is forcibly removed from their most primordial neighborhood: their home. They are extracted from participation in the intimately woven fabric of people, artifacts, vistas and ambiences, routines and associations that hitherto have constituted their daily life. Consciously and bodily they no longer directly experience the call of its demands and promptings, including those arising from its existence as a historical site of personal events, aspirations, and imaginings. Hence they can no longer presence (that is, stand forth, show themselves) as once they did, for the especial place that facilitated this is gone and the flow of immediate involvements that buoyed them through its inviting claims on them has been broken. They are left "out of place."

Overall, there is an important general sense in which one *is* in one's dealings with a world: a milieu of prompts and claims that occur in particular places and that are always partly familiar, but also always taken as partly "other." And one's inherence in a place—and hence in a world—requires that it is to some degree receptive to the self, actively accepts it, provides as it were the questions to its replies as well as the replies to its questions—as, say, when in the activity of cooking the eye's falling on a certain utensil tacitly invokes the question of a certain alternative way of proceeding that is either pursued or rejected. This is to say that a mutual anticipation (and hence invitation) of self and world is in play in which each is called forth. Such anticipation, while to some extent frequently expressed through daily routines, is not itself merely mechanical. Rather always it is a form of ecstasis, of which at times it is possible be strongly conscious as, say, with the sense of anticipation experienced when the walker sets off on a fine spring morning, open to the unknown that is to come—the challenges and surprises, the sights, the smells, the textures, and the ambiences, of say, different spots and times of day. Such anticipation speaks of a keen attentiveness that quickens life and gives a heightened sense of being. While not always as consciously salient as this, anticipation in this sense occurs at many levels and is involved in all that we do: the gardener anticipates that the soil will yield to her spade, the walker that the earth will bear her up, the reader that the text has a meaning. Such anticipation is enacted implicitly through the very movements of the limbs of the gardener and walker, in the very act of scanning the text by the reader—indeed, in the act of opening the book or envelope. It lies at the heart of the constant delicate, intelligent, adjustments both bodied and cerebral that we make within our proximate environment and through which we inhere in it—feeling in varying degree either "at home" or discomforted.

What would it be, then, for education to respect the emplaced individual? How does it affect how education should treat environmental issues when the environment is understood as something that is not simply external and susceptible to explication in purely physical—or indeed, objective—terms, but is conceived as interconnected places of ebb and flow of a wide array of sensings and significances that include prudential, aesthetic, moral, spiritual, and bodily, and that constantly reflects back one's (frequently) tacit sense of who one is and of what one is doing. Such an environment is experienced as imbued with purpose part known and part unknown, part shared and familiar, part unshared and strange. It meets us in the ways of proceeding that it invites, the attentions that it demands, the mysteries that it offers, the values that it reveals, the responses and responsibilities that it intimates. When we enter a room or look out across a valley, we can be met by a host of possibilities of thought and action. These possibilities and their fittingness arise in the very occurrence of apprehending what is there. Experience does not begin with the perception of neutral, inert, objects. The table perhaps invites our setting it for a meal, or clearing it for writing, the hovering hawk poised perfectly steady in the turbulent air perhaps invites our admiring gaze. Ultimately, there is nothing purely objective or passive about a place; it only appears so when we have lost sight of it, lost contact with its—and our own—genius.

Thus saying "I am in a world" means that "I" and "world" only exist in this relationship, not independently of it—except as abstractions. Consciousness is nothing without its things—i.e., that toward which it is directed—and these things are experienced always as there, in a world. This is true even for such phenomena as isolated color patches once taken to exemplify the raw experience of perception by sense-data theorists. It follows that in a significant sense to speak of me is to speak of my world: the places that I inhabit, my emplaced existence. And this world is normative; from the start it is shot through with intimations of fitting description and response—that is to say, of values, ethical, and other.

Yet there is an important sense in which the human self transcends its proximate environment: its anticipations outrun what is immediately present and so express intimations of a greater whole.

The Self as Part of a "Greater Whole"

Over the course of history there have been many versions of "the greater whole." Rather than attempt to recount these, here I would like to focus on the key features of some dominant modern conceptions. Perhaps the most ubiquitous of these conceptions is that of a law-governed causal network employed by classical science and related to this, the network of bio-physical interdependence employed by ecology. In addition, more recently, the systems thinking approach in which the environment as a whole is conceived as a cybernetic information feedback system has gained some prominence (see Bateson, 1972/2000). The point about all of these conceptions is that they represent highly discursive forms of structure. They involve a systematization achieved through the imposition of a super-ordinate abstract theoretical framework that delegates to each aspect of experience its character, "place," and significance. Beneath the appearance of a holism lies the reality of an atomism—in the sense that the fluid continuity of mutual anticipation has been disrupted; it is reductively broken up and ossified so that component elements and processes can be objectively specified. It is argued that such accounts, through their inattentiveness to the primary character of the native occurring of things—and in particular, the "self-arising" quality of natural things (Bonnett, 2004)—are in danger of doing violence to the phenomena that they purport to describe.

So, what might an alternative account look like? Enlarging on the theme of the preceding comment, access to a nondiscursive "whole" is facilitated by sensing the immanent in the particular and the infinite interplay of presencing and withdrawing in the occurring of things. It requires that we travel with the self-arising flow of the gathering and the dissolution of things that yet possesses a pressing reality. The self as part of a greater whole engages most authentically with that greater whole not chiefly by abstract ratiocination (although, appropriately conditioned, there is a place for this) but through the ongoing simpatico of its immediate dealings with the individuals it encounters; its attunement to its own and their own participation in the infinite interplay of presencing as it spreads out in all directions to the horizon and beyond. In other work (Bonnett, 2009b), drawing on references to ideas of "selving" and "unselving" in the poetry of Gerard Manley-Hopkins, I have explored ways in which it possible to think of the self in terms of the qualities of its presencing through its participation in the upholding of a particular neighborhood. Taking the example of the felling of a tree, I suggested that from the point of view of its selfhood:

> The nub of the issue is not that extracting living things from their natural environment will often result in physical harm or death, as, say, when a tree is removed to make way for a new road. It is that the tree, so displaced, has been withdrawn from the place that facilitates it in its occurring as the particular thing that it is.
>
> It has been withdrawn from, say, the play of sunlight on its limbs and leaves, from its movement in the breezes that stir at that spot, from the fall of its extending and diminishing shadow, from its posture in relation to its neighbors, from the sounds and sights of the birds that visit or inhabit it, from the dance of midges beneath its canopy as evening closes; that is to say from its unique and infinitely manifold contribution to the precise ambience of its neighborhood. It holds this neighborhood together, contributes to the unique and ever-changing qualities of its space, and is held by it. In other words it participates in a place-making, and is constituted as itself through this participation. (Bonnett, 2009b)

Such an example suggests that understanding of oneself as emplaced in a greater whole when viewed from the orientation given by sensitivity to presencing and place-making is not to be articulated discursively in terms of some system, but analogously (when viewed from the perspective of what is currently taken as literal truth) through the recognition of powers that are in play in attunement to presencing. I have in mind here such powers as those of lightening and darkening, sound and silence, birth and death; also, a sense of things being on their own ineffable way and of the ultimate mystery of their sheer existence. It is such poetic openness rather than scientific systemization that connects us to the cosmos. And this connection is essential to both the greater meaning that we discern in ourselves and the greater meaning that we discern in our environment. Also, as a form of attunement, it is necessarily normative, for it is not an impassive spectatorial relationship but an engagement that involves volition and emotion and that takes its cues from its sense of what is present and underway. It responds to the being (potential) yet to be fulfilled and the character of the letting be required of us in our participation in the process of the standing forth of things themselves. Hence, it is a form of measure-taking in which, through the minutiae of our manifold intimate involvements, we learn of the powers holding sway in the world and of our active place in the world: how to receive what is gifted and to gift what is received. Clearly such measure-taking has nothing to do with quantifying (or only insofar as it opens up the space in which any authentic quantification is possible). Rather it attempts to restore a sense of those suprasensory powers, including ideals, in terms of which we might yet begin to apprehend and enact the underlying intelligibility and motivation to our lives: the withdrawn source of our values. This is the hidden center that modulates our participation in the greater whole and whose exceedingly faint traces in the late modern period are the remaining conduits to authentic flourishing.

The Integrity of the Self

Elsewhere (Bonnett, 2009a), I have examined the way in which a number of recent educational texts invite us to see the self as constituted by factors that are external to it, and as possessing little or no internally maintained steady identity. For example, some influenced by Michel Foucault and Judith Butler see individual subjectivities as heavily and continuously constituted by discourse and the performative utterances and gestures of others. Again, others (e.g., Biesta, 2006) interpret Hannah Arendt and Emmanuael Levinas as seeing our subjectivity occurring by the grace of others who give us meaning. On such views the other is posited as originative of selfhood and the self becomes both decentered in the sense of losing its central epistemic position and de-nucleated in the sense that any supposed core or essence is stripped out. The interiority of the self is evacuated. It would be a mistake to think that the emphasis on relationality that informs this paper leads to a similar or parallel position; indeed, its central thrust goes in the very opposite direction, foregrounding the corrosiveness of selfhood of such accounts.

Having said this, by no means is it my intention to suggest that the self is entirely self-originative in the sense of constituting itself independently of a cultural context, including the sociopolitical. Any self (and any idea of self) arises in a cultural (including linguistic) milieu that provides the horizons of significance within which meanings are possible and issues arise. But (especially) in the late modern context this milieu is never monolithic, never homogenous through and through. Rather it consists in diverse languages, discourses, conversations, sentiments, and values. At any one time some voices are loud, some quiet yet perhaps nonetheless powerful, some fallen silent but retrievable. Hence in its continuous and fluid internal interplays culture always conditions but never determines the self in all of its potentialities. The self is never merely exclusively a creature of its cultural constitution, but possesses a responsibility for its individual expressions and engagements. It is only on such premises that my previous claim that it is the responsibility of educators to provide invitations to students to widen and deepen their sensibilities can gain a purchase.

Given the above caveat, the argument outlined in previous sections has emphasized that the integrity of the self resides in the free play of its anticipations, its mutually sustaining conserving that brings into the light things themselves in their mystery. The self's integrity is expressed in a poetic receptive-responsive agency that is imbued with a nonpossessive, noninstrumental caring, but that, nonetheless, has its own perspective for it is bodily sited and has its own intentional flow: a genealogy of meaning-giving/receiving that while in one sense is always beyond itself—with its objects—has the resident potential to be self-aware to varying degree, having some sense of its own history and a future. The meanings that it intends and for which it feels responsible are essential both to its sense of being in a world and to the character of that world. Hence the self as self-in-relationship possesses a powerful interiority that necessarily participates in the movement of place-making in which intimations of those powers that vivify the greater whole are present. The greater whole is not some system in which the self is objectively located, but, like the self, is a participant in a mutual facilitation in which senses of the "inner" and the "outer" arise and fall back.

Education

Considerations discussed in the preceding sections suggest that environmental education should be concerned with an idea of holistic education whose normative character would become salient through examining what it means for education to respect and engage the embodied person as a whole (i.e., as locally emplaced, part of a greater whole, but possessing its own integrity), and to enact the idea of its being understood as not simply contingently embedded in some greater whole but ontologically as an individual in indissoluble relationship with it, yet clearly capable of tranquillizing its awareness of this to varying degrees. On the account of this relationship developed above in which an environmental dimension is voiced, it can be seen that holistic education will be an education that derives its norms and imperatives not from some set of quasi universal rationally justified principles, but from a constantly ongoing "reading" of the emplaced individual from which emanate certain guiding values (as contrasted with discursively derived prescriptive aims). These arising values can be expressed in an infinite number of specific ways and embody a nondiscursive understanding of reality (nature).

In more general terms, this offers an invitation to view the self not predominantly in terms of, say, space-time or bio-physical/causal continuity, but in terms of the qualities of its presencing: their flow. This de-reification of the self and its worldly encounters could be important in the educational context. It demands an ongoing attentiveness to the fluid being of individual things—for example, an attentiveness to how they are from situation to situation, rather than a reliance on how they have been defined, averaged off and frozen for the convenience of easy speculation.

In addition, that normative dimension of education noted in the introduction that speaks of what it is worthwhile to participate in, seeks to develop an ever growing sensitivity to the primal powers previously identified as involved in the occurrence of meaning in the cosmos, and their instantiation in the things we encounter. Because of nature's essential autonomy, this sensitivity is likely to be experienced most strongly in relation to the natural world. And in such experience distinctions between different kinds of normative impulse can become attenuated. What feels fitting—what we are called to do or desist in—will not always be experienced as being of a distinctly moral or aesthetic or prudential character. Our sense of our own good and the good of the things that we encounter can as it were coalesce, as can the beautiful and the true. For example, the felt rightness of pausing to bear witness to the play of sunlight on water can both defy and encompass such categorization, and is not necessarily much enhanced by post hoc reflection in its terms.

It is clear that the above account suggests an understanding of the normative significance of the environment for education in a fuller sense than is often the case. The account makes it clear that what is at issue is not simply or primarily the matter of addressing our environmental predicament as gauged against the kinds of degradation to which attention is drawn by ideas of climate change, pollution, and species extinction—important as they are. Rather, what becomes central is the quality of our individual being in a world that is constituted by an ever extending and subtly nuanced interplay of significances and mutual anticipations that have sensuous as well as cerebral, aesthetic, and ethical as well as prudential, dimensions. At the heart of this mode of being is a caring that is open to nonanthropocentric impulses—a caring that lies in a recognition of the reality of the other and the responsibility for noninterference that such recognition incurs. It invites a celebration of alterity, of the sheer existential autonomy of nonhuman nature, and of the nobility of that which is truly itself. This necessarily checks the rush to action and the drive for mastery. A whole that is not regarded essentially as a resource and that therefore is not in need of definition, organization, and manipulation can be accepted in its fluidity and mystery. So apprehended it is allowed to flourish in a space in which it can gift inspiration. Only by listening for the call of what is not us can we begin to receive the norms that imbue the places (neighborhoods) we inhabit and so receive intimations of how our existence is in interplay with all else. In this way a sensing of a greater whole that is neither totalizing nor atomizing, but that embraces the ever-changing cadences of the places—the populated (but not simply peopled) neighborhoods—that constitute it can occur.

What, then, does such a view suggest for environmental education research? In order to address this we must first summarize the fundamental purposes and values of environmental education that have emerged in the preceding discussion. I have argued that environmental education properly understood is ineluctably holistic in the sense that its ultimate concern must be with understanding our place in the cosmos and the proper character of our participation in it, particularly in its natural dimension. Such understanding is not revealed in the consumerist and performative models of flourishing which dominate late modern culture, nor is it illuminated by the forms of knowing that have been recruited to their cause and are legitimated by their "success." Hence, essentially environmental education must be concerned to invite a different way of being in the world to that which currently holds sway. Given this central concern, the following are some emerging foci.

First, because we live in a predominately un-attuned world, we need to continue to employ science to monitor the effects of human behavior on the natural world. But it remains important to continue to identify the ways in which science and social/political policies and practices are implicated in the degradation of the environment—and, equally important, our relationship to it—and to investigate ways of enabling pupils to address the issues that arise. Ultimately this will involve students engaging in a metaphysical investigation in the sense of seeking to discern those deep motives that are in play in their lifeworld and the larger world in which it is located. However the spirit in which this is undertaken needs to be conditioned by a more central educational concern that it has been the business of this paper to foreground. How are we to re-engage students holistically? What contribution can and should education make to the fuller understanding of the self-in-relationship in its various dimensions? What should be its approach and upon what resources can it draw? Such questions draw attention to what remains a central research concern: to engage in an ongoing exploration and development of nonscientific understandings of the greater whole and our (internal) relationship with it.

While some aspects of such research will be empirical, a good deal of further theoretical research is also required. The following more specific questions are relevant:

How most effectively to incorporate experiences of nature into the life of the school such that the normative orientation that arises through poetic contact with the elemental is felt and allowed to matter? It is the establishment of a certain quality of ongoing everyday engagement with nature—our attunement to it as a vivifying dimension of experience—that is the greatest challenge for environmental education. In an age that is in thrall to the metaphysics of mastery and whose dominant technologies express this in increasingly purified and narcotic forms the possibilities for such everyday engagement constantly recede as increasingly we are invited, if not compelled, to inhabit places most notable for the absence of, or extreme attenuation of, nature.

This raises research questions concerning the educational possibilities of reasserting the value of knowledge by acquaintance that includes a sensual/intuitive sensing of the presence of things and a nondiscursive systemic wisdom in which the ethical is embedded and that involves an integration of thought, feeling, and action. Furthermore, insofar as knowledge is to be organized pedagogically in advance of learning, the issue is then raised as to how best to do this, for example—should it be in the form of narratives or subject disciplines, or narratives within subject disciplines, both informed by an enlarged conception of human flourishing? This in turn poses the question of how to facilitate and manage an emergent curriculum that springs from the life-world of students, respects the self-arising nature within the student, and yet incorporates a metaphysical examination of that life-world that interalia can reveal the sociopolitical cultural powers that are in play in the shaping of the self and its sensitivities. The acknowledgment of, and engagement with, the unknown and the mysterious that such ideas betoken clearly pose serious questions for the undertaking of assessment. What forms will be both acceptable to those who require that educational progress be measured and truly adequate to the progress that they should measure?

This paper has developed a normative conception of holistic education that seeks to facilitate the presencing and engagement of the whole person (including the body) in the place of the school in a way that expresses its inherent participation in the greater whole. The demands that this conception places on further theoretical and empirical research are stringent, but cannot be ducked if the more fundamental goals of environmental education are to be achieved.

References

Bateson, G. (2000). *Steps to an ecology of mind.* Chicago, IL: University of Chicago Press. (Original work published 1972).

Biesta, G. (2006). *Beyond learning: Democratic education for a human future.* Boulder, CO: Paradigm Publishers.

Bonnett, M. (2004). *Retrieving nature: Education for a post-humanist age.* Oxford, UK: Blackwell.

Bonnett, M. (2009a). Education and selfhood: A phenomenological investigation. *Journal of Philosophy of Education, 43*(3), 357–370.

Bonnett, M. (2009b). Education, sustainability and the metaphysics of nature. In M.McKenzie, H. Bai, P. Hart, & B. Jickling (Eds.), *Fields of green: Restorying culture, environment, and education* (pp. 177–186). Cresskill, NJ: Hampton Press.

Bonnett, M. (2009c). Schools as places of unselving: An educational pathology? In G. Dall' Alba (Ed.), *Exploring education through phenomenology. Diverse approaches* (pp. 28–40). Malden, MA: Wiley-Blackwell.

Dewey, J. (1902). *The child and the curriculum.* Chicago, IL: University of Chicago.

Mathews, F. (1994). *The ecological self.* London, UK: Routledge.

Naess, A. (1989). *Ecology, community and life style.* Cambridge, UK: Cambridge University Press.

Peters, R. S. (1966). *Ethics and education.* London, UK: Allen & Unwin.

Rousseau, J.-J. (1911). *Emile.* (B. Foxley, Trans.). London, UK: Dent. (Original work published 1762).

Whitehead, A. N. (1962). The aims of education. In *The aims of education and other essays.* London, UK: Ernest Benn. (Original work published 1932).

9

A Critical Theory of Place-Conscious Education

David A. Greenwood
Lakehead University, Canada

Place and Education

Describing the significance of place to human life, Clifford Geertz (1996) quipped, "[N]o one lives in the world in general" (p. 259). People, and other species, live embodied and emplaced lives, and a multidisciplinary analysis of place reveals the many ways that places are profoundly pedagogical (Gruenewald, 2003b). That is, as centers of experience, places teach us and shape our identities and relationships. Reciprocally, people shape places: places can be thought of as primary artifacts of human culture—the material and ideological legacy of our collective inhabitation and place-making. The kind of teaching and shaping that places and place-making accomplish, of course, depends on what kinds of attention we give to them and on how we respond to them. Culture, place, and identity are deeply intertwined (Basso, 1996; W. Berry, 1987, 1992; Casey, 1997), yet the formal process of education often obscures and distorts relations to place, especially with respect to the land. As Aldo Leopold (Leopold, 1949, 1968) remarked in the first half of the twentieth century, "our educational and economic system is headed away from, rather than toward, an intense consciousness of land" (p. 223). Although contemporary place-study involves more than Leopold's land ethic, care for place is in part an act of remembering people to the land from which all the complexities of culture and identity arise. This chapter argues that place-conscious education can provide environmental education research with a fluid conceptual framework that bridges cultural and ecological analysis, and that is responsive to diverse and changing cultural and ecological contexts. My purpose is to contribute to a theory of environmental education that is culturally responsive, and committed to care for land and people, locally and globally.

To appreciate "place" as a productive educational construct, one must first explore its meanings. In the simplest terms, place signifies a unique and bounded biophysical and cultural environment. Its physical scale can range from small indoor corners to sweeping outdoor expanses: mud puddles, bedrooms, backyards, prison cells, back alleys, neighborhoods, farms, watersheds, regions—all of these can be identified and experienced as places. All experience is placed. As philosopher Edward Casey (1997) wrote:

> To be at all—to exist in any way—is to be somewhere, and to be somewhere is to be in some kind of place. Place is as requisite as the air we breathe, the ground on which we stand, the bodies we have. We are surrounded by places. We walk over and though them. We live in places, relate to others in them, die in them. Nothing we do is unplaced. How could it be otherwise? (p. ix)

Places hold and shape our experiences, and the diversity of possible places, and the variety of human relationship to them across cultures, makes place-study complex. In the last several decades, place has become a niche focus for inquiry across many academic disciplines, from architecture, ecology, geography, and anthropology, to philosophy, sociology, literary theory, psychology, and cultural studies.[1] Many theories of place from these fields could inform educational studies, as most scholars who study place agree that place-consciousness can lead to vital understandings of our relationships with each other and the culturally situated, biophysical world.

Paradoxically, the recent attention to place studies across disciplines parallels contemporary interest in globalization in social, political, and educational thought. Attention to place can in part be explained as a counterpoint to the abstractness that sometimes surrounds globalization discourse, which can suggest a kind of "placelessness" (Relph, 1976), or a "geography of nowhere" (Kunstler, 1993). Philosophers have observed that throughout the modern era, ideas of place were subordinated to space and time so that place became synonymous with a location to be charted and exploited (under imperialism, colonialism, or globalization), rather than an environment to be experienced and understood on local terms (Casey, 1997; Foucault, 1980). In education, the dominant neoliberal understanding of globalization frequently leads policymakers and other leaders to disregard places and espouse the reductionist view that the chief purpose of education is preparation for competition in "the global economy." Critical perspectives on culture and education that emphasize abstract narratives of displacement, movement, and rapid change also risk neglecting the primacy of everyday placed experience as a central focus for cultural and educational thought. A frequent critique of place-focus in education is that we live in a globalized, multicultural world, and that place-study might reinforce a narrow or provincial view of global realities. However, those who study places—from environmental scientists to cultural theorists—argue that local places provide the specific contexts from which reliable knowledge of global relationships can emerge. Additionally, place-study is vital for understanding how human and other species adapt to ecological and cultural changes on a planet in continual flux.

Wendell Berry's (1992) writing on the difference between global and local thinking is helpful in determining what scale of thought is most appropriate for addressing cultural and environmental concerns:

> Properly speaking, global thinking is not possible. Those who have "thought globally" (and among them have been imperial governments and multinational corporations) have done so by means of simplifications too extreme and oppressive to merit the name of thought. Global thinkers have been and will be dangerous people. ... Global thinking can only be statistical. Its shallowness is exposed by the least intention to do something. Unless one is willing to be destructive on a very large scale, one cannot do something except locally, in a small place. (pp. 19–20)

Focusing on small places does not preclude interest in the larger world. The point, rather, is that the world is only knowable as a collection of diverse experiences with places.[2] This view corresponds with the perspective of place-based educators such as David Sobel (1996, 2005), who argues for a developmental view of learning that begins not with abstractions about the environment, but with direct, local experience.

Because of their ubiquity and their social and ecological complexity, place studies can help to frame environmental education research in three important ways. First, places provide a local focus for socioecological experience and inquiry. Such contexts are accessible and relevant to people's everyday lives and can be easily linked (ecologically and geographically) to other experiences and places elsewhere, both regionally and globally. Second, the construct "place" helps to overcome the dualism between "culture" and "environment" that some environmental educators have functioned to maintain even as the field has worked to overcome this rhetorical dichotomy. In educational literature generally, a focus on culture frequently lacks attention to environment, just as a focus on environment often neglects a deeper examination of culture. As the nexus of environment and culture, place serves as a reminder that all cultures are nested in biophysical environments, and that environments are culturally produced or experienced, as well as understood from within a culturally specific epistemology. Third, place-studies complicate the idea of socioecological context with the observation that as socially constructed contexts, places are fluid and contested terrain. Here, the field of critical geography (Helfenbein & Taylor, 2009) is helpful to explore the contested nature of places, and how a cultural politics of identity are connected to the production and experience of place (and space) in ways that function to privilege, exclude, oppress, and erase. In other words, to really know a place is to understand how environment, culture, and politics have worked to shape it, and to appreciate that any particular place may have diverse and competing meanings for different groups or individuals. As Margaret Somerville (2007) has written, places are "the contact zone of contested place stories" (p. 153) with transformative potential for culture, environment, and education.

Paradoxically then, the socioecological construct place both provides for contextual focus for environmental education research even as it problematizes the idea of focused context with the possibility of multiple and conflicting place-based experiences. Although I argue that, in terms of its value to education, place should be understood as a socioecological construct capable of remembering people to the natural environment that supports all life, it should be pointed out that much of the vast literature on place and space, including literature from the field of critical geography, entirely disregards the ecological context of culture. Critical geography has been called "spatialized" critical social theory (Soja, 1989), and like critical theory generally, its concerns are mainly cultural rather than ecological. However, critical perspectives on place, even those that are not explicitly concerned with "the natural environment," remain important to environmental education because a spatialized social analysis is sometimes necessary to understand how particular ways of knowing with regard to the environment are culturally and spatially constituted.

The conceptual range of the construct place signifies that place can inform education from a deeply philosophical and theoretical perspective. Here it is useful to make a distinction between place-based education as a movement and methodology, and place-conscious education as a philosophical and political orientation to the field. In the literature, place-based education is most often described as reform pedagogy that can transform the experience of schooling for teachers, students, and community members. Nearly synonymous with community-based learning (Hart, 1997; Melaville, Berg, & Blank, 2006), place-based education is usually presented as a schooling methodology that offers learners authentic experiences in their local communities and environments for the purpose of increasing student engagement and achievement, and for promoting democratic participation in local community processes (Gruenewald & Smith, 2008; G. Smith, 2002; Sobel, 2005).While place-based education (called also by other names) has helped to transform the experience of school in a great number of communities, the "constraining regularities" (Stevenson, 1987, 2007) of schooling as an institution limit its impact as a wider educational reform movement (see Gruenewald, 2005; G. Smith, 2007). In its implementation, however, place-based education is not limited to schooling, as a growing number of environmental groups, including those with commitments for environmental justice, embrace the term as a wider public pedagogy. When discussed as a schooling methodology alone, place-based education can become absorbed by the discourses surrounding schooling. For example, in order to demonstrate legitimacy, some advocates of place-based education, like some environmental educators before them, argue that the adoption of a place-based curriculum can improve the achievement outcomes associated with conventional schooling, such as content area test scores (Sobel, 2005). This may be a strategic way to promote attention to place in school; however, promoting place-based education (or environmental education) as an instrumental means to the problematic ends of schooling potentially reinforces the very constraining regularities that make place-study relatively rare, and that mute its political content (Gruenewald, 2004, 2005).

On the other hand, as a philosophical orientation to the field, place-conscious thinking can help build a conceptual framework that explicitly critiques problematic ends of schooling and that articulates educational purposes and possibilities that are woefully neglected in the discourse of schooling. This does not mean that place-based education and place-conscious education are in conflict. However, if place-based education is to be more than a schooling methodology, it needs to be located in a coherent theory that names political purposes and provides ethical direction for critical praxis. In short, while place-based education is often framed as a method to improve school outcomes, place-conscious education can be framed as a philosophy that challenges educators to rethink the assumptions of schooling in the context of the places we inhabit and leave behind.

Theoretical Direction for Environmental Education: A Critical Pedagogy of Place

I have previously described "a critical pedagogy of place" as a broad-spectrum theoretical framework for place-conscious education (Gruenewald, 2003a). While the phrase has stirred some controversy,[3] the naming is an attempt to frame an emergent and generative educational theory that is responsive to the cultural and ecological politics of place. The naming is also an invitation to educational researchers mainly interested in culture to think more deeply about environment, and an invitation to researchers mainly interested in environment to think more deeply about culture. In the wider educational research community, it is clear that this invitation needs continued attention, perhaps especially among those researchers who focus on environment and place.

Notably, as I write this chapter for the first American Educational Research Association (AERA) publication to focus on environmental education, an issue of *Educational Researcher* features an expanded version of William F. Tate's AERA 2008 Presidential Address, titled "'Geography of Opportunity': Poverty, Place, and Educational Outcomes" (Tate, 2008). Tate's geographical research features sophisticated maps and charts that show where businesses invest resources in communities and how the political and economic geography of urban/neighborhood planning deeply impacts the educational and life experience of adults and children. Tate's critical geography of education suggests several implications significant to the field of environmental education and the development of place-conscious educational thinking. First, the idea of place as a cultural construct has traction in a general educational research community with widespread interest in issues related to social justice and equity. Second, as in Tate's use of the term "place" in relation to urban poverty, geographically oriented environmental inquiry must include critical cultural perspectives on human inhabitation in the places and neighborhoods that are contexts for everyday living and learning. Such inquiry must also probe the larger political and economic contexts that function to construct places of privilege, oppression, and resistance. This is not necessarily new terrain for environmental education research, but the neglect of critical cultural perspectives remains an enduring problem in how the field is constituted. Put another way, Tate's presidential address on the geography of poverty should be acknowledged, at least by environmental education researchers, as an address *on* environmental education themes.

Although Tate explicitly framed his address with the language of geography and place, paradoxically, he gives no explicit attention to the link between his topic and environmental thought.[4] His discussion of huge disparities in economic investment between St. Louis neighborhoods, or of the high concentration of liquor stores and taverns around a Dallas middle school, are perfect examples of "environmental racism" or "environmental injustice," but Tate does not make explicit this connection to "environmental" themes. Further, as President of AERA, Tate promotes embracing the geographical "social, cultural, and economic" contexts of communities as a "civic responsibility" for educational researchers, but he fails to include the ecological, biophysical, or "more than human" (Abram, 1996) environment as part of how one should think about geography, cultural context, or civic responsibility. The point of this observation is not to discredit Tate or to condemn the socially constructed lack of ecological awareness of most educational researchers (including most who work from critical perspectives). The point, rather, is to suggest that all researchers who are interested in understanding lived experience in cultural context are working the terrain of place-conscious environmental education research. Such researchers, even when they employ geographical thinking, may not explicitly connect their work to the diverse fields that make up environmental studies. But it would be a missed opportunity if environmental education researchers themselves did not stretch to make this conceptual connection.[5] The issue here at a time of deepening environmental crises is one of strategy for building a broader and more inclusive theory of environmental education: in what ways will a larger movement for ecological conscience develop? A critical pedagogy of place is one construct (among others) that can help to bridge the negligent and unproductive divide between environment and culture in educational thought and practice.

Decolonization and Reinhabitation as Twin Educational Aims

Pulling from cultural and ecological traditions, a critical pedagogy of place theorizes two broad and related goals for place-conscious education: decolonization and reinhabitation. As theoretical constructs, it should be pointed out that these goals parallel other aims of environmental education research and practice; naming them is meant to be read as building inclusive rather than exclusive theory. Meanings around decolonization, for example, are in various contexts similar to meanings surrounding deconstruction, deinstitutionalization, deschooling, and other terms that signal a strong critique of cultural practices and their underlying ontological and epistemological assumptions. However, as a theoretical category, decolonization is an important term because its usage specifically problematizes colonization both as historical practice and as the ideological and political progenitor of today's socially and ecologically catastrophic globalization and development trends. Moreover, a critical pedagogy of place posits that mere critique and "the undoing of things" is insufficient theory for environmental education research, and thus the pairing of decolonization with the equally important challenge of reinhabitation. It is this pairing of decolonization—a term that suggests a strong critique of cultural practices—with reinhabitation—a term that suggests the need to reimagine and recover an ecologically conscious relationship between people and place—that gives a critical pedagogy of place its conceptual power. Though for the sake of theory-building the two terms are called out as distinct, they are most appropriately conceived as two dimensions of the same task.

Decolonization roughly corresponds to the aims of a wide range of critical educators who are primarily interested in culture, and in transforming or resisting oppressive relationships that limit people's ability to control their own life circumstances. Cultural decolonization involves learning to recognize disruption and injury in person-place relationships, and learning to address their causes. Because colonization refers also to the colonization of the mind achieved through cultural learning, decolonization refers also to the educational process of identifying and unlearning patterned and familiar ways of experiencing and knowing to make room for practices that are unfamiliar (e.g., deconstruction/reconstruction).[6] Reinhabitation roughly corresponds to the aims of a wide range of ecological educators who seek to maintain, reclaim, and create ways of living that are more in tune with the ecological limits of a place, and less dependent on a globalized consumer culture that values profits and conveniences more than people and places. Reinhabitation involves learning to live well socially and ecologically in a place, and learning to live in a way that does not harm other people and places.[7] From ontological and epistemological perspectives, reinhabitation also implies the possibility of taking a new stance toward one's own being and knowing—the possibility for reinhabiting and redirecting one's own experience of learning and becoming. Articulating decolonization and reinhabitation as twin goals for (environmental) education makes it clear that place-conscious education is not just a methodology to make learning more relevant and meaningful; it is a philosophy for personal, cultural, and ecological consciousness, renewal, and creativity.

As theoretical aims for education, the constructs of decolonization and reinhabitation take for granted or assume a global macro-context of colonization with local impacts everywhere. By colonization I refer to: (a) the historical practice, stretching from the colonial era through the present, of political entities dominating other people's territory and other people's bodies and minds for the production of privilege maintained by military, political, and economic power, and; (b) other dominant and assimilative cultural practices (e.g., schooling, consumerism, and other fundamentalisms) that tend to over-determine or restrict possibilities for people and places. As an educational theory, a critical pedagogy of place provides a frame of reference from which one can identify, and potentially resist or transform, the colonizing practices of culture and its political economy, including the ways in which cultural assumptions are transmitted through language and education (Bowers, 2001).

As broad educational aims, decolonization and reinhabitation can be described as context-dependent goals for cultural and ecological learning and renewal. They are not meant to be read as prescriptions, but generative constructs for reflection and action, constructs that are resonant with the goals of a wide range of other educators interested in culture and/or environment. Several interrelated dimensions of decolonization/reinhabitation suggest the expansive range of place-conscious educational theory: political, ecological, and epistemological dimensions. In general terms, political decolonization/reinhabitation involves the process of resisting or transforming relationships of domination and control that limit people's possibilities to direct their own life circumstances. Political decolonization/reinhabitation also implies the space needed to maintain, renew, and create ways of being and knowing that serve people and the places in which they live. Examples include the diverse traditions of critical pedagogy, critical geography, and the "decolonizing methodologies" of Indigenous people (e.g., Esteva & Prakash, 1998; Grande, 2004; L. T. Smith, 1999). Ecological decolonization/reinhabitaiton refers to the process of restoring balance between ecological and social systems. Educationally, the emphasis is on enabling a socioecological ethic to emerge from a learning relationship between cultural and natural landscapes in the context of a vision for the long-term environmental health of human and other-than-human communities. Examples include the socioecological politics of Wendell Berry and related traditions such as bioregionalism.

Connected to both cultural and ecological perspectives, epistemological and ontological reinhabitation/decolonization signifies the possibility of (a) maintaining, recovering, or creating ways of knowing and living in relation to place that are threatened or have been lost or silenced, and/or (b) unlearning patterns of thought and action that limit potential for experience and learning in relationship to places. The beliefs, values, and theories in action that make up one's way of knowing and living in large part determine one's political and ethical stance toward environment. One's way of knowing and living, however, is not culturally fixed, nor is it static in the development of one's own personal life story. Epistemology both shapes and is shaped by one's relation to place, and both can change through place-conscious learning. The subject of decolonization/reinhabitation, in other words, is both the places in which people live, and the people who inhabit and shape those places.

As Marcia McKenzie has pointed out, reinhabitation works in tandem with decolonization: it is not as if one aspect of the pairing can be worked out in isolation from the other (Greenwood & McKenzie, 2009). Many kinds of learning experiences could nurture epistemological shifts that could change one's relation to the cultural and ecological dimensions of place (McKenzie, 2008). These could include, for example, experiences with places, literature, autobiography, psychotherapy, other cultures, other human and nonhuman relationships, or any experience that focuses one's attention on one's embodied and emplaced, culturally and ecologically inscribed reality. Awareness of self and others in relation to place can either be nurtured or blocked through education, the process of schooling, or other public pedagogies such as media-sponsored consumer culture. Educational theory and practice that is negligent of the person-place relationship is, therefore, miseducative, and not without consequences for people, culture, and environment. The goals of decolonization/reinhabitation serve as guideposts to ground multiple approaches to learning in the places we inhabit, the places we impact, and the places we leave behind for others and for future generations.

Critical Questions for Place-Conscious Learning

Although decolonization/reinhabitation is not a prescriptive framework, the constructs do suggest several specific questions for place-conscious inquiry, dialog, and action anywhere. A first set of broadly applicable curricular questions focuses on an identifiable place and queries the historical, socioecological, and ethical dimensions of place-relations:

a. What happened here? (historical)
b. What is happening here now and in what direction is this place headed? (socioecological)
c. What should happen here? (ethical)

Of course, exploring such questions always involves a politics of place, as places are a meeting ground of contested place stories. What follows is an abbreviated sketch of key themes that these questions imply for place-conscious learning.

What Happened Here? Knowing something of the history of a place is fundamental knowledge lost in the processes of colonization. Who were its original inhabitants, both human and other-than-human? What are the indigenous people's stories of the place? What were—and what are—indigenous place-relationships, and how did these relationships, and the place itself, change or persist over time? Attention to indigenous inhabitation and the impact of settler society is fundamental to knowing a place itself, as well as seeing the ways in which place-relations depend on epistemology. In an article that examined educational controversies surrounding the 1999 Makah whale hunt around Neah Bay at the northwest tip of the Olympic Peninsula (Washington State, US), Michael Marker (2006) commented on the dissonant nature of epistemic encounters between settler societies and the Indigenous other:

> There is a deep insecurity within the consciousness and conscience of settler societies that, when confronted by the indigenous Other, is awakened to challenges about authenticity in relation to land and identity. There is embedded in this encounter with indigenous knowledge a challenge about both epistemic and moral authority with regard to indigenous relationships to land and the spirit of the land. Whereas other minoritized groups demand revisionist histories and increased access to power within educational institutions, indigenous people present a more direct challenge to the core assumptions about life's goals and purposes. (pp. 485–486)

To articulate the centrality of Indigenous inhabitation to place-conscious learning is neither quixotic romanticization nor pernicious appropriation, although both of these critiques offer valuable cautions (see Nespor, 2008).

Critiquing efforts to invite a more central role for indigenous experience in education as mere romanticizing is historically and politically problematic when it becomes another excuse for forgetting the past and misunderstanding present and future geopolitical relationships. Contrary to the critique that to embrace indigenous cultural and educational models is to romanticize, any honest look at indigenous experience, or at the majority of the world's population, is to reveal the extreme romanticization and denial undergirding "settler" notions of progress, which are fundamental to the project of schooling, and synonymous with the ongoing project of colonization under global capitalism. A place-conscious historical inquiry calls progress and development into question, and invites the perspective of first inhabitants. It also calls into question life's goals and purposes, and all of the assumptions of settler society on which contemporary ideas about education and place are based.[8]

Knowing something of the history of a place obviously involves more than learning about, and from, its first peoples. Because a critical pedagogy of place emphasizes human relationship to the land, however, indigenous perspectives on inhabitation and decolonization are vital. Interest in the story of indigenous inhabitation and survivance[9] over time suggests not only learning from Indigenous people, but learning through their experience more about the ongoing process of colonization and the possibilities for creative resistance. In her book, *Decolonizing Methodologies: Research and Indigenous People*, Linda Tuhiwai Smith (1999) observed that:

> Coming to know the past has been part of the critical pedagogy of decolonization. To hold alternative histories is to hold alternative knowledges. The pedagogical implication of the access to alternative knowledges is that they can form the basis for new ways of doing things. (p. 34)

Decolonization, in other words, involves access to other ways of being and knowing, which can form the basis for reinhabitation. Indigenous cultures, of course, are not the only cultures that have alternative histories worth remembering and recovering. American cultural, social, environmental, and even educational histories, for example, include many voices and movements for resistance and change. Whether it is called decolonization, remembering the past, or accessing other ways of being and knowing, place-conscious learning involves reflecting on the multicultural traditions that shaped places, and that can inform their reinhabitation now and in the long run.

What is Happening Here? Place-conscious learning depends on direct, phenomenological experience with places, the idea being that we don't just learn about places, but that we learn from them first hand. A theory of place that is concerned with the quality of human-world relationships must first acknowledge that places themselves have something to say. Human beings, in other words, must learn to listen (and otherwise perceive). Ecological theologian Thomas Berry (1988) observes that, as a species, we have gradually become "autistic" and have forgotten how to hear, communicate, and participate in meaning making with our places on the living earth. That places are alive may seem obvious to ecological thinkers and others disposed to perceive and appreciate the lives that places hold. But modern institutions, such as schools, governments, and corporations, have not demonstrated an orientation of consciousness and care toward the places that they manipulate, neglect, and destroy.[10] As in David Abram's (1996) phenomenology of perception in *The Spell of the Sensuous*, reawakening a sense of embodied and emplaced connection to the world is essential to the aim of knowing a place firsthand. Further, Abram argued, an ecological ethic might emerge,

> Not primarily through the logical elucidation of new philosophic principles and legislative strictures, but through a renewed attentiveness to this perceptual dimension that underlies all our logics, through a rejuvenation of our carnal, sensorial empathy with the living land that sustains us. (p. 69)

Beyond reawakening our senses through direct experience with place, contemporary place-knowing involves myriad culturally and ecologically specific learnings that range from the simple to the complex,[11] and can be pursued through the traditions of natural history, cultural history, and action research (Gruenewald, 2003b). As Robert Michael Pyle (2008) observed, "place-based education, no matter how topographically or culturally informed, cannot fully or even substantially succeed without reinstating the pursuit of natural history as an everyday act" (p. 156). Pyle suggests that it is only good neighborliness to demonstrate interest in and care for our nonhuman neighbors, and that we will never be able to decolonize/reinhabit our imperiled ecosystems without a functional knowledge of the system's working parts. At minimum, this means that understanding a place means knowing something about its biophysical systems, and how these systems have become politicized through cultural processes. Because the ecological contexts of places are not precultural, place-consciousness depends equally on knowledge of, and experience with, ecological and cultural systems, as well as the interactions between them. From a cultural perspective, place-consciousness means understanding that places themselves are not inevitable or predetermined, but that they are cultural products, the product of intended and unintended consequences. Further, as cultural products, places are perceived very differently by different cultural groups who hold different ways of being and knowing. Coming to know a place, therefore, means learning the diverse and competing stories told about it.[12] A critical pedagogy of place is concerned not just with the dominant story, but with all of the stories at risk of being silenced or erased, including the voice of the land itself.

What Should Happen Here? As politicized educational goals, decolonization and reinhabitation signify the need for the inhabitants of a place to become more aware of the process of place-making and the possibilities for participating in it. The ethics implied in the question—What should happen here?—will obviously be shaped by one's experience with a place, by one's assumptions about its value, and one's understanding of the competing claims to its significance and possibilities. As a starting point for critical praxis, several additional questions help to focus place-conscious inquiry and action:

a. What needs to be remembered?
b. What needs to be recovered or restored?
c. What needs to be conserved or maintained?
d. What needs to be changed or transformed?
e. What needs to be created?

These questions mirror the broad methodological concerns of a variety of critical traditions, including critical pedagogy/geography, bioregionalism, and the traditions of Indigenous peoples and of other minority or dissident cultural groups. The questions apply both to "cultural" work such as a maintaining civil rights, and to "ecological" work such as habitat restoration. Taken as a whole, this line of inquiry aims to create opportunities for responding to the ethical question—What should happen here?—as part of a participatory process of decolonization/reinhabitation.

Each of these questions prompts a different kind of cultural and ecological critique and vision, and each suggests the need to listen to multiple voices, including the voice of the land. Learning to pay attention to the land and its inhabitants, and to interpret the multiple meanings of a place, is no mere intellectual exercise, but depends on a way of being that itself needs to be created, recovered, and continually renewed. Place-conscious education aims to activate and integrate social and ecological awareness so that learning, ethics, and politics are all grounded in the enfleshed world of social and ecological experience.

Conclusion

Bridging cultural and environmental perspectives, a critical pedagogy of place proposes the twin aims of decolonization and reinhabitation as inquiry-oriented entry points into identifying and shaping place relations. In order to explore the dynamic connections between place, geography, culture, and education, and in order to envision a livable future that is authentically integrated with its past, it is necessary to look back across time, and to look beyond schooling, for the places and selves we may not yet know. Decolonization and reinhabitation are more than just political goals—they are simultaneously educational goals, in the sense of education as living and learning. It is not just places that can be decolonized and renewed. It is also people—our minds and bodies can be decolonized and reinhabited, even in this moment. What needs to be restored, maintained, transformed, or remembered is, therefore, as much a project of self-discovery as it is one of discovery of place. The point of a critical, place-conscious education is to discover/recover/reconstruct self in relation to place. Learning to listen to this complex relationship of self and other, human and nonhuman, is the ultimate educational challenge.

Notes

1. See Bruce Janz' website, *Research on Place and Space*, for comprehensive bibliographies of place studies in over a dozen academic fields: http://pegasus.cc.ucf.edu/~janzb/place/
2. Thus, instead of thinking of "place" in the singular, it is perhaps best thought of in the plural—"places."
3. In 2008, *Environmental Education Research* published a series of articles on "a critical pedagogy place" led off by Chet Bowers' critique of the term. See Bowers, 2008; Greenwood, 2008; McKenzie, 2008; G. Smith, 2008; Stevenson, 2008.
4. The same lack of obvious connection occurs whenever educational researchers use the language of "ecology" to describe their work without attending at all to the ecological contexts that make their anthropocentric work possible.
5. Further, it seems essential to the work of deepening ecological conscience that environmental education researchers collaborate with others and seek audiences outside of their own field. In terms of the literature, this may mean that environmental education researchers should look to publish their work in general education books and journals, as well as in their own specialized venues.
6. Within the dominant culture, such unlearning is frequently met with resistance, as the new learning requires time for new experiences, relationships, and concepts to develop.
7. Participation in the money economy makes it difficult to know how one's consumption is impacting other people and places. The point is that to fully reinhabit place, one must become more aware of how one's actions in one place are connected to other people and places, near and far, now and in the future.
8. For some, this emphasis on indigenous place will still seem overdone. However, from a historical perspective, settler society in the place where I live is only 150 years old, while indigenous presence goes back thousands of years. The point of remembering indigenous history and presence is not to idealize culture, but to look honestly at settler society's impact on people and places.
9. The term survivance is used in Native American Studies to describe the self-representation of indigenous people against the subjugations, distortions, and erasures of white colonization and hegemony (see, Villegas, Neugebauer, & Venegas, 2008). Survivance emphasizes continued presence over absence. For some indigenous people, therefore, it may be more appropriate to speak of inhabitation rather than reinhabitation. Though given effects of colonization on Indigenous people, the term reinhabitation may still be appropriate—at least to some extent.
10. Premodern people, of course, also manipulated and destroyed places. The scale of today's wreckage, however, along with knowledge of the consequences, legitimizes a strong critique of modern institutions.
11. This complexity of knowing around place raises the question of what kinds of knowledge ought to be made more accessible to everyone and what kinds of knowledge are so specialized that perhaps not everyone needs to know.
12. Mapping and oral history are two pedagogies of place that can begin to document the contested place stories that people and places hold.

References

Abram, D. (1996). *The spell of the sensuous*. New York, NY: Vintage.
Basso, K. (1996). *Wisdom sits in places*. Albuquerque, NM: University of New Mexico Press.
Berry, W. (1987). *Home economics*. New York, NY: North Point Press.
Berry, T. (1988). *The dream of the earth*. San Francisco, CA: Sierra Club Books.
Berry, W. (1992). *Sex, economy, freedom and community*. New York, NY: Pantheon Books.
Bowers, C. A. (2001). How language limits our understanding of environmental education. *Environmental Education Research*, *7*, 141–151.
Bowers, C. A. (2008). Why a critical pedagogy of place is an oxymoron. *Environmental Education Research*, *14*, 325–335.
Casey, E. (1997). *The fate of place: A philosophical history*. Berkeley, CA: University of California Press.
Esteva, G., & Prakash, M. (1998). *Grassroots postmodernism: Remaking the soil of cultures*. London, UK: Zed Books.
Foucault, M. (1980). *Power/knowledge: Selected interviews and other writings* (C. Gordon, Ed.). New York, NY: Pantheon Books.
Geertz, C. (1996). Afterward. In S. Feld & K. Basso (Eds.), *Senses of place* (pp. 259–262). Santa Fe, NM: School of American Research Press.
Grande, S. (2004). *Red pedagogy: Native American political thought*. Lanham, MD: Rowman & Littlefield.
Greenwood, D. (2008). A critical pedagogy of place: From gridlock to parallax. *Environmental Education Research*, *14*, 336–348.
Greenwood, D., & McKenzie, M. (2009). Context, experience, and the socioecological: Inquiries into practice. *Canadian Journal of Environmental Education*, *14*, 5–14.
Gruenewald, D. (2003a). The best of both worlds: A critical pedagogy of place. *Educational Researcher*, *32*(4), 3–12.
Gruenewald, D. (2003b). Foundations of place: A multidisciplinary framework for place-conscious education. *American Educational Research Journal, 40*, 619–654.
Gruenewald, D. (2004). A Foucauldian analysis of environmental education: Toward the socio-ecological challenge of the Earth Charter. *Curriculum Inquiry, 34*, 63–99.
Gruenewald, D. (2005). Accountability and collaboration: Institutional barriers and strategic pathways for place-based education. *Ethics, Place and Environment: A Journal of Philosophy and Geography*, *8*, 261–283.
Gruenewald, D., & Smith, G. (2008). *Place-based education in the global age: Local diversity.* Mahwah, NJ: Lawrence Erlbaum.
Hart, R. (1997). *Children's participation: The theory and practice of involving young citizens in community development and environmental care*. London, UK: Earthscan, Unicef.
Helfenbein, R., & Taylor, (Eds.). (in press). *Educational Studies* [theme issue: Critical Geography and Education]
Kunstler, J. (1993). *The geography of nowhere: The rise and decline of America's man-made landscape*. New York, NY: Simon & Schuster.
Leopold, A. (1949, 1968). *A Sand County almanac*. Oxford, UK: Oxford University Press.
Marker, M. (2006). After the Makah whale hunt: Indigenous knowledge and limits to multicultural discourse. *Urban Education*, *41*, 482–505.
McKenzie, M. (2008). The places of pedagogy: Or, what we can do with culture through intersubjective experiences. *Environmental Education Research*, *14*, 361–373.
Melaville, A., Berg, A., & Blank, M. (2006). *Community-based learning: Engaging students for success and citizenship*. Washington, DC: Coalition for Community Schools.
Nespor, J. (2008). Education and place: A review essay. *Educational Theory*, *58*, 475–489.
Pyle, R. M. (2008). No child left inside: Nature study as a radical act. In D. Gruenewald & G. Smith (Eds.), *Place-based education in the global age: Local diversity* (pp. 155–172). Mahwah, NJ: Lawrence Erlbaum.
Relph, E. (1976). *Place and placelessness*. London, UK: Pion.
Smith, G. (2002). Place-based education: Learning to be where we are. *Phi Delta Kappan*, *83*, 584–594.
Smith, G. (2007). Place-based education: Breaking through the constraining regularities of public school. *Environmental Education Research*, *13*, 189–207.
Smith, G. (2008). Oxymoron or misplaced rectification. *Environmental Education Research*, *14*, 349–352.
Smith, L. T. (1999). *Decolonizing methodologies: Research and indigenous peoples*. New York, NY: Zed Books.
Sobel, D. (1996). *Beyond ecophobia: Reclaiming the heart in nature education*. Great Barrington, MA: The Orion Society and The Myrin Institute.
Sobel, D. (2005). *Place-based education: Connecting classrooms and communities*. Great Barrington, MA: The Orion Society and The Myrin Institute.
Soja, E. (1989). *Postmodern geographies: The reassertion of space in critical social theory*. London, UK: Verso.
Somerville, M. (2007). Place literacies. *Australian Journal of Language and Literacy*, *30*, 149–164.
Stevenson, R. B. (1987, 2007). Schooling and environmental education: contradictions in purpose and practice. *Environmental Education Research*, *13*, 139–153.
Stevenson, R. B. (2008). A critical pedagogy of place and the critical place(s) of pedagogy. *Environmental Education Research*, *14*, 353–360.
Tate, W. F. (2008). "Geography of Opportunity": Poverty, place, and educational outcomes. *Educational Researcher*, *37*, 397–411.
Villegas, M., Neugebauer, S. R., & Venegas, K. R. (2008). *Indigenous knowledge and education: Sites of struggle, strength, and survivance*. Cambridge, UK: Harvard Education Review.

10

Learning From Hermit Crabs, Mycelia, and Banyan

Schools as Centers of Critical Inquiry and Renormatization

Heesoon Bai and Serenna Romanycia
Simon Fraser University, Canada

Looking for a New Home

> Most species of hermit crabs have long soft abdomens [that] are protected from predators by the adaptation of carrying around a salvaged empty seashell into which the whole crab's body can retract . . . . As the hermit crab grows in size, it has to find a larger shell and abandon the previous one. (Wikipedia, 2009b, para. 1–2)

Humans instruct, coach, model, and guide each other, especially the young, to transmit norms of a given cultural group, be it a clan, tribe, kingdom, or nation. Through this intergenerational transmission of cultural content, individuals acquire particular beliefs, values, tastes, desires, habits, behaviors, skills, and social and bodily practices that belong to a cultural group.

Schools as modern institutions of intergenerational cultural transmission have been specifically mandated to perform the formal task of shaping citizenry to fit into the modernist industrial consumer society. Through mass schooling, individuals become *normatized* and *normalized.* To normatize means to shape individuals' beliefs, values, taste, and conduct to the norms of society. Through this process, individuals come to feel "normal" about their worldview and their ways of being and doing in society.

This mass schooling process of normatization and normalization would not be an egregious problem for the present world, *if* these instrumentalist and consumerist worldviews and values were viable to the survival of our species and the well-being of the planet. But, decisively, they are not. Current literature sources are vast and well established on the indictment against our dominant global culture and civilization for their utterly damaging contribution to the biosphere and human community (e.g., Korten, 2001; McMurtry, 1998; Orr, 1994; Schor, 2004).

What are the characteristic worldviews and values of modernity? A fairly well-defined picture emerges from the works of many contemporary thinkers most notably, instrumentalism, industrialism, individualism, consumerism, and economism, all of which are backed up by militarism (e.g., Borgmann, 1992; Fromm, 1976; Taylor, 1991). There is every indication that many of our cultural habits of mind, heart, and body, constituting the modern-western and global-industrial-military-consumeristic way of life, are rapidly eroding the very carrying capacity of the planet (Harrison, 1996). Simply put, this civilization seems to be on course for extinction. Many are ignorant or in denial about the precarious planetary situation that we have put ourselves and all other beings into. Others writhe in fear about the bleak future, but their "solution" is doing more of the things that accelerate destruction. Some explode in anger and blame, and many more are simply feeling lost, hopeless, helpless, and numb (Macy & Young, 1998). The present global culture of consumeristic capitalism that has a stranglehold on almost all human societies on this earth is like a hermit crab's old shell and has become a threat to the inhabitant's viability and eventual survival.

As we indicated earlier, schooling is modern society's means of shaping its citizenry through instilling normative beliefs and values of what it is to be a functioning and successful human being within industrial-consumer culture. Should our schools continue to instill in students existing societal beliefs, values, and practices that threaten the very core of life on earth (Orr, 1994), let alone the survival of humanity, just because these are the societal norms? Or should schools instead strive to become *awakened* institutions that systematically examine how the existing norms are damaging, and deliberate what we need to do to change them? Can public and private schools become leaders in a social movement toward our current humanity practicing a more ecological life on earth?

We propose that both the content and manner of schooling—the particular norms of beliefs and values—are in need of radical revision. We believe that schools *should* indeed be committed to changing the norms of society and civilization. Hence, this chapter will focus on what we the authors have come to call the "hermit crab project"—looking for new norms of schooling. Coincidentally, Evernden (1985/1993) has also used the illustrative example of a hermit crab:

> Like hermit crabs, who protect their delicate abdomens by confiscating the shells of snails, we [the natural alien] crawl into a structure of belief and refuse to expose ourselves, even to move to a more adequate abode. This is a dangerous policy. (p. 143)

At the moment, we the authors see no other institution that has the potential to implement this large-scale project of renormatizing the culture so as to avert the modern civilization's socioecological collision course, schools. Hence, we would like to propose rededicating schools to teaching and learning ecological ways of human life. What we are suggesting for schools here is far bigger and more radical than the "Save the Earth" environmental awareness campaigns that promote recycling, tree-planting, and energy saving solutions. While these campaigns are worthy and indispensable, they fall short of addressing the real heart of the issue: we need to dismantle the exploitative cultural and economic systems that are continuously being enacted through normative institutions, especially schools (Apple, 2009). This systemic dismantling is what we suggest that schools of the future should, if not must, do. This requires that K-12 schools and postsecondary institutions become leaders in the *renormatization* of humanity.

Understanding the process of normatization is crucial here, as the proposal for renormatization requires reversing the process. The force of learning is revealed by its power to render us *unconscious* in the process of normatization. That is, the knowledge and skills transmitted by the culture and acquired by the individual go deep into the individual's psyche and body, and become "second-nature." Through normatization, we internalize various assumptions and presuppositions that have been normalized in the culture. Hence, the degree of success in schooling shows up in its citizens' tendency to unquestioningly accept and enact norms. Norms become reality for the successful learners, and this phenomenon demonstrates the ubiquitous power of mass schooling. However, today's successful schooling may be tomorrow's miseducation, leading to demise (Orr, 1994).

For our society to survive through the challenge of changes and uncertainties there has to be a critical balance between unconscious normatization and conscious undoing. Can schools play this critical role of managing the balance? Not if we continue to look to schooling to mainly carry out society's mandate of normatization. Instead, schools must be specifically dedicated to be institutions of enlightenment that serve to wake students from the unconscious force of normatization and send them to work, individually and collectively, on reconfiguring norms. Schools can then become sites of critical examination of prevailing norms and creative experiments with alternative norms. But, of course, this is easier said than done. Where can we get the help and support we need to convert our existing educational institutions into facilitators of wide-awakeness and change?

We believe that the best possibility lies with universities and their faculties or departments of education that educate—not just instruct and transmit knowledge and skills—the future generations of teachers and educational leaders. We understand that "education" is an evocative word that is used in different ways in different contexts and in different cultures and linguistic groups. As space does not allow us to unpack these meanings, we will focus primarily on schools and their latent potential at the crossroads of normatization and ecological imperative.

Higher education has the best chance of fulfilling the function of renormatization of beliefs and values through critical analysis and discourse (Rorty, 1996). Specifically, faculties of education in universities tend to infuse general culture with unconventional and innovative visions and practices of schooling through preservice and in-service teacher education and graduate education. This is particularly the case for faculties of education in universities that offer courses that are philosophical in nature. These philosophy courses in education can, and often do, play a critical and crucial role in renormatizing the culture of teacher education.

An important function of philosophy is its radical critique of "reality," showing how so-called reality known to socialized human beings is a construction of the human mind and culture, and that, for this reason, human reality can be deconstructed and reconstructed again and again (Bai, 2006). To this end, philosophical approaches in education examine the cultural assumptions and presuppositions about what educated human beings are like, what we see as the ideal of human life, and what we value and why (Martin, 1994). The details of such examination and deconstruction make the study a rich and challenging experience, as there is no aspect of human personal and collective identity and conduct that cannot be brought to the critical lens of inquiry. We therefore urge faculties of education in universities to use their academic leadership to help and guide schools to become centers of critical awareness and renormatization. Hence we also urge them to centralize philosophical studies and practices in their teaching and research. As well, there is no theoretical or practical reason whatsoever why we cannot include philosophy in K-12 curriculum and pedagogy (Moran, 2004). It is good to start philosophizing sooner than in higher education. Sow the seed sooner rather than later.

Philosophy, however, is multifaceted. We need to clarify the kind of philosophy we need for our project. Some assume that philosophy has no end other than purely theoretical or speculative inquiry. While this might be true for some domains of philosophical activity, this is not the case for our suggested philosophical examination of social norms and their internalization by students in schools. The philosophical examination of schooling (and, indeed, education) should not be neutral. Instead, it should show both intention and commitment, for the same reason that the medical examination of a patient is not neutral but is purposefully guided by the committed interest in the well-being of the patient.

But what will be the guiding hand for the philosophical examination of schools? Will it be the same as the doctor's hand? We reply, yes. Both education and medicine are formally concerned with human survival and well-being (Bai, 2004). This does not mean that all educational or medical practices result in promoting survival and health. There can be miseducation or medical malpractice, which necessitates a self-critical guidance system to reflect on and attempt to correct, or if problems are systemic, renormatize, the practice of education or medicine. The self-critical guidance system for human conduct is known as *ethics,* which is part of the study in philosophy. Ethics asks this fundamental question: Are we harming the world and ourselves through living out the existing norms? Such a question, as Weston (1992) reminds us, can only be asked and responded to in a concrete sociopolitical, economic, and ideological context of the culture that is critically examining its own values. If we, the citizens who are living out the culture's worldviews and normative values, can clearly see that the latter is not conducive to ecologically and socially sustainable life on this planet, then we would need to ask the next question: How can we change the habits of mind, heart, and body so as to reduce harm and increase well-being?

In the next section, we further elaborate the centrality of ethics as the guiding and healing hand for future schools committed to re-examination and renormatization of humanity. Most importantly, ethics is not just theoretical, but is a practice.

New Homemaking

> Almost since life began on earth, mycelia have performed important ecological roles: nourishing ecosystems, repairing them, and sometimes even helping create them. The fungi's exquisitely fine filaments absorb nutrients from the soil and then trade them with the roots of plants for some of the energy that the plants produce through photosynthesis. No plant community could exist without mycelia. (Jensen, 2008, para. 5)

We stated in the last section that ethics should guide schools (their aims, goals, objectives, and curricular and pedagogical choices). At the core of ethics is the inquiry into the "good life" or human flourishing, that is, how *well* we are living our lives, with and for each other. In our understanding of ethics, the question and the quest for the good life precedes the question of moral right or wrong because the latter requires a conception of the good life that will act as a criterion or a rubric against which particular practices are evaluated and judged to be "right" or "wrong." And the inquiry into the good life is a normative activity since it challenges us to question what constitutes as a good life, and to consider culturally and historically contingent conceptions of a good life. In much of western philosophy, the good life discourse has been limited to human welfare. Ethics has been by and large about human well-being. However, this limitation has become ecologically deadly, especially in modern times when the good life and well-being discourse is increasingly constricted to the instrumentalist and consumerist visions (Bai, 2009). Human well-being cannot be separated from planetary well-being, for humans are an integral part of the biosphere. The modern ideal of a "good life" that we have been protecting and enacting—namely, capitalistic consumerism backed up by militarism and spiced up by the entertainment industry—is decisively damaging and compromising the health of the planet. Therefore, we need a new ethic that is eco-centric, or ecological ethics (Kohák, 2000). What characterizes ecological ethics from ordinary—human-centered—ethics?

A cardinal understanding in ecology is the complex interconnectedness of all life phenomena, which is captured in the etymology of the very word "ecology" (eco+logy = study of household). In the operations of a household, in which everybody is interconnected and every element interpenetrates, a dedicated and resourceful homemaker knows, intimately and intuitively, how everything fits and flows together, and he or she can care for the household in such an intelligent, skillful way that everybody in the household flourishes. The key idea here that we need to pay attention to is how a good homemaker comes to know his or her household intimately and how to help it to flourish. The homemaker knows how every element in the household connects and fits together with everything else. This knowing is born out of the intimacy of participation (Skolimowski, 1994b), as the homemaker knows the household so intimately through being *one-bodied* with every element, participant, aspect, and detail of it. Such intimate knowing is an integral function of love and care. A devoted homemaker or householder loves his or her family and household, and embraces all aspects of it—including pains, sorrows, and toils. They care for it with the knowledge and skills that can only come from knowing something proprioceptively, like one's own body. Students and practitioners of ecological ethics are people who are in love with the earth community and its inhabitants, including the human community, which disposes them to deeply *care about* the earth's well-being and to naturally *care for* life on earth. Hence the foundation of ecological ethics is love and care *of* the earth, and love and care *for* earth's beings. Ecology as ethics is love's knowledge. Upon this foundation of ecological ethics, we can build a new normative dimension within schooling. Moving forward with this understanding, let us look at the actual practice of love and care. What is it to practice love and care? How do we learn to love and care?

Ecology as an ethic and moral education has much in common with the ethics of care that Noddings (2003), Held (2006), and Slote (2007) explored and developed. Ethics of care is also known as a branch of relational ethics because its key principle is relationship. Basically, in this ethical framework, an action is good and right when it is in the service of nourishing, fostering, repairing, and preserving relationship of intrinsic worth and regard for the other. Based on this, we can construe ecological ethics as an extended form of ethics of care in that the complex and irreducible relationships supported by intrinsic valuing include relationships with the earth community, its inhabitants, and constituent members. Earth is one large household, and ecology is an ethic of care that supports the leadership and partnership pertaining to the earth-household. Let the mycelia be our inspiration! For so long the human presence on this planet has been destructive to the ecosystems. Let us learn the ways of mycelia and join their bioecological functions of nourishing, repairing, and creating ecosystems. Our practice of ethical living is analogous to the mycelium's bioecological functions, illustrating yet again that humans and other earthly beings are consanguineous.

Love, care, knowledge, vigilance, and skillful actions are required to nurture, repair, and recreate the earth household that integrates and harmonizes everything into mutually affording and supporting relationships. Do our schools centralize such learning? Does our schooling encompass learning all about creating, nurturing, repairing the complex web of relationships that we call the earth community? We believe this is not the case as of yet. There is much we can learn and borrow from ethics of care for our task of conceptualizing ecological ethics for schools. What is particularly attractive about ethics of care for our purpose is its dimension of practice. We can tell the world all we want, that it *should be* this way and it *should act* that way, but if we cannot model and teach how this and that way can be achieved and practiced every day and every moment, ethics as practice ultimately fails. Failed ethics remain as moralistic "shouldism" that ends up being oppressive, and worse, if practiced with our future generations, may destroy the very fiber of moral autonomy and agency. In ethics of care, care as moral agency is understood as natural endowment or birthright, and the task of moral education is to create optimal conditions for its realization and extension. Noddings (1992) articulates four pillars of ethics of care as moral education: dialog, modeling, practice, and confirmation. She also speaks of organizing the school curriculum in the terms of centers of care and concern such as "caring for animals, plants, and the earth" (p. 126), "caring for strangers and distant others" (p. 126), "caring for self" (p. 126), and so on.

Optimal conditions for generating the expansive care consciousness (Bai, 1999; Chinnery & Bai, 2000) are not easy to create, given the historical entrenchment of humanity in all kinds of ideological constructions that drastically limit care consciousness. Feminists, critical theorists, postcolonial scholars, and ecophilosophers all have thoroughly critiqued these ideological constructions that have wounded and damaged cultural norms. One combination that has had the strongest hold on humanity is the belief of superiority of the human (anthropocentricism), the masculine (patriarchy), the white (western colonialism), the propertied (capitalism), and the mental (rationalism) (e.g., Bordo, 1986; Martin, 1994; Merchant, 2005). According to this normative combination, a great proportion of humanity becomes "underprivileged" and "underdeveloped" (Martin, 1994). This restrictive and oppressive worldview cannot support the ecological understanding of a systemic worldview that insists on interconnectedness and interpenetration of everything in the phenomenal world (Macy, 1991). Hence, for an ecologically expansive consciousness of care to take root and flourish in our schools and culture in general, we need to work steadily to undo and remove such patriarchal, colonial, capitalistic, and rationalistic norms from our formal and informal learning goals, policies, materials, and practices. We suggest that this very undoing and removing work, which is repair or healing work, should be an essential part of school curriculum and pedagogy. By the time students come to school, even at a primary level, they will have internalized the dominant values and views that are damaging to both the earth community and their own immediate human community. Therefore, we need an ethical curriculum and pedagogy that aims at healing and nourishing both ourselves and our relationships with all earth beings. Recall that mycelia perform both nourishing and repairing functions. The two functions go together. An ethic of care plays the role of mycelia, and when applied to ecology, both repairs the damaged relationship between humans and the biosphere, and nourishes it.

In the next section we will consider some practical strategies for rooting alternative norms. Given that the task of resisting and breaking into the existing norms and inserting new norms is likely difficult and inherently challenging, we need strategies. This time, the example of banyan tree with its amazing ability to germinate anywhere and send down its root into any available place shall be our inspiring and guiding imagery and example.

Rooting Down

> A banyan is a fig that starts its life as an epiphyte when its seeds germinate in the cracks and crevices on a host tree (or on structures like buildings and bridges). . . . The seeds germinate and send down roots towards the ground, and may envelope part of the host tree or building structure with their roots, giving them the casual name of "strangler fig." (Wikipedia, 2009a, para. 1)

For a norm to be successful as a guide to ethical living, it needs the support of stories and practices that permeate a culture and work into every aspect of personhood. Stories and practices *inform*, that is, give form to, what the world is like, what people are like as creatures among other creatures, what sense and meaning we can make of human life, who we are as individuals, and who we become through collective life-making. When these stories are told and retold in a million different variations and disguises, and practices and experiments are undertaken in innumerable ways and spaces (Weston, 1992), they infiltrate our unconscious and conscious, and sooner or later, start to permeate our perceptions, emotions, and our conduct. Currently the dominant storytelling that guides billions of people on this planet is ecologically unsustainable and even deadly. That much is clear. Look around the world today: what are the dominant stories of a good life and success for humanity? Consumerism as fulfillment; social progress through the production of goods and services; institutionalization and commodification of fundamental human needs, such as turning learning into earning grades, love into sex, bliss into entertainment, and so on and so forth. At the heart of all these unecological ways of being is the reductionist and instrumentalist mindset—as opposed to ethics of care we referenced earlier—that dichotomizes the means from the ends, and privileges the latter over the former, thereby turning the world into a vast resource base that acts as the means to human consumptive ends (Bai, 2001). The task before us is then to change the habits of the human heart and mind, thus change the major story line of human survival and flourishing. And this, we argue, is what schools can and must help to do.

True to the principle of ecology that shows that nothing is wasted and all is recycled, composted, and renewed, we may find in our midst right now a wealth of stories and practices that we can take up, and, with some intelligent labor, turn into ecological ways of feeling, perceiving, thinking, sensing, and acting. Thus we are not talking about a brand new and major invention and production of stories to live by and practices to try out. We already have stories and practices that can be revitalized and remodeled to serve ecology as a life-philosophy. We will illustrate this with some examples from our local culture, the west coast of Canada, which is home to cross-cultural confluences, especially Asian cultural influences.

Yoga studios are beginning to proliferate our culture, along with Tai Chi and Qi Gong Centers. Mindfulness meditations, originally of Buddhist origin but now secularized, are being taught in public schools by teachers who find that the practice helps children cope with stress and anxiety, and that it helps teachers manage their classrooms. Practices that would have been once thought exotic and suspicious have part of the everyday routine for countless people going to work and school. These are, essentially, healthy signs. However, in many instances, what is vitally missing in the way these practices are taken up is a substantial and comprehensive philosophy, such as eco-philosophies and care ethics, to give them a larger vision of human flourishing in partnership with all earth beings and a critical understanding of human presence in the world today. When mindfulness and yoga, or any other practices that aim to nourish and balance us, are taught under the guiding eco-ethics and eco-philosophy, these practices can go far beyond making a flexible body, stress release, and better classroom management. They can heal us from the largely objectified, alienated, and even commodified state of our being by awakening and reconnecting us to the cosmic reality of *interbeing*, to use Hanh's wonderful neologism for ecology (1998). Unfortunately, in the way these practices are usually promoted and purveyed in public culture and schools, they are no more than techniques for releasing stresses associated with the "unecological" forms of life that are harming us all. In other words, these practices are most often just a useful adjunct for the highly commodified lifestyle, rather than a catalyst for a new way of life.

The task of turning ecological ethics into life practice is really a matter of taking up some suitable everyday practices and habits of mind-body-heart, be they gardening, cooking, cleaning, exercising, walking, writing, or any number of different things we do each day, and infuse them with a clear eco-ethic and eco-philosophy, such as intrinsic valuing based on the realization of interconnection and interpenetration (Bai, 2004). Indeed, there is neither a lack of, nor limit to what we can take up from our everyday life and work into them ecological ethics as a way of being. We can be like the banyan tree whose "seeds germinate in the cracks and crevices on a host tree (or on structures like buildings and bridges)" (Wikipedia, 2009a, para. 1). Wherever we are and go, we find "cracks and crevices" of our everyday activities into which we can drop seeds of mindfulness, yoga, integral ways, or so on, and turn these activities into enlightenment practices. Here, when we say "enlightenment," we mean nothing other than freedom from alienated consciousness and full presence in *interbeing* (Hanh, 1998). In many ways, the simplest human actions, such as the act of seeing, touching, and hearing afford us profound possibilities of realizing interconnectedness and interpenetration of life—that is, the ecology. We see these and other possibilities as central to the curriculum of schools that set to change the civilizational matrix.

Here is one particularly moving example of what we were suggesting above: Skolimowski (1994a) describes four elements (rock, earth, water, and tree) that can be a source of "re-enchantment with the world" (p. 32). Skolimowski begins by inviting "us to go for a journey—if possible into a forest, or wild place. Let your mind be free. Allow time for this journey" (p. 30). However if only a park is nearby or only "a piece of solid rock" can be brought home, that is fine too. Being "in the right frame of mind" is more important in yoga (p. 30). His instruction for identifying with the rock begins thus:

> Approach the rock and embrace it as tightly and meaningfully as you can. You must not feel embarrassed or shy.

> Your bones are made from the remnants of rocks, and everything there is was once a rock. So in embracing the rock you are embracing yourselves—in your earlier states of being. Embrace the rock as a part of yourself. You are this rock. Feel its solidity, its roughness, its texture. Feel how wonderfully enduring it is, and how it is already cracked, ready to disintegrate further to give rise to other forms of being.
>
> Spend several minutes contemplating this rock. Look at it in a new way, as if you have never seen a rock before! See the forms of life which are already there—though hidden. Contemplate its origin, and what it wants to become. Feel yourself in this rock. A rock is a frozen spark out of which the tulips will grow. Then take a deep breath and wonder for a little while longer.
>
> Now you know the meaning of empathy with the rock. (p. 33)

In a similar vein, Skolimowski (1994a) describes how to "feel the heat and life in the earth":

> The earth is the same element that was once a rock. You can experience the reality of the earth.
>
> Choose a patch of ground to sit on, perhaps a bed of grass next to some soil. Submerge your fingers in the soil, and feel it. Feel it profoundly. Feel the beat of life in this earth. The entire earth is the dust of rocks transformed into life. Listen to the earth and feel its great reverberating rhythm.
>
> Take another deep breath and wonder for a minute or two. Don't rush. Time is your friend.
>
> Now you know the meaning of empathy with the earth. (p. 34)

The instruction to "submerge our fingers in the soil" (Skolimowski, 1994a, p. 34) can also be used to renormatize schooling in the key of ecological ethics through the practice of horticulture. One well-known urban school project is the "Edible Schoolyard" at Martin Luther King Jr. Middle School in Berkeley, California (Chez Panisse Foundation, 2009a, 2009b). Cofounder and chef Waters (2009) recently chronicled her fourteen-year journey in *Edible Schoolyard: A Universal Idea.* The "Edible Schoolyard" originally began in 1995 with monthly visits by students. The project now reaches nearly 1,000 students at King Middle School (Chez Panisse Foundation, 2009a). Currently interests are also picking up in metro-Vancouver's urban horticulture, which was the topic of a recent public dialog event that took place at the heart of downtown Vancouver, Canada (Morris J. Work Centre for Dialogue, 2009). All these school and public space projects are hopeful signs and offer us the opportunity to frame them as renormatization of education, thereby serving the cause of shifting, however subtly, the civilizational matrix.

Coda

Schooling has been the main vehicle of transmitting society's values and beliefs (Eisner, 1985). Students are neither trained to become deeply critical thinkers, nor to question the cultural values and beliefs that may arguably be the reason behind global crises. They are essentially ill equipped to respond to the current critical challenges facing humanity today, being blind to their own participation in global problems as well as being kept powerless to precipitate true change. We can trace our actions to our educational roots and the epistemology our culture has taught us. Ecological crises, global warming, poverty, exploitation, violence, greed—these are knowledges of some kind or another that we practice, and that have devastating effects on the earth and its inhabitants. Some argue that these destructive patterns are due to a lack of knowledge. Some also argue that if only we have better technology, if only we research more about the problems and gather more information about these problems, then we will be able to find solutions to these problems. We assert this is decidedly not so. These harmful effects are not because we lack knowledge, or technological prowess. It is because we acquire knowledge without learning to form respectful relationships with the object of our knowledge.

Meyer (2001) writes that "[as indigenous people], what we know, what we value about knowledge, how we exchange what that knowledge is, and how it endures through time are some of the most vital aspects of who we are" (p. 197). We believe this is true not only for indigenous people, but for all beings on earth. Animals, plants, and humans all pass knowledge from one form or another through time, exchanging and changing it to better the chances of survival. And if survival is not one of the most important uses of knowledge, then what is? As daily reports of the destruction and harm humans are causing to the planet increase, we can only deduce that the knowledge we are exchanging with one another is not the knowledge of survival, let alone knowledge of nourishing and flourishing. It is not the knowledge of respectful relationships, mutually benefitting cooperation and life. And if we continue to teach it, if we continue to enculturate generations with this flawed knowledge that has perpetrated so much harm, then harm will continue to happen. It is that simple. At this point in history, as we face the necessity of changing civilizational norms, we can deliberately choose to make schools an institution of renormatization and shape them to be more radically (*radix* = roots) and critically (*critus* = judgment) aware, questioning, and responsive to a changing world and a challenging reality.

Will our schools wake up and become the leaders in changing social norms? Will schools prepare citizens to encounter a critically challenging life on earth? Let us not even allow these to be questions. The earth may not be able to wait any longer, and neither can we. Schools must become awakened seekers and leaders of change in societal and civilizational norms.

References

Apple, M. (Ed.). (2009). *Global crisis, social justice, and education.* London, UK: Routledge.

Bai, H. (1999). Decentering the ego-self and releasing of the care-consciousness. *Paideusis, 12*(2), 5–18.

Bai, H. (2001). Challenge for education: Learning to value the world intrinsically. *Encounter, 14*(1), 4–16.

Bai, H. (2004). The three I's for ethics as an everyday activity: Integration, intrinsic valuing, and intersubjectivity. *Canadian Journal of Environmental Education, 9*, 51–64.

Bai, H. (2006). Philosophy for education: Cultivating human agency. *Paideusis, 15*(1), 7–19.

Bai, H. (2009). Re-animating the universe: Environmental education and philosophical animism. In M. McKenzie, H. Bai, P. Hart, & B. Jickling (Eds.), *Fields of green: Restorying culture, environment, education* (pp. 135–151). New Jersey: Hampton Press.

Bordo, S. (1986). The cartesian masculinization of thought. *Signs: Journal of Women in Culture and Society, 11*(3), 439–456.

Borgmann, A. (1992). *Crossing the postmodern divide.* Chicago, IL: University of Chicago Press.

Chez Panisse Foundation. (2009a). About us: Edible schoolyard. Retrieved from http://www.edibleschoolyard.org/about-us

Chez Panisse Foundation. (2009b). Edible schoolyard video. Retrieved from http://www.edibleschoolyard.org/video

Chinnery, A., & Bai, H. (2000). Altering conceptions of subjectivity: A prelude to the more generous effluence of empathy. In M. Leicester (Ed.), *Values and education in a pluralist society* (pp. 86–94). London, UK: Falmer Press.

Eisner, E. W. (1985). *On the design and evaluation of school programs* (2nd ed.). New York, NY: MacMillian.

Evernden, N. (1993). *The natural alien: Humankind and environment* (2nd ed.). Toronto, ON: University of Toronto Press. (Original work published 1985).

Fromm, E. (1976). *To have or to be?* New York, NY: Continuum.

Hanh, T. N. (1998). *The heart of the Buddha's teaching: Transforming suffering into peace, joy & liberation: The four noble truths, the noble eightfold path, and other basic Buddhist teachings.* New York, NY: Broadway Books.

Harrison, P. (Ed.). (1996). *Caring for the future: Making the next decades provide a life worth living.* Oxford, UK: Oxford University Press.

Held, V. (2006). *The ethics of care: Personal, political, and global.* Oxford, UK: Oxford University Press.

Jensen, D. (2008). Going underground [Electronic version]. The Sun, (386). Retrieved from http://www.thesunmagazine.org/issues/386/going_underground

Kohák, E. V. (2000). *The green halo: A bird's-eye view of ecological ethics.* Chicago, IL: Open Court.

Korten, D. (2001). *When corporations rule the world* (2nd ed.). San Francisco, CA: Kumarin Press.

Macy, J. (1991). *Mutual causality in Buddhism and general systems theory: The dharma of living systems.* New York, NY: State University of New York Press.

Macy, J., & Young, M. (1998). *Coming back to life: Practices to reconnect our lives, our world.* Gabriola Island, BC: New Society Publisher.

Martin, J. R. (1994). *Changing the educational landscape: Philosophy, women, and curriculum.* New York, NY: Routledge.

McMurtry, J. (1998). *Unequal freedom: The global market as an ethical system.* Toronto, ON: Garamond Press.

Merchant, C. (2005). *Radical ecology: The search for a livable world. Revolutionary thought/radical movements* (2nd ed.). New York, NY: Routledge.

Meyer, M. A. (2001). Acultural assumptions of empiricism: A native Hawaiian critique. *Canadian Journal of Native Education, 25*(2), 188–198.

Moran, A. (2004). *Philosophy as pedagogy* (Unpublished MA thesis) Simon Fraser University, BC, Canada.

Morris J. Work Centre for Dialogue. (2009). Growing citizens: Gardening as a catalyst for civic engagement. Heart of a citizen: Public dialogue series on civic engagement. Retrieved from Simon Fraser University, Centre for Dialogue website: http://www.sfu.ca/dialog//study+practice/workshop.html#location

Noddings, N. (1992). *The challenge to care in schools: An alternative approach to education.* New York, NY: Teachers College Press.

Noddings, N. (2003). *Caring: A feminine approach to ethics and moral education* (2nd ed.). Berkeley, CA: University of California Press.

Orr, D. W. (1994). *Earth in mind: On education, environment, and the human prospect.* Washington, DC: Island Press.

Rorty, R. (1996). Education without dogma: Truth, freedom, and our universities. In J. Portelli & W. Hare (Eds.), *Philosophy of education: Introductory readings* (2nd ed., pp. 207–218). Bellingham, WA: Temeron Books Inc.

Schor, J. B. (2004). *Born to buy: The commercialized child and the new consumer culture.* New York, NY: Scribner.

Skolimowski, H. (1994a). *EcoYoga: Practice & meditations for walking in beauty on the earth.* London, UK: Gaia.

Skolimowski, H. (1994b). *The participatory mind: A new theory of knowledge and of the universe.* London, UK: Arkana/Penguin Books.

Slote, M. A. (2007). *The ethics of care and empathy.* London, UK: Routledge.

Taylor, C. (1991). *Malaise of modernity.* Concord, Ontario: Anansi.

Waters, A. (2009). *Edible schoolyard: A universal idea.* San Francisco, CA: Chronicle Books.

Weston, A. (1992). Before environmental ethics. *Environmental Ethics, 14*, 323–340.

Wikipedia. (2009a). *Banyan.* Retrieved from http://en.wikipedia.org/wiki/Banyan

Wikipedia. (2009b). *Hermit crab.* Retrieved from http://en.wikipedia.org/wiki/Hermit_crab

11

Why We Need a Language of (Environmental) Education

Lesley Le Grange
Stellenbosch University, South Africa

> Education is the point at which we decide whether we love the world enough to assume responsibility for it and by the same token to save it from that ruin which, except for renewal, except for the coming of the new and young, would be inevitable. And education, too, is where we decide whether we love our children enough not to expel them from our world and leave them to their own devices, nor to strike from their hands their chances of undertaking something new, something unforeseen by us.
>
> Hannah Arendt (1954, p. 193)

Introduction

Environmental education emerged as a response to a growing awareness of the negative impact of humans on nonhuman nature. It is partly concerned with how humans might act so as to curtail environmental destruction or to turn the tide of destruction—it is about education's role in addressing issues of environmental deterioration. However, action is often preceded by assumptions about how humans ought to act in constructive ways toward nonhuman nature. It is in this sense that environmental education is at least partly normative in nature. It is the normative nature of "education" that makes it different to "learning." Education is inherently value-laden and associated with a positive human development process. While learning might be necessary for education, it is not sufficient. We can, for example, learn to harm others (learn to kill, steal, hate, etc.), which is not typically associated with education. In this chapter, however, I wish to take the discussion beyond a focus on just learning by devoting attention to an emerging discourse on learning that I shall refer to as a "language of learning." Furthermore, it should be noted that education is not equivalent to schooling; though, the terms are often used interchangeably by governments and also by some of the scholars I refer to in this text.

Since the term "environmental education" first appeared in a publication in the late 1960s it has been variously described as a consequence of renewed understandings of nonhuman nature as well as the interrelationships between humans and nature and education. Even though the term has taken on different meanings, it has by and large been retained in scholarly discourse. But there are some who have argued that environmental education should be redirected toward a focus on sustainability; one of these approaches is referred to as education for sustainable development (ESD). More recently we have witnessed the term "learning" being used instead of "education"—"environmental learning" instead of "environmental education." For example, "learning" features prominently in the titles of published books related to education and the environment: *Key Issues in Learning and Sustainability: A Critical Review* (Scott & Gough, 2004), *Social Learning towards a Sustainable World* (Wals, 2007) and *Participation and Learning: Education and the Environment, Health and Sustainability* (Reid, Jensen, Nikkel, & Simovska, 2008). In South Africa "environmental learning" also features in the titles of completed dissertations, for example, those by Hoffmann (2007) and Sehlola (2007). It is the shift from a language of education to a language of "learning" that I shall focus on in this chapter.

I shall argue that language matters in environmental education, because there is not a simple correspondence between language and reality—language does not simply describe what is out there. Language shapes or constitutes reality (practices) and is itself a practice. It constitutes "what can be seen, what can be said, what can be known, what can be thought, and ultimately, what can be done" (Biesta, 2006, p. 13). The idea that language has the power to determine what can be seen and thought is a strong and perhaps contentious claim. However, the point here is that language makes some things possible and others things not—the language we use could either enable or place constraints on what might be done. Considering our use of language in environmental education is important to the future of the field, because different languages are in effect different practices. This may be the reason why scholars inside and outside the field of environmental education are concerned, for example, that the loss of indigenous languages could diminish our ways of knowing.

In presenting my argument I shall divide my chapter into the following headings. First, I shall discuss the recent emergence of a language of learning. Second, I shall discuss how environmental learning has penetrated educational discourses in South Africa, including education policies. I do so for the purposes of illustration, convinced that the discussion might find resonance elsewhere. Third, I shall discuss the nature of education and why a language of education is important. Fourth, I shall argue that a language for environmental education is necessary and suggest why it is important that we reinvent a language of environmental education. Reinventing a language of education might be needed if we are to respond meaningfully to challenging environmental issues.

The Rise of a Language of Learning

Biesta (2004) argues that the emergence of a language of learning is not an expression of a single underlying agenda, but should rather be understood as the unintended outcome of a range of different developments: theories of learning, the silent explosion of learning, and the erosion of the welfare state. Theories of learning refer to developments in the field of the psychology of learning and specifically the emergence of constructivism and socioconstructivism. Constructivist theories of learning shift the emphasis from teacher to learner, since they are premised on the view that learners actively construct knowledge and understanding, and that knowledge cannot be transferred intact from teacher to learner (for a more detailed discussion on constructive theories of learning see Wals & Dillon, 2012, in this volume). By the "silent explosion" Biesta (2004, p. 73) refers to the mushrooming of nonformal kinds of learning such as "fitness centres, sport clubs, self-help therapy manuals, Internet learning, self-instructional videos, DVDs and CDs etcetera." The rise of a language of learning is also associated with the decline of the welfare state and the rise of neoliberalism. The welfare state provides all citizens (rich and poor) with health care, security, education (it claims to provide education, but what it mainly provides is schooling), and so on. Biesta (2004) argues that within the neoliberal state, "value for money" (p. 33) has become the key principle in many of the transactions between the state and taxpayers. The state's role has shifted from provider of the mentioned goods to taking on a monitoring role with tighter systems of inspection and control, and prescriptive protocols over education—what Ball (2003) refers to as a rising culture of performativity/accountability. In this context, Biesta (2004) argues, parents are viewed as purchasers of the services that education/schooling provides for their children, and the suitable name for the consumer is therefore the learner.

The developments that I have described are not all bad. Wals and Dillon (2012) additionally point to the emergence of emancipatory forms of learning in the context of high levels of complexity, contestation, and uncertainty that characterize contemporary sustainability/environmental concerns. Emancipatory forms of learning include: transdisciplinary learning, transformative learning, crossboundary learning, anticipatory learning, action learning, and social learning (for detail see Wals & Dillon, 2012). My concern is not with the (de)merits of each development, but rather with an unintended outcome that some of these developments have collectively produced, that is, the rise of a language of learning. The rise of a language of learning that I refer to has relevance to the developments I discuss above, given the dominance of neoliberalism across the globe, and would not include the emergent forms of learning that Wals and Dillon (2012) refer to. So, what is wrong then with a language of learning? Biesta (2004) argues that one of the main problems with the new language of learning is that it makes possible the redescription of the process of education in terms of an economic transaction, that is, the learner (who has the needs) is the consumer, the teacher, or education institution the provider, and education becomes a commodity. But what is the problem with the education process being understood as an economic transaction? In a typically economic transaction a consumer knows what (s)he needs and wants, and manufacturers and retailers provide for such needs and wants. However, it is questionable whether children know what it is that they want from education. Even adults do not always know what they want from education or what their needs are. People engage in education precisely to find out what are their needs and wants, and education professionals play a vital role in helping students with finding out what it is that they actually need.

Furthermore, viewing education as an economic transaction dilutes education processes to technical concerns of efficiency and the effectiveness of such processes, neglecting questions concerned with the content and purpose of education, which Biesta (2004) argues should form part of the education process—that asking questions about the content and purpose of education are important educational questions. I need to clarify, though, that a "language of learning" which leads to viewing education as an economic transaction is associated with the ascendency of neoliberalism in Western(ized) societies. An emphasis on learning in the child-centered tradition, for example, would be hostile to viewing education as an economic transaction. So too the emancipatory forms of learning Wals and Dillon (2012) refer to. Before taking the discussion on the importance of a language of education forward, I shall discuss how a language of learning generally, and a language of environmental learning specifically, has penetrated South African (environmental) education policy discourses.

A Language of (Environmental) Learning in South Africa

The developments described above provide the backdrop against which the emergence of environmental learning might be understood. Before discussing environmental learning in South Africa, I shall briefly discuss the rise of a language of learning in the country. This rise in South Africa is evidenced by the frequent use of the concept "learning" in policy documents; by the identification of life-long learning as one of the key features of the National Qualifications Framework (NQF); by increased talk of adult learning and not adult education; and more recently, by the emergence of a language of environmental learning. The expansion of neoliberalism has been strongly felt in South Africa. Since 1994 we have witnessed both the commercialization and privatization of government assets, and that the state is actively putting in place tighter systems of inspection and control. In relation to education, for example, quality assurance has become a favorite term in many of the education policy documents. These developments are not unique to South Africa. For example, in October 2007 the North American Association of Environmental Education (NAAEE) submitted environmental education standards to the National Council for the Accreditation of Teacher Education (NCATE), a body that accredits more than 600 Colleges of Education in the United States of America (http://www.naaee.org/programmes). The adoption of a language of learning in South African policy documents might be the consequence of the developments I have described, particularly the ascendency of neoliberalism. But it also needs to be acknowledged that education might be viewed by governments and corporate elites as intrinsically dangerous, since it implies at its best the development of a person's capacity to think critically, evaluate options, assess value, understand the differences between forms of knowledge, etc. These attributes do not make people easy to govern, nor do they tend toward creating mindless consumers.

Concerning theories of learning, developments that led to the emergence of a language of learning internationally have influenced the development of a language of learning in South Africa. Constructivist and socioconstructivist theories of learning have had a particular appeal in South Africa, because they offer a response to the teacher-dominant pedagogies that characterized education practices during apartheid. The post apartheid national curriculum framework was ostensibly introduced to replace content-based education with outcomes-based education, and teacher-centered education with learner centered education. As a consequence of globalization there is also a growing market for nonformal forms of learning such as the ones described earlier. This might be because of power shifts in regard to who has power in the world of schooling and accreditation. The major universities and large public school systems are finding their authority diluted, because access to instruction and information, over which they once exerted strong control, has moved to the World Wide Web, the Internet, and to social media. The effects of these developments might have both good and bad aspects.

But how has a language of learning penetrated curriculum policy more specifically? The post apartheid curriculum framework for general education and training (Grades R-9) replaced traditional school subjects with eight new integrated learning areas. Each of these learning areas has learning outcomes which learners are expected to attain. From these outcomes, assessment standards (used to distinguish phase levels of outcomes), and content specifications are derived. Furthermore, instead of designing lesson plans, teachers now have to design learning programs. Learning programs are a compilation of activities, based on a particular topic, which enable students to achieve specified critical (generic and cross-curricular) and learning outcomes (specific to learning areas).

From the above it is evident that a language of learning features prominently in post apartheid education discourses and it is this context that shapes a language of environmental learning in school curricula. Environmental concerns feature prominently in the post apartheid curriculum (at least in the intended curriculum), but it is important to notice that these concerns are framed as follows: they are included in some of the learning outcomes and/or assessment standards and/or content specifications of the eight learning areas. Moreover, teachers are expected to include these concerns in learning programs, which are essentially a compilation of learning activities. It is in this context that some officials in the national and district education departments claimed that "environment is in the curriculum" and that talk of environmental education should be abandoned in favor of environmental learning (see Le Grange, 2004). Moreover, it is also why environmental learning featured in many conversations at the Environmental Education Association of Southern Africa (EEASA) conference of 2004 (see Le Grange, 2004) and why it appears in the titles of theses produced in South Africa (see introduction).

A Word on Education

Education has traditionally been understood as an "intervention" into someone's life—a process that makes a difference to a life. Therefore education cannot simply be constituted as practices of socialization, even though novices need to be equipped with the cultural tools to participate in a particular form of life so as to ensure cultural and social continuity. For Biesta (2006) the danger of viewing education narrowly as socialization is that socialization contributes to the reproduction of inequalities. Therefore, education additionally includes individuation, that is, it focuses on the cultivation of the human person or the individual's humanity. This idea that education is about cultivating the human person might be traced back to the tradition of *Bildung*—an educational ideal that emerged in Greek society, and through its adoption in Roman culture, humanism, neo-humanism, and the Enlightenment, became one of the central notions of the modern Western educational tradition (see Biesta, 2006 for a more detailed discussion). The upshot of these developments was that the notion of human being was configured in a particular way. For example, when *Bildung* became intertwined with the Enlightenment and the particular influence of Emmanuel Kant, "human being" came to mean "rational autonomous being"—consequently the purpose of education was to develop rational autonomous beings. Critical theory/pedagogy and its derivatives followed from this understanding of education—that emancipation was a rational process—a process of conscientization in the case of Freirian pedagogy. My use of rationality, however, has reference to a particular kind of rationality that the Enlightenment period has bequeathed to us. I acknowledge that there might be other kinds of rationality. For example, Næss (2002) argues that rationality includes emotions and experiences, suggesting that it should be rescued from an erroneous turn.

Importantly, some have critiqued a particular notion of humanism (the Enlightenment idea of what it means to be human) and education based on such an understanding. For example, Heidegger points out that humanism's response to the question of what it means to be human focuses on the essence or nature of the human being. He argues that the focus should instead be on the being of this being, on the existence of the human being, on the ways in which the human being exists in the world (Heidegger 1962/2008). The problem with focusing on what it is to be human (the essence of a human being) is that it opens up possibilities for defining "human" in particular ways that declare others as less human or nonhuman. The holocaust, apartheid, genocides in Bosnia, Rwanda, and Cambodia forcefully remind us of the effects of the humanism of the Enlightenment. Levinas (1990) goes as far as to argue that the crisis of humanism began with the inhuman events of recent history:

> The 1914 War, the Russian Revolution refuting itself in Stalinism, fascism, Hitlerism, the 1939–45 War, atomic bombings, genocide and uninterrupted war . . . a science that calculates the real without thinking it, a liberal politics and administration that suppresses neither exploitation nor war . . . socialism that gets entangled in bureaucracy. (p. 279)

Biesta (2006) provides an alternative to the Enlightenment understanding of education, which is based on the idea that rational autonomous persons need to be produced—the educator takes on the role of midwife so as to release the rational potential of the human being. In contrast, Biesta's idea is not based on the educator producing or releasing anything, but that education "should focus on the ways in which the new beginning of each and every individual can come into presence" (p. 9). He goes on to argue that this is not a version of child-centered pedagogy:

> [W]e can only come into presence in a world populated by others who are not like us. The "world," understood as a world of plurality and difference, is not only the necessary condition under which human beings can come into presence; it is at the very same time a troubling condition, one that makes education an inherently difficult process. The role of the educator . . . has to be understood in terms of a responsibility for the "coming into the world" of unique, singular beings, and a responsibility for the world as a world of plurality and difference. (p. 9)

Coming into presence as singular beings involves taking responsibility for the world, or as Arendt (1954) puts it, "to love the world enough to assume responsibility for it" (p. 193). Education involves providing opportunities for the young to bring something new into the world, while not leaving them to their own devices. The role of the educator is crucial in ensuring that in the education process there is a balance between engagement with pressing matters in the world (including environmental problems) and opportunities for new imaginings. Moreover, it is important to understand that education always involves an element of risk—education processes always produce unintended outcomes. For example, in the education process there might be serendipitous moments where students might not have imagined what they are experiencing, and also instances when their beliefs, values, and notions of truth are challenged or disturbed. Biesta (2004) goes so far as to claim that education only begins when the teacher/student is willing to take a risk. He writes:

> To negate or deny the risk involved in education is to miss a crucial dimension of education. To suggest that education can be and should be risk free, that learners don't run any risk by engaging in education, or that "learning outcomes" can be know[n] and specified in advance, is a gross misrepresentation of what education is about. (p. 77)

Therefore, the teacher's role in "managing" the risk involved in the education process is crucial. Sometimes teachers have to play nurturing roles and sometimes they have to disturb students. They also have to make pedagogical judgments, for example, when the pedagogical episode should begin and when it should end. These responsibilities, Shalem (1998) argues, constitute the pedagogical authority of the teacher. A language of learning does not necessarily require teachers to take on such roles.

Both Biesta and Levinas help us in understanding why a language of education is paramount. However, they do not go far enough. Coming into presence in a world populated by others—a world of plurality and difference—should not only include the world of human beings, but the entire world, including all species and physical elements and the complex networks formed by all facets of the world. The critique of humanism should also be a critique of anthropocentrism. We must love the world in its entirety and take responsibility for it. To come into presence in the world (to be more human) must mean not to be human-centered. For this reason we need more than just a language of education, but a language of environmental education. The idea of coming into presence in relation to the natural world is supported by Bonnett (2009) who argues that the world of nature is quintessentially "other" (p. 364). Bonnett also raises an important matter that Biesta neglects in his work, that is, the possibility of coming into presence through moments of self-awareness, recognizing the interiority of the individual or taking into account the "phenomenological self" (p. 369). Moreover, Le Grange (2011, 2012) points out that the African value *Ubuntu* (humanness) concerns a condition of being that becomes/unfolds in relationship with the other (other human beings and the biophysical world).

Why a Language of (Environmental) Education?

Levinas (1990) relates the crisis of humanism to several inhuman events of the twentieth century. I wish to suggest that the global environmental crisis is as much a manifestation of the crisis of humanism. I refer here to a focus on the essence of human being (as rational autonomous being), which reinforces the distinctiveness of humans from the rest of the natural world. The view of education derived from this is that education should play the role of producing rational autonomous beings. Humanism and concomitant views of education (shaped by Western modernity) have shortcomings, and a return to a language of education based on humanism is not what I am arguing for.

We need a new language of education that involves the young coming into presence in a world of difference and plurality. But I want suggest that the world of difference and plurality does not mean engaging with other humans only, but that plurality and difference would refer to birds, bees, butterflies, ferns, lichens, whales, dolphins, ants, trees, rocks, rivers, sea, atmosphere, etc. This perspective may be common among many indigenous systems of education and acculturation, but not necessarily so in Western(ized) education systems. Therefore the need for a language of environmental education, which would entail teachers nurturing the young and "intervening" in their lives—providing opportunities for them to bring new things to the world, but also not to leave them to their own devices. In other words, it entails teaching the young to love the world (in its entirety) enough to take responsibility for it so as to save it from ruin (including environmental degradation). This might initially involve teachers identifying the principal sites that are conducive for critical reflection and deliberation, including what ought to be the right/appropriate actions (be they caring for aspects of the natural world or efforts to achieve social justice). Deliberations among students and between teachers and students are not meant to be easy but uncomfortable, as they might challenge taken-for-granted values, beliefs, and assumptions. And furthermore, teachers need to create spaces for their own and students' coming into presence though mutual dialog and actions in concert. Learning remains a necessary response but not a sufficient one in actions aimed at saving the world from ruin, because it does not require "intervention" into someone's life—learning does not require the disturbing of someone's life or the taking of risks. But let me acknowledge that the term "intervention" has a medical or intrusive overtone; however, what I mean by the term "intervention" is that the young should not be left to their own devices—they should be taught to think critically, to question and assess their assumptions and taken-for-granted values, etc.—it is in Western(ized) systems in particular that greed, consumerism and an unquestioning belief in science, technology, and progress need to be questioned.

As mentioned, a new language of education involves risk, but when environment is added the risk is compounded. Therefore, when I refer to risk there is a dual sense in which I use the word: first, it refers to the risk that a student/teacher takes to engage in education processes that will inevitably have unintended outcomes; second, it refers to the dangers or hazards that are prevalent in contemporary society. Environmental problems are complex and so are their solutions. Today's solution may be tomorrow's problem. Associated with environmental problems are risks that have become pervasive in contemporary society, so much so that Beck (1992) refers to the society of late modernity as a risk society. A risk society is characterized by the distribution of "bads" or dangers across the globe. Beck argues that a risk society is concerned with a type of immiseration of civilization. The immiseration he refers to does not involve material impoverishment, as was the case of the working masses of the nineteenth century, but rather concerns the threatening and destruction of the natural foundations of life. The ubiquity of risk is evident today when harmless things such as wine, tea, beef, pasta, etc. turn out to be dangerous (see Beck, 1992). Beck points out that in contrast to the immediacy of personally and socially experienced misery in the nineteenth century, today's civilization presents threats that are intangible, brought to consciousness chiefly in scientized thoughts. More and more the public are dependent on the knowledge of experts in the field of science to make decisions concerning risks that might affect their lives. People are therefore becoming increasingly incompetent in addressing their own afflictions. I contend that in the developing world risk associated with material impoverishment is additionally still present, thus compounding risk in such societies.

In summary, coming into presence in the world involves the risk a student takes of being challenged or disturbed by what (s)he encounters through the educational process—the risk of the unintended outcomes of education. But coming into presence in the world (in its broad sense) also involves encountering environmental risks and engaging with these to improve risk positions.

Conclusion

In a recently published article Jickling (2009) responds to a claim made at a conference by an environmental educationist that (environmental) learning is a richer concept than (environmental) education. It may be that the scholar might have meant that learning a richer concept than schooling; nevertheless, Jickling's response to the claim is important in clarifying the notion that education and learning are different in nature. He writes:

> The critical point in this case is that learning is not a richer concept than education. They are different in nature, and it is a category mistake to compare them this way, or to substitute one for the other. (p. 5)

I have argued that a language of education is imperative if students are to respond meaningfully to the challenges presented by the unprecedented levels of environmental deterioration witnessed by the present generation. Learning is an important ingredient (a necessary condition) for responding meaningfully to ameliorating environmental problems, but it is insufficient. In fact, students can learn (and have learned) that they are separate from and/or superior to the natural world/other species, leading to an exacerbation of environmental problems—they can learn not to care, to be silent and complacent concerning environmental destruction and social injustices. However, in arguing for a language of (environmental) education, I am not suggesting that we return to the languages of the past. We can't return to a language of education that is akin to the Enlightenment idea of emancipation within the humanist framework in which rationality is seen as both the essence and destiny of the human being. We need a language of education that focuses on the being of the human being, which focuses on the existence of the human being in the world—a language that will enable the young to love the world enough to take responsibility for it. If by the world we mean everything that comprises planet Earth, then the language we need might be environmental education.

Perhaps my discussion takes matters beyond the meaning that those who use the term "environmental learning" give to it. It may be that "environmental learning" is used in a more trivial way than the language of learning I have discussed in this chapter. I mean trivial in the sense that it is unquestioningly accepted by some in South Africa (and elsewhere) that all that is to be said about the environment is embodied in curriculum policy documents ("environment is in the curriculum") and all that must happen in classrooms is the learning of what is defined in the policy documents ("environmental learning"). Perhaps my discussion of an emerging language of learning can be meaningfully employed to read education policies in general and constructs such as "environmental learning" deconstructively, that is, to lay bare traces of a new language of learning.

The complex and contingent nature of environmental problems and their associated risks cannot be captured in a few learning outcomes of a National Curriculum Statement—so "environment is [not] in the curriculum." What the Revised National Curriculum Statement (RNCS) (Department of Education, 2002) for General Education and Training (GET) in South Africa does is to provide the spaces for enabling environmental education processes. But recognizing these spaces requires teachers who have an understanding of environmental education processes. To ensure this, environmental (teacher) education has a cardinal role to play. Moreover, environmental learning is dependent on teachers mediating environmental knowledge, on teachers exercising their responsibility and pedagogical knowledge. Environmental knowledge is produced in interdependent and interactive relationships between teachers and students who, in biophysical environments, engage critically with information, issues and problems, often resulting in unintended outcomes. An appreciation of this necessitates a language of environmental education and not merely a language of environmental learning. Language is loaded, so we must not dismiss lightly the importance of the terms we use: an uncritical adoption of a notion such as "environmental learning" can have potentially damaging consequences.

Some might argue that education of any worth should address the perennial questions: What does it mean to be human? What is our place in the world and our roles in relation to other beings, life forms and the planet? What is good, evil, right, wrong, true, false? How do we know and what are we capable of knowing? This raises the question: Do we need the term environmental education? Perhaps not. What we certainly need is a language of education. As mentioned, although learning is a necessary ingredient of education, it is not sufficient in raising and responding to the questions of what it means to be human, how we should live, what is good, evil, right, wrong, and so on. We need a language of education (as distinct from schooling) that is inherently concerned with these value-laden matters/questions. Given the wide consensus that the planet is on the brink of ecological disaster and that the response of formal education to the planetary crisis has been inadequate, it might do no harm to insert "environmental" before "education." A language of environmental education I am referring would raise critical questions of how we should live, how we should learn, and what knowledge is most worth learning or including in curricula/educational programs.

As languages of learning proliferate in contemporary society and as they migrate into environmental discourses, it might be important for environmental educators to reflect on these developments and their implications for the future direction of environmental education as a field. This chapter opens up the discussion for further consideration.

References

Arendt, H. (1954). *Between past and future*. New York, NY: Penguin Group.

Ball, S. (2003). The teacher's soul and the terrors of performativity. *Journal of Education Policy, 18*(2), 215–228.

Beck, U. (1992). *Risk society: Towards a new modernity*. London, UK: Sage Publications.

Biesta, G. (2004). Against learning: Reclaiming a language for education in an age of learning. *Nordic Pedagogy, 24*, 70–82.

Biesta, G. (2006). *Beyond learning: Democratic education for a human future*. London, UK: Paradigm Publishers.

Bonnett, M. (2009). Education and selfhood: A phenomenological investigation. *Journal of Philosophy of Education, 43*(3), 357–370.

Department of Education. (2002). *Revised national curriculum statement grades R-9 (schools): Overview*. Pretoria: Author.

Heidegger, M. (2008). *Being and time*. New York, NY: Harper and Row publishers. (Original work published in English 1962).

Hoffmann, P. A. (2007). *Reviewing the use of environmental audits for environmental learning in school contexts: A case study of environmental auditing processes within a professional development course* (Unpublished master's thesis). Rhodes University Grahamstown.

Jickling, B. (2009). Environmental education research: To what ends? *Environmental Education Research, 15*(2), 209–216.

Le Grange, L. (2004). Against environmental learning: Why we need a language of environmental education. *Southern African Journal of Environmental Education, 21*, 134–140.

Le Grange, L. (2011). Ubuntu, ukama and the healing of nature, self and society. *Educational Philosophy and Theory*. Retrieved from http://onlinelibrary.wiley.com.ez.sun.ac.za/journal/10.1111/(ISSN)1469–5812/earlyview

Le Grange, L. (2012). Ubuntu, ukama, environment and moral education. *Journal of Moral Education, 41*(3), 329–340.

Levinas, E. (1990). *Difficult freedom: Essays on Judaism*. Baltimore, MD: John Hopkins University Press.

Næss, A. (2002). *Life's philosophy: Reason and feeling in a deeper world*. Athens, Georgia: University of Georgia Press.

Reid, A., Jensen, B. B., Nikkel, J., & Simovska, V. (Eds.). (2008). *Participation and learning: Perspectives on education and the environment, health and sustainability*. Heidelberg: Springer.

Scott, W., & Gough, S. (Eds.). (2004). *Key issues in lifelong learning and sustainability: A critical review*. London, UK: RoutledgeFalmer.

Sehlola, M. S. (2007). *A case study of the integration of environmental learning in the primary school curriculum* (Unpublished master's thesis). University of Pretoria, Pretoria.

Shalem, Y. (1998). Epistemological labour: the way to significant pedagogical authority. *Education Theory, 49*(1), 53–70.

Wals, A. (Ed.). (2007). *Social learning towards a sustainable world*. Netherlands: Wagingen Academic publishers.

Wals, A., & Dillon, J. (2012). Conventional and emerging learning theories: Implications and choices for educational researchers with a planetary consciousness. In R. Stevenson, M. Brody, J. Dillon, & A. Wals (Eds.), *International handbook of research on environmental education*. New York, NY: Routledge and AERA.

12

Environmental Ethics as Processes of Open-Ended, Pluralistic, Deliberative Enquiry

Lausanne Olvitt
Rhodes University, South Africa

By the very nature of their work, environmental education researchers must engage with environmental philosophy and questions of values and ethics. But this terrain, despite being resourced with an apparently endless supply of typologies, anthologies, and handbooks, can remain a vast and daunting philosophical sea—at least in my experience as a newcomer to the field, and possibly for many other scholars and researchers. This essay makes no claim to altering that and instead optimistically pursues Ball's (2001, p. 89) suggestion that "there is much to be learned about, and from, the philosophical life-forms inhabiting these thickets and swamps." My intention here is to review a relatively small but growing cluster of work in environmental ethics that proposes that: "Ethical positions are always open for discussion, re-examination, and revision" (Jickling, 2004, p. 16) and are thus, by their very nature, open-ended, relational processes. My starting point in writing this essay is as an educator-researcher-environmentalist trying to explore what the field of environmental ethics has to offer in response to the question: "*As educators, how can we learn and do more with others in the face of an unprecedented socioecological crisis?*"

Environmentalism, as a new social movement, speaks into the face of unprecedented ecological collapse and human vulnerability, and recognizes that our conceptions of environment are bound up in diverse social realities and people-environment relationships. Much of the so-called "environmental crisis" originates in modern patterns of production, consumption, and domination that are out of synch with the prerequisites for the long-term integrity of Earth's natural systems. Unsurprisingly, these socio-ecological relationships have so far defied the application of a singular ethical system or the unanimous adoption of a particular value theory. It seems, therefore, that we need to seek more deliberative and transformative ways of engaging with the diversity and complexity of people-environment relationships.

To explore and contextualize the potential of these ideas, I talked with a group of Southern African educators who were at the time studying environmental education. They described how their efforts to take responsible environmental action are influenced by factors as varied as financial concerns, time frames, social conformity, the value attached to the subject, and lack of alternatives to current practices (Olvitt, 2009). When asked what values or codes guide them when faced with making decisions affecting the environment, the students referred variously to:

- *social concerns*: "I'm guided by society, e.g. what will people think about my decision";
- *self-interest*: "I am more concerned whether I'll end up in a safe side or not";
- *cultural influences*: "I respect it [nature] because it was made for me, therefore it is taboo to spoil it";
- *religion*: "As a Christian, the concept of stewardship of the earth is the most influential when it comes to my values";
- *past experiences*: "The prolonged drought I experienced in my childhood where I had to carry 25 litre bucket of water [on my head] and 5–10 litres in hands from the spring . . . forced me to modify habits in handling and use of water" (Olvitt, 2006, 2009).

Similar insights were gained in a research project with a group of South African youths identified as "at risk" due to their social context of homelessness, poverty, substance-abuse, or gangsterism. The youths' narrations of their environmental actions revealed that their social and economic identities and experiences influenced their actions more than conscious adherence to an explicitly articulated ethical code (Ayair, 2009). One of the youths, for example, stated that she willingly switched off the lights if she was the last to leave a room in the hostel, but that she was not permitted to do so at the family's rural home because her grandmother wanted all the lights in the home switched on so that others would know that the family could afford to pay the electricity bill. Another youth explained that he had in the past littered intentionally in the city streets with hopes of creating employment opportunities for garbage collectors. The same youths collectively recognized the importance of receiving more environmental knowledge in order to make better choices, and many indicated that they had, on the grounds of their educational interactions with the researcher, undertaken to act more responsibly in relation to environmental matters in future (Ayair, 2009, pp. 7–9).

These and other stories of people's authentic *in situ* struggles to live well in the world can offer insights into the diverse sociocultural, economic, and political spaces in which ethical practices are negotiated daily. They can offer glimpses into the complexity, contingency, and generative possibility of people's lived experiences. For example, hearing from South African youths who alter their environmental practices as they shift between rural and urban identities, or from the Basotho school teacher whose passion for water conservation can be traced to her childhood burden of having to collect and carry her family's daily water ration, can help environmental educators bridge philosophical propositions and lived experiences. Wals (2007) suggests that learning processes in such contexts of sustainability are "rooted in the life-worlds of people and the encounters they have with one another" and are thus open-ended and transformative. For Weston (1992, 2009, p. 27), this is a significant opening for the work of environmental ethics. Recognizing that values are socially and culturally shaped, "deeply embedded in and co-evolved with social institutions and practices," he concludes that values are, by their very nature, contingent and open to reshaping.

Similarly, Zygmunt Bauman (1993, 1995) would likely regard such stories as manifestations of the fragmented and episodic nature of life pursuits in a postmodern world.[1] His concern is that postmodern life divides not only labor, expertise, and communality, but also the whole person; it fragments not only the roles we play and the settings in which we play them, but also responsibility for their consequences. Our inherited, traditional, time-tested ethical rules are made redundant by the scale of our globalizing, technological achievements whose consequences now ripple across continents and generations like no other time in history. Bauman (1993, p. 33) proposes that we "learn again to respect ambiguity, to feel regard for human emotions, to appreciate actions without purpose and calculable rewards." Such a sense of moral responsibility is never past, never complete, and characterized instead by a "gnawing sense of unfulfilledness, by its endemic dissatisfaction with itself" (Bauman, 1993, p. 80). Ethics then becomes less of a systematic collection of generalizations set in place by external authorities to guide us in deciding whether an action is "right" or "wrong," and becomes more of an *open-ended, exploratory process of generative enquiry*, which Bauman (1993) describes as more like a string of footprints than a network of well-mapped roads.

For the purposes of this chapter, I will, following Weston (2001), consider moral values to be those values "that give rise to the needs and legitimate expectations of others as well as our selves" (p. 12), and ethics to be a more critical and self conscious process of "reflection on how best to think about moral values and clarify, prioritize, and/or integrate them" (p. 12). This view of ethics as processes of negotiation and responsiveness is a radical departure from the dominant modernist view of unambiguous "ethical rulings" which reduce moral responsibility to "a finite list of duties and obligations" (Bauman, 1995, p. 4). It is significant that people—ordinary citizens concerned with how their lives influence the world—are now compelled to be part of an open-ended conversation about how we might live better in the world. Stories have already been shared and new stories continue to emerge, all contributing something to the richness of the bigger conversation. This essay considers the significance of that conversation for environmental learning and change, and attempts to answer the question posed at the start: "How might the field of environmental ethics help us to learn and do more with others in the face of an unprecedented socio-ecological crisis?"

South African environmental philosopher, Johan Hattingh (1999), compares environmental ethics to "a large toolkit" containing many "strange and wild ideas," some of which might be more useful than others. Making a case for environmental pragmatism, he suggests that we engage with these diverse and evolving ideas as:

> genuine efforts to articulate new ideas and transform existing practices, institutions and experiences. They do not fulfil a prescriptive function within an already established framework of values and practices. They serve rather as rhetorical devices: open-ended challenges to that which already exists. They serve to open up questions, not to settle them. (p. 80)

Weston (2001) also proposes the metaphor of a "toolbox." The diversity of (sometimes contradictory) theoretical and philosophical positions is a practical tool for which we can reach when we need new, different, or challenging perspectives. These tools are as diverse and dynamic as the multiple contexts from which they arise. They may help us (generally as world citizens and more specifically as researchers, storytellers, teachers, activists, consumers, decision makers) to think more deeply about and act more reasonably and ably in the world. Hattingh (1999) urges us to keep working with these tools in practical ways so as to keep them "sharp," even developing new ones for unknown times ahead.

As we draw on various environmental ethics tools, however, we should remain sensitive and attentive to the nature of the overall project, that is, the *work* that environmental ethics can do. It can be tempting to become absorbed in a particular philosophical position or environmentalist agenda and lose the groundedness in remembering and reclarifying what the journey is really about, why we are pursuing these ideas, why we are reaching for these tools in the first place. For example, as I interact through research projects and university courses with environmental educators from around southern Africa, what work can the vast field of environmental ethics do as we face changing weather patterns, biodiversity loss, the HIV/AIDS pandemic, water-borne diseases, cultural change, land degradation, xenophobia, and so on? How might spaces be created to engage with such things in ways that account for, rather than silence, values, dignity, and future aspirations? This essay proposes a pragmatic approach that regards the diversity of ecophilosophies and ethical ideas as a "toolkit" to use in different ways at different times. The following sections consider some of the qualities that might enable us to draw on this toolkit in ways that advance the vision of environmental ethics as a necessarily uncharted journey that is both tentative and generative. In these sections I present three potentially helpful touchstones: environmental ethics as "open-ended processes" (being indeterminate and shaped over time and space by human agency), environmental ethics as "attentive and pluralistic" (seeking respect, pursuing responsibility, and being responsive to things new, different, even uncomfortable), and finally, environmental ethics as "deliberative enquiry" (asking questions and continually exploring for shared depth, nuance, and change).

Environmental Ethics as Open-Ended Processes

According to Aldo Leopold, a pioneer of environmental ethics in the North American tradition, "Nothing as important as an ethic is ever 'written'" (Leopold, 1949/1989, p. 225). Instead, ethics evolve in the minds of a thinking community and this evolution never stops. There is now a growing body of environmental philosophers, educators, and activists whose work reflects this central concern for enabling and supporting *processes* of environment-oriented ethical enquiry (Cheney, 1999; Hattingh, 1999; Jickling, Lotz-Sisitka, O'Donoghue, & Ogbuigwe, 2006; Næss, 2000; Wals, 2007; Weston, 1999). The generative possibilities that arise through approaching ethics in this way is well captured by Jickling (2004):

> [E]thics is a process of inquiry—a philosophical examination of those varied and sometimes contested stories that constitute our social reality. This is quite different from following prescribed rules or an ideologue. Rather, "ethics as process" invites individuals into an ongoing process of defining and redefining their own rules for individual and community conduct. Ethics in this sense is an everyday activity for ordinary people. And, it is the essence of citizen-based democracy. (p. 16)

Alongside this, it is important to recognize that environmental ethics, as a relatively new and emergent subfield of philosophy, is still in its originary stages and so exploration, experimentation and some "red herrings" are to be expected (Hattingh, 1999; Weston, 1992/2009). We are, after all, moving in uncharted territories; "our world is one of continuous change and ever-present uncertainty" (Wals, 2007, p. 1).

It is important to acknowledge the educational implications of this: that we cannot really teach for the future by limiting ourselves to the metaphors, discourses, priorities, and knowledge bases of the present. And seeking to find *the* right answer may prove futile and unnecessary anyway because "the future is open . . . Everything could be worse than we think, or better, but it will almost certainly be different than we think. All we can do is act on our best guesses and our hopes" (Weston, 1994, p. 176). Bauman (2001, p. 139) lays out the consequent challenge for education: to theorize "a *formative* process which is not guided from the start by the target form designed in advance . . . an open-ended process, concerned more with remaining open-ended than with any specific product and fearing all premature closure more than it shuns the prospect of staying forever inconclusive." For Cheney and Weston (1999), this is an epistemological question. They ask us to consider how things might be different if we entered into ethical relationships with others *in order* to know them better, rather than letting our predefined knowledge of them determine the nature of our ethical relations. In a world that "has barely unfolded for us," shifting from an epistemology-based ethics to an ethics-based epistemology is perhaps fundamental to seeing ethical action as "an attempt to open up possibilities, to enrich the world" (Cheney & Weston, 1999, pp. 117–118).

Environmental Ethics as Attentive and Pluralistic

Pluralism can mean different things in different settings. It could refer to diverse perspectives within and across human communities, for example, as used by Wals and Heymann (2004) when they state: "[w]e live in a pluralistic society, characterized by multiple actors and diverging interests, values, perspectives and constructions of reality" (p. 2). But it can also refer to something much wider, as we find in Abram's (1999) reference to shamans who:

> . . . readily slip out of the perceptual boundaries that demarcate his or her particular culture . . . in order to make contact with, and learn from, the other powers in the land. His magic is precisely this heightened receptivity to the meaningful solicitations—songs, cries, gestures—of the larger, more-than-human field. (p. 24)

Or, Weston's (1994, p. 113) brief experience of living "other possibilities" when, following a week-long electricity outage after Hurricane Gloria in the United States in 1985, "people would just sit for hours in someone's backyard and watch the full moon rise."

People's increasing detachment from the plurality and receptiveness alluded to in Abram's and Weston's examples is a central theme in environmental philosophy. Plumwood (2002) traces human detachment from nature to moral dualism in the neo-Cartesian tradition. A far cry from pluralism's desire to explore and invite, moral dualism divides the world into two sharply contrasting orders: humans (worthy of ethical concern) and all the rest (which may have instrumentalist value but are otherwise underqualified for ethical consideration). Plumwood (2002) notes that this kind of:

> sharp cut off or boundary for moral consideration is neither necessary nor desireable, and the forms of life that correspond to the dualisms between use and respect are unjust and diminishing both for "persons" and for the greater multiplicity of beings that make up planetary life. (p. 145)

The inherent logic of moral dualism can be dominating and all-pervasive to the extent of losing sight of other possibilities that might bring other voices, other lived experiences into our (shared) space (Evernden, 1985; Weston, 1992/2009). Consider, for example, how some architectural styles reinforce their detachment from the natural space they inhabit, shutting out natural light and views and replacing them with artificial lighting and walls-as-barriers. Fences delineate the "home" from the "wild," "safety" from "threats." A pluralistic view and its attendant values might be more attentive to the nuance of walls, windows, and doors; incorporating natural variations, accommodating small creatures, paying tribute to other histories of that space, and so on. This is the kind of care-full openness and plurality that an ethics-based epistemology (Cheney & Weston, 1999) might make realizable.

Yet, in the overlapping fields of environmentalism and education, we see how pervasive dualistic assumptions about people-environment relationships can be. Consider the concept of "ecosystem services" in the Millennium Ecosystem Assessment Report (MA, 2005), the largest international study to assess the integrity of Earth's ecosystems. Ecosystems, it notes, provide "services" for humans in numerous categories: provisioning services (e.g. food, fuel, building materials), regulating services (e.g. climate, water flow), cultural services (spiritual, aesthetic, recreational), and supporting services (e.g. soil formation, photosynthesis, nutrient cycling). Through the services framework, then, our understandings of and respect for the diversity of life forms and their intricate relationships are pre-empted and coached by a human-centered, resource-oriented view. Trees become timber; plants become food, medicine, or floral arrangements. As Evernden (1985, p. 10) cynically notes, "Nature is a conglomeration of natural resources, a storehouse of materials."

It is not my intention here to dismiss the ecosystem services lens as unhelpful or harmful, but I do want to note that it is just that: a lens, a way of seeing the world. There are, and need to be, other ways of seeing, understanding, and valuing, but if the storylines of all our textbooks, policies, research reports, advertisements, and environmental documentaries reinforce only one view of the world, then the possibilities for more exploration and creativity are radically reduced. When we encounter glimpses through different lenses, listen to different voices, we become challenged and attentive and better able to think and act beyond self-referential monologue.

But the attentiveness and plurality of dialog is relevant to more than just our relationships with the nonhuman or "more-than-human" (Abram, 1999) worlds. Human multiple-voicedness—across countries, cultures and generations, and within communities, schools, and homes—is prerequisite to ethical process. It requires listening, respect, tolerance, and equity. This does not always come easily and many social practices indeed appear designed to replicate and advance a single voice, but Naess (2000, p. 49) reminds us: "If we didn't disagree on anything of importance it means that we are getting into a kind of completely homogenous culture which is a terrible thing, so better to really dislike each other's position than to have no differences." Cheney and Weston (1999) similarly urge us to recognize ethics' pluralistic and discontinuous nature in which dissonance abounds. The work of environmental ethics in mediating this dissonance and seeking creative, generative solutions is the focus of the final touchstone in this essay: environmental ethics as deliberative enquiry.

Environmental Ethics as Deliberative Enquiry

Deliberation is the weighing up of things to achieve reasoned insight and future direction. In the light of the preceding sections' overview of ethics as dissonant, pluralistic, uncertain, and open-ended, it seems inevitable that this final section considers the role of deliberative enquiry in processes of ethical engagement.

There is a natural flow that happens: as we open ourselves up to engage with alternatives, we notice previously hidden, even inconceivable connections, and opportunities then increase for talking more and talking differently with others. And thus we learn more and can do more. Wals (2007) helps to clarify this connection between processes of ethical engagement and processes of learning with his work on dialog, deliberation, and social transformation:

> Social learning often includes a critical analysis of one's own norms, values, interests and constructions of reality (deconstruction), exposure to alternative ones (confrontation), and the construction of new ones (reconstruction). Such a change process is greatly enhanced when the learner is mindful and respectful of other perspectives. Obviously, not all participants in a social learning process display the same amount of initial openness and respect, but as they develop social relationships and mutual respect (social capital), they not only become more open towards ideas alternative to their own, they, as a group,

> also become more resilient and responsive to challenges both from within and from outside. (p. 43)

Wals and Heymann (2004, p. 2) point to a need for "facilitated cultivation of pluralism and conflict in order to create more space for social learning in nonformal and informal settings." They identify conflict not as the destructive, closing-down sort but rather as respectful explorations of divergent norms and values. In such dialogic processes, diversity and contextual depth are celebrated and conflict serves creatively to advance not debilitate transformative social learning.

Central to this process of deliberative ethical enquiry is uncovering and re-examining assumptions that lie at the heart of how we conduct ourselves in the world. How this might be achieved and what education's role might be remains loosely defined and under construction, but perhaps Jickling's (2004) call to make ethics "an everyday activity" is a bold step in a useful direction. When ethics becomes an everyday activity for ordinary people, ethical positions become, unlike static ethical codes, open to reflection, deliberation, and perhaps even revision—ethics becomes an open-ended process of enquiry.

The enquiry seems simultaneously theoretical and practical. Theoretically, environmental ethics as a recent and applied philosophical field finds itself problematizing the very foundations on which it rests. Some commentators suggest that ethical theories have achieved little more than to lead us astray, that patterns of thinking embedded in many established philosophies and theories are actually responsible for many contemporary environmental problems (Ball, 2001; Bonnett, 2004; Cock, 2007; Des Jardins, 2006). Bauman (1993) laments what he considers to be the inadequacy of traditional concepts of moral responsibility:

> Moral responsibility prompts us to care that our children are fed, clad and shod; it cannot offer us much practical advice, however, when faced with numbing images of a depleted, desiccated and overheated planet which our children and the children of our children will inherit and will have to inhabit in the direct or oblique result of our present collective unconcern. Morality which always guided us and still guides us today has powerful, but short hands. It now needs very, very long hands indeed. What chance of growing them? (p. 218)

Cheney and Weston (1999) suggest that these "moral hands" might be extended if we spend more time *questioning* the basic assumptions of the relationships between knowledge and ethical practice. They ask: "What if the world we inhabit arises most fundamentally out of our ethical practice, rather than vice versa?"—a question that reflects Weston's related work on an ethics-based epistemology (Cheney & Weston, 1999). Would our moral hands have more reach if we developed an "etiquette" which, together with knowledge, would "constitute a fluid and ongoing relationship" with other species, cultures, generations (Cheney & Weston, 1999, p. 134)?

If the answer to this question is yes, then we face a question of educational methodology which Wals (2007, p. 3) picks up on when he asks: "How can people become more sensitive to alternative ways of knowing, valuing and doing, and learn from them? How do we create spaces or environments that are conducive to this kind of learning?" He concludes, in an appropriately open-ended way, that "reflexively stumbling toward sustainability" depends on democratic and emancipatory learning processes that allow for unrehearsed dialog. This means that consensus cannot be forced on people, and that the dialog, diversity, antagonisms, and dissonance arising from such social learning processes may be radically indeterminate. As educators, we have access to a "toolkit" of environmental ethics ideas but, in negotiating its diversity, we should be guided by Plumwood's (2002) advice to adopt "methodologies and stances of openness rather than methodologies of closure" (p. 169).

It seems fitting then to close this essay with some considerations of what forms such "methodologies and stances of openness" (Plumwood, 2002, p. 169) might take in schools, colleges, communities, and workplaces. Enacting environmental ethics as processes of open-ended, deliberative enquiry is a tall order. Habituated practices within schools, organizations, and communities often (inadvertently) promote simplistic, technicist, or predetermined learning processes that can hinder reflexive deliberation. Jickling et al. (2006) emphasize the need to disrupt and move beyond superficial or taken-for-granted norms regarding our relationship with the world, and to aim for critical depth. They note that "[c]omplexity is also an ethical issue—the simplification of complex issues is both fraudulent and irresponsible" (p. 19).

Working with teachers in peri-urban and urban South African schools, Lotz-Sisitka and Schudel (2007) observed a tendency toward "empty moralising" (p. 258) in environmentally focused classroom activities. They found that the normative framework of the national curriculum does indeed provide opportunities for contextualized environmental learning (with learners and teachers facing daily challenges such as poverty, high levels of HIV/AIDS, and insecure access to increasingly degraded natural resources). However, in the absence of adequate knowledge resources, appropriate pedagogical practices, language access,[2] community support, and educational resources, the complexity of local environmental issues was often reduced to superficial fragments and an assumption "that what is said to be good will come about" (p. 258). They observed that:

> The "viability" and "reality congruence" of learners' responses were not being probed [by teachers], and solutions seemed to be at the level of short term solutions, or relatively superficial levels of awareness raising. There was also little evidence of "in-depth" engagement with the complexity of the issues. (p. 256)

In the absence of factually grounded, in-depth and coherent environmental learning, classroom discourse becomes characterized by empty moralizing and the circulation of good intentions. In such settings, reflexive deliberation of ethical concerns and the enabling of ethically grounded action are limited.

Even where adequate levels of depth, factual rigor, and critical engagement are achieved, environmental learning falls short of environmental change in the absence of practical alternatives. I have experimented with the "meat-eating versus vegetarianism" debate as a vehicle to stimulate discussion about environmental values and ethics during courses with other environmental educators. The debates are usually intense because meat-eating is central to most southern African cultures and is closely linked to other cultural practices (e.g. ritual slaughtering of livestock during important family events) and to status (cattle are traditionally indicators of wealth). Many students reflected that they had been challenged by the course readings and the environmental impacts of commercial meat production (see, e.g., Worldwatch Institute, 2004). However, as the course had not provided additional action-oriented resources such as recipes for vegetarian dishes, one student reported feeling ill-equipped to transition to a more vegetarian diet, despite his interest. Another student explained that, despite her willingness to eat less meat, she still lived with her parents and could not influence the type of meals in the home. In both cases, the students' ethical intentions could only be translated into ethical actions when certain practical circumstances changed. For educators, this small example illustrates the significance of the constructive tension between knowledge resources *and* action possibilities if people are to take up ethically deliberated practices in meaningful ways. It also reminds us of Sayer's (2000) caution that normative frameworks (such as those explicit or implicit in school and course curricula) are often naively "utopian" and unachievable given the actual conditions in which people find themselves desiring to act. Sayer (2000) explains:

> Unless the normative ideas are related to recognizable or at least imaginable kinds of social organisation and individuals, they are liable to become utopian in the bad sense—imagined felicitous communities populated by ciphers and largely irrelevant to existing and possible alternative societies. (p. 178)

Toward providing a pedagogical route toward practical, achievable ethical action-taking in South African schools, O'Donoghue and Fox (2008, 2009) developed a series of Hand Print[3] resource books. The booklets use a deliberative and action-oriented methodology to respond to sustainability concerns through practical projects such as reducing bath and shower water, clearing invasive weeds, reducing and reusing waste, and propagating indigenous tree seedlings. Each resource book's sequencing of: information-rich, orientating stories; questions to discuss; investigative tools to find out about local sustainability concerns; practical action projects to try out; and ideas to deliberate, provides opportunities for reality congruent learning processes. The ethical dimensions of local environmental issues are thus embedded in authentic, contextualized deliberations, and are less likely to slip into abstracted moralizing.

Situated, practical action-taking is equally significant in contexts of adult and community-based learning, although the scope and complexity of the processes is often greater than in school-based projects. Consider the reflections of a student environmental educator who lives and works in a poor peri-urban community in KwaZulu-Natal, South Africa. When asked about his experience of the environmental values and ethics component of the course he is studying, he responded:

> I have not encountered them [philosophical concepts and terminology] before and they just don't exist in my lifestyle and in my language. . . . When we want to segregate someone from a discussion, so then you can use these words. "Eco-centrism!" [laughs]. And then people start to say "I don't belong here," you know, whereas we need collective effort in terms of alleviating what we are doing on our environment and relating it to what the environment does to us. . . . I can only use these words to meet the requirement of the qualification, but not really at my workplace or at my professional life . . . Because our communities don't need these words. They only need action that would save their lives. (Anonymous, personal communication)

This young environmental educator has a clear and nuanced understanding of his community's development needs and of the power gradients affecting roleplayers' engagement with them. His words reflect a view that the power of environmental learning lies not in the circulation of words and good intentions, but in collaborative processes of care-filled action-taking. His perspectives can alert course developers and tutors to consider more carefully the "authority" with which certain ethical perspectives are put forward in mediational processes and tools such as curricula, text books, community forums, classroom activities, and so on (Olvitt, 2010). The pursuit of open-endness, pluralism, and deliberation can be reflected at the level of pedagogy but subtly contradicted though more tacit expressions of value judgments and moral expectations.

At the start of this essay, I posed the question: *As educators, how can we learn and do more with others in the face of an unprecedented socioecological crisis?* Collectively, a picture begins to emerge of environmental ethics that has work to do beyond asserting what *ought* to be done. Environmental ethics emerges as a dynamic, exploratory field characterized by attentiveness and deliberation, which embraces uncertainty, is never satisfied that a final destination has been reached, and is restless to learn more. Above all, it is action-centered and change-oriented.

But this identity poses a methodological challenge to educators and researchers alike: the challenge of pursuing a process that is true to the form and function of environmental ethics. It suggests that we can *learn* more by listening attentively, by reflecting, and by remaining open to new, different possibilities. And by implication, we will be able to *do* more from better-informed, situated, respectful starting points.

Notes

1. Some readers might disapprove of the apparent eclecticism of this essay which draws on social theorists (such as Bauman), environmental education researchers (such as Wals and Jickling), and environmental philosophers (such as Weston, Næss and Hattingh). (And it is likely that even this loose categorization is problematic for some of the listed authors!) My intention, however, is to reveal useful synergies between these distinct traditions and to encourage (perhaps in line with the environmental pragmatism described later in this essay) more of such interdisciplinary work in the interests of growing the young and emergent field of applied environmental ethics.
2. English is the official language of instruction in South African classrooms beyond Grade 3. Hence, teaching and learning takes place in a second or third language which significantly compromises learners' ability to access or express complex ideas, such as those associated with environmental ethics.
3. The Hand Print concept was developed by the Centre of Environment Education (CEE) in Ahmedabad, India, with various other partners around the world. It promotes individual and collective action-taking to increase our hand print while decreasing our footprint on the planet. (see www.handsforchange.org).

References

Abram, D. (1999). A more-than-human world. In A. Weston (Ed.), *An invitation to environmental philosophy*. New York, NY: Oxford University Press.

Ayair, U. (2009). *Field notes from M.Ed research project*. Grahamstown: Rhodes University.

Ball, T. (2001). New ethics for old? or, how (not) to think about future generations. *Environmental Politics, 10*(1), 89–110.

Bauman, Z. (1993). *Postmodern ethics*. Oxford, UK: Blackwell.

Bauman, Z. (1995). *Life in fragments*. Oxford, UK: Blackwell.

Bauman, Z. (2001). *The individualized society*. Cambridge, UK: Polity Press.

Bonnett, M. (2004). *Retrieving nature: Education for a post-humanist age*. Oxford, UK: Blackwell.

Cheney, J. (1999). The journey home. In A. Weston (Ed.), *An invitation to environmental philosophy* (pp.141–167). New York, NY: Oxford University Press.

Cheney, J., & Weston, A. (1999). Environmental ethics as environmental etiquette: Toward an ethics-based epistemology. *Environmental Ethics, 21*(2), 115–134.

Cock, J. (2007). *The war against ourselves: Nature, power and justice*. Johannesburg: Wits University Press.

Des Jardins, J. (2006). *Environmental ethics: An introduction to environmental philosophy*. Toronto, ON: Thomson Wadsworth.

Evernden, N. (1985). *The natural alien: Humankind and environment*. Toronto, ON: University of Toronto Press.

Hattingh, J. (1999). Finding creativity in the diversity of environmental ethics. *Southern African Journal of Environmental Education, 19*, 68–84.

Jickling, B. (2004). Making ethics an everyday activity: How can we reduce the barriers? *Canadian Journal of Environmental Education, 9*, 11–30.

Jickling, B., Lotz-Sisitka, H., O'Donoghue, R., & Ogbuigwe, A. (2006). *Environmental education, ethics and action: A workbook to get started*. Nairobi: UNEP.

Leopold, A. (1989). *A sand county almanac and sketches here and there*. New York, NY: Oxford University Press (Original work published 1949).

Lotz-Sisitka, H., & Schudel, I. (2007). Exploring the practical adequacy of the normative framework guiding south africa's national curriculum statement. *Environmental Education Research, 13*(2), 245–263.

Millennium Ecosystem Assessment. (2005). *Ecosystems and human well-being: Biodiversity synthesis*. Washington DC: World Resources Institute.

Naess, A. (2000). Deep ecology and education: A conversation with Arne Næss. *Canadian Journal of Environmental Education, 5*, 48–62.

O'Donoghue, R., & Fox, H. (2008). *Have you sequestered your carbon? A hand print resource book*. Howick: Share-Net.

O'Donoghue, R., & Fox, H. (2009). *The secret of the disappearing river. A hand print resource book*. Howick: Share-Net.

Olvitt, L. (2006). *Early experiments in engaging with others around questions of how we might live better in the world: A meaty issue*. Presentation notes for the Environmental Education Research Symposium, St. Paul, Minnesota.

Olvitt, L. (2009). Working with environmental ethics and adult education: Some experiments and reflections from southern Africa. In L. Olvitt, L. Downsborough, & H. Sisitka (Eds.), *Learning in a changing world: Selected papers from the 4th world environmental education congress* (pp. 25–30). EEASA Monograph. Howick: EEASA.

Olvitt, L. (2010). Ethics-oriented learning in environmental education workplaces: An activity theory approach. *Southern African Journal of Environmental Education, 27*, 71–90.

Plumwood, V. (2002). *Environmental culture: The ecological crisis of reason*. New York, NY: Routledge.

Sayer, A. (2000). *Realism and social science*. London, UK: Sage.

Wals, A. (2007). Learning in a changing world and changing in a learning world: Social learning towards sustainability. Background paper in support of the WEEC Panel Session on Social Learning towards a sustainable world. In L. Olvitt, L. Downsborough, & H. Sisitka (Eds.), *Learning in a changing world: Selected papers from the 4th world environmental education congress* (pp. 43–45). EEASA Monograph. Howick: EEASA.

Wals, A., & Heymann, F. (2004). Learning on the edge: Exploring the change potential of conflict in social learning for sustainable living. In A. Wenden (Ed.), *Working toward a culture of peace and social sustainability*. New York, NY: SUNY Press.

Weston, A. (2009). Before environmental ethics. In A. Weston (Ed.), *The incomplete eco-philosopher* (pp. 23–43). Albany, NY: State University of New York (Original work published 1992).

Weston, A. (1994). *Back to earth: Tomorrow's environmentalism*. Philadelphia, PA: Temple University Press.

Weston, A. (Ed.). (1999). *An invitation to environmental philosophy*. New York, NY: Oxford University Press.

Weston, A. (2001). *A 21st century ethical toolbox*. New York, NY: Oxford University Press.

Worldwatch Institute. (2004). *Meat: Now it's not personal*. Retrieved from www.worldwatch.org

Section III

Analyses of Environmental Education Discourses and Policies

Ian Robottom
Deakin University, Australia

Robert B. Stevenson
James Cook University, Australia

Introduction

Environmental education is a relatively new discipline. From its origins in the mid-seventies in the seminal projects sponsored by UNESCO with its environmental education project headed up by Professor Bill Stapp, the field has expanded and matured greatly. The profile (both quantitative and qualitative) of its research activity has changed significantly over this time (Robottom, 2005) to the current stage of complexity represented in this handbook. Similarly, as the status of the field and its research (both methodologically and substantively) has changed, the very language of the field of environment-related education has changed as well.

Within the past two decades, the defining language of the field of environment-related education has changed from "environmental education (EE)" to "education for sustainability (EfS)" or "education for sustainable development (ESD)." If we can describe the more patterned, institutionalized form that language may assume within organizations and institutions as a *discourse*, it may be argued that the field formerly known as "environmental education" is now marked by the rise of the new discourse of sustainability. This shift in discourse (and attendant changes in *policies* as the yet more institutionalized, formal, and influential from of language) is the subject of the essays presented here.

There has been some concern that this change in discourse might be a simple supplanting of one slogan with another without any real change in the educational practices that the slogans qualify (Campbell & Robottom, 2008). There is a challenge for research to demonstrate if and how environment-related educational practice conducted within the frame of the "new" discourse of EfS or ESD is qualitatively different from that conducted within the frame of the former discourse of EE. While this point may be explored with reference to the contents of this Handbook as a whole, it is important to illustrate the ways in which the discourse of the field is understood, appropriated, and interpreted at different levels (internationally, nationally, locally) and different geographical locations, and how interpretations of the discourse of the field may be a function of level and location. To what extent does the discourse of this global field remain (explicitly or implicitly) contested across educational systems? What are some of the factors shaping this contestation?

This section includes seven essays on the topic of policies and discourses in environmental education at international, national, and local levels. The essays are written by authors from Australia, Brazil, Canada, China, Hong Kong, Mexico, South Africa, Taiwan, and the United States, and consequently reflect a diversity of cultural and political influences on education and research. While providing instances of some of the more influential discourses in EE, and the recent shifts in these, the authors also provide insight into ways in which the discourses are or may be appropriated by various constituencies, the importance of social justice issues, and different relationships among discourse, policy, organization, and research. The first four chapters focus on scholarly analyses of different aspects of language and discourse, while the last three focus on research on policies in particular educational sectors (vocational and technical education) or regions (Latin America and China).

Lesley Le Grange's essay deconstructs a component of the Education for Sustainable Development (ESD) discourse that has not commonly been subject to such an analysis. The essay is both an exposition and critical analysis of the concept of needs as a component of sustainable development. It is also concerned with the role and place of sustainability as a relative of environmental education (with some acknowledged controversy over whether "parent" or "child"). In his analysis, the author draws upon an impressive array of scholarship both within and outside the field of environmental studies. In particular, the author refers to literature that addresses the problematic nature of linking sustainability to both present and future needs—a root but contestable concept of the Brundtland Commission's Report of 1972. In addition the author is sensitive to the moral dimensions of the concept of need, and distinguishes it from what he calls the humanitarian dimension. His analysis of the discourse of "needs" suggests that scholarly work could explore further how these specific views of needs may advance our thinking about EE/ESD. At the very least, Le Grange's analysis illustrates that the discourses of environment-related education are susceptible to further critical analysis.

Berryman and Sauvé's essay opens with an argument for recognizing the centrality of "language" (and its more institutionalized variant, "discourse") in environment-related education, and draws on a wide range of literature to demonstrate some of the key influential elements of contemporary environment-related educational discourse. The authors, in asserting the importance of language, gradually engage in a more critical examination of the field, indicating instances where the once-dominant discourses of environment-related education came to be challenged by other discourses in a progressively more open society of scholarly critique. The essay builds on these considerations to mount a critique of the effect of the now-dominant environment-related discourse of Education for Sustainable Development. Finally, they explore prospects for alternative languages by drawing on four sources of inspiration, from Ricoeur's philosophical essays on hermeneutics to Berger and Luckmann's sociology of knowledge and Dorothy Smith's institutional ethnography.

Bob Stevenson begins by examining recent developments in educational policy scholarship, specifically the emergence of critical sociological and sociocultural approaches, and what they reveal about recent shifts in policy processes, particularly as a result of globalization. He then draws on these perspectives to analyze international and national policy processes and discourses in environmental education (EE) and education for sustainable development or sustainability (ESD/EfS). His analysis reveals the underlying environmental/sustainability and educational ideologies that are involved in shaping policy formation, including the struggles and negotiations around different values and ideological interests. After highlighting the converging and competing policy discourses in national contexts, Stevenson concludes by arguing that policy discourses and processes should be democratized and a more dialectic relationship between policy and practice should be constructed.

Ian Robottom discusses some trends in recent discourses in environment-related educational work, arguing that the field is marked by a policy context based on shifting discourses at national and international levels and a language that is slogan-like in effect, creating the conditions for changeless reform. The essay describes some instances of environment-related educational practice, considering these in relation to the recent discourse shift in the field, and presents an approach to professional self-development that takes into account the issues associated with the relationship of discourse to practice within particular educational contexts.

Arenas and Londoño focus on sustainability educational initiatives within vocational education in secondary schools. They address the intersection of the ESD discourse with that of vocational and technical education (VTE) in asserting that vocational educators are becoming increasingly interested in addressing the need to protect the natural environment. They speak of the difficulties in moving from agreement at the level of discourse to institutional policy, organization, and, above all, educational practice. They make the point that moving from agreement in language to effective informed action is less a technical challenge and more a "political" matter of engaging professional contextual factors—especially those with the potential to oppress, shape and constrain classroom practice. Their chapter highlights the limited attention given to the place of sustainability education in vocational and technical education, although its place in universities has been widely conceptualized and examined in and across many geographical and cultural contexts. In contrast, however, the current and potential contribution of sustainability education to the preparation of individuals for technical and trade-related careers has been under-theorized and under-studied.

The contribution of Edgar González Gaudiano and Leonir Lorenzetti focuses on the Latin American region, with reference to Mexico, Brazil, Colombia, Cuba, Chile, Ecuador, and Venezuela. They provide a regional overview of newly evolving policies, discourses, and practices in these developing or newly industrialized countries. Of these, Colombia has adopted an interesting approach to environmental education sponsored by government but retaining a fairly critical academic edge (see Torres, 2010), and Mexico has significant environment-related programs producing and supporting doctoral candidates focusing on EE/ESD. Despite the emergence of ESD as a new regional discourse, the authors report limited progress in other Latin American countries, concluding with an asserted need for major institutional change before more rapid development is possible.

Chinese researchers Lee, Wang, and Yang focus on three Chinese communities (mainland China, Taiwan, and Hong Kong) in which there has been little EE research, enabling the authors to draw from research in other contexts to examine discourse and policy issues. Their essay presents accounts of EE policies and practices in three Chinese community states. The approach identifies a number of interesting issues such as the effect of journal rankings on developments in the field. The authors also point to some absences and silences in EE in the countries embraced in their review.

This diverse range of essays, related through their interest in discourse and policy of the field, illustrate the kind of contestation existing at the level of language within EE/ESD, exposing such diverse issues are:

- the importance of language and the dialectic relationship between policy and practice in particular cultures;
- the importance of context, and within context the effect of cultural, economic, political, environmental, and social justice issues on EE policies and practices;
- the necessity of multiple languages to express encounters and relationships with the environment, given the unlikelihood of an all-embracing metalanguage;
- the ill-defined intergenerational equity "needs" principle that is central to the ambiguous and contested concept of sustainable development and therefore also to ESD;
- the scarcity of sustainability education in policy and practice in the preparation of individuals for technical and trade-related careers in vocational and technical colleges;
- the contribution of the increasingly extensive research on regional perspectives by Latin American and Chinese researchers and research students working or studying in universities in other countries (and the relationship of this with research conducted by scholars actually based in the Latin American and Chinese countries); and
- the potential to inform EE research of environment-related research that is conducted within a social science context that may not be commonly recognized as EE. For example, community forestry management is pushing the boundaries on social learning and participatory extension projects. Such research is not usually viewed as EE but may be part of a broader conception of EE research that encompasses efforts to understand the role of learning and human agency in natural resource management and human-environment relations more generally.

These issues and others identified in the following chapters suggest some future directions for research that might include such questions as:

- What might the theoretical and methodological frameworks employed in environment-related social science fields offer EE research?
- In what ways is the discourse of ESD appropriated in support of a disparate range of environment-related activity and how does this appropriation shape ESD practices?
- What can alternative conceptions of "needs" and alternative languages and discourses contribute to EE/ESD practice and research?
- What is the influence of other common, often competing, educational discourses on EE practices, and how might coherent and compatible policies and practices be established across these discourses?
- Given the historical tendency of discourses in the field of environment-related education to become slogan-like in nature, how may a critical perspective on the relationships among discourse, organization, and practice mitigate the scenario?

These last points relate to a challenge identified at the international launch of the Decade of Education for Sustainable Development in New York in March 2005, by the UNESCO Director General Koichiro Matsuura who addressed specifically the issues of slogans in educational reform:

> The ultimate goal of the decade is that education for sustainable development is more than just a slogan. It must be a concrete reality for all of us—individuals, organizations, governments—in all our daily decisions and actions, so as to promise a sustainable planet and a safer world to our children, our grandchildren and their descendants . . . Education will have to change so that it addresses the social, economic, cultural and environmental problems that we face in the twenty-first century (UNESCO, 2005, p. 2).

As stated earlier, the challenge requires further research on these shifting discourses, but not just in isolation; the relationship among discourse, organization, and educational policy, and practice remains a highly justified arena of research in light of the essays in this section.

References

Campbell, C., & Robottom, I. (2008). What's in a name? Environmental education and educational for sustainable development as slogans. In E. Gonzalez-Gaudiano & M. Peters (Eds.), *Environmental education: Meaning and constitution. A handbook*. Rotterdam, The Netherlands: Sense.

Robottom, I. (2005). Critical environmental education research: Re-engaging the debate. *Canadian Journal of Environmental Education, 10*, 62–78.

Torres, C. M. (Ed.). (2010). *Investigacion y educacion*. Medellin, Colombia: Ambiental. Corantioquia.

UNESCO. (2005). *Initiating the United Nations decade of education for sustainable development in Australia*. Report of a National Symposium, Australian National Commission for UNESCO, Paris.

13

The Politics of Needs and Sustainability Education

Lesley Le Grange
Stellenbosch University, South Africa

Introduction

The term "sustainability" was first used in eighteenth-century German forestry management practices. However, according to the 1986 supplement of the *Oxford English Dictionary*, in English the use of the word "sustainability" dates from only 1972. The construct "sustainability" was introduced into popular discourse by the Brundtland Commission Report (World Commission on Environment and Development [WCED], 1987). It defined sustainable development as "development which meets the needs of the present without compromising the ability of future generations to meet their own needs" (WCED, 1987). As an adjective, sustainability has been combined with varied entities such as yields of renewable natural resources, crop yields, agricultural practices, development, ecosystems, communities, societies, living, and even the entire planet (Wensveen, 2001). In this chapter I shall use the terms "sustainability" and "sustainable development" interchangeably.

Sustainable development is a contested term. Some criticisms leveled against the term are: it has internal contradictions, it manifests epistemological difficulties, it reinforces a problematic anthropocentric stance, it has great appeal as a political slogan, it is a euphemism for unbridled economic growth, and it does not take into consideration the asymmetrical relation between present and future generations (for a more detailed discussion see Bonnett, 1999, 2002; Goodwin, 1999; Le Grange, 2007; Stables & Scott, 2002). Furthermore, environmental educationists have widely differing views on the relationship, between sustainability and environmental education. As Sauvé (1999) points out, for some people sustainable development is the ultimate goal of environmental education, hence the term *environmental education "for" sustainable development* (EEFSD). For others, sustainable development encompasses specific objectives that should be added to those of environmental education, hence the expression *education for environment "and" sustainable development* (EFE & SD). For others still, environmental education inherently includes *education for sustainable development*, and thus the use of both terms is tautological.

Moreover, some scholars have troubled the very idea of "educating for" sustainability, arguing that such an approach suggests an instrumentalist view of education. Some have gone as far as to say that the approach is anti-educational and tantamount to indoctrination. As Jickling (1997, p. 95) writes:

> When we talk about "education for" anything we imply that education must strive to be "for" something external to education itself. We may argue, in an open sense, in favour of education for citizenship or character development. However, as prescriptions become more specific, interpretations of education become more loaded and more problematic . . . .

I shall not discuss the contested nature of *education for sustainability* in detail because this has been reviewed extensively elsewhere (see, e.g., Campbell & Robottom, 2008; Huckle, 1999; Jickling, 1995, 1997; Jickling & Spork, 1998; Le Grange, 2008; Sauvé, 1999). My interest is rather in focusing on the Brundtland Commission Report's definition of sustainable development and in particular on the notion of needs, which is a largely unexplored area in environmental education discourses. The Brundtland Commission Report's definition is popularly invoked in policies of supranational organizations and national governments, but it is rarely discussed critically.

In exploring the concept of needs in relation to sustainable development, I shall divide my chapter is into three main sections. First I shall discuss the notion of needs in relation to environmental theory and justice theory. Secondly, I shall discuss the emergence of needs discourses in late capitalist societies and in so doing shift the angle of vision onto the politics of needs. Thirdly, I shall attempt to show how the emergence of needs discourses in late capitalist societies might relate to the Brundtland Commission's definition of sustainable development. Fourthly, I shall examine the implications that "needs talk" (in relation to sustainable development) might have for education and explore whether the needs lens might offer an appropriate point of departure for students to engage matters related to sustainable development.

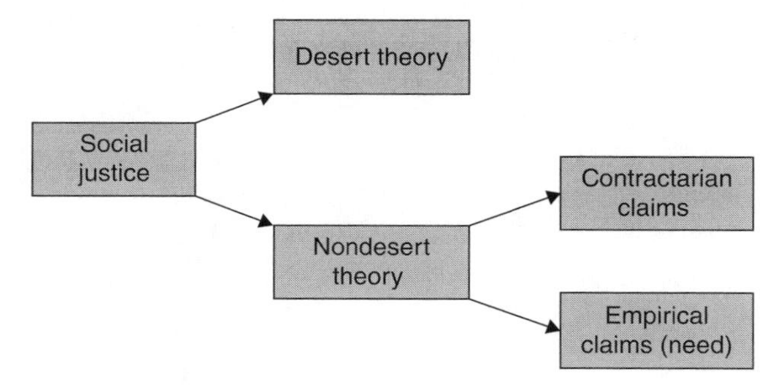

Figure 13.1 Need and social justice theory.

Locating Need Within Environmental Theory and Justice Theory

As a first level of analysis I begin with a brief discussion of how needs might be located in environmental theory and justice theory. The reason for doing so is that the term "sustainable development" is informed by both environmental theory (the interest in conserving the earth's resources) and justice theory (a concern for the needs of present and future generations). Vincent (1998) offers a useful typology of both environmental theory and justice theory to inform our discussion. I begin with justice theory—with social justice or distributive justice theory in particular as it is pertinent to the discussion here. Distributive justice/social justice might be understood in at least two senses: the fair allocation of burdens/benefits of society; to each according to what is their due. Concerning the latter understanding, Vincent (1998, pp. 124–125) suggests that there are two broad categories of theory that determine what is due: desert theory and nondesert theory. Desert theory holds that a person is rewarded based on what they deserve, whereas nondesert theories posit that what is due is based on either rational reasoning (contractarian claims) or on need (empirical claims) (Figure 13.1). Contractarian claims can either be determined through mutual bargaining (proponents of this position are, for example, Buchanian and Gauthier) or through rational agreement (proponents of this position are, for example, Rawls, Barry, and Scanlon). In short, within justice theory "need" is an empirical claim forming part of nondesert theory. Matters of justice, however, always relate to human beings—it is about what humans deserve or need. Mutual bargaining or rational agreement takes place between human beings. It is only human conduct that can be called just or unjust—nature cannot be just or unjust. What this points to is that the invocation of the notion of "needs" in the Bruntland Commission's definition of sustainable development makes the concept decidedly anthropocentric—it refers to the needs of present humans and future generations of humans. But is anthropocentrism antagonistic or integral to environmental theory?

Vincent (1998, pp. 122–123) argues that environmental theory comprises three categories: pliant anthropocentrism, intermediate axiology, and ecocentrism (Figure 13.2). These three categories are distinguishable based on the extent to which nature is the criterion of value. The pliant anthropocentric position holds that nature has a quasi-instrumental character, that is, that the biophysical world (nonhuman nature) has value for humans. This does not mean that nature is low on the value priority scale, but rather that nature without human beings is valueless. On the other extreme, the ecocentric view holds that the whole ecosphere is the locus of value. In other words, the whole ecosphere has intrinsic value and should not be used instrumentally to serve human ends. This position is held by deep ecologists such as Arne Naess and Gaianists such as James Lovelock. Between pliant anthropocentrism and ecocentrism there is an intermediate category which does not fully accept either anthropocentrism or ecocentrism. Two categories of intermediate axiology can be distinguished: moral extensionism and reluctant holism. Moral extensionists such as Peter Singer and Tom Regan argue that rights enjoyed by human beings should be extended to all sentient beings—animals therefore also have rights. Reluctant holists such as Baird Callicot and Homes Rolston go further than moral extensionists by extending value to, for example, whole biotic communities (see Vincent, 1998, p. 124). However, they do not go as far as to suggest that mountains or rocks can think. From Vincent's typology we might conclude that all environmental theories place the value of nature centrally. In other words, environmental theories extend value beyond only human interests. This is the reason why deep or hard anthropocentricism (nature has no value, neither intrinsic nor instrumental) would not be included in environmental theory. However, the fact that "need" is by definition anthropocentric (at least according to justice theory) does not necessarily make it antagonistic to environmental theory because environmental theory does incorporate pliant/weak anthropocentrism. One might therefore argue that the term "sustainable development" (at least the Brundtland Report definition of it) is informed by aspects of both justice theory and environmental theory—nonhuman nature ought to be conserved and the needs of present and future generations ought to be met. However, the anthropocentric leaning of the Brundtland Commission's definition of sustainability, with its emphasis on human needs, might dominate the interest in conserving nonhuman nature if the focus on needs in this definition is understood in the context of the emergence of needs talk as a major vocabulary in political discourses in welfare states. I now turn to this discussion.

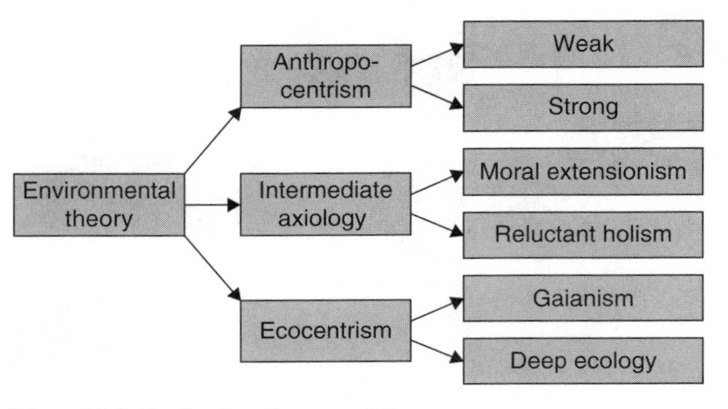

Figure 13.2 Need and environmental theory.

The Emergence of Needs Talk as a Major Vocabulary in Political Discourse

In my view, the angle of vision of much of the critique of sustainable development and education evident in the literature might need to shift to a focus on the analysis of needs. "Needs talk" has been given scant attention in the proliferation of the literature on education and sustainable development over the past twenty-five years, despite the fact that word "needs" features strongly in the most widely quoted definition of sustainable development. With a few exceptions (e.g., Hamilton, 2003; Miller, 1999), even in political philosophy there is little theorization of needs, and this is so despite the fact that need has become "institutionalised as a major vocabulary of political discourse" (Fraser, 1993, p. 162).

In the Brundtland Commission Report (WCED, 1987) two sets of needs are mentioned: the "needs of present generations" and the "needs of future generations." Before referring to these, I shall first focus on what is meant by needs by generating questions on the concept. For example, are needs the distribution of satisfactions; a principle of social justice; or a variant of desires and wants? What is meant by "needs?" Answers to these questions (and many other related questions) are complex. For one thing, the term "distribution of satisfactions" would have to be interrogated. For example, what is meant by "satisfactions": are they individual or group satisfactions, and how can competing satisfactions be met if there are not sufficient resources available? Many needs may qualify on the basis of principles of humanitarianism but not necessarily principles of social justice. Distinguishing when needs claims are claims of justice or claims of humanity/benevolence becomes crucial—the boundaries between the two are often blurred (see Miller, 1999 for a detailed discussion). Needs could be distinguished from desires and wants in that if the former are not met, the individual or group suffers. But this raises the question of what constitutes harm. There are many more questions concerning what is meant by need(s) that could be generated. However, suffice it to say that questions such as these belie the apparent simplicity of the definition of sustainability as "development which meets the needs of the present without compromising the ability of future generations to meet their own needs" (WCED, 1987). I shall return to particular difficulties with respect to meeting the "needs of future generations."

Fraser (1993), however, introduces another dimension to the analysis of needs which focuses on the politics of needs. Talk about needs has not always been central to Western political culture. In the past this has often been relegated to the margins and considered antithetical to politics. So why has talk about needs become so prominent in the political culture of contemporary welfare states societies? Fraser (1993, p. 162) raises several other questions of which I shall mention two. Firstly, does the emergence of the needs idiom presage an extension of the political sphere or, rather, a colonization of that sphere by newer modes of power and social control? Secondly, what are the varieties of needs talk and how do they interact polemically with one another? In responding to these questions, Fraser does not offer definitive answers but rather outlines an approach to thinking about such questions. I shall elaborate on this, arguing that her approach could provide a more nuanced understanding of needs (talk) in relation to sustainable development.

The central focus of Fraser's inquiry is not on needs but rather on discourses about needs and in so doing she shifts the angle of vision to the politics of needs. Put another way, she shifts the focus from the usual understanding of needs, which pertains to the distribution of satisfactions, to the politics of needs interpretation. She sharpens the focus on the contextual and contested character of needs claims so that the interpretation of people's needs is not seen as simply given and unproblematic; the politics of needs concerns a struggle over needs. Fraser (1993, p. 164) goes on to suggest that the politics of needs comprises three moments that are analytically distinct but interrelated in practice. I summarize them as follows:

1. The struggle to validate a given need as a matter of legitimate political concern or to enclave it as a nonpolitical matter;
2. The struggle to interpret the need—the struggle for the power to define it and to determine what would satisfy it;
3. The struggle to satisfy the need—the struggle to secure or withhold provision.

Fraser's inquiry into the politics of needs led to a social discourse model which maps three major kinds of needs discourses in late capitalist societies: "oppositional" discourses, "reprivatization" discourses, and "expert" needs discourses. Oppositional discourses arise when needs are politicized from the bottom, which lead to the establishment of new social identities on the part of subordinated groups. These discourses arise when needs become politicized, such as when women, people of color, or workers contest the subordinate identities and roles they have been assigned or that they have embraced themselves. Among other things, oppositional discourses create new discourse publics and new vocabularies and forms of address. Fraser (1993) points out, for example, that the wave of feminist ferment established terms such as "sexism," "sexual harassment," "date rape," and "wife battering." In her view, reprivatization discourses have emerged in response to the oppositional discourses. They articulate entrenched needs that would previously have gone without saying. Institutionally, "reprivatization" initiatives are aimed at dismantling or cutting back social welfare services, selling off nationalized assets, and deregulating private "enterprise"; discursively it means depoliticization. Advocates of reprivatization may insist that "wife battering" is a domestic rather than a political issue.

For Fraser (1993) expert needs discourses link popular movements to the state. They are best understood in the context of "social problem solving," institution building and professional class formation. They are closely connected with institutions of knowledge production and utilization, and they include social science discourses generated in universities and "think tanks," legal discourses generated in judicial institutions, journals, professional associations, and so on. Expert discourses tend to be restricted to specialized public discourses associated with professional class formation, institution building, and "social problem solving." However, sometimes expert rhetorics are disseminated to a wider spectrum of educated lay persons—expert public discourses sometimes acquire a certain porous quality—and become the bridge discourses linking loosely organized social movements with social states. It is the polemical interaction of these three kinds of needs talk that structures the politics of needs in late capitalist societies. The interaction between these three kinds of needs talk could provide a basis for reflecting on the idea of sustainable development, particularly in view of the definition of sustainable development which appeared in the Brundtland Commission Report.

Needs Talk and Sustainable Development

Fien's (1993) typology provides a useful starting point for critical reflection on sustainable development in terms of *people to nature* values/principles and *people to people* values (Table 13.1), as does his placement of needs as a value/principle of social justice. His two broad categories—ecological sustainability and social justice—are in tension with one another, or they could be perceived to be. A primary focus on ecological sustainability may be described as biocentric/ecocentric whereas viewing social justice as central to sustainable development could be described as anthropocentric. People who argue from liberal and Gaianist (influenced by deep ecological perspectives) positions favor values related to ecological sustainability, but view values associated with social justice as being anthropocentric (human-centered). Those who take up critical and/or ecosocialist positions emphazise issues related to social justice in preference to those associated with ecological sustainability. The first group, which favors values related to ecological sustainability, might extend the notion of needs to nonhuman nature and refer to the needs of nature. Some of them would restrict needs to sentient beings only, arguing that animals, which have rights, also have needs. As a whole, Gaianists and deep ecologists might contest the idea that "needs" are endemic to human beings exclusively.

But the value/idea of "basic human needs" is contested and controversial. In her cogent argument Fraser (1993) presents the view that thin needs are uncontroversial, but when one descends to a lower level of generality the needs claims become controversial. For example, let us assume that shelter is a basic human need in nontropical climates—at this level of generality such a claim would be uncontroversial. However, as soon as we become more specific and ask what homeless people need in order to be sheltered from the cold, the needs claim becomes more controversial. Fraser (1993, p. 163) illustrates this when she asks: ". . . [should they] sleep undisturbed next to a hot-air vent on a street corner, in a subway tunnel or bus terminal . . . a bed in a temporary shelter . . . a permanent home? And we can go on to proliferate such questions—in doing so, we will proliferate controversy." Furthermore, "needs" is not necessarily a value/principle of social justice. Miller (1999, p. 223) argues that an individual's satisfaction or relief is based on the moral imperative of benevolence or humanity rather than justice. Justice, he argues, is concerned with the fair allocation of resources

TABLE 13.1
Core Values Central in Sustainable Development

People and Nature: Ecological Sustainability	People and People: Social Justice Principle
Interdependence: people are part of nature and are dependent on it.	*Basic human needs*: the needs of all individuals and societies should be met, within the constraints of the planet's resources.
Biodiversity: every life form warrants respect independently of its perceived worth to humans.	*Inter-generational equity*: future generations should be left with a planet that has at least similar benefits to those enjoyed by present generations.
Living lightly on the earth: all persons should use biophysical resources carefully and restore degraded ecosystems.	*Human rights*: all persons should enjoy the fundamental freedoms of conscience, religion, expression, etc.
Interspecies equity: people should treat all life forms decently and protect them from harm.	*Participation*: all persons in communities should be empowered to exercise responsibility for their own lives.

Adapted from Fien (1993)

to meet satisfactions. Also, much of the discussion here focuses on the needs of present generations. Thinking about the needs of future generations further complicates matters. Can and should present generations determine the needs of future generations? How would they determine what such needs might be? Yet at the same time the decisions that present generations make could place future generations in very vulnerable/needy positions. In short, I reiterate the complex nature of needs. But the discussion should be taken further and so I turn to the politics of needs, that is, to discuss some needs discourses and how they might relate to sustainable development.

Oppositional discourses in relation to sustainable development are evident in the contemporary era. For example, the NGO forum that met at the Rio de Janeiro Earth Summit in 1992 formulated alternative principles on sustainable development to those of governments. As far as new vocabularies are concerned, "environmental justice" is an example of a new term constructed within oppositional discourses—the environmental justice movement is led mainly by women of color in the United States. More recently we have also witnessed oppositional voices from what is referred to as the new social movements. As Irwin (2003, p. 329) writes:

> Contemporary anti-globalisation protest is a remarkable "rhizome" of radical groups, upstanding citizens, charities, long standing emancipatory organisations, environmental groups, right-wing organisations, anarchists, communists and so forth, who have all found a common thread which weaves together their disgust as the solidified locus of financial, discursive and policy flows which have coagulated in supra-national organisation such as the WTO [World Trade Organisation], World Bank, IMF [International Monetary Fund], and various events such as the recent United Nations Earth Summit at Johannesburg.

Referring to reprivatization discourses, Irwin (2003, p. 329) argues that most of the nations of the world currently adhere to neoliberal policies of privatization and devolvement promoted by the World Bank, IMF, and WTO. As far as expert discourses are concerned, over the past two decades we have witnessed several conferences on sustainable development as well as education for sustainable development; journals on sustainable development (e.g., *Journal of Sustainable Development in Higher Education*) have been established and several special issues of journals have been published on education and sustainable development (e.g., *The Trumpeter*, *Philosophy and Theory of Education*, *Environmental Education Research*); and intergovernmental conventions such as the World Summit in Johannesburg (2002) have been held. Fraser's social discourse model is pertinent to sustainable development. The struggle over needs is evident in the oppositional discourses of the antiglobalization movements, which struggle against reprivatization discourses produced by organizations such as the WTO, IMF, and the World Bank as well as national governments. Expert discourses influence both oppositional and reprivatization discourses and are also influenced by them. However, the issue here is what implications these may have for education.

Some Implications for Education

As I said earlier, need is a complex construct, easily invoked in political speeches and social policies. However, it is a controversial and contested idea. In this chapter I briefly discussed why need is such a complex issue and considered the importance of understanding the politics of needs, that is, how needs are constructed or produced within different political discourses. Need in this instance has been constructed within what might be viewed as an environmental discourse—in the Brundtland Commission definition of sustainable development—which has penetrated a myriad social and education policies across the globe and will continue to do so during the UNO decade of education for sustainable development (2005–2014). However, need in relation to sustainable development should be understood as part of a political struggle over needs in contemporary society. As Foucault (1977, p. 26) writes: "need is also a political instrument, meticulously prepared, calculated, and used." So we can't simply invoke sustainable development in education programs without opening up its complexity and its political dimension to students. The complexity of the construct "need" ("needs"), including its production in different political discourses, as well as the anthropocentric nature of the term "sustainable development" raises questions for consideration within the field of environmental education. If sustainability is to be a focus of environmental education, and given environmental education's interest in promoting caring for nonhuman nature (or seeing intrinsic value in nonhuman nature), we might ask what might be the lines of escape from the dominance of anthropocentrism? Can we extend the notion of needs at least to sentient beings, as in the case of rights (animal rights), or is it—as Vincent (1998) suggests—a category mistake to ascribe animistic or anthropocentric attributes to animals or biotic communities? Or might we benefit from the knowledges of indigenous communities who view the cosmos holistically rather than atomistically?

At a time of increasing concern about the planet's biophysical base that is rapidly being eroded, the perennial, existential question of *how should we live* emerges strongly. Related to this are questions of how and what should be learned and taught in educational institutions. In response to these questions, notions such as sustainable development and *education for sustainable development* have great appeal, presumably because they purport to serve multiple and often disparate aspirations. However, as pointed out, sustainable development generally and "needs" more specifically are complex and contested ideas. It is very difficult to determine what is meant by needs for needs form part of, and are constituted by, much of the political struggles that prevail in contemporary societies. Viewing sustainable development through the needs lens helps to make sense of the complex, contested, and controversial nature of the term. Such a perspective could provide students with the opportunity to learn about what Wals and Jickling (2002, p. 123) describe as "a highly relevant, controversial, emotionally charged, and debatable topic at the crossroads of science, technology, and society."

In South Africa's new National Curriculum Statement (NCS) sustainable development forms part of the knowledge foci/content prescriptions of subjects such as Geography and Life Orientation. *Environmental and social justice* is also one of principles which underpin the NCS for Further Education and Training (FET). Sustainable development will therefore form part of school learning programs. However, because of its popular appeal there is a danger that it will be simplistically defined and formulated in terms of narrowly defined outcomes. It is vital that sustainable development should form part of classroom conversations. However, these conversations must recognize and initiate critical debate on the complex, controversial, and contested nature of the term. For this to happen, teachers might have to understand the complexity of sustainable development and engage critically with the construct. They might also have to understand learning outcomes as being dynamic and not static. Looking through the "needs" lens makes possible an appreciation of the complex nature of the term sustainable development and, furthermore, reference to needs in classrooms may enable better understanding of sustainable development because "needs" is a term learners might be able to relate to more easily. This raises another question for further exploration: if learners relate easily to the concept of need, can caring for the self (a concern with personal needs) serve as a catalyst for caring for others and the biophysical world, as Guattari (2001) seems to suggest in his book *The Three Ecologies*?

The introduction of aspects of sustainable development into South African education coincides with, and in a sense is integral to, the curriculum and school reform currently taking place. However, Popkewitz (1991, p. 244) cautions against accepting reform as automatically truth producing and progressive, referring to what he terms the "dangers of an epistemology of progress." Dominant discourses on sustainable development (produced through supranational bodies) and curriculum reforms (influenced by globalization), such as outcome-based education, are underpinned by an "epistemology of progress." Progress stories embedded in both global discourses on sustainability and curriculum reforms threaten to narrow democracy by thwarting attempts to achieve social justice and develop a critical citizenry that is reflexive in a rapidly growing consumerist society. Shifting the angle of vision on sustainability to a focus on needs could shift the understanding of reform (social, political, and educational) as "truth producing and progressive" to understanding it as an "object of social relations" (Popkewitz, 1991, p. 244). This opens up the possibility for challenging dominant forms of power (as captured in Western discourses of progress) by empowering persons (including teachers) to act in concert in attempts to change the world (or their world) as captured in Nyberg's book *Power over Power* (1981). Such a move might provide pedagogical space for critiquing/deconstructing education programs embedded in Western (enlightenment) progress stories and open up opportunities for alternative actions—actions that are empowering.

References

Bonnett, M. (1999). Education for sustainable development: A coherent philosophy for environmental education? *Cambridge Journal of Education, 29*(3), 313–324.

Bonnett, M. (2002). Sustainability as a frame of mind-and how to develop it. *The Trumpeter, 18*(1), 1–9.

Campbell, C., & Robottom, R. (2008). What's in a name? Environmental education and education for sustainable development as slogans. In E. Gonzalez-Gaudiano & M. A. Peters (Eds.), *Environmental education: Identity, politics and citizenship* (pp. 195–206). Rotterdam, The Netherlands: Sense publishers.

Fien, J. (1993). *Education for the environment: Critical curriculum theorising and environmental education*. Geelong, VIC: Deakin University Press.

Foucault, M. (1977). *Discipline and punish: The birth of the prison* (A. Sheridan, Trans.). England, UK: Penguin Books.

Fraser, N. (1993). *Unruly practices: Power, discourse and gender in contemporary social theory*. Cambridge, UK: Polity Press.

Goodwin, R. (1999). The sustainability ethic: Political, not just moral. *Journal of Applied Philosophy, 16*(3), 247–254.

Guattari, F. (2001). *The three ecologies* (I. Pindar & P. Sutton, Trans.). London, UK: The Athlone Press.

Hamilton, L. (2003). *The political philosophy of needs*. Cambridge, UK: Cambridge University Press.

Huckle, J. (1999). Locating environmental education between modern capitalism and postmodern socialism: A reply to Lucie Sauvé. *Canadian Journal of Environmental Education, 4*, 36–45.

Irwin, R. (2003). *Education and Guattari's philosophy of the environment*. Paper presented at the annual conference of the Philosophy of Education Society of Great Britain. Oxford, UK: New College, Oxford University.

Jickling, B. (1995). Sheep, shepherds, or lost? *Environmental Communicator, 26*(6), 12–13.

Jickling, B. (1997). If environmental education is to make sense for teachers, we had better rethink how we define it. *Canadian Journal of Environmental Education, 2*, 86–103.

Jickling, B., & Spork, H. (1998). Education for the environment: A critique. *Environmental Education Research, 4*(4), 309–327.

Le Grange, L. (2007). An analysis of 'needs talk' in relation to sustainable development and education. *Journal of Education, 41*, 1–14.

Le Grange, L. (2008). Towards a language of probability for sustainability education (South) Africa. In E. Gonzalez-Gaudiano & M. A. Peters (Eds.), *Environmental education: Identity, politics and citizenship* (pp. 208–217). Rotterdam, The Netherlands: Sense Publishers.

Miller, D. (1999). *Principles of social justice*. London, UK: Harvard University Press.

Nyberg, D. (1981). *Power over power*. Ithaca, NY and London, UK: Cornell University Press.

Popkewitz, T. (1991). *A political sociology of educational reform: Power/knowledge in teaching, teacher education, and research*. New York, NY: Teachers College Press.

Sauvé, L. (1999). Environmental education between modernity and post-modernity: Searching for an intergrating educational framework. *Canadian Journal of Environmental Education, 4*, 9–35.

Stables, A., & Scott, W. (2002). The quest for holism in education for sustainable development. *Environmental Education Research, 8*(1), 53–60.

Vincent, A. (1998). Is environmental justice a misnomer? In D. Boucher & P. Kelly (Eds.), *Social justice: From Hume to Walzer* (pp.120–140). London, UK: Routledge.

Wals, A., & Jickling, B. (2002). "Sustainability talk" in higher education: from doublethink and newspeak to critical thinking and meaningful learning. *Higher Education Policy, 15*, 121–131.

Wensveen, L. (2001). Ecosystem sustainability as a criterion for genuine virtue. *Environmental Ethics, 23*(3), 227–241.

World Commission on Environment and Development. (1987). *Our common future.* Oxford, UK: Oxford University Press.

14

Languages and Discourses of Education, Environment, and Sustainable Development

Tom Berryman and Lucie Sauvé
Université du Québec à Montréal, Canada

Language and discourse are crucial dimensions of environmental education though they are not so frequently recognized as such. One needs only to look at the various research journals, textbooks, guidelines, policy documents, and this handbook, to realize the centrality of language and discourse in education. Browsing research journals in education and environmental education reveal the weight of language as part of a research problematic or as an object of research, directly as "language" or "discourse," or through associated themes such as "voice," "story," "metaphor," "rhetoric," "literacy," "text," "writing," "reading," "defining" and "debating." Language and discourse are also central in research methodologies through "literature review," "interview research," "narrative research" or through the study of "discussions" or "conversations," through "content analysis," "discourse analysis," "linguistics," "hermeneutics" and "language research." In various settings, exchanges and communications between educators, learners, and researchers are mediated through languages and our understandings are also shaped through them. However central to our experience, language, and discourse are not spontaneous givens. They are historical and contextual constructs that we learn, being shaped and educated by them and educating through them. They also change, some rising to ascendancy, some losing ground, some exerting more power than others, and some being culturally or politically silenced or amplified. Finally, they play crucial roles in our ecological encounters and experiences, especially when we try to name, express, or communicate the various components and relations of such encounters. All of the above stress the immense importance and complex intricacies of language and discourse in our lives. This chapter thus examines successively the languages of the environment, of education, of environmental education, and sustainable development. It opens with a story about the use of a certain language to legitimize a position and concludes with some prospects for reflexivity and further research on issues of language and discourse in environmental education.

"The Same Old Story?"

Environmental education arising conceptually, formally, and institutionally in the end of the 1960s and the early 1970s, we can now interpret more easily, with the distance of years, some of the educational and environmental languages and the dominant discourses that produced it and that environmental educators reproduced or adapted, and sometimes contested or strived to emancipate from.

Before environmental education was constructed, nature education, outdoor education and conservation education were some of the major educational voices about relationships with the surrounding world. One of the key components of the dominant discourse striving to legitimize environmental education was inclusiveness and broadness (with regards to the complexity of environmental realities and problems). It was then articulated in institutional discourses, notably within UNESCO, in a similar fashion that we are now hearing and reading with the discourse of education for sustainable development. Matthew Brennan, who wrote "*Total Education for the Total Environment*" in 1964 and worked in the interstices between conservation education and environmental education, expressed these tensions sharply in a 1972 UNESCO paper and in a 1976 "*Journal of Environmental Education*" editorial as executive director.

> Some are naïve enough to believe that changing the name from conservation education to environmental education will bring instant results, success where there was failure, support where no support was available. (Brennan, 1972, p. 472)

> However, when early conservation efforts failed, those of us interested in education for the preservation of planet

earth and its unique systems took the easy way out. We created a new program called "environmental education," which would be more saleable to our "apathetic" public than "conservation" had been. We said that our new program would embrace all of the various conservation education efforts; but it didn't happen that way. Nature study, outdoor education, and conservation education were left out, and proposal writers soon learned that to mention them was an open invitation to failure at getting funding grants. (Brennan, 1976, p. 65)

With inclusiveness and broadness, another key component of the legitimizing discourse of environmental education as it appeared in early documents such as the Belgrade Charter (UNESCO-UNEP, 1976) and the Tbilisi Declaration (UNESCO-UNEP, 1977), was a conflation of crisis or impeding catastrophe, with urgency and with a salvation or a redemption narrative through problem solving. Brennan opens his 1972 paper with a caustic interpretation of this story:

It would be funny if it were not serious. Faced with a national crisis of rapid deterioration of the quality of the environment, the people of the United States of America have turned to education for solutions. And education has no solutions [. . .] we have run off in all directions in our haste to develop an "instant" education programme to match our instant replay of television action, instant rice and mashed potatoes and answers to questions with no answers, only possible alternatives for action. (Brennan, 1972, p. 472)

Such a language with themes such as inclusiveness, broadness, crisis, emergency, impending gloom or catastrophe, problem solving, salvation, and redemption has shaped discourses in education and in environment, and again in their coupling through either environmental education in the 1970s and 1980s or education for sustainable development in the 1990s and 2000s.

So what is the problem with the aggressive re-badging of EE? It is the problem spoken of by Popkewitz (1982)—that the slogans can be used to justify a lot of activity at the level of language and organization without actually leading to any real or lasting change at the important level of practice. There is a danger that ESD will not lead to an improvement of environmental-related education in schools. This is the lesson from environmental education—that where there is a slogan system operating, there is every chance that change will be symbolic only. (Campbell & Robottom, 2008, p. 205)

However, the point here is not the issue of change in itself, nor is it the issue of language and discourse in itself, but the dynamic and dialectic relationships that are being played out in environmental education between language, discourse, and practice. Language and discourse bear some meaning, direction, intent and they orient or focus attention, perception and action that also feedback on language and discourse.

Languages of Environment

The notion of environment gained importance in the 1960s, slightly before the idea of environmental education. The language and discourse of environment is also framed in the web of "disciplinary matrixes," following Thomas Khun's (1970, p. 182) language for paradigms, to which it is dialectically related. And, once again, such disciplinary matrixes do not arise in a social void. Environmental languages and discourses in the sciences are linked to issues, language, and discourse in the broader social world, the broader social matrix. The ascension of the "environmental" in the 1960s can be linked to synergies in: (a) the rising of global or planetary perspectives, notably stemming from spatial exploration, (b) the emergence of cybernetics and systems thinking, (c) the important developments in the science of ecology, (d) the fabulously increasing industrial powers to exploit and transform matter, (e) the importance of newer and larger pollutions or disturbances in environmental systems, (f) the depletion of resources linked to the after war economic boom, and (g) a sense of the finitude or the limits of the earth. All of these contributed to a first anchoring of the "environmental" within the confines of the sciences in university structures and languages, and more broadly in education. The "environmental" also gained importance in the broader social world, in the public and in the governing of collective life through the creation of environmental departments in many levels of governments during the 1960s and 1970s. The salvation discourse was thus also active: sound environmental science would lead to sound environmental solutions and policies. The "environmental" would be salvational. However, things proved to be much more complex and more complicated. Problems did not recede and further exploring of the "environmental" moved us out of the strictures of sciences, even if it continued to develop in the sciences. The "environmental" thus found its way in more and more disciplines within universities, each coloring the "environmental" with its own language and discourse. A search in a larger university course catalog with the words environment, environmental, and some synonyms or quasi-synonyms discloses a vast spectrum of angles, approaches, preoccupations, types of languages, discourses and other practices in many faculties, departments, programs, and courses.

This invites to border crossings and to reflect on "the environmental in environmental education," as aptly titled by Rob O'Donoghue (2006) or what Paul Hart (2003, p. 3) mentions, "might be described more cautiously as environment-related education." Within the different practices, voices, and discourses about environment-related education, it is not necessarily the same "environmental" as the one identified in early definitional and policy document such as the Belgrade Charter and the Tbilisi Report of 1975 and 1977. Some researchers are trying to identify and classify the various meanings and changes that stand under the name "environment." Lucie Sauvé thus continues to try mapping environmental education as it changes in time within various currents, and one of the keys used for such mapping are the social representations of the environment such as "nature," "problem," "resource," "territory," "place," "community project," "biosphere," "system," "object of study," "field of values," "locus of action," "object of transaction," "locus for identity construction" (Sauvé, 2005, pp. 33–34). Robert Brulle (2000, p. 98) has mapped the "discourses of the US environmental movement," notably analyzing the movement through Habermas' lense of "communicative action" and identifying "discursive frames" presented in a somewhat historical sequence: "manifest destiny," followed by the "early development of the movement" with the "wildlife management," "conservation" and "preservation" discourses, then followed by "reform environmentalism" where "actions can be developed and implemented through the use of natural sciences" and finally the "alternative voices" such as "deep ecology," "environmental justice," "ecofeminism," and "ecotheology." Across the Atlantic, Kerry Whiteside (2002, p. 12) has mapped "varieties of French ecologism" differentiating them sharply from "English-speaking counterparts." This highlight the contextual, cultural, and political dimensions of the various constructions of the "environmental." Also from a cultural perspective, Roger Cans (1997) compares what he names Latin, Germanic, and Anglo-Saxon ecologies. William Cronon (1995, p. 34) proposes "a guided tour of the several versions of nature," while Holmes Rolston (1994, p. 134) presents a series of values in nature, including fourteen "human values carried by nature": life support, economic, recreational, scientific, aesthetic, biodiversity, historical, cultural symbolization, character-building, diversity-unity, stability and spontaneity, dialectical, life, and finally philosophical and religious values. In the special issue of *Environmental Education Research* on the metaphor of natural capital, Maria Åkerman (2005, p. 45), displays how this capital metaphor emphasizes some "attributes of nature" (ecosystem, sink, resource, and raw material) and "suppressed some other attributes" (unique, life support, lived environment, space, wilderness), whereas Derek Bell (2005, p. 59) revisits the series of nature metaphors by Lakoff and Johnson (1980): organism, mechanism, community, "red in tooth and claw," work of art, work in progress, garden, garden of Eden, mother, adversary, resource, and capital.

The above is just a sampler of the vastness of what lies under "nature," "environment," or "ecology" and that should draw the attention of educators (also as researchers) and researchers (also as educators). Attending to the meaning of nature, environment, and ecology convenes into a maelstrom with a long, profound, and varied history in different cultures. Environment and nature will thus not be so easily enclosed, "circumscribed," or delineated through writing. Just to slightly expand the above sample, one could explore issues of language and discourse around the ideas of "nature," "environment," or "ecology" and how they can act in environmental education through a variety of lenses such as anthropology, cultural studies, and aboriginal perspectives (Cole, 2002; Descola, 2005), history in general and history of western thought (Coates, 1998; Harrison, 1992; Pepper, 1989; Simmons, 1993), environmental thought (Evernden, 1985, 1992; Livingston, 1981), environmental or ecological literacy (Curthoy and Cuthbertson, 2002; Stables, 1996, 1998, 2007), social ecology (Bookchin, 1990), social ecology and ecofeminism (Heller, 2003), gender studies and issues in environmental education (Gough, 1997), critical and emancipation theories (Bowers, 2001; Leiss, 1994; Pepper, 1993, Vogel, 1996), social philosophy of nature (Haber, 2006; Moscovici, 2002), phenomenology (Abram, 1996; Coquet, 2007; Grange, 1997; Merleau-Ponty, 1969), hermeneutics (Mugerauer, 1995) and of course, in various environmental education context (Ferguson, 2008; Hammond, 1998; Lake, 2001; Lousley, 1999; Meisner, 1993; Scott & Gough, 2003).

The above certainly does not pretend to any completeness or systematic inventory. It neither pretends to identify key texts within each specific lens. It simply wishes to point, with insistence, to potential analytical lenses in order to loosen overconfident certainties so as to infuse research and educational endeavors with prudence and a renewed attention to the language we use and the discourse we shape. Italian novelist and essayist Italo Calvino captures it beautifully.

> I think our basic mental processes have come down to us through every period of history, ever since the times of our Paleolithic forefathers, who were hunters and gatherers. The word connects the visible trace with the invisible thing, the absent thing, the thing that is desired or feared, like a frail emergency bridge flung over an abyss. For this reason, the proper use of language, for me personally, is the one that enables us to approach things (present or absent) with discretion, attention, and caution, with respect for what the things (present or absent) communicate without words. (Calvino, 1988, p. 77)

Languages of Education

As for the environment, education is also shaped into languages and discourses. The histories of educational languages and discourses is however much longer. Israel Scheffler (1960) and Olivier Reboul (1984) have both written on the language of education, focusing on issues of "definitions," "slogans," "metaphors," "discourses," "teaching and telling" and the "rhetoric of pedagogical discourses." As for the environment, exploring language and discourse in education, conveys to prudence, attentiveness and a loosening of easy certainties. Neil Postman captures a sense of such issues by addressing metaphors.

> . . . it has always astonished me that those who write about the subject of education do not pay sufficient attention to the role of metaphor in giving form to the subject.

> In failing to do so, they deprive those studying the subject the opportunity to confront its basic assumptions. Is the human mind, for example, like a dark cavern (needing illumination)? A muscle (needing exercise)? A vessel (needing filling)? A lump of clay (needing shaping)? A garden (needing cultivation)? Or, as many say today, is it like a computer that processes data? And what of students? Are they patients to be cared for? Troops to be disciplined? Sons and daughters to be nurtured? Personnel to be trained? Resources to be developed? (Postman, 1995, p. 174)

Systematic analysis of educational languages and discourses is challenging, especially in a comparative perspective striving to be contextually and culturally sensitive. A fruitful approach is to try discerning how the discourse focuses either on educators, on learners, on knowledge, on society or on the weight of their various relationships and the dominant perspectives on a typical spectrum between reproduction and emancipation. Out of such analysis have emerged various typologies. However, Chet Bowers (2001) and Ilan Gur-Ze'ev (1998, 1999, 2003) both capture the slipperiness and messiness of our too easy circumscribing of educational languages and discourses within sharp and discrete categories. Sometimes, we can see a common thread uniting discourses that are generally seen as radically different or opposed.

> Both the Judeo-Christian tradition of religious redemption and the tradition of utopianism needed the idea of "progress" as well as the optimistic historical conscience in general. From the prophets of Israel to Rabbi Cook, from St. Augustine to Hegel and Marx, the idea of progress ensured confidence in the possibility of redemption, or in the realization of the future utopia. (Gur-Ze'ev, 1998)

We find here an unusual and challenging association of reproduction (tradition, conservation) and emancipation, under the umbrella of a salvation discourse. Thomas Popkewitz (2004, p. 244) also notices "the themes of salvation and redemption are not recent, but rather belong to the institutionalization of schooling at the world scale that started in nineteenth century." Focusing on educational reforms, an element of discourse that environmental education and education for sustainable development have insisted upon, Popkewitz adds, "narratives of salvation through school reform are also narratives of redemption" (p. 243) and in "the common narratives of salvation," "preservation does not come from a return to the past, but rather from the renovation of the nation by its children" (p. 252).

Under such a transversal salvation discourse, rooted in a specific culture and worldview, which we should ponder a little more before universalizing it, the educational in environmental education and more recently in education for sustainable development or education for sustainability has been acted upon by changing languages of education according to their ascent and descent. In education—more specifically in North America—the short spell of humanist education (epitomized e.g., in Rogerian approaches) diminished under the rise, ascendancy and domination of programmed teaching and the various taxonomies of development (cognitive, moral, psychomotor, etc.) with their typical action verbs translating into programs and curriculum with their statement of goals, general educational objectives, terminal objectives, and intermediate objectives. The categories of objectives for environmental education proposed at the Belgrade and Tbilisi conferences are essentially bounded and framed in the language and discourse of Benjamin Bloom's 1956 taxonomy of the cognitive domain. These taxonomies gained ascendancy in the world of formal education in North America and also influenced the language and discourse of organizations striving to work in collaboration with schools: nature centers, interpretation centers, parks, museums, etc. In his 1976 doctoral thesis, Gary Harvey proposes a "substantive structure" for environmental education that is essentially framed in Bloom's, Kohlberg's, and Krathwohl's common language and discourse for cognitive, affective, and moral development. Such a language finally formed the backbone of environmental education's educational discourse that was adopted in United Nations conferences, declarations, and reports. It was thus generalized and pushed, unproblematically, as a universal.

What would the world of formal education look like if it would have been the educational languages, discourses, and practices of Carl Rogers, or Paulo Freire, or Celestin Freinet, or Maxine Greene, or Maria Montessori that would have gained ascendancy or if more attention to the diversity of educational visions and practices had been a stated pursuit?

As for the environmental, the educational also became rapidly more complex, more complicated and more diverse. Early questioning of the Belgrade and Tbilisi discourse stimulated a defense of environmental education delineated in these international and intergovernmental meetings. Harold Hungerford, Ben Peyton, and Richard Wilke could quite confidently assert "*Goals for Curriculum Development in Environmental Education*" (1980) based on the Belgrade and Tbilisi reports and could then answer to a perplexed that "*Yes, EE Does Have Definition and Structure*" (1983). Later, Ian Robottom (1987) could somehow reply about "*Contestation and Consensus in Environmental Education.*" In a similar sense of questioning conventions, Bob Jickling (1991) could write "*Environmental Education, Problem Solving, and Some Humility Please*" and then Peter Corcoran and Eric Sievers (1994) "*Reconceptualizing Environmental Education: Five Possibilities.*" The early educational language and discourse and the accompanying practices that were built into environmental education were thus being questioned from various standpoints. Ian Robottom and Paul Hart (1993) provided a good synthesis of the issues and positions with their "*Research in Environmental Education: Engaging the Debate.*" From an educational perspective then, throughout forty years of exploring the variety of meanings and practices in environmental education in a diversity of contexts, the easy certainties about education progressively eroded.

As an example of changes in the educational languages and discourses, the increasing presence of "learning" as a core or central idea in texts is remarkable (Table 14.1). While recognizing the great importance of research on learning, such as Michael Brody's (2005), we also need to question the language of "learning" as stressed by Lesley Le Grange (2004), especially with the rise of the "free-choice learning" (see the various papers in *Environmental Education Research, (11)*3, special issue on "Free-choice learning and the environment"). While being more attentive to the learning processes and the learners activities is essential for environmental educators, Lesley Le Grange (2004, p. 135) deplores that "learning can involve the ability to access information rather than being a process that engages deliberation among learners and between learners and teachers." He stresses "that education should not be viewed as an economic transaction" (p. 137) where knowledge becomes reducible to objectified and reified merchandise exchanged in a so called knowledge based economy. However, the notion of "social learning" introduces another vision of learning. Fritjof Capra (2007, p. 15) explains such "learning is called social learning to emphasize the importance of relationships, collaborative learning, and the roles of diversity and flexibility in responding to challenges and disturbance" while Arjen Wals and Tore van der Leij (2007, p. 18) draw attention to "learning that takes place when divergent interests, norms, values, and constructions of reality meet in an environment that is conductive to learning." Thus, while a greater attention to learners and learning can be a move toward more interpretative or humanistic perspective, it can also be a move toward more socially oriented perspective and it can also be a move toward more behaviorist perspective as a device to better shape learner's behaviors. Certainly then, the divergent languages and discourses of learning should draw the attention of researchers and it also invites to attend very closely to meanings and finalities in specific educational texts and contexts.

Thus, more recently, with the rise of globalization and the increasingly rapid pace of changes, notably those stemming from sciences and technologies, and influencing knowledge, the language of education is shifting. Competitiveness and adaptability to changes are often highlighted in policy documents and in curriculum rationale. Robert Stevenson witnesses, along with many others, the rise of "emphasis on literacy and numeracy" (Stevenson, 2006, p. 286). Elsewhere, reflecting on "discourses of policy and practice," he notes how,

> Most governments have focused on the role of schools in preparing workers to compete in this new global knowledge-based economy through centrally defined curricula; an emphasis on mathematics, science and technology; and an increased reliance on standard measures of student's performance, including international comparisons. (Stevenson, 2007, p. 270)

The language of education would seem to be undergoing metamorphosis under a sort of cross pollination or synergy between various ideas, movements, and social forces: competence based education and employment to produce generalist problem solvers, adaptability, transferability, mobility, expanding human capital, and lifelong learning. Salvation would now come to the nation by the strivings of its individual members, "human resources," in a worldwide competition, typically quantified and universalized in the "Program for International Student Assessment" (PISA). Promoting the notion of human capital, Brian Keeley (2007, p. 87) thus asserts, unpoblematically, as if we were similar to rapidly obsolete computers, "we'll also have to go on upgrading our skills, education, abilities—our human capital—throughout our lives," and "to stay relevant at work, people will need to go on continually upgrading their education."

The pressures of a changing world affect the world of education, how it is conceived, the goals it should pursue, and of course its languages and discourses (Lamarche, 2006; Laval et al., 2002; Le Goff, 2003; Petrella, 2000). The changes are paradoxical to say the least, opening possibilities for people, thus offering an emancipation-salvation potential, but often framing these possibilities in a strictly personal or individual challenge and lifelong race to achieve, to survive, and to thrive in a menacing and a rapidly changing world. Even though these evolutions are socially constructed, it seems to be more and more

TABLE 14.1
The Rise of Learning in the Languages of Education and Environmental Education

	Percentage of Documents in the ERIC Collection with Lear*: Wildcard for Learn, Learning, Learners…			
	… in the Totality of the Collection (Any Field)	… in the Collection (Any Field) with Document Using the "Environmental Education" Descriptor	… in the Title Field in the Totality of the Collection	… in the Title Field for Documents Using the "Environmental Education" Descriptor
1970–1979	19.5%	28.7%	3.6%	2.3%
1980–1989	22.5%	36.2%	4.4%	2.3%
1990–1999	28.8%	36.6%	6.8%	5.6%
2000–2009	34.0%	43.9%	10.0 %	9.6%

difficult to face them socially, collectively. In education, the salvation of the world is subsumed under a personal quest and unending race for salvation.

> The soul of modern school is not the religious soul of the Christian afterlife, but rather the one of an individuality ordered by dispositions, sensibilities and a consciousness that make the civilized actor a person able to evolve in a global culture and economy. The new child in pedagogy is analogous to the manager in the Swedish State and his capacities. Both embody characteristics of an active adaptability that resolves problems in contexts of uncertainty without fixed rules of application, without prescribed solutions to the problems of education. But the child not only resolves problems, because, in the language of pedagogical reform, the child actively constructs his knowledge in collectivities. One of the corner stones of reforms is psychological and sociopsychological constructivism. Learning mathematics or science, as examples, is developing flexible approaches and answering new contingencies since there is not any longer a single correct answer. The child, in the language of contemporary reforms, is active, resolves problems and learns through lifetime, evolving without a rigid center and within fluid frontiers. (Popkewitz, 2004, p. 253)

In the world of education, work needs to be pursued to recall and highlight how the language of the individual is a contextual, social, and historical construct that hides the social matrix producing it, thus reifying this notion and way of being, naturalizing the calculative and almost anti-social individual. But again, the language of the individual can help emancipate, depending on the context. As an example, biographical and autobiographical approaches in education and educational research have historically been seen by many as critical endeavors to emancipate from rigid structures and determinations, or at least to highlight them (see as examples, Antikainen, Houtsonen, Huotelin, & Kauppila, 1996; Bullough & Gitlin, 1995; Pinar, 1994). However, with the rise of individualism or pseudoindividualism, some contextual injunctions to "produce oneself," to "autopresentation" can force people into predefined format and files and thus act as processes of "auto-reification," turning oneself into an object, and thus dominate rather than emancipate (Delory-Momberger, 2009, p. 85, see also Hesford, 1999). The above highlight the dynamic and highly complex, contextual and historical relationships, tensions, determinations, and conflicts between what are sometimes called the individual and the social. Providing adequate language to face such tensions is challenging. Whereas the individual quest to perform for one's salvation is now an important social discourse, Miguel Benasayag writes about "the myth of the individual" (Benasayag, 1998) and instead of shunning from various conflicts and tensions, writes a book titled *Praise to Conflict* (Benasayag & del Rey, 2007). In various cultural contexts, environmental educators and environmental education researchers are not blindly abiding to a dominant discourse of the individual educational quest for a more personal salvation in a fatalistic global capitalism but include, at the core of their work, social, political, and collective dimensions (Caride Gómez, 2007; González-Gaudiano, 2007; Naoufal; 2009, Taleb, 2008; Torres de Oliveira, 2008).

In environmental education, by focusing educational attentiveness on the environment, or to be more precise, on our relationships to our surroundings—the ecological dimensions of our life—we draw attention to an "object" that offers a potential to pull us outside or beyond the strictly individual quest and highlight inescapable and vital connections. There is a joyous potential lying there that could move us beyond the confines of the environmental problem solving approach to which environmental education has so often been associated and reduced. It could also move us beyond the purely personal quest for salvation through economic gains. Contrasting with the narrative of a national salvation through the adaptability and competitiveness of its individual members striving for their human capital, as lubricants for growth, Nel Noddings (2003, p. v) proposes to connect "happiness and education" and proposes some lines around which curriculum could be redrawn. First, "educating for personal life" could be reconceptualized around "making a home," "places and nature," "parenting," "character and spirituality" and "interpersonal growth." Second, "educating for public life" could be reconceptualized around "preparing for work," "community, democracy and service" and an explicit strive for "happiness in schools and classrooms."

Of course, Nel Noddings is not the sole source of alternative languages and discourses in education. As one of various outlooks on education, her proposal contrasts however with a dominant discourse by focusing on happiness rather than achievement as a solely economic and individual pursuit, and it helps highlight the importance of a critical vigilance and a dedicated search for options in education. Once again, critical pedagogy provides another very important contribution by highlighting the social, collective, and political dimensions of education.

Discourse analysis and critical discourse analysis in education (Rogers, 2003, 2008; Rogers Malancharuvil-Berkes, Mosley, Hui, & O'Garro Joseph, 2005; Warriner, 2008) could further help environmental educators avoid blindly reproducing hegemonic educational languages and continue exploring the immense possibilities and diversity of the educational in environmental education. Especially in the education of educators, it seems important to explore such a diversity and look at the various discourses within a curriculum (Eisner, 1994). Language and discourse are dialectically shaping and being shaped by the whole education spectrum: policy, theory, curriculum, school and classroom organization, pedagogy, evaluation, and the daily relationships and exchanges between educators and learners in any educational setting.

Languages of Environmental Education and of Sustainable Development

The above sections drew attention to salvation as a key dimension of narratives in environmental languages and discourses, as well as in education. In such a context, salvation was somehow bound to be central when the languages and discourses of education and environment met in the various rhetoric of environment related education.

This theme of salvation is one of those to appear in our discourse analysis of United Nations (UN) documents touching upon environment related education from 1972 to 2009 (Sauvé, Berryman, & Brunelle, 2007). The research, by focusing on the representations of education, environment, and development, also draws attention to other key features of the discourse in UN policy documents. Education is most often presented as an instrument to implement a predefined political agenda, where environment is considered a reservoir of resources needed for development and development confused with economic growth. The language of sustainable development goes further in this direction: economy is considered as an autonomous entity, extracted from society that somehow imposes its rules on society-environment relationships. Such a framing and reduction of the environmental in the strictures of the developmental needs to be questioned. Once again, in the context of language and discourse, the various critical examinations of development and economy and of their joining in a reformed salvation narrative need to be stressed (Latouche, 2004, 2005; Laval, 2007; Livingston, 1973; Polanyi, 1983; Rist, 2001; Sachs, 1999).

However, it is important to recognize that seeds of the sustainable development discourse were clearly manifest in UN policy documents since the 1970s. As an example, sustainable development lies at the core of the *World Conservation Strategy* (WCS) published in 1980, as clearly illustrated by the report's subtitle and the title of it's first chapter: "Living resource conservation for sustainable development" (International Union for the Conservation of Nature [IUCN], 1980). The outlook is clearly framed in a resourcist perspective even if the authors still refer to environmental education when they discuss educational issues. All along the document, the basic project is to integrate conservation and development. This also forms the backbone of *Our Common Future*, (World Commission on Environment and Development, 1987), the Brundtland Report, often and mistakenly presented as the origin of the notion of sustainable development. It also forms the backbone of *Agenda 21* whose section on the environment is titled "Conservation and Management of Resources for Development" (UNCED, 1993). All these documents originated from negotiations and compromises between delegates of different countries. In such international negotiations, the quest for environmental conservation, one of the features of environmental education, encountered the quest for a better life in what is designated as undeveloped or underdeveloped countries. The compromise often meant that developed countries should reduce their pressure on resources so the underdeveloped could develop. Such an outlook finally lead to the now classic figure of the three interlocked spheres: environment, society, and economy. Out of this also came an often heard discourse that sustainable development is broader than the concern for the environment and that it includes it. However legitimate is the sustainable development outlook, it clearly orients and focuses approaches to the environment in specific directions that do not encompass all the diversity of environmental perspectives. Corollary, not every environmental perspectives encompass or address sustainability or sustainable development issues. Finally, sustainable development is not limited to the environment and, conversely, the environment is not limited to sustainable development.

How then does this play out in educational discourses: policies, curriculum, educational material, and various educational encounters? This chapter of the handbook opened with the question about the replay of the same old story of a legitimizing discourse built around ideas of inclusiveness, broadness, impeding catastrophe, urgency, and salvation. As for the discourse of environmental education interpreted by Brennan (1972), the discourse of education for sustainable development somehow strives to posit and legitimize it as a broad and inclusive, generous endeavor, and as a salvation. Bob Stevenson and Edgar González-Gaudiano also notice this salvation discourse.

> A view of sustainable development as a "salvation narrative" that represents the way for society to be rescued from environmental and social destruction, along with its prominence in international conferences, risks reifying ESD policy. The reification of international policy discourse can imply an unquestioning faith in so-called "experts" and authorities, in centralized global institutions and intergovernmental agreements, and in top-down approaches to educational reform. (Stevenson, 2006, p. 287)

> I am inclined to follow Lyotard and think that sustainable development is a case of an empty signifier which operates like a huge myth with pretensions of being a salvation grand narrative. (González-Gaudiano, 2006, p. 297)

And once again, with education for sustainable development, we could start with the same caustic line Mathew Brennan (1972, p. 472) used to open his paper on education and the environment: "It would be funny if it were not serious." Read on!

> Emerging from this discussion is an agreement amongst scholars and researchers that environmental education in the coming decade must reorientate itself towards improving the quality of life for all citizens under the focus of environmental education for sustainability. (Tilbury, 1995, p. 197)

> . . . the path to sustainability and the end place are unknown. (McKeown & Hopkins, 2003, p. 126)

Yes, "it would be funny if it were not serious" but these are two typical lines and often repeated claims about sustainable development and education for sustainable development. We do not know what it is, what the goals are, nor the path, but it is broader than environmental education and education must be reformed toward this end or path.

The discourse of education for sustainable development often strives to legitimate and position itself by claiming such an encompassing perspective toward environmental education in a similar fashion that the discourse of environmental education often did toward nature education, conservation education and outdoor education. Since sustainability is in such a rise in the language of education, as illustrated in Table 14.2, the need for various critical analysis of education for sustainable development will have to be pursued as are critical analysis of environmental education. Of course there are different interpretations of the "sustainable" as there are of the "environmental" and the "educational." As an example, some strategically use "sustainability" and thus deliberately avoid using "development" for various motives. Yet, others will circumvent the issue and the sustainable development perspective by systematically avoiding both words. Others will use them uncritically and yet others criticize this language and discourse.

The future will tell how the world of education has treated the language and discourse of sustainability. For now however, since the discourse about the inclusiveness, the generosity, and the broadness of environmental education is close to forty years old, we can look back and try to see how this part of the story turned out.

Lucie Sauvé's (2005) ongoing strivings to map "currents in environmental education" provide a view of the diversity of practices, languages, and discourses. Some currents meet or intersect with sustainability and sustainability itself interpreted as a current to which some environmental education practices identify. The fifteen currents she characterizes are: naturalist, conservationist/resourcist, problem-solving, systemic, scientific, humanist/mesological, value-centered, holistic, bioregionalist, praxic, socially critical, feminist, ethnographic, eco-education, sustainable development/sustainability. This is not exhaustive and is evolving. As an example of more recent changes, in universities, often in language departments, "literature and environment," sometimes named "ecocriticism," is on the rise as a form of environment related education and as a field of research. Productions in this academic field will clearly contribute to research on the language and discourse of the environment.

In a way then, environmental education has emancipated from early UN "policing" by exploring, adapting, or constructing diverse avenues to approach and outline its own object, which is the network of our relations to the environment, rather than focusing solely on the early "project" defined in international policy. There is more than a pun at work here with the question about governance of the city, the Greek *polis*, the political, policy, and police. A policy wishes to orient and to police conduct. Policy clearly addresses and strives to govern, to direct, to shape. We cannot avoid issues of governing our collective life and our environmental circumstances, but we cannot be totally reduced to these dimensions. The major strength of environmental education is the various ways of exploring its "object" that provide numerous opportunities for decentring. Neil Evernden provides a compelling view of how language structures our worldview and the importance of returning to experience the world.

> Obviously we cannot avoid the creation of categories, any more than we can avoid the social construction of reality. The inclination to tell the story of "how the world is" seems basic to being human. Indeed, the formation of categories may be important to the development of human capacities. We can only hope that when the story turns out to be too far removed from actual experience to be reliable, we still have the skill to return to the world beneath the categories and reestablish our connection to it. (Evernden, 1985, p. 56)

The above highlights issues for environmental education practice and research. Since there is such a web of relationships between languages, discourses, policies, curriculum, textbooks, practices, experiences, and worldviews, we should strive to shuttle, navigate, circulate, dance, and move back and forth between these points. Robert Stevenson (2007, p. 267) justly decries the fact "there as been a continuity in the power of certain groups in maintaining control of the discourse." Policy should

TABLE 14.2
The Rise of Sustainability in the Languages of Education and Environmental Education

	Percentage of Documents in the ERIC Collection with Sustainab*: Wildcard for Sustainable, Sustainability...		
	... in the Totality of the Collection (Any Field)	... in the Collection (Any Field) with Documents Using the "Environmental Education" Descriptor	... in the Title Field for Documents Using the "Environmental Education" Descriptor
1970–1979	0%	0%	0%
1980–1989	0%	1.2%	0.3%
1990–1999	0.3%	8.7%	2.8%
2000–2009	0.9%	24.0%	12.3%

not be left solely to specialized policy writers, curriculum solely to curriculum writers and then teaching the policies and the curriculum solely to teachers as agents. We need to attend to the many languages and discourses of environmental education. In terms of diversity, Leigh Price thus mentions:

> Different cultures will provide different language resources, different histories, different geographical potentials, constraints and evocative imagery; thus the same phenomenon, mobilized by people from different cultural, geographical, and historical heritages may have significantly different characteristics. (2005, p. 95)

He thus draws attention to the issue of diversity and hegemony, and he adds an ecological layer of complexity by suggesting that "we should be seeking a fit between what we say, what others say and our experience of *the world*, not just a fit between what we say and what other people say" and also that "for the 'other' which is not human, to allow its 'voice' to be heard we may need to actively research and mobilize information" (Price, 2005, p. 99). Here, his voice meets those of Anne Bell and Constance Russell:

> The belief that language is an exclusively human property that elevates mere biological existence to meaningful, social existence. Understood in this way, language undermines our embodied sense of interdependence with a more-than-human world. Rather than being a point of entry into the webs of communication all around us, language becomes a medium through which we set ourselves apart and above. (Bell & Russell, 2000, p. 193)

This is part of the decentering that closer attention that environmental encounters can foster. In a way, emancipating from early policy allowed to reconnect to some of the historical roots of environmental education such as encounters with nature, as in nature education, and encounters and various experiences of the outdoors, as in outdoor education. And, to make the point more clear, these encounters can be educationally relevant without resorting to sustainability or sustainable development frameworks.

However, from a critical vigilance stance, wondering about tensions between the individual and the collective, Rob O'Donoghue (2006, p. 348) is perplexed by the "ways in which the mediating dialectic associated with a maturing liberal democracy and an increasing concern for the individual were shaping environment as a more personal (diverse) and culturally relative (provisional) phenomenon in environmental education research."

The observation is crucial but we could find ourselves once again in the opposition of the individual way "or" the collective way of facing our ecologies, considering the economy of our ecologies or not. Maybe a future challenge for environmental education is to continue to explore ways of replacing "or" with "and." Phillip Payne is one of those exploring such prospects and he proposes a map "to explore the conceptual relations between embodied and textual selves" (Payne, 2005, p. 427) where the "social" is closely related to the "personal." This is a crucial component of the "critical ecological ontology" out of which he draws "sensitizing questions" for educators that invite to formally link, within a curriculum, the ecological, the social, and the personal, through an embodied presence, attention or vigilance (Payne, 1999, 2003a, 2003b, 2006).

Finally, the ways and languages for approaching the environment in educational contexts are numerous and increasing in diversity, echoing the diversity and complexity of the relationships which support us and the others. There is a broad spectrum of approaches and languages that can also be combined, from the variety of art-based approaches with their language to the place-based, from the foodshed to the ecological footprint, from eco-justice to nature writing and from environmental ethics to ecological economics.

Attending to Language and Discourse: Some Prospects

There are many possible venues, angles, approaches, and sources of inspiration to address the question of language and discourse in environmental education and in environmental education research. For such purposes, there are in fact various research traditions with their paradigms, their epistemology, their methodology, their school of thought, and of course, their own language and discourse that are more or less broadly shared. It is not quite possible, nor desirable to invent a metalanguage to produce a metadiscourse that would strive to contain all the others.

The same holds for another more immediate notion of language. Catalan, English, French, Gaelic, Mandarin, Occitan, Peul, Quechua, Soussou, Spanish, Wolof, Zoulou are all languages that bear upon intersubjective communications between people in particular cultural and environmental settings. Language and discourse are dialectically related to action and history, giving shape to a culture and being shaped by a culture. We cannot imagine a metalanguage that would accommodate all the languages. However, we can easily imagine a language rising to prominence and exerting dominion and hegemony over the others and eventually the collapse of other languages and the cultures, the worldviews and the practices they framed linguistically.

The above points to key dimensions of language and discourse that need to be stressed. Issues of language and discourse should never be downplayed as cumbersome blah blah that must be set aside in order to act urgently to face some environmental—socioecological—circumstances. Language, discourse and action are intertwined and they feedback upon each other. The current context of globalization highlights even more sharply and urgently the importance of being much more attentive to the cultural and the ecological diversity of contexts, practices, and languages (Fairclough, 2006).

Finally, we have to wonder about the future of the written text as the most important language and bearer of meaning and of rules. For many, we are living in an era where the textual was democratized. Before, the textual was more marginal, reserved for the elites, for law, and for some professions. The increasing presence and reliance on the pictorial (photo and video) and the audiovisual as sources of information and as forms of languages and discourses is raising new challenges. Are we now in changing era? Are we moving from an era where the textual is(was) central and where audiovisual is(was) in the hands of the more powerful to an era where the audiovisual is more present in ruling and is also increasingly democratized? This raises new challenges to address issues of language and discourse. As an example, papers about neurosciences are given more credibility by readers when they are accompanied by images of the brain (Sicard, 2008).

Attending to issues of language and discourse in environmental education opens various prospects to further explore such educational endeavors. From the various traditions to approach issues of language and discourse, we will focus here on four sources of inspiration to approach environmental education. More than inspiration, these sources point to the importance of language and discourse in relation to practice and they point to some methodological considerations. Taken together, they allow a more critical attentiveness to the world in which we are invited to dwell, act, teach, and learn as environmental educators and environmental education researchers when we read a text and write one, when we listen and talk, even when it is about our acting.

First, we find inspiration and some grounding in Paul Ricœur's philosophical essays about interpretation or hermeneutics, particularly, on the one hand, when he examines the "remarkable connections between the theory of text, the theory of action and the theory of history" and, in second hand, when he asserts, "The question is not anymore to define hermeneutics as an inquiry into the psychological intents that would be hidden under the text but rather the clarification of the being-in-the-world shown by the text. What is to be interpreted in a text is the proposition of a world, the project of a world I could inhabit. . ." (Ricœur, 1986, p. 58). Elsewhere he invites readers of texts to attend to "the kind of world opened up by the text" (Ricœur, 1986, p. 407). Attending to the world opened up by a text, being attentive to the one constructed and transformed through our action and finally being attentive to the world represented through our educational and our research activities have a special importance in environmental education. Our research on the discourse of UN policy documents and our critical understanding of the world opened up by such a discourse and the one closed by it (Sauvé et al., 2007) is sustained by Ricœur's proposal and his shrewd linking of the theory of text, the theory of history, and the theory of action. Analyzing policy documents, Stevenson (2006, 2007) and González-Gaudiano (2006) also attend to the interplay of text and action. In a similar vein, analyzing international environmental agreements on their representation of education and public participation, Elin Kelsey (2003) also depicts a series of sad and reductive tendencies in these "mechanism of global environmental governance" (p. 403): the weight of "management bureaucracies" (p. 410), a discourse where "information is synonymous with knowledge" (p. 412), "education is a vague concept" (p. 415), the dominance of "administrative rationalism" (p. 416) and "answer culture" where "the answer culture is a metaphor for the expectation that knowledge will save us" and where "language functions like a conduit, transferring thoughts directly from source to recipient" (p. 419). To help compensate the possible domination exercised by policy documents, such as curriculum policies, Michael Singh (1998) proposes various "critical literacy strategies" for reading, counter reading, and rewriting policy discourses in environmental education. These can be empowering and help educators and researchers alike avoid blindly reproducing a dominant discourse. We cannot just wish away policy documents and should be attentive to their role in transforming and otherwise influencing language, discourse, and practice at various levels and locations.

Second, closely related to Ricœur's invitation to an attentiveness to the relation between text, action and history, we find inspiration in Peter Berger's and Thomas Luckmann's (1966) classic in sociology of knowledge. When they analyze the "dialectical moments" (p. 61) in "the social construction of reality" (also title of the book), or in institutionalization, they highlight the role of language in the linkages between "society as an objective reality," produced by "externalization" through processes of "habitualization," "typification," "sedimentation," and "reification," and "society as a subjective reality" constructed by "internalization" through socialization. Here, language plays an important role as a kind of shuttle between the more internal and the external, or between the subjective and the objective: "Language provides the fundamental superimposition of logic on the objectivated social world. The edifice of legitimization is built upon language and uses language as its principal instrumentality" (p. 64). For educators then, as well as for educational researchers, Berger and Luckmann clearly stress the importance of the language and discourse we use. Once again, somehow along Ricœur's line of thinking, we are invited to be more attentive to the powerful role of language in constructing a world and a worldview, objectifying them, legitimizing them, institutionalizing them, and sometimes reifying them, leading one to forget the "social construction of reality." This can be linked to Jean Piaget's constructivist work and his observations on the representations of the world in children: "every language contains a logic and a cosmology and the child, learning to talk at the same time or before he learns to think, thinks according to the adult social world" (Piaget, 1947, p. XL–XLI).

Thus, moving from the interpretations of texts to all sorts of exchanges through language, we are invited to consider the world we are creating and into which we accompany educators and learners alike.

Third, easily related to Ricœur as well as to Berger and Luckmann, we also find inspiration from Dorothy Smith's institutional ethnography, specially as it points to "how people are putting our world together daily in the local places of our everyday lives and yet somehow constructing a dynamic complex of relations that coordinates our doing translocally" (2005, p. 2) and as institutional ethnography invites to examine "the textual realities that are essential to the existence of institutions and of the ruling relations in general" (2005, p. 27). Marjorie DeVault and Liza McCoy summarize by stressing the "linkages among local settings of everyday life, organizations, and translocal processes of administration and governance" and "the increasingly textual forms of coordination" (2002, p. 751). Quoting Dorothy Smith, they add that these textual forms are "forms in which power is generated and held in contemporary societies" (p. 751) and finally they emphasize that "a central feature of ruling practice in contemporary society is its reliance on text-based discourses and forms of knowledge." (p. 753). In environmental education, institutional ethnography provides a compelling invitation to attend much more closely and much more critically to the whole series of strings linking global policy and educational encounters between educators and learners. Institutional ethnography draws attention to power and ruling through various types of texts and strives to emancipate from subtle and not so subtle forms of domination.

Fourth, we also find inspiration in Chaïm Perelman and Lucie Olbrechts-Tyteca (1988) writing on argumentation and their use of the dissociation of notions to better highlight, in a text, what is explicitly or implicitly qualified or valued from what is disqualified or devalued, once again explicitly or implicitly. Dissociation of notions, by naming, pair by pair, the qualified and disqualified opposites in a text, is thus one of the analytical techniques that can help identify what is highlighted and overshadowed in language and what is more clearly shaped and formed into a specific discourse, in our case, discourse about formation and environment. Perelman and Olbrechts-Tyteca dissociation of notions (1988) helped us identify what UN policy documents tended to highlight and what they tended to downplay.

All of the above thus stress the importance of: (1) the world opened up or disclosed in texts or by texts, (2) the powerful links between texts, actions, and histories, (3) the importance of texts in ruling daily life, (4) a way to better highlight the world opened up and the one that is closed, veiled by a specific or chosen or dominant language, text, discourse, and finally, (5) an invitation to try to lessen domination in the relationships between people, between cultures and between us and the surrounding world, the environment and nature.

In this chapter, we stressed the importance of attending to language and discourse in environmental education. In research and in educational setting, language is too often an invisible agent, shaping and structuring theories and practices, visions, and actions. Being more attentive to language and discourse and their influences upon theoretical and methodological frameworks is helpful, noticing how and what they can open and what they can close. Finally, critical education into the various languages and discourses of education, of the environment, of environmental education, of sustainable development, and finally of educational research can greatly improve researchers culture with regards to diversity in these areas and such a critical education can allow one to make more conscious and deliberate choices. As the body of written material in environmental education increases and is now belonging to longer historical horizons, there are improved opportunities to undertake research into the action of language and discourse in shaping environmental education.

References

Abram, D. (1996). *The spell of the sensuous: Perception and language in a more-than-human world*. New York, NY: Pantheon.

Åkerman, M. (2005). What does "natural capital" do? The role of metaphor in economic understanding of the environment. *Environmental Education Research, 11*(1), 37–52.

Antikainen, A., Houtsonen, J., Huotelin, H., & Kauppila, J. (1996). *Living in a Learning Society: Life-Histories, Identities and Education*. Coll. Knowledge, Identity and School Life Series, No 4. Londres: Falmer.

Bell, A. C., & Russell, C. L. (2000). Beyond human, beyond words: Antrhopocentrism, critical pedagogy, and the poststructuralist turn. *Canadian Journal of Education, 25*(3), 188–203.

Bell, D. (2005). Environmental learning, metaphors and natural capital. *Environmental Education Research, 11*(1), 53–69.

Benasayag, M. (2004). *Le mythe de l'individu*. Translated and adapted from Spanish by Anne Weinfeld. Coll. Sciences humaines et sociales No 168. Paris, France: La Découverte. Originally published in 1998 at Éditions La Découverte in the Coll. Armillaire.

Benasayag, M. (2006). *Connaître est agir: paysages et situations*. Paris: La Découverte.

Benasayag, M., et del Rey, A. (2007). *Éloge du conflit*. Coll. Armillaire. Paris, France: La Découverte.

Berger, P. L., & Luckmann, T. (1966). *The social construction of reality: A treatise in the sociology of knowledge*. New York, NY: Doubleday & Anchor.

Bookchin, M. (1990). *The philosophy of social ecology: Essays on dialectical naturalism*. Montreal, QC: Black Rose.

Bowers, C. A. (2001). *Educating for eco-justice and community*. Athens, GA: University of Georgia Press.

Brennan, M. J. (1964). *Total education for the total environment*. Paper presented at the 131st annual meeting of the American Association for the Advancement of Science, 26–31 December 1964. Montreal. Reprinted in *Journal of Environmental Education, 6*(11), 1974, 16–19.

Brennan, M. J. (1972). Environmental conservation education in the United States of America. *Prospects, II*(4), 472–476. UNESCO Quarterly Review of Education, Winter 1972.

Brennan, M. J. (1976). Editorial. *Journal of Environmental Education, 7*(4), 65.

Brody, M. (2005). Learning in nature. *Environmental Education Research, 11*(5), 603–621.

Brulle, R. J. (2000). *Agency, democracy and nature: The U. S. environmental movement from a critical theory perspective*. Cambridge, MA: The MIT Press.

Bullough, R. V., & Gitlin, A. (1995). *Becoming a student of teaching: Methodologies for exploring self and school context*. Coll. Critical Education Practice. New York, NY: Garland.

Calvino, I. (1988). *Six memos for the next millennium*. Cambridge, MA: Cambridge University Press.

Campbell, C., & Robottom, I. (2008). What's in a name? Environmental education and education for sustainable development as slogans. In E. González-Gaudiano & M. A. Peters (Eds.), *Environmental education: Identity, politics and citizenship* (pp. 195–206). Netherlands: Sense Publishers.

Cans, R. (1997). Les trois sœurs de l'écologie. In J. M. Besse & I. Roussel (Eds.), *Environnement: Représentations et concepts de la nature*. (pp. 209–211). Coll. Les Rendez-Vous d'Archimède. Montréal, QC & Paris, France: L'Harmattan.

Capra, F. (2007). Foreword. In A. Wals (Ed.), *Social learning towards a sustainable world*. Wageningen, The Netherlands: Wageningen Academic Publishers.

Caride Gómez, J. A. (2007). Educación ambiental, desarrollo y pobreza: Estrategias para otra globalización. (pp 165–178). In Collectif. (2007). *Reflexiones sobre educación ambiental ll*. Centro Nacional de Educación ambiental. Spain.

Coates, P. (1998). *Nature: Western attitudes since ancient times*. Berkeley, CA: University of California Press.

Cole, P. (2002). Land and language: translating aboriginal cultures. *Canadian Journal of Environmental Education, 7*(1), 67–85.

Coquet, J.-C. (2007). *Phusis et Logos: Une phénoménologie du langage*. Paris, France: Presses universitaires de Vincennes.

Corcoran, P. B., & Sievers, E. (1994). Reconceptualizing environmental education: Five possibilities. *Journal of Environmental Education, 25*(4), 4–8.

Cronon, W. (1995). Beginnings—introduction: In search of nature. In W. Cronon (Ed.), *Uncommon ground: Rethinking the human place in nature* (pp. 23–56). New York, NY: W. W. Norton.

Curthoy's, L., & Cuthbertson, B. (2002). Listening to landscapes. *Canadian Journal of Environmental Education*, *7*(2), 224–240.

Delory-Momberger, C. (2009). Enjeux et paradoxes de la société biographique. In D. Bachelart & G. Pineau (Ed.), *Le biographique, la réflexivité et les temporalités: Articuler langues, cultures et formation* (pp. 75–85). Coll. Histoire de vie et formation. Paris, France: L'Harmattan.

Descola, P. (2005). *Par-delà nature et culture*. Coll. Bibliothèque des sciences humaines. Paris, France: nrf Gallimard.

DeVault, M. L., & McCoy, L. (2002). Institutional ethnography: using interviews to investigate ruling relations. In J. F. Gubrium & J. A. Holstein (Eds.), *Handbook of interview research: Context and method* (pp. 751–776). Thousand Oaks, CA: Sage.

Eisner, E. W. (1994). *The educational imagination: On design and evaluation of school programs*. New York, NY: Macmillan.

Evernden, N. (1985). *The natural alien: Humankind and environment*. Toronto, ON: University of Toronto Press.

Evernden, N. (1992). *The social creation of nature*. Baltimore, MD: Johns Hopkins University Press.

Fairclough, N. (2006). *Language and globalization*. New York, NY: Routledge.

Ferguson, T. (2008). 'Nature' and the 'environment' in Jamaica's primary school curriculum guides. *Environmental Education Research, 14*(5), 559–577.

González-Gaudiano, E. (2006). Environmental education: A field in tension or transition. *Environmental Education Research, 12*(3–4), 291–300.

González-Gaudiano, E. (2007). Educación, globalización y consumo: una mirada crítica. *Anales de la educación común*. No 8 (Educación y ambiente), 16–23.

Gough, A. (1997). Founders of environmental education: Narratives of the Australian environmental education movement. *Environmental Education Research, 3*(1), 43–57.

Grange, J. (1997). *Nature: An environmental cosmology*. Albany, NY: State of New York University Press.

Gur-Ze'ev, I. (1998). Walter Benjamin and Max Horkheimer: From Utopia to Redemption. *Journal of Jewish Thought and Philosophy, 8*, 119–155.

Gur-Ze'ev, I. (1999). Max Horkheimer and Philosophy of Education. *Encyclopedia of Philosophy of Education*. 8 août 1999. Retrieved from http://construct.haifa.ac.il/~ilangz/horkheimer2.pdf.

Gur-Ze'ev, I. (2003). Critical theory, critical pedagogy and the possibility of counter-education. In M. Peters, C. Lankshear, & M. Olssen (Eds.), *Critical theory and the human condition: Founders and praxis* (pp. 17–35). Coll. Studies in the Postmodern Theory of Education, No 168. New York, NY: Peter Lang.

Haber, S. (2006). *Critique de l'antinalturalisme: Études sur Foucault, Butler, Habermas*. Coll. Pratiques théoriques. Paris, France: PUF.

Hammond, W. F. (1998). *The earth as a problem: A curriculum inquiry into the nature of environmental education*. Doctoral dissertation. Simon Fraser University.

Harrison, R. (1992). *Forêts: Essai sur l'imaginaire occidental*. Translated from English by F. Naugrette. Coll. Champs. no 218. Paris, France: Flammarion.

Hart, P. (2003). *Teachers' thinking in environmental education: consciousness and responsibility*. Coll. Rethinking Childhood, No 29. New York, NY: Peter Lang.

Harvey, G. D. (1976). *Environmental education: A delineation of substantive structure*. Doctoral dissertation. Southern Illinois University—Carbondale. ERIC ED 134 451 or *Dissertation Abstracts International*, 38:611.

Heller, C. (2003). *Désir, nature et société: L'écologie sociale au quotidien*. Montréal: Écosociété. Originally published in English in 1999 under *Ecology of Everyday Life: Rethinking the Desire for Nature*.

Hesford, W. S. (1999). *Framing Identities: Autobiography and the Politics of Pedagogy*. Minneapolis, MN: University of Minnesota Press.

Hungerford, H., Peyton, R. B., & Wilke, R. J. (1980). Goals for curriculum development in environmental education. *Journal of Environmental Education, 11*(3), 42–47.

Hungerford, H., Peyton, R. B., & Wilke, R. J. (1983). Yes, EE does have definition and structure. *Journal of Environmental Education, 33*(3), 5–9.

International Union for the Conservation of Nature. (1980). *World Conservation Strategy: Living Resource Conservation for Sustainable Development*. International Union for the Conservation of Nature, United Nations Environment Programme (UNEP) and World Wildlife Fund (WWF). Switzerland.

Jickling, B. (1991). Environmental education, problem solving, and some humility please. *The Trumpeter, 8*(3), 153–155.

Keeley, B. (2007). *Le capital humain: Comment le savoir détermine notre vie*. Coll. Les essentiels de l'OCDE. Paris, France: Les éditions de l'OCDE.

Kelsey, E. (2003). Constructing the public: Implications of the discourse of international environmental agreements on conceptions of education and public participation. *Environmental Education Research, 9*(4), 403–427.

Khun, T. (1970). *The Structure of Scientific Revolutions* (2nd ed.), enlarged. Chicago, IL: University of Chicago Press.

Lake, D. (2001). Waging the war of the words: Global warming or heating. *Canadian Journal of Environmental Education, 6*, 52–57.

Lakoff, G., & Johnson, M. (1980). *Metaphors we live by*. Chicago, IL: University of Chicago Press.

Lamarche, T. (Ed.). (2006). *Capitalisme et éducation*. Coll. Nouveaux Regards/Syllepse. Paris, France: Nouveaux Regards.

Latouche, S. (2004). *Survivre au développement: de la décolonisation de l'imaginaire économique à la construction d'une société alternative*. Coll. Les petits libres, No 55. Paris, France: Mille et une nuits.

Latouche, S. (2005). *L'invention de l'économie*. Coll. Bibliothèque Albin Michel Économie. Paris, France: Albin Michel.

Laval, C. (2007). *L'homme économique: Essai sur les racines du néolibéralisme*. Coll. nrf essais. Paris, France: Gallimard.

Laval, C., et Weber, L. (Eds.), Baunay, Y., Cussó, R., Dreux, G., & Rallet, D. (2002). *Le nouvel ordre éducatif mondial: OMC, Banque mondiale, OCDE, Commission Européenne*. Institut de Recherches Historiques, Économiques, Sociales et Culturelles (IRHESC) de la Fédération syndicale unitaire (FSU). Paris, France: Éditions Nouveaux Regards et Éditions Syllepse.

Le Goff, J.-P. (2003). *La barbarie douce: La modernisation aveugle des entreprises et de l'école*. Coll. Sur le vif. Paris, France: La Découverte. Originally published in 1999.

Le Grange, L. (2004). Against environmental learning: Why we need a language of environmental education. *Southern African Journal of Environmental Education, 21*, 134–140.

Leiss, W. (1994). *The Domination of Nature*. Montréal, QC et Kingston, ON: McGill-Queen's University Press. Originally published in 1972.

Livingston, J. A. (1973). *One Cosmic Instant: A Natural History of Human Arrogance*. Toronto, ON: McClelland and Stewart.

Livingston, J. A. (1981). *The Fallacy of Wildlife Conservation*. Toronto, ON: McClelland and Stewart.

Lousley, C. (1999). De(Politicizing) the Environmental Club: Environmental Discourses and the Culture of Schooling. *Environmental Education Research, 5*(3), 293–304.

McKeown, R., & Hopkins, C. (2003). EE ≠ ESD: Defusing the worry. *Environmental Education Research, 9*(1), 117–128.

Meisner, M. (1993). Wild words: Nature, language and outdoor education. *Pathways: The Ontario Journal of Outdoor Education, 5*(6), 5–11.

Merleau-Ponty, M. (1969). *La prose du monde*. Text established and presented by C. Lefort. Coll. Tel, No 218. Paris, France: Gallimard. Text by Merleau-Ponty written in 1952.

Moscovici, S. (2002). *De la nature: Pour penser l'écologie*. Paris, France: Éditions Métailié.

Mugerauer, R. (1995). *Interpreting Environments: Tradition, Deconstruction, Hermeneutics*. Austin, TX: University of Texas Press.

Naoufal, N. (2009). Éducation relative à l'environnement, dialogue intercommunautaire et apprentissage du vivre-ensemble. *Éducation et Francophonie, XXXVll*(2), 186–203.

Noddings, N. (2005). *Happiness and Education*. Cambridge, UK: Cambridge University Press. Originally published in 2003.

O'Donoghue, R. (2006). Locating the environmental in environmental education research: A review of research on nature's nature, its inscription in language and recent memory work on relating to the natural world. *Environmental Education Research, 12*(3–4), 345–357.

Payne, P. (1999). Postmodern challenges and modern horizons: Education "for being for the environment." *Environmental Education Research, 5*(1), 5–34.

Payne, P. (2003a). Postphenomenological enquiry and living the environmental condition. *Canadian Journal of Environmental Education, 8*, 169–190.

Payne, P. (2003b). The technics of environmental education. *Environmental Education Research, 9*(4), 525–541.

Payne, P. (2005). Lifeworld and textualism: Reassembling the researcher/ed and "others." *Environmental Education Research, 11*(4), 391–400.

Payne, P. (2006). Environmental education and curriculum theory. *Journal of Environmental Education, 37*(2), 25–35.

Pepper, D. (1989). *The Roots of Modern Environmentalism*. Londres: Routledge. Originally published in 1984.

Pepper, D. (1993). *Eco-socialism: From deep ecology to social justice*. London, UK: Routledge.

Perelman, C., et Olbrechts-Tyteca, L. (1988). *Traité de l'argumentation: la nouvelle rhétorique*. Bruxelles: Éditions de l'Université de Bruxelles.

Petrella, R. (2000). *L'éducation victime de cinq pièges: À propos de la société de la connaissance*. Coll. Les grandes conférences. Montréal, QC: Fides.

Piaget, J. (1947). *Les représentations du monde chez l'enfant: avec le concours de onze collaborateurs*. Paris, France: Presses universitaires de France. Originalement publié en 1926.

Pinar, W. F. (1994). *Autobiography, politics and sexuality: Essays in curriculum theory 1972–1992*. Coll. Counterpoints. New York, NY: Peter Lang.

Polanyi, K. (1983). *La Grande Transformation: Aux origines politiques et économiques de notre temps*. Translated from English by Catherine Malamoud. Coll. Bibliothèque des Sciences Humaines. Paris, France: Gallimard. Originally published in 1944.

Popkewitz, T. (1982). *The myth of educational reform: A study of school responses to a program of change*. Madison (Wisconsin), WI: University of Wisconsin Press.

Popkewitz, T. (2004). Une perspective comparative des partenariats, du contrat social et des systèmes rationnels émergents. In M. Tardif et C. Lessard (Eds.), *La profession d'enseignant aujourd'hui: évolutions, perspectives et enjeux internationaux* (pp. 243–264). Coll. Formation et profession. Saint-Nicolas Presses de l'Université Laval.

Postman, N. (1995). *The end of education: Redefining the value of school*. New York, NY: Vintage Press.

Price, L. (2005). Social epistemology and its politically correct words: Avoiding absolutism, relativism, consensualism, and vulgar pragmatism. *Canadian Journal of Environmental, 10*(1), 94–107.

Reboul, O. (1984). *Le langage de l'éducation: Analyse du discours pédagogique*. Coll. L'éducateur, no. 90. Paris, France: Presses Universitaires de France.

Ricœur, P. (1986). *Du texte à l'action: Essais d'herméneutique II*. Coll. Essais, no. 377. Paris, France: Seuil.

Rist, G. (2001). *Le développement: Histoire d'une croyance occidentale*. Paris, France: Presses de la Fondation nationale des sciences politiques.

Robottom, I. (1987). Contestation and consensus in environmental education. *Curriculum Perspectives, 7*(1), 23–27.

Robottom, I., & Hart, P. (1993). *Research in environmental education: Engaging the debate*. Geelong, Australie: Deakin University Press.

Rogers, R. (Ed.). (2003). *An introduction to critical discourse analysis in education*. Mahwah, NJ: Lawrence Erlbaum.

Rogers, R. (2008). Critical Discourse Analysis in Education. In M. Martin-Jones, A. M. de Mejia, & N. H. Hornberger (Eds.). *Encyclopedia of language and education. 2nd Edition, Volume 3: Discourse and Education*. (pp. 53–68). New York, NY: Springer.

Rogers, R., Malancharuvil-Berkes, E., Mosley, M., Hui, D., & O'Garro Joseph, G. (2005). Critical discourse analysis in education: A review of the literature. *Review of Educational Research, 75*(3), 365–416.

Rolston, H. (1994). *Conserving natural value*. Coll. Perspectives in Biological Diversity. New York, NY: Columbia University Press.

Sachs, W. (1999). *Planet dialectics: Explorations in environment and development*. Halifax, UK: Fernword.

Sauvé, L. (2005). Currents in environmental education: Mapping a complex and evolving field. *Canadian Journal of Environmental Education, 10*, 11–37.

Sauvé, L., Berryman, T., & Brunelle, R. (2007). Three decades of international guidelines for environment-related education: A critical hermeneutic of the united nations discourse. *Canadian Journal of Environmental Education, 12*, 33–54.

Scheffler, I. (1960). *The language of education*. Springfield, UK: Charles C Thomas.

Scott, W., & Gough, S. (2003). *Sustainable development and learning: Framing the issues*. New York, NY and London, UK: RoutledgeFalmer. 2- Policy context 3- language and meaning.

Sicard, M. (2008, June, July, & August). Images de l'invisible. *Sciences Humaines—Les grands dossiers No 11. Entre image et écriture: La découverte des systèmes graphiques*. 71–73.

Simmons. I. G. (1993). *Interpreting nature: Cultural constructions of the environment*. Londres, Grande-Bretagne: Routledge.

Singh, M. (1998). Critical literacy strategies for environmental educators. *Environmental Education Research, 4*(3), 341–354.

Smith, D. (2005). *Institutional ethnography: A sociology for people*. Lanham, MD: AltaMira Press.

Stables, A. (1996). Reading the environment as text: Literary theory and environmental education. *Environmental Education Research, 2*(2), 189–195.

Stables, A. (1998). Environmental literacy: Functional, cultural, critical. The case of SCAA guidelines. *Environmental Education Research, 4*(2), 155–164.

Stables, A. (2007). Is nature immaterial? The possibilities for environmental education without an environment. *Canadian Journal of Environmental Education, 12*, 55–67.

Stevenson, R. (2006). Tensions and transitions in policy discourse: Recontextualizing a decontextualized EE/ESD debate. *Environmental Education Research, 12*(3–4), 277–290.

Stevenson, R. (2007). Schooling and environmental/sustainability education: From discourses of policy and practice to discourses of professional learning. *Environmental Education Research, 13*(2), 265–285.

Taleb, M. (2008). L'éducation relative à l'environnement contre la modernité capitaliste—Une contribution au réenchantement du monde. *Éducation relative à l'environnement—Regards, Recherche, Réflexions, 7*, 277–290.

Tilbury, D. (1995). Environmental education for sustainability: Defining the new focus of environmental education in the 1990s. *Environmental Education Research, 1*(2), 195–212.

Torres de Oliveira, H. (2008). Popular education and environmental education in Latin America: Converging paths and aspiration. In E. González-Gaudiano & M. A. Peters (Eds.), *Environmental education—identity, politics and citizenship* (pp. 219–230). Rotterdam: Sense Publishers.

UNCED. (1993). *Agenda 21: Programme of Action for Sustainable Development. United Nations Conference on Environment and Development*. Rio de Janeiro, 3–14 June 1992.

UNESCO-UNEP. (1976). *Belgrade charter: A global framework for environmental education*. International Environmental Education Workshop. Belgrade, 13–22 October 1975. *Connect, 1*(1), 1–9.

UNESCO-UNEP. (1977). *Tbilisi Declaration and Final Report—Intergovernmental Conference on Environmental Education*. Tbilisi, 14–26 October 1977.

Vogel, S. (1996). *Against nature: The concept of nature in critical theory*. Albany, NY: State University of New York Press.

Wals, A. E. J., & van der Leij, T. (2007). Introduction. In A. Wals (Ed.), *Social learning towards a sustainable world*. Wageningen, The Netherlands: Wageningen Academic Publishers,.

Warriner, D. (2008). Discourse analysis in educational research. In K. A. King & N. H. Hornberger (Eds.), *Encyclopedia of language and education. 2nd Edition, Volume 10: Research Methods in Language and Education* (pp. 203–215). New York, NY: Springer.

Whiteside, K. H. (2002). *Divided natures: French contributions to political ecology*. Cambridge, MA: The MIT Press.

World Commission on Environment and Development. (1987). *Our common future: The brundtland report*. New York, NY: Oxford University Press from the World Commission on Environment and Development.

15

Researching Tensions and Pretensions in Environmental/Sustainability Education Policies

From Critical to Civically Engaged Policy Scholarship

Robert B. Stevenson
James Cook University, Australia

Introduction

Although educational policies have traditionally emanated from national or sub-national governments and their agencies, international organizations (e.g., UNESCO, IUCN) and intergovernmental conferences have played a major role in environmental education (EE) and education for sustainable development (ESD) or sustainability (EfS) policies. This change accompanies a more recent broader trend in which educational policies are no longer exclusively developed with a national system but have come, under the forces of globalization, to be "now framed, produced, disseminated and implemented differently" (Rizvi & Lingard, 2010, p. 14). Over the past four decades there has been much activity in the production of policy statements for EE and, more recently, ESD or EfS. These statements, which were developed at an international level at meetings and conferences sponsored by UNESCO in Stockholm (1972), Belgrade (1975), Tbilisi (1977), Moscow (1987), Rio de Janeiro (1992), Thessaloniki (1997), Johannesburg (2002), and Bonn (2009), have been very influential in shaping national policies. Such policies have proliferated around the world, initially mainly in advanced industrialized countries, but in recent years more globally in response to the emerging processes of globalization and an emerging focus on education for sustainable development or sustainability. Many scholars have suggested that policies are more common and more developed than actual practices in formal education systems (Gough, 1997; Scott & Gough, 2003; Stevenson, 1987, 2007), and some have argued that the discourse of policy differs in important ways from the discourse of practice (Gough, 1997; Stevenson, 2007).

This chapter examines the role of policy processes and discourses in shaping the formation of policy statements for environmental and sustainability education. I begin by reviewing different approaches to the study of educational policy in general with a focus on more recent policy scholarship, specifically critical sociological and sociocultural perspectives. I then draw on these perspectives for analyzing international and national policy processes in environmental education (EE) and education for sustainable development or sustainability (ESD/EfS), particularly to reveal the underlying environmental/sustainability and educational values and ideologies that are involved in policy formation, including the tensions, struggles, and negotiations around these ideological interests that are masked in policy texts by a pretension of consensus around a claimed common good. Converging and competing EE/ESD/EfS policy discourses in national contexts and educator responses are examined before concluding with a call for democratizing policy discourses and processes and moving toward a more balanced and dialectic relationship between policy and practice.

Educational Policy Scholarship

Policy can be viewed as a process as well as a product or text. Traditionally, however, scholarship on educational policy has treated the policy process as simply two separate phases of policy formulation and policy implementation. Accordingly, there were in the past essentially two distinct bodies of policy study work: one on the generation of policy, mainly macroanalyses of policy documents; and the second more micro-level analyses of the implementation and impact of policies on practices and situations in educational institutions (i.e., on the consequences of policy enactments). This separation of the studies of formulation and implementation assumes a linear conceptualization of the policy process and implicitly a top-down one-way managerial model (Bowe & Ball, 1990; Levinson & Sutton, 2001) of the "relationships between setting the policy agenda, the production of the policy text and its implementation into practice" (Rizvi & Lingard, 2010, p. 6). This perspective, labeled technical-rational by critics, tends to focus in the case of the study of policy implementation on "the administrative and procedural aspects. . . . while ignoring or downplaying the influence of context" (Datnow & Park, 2009, p. 348).

Scholars concerned with local responses to federal educational policy initiatives in the 1960s in the United States engaged in a more substantive analysis of policy implementation and drew attention to the importance of the local context and teachers' and administrators' interpretations in shaping how policy is enacted. For example, the "Rand Change Agent" study in the United States revealed a process of mutual adaptation, rather than one of uniform implementation, that takes place at a local school level (McLaughlin, 1987). These studies, especially the work of Elmore and McLaughlin (1988), began to challenge a simple linear model of the policy process and brought increasing attention to "implementation as an integral part of the policy process, rather than what happened after policy is made" (Levinson & Sutton, 2001, p. 6).

Scholarship in policy sociology, centered in the UK and Australia, expanded on this sensitivity to context and responded to the above criticisms by focusing on the various contexts that shape the policy process. For example, in the UK Stephen Ball and his colleagues proposed a tripartite framework or "policy cycle" for examining policy: the context of influence where policy agendas are established; the context of policy text production; and the context of practice or implementation, the arena to which policy is aimed (Bowe & Ball, 1990). This research, particularly from a critical sociology orientation, also began to address one of the criticisms of the mutual adaptation perspective in neglecting the issue of power and the asymmetry of relationships among the different actors and agencies in the policy process (Datnow & Park, 2009). Simply put, it was argued that policy can be viewed as an instrument—and at times a blunt instrument—for expressing and enacting power through one, usually elite, group attempting to impose their ideas on others.

A similar orientation emerged in the form of a sociocultural approach to policy which reconceptualizes policy as a complex social practice—and analyzes the meaning of policy in practice, particularly as a practice of power (Levinson & Sutton, 2001). This conception of policy does not "privilege official governing bodies only, and includes unofficial and occasionally spontaneous normative guidelines developed in diverse social spaces" (Levinson & Sutton, 2001, p. 2). Introducing the notion of practice further allows for examining the way individuals and groups "engage in situated behaviors that are both constrained and enabled by existing structures" (p. 3) but while exercising agency in these situations. The ongoing nature of practice also captures the process in which policies are formulated and reformulated, often being revised in light of reactions by those affected and by changes in the policy actors. In other words, educators can also be viewed "as policy makers or potential makers of policy, and not just the passive receptacles of policy" (Ozga, 2000 cited in Rizvi & Lingard, 2010, p. 5).

Both critical sociological and sociocultural approaches conceptualize policy processes as complex, "messy, often contested and nonlinear relationships that exist between aspects and stages of policy processes" (Rizvi & Lingard, 2010, p. 6). These stages include, in addition to the production of texts, agenda setting, the representation of the text and the interpretation and implementation of the policy in a recursively dynamic interrelationship that continues over time—rather than a simple linear process, charter, or political act (Levinson & Sutton, 2001; Rizvi & Lingard, 2010). These approaches suggest the need to examine not only the content of EE/ESD/EfS policy statements, but also the processes and factors influencing the formulation of these statements in order to determine the underlying assumptions, values and ideologies (Codd, 1998).

Environmental and Sustainability Education Policy Processes

These new perspectives over the last couple of decades have reframed how policy studies are conceptualized and investigated. These approaches have contributed to revealing how the processes and pressures of globalization have produced a number of shifts in education policy processes; for example, policy is now "multidimensional and multilayered and occurs at multiple sites" (Rizvi & Lingard, 2010, p. 14). Given that there are multiple actors involved at multiple sites, there are multiple influences on policy formulation (as well interpretation and implementation). As a form of normative decision-making practice, policy is influenced by the values and ideologies of these actors, which in turn are shaped in part by the contexts within which those actors are located. These influences represent different and often competing political and ideological interests which result in struggles over discourses and purposes that continually need to be resolved in the formulation of policy. This suggests that attention should be paid not only to the content and contexts of policy production but also to the contests over values and ideals among those engaged in policy formulation.

In this section, the influence of different political and ideological interests of the actors involved in producing policy statements on EE/ESD/EfS at international conferences and meetings sponsored by UNESCO is examined, as well as how struggles over different interests have played out in the resultant policy texts. I draw on the work of Annette Gough (1997) and Masahisa Sato (2006) who have provided descriptions and analyses of the policy processes and statements developed at UNESCO conferences and meetings.

The Belgrade Workshop (1975), which identified a goal, objectives, and guiding principles for EE, was a working meeting of mainly people actively involved in EE. The content of the Belgrade Charter, which came to be widely cited by policymakers and EE scholars, was not surprisingly influenced by the first director of the UNESCO-UNEP International Environmental Education

Program (IEEP), a well-known academic from the School of Natural Resources at the University of Michigan, Professor Bill Stapp. Its most notable inclusion was the addition to previous conceptualizations of EE of an action component, expressed as part of the goal statement as a "commitment to work individually and collectively toward solutions of current problems and prevention of new ones." Stapp and his colleagues (1969) had gestured toward an action dimension in identifying an objective for EE of helping individuals acquire "attitudes of concern . . . which will motivate citizens to participate in biophysical environmental problem-solving" (p. 31), but the Belgrade Charter statement was a more explicit call for taking action as a goal of EE. This reflected the ideological commitment of Stapp and many others working in EE at the time as scholars or state policymakers to foster citizen participation in natural resource management and the resolution of environmental issues.

However, as already mentioned, there are multiple influences on and sites of policy formulation and text production. Recognizing that one of the most important policy sites is at the level of the state or national governments, two years later UNESCO organized the Tbilisi (1977) intergovernmental conference on EE to create a commitment among governments to establish EE as a national policy priority (Fensham, 1978, cited in Gough, 1997). In contrast to the Belgrade workshop, high level government representatives were the main participants. It was carefully structured, according to an Australian representative Professor Peter Fensham, to produce a consensus document, the Tbilisi Declaration, which was seen as providing an agreed framework for translating goals and policies into national action plans (Gough, 1997).

Gough reports Fensham's observation that the evaluation and participation objectives from the Belgrade Charter were deleted at the Tbilisi conference as not serving the political interests of the government participants in Tbilisi. For example, the evaluative ability objective could be seen as potentially supporting critiques of government policies and programs and was noticeably omitted from the categories of objectives in the Tbilisi Declaration (Gough, 1997). The participation objective was "softened" from "to help individuals and social groups develop a sense of responsibility and urgency regarding environmental problems to ensure appropriate actions to solve those problems" (Belgrade Charter, 1975) to "provide social groups and individuals with an opportunity to be actively involved at all levels in working toward resolution of environmental problems" (Tbilisi Declaration, 1978). That the participation objective was modified rather than deleted seemed to reflect a compromise to accommodate the representatives present from the Belgrade meeting who worked in EE. Although evaluation was viewed by most environmental educators as important for developing the ability to make decisions on complex environmental issues, participation and action were generally seen as not only important but as an essential characteristic if EE was to serve the purpose of influencing environmental amelioration. And encouraging participation was probably viewed by the government representatives as less threatening than supporting evaluation. These positions suggest there was more room for negotiation to include participation in some form than for retaining the evaluation objective. This compromise illustrates the kind of ideological strategies used by political elites to formulate policy (Eagleton, 1991 cited in Levinson & Sutton, 2001).

Policy formulation typically involves these kinds of political and ideological tensions and conflicts and policy text represents the outcome of those struggles—usually in the form of compromises. As Stephen Ball and his colleagues have argued, the context of influence is often related to the articulation of specific narrow interests and ideologies, whereas policy text is usually framed in the language of the general public good and masks underlying ideologies as well as conflicts and compromises. The result is a pretension of consensus around the common good. Yet, important questions need to be asked about the context of influence, including: Whose voices and power/interests are represented? What and whose knowledge and discourse are represented? What informs the policy? Who benefits from policy?

Tensions and Struggles Over Environmental and Sustainability Ideologies

Two other major struggles over different political and ideological interests illuminate some of these questions in relation to environmental and sustainability policies. The first concerns the differing political interests and economic concerns of the developed and developing countries which were reflected in struggles over language and discourse.

The World Conservation Strategy (1980), which was produced by UNEP in conjunction with two other international organizations, the International Union for the Conservation of Nature (IUCN) and the World Wildlife Fund (WWF), addressed for the first time the conflict between environmental conservation and development and the notion that development was an important means of achieving the goals of conservation which led to the introduction of the concept of sustainable development (Sato, 2006, p. 3). This concept was formulated at intergovernmental conferences that included representatives from developed, or countries of the North, and developing or countries of the South. It was pressures from the latter countries concerned about their need for further development of their economies and infrastructures which resulted in the emergence of sustainable development in the discourse and the substitution of "environment and development" in Agenda 21 (the product from the UNESCO-UNEP conference in Rio de Janeiro in Brazil in 1992) for "environment" in the policy statements from Tbilisi (Gough, 1997). With the exception of this substitution, the principles enshrined in the Tbilisi Declaration remained the basis for the Agenda 21 chapter on promoting education, public awareness, and training.

By the late 1980s, education for sustainable development (ESD) began to replace EE as the dominant discourse in international policy circles. A sociocultural approach to analyzing policy as practice attempts to illuminate the social networks that are in play in both the policy production and appropriation process. For example, an international network of policymakers and scholars associated with IUCN, which was one of the authors of the World Conservation Strategy that introduced the concept of sustainable development, was influential in the resulting textual discourses. Members of the IUCN Commission on Education and Communication were involved in the UNESCO conferences and dialogs and had an interest in the replacement of EE with ESD. A sociocultural approach, drawing on ethnographic methods, might interrogate the ways this network contributed to shaping the policy texts.

Five years after Agenda 21, Declaration 10 from the UNESCO-UNEP conference in Thessaloniki (1997) introduced the concept of sustainability to encompass "not only the environment but also poverty, population, health, food security, democracy, human rights and peace" (p. 6). Subsequently, the text of Declaration 11 stated that EE has evolved from its conceptualization at the Tbilisi conference to address all of these global issues included in Agenda 21 and has also now become recognized as education for sustainability (Sato, 2006). Interestingly, Sato argues that in order to make clear that the roots are planted in environmental education, it was also recognized that EE may be referred to as education for environment and sustainability (Sato, 2006). This addition of the word "environment" could be viewed as responding to concerns voiced in debates at conferences and in the literature about the ESD/EfS discourse deflecting attention from environmental issues (see, e.g., Jickling, 1992).

A related tension and struggle over ideological interests is evident in the ways in which the discourse has been appropriated. For example, one UNESCO observer commented recently that "I find that 'sustainability' is used more and more in the context of business and innovation (in the sense of sustaining profit) and stripped from any planetary conscience, whereas sustainable development at least still has something do with the planet" (Wals, personal communication). Yet, ironically the use of the term "sustainabililty" instead of "sustainable development" has been suggested as being preferred in order to avoid the term "development," while others, as mentioned, have expressed concerns that both terms—sustainability and sustainable development—leave out "environment" both literally and in practice. This omission creates the space in which the terms can be (re)presented or appropriated by those not involved in official policy formation in ways that are more favorable to their own interests. Others will represent or frame policies in discourses with which they are familiar: for example, in Australia, the UK, and North America teachers still tend to use environment and environmental when discussing issues that others would classify as being issues of (ecological) sustainability.

These struggles over language and discourse illustrate how there is confusion and contestation within and across contexts of influence.

> In the production of policy texts . . . there are attempts to appease, manage and accommodate competing interests. . . . policy texts often seek to suture together and over competing interests and values. At the same time, policies usually seek to represent their desired or imagined future as being in the public interest, representing the public good. As a result, they often mask whose interests they actually represent. (Rizvi & Lingard, 2010, p. 6)

The development of the concepts of sustainable development and sustainability and their educational equivalents of ESD and EfS as described is a prime example of accommodation of competing interests. The policy texts that are produced leave out the context in which they were formulated, including any conflicts over discourse and thereby do not reflect struggles over underlying values and ideologies. Instead they connote a consensus, although frequently the text represents only a pretension of consensus. Ideological interests can play out or result in compromise or changes in the discourse. The consequence can be what has been described as "a paradoxical compound policy slogan" (Stables & Scott, 2002) "that enables people with widely different views to accept it to some degree without agreeing on any of the underlying philosophical and political issues which remain obscured and in tension" (Stevenson, 2006, p. 278).

In this respect, the word "sustainability" can be viewed, like democracy and equity, as "a floating signifier" that attempts to establish a new hegemony around a common agenda for change (Laclau, 1990). Couched in language that suggests the agenda is in the public good, its use in policy text seeks to create universal appeal and influence while enabling the masking of underlying ideologies, conflicts, and compromises (Bowe & Ball, 1990). Yet what constitutes the public good and in what time frame (e.g., short-term or long-term, present or future generations) is a contested issue, as is what policy makers see as representing the public good or the interests of society-at-large and the interests of particular groups (Scott & Gough, 2003). The common conception of sustainability, specifically intergenerational equity, calls for addressing the interests of future generations, but this can involve costs for individuals and families of the present generation. The important question that underlies these issues is: Whose interests are being served? Consideration of this question invariably reveals that policies not only have costs as well as benefits to people, the environment and the economy, but also can result in inequities arising from the hegemony of particular policy actors.

Tensions and Struggles Over Educational Ideologies

A second set of tensions and struggles over policy statements involves different educational ideologies among scholars in the field and policymakers inside and outside the field. The agenda for the UNESCO-UNEP Congress on Environmental Education and Training in Moscow (1987) sought to produce an international strategy for action based on the assumption that the Tbilisi Declaration offered a basic framework for EE at all levels, inside and outside schools (Gough, 1997) and therefore there was no need for further work on conceptualizing the field. Yet scholarly debates had taken place over the guiding principles for EE from Tbilisi. Some argued that there was now a consensus about the goals, most notably Hungerford, Peyton, and Wilke (1980) who had developed a set of goals for curriculum development in EE and therefore had an interest in truncating debate and moving the research and evaluation agenda to curriculum and professional development based on their conceptualization of EE (Gough, 1997). Other writers, however, had challenged such assumptions (e.g., Hart, 1981; Robottom, 1987) with many Australian writers arguing for a socially critical approach to EE—an approach that emphasized the evaluation and participation objectives that were deleted or modified from the Tbilisi Declaration. They also questioned the human interests and ideologies underlying a positivist and technical-rational approach, as well as the influence of historical and political antecedents of contemporary practices in the field.

Gough (1997) argues that the UNESCO and IUCN statements represent a managerial view of policy in that the "subjects of the policy product are seen as passive recipients of outcomes of the policy process rather than actors in the process" (p. 41) and the discourses are assumed to be institutionalized in national and regional policies. Furthermore, statements of objectives for EE from Tbilisi and for education in Agenda 21 tend to assume an instrumental role in achieving the goals of sustainable development. For example, Chapter 36: "Education, training and awareness" appears in Agenda 21 in the section "Means of Implementation" of sustainable development. This perspective is supported by Sato (2006) who argues that a top-down RDDA (research, development, dissemination, and adoption) approach was taken by IEEP to the promotion of environmental educational activities from its establishment by UNESCO-UNEP in 1975 to when the program ended in 1995. He also points out that the Brundtland (1987) report, "Our Common Future," in calling for raising awareness of environmental issues, made the assumption that greater awareness would change attitudes and behaviors—an assumption that has been challenged by many scholars as not representing the complex nature of behavior change (e.g., Agyeman & Kollmuss, 2002; Trainer, 1994) and premised on the questionable belief that only individual behavior change is necessary for addressing environmental issues.

Furthermore, as Annette Gough points out, this instrumental approach to education is challenged by the issues raised by Noel Gough (1987) in questioning who has the right to declare what is good for the environment and Bob Jickling (1992) who articulated a concern that education should be about building people's capacity to think for themselves rather than being "for" something, such as sustainable development. Yet accompanying the emergence of ESD, according to a participant in UNESCO conferences, has been a claimed focus on a bottom-up approach of "participatory learning, higher-order thinking and action research" (Sato, 2006, p. 10). Developing such an approach is a challenge given that, as indicated, the construction of international policy statements on ESD and EfS has only included certain privileged participants, including government representatives from both the environment and education policy making arenas, and excluded environmental education practitioners. Sato, however, argues that there has been a move toward more regionally based conferences and meetings (e.g., South East Asia) enabling the involvement of a greater number of participants, but these still generally fail to include school and community-based workers.

Most recently, the Bonn Declaration (2009) from the UNESCO World Conference on Education for Sustainable Development, which followed and referenced the global financial crisis, draws attention to "the risks of unsustainable economic development models and practices" and the need for "new economic thinking" as well as "stronger political commitment and decisive action" (www.esd-world-conference-2009.org). The implication being that environmental and sustainability issues are more than technical and demand attention not only to individual behaviors but also to our social and economic systems. Yet there simultaneously remains an apparent faith in a positivist ontological worldview and externally produced (science and social science) knowledge that is applied in a technical rational approach to change as expressed in the statement: "We now need to put this knowledge into action." The seemingly contradictory messages continue with the call for action at a policy level emphasizing the promotion of awareness and understanding through disseminating information and providing resources and training in ways that could continue a top-down linear approach, while actions at a practice level represent efforts to engage all sectors of society in order to stimulate their participation in and "ownership of ESD questions and issues."

National Policy Contexts and Converging and Competing Discourses

Public policy remains an activity of the nation state, but educational policymakers within the state today "are also networked with policymakers in agencies beyond the state, including international organizations such as the OECD, UNESCO and supranational organizations such as the EU, resulting in the creation of an emergent global education policy community" (Rizvi & Lingard, 2010, p. 16). The emergence, or perhaps more accurately heightened awareness, of global environmental issues has further enhanced the participation of nongovernment and international agencies in environment and environment-related education matters. These trends have contributed to a global convergence of policy ideas and ideologies which are circulated, in the case of EE/ESD/EfS, through conferences, meetings and electronic media among a relatively small group of policymakers in international organizations, primarily UNESCO and IUCN, as well as nongovernment agencies, such as World Wide Fund for Nature/World Wildlife Fund (WWF), and national governments.

There are three related consequences of this situation. First, policy support, at least in terms of rhetoric, now exists for the inclusion of ESD/EfS in school systems in numerous countries. A second is that national policies in ESD/EfS are largely borrowed from and been shaped by the policy statements that emerged from the international conferences that have been described. The third result is that, although a few local academics are often engaged as consultants to varying degrees in different countries in the process of producing national policy texts, there is considerable convergence in the policies across different historical, cultural, political, and economic contexts.

National policies convey an apparent global consensus on ESD or EfS replacing EE as the preferred discourse. So despite the struggles over environmental and educational ideologies that have been discussed, national policies converge around a similar discourse and (ambiguous) conception of ESD/EfS. The political rationality that emerged from conflicts and negotiations at international meetings and resulted in the creation of the sustainable development discourse has parallels in policy formulation at the national level. For example, Annette Gough (Greenall, at the time) noted that "some Australian states felt that statements about environmental education needed to be positive rather than negative and an emphasis on environmental problems was seen as negative" (Greenall, 1981 cited in Gough, 1997, p. 20). This concern was more widespread among governments than just in Australia and foreshadowed some of the arguments for the later move to ESD which was viewed as having a more positive futures-orientation than a perceived negative connotation of a problem orientation of EE (see Smyth, 1995). There was also a shift in the policy discourse in Agenda 21 from a focus on the resolution of environmental problems or a current problem orientation to a futures orientation in focusing on the creation of a sustainable future (Smyth, 1995).

However, although the international networking situation has significantly contributed to national educational policy contexts which could now be regarded as very supportive of sustainability education, the nature of this support should be examined. For example, the draft national curriculum in Australia includes such statements as "complex environmental, social and economic pressures such as climate change that extend beyond national boundaries pose unprecedented challenges." If this is the extent of statements about the climate change, then it illustrates how policies of the state emphasize safe knowledge that is unlikely to create community dissent or anxiety. This also means that knowledge is likely to be treated as devoid of complexities, nuances, and uncertainties.

Negotiations over discourse in EE/ESD/EfS policy arenas serve as a reminder that power and knowledge are contested along other axes of social and institutional contexts (Carlson, 1992) such that these policy discourses compete with other more dominant policies and discourses both outside and within education. For example, educational policies in most English-speaking advanced industrial nations (e.g., Australia, Britain, Canada, United States) have been tied to economic policies by emphasizing performativity and accountability which have resulted in a narrowing of the curriculum to literacies and skills that are seen as enabling these countries to compete in the global economy. Coupled with the emergence of international comparisons of student achievement that has also shaped the educational policy arena on a global scale and altered the priorities and values at national and local levels, the consequence has been that subjects outside the core tested areas of literacy, math and science are being de-emphasized and the purposes of schooling and teaching and learning are being narrowed. Thus EE/ESD/EfS discourses and policies have to compete for resources with other educational (and environmental) policies that usually receive higher priority.

In some regions and countries, such as in Latin America, the discourse of EE competes with other related discourses associated with such progressive traditions as emancipatory education and critical pedagogy (Gonzalez-Gaudiano, 2007). As a result, Gonzalez-Gaudiano argues there has been a hybridization of the field with the emergence of ESD and connections made to adult/community and emancipatory educational and political traditions. He argues this has created a problem of splitting the discursive identity and configuration of the field in Latin America.

However, even if EE/ESD/EfS discourses and policies receive greater attention, schools are widely seen as instruments for changing behaviors toward those governments and experts have determined are desirable. In other words, political history suggests that the state will continue to try to use policy to dictate how education reforms such as EE/ESD/EfS will be defined and implemented. The likely result is that "weak conceptions of sustainability will be invoked in which environmental issues are not portrayed as open to interpretation" (Stables & Bishop, 2001; Stevenson, 2007b, p. 272).

This reflects one of the tensions in policy processes between discourses of prescription, in the form of mandated policies and/or specified courses of action, and discourses of empowerment that encourage participation and support self-initiated responses by individuals, groups, and communities. Empowerment should start with engaging educational practitioners in policy formation—an issue I address in the final section of this chapter.

Policy Interpretations and Responses

Policy is about change, either desired or imagined, as it offers an imagined future state of affairs, but in articulating desired change always offers an account somewhat more simplified than the actual realities of practice. In many ways policies eschew complexity . . . They are designed to provide a general overview, leaving a great deal of room for interpretation . . . a policy is designed to steer understanding and action but without ever being sure of the practices it might produce (Rizvi & Lingard, 2010, p. 5).

In addition to concealing the complexities and uncertainties associated with the foundational concepts of sustainability, policy is also not sensitive to complex structures such as teaching and learning. Perhaps it is not surprisingly then that EE/ESD/EfS policy texts have little to say about the curriculum and pedagogical tensions and challenges in enacting the goals in local settings.

These international and national policy discourses can offer a framework for local initiatives but must be mediated through other national, regional, and local educational policies (Stevenson, 2006). Principles that frame the sustainability discourse (e.g., biodiversity, intergenerational equity) also need to be translated into curriculum and pedagogical practices, so "policy discourse must be (re)contextualized and transformed by teachers into their own discourses of practice, and most importantly, into pedagogical actions" (Stevenson, 2007b, p. 269). Educators have developed, if not always articulated or made explicit to themselves or others, their own practical theories or theories of practice which encompass their assumptions, beliefs and values regarding schooling, teaching, learning, and students (as well as their perceptions of the norms and expectations of their school and community) and largely determine their responses to educational policies. Recontextualization involves educators in interpreting what the policy as an intervention means for (changes in) their own practice and is filtered through these practical theories and shaped within educators' understandings of other policies and the constraints and possibilities of the context in which they work. Teachers' responses to local policies also have been found to be related to the degree to which they perceive the policies as threats to their professional autonomy (Agnostopoulous, 2003).

Interrogating how educators appropriate the meaning of the discourse and the institutional power and resources of a policy in light of their own interests, intentions, and theories of action is a focus of sociocultural analysis. Such a policy analysis includes how local actors affected by a policy interpret its meanings in relation to their everyday practices and respond or exercise agency, which can include adopting or applying, adapting, or contesting a particular policy. Given the widely recognized gap between policy discourses and implementation or responses in schools (Stables & Scott, 2002; Stevenson, 2007a), an important question for environmental education researchers is: How are EE/ESD/EfS policies represented, interpreted, appropriated, and implemented at the local level?

A response of many educators and some of their professional associations (e.g., North American Association for Environmental Education) has been to package EE "in the language of student achievement in the basic content areas" by aligning its goals with state standards (Gruenwald & Manteaw, 2007, p. 176). These authors argue that this response of accommodation or "playing the achievement game," while at times being necessary, mutes or distorts the purposes of environmental education.

This response reinforces the need for policy discourses to be informed by research and practice and continually and reflexively reconceptualized (Plant, 1995). I would argue that this process should be democratized to engage not only policymakers and academics but also educational practitioners and other community stakeholders. Such a proposal, of course, raises many questions about challenging power relations and ideological interests, including issues of empowerment and disempowerment and resistance to loss of power (Scott & Gough, 2003).

Civically Engaged Policy Research and Democratizing Policy Processes

A sociocultural approach to educational policy should go beyond analysis to democratization by stimulating and supporting more participatory approaches to policy processes. This demands reforming "the cultures of policy formation" (Rizvi & Lingard, 2010, p. 7) which are more complex in situations where "international as well as national actors and institutions are involved" (p. 9). It also demands reforming the cultures of academia and scholarship. A civically engaged scholarship is not only concerned with the production of knowledge but also with reflecting on its use and impact and expanding the (public) dialog or scope of conversation beyond the (walls of) academy to all the many publics in any society (Nozaki, Weis, Granfield, & Olsen, 2006). A democratizing of the policy process demands opening up participation in policy production and enabling multiple voices, especially from traditionally disenfranchised groups, to be heard so that diverse perspectives are incorporated into the discourse. Scholarly discourses inform policy discourses when academics are involved in policy production and the discourses of policy texts inform scholarly discourses. However, neither tends to be regularly informed by discourses of practice. Yet policy should be informed by both discourses of scholarship and practice. In the case of EE/ESD/EfS, there is "a need for identifying and creating spaces for engaging educators in the discourse so it is constructed *with* them rather *for* them and is contextualized historically, pedagogically, and politically" (Stevenson, 2006, p. 288).

Rather than calls for evidence-based policy, which tends to only address what works and to ignore or de-emphasize other factors that shape educators' decisions about their practice, what is needed is evidence-informed policy which recognizes the place of professional knowledge and values, alongside research-generated knowledge or evidence to inform decision making in practice. The Bonn Declaration recommends promoting evidence-informed policy dialog on ESD "drawing upon relevant research, monitoring and evaluation strategies, and the sharing and recognition of good practices." Evidence-informed policy (and practice) is consistent with Robinson's (1998) argument that educational policies and practices both can be viewed as activities intended to solve problems about what to do in particular situations. Most importantly, she further argues the need to understand the reasoning of educators in formulating the problem and responding to any tensions in the various constraints surrounding, in this case, a pedagogical response to a policy.

Policy often tends to serve more symbolic and political purposes than educative ones, although it can serve as leverage for educational reform. Although usually used as a (blunt) instrument for prescribing reform, policy can also be used to leverage learning and participation by educators to imagine and enact authentic reform. In order to improve the quality of normative decision making, and thereby the quality and influence of policy, policy practices could also be treated as a learning process—just as sustainability itself should be viewed as a learning process (Scott & Gough, 2003).

One possible approach is through constructing a new discourse of professional learning which, instead of treating policy and practice as a linear top-down relationship, embraces a dynamic and dialectic relationship between, on the one hand, goals and conceptual characteristics (typically encapsulated in policy discourse) and, on the other hand, curriculum and pedagogical practices (as represented in discourses of practice) such that learning about one informs learning about the other (Stevenson, 2007b). This discourse could occur in communities of practice within schools as well as in networks and partnerships linking educators with local, regional, national and even global environmental groups and organizations with specific expertise and resources in environmental and sustainability issues. A discourse of professional learning could shift the focus of policy processes from implementation of EE/ESD/EfS as framed in traditional policy discourse to building normative and technical capacity, of both individuals and organizations, for developing meaningful practices in local contexts.

Learning could be further fostered by including in policy documents references to the debates, tensions, and conflicts that took place in formulating policy text. This could encourage different perspectives to be aired and legitimize disagreements and debates among those expected to respond to the policy and provide them with a sense of agency. Too often critical reactions to policy are silenced and the agency of respondents is not acknowledged or supported.

There will be continuing challenges and tensions in policy processes in relation to EE/ESD/EfS even with a democratization of these processes, but these tensions are more likely to be explicated and examined in a policy democracy. Such democratization of policy does not involve displacing those people currently making policy decisions but creating wider representation. Besides responsiveness to the diversity of stakeholder perspectives, it also demands being cognizant of and responsive to the diversity of contexts in which a policy is to be enacted. A pretension of a consensus or a common context deprives a policy of integrity. Hopefully, pretensions of consensus around meanings and understandings and guidance for practice will also thereby be reduced.

References

Agnostopoulous, D. (2003). The new accountability, student failure and teachers' work in urban high schools. *Educational Policy, 17*(3), 291–316.

Agyeman, J., & Kollmuss, A. (2002). Mind the gap: Why do people act environmentally and what are the barriers to pro-environmental behavior? *Environmental Education Research, 8*(3), 239–260.

Bowe, R., & Ball, S. (1990). *Reforming education and changing schools: Case studies in policy sociology*. London, UK: Routledge.

Carlson, D. (1992). Review essay: Postmodernism and educational reform. *Educational Policy, 6*(4), 444–456.

Datnow, A., & Park, V. (2009). Conceptualizing policy implementation: Large-scale reform in an era of complexity. In G. Sykes, B. Schneider, & D. Plant (Eds.), *Handbook of education policy research*. Philadelphia, PA: Routledge.

Elmore, R., & McLaughlin, M. (1988). *Steady work: Policy, practice and the reform of American education*. Santa Monica, CA: Rand Corporation.

Gonzalez-Gaudiano, E. (2007). Schooling and environment in Latin America in the third millennium. *Environmental Education Research, 13*(2), 155–169.

Gough, A. (1997). *Education and environment: Policy, trends and the problems of marginalization*. Melbourne, VIC: Australian Council for Educational Research.

Gruenwald, D., & Manteaw, B. (2007). Oil and water still: No child left behind limits and distorts environmental education in US schools. *Environmental Education Research, 13*(2), 171–188.

Hart, P. (1981). Identification of key characteristics of environmental education. *Journal of Environmental Education, 13*(1), 12–16.

Hungerford, H., Peyton, B., & Wilke, R. (1980). Goals for curriculum development in environmental education. *Journal of Environmental Education, 11*(3), 42–47.

Jickling, B. (1992). Why I don't want my children to be educated for sustainable development. *Journal of Environmental Education, 23*(4), 5–8.

Levinson, B., & Sutton, M. (2001). Introduction: Policy as practice. In M. Sutton & B. Levinson (Eds.), *Policy as practice: Toward a comparative sociocultural analysis of educational policy*. Santa Barbara, CA: Praeger.

McLaughlin, M. (1987). Learning from experience: Lessons from policy implementation. *Educational Evaluation and Policy Analysis, 9*(2), 171–178.

Noel Gough (1987). Arts of anticipation in curriculum inquiry. Paper presented at the annual meeting of the American Educational Research Association, Washington, April 20–24.

Nozaki, Y., Weis, L., Granfield, R., & Olsen, N. (2006). A call for civically engaged educational policy related scholarship. *Comparative & Global Studies in Education Newsletter, 8*(1), 8–10.

Plant, M. (1995). The riddle of sustainable development and the role of environmental education. *Environmental Education Research, 1*(3), 253–266.

Rizvi, F., & Lingard, B. (2010). *Globalizing education policy*. London, UK: Routledge.

Robinson, V. (1998). Methodology and the research-practice gap. *Educational Researcher, 27*(1), 17–26.

Robottom, I. (1987). Contestation and consensus in environmental education. *Curriculum Perspectives, 7*(1), 23–27.

Sato, M. (2006). *Evolving environmental education and its relation to EPD and ESD*. Paper presented at the UNESCO Expert Meeting on Education for Sustainable Development (ESD): Reorienting education to address sustainability. Kanchananaburi, Thailand, May 1–3, 2006.

Scott, W., & Gough, S. (2003). *Sustainable development and learning: Framing the issues*. London, UK: RoutledgeFalmer.

Smyth, J. (1995). Environment and education: A view of a changing scene. *Environmental Education Research, 1*(1), 3–20.

Stables, A., & Bishop, K. (2001). Weak and strong conceptions of environmental literacy: Implications for environmental education. *Environmental Education Research,* 7(1): 89–97.

Stables, A., & Scott, W. (2002). The quest for holism in education for sustainable development. *Environmental Education Research, 8*(1), 53–61.

Stapp, W., Bennett, D., Bryan, W., Fulton, J., McGregor, J., Nowak, P., . . . Havlick, S. (1969). The concept of environmental education. *Journal of Environmental Education, 1*(1), 30–31.

Stevenson, R. B. (1987). Schooling and environmental education: Contradictions in purpose and practice. In I. Robottom (Ed.), *Environmental education: Practice and possibility* (pp. 69–82). Geelong, Victoria: Deakin University Press. (Reprinted in 2007 in *Environmental Education Research, 13*(2), 139–153.)

Stevenson, R. B. (2006). Tensions and transitions in policy discourse: Recontextualizing a decontextualized EE/ESD debate. *Environmental Education Research, 12*(3–4), 277–290.

Stevenson, R. B. (2007a). Editorial (Special issue: Revisiting schooling and environmental education: Contradictions in purpose and practice.) *Environmental Education Research, 13*(2), 129–138.

Stevenson, R. B. (2007b). Schooling and environmental/sustainability education: From discourses of policy and practice to discourses of professional learning. *Environmental Education Research, 13*(2), 265–285.

Trainer, T. (1994). If you really want to save the environment. *Australian Journal of Environmental Education, 10*, 59–70.

16

Changing Discourses in EE/ESD

A Role for Professional Self-Development

Ian Robottom
Deakin University, Australia

Introduction

This chapter will discuss some trends in recent discourses in environment-related educational work, arguing that the field is marked by a policy context based on shifting discourses at national and international levels and a language that is slogan-like in effect, with the attendant risk of creating the conditions for changeless reform.

The chapter will describe some instances of environment-related educational practice, considering these in relation to the recent discourse shift in the field. An approach to professional development will be presented that takes into account the issues associated with the relationship of practice and discourse within particular educational contexts.

Recent Discourse Trends in Environment-Related Educational Work

In the field of environment-related education, the period from the early '70s to the present is marked by continuity and contestation. There has been a remarkable continuity of environment-related practice; and there has also been contestation in the language of the field, with terms like ecology education, environmental education, and education for sustainable development becoming highly visible at different times. Presently, the environment-related work formerly known as "environmental education" (EE) is being aggressively and extensively "rebadged" as "education for sustainable development" (ESD) (Campbell & Robottom, 2008). The United Nations has taken the significant step of establishing an international Decade of Education for Sustainable Development (UNDESD) (UNESCO, 2003), which has encouraged an impressive array of research and development projects around the world. UNESCO defines ESD as follows: "ESD involves learning how to make decisions that balance and integrate the long-term future of the economy, the natural environment, and the well-being of all communities, near and far, now and in the future" (UNESCO, 2005, p. 1). ESD has emerged as an internationally mandated educational movement concerned with responding to the rise of sustainability issues worldwide (UK Department for Education and Skills [DfES], 2006; Selby, 2006). That the rise of ESD represents a shift in discourse from EE is clear in the minds of key participants in the field:

> There is clearly a shift taking place between conservative approaches to informing people and students about the environment (commonly practised as environmental education) towards educating to think more critically and reflectively about change and how to engage in change for sustainability, which underpins ESD approaches. (Tilbury, 2004, p. 104)

This internationally mandated shift from a discourse of Environmental Education to one of Education for Sustainable Development is also clearly evident in Australia. Australian Government is developing a National Action Plan for Education for Sustainable Development (NAPESD). The discussion paper relating to the new national plan indicates that the shift in thinking and ESD terminology will be incorporated in the plan. The broader concepts of sustainability are included in many government documents such as "Caring for our Future" which is the Australian Government Strategy for the United Nations Decade of Education for Sustainable Development (2005–2014). This document defines ESD in this way:

> Education for sustainable development aims to equip individuals, organizations and communities to deal effectively with the complex and inter-related social, economic and environmental challenges they encounter in their personal and working lives, in a way that protects the interests of future generations. (Australian Government Department of the Environment and Heritage [DEH], 2005)

The discussion paper indicates that ESD is seen as transformative and is about managing change such that people are provided with not only knowledge and understanding, but also skills and capacity to administer change while recognizing the relationships between environmental protection, economic prosperity, and social cohesion. The paper specifies that ESD terminology "encompasses the activities of environmental education, learning for sustainability and change for sustainability." It is also clear that the new discourse of ESD has been institutionalized into a more powerful form—that of official intergovernmental and national policy. For example in Australia, the Australian Vice-Chancellors' Committee (AVCC) (the peak body of Australian university leaders, now Universities Australia) has declared a commitment to ESD in its formal statement that:

> By 2020, the university sector in Australia will be playing a key role in promoting sustainability in the community through research and building capacity to achieve change for sustainability . . . The AVCC will further promote sustainability by supporting its members and through the creation of strategic linkages with government . . . and encourages its members to engage with schools, industry and communities in partnership and projects which promote sustainable development. (AVCC, 2006)

Right from the start of UNESCO's public commitment to ESD, there have been allusions to a potential risk that then new ESD policy suite in environment-related work could lead to an expression in practice that is merely symbolic. Speaking at the international launch of DESD in New York in March 2005, UNESCO Director General Koichiro Matsuura set out a clear challenge for ESD:

> The ultimate goal of the Decade is that education for sustainable development is more than just a slogan. It must be a concrete reality for all of us—individuals, organizations, governments—in all our daily decisions and actions, so as to promise a sustainable planet and a safer world to our children, our grandchildren and their descendants . . . Education will have to change so that it addresses the social, economic, cultural and environmental problems that we face in the 21st century. (UNESCO, 2005, p. 2)

So if the discourse of ESD represents a "clear shift" from that of EE, and if ESD is to be "more than just a slogan," what does this change in language really mean, what lessons can be learned from the "environmental education" experience for the proponents of the newer term "education for sustainable development," and what are the implications of these lessons for professional development of practitioners in this field? It is timely now to ask whether this shift in discourse—this evolution in the language of the field—has been accompanied by real change in educational practices beyond the changes in descriptors, that is beyond mere language (including its more institutionalized forms of discourse and policy) to levels of organization and, especially, curriculum practice. As indicated by Matsuura, the real test of efficacy for a new policy is whether it occasions a significant change in everyday actions and (from an institutional education perspective) in educative practices concerned with addressing sustainability problems in the twenty-first century. Putting this another way, can contemporary educational practice conducted under the ESD discourse be differentiated from practices conducted under the EE discourse? The broader issue is whether the ESD discourse is governing environment-related educational reform in name only—reform that may entail no actual change in the practice of EE. This issue is of particular relevance to the matter of professional development of practitioners in environmental education. We will now look at recent instances of environmental education practice, and then consider implications for professional development.

An Instance of Contemporary Environment-Related Educational Work

The Southeast Asian Ministers of Education Organization (SEAMEO) *SEAMEO Search for Young Scientists (SSYS) Congress* is a regional science and ESD talent search held every two years at the Regional Center for Education in Science and Mathematics (RECSAM), in Penang, Malaysia. Participating in this congress are students and teachers from eleven countries in the Southeast Asian region, including a number of developing countries. Thus the SSYS Congress provides a wide-angled "window" into science education and ESD in much of the Southeast Asian region. A feature of many of the students' presentations at SSYS is that they are based on authentic local community problems. The approach has particular relevance for students in those communities in Southeast Asia where experience of ESD may otherwise be limited by their relative isolation and limited economic development—it provides an alternative source of ESD activity to the traditional "curriculum package" employed in well-resourced settings in developed countries. Thus this approach may have the capacity to inform innovation in ESD beyond the Southeast Asian region.

In 2010 the SSYS Congress adopted as its key theme "Sustainable Solutions for the Local Community." The SSYS Congress's general guidelines include the statement that, "Projects need to focus on the nature of the concept of sustainable community development and the three pillars of sustainable development: environment, economy and society." The SSYS theme is thus directly related to the discourse of ESD as can be seen in a comparison with the UNESCO Director-General's statement cited above that, "Education will have to change so that it addresses the social, economic, cultural and environmental problems that we face in the 21st century" (UNESCO, 2005, p. 2).

The following list of titles of students' sustainability projects provides an indication as to their focus orientation:

While it is evident even from these titles that there is a range of interpretations of the SSYS sustainability theme,

Country	Project Title
Brunei	Sustainability of stream quality in Lumapas
Cambodia	Design of a wind generator with low speed and low cost
Lao PDR	Introducing an incredible curtain which can save and produce electric energy
Lao PDR	Culture-based fisheries in the local community
Malaysia	Eco-friendly household cleaner from fruit waste
Malaysia	Sustainable flood warning system
Philippines	A novel glass from rice hulls and oyster shells
Singapore	Evaluating different methods of synthesizing biodiesel
Thailand	Soil improvement with soil-borne bacteria and EM microorganism
Thailand	Catfish command-feeding machine
Vietnam	Reuse waste paper to grow mushrooms: reuse paper napkins to grow mushrooms and vegetable sprouts.

it is also clear that local community issues form a rich source of topics for school ESD. I will now fill out in more detail one instance of the community-based sustainability projects presented at SSYS.

Ambiguities in Environment-Related Educational Work

Educational activity of this kind is clearly valuable in its own right; it is also valuable in terms of illustrating a basic ambiguity in the shift in discourse in environment-related educational work. The teachers and students associated with this instance explicitly position the activity within the discourse of ESD and it certainly appears to tick the ESD boxes, as indicated above. But in what sense may it not also, and equally justifiably, be represented as an instance within the discourse of environmental education? The Thai catfish-feeding project appears to comply with the policy statements emanating from UNESCO in the late seventies:

> Whereas it is a fact that biological and physical features constitute the natural basis of the human environment, its ethical, social, cultural and economic dimensions also play their part . . . A basic aim of environmental education is to succeed in making individuals and communities understand the complex nature of the natural and the built environments resulting from the interaction of their biological, physical, social, economic and cultural aspects, and acquire the knowledge, values, attitudes, and practical skills to participate in a responsible and effective way in anticipating and solving environmental problems, and the management of the quality of the environment.—
>
> Special attention should be paid to understanding the complex relations between socio-economic development and the improvement of the environment. (UNESCO, 1978)

In one sense it may not matter in terms of which theoretical discourse this kind of valuable practice is justified. However the task of thinking critically about the relationship between the theoretical positioning of valuable practice is useful in informing both theory (the respective adequacies of and relationships among alternative discourses) and practice (including its congruence with sanctioning policies of the day); such critical reflection on the relationship of theory and practice is a justifiable approach to research aimed at the enhancement of the coherence of educative work (Robottom, 1992, 2000). It may also form the core of effective approaches to professional development in the field of environment-related educational work in this time of shifting discourses.

Responding to the Challenges of EE/ESD: A Case for Professional Self-Development

Teacher education (teacher professional development) is identified as the "priority of priorities" in EE/ESD (Tilbury, 1992). If EE/ESD has the responsibility to educatively engage environmental sustainability problems as exhorted by UNESCO Director General Koichiro Matsuura, if such issues are socially, politically, and contextually constructed around contending social, economic, and environmental interests as indicated in the current ESD discourse, and if (as argued above) educators are operating within an ambiguous discourse context, then we need access to a form of research-based professional development that is capable of exploring practically the relationships among theories, practices, and the contexts within these have meaning.

The issue being addressed here is how best to respond to a situation in which educators find themselves located within a changing discursive context—the shift in dominant language of the environment-related field from that of "environmental education" (EE) to that of "education for sustainable development" (ESD)—marked by a lack of definition about how educational *practice* under the "new" discourse of ESD may differ from educational practice established under the "old" discourse of EE. How may educators develop a greater understanding of the complexity of their own professional circumstances? What are the implications of this pedagogical/curricular situation for professional development of educators? We will now consider some themes that may constitute a coherent approach to professional development for educators in ESD.

Specific Example: Catfish Command-Feeding Machine (Thailand)

This example of community-based EE/ESD took place in a rural community in Surattani Province of Thailand. In some traditional communities in the Southeast Asian/Indo-China region, sustainability of social life depends in part on the economic viability of traditional fish-farming practices within a context of ecological sustainability.

The background to the project, conducted by students at a local secondary school, is that catfish aquaculture is an important component of the community's local economy. Thai channel catfish have the capacity to grow very rapidly if appropriate levels of food are available, and they are amenable to aquaculture. One interesting aspect is that Thai channel catfish have the capacity to live and walk on land, though of course they are more usually found in large reservoirs, lakes, ponds, and some sluggish streams. Feeding can occur by day or by night, and they can eat a variety of plant and animal material.

Thai channel catfish are capable of gorging themselves with food through vigorous, even voracious, feeding characterized by high physical activity and rapid movement. In traditional aquaculture, catfish were historically fed by hand. In this method, it is easy to underfeed or overfeed the fish. Underfeeding obviously retards growth of the population of fish in the reservoir; overfeeding leads to residual food accumulating on the water, causing bacterial growth and disease among the fish. It is in the interests of an economically more successful aquaculture operation (and hence greater sustainability of the local environment and traditional community/social life) if a balanced quantity of food is delivered to the fish population in a given reservoir.

With a view to managing the feeding program through delivery of the "correct" amount of food (enough to lead to maximum growth of the fish leaving minimal residual food that may otherwise pollute the water leading to increased incidence of disease among the fish), the students conducted a number of experiments to develop a fish-feeding machine capable of resolving this problem resulting in a *catfish self-feeder.* Essentially, the food supply was located in a floating platform on a large truck tire tube, and release of the food was controlled by a servo whose switch was activated by a slender paddle extending into the water next to the floating platform. When the catfish are hungry, they swim about vigorously, disturbing the paddle and activating release of food. As their appetite is satisfied, they lose interest in food, swim less vigorously or swim away from the platform, and hence the paddle is disturbed less, resulting in a slowing and cessation of food release.

This example illustrates a number of points about the way teachers and students in the region are interpreting ESD. Issues in the local community that are real, current and relevant appear to serve as effective topics for school project work. These issues have discernible social, economic, and environmental interests that are to a greater or lesser extent aligned. And the resolution of the intersection of interests is essential to the cultural survival of the community within which the issue is constructed and has meaning. For example, catfish aquaculture is part of the social/cultural history of certain Southeast Asian and Indo-Chinese communities; greater economic productivity is being sought; and the longer-term viability of the community is dependent on present good management. Any shift in the balance of social, economic, and environmental interests has the potential to affect an aspect of local community identity.

Personal Professional Theories and Practice as a Starting Point in Professional Development Perhaps the first point to make is that since the challenge of ESD as outlined above is directly related to practice (that is, avoidance of the pitfalls of a slogan system entails real change in the field which is achieved only when a movement at the level of discourse is matched by a shift in actual educational practice), personal professional theories and practice may be a useful starting point. This is itself a departure from the more conventional, instrumentalist approach to professional development in which participants engage in professional development activities organized for them by operatives outside their own professional context. It is also a starting point that recognized the complexity of teacher change, as discussed by Sikes (1992) who asserts that change:

> Is not a one-way process, for the implementation of change is influenced by teachers' ideologies: in other words, by the beliefs and values, the body of ideas which they hold about education, teaching, the schooling process in particular and life in general. This means that it is not possible to attempt to change one aspect without affecting all the others. (p. 38)

The importance of personal professional theories and practice as a starting point in thinking about an appropriate approach to professional development is evident in the work of the well-known curriculum theorist Robert Stake, who proposes that:

> The kind of knowings generated by experiencing, whether direct of vicarious, are different from the knowings which come from an encounter with [already] articulated propositions of knowledge. The knowing which arise from experience are more tacit, contextual, personalistic. They are self generated knowings, *naturalistic generalizations,* that come when, individually for each practitioner, new experience is added to old. (Stake, 1987, p. 60)

Stake is suggesting that educators take an active role in their own professional development with a view to generating their own "naturalistic generalisations" about their professional work through professional *self*-development. It is evident that for Stake, careful reflection on personal, practical experience is the key to this form of professional development; this gives rise to the second theme in this discussion about professional development in ESD.

A Commitment to Reflection on Practice Stake's assertion of the importance of the knowings that arise from experience, as a starting point in professional *self*-development, may also be applied to effective learning communities that acknowledge the significance of reflective practice as a means of professional development. In ESD, as elsewhere in other disciplines, educational work is shaped by certain key features: the educators themselves; the learners for whom they have responsibility; the subject matters under consideration; and the context within which the professional work is being conducted. Without trying to oversimplify what Sikes and Stake and others have indicated is a very complex situation, most educational situations may be defined in terms of these four elements and their interactions. A better (more complex?) understanding of these interactions requires careful reflection on the nature of these elements and their interactions within particular contexts. This may be particularly the case in ESD, when there is arguably an overlaying shift in the discourse within which these elements and their interactions are given meaning.

Put another way, it is only those participants whose professional practice is implicated in these shifts that can properly articulate what it's like to be caught up in these changes and therefore what their professional development needs really are. It is important in professional development that the educators be given the opportunity to themselves engage in research aimed at better understanding the relationships among the features making up their professional work (Louden, 1992), and how these may change within a shifting overarching discourse like ESD.

Recognizing Context and Culture in Professional Development While recognizing that individual ESD practice may serve as an appropriate starting point in professional development in this field, and that reflection on that personal practice may be effective in the generation of new understandings about the complexity of professional work, it is also argued that it is important to recognize the influence of context and culture in shaping individual and collective professional work and how these are understood.

Patterns of work are changing rapidly, partly owing to the impact of globalization on the educational sector. One aspect of this is the increasing cultural diversity of classrooms as universities pursue vigorously their policies of internationalization, resulting in the likelihood of a range of differing (culturally embedded) interpretations of key substantive concepts such as sustainability and the relative significance of social, environmental, and economic interests. While increasing cultural diversity is a clear feature of formal primary and secondary classrooms, the postschool sector of education and training is possibly even more diverse and complex. Postschool education takes place in a variety of workplaces, from hotels to hospitals to shop floors in the manufacturing industry. Adult community education takes place in community halls in programs organized by local government, in neighborhood centers, local libraries, and technical and further education institutes, and in all these there is likely to be participants from any and all parts of the world, interpreting the subject matters they engage (including ESD) through a range of culturally derived lenses, conferring a necessary contextuality to learning experiences.

To be effective, professional development in ESD needs to recognize not only this source of contextuality of professional work, but also the contextuality of the substantive topic of ESD itself. ESD seeks to be educative about sustainability issues, and sustainability issues themselves are strongly embedded in local communities, as illustrated by the Thai fishfeeding project outlined above. The meaning that learners construct from engagement with sustainability issues is shaped by their own biographies (their cultural identities) and the ways in which the issues are played out in their local communities. Even for seemingly similar topics (e.g., sustainable aquaculture), this will differ from site to site.

Recognizing Social Politics in Professional Development So far I have argued for an approach to professional development in ESD that regards the practical experience of ESD educators as the appropriate starting point in a process of professional *self*-development mediated by processes of reflection upon such practice, while recognizing the impact of context and culture on the forms of professional work at the center of professional development endeavors. In a sense, this approach is related to views regarding what it is to be a professional educator. The concept of "professional" is itself a contested one. One may draw a distinction between a notion of "professional" as associated with the self-directedness and autonomy of educators, as against those who see the term as antagonistic to the legitimate interests of central authority. The latter might entail an understanding of the professional as "an efficient deliverer of a predetermined product," and a greater emphasis on skills and discrete competencies with a primacy accorded to skills over understanding and judgment. The issue is one of power and control, about who gets to set the professional development agenda, and how we construct our responsibilities toward students, government authority, and our own professional and personal identities. This point has relevance in the United Nations Decade of Education for Sustainable Development which is marked by vigorous attempts to impose centrally developed curriculum packages seemingly designed for universal implementation.

Recognizing That Professional Development Is Located Within a Broader Ambit of Organizational Change The field of professional development is also shaped by the tension between the "ideal" of autonomous professional self-development referred to earlier and the opportunities and demands of the broader educational organization with its higher-order collective strategic goals, which are often market-driven. The presence of environmental education is frequently an outcome of the endeavors of a committed individual, occasionally located toward the margins of the school curriculum. In such cases, the independence of professional self-development is a valuable commodity. However, educational organizations taken as a whole tend to reflect the society and culture within which they are located and from which they derive their financial support. So the individual environmental educator will almost inevitably need to balance opportunities, constraints and competing forces: individual philosophical commitment will encounter a potentially unsympathetic policy-led organization responding to conflicting ideologies within a market economy. In professional development terms, a possible challenge for the salaried professional educator is to determine a justifiable balance between (environmental) self-interest and a seemingly indifferent organization (Henry, 2003).

The approach to professional development in EE/ESD advanced above is based on a similar perspective on the relationship of theory and practice to that of action research in the broader field of education. Within EE/ESD, research has evolved to the point where it is possible to undertake systematic inquiry (and to have it constructed and accepted as proper research) that recognizes a special relationship between theory (in its various forms) and practice (in its various forms). Key figures in Australia in reconceptualizing educational research along these lines include Wilfred Carr and Stephen Kemmis (1987), and Robin McTaggart (1991). Essentially (and oversimplistically) action research in particular came to respect and embrace the relationships among the theories held by participants in educational settings, those participants' practices, and the characteristics of the professional settings in which these theories and practices have meaning. A consequence of this research perspective is arecognition that practitioners themselves, including the range of educational stakeholders, are properly the central figures in the conduct of these inquiries, and that their differing perspectives ought to be acknowledged, explored and tested. These considerations support a form of professional *self*-development in EE/ESD.

Conclusions

In this chapter I began with an overview of recent discourse trends in environment-related educational work, making the point that there has been a significant recent shift in the discourse framing this work. I then described some recent environmental educational activity of obvious value in its community that could as readily be described as an instance of EE and one of ESD, pointing to the contentious relationship between (changing) discourse and educational practice in this field. Finally, I proposed a form of professional development that takes into account the challenges for EE/ESD practitioners within a context of shifting discourses.

The work of Tom Popkewitz on "slogan systems" in educational reform may be useful in seeking to understand the relationship of powerful educational discourse on the one hand and continuity of environment-related educational practice on the other. In exploring instances of school reform and institutional life, Popkewitz (1982) refers to the "myth of educational reform" and proposes the role of "slogan systems" as one key agent in the circumstances of changeless reform:

> In many cases reform activities take on ceremonial or symbolic functions. The rational approach offered by reform program demonstrates to the public that schools are acting to carry out their socially mandated purpose, and that the procedures and strategies of reform offer dramatic evidence of an institution's power to order and control change. But the ceremonies and rituals of the formal school organization may have little to do with the actual schoolwork or with the teaching and learning that goes on in the classroom . . .
>
> The legitimizing function of reform can be clarified by examining the symbolic nature of slogans. The terms "individualization," "discovery approaches," and "participation" are slogans, each of which symbolizes to educators a variety of emotions, concepts and values, just as terms like "democracy" and "national security" symbolize the values and aspirations of political groups. Slogans, however, are symbolic, not descriptive: they do not tell us what is actually happening . . . Reform ban be a symbolic act that conserves rather than changes. (Popkewitz, 1982, p. 20)

The slogan system notion was originally proposed to illuminate changeless reform (adoption of a new and high-impact name in absence of any real change in practice). For Popkewitz, adoption of an active, high-profile slogan has at times been associated with a process whereby practitioners seek the benefits accompanying a concept that carries contemporary popularity (and an instantly recognizable name-as-slogan) by simply adopting the slogan symbolically, while retaining practice in largely unchanged form. Slogans can be used to justify activity at the levels of language and organization without actually leading to any real or lasting change at the important level of practice. The slogan enables a continuity of established practice: resources will be expended, careers developed, associations formed, journals filled, yet environment-related practice will not necessarily change for the better. This is the concern identified in 2005 by no less than the UNESCO Director-General when he issued the challenge that ESD needs to be "more than just a slogan."

If ESD is to be more than a slogan the challenge is to promote ESD practice in schools and elsewhere that is qualitatively different from established practice already conducted successfully within the discourse of environmental education. Given the stated importance of supporting practitioners in the field (a priority of priorities for some commentators) we need to give thought to ensuring that adopted approaches to professional development are designed with the nature of the challenge in mind. A form of professional development with the following themes has been suggested here:

- Personal professional theories and practice as a starting point in professional development;
- A commitment to reflection on practice;
- Recognizing context and culture in professional development;
- Recognizing social politics in professional development; and
- Recognizing that professional development is located within a broader ambit of organizational change.

References

Australian Government Department of the Environment and Heritage. (2005). *Educating for a sustainable future: A national environmental education statement for Australian schools.* Author. Retrieved from www.deh.gov.au/education

Australian Vice-Chancellors' Committee. (2006). *AVCC policy on education for sustainable development.* Canberra, ACT: Author Retrieved from www.avcc.edu.au/sustainability

Campbell, C., & Robottom, I. (2008). What's in a name? Environmental education and educational for sustainable development as slogans. In E. Gonzalez-Gaudiano & M. Peters (Eds.), *Environmental education: Meaning and constitution. A Handbook.* Rotterdam, The Netherlands: Sense.

Carr, W., & Kemmis, S. (1986). *Becoming critical: Education, knowledge and action research.* Geelong, Victoria: Deakin University Press.

Henry, J. (2003). *Advancing professional development.* Geelong, Victoria: Deakin University.

Louden, W. (1992). Understanding reflection through collaborative research. In M. Fullan & A. Hargreaves (Eds.), *Teacher development and educational change.* London, UK: The Falmer Press.

McTaggart, R. (1991). *Action research: A short modern history.* Geelong, Victoria: Deakin University Press.

Popkewitz, T. (1982). *The myth of educational reform: A study of school responses to a program of change.* Madison, WI: University of Wisconsin Press.

Robottom, I. (1992). Matching the purposes of environmental education with consistent approaches to research and professional development. *Australian Journal of Environmental Education, 8,* 80–90.

Robottom, I. (2000). Environmental education and the issue of coherence. *Themes in Education, 1*(3), 227–241.

Selby, D. (2006). The firm and shaky ground of education for sustainable development. *Journal of Geography in Higher Education, 30*(2), 351–365.

Sikes, P. (1992). Imposed change and the experienced teacher. In M. Fullan & A. Hargreaves (Eds.), *Teacher development and educational change.* London, UK: The Falmer Press.

Stake, R. (1987). Imposed change and the experienced teacher. In M. Wideen & I. Andrews (Eds.), *Staff development for school improvement.* Philadelphia, PA: The Falmer Press.

Tilbury, D. (1992, October–December). Environmental education within pre-service teacher education: The priority of priorities. *International Journal of Environmental Education and Information, 11*(4), 267–280.

Tilbury, D. (2004). Rising to the challenge: Education for sustainability in Australia. *Australian Journal of Environmental Education, 20*(2), 103–114.

UK Department for Education and Skills. (2006). *Sustainable schools: For pupils, communities and the environment.* Nottingham, UK: Author. Retrieved from www.teachernet.gov.uk/publications

UNESCO. (1978). *Tbilisi Intergovernmental Conference on Environmental Education organised by UNESCO in cooperation with UNEP Tbilisi (USSR), 14–26 October 1977.* Final Report. Paris: Author. Retrieved from http://unesdoc.unesco.org/images/0003/000327/032763eo.pdf

UNESCO. (2003). *United Nations decade of education for sustainable development (2005-2014): Framework for the international implementation scheme.* Paris: Author.

UNESCO. (2005). *Initiating the UNITED NATIONS decade of education for sustainable development in Australia. Report of a National Symposium.* Paris: Australian National Commission for UNESCO.

17

Connecting Vocational and Technical Education With Sustainability

Alberto Arenas and Fernando Londoño
University of Arizona, USA

Introduction

Vocational and Technical Education (VTE), also known as career and technical education, productive education, work education, or technology education, is the form of education most directly tied to the world of work because it provides students with the knowledge and skills to become economically productive members of society. VTE has generally been perceived as a less prestigious form of education relative to general, academic education, in great part because it was created to serve youth from lower socioeconomic backgrounds and because the pecuniary and status rewards derived from it have been, especially in developing countries, lower than those for graduates of the academic strand. In terms of the role it has played in industrial societies, VTE has sought to train students to produce goods and services that maximize economic growth while yielding individual benefits (as measured by salary earnings and occupational opportunities) and societal ones (as measured by GDP and other macro-economic indicators). The conventional discourse and policies surrounding VTE have focused mostly on increasing personal and social rates of return and seldom have they sought to present a critical analysis of the social relations inside VTE programs—related to power differences around sociological categories such as class, ethnicity, gender, physical ability, or sexual orientation—and much less on the social or environmental consequences of the production of the targeted goods and services (Anderson, 2009; Arenas, 2008; Kincheloe, 1995).

In the last two decades, however, a VTE counter-discourse has arisen that focuses on the intersection among economic, social, and environmental sustainability. Some of the main concepts behind each of these three partially overlapping dimensions are as follows (UNESCO & UNEVOC, 2004):

1. *Economic Sustainability*: This dimension promotes economic literacy (how to ensure that a business remains alive over the long term) in tandem with sustainable production (how to provide goods and services that address basic human needs and offers a better quality of life while being accountable for social and environmental costs of these goods and services) and sustainable consumption. This latter theme is especially true for the developed world, which has only 20 percent of the world's population but consumes 75 percent of the world's energy and 80 percent of its resources, while generating 75 percent of its pollution.
2. *Social Sustainability*: This dimension advocates, first and foremost, poverty alleviation, with an emphasis on access to basic needs such as clean water, air, and soil; decent housing; universal health care; and dignified work. It also calls for respect for cultural diversity, gender equality, and inclusion of groups that have been historically marginalized. And finally, it pushes for fair and safe working conditions inside the workplace.
3. *Environmental Sustainability*: This dimension calls for the conservation of natural resources, and for a use that minimizes waste and pollution by engaging in life-cycle analysis of all products and services. At the same time it stresses a new set of attitudes, skills, and values that cultivate a respect for the earth in all its diversity, and a sense of care for the earth's community in a way that secures its bounty and beauty for present and future generations.

In addition to analyzing the shift from conventional VTE toward VTE for sustainability, this chapter examines how it is currently enacted in secondary schools. It ends with the presentation of several areas that need further exploration to advance the cause of VTE in the context of sustainability. These areas (i.e., curriculum development; teacher education; facilities, equipment, and maintenance; and assessment strategies) harbor the reforms that need to occur in academic institutions at the secondary level to ensure a fundamental altering of purposes, structures, and roles.

Debates About VTE and the Rise of a New Discourse

VTE started to gain traction worldwide in the 1960s, thanks to national and international education policies that emphasized the development of human capital to increase economic productivity in both rural and urban areas (Heyneman, 1986). It was assumed that through VTE youth and adults would acquire a new set of skills and attitudes toward work (and modernity in general) that could help impoverished societies attain a higher standard of living. International organizations and ministries of education and labor became engaged in labor forecasting that would help mold VTE programs in accordance with the economic needs of the country five, ten, or fifteen years down the road. But as soon as implementation of these policies started, several publications emerged criticizing the alleged benefits of VTE (e.g., Blaug, 1973; Foster, 1963). Essentially, these studies pointed out that VTE did not confer the pedagogical, social, or economic benefits that national education planners were promising: Students' attitudes toward manual work did not improve; teachers often used manual work as a form of punishment; students, parents, and teachers perceived the vocational track as inferior to the academic one; VTE did not halt the migration to urban areas or prevent vocational education graduates from pursuing a university degree; and it did not alleviate unemployment or increase the salaries of vocational graduates. Moreover, it was concluded that needs for human power could not be forecasted with any degree of certainty; in fact, VTE programs based on the employment predictions put forth by planning agencies ended up yielding widely inaccurate results because graduates from these programs were pursuing either a higher status university degree or a different line of work altogether (Heyneman, 2003).

Research on VTE continued in the 1980s and 1990s with a focus on cross-country analyses of the social rates of return of VTE (World Bank, 1991, 1995). These studies compared VTE and general academic education in terms of internal efficiency (i.e., the costs of both tracks), and of external efficiency (i.e., the amount of time needed to find employment after graduation and the graduates' earning patterns). They concluded that VTE, although it cost at least twice as much, delivered lower economic benefits in comparison to general academic education. The low rates of return for VTE led organizations such as the World Bank to conclude that it should be pushed out of secondary schools and moved into postsecondary institutions and the workplace (World Bank, 1995). Indeed, as a result of these studies, international funding for VTE was slashed and many programs were discontinued, despite the fact that other studies contradicted the low rates of return attributed to VTE (e.g., Bennell & Segerstrom, 1998). More recent research on the role of VTE in specific countries (e.g., India: World Bank, 2008; new EU members: Canning, Godfrey, & Holzer-Zelazewska, 2007) or globally (World Bank, 2007) have supported the premise that the social rates of return on VTE are too low to warrant its continuance as an independent curriculum strand in secondary schooling.

A collective analysis of these various studies reveals that the bulk of the research and discourse on VTE has concentrated on how to increase the employment prospects and lifetime earnings of individual graduates and, simultaneously, to improve the economic productivity of the society as a whole. These raisons d'être of VTE have been inscribed directly into a development paradigm that supports industrialization, rationality, homogenization, and growth economics, while paying scant attention to developing local economies that reaffirm local culture and foster self-sufficient communities that protect the environment (Anderson, 2009). To be fair, the impetus behind vocational education has always been to reduce poverty, but the strategies have been mostly focused on introducing students to what Damon Anderson (2009, p. 36) called "productivism"; that is, a geopolitical space in which "economic growth and [labor] are [viewed as] permanent and necessary features of human existence, regardless of their adverse impact and consequences, social, cultural and environmental." In general, the practice of VTE and its mainstream discourse have had little regard for how they can address issues of social and economic justice, and far less for how they can improve ecological integrity. Productivism has been a key rationale behind VTE since its inception, and has been and continues to be vital in its reproduction.

Countering productivism, a new discourse in VTE has emerged that strongly advocates a form of development that fosters a responsible economic system that is highly responsive to social and environmental issues, what Anderson (2009) called "ecologism." This trend became particularly evident starting in 1999 with the UNESCO-sponsored Second International Congress on Technical and Vocational Education, held in Seoul, and the subsequent establishment in 2000 of the UNESCO-UNEVOC International Centre for Technical and Vocational Education and Training in Bonn. The Final Report of the 1999 Congress made explicit statements regarding the relationship of VTE to social and environmental stewardship (UNESCO, 1999, p. 1, 27):

> The VTE of the future must not only prepare individuals for employment in the information society, but also make them responsible citizens who give due consideration to preserving the integrity of their environment and welfare of others. An integral component of life-long learning, [VTE] has a crucial role to play in this new era as an effective tool to realize the objectives of a culture of peace, environmentally sound sustainable development, social cohesion and international citizenship.

Other international organizations have issued statements that echo this new orientation, as demonstrated by these joint recommendations from UNESCO and ILO on the purposes of VTE (2002, p. 9):

1. Contribute to the achievement of the societal goals of greater democratization and social, cultural, and economic development;

2. Lead to an understanding of the scientific and technological aspects of contemporary civilization . . . while taking a critical view of social, political, and environmental implications of scientific and technological change;
3. Empower people to contribute to environmentally sound, sustainable development though their occupations and other areas of their lives.

Implicit in these statements is the rise of a new philosophy surrounding the notion of development. This new social and environmental discourse surrounding VTE was captured by Rupert Maclean, UNESCO-UNEVOC's founding director, when he wrote (2005, p. 270),

> All countries want development, since this implies improvement; and they also want development that is long-term, and therefore sustainable. But communities increasingly want development that not only stresses economic matters but which also pays greater attention to important social, cultural, political and environmental considerations. Increasingly, countries are not willing to accept economic development at any cost and expect the benefits of development to reach all sections of the community.

This new discourse, based in a human-in-nature focus, has several common themes across sociocultural, environmental, and economic strands (Fien & Wilson, 2005). From a sociocultural perspective, it includes human rights, peace, gender equality, cultural diversity, health, and democracy. From an environmental perspective, it addresses natural resource conservation, climate change, urban and rural sustainable development, and disaster prevention and mitigation. And from an economic perspective, it involves poverty reduction, corporate responsibility and accountability, a critique of unregulated market economies, and sustainable production and consumption.

One reform common to both forms of discourse is a push for a comprehensive curriculum that combines academic and vocational education. Advocates of both forms of discourse realize the importance of having all students, regardless of their socioeconomic background, engage in VTE as a means of providing relevance, context, and concreteness to academic learning. The convergence of academic and vocational education has the potential to challenge the individualistic nature of today's schooling and to give students the opportunity to learn collectively for a reward other than a grade. Many schools worldwide have come to recognize that curricular convergence is worth considering, as a 2005 World Bank report concluded:

> Curriculum-based reform of secondary education in the twenty-first century is prioritizing skills and competencies that go beyond and cut across the traditional general-vocational divide. The frontier between general and vocational curricula is shifting and fading, and the heretofore hard-to-strike balance between vocational and general education is becoming increasingly irrelevant (p. xxi).

The Challenges of Translating Discourse Into Policy

As explained in the previous section, two radically different approaches to the discourse and research surrounding VTE have developed: productivism and ecologism (Anderson, 2009). How these discourses and research become translated into actual policy is a question with an evolving answer. In general terms it is fair to say that the productivist discourse has had more influence and overall appeal to national-level policymakers, whereas ecologism's influence has been more manifested at the local level, and even then, its presence has depended on the idiosyncrasies of specific sites and educators. In a case study looking at the national level, Heila Lotz-Sisitka and Lausanne Olvitt (2009) analyzed the influence, or lack thereof, of a sustainability agenda in South Africa's National Qualifications Framework (NQF). The NQF is a set of national educational and work standards based on quality, equity, and redress, agreed to by education and human resource development experts throughout South Africa. The NQF, jointly administered by the Ministries of Education and Labor, is meant to be inspired by the post apartheid 1996 South African Constitution which enshrined the right to a healthy environment for all citizens and the sustainable use of natural resources. However, according to Lotz-Sisitka and Olvitt (2009, pp. 320), the NQF failed to live up to its sustainability expectations: "The NQF was not creating adequate opportunities for environmental and sustainability learning, as few qualifications were being designed. The qualifications design was being driven mainly by industry's immediate needs for skills development, rather than the requirements of the new national policy guiding sustainable development."

The researchers attributed this lack of success to two main problems: (1) the NQF failed to include a set of generic standards that addressed education for sustainable development; and (2) the Ministry of Labor, which is in charge of developing, funding, and implementing vocational programs, has been more concerned with the provision of skills and competencies to increase economic output and employability than with the other two legs of the sustainability stool. To address these issues, Lotz-Sisitka and Olvitt (2009, pp. 325–326) advocate for a concerted dialog between the Ministries of Education and Labor to create a new vision and set of tools to assist groups at the national and municipal level to implement lessons related to sustainability. Such dialog could encompass the following six categories: (1) environmental; (2) management/planning and administrative; (3) legislative; (4) communications; (5) social justice and ethics; and (6) monitoring, evaluation, and research.

South Africa is not alone in grappling with the predicament of translating idealist-sounding documents into national educational standards, particularly at the local site level. Canada, Germany, Australia, India, and Azerbaijan are among a long list of other countries that are traversing a similar quagmire (Fien, Maclean, & Park,

2009), and just as in South Africa's case, research in some of these countries has yielded a set of generic indicators focused on sustainable development that could be transferred to VTE and the workforce (e.g., for Canada, see Chinien, Boutin, McDonald, & Searcy, 2004). How well sets of generic indicators, or even occupational-specific ones, can be translated into actual practice remains to be seen. What is clear, however, is that many manifestations of VTE related to sustainability can be found at the local level but not in the national sphere. That is the topic we turn to in the next section.

Expressions of VTE and of Sustainability

Although it has become fashionable to talk about "environmental jobs" (Renner, 2000) or "employment for sustainability" (Edwards, 2005), the truth is that all forms of employment, regardless how "green" they purport to be, include expressions that could be considered unsustainable. Such is the intrinsic condition of living in a modern society. For example, a person installing a wind turbine is considered to have an environmentally friendly job, yet a wind turbine consists of more than 8,000 parts, everything from screws to generators to blades, parts that tend to be manufactured under less-than-sustainable conditions. Or consider the receptionist working at a solar panel manufacturer who is paid a miserly wage or the solar panel distributor who discriminates against employees of a minority background. Clearly, many jobs wear many hats, and although the push should be toward truly sustainable employment that simultaneously addresses the needs of people, nature, and the economy, a dogmatic opposition to industrialism or to livelihoods that fall short of green or sustainable ideals would be counterproductive (Lehmann, 2008).

VTE, the form of education most directly connected to employment, faces a similar predicament. Take agricultural education, for instance. The best-case scenario for VTE would be to teach organic farming and all that it implies—using minimal tillage, crop rotation, green manure, compost, and biological pest controls—but the reality is that many agricultural VTE schools in rural areas teach conventional methods of farming that use synthetic pesticides and fertilizers, plant growth regulators, and even genetically modified organisms. Nonetheless, even in such schools, students are learning skills and values related to social and environmental sustainability, including the importance of manual labor, of working in nature, of producing a basic need, and of becoming self-reliant. Thus, the reality of the day-to-day practices of VTE carries within it a combination of sustainable and unsustainable practices.

Having said this, VTE practitioners and researchers would do well to adopt a set of sustainability principles that can guide the creation of new VTE programs that address social and environmental costs. Such sets of principles already exist, four of which are the Ontario Round Table on Environment and Economy Model Principles (to assist local communities to define their sustainable development goals); the Natural Step (created by Karl-Henrik Robert and others in Sweden, and based on systems principles); the Principles of Ecological Design (by John and Nancy Todd, which provides a biological framework that places nature at the center of the design process); and the Earth Charter (completed in 2000 and highlighting basic values of sustainability such as respect for life, protection of the environment, social justice, and democracy; for a review of these and other sets of sustainability principles, see Edwards, 2005).

Attempts to operationalize these principles in the context of VTE at the local level can be found in the work of Anderson (2009), Arenas (2001–2003, 2008), and UNESCO (1999). Some examples of principles include (Arenas, 2001–2003, p. 82),

In terms of social responsibility, VTE programs should:

- Ensure an equal concern for imparting adequate skills alongside a critical analysis of the social and political history of vocational education.
- Promote horizontality and ample dialog in the decision-making process regarding the process and purpose of production.
- Ensure that the product or service addresses a social or environmental need. Do not create a product that intentionally harms humans or the environment, such as bombs; or that has a built-in obsolescence; or that uses resources obtained by exploiting the labor of others.
- Ensure that students are exposed to alternative forms of economic production (e.g., worker-owned businesses; cooperatives).

In terms of environmental responsibility, VTE programs should:

- Establish meaningful and productive relationships with firms that engage in "green" practices.
- Assess the needs of the locality first, instead of establishing standardized vocational models with prepackaged answers for a whole country or state.
- Strive for usage of local, native, and organically grown resources.
- Emphasize products that are durable, repairable, refurbishable, and that make use of renewable energy and/or that are energy efficient.
- Push for product life-cycle analysis at school to help unveil hidden externalities.

Expressions of these principles in secondary schools can be found in the following examples (e.g., Arenas, 2008; Delisio, 2000; Schreuders, Salmon, & Stewardson, 2007; Soledi, Sorial, McNerney, & Husting, 2007; Wolf, 2001):

- Green construction, including integrated design systems, low-impact and energy-efficient materials, rainwater harvesting, and compost toilets.

- Organic and integrated farming and animal husbandry, organic urban gardens, and native-plant landscaping.
- Environmental engineering, including the prevention of water, air, and soil pollution and restoration of polluted areas.
- Repairing automobiles and home appliances, such as old computers.
- Furniture-making with sustainably harvested materials.
- Creating and rehabilitating urban and rural parks.
- Rescuing noncommodified and ancestral knowledge and skills, such as those of indigenous groups, related to healing, construction, food, clothing, recreation, and other basic needs.
- Providing services to needy populations, such as the elderly and the homeless.
- Tapping onto alternative and low-impact energy sources, such as solar, wind, and biodiesel.

The actual manifestation of these types of education will vary from school to school in terms of who benefits from the vocational program (do all students take the courses or only those in a vocational track?); the school's organization (is the school organized around vocational clusters or limited to a vocational course that students take, or is it an after-school or summer program?); or the school's curriculum (is there a vocational track per se or is it part of the regular academic courses, or is it considered part of service-learning or community service?). These differences are less important than the fact that students have the opportunity to provide a product or service that benefits the community as a whole and the environment.

Areas for Further Exploration

Curriculum Development Since the late 1980s, several scholars have singled out VTE as being silent at the social level about patterns of exploitation and discrimination in terms of the population it serves and the outcome of its programs (Kincheloe, 1995; Shor, 1988; Simon, Dippo, & Schenke, 1991). These researchers have argued that VTE has historically paid scant attention to societal pathologies that reproduce an unequal distribution of wealth; racist, sexist, and homophobic practices in the workplace; and dull and degrading forms of work. As Kincheloe (1995, p. 25) observed, VTE programs have been far more concerned with teaching job skills, leadership strategies, public relations, and program evaluation than with the socioeconomic realities of industrial practice and the nature of dignified work.

These criticisms have been coupled with those put forth in the 1990s by researchers who observed the lack of connection between VTE and the conservation of natural resources and biodiversity protection (UNESCO & UNEVOC, 2004). They analyzed how VTE was following in the footsteps of an industrial economy that showed little regard for environmental integrity, and they proposed that the outcome of the productive process (the effects on the environment as a result of the extraction, manufacturing, distribution, and delivery of goods and services) was just as important as the process of production itself.

Despite these critiques, national, or regional curricula that integrate VTE studies with principles of social justice and environmental integrity are hard to come by. To be fair, many curricular units related to sustainable development have been created, but only a few of these have been translated into the field of VTE (Pavlova, 2009). Nonetheless, three key strategies for curriculum development that would counteract these trends and history are (1) the application of sustainability principles in all courses, vocational and nonvocational alike; (2) identify specific curricular implications for occupationally relevant areas; and (3) whenever possible, match students with employers to ensure students receive actual work experience (UNESCO & UNEVOC, 2004, p. 23).

A related issue is the creation of curricular units related to noncommodified and low-status knowledge and skills that can help communities to become more self-reliant for many basic needs. Thus, if a particular vocational program that focused on culinary arts wished to gain sustainability credentials it would behoove it to offer meals that employ not only local, organic, and fair-trade produce, but also produce that is native to the area and is in danger of becoming extinct. This would ensure that heritage crops and localized forms of preparation remain alive and ultimately keep intergenerational networks of support vibrant (Bowers, 2001). One example of rescuing noncommodified knowledge is found in the work of Arenas (2003), who studied secondary schools in Colombia that have created curriculum around productive education. In one case, sixth grade students established an ethnobotanical garden and learned the various cultural uses of the plants; several times of year, students would go out into the community to sell specimens of the plants and teach the local population about the various uses of each plant. The ethnobotanical garden has become a centerpiece for teachers of various subjects who have created a truly cross-disciplinary curriculum. For other examples working with youth inside and outside schools, see Hautecoeur (2002).

Teacher Education Just as with curriculum development, the field linking teacher education, VTE, and sustainability is quite incipient. Given the novelty of the field and the ongoing emergence and re-emergence of new and old forms of knowledge that lower humans' ecological impact, it becomes imperative to transform teacher education from a paradigm that views the teacher as the expert who imparts knowledge to a paradigm wherein the teacher is knowledgeable but learns alongside the student. Probably the person who has conducted most research in this field is Margarita Pavlova, who has studied the sustainability attitudes and knowledge of technology education teachers in western and eastern European countries and Australia. She concluded (Pavlova, 2009, p. 87),

> Most preservice teachers and practicing teachers . . . were not familiar with what sustainable development or ESD [education for sustainable development] means. The exceptions were teachers who were personally engaged in activities relevant to sustainable development outside their work, or those who had studied environmental sciences at university.

On a hopeful note, Pavlova found that teachers did express interest in being able to incorporate sustainability lessons into their daily work but would require extensive professional development in order to do so (2009, p. 106). Clearly then, both preservice and in-service education are indispensable for successful integration of sustainability in teacher education. During preservice preparation teachers-to-be would acquire the concepts and methodologies to reformulate old patterns related to VTE, whereas in-service education would allow previously trained VTE teachers to learn a whole new set of concepts and practices to integrate sustainability into their daily work. What are some of the practices that technology education teachers could be exposed to (Pavlova, 2009, p. 118)?

- They could learn to extend the life of household products by repairing, reusing, refurbishing, or remanufacturing them.
- They could learn how to make goods that are more energy-efficient.
- They could learn to harness renewable resources.
- They could help in the rebirth of traditional crafts as a way of supporting social and cultural revitalization.

Clearly, none of these topics is easily to develop. Current teachers were not exposed to them while in preparation, and colleges of education preparing future teachers are currently not equipped to do the job adequately (or at all). The task becomes even more daunting with the last topic, the rebirth of traditional crafts. Not only were most teachers not exposed to these practices when growing up, but they were also trained to master and admire high-status knowledge and to ignore or despise low-status one, which ultimately is the basis of community life for many poor communities around the world (Bowers, 2001, p. 155). Two strategies will become crucial: First, exposing students to the effectiveness and the beauty of many of these noncommodified practices to ensure adequate buy-in from preservice and in-service teachers; and second, stepping outside the field of technique and entering into the field of moral philosophy to attune students to the importance of such values as humility and appreciation for alternative epistemologies that offer a different set of answers than those provided by positivist science and methodology.

Facilities, Equipment, and Maintenance Conventional VTE has been eliminated from many schools because of its higher cost per student relative to general academic education. The problem of expense is compounded further when one adds sustainability considerations, given that new materials and tools would be needed for such activities as manufacturing photovoltaic panels, installing rainwater harvesting systems, or repairing hybrid vehicles. Government programs that pay for the costs of procurement could offset some of the initial costs, but once the purchases are made, ongoing funding would still be needed for maintaining equipment and upgrading or replacing obsolete equipment.

For strictly vocational programs, establishing an apprenticeship model involving internships at appropriate companies may allow schools to save on some expenses, but such a strategy would not work for comprehensive schools that want to expose all students to technical and vocational skills. For such schools a continued financial commitment to the VTE component would be indispensable to ensure its permanence as a viable program.

One example of how effective teacher education can be implemented at both the preservice and in-service levels as long as adequate materials and facilities are available is seen in a module entitled "Product Design (Plastics)" (Pavlova, 2009, p. 112). This thirteen-week module asks preservice or in-service teachers to make a board game for children. They are expected to design and manufacture the board game, including the packaging, applying concepts of sustainability. The materials used in the production of the board game need to be made from reused materials or have a high recycled content. The actual design and manufacturing of the board game is accompanied by a series of lectures on plastics for a sustainable future, recycling, life-cycle analysis, design and manufacturing techniques, and safety issues. To ensure a successful course, adequate facilities and proper tools are indispensable. Unique courses like this one will need to be implemented to ensure that VTE and sustainability go hand in hand.

Assessment Strategies In the context of sustainability, quantitative and qualitative assessment methods using indicators are becoming the most commonly applied strategies. Indicators seek to measure an aspect of sustainability that shows a trend over time and space. For an indicator to be useful, data need to be available, reliable, and valid; collected on a regular basis; meaningful to the various interested groups; and easy to communicate to the community at large. Most currently available sets of indicators have been developed (or are in the process of being developed) at the national and international levels, including in the UK, Germany, Scandinavian countries, the Asia-Pacific region, and the UNECE region (fifty-five countries from Europe and North America) (for a review of these indicators, see Tilbury & Janousek, 2006). The following are the main types of indicators (Sollart, n.d.):

- *Baseline indicators*, which identify starting points and help to ascertain realistic impact indicators. Examples of questions are: What are current levels of understanding and support for sustainability? What are the opportunities for promoting it? What are factors that will act as obstacles against it?

- *Outcome indicators*, which enable researchers to ascertain anticipated and unanticipated learning results. Examples of questions are: What are learners able to understand and do as a result of the change? In what ways have learners' values and attitudes changed?
- *Performance indicators*, which show whether plans were carried out and how effectively. Examples of questions are: Do the learning methodologies communicate the issues and facilitate the learning process? Do practitioners share good practice and positive working relationships?
- *Impact indicators*, which assess progress toward goals, including long-term impacts on practice at different levels, such as changes in classroom practice, learning methodologies, and schemes of work and curriculum contents, all the way to national-level impacts such as change in accreditation systems and education policies. Examples of questions are: Have there been changes in curriculum content or methodologies? Have there been changes in institutional policy and practice?

Initial research on these sets of indicators has revealed several trends worthy of consideration (Tilbury & Janousek, 2006): First, users have focused mostly on restating the main components related to sustainability and to some extent have identified areas of progress, but very little has been done to build on such progress. Second, indicators are being adopted wholesale with little involvement on the part of intended beneficiaries. This has prevented them from being flexible enough to accommodate local needs and from developing a common language that is accessible to all. Third, whereas both quantitative and qualitative indicators are being used, researchers have found out that quantitative indicators have been less useful than qualitative ones in measuring sustainability. The reason for this is that quantitative forms of measurement have been more concerned with hitting performance targets than with the social processes that produce sustainability in the first place, which ultimately are more difficult to measure. And fourth, there has been a tendency to focus on single indicators when in fact there should be a push toward using multiple ones that capture the complexity of sustainability—as long as the greater number is still manageable and realistic.

Conclusions

The emergence of a new discourse related to VTE has been a long time in the making. With a greater presence of environmental concerns in both the popular media and in policy-making circles, an increasing number of academic and vocational educators are becoming attuned to the importance of imbuing their programs with a concern for protecting the natural environment. However, translating such a discourse into effective and comprehensive educational policies is a task that remains elusive. To ensure a greater effectiveness in enacting new policies and to do justice to the complexity of connecting sustainability to VTE, several issues are worth highlighting: First, in an effort to "green" VTE programs, there is a danger of displacing or not giving equal import to social and cultural concerns. Becoming "green" has become fashionable in contemporary society, and firms, big and small, have jumped on the environmental bandwagon. It is easy to fall into the belief that the solution to environmental problems is simply a matter of improved technologies and better pricing systems. Although these strategies are often part of the solution, just as important is the need to address issues of oppression and injustice inside schools and places of work. Thus, it is imperative that educators do not lose sight of the importance of working on both environmental and social concerns simultaneously.

Second, academic educators sensitive to sustainability issues are becoming increasingly aware of the importance of joining forces with vocational educators. They are realizing that it is no longer enough simply to talk about the importance of organic farming if the student (or the teacher) does not know how to farm organically. This issue calls for new curriculum development and teacher education (both pre and in-service) that imparts new knowledge and skills to all teachers and includes a permanent dialog among teachers to create cross-disciplinary and experiential learning. It also requires the constant support of school administrators and of new educational policies that allow for a more flexible school schedule and a performance-based system for grading students.

Third, in times of financial shortfalls VTE programs are often among those that suffer the most. Many shut down entirely, or the equipment they rely on is not replaced or repaired when needed. Creating the conditions described here, especially in terms of the convergence of academic and vocational education, will require added financial expenditures by schools, and a cost-benefit analysis of the personal and social rates of return of VTE should not focus narrowly on expenditures per student or student earning potential post graduation. Equally important are such considerations as revitalizing local economies, protecting natural resources, reclaiming traditions rich in cultural heritage, and enabling individuals to be more self-sufficient, all of which may bring forth much greater individual and collective economic well-being in the long run.

And fourth, educators must be aware that this new form of VTE clashes head-on with industrial economies that are characterized by incessant material growth that is undifferentiated and unqualified. As Schumacher wrote, in the context of modern economies it has become nefarious to consider "the idea that there could be pathological growth, unhealthy growth, disruptive or destructive growth" (1989/1973, p. 51). In rejecting this notion, educators must be steadfast in the belief that the goal of VTE should be to maximize human satisfaction through the optimal amount (not the maximum possible amount) of production and consumption. The opposite of growth should not be non growth, but rather sufficient growth, less material growth with regard to wants but more with regard to basic needs (including services such as health, housing, mass transportation, clean water, and so on). In essence, there needs to be a recognition that the new approach to VTE redefines the current concept of well-being, from one that stresses material accumulation to one of a development of solidarity in a harmonious relationship with nature.

References

Anderson, D. (2009). Productivism and ecologism: Changing dis/courses in VTET. In J. Fien, R. Maclean, & M. Park (Eds.), *Work, learning and sustainable development: Opportunities and challenges* (pp. 35–57). Dordrecht, The Netherlands: Springer.

Arenas, A. (2001–2003). Vocationalism for social and environmental responsibility. *La educación, 136–138*, 77–92.

Arenas, A. (2003). School-based enterprises and environmental sustainability. *Journal of Vocational Education Research, 28*(2), 107–124.

Arenas, A. (2008). Connecting hand, mind, and community: Vocational education for social and environmental renewal. *Teachers College Record, 110*(2), 377–404.

Bennell, P., & Segerstrom, J. (1998). Vocational education and training in developing countries: Has the World Bank got it right? *International Journal of Educational Development, 18*(4), 271–287.

Blaug, M. (1973). *Education and the employment problem in developing countries*. Geneva, Switzerland: International Labor Office.

Bowers, C. A. (2001). *Educating for eco-justice and community*. Athens, GA: University of Georgia Press.

Canning, M., Godfrey, M., & Holzer-Zelazewska, D. (2007). *Vocational education in the new EU member states: Enhancing labor market outcomes and fiscal efficiency, World Bank Working Paper No. 116*. Washington, DC: World Bank.

Chinien, C., Boutin, F., McDonald, C., & Searcy, C. (2004). *Skills to last: Broadly transferable sustainable development skills for the Canadian workforce*. Ottawa, ON: Human Resources Development Canada.

Delisio, E. R. (2000). School gardens sprouting in Chicago. *Education World*. Retrieved from http://www.educationworld.com/a_curr/curr313.shtml

Edwards, A. (2005). *The sustainability revolution: Portrait of a paradigm shift*. Gabriola Island, Canada: New Society Publishers.

Fien, J., Maclean, R., & Park, M. (Eds.). (2009). *Work, learning and sustainable development: Opportunities and challenges*. Dordrecht, The Netherlands: Springer.

Fien, J., & Wilson, D. (2005). Promoting sustainable development in VTET: The Bonn declaration. *Prospects, 35*(3), 273–288.

Foster, P. J. (1963). The vocational school fallacy in development planning. In C. A. Anderson & M. J. Bowman (Eds.), *Education and economic development* (pp. 142–166). Chicago, IL: Aldine.

Hautecoeur, J. P. (2002). *Ecological education in everyday life: Alpha 2000*. Toronto, ON: University of Toronto Press.

Heyneman, S. P. (1986). *Investing in education: A quarter century of World Bank experience*. Washington, DC: World Bank.

Heyneman, S. P. (2003). The history and problems of making education policy at the World Bank, 1960–2000. *International Journal of Education Development, 23*(3), 315–337.

Kincheloe, J. L. (1995). *Toil and trouble: Good work, smart workers, and the integration of academic and vocational education*. New York, NY: Peter Lang.

Lehmann, C. (2008). What about the dirty jobs? *Mother Jones*, 55–56.

Lotz-Sizitka, H., & Olvitt, L. (2009). South Africa: Strengthening responses to sustainable development policy and legislation. In J. Fien, R. Maclean, & M. Park (Eds.), *Work, learning and sustainable development: Opportunities and challenges* (pp. 319–328). Dordrecht, The Netherlands: Springer.

Maclean, R. (2005). Orienting VTET for sustainable development: Introduction to the open file. *Prospects, 35*(3), 269–272.

Pavlova, M. (2009). *Technology and vocational education for sustainable development: Empowering individuals for the future*. Dordrecht, The Netherlands: Springer.

Renner, M. (2000). *Working for the environment: A growing source of jobs*. Washington, DC: Worldwatch Institute.

Schreuders, P. D., Salmon, S. D., & Stewardson, G. A. (2007). Temporary housing for the homeless: A pre-engineering design project. *The Technology Teacher, 67*(4), 4–10.

Schumacher, E. F. (1989). *Small is beautiful: Economics as if people mattered*. New York, NY: HarperPerennial. (Original work published in 1973).

Shor, I. (1988). Working hands and critical minds: A Paulo Freire model for job training. *Journal of Education, 170*(2), 102–121.

Simon, R. I., Dippo, D., & Schenke, A. (1991). *Learning work: A critical pedagogy of work education*. New York, NY: Bergin & Garvey.

Soledi, S. W., Sorial, G., McNerney, P., & Husting, C. (2007). *Work in progress: Creating high school student environmental engineers*. Paper presented at the 37th ASEE/IEEE Frontiers in Education Conference, Milwaukee, WI.

Sollart, K. (n.d.). *Framework on indicators for education for sustainable development: Some conceptual thoughts*. Unpublished manuscript. The Netherlands: The Netherlands Environmental Assessment Agency.

Tilbury, D., & Janousek, S. (2006). *Development of a national approach to monitoring, assessment and reporting on the decade of education for sustainable development: Summarising documented experiences on the development of ESD indicators*. Sydney, NSW: Australian Research Institute of Education for Sustainability.

UNESCO. (1999). *Final report of the second international congress on technical and vocational education, Seoul*. Paris, France: Author.

UNESCO & ILO. (2002). *Technical and vocational education for the twenty-first century: UNESCO and ILO recommendations*. Paris & Geneva: Authors.

UNESCO & UNEVOC. (2004). *Orienting technical and vocational education and training (VTET) for sustainable development: A discussion paper*. Bonn: Authors.

Wolf, J. E. (2001). Building green: Creating environmentally friendly trade programs in vocational-technical schools. *Green Teacher, 66*, 14–18.

World Bank. (1991). *Vocational and technical education and training: A World Bank policy paper*. Washington, DC: Author.

World Bank. (1995). *Priorities and strategies for education: A World Bank review*. Washington, DC: Author.

World Bank. (2005). *Expanding opportunities and building competencies for young people: A new agenda for secondary education*. Washington, DC: Author.

World Bank. (2008). *Skill development in India: The vocational education and training system, Report No. 22*. Washington, DC: Author.

18

Trends, Junctures, and Disjunctures in Latin American Environmental Education Research

Edgar González Gaudiano
Universidad Veracruzana, Mexico

Leonir Lorenzetti
Universidade do Vale do Rio do Peixe, Brazil

Introduction

The environmental education research field in Latin America is practically virgin. The few attempts made to find out what is currently being done have come from Mexico and Brazil, where efforts have been and are being made to promote educational research in the field on an institutional level. In other countries, this effort is limited to a few isolated institutions and researchers who, with rare exceptions, carry out research projects of limited viability due their policies not being specifically oriented toward the discipline. There is also the case of research done on the regional situation by Latin-American environmental educationalists working out of universities in developed countries.[1]

In the cases of Brazil and Mexico, the fact that increasing numbers of students are now successfully completing postgraduate programs in environmental education has contributed notably to the growth of research in the field.[2] The same can be said of Colombia, Venezuela, and Cuba, although there are a fewer number of graduates. However, in Mexico and Brazil, conditions have been more favorable than in the other countries. For example, national environmental education research meetings have been held in Mexico (1999 and 2011) and Brazil (2001), and both countries have a group of promoters with considerable influence in the environmental education community. In Mexico, this group has grown around the National Academy of Environmental Education, founded in 2000, and the magazine *Tópicos en Educación Ambiental* (*Topics in Environmental Education*). There is greater diversification in Brazil, but the main effort gravitates around the research group "The Environmental Thematic and the Education Process," which came into existence in 2001 with the integration of several universities from the state of São Paulo. This group holds a research conference every two years and publishes the magazine *Pesquisa em Educação Ambiental* (*Environmental Education Research*).

It is not possible to speak of environmental education research in Mexico without taking into account the institutional positioning and policies of the broader field of environmental education. It is therefore important to recognize the role played in this process by Ibero-American environmental education congresses and other domestic events, in addition to the existence of postgraduate master's programs that were first offered by several universities in the country during the last decade. The most relevant factors have been the increasing participation of environmental education researchers in the Mexican Council for Educational Research (*Consejo Mexicano de Investigación Educativa*: COMIE) and the biannual congresses that they hold. A study of the state of knowledge of each of the areas recognized by the COMIE is undertaken every ten years. Environmental education was first accepted by this organism as a bona fide area of research in 2005, which has also made it possible to strengthen dialog and reciprocation with researchers in other areas of education.

In Brazil, the momentum generated by the passing of the *Environmental Education Law*, and the creation of its governmental task manager, gave greater expression to the educational component of the growing environmentalist movement born in Rio 92. Environmental Education (EE) in Brazil can be analyzed from several perspectives; worthy of note in this respect is the production of academic papers, theses, and dissertations (known as state-of-the-art research), which are important because they contribute to the process of understanding the consolidation of this field of knowledge and the way it has changed over time. Through these studies, it is possible to identify the trends in daily EE research practices that are occurring in school settings.

It is important to emphasize that environmental education research in Brazil started in the 1980s, when the first master's thesis was presented (Lorenzetti, 2008). It is not easy to identify the exact moment that EE practices were adopted in the country, but there is a consensus that concern for environmental issues was commonplace in Brazilian education. However, the critical, emancipatory, and transformative focus that EE requires was largely absent.

Mexico

In the specific case of Mexico, the state of knowledge on environmental education research for the period 1992–2002 reported a series of important advances with respect to the previous period. In the first place, because environmental education had not been recognized as a specific area of educational research by the COMIE, an independent state of knowledge report was written and published as part of the "Education, Social Rights and Equity" area (Bertely, 2003). It was incorporated into this area's Volume 1 (pp. 237–463) and comprises several sections:

1. A critical discussion on the conceptual construction process of environmental education in Mexico;
2. A general report on the stages environmental education research has gone through in the country, from 1984 to 2002;[3]
3. The most noteworthy events held in Mexico and the organizational strategy of the working party that compiled and interpreted the studies that were found, using classification categories designed for the purpose;
4. An explanation of the database used to organize the relevant information;
5. A set of final remarks aimed at describing the trends predicted for the environmental education research field in the coming years.

What can be taken from this incipient process is the need to make the environmental education research field more readily accessible, not only by means of researcher training programs—to ensure that interested environmental educators have the theoretical and methodological toolkit that will allow them to go beyond the systematization of their experience—but by trying to establish a research agenda designed to more clearly identify priority issues and problems, with the intention of coordinating better collective efforts.

In this sense, the issue of education and climate change, which really must be included, growing institutional support and growing political relevance is starting to spark interest within the academic community and is becoming an educational research field with great potential (González Gaudiano, 2009; Meira & González Gaudiano, 2009). It is desirable that this issue be used as a meeting place for other issues that have remained relatively isolated in spite of the United Nations' Decade for Education for Sustainable Development (2005–2014), such as education for sustainable consumption, multicultural education, and the reduction of social and political inequality.

The aspect of research that has received most attention has to do with different forms and levels of schooling, which is paradoxical since the environmental education field in Mexico and other Latin-American countries, *mutatis mutandis*, is barely institutionalized (González Gaudiano, 2007a). A recent characteristic of this environmental education research trend in elementary education is the analysis of teachers' and students' social representations on the environment and environmental education (Calixto, 2007; Fernández, 2002; Terrón, 2009), and studies based on discourse analysis (Andrade & Ortiz, 2004; Fuentes, 2008) from the viewpoint of diverse theoretical approaches. However, a problem often remarked upon in Mexico is that numerous environmental education experiences are poorly documented and, in consequence, an effort is being made to keep records, at least in the most consolidated cases.[4]

It is clear that in Mexico the practice of environmental education and related research projects are fraught with interest and significance when it comes to exploring certain problems and substantive matters related to the quality of schooling in general, and to genuine social participation with heightened public awareness of personal decisions affecting the lives of others. This is more evident at times, like the present, of national elementary school curriculum reform.

Twelve characteristics in the environmental education research field in Mexico have been identified (de Alba, 2007):

1. It continues to be an emerging field in the throes of constitution;
2. It has a marginal nature in both the education and environment fields;
3. It has an incipient structure, especially due to of its lack of defined centrality;
4. It is characterized by a weak, limited autonomy due to the lack of clear rules of entry and permanence;
5. The collaboration of researchers from different disciplines is complex and conflictive;
6. It suffers exigencies and pressures to become interdisciplinary, multireferential, interscientific, and interprofessional;
7. Its symbolic goods market is incipient and emerging;
8. Its capacity for explanation is precarious and unstable;
9. Its interior and exterior positioning is ambiguous and weak;
10. Its identity is vague;
11. It enjoys scant, limited prestige, and recognition;
12. There is a tense intersection of the real (in a Lacanian sense) with our symbolic and imaginary reality.

Brazil

A catalog of scientific research (Megid, 1998) reveals that, from 1972 to 1995, only 36 of 572 studies were on environmental education in Brazil. The subject matter of this research is likely to spark the development of: curriculum and programs, content-method, teacher development, didactic materials, teacher characteristics, public policies, institutional organization/nonschool learning programs, learner characteristics, school organization, the history of science teaching, the philosophy of science, and concept formation. It is clear that these studies are especially aimed at promoting EE in schools and developing pedagogical practices.

Reigota (2002, 2005) analyzed the titles of theses and dissertations, placing emphasis on their cartography and their primary pedagogical and policy features. Four categories were identified: thematic environmental issues, pedagogical characteristics, theoretical and methodological context, and policy characteristics.

The scientific conclusions of the Environmental Education Research Conference (*Encontro de Pesquisa em Educação Ambiental*—EPEA) have inspired further research. Valentin (2004) points out that most studies are a collection of EE concepts, practices, procedures, and goals related to teachers and learners, with a great diversity of study topics. Few studies examine environmental education teaching and learning processes in schools.

In papers presented at the EPEA conference, Freitas and Oliveira (2006), Cavalari, Santana and Carvalho (2006), and Avanzi and Silva (2004) identified methodological trends in research, education, and environmental education concepts, and research trends in this field respectively. Lorenzetti and Delizoicov (2006) researched theses and dissertations from three Brazilian universities; they found that research on environmental education in schools involved initial and in-service teacher education, reflections on the importance of the curriculum, the use of didactic materials in the development of environmental education, and general issues like sexuality, art, and philosophy. Research projects on out-of-school environmental education are related to the debate on natural factors, communities and groups of people, institutions, and the media among others.

Thinking Styles in Environmental Education

The analysis of scientific papers produced by graduate programs in Brazil, concerning studies set by teachers at different levels, comprised seventy-seven theses and dissertations, which dealt with the following research topics:

The data collected evidenced that, from 1981 to 2003, emphasis was placed on aspects related to teaching and learning processes. It can also be noted that there was concern for teacher education and professional development: a need for teachers to be prepared to take effective and meaningful action in order to promote environmental education.

Using the epistemology of doctor and epistemologist Ludwig Fleck (1986) as a reference, analyses were done on the environment, social representations of environmental education, practical applications of environmental education developed in schools, the stylistic language used by teachers and the authors of theses and dissertations, as well as the references cited by the authors of seventy-seven theses and dissertations. Two of Fleck's (1986) thinking styles, or thinking collectives, were found in the researched data: the ecological thinking style and the critical-transformative environmental thinking style.

Ecological Thinking Style

The ecological thinking style began its journey in Brazil associated with the environmentalist movement. Among its main characteristics, it is important to mention concern for the destruction of natural resources and the conservation and preservation of the natural environment through the promotion of ecology. It presents a strong tendency toward behaviorism and technicism in the teaching of ecology and the resolution of environmental problems, which are usually restricted to the world of ideas.

Discourse is informative in nature and only serves to challenge the issues under discussion; it does not answer questions related to the environment, quality of life and citizenship. Currently, the presence of the ecological thinking style in schools is a result of poor theoretical-epistemological training of the professionals that work in environmental education development.

TABLE 18.1
Number of Studies by Research Topic and Period

Research Topics	1st Period (1981–1991)	2nd Period (1992–1996)	3rd Period (1997–2003)	Total
Content-Method	1	6	23	30
= and Development	-	3	20	23
Didactic Resources	-	1	7	8
Social Representation	-	3	9	12
Relationships in Environmental Education	-	2	2	4
Total	**1**	**15**	**61**	**77**

The teachers that share this thinking style restrict their teaching to naturalistic aspects of the environment and have made no progress in terms of understanding the interrelationships and interdependence among human beings. Therefore, these teachers use stylistic language related to ecology. They do not understand environmental education as an integral part of education as a whole. Moreover, they do not appreciate the role reflective and political components of environmental education play in the formation of active, conscientious citizens who recognize and exercise their rights and duties in modern society.

Critical-Transformative Environmental Thinking Style

The critical-transformative environmental thinking style involves a broader view of the educational process in terms of comprehending and analyzing environmental problems within multiple dimensions: natural, historical, cultural, social, economical, and political. This thinking style presents a global approach to the environment, which develops under a critical, ethical, and democratic perspective. Thus, it prepares citizens to search for a better relationship with the world by questioning the causes of environmental problems. It also leads to a concern for environmental aspects, their specificities and interactions, in the way that learners establish observable and imperceptible networks for them. Transdisciplinarity is essential to environmental education because of its holistic perspective, its multiple action networks, and its propensity for individual, and collective participation.

In the studies analyzed, it was verified that critical and transformative perspectives intertwine with educational practices. It is not enough to raise learners' awareness of environmental problems and their consequences in human beings; it is necessary to establish concrete actions for comprehension and decision making in order to face up to these problems, to reflect upon effective attitudes in society and to become an instrument for building citizenship.

In addition to the ecological and the critical-transformative environmental thinking styles, the results revealed that there is a group of teachers in transition from the former to the latter. This transition is important as it indicates that ideas are circulating, that teachers are aware of the complications of these issues and are incorporating them into their thinking styles.

Since 2000, the environmental education research field in Brazil has risen through the domestic ranks to become a bona fide research area. EE researchers have become organized, constituted research groups and information has been exchanged among Brazilian universities. Furthermore, through the EE directory, a partnership with the Ministry of the Environment has been established in recent years. Due to this partnership, an articulated work with the core premises of the critical-transformative environmental thinking style has been developed.

Researchers and teachers have been attending scientific events, as well as promoting the spread of knowledge produced through research, which contributes to the process of broadening the research in the area. Some of these events are: Brazilian South Environmental Education Symposium (*Simpósio Sul Brasileiro de Educação Ambiental*), Environmental Education Research Conference (*Encontro de Pesquisa em Educação Ambiental*), National Science Education Research Conference (*Encontro Nacional de Pesquisa em Educação em Ciências*), Brazilian South Symposium for the Teaching of Science (*Simpósio Sul Brasileiro do Ensino de Ciências*), and the Annual ANPEd Meeting, with the creation of the "Working with Environmental Education" group. Moreover, informative science magazines also contribute to the spread of research in the field. Environmental education articles are often found in magazines related to the teaching of science and education. Thus, this area of research has gained a lot in terms of more consistent theoretical publications with the production of higher quality materials that can influence pedagogical practice. It can therefore be concluded that while environmental education has embarked upon a journey of gradual development, there is still a long road to travel.

The Situation in Other Countries in the Region

In Colombia, the environmental education field has shifted progressively towards education for sustainable development. School environmental projects (*proyectos ambientales escolares*: PRAES) supported by the Colombian Ministry of Education have followed this shift. Very little is done in the environmental education research field. Even though the implementation of the decennial plan for environmental education, legally in effect from 2005 until 2014, has contemplated a proposal for a diagnostic research project on the progress of environmental education since its inception in Colombia, the trend experienced by the PRAES has also taken hold with education for sustainable development being included in the plans. Any other studies start with the premise that environmental education is for sustainable development, with the exception of isolated research projects such as that carried out at the National University of Colombia, particularly the Institute of Environmental Studies, which takes a very critical look at the way environmental education had been handled in Colombia (Ángel, 2000; Noguera, 2002).[5]

Cuba has received international recognition for the quality of its scientific research, but the environmental education field *as a subject of research* has not been granted the same status and has lagged behind for years. For this reason, at its national conference in 2005, the Cuban Environmental Education Network (*Red Cubana de Formación Ambiental*) approved an agreement to promote the development of research in this field. As of that year, progress has been swift. The Ministry of Education is currently working on its national program

(*Programa Ramal Nacional*) and two study centers have been set up with this in mind. At the same time, there has been a significant increase in the number of master's and doctoral theses on the subject at the majority of universities, especially so since the incorporation of a master's degree in educational sciences offering accessible study opportunities.[6]

Environmental education research sensu stricto in Chile is virtually nonexistent. An examination of published materials reveals informative articles and opinions, some of which are thought-provoking, but always on the same theme (e.g. diagnoses, concepts, relevant experiences). This vacuum is the result of neoliberal policies of intense natural resource exploitation combined with government disinterest in the environmental variable and public apathy about the topic (except for spontaneous responses to catastrophic environmental events). There are no groups of researchers analyzing the priorities for the development and application of formal and nonformal environmental education (for exceptions, see Moyano Encina, & Vicente, 2007). The scenario has not changed, nor will it change, while the above-described variables exist (see Lagos, 2005; Ruthenberg, 2001; Squella, 2001).[7]

As is the case throughout the region, environmental education research in Ecuador has not been one of the strongest fields, nor has it figured among the country's top priorities. This has been true in schools and in nonformal education, such as open or informal unsystematic education. It may be assumed that this is a reflection of the country's general education system's priorities and is a consequence of the permanent economic, academic, technical, and political crisis that the country has suffered throughout its history. That is to say, education research has been almost completely neglected, with a few exceptions mainly as a result of cooperation between the government and the civil society. Issues worthy of note are the scope, limits, and approaches of the transversality of environmental education with respect to Ecuador's official curriculum (Oikos, 1996; Oikos-MEC-UNESCO-UNICEF-CONAMU, DNI, INFFA., 1998), the state of information and attitudes concerning the environmental problematic and the training of leading actors in the education community, especially teachers and parents (Oikos-MAE-Comunidad Europea, 2007), and the need for environmental education in the rural education system (Oikos-CRM, 2006). Research on the effectiveness of certain conceptual education approaches has been very limited (Oikos-Grupo Interinstitucional de Educación Ambiental, 2006), but even more so research into appropriate methods, techniques, and pedagogical tools for environmental education.

As far as informal education is concerned, there has been a tendency to embark upon formative research in view of the need to plan specific programs of a fixed length, which is predominantly the way things are done (Corporación Oikos-HCPP, 1998). From this perspective, since 1979 certain circles of the civil society have shown interest in research on the environmental education needs of different sectors of society, on the variety of current public perceptions, attitudes, and practices, taking into account the specific objectives of this kind of education, and on the best education methods and strategies for dealing with social groups that have a high propensity to become agents or victims of environmental deterioration (AED-Oikos, 1996; Corporación Oikos, 2000). This last consideration has taken on greater interest in the last ten years with reference to problems related to the problems of climate change, and particularly the vulnerability of the public to its effects. However, there persists a lack of interest in subsidizing farther-reaching research into the economy of education, and the efficiency and effectiveness of environmental education strategies, methods, and techniques (Encalada, 2003). Here, the lack of interest in investing in education in general and environmental education in particular is also relevant. Few institutions are genuinely interested in determining what happens to sustainable development processes in the absence of environmental education, or how much the country loses without environmental education.[8]

In Venezuela, as yet only one piece of research has been done that includes some information on environmental education research (Ruíz, Álvarez, & Benayas, 1999). Published materials related to the field were used as an indicator of the development of EE research, in total twenty-seven editions from the years preceding 1998, mostly from 1980 to 1988. The vast majority of these were informative books principally published by the national government and universities. Also, a small number of articles were found in periodic scientific magazines (only 6% of the total).

A process of information updating is currently under way (Álvarez Iragorry, in prep.). Preliminary results indicate a significant increase, in quantitative and qualitative terms, in the number of published materials in the country. The preliminary investigation uncovered forty-three works published between 1999 and 2008, the number increasing on a yearly basis from 2003 until at least 2007 when ten were produced. Furthermore, most of the publications were generated by universities (65.1% of the total) and the number of articles appearing in periodic scientific publications grew significantly (72%). After these preliminary results, which still have to be considered in greater depth, it will be necessary to analyze the effect of the increasing number of national graduates on environmental education, the development of incentives on the part of university institutions, and the creation of specialized magazines on educational matters.

Closing Remarks

As can be seen, there is a positive trend of slow but gradual growth in the environmental education research field in Latin America, and it has become an area of opportunity for both formal and nonformal education. It is obvious that there is a need for more institutional promotion through policies and funding, in addition to placing more importance on the training of researchers to broaden the scope of methodological approaches and perspectives, with a more inclusive and pragmatic attitude in the choice of research topics (Reid & Scott, 2009), the better to respond to the priorities of the field instead of clinging to theoretical approaches.

To this end it would perhaps be necessary to breathe some new life into the ontological, epistemological, paradigmatic, and methodological debate such as took place in the developed world during the nineties (See Mrazek, 1993; Reid, 2003; Robottom, 1993; Robottom & Hart, 1993), given that empiricist and positivist discourse and viewpoints are still dominant, especially where specialists in science education research play a part or where an instrumental vision of environmental education prevails, as is the case of many educational projects set up in protected natural areas. A debate would therefore serve to challenge the empirical-analytical tradition that is embedded in our research conceptions and practices, and make it possible to get the most out of theoretical and methodological approaches that improve the quality of different ways and genres of knowing.

No field of knowledge about any area of reality may be consolidated without research. The absence of an environmental education research program, sustained over time, with lines of research defined according to priorities, may be one of the explanations for the faltering progress made in environmental education in the region. However, the winds of change do not seem to be picking up very quickly in the environmental education field. Along with other negative consequences, this means that the dominant discourses on education and on the environment continue to hold the floor without sufficient well-informed opponents to impugn and challenge them.

Notes

1. To mention just three, Isabel Orellana from Chile, who works at the University of Quebec in Montreal, Canada; Alberto Arenas from Colombia, who works at the University of Arizona in the United States and Germán Vargas Calleja from Bolivia, who is currently at the University of Santiago de Compostela, Spain.
2. The same is true of Spain, where six volumes have been published containing the summaries of 113 doctoral theses from the Interuniversity Doctorate Program in Environmental Education, jointly run by nine Spanish universities, which is excellent material for course planning on the subject. See: Benayas, Gutiérrez and Hernández (2003), Barroso, Benayas and Cano (2004) and Sureda and Cano (2007), Gutiérrez and Cano (2008), Meira, Cano, Iglesias, and Vargas (2009), and Junyent and Cano (2010).
3. Three stages were identified: the first was called Birth of the Field–Early Research (1984–1989); the second was the growth and diversification of environmental education research (1990–1994); and the third was the consolidation process of the environmental education research field (1995–2002). In 2012 we started the compilation of information for the state of knowledge in EE research (2003–2012).
4. See González Gaudiano (2007b); González Gaudiano and Peters (2008); and Castillo and González Gaudiano (2010).
5. The information on Colombia was provided by Dr. Ana Patricia Noguera of the National University of Colombia, Manizales campus.
6. The information on Cuba was provided by Dr. Martha Roque of the Ministry of Science, Technology and the Environment.
7. The information on Chile was provided by Dr. Andrés Muñoz-Pedreros of the School of Environmental Sciences, Catholic University of Temuco, Chile.
8. The information on Ecuador was provided by Dr. Mario Encalada of the Oikos Corporation.

References

Academia Nacional de Educación Ambiental (ANEA). Retrieved from http://www.anea.org.mx/

Andrade, B., & Ortiz, B. (2004). *Semiótica, Educación y Gestión Ambiental*. Puebla, México: UIA/BUAP.

Ángel, M. A. (2000). *La aventura de los símbolos. Una visión ambiental de la historia del pensamiento*. Bogotá, Colombia: Ecofondo.

Avanzi, M. R., & Silva, R. L. F. (2004). Traçando os caminhos da pesquisa em educação ambiental: uma reflexão sobre o II EPEA. *Quaestio—Revista de estudos de Educação, 6*(1), 123–132.

Barroso, J. C., Benayas del Álamo, J., & Cano Muñoz, L. (Eds.). (2004). *Investigaciones en educación ambiental. De la conservación de la biodiversidad a la participación para la sostenibilidad*. Madrid, Spain: Organismo Autónomo Parques Nacionales. Ministerio de Medio Ambiente. (Naturaleza y Parques Nacionales. Serie educación ambiental).

Benayas, J., Gutiérrez, J., & Hernández, N. (Eds.). (2003). *La investigación en educación ambiental en España*. Madrid, Spain: Ministerio de Medio ambiente.

Bertely Busquets, M. (Ed.). (2003). *Educación, derechos sociales y equidad. La investigación educativa en México 1992–2002*. México: COMIE. (Vol. 1: Educación y diversidad cultural/Educación y medio ambiente).

Calixto, R. (2007). *Representaciones del medio ambiente de los estudiantes de la licenciatura en educación primaria*. (Doctoral dissertation). Facultad de Filosofía y Letras-UNAM.

Castillo, A., & González Gaudiano, E. (Eds.). (2010). *Educación ambiental y manejo de ecosistemas en México*. México: Instituto Nacional de Ecología.

Cavalari, R. M. F., Santana, L. C., & Carvalho, L. M. (2006). Concepções de educação e educação ambiental nos trabalhos do I EPEA. *Pesquisa em Educação Ambiental, 1*(1), 141–173.

Consejo Mexicano de Investigación Educativa (Comie). Retrieved from http://www.comie.org.mx/v1/sitio/portal.php

Consorcio Mexicano de Programas Ambientales Universitarios para el Desarrollo Sustentable (Complexus). Retrieved from http://www.complexus.org.mx/

Corporación Oikos-HCPP. (1998). Diagnóstico sobre conocimientos, actitudes y prácticas ambientales en las comunidades del Occidente de la Provincia de Pichincha. *Plan de Educación Ambiental del Honorable Consejo Provincial de Pichincha*. Documento de trabajo. HCPP y Banco Interamericano de Desarrollo.

Corporación, Oikos. (2000). *Investigación diagnóstica sobre conocimientos, actitudes y prácticas de la comunidad en la Biorreserva del Cóndor*. Programa Educar Oikos—USAID 1993–2000.

de Freitas, D., & de Oliveira, H. T. (2006). Pesquisa em educação ambiental: um panorama de suas tendências metodológicas. *Pesquisa em Educação Ambiental, 1*(1), 175–191.

De Alba, A. (2007). Investigación en educación ambiental en América Latina y el Caribe. Doce tesis sobre su constitución, en González Gaudiano, E. (2007b). *Op. Cit.* 277–287.

Encalada, M. (2003). Optimizing the use of research in order to consolidate communications planning and practice for protected areas. *International workshop on communication as a means to create support for the protected areas*. Commission of Education and Communication of UICN. World Park Congress, Durban, South Africa.

Fernández, C. A. (2002). *Análisis del modelo de educación ambiental que transmiten los maestros de primaria del municipio de Puebla (México)*. (Doctoral dissertation). Universidad Autónoma de Madrid.

Fleck, L. (1986). *La génesis y el desarrollo de un hecho científico*. Madrid, Spain: Alianza Editorial.

Fuentes, A. S. (2008). *Sujetos de la Educación: Identidad, ideología y medio ambiente*. México: UPN.

González-Gaudiano, E. (2007a). Schooling and environment in Latin-America in the Third Millennium. *Environmental Education Research*, Special Issue: Revisiting "Schooling and EE: Contradictions in purpose and practice," *13*(1), 155–169.

González-Gaudiano, E. (Ed.). (2007b). *La educación frente al desafío ambiental global. Una visión latinoamericana*. México: Plaza y Valdés-Crefal.

González-Gaudiano, E. (2009). Education against climate change: Information and technological focus are not enough. In R. Irwin (Ed.), *Climate change and philosophy; Transformational possibilities*. London, UK: The Continuum International.

González-Gaudiano, E., & Peters, M. (Eds.). (2008). *Environmental education. Identity, politics, and citizenship*. Rotterdam, The Netherlands: Sense Publishers.

Gutiérrez Pérez, J., & Cano Muñoz, L. (Eds.). (2008). *Investigar para Avanzar en Educación Ambiental*. Madrid, Ministerio de Medio Ambiente y Medio Rural y Marino. Organismo Autónomo Parques Nacionales (Serie Educación Ambiental).

Junyent, M., & Cano Muñoz, L. (Eds.). (2010). *Investigaciones en la eDécada de la Educación para el Desarrollo Sostenible*. Madrid, Ministerio de Medio Ambiente y Medio Rural y Marino. Organismo Autónomo Parques Nacionales (Serie Educación Ambiental).

Lagos, D. A. (2005). *Tendencias en los objetivos de los programas de educación ambiental en Chile entre los años 1994 al 2002*. Chile, Facultad de Universidad de Chile. Memoria de Título.

Lorenzetti, L., & Delizoicov, D. (2006). Educação Ambiental: um olhar sobre dissertações e teses, em: *Revista Brasileira de Pesquisa em Educação em Ciências, 6*, 25–56.

Lorenzetti, L. (2008). *Estilos de pensamento em educação ambiental: Uma análise a partir das dissertações e teses*. Tese (Doutorado em Educação Científica e Tecnológica). Universidade Federal de Santa Catarina.

Megid, J. (coord.) (1998). *O ensino de Ciências no Brasil: Catálogo analítico de teses e dissertações (1972–1995)*. Campinas, Brazil: UNICAMP/CEDOC.

Meira Cartea, P. Á., & Andrade, M. (coord.) (2008). *Formación e investigación en educación ambiental. Novos escenarios e enfoques para un tempo de cambios*. A Coruña: CEIDA.

Meira Cartea, P. Á., Cano Muñoz, L., Iglesias da Cunha, L., & Vargas Callejas, G. (Eds.). (2009). *Educación Ambiental: Investigando sobre la Práctica*. Madrid, Ministerio de Medio Ambiente y Medio Rural y Marino. Organismo Autónomo Parques Nacionales (Serie Educación Ambiental).

Meira, P. Á., & González Gaudiano, E. (2010). Climate change education and communication: A critical perspective on obstacles and resistances. In D. Selby & F. Kagawa (Eds.), *Education and climate change. Living and learning in interesting times*. New York — Oxon, UK, Routledge.

Moyano, E., Encina, Y., & Vicente, D. (2007). *Evaluación del Sistema Nacional de Certificación Ambiental de Establecimientos Educacionales (SNCAE) en Chile: Operatoria e impacto*. Psicología Ambiental Latinoamericana, 10.

Mrazek, R. (Ed.). (1993). *Alternative paradigms in environmental education research*. Troy, MI and OH: NAAEE.

Noguera, A. P. (2002). *El reencantamiento del mundo*. Manizales, Colombia: PNUMA-UNCIDEA.

Noguera, A. P. (2004). *El reencantamiento del mundo*. Manizales, Colombia: PNUMA-UNC-IDEA.

Oikos. (1996). *Estudio exploratorio sobre la temática socio-ambiental en los contenidos de la educación básica ecuatoriana*. Programa Educar. OIKOS-USAID.

Oikos-CRM. (2006). Diagnóstico de comunicación sobre el PIGSA, primera etapa. *Plan Integral de Gestión socio-Ambiental*. Corporación Reguladora de Manejo Hídrico de Manabí. Julio 2006.

Oikos-Grupo Interinstitucional de Educación Ambiental. (2006). Breve diagnóstico de la educación ambiental en el Ecuador. *Plan Nacional de Educación Ambiental para la Educación Básica y el Bachillerato*, elaborado por el Ministerio de Educación, Ministerio de Turismo, Ministerio de Defensa, Ministerio de Salud Pública y CEDENMA (organización de ONGs) con el apoyo técnico de UNESCO Quito-UNESCO BRASIL.

Oikos-MAE-Comunidad Europea. (2007). *Diagnóstico de conocimientos, actitudes y prácticas sobre educación ambiental de docentes de las provincias de Carchi, Imbabura y Esmeraldas, con miras a formular proyectos de innovación curricular de educación ambiental en la educación básica y el bachillerato*. Programa PRODERENA. Quito: Gobierno del Ecuador.

Oikos-MEC-UNESCO-UNICEF-CONAMU, DNI, INFFA. (1998). *Hacia políticas y estrategias de aplicación de los ejes transversales y de los temas de interés social en la educación básica del Ecuador*. Septiembre 1998.

Reid, A. (2003). Sensing environmental education research. *Canadian Journal of Environmental Education, 8*, 9–30.

Reid, A., & Scott, W. (Eds.). (2009). *Researching education and the environment. Retrospect and prospect*. London, UK and New York, NY: Routledge.

Reigota, M. (2002). El estado del arte de la educación ambiental en Brasil. *Tópicos en Educación Ambiental, 4*(11), 49–62.

Reigota, M. (2005). O estado da arte da educação ambiental no Brasil, em: *III Encontro de Pesquisa em Educação Ambiental*. Ribeirão Preto—SP. (mimeo).

Robottom, I. (1993). Towards a meta-research agenda in science and environmental education. *International Journal of Science Education, 15*(5), 591–605.

Robottom, I., & Hart, P. (1993). *Research in environmental education: Engaging the debate*. Geelong, Victoria: Deakin University Press.

Ruíz, D., Álvarez, A., & Benayas J. (1999). Contrastes y expectativas: Una mirada a la situación de la educación Ambiental en Venezuela. *Tópicos en Educación Ambiental, 1*(3), 31–45.

Ruthenberg, I. M. (2001). *A Decade of Environmental Management in Chile*. The World Bank, Environmental Economics Series. Paper No. 82.

Squella, M. P. (2001), Environmental education to environmental sustainability. *Educational Philosophy and Theory, 33*(2), 217–230.

Sureda, J., & Cano, L. (Eds.). (2007). Tendencias de la investigación en educación ambiental al desarrollo socioeducativo y comunitario. Madrid, Spain: Organismo Autónomo Parques Nacionales. Ministerio de Medio Ambiente. (Naturaleza y Parques Nacionales. Serie educación ambiental).

Terrón, E. (2009). *Educación ambiental. Representaciones sociales de los profesores de educación básica y sus implicaciones educativas* (Doctoral dissertation). Facultad de Filosofía y Letras-UNAM.

The Academy for Educational Development-Oikos. (1996). *Estudio exploratorio sobre las actitudes y prácticas de las mujeres en el programa de administración comunitaria de residuos sólidos en barrios del sur de Quito*. Quito, Oikos.

Valentin, L. (2004). Tendências das pesquisas em educação ambiental no Brasil: algumas considerações. *27ª Reunião Anual da ANPEd*, Caxambu—MG, 21 a 24 de novembro.

19

EE Policies in Three Chinese Communities

Challenges and Prospects for Future Development

Lee Chi Kin John
Hong Kong Institute of Education, China

Wang Shun Mei
National Taiwan Normal University, Taiwan

Yang Guang
Capital Normal University, China

Origins and Development of Environmental Education in Three Chinese Communities Since the 1980s

Environmental education continues to be prominent in educational reform agendas in the Chinese communities as well as internationally. In this chapter, we focus on the Chinese communities of Taiwan, the Hong Kong Special Administrative Region (SAR) and the People's Republic of China (PRC) (the Chinese mainland). While these communities share many aspects of a common Chinese cultural heritage and related historical development, it should be noted that Hong Kong became a British colony in 1897 and was returned to Chinese sovereignty in 1997 and, since the 1940s Taiwan has had a contentious history. Both Hong Kong and Taiwan currently have significantly different relationships with the mainland. In terms of spoken Chinese, standard Mandarin is the official spoken language of Taiwan and the PRC (known as Putonghua) while Cantonese is the common spoken language of Hong Kong. For written Chinese, traditional Chinese characters are officially used in Taiwan and Hong Kong while simplified Chinese characters are officially used in the PRC (http://en.wikipedia.org/wiki/Chinese. Accessed on October 15, 2008).

The purpose of this chapter is to provide an overview of the status of environmental education in the schools of these Chinese communities highlighting governmental policies and more decentralized programs, including green school development as well as discussing the challenges and prospects for their future development. In each community, evidence of developments in environmental education can be found in government environmental policy documents, government educational reform documents, publications of NGOs, academic papers and books, and school websites. All of this evidence needs to be read carefully, since some of it is largely rhetorical in character and some of it is heavily practical. Also, in the Chinese communities there is a dearth of English language publications on EE and a lack of empirical studies on EE. There is an obvious need for critical reflection in interpreting progress made toward implementing environmental education in the Chinese communities.

Development of EE Policies in Taiwan As early as 1987 in Taiwan, an Environmental Protection Administration Executive Yuan was established followed in 1990 by the creation of the Division of Environmental Protection Education in the Ministry of Education. "Essential Components of Environmental Education" was a highly influential document published in 1992 by the Environmental Protection Administration Executive Yuan (http://ivy3.epa.gov.tw/cp/education-announce/page/2-sch-ok.htm. Accessed on August 8th 2008). This stipulated that primary schools should be encouraged to promote students' understanding of the places in which they live, love of their home environment, history, and culture through direct contact with the natural environment and historical antiquities so as to cultivate correct life attitudes and habits (Teng & Liu, 1995). The secondary schools were encouraged to focus on society, politics, economics and the global ecosystem to build up students' evaluation and critical thinking skills and to encourage them to demonstrate environmentally friendly behavior.

Min-Hwang Liang (2007) has identified six stages in the evolution of EE policy in Taiwan:

- 1990–1999: establishment of EE Centers in nine teacher colleges;
- 1995–1998: establishment of the Nature Education Centers through the national park system and the nature

reserve areas system to implement outdoor education and hands-on activities;

- 1998–today: establishment of the Environmental Security and Hygienic Center;
- 2000–today: Taiwan Greenschool Partnership and Network Project;
- 2003–today: Sustainable School Campus Project; and
- 2000–today: County EE Project.

He further pointed out that EE tended to highlight local governance, collective school action, EE as an interdisciplinary theme, a sustained stimulus, a whole school approach, cooperation and partnership (p. 5). All of these are very ambitious, requiring considerable changes in school organization and teacher behavior. Both the Environmental Protection Administration (EPA) and the Ministry of Education (MOE) launched in 1995 a "three-year implementation plan for strengthening environmental education in schools" and this was revised and extended to 2009 (Chou & Gau, 2007; Wang, 2004a, p. 87). Each year the EPA chose ten schools that demonstrated outstanding performances in EE which encompassed the priority areas of environmental cleanliness, recycling, and resource conservation (Wang, 2009a). Modeling practices evident in the United States and Japan, there has been an advocacy of enacting EE laws, leading to an environmental education law, stipulating four hours of professional development of EE per year for personnel in many public-funded organizations being enacted in May 2010.

Development of EE Policies in China It is possible to identify four phases in the development of environmental education policies in China. First, there was the expert phase extending from 1973 to 1982. The Standing Committee of the fifth National People's Congress passed the Law of Environmental Protection of the People's Republic of China (for trial use) in 1979 and this adopted an environmental science rather than an environmental education approach, emphasizing knowledge and the training of experts as the panacea for environmental problems (Lee & Tilbury, 1998).

Secondly, there was the "red" phase from 1983 to 1992 when environmental education was promoted as a component of state directed, ideologically driven, civic, and political education. In the Second National Environmental Protection meeting in 1983, Vice-premier Li Peng announced that environmental protection was to be one of the major national policies in China, advocating that, from 1990, national defense education, and environmental education were to be included as part of extracurricular activities and it should also permeate related school subjects. Being individually responsible for the natural environment was viewed as a national duty and a patriotic act. Environmental education was gradually moving from environmental science toward a "red" approach that included propaganda directed at raising public consciousness of the need for environmental protection rather than emphasizing social transformation as the critical, socialist approach in most Western societies.

Thirdly there was the "green education" phase extending from 1992 to 2000. In 1992, the Ministry of Education issued the Nine-Year Curriculum Plan for Full-time Compulsory Education in Elementary, Middle, and High Schools (trial implementation). Students were expected to understand basic aspects of such topics as population, resources and environment in China. In 1994, following the 1992 United Nations Conference on Environment and Development in Rio de Janeiro, the Chinese government drew up its own China's "Agenda for the Twenty-first Century" and then published a White Paper on "China's Population, Environment, and Development in the Twenty-first Century." This called for "strengthening the indoctrination of the concept of sustainable development for people receiving education . . . [and] cultivating students' affection for the environment and responsibility for the society, which in turn change unsustainable behavior and lifestyle towards the environment" (Ministry of Education, PRC, 2003, p. 5).

In 1996, the Chinese government advocated the national development strategy "Invigorate China through Science, Technology and Education as well as Sustainable Development." Despite the efforts to promote it, environmental education has tended to emphasize environmental knowledge while the cultivation of environmental skills and values have been relatively ignored (Ministry of Education, PRC, 2003, p. 6).

In 1997, the concept of sustainable development was formalized in the UNESCO conference in Slovenia, and since then it has become much more familiar in Chinese central policy-making. Sustainable development refers to meeting the needs of the present without compromising those of future generations and education for sustainable development (ESD) and highlights helping "people to develop the attitudes, skills and knowledge to make informed decisions for the benefit of themselves and others, now and in the future, and to act upon these decisions" (http://www.unesco.org/en/esd/. Accessed on 1st May 2010).

Since 2001, a fourth phase has been inaugurated. This can be termed the "sustainability" phase that brought into alignment the national curriculum reform, and later the United Nations Decade of Education for Sustainable Development (UNDESD) (2005–2014). Despite UNESCO's rhetoric scholars have been divided in their interpretations of SD and ESD generating new tensions within the global EE community. As Sauvé (2002, p. 3) argued, ". . . environmental education is reduced to a mere instrument in the service of sustainable development. Moreover EE ceases to be seen as a setting for interdisciplinarity and the dialogue of knowledge systems . . . It is possible however to conceive of an EE that considers the sustainable development proposal (as a sociohistorical phenomenon), but that is not locked into it." Of course, as with governmental documents in some other parts of the world, these policy statements need to be treated with caution with regard to the implementation of changes at school and teacher level. We shall return to practical implementation issues later in this chapter.

Development of EE Policies in Hong Kong Turning to Hong Kong, until the mid-1980s, environmental education (EE) had not been given a high priority. In Hong Kong, the Government White Paper Pollution in Hong Kong—*A Time to Act* was published in 1989 and one of its responsibilities was to "encourage the development, through the formal education system, of a well informed, environmentally aware and responsible community" (p. 43). In early 1990, the Environmental Campaign Committee was established to coordinate the distribution of resources for promoting environmental activities followed by the issuing of the Guidelines on Environmental Education for Schools in 1992 (revised in 1999) (Lee, 1997, 2000). Here, echoing changes taking place in countries like Australia and England, environmental education was viewed as encompassing education about, in and for the environment. Following the changeover to Chinese sovereignty in 1997, the chief executive highlighted the importance of sustainable development in his Policy Address in 1999 (http://www.susdev.gov.hk/html/en/sd/index.htm. Accessed on June 16, 2007). Consequently, a Sustainable Development Unit (SDU), the Council for Sustainable Development, and the Sustainable Development Fund were set up to promote sustainability in Hong Kong. These bodies have alerted the educational community to the need to develop programs and activities that focus on education for sustainable development.

In these three Chinese communities, it can be seen that EE has received increasing attention at central government level since the 1980s. The recommendations from the Rio Summit in 1992 had a significant impact on the language and recommendations for EE practice in schools exemplified by the issue of EE guidelines or documents. In the late 1990s, the concept of education for sustainable development as reflected in UN EE-related conferences was increasingly incorporated into the rhetoric of EE in these Chinese communities. The Chinese mainland tended to follow UNESCO's EE rhetoric more closely probably because the PRC was an active participant in UNESCO activities.

In Hong Kong, as a British colony, reference was made by the education authorities to the UK's Sustainable Development Education Panel's seven domains of ESD: interdependence, citizenship and stewardship, needs and rights of future generations, diversity, quality of life, equity and justice, sustainable change, and precaution in action in Curriculum Development Institute (CDI) of Education Bureau's Primary School Teachers' Handbook on EE in 2002. However, no specific actions were made to help incorporate these domains into the school curriculum.

While there was a strong official rhetoric concerning EE in government pronouncements in these Chinese communities in the 1990s, the related EE guidelines for schools, were noncoercive and there have been little, if any, negative consequences for those schools that failed to comply with the official proposals. It was not until the new century that EE was implemented more extensively in the three Chinese communities as they engaged in broader curriculum reform, to which we now turn.

Environmental Education Guidelines and EE Curricula In 1999 the Ministry of Education in Taiwan launched The Action Program for Education Reform and initiated the Nine-year Articulated Curriculum Guideline, which included the promotion of a school-based curriculum and an integrated curriculum including themes such as green school, school campus, water, and energy (Hwang, Yu, & Chang, 2006). The curriculum guidelines covered the whole curriculum and identified not only ten core competencies (self-understanding and exploration of potentials; appreciation, representation, and creativity; career planning and lifelong learning; expression, communication, and sharing; respect, care, and team work; cultural learning and international understanding; planning, organizing, and putting plans into practice; utilization of technology and information; active exploration and study; and independent thinking and problem solving) and seven key learning areas (language arts, health and physical education, social studies, arts and humanities, mathematics, science and technology, and integrative activities) but also six key issues, one of which was environmental education (Ministry of Education, (Taiwan), undated). Taking the competency indicator of, "Participate in school association's and community's environmental protection related activities" (Third stage: Grades 6–9 of Domain 5) of environmental action experience as an example, it could be infused with the following core competencies and key learning areas: core competencies of "cultural learning and international understanding" and "independent thinking and problem solving" and with the learning area of Integrative Activities; core competency of "planning, organizing and putting plans into practice" and with the learning area of science and technology; core competency of "independent thinking and problem solving" and with the learning area of health and physical education [Lai & Lee, 2009; Ministry of Education, R.O.C. (Taiwan), 2000].

Chang Tzuchau (2001) suggested that by infusing environmental education into the nine-year curriculum in Taiwan, an emphasis on environmental awareness and environmental sensitivity could be introduced into outdoor teaching and experiential activities. The emphasis on environmental values and attitudes could make use of values analysis and values clarification. The emphasis on subject integration could use the "environment" as an issue for curriculum and instructional design. The dual attention paid to a global perspective and local consciousness could make use of local environmental teaching. An emphasis on environmental action could start from issues in the students' daily lives.

In China, elements of environmental education have been incorporated into many school subjects in the context of curriculum reform since 2001. Environmental education has been included as one of the themes in the integrated activity curriculum (Ministry of Education, PRC, 2003). In 2003, the Ministry of Education published the "Framework for Primary and Secondary School Environmental Education Special Topic Education." In 2003, the National Environmental Education Guidelines (NEEG) had been formulated by the Ministry of Education in collaboration with the WWF-China through the Environmental Educators' Initiatives (EEI) project. The overall objective of EE in the context of China, as expressed in the NEEG, is (Ministry of Education, PRC, 2003):

> . . . guiding students to concern about problems faced by the family, the community, the nation and the globe and know correctly the interdependent relationships among person, society and nature; to help students acquire knowledge and skills required for maintaining person-environment harmony; to foster affection, attitudes and values conducive to the environment; to encourage students to participate actively in decisions and actions for sustainable development so as to become citizens with social practices ability and responsibility. (p. 11)

As referred to above, the Guidelines on Environmental Education in Schools was published in Hong Kong in 1992 and then revised in 1999. In the revised document, a model for environmental education toward sustainable development was proposed. Here it was suggested that students "should develop these qualities of environmental citizenship, including values, attitudes, competence, belief, and behavior which enable them to become informed decision makers through active interaction with the natural and social environments in the local context while bearing the global context in mind" (Curriculum Development Council, 1999, p. 7). However, in Hong Kong EE had become marginalized in the curriculum reform agenda of the new century. In 2001, the Curriculum Development Council published a curriculum reform blueprint, "Learning to Learn." While this document encouraged school-based curriculum development, environmental education was mentioned only as one kind of moral and civic education, which was promoted as one of the four key tasks for curriculum reform. Nonetheless, the Curriculum Development Institute of the Education Department issued a teacher handbook on sustainable development education for primary school teachers and another handbook on life-wide learning for schoolteachers to assist them in implementing environmental education.

While there were official EE guidelines or policies revealing an orientation toward education for sustainable development and an integration of EE with curriculum reform in each place, the rhetoric, and theoretical underpinnings have different emphases. The EE objectives in Taiwan tend to reflect the Environmental Literacy Model of Hungerford and Peyton (1986). This contained the following components: ecological foundations, the awareness of issues and human values, the investigation and evaluation of issues and actions, and citizenship action (Chang, 2001). However, there were more emphases on awareness of environmental issues and values and less success was achieved in enhancing students' environmental skills and actions. In Hong Kong, the triad components of education about, in and for the environment in EE were emphasized (Robottom, 1987). In China, the EE objectives, as expressed in government documents, are divided into three categories, in order of emphasis: knowledge and capability; process and methods; and affection, and values.

Institutional and Professional Development To introduce successfully a curriculum innovation such as EE or a whole school approach to EE such as a green school, schools as organizations and teachers in their classrooms need to be carefully and systematically prepared for the innovation. Universities and associated professional associations should also be involved in EE research to support, monitor, and evaluate the changes. There is a need for research and development projects to examine issues such as school environmental management and curriculum integration and individual and organizational capacity including professional development of teachers and principals, to which we now turn.

Since 1993, five graduate institutes of environmental education have been established in north, south, east, and central Taiwan. These institutes have played an important role in cultivating well-educated environmental educators. Recently, more students have been working as program coordinators in nature centers. However, because of the lack of funding and manpower, there has been no research to monitor the graduate students' EE service and their impact on raising public environmental awareness. In China, postgraduates of EE face similar problems of career development and many of them have been employed in the publishing industry or as school teachers.

Besides cultivating professionals, these institutions promote EE in the real world. The EE professors, with their students, are deeply involved in projects supported by various central and local government agencies. These projects are concerned with biodiversity, protected areas, natural resources, pollution, health, animal welfare, disasters, and community development. However, EE professors have been constrained by the emphasis on social science or science citation index journal publications which have marginalized or even ignored EE research and development endeavors because international or national EE journals published in China or Taiwan are not on these lists and do not carry impact ratings.

The Chinese Society for Environment Education (CSEE) in Taiwan was established in 1993 with the publications of the *Chinese Journal of Environmental Education Research* and *Green Bud Teacher*. It also promoted the enactment of the laws of EE in 1998 and 2005.

As the CSEE was a professional association, it was not fully accepted by grassroots EE nongovernmental organizations. In the future, it would be desirable for the CSEE to engage in dialoge with EE NGOs in promoting EE.

As regards in-service teachers' professional development in EE, Hsu and Roth (1998) found that for secondary teachers in the Hualien area of Taiwan, their perceived knowledge of and skill in using environmental action strategies were moderate to low.

In the case of China, EE centers have been established in many key or provincial normal universities, providing training for teachers and supporting school-based curriculum development for EE. It is notable that in the case of Hong Kong, in addition to the government, nongovernmental organizations such as the World Wide Fund (Hong Kong), the Friends of the Earth (Hong Kong), the Green Power, and the Conservancy Association have made significant contributions to the promotion of EE in schools by disseminating environmental messages through seminars, exhibitions, and bulletins and providing environmental activities for school children. Such support is generally lacking in Taiwan and the PRC (Zhu & Pan, 2004). Nonetheless, there has been less attention given to preservice teacher education for EE in the three Chinese communities. A study of Hong Kong teacher trainees' environmental attitudes, behavior, and knowledge by Lui, Tsang, and Chan (2000) found that teacher trainees tended to reveal positive attitudes toward environmental protection but their behavior and habits were not totally consistent with their stated attitudes. In addition, their understanding of environmental protection and conservation concepts tended to be deficient. A recent study of in-service and preservice primary teachers in Hualien in Taiwan found that in-service teachers had higher scores on environmental action and other environmental literacy variables than their preservice counterparts but both groups were weak in aspects political action and legal action in the domain of citizenship action (Hsu & Chang, 2005). There are lessons to be learned from teacher education practice elsewhere, such as establishing partnerships between geography and science tutors and student teachers to develop personal understanding of sustainable development and EE pedagogy, which could then be extended collaboratively between geography and science teachers/mentors in schools (Summer, Childs, & Corney, 2005).

EE Projects or Programs There are many EE projects in these three Chinese communities. Many of them are government funded national initiatives while some of them are funded by nongovernmental organizations or private donations. By far the largest completed project, the Environmental Educators' Initiative (EEI), was initiated in China. It was scheduled to take ten years, ending in 2007. It was an ambitious project that involved twenty-one EE/ESD centers and eighty-six pilot schools in thirty-one provinces (including municipal cities and autonomous regions). Over the decade, it has enhanced capacity building through the training of 5,000 teacher advisors and 160 university educators who have helped to promote EE/ESD in schools. The most important success of this project, that was sponsored by a partnership of the Chinese government, British Petroleum (BP) and the World Wide Fund (WWF), was the establishment of an education for sustainable development (ESD) network.

The design of the EEI, especially in the first phase, had been influenced by world trends in EE/ESD in general. One of the EEI activities was the training of teacher advisors and university teacher educators to grasp the essence and ethos of the EEI approach, which was characterized by an "emphasis on critical awareness, participative method, and social responsibility as keys to social change and environmental conservation" (Sterling, 2004, p. 12). The EEI also promoted innovative approaches to teaching and learning in primary schools, including: "From education about the environment to education for sustainability"; "Learning through experience"; and "Cross-curriculum mode" (Shi, Hutchinson, & Yu, 2000, pp. 209–210). This is to some extent congruent with the EE in-service education and training in South Africa where participants appreciated the guidance on the use of cross-curricular approach in EE as well as the communicative and interactive approach used by the presenter (Roux & Ferreira, 2005).

However, despite these successes, the EEI project encountered significant constraints. Of these, the most important were related to the difficulties in organizing team teaching and in using a theme for integrating content from various subjects and developing EE/ESD programs or activities at senior secondary levels where teachers and students encountered examination pressures. The challenge for the next decade is continuing the work of the project without the level of funding.

Also influential is the implementation of Environment, Population and Development (EPD) project by the Chinese National Commission for UNESCO, which echoed the UN Decade of Education for Sustainable Development (UNDESD). This project was initiated in.... and is/was a.... year project. The objectives of the project are to establish a new education outlook on environment, population and sustainable development in all Chinese educators and construct a new pattern of quality education mode; to help students acquire knowledge, capability, and awareness of environment, population, and sustainable development in schools; to train a new generation with a sense of responsibility and relevant capabilities for sustainable development; and participation from the youth and all social members, to make improvement in environment, population education, and sustainable development (http://www.eepdinchina.com/english/index.htm. Accessed on November 1, 2008.

In Taiwan, a recent initiative has been the Taiwan Sustainable Campus Program (TSCP) launched in 2004. This government funded initiative consisted of three components: campus ecology, sustainable technology, and environmental management (http://140.135.10.174/esdtaiwan_english/intro.php. Accessed on August 11, 2008). The program was intended to be participatory, eco-designed, innovative, having a new curriculum, and having a campus-community connection (Chang, 2007, pp. 3–4). Each school in the program was encouraged to introduce such changes to the school campus as: the use of energy-saving appliances, the introduction of a water recycling and reuse system, improving the permeable ground surface, creating an artificial wetland, organizing green design for CO_2 reduction and conserving biodiversity, producing compost from foliage and kitchen waste, and setting up an educational organic farm or eco-pond. This program had an advisory committee, comprising experts from many disciplines, established by the Ministry of Education. It was charged with: writing general guidelines; providing the necessary knowledge and technological support; and undertaking essential supervision and evaluation.

The program had little impact on the school curriculum since most schools devoted their energies to physical improvement, rather than environmental teaching. Moreover, the experiences and processes of school campus development were not publicly shared and only a summary was distributed until the project was completed. It would have been much better if the TSCP had established a partnership with the Greenschool Partnership Project in Taiwan (GPPT) in which good examples of school-based environmental education were displayed on a green school website (http://www.greenschool.org.tw/eng/about.htm. Accessed on November 1, 2008). A research study of the sustainable campus of elementary schools in central Taiwan has revealed that teachers frequently conducted activities on knowing the organisms in school and providing knowledge about the natural ecological environment. In addition, there were great difficulties in developing educational resources for a sustainable campus and these difficulties were related to insufficient funding, lacking specialists, and having no suitable places in school for building environmental trails, ecological pools, and other learning sites (Lin & Yeh, 2005).

Another recent Taiwan initiative was the establishment of Environmental Learning Centers (http://refuge.wwg.org.tw/. Accessed on November 1, 2008), that shifted recreation-oriented outdoor education to environmental learning guided by EE professionals. It is interesting to note that while outdoor environmental learning, such as wetland studies, could enhance elementary students' cognitions, attitudes, and behavioral intentions toward wetland conservation, lower graders tended to show better learning effects than higher graders (Lin & Wang, 2006).

In Hong Kong, a project titled "Education for Sustainable Development in Primary Schools" (funded by the Sustainable Development Fund from 2003 to 2004) had the following aims:

> To develop web-based teaching and learning resources to support the school-based curriculum development framework for sustainable development education proposed by the Curriculum Development Institute of the Hong Kong Education and Manpower Bureau (now called the Education Bureau); to heighten the awareness and increase the knowledge of teachers, students, parents and community helpers on issues related to sustainable development education; and to empower primary school students in community problem solving and environmental improvement.

Project deliverables included a website, workshops for teachers, parents, and student leaders, visits as well as school-based programs and activities. The project was introduced to twenty-five primary schools (students aged five to eleven years). The more successful schools integrated EE with project-based learning and general school life (e.g., mounting energy-saving campaigns). Sustainable development was seen as a scenario for developing and practicing problem-solving skills, since students were encouraged to explore the environment, the economy, and society so as to come up with feasible solutions to real environmental problems. Less successful schools perceived education for sustainable development as yet another optional theme that teachers might introduce for cross-curricular learning or thematic teaching. EE usually turned out to be a short term and a stand-alone learning activity. From the experience of the project it appears more feasible and desirable for primary schools in Hong Kong to utilize existing subjects (e.g., general studies) or a recent government curriculum initiative (e.g., project learning) as a platform for EE activities.

It should be noticed that in Hong Kong, as in China, Taiwan, and elsewhere, the school timetable is tightly packed and there are many pressures on limited amounts of time. The introduction of EE activities must, inevitably, be achieved in a gradual and small-scale manner. Further, much more attention must be paid to the leadership of school principals alongside the development of a committed core-group of influential teachers in each school. In Hong Kong, where school principals are the keys to school-based changes, these are critical to the smooth implementation of EE programs and activities.

Recently, the Ministry of Environmental Protection's Center for Environmental Education and Communication in China has advocated the establishment of "green communities" through the project "old communities, new greenery" which involved environmental protection activities such as increasing rainwater collection devices, planting green plants, introducing composting to enhance the environmental awareness of community residents. (http://big5.ce.cn/gate/big5/finance.ce.cn/rolling/201005/10/t20100510_15753963.shtml. Accessed on May 14, 2010).

China also joined the eco-schools program of the Foundation for Environmental Education (FEE) and one of the recent initiatives is the Hong kong and Shanghai Banking Corporation (HSBC) Eco-schools Climate Initiative which involves the participation of HSBC employees in

China and the reduction of carbon footprints through the environmental audit by students and consultation with the wider community (http://www.eco-schools.org/climateinitiative/index.php?lang=cn Accessed on May 17, 2010). As regards the future development of "green communities," more could be done to evaluate the impact of such environmental activities on students and community residents' environmental knowledge, awareness, and behavior. Environmental education policymakers in the Chinese communities could also reflect on the Dutch experience and evaluate the value and feasibility of both instrumental and emancipator approaches to EE (Wals et al., 2008).

Green School Development The green school concept was introduced in Europe in the 1990s and has rapidly spread from country to country. Green schools worldwide, sometimes referred to as eco-schools, share the following features (Huang, 2003, pp. 25–26):

- a curriculum that highlight the importance of people-environment harmony and sustainable development;
- an environmentally friendly educational climate/ethos;
- encouragement of whole-school participation (teachers, students, and staff);
- an emphasis on the use of the immediate school environment and the local environment as resources for environmental education;
- mutual openness of community and school; and
- the promotion of student-centeredness.

It has been argued that, to realize the green school ideals, concepts such as sharing and cultivating a harmonious interpersonal environment, as well as respecting diverse cultures and values approaches needed to be reinforced (Huang, 2003, p. 27).

As early as 1996, the Ministry of Education of The People's Republic of China initiated the Green School Project, funded by the State Environmental Protection Administration. This Green School program, informed by ISO14000 and the European green schools, has been administered by the Center for Environmental Education and Communications (CEEC) in China since 2000 (Henderson & Tilbury, 2004, p. 13). In 2003, the Center for Environmental Education and Communication (CEEC) of the Ministry of Environmental Protection, (MEP since 2008 and formerly the State Environmental Protection Administration) issued the Guide to China Green Schools. This emphasized the following principles: objectivity; openness, and fairness; localized; and encouragement-oriented (pp. 19–20).

The number of green schools in China has increased from more than 3,200 schools in 2000 to more than 35,800 schools in 2006. China has a three-tier structure (national, provincial, and city) for managing green schools and those at the national level are recognized by the both Ministries of Environmental Protection and Education of the national government every two years and reassessed every four years. Green schools have also been promoted through some sub-projects named Young Masters Program in China (YMPiC) and Profitable Environmental Management (PREMA), in collaboration with Lund University of Sweden and the Heinrich-Boell-Foundation of Germany respectively, to help green schools to develop sustainably (http://www.iiiee.lu.se/site.nsf/AllDocuments/3FF88126AF2B70EDC1256F6B00489555 and http://www.jeef.or.jp/EAST_ASIA/programs/view/3795. Accessed on November 1, 2008). The main features of the green schools in China are the dual emphasis on government's directives and localized school-based uniqueness as well as on the balance of school campus construction and curriculum permeation in order to enhance the overall school quality and prestige (Zeng, Yang, & Lee, 2009).

In Hong Kong, a Green School Award was introduced by the Environmental Campaign Committee, the Environmental Protection Department and the Education Department in 2000 to promote comprehensive green management in primary and secondary schools. In the Green School Award, the participating schools were required to submit a self-assessment plan that matched the criteria set out by the organizers. These criteria encompassed environmental infrastructure, environmental management, environmental education, effectiveness of environmental education, and school members' participation in environmental activities. The award scheme attracted 186 primary and secondary schools in 2003 (http://www.epd.gov.hk/epd/misc/er/er2004/tc_chi/chapter5_5.html. Accessed on November 24, 2008). By 2005, the figure had increased to 467 primary and secondary schools (Environmental Protection Department, 2007). Lee (2009) studied winning schools of the Green School Award and found that factors instrumental in launching a school to take the initiative of evolving into a green school were related to the leadership of the environmental education coordinator/core group or the vice-principal and empowerment of the principal. This is in line with Gough's (2005) study of sustainable schools in Victoria, Australia that highlighted such success factors as the support of the school leadership team and enthusiastic and committed staff. In addition, while these green schools have been successful in such aspects as curriculum integration, school ground enhancement, community-based education, school sustainability, and administrative support, there was room for improvement. Improvements included reinforcing the role of Student Environmental Protection Ambassadors who help disseminate the EE messages enhancing the quality of individual environmental education activities for students, and forging greater shared understanding of the green school concept among teachers. Moreover, it may be desirable to consider how a socially critical orientation to EE in which student-teacher relationships are based on equality and various stakeholders could work together for setting social and educational agendas for environmental betterment.

The Green School Partnership Project in Taiwan (GPPT) was initiated and implemented by the Graduate Institute of Environmental Education, National Taiwan Normal University (NTNU) in 1999. It was funded and promoted by the Ministry of Education in Taiwan (http://www.greenschool.moe.edu.tw/eng/home.html. Accessed on May 17, 2010). The GPPT is characterized by school autonomy, connections among green school partners, and having a reward and evaluation system (Wang, 2009a). At the end of 2005, the elementary school partners comprised more than 60 percent of all elementary schools in Taiwan. Now there are a total of 3,672 schools in the GPPT. While the project has similar names in Hong Kong and China, the nature of Taiwan's GPPT is different, especially because it has a volunteering orientation, which invites schools to become partners in a web-based community so long as participating schools share the same vision with GPPT and implement appropriate activities.

While green schools in Hong Kong and China need external review and recognition, Taiwan green schools have a more equal relationship with the central office of GPPT. As regards the development of green school projects in China, Hong Kong, and Taiwan, there have been important changes at different national, provincial, and city levels, especially in China. Some frontline practitioners in China have expressed concern that, in their schools, the quest for commendation or having a green school badge became the means to an end rather than a search for the successful fulfillment of ends. In Taiwan, Jenq's (2004) research revealed that there was a lack of team work and internal support for GPPT within the schools. A research study, focused on the green school project in Taiwan, showed that there was a need to establish a professional membership list in the Green School network, provide advice and support to help schools resolve their problems, organize seminars for experience sharing, and to offer recognition for green school development (Liu, 2002). There was also a need for systematic and rigorous program evaluation of its impact not only on students, teachers, and schools but also on local communities (Henderson & Tilbury, 2004, p. 48).

Recently, Wang (2009b, 2009c) identified three sets of criteria for a Taiwan green school (participation and partnership, reflection and learning, and ecological consideration) and developed three operational dimensions (learning context, administration, and teaching) and related indicators which had good validity and reliability. These criteria and dimensions not only echo international trends of green school (Henderson & Tilbury, 2004) but also emphasize the process instead of predetermined outcomes, which are in line with Wals and Jickling's (2000) process-based environmental education and constructive standards and Gough's (2005) study of sustainable schools in Victoria, Australia that highlights the essential educational process of development in different school contexts.

Challenges and Prospects for Future Development These projects and endeavors in different contexts in the Chinese communities provide, on the one hand, success stories of EE and, on the other hand, highlight the difficulties encountered in implementing EE on a large-scale and in building a sustainable culture in schools. Studies of significant life experiences affecting environmental awareness of adults in the Hualien Area of Taiwan and Hong Kong showed that, contrary to findings elsewhere in English-speaking countries, "education" as a factor was rarely mentioned (Hsu, 2003; Palmer et al., 1998). This could be partly attributed to protective and conservative parenting styles, social conventions, and the emphasis on academic excellence rather than on the development of attitudes and values in education (Huang & Yore, 2003; Kennedy & Lee, 2008). In addition, an emphasis on education about the environment rather than education in and for the environment remains an entrenched challenge for environmental education and ESD in the Chinese communities. However, based on interviews with school teachers, Wang (2004b) reported that many teachers were not often close to nature and were not concerned about the environment and they lacked environmental knowledge and skills. They considered teaching about environment protection or sustainable development from textbooks a burden. Although environmental education modules that matched the textbooks were provided, the complete modules were not distributed widely for schools and inadequate workshops were provided for teachers. This is in line with findings from a study by Summers, Corney, and Childs (2003) that showed that teachers' personal knowledge and understanding of sustainable development could be inhibiting factors for implementation.

In China, Yang (2006) conducted a quantitative study through the development of a new instrument to assess secondary geography teachers' beliefs about ESD. The finalized instrument comprised two components of sustainability values (VSD) and teaching beliefs of ESD (TESD), with satisfactory reliability indices. VSD consists of four subscales: respect and care for the community of life; ecological integrity; social and economic justice; and democracy, nonviolence, and peace. TESD consists of three subscales: relevance to daily life; students' needs in the future; and integrated teaching. The findings revealed that while geography teachers were receptive to the directions and values of ESD, they had no clear definitions of ESD in teaching practices.

One should not underestimate the strength of the constraints in, for example, China where teachers have been busily occupied on tasks stipulated by the educational authority and had "no time to research the methods of environmental education in the relevant subjects' and have had "no time to consider how to educate their students for the environment and in the environment" (Zhu & Dillon, 1999/2000, p. 41). Unreformed examinations and insufficient or nonexistent teacher education are other difficulties encountered by advocates of EE in the Chinese communities.

Stimpson (2000, p. 72) observed for Hong Kong and Guangzhou in China: "A broad commitment to cross-curricularity was lacking among subject oriented teachers. Informal, extracurricular activities have, thus, proved an important avenue for environmental education in both cities." Moreover, most teachers did not teach much environmental content because of the following constraints (Wasmer, 2005, p. 22): lacking time; lacking teaching materials; lacking recognition from the school head; lacking financial support; and having difficulties with implementation. Some of these constraints echo the factors hindering the promotion of ESD in schools elsewhere which encompass the lack of policy, the lack of time as a result of an overcrowded curriculum, the lack of opportunities for ESD within the curriculum and limited funds (Zachariou & Kadji-Beltran, 2009).

In terms of approaches to and the design of EE activities, a study of junior high schools in Taiwan concluded that a combination of classroom teaching, school activities (e.g., recycling and an environmental slogan competition), and family participation (e.g., a family forum and pamphlets to families) could lead to improvement in environmental knowledge and behavior of grade seven students although its short-term effectiveness (after two months) was insignificant as compared with schools either with classroom instruction or school activities and family participation (Tung, Huang & Kawata, 2002). These results suggest that a combined approach of classroom instruction, school activities, and family participation would be ideal for environmental education.

In another study, the Delphi technique was used to construct a "school environmental education indicator system" in Taiwan. This identified the main indicators: "team power" (under the domain of organizational operation), "systematic reform," "pollution prevention," "resource management," "landscaping" (under the domain of environmental management), as well as "activities planning and participation" (under the domain of promotion activities), and "student literacy" (under the domain of "environmental literacy") for universities, high schools, and elementary schools (Yen, Ferng, & Liu, 2006). These indicators suggested that teamwork could lead to the promotion of environmental education, particularly in aspects related to environmental management. More emphasis could be placed on leadership and teacher development for ESD encompassing the components that echo the UNDESD: interdisciplinary and holistic; values-driven; critical thinking and problem-solving; multi-methods; participatory decision making; applicability; and locally relevant) (UNESCO, 2005, p. 6).

EE has been shifting to ESD since the 1990s and there have been emphases on ESD in policy discourses in the three Chinese communities. Nonetheless, the schooling systems in all three communities has been influenced by their ongoing national curriculum reforms, which, on the one hand, provide "space" or opportunities for integrating EE with school-based curriculum initiatives but, on the other hand, enhance the competition between EE and other curricular agendas.

In addition, schools in the Chinese communities have experienced the impact of globalized educational reforms and the accountability mechanisms which cause overloads for schools and teachers as well as marginalizing EE or ESD. Teachers simply do not have adequate time or knowledge to handle ESD, often seen by them as an ambiguous concept, both in theory and practice (Lee & Williams, 2009; Yang & Lam, 2007).

Under these constraints, the professional development of teachers and principals (head teachers) takes on much greater significance since teacher and administrative leaders are known to be the key factors in facilitating processes of student learning, where students are encouraged to think globally and act locally.

There are positive lessons to be learned from the Environment and School Initiatives Project (sponsored by the Organization for Economic Cooperation and Development/Centre for Educational Research and Innovation OECD/CERI) where the participatory (action) research approach to professional development is highlighted. There are also lessons to be learned from the Australia/South Africa Institutional Links (AusLinks) Project where professional development adheres to the following major principles (Robottom, 2000, pp. 254–255): contextual (related closely to workplaces and workplace issues); responsive (related to the interest and concerns of participants); emergent (professional knowledge emerging from the case study work at the center of the professional self-development process); participatory (direct and equitable involvement of participants); critical (looking beyond the surface of activities); and praxiological (mediated by praxis comprising reflective interaction between personal professional theory and practice and the professional settings). This partly echoes Stevenson's (2007, p. 275) suggestion of "new approaches to teacher professional development that recognize the importance of teacher agency and professional communities and incorporate more constructivist and collaborative learning opportunities."

In recent years environmental education in Taiwan has encountered a crisis, as the enrolment numbers of the five specialist EE graduate institutes have fallen. In 2005, the NAAEE (North America Association for Environmental Education) certification system was introduced to Taiwan to enhance the visibility of EE. A survey by Chang, Yeh, and Liang (2007) showed that there was a diversity of opinions about the introduction of certification systems among governmental agency staff, graduate students and university professors, and experienced teachers. It seems necessary to strengthen the communication and dialoge between influential groups so as to build consensus and partnership. A better policy and support infrastructure is also essential for the successful promotion of EE. In China, the successful partnership between the government

(Ministry of Education), the nongovernmental organization (NGO) (WWF-China), and the multinational business organization (British Petroleum) as well as collaboration with universities provide insights for future collaboration between government, NGOs, and external community groups.

References

Chang, T. (2001). Nine-year articulated curriculum reform and infusion of environmental education. In T. Chang (Ed.), *Environmental education curriculum design* (pp. 109–127). Taiwan: National Taiwan Normal University. [in Chinese]

Chang, T. (2007). *The enriching subject matters of learning through "Taiwan Sustainable Campus Program."* Paper presented in the Fourth World Environmental Education Congress-Learning in a changing world, July 2–6, 2007, South Africa.

Chang, M-H., Yeh J-C., & Liang, M-H. (2007). Initial study on the implementation of certification system for environmental education personnel in Taiwan. In *Proceedings of Taiwan Chinese Society for Environmental Education* (pp. 1038–1045), 28–30 September 2007. [in Chinese]

Chou, J., & Gau, T. S. (2007). Analysis and development of environmental education strategies of Environmental Protection Administration Executive Yuan. Collaborative project of Environmental Protection Administration Executive Yuan and Taiwan Normal University. [in Chinese]

Curriculum Development Council (1992/1999). Guidelines on environmental education in schools. Hong Kong: Education Department Environmental Protection Department (2007). Proposed capital injection into the Environment and Conservation Fund CB(1)283/07-08(05). Retrieved November 24, 2008, from http://legco.hk/yr07-08/english/panels/ea/papers/ea1126cb1-431-1-e.pdf.

Gough, A. (2005). Sustainable schools: Renovating educational process. *Applied Environmental Education and Communication, 4*, 339–351.

Henderson, K., & Tilbury, D. (2004). Whole-school approaches to sustainability: An international review of sustainable school programs. Report prepared by the Australian Research Institute in Education for Sustainability (ARIES) for the Department of Environment and Heritage, Australian Government.

Hsu, S-J. (2003). Significant life experiences affecting the environmental action of active members of environmental organizations in the Hualien area. *Chinese Journal of Science Education, 11*(2), 121–139. [in Chinese]

Hsu, S-J., & Chang, N-C. (2005). An assessment o environmental literacy and analysis of predictors of environmental action held by in-service and pre-service primary teachers in Hualien county. *Environmental Education Research, 2*(2), 91–123.

Hsu, S-J., & Roth, R. E. (1998). An assessment of environmental literacy and analysis of predictors of responsible environmental behavior held by secondary teachers in the Hualien area of Taiwan. *Environmental Education Research, 4*(3), 229–249.

Huang, Y. (2003). Development of international environmental education and China's "Green School." *Comparative Education Review, 1*(152), 23–27. [in Chinese]

Huang, H-P., & Yore, L. D. (2003). A comparative study of Canadian and Taiwanese Grade 5 children's environmental behaviors, attitudes, concerns, emotional dispositions, and knowledge. *International Journal of Science and Mathematics Education, 1*, 419–448.

Hungerford, H. R., & Peyton, R. B. (1986). *Procedures for developing an environmental education curriculum*. Paris: UNESCO.

Hwang, J. J., Yu, C-C., & Chang, C-Y. (2006). School improvement in Taiwan, 1987–2003. In J. C. K. Lee & M. Williams (Eds.), *School improvement: International perspectives* (pp. 201–212). New York, NY: Nova Science Publishers, Inc.

Jenq, C. S. (2004). *A study of the process and obstacles of the green elementary school in central Taiwan*. Unpublished master's thesis, National Taichung University, Taichung, Taiwan. [in Chinese]

Kennedy, K. J., & Lee, J. C. K. (2008). *The changing role of schools in Asian societies: Schools for the knowledge society*. London and New York: Routledge.

Lai, K. C., & Lee, J. C. K. (2009). Assessment and evaluation. In J. C. K. Lee & M. Williams (Eds.), *Schooling for sustainable development: Chinese experience with younger children* (pp. 53–75). Dordrecht, The Netherlands: Springer.

Lee, J. C. K. (1997). Environmental education in schools in Hong Kong. *Environmental Education Research, 3*(3), 359–371.

Lee, J. C. K. (2000). Teacher receptivity to curriculum change in the implementation stage: the case of environmental education in Hong Kong. *Journal of Curriculum Studies, 32*(1), 95–115.

Lee, J. C. K. (2009). Green primary schools in Hong Kong. In J. C. K. Lee & M. Williams (Eds.), *Schooling for sustainable development: Chinese experience with younger children* (pp. 195–212). Dordrecht, The Netherlands: Springer.

Lee, J. C. K., & Tilbury, D. (1998). Changing environments: The challenges for environmental education in China. *Geography, 83*(3), 227–236.

Lee, J. C. K., & Williams, M. (Eds.). (2009). *Schooling for sustainable development: Chinese experience with younger children*. Dordrecht, The Netherlands: Springer.

Liang, M-H. (2007). *Environmental education policy and institutional transformation in Taiwan*. Paper presented at the Fourth World Environmental Education Congress (WEEC), July, Durban, South Africa. Retrieved March 3, 2008, from http://www.weec2007.com/papers/files/2007-08-04_33/47%20Liang.doc?PHPSESSID=2b6d7804b5bbfe278ffe6216b7bea2f4.

Lin, M. R., & Yeh, M. S. (2005). Research of connotation of sustainable campus of elementary schools in central Taiwan: Research from the natural environmental point of view. *Journal of Environmental Education Research, 2*(2), 1–24. [in Chinese]

Lin, M. R., & Wang, S. H. (2006). A experimental teaching study of conservation curricula of the Fubow Wetland on the influences for third to sixth graders' cognitions, attitudes of wetland conservation. *Journal of Environmental Education Research, 4*(1), 103–146. [in Chinese]

Liu, C. C. (2002). *The research of administrators participate in Project Taiwan Green School–the motivation of promoting environmental education in the school*. Unpublished master's thesis, National Taiwan Normal University, Taipei, Taiwan. Retrieved August 11, 2008, from http://etds.ncl.edu.tw/theabs/site/sh/detail_result.jsp?id=090NTNU0587001. [in Chinese]

Lui, K. C. W., Tsang, E. P. K., & Chan, S. L. (2000). A key to successful environmental education: Teacher trainees' attitude, behavior, and knowledge. *New Horizons in Education, 41*, 1–18.

Ministry of Education (PRC). (2003). The National Environmental Education Guidelines (Trial). Beijing: Beijing Normal University. [in Chinese] Ministry of Education (Taiwan) (undated). General Guidelines of Grade 1–9 Curriculum of Elementary and Junior High School Education. Retrieved May 18, 2010, from http://english.moe.gov.tw/public/Attachment/66618445071.doc.

Ministry of Education, R.O.C. (Taiwan) (2000). *The nine-year articulated curriculum guidelines for compulsory education*. Taipei, Taiwan: Ministry of Education.

Palmer, J. A. et al. (1998). An overview of significant influences and formality experiences on the development of adults' environmental awareness in nine countries. *Environmental Education Research, 4*(4), 445–464.

Robottom, I. (1987). Contestation and consensus in environmental education. *Curriculum Perspectives, 7*(1), 23–27.

Robottom, I. (2000). Recent international developments in professional development in environmental education: Reflections and issues. *Canadian Journal of Environmental Education, 5*, 249–267.

Roux, C. L., & Ferreira, J. G. (2005). Enhancing environmental education teaching skills through in-service education and training. *Journal of Education for Teaching, 31*(1), 3–14.

Sauvé, L. (2002). Environmental education: Possibilities and constraints. *Connect, XXVII*(1/2), 1–4.

Shi, C., Hutchinson, S. M., & Yu, L. (2000). Moving beyond environmental knowledge delivery: Environmental educators' initiative for China. *International Journal of Environmental Education and Information, 19*(3), 205–214.

Stevenson, R. B. (2007). Schooling and environmental/sustainability education: From discourses of policy and practice to discourses of professional learning. *Environmental Education Research, 13*(2), 265–285.

Sterling, S. (2004). *Investing in the future: WWF China Education Programme Evaluation 2004*. WWF China, Beijing: Author.

Stimpson, P. (2000). Environmental attitudes and education in southern China. In D. Yencken, J. Fien, & H. Sykes (Eds.), *Environment, education and society in the Asia-Pacific* (pp. 51–74). London, UK: Routledge.

Summers, M., Childs, A., & Corney, G. (2005). Education for sustainable development in initial teacher training: Issues for interdisciplinary collaboration. *Environmental Education Research, 11*(5), 623–647.

Summers, M., Corney, G., & Childs, A. (2003). Teaching sustainable development in primary schools: An empirical study of issues for teachers. *Environmental Education Research, 9*(3), 327–346.

Teng, M., & Liu, C-C. (1995). The current status of environmental education in Taiwan, R.O.C., *Bulletin of the National Institute of Educational Resources and Researches, 20*, 161–183. [in Chinese]

Tung, C-Y., Huang, C-C., & Kawata, C. (2002). The effects of different environmental education programs on the environmental behavior of seventh-grade students and related factors. *Journal of Environmental Health, 64*(7), 24–29.

UNESCO. (2005). United Nations decade of education for sustainable development 2005-2014: Draft international implementation scheme. Retrieved November 21, 2008, from http://portal.unesco.org/education/en/file_download.php/e13265d9b948898339314b001d91fd01draftFinal+IIS.pdf.

Wals, A. E. J., Floor, G-E., Hubeek, F., van der Kroon, S., & Vader, J. (2008). All mixed up? Instrumental and emancipator learning toward a more sustainable world: Considerations for EE policymakers. *Applied Environmental Education and Communication, 7*(3), 55–65.

Wals, A. E. J., & Jickling, B. (2000). Process-based environmental education: Seeking standards without standardizing. In B. B. Jensen, K. Schnack, & V. Simovska (Eds.), *Critical environmental and health education: Research issues and challenges*. Copenhagen, Denmark: Research Centre for Environmental and Health Education, Danish University of Education.

Wang, S. M. (2004a). The exploration of the environmental education with whole school approach in the middle schools of Taiwan. *Journal of Taiwan Normal University: Mathematics & Science Education, 49*(2), 87–106. [in Chinese]

Wang, S. M. (2004b). Environmental education and social change—the Green School project. *Journal of Taiwan Normal University Education, 49*(1), 159–170. [in Chinese]

Wang, S. M. (2009a). The Greenschool project in Taiwan. In J. C. K. Lee & M. Williams (Eds.), *Schooling for sustainable development: Chinese experience with younger children* (pp. 213–231). Dordrecht, The Netherlands: Springer.

Wang, S. M. (2009b). The development of indicators and their evaluation instrument for green schools in Taiwan. *Journal of Environmental Education Research, 6*(1), 119–160. [in Chinese]

Wang, S. M. (2009c). The development of performance evaluation for green schools in Taiwan. *Applied Environmental Education and Communication, 8*(1), 49–58.

Wasmer, C. (2005). Towards sustainability: Environmental education in China. Can a German strategy adapt to Chinese schools? *Duisburg working papers on East Asian economic studies. No. 73*. Universitat Duisburg-Essen, Germany.

Yang, G. (2006). Secondary teachers' beliefs about Education for Sustainable Development: An exploratory study to develop an instrument for measuring ESD beliefs. (Doctoral thesis, The Chinese University of Hong Kong).

Yang, G., & Lam, C. C. (2007). What kind of education for sustainable development do we need? *Geography Teaching, 4*, 9–12. [in Chinese]

Yen, H-W., Ferng, J-Y., & Liu, C-H. (2006). The construction of school environmental education indicator system in Taiwan. *Journal of Taiwan Normal University: Education, 51*(1), 85–102. [in Chinese]

Zachariou, A., & Kadji-Beltran, C. (2009). Cypriot primary school principals' understanding of education for sustainable development key terms and their opinions about factors affecting its implementation. *Environmental Education Research, 15*(3), 315–342.

Zeng, H. Y., Yang, G., & Lee, J. C. K. (2009). The Green Schools in China. In J. C. K. Lee & M. Williams (Eds.), *Schooling for sustainable development: Chinese experience with younger children* (pp. 137–156). Dordrecht, The Netherlands: Springer.

Zhu, H. X., & Dillon, J. (1999/2000). Environmental education in the People's Republic of China: Features, factors and trends. *Australian Journal of Environmental Education, 15/16*, 37–43.

Zhu, H-X., & Pan, H-P. (2004). Analysis of the policy and practice of environmental education in Hong Kong. *Comparative Education Review, 2004*(1), 34–38. [in Chinese]

Part B

Research on Environmental Education Curriculum, Learning, and Assessment

Processes and Outcomes

Section IV

Curriculum Research in Environmental Education

Heila Lotz-Sisitka
Rhodes University, South Africa

Introduction

Curriculum research is widely practiced in environmental education. It is located in, but also departs from the wider field of curriculum research as found in the broader field of educational research. Because environmental education is a newly emerging subfield of educational research it is possible to trace the linkages. However, environmental education research is also concerned with the environment, and the relationships that exist between the environment, learning and social-ecological change, sustainability, and resilience. Thus it cannot simply be reduced to be a replica or copybook variety of mainstream educational research. This section of the handbook also, however, maps out and presents chapters that show the divergences, and point to "new niches" for environmental education research within the wider curriculum research landscape.

It deals with the question of curriculum research in three distinct ways. First, environmental education curriculum research is discussed historically. Second, it is also discussed methodologically, showing trends in methodological direction and orientation. And third, it deals with the educational knowledge interests of environmental education curriculum research such as participation by learners in the curriculum process; cultural change in and through curriculum, teaching and learning; and place-based environmental education. The paper by Lotz-Sisitka et al., maps out various new niches that are emerging and seem to be of particular interest to environmental education curriculum researchers. Their analysis shows that only a small portion of such interests can be addressed in a publication such as the IHREE, and the section therefore remains "essentially incomplete."

The first paper in this section by Lotz-Sisitka, Fien, and Ketlhoilwe (writing together across country and continental borders of South Africa, Botswana, and Australia), presents a detailed "mapping" or overview of "Traditions and New Niches" associated with environmental education curriculum research. The paper begins with an orientation to the history of education in modernity, noting dominant forms of curriculum and learning as these have come to be practiced in modern education systems today. This opening provides a backdrop for a "reading" of environmental education research. The authors continue with this historical trajectory noting that it is "important to locate these [environmental education research traditions] within a much longer and more established research landscape." From this perspective they argue that environmental education research is "young" with a relatively short history of some forty years, making it essentially a post '60s phenomenon. They authors trace the emergence of different research traditions that appear to be substantively influential in shaping environmental education curriculum research, vis-à-vis an empirical analytical; interpretivist or constructivist and a critical research tradition, which, they argue reflect Habermasian views on knowledge interests. They note too that environmental education curriculum research has also been shaped and influenced by poststructuralism, and more recently by critical realism. The authors show that despite this "methodological ferment" and despite years of critique of some of these research traditions, they continue to be practiced. The authors show the gradual emergence of an interest in critical research traditions, showing too how these are being refined and deliberated. The authors do not want to leave the readers with the impression that "methodology" is the only characterizing feature of environmental education curriculum and learning research, noting that "research can suffer from too great an emphasis on methodology, without giving adequate attention to theory development." Thus, they urge the reader to consider the educational knowledge "spaces" that seem to be emerging in environmental education curriculum research, where such theory development is taking place. They point to various thematic or niche areas, such as research on action and action competence;

and research on constructivist or phenomenological forms of place-based research; and sociocultural learning. The chapter provides a useful "landscape" for further literature review work, and for further critical analysis of a developing field, from historical, methodological, and educational knowledge vantage points.

The chapter by Shallcross and Robinson, both writing from the UK, follows. It provides a methodological deliberation on how one might observe "action" in environmental education research. They argue that environmental education (or education for sustainable development) research is oriented to understanding the role that education can play in changing lifestyles and actions, seeing this research as primarily a transformative research "type." They raise the problem that while education tends to equip learners to conceptualize problems, it seldom supports them to act out solutions. This, they see as being an important area for innovation in environmental education research. Their dilemma, however, is *how to observe* this kind of curriculum research, and they make a compelling case for use of visual methods in research. They explain, through the use of meta-analysis of case study research, some of the dilemmas and new possibilities for such methods, providing a sophisticated and interesting perspective on what one may ordinarily and perhaps naively think of as being a "straightforward process"—after all it is not difficult to use a video camera or photographs in research—or is it—especially when one begins to think carefully about what is involved in mixed method and multimodal research?

Hoeffel, Fadini, Machado, Reis, and Lima, writing from Brazil, raise another methodological issue in and for environmental education curriculum work, and the associated research processes. They raise the question of participation in environmental decision making, and point out how research can be constituted in ways that support and enable such decision making in rural community contexts. Their interest is political, but also social, ecological, and educational. They point to the affects and impacts of colonial intrusion on traditional societies in Brazil, and seek out and test a methodological process that is less intrusive and violent. The methodology that they describe, developed in the context of a catchment management program with local communities, is a patient methodology; it requires time, consultation, deliberation, and slow movements in support of emergent changes. This, they argue is necessary if research is to contribute to transformation of local reality. They are careful too to distinguish between *this* variety of participatory research, and popular participation, as appropriated by development organisations and governments seeking to co-opt people into governance agendas of the day. Their interest is emancipatory and supportive of autonomy, social critique and transformation, and recognizes communities of practice as a unit of engagement.

The chapter on place-based education by Smith, writing from the United States, considers a similar "locatedness" or "situational" theme in environmental education curriculum research, but focuses more on the curriculum "value" of such approaches to education and learning. Smith notes that this approach fosters community and environmental renewal. Through an extensive review of what he claims to be a limited body of research (normally undervalued and underfunded) he begins to point toward the relationship between place-based learning, learner motivation and participation, construction of knowledge and experience, and learner achievement. He describes two place-based educational projects in detail presenting and showing the significance of these as modalities for and of citizenship participation and stewardship. The chapter also argues that place-based curriculum approaches help to "bridge the divide" between everyday knowledge and experience, and more abstract forms of knowledge presented in schools, linking this to an ever-present discussion in learning theory research, since the time of Dewey.

Chapter 21 provides another perspective on what counts as valuable contemporary curriculum theory research in environmental education. Zeyer and Kelsey (writing from Switzerland and the United States), focus on the cultural context of environmental education curriculum and learning. They consider the significance of emotions, and their embedding in norms, values, and beliefs in culture. They consider this particularly in the context of science education, and they argue that a "clash of cultures" exists between students' life world culture and the culture of environmental education represented in schools. Through analysis of case data, they show this disjuncture, representing students' disempowerment, disinterest, "depression," and beliefs in technical fixes, commitments to consumerism, and their dislike for government actors and green politics as critical emerging factors shaping and influencing how environmental education is being viewed and responded to (or not) by young people today (at least in the contexts and literature that they discuss). In the final analysis they reason that this represents a significant clash between school culture and life world culture, and that environmental educators should "take culture seriously" in their curriculum and pedagogical deliberations.

While remaining essentially complete as there are many other innovative areas of curriculum research currently taking place in environmental education, such as those pointed to in the Lotz-Sisitka et al. chapter on trends and niches, this section of the IHREE presents readers with the tools to engage critically with environmental education curriculum research by tracing the histories of how environmental education curriculum research has become constituted; by engaging critically with methodological questions embedded in different research traditions and orientations; and by critically considering how the research is contributing to theory building in this young, but clearly dynamic, subfield of educational research.

20

Traditions and New Niches

An Overview of Environmental Education Curriculum and Learning Research

Heila Lotz-Sisitka
Rhodes University, South Africa

John Fien
RMIT University, Australia

Mphemelang Ketlhoilwe
University of Botswana, Botswana

Introduction: Educational Change Landscape

The seventeenth to twentieth centuries were characterized by the rise of modernity, industrial societies, and the expansion of colonial intrusions. The associated modern education project was oriented primarily to the development of scientific, linguistic, and technical capabilities necessary for establishing, maintaining and expanding production systems, resource flows, economies of scale, modern life styles, and competitive capital advantage. Formal education institutions were expanded by nation states to carry this agenda forward. Early curriculum and learning practices were transmission oriented, and structural functionalist in nature. These approaches were valorized by educational theorists such as Tyler (1950) and Popham (1972), who proposed objectives-driven approaches to curriculum. In the latter parts of the twentieth century, this education project was critiqued for its behaviorist and deterministic character, coming under fire for creating, maintaining and exacerbating exclusions, and narrowing learning possibilities. Indeed, learner-focused and social reconstructionist approaches, respectively based upon based upon Rousseauan and Dewian ideals existed as counter discourses of education. Despite this contestation, Bernstein (1996) and other curriculum theorists have shown that this "performance model" retains its status as the dominant curriculum and learning model in education systems. The performance model "... clearly emphasizes marked subject boundaries, traditional forms of knowledge, explicit realization and recognition rules and the designation and establishment of strong boundaries between different types of students" (Scott, 2008, p. 4). Within this structural frame, it is, and has been, difficult to insert new curriculum innovations such as environmental education (Stevenson, 2007).

As the developmental and welfare states expanded after World War II, so did the call for "Education for All," and in the post war period, a massive international effort was put in place to enable access to education. This expansion of education has shown continuities with earlier objectives driven curriculum models and associated learning assumptions, as discussed above. There are, however also visible influences on curriculum and learning emerging from a decade of psychological and sociological research into human learning, education and schooling. The most prominent influences here include the work of Dewey, and the work of cognitive and social psychologists Piaget and Vygotsky; and critical cultural sociologists such as Bourdieu, Gramsci, and Bernstein, even though the effects of their critical scholarship remains "on the margins." More recently feminist theory and postmodern perspectives on education (after Foucault) have raised interest in critical issues such as inclusivity, knowledge, and power. Also increasingly powerful in this area, is the work of critical post colonial scholars, such as Said (1993), Kapoor (2009), Bhabha (1994), who have helped to identify ideo-cultural "grand narratives" situated in modern forms of education and associated implications for identity formation, oppression, and cultural reproduction. Examples of this include the significant political influence of the written word and its pretext for the project of colonization and so-called "civilizing" of the "backward" [sic] areas of Africa, Asia, Latin America, and other locations of the same genre (Abdi, 2008). This critique of mainstream curriculum and learning theories extends to researchers working in the East, who argue that fundamentally different philosophical traditions shape education in different parts of the world, a point which has also been largely ignored in the rise of modern forms of education (Eppert & Wang, 2008).

These influences have created the space for a more "open" curriculum in which acquirers have some control over the selection, pacing, and sequencing of their curriculum and thus their learning. Bernstein describes such modes as "competence" modes which "may be seen as interrupts or resistances to this normality [of performance modes]." He goes on to say that such "competence modes" were generally appropriated by specific (e.g. environmental education) or local purposes and tended to mostly characterize early years education or education in "repair sections" (Bernstein, 1996, p. 65). Given the affinity to more critical orientations (see below), it is not surprising that most environmental education curriculum and learning research tends toward favoring competence modes of curriculum over performance modes, some explicit examples being the work on action competence (Jensen & Schnack, 2006), critical learner-centered pedagogies (Fien, 1993; Gough & Robottom, 1993), deliberative participatory models (Wals, 2007; see also Hoeffel et al., this section); and ecophenomenological, cultural, and place-based modes of curriculum theorizing (e.g. Gruenewald & Smith, 2008; Payne, 2010; see also Smith; Zeyer and Kelsey, this section). Models of curriculum are closely related to views of learning, and the interests of environmental educators in critical competence models of curriculum are mirrored by an interest in participation metaphors of learning (Sfard, 1998; see Hoeffel et al.; Smith; Zeyer and Kelsey, this section). Rickinson (2001, 2006), in a critical commentary on learning research in environmental education notes, that despite an emerging interest in participatory metaphors or more cognitive and situative forms of learning in environmental education, learning process research in environmental education fails to engage deeply with learning theory, learning processes, and learning *per se*.

As the twentieth century saw the foregrounding and expansion of education to include more people everywhere, inclusivity and participation discourses expanded to replace early civilizational rhetoric, even though these may not have been engaged as deeply in environmental education as Rickinson (2006) expected. Critics have noted that despite the inclusivity discourses shaping recent curriculum and learning theories, modern forms of education continue to paradoxically result in alienation and exclusion (Popkewitz, 2008). Today large bodies of educational research are oriented toward strengthening inclusivity and the efficacy of this cosmopolitan agenda. Among the subfields of educational research operating in this terrain (perhaps somewhat blindly) is environmental education research. Popkewitz comments that the cosmopolitan agenda of education and associated research, seeks mostly to shape a local and global [environmental] citizen with capabilities to "use reason and science to perfect the future" (2008, p. 7). This, Popkewitz notes, is an historically recent idea, and to fully understand this tendency, he urges us to critically understand the social epistemologies of pedagogy which have been established in the modern education project to "cultivate, develop and enable the reason necessary for human agency and progress" (2008, p. 17). It is against this broader educational research landscape that environmental education curriculum and learning research can be interpreted.

Environmental Education Research

Along with globalization and progress narratives, the late twentieth century brought various socioecological issues into global/local focus, with a greater realization that human beings are living unsustainably with each other, and within the earth's support systems, adding these issues to the cosmopolitanist agenda. These insights have raised significant questions about the future of human development. In the past three years, the evidence of anthropogenic impacts on the environment have shot to the top of the global agenda, with reports on the radical projected impacts of climate change on human existence beginning to shape educational research in areas such as curriculum and learning (Lister, 2010; Selby & Kagwa, 2009).

While curriculum and learning research in this direction remains underdeveloped in the broader field of education, there is an ever increasing recognition that education can no longer be about reproduction of knowledge and skills for production or consumption patterns that were the focus of nineteenth and twentieth century aspirations, as evidenced by the international UN Decade on Education for Sustainable Development (United Nations Educational, Scientific and Cultural Organization [UNESCO], 2005). Despite UN attention, there is inadequate engagement with the question of whether society can afford to continue to educate young people around the world for an unsustainable future. Biersta (2009), in a discussion on the value of educational research in the UK, draws on De Vries's work to explain that educational research can serve two important roles in relation to educational practice and society, namely it can provide technical support, guidance and contributions to changing practices, or it can serve *a cultural innovation role*, a perspective which is helpful to shed light on the role of environmental education research within the wider landscape of educational research. He argues that these roles are not mutually exclusive, but that the cultural innovation role is particularly significant for furthering transformation and democracy. Thus, Lotz-Sisitka recently commented on a review of environmental education research that

> . . . environmental education research (as cultural practice) is reflexive: it is shaped by the nature of its contestability through the social practices of actors. Environmental

> et al. education research provides actors with resources for the project of educational, social and environmental change, while transforming these projects at the same time. (2009, p. 166)

In this chapter we consider the traditions of environmental education curriculum and learning research, their relationship to wider education research traditions, and point to new niches for curriculum and learning research, as opened up (in part)[1] through the contributions in this section of the IRHEE. The chapter points to the fact that environmental education research seems to primarily be seeking to fulfill a "cultural innovation role" in the wider education research landscape, carving out niches and spaces that speak to educational innovation/transformation and change. This may in part be due to its "youthfulness" within the more established and traditional education research landscape and trajectory, but also to its transformative intent. Environmental education researchers such as Stevenson (2007) continue to lament the "marginal" or "permanently peripheral" status of environmental education and environmental education research, noting that it is almost impossible to situate effectively within modernist educational paradigms oriented mostly toward reproduction of existing cultures and practices, traditions which continue to characterize formal education institutional settings.

Environmental education curriculum and learning research traditions. In describing environmental education curriculum and learning research "traditions," it is important to locate these within a much longer and more established research landscape (as broadly outlined above). Seen within a longer term social and educational history, environmental education research is young (with a relatively short history of some forty years—making it essentially a post-'60s, or even more accurately, a post-'80s phenomenon, with most environmental education journals only being in existence for the past ten to fifteen years). "Traditions" when—located within this short historical configuration, and in relation to the much longer history of educational research and praxis, may therefore not be an adequate descriptor, and it may be more accurate to discuss trends or streams of work emerging in the field of environmental education curriculum and learning research. Added to this is the time-space configuration of these "traditions," with most established environmental education journals being published in the English speaking world, and then mostly in the European-American nexus, a fact which influences knowledge generation. This descriptor therefore remains open to deliberation. Our criteria for naming various "traditions" here was in some ways empirical, in that we considered those traditions that were appearing to influence the field of curriculum and learning, and which were "gaining ground" in the curriculum and learning research arena in environmental education (at least insofar as these are reflected in journals and other published media in the English speaking world).

Influential Research Traditions Three major influential approaches in environmental education curriculum research, with their roots in Habermasian views on human knowledge interests (Habermas, 1972) can be discerned. They are briefly reviewed here as they appear to be particularly prominent forms of environmental education curriculum and learning research. What is of interest is the manner in which the research tradition or approach shapes the type of knowledge produced about curriculum and/or learning. Examples are included by way of illustration, and these are by no means representative of the full scope of work undertaken within these broadly defined research traditions. They merely provide openings for further literature review work by scholars interested in this arena.

Habermas (1972) argued that humans have three distinct categories of needs and interests ("knowledge constitutive interests"). The first of these, the technical interest, involves the need for mastery and control over the physical world. This gives rise to the need for instrumental knowledge that can satisfy physical and economic needs and allow one to fit into society as it presently constructed. The technical interest underlies functionalist, positivist or behaviorist approaches to environmental education in which predetermined knowledge, attitudinal, and/or behavioral changes are sought through carefully structured curriculum and instruction, showing continuities with early Tylerian thinking in the wider educational field, and with performance models of curriculum. Such curriculum and learning research interests in environmental education are often reflected in empirical-analytical approaches to research using quasi-experimental research designs, survey data, or phenomonographic methods. These study designs, despite various forms of critique, remain popular. A typical (more recent) example of such research is a study by Rideout (2005) who tested the increase in, and retention of knowledge of a group of students exposed to a three week environmental problem module, against a control group (who had not been exposed to the same module) to establish effectiveness of the teaching intervention (see also examples from Aivazidis, Lazaridou, & Hellden, 2006; Smith-Sebasto & Cavern, 2006). In 1993 Robottom and Hart critiqued these approaches to curriculum and learning research for assuming that control of the flow of knowledge or efficacy of the pedagogy can lead to proenvironmental behavior; making proenvironmental behavior the key objective of environmental education curriculum and learning. The approaches were also critiqued for their individualistic assumptions about social change, and a failure to address structural underpinnings of existing practices (Robottom & Hart, 1993).

A second knowledge interest, according to Habermas, is important because we, as humans, inhabit a social world as well as a physical one and, thus, have a practical interest in understanding the symbols, and participating in the cultural traditions, that shape social life and experiences of the environment. The liberal-arts approach to planning the curriculum and instruction reflects this paradigm, and includes negotiating with students what they would like to learn, based upon the various need and interests, and planning learning experiences through child-centered and place-based pedagogies. Curriculum based upon this practical interest in environmental education has generally been developed upon research that focuses on the lived experiences of planning, teaching, and learning and, thus, relies upon interpretive or constructivist approaches to research and learning. An example of such research is a study focusing on how teaching through modeling sustainability practices in schools promoted environmental learning and adoption of sustainability oriented behavior in schools, often using qualitative or constructivist approaches (e.g., Higgs & McMillan, 2006). Other examples are a study by Corney and Reid (2007) focusing on teachers learning about subject matter and pedagogy in education for sustainable development and Carew and Mitchell's (2006) study focusing on metaphors used by university engineering lecturers to explain sustainability. These types of studies have also been critiqued for privileging understandings and individual meaning making over structural conditions and critical engagement with the "real causes" of environmental issues (Robottom & Hart, 1993).

The third human interest, according to Habermas, is a goal of emancipation that derives from a desire to be free of the constraints of ignorance, authority, and tradition upon human reason. Such constraints impede the freedom of teachers and students to determine their plans and actions on the basis of their self-identified needs. This has led to an emphasis on socially critical thinking, media literacy, and community problem solving through action research as approaches to teaching and learning in environmental education. In research, this emancipatory interest is satisfied by critical social science studies that seek to uncover the ideological origins of curriculum practices; address the inequalities and related problems created by unequal power relations in schooling; and, where appropriate, empower teachers and students to think and act in the interests of social justice, environmental sustainability, and democracy. Typical examples of curriculum and learning studies in this tradition are a study by Mortari (2003) which seeks to understand how to foster youth civic participation through youth town councils in Europe; a study by O'Donoghue (2003) seeking to foreground indigenous knowledge methodologies and materials that respond to social processes of marginalization in Eastern Southern Africa, and a study focusing on learners' engagement with conflicts of interest in action competence approaches to pedagogy by Lundegård and Wickman (2007). These studies are also being critiqued, this time because their assumptions of empowerment are primarily located in a "philosophy of consciousness" with inadequate attention being given to the historical antecedents and feasible and practical responses to the structural problems being uncovered in curriculum and learning situations (Popkewitz, 2008), and for inadequately taking account of embodied experiences, habitus, and other forms of knowledge (N. Gough, 1990; Naess & Jickling, 2000; Payne, 2010)

Robottom and Hart's earlier (1993) critique of approaches that privilege "responsible environmental behavior" as focus for curriculum and learning research, led to an interesting body of work that has extended the scope of environmental education curriculum and learning research to include the ecophilosophical, place-based, culturally situated, and phenomenological perspectives that underpin constructivist and socially critical environmental education approaches. These approaches take account of the contextual nature of environmental issues and risks, and also consider the cultural fabric and ontological dynamics of environmental education praxis. This ontological perspective, rather than seeing humans as separate to nature or environment (with capacity to control it with knowledge), sees humans as part of nature or the environment, where values and practices emerge as being as significant as knowledge. Place-based education approaches, action oriented, participatory and cultural education approaches (as reflected in the contributions in this section of the IRHEE) give rise to different approaches to environmental education curricula and learning to those privileged by performance orientations to curriculum. On a methodological level, the ecophilosophic orientation holds that knowledge (scientifically constituted) is not primarily the domain of the expert educator who is able to successfully "transfer" and "plan for" transmission of this knowledge to students or scholars. Instead, it sees knowledge through pedagogical constructivist lenses, as a process that accumulates through experiences with others, with society, within and through mutually constitutive relations with the environment. Epistemologically, the ecophilosophical orientation treats knowledge as subjective and maintains that valid knowledge can be both rational and nonrational unlike the empiricism of the scientific worldview, and the "hard cosmopolitanism" described by Popkewitz (2008). Scott and Gough (2004) argue that learning in environment and sustainability education needs to take account of complexity and uncertainty, and that not everything can be rationally controlled through pedagogy and/or scientific thought. Ontologically, the ecophilosophical orientation recognizes that humans are mutually constituted through and in environments, and that there is a reality that exists outside of, but not always separate from humans (see Payne, 2005, 2010; Price, 2007; Scott & Gough, 2004). Perhaps it is the ecophilosophical orientation that integrates fact, experience, practice, value and reflexivity in environmental education that differentiates it from performance models of curriculum, locating the research terrain more within, or even "beyond" the realm of Bernstein's competence models that seek to "interrupt" the dominance of performance models of modern curricula. The difficulty of this interruptive process creates some critical frustration; and an ongoing search for theory, methods, and approaches among environmental education researchers who are anxious for environmental education to be constituted within critical, democratic, phenomenological or values-centered orientations to education and learning (see, e.g., Mogensen & Schnack, 2010; Payne, 2010; Rudsberg & Öhman, 2010; Wals, 2010).

New Emerging "Traditions" This early work on research traditions in environmental education has proved to be highly influential not only in catalyzing debates about environmental education curriculum and learning (e.g., Jensen, 2002; Jensen & Schnack, 2006; Huckle, 1993) but also in helping to frame alternative approaches to curriculum policy, teaching, and learning, and approaches to teacher professional development in environmental education in different socioecological and educational contexts (e.g., Elliott, 1999; Huckle, 1993; Kyburz-Graber& Robottom, 1999; LeGrange, Makou, Neluvhalani, Reddy, & Robottom, 1999; Lupele, 2008; Posch, 1996; Rauch, 2002). Despite the value of articulating traditions, too fixed a rendition of "traditions" in environmental education research can have the effect of neglecting a rich array and diversity of emerging forms of research, and the important point that different approaches to research may be used to answer different sorts of research questions. Biersta (2009) also warns against too heavy an emphasis on methodology, and encourages theory development. Thus, if we return to the point made by Biersta about research being a cultural innovation process, it would also be important to continue to see environmental education research as an open-ended, reflexive process, not too fixed by the boundaries of research traditions and methodological frameworks, even though these are helpful to provide a mapping of an emerging field.

As noted briefly above, ecophilosophic orientations and their associated critical research traditions, and constructivist learning assumptions, while valid and important, can be critiqued for neglecting important ecocentric aspects of ecophilosophic viewpoints. Researchers working on deep ecology, ecofeminism and socioecological justice perspectives who take equal account of humans and the more than human in curriculum practices and research approaches, have tended to draw more on poststructuralist forms of enquiry (Barret, 2005; Greenall Gough, 1993) and more recently on critical realist perspectives (Lupele, 2008; Price, 2007).

This introduces new research traditions that are increasingly becoming influential in environmental education research—that of poststructural enquiry and its counter-movement critical realist enquiry. The emerging tradition of poststructural inquiry in curriculum and learning research focuses on the deconstruction of the disciplinary and bio-power of discourses that become powerful, taken for granted and normalized in society, often leading to forms of governmentality (self governance) (Ferreira, 2007; Lather, 1992; Popkewitz, 2000). Lather's early feminist educational research using poststructural enquiry influenced environmental education research and introduced a reflexive orientation to existing research traditions (Janse van Rensburg, 1995). For curriculum and learning research this signaled an understanding that knowledge could not be taken for granted in its current form, and that significant knowledge-power relations shape curriculum, teaching and learning. Noel Gough never fails to remind us that curriculum is the stories we choose to tell our children, highlighting the powerful nature of the narratives that are selected for representation in and as curriculum, and in their globalized form, how such narratives tend to exclude or turn local knowledge forms into abjection and exclusion (N. Gough, 1999, 2000, 2002, 2008). Not unlike Popkewitz (2008) he notes that curriculum enquiry is a "cultural production" through which "the transnational imaginary of globalization is being envisaged, negotiated and materialized" (N. Gough, 2002, p. 174). Poststructural research that lays power-knowledge relations bare, has been most powerful in its capacity for deconstruction, or making visible that which is taken for granted. Like other critical research traditions, its full effect and impact on mainstream education systems is yet to be determined. Despite the growing number of research-orientated papers advocating for poststructuralist, feminist, and queer research in environmental education, or critiquing other research from such a perspective (e.g., A. Gough 1999; A. Gough, & N. Gough, 2003; A. Gough & Whitehouse, 2003; N. Gough 1999, 2000, 2008; Walker, 1997), there are fewer that actually display the practice of such research (e.g., Bennett, 1996; Ferreira, 2007; N. Gough, 1996, 1997; Hardy, 2006; Ket lhoilwe, 2007) and far fewer still that actually display poststructuralist, feminist and/or queer curriculum or educational practice in environmental education (e.g., A. Gough, 1999b; Hardy, 1998).

Ketlhoilwe's (2007a, 2007b) study is a good example of the potential value of such research. Through poststructural analysis of the Botswana environmental education curriculum policy, he was able to illuminate and show how curriculum policies are actually interpreted by teachers through various dominant discourses that maintain continuities despite transformative intentions, and also how various practices of self-governance become established in schools by teachers and learners as a result of power-knowledge relations that circulate in education systems. His research also demonstrated the effects of the normalization of such discourses, showing how environmental education policy can simply result in normalized "cleaning regimes" in schools where children take on the role of cleaners in/as environmental education praxis, as a result of the complex discourse-power-knowledge relations that are embedded in policy transformations. His research is being followed by a study on reoriented praxis in schools to address the normalized "cleaning regimes" that became evident (Silo, 2009).

Critical of the ontological relativism embedded in poststructural research designs and their overemphasis on language as constituting element of society and environments, Price (2007) and other critical curriculum and learning researchers (e.g., Lotz-Sisitka, 2009; Lupele, 2008; Mukute, 2009; Pesanayi, 2009) seeking to provide robust critical explanatory critiques are turning to critical realist ontological perspectives and research tools to probe curriculum issues. Price (2004) for example, argues that it is important for environmental educators to take account of transitive and intransitive dimensions of reality in curriculum design, and if this is done adequately, all participatory processes may not be equally valid, and there may be good reason (e.g. if the environment is being severely degraded or if someone is being hurt) for an educator to go against the popular views of "participants" who are constructing meanings. She argues that it may well be an ethical responsibility of educators to "go against the grain" and thus challenge conventional perspectives on participation, learning, constructivism, and critical traditions that privilege the philosophy of consciousness of the learners involved. This research provides an example of why it is important to continue to engage with methodological experimentation and development in environmental education curriculum and learning research.

Table 20.1 provides a "brief overview" of how a picture of changing research traditions can be simplistically represented, for heuristic purposes, and to help "map" a changing field. Care should be taken *not* to use this table as "fixed paradigms" to be slavishly followed. Relegating possibilities for environmental education curriculum research to choices that reside within the "established traditions" is likely to close in and narrow research potential, new knowledge generation, and the cultural innovation potential of environmental education research. This would be a tragedy in a newly emerging subfield of education, hence Table 20.1 should be approached with the utmost caution and skepticism.

New Niche Areas

As mentioned by Biersta (2009), research can suffer from too great an emphasis on methodology, without giving adequate attention to theory development. Environmental education curriculum and learning research, when considered in relation to the broader field of educational research can also be defined, not only for its engagement with methodological innovations, but also for some areas of theoretical innovation, although the latter remains youthful. With further work and greater maturity, some of these fields of curriculum and learning research promise to provide interesting "cultural innovations" in educational research. As shown above, some of these developing areas

TABLE 20.1
Ontological, Epistemological, and Methodological Aspects of Identified Research "Traditions" in Environmental Education Curriculum Research

Research Orientation or Tradition	Ontology (What is the Nature of Reality?)	Epistemology (What is the Nature of Knowledge?)	Methodology (How is Knowledge Developed?)	Typical Research Methods	Sample Research Themes	Sample Studies
Empirical-Analytical	• Reality is 'real', "concrete, material, out there"—independent of human thoughts and feelings. • Generalizations about "reality" can be made free of context.	• Objectivity is the ideal goal. • Values and other factors can produce some bias if not regulated or controlled for.	• Knowledge grows from the gradual accumulation of findings and theories and testing the significance of relationships. (Deductive modes of inference)	• Experiment • Pretest/posttest • Questionnaire • Survey	• Factors affecting responsible behavior • Surveys of environmental knowledge and/or attitudes • Impacts of particular instructional approaches	• Culen and Volk (2000) • Orams (1999) • Wesley (1999) • Rideout (2005)
Interpretivism/ Constructivism	• Reality is not "out there." It exists in the human mind, and is conditional upon human experiences and interpretation. • Reality is not independent but socially constructed and can have varied meanings.	• Knowledge is not objective but subjective. • Knowledge is constructed through the interaction of people (including the researchers) and the objects of inquiry.	• Identification and analysis of individual and group constructions or interpretations of reality and an attempt to recognize patterns in them or bring them into some consensus.	• Ethnography • Case study • Phenomeno-graphy • Interview • Focus group • Life-history • Narrative enquiry	• Student or teacher experiences of a particular instructional approach or encounter with nature. • Development or review of a course or teaching / learning episode.	• Fien and Rawling (1996) • Ballantyne, Fien, and Packer (2001) • Wals and Alblas (1997)

Research Orientation or Tradition	Ontology (What is the Nature of Reality?)	Epistemology (What is the Nature of Knowledge?)	Methodology (How is Knowledge Developed?)	Typical Research Methods	Sample Research Themes	Sample Studies
***Critical Research**	• Reality is "out there"; it is material but interpretations of it can be controlled by human power relations.	• Knowledge is not objective but subjective. • Values and power play a pivotal role in shaping what counts as knowledge. • Knowledge and issues of equity and power are closely intertwined.	• Research seeks to understand the practices and effects of power and inequality, and to empower people to transform environmental and social conditions.	• Critical ethnography • Action research	• Addressing inequality in society through education • Improving educational or environmental conditions	• O'Donoghue and McNaught (1991) • Hart, Robottom, and Taylor (1994) • Malone (2006)
***Poststructural Research**	• Multiple, representations constituted in and through language and discourse	• Events are understood in terms of powerful and subordinated discourses which constitute social realities	• Deconstruction: Exposing how dominant interests constructed through language and discourse preserves social inequities	• Discourse analysis • Geneology	• Critique of assumptions and knowledge-power relations in environmental education practice • Identification of processes of governmentality	• N. Gough (1997) • N. Gough and A. Gough (2003) • Hardy (2006) • Ketlhoilwe (2007) • Ferreira (2007)
***Critical realist research**	• Recognition of transitive and intransitive reality • Depth ontology distinguishing between empirical experiences, actual events and the real (that which exists but may not be actualized or visible).	• Knowledge is based on the "best possible truth" claim, but is recognized as being fallible. • Not everything can be reduced to language. • Events and empirical experiences are understood as being influenced by causal mechanism that may not always be visible.	Explanatory critique – providing complex descriptions of reality with a view to identifying emancipatory options for absenting causal factors constraining actions, or enabling actions.	• Follows a process involving description; analytical resolution; abduction or theoretical redescription; retroduction; comparison between different theories and abstractions; and concretization and contextualization. • Mixed methods can be used, but are not "compulsory."	• Identifying causal mechanisms influencing circumstances and practices • Structure agency relations analysis and identification of morphogenesis (change) or morphostasis (no change) • Identification of possibilities for action and inaction through absenting	• Price (2007) • Lupele (2007) • Mukute (2009) • Lotz-Sisitka (2009)

*All three of the above are critical research traditions, showing that ongoing clarification of various critical orientations to research is an interesting feature of environmental education curriculum research

Sources: Adapted from Robottom and Hart (1993); Danermark, Ekström, Jakobsen, and Karlsson (2000); Price (2007)

Note: A number of years ago, Robottom and Hart (1993) analyzed the *ontological*, *epistemological*, and *methodological* characteristics in the first three environmental education curriculum research traditions dealt with in Table 20.1. Table 20.1 summarizes some of these perspectives, and includes discussion on more recent research traditions that have, and are influencing environmental education curriculum research. Mrazek's 1993 edited collection of essays and Hart's 2007 review of approaches to environmental education research provide further references to, and examples of the many studies that have been conducted within each of these research traditions. Price's 2007 study provides insight into the latter two "traditions" that are included in Table 20.1. See also Section XXX of the IHREE).

are shaped by the methodological, ontological, and epistemological processes embedded in the research process, but are equally related to working with theory in environmental education. To comment briefly on other defining features of environmental education curriculum and learning research, we briefly consider an introductory scoping of "thematic areas" that are prominent in environmental education research (broadly clustered) and provide a brief discussion of the various theoretical influences and developments that are taking place. Again the examples are illustrative, and mark out an area for further literature review work.

These "niche areas," are visible across the range of environmental education journals, and a careful reading of most of the curriculum and learning research papers can be included within these broad areas, although new

TABLE 20.2
Thematic Areas in Environmental Education Curriculum, and Learning Research, and Theoretical Influences (Broadly Analyzed)

Thematic / Niche Areas	Broad description with some examples cited
• Knowledge, attitudes, values, concepts, and perceptions research	This research works on "measuring" knowledge, values, attitudes, and perceptions of people relating to environmental issues and questions. This research has been critiqued for empirical analytical assumptions that are positivist, and a lack of engagement with "what can't be measured" but which still exists (i.e. for including some variables, and excluding others). Influences are typically from the behavioral and/or cognitive sciences. **Typical examples include:** Andrews, Tressler, and Mintzes (2008); Bright and Tarrant (2002); Kollmus and Agyeman (2002); Malandrakis (2008)
• Research on curriculum programs, courses, course design, course impacts, and learning outcomes	This research works on establishing effectiveness and value of curriculum programs, courses, course design, and course impacts and seeks to consider the value of educational interventions, or learning outcomes. As indicated by Rickinson's (2003) review, such research has tended to neglect a focus on learning, the learning process, and learning theory. Influences are typically from objectives oriented or reflexive curriculum models. **Typical examples include:** Mokuku et al., (2005); Kowalewski (2002); Shepardson et al. (2002)
• Research on action, action competence, social learning, and participation in change processes	This research works on developing insight into pedagogical approaches, philosophies, and methodologies to include learners in decision making about environmental issues and actions. Some examples of this research fall victim to behavioral assumptions, while other are careful to emphasize the open process and philosophy of participation. Participatory approaches have also been critiqued for relying too heavily on "philosophies of consciousness" or the "individual agent" and some researchers are beginning to focus more on cultural practices and historical influences. Influences are typically from constructivism and liberal education. **Typical examples include:** Dyment and Reid (2005); Jensen and Schnack (2006); O'Donoghue (2007); Wals (2007)
• Constructivist and phenomenological forms of place-based research which may or may not include sociocultural dynamics	This research works on presenting an integrated embodied and experiential theoretical and philosophical foundation for environmental education curriculum and learning. Some of this research tends to focus more on the human-nature relationship, discounting sociocultural dynamics that influence and shape such relationships. Influences are typically from phenomenology and deep ecology. Some of this research also includes a focus on "worldviews" particularly ecological worldviews. **Typical examples include:** Andrews, Tressler, and Mintzes (2008); Gruenewald and Smith (2008); Guenther (2002); Payne (2005, 2010); Sandell and Öhman (2010)
• Sociocultural learning, cultural critiques, and curriculum research	This research works on developing knowledge of pedagogical and social change processes that are embedded in sociocultural understandings and practices. It tends to emphasize history and context, and at times emphasizes place-based pedagogies but may focus more on the sociocultural dynamics of learning and change than on place-based influences per se. Influences are typically from post-Vygotskian social constructivist theories, historical materialism and post-Marxian perspectives; as well as historical and other critical process sociologies. Postmodern perspectives on knowledge and culture are also influential in this stream, more from the perspective of cultural critique than culture and agency. This niche area also sometimes includes reference to indigenous knowledge research. **Typical examples include:** Foster (2005); Gough (2005); Hogan (2008); Krasny and Tidball (2009); Laesse (2010); Mukute (2009); O'Donoghue and Neluvhalani (2002)
• Values, ethics, and normative deliberations in curriculum and learning research	This research works on developing democratic practices that take account of explicit normative engagements with values and ethics within deliberative communicative engagements. It tends to emphasize communicative rationality and democratically constituted pedagogical deliberations, and also takes account of diversity, difference, and negotiating conflicts or tensions in social settings related to diversity in ethical or normative perspectives. Influences tend to be Kantian ethics; deep ecology; environmental ethics theory, Habermasian communicative rationality, and /or theories of language and pragmatism after Dewey, Rorty, and Wittgenstein. **Typical examples include:** Jickling (2007); Östman (2010); Rudsberg and Öhman (2010); Wals (2010)

niche areas continue to emerge, and a framework such as that presented in Table 2 will continue to remain indicative rather than definitive, and open to redescription and challenge. It is included here more as an heuristic to encourage a critical analysis of the emerging niche areas as *areas of theory development* in the wider field of educational research, in order to fully establish the potential cultural innovation role of environmental education research. Suffice to say that the field remains "too young" to complete a definitive analysis such as the one pointed to above. It is also possible to discern that the environmental education curriculum and learning "niche areas" identified above, are in some ways linked to the description of research "traditions" outlined above, but that they also deviate from these broad framings of research traditions. While not all of these niche areas are "theoretically innovative" in the sense of being culturally innovative within the wider field of educational research, some of them are showing promise to become theoretically innovative with stronger forms of dialog, and ongoing deliberation on the research patterns in environmental education. In-depth reviews such as the one provided by Rickinson (2001, 2006) into learning research in environmental education are important for "marking out the terrain" and for assisting environmental education researchers to see their own body of work within a wider landscape of research in environmental education and in the wider field of educational research. Border crossings into other field s of research, exemplified by the interest in social learning in both environmental sciences and in environmental education (see Scott & Gough, 2004; Wals, 2007) is likely to make environmental education researchers more reflexive of their own research trajectories and contributions to theory development (as evidenced in a recent social learning seminar held at the Stockholm Resilience Institute).

Conclusion

In closing this chapter, it is difficult to feel or capture any form of "consolidation," particularly in the light of the point made above that the environmental education curriculum and learning research remains "young" and experimental within a wider and much longer and more established tradition/s of educational research. It is, however, possible to trace the relationship between these longer, more established educational research traditions and environmental education curriculum and learning research, and to begin to mark out the deviations, the niche areas, and the emergence of culturally innovative research patterns. The chapters that follow in this section of the IHREE mark out some of this terrain by focusing on specific contemporary issues of interest to environmental education curriculum and learning research (e.g. participation; cultural context; place-based education; and methodological tools for observing action), and researchers entering the field of environmental education curriculum and learning research will be faced with the choice of extending further cultural innovation oriented work, and or consolidating environmental education curriculum and learning traditions within the "mainstream" of educational research and practice. All of the authors contributing to this section of the IHREE have chosen the former path. As indicated above, more robust debates about the emerging and changing form and focus of environmental education curriculum and learning research are needed in future.

Note

1. It was not possible to include the full scope of "new niches" that are opening up in and for environmental education curriculum and learning research. The papers referred to in this section are therefore exemplary and demonstrative, rather than conclusive.

References

Abdi, A. (2008). Oral societies and colonial experiences: Sub-Saharan Africa and the *de facto* power of the written word. In D. Kapoor (Ed.), *Education, decolonization and development. Perspectives from Asia, Africa and the Americas*. Rotterdam: Sense Publishers.

Aivazidis, C., Lazaridou, M., & Hellden, G. F. (2006). A comparison between a traditional and an online environmental education program. *The Journal of Environmental Education, 37*(4), 45–55.

Andrews, K., Tressler, K., & Mintzes, J. (2008). Assessing environmental understanding: An application of the concept mapping strategy. *Environmental Education Research, 14*(5), 519–536.

Ballantyne, R., Fien, J., & Packer, J. (2001). School environmental education programme impacts upon student and family learning: A case study analysis. *Environmental Education Research, 7*(1), 23–37.

Barret, M. J. (2005). Making [some] sense of feminist poststructuralism in environmental education research and practice. *Canadian Journal of Environmental Education, 10*, 79–107.

Bennett, S. (1996). Discourse analysis: A method for deconstruction. In M. Williams (Ed.), *Understanding geographical and environmental education: The role of research* (Chapter 13). London, UK: Cassells Education.

Bentz, V. M., & Shapiro, J. J. (1998). *Mindful inquiry in social research*. Thousand Oaks, CA: Sage.

Bernstein, B. (1996). *Pedagogy, symbolic control and identity: Theory, research and critique*. London, UK: Taylor and Francis.

Bhabha, H. K. (1994, March). *The location of culture*. London, UK: Routledge.

Biersta, G. (2009). *Educational research, democracy and TLRP*. Invited lecture presented at the TLRP event "Methodological Development, Future Challenges," London, March 19, 2009.

Bright, A., & Tarrant. M. (2002). Effect of environment-based coursework on the nature of attitudes towards the endangered species act. *The Journal of Environmental Education, 33*(4), 10–19.

Carew, A. L., & Mitchell, C. A. (2006). Metaphors used by some engineering academics in Australia for understanding and explaining sustainability. *Environmental Education Research, 12*(2), 217–231.

Corney, G., & Reid, A. (2007). Student teachers' learning about subject matter and pedagogy in education for sustainable development. *Environmental Education Research, 13*(1), 33–54.

Culen, G., & Volk, T. (2000). Effects of an extended case study on environmental behavior and associated variables in seventh- and eighth-grade students. *Journal of Environmental Education, 31*(2), 9–15.

Danermark, B., Ekström, M., Jakobsen, L., & Karlsson J. (2000). *Explaining society. Critical realism in the social sciences*. London and New York: Routledge.

Dyment, J., & Reid, A. (2005). Breaking new ground? Reflections on greening school grounds as sites of ecological, pedagogical, and social transformation. *Canadian Journal of Environmental Education, 10*, 286–301.

Elliott, J. (1999). Sustainable society and environmental education: future perspectives and demands for the educational system, *Cambridge Journal of Education, 29*(3), pp. 325–340.

Eppert, C., & Wang, H. (2008). *Cross-cultural studies in curriculum. Eastern thought, educational insights*. New York, NY: Lawrence Erlbaum Associates.

Ferreira, J. (2007). *An unorthodox account of failure and success in environmental education* (Unpublished PhD thesis). Griffith University, Australia.

Fien, J. (Ed.). (1993). *Environmental education: A pathway to sustainability*. Geelong, Victoria: Deakin University Press.

Fien, J., & Rawling, R. (1996). Reflective practice: A case study of professional development for environmental education. *Journal of Environmental Education, 27*(2), 11–20.

Foster, J. (2005). Options, sustainability policy and the spontaneous order. *Environmental Education Research, 11*(1), 115–136.

Gough, A. (1999a). The power and the promise of feminist research in environmental education. *Southern African Journal of Environmental Education, 19*, 28–39.

Gough, A. (1999b). Recognising women in environmental education pedagogy and research: Towards an ecofeminist poststructuralist perspective. *Environmental Education Research, 5*(2), 143–161.

Gough, A., & Robottom, I. (1993). Towards a social critical environmental education: Water quality studies in a coastal school. *Journal of Curriculum Studies,* 25, 301–316.

Gough, A., & Whitehouse, H. (2003). The "nature" of environmental education research from a feminist poststructuralist standpoint. *Canadian Journal of Environmental Education, 8*, 31–43.

Gough, N. (1990). Healing the earth within us: Environmental education as cultural criticism. *Journal of Experiential Education, 13*(3), 12–17.

Gough, N. (1993). Narrative inquiry and critical pragmatism: Liberating research in environmental education. In R. Mrazek (Ed.), *Alternative paradigms in environmental education research* (pp. 175–197). Troy, OH: North American Association for Environmental Education.

Gough, N. (1996). Virtual geography, electronic media culture, and the global environment: Post-modernist possibilities for environmental education research. *Environmental Education Research, 2*(3), 457–467.

Gough, N. (1997). Weather incorporated: Environmental education, postmodern identities, and technocultural constructions of nature. *Canadian Journal of Environmental Education, 2*, 145–162.

Gough, N. (1999). Surpassing our own histories: Autobiographical methods for environmental education research. *Environmental Education Research, 5*(4), 407–418.

Gough, N. (2000). Interrogating silence: Environmental education research as postcolonialist textwork. *Australian Journal of Environmental Education, 15/16*, 113–120.

Gough, N. (2002). The long arm(s) of globalization: Transnational imaginaries in curriculum work. In W. E. Doll Jr & N. Gough (Eds.), *Curriculum visions*. New York, NY: Peter Lang.

Gough, N. (2008). Narrative experiments and imaginative inquiry. *South African Journal of Education, 28*(3), 335–349.

Gough, N., & Gough, A. (2003). Tales from Camp Wilde: Queer(y)ing environmental education research. *Canadian Journal of Environmental Education, 8*, 44–66.

Gough, S. (2005). Rethinking the natural capital metaphor: Implications for education and learning. *Environmental Education Research, 11*(1), 95–114.

Greenall Gough, A. (1993). *Founders of environmental education*. Geelong, Victoria: Deakin University Press.

Gruenewald, D. A., & Smith, G. (Eds.). (2008). *Place-based education in the global age*. New York and London: Routledge Taylor and Frances Group.

Guenther, L. (2002). Towards a phenomenology of dwelling. *Canadian Journal of Environmental Education, 7*(2), 38–46.

Habermas, J. (1972). *Knowledge and human interests*. London, UK: Heinemann.

Hardy, J. (1998). Chaos in environmental education. *Environmental Education Research, 5*(2), 125–142.

Hardy, J. (2006). 'In the neighbourhood of': Dialogic uncertainties and the rise of new subject positions in environmental education. *Mind, Culture, and Activity, 13*(3), 257–274.

Hart, P. (2003). *Teachers' thinking in environmental education: Consciousness and responsibility*. New York, NY: Peter Lang Publishing.

Hart, P. (2007). Environmental education. In S. K. Abell & N. G. Lederman (Eds.), *Handbook of research on science education*. Mahwah, NJ: Lawrence Erlbaum Associates.

Hart, P., Robottom, I., & Taylor, M. (1994). Dilemmas in participatory enquiry: A case study of method-in-action. *Assessment and Evaluation in Higher Education, 19*(3), 201–214.

Higgs, A. L., & McMillan, V. M. (2006). *The Journal of Environmental Education, 38*(1), 39–56.

Hogan, R. (2008). Contextualising formal education for improved relevance: A case from the Rufiji Wetlands, Tanzania. *Southern African Journal of Environmental Education, 25*, 44–58.

Huckle, J. (1993). Environmental education and sustainability: A view from critical theory. In J. Fien (Ed.), *Environmental education: A pathway to sustainability* (Chapter 13). Geelong, Victoria: Deakin University Press.

Hungerford, H., & Volk, T. (1990). Changing learner behavior through environmental education. *Journal of Environmental Education, 21*(3), 8–21.

Janse van Rensburg, E. (1995). *Research in environmental education in southern Africa. A landscape of shifting priorities* (Unpublished PhD thesis). Rhodes University, South Africa.

Jensen, B. (2002). Knowledge, action and pro-environmental behaviour. *Environmental Education Research, 8*(3), 325–334.

Jensen, B., & Schnack, K. (2006). The action competence approach in environmental education. *Environmental Education Research, 12* (3–4), 471–486.

Jickling, B. (2007). Researching in a changing world: Normative questions and questions that matter. *Southern African Journal of Environmental Education, 24*, 108–118.

Kapoor, D. (Ed.). (2009). *Education, decolonization and development. Perspectives from Asia, Africa and the Americas*. Rotterdam: Sense Publishers.

Ketlhoilwe, M. J. (2007a). *Genesis of environmental education policy in Botswana: Construction and interpretation* (Unpublished PhD thesis). Rhodes University, South Africa.

Ketlhoilwe, M. J. (2007b). Environmental education policy interpretation challenges in Botswana schools. *Southern African Journal of Environmental Education, 24*, 171–184.

Kollmus, A., & Agyeman, J. (2002). Mind the gap: Why do people act environmentally and what are the barriers to pro-environmental behavior. *Environmental Education Research, 8*(3), 239–260.

Kowalewski, D. (2002). Teaching deep ecology: A student assessment. *The Journal of Environmental Education, 33*(4), 20–27.

Krasny, M., & Tidball, K. (2009). Applying a resilience systems framework to urban environmental education. *Environmental Education Research, 15*(4), 465–482.

Kyburz-Graber, R., & Robottom, I. (1999). The OECD-ENSI project and its relevance for teacher training concepts in environmental education. *Environmental Education Research, 5*(3), 273-291.

Laesse, J. (2010). Education for sustainable development, participation and socio-cultural change. *Environmental Education Research, 16*(1), 39–58.

Lather, P. (1992). Critical frames in educational research: Feminist and post-structural perspectives. *Theory Into Practice,* 31(2), 87–99.

LeGrange, L., Makou, T., Neluvhalani, E., Reddy, C., & Robottom, I. (1999, September). *Professional self-development in environmental education: The case of the "Educating for Socio-Ecological Change Project."* Paper presented at the annual conference of the Environmental Education Association of Southern Africa, Grahamstown.

Linke, R. (1984). Reflections on environmental education: Past development and future concepts. *Australian Journal of Environmental Education, 1*(1), 2–4.

Lister, R. (Ed.). (2010). *Climate change and philosophy*. London, UK: Continuum Publishing Group.

Lotz, H., & Robottom, I. (1998). Environment as text: Initial insights into some implications for professional development in environmental education. *Southern African Journal of Environmental Education, 18*, 19–28.

Lotz-Sisitka, H. (2009). Why ontology matters in reviewing environmental education research. *Environmental Education Research Journal, 15*(2), 165–175.

Lundegard, I., & Wickman, P. (2007). Conflicts of interest: An indispensable element of education for sustainable development. *Environmental Education Research, 13*(1), 1–15.

Lupele, J. (2008). Underlying mechanisms affecting institutionalization of environmental education courses in southern Africa. *Southern African Journal of Environmental Education, 25*, 113–131.

Malandrakis, G. (2008). Children's understandings related to hazardous household items and waste. *Environmental Education Research, 14*(5), 579–601.

Malone, K. (2006). Environmental education researchers as environmental activists. *Environmental Education Research, 12*(3–40), 375–389.

Mogensen, F., & Schnack, K. (2010). The action competence approach and the 'new' discourses of education for sustainable development, competence and quality criteria. *Environmental Education Research, 16*(1), 59–74.

Mokuku, T., Jobo, M., Raselimo, M., Mathafeng, T., & Stark, K. (2005). Encountering paradigmatic tensions and shifts in environmental education. Canadian Journal of Environmental Education, *10*, 157–172.

Mortari, L. (2003). Fostering civic participation through youth town councils in Europe. *Australian Journal of Environmental Education, 19L*, 47–56.

Mrazek, R. (Ed.). (1993). *Alternative paradigms in environmental education research*. Troy, OH: NAAEE.

Mukute, M. (2009). Cultural historical activity theory, expansive learning and agency in permaculture workplaces. *Southern African Journal of Environmental Education, 26*, 150–166.

Naess, A., & Jickling, B. (2000). Deep ecology and education: A conversation with Arne Naess. *Canadian Journal of Environmental Education, 5*, 48–62.

O'Donoghue, R. (2003). Indigenous knowledge: Towards learning materials and methodologies that respond to social processes of marginalisation and appropriation in eastern southern Africa. *Australian Journal of Environmental Education, 19*, 57–68.

O'Donoghue, R. (2007). Environment and sustainability education in a changing South Africa: A critical historical analysis of outline schemes for defining and guiding learning interactions. *Southern African Journal of Environmental Education, 24*, 141–157.

O'Donoghue, R., & McNaught, C. (1991). Environmental education: The development of a curriculum through 'grass-roots' reconstructive action. *International Journal of Science Education, 13*(4), 391–404.

O'Donoghue, R., & Neluvhalani, E. (2002). Indigenous knowledge and the school curriculum: A review of developing methods and methodological perspectives. In E. Janse van Rensburg, J. Hattingh, H. Lotz-Sisitka, & R. O'Donoghue (Eds.), *EEASA monograph: Environmental education, ethics and action in southern Africa* (pp. 121–134). Pretoria: EEASA / HSRC.

Orams, M. (1999). The effectiveness of environmental education: Can we turn tourists into "greenies"? *Progress in Tourism and Hospitality Research, 3*(4), 295–306.

Östman, L. (2010). Education for sustainable development and normativity: A transactional analysis of moral meaning-making and companion meanings in classroom communication. *Environmental Education Research, 16*(1), 75–94.

Payne, P. (2005). "Ways of doing." learning, teaching, and researching. *Canadian Journal of Environmental Education, 10*, 108–124.

Payne, P. (2010). The globally great moral challenge: Ecocentric democracy, values, morals and meaning. *Environmental Education Research, 16*(1), 153–171.

Pesanayi, T. (2009). A case of exploring learning interactions in rural farming communities of practice in Manicaland, Zimbabwe. *Southern African Journal of Environmental Education, 26*, 64–73.

Piaget, J. (1971). *The science of education and the psychology of the child*. London, UK: Routledge and Kegan Paul.

Popham, W. J. (1972). *An evaluation guidebook: A set of practical guidelines for the educational evaluator*. Los Angeles, CA: The Instructional Objectives Exchange.

Popkewitz, T. (2000). Reform as the social administration of the child: Globalization of knowledge and power. In N. Burbeles & C. Torres (Eds.), *Globalisation and education: Critical perspectives*. New York and London: Routledge.

Popkewitz, T. (2008). *Cosmopolitanism and the age of school reform. Science, education, and making society by making the child*. New York and London: Routledge Taylor and Francis Group.

Popkewitz, T. S. (Ed.). (2000). *Educational knowledge. Changing relationships between the state, civil society, and the educational community*. New York, NY: State University of New York Press.

Posch, P. (1996). Curriculum change and school development. *Environmental Education Research, 2*(3), 347–362.

Price, L. (2004). Applied methodological lessons from A.S. Byatt's book "The Biographer's Tale." *Environmental Education Research, 10*(3), 429–442.

Price, L. (2007). *A transdisciplinary explanatory critique of environmental education. (Volume 1), Business and Industry* (Unpublished PhD thesis). Rhodes University, Grhamstown, South Africa.

Rauch, F. (2002). The potential of education for sustainable development for reform in schools. *Environmental Education Research, 8*(1), 43–51.

Reason, P. (2002). Justice, sustainability, and participation: Inaugural professorial lecture. *Concepts and Transformations, 7*(1), 7–29.

Rickinson, M. (2001). Learners and learning in environmental education: A critical review of the evidence. *Environmental Education Research, 7*(3), 207–320.

Rickinson, M. (2003). 'Reviewing research evidence in environmental education: some methodological reflections and challenges', *Environmental Education Research, 9*, 2, 257-271.

Rickinson, M. (2006). Researching and understanding environmental learning. *Environmental Education Research, 12*(2–3), 445–457.

Rideout, B. E. (2005). The effect of a brief environmental problems module on endorsement of the new ecological paradigm in college students. *The Journal of Environmental Education, 37*(1), 3–12.

Robottom, I. (1989). Social critique or social control: Some problems for evaluation in environmental education. *Journal of Research in Science Teaching, 26*(5), 435–443.

Robottom, I. (2005). Critical environmental education research: Re-engaging the debate. *Canadian Journal of Environmental Education, 10*, 62–78.

Robottom, I. (2007). Emerging methodological issues in environmental education research. In M. Salomone (Ed.), *Educational paths towards sustainability*. Proceedings of 3rd World Environmental Education Congress, Torino, 133–142.

Robottom, I., & Hart, P. (1993). *Research in environmental education*. Geelong, Victoria: Deakin University Press.

Robottom, I., Malone, K., & Walker, R. (2000). *Case studies in environmental education: Policy and practice*. Geelong, Victoria: Deakin University Press.

Rudsberg, K., & Öhman, J. (2010). Pluralism in practice—Experiences from Swedish evaluation, school development and research. *Environmental Education Research, 16.1*, 95–111.

Said, E. W. (1993). *Culture and imperialism*. New York, NY: Vintage Books.

Sandell, K., & Öhman, J. (2010). Educational potentials of encounters with nature: Reflections from a Swedish outdoor perspective. *Environmental Education Research, 16*(1), 113–132.

Scott, D. (2008). *Critical essays on major curriculum theorists*. Abingdon, UK: Routledge.

Scott, W., & Gough, S. (2004). *Sustainable development and learning: Framing the issues*. London and New York: Routledge Falmer.

Scott, W., & Oulton, C. (1999). Environmental education: Arguing the case for multiple approaches. *Educational Studies, 25*(1), 89–97.

Selby, D., & Kagawa, F. (Eds.). (2009). *Education and climate change. Living and learning in interesting times*. London, UK: Routledge.

Sfard. A. (1998). On two metaphors for learning and the dangers of choosing just one. *Educational Researcher, 27*(2), 4–13.

Shepardson, D., Harbor, J., Cooper, B., & McDonald, J. (2002). The impact of a professional development program on teachers' understanding about watersheds, water quality and stream monitoring. *The Journal of Environmental Education, 33*(3), 34–40.

Silo, N. (2009). Exploring learner participation in waste management activities in a rural Botswana primary school. *Southern African Journal of Environmental Education, 26*, 176–192.

Smith-Sebasto, N. J., & Cavern, L. (2006). Effects of pre-and post-trip activities associated with a residential environmental education experience on students' attitudes toward the environment. *The Journal of Environmental Education, 37*(4), 3–18.

Stevenson, R. B. (2007). Schooling and environmental education: Contradictions in purpose and practice. *Environmental Education Research, 13*(4), 139–153.

Tyler, R. (1950). *Basic principles of curriculum and instruction*. Chicago, IL: University of Chicago Press.

United Nations Educational, Scientific and Cultural Organization. (2005). *United Nations Decade of Education for Sustainable Development (2005–2014). Framework for the International Implementation Scheme*. Retrieved from www.unesco.org/desd

Vygotsky, L. (1978). *Mind in society*. Cambridge, MA: Harvard University Press.

Walker, K. (1997). Challenging critical theory in environmental education. *Environmental Education Research, 3*(2), 155–162.

Wals, A. (2007). Learning in a changing world and changing in a learning world: Reflexively fumbling towards sustainability. *Southern African Journal of Environmental Education, 24*, 35–45.

Wals, A. (2010). Between knowing what is right and knowing what is wrong to tell others what is right: On relativism, uncertainty and democracy in environmental and sustainability education. *Environmental Education Research, 16*(1), 143–152.

Wals, A., & Alblas, A. (1997). School-based research and development of environmental education: A case study. *Environmental Education Research, 3*(3), 253–267.

Wesley, P. (1999). Changing behavior with normative feedback interventions: A field experiment on curbside recycling. *Basic and Applied Social Psychology, 21*(1), 25–36.

21

Environmental Education in a Cultural Context

Albert Zeyer
University of Zurich, Switzerland

Elin Kelsey
Royal Roads University, Canada; James Cook University, Australia

Introduction

In this chapter, we contend that emotions are of primary practical importance in environmental education. They affect students' and teachers' interest, engagement, and achievement, as well as their personality development, health and well-being. By implication, they can profoundly influence the productivity and quality of life in educational institutions and in society at large. Further, we adopt a social constructivist stance, suggesting that emotions are embedded in and part of a student's introduction to the beliefs, norms, values, and expectations of his or her culture (Harré, Armon-Jones, Lutz, & Averill, 1986). In fact, the concept of cultural border crossing is well known in science education (Andree, 2005). Within this concept, problems of science learning are interpreted in terms of a clash between the culture of Western science (as represented by school science) and the way students experience science in their everyday lives; what Costa (1995) has coined "life-world culture." A worrisome result of the lack of alignment between these cultures is student alienation from Western science. As Aikenhead (1996) puts it: "For the vast majority of students of any culture, their cultural identities are at odds with the culture of science" (p. 1).

We argue that a similar clash exists between students' life-world culture and the culture of environmental education represented in schools, and raise this as an important focus of environmental education curriculum research. We believe that a cultural perspective has much to offer environmental education researchers and practitioners. It can explain why well-intentioned modernist curricular approaches to environmental education may lead to unintended and unforeseen consequences like environmental depression and ecological passivity. We argue that such findings can be extended to a broader cultural perspective because they are supported by other empirical findings of research in environmental education. Insofar, the classroom can be conceived as a societal laboratory (Zeyer & Roth, 2009) in order to understand cultural mechanisms and to improve strategies of embedding new value orientations in society, institutionalizing transformative learning approaches, and traditional and institutionalized knowledge structures.

In this chapter we firstly briefly locate our approach in research on environmental emotions and in the concept of cultural border crossing in science education. We then explore the implications of cultural frames, such as post-ecologism, on environmental education, and draw on research in Swiss schools to demonstrate how a cultural approach can indicate problems with the teaching process that otherwise might remain undiagnosed. We close the chapter by outlining possible strategies to support the recentering of learner and teacher experience and dialog in socioecological contexts.

Environment and Emotion

Interviewer: Fifty percent of you said that we are destroying the world. This is pretty extreme. Do you really think that way?

Student A: Well, you can see this everywhere. Of course, you always have to be optimistic. But it is a simple fact. You have to see it. It just is so.

Student B: Well, I think it was us that have made the environment as it now is. It really is our own fault. If we want to change anything about it now, then we have to do it ourselves. We cannot blame someone else for it.

Interviewer: But would you count yourself among those who say that if we were to do the right thing, it would turn out well?

Student B: That I cannot say.

This excerpt from a class discussion between fifteen- and sixteen-year-old Swiss high school students and an education researcher convey the tone of students' feelings about the state of the planet. Rather than outrage or despair, they respond in a matter-of-fact and taken-for-granted manner and with apparent resignation to the bleak forecast that we are destroying the world.

Much of what currently passes for environmental education in schools is better classified as environmental information (Coyle, 2005). On face value, the students in this Swiss classroom can accurately reproduce the required facts. Yet their feelings echo a growing recognition from a range of disciplines—environmental sociology (Kervorkian, 2004) political theory (Blühdorn, 2002, 2008); environmental education (Kool & Kelsey, 2006); and environmental philosophy (Macy, n.d.) that people feel overwhelmed and hopeless in response to the state of the environment, and their own futures.

A careful analysis of the mentioned classroom discussions shows that the students' statements share important traits in common with the medical diagnosis of depression: the bleak picture of the future; the pessimistic mood; the motive of guilt; the lack of feelings of control. Do these pupils have a form of latent environmental depression? More than a decade ago, Sobel (1996) coined the phrase "ecophobia" to describe "a kind of despondency among the children" and "a submerging of children's natural interests in a sea of problems" among elementary school students who had participated in Earth Week curriculum activities focused on rainforest destruction. Ecophobia, he writes, is a fear of ecological problems and the natural world.

> Fear of oil spills, rainforest destruction, whale hunting, acid rain, the ozone hole, and Lyme disease. Fear of just being outside. If we prematurely ask children to deal with problems beyond their understanding and control, then I think we cut them off from the possible sources of their strength. In response to physical and sexual abuse, children learn distancing techniques, ways to cut themselves off from the pain. My fear is that our environmentally correct curriculum will end up distancing children from, rather than connecting them with, the natural world. The natural world is being abused, and they just don't want to have to deal with it. (Sobel, 1996, p. 5)

Sobel's findings echo an earlier study described to him by George Russell of Adelphi University that was conducted in West Germany during the 1980s when the Germans implemented a conscientious national curriculum intended to raise the consciousness of elementary school students across the country regarding environmental problems. As Sobel explains:

> By informing students about the problems and showing them how they could participate in finding the solutions, the education ministry hoped to create empowered global citizens. Follow-up studies conducted some years after implementation indicated just the opposite had occurred. As a result of the curriculum initiative, education officials found that students felt hopeless and disempowered. The problems were seemingly so widespread and beyond the students' control that their tendency was to turn away from, rather than face up to, participating in local attempts at problem solving. (Sobel, 1996, p. 20)

Awareness of the problem of environmental despair among students is not new. Yet, environmental education research is strangely silent about dealing with the emotional implications of the environmental crisis on children and young people. There is little in the literature addressing appropriate ways to deal with the emotions associated with environmental degradation. Words like grief, despair, or anger rarely appear in our writings (Kool & Kelsey, 2006). In part, this may be due to a wider tendency to omit emotions from educational inquiry (Pekrun, Goetz, Titz, & Perry, 2002). The one notable exception is investigations of Students' test anxiety. According to Schutz and Pekrun (2007, p. 3):

> From more than 1,000 empirical studies conducted over a span of more than five decades, we have evidence on the structures, antecedents, and effects of this emotion, as well as on measures suited to prevent excessive test anxiety by changing education and to treat this emotion once it occurred. However, what about student emotions other than test anxiety? And what about teachers' emotions other than anxiety, such as anger, hopelessness, shame, or boredom; and what do we know about pleasant emotions; such as enjoyment, hope or pride in educational settings?

Cultural Border Crossing

Aikenhead (2000) describes the impact of a cultural border crossing on science teaching. Following Geertz (1973), he defines a culture to be an ordered system of meaning and symbols, in terms of which social interaction takes place. This includes norms, beliefs, expectations, values, and conventional actions of the group of people involved (Phelan, Davidson, & Cao, 1991). Students are seen as living simultaneously in different cultures. The culture of western science is only one of many. Others include the culture of school science (transmitting the dominant culture of the school and its community) and, of course, students' life-worlds including family, peers, community, and so on. According to Costa (1995) very few pupils have identities and abilities that harmonize closely enough with the culture of western science to be easily enculturated. Costa labels this group "Potential Scientists" and compares them to other empirical categories of students who deviate further from a western scientific worldview. There are, for example, "Other Smart Kids," a term Costa borrows from Tobias (1990). These are pupils who do well at school, even in science, but for whom science is neither personally meaningful nor believed to be useful to their everyday lives. There are also "I Don't Know" pupils who are labeled for their ubiquitous response to a host of questions about science and about school, and for their overall noncommittal attitude toward school science (Aikenhead, 1996). If the subculture of science is generally at odds with a student's life-world culture, and science disrupts the student's view of the world by trying to replace it or marginalize it ("assimilation"), then "Other Smart Kids" and "I Don't Know Students" quite often react by playing the "school game" of passing science classes without really being involved in meaningful learning and even without understanding the content. Larson (1995) captured this phenomenon as "Fatima's Rules," named after an articulate student in a high school chemistry class. We now will report results from research in Swiss schools (Zeyer, 2007, 2008) to flesh out the described cultural framework and provide a local basis for the unfolding of our further argumentation.

A Swiss School Culture of Ecoscientism

In the lower secondary schools of Central Switzerland, environment and environment protection play a crucial role in the science curriculum. *Acting responsibly in environment and society* is one of the four explicitly formulated main teaching goals of science education. The contradictory impact of science on environment is also emphasized. Subject areas that explicitly involve environmental topics are for example: "our world we live in," "our world, a network system," "energy," "water as a basis of life," "green plants and their life," "soil, basis of our food" (IEDK, 1997).

The Swiss science curriculum was created in 1986 prior to the uptake of the concept of sustainability. Progressive for its time, it forwards the assumption that scientific understanding of nature dictates environmentally friendly behavior. Each subject area begins with the description of the scientific background of an environmental issue and ends with possible consequences of ignoring it, with norms for environmentally friendly behavior, and with appeals to personal commitment and societal engagement. The description of the nurture chain, for example, is followed by the assertion that our consumption of meat has a global impact. The analysis of ecosystems—another issue in the curriculum—is focused on scientific aspects like feedback loops and biological equilibrium, followed by conservationist concepts of environment and nature protection. Similarly, the interpretation and comparison of energy data initiates the discussion of energy saving, efficiency factors and green technologies.

Qualitative content analysis (Berg, 2004) has shown that the Swiss science curriculum operates within a culture of Ecological Scientism or Ecoscientism (Zeyer, 2007). Ecoscientism assumes behavioral rules and norms to be the direct output of scientific insight. It further assumes that students who do not act on behalf of the environment lack understanding of environmental problems.

This curriculum influences the school culture as a whole. For example, when asked about important environmental issues, students and teachers reproduced a broad spectrum of topics provided by the curriculum. Teachers reported to be influenced by the curriculum in their teaching (motivation, choice of topics, amount of lessons dedicated to these topics) and also in their behavior as a role model. Additionally, these schools were sensitive to environmental issues in many organizational and infrastructural aspects. Waste separation was institutionalized in almost all cooperation schools. Many schools realized reasonable light and climate management and provided natural outdoor areas, such as a pond or a hedge, where students could enjoy and study nature and biota.

A Life-World Culture of Postecologism

While the school culture was strongly oriented toward ecoscientism, the students lived in a completely different culture. In fact, discourse analysis of three whole class interviews, as well as in-depth interviews with twelve students revealed that students' life world cultures were much more closely aligned to what Blühdorn (2005) calls postecologism (Zeyer, 2008). Postecologism argues that individuals on face value understand that the status quo is unsustainable and that radical change is necessary but when it comes to the realization of environmental concerns, they forward many reservations and express serious doubts that such change will ever occur. They tend to defend the status quo, often through arguments of materialism and consumerism. They "normalize" the environmental crisis by adopting the paradox assumption that unsustainability is here to stay, and that they, as individuals, are powerless to do anything about it. They no longer believe social movement campaigns that "another world" is possible. In the following section, we provide a short description of our students' responses through the lens of postecologism in order to be able to demonstrate how these findings can be interpreted in the light of cultural border crossing.

Apathy About Their Own Influence Students in the Swiss study did not believe that they could personally influence what happens to the environment. They said that young people had no influence at all on ecological matters. They believed that most people were apathetic and would not join environmental-friendly initiatives, and this was the main reason that environmental conditions would not improve. As a result, many of the students did not see a point in engaging in environmental-friendly actions. Some even found that this would be a mere waste of time.

A Rhetoric of Othering Discourse analysis reveals a tendency for the students to adhere to what Riggins (1997) calls "the rhetoric of othering" when they spoke about people who abstained from environmental-friendly actions. Students talked about "the others," when they wanted to legitimize their own lack of environmental-friendly engagement. "The others" were those people who did not care about environment, who broke the environmental rules, who did not understand the importance of environmental issues, and thus, compromised environmental-friendly actions. In the students' discourse "the others" were often part of "the industry" that was supposed to dodge environmental law and to generally conspire against environmental progress. The motorcar lobby, for example, was suspected of systematically blocking the development of pollution-free car engines. Accusations like these were never concretized; they always reduced to a mystifying, fiction-like portrayal of the evil.

Sometimes "the others" were much bigger than a single branch of "the industry." They could be, for example, other countries. Many pupils thought that "other," bigger countries were less concerned about the environment than their own country. Since Switzerland is very small, they suspected their own country's efforts to be spoiled through international carelessness.

A Belief in Technological Fixes The students appreciated innovation and technical progress generally as a grant toward prosperity and economic growth. They believed that technical solutions were more likely to be successful in protecting the environment than social and behavioral changes. Their hope for solutions to the environmental crisis rested with science and technology and credited technological progress with the largest potential to improve environmental protection. Only a few acknowledged that behavioral change could be more important than technological progress. Generally it was suggested that technology was most promising and trustworthy in this context. The vision of a technology-free world in harmony with nature scarcely emerged. If somebody alluded to concepts such as "less technology" or "more nature," it was not without asserting that changes in this direction should of course not compromise a consumption-oriented life style.

A Commitment to Consumerism The students considered consumerism to be normal; a natural part of their life-world. Consumerism was not criticized but seen as a source of pleasure and satisfaction, and of jobs and goods. Economic growth, globalization and the consequences of both were taken-for-granted and scarcely contested. They did not criticize the role of the global market nor its potential impact on the environment.

Students were quite critical of certain aspects of green consumerism or services, labeling them as "idealistic" or "unrealistic." For example, they cited organic goods as being more expensive than conventional goods without providing more quality. Organic agronomy, which is very prominent in Switzerland, was categorized as "expensive" and "illusory." Public transport, which is also highly approved in Switzerland by both the public and the media, was accused of being unable or unwilling to fully connect all parts of the country.

A deep reaching skepticism toward comprehensive and normative worldviews ("Weltanschauung") and an obvious commitment to consumerism flavored the whole discussion about social conceptions. Any credited draft for the future had to guarantee materialistic desires. Social visions of the future world therefore were always pragmatic and combined with a slight regret that there was no proper room left for environmental issues.

Environmental Issues Shouldn't Be Overstated Students in the Swiss study did not deny the importance of environmental problems. On the contrary, they underlined their crucial impact on the future of the world. They considered environmental issues to be important, however, they also believed them to be less important than other topics like peace (meaning personal and social peace as well as world peace), racism, or immigration problems. In this they mirror the societal worries of Switzerland as a nation, which during the last years has been confronted (as have many other countries) with decreased economic growth, increasing levels of unemployment, a new problem of poverty (especially of young single mothers), and societal restlessness in the context of social tensions, immigration, and financial problems across the nation's households.

A Dislike for Nongovernmental Actors and Green Politics Because of their clear recognition of environmental problems, one might assume that these students would be sympathetic to environmentalists. This was not the case. The students said that nongovernmental organizations (NGOs) would hinder economic progress and the development of the countryside, especially when they tried to impede the installation of new sport and leisure time arenas. They felt that nongovernmental organizations (NGOs) like World Wildlife Fund (WWF) and Greenpeace were "unrealistic" and "naïve." They respected the political intentions of these organizations and their role in modern society. However they did not admire environmental activists as modern heroes but rather suspected them to be idealistic troublemakers that did not notice societal reality. Many of the pupils disliked the public interventions by Greenpeace. They found their actions dangerous and overdone. They said that it was not sensible to provoke other people and to disturb their daily life. They especially abhorred the obstruction of economical processes. Actions like the boarding of ships or the blocking of truck transport were completely inexcusable in their view. In this context they also criticized left wing and Green politics as "exaggerated," "extremist," and "economically destructive."

In summary, students in the Swiss study considered environmental issues to be important, but other issues to be at least equally important. They believed that environmental issues should not be overstated, and thought that environment protection must be realized by the political system as a whole and must be part of continuous systemic innovation and change. They considered the possible contribution of one person alone to environmental-friendly solutions to be marginal. Their personal life style was materialistic and consumption-oriented. Economic growth and political globalization were believed to be necessary to grant quality of life, and technical progress was seen as crucial and the only reasonable approach to solve environmental problems. They also believed that official politics and nongovernmental organizations (NGOs) as well should respect economic constraints of an environmental-friendly attitude.

A Clash Between School Culture and Life World Culture

These findings revealed a postecologist life world culture among the students in this study that is in stark contrast to the ecoscientistic culture found in their schools. The school culture reproduced the intentions of the curriculum, but the students did not. Thus, the students experience a cultural clash between the school culture and their own home and peer culture.

Using interpretive repertoires to analyze the students' discourse (Zeyer & Roth, 2009), it appears that in the postecological discourse, students are locked into a framework of "facts" provided by ecoscientism. Students take these facts as intrinsic, unchangeable, and indisputable. Students thus come to be stuck in a world full of unchangeable "facts" that confine their room to maneuver. This confinement is the result of their feelings of sheer personal helplessness in the face of an anonymous mass of "others." Consequently they delegate agency to impersonal, institutional goals such as innovation, growth, and progress. Postecologism is not an active orientation toward materialism and hedonism, a treason of the ideals of a green culture. It is first and foremost a reaction to a lost locus of self-control in a "fact-"oriented inner and outer world. This is a severe diagnosis, because "the perception of personal control is one of the most significant aspects of an individual's self-perception" (Falomir, Mugny, Quiamzalde, & Butera, 2000, p. 443). Control beliefs are regarded as primary in determining a person's decisions to act. They are prerequisites for the planning, initiation, and regulation of goal-oriented actions, and they are part of the self-concept. As such, they largely determine feelings of self-esteem, causing such emotional states as pride, shame, or depression (Flammer, 1995). We came to the conclusion that postecologism might reflect a form of environmental depression, based on the loss of articulated agency with respect to issues of control belief concerning the environment and environmental protection.

Thus, the concept of cultural clash helps to explain some important problems of environmental education: the gap between a verbal attribution of great importance to environment protection and a substantial lack of willingness to sacrifice personal goods to this aim; low belief in personal influence on environmental problems; and the lack of transfer of scientific knowledge gained in school to personal behavior and orientation in daily life and lifestyle, and an emotional disposition to environmental depression.

So Where Do We Go From Here?

Interpreting findings like these in terms of a cultural clash, in the theoretical framework of "cultural border crossing," means moving beyond more familiar approaches of teaching and learning environmental education content. It can reveal a deeper awareness of the cultural divides that exist within classrooms, and how these may compromise the embedding of new value orientations in society, and the institutionalization of transformative learning approaches in environmental education.

For example, we believe that the findings in the Swiss study are paradigmatic not only for Swiss classrooms but for a more general cultural constellation that can be identified in a variety of environmental contexts. In fact, post-ecologism seems to be widespread at least in German speaking European countries (Blühdorn, 2000). Moreover results reminiscent of this constellation can also be found in other research studies on students' attitudes toward the environment and environmental protection, although they are not made explicit as such. For example, a study analyzing environment and environmental protection in "The Relevance of Science Education (ROSE)" project (Schreiner & Sjøberg, 2004) concludes that students found environmental problems important but were unwilling to sacrifice many personal goods to solve or alleviate environmental problems (Jenkins & Pell, 2006). Only a minority of students agreed that they could personally influence what happens to the environment. The same pattern can essentially be seen in a body of studies in a variety of cultural contexts (e.g., Chu et al., 2007).

A cultural approach as proposed reveals the core of this constellation: it is a cultural clash between an ecoscientist approach of environmental education and the students' life world. Notice that thereby students' life worlds may considerably differ, depending on the world region, the socioeconomical background etc. In a wealthy region for example, the clash with an ecoscientist educational approach may be informed by consumerist and hedonist lifestyle arguments (as it is in our example). In a poor region, the urge of personal needs may lead to the same clash pattern in a quite different context. However, the cultural point of view shows the common backdrop, namely the rejection of an ecoscientist educational approach, which assumes behavioral rules and norms to be the direct output of scientific insight.

There are implications of this type of interpretation. One is for example, that successful environmental education should take cultural issues seriously. As our example shows, a cultural clash may in fact compromise the best intentions of curriculum, teachers, and school to establish environmental consciousness and behavior in students. The concept of cultural border crossing explains such "failures" in terms of an unsuccessful assimilation of students into an unfamiliar culture—in this case the eco-scientistic culture.

In science education, cultural border crossing proposes an alternative to assimilation (with students playing Fatima's rules as a reaction). It is called autonomous acculturation. Students reconstruct some content from science culture in their own cultural framework because the content appears useful. Thereby they replace some former ideas of their own culture. Their everyday thinking becomes an integrated combination of common sense thinking and useful science/technology thinking. The teacher's role is to be a culture broker. She/he makes border crossings explicit for students. She/he acknowledges the validity of students' cultural identities and cultural environment and, at the same time, helps them feel at ease in the culture of science, and to move back and forth between their everyday cultures and the culture of school science.

Scientism—the myth "that scientific knowledge deserves unquestioned epistemic privilege" (Cobern & Loving, 2007, p. 427)—is one important reason for a cultural clash in science education. Science teachers as cultural brokers should therefore accept a proviso, for example in a Rawlsian sense, i.e. they should not support scientistic practice in the classroom—neither on the epistemological level nor on any other level of theoretical and practical reasoning (Zeyer, 2009).

In analogy, a cultural approach suggests that environmental education teaching should be understood as a form of "cultural brokering" in which the teacher attempts to understand the life world cultures of his or her students, and to effect change by bridging, enlinking, or mediating between those cultures and the dominant culture of environmental education inherent within the school curriculum. Students would not be assimilated to ecoscientistic culture, but rather the teacher would abstain from teaching ecoscientism as a comprehensive doctrine of environmental education. Here we see a close link to other approaches like the fostering of informal reasoning on socioscientific issues (Sadler, 2004), which could enrich the talk about environmental protection with conversations about the nature of science, about complexity, and applied ethics.

Another implication of a cultural approach is that further research into the environmental life-world cultures of students, of teachers, and of curriculums and schools is warranted. Only an intimate knowledge of these cultures can facilitate successful border crossing and avoid the playing of Fatima's rules by alienated students. Such investigations could be conducted by teachers themselves in the form of case studies as proposed in reflective teaching (Kyburz-Graber, 2003). However long term research by professional researchers should also be undertaken as cultural characteristics evolve over time, especially when it comes to complex issues like the environment. It would be interesting, for example, to repeat the study described in this paper in light of the recent assessment report of the Intergovernmental Panel on Climate Change (2007) and the enormous scale of the media response. (The students in this study often asserted that media events play an important role in their thinking and their emotions toward environment.)

The growing body of research exploring environmental emotions, such as despair and hope, exists within the fields of sustainability, deep ecology, conservation biology, conservation psychology, and other forms of ecopsychology (Kelsey & Armstrong, 2012). See for example, the work of Dillon, Kelsey, and Duque-Aristizabal (1999) on culture and environmental identity. The findings described in this chapter argue in favor of broadening the fields of inquiry to include cultural contexts, such as post-ecologism, as well.

References

Aikenhead, G. (1996). Science education: Border crossing into the subculture of science. *Studies in Science Education, 27*, 1–52.

Aikenhead, G. (2000). Renegotiating the culture of school science. In R. Millar, J. Leach, & J. Osborne (Eds.), *Improving science education: The contribution of research* (pp. 245–264). Philadelphia, PA: Open University Press.

Andree, M. (2005). Ways of using "everyday life" in the science classroom. In K. Boersma, M. Goedhart, O. de Jong, & H. Eijkelhof (Eds.), *Research and the quality of science education* (pp. 107–116). Dordrecht: Springer.

Berg, B. L. (2004). *Qualitative research methods* (5th ed.). Boston, MA: Pearson Education.

Blühdorn, I. (2000) *Post-ecologist politics. Social theory and the abdication of the ecologist paradigm*. London, UK: Routledge.

Blühdorn, I. (2002). Unsustainability as a frame of mind and how we disguise it: The silent counter-revolution and the politics of simulation. *The Trumpeter, 18*, 1–11.

Blühdorn, I. (Ed.). (2005). Social movements and political performance. Niklas Luhmann, Jean Baudrillard and the Politics of Simulation. Würzburg: Königshausen & Neumann.

Blühdorn, I., & Welsh, I. (Eds.). (2008). *The politics of unsustainability. Eco-politics in the post-environmental era*. New York, NY: Routledge.

Chu, H.-E., Lee, E. A., Ko, H. R., Shin, D. H., Min, B. M., & Kang, K. H. (2007). Korean year 3 children's environmental literacy: A prerequisite for a Korean environmental education curriculum. *International Journal of Science Education, 29*(6), 731–746.

Cobern, W., & Loving, C. C. (2007). An essay for educators: Epistemological realism really is common sense. *Science & Education, 17*, 425–447.

Costa, V. B. (1995). When science is "another world": relationships between worlds of family, friends, school and science. *Science Education, 79*, 313–333.

Coyle, K. (2005). *Environmental literacy in America: What 10 years of neetf/roper research and related studies say about environmental literacy in the U.S.* Washington, DC: The National Environmental Education & Training Foundation.

Dillon, J., Kelsey, E., & Duque-Aristizabal, A. M. (1999). Identity and culture: Theorising emergent environmentalism. *Environmental Education Research, 5*(3), 395–405.

Falomir, J. M., Mugny, G., Quiamzalde, A., & Butera, F. (2000). Social influence and control beliefs in identity threatening contexts. In W. J. Perrig & A. Grob (Eds.), *Control of human behavior, mental processes and consciousness: essays in honor of the sixtieth birthday of august flammer*. New York, NY: Lawrence Earlbaum Associates.

Flammer, A. (1995). Development analysis and control beliefs. In A. Bandura (Ed.), *Self-efficacy in changing societies*. New York, NY: Cambridge University Press.

Geertz, C. (1973). *The interpretation of cultures*. New York, NY: Basic Books.

Harré, R., Armon-Jones, C., Lutz, C., & Averill, J. (1986). *The social construction of emotions*. Oxford, UK: Blackwell.

IEDK, Innerschweizer Erziehungsdirektorenkonferenz (1997). *Lehrplan Naturlehre für die Orientierungsstufe*. Ebikon. [Central Swiss Conference of Cantonal Ministers of Education (1997). Science Curriculum for Lower Secondary Schools. Ebikon (CH)].

Jenkins, E. W., & Pell, R. G. (2006). "Me and the environmental challenges": A survey of English secondary school students' attitudes towards the environment. *International Journal of Science Education, 28*(7), 765–780.

Kelsey, E., & Armstrong, C. (2012). Finding hope in a world of environmental catastrophe. In A. Wals & P. B. Corcoran (Eds.), *Learning for Sustainability in Times of Accelerating Change*. Wageningen Academic Publishers' Education and Sustainable Development Series, Newfoundland.

Kervorkian, K. (2004). Environmental grief: Hope and healing. PhD diss., Union Institute and University, Cincinatti, Ohio.

Kool, R., & Kelsey, E. (2006). Dealing with despair: The psychological implications of environmental issues. In W. L. Filho & M. Salomone (Eds.), *Innovative approaches to education for sustainable development*. Frankfurt: Peter Lang Publishing.

Kyburz-Graber, R., Posch, P., & Peter, U. (Eds.). (2003). Challenges in Teacher Education—Interdisciplinarity and Environmental Education. Innsbruck/Wien: Studienverlag.

Larson, J. O. (1995). *Fatima's rules and other elements of an unintended chemistry curriculum*. Unpublished manuscript, San Francisco, CA.

Macy, J. (n.d.) *The Great Turning*. Retrieved February 24, 2009, from http://www.joannamacy.net/html/great.html

Pekrun, R., Goetz, T., Titz, W., & Perry, R. P. (2002). Academic emotions in students' self-regulated learning and achievement: A program of qualitative and quantitative research. *Educational Psychologist, 37*, 91–105.

Phelan, P., Davidson, A. L., & Cao, H. T. (1991). Students multiple worlds—Negotiating the boundaries of family, peer, and school cultures. *Anthropology & Education Quarterly, 22*, 224–250.

Riggins, S. H. (Ed.). (1997). *The language and politics of exclusion: Others in discourse*. London, UK: Sage Publications.

Sadler, T. D. (2004). Informal reasoning regarding socioscientific issues: A critical review of research. *Journal of Research in Science Teaching, 41*(5), 513–536.

Schreiner, C., & Sjøberg, S. (2004). ROSE (The relevance of science education)—a comparative study of students' views of science and science education. *Acta Didactica, 4*, 1–126.

Schutz, P. A., & Pekrun, R. (2007). *Emotion in education*. San Diego, CA: Academic Press.

Sobel, D. (1996). *Beyond ecophobia: Reclaiming the heart in nature education*. Nature Literacy Monograph Series #1, Great Barrington, MA: Orion Society.

Sobel, D. (n.d.) *Beyond Ecophobia*. Retrieved September 7, 2010, from http://www.eenorthcarolina.org/certification/beyond_ecophobia.pdf

Tobias, S. (1990). *They're not dumb. They're different*. Unpublished manuscript, Tucson, AZ.

Zeyer, A. (2007). *The impact of a secondary STSE science curriculum on students, science teachers, and their schools*. Full paper presented at the ESERA Conference Sweden 2007, Malmö.

Zeyer, A. (2008). *Students' post-ecological discourse in a secondary one STSE (science-technology-society-environment) education*. Full paper presented at the NARST Annual International Conference, San Francisco.

Zeyer, A. (2009). Public reason and teaching science in a multicultural world: A comment on cobern and loving. "An essay for educators . . ." in the light of John Rawls' political philosophy. *Science & Education, 18*(8), 1095–1100.

Zeyer, A., & Roth, W. M. (2009). A mirror of society: A discourse analytic study of fifteen- to sixteen-year-old Swiss students' talk about environment and environmental protection. *Cultural Studies of Science Education, 4*(4), 961–998.

22

Place-Based Education

Practice and Impacts

Gregory A. Smith
Lewis & Clark College, USA

In the mid-1990s, educators in different locations across the United States and elsewhere began to experiment with a vision of teaching and learning aimed at fostering both community and environmental renewal. Claiming the local as the focus of its attention, this approach has come to be known as place-based education. Its adherents have been drawn to this work for a variety of reasons: the development of more engaging forms of instruction, the cultivation of involved citizens, the development of people committed to the wise stewardship and protection of natural resources and areas.

Place-based education cannot be slotted into specific curricular domains such as science or social studies. It is instead an approach to curriculum development and instruction that acknowledges and makes use of the places where students live to induct them into the discourses and practices of any and all school subjects. More than anything else, teachers who use this approach share a perspective about teaching and learning that alerts them to the educational potential of phenomena outside the classroom door. For them, community and place become additional "texts" for student learning.

Closely related to project-based learning, place-based education addresses one of John Dewey's central concerns about public school classrooms, their separation from the lives children lead when they are not in school. As Dewey observed more than a century ago,

> From the standpoint of the child, the great waste in the school comes from his inability to utilize the experience he gets outside the school in any complete and free way within the school itself; while on the other hand, he is unable to apply in daily life what he is learning at school. That is the isolation of the school, its isolation from life. When the child gets into the school he has to put out of his mind a large part of the ideas, interests, and activities that predominate in his home and neighborhood. So the school, being unable to utilize this everyday experience, sets painfully to work, on another tack and by a variety of means to arouse in the child an interest in school studies. (1959, p. 76–77)

When place is incorporated into the act of curriculum development, children's everyday experiences become one of the foundations upon which learning is constructed. When teachers do this, students more easily understand why they are learning what is being taught.

But an exploration of place can do more than simply make school meaningful. It can also act as an antidote to the experience of disconnection or alienation that sociologists from Emile Durkheim (1951) to Robert Bellah (1985) and Robert Putnam (2000) have associated with life in industrial and postindustrial civilizations. When electronic media and schools together direct children's attention away from their own lived experience, it is not surprising that they find it difficult to become attached to and responsibly involved with their communities. Nor is it surprising that they are spending less and less time getting to know natural places within or beyond their neighborhoods. As Richard Louv (2005) has made apparent, we live in a culture that is preventing many children from establishing the relationships with nature that may undergird both emotional health and environmental preservation. Place-based education can address these gaps by providing young people with opportunities to experience in visceral ways both the human and natural communities that surround them.

During an era when policymakers and foundations have been primarily preoccupied with raising achievement scores, place-based education advocates have found it necessary to demonstrate the link between learning experiences grounded in the local and improved academic performance. To justify their own claims about enhanced citizenship and stewardship, they have also been faced with the challenge of assessing how these dispositions and the activities associated with them are affected by learning that uses as its starting point student encounters with their community and place. Given the recent and small-scale nature of most place-based education endeavors, the number of these studies is small and minimally funded. Their results, however, have generally been encouraging and suggest the need for more extensive investigation.

In what follows, I will present a description of two successful place-based education projects, both located in the United States. Similar projects are emerging in places as diverse as the Scottish highlands, Norwegian coastal communities, and the Australian outback as people seek to make the boundary between schools and their communities more permeable and overcome the division between school and life that worried Dewey in the late-1800s. Following these examples of practice, I will then discuss some common principles that underlie this approach to teaching and learning and then review research studies that provide some preliminary insights into its potential impact on academic engagement and achievement as well as on students' perception of themselves as citizens and stewards.

Place-Based Education in Practice

The Llano Grande Center grew out of the work of a high school teacher (now university professor) named Francisco Guajardo. Guajardo had grown up in the Edcouch-Elsa region of South Texas. The son of Mexican immigrants, he had become a successful student who earned a university degree in education and returned to his home community to teach. Interested in helping more students become successful in school but also committed to improving their community, he designed a curriculum around the collection and dissemination of oral histories. He hoped that listening to their elders would give young people an opportunity to learn about the qualities of persistence, care, and generosity that are so essential to the maintenance of positive social environments. In the early years, students would gather their interviews, transcripts, photographs, and video recordings and present them to community members in a museum format.

In the mid-1990s, Guajardo successfully wrote a grant proposal to the Annenberg Rural Challenge—a national initiative aimed at improving rural education—to enhance student achievement, community development, and cultural pride through the creation of the Llano Grande Center. In the years since Guajardo began teaching in his hometown, over eighty students (from a high school whose student population numbers 1,570, 99 percent of whom are Latino and 83 percent, economically disadvantaged) have gone on to Ivy League or other competitive colleges and universities to pursue higher education. They are among the 70 percent of students from Edcouch-Elsa High School who pursue postsecondary studies (F. Guajardo, personal communication, November 27, 2008), a number that is remarkably high given the demographic characteristics of this community.

Among projects taken on by students and teachers as this work has evolved are regular television productions featuring interviews with local residents and a Spanish-language school for adults marketed to people from outside the region and state. During the 2007–2008 academic year, students investigated the history of a chemical plant in town that had once produced fertilizers but was now abandoned. They interviewed people who lived close to the plant and discovered repeated stories about cancer. In the spring of 2008, they presented their findings through a film entitled *Toxic Nopales* that explored the effects of the chemical company on their community. Students had hoped that their presentation would motivate elected officials to take action on this issue. Although city leaders in Elsa did not choose to do so, leaders in a neighboring community saw the film and embarked on an effort to get property owners to clean up a similarly contaminated site (F. Guajardo, personal communication, February 12, 2010).

A second example is from the Young Achievers Science and Mathematics Pilot School in Boston. Young Achievers is one of Boston's twenty pilot schools, schools that are part of the public system but that have more autonomy with regard to budget, hiring, and curriculum decisions than conventional schools. Young Achievers primarily serves low-income kindergarten through eighth-grade black and Latino children. It demonstrates a strong social justice commitment and collaborates closely with the Dudley Street Neighborhood Initiative, one of the nation's more successful grassroots community development organizations (Sklar, 1999).

Since 2003, Young Achievers has worked with the CO-SEED (Community-based School Environmental Education) program at Antioch New England University to situate its educational program even more firmly in the neighborhoods that surround the school. Place-based education is becoming a central feature of its offerings. Students commonly participate in environmental field studies in a nearby cemetery as well as the Harvard Arboretum. When fourth-grade students study American history, they visit Plymouth Rock and then imagine what it would have been like to start their lives over again in a new place completely cut off from everything they had once known. The spring I visited the school (2007), part of their classroom was filled with a village built to scale consisting of log cabins. A large mural of Boston Harbor before European settlement graced a wall outside the classroom.

During the 2007–2008 academic year, second graders participated in a place-based learning project typical of the school. Community studies are central to what students do at Young Achievers, even when they are no more than seven years old. Among other topics, students decided to investigate the experience of residents of Boston's Chinatown, the impact of air pollution on local asthma rates, the value of public art such as murals, and the need for more academic learning space at their school. In teams of five to seven, they researched these issues and developed presentations based on their findings that were presented on Con Salsa, WBUR's weekly Saturday night radio program, during the spring. Describing the academic outcomes of this work, Robert Hoppin (personal communication, June 5, 2008), the school's place-based education coordinator, wrote:

> The entire initiative held the second grade students to a very high standard. It asked them to apply, understand and use the power of proficient literacy skills to explain their findings from the neighborhood investigations. The program demanded the use of formal oral language. There is not a casual word on their broadcast. You can distinguish the kids who read above grade level and below grade level, but because they worked with a professional sound technician from WBUR, all students sound professional in their speaking and reading skills. Each student has the opportunity to reflect on their own literacy skills by listening to their professionally recorded voice. The whole process sends the message "you are all expected to read proficiently and speak to sound important." The boosting of self-esteem for lower level readers is profound. The use of this audio recording next year will set a high bar for the incoming second grade students to surpass. (2008)

In these two examples can be seen some common principles that are often encountered in other place-based efforts, as well (Smith & Sobel, 2010). First, teachers took full advantage of local possibilities for curriculum development. They found ways to teach some of the central skills associated with language arts or social studies or science by directing students to people or issues in their own communities worth investigating and reporting on.

Second, learning experiences required students to become knowledge producers rather than consumers and gave them the chance to exercise their voices beyond the classroom. In Boston, students' findings were based on their own investigatory activities and were then aired on a popular radio show. In Edcouch-Elsa, creating a film involved reviewing legal documents, interviewing numerous people, and then synthesizing what had been learned into a factually accurate story. When the film was shown to the public, a large crowd provided students with an opportunity to share their concerns and perspectives with families, neighbors, and decision makers. Important in both of these examples, as well, is the fact that work students were producing was of interest to the broader community, leading them to want to put in their best effort.

Third, students at Ed Couch-Elsa High School and the Young Achievers School were able to learn from other community members as well as their teachers. To make *Toxic Nopales*, student went far beyond the classroom as they interviewed elected officials, agency personnel, and adults who had been affected by the chemical plant. In Boston, second graders got to interact with media professionals and learn some of the in's and out's of broadcasting when they shared their research on the air. Drawing upon this outside-of-school expertise is often central to the success of place-based learning experiences.

Finally, place-based educators often design learning activities that could potentially engender a sense of appreciation or positive regard for students' home communities and regions. Even when problems are surfaced and students examine difficult aspects of their common life, participating in efforts to make things better can lead to the forms of affiliation and empowerment that underlie both stewardship and citizenship. This development of connection and the capacity to make change underlies the cultivation of the kind of people needed to support or create socially healthy and ecologically sustainable communities.

Impact of Place-Based Education on Academic Engagement and Achievement

One of the challenges in justifying grassroots and small-scale innovations like place-based education is that much of the evidence supporting the value of these approaches will be limited and anecdotal. Furthermore, the costs of engaging in cross-institutional studies capable of overcoming this problem can be prohibitive. A solution that has been adopted by some advocates of place-based education involves the use of logic models which assert that if a similar approach has resulted in improved academic engagement and performance, one can hypothesize that comparable instructional strategies will lead to similar results (Duffin, Chawla, Sobel, & PEER Associates, 2005; Powers, 2004).

Place-based education, for example, has much in common with the forms of authentic instruction that were the focus of a large-scale study conducted at the National Center on School Organization and Restructuring during the 1990s. Although this important work was quickly upstaged by the accountability movement following the passage of the No Child Left Behind legislation, its findings point to the powerful effect that learning activities involving the mastery of complex tasks and information connected to the world beyond the classroom can have on student achievement (Newmann, Marks, & Gamoran, 1995).

Researchers involved with this project looked for classrooms where students participated in inquiry activities that incorporated original research, the consideration of alternatives, and the application to real world problems of the forms of inquiry and communication associated with academic disciplines. As in the Llano Grande Program and the Young Achievers School, students in these classrooms were expected to be knowledge producers, not merely consumers. They were asked to write reports and speeches and convey their findings and ideas to people outside the school. Public presentations were central to the experience of authenticity. Data collected regarding student achievement in twenty-four different elementary, middle, and high schools across the United States revealed that students in the classrooms of teachers who ranked high in their use of authentic instruction performed more successfully on tests based on items from the National Assessment of Educational Progress. Place-based education, grounded as it is in local phenomena and issues, is by its very nature authentic.

There is also a logical relationship between the integration of school and community encountered in classrooms where place undergirds instructional decisions and findings from Coleman and Hoffer's 1987 study, *Public and Private High Schools: The Impact of Communities.* One of the central discoveries reported in this volume is that when the messages conveyed to children and youth from their homes, communities, and schools all converge, academic engagement, and achievement improve. As one might expect, given the widely recognized relationship between income and achievement, student performance is higher in private than public schools; what was of interest to Coleman and Hoffer is the way that less affluent and more diverse students in Chicago's Catholic schools frequently outperformed their wealthier counterparts in the city's secular private schools.

The critical difference between these two populations was that students in Catholic schools were generally more connected to their neighborhoods and families than was the case for the primarily white students in non-Catholic private schools. The lives of students in the Catholic schools were characterized by a set of common understandings that crossed institutional boundaries. In contrast, students in secular private schools, although wealthier, lived social lives that were much less consistent. Growing up in homes with parents whose work lives left little time for family or for community, they experienced a kind of isolation and disconnection not found in Catholic schools. For Coleman and Hoffer, this affected academic motivation and performance. Although place-based education is unlikely to have any impact on the amount of time parents spend with their children, it can enhance students' experience of the different domains of their lives as integrated rather than separate and lead them to more easily internalize the purpose behind the work they are expected to do in school.

From a logic model standpoint, the relationship between studies of the academic impact of environmental education—especially when it focuses on local observations, projects, and issues—and place-based learning is even stronger. One of the few large-scale studies investigating environmental education and academic achievement is Oksana Bartosh's *Environmental Education: Improving Student Achievement* (2003). It is one part of a broader research effort conducted by Washington State's Environmental Education Consortium. Such a study is possible in Washington because of the state's inclusion of environmental and sustainability issues in its mandated standards. Schools, however, vary in the degree to which they have implemented environmental education into their curriculum. This allowed Bartosh to conduct a comparative study of 144 schools, half of which featured strong environmental education programs and half of which had either not yet integrated these approaches into their curriculum or were just starting to do so. In schools that exhibited strong programs, students were encouraged to construct their own knowledge, and efforts were made to link learning activities to the community—initiatives also characteristic of schools where place-based education is encountered.

Student achievement measures include scores on the Washington Assessment of Student Learning (WASL) and the Iowa Test of Basic Skills (ITBS). In addition, a survey assessing the degree to which the practices described above were being implemented was made available on the Internet for personnel in participating schools. Bartosh reports that:

> According to this research, schools that undertake systemic environmental education programs consistently have higher test scores on the state standardized tests over comparable schools with "traditional" curriculum approaches. The mean percentage of students who meet standards on WASL and ITBS tests are higher in all six areas (i.e., WASL-Math, WASL-Reading, WASL-Writing, WASL-Listening, IT-Reading, IT-Math) in the schools with environmental programs. According to the statistical analysis, schools with EE programs performed significantly better compared to non-EE schools on the state standards tests. (2003, p. 117)

Bartosh is careful to acknowledge that these effects are correlational rather than causal, but she still believes the adoption of these approaches provides "tremendous opportunities for schools, teachers, and students" (p. 118).

> It improves students' behaviors and motivation to learn. It encourages parents and members of the community to take part in school learning activities. Also, students have a unique opportunity to participate in real-life projects and try to solve issues and problems in their communities. They see the relationship between knowledge and skills they receive in the classrooms and the real world around them. (p. 118)

These three robust studies, related as they are to many of the practices encountered in place-based educational settings, demonstrate that contextualizing learning and inviting young people to be knowledge creators can have a powerful impact on students' willingness to attend to educational tasks and demands. But what about research that explicitly addresses place-based educational reforms? Even though they may be smaller scale and more anecdotal, what do they show about the relationship between learning that addresses local phenomena and issues and academic engagement and achievement?

Gerald Lieberman and Linda Hoody's 1998 study, *Closing the Achievement Gap: Using the Environment as an Integrating Context for Learning*, is widely cited as one of the first efforts to link what in effect is place-based learning to a variety of student outcomes including academic engagement and achievement as measured on standardized tests. The environment referred to in the title of the study includes both natural and social settings; at issue for the authors is the way teachers seek to contextualize learning through the use of local lived experiences in both natural and social settings.

The study involved site visits to forty schools where teachers were using the environment as an integrating context (EIC), interviews with more than 400 students and 250 teachers and administrators, and the completion by school personnel of four instruments aimed at uncovering additional information about student and teacher participation, engagement, instructional practices, and effects of the program on students' knowledge, skills, retention, and attitudes (Lieberman & Hoody, 1998).

What Lieberman and Hoody report parallels research from the Center on School Organization and Restructuring about the impact on engagement of learning activities that have real world applications. They discovered, for example, that 98 percent of educators noted an increase in student enthusiasm and engagement in school activities after the adoption of EIC approaches. Eighty-nine percent reported that students were more willing to stay on task when learning in this way. Especially important for respondents was the way that EIC was more responsive to students with varied learning styles, particularly those considered at risk of school failure. Educators at all of the forty schools involved in the study reported that learning in outdoor settings provided opportunities for many of these students to become more successful learners.

Although scores on standardized achievement tests were available from only ten of the forty schools, available data suggested a positive link between place-based learning experiences and higher scores on reading, writing, math, science, and social studies tests. Studies of a smaller number of schools in California conducted in 2000 and 2005 (State Education and Environment Roundtable, 2000, 2005) showed similar trends.

Another place-based educational reform effort important to consider is the Annenberg Rural Challenge, mentioned earlier with regard to the Llano Grande project. Started in the late 1990s, this $50 million initiative paralleled the Annenberg Challenge in urban schools and provided support over a five-year period for educators and activists in the nation's heartland who sought to strengthen rural schools. Vito Perrone, then chair of the Department of Teaching, Curriculum, and Learning Environments at Harvard University, took on the task of overseeing the evaluation of the Rural Challenge (1999). This work resulted in a publication entitled *Living and Learning in Rural Schools and Communities* in 1999 following the first year evaluation. Much of this publication is focused on a description of projects and their positive impact on student and community engagement. Research regarding student achievement for all of the Rural Challenge sites, however, was never completed after Perrone suffered a massive stroke in 2000. Studies of Rural Challenge projects in Louisiana and Alaska, however, did investigate the impact of place-based education on student achievement. As with Lieberman and Hoody's work, they are also suggestive of the positive effect this approach can have on student learning.

Emekauwa's 2004 article, *"They Remember What They Touch: The Impact of Place-based Learning in East Feliciana Parish"* describes changes in student performance over a three-year period in three Louisiana elementary schools. Students in this parish are largely low-income and African American. Before their teachers had participated in three summers of professional development focused on integrating the study of local natural resources into the teaching of science, math, and technology, a significant proportion of fourth graders scored unsatisfactorily on state tests. In 1999–2000, for example, 32.6 percent failed the English-language arts test; 39.0, the mathematics test; 27.5, the science test; and 39.4 the social studies test. In 2002, the number of failing students had been reduced to 18.4 percent in English-language arts; 24.9 in math; 19.4 in science; and 29.1 in social studies. (Emekauwa, 2004b, p. 5), bringing student scores in East Feliciana Parish much closer to state averages. A district director for place-based learning maintained that the incorporation of local phenomena into classroom teaching served "as a hook to get students excited by learning" (Emekauwa, 2004b, p. 8).

Emekauwa (2004a) also studied the impact of place-based learning in Alaska where the Rural Challenge initiative was linked to a National Science Foundation supported project. Together, both initiatives aimed at integrating indigenous values, knowledge, and skills into the teaching of science, math, and the humanities as they worked with twenty rural districts that included 176 schools serving 19,000 students, the majority of whom were Alaska Natives. The culminating report of this initiative published in 2006 indicated that between the 1999–2000 and 2004–2005 academic years the proportion of students who scored at the advanced/proficient level on the eighth-grade mathematics benchmark increased from 21 percent to 42 percent; tenth-grade scores on the high school graduation qualifying exam, from 20 percent to 50 percent (Alaska Rural Systemic Initiative, 2006, p. 23). As in East Feliciana, many students continued to struggle academically, but the opportunity to learn in ways that were grounded in students' local experience and culture appeared to have a positive impact on student achievement as measured by state tests.

One final set of research studies regarding student engagement and achievement bears mentioning. In 2002, a group of nonformal and university place-based educators in New England formed a collaborative to engage in more systematic assessments of their programs. Faced with foundations that were demanding quantitative data regarding the impact of their work on student engagement and achievement, citizen participation, and stewardship attitudes and behaviors, they hoped that the jointly supported Place-Based Education Evaluation Collaborative (PEEC) would provide the evidence need to justify additional funding. Since its founding, PEEC has produced nearly fifty reports about a wide range of place-based educational projects. Defining the relationship between stream studies or the collection of oral histories and student performance on standardized tests, however, has generally proven to be difficult (Duffin, Powers, Tremblay, & PEER Associates, 2004b). Two studies of schools that had worked closely with the CO-SEED program (mentioned earlier with regard to the Young Achievers School) are again suggestive of the positive link between this approach and enhanced student learning.

The Beebe Environmental and Heath Science Magnet School in Malden, Massachusetts, serves a primarily low-income and minority student population. CO-SEED worked for three years with this school encouraging teachers to incorporate more hands-on and real world learning experiences with their students. Students who had had more exposure to CO-SEED influenced instruction demonstrated higher scores on the Massachusetts Comprehensive Assessment System (MCAS) tests of Life Science and Math (Duffin, Phillips, Tremblay, & PEER Associates, 2006a). A similar study was conducted at the Gorham School in Gorham, New Hampshire (Duffin, Phillips, Tremblay, & PEER Associates, 2006b). Data were collected for eight cohorts of Grades 3–6 students from before the introduction of CO-SEED initiative to two years afterward. Student scores on the New Hampshire Educational and Improvement and Assessment Program (NHEIAP) tests of math and language arts showed more Gorham students in the passing category in comparison to the novice category as they moved through the grades. The reverse happened with regard to state average scores for these tests, with more students in the novice category as they progressed from Grades 3 to 6, and correspondingly fewer in the passing category. Because the trend predated the introduction of CO-SEED, the researchers acknowledge that other factors must be at work in this school despite the belief on the part of teachers and administrators that CO-SEED was a contributing factor. At the same time, the researchers suggest that CO-SEED did not interfere with this trend.

It is conceivable that the kind of learning encouraged by place-based education is more likely to become evident in authentic demonstrations of student mastery as seen in reports, exhibits, speeches, films, or the completion of community projects than on multiple choice tests. Standardized measures may not be the best way to evaluate the intellectual achievements associated with this kind of learning. Portfolios and assessments by family and community members during performances and exhibitions could provide a more reliable indication of the degree to which students are mastering skills and knowledge important to their communities.

Still, academic achievement, however measured, should be seen as only one of the many outcomes of any educational process. Schools need to prepare individuals who also perceive their membership in broader social and natural communities, and then act accordingly.

Citizen Participation and Stewardship

As with academic achievement, research studies regarding place-based education's impact on citizen participation and stewardship remain largely small-scale and anecdotal. Once more, drawing upon studies of similar approaches is useful. Service learning—another approach closely related to place-based education but without its emphasis on the cultivation of students' membership in natural as well as human communities—has been widely studied. Billig, Root, and Jesse (2005) report that a

> . . . gradually accumulating body of evidence suggests that service-learning helps students develop knowledge of community needs, commit to an ethic of service, develop more sophisticated understandings of politics and morality, gain a greater sense of civic responsibility and feelings of efficacy, and increase their desire to become active contributors to society. (p. 4)

They speak specifically of increases in civic-related knowledge, civic-related skills, civic attitudes, service behavior, and the cultivation of social capital, ". . . including increased connections to schools and other organizations and increased social networks" (p. 4). Given the way that many place-based educational experiences mirror the forms of applied learning encountered in well-implemented service learning programs, one could expect comparable gains in citizenship behavior on the part of students who participate in these programs, as well.

Findings from the Place-Based Education Evaluation Collaborative (PEEC), suggest that this is the case. During a 2003–2004 investigation of nine rural and urban schools that were working closely with the CO-SEED program, evaluators discovered that the most notable impacts of this approach were increases in students' civic engagement and their environmental stewardship. Students of teachers who incorporated the local place and people into their curriculum, made use of service-learning opportunities, took children outside on a regular basis, and in general embedded place-based education in their lessons were more likely to agree than disagree with the statements listed below:

- I feel like I am part of a community.
- I pay attention to news events that affect the community.
- Doing something that helps others is important to me.
- I like to help other people, even if it is hard work.
- I know what to do to make the community a better place.
- Helping other people is something everyone should do, including myself.
- I know a lot of people in the community, and they know me.
- I feel like I can make a difference in the community.

- I try to think of ways to help other people.
- In the last two months, I have done something *with my classmates* to take care of my neighborhood or community.
- In the last two months I have done something *on my own time* to take care of my neighborhood or community.
- I enjoy learning about the environment and my community. (Duffin, Powers, Tremblay, & PEER Associates, 2004a, p. A-19)

In her study of the Maryland Bay School Project, a regional effort that used the environment as an integrating context (EIC), Von Secker (2004) similarly found that students in the classrooms of teachers who emphasized project- and place-based learning reported higher levels of stewardship behaviors than their peers in classrooms where teachers did not incorporate these practices to the same extent.

A 2008 summary of research findings from New England affirmed once again the persistence of the impact of place-Based approaches on civic engagement and stewardship (Place-Based Education Evaluation Collaborative, 2008).

Central to the experience of such participation is the belief that the activities of average citizens can actually make a difference. For many young people and Americans, in general, the idea that their own efforts could lead to social betterment is often viewed as fanciful. The result is that people assume "you can't fight City Hall" and withdraw from a community's public life altogether. When children are shown that their efforts can influence decision making and lead to improvements in local social settings and the natural environment, they often develop a sense of their own capacity as change agents, something that seems likely to set a life trajectory characterized more by activism than passivity. After participating in a course on environmental justice that played a major role in raising public awareness about the relationship between declining air quality and rising asthma rates, a student in Boston's Greater Egleston Community High School made the following observation about her induction into the process of public decision making.

> I am proud of my accomplishments in environmental justice this trimester. Most importantly, I have been able to gain confidence to speak in front of large groups of people. Before presenting to the City Council I was very nervous. But after watching them and my classmates somewhat debate I realized they are regular people just like my family, my teachers, and my friends, and I should not be nervous when it comes to speaking my mind. (Senechal, 2008, pp. 100–101)

Statements like this point to the way that learning experiences grounded in local communities and issues can positively affect the development of the capacities and dispositions associated with active citizenship.

In a similar vein, Buxton (2010) demonstrates how adopting the more critical approach to social and environmental analysis proposed by Gruenewald (2003) can lead students to transformative shifts in worldviews as well as an increased willingness to act. Gruenewald argues for a critical place-based pedagogy aimed at helping students understand the nature of social inequities and power dynamics (decolonization) coupled with efforts to draw them into activities aimed at improving their own places (reinhabitation). After being engaged in a week-long investigation of water issues with the aim of using science to promote social justice, Buxton found that middle school students from diverse socioeconomic backgrounds developed a much more complex understanding of the way access to safe water is tied to social position. At the same time, by associating science with social problem solving, he found that students became motivated to master more content since they were given the opportunity to consider the questions of "why here" and "so what."

Lim and Calabrese Barton (2006) have also studied the way that situating the study of science in students' own life worlds can enhance their engagement as they explore the relationship between content that can often seem abstract and decontextualized, and the needs and possibilities of their own community. Although Lim and Calabrese Barton do not explicitly address issues of decolonization and reinhabitation, the park design project they describe could well have evolved in this direction in the hands of a teacher committed to social action as well as the incorporation of students' sense of place in his instructional decisions. A slowly growing collection of studies strongly suggests that situating learning experiences in the places where students live their lives increases their willingness to become engaged academically, civically, and environmentally.

Preparing Students to Compete and Cooperate

In many respects, one of the central aims of place-based education is to enlist the intelligence, energy, and skills of young people in the process of community and environmental revitalization and restoration. The hope is that by inviting students into this work, they will find reasons to learn and that the social and natural communities that surround them will become both healthier and more sustainable as a result of their involvement. The motto of the Rural School and Community Trust, one of the most significant national supporters of this approach, is "Schools and Communities Getting Better Together." This perspective about the aims of education are to some extent at odds with our current system with its focus on the preparation of individuals who are seen primarily as competitors for jobs in an increasingly unstable and threatening global economy. One of the challenges facing advocates of this approach is demonstrating the way schools and teachers who have embraced its practices can prepare children to both compete and cooperate.

Research done to this point in time suggests strongly that something like this may be happening. Because students are able to connect academic learning to their own lives and the needs of their communities, they see the value of the work associated with becoming well educated. And given the opportunity to direct their energies to authentic needs and concerns, many develop a taste for participation and action. More studies will be needed to demonstrate the conditions that support such learning and involvement, but those that have been completed so far indicate that place-based education could well provide a way for overcoming the division between the classroom and community Dewey identified over a century ago. In doing so, students, communities, and the environment could all be beneficiaries.

References

Alaska Rural Systemic Initiative. (2006). *AKRSI final report*. Fairbanks, AK: Alaska Native Knowledge Network, University of Alaska Fairbanks.

Bartosh, O. (2003). *Environmental education: Improving student achievement* (Unpublished master's thesis). Olympia, WA: The Evergreen State College.

Bellah, R. N., Madsen, R., Sullivan, W. M., Swidler, A., & Tipton, S. M. (1985). *Habits of the heart: Individualism and commitment in America*. Berkeley, CA; University of California Press.

Billig, S., Root, S., & Jesse, D. (2005). *The impact of participation in service-learning on high school students' civic engagement*. Denver, CO: RMC Research Corporation.

Buxton, C. (2010). Social problem solving through science: An approach to critical, place-based, science teaching, and learning. *Equity and Excellence in Education, 43* (1), 120–135.

Coleman, J., & Hoffer, T. (1987). *Public and private high schools: The impact of communities*. New York, NY: Basic Books.

Dewey, J. (1959). School and society. In M. Dworkin (Ed.), *Dewey on education*. New York, NY: Teachers College Press.

Duffin, M., Chawla, L., Sobel, D., & PEER Associates. (2005). Place-based education and academic achievement. Retrieved from http://www.peecworks.org/PEEC/PEEC_Research/S0032637E

Duffin, M., Phillips, M. Tremblay, G., & PEER Associates. (2006a). CO-SEED and test scores in Malden, Massachusetts. Retrieved from http://www.peecworks.org/PEEC/PEEC_Reports/S01469109-01469165

Duffin, M., Phillips, M., Tremblay, G., & PEER Associates. (2006b). CO-SEED and test scores in Gorham, New Hampshire. Retrieved from http://www.peecworks.org/PEEC/PEEC_Reports/S010DFF34-014690A4

Duffin, M., Powers, A., Tremblay, G., & PEER Associates. (2004a). *An evaluation of Project CO-SEED: Community-based school environmental education, 2003–2004 Final Report*. Retrieved from http://www.peecworks.org/PEEC/PEEC_Reports/S0019440A-003A00C7

Duffin, M., Powers, A., Tremblay, G., & PEER Associates (2004b). *Report on cross-program research and other program evaluation activities, 2003–2004*. Richmond, VT: PEER Associates.

Durkheim, E. (1951). *Suicide: A study in sociology*. New York, NY: New Press

Emekauwa, E. (2004a). *The star with my name: The Alaska Rural Systemic Initiative and the impact of place-based education on native student achievement*. Washington, DC: Rural School and Community Trust.

Emekauwa, E. (2004b). *They remember what they touch. The impact of place-based learning in E. Feliciana Parish*. Washington, DC: Rural School and Community Trust.

Gruenewald, D. (2003). The best of both worlds: A critical pedagogy of place. *Educational Researcher, 32*(4), 3–12.

Lieberman, G., & Hoody, L. (1998). *Closing the achievement gap: Using the environment as an integrating context for learning*. San Diego, CA: State Education and Environment Roundtable.

Lim, M., & Calabrese Barton, A. (2006). Science learning and a sense of place in an urban middle school. *Cultural Studies of Science Education, 1*, 107–142.

Louv, R. (2005). *Last child in the woods: Saving our children from nature deficit disorder*. Chapel Hill, NC: Algonquin Books of Chapel Hill.

Newmann F., Marks, H., & Gamoran, A. (1995). Authentic pedagogy: Standards that boost student performance. *Issues in Restructuring Schools* (Issue Report No. 8, 1–4).

Place-Based Education Evaluation Collaborative. (2008). *The benefits of place-based education: A report from the Place-Based Education Evaluation Collaborative*. Richmond, VT: Author.

Power, A. (2004). An evaluation of four place-based education programs. *The Journal of Environmental Education, 35*:4 (Summer), 17–32.

Putnam, R. (2000). *Bowling alone: The collapse and revival of American community*. New York, NY: Simon & Schuster.

Rural Challenge Research and Evaluation Program. (1999). *Living and learning in rural schools and communities*. Cambridge, MA: Harvard Graduate School of Education.

Senechal, E. (2008). Environmental justice in Egleston Square. In D. Gruenewald & G. Smith (Eds.), *Place-based education in the global age: Local diversity*. New York, NY: Erlbaum/Taylor & Francis.

Sklar, H. (1999). *Chaos or community: Seeking solutions, not scapegoats, for bad economics*. Boston, MA: South End Press.

Smith, G., & Sobel, D. (2010). *Place- and community-based education in schools*. New York, NY: Routledge.

State Education and Environment Roundtable. (2000). *California student assessment project: The effects of environment-based education on student achievement, phase one*. Poway, CA: Author.

State Education and Environment Roundtable. (2005). *California student assessment project, phase two: The effects of environment-based education on student achievement*. Poway, CA: Author.

Von Secker, C. (2004). *Bay Schools Project: Year three summative evaluation*. Annapolis, MD: Chesapeake Bay Foundation.

23

Getting the Picture

From the Old Reflection—Hearing Pictures and Telling Tales, to the New Reflection—Seeing Voices and Painting Scenes

Tony Shallcross
University of Hull, UK

John Robinson
Robinson Consulting, UK

Introduction

Research into environmental education (EE) and education for sustainable development (ESD), in their aspiration to change lifestyles, are often caught in a tension between the need to investigate the outcomes of EE/ESD interventions on one hand and on the other to understand the educational processes that underpin the achievement of these outcomes. Furthermore ESD research in particular, because it concerns environmental as well as economic and social outcomes and processes (UNESCO, 2005) must remain open to the influence of the findings of positivist, scientific research, while educational research, in general, is bound up in an orthodoxy of qualitative research. This is not to say that research paradigms are the only issue for EE/ESD research but to argue that these tensions illustrate the deep dilemmas that fracture and to some extent marginalize educational research.

The orthodoxy of qualitative/interpretivist research is often underpinned by postmodern and poststructuralist thinking. This new scholarship focuses more on the reflectivity and reflexivity of the researcher and the problematics of research in a pluralized world and rarely arrives at conclusions that have anything other than local, temporary implications. It is a research trend that eschews vision as Utopian, that places deconstruction before reconstruction and regards number as a form of viral infection. Meanwhile EE/ESD research grapples with understanding the role that education can play in promoting and fostering actions that have the potential to contribute to the resolution of global problems such as climate change, economic inequalities, and the loss of biodiversity. EE/ESD is part of the response to the global emergency of unsustainable lifestyles, which necessarily engage EE/ESD with scientific, positivist research that indicates that planetary systems are rapidly reaching a tipping point beyond which for example climate change will be irreversible (Lovelock, 2007, 2009). Deconstruction sits well with understanding the origins of these global issues but does little to provide the professional knowledge that will promote educators' engagement with EE/ESD as transformatory processes.

> Students in schools, I am quite sure, could not care less about our methodological quibbles; they would like us to forget our concerns about our "selves," so we can fully listen to them to help us understand how to turn schools into better places to be (Heshusius, 1994, p. 20).

By rejecting the visioning associated with reconstitutive postmodernity (Des Jardins, 1993; Zimmerman, 1994) new scholarship can result in the deskilling and disempowerment of educators and learners because it conveys an understanding of fracture and fragmentation but cannot engage with synthesis and convergence. While deconstruction can conceptualize problems it rarely acts out solutions. A more likely outcome is that deconstruction will lead practitioners to ignore new scholarship research because it does not provide the evidence to inform practice. The combination of scientific research that identifies negative trends with educational research that eschews answers will widen the gulf between research evidence and practice that is already more evident in education than most other professions.

What is more surprising methodologically is that the infatuation with postmodernity has not been accompanied by a more vigorous interest in and advocacy of visual methods in educational research. Hall has suggested that in a postmodern world visual representation increases in importance (S. Hall, 1990) yet the same author argues that visual images have been neglected in educational research (Evans & Hall, 1999; Fischman, 2001). However, even when visual sources are used, the findings derived from them are usually reported in text and number. Yet, technological developments in webcams and video cameras for example, have made visual recordings much more accessible for research purposes. The advantages for educational research of deriving data from visual recordings are being recognized. The Board on International Comparative Studies in Education recommends the use of visual recordings as a first step toward data collection in research studies (Ulewicz & Beatty, 2001).

Throughout this chapter we will try to address three intermingling sets of issues. The first relates to the problem of capturing direct action in EE/ESD research. We will argue that too often in such research we are faced with reports of intentions, opinions, or impacts rather than images of the actions themselves. An extension of this issue is the potential gap between espoused and actual actions. The second of our intermingled concerns relates to the relative lack of engagement with visual images. The first issue raises the questions of memory, recollection, presence, and absence. The second issue raises questions of (re)presenting. One of the legacies of poststructuralist/postmodern thinking is the spread of "anxiety into a more general worry about the representation of 'The Other'" (Geertz, 2000, p. 95). What we are focusing on here is an anxiety about the representation of the self. The third issue is the need for a research paradigm in EE/ESD that addresses the complexity of the research agenda each faces by adopting a mixed method approach that uses visual information gathering where possible and feasible. Multi-modal analysis would seem to offer the prospect of a unified approach to this visually enriched mixed method analysis.

Haunting this third debate is the specter of increased practitioner involvement in the development of pedagogical insights, which in turn raises questions about the transferability of policies and practices from one educational context to another. Using visual recordings to share applied professional knowledge creates friction between UNESCO's notions of best practice and the more appropriate concept of contextually sensitive good practice. Reflection becomes situated in context rather then focusing on the mirror of grand pedagogical theory. After all reflection is a metaphor and theorizing is a form of visualizing, which, when supported by visual images, can help to reposition theory and ground it so that it becomes a more concrete representation of practice in context.

There are a number of reasons why researchers in EE/ESD should consider redressing the neglect of visual research methods. Firstly self-evaluation of classroom performance is generally considered to be the weakest feature of trainee teachers and newly qualified teachers' practice. Evaluation is clearly relevant to all branches of the educational process as well as being a key component of the whole school approaches associated with EE/ESD (Shallcross, 2008; Shallcross & Robinson, 2008). Indeed EE researchers have written about the need for more ethnographic, situated approaches to organizational self-evaluation. Secondly, there is the gap between espoused values and values in action that leads to the socialization of hypocrisy when schools, teachers and other adult staff fail to practice what they teach, which Murthy (2008) refers to as a form of cognitive dissonance. While this discrepancy may occasionally be deliberate, the usual pattern is that schools and teachers frequently overestimate the degree of democracy, participation, and transformative pedagogy that occurs in their classrooms. Thirdly, while EE/ESD are both concerned with transforming lifestyles and actions there is a problem in observing and recording direct actions (Jensen, 2002) that illustrate these transformations. What researchers are left with are statements of intent, or reports of what educators or learners say they, have done. If such research is concerned with assessing outcomes then reporting supported by researcher observation may meet the requirements of professional evidence. However, for those who pursue research as investigating, reporting, and interpreting the dynamics of the EE/ESD process the surrogacy of text, impression, or postnatal commentaries is a limitation. Fourthly, there is a need to question the mantra of reflection, which has come to mean little more than "thinking about" in the modern educational lexicon. If reflection is to be improved trainees and teachers need not only the mirrors of theory, context, and personal experience against which to make judgments; reflection will be enhanced if they can see physical images in which their propositional knowledge can be grounded. Kuit, Reay, and Freeman summarize the new reflection succinctly if one incorporates their precepts with a mixed method multimodal methodology:

> [a] reflective teacher is one who compares their teaching against their own experience and knowledge of educational theory that predicts what might happen. Invariably these comparisons highlight differences between theory and practice, and the reflective process readjusts the theory until it accurately describes the practice. (2001, p. 130)

It is often the case, especially with trainees and novice teachers, that it is the practice that is reformed to match theory, particularly if the theory is the intellectual property of powerful stakeholders. Ball (1994) claims that feedback is being used to affect teaching not to cultivate the reflection that should inform teaching. How much greater is this danger now with the advent of competences than it was in the early 1990s? How can we use a visual metaphor to describe what is essentially a text based process? Towers (2007), drawing on Pimm (1993), suggests that often the relationship between trainees and novice teachers and their mentors and coaches can often be intimidating. What digital video images of classroom practice, particularly web-based images, may allow is for trainee and novice teachers to see others at the same stage of their career development. Fifthly as debates about research design in educational research become ever more occluded by the plethora of research paradigms that exist within the orthodoxy of interpretive research, there is a flickering engagement with mixed method approaches. One of the more interesting improvisations on this theme is the interest in mutlimodal research because of its integration of different modes of representation. The difficulty is that multimodality, including visual methods, while having significant influences on the collection of data, rarely impacts on modes of reporting those data, which are still largely dominated by written text. We will return to this point later. Finally, the use of multimodality, including video capture of classroom practices and processes opens up the space for the breaking down of the distinction between "researchers" and "teachers" (Armstrong & Curran, 2006) so that theory is modified in practical contexts (Hennessy & Delaney, 2009). This is not meant to imply that such techniques set out to identify best practice, rather that they allow for the development of intermediate theory through the dialog that takes place focused on visual excepts (Hennessy & Delaney, 2009, p. 1756) drawing upon both scholarly or academic knowledge and professional or craft knowledge (Hennessy & Delaney, 2009, p. 1765).

Trust Me, I am a Researcher: Mixed Method, Multi-Modal Research in EE/ESD

In terms of analyzing visual recordings Jacobs, Hollingworth, and Givvin (2007) recommend a multi-perspectival approach. Visual recordings of school activity enrich both young people's and adult's voices in ESD. They draw on a case study approach based on situated learning in whole institution approaches to school improvement. We draw on data from different case studies to examine how learners and educators perceive their participation in school engagement with sustainable development.

This chapter seeks a meta-analytical rather than descriptive account of the case studies as a way of identifying commonalities across the cases relating to participation in decision making. This analysis is then used as a backdrop to explore the contribution of visual methodologies to multi-modal analysis (MMA) in EE/ESD. By juxtaposing multi-level data forms which draw upon transcription, sound and dynamic visual images MMA examines how each of these data sets give rise to different, perhaps competing, but authentic stories about the data. This approach allows young people and adults, who have collected and analyzed the data, not only to explore more sustainable lifestyles but also to interrogate their own roles as researchers and researched more fully. Representations are constructed as data through processes of selection and exclusion that privilege different modes of communication. Making principled decisions that guide the choice of representation ensures that this process is explicit (Plowman & Stephen, 2008, p. 542). All too frequently research papers, of whatever hue, fail to indulge in this level of procedural detail. This is a problem facing all research paradigms because, as Plowman and Stephen (2008), Fischman, (2001), Pink (2001) and Rose (2001) all observe, selecting ways of representing data that are significant methods of data gathering that sustain the manifestation of verisimilitude is a central issue. It is embedded in the way that researchers gather and represent information as data, regardless of the medium.

R. Hall (2000) and Erickson (2006) point out that despite what participants say, visual data gathering is neither objective nor theory neutral. A recording is not data and it cannot be assumed that the picture does not lie or perhaps more accurately that the visual image is not tinged with ignorance. Conversely the camera is less likely to censor stories than verbal forms of communication (Fiske, 2002). When researchers refer to video capture (Pea, 2006) there is a connotation that the camera can appropriate reality. We refute this notion for both technical and epistemological reasons. Camera lenses are not manufactured with the field of view to gather all the information that is found in the classroom.

The selection of the camera position and which learning episodes to record are partly ideological decisions. Gathering complete information from any social interaction is impossible but the threat of leakage does not end at the gathering stage. The decision about how to represent information is partly about simplifying complex processes in order to sort, describe and analyse data. What we argue is that gathering information using multimodal approaches and then representing and analyzing the "data" that emerge as text represents a major leakage. When such approaches are not accompanied by a principled outline of the research design leakage can become a torrent, because the reader does not know the location or reason for the initial leak. Such leakage is unsustainable not only in research but also environmental terms.

The standard practice of transcribing interviews loses important elements of social interactions and the connections between oral/aural—emotion, intonation, context, sound—and body language does not display the fullness of that language. Representation needs to acknowledge that "voice" includes emotion, gesture, person, and place. Elements of voice are necessarily lost when a visual tapestry is translated into a transcribed text particularly if the transcript is itself a translation from a signed language:

> [s]imilar to a silhouette, the texts in front of me were a manifestation, a reproduction of the visual and visceral experience, but it appeared featureless and lacking the important nuances of the performed texts. (Hole, 2007, p. 704)

Data analysis and reporting reduce the thickness of research descriptions and consequential analysis of information collected. As Kundera (2002) remarks, understanding is based on a knowledge of what is absent as well as what is present. Our problem, as researchers is that we cannot know what is missing or absent only what has been consciously omitted and so understanding will always remain partial and sequential. What we are not suggesting here that video recorded data will rectify absence nor will they allow complete understanding—both claims would be fallacious—what we do want to suggest is that video recorded data will gather more complete understanding because recordings capture *more* of the processes and direct collective actions and their interactions. As Evans and Hall maintain the image is "multi-vocal" (1999, p. 309) and its meaning cannot be conveyed solely through the medium of written text.

In trying to achieve a more complete understanding—by drawing upon and juxtaposing multiple sources of data—researchers need to open up to the possibilities of new forms of representation. Disputes along lines of cleavage relating to paradigmatic allegiances are well reported and we do not intend to explore them further here; however, drawing on these arguments what we do notice is a (self-imposed) tyranny of conventional modes of research reporting. Breaking these conventions through a synthesizing of reflection and reflexivity that draws on video-recorded data, allows both a more complete understanding of the data but also, as our case study analyses will show, opens up the space for challenges to asymmetric power relationships (Bourdieu, 1991) between learners and teachers and between preservice teachers and their mentors. Armstrong and Curran (2006) argue that the use of the visual image gets teachers to "learn to notice and to develop new ways of seeing what was happening in their classrooms" (p. 342).

These challenges are based on the possibilities opened up by video recordings, for as Daniel-McElroy and Dalton observe, "[t]he human eye does not have the capacity to selectively frame and compose objects or focus on the minutiae of detail in an image, nor does it travel slowly over surfaces or use differential focus to make an emphasis" (2002, p. 2). Repeated watchings of data overcome this gap in our abilities. Memory does not equate to recollection. It is more an act of remembering which embodies conjuring up, selecting, rejecting, and accepting created recreations. This style of remembering is the basis on which repeated watchings should be reread. Meaning is not in the repeated visuals and words, but in the structural act, or enactment, of repetition. Meaning is in the structure of the piece, and this piece does not move forward with a false linearity, but it moves slowly and repeatingly. This is where the meaning is.

Despite the initial association of visual-recordings with CCTV style surveillance and control, the use of webcams in self-evaluation projects progressively fosters connections with "communication and democracy." It gives more power to learners/trainees and thereby develops a more empowering approach to classroom research and practice. As situated learning emerges as a favored approach to learning in the UN Decade of ESD, gathering visual information enriches the research database that is needed to understand the complexity of the relationships between the economic, social, and environmental strands of sustainable development. The facility to see other educational contexts not only augments research, it enables collaboration and the sharing of practitioner knowledge which, if it is rooted in a strong practitioner base, can form the basis of a more democratic professional educational knowledge base (Hiebert, Gallimore, & Stigler 2002). Cameras provide recorded packages that make teaching more visible and facilitate the sharing of practice across latitudinal studies which avoid some of the ambiguities associated with conventional reporting (Hiebert, Gallimore, & Stigler, 2003). Visual records, then, not only facilitate the collaboration that will allow practitioners to make contributions to professional knowledge in education, they also situate this knowledge and shift the locus of evaluation from writer to viewer.

Process and Principles

We referred earlier to the need to share the principled selection of methods and methodology with the reader. Such a principled selection involves three steps. Stating an explicit rationale for the approach adopted, maintaining the alignment between theory and empirics and ensuring fitness for purpose (Plowman & Stephens, 2008). In this section we want to explore some methodological issues that have arisen from our use of visual images as data and multi-modal analyses. In doing so we are drawing on a series of research reports by ourselves and our colleagues (Robinson, Shallcross, & McDonnell, 2008; Shallcross & Robinson, 2007, 2008; Tamoutseli & Pace, 2008) which space does not allow for a full description here (for a summary of these case studies see Shallcross & Robinson, 2008).

The presence of the physical image can assist reflection regardless of whether the route is an action research, a narrative or other approaches. The old paradigm is one in which reflection and research are based on observation, tape recording or the use of logs. In its analysis of situated social practices, the theoretical basis of a sociocultural perspective leads to a focus on activity and mediation rather than a focus on individuals' internalized learning. The multimodality of the ways in which this activity is enacted and the ways in which learning is mediated by people and artifacts are not necessarily represented, however, leading to a disjunction between the theoretical orientation and its empirical basis. This makes the task for visual ethnography one which sees the visual as a medium through which knowledge and critiques can be created outside of text (Mason, 2005, p. 329).

Although there are different approaches to visual methods research the type that we are suggesting here is a combination of two main trends. The first is the study of the social by producing images, the second is the collaboration between researchers and participants in the production of dynamic visual representations (Banks, 1995) of EE/ESD activities involving children and adults. The collection of visual images in order to analyze them as written text is in part a waste of the richness of these images. On the other hand if we are analyzing these images as different modalities, does this not suggest that researchers need training in approaches to the analysis of visual information in order to convert it into data. Banks (1995) suggests that researchers, be they insiders or outsiders, can look at conversations as well as hearing them and also observe the kinesics (body language) and proxemics (personal spatial behavior i.e., positioning). It is in this melange of words and their oral and visual presentation, the interlocution between the spoken, the heard and the seen, that the art of the educator lies (Hiebert et al., 2002).

Visual recordings represent the most comprehensive technology we have for gathering and generating thick descriptions (Geertz, 1973) created by complex educational interactions. Teaching and participation are forms of performance whose quality does not lie solely in the script, both are enhanced by the quality of speaking and the art forms of posture and gesture. However, in using this metaphor do we cast the educator as director or lead actor? Starting with a good script (lesson plan) is an important foundation but just as good scripts can be murdered by poor direction and/or acting so too can lesson plans. The visual richness of teaching as a form of performance is still overlooked in much educational training and research.

One question that emerges from this advocacy of visual information within a MMA research approach is whether researchers and perhaps more significantly practitioners have the data reduction, display and summarizing skills (Miles & Huberman, 1994) to convert the rich and diverse information gathered by MMA into meaning. Rose (2001) sees the translation from image to meaning to text as key and representation clearly has an important impact on this translation process but equally is not of itself translation. Even if the end result is still reported as text, the employment of principled approaches brings reflective benefit:

> By the end of the process, even if we do ultimately resort to language, [researchers] will have developed a set of responses which may be quite different to what their initial "gut reaction" may have been. (Gauntlett, 2004, p. 2)

So what form might an MMA observation take? Hopkins (1993) argues that there are three clear focuses within MMA: agreed criteria for what is to be collected and observed; observational skills; and feedback. In order to extract the greatest amount of learning from the visual data Hopkins recommends that researchers should ensure that a recording has been watched by the observed and preferably within twenty-four hours of the event so that initial comments from the observed can be collected. This allows for a systematic data recording based on the agreed and applied criteria. Observations (Hopkins, 1993) can be open, structured or systematic or a combination, so long as they meet with the agreed criteria, which establish the recording and access protocols.

> Grace: "I'm not experienced enough to make the most out of the situation while I'm in it right then. [I get a lot out of watching my tapes with you] but I really need your feedback. Because there's tons I would have missed, really, without you. . . . [I'm beginning to] feel more comfortable . . . because each time I read the story I see a little bit more." (Hiebert et al., 2002, p. 5)

Prosser (1998) argues that visual methods need to be open to the same methodological issues as any other source to ensure that researchers guard against interfering with subjects that may distort their behavior; that judgments are made about what is selected and the interpretation which is derived from its use; and that any narrative produced is based upon such a selection. However, it is true that critical and reflexive discussions are rare in much video-based research.

There needs to be an appreciation of context when visual images are used in research. This means that the reporting and analysis of data derived from visual images requires a discussion of the use of images in the research that focuses on the importance of the physical context that produced the images and the social context in which images are viewed as they impact the narrative of research reporting and researchers' reflexivity. Visual recordings allow for the potential impacts of physical settings to be explored as they allow nonverbal data to be observed. While it is difficult to generate a verbal/written account of a complex audiovisual experience on demand, researchers need to make every effort to do so. While this may be an overly positivistic interpretation of visual data as it reinscribes the dataistic (Alvesson & Skoldberg, 2000) credo of grounded theory work, what it does raise for the research community (including editors and publishers) is the urgent need to explore alternative and more appropriate ways of reporting visual data, as explored by Goldman-Segall and her colleagues.

In an analysis of digital ethnographies of children's thinking Goldman-Segall (1990, 1991, 1998)

> . . . explored the tenuous, slippery, and often permeable relations between creator, user, and media artifact through an online environment for video analysis. A video chunk . . . became the representation of a moment in the making of cultures. This video chunk became both cultural object and personal subject, something to turn around and shape. And just as we, as users and creators (readers and writers) of these artifacts, change them through our manipulations, so they change us and our cultural possibilities. (Goldman-Segall & Maxwell, 2003, p. 407)

Goldman-Segall (1995) argued that technologies and video analysis allow the layering of views and perspectives into new theories with configurational validity in the form of thick description by drawing upon perspectivity theory. Using perspectivity theory technologies provide lenses for analysis "that enhance, motivate, and provide opportunities for learning, teaching and research because they address how the personal point of view connects with evolving discourse communities" (Goldman-Segall & Maxwell, 2003, p. 407). Perspectivity theory is a partnership between viewer, author, and media text, "a set of partnerships that revolve around and is revolved around, the constant recognition of cultural connections as core factors in using new-media technologies" (Goldman-Segall & Maxwell, 2003, p. 407). Goldman-Segall (1990) developed Learning Constellations, a computer-based video analysis, drawing on Geertz's (1973) concept of "thick description." Using such technological tools for analysis, according to Goldman-Segall, creates a "partnership of intimacy and immediacy" (1998, p. 33, (see http://mf.media.mit.edu/pubs/detail.php?id=1538)) by supplementing written analyses of data with visual representations available on the web (Goldman-Segall, 1998 (see http://www.pointsofviewing.com)).

Beers and Beers (2001) and Goldman-Segall (2001) have also explored how preservice MFL teachers have created video artefacts as representations of their various cultures.

Where investigators produce the visual record there is a danger of content and practice issues becoming more important than context. Camera location is crucial and we need to be aware of how camera shots are established and reflect upon the social and cultural construction of vision as "common cultural conventions of visuality are something every researcher brings to the analytic work behind the desk" (Sparrman, 2005, p. 252). So analyses of images are not only temporally, spatially, and culturally located they are also subject to "preferred readings" (S. Hall, 1997). In the collaborative ethnographic approach the researcher is actively involved with the researched. Context is central for images are located not only in time, space and physical settings but also in a set of culturally loaded historical conditions. "Or, perhaps most humanistically as well as most interestingly, it may involve working together on a project that simultaneously provides information for the investigator while fulfilling a goal for the subjects" (Gauntlett, 2004, p. 3). In viewing the data generated by visual recording methodologies the researcher needs also to be aware of the possible Hawthorne effect (Mayo, 1987). Familiarity with the recording equipment can alleviate some of these concerns.

Of more concern in most journals is the adequate balance between words and numbers in research reporting. There is a tendency to translate visual data into words and numbers. Images are expensive to use in journal papers. In exploring the extant knowledge in a field of study scholars need to be aware of and know why these blind spots exist in particular disciplines. Although, as we have previously indicated, the presence of and presenting of visual data do not mean that the whole of a picture has been revealed, what they do mean is that less of what was absent remains absent from the picture.

Gauntlett (2004) argues that when using images in research we need to be aware of the internal and external *narrative* of the image. The *internal* narrative is the content (or "story") of the image as it is read by the viewer, rather than necessarily the story that the image maker intended to convey. The *external* narrative is the context which produced the image, and the context in which the image is being viewed. This is closer to the approach of Pink (2001) who is less concerned with the claims of "validity" and more with principles for the generation of ethnographic knowledge which underpin such research, including that with images. She argues that it is impossible to record an image without some interference; what is crucial is reflexivity and a recognition of the context which produces an image and which produces ethnographic knowledge. It is the reflexivity of the researcher—in the case of visual ethnography their awareness of the production and use of the image throughout the research process—which is crucial for Pink (Mason, 2005).

Visual material is important because it represents a source of data that might lead to the learning of different things through different media.

Conclusion

What our analysis of the cases we have referred to here suggests is that significant understanding may fall through the gap between actions and their representations. As researchers we face the dilemma of what is it that can be said about what it is that we see when we see something. What is it that can be said to be true? To what does that which is said to be true refer? What are its referrants? Dataistic approaches to analysis, particularly those which rely on a grounded theory approach often rely on the transcription of data and tend to impose a positivist frame of reference (Alvesson & Skolberg, 2000) that fixes meaning as more or less certain. Transcribing an idea is not the same as the idea, it is part way between the thing (the idea) and the work (the transcription), which is the mimesis of the thing (idea) (Hobson, 2001, p. 147). Transcription is a form of reflexivity, for it is a conscious act on behalf of the researcher, and, as such the transcript is not itself theory neutral. Hobson argues that undecidability is a positive inscription (2001, p. 136). Derrida (1994) points to the unstable distinction between effect/real and the phantasmic or phantasmal. All becomes in/undistinct—model/copy/simulacra, act/its memory, repetition (both representation (mimesis) and representation), true/false appearance. At what point does the representation cease to be imitation and inscribe itself (phantasmically—"as a ghost coming forward" (Kundera, 2002, p. 97)) as something else (real?)—for example tribute bands or forged paintings? Film repeats, remakes, tribute bands all call in to question what is original and what is a simulacre and what is a reminiscence and what is a nostalgic moment (Kundera, 2002). Which is the past, and which is the present? Consider two brief examples, The Bootleg Beatles (a tribute band formed in 1979 (www.bootlegbeatles.com/)) and John Wyatt.

The Bootleg Beatles are renowned for having all the right equipment, playing every tune note and pitch perfect. When Neil Harrison (the John Lennon "character") begins to play *Imagine* on the piano the crowd joins in and Neil/John stops and says "I haven't written that yet" (Holman, 2005). Also John Wyatt, the convicted art forger, now markets his copies as "genuine fakes" and who is micro-chipping his paintings because his copies are being copied by a faker. These examples might be seen as examples of "hyper-reality" (Baudrillard, 1994), but what they do is to cause us to ask, "Are the 'Bootleg Beatles' a copy of the Beatles or are they themselves?" (the past repeated as the present or the present present or the present future (Deleuze, 2004)). At what point does a happening of truth take place? This point applies to all research data, because they have been selected and sorted by human beings who, wittingly or otherwise, are subject to prejudgments and biases. The fact that video recordings of social interactions are available to be interrogated to generate research data that may have a stronger case to be representative of reality does not erase the problem of "truth happenings."

"The 'undecidable' can be pondered as a thematic thread but must also be understood in the relation in which it is proffered, as 'the operation of a certain syntax' . . . In representation the presentation comes back as a double, a copy, and the idea as the picture (or writing) of the thing which is at our disposal . . . Representation and what it had substituted are indissoluble" (Hobson, 2001, p. 139 and 149, speech marks in the original). As Blackmore suggests the repeated, but changed repetition of an idea hotwires into the brain (Blackmore, 2006) for repetition is not possible—"[t]he only thing repeated (is) the impossibility of repetition" (Kierkegaard, 1989, p. 56). What recordings of visual data allow is the repeated watching of the events recorded. What the analysis above suggests, however, is that each time we watch a recorded event we are seeing it anew, because we cannot expunge the impact the previous viewings have had on us to add to the context of the present viewing.

What we are not suggesting here is that video recording (whether using a webcam or other moving image capture techniques) and moving picture data sharing "solves" the problems faced by researchers and practitioners in ESD/EE. Rather, what we are saying here is that it can provide a useful technique within the repertoire of approaches to research. Logocentrism encourages and legitimates the idea that there are fixed truths, things in themselves (Novitz, 1985). In saying this we are mindful of Anscombe's (1958) concept of "brute facts" or what Novitz (1985) refers to as "brute, extra-lingual facts," which are to be distinguished from "mental facts" (Searle, 1995). Derrida (Borradori, 2003) notes, in relation to the attack on the Twin Towers in the United States that came to be known as 9/11, that video is particularly good at capturing the brute facts of an event, but that these facts need to be distinguished from the interpretation of them and from the ascription of meaning to them (the mental facts). As we indicated above, we are not proposing that video expunges ideologically driven choices that researchers consciously or unconsciously draw upon, but that such techniques may illuminate more of the darker elements of the chiaroscuro of the researched landscape of ESD/EE and so redress the balance in relation to an over-emphasis on written textual interpretations of truth that may influence our and others actions.

Our work in this field to date suggests that the collection of visual data does not just improve reflection by augmenting standard approaches to data collection, it has the capacity to reveal different information. This is one of the perceived advantages of digital ethnography. In opening up the epistemic spectrum of another culture, it can lead participants into more reflexive thoughtful observation by the researcher. The visual data are captured using webcam technology because this approach gives the most naturalistic approach to data collection about the context and situation in which learning takes place. These data can illuminate the physical, cultural, and social dynamics of the learning spaces. The physical data allow a lens on the characteristics of the physical space in which learning takes place and can be influenced by class dynamics and interactions, for example classroom features such as color and decoration of walls, use of displays, layout of classroom furniture, and lighting. The links between these features and social interactions in classrooms will rarely be directly observable from webcam-based recording but these links can be investigated or derived from follow-up discussions or interviews as tentative hypotheses for corroboration or falsification. The data also open up the possibilities of exploring the cultural and social dynamics involving the nature of the social interactions between groups of pupils, the body language, and positioning of teachers in classrooms, the differences in the quality and length of interactions between different groups in the class, the extent to which pupils stay on task are all observable. Furthermore, as Hennessy and Delaney (2009, p. 1754) suggest, such approaches respect a variety of voices and can promote co-learning. These data can then be investigated for the light that they throw on the perennial research analysis gaps between knowledge, attitudes and action.

Perhaps one crucial gap that may be filled is the space left by a lack of sufficient attention given to the perspectives of students in classroom settings. As Murthy (2008, p. 839) suggests the larger and more exciting array of methods that researchers draw upon then the more realistic views of interactions in classrooms that are potentially created.

We have drawn upon multimodal analysis (MMA) and dynamic visual data because, in combination, they make complex data accessible by offering new insights into situated, sociocultural learning by considering a spectrum of modes of making meaning rather than a unimodal approach through the study, for example, of transcripts of the spoken. In a multi-vocal conversation—a polylogue—the voices cannot be distinguished from each other, other than typographically (Hobson, 2001, p. 140). Erickson refers to the way in which this form of transcription can overprivilege the representation of speech at the cost of nonverbal activity and lead to the concept of social interaction as a ping-pong match (2006, p. 548). By drawing upon video-based data different voices become clearer. The videos raise issues of agency and participation in pupils' and teachers' choices of different modes during classroom interactions. They focus on the processes of forging meaning rather than outcomes of classroom interaction. Rather than focusing on what pupils are learning the MMA approach focuses on, for example, the meaning of activities to pupils and their engagement with these activities. No one research tool can eradicate all the problems of understanding being lost in translation. What video data and MMA do allow, however, is a more "sensitive translation" (Phillips & Ochs, 2004). They can lead to a transformation in what teachers consider to be evidence of learning with a movement away from a narrow focus on the written or spoken word as the arbiter and assessor of learning toward an approach that captures the richness and diversity of contexts. This does, however, raise the issue of the development of the technical skills of analyzing visual data to which Jacobs et al. (2007) refer.

> Attempting to incorporate problems and methods of data collection, interpretation and representation related to images and visuality, including new questions and new actors, the revisioning of old arguments and imagining of comprehensive and socially relevant projects are what is called for. (Fischmann, 2001, p. 33)

In closing we wish to draw to draw on Hiebert, et al. (2002) in their vision for a future of professional development supported by visual recordings of practice.

> Indeed, a final reason for optimism is that Internet accessible digital libraries of lesson videos with teacher commentary could provide tools and resources needed to address at least two challenges faced by teachers as they transform personal knowledge into a professional knowledge base. One challenge is to envision alternatives to current practice. Earlier we mentioned expertise as one source of new ideas, but easily accessible digital video libraries that contain examples of other teachers teaching similar topics can provide another source. A second challenge for teachers is communicating what they have learned by trying out a particular lesson or teaching approach and coordinating multiple trials of similar lessons across different sites. Again, web-based video libraries can help. Lesson videos provide enough detail that multiple trials can be conducted with each test site enacting the same approach. (Hiebert et al., 2002, p. 9)

What we take from this is that in reflecting on or about practice and reflexively approaching visual data within an MMA approach the mirror provided by others' practice can be a useful tool. As Wren suggests the value of the visual in professional development is considerable:

> [s]imply assessing course participants' views on the value of the training without investigating their subsequent knowledge of useful principles and strategies is insufficient . . . (Course participants) suggest that observation is more valid than self-testimony when assessing the value of workplace learning (Wren, 2003, p. 117 and 116).

In addressing issues relating to research within the scholarship fields of EE/ESD what we have suggested here is that blind spots need to be acknowledged and that visual images need to be given a greater emphasis. As we have indicated, however, portraying the full picture can never be possible, what visual images do is to help the painting by providing more of the pigment.

Acknowledgments

We are grateful for the supportive comments of the anonymous referees and thank them for pointing us in the direction of some thoughts about our work and further references that have helped us to draw the conclusions reached here.

References

Alvesson, M., & Skolberg, K. (2000). *Reflexive methodology,* London, UK: Sage.

Anscombe, G. E. M. (1958). On brute facts. *Analysis, 18,* 69–72.

Armstrong, V., & Curran, S. (2006). Developing a collaborative model of research using digital video. *Computer & Education, 46,* 336–347.

Ball, S. J. (1994). *Education reform: A critical and post-structural approach.* Buckingham, UK: Open University Press.

Banks, M. (1995). *Social research update,* 11 (Winter 1995). University of Surrey, Guilford, CT. Retrieved from http://sru.soc.surrey.ac.uk/SRU11/SRU11.html accessed 24/08/2007

Baudrillard, J. (1994). *Simulacre and simulation* (S. F. Glaser, Trans.). Ann Arbor, MI, University of Michigan Press.

Beers, M. (2001). *Subjects-in interaction version 3.0: An intellectual system for modern language students to appropriate multiliteracies as designers and interpreters of digital media texts* (Unpublished doctoral dissertation). University of British Columbia, British Columbia, Canada.

Beers, M., & Goldman-Segall, R. (2001). *New roles for student teachers becoming experts: Creating, viewing, and critiquing digital video texts.* Paper presented at the American Education Research Association Annual Meeting, Seattle, WA.

Blackmore, S. (2006). *Darwin's meme: Or the origin of culture by means of natural selection.* The British Humanist Association Darwin Day Lecture, London, UCL (http://www.humanism.org.uk/site/cms/ and http://www.susanblackmore.co.uk/

Borradori, G. (2003). *Philosophy in a time of terror: Dialogues between Jurgen Habermas and Jacques Derrida.* Chicago, IL: University of Chicago Press.

Bourdieu, P. (1991). *Language and symbolic power.* Harvard, MA: Harvard University Press.

Daniel-McElroy, S., & Dalton, A. (2002). *Real life* (Broadsheet to accompany the Real Life Film and Video Art Exhibition, Tate St. Ives, UK, 26-10-2002 to 26-01-2003).

Deleuze, G. (2004). *Difference and repetition.* London, UK: Continuum.

Derrida, J. (1994). *Spectres of Marx: The state of the debt, the work of mourning and the new international* (Peggy Kamuf, Trans.). London, UK & New York, NY: Routledge.

Des Jardins, J. R. (1993). *Environmental ethics: An introduction to environmental philosophy.* Belmont, CA: Wadsworth.

Erickson, F. (2006). Definition and analysis of data from videotape: some research procedures and their rationales. In J. L. Green, G. Camilli, P. B. Elmore, A. Skukuaskaite, American Educational Research Association., & E. Grace (Eds.), *Handbook of complementary methods in education research.* New York, NY and Abingdon, UK: Routledge.

Evans, J., & Hall, S. (Eds.). (1999). *Visual culture: The reader*. London, UK: Sage.

Fischman, G. E. (2001). Reflections about images, visual culture, and educational research. *Educational Researcher*, *30*(8), 28–33.

Fiske, J. (2002). Videotech. In N. Mirzoeff (Ed.), *The visual culture reader.* London, UK: Routledge.

Gauntlett, D. (2004). *Using new creative visual research methods to understand the place of popular media in people's lives.* Paper presented at the annual conference of the International Association for Media and Communication Research (IAMCR), Porto Alegre, Brazil, 25–30 July 2004:1–17.

Geertz, C. (1973). *The interpretation of cultures: Selected essays.* New York, NY: Basic Books.

Geertz, C. (2000). *Available light: Anthropological reflections on philosophical topics*. Princeton, NJ: Princeton University Press.

Goldman-Segall, R. (1990). *Learning constellations: A multimedia ethnographic research environment using video technology to explore children's thinking* (Unpublished doctoral dissertation). MIT, Cambridge, MA.

Goldman-Segall, R. (1991). Three children, three styles: A call for opening the curriculum. In I. Harel & S. Papert (Eds.), *Constructionism* (pp. 235–268). Cambridge, MA: MIT Press.

Goldman-Segall, R. (1995). Configurational validity: A proposal for analyzing ethnographic multimedia narratives. *Journal of Educational Multimedia and Hypermedia*, *4*(2/3), 163–182.

Goldman-Segall, R. (1998). *Points of viewing children's thinking: A digital ethnographer's journey.* Mahwahn, NJ: Erlbaum (accompanying video cases retrieved from http://www.pointsofviewing.com).

Goldman-Segall, R., & Maxwell, J. W. (2003). Computers, the internet, and new media for learning. In W. R. Reynolds & G. E. Miller (Eds.), *Handbook of psychology* (Vol. 7) *Educational Psychology* (pp. 393–427). Hoboken, NJ: John Wiley & Sons.

Hall, R. (2000). Video recording as theory. In D. Lesh & A. Kelly (Eds.), *Handbook of research design in mathematics and science education*. Mahweh, NJ: Lawrence Erlbaum.

Hall, S. (1990). Cultural identity and Diaspora. In S. Hall & J. Rutherford (Eds.), *Identity, community, culture, difference.* London, UK: Lawrence & Wishart.

Hall, S. (1997). *Representation: Cultural representations and signifying practices*. London, UK: Sage.

Hennessy, S., & Delaney, R. (2009). "Intermediate theory" building: Integrating multiple teacher and researcher perspectives through in-depth video analysis of pedagogic strategies. *Teachers College Record*, *111*(7), 1753–1795.

Heshusius, L. (1994). Freeing ourselves from objectivity: Managing subjectivity or turning toward a participatory mode of consciousness. *Educational Researcher, 23*(3), 15–22.

Hiebert, J., Gallimore, R., & Stigler, J. W. (2002). A knowledge base for the teaching profession: What would it look like and how can we get one? *Educational Researcher, 31*(5), 3–15.

Hiebert, J., Gallimore, R., & Stigler, J. W. (2003). The new heroes of teaching. *Education Week, 23*(10), 56, 42. http://www.edweek.org/ew/ew_printstory.cfm?slug=10hiebert.h23

Hobson, M. (2001). Derrida and representation: mimesis, presentation, and representation. In T. Cohen (Ed.), *Jacques derrida and the humanities: A critical reader*. Cambridge, NY, Port Melbourne, Madrid and Cape Town: Cambridge University Press.

Hole, R. (2007). Representation, voice, and authority intensified working between languages and cultures: Issues of representation, voice, and authority intensified. *Qualitative Inquiry*, *13*, 696–711.

Holman, S. (Ed.). (2005). *Access all eras: Tribute bands and global pop culture*. London, UK: Open University Press.

Hopkins, D. (1993). *A teacher's guide to classroom research.* Milton Keynes, UK: Open University Press.

Jacobs, J. K., Hollingworth, H., & Givvin, K. B. (2007). Video-based research made "easy": Methodological lessons learned from the TIMSS video studies. *Field Methods, 19*(3), 284–299.

Jensen, B. B. (2002). Knowledge, action, and pro-environmental behavior. *Environmental Education Research, 8*(3), 325–334. http://www.ingentaconnect.com/content/routledg/ceer;jsessionid=7dqbcdgd4i8ca.alice

Kierkegaard, S. (1989). *The sickness unto death: A Christian psychological exposition of edification and awakening by anti-climacus* (Vol. 19, S. Kierkegaard , Ed., Alistair Hannay, Trans.). London, UK: Penguin Books.

Kuit, J. A., Reay, G., & Freeman, R. (2001). Experiences of reflective teaching. *Active Learning in Higher Education, 2,* 128–142.

Kundera, M. (2002). *Ignorance* (L. Asher, Trans.). London, UK: Faber and Faber.

Lovelock, J. (2007). *The revenge of gaia: Why the Earth is fighting back—and how we can still save humanity.* London, UK: Penguin.

Lovelock, J. (2009). *The vanishing face of gaia: A final warning*. London, UK: Penguin.

Mason, P. (2005). Visual data in applied qualitative research: Lessons from experience. *Qualitative Research, 5*(3), 325–346.

Mayo, E. (1987). *Hawthorne and the western electric company*. London, UK: Random House.

Miles, M. B., & Huberman, A. M. (1994). *Qualitative data analysis: An expanded sourcebook*. Thousand Oaks, CA: Sage.

Murthy, D. (2008). Digital ethnography: An examination of the use of new technologies for social research. *Sociology*, *42,* 837–854.

Novitz, D. (1985). Metaphor, Derrida, & Davidson. *The Journal of Aesthetic and Art Criticism, 44*(2), 101–114.

Pea, R. (2006). Video-as-data and digital video manipulation techniques for transforming learning sciences research, education, and other cultural practices. In J. Weiss, J. Nolan, J. Hunsinger & P. Trifonas (Eds.), *The international handbook of virtual learning environments*. Dordrecht, The Netherlands: Springer.

Phillips, D., & Ochs, K. (2004). Researching policy borrowing: Some methodological challenges in comparative education. *British Educational Research Journal*, *30*(6), 773–784.

Pimm, D. (1993). From should to could: Reflections on possibilities for mathematics teacher education. *For the Learning of Mathematics*, *13*(2), 27–32.

Pink, S. (2001). More visualizing, more methodologies: On video, reflexivity and qualitative research. *The Sociological Review*, *49*(4), 586–599.

Plowman, L., & Stephen, P. (2008). The big picture? Video and the representation of interaction. *British Educational Research Journal*, *34*(4), 541–565.

Prosser, J. (1998). *Image-based research: A sourcebook for qualitative researchers*. Oxford, UK: Routledge Falmer.

Robinson, J., Shallcross, T., & McDonnell, P. (2008). Being sensible and social sensibility: Smallwood supporting Somaliland. In T. Shallcross, J. Robinson, P. Pace & A. Wals (Eds.), *Creating sustainable environments in our schools.* Stoke-on-Trent and Sterling, US: Trentham Books.

Rose, G. (2001). *Visual methodologies*. London, UK: Sage.

Searle, J. (1995). *The construction of social reality.* New York, NY: The Free Press.

Shallcross, T. (2008). Whole school approaches, forging links, and closing gaps between knowledge, values and actions. In T. Shallcross, J. Robinson, P. Pace & A. Wals (Eds.), *Creating sustainable environments in our schools*. Stoke on Trent, UK and Sterling, US: Trentham Books.

Shallcross, T., & Robinson, J. (2007, September). *Getting the picture: Using ICT to support teacher reflection.* Paper presented to the British Educational Research Association Annual Conference, London, UK.

Shallcross, T., & Robinson, J. (2008). Sustainability education, whole school approaches, and communities of action. In A. Reid, B. B. Jensen, J. Nikel & V. Simovska (Eds.), *Participation and learning: Perspectives on education and the environment, health and sustainability*. Dordrecht, The Netherlands: Springer.

Sparrman, A. (2005). Video recording as interaction: Participant observation of children's everyday life. *Qualitative Research in Psychology. 2*, 241–255.

Tamoutseli, K., & Pace, P. (2008). Redeveloping a school's physical environment. In T. Shallcross, J. Robinson, P. Pace & A. Wals (Eds.), *Creating sustainable environments in our schools*. Stoke on Trent, UK and Sterling, US: Trentham Books.

Towers, J. (2007). Using video in teacher education. *Canadian Journal of Learning and Technology/La revue canadienne de l'apprentissage et de la technologie, Spring/Printemps, 33(2).* http://www.cjlt.ca/index.php/cjlt/article/view/7/7

Ulewicz, M., & Beatty, A. (2001). *The power of video technology in international comparative research in education.* Washington, DC: National Academy Press.

UNESCO. (2005). *UN decade of education for sustainable development draft strategy*. Paris, France: Author.

Wren, Y. (2003). Using scenarios to evaluate a professional development program for teaching staff. *Child Language Teaching and Therapy, 19,* 115–134.

Zimmerman, M. E. (1994). *Contesting earth's future: Radical ecology and postmodernity*. Berkeley, CA and Los Angeles, CA: University of California Press.

24

Moinho D'Água

Environmental Education, Participation, and Autonomy in Rural Areas

João Luiz de Moraes Hoeffel
FAAT College (NES/FAAT), Brazil; São Francisco University, Brazil

Almerinda B. Fadini
São Francisco University, Brazil

M.K. Machado
FAAT College (NES/FAAT), Brazil

J.C. Reis
São Francisco University, Brazil

F.B. Lima
São Francisco University, Brazil

Introduction

In contributing to international discourse on environmental education research, this chapter presents an environmental education project that used as study area the rural region known as Moinho located in the town of Nazaré Paulista, Sao Paulo, Brazil. It does this to foreground rural areas and participation as significant in environmental education research on participation and learning. Rural areas, participation, and participatory forms of research have been the subject of environmental education research in a number of contexts, including developing country contexts, where large majorities of people continue to depend on natural resources, and their traditional ecological knowledge practices for livelihoods and survival (see, e.g., Mukute, 2009; Pesanayi, 2009).

Like many rural areas in formerly colonized countries, Moinho is suffering diverse social environmental transformations resulting largely from the development of nonparticipatory projects that caused changes to the landscape and disorganization of economic and cultural activities. These have brought about a rural exodus and a loss of local traditions and ways of life. The chapter reports on how, through a process of environmental education participatory research and reflection, the project has encouraged learning and social transformations among the residents of Moinho in relation to their environment. It describes the participatory research methodologies of engagement and reflection applied in this context, shedding light on researcher-community links, cooperative engagements, and community autonomy.

The methodology described in the chapter includes the following phases: (1) *Social Environmental Prediagnosis*: based on the involvement of researchers in the reality of local communities; (2) *Social Educational Intervention*: based on this prediagnosis the project has developed educational actions aimed to awaken the interest of local inhabitants in the social environmental questions; (3) *Social Environmental Diagnosis*: Through the integration of researchers with the local community the researches have acted as facilitators encouraging the participation of the population in process; and (4) *Community Autonomy*: This phase that involved a process of reflection and participatory action had resulted in the creation of The Moinho Residents Association—Agua Viva (Living Water). Environmental education, present in all phases of the process, is reflected through the association and the associate's goal to bring about improvement in the quality of their lives, financial autonomy, and environmental sustainability. The results of this work have demonstrated that the insertion of environmental education in projects of participatory intervention in rural communities located in diverse contexts contributes to the transformation of local reality when the local social actors recognize the importance of their local traditional knowledge and promote their own autonomy.

Popular Participation in the Social Environmental Context

Popular participation can be understood as a way of satisfying essential human needs and it is also an important way of bringing about social inclusion (Alexander, 1998; Gayford, 2003). The principal means of participation in society have been social movements that allow popular participation in decision-making processes. Historically, this participation has shown itself to be more related to the capacity of people to demand the right to participate than the opening allowed by their governments.

The beginning of the 1960s saw the intensification of the worker, student, ecology, and union movements in diverse Latin America countries, including Brazil. This was due to a series of factors among which, as pointed out by Peruzzo (1998), were related to perception of a need for collective action necessary to interrupt the decision-making processes of public power and private industry. Support for this was found in civil society, especially in institutions sensitive to human rights (Castro & Canhedo, 2005; Peruzzo, 1998).

The growth of popular movements provoked various reactions among the privileged classes and among the still powerful oligarchies. As expressed by Cardoso (1994), these movements contributed to a change in the political culture. This was a result of the autonomy of these movements, breaking a relationship of dependence and changing the role of the traditional political system.

Faced with this reality, politicians, public institutions, and other institutions of civil society began to use the term *participation* representing a range of differing interests, and manifesting in a range of forms of participation—from mere assistance to full active participation (Peruzzo, 1998). In this sense it is essential to question the possibilities presented to the population, in other words, to question if in fact there is a real conquest made or only a game played out by political and economic interests and if participation only truly exists when it is fought for and won (Demo, 1996).

This entire historical context reveals that society is always changing and that because of its pluralistic characteristics it is in constant movement. This dynamic situation stems from its contradictions and diversity and the conflict between opposing forces that sometimes move in harmony and at other times enter into shock, and through this new social realities may be initiated (Peruzzo, 1998).

The historical context also allows verification of the current ways and means of participation that truly permit the practice of citizenship and identifies the best ways of involving all social actors in a process based on environmental issues. This question (involving social actors based on environmental issues), according to Sachs (2000), already includes all of the interactions between nature and human society—inclusive of the ecological, social, economic, and political realms. As already stated, various segments of civil society together with the state are seeking to foster social inclusion and today environmental participatory planning is especially prevalent in legally protected conservation sites, providing good sites for examining such participatory practices.

In Brazil involvement of the population in environmental planning is encouraged by the state in Conservation Areas where human presence is permitted and sustainable usage of natural resources is allowed. This is the case in Environmental Protected Areas (EPA). This has been due to the creation of another category of environmental protected areas known as Integrally Protected Conservation Areas where human habitation is prohibited.

This systematization of protected areas created by the Conservation Areas National System—CANS (Brasil, 2000, 2001), has, however, been criticized by certain authors such as Diegues (1996), Cabral and Souza (2002), and Ferreira (2004). The reason for this critique is linked to the fact that CANS was elaborated based on the North American model, especially the Integrally Protected Conservation Areas, without taking into consideration the fact that the majority of Brazilian forests are already inhabited by traditional societies and that the country already faces serious social, political, and economic problems (Diegues, 1996).

Although the CANS has been greatly criticized, the category EPA possesses certain important peculiarities, especially with regards to participation. According to Cabral and Souza (2002), because this type of conservation area is defined as an area allowing sustainable development and sustainable use of natural resources, it is important from the social economic point of view in that human activities must be carried out responsibly allowing the continuity of environmental quality and integrity of the conservation area on a local and regional dimension.

It is understood that participation of the local community and the various social actors especially in the area of environmental conservation is extremely relevant. Authors like Diegues (1996) and Arruda (1997), have presented studies showing that part of the loss in biodiversity in natural areas is directly related to the restrictions imposed on the activities of the local community. Pimbert and Pretty (2000, p. 196) point out that "for these reasons the efforts of conservation must identify and promote social processes that permit local communities to conserve and even increase biodiversity as part of their way of life."

According to Pimbert and Pretty (2000), this vision of popular participation in conservation is recent. They argue that until the 1970s participation was seen as a way of achieving voluntary submission of the population to conservation plans, that in the 1980s participation was defined as an interest in the protection of natural resources and that at the beginning of the 1990s participation became a means of involving the population in the management of protected areas.

A question to be raised is, however, whether community participation in the planning of environmental protected areas implies changing participation patterns from consultative participatory practices (with limited transfer of power), to interactive participatory practices that give greater access to power. The challenges associated with this shift in participatory practice creates new roles for conservation professionals and other outsiders (Pimbert & Pretty, 2000).

The involvement of communities in decision making, responsibilities, and autonomy is a task that demands time, especially when it involves an already existing environmental protected area. This is because the local population often has little knowledge of their role in environmental conservation and improvement to the quality of life, especially when this reality poses great restrictions in the area where they live, restrictions that have been created, implemented, and imposed from outside the community.

The process of the creation of the conservation areas—the Piracicaba and Juqueri-Mirim River Basin Area II (Piracicaba Environmental Protected Area) and the Cantareira System (Cantareira Environmental Protected Area) began within the historical context of nonparticipation and Moinho, the area of study presented in this chapter and located in the municipality of Nazare Paulista/SP, lies within these conservation areas.

According to Fadini and Carvalho (2004) the sole creation of these conservation areas is not enough to guarantee social and environmental integrity. These areas are still not regulated and there has been no effective participation of the community in a series of decision-making processes, factors that have damaged the ability to involve the local population in environmental conservation.

The Historical Process of Social Environmental Transformation in Moinho

The area of study discussed in this chapter presents a unique environmental problem because it is located in an area vital to water conservation. It is home to the sources of rivers and important watershed areas. Despite its proximity to the Metropolitan Area of the city of Sao Paulo important remnants of the Atlantic Forest are found.

In the state of São Paulo this question determined the creation of the environmental protected areas of-the Piracicaba and Juqueri-Mirim River Basins—Área II (EPA Piracicaba) and the Cantareira Reservoir System Environmental Protected Area (EPA Cantareira). The protection of the many river sources found here and their economic importance provided the rationales for defining these areas as environmental protected areas (Brasil, 2001; São Paulo, 2000, 2001).

The construction of the Canteira Reservoir System occurred in the 1960s when Brazil was going through intense population redistribution caused by a combination of intense demographic growth and the modernization of production that was aided by investments in infrastructure improvements, especially transportation and communication infrastructure. This migratory movement resulted in a rural exodus toward the agricultural frontier and an ongoing process of urbanization (Taschner, 1998). The urbanization of the state of São Paulo and the rapid demographic growth provoked by industrialization created the need to search for water sources protected and far form urban centers (Whately & Cunha, 2007).

In order to supply a growing demand and resolve the water problem in the metropolis of São Paulo the building of the Cantareira Reservoir Water System took place. This system is located north of the metropolis of São Paulo over a great area of the Cantareira Mountain Range and the technology used to create it is considered a historical step in the creation of public hydraulic infrastructure in the region and in Brazil. The Cantareira Reservoir Water System is made up of a group of large scale dams planned specifically to attend the water demands of the metropolis and the areas adjacent to the reservoirs. This system is known as the Cantareira System and made up of the Jaguari-Jacarei River Reservoirs, the Juquery River Reservoirs (Paiva Castro), the Cachoeira River Reservoir, and the Rio Atibainha Reservoir (São Paulo, 1989).

The construction of the Cantareira System along with the creation of the Cantareira and Piracicaba-Juqueri-Mirim Area II Environmental Protected Areas imposed significant economic restrictions on local communities in relation to land use and agriculture. Within this context the area of study presented in this chapter is marked by intense social environmental transformations during diverse historical periods, especially during the 1960s with construction of the Rio Atibainha Reservoir and the creation in the 1990s of the referred Environmental Protected Areas already mentioned and the expansion of the Dom Pedro I and Fernão Dias regional highways.

The result of these undertakings have been an altered landscape, the emergence of environmental conflicts that revolve around the question of water and environmental legislation and the disorganization of economic activities that has significantly stimulated a process of rural exodus and the disfiguration of the traditional way of life. The economic difficulties suffered by the regional population represent a possible threat to conservation of natural areas and water resources. At present many landowners are selling their properties to real estate speculation ventures and weekend tourists. This has resulted in a considerable increase in damage to the environment (Hoeffel, Machado, & Fadini, 2005).

Within this context is the rural settlement of Moinho located along the margins of the Rio Atibainha Reservoir. Moinho has approximately 450 habitants and originated along Moinho Creek, an affluent of the Atibainha River. Moinho has social characteristics reminiscent of a rustic culture known in Portuguese as *caipira* (Fadini & Carvalho, 2004). The settlement presents a social structure marked by collectivity. This was very evident in the organization of labor which was carried out communally through the coming together of the inhabitants living in close proximity to help each other with the strenuous tasks of planting, harvesting, and building. After work was completed the owner of the house organized a feast with music and handcrafted rum (*pinga*) turning work into an opportunity for festive pleasure (Machado et al., 2007).

The study area shows a breakdown in society marked by authoritarian economic development that greatly limits community involvement in collective actions that relate to social environmental conservation (Fadini & Carvalho, 2004; Lima et al., 2003). Since 2002 the authors have been working with this reality in a project entitled "Moinho D'Água (Water Wheel)." The Moinho D'Água project has aimed to encourage the organization and involvement of the inhabitants of Moinho in a participatory process and reflection based on the assumptions of environmental education.

Moinho D'Água: Environmental Education, Participation, and Community Autonomy

The methodology used in development of the project with the community of Moinho was elaborated taking into consideration the social environmental characteristics of the local and how to create ways that lead to ownership of the educational process by the community in order that the community itself is able to transform a reality that up till now has been the result of a historical process of nonparticipation.

The methodology is made up of the following phases (described in more detail below): *Social Environmental Prediagnosis, Social Educational Intervention, Participatory Social Environmental Diagnosis, and Community Autonomy.*

Social Environmental Prediagnosis During this phase of the project the researchers elaborated a prediagnosis by involving themselves with the local reality and with the community. Field work was carried that involved participation in religious festivals, informal conversations, and interviews with the inhabitants, interpretative walks and bibliographical elaboration. The methodology of participatory research and observation was based on the work of Robottom and Sauvé (2003), Gayford (2003), Thiollent (2003), Brandão (1990), Richardson (1999), Laville and Dionne (1999), and Gil (1999).

Thiollent (2003) and Borda (1990) have classified participatory research as a project where there is actual participation of the poorest classes in contemporary social structures and takes their aspirations and potential learning capacity and ability to act into consideration. For Brandão (1990) it is important that the community be actively involved in research. For Richardson (1999) in participatory observation the researcher is not only a spectator of facts that are being studied, he/she places him/herself in the position and on the level of the other human elements that make up the phenomena that are being observed.

It is important to emphasize that the prediagnostic phase not only provides knowledge about the local reality, but it helps the researcher to gain greater proximity to the community facilitating the development of the next phase.

Social Educational Intervention Based on the prediagnosis, educational actions were developed to awaken interest and involvement of the inhabitants in the process or research action. In the Moinho D'Água project the educational actions were carried out with students from Moinho's two rural schools (Rural Moinho School and Rural Caraça School). These actions sought a learning and reflection context through play in ways that took the historical, social, cultural economic and natural aspects of the local into consideration.

Educational actions were divided up into eight assemblies or meetings per school, each one lasted for four hours. The meetings utilized methodology derived from the work of Joseph Cornell (1996, 1997, 2005) on sequential learning. Sequential learning is divided up into four phases: the awakening of enthusiasm, concentration of attention, direction of experience and sharing of inspiration. From this methodology the Environmental Education Activities Group (EEAG) was elaborated. This group elaborated inspirational themes related to the social environmental reality identified in the prediagnostic phase which included among others water resources and *caipira* culture. In this way each encounter approached an inspirational theme that was developed into four basic activities.

In the social educational intervention phase approximately fifty children from the two participating schools were involved. It is interesting to observe that during the carrying out of the activities many parents demonstrated interest and curiosity in knowing what was happening at the schools. In this way, the proximity of the researches to the community was refined.

Four months after this phase ended an evaluation was undertaken with the student's parents. This interval of time was important in order to verify if the subjects taken from the themes were learned by the children and shared with their families; and to see if they resulted in a multiplying effect. This evaluation took place through interviews conducted in the homes of the inhabitants. The main result from this was insight into the interest the adult population showed in participating in a process similar to the one the children were involved in at the schools. Based on these results the inhabitants were invited to participate in the participatory social environmental diagnosis which was the next phase of the methodology elaborated for the Moinho D'Água project.

Participatory Social Environmental Diagnosis After the integration of the researchers into the community in the previous phases; the social environmental prediagnosis and the social-educational intervention; a participatory social environmental diagnosis was carried out with the researchers acting as the facilitators of the process and encouraging the effective participation of all those involved.

In the Moinho D'Água project this phase took place in ten 2-hour encounters with the community that took place once a month at the Caraça Rural School.

The methodology used encouraged the participation of those involved through a range of related and complementary research methods including:

- **Informal conversation** with the inhabitants using nonstructured interview techniques proposed by Laville and Dionne (1999), where the researcher supported by one or various themes preserves the spontaneity and personal character of the interactions and the answers obtained from the participants.
- **Focal Groups** based on the work of Iervolino and Pelicioni (2001), Laville and Dionne (1999), and Gaskell (2002), where the principal interest is to recreate a context or social environment where participants can interact, defend their opinions and contest each other. The debate is open and accessible to all and the subjects in questions are of common interest; the different status among the participants is not a consideration.
- A **Dynamic "Wailing Wall" strategy** where the participants construct a wall that represents the problems where they live. They also construct a dynamic "Tree of Hope" with dreams they have for the place where they live (Ecoar, 1999).
- **Interpretive Walks** based on the works of Lima et al. (2003) and Hoeffel et al. (2003) with the purpose of verifying the environmental perception of the participants of the social environmental reality of Moinho.
- **Debate Week—Planting Citizenship**. This strategy seeks to organize a week of debates that are concerned with guaranteeing the community an opportunity to know, reflect, and question the proposals made by each candidate for mayor of their city of Nazaré Paulista. This provides the inhabitants a more participatory alternative to the actual political activities that happen during political campaigns where there is no encouragement to effectively participate in the community.

The results obtained from these participatory methods have shown that: (1) for the inhabitants no community work exists and the reason for its disappearance is the fact that people are working increasingly farther away and as a result don't have the close relationships they had in the past; (2) The inhabitants perceive that population has grown significantly; (3) The volume and quality of water in the creeks and streams has gradually decreased; (4) Numerous improvements to infrastructure have been brought to Moinho, including asphalt, power lines, and a bus system; (5) There has been an increase in violence, robberies, and drug consumption among youth; (6) There has been an increase in the number of tourists, visitors, weekend homes, and condominiums; (7) The community has also seen an increase in the amount of garbage and trash thrown on the streets and of sewers flowing into the streams and creeks; (8) The inhabitants of Moinho practically carry out no subsistence activities, in other words agriculture or the confection of objects for day-to-day use; (9) The community was somewhat divided in its opinions about the benefits brought by the increase in tourism as they recognized the improvements to infrastructure they had received; on the other hand they affirm that there has been an increase in the lack of security which is due to an increase in the number of people coming from the outside to visit the community; (10) The activity Wailing Wall demonstrated the inhabitants principal complaints. These were the degradation of nature, the end of community social relations, an increase in the use of drugs and in violence; (11) The Tree of Hope demonstrated that the most important dreams of the inhabitants were improvements to infrastructure, a return of solidarity among them, a need to build factories to bring employment, the desire to increase the amount of people going to church and to resolve the problems of violence; (12) During the Interpretive Walk the inhabitants noted that agricultural activities like sugar cane and eucalyptus disfigured the natural landscape, and decreased the volume of water. In relation to cultural aspects the community activities developed in the farm manors and farm chapels were greatly missed by the inhabitants and there was sadness that these manifestations had much diminished in their community. The inhabitants also showed much knowledge of local plants for medicinal and mystical purposes and plants used as raw material for the construction of domestic tools; (13) During the week of debates Planting Citizenship, the inhabitants showed their main concerns in relation to improvements to infrastructure in the rural communities. It is interesting to observe that participants organized themselves so as to be able to question all of the political candidates about the same concerns in order to be able to compare the proposals the candidates presented and from this analysis choose the best candidate.

From these activities diverse aspects were discovered that supported the data that had been gathered in the social environmental prediagnosis that showed that participatory methodology based on the assumptions of environmental education can guarantee a vision closer to the reality of communities involved in research projects.

Community Autonomy The last phase of the methodology proposed in the *Moinho D'Água* project is more complex and seeks to create a foundation on which the community can act with autonomy to resolve its problems and consolidate its dreams revealed in the other phases of the project. In the phase Community Autonomy, the community exercises practice and seeks to carry out participatory actions that bring improvement to the problems they raised and value the potentials that were identified.

In the case of Moinho and based on the results in previous phases the inhabitants were invited to think about what the principal cause of the problems they identified was and how they could begin a process of resolving or minimizing these problems. With the aim of encouraging and facilitating this reflection a workshop was created entitled the "Workshop of the Future" which was carried out over ten meetings and was based on the work of Agenda 21

(Rio de Janeiro, 1996), and on insights from the Ecoar Institute for Citizenship (Ecoar, 1999) and Matthaus (2001). These studies described this technique as a tool that stimulated systemic analysis and sought unconventional solutions. For Matthäus (2001), in "Workshop of the Future," participants from the various social classes and/or of different professional backgrounds discussed themes and problems in a democratic space that permitted expression of fundamental opinions, technical arguments, and scientific studies. Local traditional wisdom was welcome.

This workshop was divided into three moments: *The Critical Phase*: where all the dissatisfaction, problems, and criticism were allowed expression; *The Utopic Phase*: or the dream phase, where creative solutions were sought and recognized as "Utopias" or "dreams." These Utopias didn't need to be related to reality. The group should free themselves in an attempt to set aside conventional roads and conditioning that limits reality (Matthäus, 2001, p. 127); *The Realization Phase*: This phase can be described as a return to reality. In this phase there is an attempt to define actions that can improve problems that can be carried out by those present. In the realization phase it is possible to arrive at a plan of action (Matthäus, 2001).

The "Workshop of the Future" carried out with the community showed worry in relation to lack of security, lack of investment in infrastructure and in sports in the rural area, alcoholism, and a lack of respect for nature. The inhabitants identified the principal causes of these problems as being a lack of solidarity and information in the community. From this analysis the participants concluded that before any action related to resolving or minimizing problems they had identified could be taken they first needed to bring more people to participate in this process. Within this perspective some alternatives to bring together the inhabitants of their community were proposed. These alternatives included workshops, talks, seminars, and exhibitions.

Over a two-year period, the inhabitants planned and organized these actions in the community. This resulted in a greater sense of closeness among community inhabitants. Some of the meetings held involved more than sixty people. This process had the direct help of the researchers involved in the *Moinho D'Água* project who interfered or guided the discussions less and less in order for the community to move with true autonomy. After some time it became clear that certain inhabitants always took the lead and were more intensely involved. This characteristic and the activities lead to the formation of the Moinho Community Association "Água Viva (Living Water)" and as a result the community began to decide on its own how to solve the problems where they lived.

It is important to mention, however, that during the entire process of forming the association as well as its activities researchers gave educational and facilitation support. This process continues, but with less intervention and intensity. It is important to keep in mind that community autonomy is a long process that needs support in order to achieve good results.

Final Considerations

The work carried out in the Moinho D'Água project was based on two fundamental environmental education assumptions defined by Philippi Jr. and Pelicioni (2002) as a teaching learning process to achieve the exercise of citizenship, political and social responsibility that results human beings building new values in relation to nature and that improve the quality of life for all living creatures.

The process of this study has shown that in order for the objectives of environmental education to be fully reached participation is essential. In this manner the methodology adopted in this work and its four phases: *Social Environmental Prediagnosis, Social—Educational Intervention, Participatory Social Environmental Diagnosis, and Community Autonomy,* ensured that the essential social actors were involved in their own change instead of being the objects of change (Pilon, 2005).

During the process much difficulty was encountered, an example of this was the lack of adequate infrastructure in order to carry out the activities, many times the participants didn't carry out the activities they had committed themselves to. Researchers also had difficulty in maintaining a truly participatory proposal and were challenged not to take the lead in the decision-making process. There was also a decrease in the number of participants—at the beginning of the activities proposed by the community there were an average of forty participants, which was later reduced to an average of fifteen when the formation of the community association began to take place. Another issue was experiences of discouragement at not obtaining quick results, as the process did not offer any "quick fixes."

It is important to realize, however, that difficulties are part of a collective building process that results in much learning. Taking these questions into consideration supports a true educational process that according to Freire (1979) can't be built on an understanding of human beings as empty vessels to be filled with content.

It can also be observed that the methodology adopted by the *Moinho D'Água* project collaborates for those involved to become part of the process, taking upon themselves the responsibility of transforming their reality and becoming effective participants in building community autonomy.

Thus, the research presented here, although describing a project regionally located, once again emphasizes the role of environmental education in solving social and environmental issues. Another point to highlight is that the adopted methodology can be adapted and used in different contexts and realities that have similar situations and emphasizes the importance of effective participatory processes and their relevance and implications for the future development of environmental education. It is encouraging to notice that other communities of practice concerned with issues and community autonomy in rural areas are experimenting with similar types of participatory methodology in environmental education (e.g., Mukute, 2009). Further work and dialog on these participatory processes across continents can enrich understandings of the processes involved.

References

Alexander, J. C. (1998). Ação coletiva, cultura e Sociedade. *Revista Brasileira de Ciências Sociais, 13,* 3–51.

Arruda, R. V. (1997). "População Tradicional" e a proteção dos recursos naturais em Unidades de Conservação. In RNPUCs (Ed.), *Congresso Brasileiro de Unidades de Conservação* (pp. 262–275). Curitiba: RNPUCs.

Borda, O. F. (1990). Aspectos teóricos da pesquisa participante: considerações sobre o significado e o papel da ciência popular. In C. R. Brandão (Ed.), *Pesquisa Participante* (pp. 42–62). São Paulo: Brasiliense.

Brandão, C. R. (1990). Pesquisar participar. In C. R. Brandão (Ed.), *Pesquisa participante* (pp. 9–16). São Paulo: Brasiliense.

Brasil (2000). *Lei nº 9.985, de 18 de julho de 2000. Instituiu o Sistema Nacional de Unidades de Conservação da Natureza.* Brasília: MMA.

Brasil (2001). *Roteiro metodológico para gestão de Área de Proteção Ambiental.* Brasília: IBAMA/MMA.

Cabral, N. R. A. J., & Souza, M. P. (2002). *Área de Proteção Ambiental: planejamento e gestão de paisagens protegidas.* São Carlos: RiMa.

Cardoso, R. C. L. (1994). A trajetória dos movimentos sociais. In E. Dagnino (Ed.), *Anos 90: política e sociedade no Brasil* (pp. 81–90). São Paulo: Brasiliense.

Castro, M. L., & Canhedo Jr., S. G. (2005). Educação Ambiental como instrumento de participação. In A. Philippi Jr & M. C. F. Pelicioni (Ed.), *Educação Ambiental e Sustentabilidade* (pp. 401–411). Barueri: Manole.

Cornell, J. (1996). *Brincar e aprender com a natureza: um guia sobre a natureza para pais e professores.* São Paulo: Companhia Melhoramentos.

Cornell, J. (1997). *A alegria de aprender com a natureza: atividades na natureza para todas as idades.* São Paulo: Companhia Melhoramentos.

Cornell, J. (2005). *Vivências na a natureza: guia de atividades pais e educadores.* São Paulo: Aquariana.

Demo, P. (1996). *Participação é conquista.* São Paulo: Cortez.

Diegues, A. C. (1996). *O mito moderno da natureza intocada.* São Paulo: Hucitec.

Ecoar—Instituto Ecoar para a Cidadania. (1999). *Desafio das Águas—Agenda 21 Do Pedaço.* São Paulo: Instituto Ecoar Para A Cidadania.

Fadini, A. A. B., & Carvalho, P. F. (2004). Os usos das águas do Moinho – Um estudo na Bacia Hidrográfica do Ribeirão do Moinho—Nazaré Paulista/SP. In ANPPAS (Ed.), *Anais do II Encontro da ANPPAS.* Indaiatuba: ANPPAS.

Ferreira, L. C. (2004). Dimensões Humanas da Biodiversidade: mudanças sociais e conflitos em torno de áreas protegidas no Vale do Ribeira, SP, Brasil. *Ambiente e Sociedade, 7,* 47–66.

Freire, P. (1979). *Pedagogia da Autonomia: saberes necessários à prática educativa.* São Paulo: Paz e Terra.

Gaskell, G. (2002). Entrevistas individuais e grupais. In M. W. Bauer & G. Gaskell (Ed.), *Pesquisa qualitativa com texto: Imagem e som: um manual prático* (pp. 64–89). Petrópolis: Vozes.

Gayford, C. (2003). Participatory methods and reflexive practice applied to research in education for sustainability. *Canadian Journal for Environmental Education, 8,* 129–142.

Gil, A. C. (1999). *Métodos e técnicas de pesquisa social.* São Paulo: Atlas.

Hoeffel, J. L., et al (2003). Interpretative walks and environmental education. In *Proceedings 32º Annual Conference North American for Environmental Education.* RokSpring: NAAE.

Hoeffel, J. L., Machado, M. K., & Fadini, A. B. (2005). Múltiplos Olhares, Usos Conflitantes, Concepções Ambientais e Turismo na APA do Sistema Cantareira. *OLAM—Ciência e Tecnologia, 5,* 119–145.

Iervolino, A. S., & Pelicioni, M. C. F. (2001). A utilização do grupo focal como metodologia qualitativa na promoção da saúde. *Revista da Escola de Enfermagem da USP, 35,* 115–121.

Laville, C., & Dionne, J. (1999). *A construção do saber: manual de metodologia da pesquisa em ciências sociais.* Porto Alegre: UFMG.

Lima, F. B., et al. (2003). Caminhos do Moinho—Processos Históricos e Educação Ambiental. Um Estudo no Bairro do Moinho, Nazaré Paulista, São Paulo. In *Anais da 55ª Reunião Anual da SBPC.* Recife: SBPC.

Machado, M. K., et al. (2007). Educação Ambiental em comunidade rural. Um estudo no Bairro Rural do Moinho, Nazaré Paulista/SP. In M. C. F. Pelicioni & A. Philippi Jr (Ed.), *Educação Ambiental em Diferentes Espaços* (pp. 139–154). São Paulo: Signus.

Matthäus H. (2001). Oficina do Futuro como metodologia de planejamento e avaliação de projetos de desenvolvimento local. In M. Brose, (Ed.), *Metodologia participativa: uma introdução a 29 instrumentos* (pp. 121–130). Porto Alegre: Tomo Editorial Ltda.

Mukute, M. (2009). Cultural historical activity theory, expansive learning, and agency in permaculture workplaces. *Southern African Journal of Environmental Education, 26,* 150–166.

Peruzzo, C. M. K. (1998). *Comunicação nos movimentos populares: a participação na construção da cidadania.* Petrópolis: Vozes.

Pesanayi, T. (2009). Sigtuna think piece 6. A case of exploring learning interactions in rural farming communities of Practice in Manicaland, Zimbabwe. *Southern African Journal of Environmental Education, 26,* 64–73.

Philippi, Jr., A., & Pelicioni, M. C. F. (2002). Alguns Pressupostos da Educação Ambiental. In A. Philippi, Jr & M. C. F. Pelicioni (Ed.), *Educação Ambiental: desenvolvimento de cursos e projetos* (pp. 3–5). São Paulo: Signus.

Pilon, A. F. (2005). A ocupação existencial no mundo, um proposta ecossistêmica. In A. Philippi, Jr & M. C. F. Pelicioni (Ed.), *Educação Ambiental e Sustentabilidade* (pp. 305–352). Barureri: Manole.

Pimbert, M. P., & Pretty, J. N. (2000). Parques, comunidades e profissionais: incluindo "participação" no manejo de áreas protegidas. In A. C. Diegues (Ed.), *Etnoconservação: novos rumos para a proteção da natureza nos trópicos* (pp. 43–65). São Paulo: AnnaBlume.

Richardson, R. J. (1999). *Pesquisa social: métodos e técnicas.* São Paulo: Atlas.

Rio de Janeiro. (1996). *Agenda 21 Local, Construindo Nosso Futuro Comum—Guia do Cidadão.* Rio de Janeiro: ISER.

Robottom, I., & Sauvé, L. (2003). Reflecting on participatory research in environmental education: Some issues for methodology. *Canadian Journal for Environmental Education, 8,* 111–128.

Sachs, W. (2000). Sustainable development. In M. Redclift & G. Woodgate (Ed.), *The international handbook of environmental sociology* (pp. 71–82). Cheltenham, UK: Edward Elgar.

São Paulo. (1989). *Sistema Cantareira.* São Paulo: Sabesp.

São Paulo. (2000). *Atlas das Unidades de Conservação Ambiental do Estado de São Paulo.* São Paulo: SMA.

São Paulo. (2001). *APAs—Áreas de Proteção Ambiental Estaduais: proteção e desenvolvimento em São Paulo.* São Paulo: SMA.

Taschner, S. P. (1998). *São Paulo: moradia da pobreza e o redesenho da cidade.* São Paulo: FAUUSP.

Thiollent, M. (2003). *Metodologia da Pesquisa-Ação.* São Paulo: Cortez.

Whately, M., & Cunha, P. (2007). *Cantareira 2006: Um olhar sobre o maior manancial de água da Região Metropolitana de São Paulo.* São Paulo: ISA.

Section V

Research on Learning Processes in Environmental Education

Justin Dillon
King's College London, UK

Joe E. Heimlich
Ohio State University, USA

Elin Kelsey
Royal Roads University, Canada; James Cook University, Australia

Introduction

Bridging the Self Across Relational Inquiry to Action with Culture in Mind The context for this section is well elaborated by Cecilia Lundholm, Nick Hopwood, and Mark Rickinson in the introduction to their chapter. Rickinson's (2001) review of environmental education research from 1993 to 1999 had identified that much of the research prior to 2000 had focused on knowledge, attitudes, and behavior. Writing in response to Rickinson's review, one of us (JD), argued that whereas science education research frequently focused on learning and on developing models of learning, that was not the case for research in environmental education. Rickinson later noted that in the cases where learning had been studied it had tended to be in relation to educational interventions rather than as a process in its own right (Rickinson, 2006).

There are several possible explanations for this state of affairs. Many environmental education researchers in the early years of the field had backgrounds in environmental science. They were scientists rather than social scientists and, as such, may have been more comfortable with an instrumentalist view of learning. That is, if you could only find the right bit of knowledge to change someone's attitudes, then their behaviors would change. Another related reason, is the dominance of psychology in education departments in US universities for many years. The type of psychology that was dominant in the 1970s and 1980s was much narrower than that prevalent today. The rise of sociocultural theories of learning in recent years has edged out behaviorist and purely cognitivist views which, while still offering insight into how people learn, are only part of the story.

Another reason might be because environmentalists have tended to focus on the relationships between the environment and human beings rather than on how people make sense of their surroundings. That is, there has been a focus on the experience of engagement rather then on the processes of learning. As Philip Payne wrote near the turn of the twenty-first century there appeared to be "a lack of consideration in environmental education theory and research practices about the children who are the subjects of environmental education" (1998, p. 20). Again, the balance has now shifted somewhat.

For many years, a false dichotomy was promoted with school education being seen as "bad" and out-of-school education being seen as "good." The result was that what we knew about learning in classrooms was not seen to be relevant to what was learned beyond the classroom walls. Fortunately we have grown to realize that learning is learning and that the terms "formal" and "informal" have increasingly limited utility. That is not to say that formal and informal should not be used; rather, if they are used it is better to talk about learning in formal or informal contexts, situations, or environments rather than to refer to informal or formal learning.

In the first chapter in this section, Lundholm, Hopwood, and Rickinson describe studies that focus on the ways in which learners make sense of their environmental education in formal settings. The context for the chapter is studies carried out in high school and university contexts in England and Sweden (Rickinson, Lundholm, & Hopwood, 2009). The studies point to three key issues which together result in complexity in the student learning experience: dealing with emotions and values; questioning relevance; and negotiating viewpoints among students and teachers.

Lundholm et al. note that in recent years studies have tended to see learners as "active agents, rather than passive objects, in the learning situation." Qualitative studies of learning are far more common now than they were, say, in the 1980s. Indeed, Lundholm, Hopwood, and Rickinson all use qualitative methods in the work that they draw on in their chapter. Lundholm's research focuses on Swedish university students' learning about environmental issues as part of undergraduate and graduate programs; Hopwood followed the experiences of two students (aged 13–14) in geography lessons for three months per class; and Rickinson examined how three teachers from UK secondary schools and twelve of their students (aged 13–15) dealt with controversial environmental issues. What is particularly interesting about the chapter is that the three authors have spent a considerable time comparing and further analyzing their collective data.

Lundholm et al. examined environmental learning as a process. They did this by addressing a series of questions including "how do students view particular aspects of subject matter?" and "How do they feel about that subject matter, particular tasks, and their learning more generally?" Their findings, as was mentioned above, fall into three themes. What they have in common is an awareness and a sensitivity to the challenges that learners face in the classroom. The content of some lessons about environmental topics can be controversial and this may provoke strong emotional reactions from learners—something that is not commonly acknowledged or investigated in formal education.

The personal response to environmental education is just that, personal. Lundholm et al. identify the individual responses to learning and the range and diversity of interpretations that can be made of environmental topics. The implications of their study is that learning is a very personal cognitive and emotional process and that research, so far, has only begun to identify what learning is and what it feels like. They leave us with a challenging and somewhat disconcerting question:

> Are [. . .] students' difficulties due to their views of knowledge, so-called "epistemological beliefs" [. . .] Or, are contradicting and opposing views challenging to engage with because they become emotionally difficult . . .?

And their challenge for environmental educators is equally incisive: "Can teaching create an atmosphere where different views are shared and discussed comfortably despite power imbalances between students and their teachers?"

In the second chapter of the section, Arjen Wals and Justin Dillon ask the question, "What do theories of learning offer to an understanding of environmental education research and to those who use the research?" Staring from a similar point to Lundholm et al., that is, the lack of research into learning in environmental education, Wals and Dillon first examine what learning theories tell us about effective practice. Stella Vosniadou's (2001) summary of how people learn chimes with the messages from Ludholm et al., particularly when she states that: "People learn best when their individual differences are taken into consideration." Though she does consider motivation, Vosniadou does not spend much time on the emotional dimension of teaching *per se*.

After looking at behaviorist and cognitivist theories of learning and discussing motivation in more depth, Wals and Dillon choose as their fundamental question:

> What are or should we be changing or developing in learners? Or, alternatively, how can we create optimal conditions and support mechanisms which allow citizens, young and old, to develop themselves in the face of change?

As Wals and Dillon note, the first question has instrumental connotations, whereas the second one has emancipatory ones. They describe how environmental educators often focused on influencing people's environmental behaviors in predetermined ways. This strategy, they argue is inappropriate because it contradicts "the very foundation of education and borders on indoctrination." Even more problematic is that there is no certainty about what counts as the "best" behavior from an environmental or sustainability point of view. Drawing on the work of Ulrich Beck, Wals, and Dillon note that globalization and individualization have led to increased insecurity and unpredictability and, as a result, new forms of education are required, particularly education that is emancipatory.

Such forms of learning include transdisciplinary learning, transformative learning, crossboundary learning, anticipatory learning, action learning, and social learning. What these forms of learning have in common is a range of factors including the fact that they consider learning as more than merely knowledge-based, and that they "view learning as inevitably transdisciplinary, 'transperspectival,' and transboundary in that it cannot be captured by a single discipline and by a single perspective." Such cross-boundary learning challenges the assumptions on which more traditional models of learning have been based. Done well, the learning becomes transformative of the individual and of the society as a whole.

Wals and Dillon note that we can learn from nature about the learning process and about sustainability. Ecosystems display evidence of resilience and systems thinking allows us to examine how communities depend on each other to survive and to develop in the face of challenging circumstances. They conclude that:

> Learning in the context of environment and sustainability then becomes a means for working towards a "learning system" in which people learn *from* and *with* one another and collectively become more capable of withstanding setbacks and dealing with insecurity, complexity and risks.

In order to do this well, people need to be reflexive, that is, to be self-critical and willing to change beliefs and attitudes in the light of feedback and/or new information. Wals and Dillon argue that these new forms of learning are difficult to realize without what they term, "postnormal environmental education." There is, then, a need for postnormal research "that is capable of 'handling' multiple causation, interactions, feedback loops, inevitable complexities, uncertainties and contestations." So whereas Lundholm et al. focus on the personal response to conventional education, Wals and Dillon call for a response from educators and researchers to provide education that is personally and socially transformative and reflexive.

In the third chapter in this section, Joe Heimlich, Preethi Mony, and Victor Yocco note that we all hold beliefs about the environment and specific environmental issues. Researching beliefs, then, is important but very difficult because they "exist in the complex realities of an individual's life." Research into attitudes and beliefs has been central to environmental education for decades. Over the years, a number of instruments have been devised to measure people's opinions about a range of topics. This type of research has its critics as is pointed out in the first two chapters in this section. What Heimlich et al. do is to show how research in this area has developed and at how researchers have addressed the complexity of the issues involved. The shift toward seeing learners and learning as individual and personal identified in Lundholm et al.'s chapter finds an echo in the writing of Heimlich and his coauthors when they write: "Thus, some researchers believe that understanding concern, caring, empathy, or stewardship may reveal the affect support and belief systems that can be actuated within individuals."

Heimlich et al.'s chapter also resonates with Wals and Dillon's in that it offers a challenge to traditional education. Tellingly, they write that "studies have shown that knowledge, in the form of fact-acquisition, does appear to be significant in the model, but that it is not a predictor of behavior, suggesting knowledge as a component of conation and intent [. . .] or that systemic knowledge exerts only mediated influence on behaviors." The implication is that simply giving people the facts doesn't work and that education has to be much more sophisticated if it is to be effective and empowering. Heimlich et al. suggest that "a potential role of environmental education is in guiding individuals to see how proenvironmental behaviors may align with beliefs," rather than trying to shape beliefs directly. In terms of environmental education research, they suggest a need to study "how people learn, how learning shapes and relates to beliefs, and how beliefs are ultimately tied to an individual's choice to act."

In the fourth and final chapter of the section, Carol Brandt examines the use of landscape as a context for learning in environmental education. Brandt uses a transdisciplinary approach, drawing on the ecology, natural resource management, human geography, and environmental education. It is such an approach that Wals and Dillon advocate in their chapter and meets their call for postnormal environmental education research that is empowering and reflexive.

Brandt's concern is with the relationship between the learner and the context in which learning takes place and she focuses on the link between place and identity through the lens of landscape. She writes, "This connection of landscape to affect or human emotion is also caught up in the ways in which we reflect upon our individual and collective identity and the meaning one attaches to landscape." Brandt notes that some authors have seen the process of looking out, at a landscape, as allowing an individual to look inward, reflexively. Such a process opens up a space for critical reflection, another feature of postnormal learning.

Brandt touches on the pedagogic opportunities that places and spaces provide. Landscapes can be read in terms of what they tell us about the history of people—they provide evidence for interpretation and opportunities for developing imagination as well as empathy for the people and wildlife who live, have lived, and will live in an area. They provide an opportunity to see power and control, influence and politics writ large on the environment. Brandt concludes that: "Landscape as a context for learning in environmental education demands that we continually reposition our perspective as we come to understand our place in the world."

Like Lundholm et al., Brandt brings in young people and teachers, and examines studies of educational processes and their impacts on them. She captures the rich opportunities that engaging with landscapes offers to individuals whether they are younger or older. As well as looking at a range of strategies that have been used to facilitate learning, Brandt examines how landscape is viewed in terms of its physical, biological, and cultural features and at how "learning and education are also considered to be fundamental to the continued existence of these particular landscapes and the resiliency of particular ecosystems." Through a series of examples taken from around the world, Brandt shows how environmental education can help people, as Heimlich et al. suggested, to reflect on their own behaviors and their own choices both now and in the future.

Finally Brandt identifies some gaps in the literature: longitudinal studies of the impact that learning in landscapes might have on participation in environmental conservation and studies of environmental curricula that incorporate the political and economic processes that have shaped and continue to shape landscapes.

Taken together, the chapters provide several tools for researchers: an historical context for developments in environmental education; a set of models and theories of learning with examples of their implications for teachers and learners; a set of challenges for future research; and, more importantly, a sense that the field is moving toward broader and deeper views of the role of learning in empowering individuals to be reflexive and socially critical.

References

Rickinson, M. (2001). Learners and learning in environmental education: A critical review of the evidence. *Environmental Education Research, 7,* 207–317.

Rickinson, M. (2006). Researching and understanding environmental learning: Hopes for the next 10 years. *Environmental Education Research, 12,* 239–246.

Rickinson, M., Lundholm, C., & Hopwood, N. (2009). *Environmental learning: Insights from research into the student experience.* Dordrecht: Springer.

Vosniadou, S. (2001). *How children learn.* Brussels: International Academy of Education.

25

Environmental Learning

Insights From Research Into the Student Experience

CECILIA LUNDHOLM
Stockholm University, Sweden

NICK HOPWOOD
University of Technology, Sydney, Australia; Oxford University, UK

MARK RICKINSON
Oxford University, UK

Introduction

There is growing interest in the role of learning within debates about the environment and sustainability. This is reflected in the titles of several recent publications in the field: *Social Learning Towards a Sustainable World* (Wals, 2007); *Participation and Learning: Perspectives on Education and the Environment, Health and Sustainability* (Reid, Jensen, Nikel, & Simovska, 2008); *Environmental Education: Learning, Culture and Agency* (Stevenson & Dillon, 2010); and *Resilience in Social-Ecological Systems: the Roles of Learning and Education* (Krasny, Lundholm, & Plummer, 2011). It is also clearly articulated in Scott and Gough's (2003) volume on *Sustainable Development and Learning* which works from the basis that:

> It is not enough to say that sustainable development and learning need to go hand in hand; rather sustainable development itself needs to be understood as a learning process. (p. xiv)

Until recently, however, the field of environmental education research has had surprisingly little to say about environmental *learning*. A review of research from 1993 to 1999 found that while there were many studies that had investigated the environmental attitudes or knowledge of students, few had explored the process of their environmental learning (Rickinson, 2001; see also Hart & Nolan, 1999). Furthermore, where learning has been studied or discussed, it has tended to be in relation to educational interventions rather than as a process in its own right (Rickinson, 2006). In other words, exploration of environmental learning as "an integral part of our everyday lives" (Wenger, 1998, p. 8) and of "what learners learn, not with what teachers teach" (Scott & Gough, 2003, p. 38) has been all too rare. There have also been few attempts to "develop new models of [environmental] learning" due to an apparent reluctance by environmental education researchers "to engage with learning theories" (Dillon, 2003, p. 217).

With this situation in mind, this chapter draws together the findings of recent studies that have taken seriously the ways in which learners make sense of their environmental education in formal settings. This focuses primarily on work undertaken by ourselves on students' environmental learning experiences in secondary school and university contexts in England and Sweden (Rickinson, Lundholm, & Hopwood, 2009). Our findings highlight the complexity of the student learning experience in environmental education in terms of three key issues: dealing with emotions and values; questioning relevance; and negotiating viewpoints among students and teachers. Each of these themes is discussed using examples from our research, after which we consider the possible implications emerging from this work for research and practice in environmental education.

Research on Environmental Learning Experiences

Writing in the late 1990s, Payne (1998, p. 20) drew attention to "a lack of consideration in environmental education theory and research practices about the children who are the subjects of environmental education." Since that time there have been signs of various efforts to address this gap as researchers have started to explore the process aspects of environmental learning. If research on environmental learning is characterized in terms of three different approaches (Figure 25.1), then what has been seen over recent years has been an increasing emphasis on work concerned with exploring environmental learning processes.

Researching learners
This is where researchers have sought to discover more about some aspect of young people's environmental knowledge, attitudes, and/or behaviors (e.g., Kuhlemeier, Van Den Bergh, & Lagerweij, 1999). Studies in this category have, therefore, not specifically been about understanding environmental learning itself, but rather investigating the environmental characteristics of young people.

Measuring outcomes
This second category moves beyond studies of the characteristics of learners into research on learners in relation to environmental education programs. The focus here is on the outcomes of educational interventions in terms of the extent to which they bring about changes in students' environmental knowledge, attitudes, and/or behaviors (e.g., Bogner, 1998).

Exploring processes
This final category, which has become more prominent over recent years, is focused on exploring the processes of environmental learning as opposed to its outcomes. There is a strong emphasis in this work on the voice of the student and seeking to understand how learners themselves make sense of environmental education (e.g., Lai, 1999).

Figure 25.1 Different approaches to researching environmental learning.

One strand of work within the "exploring processes" category has been studies of young people's responses to and experience of various forms of environmental learning. Examples include investigations of:

- nature-based excursions (Ballantyne & Packer, 2002)
- field trips (Lai, 1999)
- secondary/high school environmental geography lessons (Hopwood, 2007a, 2007b, 2007c, 2008, 2009, 2011, in press; Rickinson, 1999a, 1999b)
- secondary school science classes (Osterlind, 2005)
- primary school environmental education (Nagel & Lidstone, 2008)
- school-based social/environmental change programs (Mackenzie, 2006)
- university environmental modules/courses (Lundholm, 2004b, 2005, 2007, 2008)
- national park visits (Brody & Tomkiewicz, 2002).

A common thread running through these studies is a conceptualization of learners as active agents, rather than passive objects, in the learning situation. In keeping with this, they have tended to be qualitative in nature, involving semistructured interviews and participant observation of learners. The sorts of arguments that have underpinned these types of studies have been that: (i) the voice of the learner is a severely neglected one in environmental education research and curriculum development (Payne, 1998; Rickinson, 1999b); (ii) understanding learning necessitates a "focus on the multiple ways in which [learners] make sense of the information they encounter, rather than whether [they] 'get the message' the provider intended to convey" (Ballantyne & Packer, 2005, p. 283); and (iii) learning is an active process influenced by emotional factors, and needs to be researched as such through naturalistic consultation with students about their experience and learning (Lai, 1999).

Our Research

We see our own research in environmental education as part of this emerging work on students' environmental learning experiences. Lundholm's research examined Swedish university students' learning about environmental issues as part of undergraduate and graduate programs in civil engineering, biology, and economics (Lundholm, 2003, 2004a, 2004b, 2005, 2007, 2008). Hopwood (2007a, 2007b, 2007c, 2008, 2009, 2011, 2012) chose a Year 9 (age 13–14) class in three UK secondary schools, and followed the experiences of two students from each in geography lessons for a period of three months per class. Rickinson's (1999a, 1999b) work focused on the ways in which three teachers from three UK secondary schools and twelve of their students (age 13–15) dealt with controversial environmental issues within geography.

The findings we present below reflect a cross-cutting analysis drawing on common reference points across our three studies identified through a grounded rather than theoretical analysis. However it is worth noting briefly the different theoretical frameworks adopted in these studies. These reflect different motivations for and positions adopted in seeking to better understand environmental learning.

Lundholm's research was conducted from a constructivist perspective, which takes into account the social and cultural context of learning. In so doing, it focuses on students' "intention in action" when making inferences about conceptions and learning, therefore paying attention to the way the student interprets tasks, the situation (classroom, peers), and the cultural setting (Halldén, 1988; Halldén, Scheja, & Haglund, 2008; Lundholm, 2004a, 2008, 2010, 2012; Sternäng & Lundholm, 2011, 2012). Halldén (1988) defines a "task" as "what is presented to the pupils by the teacher with the intention that they are to do something and/or to learn something" (p. 125). "Task" and "problem" distinguish what the teacher has presented to the students, in writing or orally, from students' interpretation of what has been presented to them. Looking at learning from an *intentional* perspective also means taking account of students' aims and considering these when analyzing their interpretations. This means that a "problem" a student is trying to solve can be interpreted in a wider sense by taking into account goals, where a far stretching goal is defined as a "project" the student is trying to realize (Lundholm, 2004a, 2005, 2008).

Hopwood's study was framed by the New Social Studies of Childhood or NSSC (James & Prout, 1997), reflecting a focus on younger secondary school-aged learners. The NSSC understands childhood as a social construct, deems children's relationships, cultures, and perspectives as worth of study in their own right, acknowledges a multiplicity of childhoods, and sees children as active in the construction of their own lives. Applied to the context of environmental learning, this framework underpinned questions that sought to explore how children made sense of environmental learning, constructing meaning from it in relation to their own lives and experiences. A focus on a small number of individual pupils attended to the notion of multiplicity—exploring differences as they were found from person to person, rather than looking for aggregate patterns. Although concepts from adult-derived literature on environmental education were used in analysis, the purpose was not to compare child views with (correct or informed) adult views, rather to respect the former as valid and interesting *per se*. Studies that take young people's ideas seriously can be seen as a corrective to a historical bracketing of children's voices, replacing an "outdated view of childhood that fails to acknowledge children's capacity to reflect on issues affecting their lives" (Rudduck, Day, & Wallace, 1997, p. 89). Traces of these theoretical commitments can be seen through the three themes discussed below—in the values learners bring to their experience and the emotional challenge of dealing with values-rich content, in the negotiations between teachers' and learners' points of view, and particularly in the exploration of how relevance is constructed by students as a feature of environmental learning.

Rickinson's work was underpinned by a particular conceptualization of the classroom curriculum. This was informed by previous studies of classroom teaching and learning conducted beyond the field of environmental education (e.g., Cooper & McIntyre, 1996; Doyle, 1992; Erickson & Shultz, 1992; Nespor, 1987). Such studies see the curriculum as becoming manifest in classrooms through teachers constructing and enacting particular combinations of subject matter and learning tasks that are interacted with, and experienced, by students. The process by which this happens is viewed as dynamic and interactive, that is, subject matter-tasks come into being through interactions between the teacher and the students during a lesson. As Doyle (1992, pp. 507–508) argues, it is teachers and students together who "author" the curriculum through "a dynamic process in which content is produced and transformed continuously." This understanding of the classroom curriculum helped frame the study's emphasis on exploring school-based environmental learning in terms of:

(i) *the actualities of everyday practice*—"It is the moment-to-moment character of engaging the manifest curriculum that is perhaps the least well understood by researchers, by teachers and by the students themselves" (Erickson & Schultz, 1992, p. 468)
(ii) *the perspectives of students and teachers*—"Past research on teacher-pupil interactions has largely been teacher-centred [. . .] Pupils' perceptions, intentions and evaluations have only rarely been the focus of analysis" (Hanke, 1990, p. 189)
(iii) *the interplay between curriculum and pedagogy*—"We find few descriptions of analyses of teachers that give careful attention not only to the management of students in classrooms, but also to the management of ideas within classroom discourse" (Shulman, 1987, p. 1).

This chapter is based on a collaborative synthesis of the findings of these pieces of work, emphasizing key shared empirical findings, over theoretical distinctions. Over the course of several years, we have been working through a process of sharing, analyzing, and integrating the findings of our original studies (e.g., Lundholm, Hopwood, & Rickinson, 2008; Lundholm & Rickinson, 2006; Rickinson & Lundholm, 2008). The purpose has been to generate deeper and more powerful insights into students' learning experiences which hopefully contribute to richer understandings of environmental learning (Rickinson et al., 2009).

It is important to note that, although all the studies were carried out independently, they shared a lot of common ground in terms of their:

- conceptualization of learning—a view of learners as active agents in constructing understanding rather than passive recipients of knowledge.
- conceptualization of curriculum processes—drawing on the distinction between task as presented by the teacher (the enacted curriculum) and problem (or "project") as understood by an individual learner (the experienced curriculum) (Entwistle & Smith, 2002; Erickson & Schultz, 1992; Halldén, 1988)
- methodology—shared case study approaches, employing qualitative techniques including observation/recording of classroom talk, documentary analysis (students' work), interviews/focus groups about particular learning experiences.

Each of our studies explored the process of environmental learning by asking questions such as how do students view particular aspects of subject matter? How do they feel about that subject matter, particular tasks, and their learning more generally? What do they do in classrooms, and why do they do these things? Our interest in understanding students' experiences can be described as rooted in social constructivism, in that we view learning as a process and product of the individual interacting with others in a social and institutional setting, that is, among peers and teachers. Having stressed the social and cultural aspect, our interest is still in the ways in which the individual student comes to engage, disengage, and (re)act to social and institutional norms as, for example, in feeling obliged to have an opinion or, alternatively, not feeling comfortable in expressing one. We view learning as a change in the way we look upon the world—our thoughts, feelings and actions. We see it as a process that is dependent on the learner, the object of learning, and the physical/ecological, social, cultural, and economic situation and setting.

Our Findings

The integration of our work has led to three themes that bring different aspects of students' environmental learning experiences into focus. These concern:

- students dealing with emotions and values—the challenges for students of having to deal with their own values and emotions in the environmental learning situation
- students' views on relevance—the extent to which students' see environmental learning as relevant to particular curriculum subjects, to themselves as learners and to their personal/professional futures
- negotiating viewpoints among students and teachers—the ways in which differences between teachers' and students' viewpoints can present challenges for student engagement and learning.

Dealing With Emotions and Values The fact that environmental issues are value laden and emotionally charged is very clear when we look at students' views and ways of dealing with emotions and values. This is evident in two ways: (i) students' emotions and values as part of the learning process and (ii) students' conceptions of values in subject matter and in relation to themselves.

Emotions and values are terms that encompass a range of different meanings, often given the generic label of the affective domain. Emotions can refer to students' anxiety with schoolwork and in test situations, aggressive behavior and self-control, and metacognitive skills for dealing with emotions as well as thinking (Efklides & Volet, 2005). The emotional aspect of learning also relates to motivation and students' engagement in schoolwork and instruction. With regard to the latter, there has been an increasing interest in understanding the ways motivation, emotions and values are an important part of the process of conceptual development and change (Pintrich, Marx, & Boyle, 1993; Sinatra, 2005; Sinatra & Pintrich, 2003; Watts & Alsop, 1997).

Students' Emotions and Values as Part of the Learning Process Emotions that featured in the learning processes documented in our studies include reactions of distaste and discomfort with particular learning activities, as well as engagement in learning sparked by learners' emotional responses toward particular issues. An example of the former was a secondary school student Melanie's reactions when watching a video on rainforests: "The second video [. . .] didn't seem very interesting—cutting down trees, I don't think that's very my sort of thing, I don't like things like that. I don't like cutting down trees, I don't like animals being hurt or moved or anything."

Melanie's response can be compared with evidence from other studies indicating that students may disengage from learning activities and tasks if they experience dislike or discomfort with what is being learned (e.g., Watts & Alsop, 1997). This is important when understanding students' emotional reactions to environmental issues, like global climate change, that are not only contested but also potentially alarming.

Other examples of the role of emotions and values come from undergraduate biology students describing learning challenges in the subject of economics (Lundholm, 2007, 2008). Some of the biologists struggled with the discipline's approach to the inclusion/exclusion of nature (Figure 25.2), while others found pricing itself to be problematic as "everything has to be shown in dollars and cents when a decision is to be made, and my world view really opposes to that."

As the students related pricing to natural goods (air, ocean) and services (ecosystems services as in biodiversity) they found the subject increasingly problematic. Previous studies have shown that students can retreat from economics courses because of reluctance toward the idea of monetary values being assigned to nature "despite students agreeing that such approaches can serve to achieve objectives consistent with their original beliefs (such as wildlife preservation)" (Shanahan & Meyer, 2006, p. 104). In explaining why the biology students in our study succeeded, it seems that despite emotions of "despair," they also saw strategic benefits in pricing nature in terms of their environmental concerns and professional goals: gaining economical knowledge with the purpose of improving the environment. Thus, they engaged with views to which they in fact opposed.

Students' conceptions of values in subject matter and in relation to themselves Both school and university students share a view that opinions are part of subject matter relating to environmental issues. To quote Jenie (a secondary school geography student): "I think it's a kind of opinion subject because people have different opinions about things." Similarly, Patrick (a university engineering student), while complaining about a lack of discussion about environmental solutions as opposed to environmental problems, stressed that: "When it comes to solutions, it's all about politics. I mean nuclear power, it's all politics."

Diana

The difference between biologists and economists, is that they [economists] often think there is a "base" in the economic world, or possibly in the social world, while we see the economic and social world depending on ecology as a foundation. So, it's the opposite really, the world upside down. [. . .] Nature constitutes the base, we are part of nature and through nature we get economy, that means, what is traded is taken from nature, it's not something that's just there like magic in a factory, it has its origin in nature. I think that's why I found it very difficult in economics; they picked out the environment part, and sort of put it aside, like an appendage, instead of the way I see it: as the foundation for everything. If nature wasn't there we wouldn't be here ourselves.

Figure 25.2 Challenging worldviews in economics.

Our findings also show that students' responses to learning activities, and challenges they are faced with, concern their values and views of science. Students raise questions like: How do we distinguish and separate our viewpoints and values from "factual" knowledge? If we don't, are we not being "scientific?" These questions can be seen in the context of work on scientific literacy including studies in which learners were found to distinguish the subjectivity of opinion with the objectivity of scientific knowledge (Ziedler, Walker, Ackett, & Simmons, 2002) and studies showing students distinguishing between personal value judgments and scientific descriptions, hence developing "contextual awareness" (Lundholm, 2001, 2004a, 2004b, 2005).

Questioning Relevance Our second theme investigates why learners judge environmental learning to be relevant or not, and to whom and what they see such learning as being relevant. In doing so, we distinguish between (i) relevance to the individual learner and (ii) relevance to curricular context.

Why should we look for relevance? Relevance is a much-cherished quality of learning in general and environmental learning in particular. A significant body of evidence and theoretical work suggests that learners' views as to the relevance of learning are important because they influence decisions to engage or not in learning and the value they then place on that learning. Postman and Weingarter (1969) express the core idea succinctly:

> Unless an inquiry is perceived as being relevant by the learner, no significant learning will take place. No one will learn anything he [sic] doesn't want to know. (p. 52)

Many argue that subject matter should focus on things that interest, are salient or significant to learners (e.g., Pintrich et al.,1993). Real-life, contemporary issues are often suggested as appropriate, stimulating, and relevant foci (Beane, 1997). Within environmental education a desire for relevance to immediate local contexts, real or everyday life, and contemporary issues has been articulated by researchers, educators, and students (e.g., Barratt & Barratt Hacking, 2008; Battersby, 1999; Connell, Fien, Lee, Sykes, & Yencken, 1999; Nagel & Lidstone, 2008; Smith-Sebasto & Walker, 2005).

Evidence from studies in a number of subject areas also suggests that learners' ideas about what counts as relevant to particular subjects or courses may also influence what they pay attention to, work on, and try to achieve in formal learning settings (e.g., Brook, Briggs, & Driver, 1984; Buehl & Alexander, 2005 ; Driver, Leach, Millar, & Scott, 1996).

Despite the strong evidence that learners' perceptions of relevance matter in learning, relatively little is known about this in the specific context of environmental education, and even less is known about how such perceptions play out in and affect the actual experience of learning itself. We have learned much in this regard by taking the data from our separate studies and looking at or for issues to do with relevance.

Relevance to the Individual Learner Frequently it is assumed that learners associate relevance with local or immediate contexts, their own experience, or current issues. Across our studies we found examples of this, but also subtly different notions of how environmental learning might be deemed "relevant to me, now." Three examples of learners in English secondary school geography classrooms are provided in Table 25.1.

Jenie's desire to learn about things that relate closely to her own experiences contrasts with Sara's notion that relevant learning actually focuses on the unfamiliar (she questioned the purpose of learning about things she already knew about, invoking a sense of relevance to me as a learner). Lisa's ideas focus not on immediacy, locality, or (un)familiarity. Rather she interprets her learning experiences with respect to a general passion for what she describes as "green" issues—protecting "nature" from "destruction by people."

Across our studies, we also found instances of learners considering environmental education with respect to (imagined) future selves. Ryan (UK secondary school geography) felt that learning about sustainable development made important contributions to his potential in the future, both in the world of work and as a global citizen. A group of Swedish undergraduates responded in varied ways to an ecology course taken as part of a vocationally oriented degree in engineering. Martin adopted a bare-minimum approach to his work, considering it irrelevant to his future as an engineer. Patrik and Sara were similarly not convinced of the relevance of ecology to professional engineering, but did think ecological learning was relevant

TABLE 25.1
Illustrating Different Views of Relevance to the Present Self

Learner	Environmental Learning is Most Relevant to Me When . . .	Relevant Learning Conceptualized As
Jenie	. . . it is about places where I live or where I have been, issues I am directly involved in	Learning that coincides with personal experiences
Sara	. . . it involves studying people, environments, and places of which I have little knowledge or experience myself	Learning about the unfamiliar
Lisa	. . . it addresses issues of human impacts on the natural environment, something I feel strongly about	Learning about issues I care about

in that it formed part of an understanding that everyone in society should share (encapsulated in the Swedish concept of allmanbildning).

Relevance to Curricular Context So, learners' views as to whether particular environmental learning experiences are relevant to them may take on a range of forms, may focus on the familiar or unfamiliar, particular passions and interests, or ideas about future selves as workers or citizens. However, learners across our studies also made reference to (ir)relevance of a different kind, focusing on many occasions on the idea that certain learning is or is not relevant to a particular course, subject, or unit of study.

Some learners deemed either physical (relating to nature, the physical environment) or human (relating to the built environment, people, culture, and human activity) phenomena to be relevant to a particular curricular context but not both. Such dualisms have been questioned in academic literature, but appear to be important in learners' conceptualizations of subjects or courses such as geography or ecology. Matt thought human aspects were not so relevant to geography, choosing to focus on physical processes and features rather than human experiences (as when he wrote a report about Hurricane Ivan); Aiden thought learning about people who live in tropical rainforests was not relevant to school geography, expressing frustration and dissatisfaction with a lesson about Kayapo Indians. In contrast, Tobias found the focus on physical aspects in his ecology course to deviate from the emphasis on human problems and solutions he expects in his learning as an engineering undergraduate. Other learners tended to view both human and physical as relevant, but again there was variation: Jenie thought they might be studied separately; Lisa and Sara thought learning was most relevant to geography when it dealt with both in interaction.

Negotiating Viewpoints Among Students and Teachers Our third theme frames environmental learning situations in terms of the interactions between the beliefs and values of students and the beliefs and values of teachers. It draws attention to the ways in which differences between teachers' and students' viewpoints can present challenges for student engagement and learning. Within our studies, there were several cases of students talking about their environmental learning activities in terms of some kind of conflict or tension between themselves and the teacher. What is important, though, is that these were not general interpersonal conflicts or tensions. Instead they related specifically to values and beliefs connected with the environmental content/learning activity. More specifically, they concerned students and teachers having differing viewpoints on: (i) an environmental issue; (ii) what is controversial; (iii) what is relevant; and (iv) empathy tasks.

Different Views on an Environmental Issue One area of difficulty for students within our studies occurred when they perceived there to be a significant difference between their own perspectives on an environmental issue and those of their teacher. These situations of perceived differences of viewpoint are significant for two reasons. Firstly, they very rarely get expressed by students within the learning situation but instead remain hidden and tacit. Secondly, as we will see in the example below, they can have a very real affect on the nature and extent of a student's engagement with environmental learning activities.

This kind of scenario was clearly seen among a group of engineering students in Lundholm's Swedish study. In the interviews about their ecology course, several students brought up the issue of the lecturer's perspective. Tobias, for example, felt the course had been "angled from an ecological perspective [i.e.] everything that humans do has an impact on nature and if you affect nature, it is bad." Another member of the class, Ola, commented that "Our dear ecology lecturer has the viewpoint that man was God's biggest mistake." For Ola, this perception of the lecturer's perspective had a real affect on his willingness to actively engage in discussions during the course: "I think you quickly kill all interest in discussing [. . .] You don't want to become an enemy to someone who is going to correct your exam."

What we see here is an example of how perceptions of the relationship between students' and teachers' environmental viewpoints can profoundly influence the learning experiences and responses of individual students.

Different Views on What is Controversial A second source of difference between students and teachers occurred when they held contrasting views on whether the subject matter was controversial. This is not about students and teachers holding different views on an environmental issue, as we have just seen with the Swedish students and their ecology lecturer. Instead, it concerns students and teachers having different views of whether a particular environmental issue is controversial, topical, and worth engaging with.

A clear illustration of this can be seen with Laura, a fourteen-year-old student, in a geography lesson about nuclear power. For Laura nuclear power is something she's "not bothered about" and "wouldn't sit around for an hour thinking full-on whether [it] is a good idea," while for her teacher it is something that is "really affecting us, and we all need to sit up and take notice [and] consider the possibility that [a Chernobyl-type accident] could happen here." This divergence of viewpoints between Laura and her teacher has a marked effect on the way in which Laura responds to the task of the lesson (Figure 25.3).

We see this as an example of a student engaging with a learning activity in a superficial or "going through the motions"-type way because of her belief that the topic of nuclear power is not worth thinking too hard about. In other words, a task that is designed by the teacher to encourage students to consider their own views on the issue of nuclear power is actually carried out by the student in terms of "spur of the moment thinking for what would

The students in this lesson had been asked to answer the question "Do you think nuclear power is a good thing?" after watching a video about the Chernobyl disaster. Laura, a fourteen-year-old student, explained her difficulty with this task as follows:

I could put "No nuclear power is not a good idea," but I'd rather just write "No" full stop. [. . .] It's not exactly hard to write what you think, but sometimes you're thinking, you don't write what you think, but you're thinking of something, but it's not what you think if you sat down and thought about it for ages. It's just a spur of the moment thinking for what would be right to put down.

Figure 25.3 Laura's difficulties with the nuclear power lesson.

be right to put down." For Laura, it was "just the work we have to do" and not (as the teacher felt) something "we"ve really got to grapple with."

Differing Views of What Is Relevant We have already seen how learners can have strong beliefs about what counts as relevant both to them now and in the future and to particular subjects or courses. Given that teachers can also hold strong views on what is and is not relevant to study in a particular subject, it is not surprising that diverging views on relevance was another example of student-teacher differences within our studies.

A secondary school geography lesson about the Kayapo Indians of the Amazonian rainforest provides a helpful example. For the teacher, it was very clear that learning about the indigenous peoples of the rainforest was appropriate and useful for a Year 8 geography lesson. When the experiences of individual students in this lesson were probed, however, it became clear that the teacher's views on the importance of learning about Amazonian people as well as Amazonian vegetation/climate were not necessarily shared. One boy in the class, Aiden (13 years old), felt that learning about the peoples of the rainforest was "not really geography." When interviewed shortly after the Kayapo lesson, he said: "In geography today I did not learn anything to my benefit. Yesterday's lesson on climate was much more interesting because you were actually learning something about the rainforest, not the people who live there." Far from being geography, Aiden felt that, "You probably would have learnt about the Kayapo in history as history is about things that happened and people in many ways."

This and other similar examples (see Rickinson et al., 2009) highlight the need for teachers and curriculum developers to think carefully about the ideas that students bring to the learning situation about what is and is not relevant to study in different curriculum contexts.

Different Views on Empathy Tasks Empathy tasks, such as role plays or empathetic writing activities, are commonly used by teachers wishing to present students with a balanced picture of controversial environmental issues (e.g., Cotton, 2006). When these kinds of tasks are looked at from the perspective of teacher-student negotiation, it becomes clear that they can be a source of difficulty for students. Empathetic role play, for example, can be challenging if you are very keen on expressing your own perspective in debates but are asked to play the part of a perspective that is quite different from your own (Figure 25.4). By contrast, an empathetic writing exercise like writing a poem from the perspective of someone very different to yourself can easily turn into an exercise in creative, but not empathetic, writing if you are a student who really likes writing poems that rhyme (see Rickinson, 1999b).

Picture the scene: the students, sitting in groups, have just received their role cards representing differing perspectives on rainforest development. The teacher is moving around the room from group to group when the following interaction with a male student, Simon, occurs:

Simon: I don't want to be someone against the development, I want to be someone for it
Teacher: Well see how you do with that, it's important to be able to look at another viewpoint Simon
S: I'm against stopping it, so how come?
T: I know you are, but it's a chance to see another viewpoint
S: I can see it, but I can't agree with it
T: I'm not asking you to agree with it, I'm asking you to put across that view
S: I'm not saying anything then
T: I think you're going to find you're going to have to
[Teacher moves away to deal with other student groups]

On the face of it such an interaction could well be seen as a fairly regular part of classroom life where students and teachers are continually challenging each other in any number of ways. When the teacher and student were interviewed after the lesson; however, it became clear that this interaction was more than a passing spat. Instead it represented a genuine difference of perspective between the student and teacher.

Figure 25.4 Simon's difficulties with the rainforest role play.

This example provides a glimpse of the complexity that can be involved in undertaking empathy tasks such as role play exercises. This is just one student's response to one part of one lesson but it serves to highlight the way in which these kinds of tasks can present very particular challenges for students' environmental learning and can result in learning processes that differ considerably from teachers' original intentions.

Discussion

A number of issues emerge across the three themes we have discussed. All of them point in some way to a range of challenges or difficulties learners may encounter in their experiences of environmental education in formal settings. Subject matter often provokes strong emotional reactions and may challenge learners' closely held values. Students thus have to grapple with their own affective responses and their ideas of what role these could or should play in a formal learning environment such as a school classroom or university lecture course. Across our studies we found examples of students struggling with themselves and with each other and their teachers as to how to proceed or work appropriately on particular tasks. We have also found students challenged by tensions between what is presented to them in their classrooms and what they perceive to be relevant to their own lives, their learning, or the particular course or subject at hand. This tension may play out in different ways, leading to disengagement, frustration, or sometimes confusion as to what is expected or deemed appropriate by the teacher.

All three themes also point to the marked variety in learners' experiences and responses even within very specific settings such as a particular school class or a small group of university students working on a joint project. The same lesson or projects are interpreted and experienced in different ways. These varying responses appear to be influenced by the diverse values learners hold, their personal affective reactions, and different views held about the role of students and teachers (and potential for negotiation between them). The process of environmental learning thus appears to be a highly personal one. It is reasonable to suggest that this personal dimension may be particularly significant in the environmental context in which, as we discussed above, the subject matter itself often strikes a chord with learners' values, emotions, and sense of themselves and their surroundings (social, environmental) now and in the future.

Reflecting on our shared assumptions about the nature of learning (as involving active participation rather than passive receipt of knowledge), we find indeed that learners play a significant role in shaping the process of environmental learning. They bring with them personal histories, views, and values, and these then influence a range of complex judgments they make about relationships between themselves, their peers, their teachers, subject matter, tasks, and learning outcomes.

Implications for Research and Practice

Our research has thus revealed a number of important insights into the process of environmental learning. However it has also become clear to us that much of what we have learned would normally remain hidden from the view of teachers and curriculum planners and others who share responsibility for setting up learning experiences in formal settings. As we have seen, students may well have very good reasons for wishing to keep their differences of opinion or views about the subject matter hidden from their teacher or lecturer. Equally, much of what we have found has remained hidden from the gaze of researchers who, as we discussed previously, have tended to focus on other aspects of environmental education. A key challenge is thus to procure empirically robust insights into the process of environmental learning in a wider range of settings than those that formed our focus (which we acknowledge are limited in scope). We know little about what lies beneath the surface of environmental learning in other formal settings (different countries, age ranges, subject, or curricular contexts), and there remains a wide realm of nonformal settings which could usefully be investigated in this way.

In seeking to investigate such matters it is critical that environmental learning research connects with wider research within and beyond education and draws upon learning theory. Indeed, it is clear from the editorial of a special issue of the journal *Learning and Instruction* that environmental education is not alone in that "the role of feelings and emotions in the learning process [is] . . . a new and largely unexplored area" (Efklides & Volet, 2005, pp. 377–379). Moreover, it is clear from the field of teacher education that "there has been a lack of scholarly attention to the practical work of helping students build emotional, as well as cognitive, relations to what they are learning" (Rosiek, 2003, p. 140).

In current debates concerning learning goals and aims of environmental education, we find views that stress the importance of enhancing students' capacity to be reflective about one's knowledge and ideas, engage with opposing views and critically examine proposals for change (e.g., Vare & Scott, 2007). In our findings we see students struggling with such aspects of learning. They are challenged by and uncomfortable with opposing views and find difficulties in setting their own views aside. This raises questions in relation to research findings on learning, views of knowledge, and identity. Are, for example, students' difficulties due to their views of knowledge, so-called "epistemological beliefs" (for a discussion of research on epistemological beliefs and influence on learning, see Mason, 2003)? Or, are contradicting and opposing views challenging to engage with because they become emotionally difficult (see Limon Luque, 2003, for a discussion of research on emotions, identity, and learning)?

In relation to practice, further significant questions emerge such as: Within the institutions of formal education students will always be hesitant in expressing opposing views because the aspect of power in relation to their teachers? When considering communication among peers from a social perspective, is it unavoidable that students feel uncomfortable in expressing opposing views? Can teaching create an atmosphere where different views are shared and discussed comfortably despite power imbalances between students and their teachers? Here, we wish to point out the importance of recognizing the difference between learning challenges that are *domain specific* and learning challenges that are *domain general* when furthering our understanding of environmental learning in future research (Vosniadou, Vamvakoussi, & Skopelti, 2008). By acknowledging this difference, we can develop clearer insights of the specificity of challenges that are part of students' environmental learning as compared to those encountered in other subjects and contexts (Lundholm, 2011; Lundholm & Davies, in press).

Progress on these and other questions will require more collaborative research that brings together researchers, students, teachers, educational authorities, and other stakeholders. In this way an enhanced understanding of "what is" might be closely coupled with a potential to shape "what might be." Researchers might support teachers and students in making hidden aspects of their experience more explicit, and practitioners might work with researchers in developing more research-informed learning and teaching practices. This chapter provides some early insights into the subtleties and complexity of environmental learning in school and university classrooms, but their further development will require much closer collaboration between researchers, educators, and learners in environmental education.

References

Ballantyne, R., & Packer, J. (2002). Nature-based excursions: School students' perceptions of learning in natural environments. *International Research in Geographical and Environmental Education, 11*(3), 218–236.

Ballantyne, R., & Packer, J. (2005). Promoting environmentally sustainable attitudes and behaviour through free-choice learning experiences: What is the state of the game? *Environmental Education Research, 11*(3), 281–295.

Barratt, R., & Barratt Hacking, E. (2008). A clash of worlds: Children talking about their community experience in relation to the school curriculum. In A. Reid, B. B. Jensen, J. Nikel, & V. Simovska (Eds.), *Participation and learning: Perspectives on education and the environment, health and sustainability* (pp. 285–298). Dortrecht: Springer.

Battersby, J. (1999). Does environmental education have 'street credibility' and the potential to reduce pupil disaffection within and beyond their school curriculum? *Cambridge Journal of Education, 29*(3), 447–459.

Beane, J. A. (1997). *Curriculum integration: Designing the core of democratic education*. New York, NY: Teachers College Press.

Bogner, F. X. (1998). The influence of short-term outdoor ecology education on long-term variables of environmental perspective. *Journal of Environmental Education, 29*(4), 17–29.

Brody, M., & Tomkiewicz, W. (2002). Park visitors' understandings, values and beliefs related to their experience at Midway Geyser Basin, Yellowstone National Park, USA. *International Journal of Science Education, 24*(11), 1119–1141.

Brook, A., Briggs, H., & Driver, R. (1984). *Aspects of secondary students' understanding of the particulate nature of matter*. Leeds: Children's Learning in Science Project, Centre for Studies in Science and Mathematics Education, University of Leeds.

Buehl, M. M., & Alexander, P. A. (2005). Motivation and performance differences in students' domain-specific epistemological belief profiles. *American Educational Research Journal, 42*(4), 697–726.

Connell, S., Fien, J., Lee, J., Sykes, H., & Yencken, D. (1999). 'If it doesn't directly affect you, you don't think about it': A qualitative study of young people's environmental attitudes in two Australian cities. *Environmental Education Research, 5*(1), 95–113.

Cooper, P., & McIntyre, D. (1996). *Effective teaching and learning: Teachers' and students' perspectives*. Buckingham: Open University Press.

Cotton, D. R. E. (2006). Teaching controversial environmental issues: Neutrality and balance in the reality of the classroom. *Educational Research, 48*(2), 223–241.

Dillon, J. (2003). On learners and learning in environmental education: Missing theories and ignored communities. *Environmental Education Research, 9*(2), 215–226.

Doyle, W. (1992) Curriculum and pedagogy, In P. W. Jackson (Ed.), *Handbook of research on curriculum*. New York, NY: Macmillan.

Driver, R., Leach, J., Millar, R., & Scott, P. (1996). *Young people's images of science*. Buckingham: Open University Press.

Efklides, A., & Volet, S. (2005). Editorial. Emotional experiences during learning: Multiple, situated and dynamic. *Learning and Instruction, 15*(5), 377–380.

Entwistle, N., & Smith, C. (2002). Personal understanding and target understanding: Mapping influences on the outcomes of learning. *British Journal of Educational Psychology, 72*, 321–342.

Erickson, F., & Shultz, J. (1992). Students' experience of the curriculum, In P. W. Jackson (Ed.), *Handbook of research on curriculum*. New York, NY: Macmillan.

Halldén, O. (1988). Alternative frameworks and the concept of task: Cognitive constraints in pupils' interpretations of teachers' assignments. *Scandinavian Journal of Educational Research, 32*, 123–140.

Halldén, O., Scheja, M., & Haglund, L. (2008). The contextuality of knowledge: An intentional approach to meaning making and conceptual change. In S. Vosniadou (Ed.), *International handbook of research on conceptual change* (pp. 509–532). London, UK: Routledge.

Hanke, U. (1990) Teacher and pupil cognitions in critical incidents. In C. Day, M. Pope, & P. Denicolo (Eds.), *Insights into teachers' thinking and practice*. London, UK: Falmer Press.

Hart, P., & Nolan, K. (1999). A critical analysis of research in environmental education. *Studies in Science Education, 34*, 1–69.

Hopwood, N. (2007a). Environmental education: Pupils' perspectives on classroom experience. *Environmental Education Research, 13*(4), 453–465.

Hopwood, N. (2007b). Pupils' conceptions of geography: Issues for debate. In J. Halocha & A. Powell (Eds.), *Conceptualising geographical education* (pp. 49–65). London, UK: International Geographical Union Commission on Geographical Education/Institute of Education.

Hopwood, N. (2007c). Researcher roles in a school-based ethnography. In G. Walford (Ed.), *Studies in educational ethnography, Vol. 12: Methodological developments in ethnography* (pp. 51–68). Oxford, UK: Elsevier.

Hopwood, N. (2008). Values in geographic education: The challenge of attending to learners' perspectives. *Oxford Review of Education, 34*(5), 589–608.

Hopwood, N. (2009). UK high school pupils' conceptions of geography: Research findings and methodological implications. *International Research in Geographical and Environmental Education, 18*(3), 185–197.

Hopwood, N. (2011). Young people's conceptions of geography. In G. Butt (Ed.), *Geography, education and the future* (pp. 30–43). London, UK: Continuum.

Hopwood, N. (2012). *Geography in secondary schools: Researching pupils' classroom experiences*. London, UK: Continuum.

James, A., & Prout, A. (1997). Introduction. In A. James & A. Prout (Eds.), *Constructing and reconstructing childhood: Contemporary issues in the sociological study of childhood* (2nd ed., pp. 1–6). London, UK: Falmer Press.

Kuhlemeier, H., Van Den Bergh, H., & Lagerweij, N. (1999). Environmental knowledge, attitudes, and behavior in Dutch secondary education. *Journal of Environmental Education, 30*(2), 4–14.

Krasny, K., Lundholm, C., & Plummer, R. (2011). *Resilience in social-ecological systems: The roles of learning and education*. London, UK: Routledge.

Lai, K. C. (1999). Freedom to learn: A study of the experiences of secondary school teachers and students in a geography field trip. *International Research in Geographical and Environmental Education, 8*(3), 239–55.

Limón Luque, M. (2003). The role of domain-specific knowledge in intentional conceptual change. In G. Sinatra & P. Pintrich (Eds.), *Intentional conceptual change* (pp. 33–70). Mahwah, New Jersey: Lawrence Erlbaum Associates.

Lundholm, C. (2001). *Att forska och lära om miljö. en intentionell analys av forskarstuderandes miljöforskning och tolkning av begreppet 'miljö'. Researching and learning about environmental issues. An intentional analysis of graduate students' environmental research and interpretation of 'environment'*. Paper presented at the 29th Nordic Education Research Association Congress, Stockholm.

Lundholm, C. (2003). *Att lära om miljö. Forskar-och högskolestuderandes tolkningar av ett miljöinnehåll i utbildningen/Learning about environmental issues. Undergraduate and postgraduate students' interpretations of environmental content*. (Unpublished Ph.D. thesis). Department of Education, Stockholm University, Sweden.

Lundholm, C. (2004a). Case studies: Exploring students' meaning and elaborating learning theories. *Environmental Education Research, 10*(1), 115–124.

Lundholm, C. (2004b). Learning about environmental issues in engineering programmes: A case study of first-year civil engineering students' contextualisations of an ecology course. *International Journal of Sustainability in Higher Education, 5*(3), 295–307.

Lundholm, C. (2005). Learning about environmental issues. Undergraduate and postgraduate students' interpretations of environmental content. *International Journal of Sustainability in Higher Education, 6*(3), 242–253.

Lundholm, C. (2007). Pricing nature at what price? A study on undergraduate students' conceptions of economics. *South African Journal of Environmental Education, 24*, 126–140.

Lundholm, C. (2008). Contextualisation and learning in economics. An intentional perspective on the role of values. Paper presented at the symposium Exploring "Hot" Conceptual Change: Affect, Emotions, Values, Self-Efficacy and Epistemic Beliefs at the *6th International Conference on Conceptual Change, European Association for Research in Learning and Instruction*, Turku, Finland.

Lundholm, C. (2010). Lärandets rationalitet och komplexitet / The rationality and complexity of learning. In C. Lundholm, G. Petersson & I. Wistedt (Eds.), *Begreppsbildning i ett intentionellt perspektiv/Conceptual development from an intentional perspective* (pp. 13–21). Stockholm: Stockholm Universitets Förlag.

Lundholm, C. (2011). Society's response to environmental challenges: Citizenship and the role of knowledge. *Factis Pax, 5*, 80–96.

Lundholm, C. (forthcoming). Environmental learning from a constructivist perspective: Content, context and learner. In C. Russell, J. Dillon & M. Breunig (Eds.), *Environmental education reader*. New York, NY: Peter Lang Publishing.

Lundholm, C., & Davies, P. (in press). Conceptual change in the social sciences. In S. Vosniadou (Ed.) *International handbook of research on conceptual change* (2nd ed.). London, UK: Routledge.

Lundholm, C., & Rickinson, M. (2006). I didn't really feel sorry for the trees, it was just something to write: Exploring students' learning in environmental education. Paper presented at the Annual Meeting of the American Educational Research Association, San Francisco.

Lundholm, C., Hopwood, N., & Rickinson, M. (2008). Developing lenses for understanding environmental learning. Paper presented at the Annual Meeting of the American Research Association, New York.

Mason, L. (2003). Personal epistemologies and intentional conceptual change. In. G. Sinatra & P. Pintrich (Eds.), *Intentional conceptual change* (pp. 199–236). New Jersey: Lawrence Erlbaum Associates.

McKenzie, M. (2006). Three portraits of resistance: The (un)making of Canadian students. *Canadian Journal of Education, 29*(1), 199–222.

Nagel, M., & Lidstone, J. (2008). *Green doesn't always mean go! Children's conceptions of environmental education*. Saarbrucken: VDM Verlag Dr Müller Aktiengesellschaft & Co KG.

Nespor, J. (1987). Academic tasks in a high school English class. *Curriculum Inquiry, 17*(2), 202–228.

Osterlind, K. (2005). Concept formation in environmental education: 14-year-olds' work on the intensified greenhouse effect and the depletion of the ozone layer. *International Journal of Science Education, 27*(8), 891–908.

Payne, P. (1998). Children's conceptions of nature. *Australian Journal of Environmental Education, 14*, 19–26.

Pintrich, P. R., Marx, R. W., & Boyle, R. A. (1993). Beyond cold conceptual change: The role of motivational beliefs and classroom contextual factors in the process of conceptual change. *Review of Educational Research, 63*(2), 167–199.

Postman, N., & Weingartner, C. (1969). *Teaching as a subversive activity*. New York, NY: Dell.

Reid, A., Jensen, B. B., Nikel, J., & Simovska, V. (Eds.). (2008). *Participation and learning: Perspectives on education and the environment, health and sustainability*. Dordrecht: Springer.

Rickinson, M. (1999a). *The teaching and learning of environmental issues through geography: A classroom-based study* (Unpublished D.Phil. thesis). University of Oxford, Oxford, UK.

Rickinson, M. (1999b). People-environment issues in the geography classroom: Towards an understanding of students' experiences. *International Research in Geographical and Environmental Education, 8*(2), 120–139.

Rickinson, M. (2001). Learners and learning in environmental education: A critical review of evidence. *Environmental Education Research, 7*(3), 207–317.

Rickinson, M. (2006). Researching and understanding environmental learning: Hopes for the next ten years. *Environmental Education Research, 12*(3/4), 445–458.

Rickinson, M., & Lundholm, C. (2008). Exploring students' learning in environmental education. *Cambridge Journal of Education, 38*(3), 341–353.

Rickinson, M., Lundholm, C., & Hopwood, N. (2009). *Environmental learning: Insights from research into the student experience*. Dordrecht: Springer.

Rosiek, J. (2003). Emotional scaffolding: An exploration of the teacher knowledge at the intersection of student emotion and the subject matter, *Journal of Teacher Education, 54*(5), 399–412.

Rudduck, J., Day, J., & Wallace, C. (1997). Students' perspectives on school improvement. In A. Hargreaves (Ed.), *Rethinking educational change with heart and mind: 1997 ASCD yearbook* (pp. 73–91). Alexandria, VA: Association for Supervision and Curriculum Development.

Scott, W., & Gough, S. (2003). *Sustainable development and learning: Framing the issues*. London, UK: Routledge.

Shanahan, M., & Meyer, J. H. F. (2006). The troublesome nature of a threshold concept in Economics. In J. Meyer & R. Land (Eds.), *Overcoming barriers to student understanding: Threshold concepts and troublesome knowledge* (pp. 100–114). Great Britain: Routledge.

Shulman, L. (1987). Knowledge and teaching: Foundations of the new reform. *Harvard Educational Review, 57*(1), 1–22.

Sinatra, G. (2005). The 'warming trend' in conceptual change research: The legacy of Paul R. Pintrich. *Educational Psychologist, 42*(2), 107–115.

Sinatra, G., & Pintrich, P. (Eds.). (2003). *Intentional conceptual change*. New Jersey: Lawrence Erlbaum Associates.

Smith-Sebasto, N. J., & Walker, L. M. (2005). Toward a grounded theory for residential environmental education: A case study of the New Jersey school of conservation. *The Journal of Environmental Education, 37*(1), 27–42.

Sternäng, L., & Lundholm, C. (2011). Climate change and morality: Students' conceptions of individual and society. *International Journal of Science Education, 33*, 1131–1148.

Sternäng, L., & Lundholm, C. (2012). Climate change and costs: Investigating chinese students' conceptions of nature and economic development. *Environmental Education Research, 18*, 417–436.

Stevenson, B., & Dillon, J. (Eds.). (2010). *Environmental education: Learning, culture and agency*. Rotterdam: Sense Publishers.

Vare, P., & Scott, B. (2007). Learning for a change: Exploring the relationship between education and sustainable development. *Journal of Education and Sustainable Development, 1*(2), 191–198.

Vosniadou, S., Vamvakoussi, X., & Skopelti, I. (2008). The framework theory approach to the problem of conceptual change. In S. Vosniadou (Ed.), *International handbook of research on conceptual change* (pp. 3–34). London, UK: Routledge.

Wals, A. (Ed.) (2007). *Social learning towards a sustainable world*. Wageningen: Wageningen Academic Publishers.

Watts, M., & Alsop, S. (1997). A feeling for learning: Modelling affective learning in school science. *The Curriculum Journal, 8*(3), 351–365.

Wenger, E. (1998). *Communities of practice. Learning, meaning and identity*. Cambridge, UK: Cambridge University Press.

Ziedler, D. L., Walker, K. A., Ackett, W. A., & Simmons, M. L. (2002). Tangled up in views: Beliefs in the nature of science and responses to socioscientific dilemmas. *Science Education, 86*(3), 343–367.

26

Conventional and Emerging Learning Theories

Implications and Choices for Educational Researchers With a Planetary Consciousness

Arjen E.J. Wals
Wageningen University, The Netherlands; Cornell University, USA

Justin Dillon
King's College London, UK

Introduction

In 2001, the journal *Environmental Education Research* published a special issue which, unusually, contained only one article—a review of empirical studies of learners and learning in environmental education (EE) by Mark Rickinson. This seminal review has been cited more than 200 times, according to Google Scholar, making it one of the most cited papers in the history of the field. Rickinson focused "specifically on the nature and quality of the *evidence* generated by the work in this area" (2001, p. 207, emphasis in original). In critiquing Rickinson's review, one of us [JD] made the point that the fact that "researchers, particularly those engaged in empirical research, have failed to engage with learning theories to any depth is a concern and a weakness" (2003, p. 217).

So, what do theories of learning offer to an understanding of environmental education research and to those who use the research? In this chapter we will attempt to show how different theories of learning in a range of contexts can inform and illuminate environmental education research. We will begin by outlining some basic classical perspectives on learning mostly from educational psychology which can be recognized when analyzing different strands of environmental education that have evolved in the last forty years or so. We will then argue that there are "postnormal" currents in science and society which raise into question some often taken-for-granted but rather persistent, assumptions with respect to the nature of knowing, the status of scientific understanding, the role of uncertainty and complexity, and the degree to which they can be managed and controlled. These postnormal currents are, at least in part, triggered by sustainability issues such as runaway climate change, loss of biodiversity, rising inequality, and so on, which are characterized by multiple causation, interactions, feedback loops, and the inevitable complexity, uncertainty, and contestation (Davidson-Hunt & Berkes, 2003; Wals, 2010). We will then introduce what we refer to as "postnormal environmental education" which takes advantage of new forms of learning that do not fit the classical categories outlined below. These new forms tend to be both transformative and transboundary. Finally, we will discuss the implications of postnormal environmental education for environmental research.

Classic Theories of Learning

Perhaps a good place to start to look at what we know about learning can be found in Stella Vosniadou's summary of research into "How Children Learn" carried out on behalf of UNESCO (Vosniadou, 2001). The review, according to its preface, "focuses on aspects of how children learn that appear to be universal in much formal and informal schooling" (Walberg, 2001). Synthesizing a range of research studies, Vosniadou provides a succinct and authoritative guide (Table 26.1).

The pamphlet is designed to give a general overview of learning; however, there is one example that specifically refers to the environment. In the section on how to make learning more meaningful, Vosniadou writes: "Students can learn science by participating in a community or school environmental project" (p. 11). Although Vosniadou mainly speaks about children's learning, it can be argued that the characteristics in the table also apply to adult learning.

Reducing a huge body of empirical and theoretical work is invariably going to produce oversimplifications and omissions. There is not much in Vosniadou's pamphlet about the importance of the affective domain and not every learner fits the pattern. Nevertheless, much of what Vosniadou presents would seem to be common sense. However, the evidence

TABLE 26.1
How Children Learn (Adapted From Vosniadou, 2001)

Learning requires the active, constructive involvement of the learner.
Learning is primarily a social activity and participation in the social life of the school is central for learning to occur.
People learn best when they participate in activities that are perceived to be useful in real life and are culturally relevant.
New knowledge is constructed on the basis of what is already understood and believed.
People learn by employing effective and flexible strategies that help them to understand, reason, memorize, and solve problems.
Learners must know how to plan and monitor their learning, how to set their own learning goals and how to correct errors.
Sometimes prior knowledge can stand in the way of learning something new. Students must learn how to solve internal inconsistencies and restructure existing conceptions when necessary.
Learning is better when material is organized around general principles and explanations, rather than when it is based on the memorization of isolated facts and procedures.
Learning becomes more meaningful when the lessons are applied to real-life situations.
Learning is a complex cognitive activity that cannot be rushed. It requires considerable time and periods of practice to start building expertise in an area.
People learn best when their individual differences are taken into consideration.
Learning is critically influenced by learner motivation. Teachers can help students become more motivated learners by their behavior and the statements they make.

provided by much of the environmental education literature is that this knowledge is neither common nor accepted yet as "sense." In terms of specific theories that might benefit environmental researchers I have picked three that we think provide some affordances: behaviorism, constructivism, and social constructivism.

Hohenstein and King conceptualize learning as "a relatively permanent change in thought or in behavior that results from experience" (Hohenstein & King, forthcoming). Not all learning is conscious—many of our actions are learned subconsciously. In this chapter, though, the focus is on conceptual and motivational development as well as on the transformative potential of learning, rather than on unconscious changes in our behaviors.

Theories of Learning

Most theories of learning traditionally fall into one of two camps: behaviorist or cognitivist. Crudely put, the focus of behaviorist theories is the relationship between external influences on behavior whereas cognitivist theories focus much more on the active construction of knowledge inside the brain. Behaviorist approaches have, to some extent, fallen out of favor because they can be perceived as oversimplifying human behavior.

Behaviorist Theories of Learning Most people will be familiar with Pavlov's experiments involving bells, dogs, and food. Training dogs to salivate when they receive a stimulus, such as hearing a bell is a, perhaps, an extreme example of classical conditioning. The stimulus (the bell) leads to an unconscious response (salivation). Operant conditioning, favored by Skinner (1974), involves a more sophisticated approach to training. There are numerous examples of animals being trained using punishments and rewards (such as food) in order to behave in particular ways. Behaviorist approaches to teaching humans may also involve the use of punishments and rewards and, at their most extreme, tend to treat people as though they were a blank slate to be written on or an empty vessel to be filled.

Cognitivist Theories of Learning Behaviorist theories held sway for many years and still influence classroom practices today. However, growing dissatisfaction with behaviorism led to the development of alternative theories many of which come under the umbrella of cognitivist theories. Possibly the two most well-known theorists that contributed to the development of alternative approaches to learning are Jean Piaget and Lev Vygotsky.

Piaget's body of work, developed with Inhelder, involves a recognition that there are different and progressive stages of development of the brain which are related to the types of thinking that can be undertaken. Piaget used the term genetic epistemology to describe the theory linking human development and the ways in which we build knowledge. The first stage (sensorimotor) begins at birth and lasts for around two years. Children make sense of the world through their senses, in particular sight and touch. Between the ages of two and seven most children go through the preoperational stage and then between seven and twelve are in the concrete operational stage where the first evidence of logical thinking is evident. From twelve onward, most children are in the formal operational stage and become increasingly able to engage in abstract thought. Piaget's ideas have been criticized many times although they are still widely regarded as offering value to educators. A major criticism is that children are capable of ways of thinking much earlier than Piaget's theories suggest (Adey, Robertson, & Venville, 2002).

In terms of how people learn, two processes are critical to the Piagetian model, accommodation and assimilation. Assimilation involves little change to existing ways of thinking (schema) when faced with new phenomena or situations. Accommodation involves changing the schema in the light of new information or ideas. So, for example, testing multiple water samples which are identical doesn't involve much learning. Trying to explain an anomalous result might require some accommodation. The point is that the stage of development affects what can and cannot be learned.

A critical term for educators, cognitive challenge, explains how learners are helped to move from one level to another. If children are faced with information that is too easy or too complex, learning is unlikely to take place. If, instead, they are faced with something that is puzzling but understandable with some thought and, perhaps, some help, then the level of cognitive challenge is just right. The educator's skill is to match the task to the capability of the learner.

The idea that children are involved in actively building their ideas, through being cognitively challenged and through the processes of accommodation and assimilation are the foundations of constructivist theories of learning. Such theories recognize the importance of motivation—if students aren't motivated to learn then they won't make much progress. There is some overlap with behaviorist theories here in that rewards and punishment affect motivation to learn.

Lev Vygotsky's contribution to constructivist theories of learning was a recognition of the value of language in developing ideas. Vygotsky recognized that children encounter ideas in play and discussion with others—perhaps by being read to, for example. As children develop their understanding of what these ideas are, they become part of their internal ways of thinking. It is as though the ideas exist in a shared space before they are internalized (Vygotsky, 1978). What Vygotsky offers is a view of the value of other people including peers and teachers—the term "social constructivism" is usually used to describe such theories. Consciously or unconsciously other people bring new ideas to a child—by asking questions, by pointing things out, by telling stories, etc. Wood et al. describe the active process of supporting the building of ideas as "scaffolding" (1976). Vygotsky's term, the Zone of Proximal Development (ZPD), is used to describe the potential learning that might occur with the aid of another person. An effective teacher can help a student to understand something that they would not have done without being probed, provoked, or challenged. Effective teachers, then, are those who appreciate the value of scaffolding and cognitive challenge in maximizing a learner's development in their ZPD.

A growing awareness that culture affects learning and development is recognized by sociocultural theories (Rogoff, 2003). Different cultures have different ways beliefs about learning and use different strategies. For example, a wood carver in Japan learns in a very different manner than a wood carver in England.

Learning and Motivation Earlier we mentioned that motivation is critical if people are to learn. Whereas behaviorists might favor extrinsic motivation (the rewards and punishment mentioned above), cognitivists might favor intrinsic motivation, that is motivation from within the learner. There are many examples of both types of motivation working in classrooms and in the outdoor classroom. Understanding what motivates people is critical to being an effective educator—hence Vosniadou's focus on meaningful learning above.

So, in summary, we do know quite a lot about how people learn. A pedagogy that involves cognitive conflict, discussion and an awareness of where the learner is developmentally, intellectually, and culturally is likely to be more successful than more traditional methods that treat all learners as empty vessels into which educators pour knowledge. In the next section we look at the relationship between education, learning, and the environment.

Current Developments in Education and the Environment

Education and learning have always played a role in responding to the loss of nature, environmental degradation, natural resource depletion and, indeed, the current sustainability crisis. Over the last one hundred years or so we have seen the sequential emergence of nature conservation education to environmental education to education for sustainability, along with a number of related educations such as: outdoor education, place-based education, biodiversity education, climate change education, health education, citizenship education, peace education, and the list goes on. A critical question that is continuously asked in these educations, including in EE, is what are or should we be changing or developing in learners? Or, alternatively, how can we create optimal conditions and support mechanisms which allow citizens, young and old, to develop themselves in the face of change? The first question has instrumental connotations, whereas the second one has emancipatory ones. The difference between the questions may appear small but, as we will see, speak to a large issue.

In terms of educational process or the type of learning promoted or used, there are conventional approaches that tend to focus on expanding knowledge and understanding, developing values and attitudes, and shaping environmental behavior through forms of instruction and more exploratory activities designed to achieve predetermined goals and objectives. Wals and Jickling have referred to these conventional approaches as instrumental forms of education learning (Jickling & Wals, 2008; Wals & Jickling, 2002). Much EE around the world was aimed—and still is aimed—at changing learner behavior. Learner behavior tends to be broadly defined to include awareness, attitudes, beliefs, and values (see for instance: Hungerford & Volk, 1990). Early EE was informed by insights from behaviorist sociopsychology that assumed a more or less linear causality between environmental awareness, attitudes and environmental behavior (Fishbein & Azjen, 1980). In other words, an increase in environmental awareness and a change in attitudes would lead to more responsible environmental behavior. More recently we have come to know that these models represent an oversimplification of reality and incorrectly assume a linear correlation between knowledge, awareness, and behavior (Hannigan, 1995). People's environmental behaviors are far too complex and contextual to be captured by a simple causal model. Glasser points out that even though people have a familiarity with problems related to, what he calls, eco-cultural unsustainability, they still choose not to respond or respond ineffectively (Glasser, 2007).

It is no surprise that much of the research taking place within EE during the 1970s and 1980s was of an empirical analytical nature, aimed at understanding attitudes, values and behavior and trying to assess the extent to which a predesigned learning program or educational activity influenced those attitudes, values, and behaviors. In terms of the classic theories of learning presented earlier, we can see both behaviorist and cognitivist perspectives in conventional EE. Another observation that is relevant here is that most environmental educators and environmental educators in the early years (roughly the period 1970–1990) had a natural science background or more specifically a background in forestry, natural resource management, and environmental science and not a social science background or more specifically a background in education, pedagogy, or human development. This fact explains why a concern about loss of nature and environmental degradation tended to outweigh a concern about how people live, learn, and develop.

From the mid-1980s onward we saw a number of environmental educators and educators from neighboring fields, particularly those with a strong pedagogical background, challenge the instrumental focus of EE on behavioral change. They argued that education should above all be formative and focus on the kind of learning, capacity building, and critical thinking that will allow citizens to understand what is going on in society, to ask critical questions, and to determine for themselves what needs to be done (Jickling & Wals, 2008; Mayer & Tschapka, 2008). The idea of influencing people's environmental behavior in a predetermined way, they maintain, contradicts the very foundation of education and borders on indoctrination. Doing so becomes even more problematic when there is no certainty about what the "best" behavior is from an environmental or sustainability point of view and when citizens who are taught to behave in a particular way, find out later that this was not the "right" way after all.

This counter-sentiment gained momentum when Ulrich Beck introduced the idea of the "risk society" (Beck, 1992, 2008). Beck argues that industrial growth, fueled by unbridled consumerism, is beginning to turn against us because it is leading to more and more risks. Trends such as globalization and individualization are having an enormous impact on the complexity of society, resulting in increased insecurity and unpredictability. What is typical of the risk society is that this insecurity and unpredictability stem from unintentional and (in part) unforeseen changes to (eco)-systems. Society is constantly in motion and citizens are facing problems and challenges for which there are no ready-made solutions. Questions such as: is the fatigue syndrome ME a consequence of the many hazardous substances in our environment or not? Is the increasing infertility in men a consequence of our modern lifestyles and eating habits? Is fertility in fact decreasing? Is there an ADHD epidemic or not? Has the Deepwater Horizon oil spill led to irreversible ecological damage or will nature take care of itself and self-heal? We do not have single, simple and agreed-upon answers which lead to the associated desired behaviors that can be confidently prescribed.

In such a "risk society" people need to become capable of handling uncertainty, poorly defined situations and conflicting, or at least diverging, norms, values, interests, and reality constructions. Posch writes in an OECD-ENSI publication:

> Professional, public and private life has become increasingly complex, with divergent and even contradictory demands on the individual [who lives] within an increasingly pluralistic value system. Above all, it is necessary to look beyond everyday normalities and to search for ethically acceptable options for responsible action. (Posch, 1991, p. 12)

This "emancipatory perspective" (Wals & Jickling, 2002, p. 5) has major implications for the conceptualization of education, learning, and research as it makes prescribing and transferring particular lifestyles or (codes of) behavior problematic. From an emancipatory vantage point, education, learning, and research in a risk society should foster in learners/participants a critical stance toward the world and oneself by promoting discourse, debate, and reflection. It is through discourse that participants engage in a process of self-reflection on the relationship between their own guiding assumptions and interpretations and those of others. Furthermore, engaging in the types of ambivalent and complex questions and working toward their resolution, if only temporary, requires pluralism of thought, creativity, divergent thinking, and high degrees of participation, interactivity, and self-determination.

One could also argue that the deeper the planetary sustainability crisis, the more tempting it will be to adopt more instrumental approaches as people, policy makers, and legislators will increasingly come to think that we are running out of time and need to act now (Wals, 2010). This might be a dangerous response because a flight to the instrumental might keep us from developing a more resilient society better able to cope with risk and stress. In the next section we will expand the emancipatory perspective by introducing new forms of learning and corresponding approaches to research that appear particularly appropriate in a risk society and an era of affiliated postnormal science. The new forms of learning also show that the conventional categories of behaviorist and cognitivist theories of learning need to be complemented with so-called transformative theories of learning.

Postnormal Environmental Education (Research)

As suggested earlier, the nature of the sustainability crisis—characterized, among other things, by high levels of complexity, contestation, and uncertainty—suggests that people will need to develop capacities and qualities that will allow them to contribute to alternative behaviors, lifestyles, and systems both individually and collectively. In addition to appropriate forms of governance, legislation, and regulation, alternative forms of education and learning that can help develop these capacities and qualities will be needed. Peters and Wals (in press) refer to a range of associated forms of learning that all have emancipatory characteristics, including: transdisciplinary learning (e.g. Klein, 2000; Somerville & Rapport, 2000), transformative learning (e.g. Cranton, 2006; Mezirow & Taylor, 2009), cross-boundary learning (e.g., Levin, 2004), anticipatory learning (e.g., Tschakert & Dietrich, 2010) action learning (e.g., Cho & Marshall Egan, 2009; Marquardt, 2009) and social learning (e.g., Keen, Brown, & Dyball, 2005; Pahl-Wostl & Hare, 2004; Wals, 2007). These forms of learning show a high family resemblance in that they:

- consider learning as more than merely knowledge-based,
- maintain that the quality of interaction with others and of the environment in which learning takes place is crucial,
- focus on existentially relevant or "real" issues essential for engaging learners,
- view learning as inevitably transdisciplinary, "transperspectival," and transboundary in that it cannot be captured by a single discipline and by a single perspective and that it requires "boundary crossing,"
- regard indeterminacy a central feature of the learning process in that it is not and cannot be known exactly what will be learned ahead of time and that learning goals are likely to shift as learning progresses,
- consider such learning as cross-boundary in nature in that it cannot be confined to the dominant structures and spaces that have shaped education for centuries.

(See also Wals, 2010; Wals, van der Hoeven, & Blanken, 2009.)

Wals (2010) suggests that learning in a risk society requires "hybridity" and synergy between multiple actors and the blurring of education in formal and informal contexts. Opportunities for this type of learning expand with an increased permeability between units, disciplines, generations, cultures, institutions, and sectors. Through this hybridity and synergy, new spaces might open up that will allow for transformative learning (TL) to take place. Such space includes: space for alternative paths of development; space for new ways of thinking, valuing, and doing; space for participation minimally distorted by power relations; space for pluralism, diversity, and minority perspectives; space for deep consensus, but also for respectful disagreement and differences; space for autonomous and deviant thinking; space for self-determination; and, finally, space for contextual differences. "Transformative" here refers to a shift or a switch to a new way of being and seeing (see also: O'Sullivan, 2001). Mezirow describes TL as a process of "becoming critically aware of one's own tacit assumptions and expectations and those of others and assessing their relevance for making an interpretation" (Mezirow & Taylor, 2009, p. 4) which "enables us to recognize, reassess, and modify the structures of assumptions and expectations that frame our tacit points of view and influence our thinking, beliefs, attitudes and actions" (Mezirow & Taylor, 2009, p. 18). Table 26.2 shows the steps that Mezirow distinguishes in a transformative learning process.

Recently, scholars using social learning theory have argued that pluralism and heterogeneity can be catalysts of transformative learning as long as there is sufficient social cohesion between all actors involved for the differences among them to become constructive rather than destructive (Page, 2007; Wals, 2010). Support for this assumption comes also from system thinkers with a natural science background. Using nature as guide, they argue that it is very important that we understand systems of communities and that we begin to think (again) in terms of relations and connections (Ramage & Shipp, 2009). Systems thinking—seeing connections, relating functions to one another, making use of diversity, and creating synergy—may offer support in realizing a society that is more sustainable than is presently the case.

TABLE 26.2
Phases of Transformative Learning (Mezirow, 1978, 1997)

1. A disorienting dilemma
2. Self-examination with feelings of fear, anger, guilt, or shame
3. A critical assessment of assumptions
4. Recognition that one's discontent and process of transformation are shared
5. Exploration of options for new roles, relationships, and actions
6. Planning a course of action
7. Acquiring knowledge and skills for implementing one's plans
8. Provisional trying of new roles
9. Building competence and self-confidence in new roles and relationships
10. A reintegration into one's life on the basis of conditions dictated by one's new perspectives
11. Altering present relationships and forging new relationships

They suggest we can learn a lot from how ecosystems work as networks: mutual dependency, flexibility, resilience, and, indeed, sustainability tend to be key characteristics of such systems.

According to Fritjof Capra, one essence of sustainability can be found in resilience or the manner in which ecosystems are organized and can deal with disruptions (Capra, 1994). It is not about the individual principles and elements, but rather about the system as a whole that is constantly in motion and developing and that, as a whole, makes up more than the sum of its parts. "Healthy" ecosystems are actually learning systems Capra (2007) argues. The question is whether people, too, are capable of forming a learning system that can cope with the challenges that we face in a risk society. Learning in the context of environment and sustainability then becomes a means for working toward a "learning system" in which people learn *from* and *with* one another and collectively become more capable of withstanding setbacks and dealing with insecurity, complexity, and risks. Such learning requires that we not only accept one another's differences but are also able to put these to use.

A sustainable society is, in its essence, a reflexive society in which creativity, flexibility, and diversity are released, a society that has the capacity to lay existing routines, norms, and values on the table, but also one that has the ability to correct itself (Wals et al., 2009). Such a society cannot exist without reflexive citizens who critically review and alter everyday systems that we live by and that we often take for granted. This process entails what Argyris (1990) refers to as second order or "double loop" learning, which, in line with Mezirow's ideas, calls for reflection and deliberation on the relevance and tenability of underlying background theories and normative considerations. This process does not automatically occur in practice as people unconsciously use defence mechanisms (defensive routines) (Argyris, 1990) to prevent themselves from losing face with their peers or to avoid the uneasy feelings of doubt that long fostered assumptions are perhaps not correct. And so we often ignore (unwelcome) information that collides with our views and expectations or we dismiss this information as irrelevant or false (Glasser, 2007).

Using the emergence of the risk society as a focal point, we have introduced transformative learning and reflexivity as key ingredients of learning in the context of environment and sustainability. We argue that the characteristics of the risk society are basically the same as the ones that characterize sustainability issues (complexity, uncertainty, contestation, emergence, etc.) and that conventional instrumental theories of learning which have historically been used in (environmental) education, need to be complemented, if not replaced, by ones that are more emancipatory and transformative. In the final section of this chapter we will turn to the implications of such a shift for environmental education research.

Emerging Postnormal Research

If transformative learning is essential for understanding sustainability issues such as runaway climate change, loss of biodiversity, rising inequality, and so on, then we also need to explore emerging strands of postnormal transformative *research*: research that is capable of "handling" multiple causation, interactions, feedback loops, inevitable complexities, uncertainties, and contestations. Such research also assumes different boundaries between science and society, researcher and researched, theory and practice, and other, what we might regard as false dichotomies. Boundary-crossing and the blurring of boundaries appear essential to allow for new forms of understanding and knowledge creation to emerge (Davidson-Hunt & Berkes, 2003; Wals, 2010).

A question for us as academics working in a globalizing university system, is whether our institutions provide sufficient space to develop and strengthen these "postnormal" transformative niches in both research and education. Today's universities, indeed around the world, are somewhat caught in between two trends—an hegemonic one and a marginal but emerging one. The hegemonic one very much builds upon a model of fragmentation, management, control, and accountability. Elsewhere Peters and Wals (in press) refer to this as "science as commodity." The alternative conceptualization of "science as community" is based on integration, autonomy, learning, and reflexivity. The "science as community" perspective is gaining momentum in part because of the nature of the types of issues that are currently at hand and the recognition that more of the same won't do the trick.

In some ways the single focus on academic research output as expressed by the phrase "science for impact factors" appears to be hitting a wall as the number of journals and publications is growing rapidly, but the time for academics to sit down and read or review has gone down dramatically. In a tongue-in-cheek way it can be said that "everybody is writing for nobody." The peer-review system appears to be standing on its last legs since the system only rewards one side of the equation: the publishing side. This single emphasis on academic output arguably leads to a disconnect between universities and the communities that support them and that they are supposed to serve (Unterhalter & Carpentier, 2010). Not surprisingly, a counter-movement is gaining strength as illustrated by the revival of university science shops as conceptualized in the 1970s (www.livingknowledge.org) but also by the emergence of a number of new networks of community-engaged universities (e.g. Centro Boliviano de Estudios Multidisciplinarios, Commonwealth Universities Extension and Engagement Network, Global Alliance of Community Engaged Research, Global Universities Network for Innovation, PASCAL International Observatory, Participatory Research in Asia, and the Talloires Network) and centers of expertise focusing on sustainability issues such as the Regional Centre's of

TABLE 26.3
Two Simplified Representations of Environmental Education Research (see also Peters & Wals, in press)

	Conventional EE (Type I) Research	Postnormal EE (Type II) Research
Dominant view of learning	*Instrumental*—transfer of predetermined and relatively fixed outcomes	*Emancipatory*—high degrees of self-determination, space for transformation and cocreated and emergent outcomes
Dominant epistemological orientation	*Empirical rationalism* Finding an objective truth. Establishing causality. Single truth exists and can be known. Maximize predictability, management, and control. Minimize uncertainty.	*Socioconstructivism* Cocreation of knowledge, intersubjectively validated. Not one single "truth" but many subject to interpretation. Uncertainty as a given. Facts and values are inseparable.
Type of knowledge generated	*Scientific and technical* knowledge that can be generalized across contexts to inform attempts by various social actors to predict, control, and/or intervene for specific instrumental ends.	*Phronesis*—ethically practical knowledge that is indispensible for the work of making context specific value judgments about ends and means.
Dominant modus of understanding	Empirical analytical	Socially critical Hermeneutic-interpretive Holistic-descriptive
Locus of impact	Universal-prescriptive	Contextual-transformative
Main researcher modes	Passively detached Neutral objective Scientific expert	Actively committed Explicitly supportive or biased Colearner (representing one kind of expertise)
Role of the researched	Sources of data	Active informants Change agents Colearners
Desired outcomes include	Explanatory models Tests of hypotheses Definitive answers Engineering (tools, policies, and instruments)	Improved understandings Thick descriptions Increased (self) awareness Enlightenment (alternative ways of seeing things, new ideas) Transformation (Systemic) Change Emancipation

Expertise (RCE's) in which universities are partners in a network of NGOs, civil society organizations community groups, schools, etc. Within these networks and centers a range of participatory forms of research can be found such as: action research (Reason & Bradbury, 2001), citizens science (Irwin, 1995), and transdisciplinary research (Hirsch Hadorn et al., 2008). Some universities and university systems are also beginning to emphasize societal relevance as a key quality criterion and are extending peer review to include not only scientific peers but also user groups. Furthermore new forms of so-called reflexive monitoring and evaluation are sprouting across the globe that can be labeled "postnormal" and transformative. Examples include reflexive monitoring and evaluation (RMA) developed by Wageningen University and the Free University of Amsterdam (van Mierlo et al., 2010) and so-called "Realist Evaluation" developed by Pawson and Tiley (1997).

We have argued elsewhere that the field has seen a shift from research in EE that framed by instrumental theories of learning (what we might call Type I EE research) to research that is more emancipatory and transformative (Type II) (Dillon & Wals, 2006). This shift requires a rethinking of ontological, epistemological, and axiological assumptions. There is a need and, indeed, a responsibility, for environmental education researchers to question taken-for-granted but rather persistent, assumptions with respect to the nature of knowing, the status of scientific understanding, the role of uncertainty and complexity, and the degree to which they can be managed and controlled. For example, we might see the purpose of EE research as moving from identifying gaps in knowledge to helping participants to identify what knowledge was most useful to them in meeting their own needs. Type II research shows high degrees of compatibility with the "postnormal" trends in science and society that we have described in this chapter (Table 26.3).

There is a danger of oversimplifying the differences between Types I and II research. There is also a danger of setting up a model that does not fully represent any overlapping between the different methodologies. Nevertheless, we do believe that the juxtaposition in Table 26.3 can be generative in opening up discussion among researchers in the field which can inform some critical choices that will need to be made in the coming years.

References

Adey, P., Robertson, A., & Venville, G. (2002). Effects of a cognitive stimulation programme on Year 1 pupils. *British Journal of Educational Psychology, 72*(1), 1–25.

Argyris, C. (1990). *Overcoming organizational defenses. Facilitating organizational learning*. Boston, MA: Allyn and Bacon.

Barth, M., Godemann, J., Rieckmann, M., & Stoltenberg, U. (2007). Developing key competencies for sustainable development in higher education. *International Journal of Sustainability in Higher Education, 8*(4), 416–430.

Beck, U. (1992). *Risk Society: Towards a new modernity*. London, UK: Sage.

Beck, U. (2008). *World at risk*. Cambridge, UK: Polity Press.

Berlyne, D. E. (1965). Curiosity and education. In J. D. Krumbolts (Ed.), *Learning and the educational process*. Chicago, IL: Rand McNally and Co.

Capra, F. (1994). *Ecology and community*. Berkeley, CA: Center for Ecoliteracy. Retrieved from http://www.ecoliteracy.org/publications/pdf/community.pdf

Capra, F. (1996). *The web of life*. New York, NY: Anchor/Doubleday.

Capra, F. (2007). Foreword. In A. E. J. Wals (Ed.), *Social learning towards a sustainable world* (pp. 13–17). Wageningen: Wageningen Academic Publishers.

Cho, Y., & Marshall Egan, T. (2009). Action learning research: A systematic review and conceptual framework. *Human Resource Development Review, 8*(4), 431–462.

Cranton, P. (2005). *Understanding and promoting transformative learning*. San Francisco: Jossey-Bass.

Davidson-Hunt, I., & Berkes, F. (2003). Learning as you journey: Anishinaabe perception of social–ecological environments and adaptive learning. *Conservation Ecology, 8*(1), 5.

de Waal, F. (2009). *The age of empathy: Nature's lessons for a kinder society*. New York, NY: Crown Publishing Group.

Dillon, J. (2003). On learners and learning in environmental education: Missing theories, ignored communities. *Environmental Education Research, 9*(2), 215–226.

Dillon, J., & Wals, A. (2006). On the dangers of blurring methods, methodologies, and ideologies in environmental education research. *Environmental Education Research, 12*(3/4), 549–558.

Festinger, L. (1957). *A theory of cognitive dissonance*. New York, NY: Harper and Row.

Fishbein, M., & Ajzen, I. (1980). *Understanding attitudes and predicting social behavior*. Englewood Cliffs, NJ: Prentice Hall Inc.

Glasser, H. (2007). Minding the gap: The role of social learning in linking our stated desire for a more sustainable world to our everyday actions and policies. In A. E. J. Wals (Ed.), *Social learning towards a sustainable world* (pp. 35–62). Wageningen: Wageningen Academic Publishers.

Green, D., & McDermott, F. (2010). Social work from inside and between complex systems: Perspectives on person-in-environment for today's social work. *British Journal of Social Work*, 1–17. doi:10.1093/bjsw/bcq056.

Hadorn, G. H., Hoffman-Riem, H., Biber-Klemm, S., Grossenbacher-Mansuy, W., Joye, D., Pohl, C., Wiesmann, U., & Zemp, E. (Eds.). (2008). *Handbook of transdisciplinary research*. Springer: Frankfurt.

Hannigan, J. A. (1995). *Environmental sociology: A social constructionist perspective*. London, UK: Routledge.

Hesselink, F., van Kempen, P. P., & Wals, A. E. J. (2000). *ESDebate: International online debate on education for sustainable development* (p. 98). Gland, Switzerland: IUCN.

Hohenstein, J., & King, H. (forthcoming). Learning: Theoretical perspectives that go beyond context. In J. Dillon & M. Maguire (Eds.), *Becoming a teacher* (4th ed.).

Hungerford, H., & Volk, T. (1990). Changing learner behavior through environmental education. *Journal of Environmental Education, 21*(3), 8–21.

Irwin, A. (1995). *Citizen science: A study of people, expertise, and sustainable development*. London: Routledge.

Jickling, B., & Wals, A. E. J. (2008). Globalization and environmental education: Looking beyond sustainable development. *Journal of Curriculum Studies, 40*(1), 1–21.

Kagawa, F., & Selby, D. (Eds.). (2010). *Education and climate change. Living and learning in interesting times*. London, UK: Routledge.

Keen, M., Brown, V. A., & Dyball, R. (2005). *Social learning in environmental management. Towards a sustainable future*. London, UK: Earthscan.

Kellstedt, P. M., Zahran, S., & Vedlitz, A. (2008). Personal efficacy, the information environment, and attitudes toward global warming and climate change in the United States. *Risk Analysis, 28*(1), 113–126.

Klein, J. T. (2000). Integration, evaluation and disciplinarily. In M. Somerville & D. Rapport (Eds.), *Transdisciplinarity: Recreating integrated knowledge*. Oxford, UK: EOLSS Publishers.

Levin, M. (2004). Cross-boundary learning systems—Integrating universities, corporations, and governmental institutions in knowledge generating systems. *Systemic Practice and Action Research, 17*(3), 151–159.

Marquardt, M. (2009). *Action learning for higher education institutions*. Kuala Lumpur: AKEPT Press.

Marquardt, M. J. (1999). *Action learning in action*. Palo Alto, CA: Davies-Black.

Mayer, M., & Tschapka, J. (Eds.). (2008). *Engaging youth for sustainable development: Learning and teaching sustainable development in lower secondary schools*. Retrieved from www.ensi.org/Publications/media/downloads/223/Engaging_Youth_08_internet.pdf on

Mezirow, J. (1978). *Education for perspective transformation: Women's re-entry programs in community colleges*. New York, NY: Teacher's College, Columbia University.

Mezirow, J. (1997). Transformative learning: theory to practice. In P. Cranton (Ed.), *Transformative learning in action. New directions* in *adult* and *continuing education*, no. 74 (pp. 5–12). San Francisco, CA: Jossey-Bass.

Mezirow, J., & Taylor, E. W. (Eds.). (2009). *Transformative learning in practice: Insights from community, workplace, and higher education*. San Francisco, CA: Jossey-Bass.

O'Sullivan, E. (2001). *Transformational learning*. London, UK: Zed Books.

Page, S. (2007). *The difference: How the power of diversity creates better groups, firms, schools, and societies*. Princeton, NJ: Princeton University Press.

Pahl-Wostl, C., & Hare, M. (2004). Processes of social learning in integrated resources management. *Journal of Community and Applied Social Psychology, 14*, 193–206.

Pawson, R., & Tilley, N. (1997). *Realist evaluation*. London, UK: Sage.

Peters & Wals (*in press*)

Piaget, J. (1952). *Origins of intelligence in children*. New York, NY: International Universities Press.

Piaget, J. (1964). Development and learning. *Journal of Research in Science Teaching, 2*, 176–186.

Posch, P. (1991). Environment and school initiatives. In K. Kelly-Laine & P. Posch (Eds.), *Environment, schools and active learning*. Paris: OECD.

Ramage, M., & Shipp, K. (2009). *System thinkers*. London, UK: Springer

Reason, P., & Bradbury, H. (2001). *Handbook of action research: Participative inquiry and practice*. London, UK: Sage.

Rickinson, M. (2001). Learners and learning in environmental education: A review of recent research evidence. *Environmental Education Research, 7*, 207–317.

Rogoff, B. (2003). *The cultural nature of human development*. Oxford, UK: Oxford University Press.

Scheffer, M. (2009). *Critical transitions in nature and society*. Princeton, NJ: Princeton University Press.

Skinner, B. F. (1974). *About behaviorism*. New York, NY: Alfred A. Knopf.

Somerville, M., & Rapport, D. (Eds.). (2000). *Transdisciplinarity: Recreating integrated knowledge*. Oxford, UK: EOLSS

Tschakert, P., & Dietrich, K. A. (2010). Anticipatory learning for climate change adaptation and resilience. *Ecology and Society, 15*(2), 11.

Unterhalter, E., & Carpentier, V. (2010). *Global inequalities and higher education: Whose interests are we serving?* London: Palgrave Macmillan.

van Mierlo, B. C., van Regeer, B., Amstel, M., van Arkesteijn, M. C. M., Beekman, V., Bunders, J. F. G., . . . Leeuwis, C. (2010). *Reflexive monitoring in action. A guide for monitoring system innovation projects*. Wageningen/Amsterdam: Communication and Innovation Studies, WUR/ Athena Institute, VU.

Vosniadou, S. (2001). *How children learn*. Brussels: International Academy of Education.

Vygotsky, L. (1978). *Thought and language*. Cambridge, MA: MIT Press.

Walberg, H. J. (2001). Preface. In S. Vosniadou (Ed.), *How children learn* (p. 2). Brussels: International Academy of Education.

Wals, A. E. J., & Jickling, B. (2002). "Sustainability" in higher education from doublethink and newspeak to critical thinking and meaningful learning. *Higher Education Policy, 15*, 121–131.

Wals, A. E. J. (1994). Action research and community problem solving: Environmental education in an inner-city. *Educational Action Research, 2*(2), 163–182.

Wals, A. E. J. (2010). *Message in a bottle: Learning our way out of un-sustainability*. Inaugural lecture held on May 27, 2010, accepting the endowed Chair and UNESCO Chair of Social Learning and Sustainable Development, Wageningen University, Wageningen.

Wals, A. E. J., & Blewitt, J. (2010). Third wave sustainability in higher education: some (inter)national trends and developments. In P. Jones, D. Selby, & S. Sterling (Eds.), *Green infusions: Embedding sustainability across the higher education curriculum* (pp. 55–74). London, UK: Earthscan.

Wals, A. E. J., Alblas, A. H., & Margadant-van Arcken, M. (1999). Environmental education for human development. In A. E. J. Wals (Ed.), *Environmental education and biodiversity*. Wageningen: National Reference Centre for Nature (IKCN).

Wals, A. E. J. (Ed.). (2007). *Social learning towards a sustainable world*. Wageningen: Wageningen Academic Publishers.

Wals, A. E. J., van der Hoeven, N., & Blanken, H. (2009). *The acoustics of social learning: Designing learning processes that contribute to a more sustainable world*. Wageningen/Utrecht: Wageningen Academic Publishers/SenterNovem.

Wals, A. E. J., Beringer, A. R., & Stapp, W. B. (1990). Education in action: A community problem solving program for schools. *Journal of Environmental Education, 21*(4), 13–19.

27

Belief to Behavior

A Vital Link

JOE E. HEIMLICH
Ohio State University, USA

PREETHI MONY
Columbus, USA

VICTOR YOCCO
Ohio Department of Public Safety, USA

What one believes affects how one acts. Consider the act of separating recyclables from waste. For a child, this act appears simple—put different items into different bins. For an adult, the act is complicated by issues such as having curbside collection or taking the material somewhere, convenience, cost of participation, proximity to a center, balancing effort versus direct cost, and the like. Both the child and the adult may believe recycling is a desirable behavior, but age and conditions alter the action toward the belief. As people age, the relationships between belief and behavior become convoluted with the inclusion of social norms, expectancy, prior experiences, and a host of other affective responses that serve to mask the degree to which belief and behavior are closely tied. For environmental education, beliefs themselves, and the connection between belief and behavior provide important insight into constructing programs built toward behavioral outcomes (for critical discussions on behavior and purposes of education, see Kollmuss & Ageyman, 2002; Stevenson, 2007). If, as Rickinson, Lundhom, and Hopwood (2009) argue, environmental learning and learners' experiences are intractably related, it is important to understand the structure of beliefs and how these tie to cognition and, inevitably, to behavior.

The study of how what one believes relates to how one acts is central to those areas of environmental education where behavior is a desired outcome. Is it possible to truly affect what one chooses to believe or how someone chooses to act? Are there ways in which we can facilitate or manage how one takes on new actions and integrates those as habits embedded within routines (Heimlich & Ardoin, 2008)?

Mainieri, Barnett, Valder, Unipan, and Oskamp (1997) found environmental beliefs were the strongest predictors of environmental behavior, but beliefs can change over time, and common beliefs do not necessarily lead to common actions between and among people. A cursory examination across behavioral and cognitive psychology and environmental education research presents a framework for a future environmental education research agenda related to beliefs and their necessary connection with behavior.

This chapter starts with an exploration of what beliefs are and how they reside within individuals. The second part considers what has been learned regarding the relationship of beliefs with behavior. The third section explores the centrality of beliefs to environmental behavior. The fourth section addresses the connection between environmental concerns and beliefs. The fifth section covers briefly the contentious topic of what are proenvironmental behaviors. Lastly, section five deals with connecting beliefs to behavior, and understanding the role of intention to behavioral outcomes.

Section 1: The Foundation of Beliefs

Rokeach (2000) stated that beliefs, attitudes, and values were all organized within the affect to form a somewhat integrated cognitive system linking affect with cognition. At their purest level, Rokeach suggested, beliefs are inferences about underlying states of expectancy—what one expects to happen in a certain context. Belief, then, is a foundational component of knowledge; what one believes to be true is then compared to what is held by other evidence to be true. There are propositional aspects to the structure of beliefs. For every belief, there is a subject—the holder of the belief—and an object, or the proposition of the belief.

The Rokeach centrality model (1968) is constructed of five layers of beliefs:

- *Central core* beliefs are formed from direct experience and reinforced by social consensus
- *Next to core* are direct experience beliefs that are not dependent on consensus
- *Intermediate-authoritative* beliefs are those held based on what we think people we believe should have answers to our questions would hold

- *Derived* beliefs are based on authoritative worldviews we trust
- *Peripheral-inconsequential* beliefs tend to be individual with no need for consensus.

Rokeach (1968) further posits that age, experience, and acquired knowledge those people held as referents effect how one determines what one believes to be true.

Bem (1970), however, adds an additional layer of depth to individuals' beliefs. He suggested there are also zero-order beliefs which are nonconscious and experientially based, as opposed to first-order beliefs of which a person is consciously aware. First order beliefs are very much dependent on the use of external authority—less as a child, but increasing as people get older and determine those authorities they accept or reject. Such determinations use generalizations and stereotypes based on experiences over time, and tie to the centrality model in the consensus at the core, and in the intermediate-authoritative beliefs. In terms of determining what one believes to be factual, individuals use disconfirmation on new information to determine if it fits her/his belief system (Reeder & Brewer, 1979). The important difference between Rokeach and Bem lies in the possibility for the belief to be influenced by the consensus of others, which could be environmental evidence or fact. Rokeach would hold that a shift in scenario leading to unanimous social consensus around a different belief would thereby influence the individual's belief. On the other hand, Bem would suggest that the belief would remain static due to the subconscious nature of the belief. For example, current research in the United States shows that the belief toward or against climate change and its causes is not solely related to environmental scientific data, but is tied directly to political affiliation (Dunlap & McCright, 2008; McCright & Dunlap, 2010).

Situational and dispositional factors have long been believed to interact with a belief structure to influence perceptions and behaviors (e.g., Harvey, Hunt, & Schroder, 1961; Withers & Wantz, 1993). Harvey (1986) framed a belief structure concept in which four systems reside on a continuum from belief concreteness to belief abstractness. He found people tended to have predominance in one of the four systems on the continuum where they exhibit traits related to belief discrimination, belief integration, judgment, and openness.

The literature on beliefs refers to belief structures, larger systems that link different attitudes held by an individual (e.g., Bem, 1970; Eagly & Chaiken, 1993). From these structures, symbolic beliefs (those not readily testable and presupposed) and instrumental beliefs (where there is a correlation between stated beliefs and observed behaviors), the dissonance within an individual regarding competing beliefs can emerge. Many environmental beliefs seem to be symbolic in nature where the behaviors serve to maintain consistency and relieve dissonance at the instrumental level of action (Jurin, 1995).

All people hold beliefs about the environment and specific environmental issues, their relationship to the issue and the environment, and toward the efficacy of their own actions on the issue. Beliefs exist in the complex realities of an individual's life. Researching beliefs, however, is critical for better understanding the varied ways people frame and enact environmental beliefs.

Section 2: Beliefs and Behavior

Beliefs are closely aligned with other elements of affect, but are an important bridge to behavior. The cognitive connection of how beliefs are supported by knowledge of how behaviors reflect a belief nearer to the core is important, and is strongly related to how the beliefs enhance or deflate the ego. Beliefs that support the ego, when close to core, lead to a sense of belongingness. Because belief structures are constantly being compared against new information, especially information necessary to make complex decisions and solve problems, it is posited that beliefs themselves are structured cognitively with the belief structure a template so the cognitive message has form and meaning (Tanner, 1999).

In many models of human behavior, belief plays an important role. Corraliza and Berenguer (2000) found that environmental behavior depends on the interaction of personal and situational variables. They also found that when there is high degree of conflict between personal dispositions and situational conditions, attitudes are weak predictors of behavior as the situational variables were dependent on the specific environmental behavior considered. For example, an individual may hold a negative attitude towards the use of fossil fuels. Simultaneously, the individual may choose to live in a suburb and commute via an automobile with an internal combustion engine. A common structure across behavior prediction models is the incremental change from reactions, to attitudes, to knowledge, to behavior (Danter, 2005). The Theory of Planned Behavior and its predecessor, the Theory of Reasoned Action, guide many of the models presently used (Zint, 2002). One consistent note of caution emerges across studies—individuals do, and will continue to, take action and behave in certain ways that they feel reflect their beliefs. Those ways may not always reflect the choices that would be considered rational or most reflective of the belief by outsiders, even when the individual's belief would suggest it does (Clover, 2002).

The importance of beliefs in an individual's cognition and behavior is evident throughout the research in environmental education. A short listing of some of the topical areas within EE for which belief has been a major element of study cross formal, free-choice, and community education, policy, and teacher education. Some of the broader categories of studies include **examinations of traditional beliefs** (Anthwal, Gupta, Sharma, Anthwal, & Kim, 2010); **environmental responsibility** (Jagers & Matti, 2010; Mayes & Richins, 2009; Mobley, 2010);

environmental protection (Dietz & Stern, 2002); **conservation programming** (Morgan-Brown, Jacobson, Wald & Child, 2010); **environmental risk** (El-Zein, Nasrallah, Nuwayhid, Kai, & Makhoul, 2006); **environmental policy** (Parag & Eyre, 2010; Vazquez Brust & Liston-Heyes, 2010); **environmental health** (Severtson, 2009); **environmental literacy** (Wright, 2008); and **environmental concern** (Olofsson & Ohman, 2006). Topics within the large scope of environment include topics such as **agriculture** (Muma, Martin, Shelley, & Holmes, 2010; Vignola, Koellner, Scholz, & McDaniels, 2010); **climate change** (McCright & Dunlap, 2010; Nolan, 2010; Paton & Fairbairn-Dunlop, 2010); **energy** (Delshad, Raymbnd, Sawicki, & Wegener, 2010; Jurin & Fox-Parrish, 2008; Spence, Poortinga, Pidgeon, & Lorenzoni, 2010; Wolsink, 2010); **water and oceans** (Fritsch, 2009; Needham, 2010; Rundblad 2008); off-road vehicle use (Smith, Burr, & Reiter, 2010); **litter** (Brown, 2010); **wildlife** (Blanchard, 1995; Browne-Nunez & Jonker, 2008; Kaltenborn, Bjerke, Nyahongo, & Williams, 2006); **biodiversity** (Dervisoglu, 2009; Menzel & Bogeholz, 2010); **ecotourism** (Donohoe & Needham, 2006; Fletcher, 2009); **sustainable development** (Qablan, Al-ruz, Khasawneh, & Al-Orami, 2009); **forestry** (Egan & Jones, 1993; Vaske, Donnelly, Williams, & Jonker, 2001); **tourism** (Dolnicar & Leisch, 2008; Zhang & Wang, 2009); **invasive species** (Bremner, 2007); and more. Beliefs have also been explored within the environmental education literature as **antecedents for behavior** (Cordano, Welcomer, Scherer, Pradenas, Parada, 2010); **components of attitudes** (Milfont & Duckitt, 2010) and **attitude development** (Farmer, Knapp, & Benton, 2007); and for **teacher efficacy** (Kusmawan, O'Toole, Reynolds, & Bourke, 2009; Moseley, Huss, & Utley, 2010). The findings of these studies have consistently suggested that beliefs impact actions an individual chooses to undertake; individual belief systems need to be considered when predicting or explaining behavior. There is clearly a relationship between beliefs and behaviors, but before that relationship can be explored, it is helpful to understand how central beliefs are toward individual adoption of environmental behaviors.

Section 3: Centrality of Beliefs to Environmental Behavior

Beliefs are central to many studies in the affective domain in environmental education. This is due, in great part, because of well-established relationships between beliefs and behavior (e.g., Ajzen & Fishbein, 1980). In 1978, Dunlap and Van Liere created the New Environmental Paradigm (NEP) scale to categorically determine individuals' beliefs toward the environment, developed from the argument that environmentalism challenged the Dominant Social Paradigm which was human-centric and consumption based. In the decades since, this belief scale has been tested to examine the centrality of the beliefs in generic environmental dispositions and primitive beliefs (Edgell & Nowell, 1989) and to determine whether the NEP is a source of coherence between environmental beliefs and attitudes (Xiao & Dunlap, 2008). The NEP has been used to find correlations with beliefs about money (Hodgkinson & Innes, 2000), preferences for social development (Raudsepp, 2001), correlation to ecologically responsible consumer behavior (Roberts & Bacon, 1998) and cross-cultural characteristics (Bechtel, Verdugo, & Pinheiro, 1999). Cordano, Welcomer, and Scherer (2010) tested all versions of the NEP and some NEP-derived scales and found a significant amount of variance in the measure of intention to engage in proenvironmental behaviors with all the scales.

Dunlap, Van Liere, Mertig, and Jones (2000) suggest that rather than measuring individual beliefs, the original NEP identified "primitive beliefs" that were indicative of a proenvironmental orientation or worldview. They restructured the NEP and renamed it as the New Ecological Paradigm Scale by creating a new fifteen-item scale which updated items, broadened the content, and provided a better balance between pro- and anti- NEP items. Manoli, Johnson, and Dunlap modified the New Ecological Paradigm Scale for use with children in 2007. And in 2008, Corral-Verdugo, Carrus, Bonnes, Moser, and Sinha proposed an alternative to the older NEP worldview with the New Human Interdependence Paradigm. This more recently proposed paradigm which attempted to create a belief measure that would overcome, but also partially include, the environmental beliefs and worldviews of both the old Human Exceptionalist Paradigm (old NEP) and the more recent ecological paradigm. This type of discussion leads to a need to understand the affective domain in which the individual's beliefs exist.

Section 4: Environmental Concerns and Beliefs

Various studies have examined environmental concern (Best, 2010; Borden & Powell, 1983; Fransson & Gärling, 1999; Schultz, 2000, 2001; Stern & Dietz, 1994), as well as some related but distinct constructs of environmental concern such as caring for environment/animals (Allen & Ferand, 1999; Geller, 1995; Teel, Manfredo, & Stinchfield, 2007; Vining, 2003), empathy toward nature (Hills, 1993; Myers, Saunders, & Bexell, 2009), and stewardship (Carr, 2002; Esty, Levy, Srebotnjack, & Sherbinin, 2005). Despite this there has been a lack of consistency in defining the concept of environmental concern (Takacs-santa, 2007). Weigel and Weigel (1978) define environmental concern as "individuals' relatively enduring beliefs and feeling about ecology such that predispositions to engage in pro or antienvironmental actions could be anticipated" (p. 4). While Maloney and Ward (1973) characterize environmental concern as an individual's emotionality toward ecological issues, Schultz (2000, 2001) suggests environmental concerns are strongly linked to underlying value orientations based on Schwartz's (1994) more general values inventory. More recently, theorists posit that all individuals have concern for the future of the environment and preventing catastrophic environmental degradation (Schultz, 2000, 2001; Schultz et al., 2005; Stern & Dietz, 1994; Stern, Dietz, & Kalof, 1993); it is the impetus for concern that differs, i.e., an egoistic, social-altruistic, or biospheric orientation.

Given the varying definitions of environmental concern, Xiao and Dunlap (2007) conducted a study to determine whether it is meaningful to conceptualize and attempt to measure environmental concern. They found that in United States and Canada populations, attitudes toward environmental issues are not fragmented and unorganized. Instead, public attitudes are sufficiently well organized and coherent to justify the construct of environmental concern. Studies have demonstrated the complexity of components that comprise and explain environmental concern. Wall (1995) examined general versus specific concern and found that education and political identification were important factors in widespread environmental concern. Stern et al. (1993) found that gender appears to play a role when examining consequences for self (egoistic), others (social-altruistic), and the biosphere (biospheric); women were found to hold stronger beliefs about these consequences, Xiao and McCright (2007) later showed that age, level of education, and political ideology were better predictors of environmental concern than gender and income.

Many environmental behavior models build a causal chain from attitudes to behaviors. Most of these models are based on studies that reveal attitude or awareness outcomes with a weak claim, if any, to actual behavior change, though concern remains an important predictor. Some findings suggest that egoistic concerns are negatively correlated with proenvironmental behavior, while biospheric concerns are positively correlated with proenvironmental behavior (Schultz et al., 2005). Concern also plays a role in an individual's willingness to pay for an environmental action (Nixon, Saphores, Ogunseita, & Shapiro, 2009; Van Birgelen, Semeijn, & Keicher, 2009). Beliefs about consequences of environmental conditions can also predict willingness to take political action (Stern et al. 1993).

Environmental behaviors cover a broad spectrum of actions, including individual, collective, simple, complex, passive, active, one-shot, and ongoing actions. As a result there is a challenge in assuming any one study can accurately describe how environmental concern will transfer between and among actions and issues. Stern (2000) proposed that environmental behaviors could be classified into intent-oriented (based on motivation of the individual) and impact-oriented (based on environmental impact of the action) behaviors. Building on this research, Poortinga, Steg, and Vlek (2004) found that the motivational determinants of environmental behavior could be used to further sub-divide intent-oriented behaviors into intent-oriented with direct environmental impact behaviors and intent-oriented with indirect environmental impact behaviors.

Regardless, the affective frames of environmental concern—caring, stewardship, and empathy—are components of belief, and determine what actions the individual may undertake are congruent with the beliefs that are core or close to core. Without subsequent action, changes in these forms of concern are not realized. Thus, some researchers believe that understanding concern, caring, empathy, or stewardship may reveal the affect support and belief systems that can be actuated within individuals.

Section 5: Proenvironmental Behaviors

Environmental concern including stewardship, caring, and empathy, suggests a strong orientation that then supports or counters an individual's behaviors. As a result, actions considered to be "proenvironmental" vary from individual to individual and context to context. Proenvironmental behaviors are ultimately directed at solving a problem and determined by those who will carry them out (Jensen, 2002). They do not describe a finite set of specific behaviors, but instead represent those that are understood by the individual to be proenvironmental, even when there might be better, more environmentally beneficial options (Heimlich & Harako, 1994). Consider the example of use of cloth rather than disposable diapers. In a desert environment the reusable option wastes water, which is, in that context, the more valuable natural resource. In this case, supporting the appearance of what seems to be the more proenvironmental behavior is not necessarily what science would suggest is the more beneficial behavior for the environment. This example also serves to highlight the sometimes contentious nature of what constitutes a proenvironmental behavior. If one is acting on the information at hand, and their belief is that they are conserving resources, how can environmental education present information in nonconfrontational ways that may show one's current behavior does is not truly reflecting a belief in conserving resources? This suggests that there are two proenvironmental behavior categories, the first is those behaviors in which an individual engages under the assumption the action in line with the underlying, proenvironmental beliefs. The second is those which empirical data suggest most reduce human impact on the environment. For example, the concept of recycling is, for many, a proenvironmental action based on the belief that recycling reduces impact of consumption; yet, those who study waste and ecology consistently note that consumption is the activity that needs to be addressed to lower the environmental impact. The first set of behaviors may not be *wrong* in the way the individual frames the issue and the impact; indeed, these behaviors are often normed by social pressures, media, and well-intentioned educators. The latter set are those that are from a more global perspective, context, and condition dependent, and more sensitive to change in understanding that is a necessary part of the nature of science (National Research Council [NRC], 2009).

Adopting new or different proenvironmental behavior is generally limited by a large number of constraints. Tanner (1999) frames subjective factors such as affective preference for proenvironmental behavior including responsibility and perceived barriers within the objective conditions that inhibit performance. In this study, subjective constraints also explained a significant amount of variance, with structural constraints contributing as well. Comparing three models of measurement, Bamberg and Schmidt (2003) found that role beliefs and use habits were explanatory and predictive to planned behavior. How one perceives their behavior reflects who they are and may explain individual behavior; what one has done before is a strong predictor of future behaviors.

In 2000, Stern examined various theories of environmentalism and suggested that rather than individual variables, groups of causal variables have a more predictive value explaining what Stern called "environmentally significant behaviors," comparable to responsible environmental behavior (Hines, Hungerford, & Tomera, 1987) or proenvironmental behavior (Kollmuss & Agyeman, 2002). The four types of causal variables Stern identified include attitudinal factors such as values, beliefs, and norms; external or contextual forces; personal capabilities; and habits or routines. The extent to which each type of causal variable is predictive of behavior depends on the constraints related to the behavior. Across these studies, the reality of context and individual holdings are important barriers or enablers to actions. Proenvironmental behaviors are complex and research can continue to inform practice by revealing ways to connect beliefs with the desired sets of behaviors.

Section 6: Connecting Beliefs to Intention and Behavior

For decades, a number of researchers and practitioners of environmental education acted under an assumption that knowledge leads to attitudes, which in turn lead to behaviors (Peyton & Miller, 1980; Ramsey & Rickson, 1977). Such an assumption led, logically, to a focus on cognitive learning with hopes and aspirations that attitudes and behaviors would change. This one-way linear model (Fig. 27.1) assumed cause and effect between variables and was never validated by EE research (Hungerford & Volk, 1990).

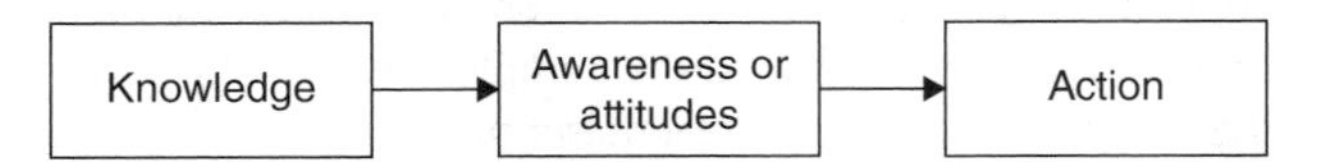

Figure 27.1 Behavioral change system, oversimplified model of behavior. (From Kollmuss & Agyeman, 2002)

In the seminal meta-analysis of 128 published environmental education studies, Hines et al. (1987) developed a model of responsible environmental behavior (Fig. 27.2). The model is predicated upon a number of factors that precede an individual's intention to engage in such behaviors, with the assumption that intention leads to behavior. Knowledge of the issues, knowledge of the action strategies, and action skills interact with internal and external locus of control, attitudes, and personal responsibility to form the intention. External variables such as finances, social group norms, and other choices can reinforce or weaken factors of the model.

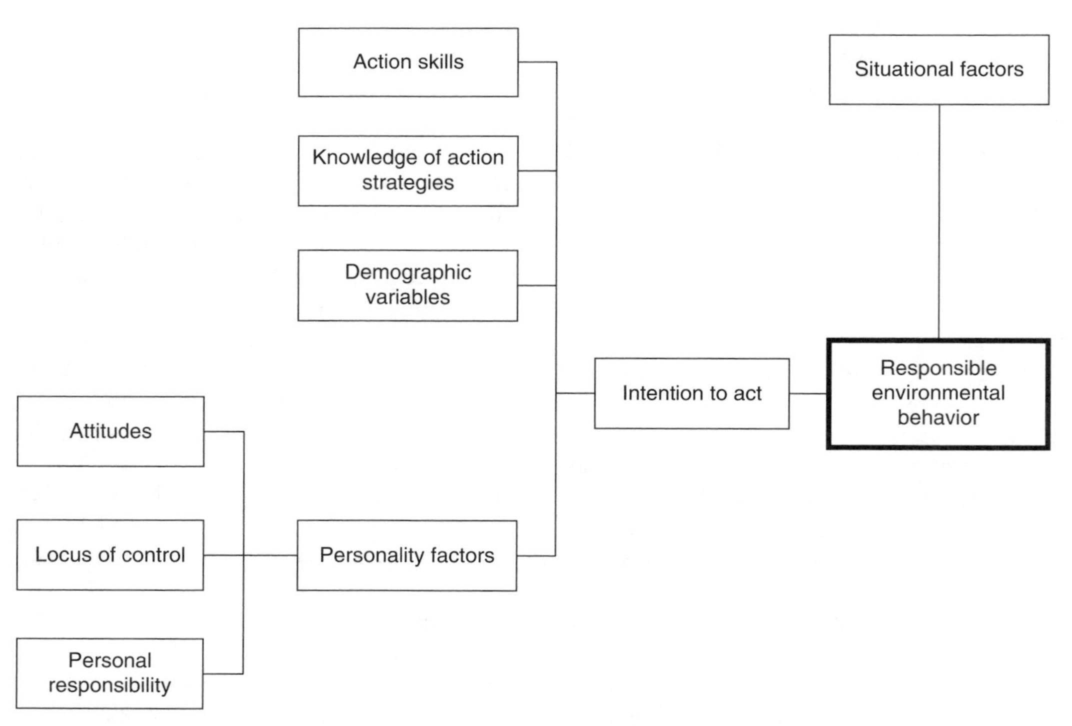

Figure 27.2 Responsible environmental behavior model. (From Hines, Hungerford, and Tomera, 1987)

Additional researchers of the period also looked at factors contributing to behavior (Borden, 1984–1985; Borden & Powell, 1983; Holt, 1988; Hungerford, Tomera, & Sia, 1985–1986; Koslowsky, Kluger, & Yinon, 1988; Marcinkowski, 1989; Ramsey, 1989; Simpson, 1989; Sivek, 1989). Some studies since have also looked at instructional strategies and their impacts on responsible environmental behaviors and have consistently found connections between instruction and behaviors (Culen & Volk, 2000; Hsu, 2004; Hsu & Roth, 1999; Ramsey, 1981; Smith-Sebasto, 1995; Tung, Huang, & Kawata, 2002; Zelezny, 1999).

In a third model developed by Hungerford and Volk (1990), predictor variables were organized into three categories (entry level, ownership, and empowerment) that interact to lead to behavior. These categories, as shown in Figure 27.3, suggest the importance and role of each predictor variable in shaping the intention to act.

No common set of variables of responsible environmental behavior predict a wide range of proenvironmental behaviors, but there do appear to be differing predictive factors for different specific behaviors (McKenzie-Mohr, Nemiroff, Beers, & Desmarais, 1995). Kaiser, Wolfing, and Fuhrer (1999) found attitude to be a powerful predictor of environmental behavior. They claimed that prior studies finding knowledge or other factors to be stronger, failed to hold a unified concept of attitude, lacked measurable correspondence between attitude and behavior on a more broad level, and did not adequately consider the external constraints on individuals' behaviors. In general, studies have shown that knowledge does appear to be significant in the model, but that it is not a predictor of behavior, suggesting knowledge as a component of conation and intent (Kaiser et al., 1999) or that systemic knowledge exerts only mediated influence on behaviors (Frick, Kaiser, & Wilson, 2004). Courtenay-Hall and Rogers (2002) identified what they saw as seven gaps between knowledge and behavior: (1) critical thinking/behavior change; (2) reflective practice/researcher as authority; (3) two literatures; (4) conscious/nonconscious; (5) direct/indirect action; (6) gender; and (7) moral agency (internal)/cultural identity (external). Research on how these gaps can be bridged will provide useful data in moving forward environmental education research, and improving the effectiveness of environmental education programming.

Several studies reveal a variety of affect factors as strong predictors of behavior (Finger, 1994; Grob, 1995; Hwang, Kim, & Jeng, 2000; Kuhlemeier, Van den Bergh, & Lagerweij, 1999). As examples, environmental beliefs were found by Mainieri et al. (1997) as the strongest predictor of behavior. Attitude (Hwang et al., 2000; Kuhlemeier et al., 1999) and locus of control (Hwang et al., 2000) were found to be more important than knowledge. Personal-philosophic values and emotions were put forth by Grob (1995) as being useful for prediction. Finger (1994) found environmental experiences and the reaction to the experiences as important predictors of individuals' environmental behaviors.

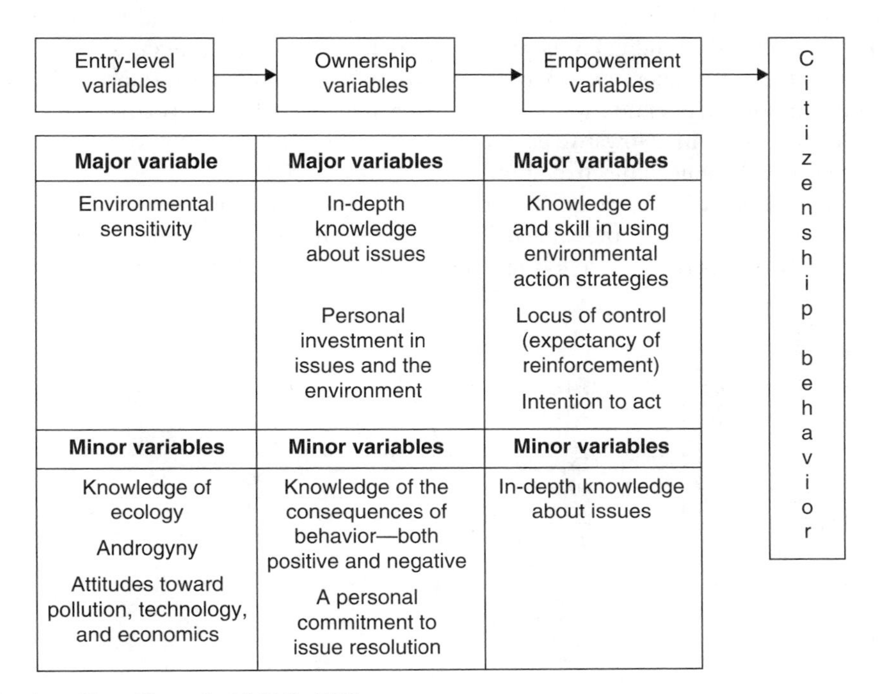

Figure 27.3 Behavior flowchart. (From Hungerford & Volk, 1990)

Another set of factors in understanding behavior is the situation or context in which the learner is engaged. The unique conditions of learning in free-choice environmental settings create a differing set of variables than does formal education (Falk & Heimlich, 2009). Free-choice learning settings such as zoos, nature centers, and natural history museums are contexts in which visitors often encounter programming on how human behavior impacts the environment (Association of Zoos and Aquariums, 2010). In 2008, Zeppel offered a model for behavior based on visitor experiences in free-choice learning environments. Based on a meta-analysis of eighteen studies conducted at marine tourism facilities, Zeppel suggested that experience, learning, and action do appear to follow; however, as Zeppel noted, this was not a causal study, so the conclusion may unintentionally imply a linear, rather than a correlational, relationship. Evidence for Zeppel's model included work from Ballantyne, Packer, and Bond (2007) showing persistence of behaviors after six-month postvisit, and visitor willingness to pledge during or concluding a visit (Mayes, Dyer, & Richins, 2004).

The Theory of Reasoned Action, introduced by Fishbein in 1967 (Ajzen & Fishbein, 1980), was based on the assumption that human behavior was grounded in rational thought and systematic use of information and included internal variables of behavioral beliefs and normative beliefs.

To move from the belief system to behavior, determinants of intention were hypothesized; intention has repeatedly and consistently been shown to be the direct antecedent to behavior. The two determinants of intention identified were (1) attitude toward the behavior and (2) the perception of social pressures surrounding the behavior (subjective norm) which roughly parallel the situational and dispositional constructs of belief identified above. In the model, the relative weights of attitudinal or normative factors were key to the value of the theory as they provide an understanding of the reason the intention was formed.

Attitudes—ones' learned disposition toward a person or thing (Ajzen & Fishbein, 1980)—and subjective norms—perceived social pressure to engage in a behavior (Ajzen & Fishbeing, 1980)—are both a function of beliefs. The underlying beliefs of the two constructs serve to improve the understanding of the intention formation and build on the models offered by Rokeach (1968) and Bem (1970). Expectancy Theory (Bandura, 1977) suggests individuals have expectations for the effects of outcomes of behaviors. Attitudes toward a specific behavior are comprised of the attitudes toward the expected outcome to the individual of performing that behavior. We would therefore expect that an individual who holds a belief that the environment is sacred to expect that personal behaviors towards the environment would lead to outcomes conserving the environment. Expectations of a positive outcome most likely lead to a positive attitude. Someone close to the actor may disapprove or approve of the behavior which, depending on whether the belief is core, next to core, or intermediate-authoritative (Rokeach, 1968), may shift the expected outcome.

The Theory of Reasoned Action was expanded by Ajzen (1991; Ajzen & Madden, 1986) to include prediction of behaviors for which one does not necessarily have agency (Heimlich & Ardoin, 2008). This "perceived behavioral control" construct (Fig. 27.4) shifts the model to suggest that behavior is influenced by three belief constructs rather than the two of attitude toward the behavior and perception of social pressures surrounding the behavior (Ajzen, 1991; Armitage & Conner, 1999; Terry & O'Leary, 1995). Ajzen (2002) noted that perceived behavioral control is similar to Rosentock's barriers construct (1966), Triandis's facilitating conditions (1977), and Bandura's concept of self-efficacy (1977). Trafimow, Sheeran, Conner, and Finlay (2002) divided perceived behavioral control into two parts: the extent to which people consider the behavior within their control; and whether people consider a behavior to be easy or difficult to perform. Behavioral beliefs and beliefs about consequences to one's behavior support the attitude held toward the behavior, normative beliefs emerging from ones' beliefs about expectations of important others lead to the subjective norm, and control beliefs related to control over what may support or prevent the behavior lead toward perceived behavioral control (Ajzen, 2002).

Several revisions to the Theory of Reasoned Action have been offered, e.g., Project SAFER (Montano, Kasprzyk, von Haeften, & Fishbein, 2001; von Haeften, Fishbein,

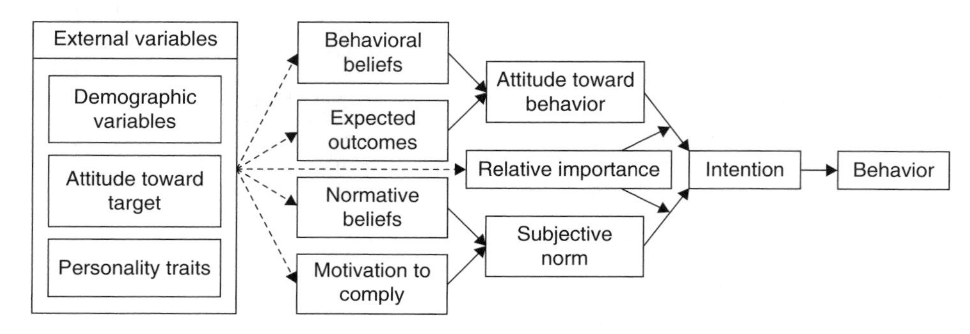

Figure 27.4 The theory of reasoned action. (Adapted from Ajzen & Fishbein, 1980, p. 84)

Kasprzyk, & Montano, 2001), Project RESPECT (Fishbein, 2000), and the Integrative Model of Behavior Change (Fishbein, Hennessy, Yzer, & Douglas, 2003). Taylor and Todd (1995) conducted a test of an integrated model related to household garbage reduction behavior and Cordano and Frieze (2000) applied the Theory of Planned Behavior against pollution reduction preferences of environmental managers. In both studies, the model was supported and the role of attitude, as defined by Hines et al. (1987), was reinforced.

Many studies have used the Theory of Planned Behavior related to proenvironmental behaviors. Kaiser, Hubner, and Bogner (2005) found that the Theory of Planned Behavior intention accounted for 95 percent of German university students' conservation action, and Value-Belief-Norm accounted for 64 percent. In 2006, Oreg and Katz-Gerro used twenty-seven countries to test a new model building on TPB and VBN along with Schwartz's (1994) country-level harmony dimension and Inglehart's (1995) postmaterialism index and contextual antecedents expected to affect environmental concern. It appears the Theory of Reasoned Action model holds in part because the Integrative Model of Behavioral Predication allows variables to be added to the theory of planned behavior depending on the purpose of the study (Conner & Armitage, 1998) and its use in environmental education will likely continue.

Across belief/behavior models, intentions to act are influenced by attitudes, social norms, context, and a host of individual factors. Even so, there are elements within the behavioral model that clearly are strong predictors of behavior: intentions.

As complex as they are, behaviors are congruent with a person's beliefs *at some level,* even if the individual does not know what conditions are determining the action (Ajzen, 1991; Stern, 2000). Self-regulation theory and its applications are concerned with how individuals meet and overcome obstacles to achieving their behavioral goals (Danter, 2005). Cottrell (2003) found verbal commitment (to act) as the strongest predictor in a structural equation model using the Responsible Environmental Behavior model discussed above. Rise, Thompson, and Verplanken (2003) introduced what they called implementation intentions as a step between intentions and behavior, as they felt planning, along with motivation to act, are both necessary. Barriers to action (Rosenstock, 1966), facilitating conditions for action (Triandis, 1977), and self-efficacy (Bandura, 1977) provide insights into moving from intention to action. Indeed, the Transtheoretical Model of Behavior Change as presented by Prochaska (1979) includes five stages, from precontemplation through maintenance of a behavior. In this model, the stages describe readiness to pursue new or changes to behaviors with key predictors being self-efficacy and the decisional balance, or pros and cons, associated with a particular behavior (Armitage, Sheeran, Conner, & Arden, 2004).

In a meta-analysis of environmental education programs, Bamberg and Moser (2007) used fifty-seven sample studies to examine psycho-social determinants and proenvironmental behavior using Hines et al.'s model and found similar mean correlations as in the original study. In addition, they postulated structural relationships among eight determinants of proenvironmental behavior including problem identification, attribution, social norm, guilt (a construct not previously included in models), perceived behavioral control, attitude, moral norm, and intention. They found intention mediates all other psycho-social variables and explained 27 percent of the variance. Across all studies, intention, when measured as a component of conation, is the single, strongest predictor of behavior.

Conclusion

There is a questioned assumption of logic underlying belief-behavior that people act rationally and in accordance with what would consistently be "true" to their beliefs. For those engaged in environmental education, it may sometimes appear this logic does not hold. For example, witness individuals who drive overly large, gas-consuming vehicles long distances in order to "be in nature," or those who engage in environmentally degrading recreational activities because they feel the activities connect them to their physical environment. There are some in the simple-living movement who forgo energy efficiency such as replacing older, more energy intensive products with newer energy efficient products in order to "not consume." This may result in higher energy consumption by the individual, while simultaneously reflecting the belief that the behaviors are favorable toward conservation. While there are no universally right ways to behave with respect to the environment as conditions, contexts, and scientific understanding change, these examples serve to highlight situations where an outside observer would question the proenvironmental nature of the individual's behavior, yet the individual is acting in accordance with personally held, proenvironmental beliefs. What one believes, and how those beliefs translate to behavior, have been and will continue to be a major focus of research in psychology as well as environmental education.

The key theories related to belief reviewed in this chapter do not suggest that environmental education or any type of education necessarily shapes one's beliefs. Rather, they hold that beliefs are either located in the subconscious and therefore relatively inaccessible, or that beliefs are formed and reinforced based on the social norms surrounding the individual. Thus, a potential role of environmental education is in guiding individuals to see how proenvironmental behaviors may align with beliefs (e.g., Ajzen, 1991; Ajzen & Fishbein, 1980, Bem, 1970, Rokeach, 1968). Although the connection between beliefs and behavior exists in the cognitive realm, the literature holds that when operating congruently or authentically, one's behaviors reveal how one interprets behavior as a reflection of an underlying belief. To the self, one does not act in opposition of beliefs.

We believe people act in congruence with their beliefs; this is the ultimate belief-behavior connection. However, beliefs are complex and structurally difficult to disaggregate. Beliefs are held in constellations, have different intensities depending on surrounding operationalized beliefs (contexts), and will always defer to more core beliefs (Bem, 1970). Thus, from the viewpoint of others it may appear that beliefs and behaviors fail to match, the mismatch is in our understanding of how they are linked. Beliefs surrounding environmental issues are tremendously complicated as there are personal, social, and economic concerns of varying value to the individual, confounded with the ability and intention of sustaining a related action. Research on how the affect, cognition, and behavior link for a person related to complex environmental issues needs to continue to better reveal how people learn, how learning shapes and relates to beliefs, and how beliefs are ultimately tied to an individual's choice to act.

References

Ajzen, I. (1991). The theory of planned behavior. *Organizational Decision and Human Decision Process, 50*, 179–211.

Ajzen, I. (2002). Perceived control, self-efficacy, locus of control, and the theory of planned behavior. *Journal of Applied Social Psychology, 32*, 107–122.

Ajzen, I., & Fishbein, M. (1980). *Understanding attitudes and predicting social behavior*. Englewood Cliffs, NJ: Prentice-Hall.

Ajzen, I., & Madden, T. J. (1986). Prediction of goal-directed behavior: Attitudes, intentions, and perceived behavioral control. *Journal of Experimental Social Psychology, 22*, 453–474.

Allen, J. B., & Ferrand, J. L. (1999). Environmental locus of control, sympathy, and pro-environmental behavior. *Environment & Behavior, 31*(3), 338–353.

Anthwal, A., Gupta, N., Sharma, A., Anthwal, S., & Kim, K. (2010). Conserving biodiversity through traditional beliefs in sacred groves in Uttarakhand Himalaya, India. *Resources, Conservation and Recycling, 54*(11), 962–971.

Armitage, C. J., & Conner, M. (1999). The theory of planned behavior. Assessment of predictive validity and perceived control. *British Journal of Social Psychology, 38*, 35–54.

Armitage, C. J., Sherran, P., Conner, M., & Arden, M. A. (2004). Stages of change or changes of stage? Predicting transitions in transtheoretical model stages in relation to healthy food choice. *Journal of Consulting and Clinical Psychology, 72*(3), 491–499.

Association of Zoos and Aquariums. (2010). *The accreditation standards and related policies 2010 edition*. Silver Spring, MD: Author.

Ballantyne, R., Packer, J., & Bond, N. (2007). *The impact of a wildlife tourism experience on visitors' conservation knowledge, attitudes and behavior: Preliminary results from Mon Repos Turtle Rookery, Queensland.* Paper presented at CAUTHE Conference, Sydney, Australia.

Bamberg, S., & Möser, G. (2007). Twenty years after Hines, Hungerford, and Tomera: A new meta-analysis of psycho-social determinants of pro-environmental behavior. *Journal of Environmental Psychology, 27*(1), 14–25.

Bamberg, S., & Schmidt, P. (2003). Incentives, morality, or habit? Predicting students' car use for university routes with the models of Ajzen, Schwartz, and Triandis. *Environment and Behavior, 35*(2), 264–285.

Bandura, A. (1977). Self-efficacy: Toward a unifying theory of behavior change. *Psychological Review, 84*(2), 191–215.

Bechtel, R. B., Verdugo, V. C., & Pinhiero, J. Q. (1999). Environmental belief systems: United States, Brazil, and Mexico. *Journal of Cross-Cultural Psychology, 30*(1), 122–128.

Bem, D. J. (1970) *Beliefs, attitudes and human affairs.* Belmont, CA: Brooks/Cole Pub.

Best, H. (2010). Environmental concern and the adoption of organic agriculture. *Society and Natural Resources, 23*(5), 451–468.

Blanchard, K. (1995). Seabird conservation on the north shore of the Gulf of St. Lawrence, Canada: The effects of education on attitudes and behavior towards a marine resource. In J. Palmer, W. Goldstein, & A. Curnow (Eds.), *Planning education to care for the earth* (pp. 39–50). Gland, Switzerland: International Union for Conservation of Nature.

Borden, R. (1984–85). Psychology and ecology: Beliefs in technology and the diffusion of ecological responsibility. *Journal of Environmental Education, 16*(2), 14–19.

Borden, R., & Powell, P. (1983). Androgyny and environmental orientation: Individual differences in concern and commitment. In A. Sacks (Ed.), *Current issues in environmental education and environmental studies volume VIII* (pp. 261–275). Columbus, OH: ERIC/SMEAC.

Bremner, A. (2007). Public attitudes to the management of invasive non-native species in Scotland. *Biological Conservation, 139*(3–4), 306–314.

Brown, T. J. (2010). Picking up litter: An application of theory-based communication to influence tourist behavior in protected areas. *Journal of Sustainable Tourism, 18*(7), 879–900.

Browne-Nuñez, C., & Jonker, S. A. (2008). Attitudes toward wildlife and conservation across Africa: A review of survey research. *Human Dimensions of Wildlife, 13*(1), 47–70.

Carr, A. (2002). *Grass roots and green tape: Principles and practices of environmental stewardship.* Sydney, Australia: The Federation Press.

Clover, D. (2002). Traversing the gap: Conceientizacion, educative-activism in environmental adult education. *Environmental Education Research, 8*(3), 315–323.

Conner, M., & Armitage, C. J. (1998). Extending the theory of planned behavior: A review and avenues for further research. *Journal of Applied Social Psychology, 28*(15), 1429–1464.

Cordano, M., & Frieze, I. H. (2000). Pollution reduction preferences of US environmental managers: Applying Ajzen's theory of planned behavior. *Academy of Management Journal, 43*(4), 627–641.

Cordano, M., Welcomer, S., & Scherer, R. (2010). An analysis of the predictive validity of the New Ecological Paradigm Scale. *The Journal of Environmental Education, 34*(3), 22–28.

Cordano, M., Welcomer, S., Scherer, R., Pradenas, L., & Parada, V. (2010). Understanding cultural differences in the antecedents of pro-environmental behavior: A comparative analysis of business students in the United States and Chile. *Journal of Environmental Education, 41*(4), 224–238.

Corraliza, J. A., & Berenguer, J. (2000). Environmental values, beliefs, and actions: A situational approach. *Environment and Behavior, 32*(6), 832–848.

Corral-Verdugo, V., Carrus, G., Bonnes, M., Moser, G., & Sinha, J. B. P. (2008). Environmental beliefs and endorsement of sustainable development principles in water conservation. *Environment and Behavior, 40*(5), 703–725.

Cottrell, S. P. (2003). Influence of socio-demographics and environmental attitudes on general responsible environmental behavior among recreational boaters. *Environment and Behavior, 35*(3), 347–375.

Courtenay-Hall, P., & Rogers, L. (2002). Gaps in mind: Problems in environmental knowledge-behavior modeling research. *Environmental Education Research, 8*(3), 283–297.

Culen, G., & Volk, T. (2000). The effects of an extended case study on environmental behavior and associated variables in seventh and eighth grade students. *Journal of Environmental Education, 31*(2), 9–15.

Danter, E. H. (2005). *The intention-behavior gap: To what degree does Fishbein's integrated model of behavioral prediction predict whether teachers implement material learned in a professional development workshop* (Unpublished dissertation). The Ohio State University, Columbus, OH.

Delshad, A. B., Raymond, L., Sawicki, V., & Wegener, D. T. (2010). Public attitudes toward political and technological options for biofuels. *Energy Policy, 38*(7), 3414–3425.

Dervisoglu, S. (2009). Influence of values, beliefs, and problem perception on personal norms for biodiversity protection. *Hacettepe University Journal of Education, 37*, 50–59.

Dietz, T., & Stern, P. C. (2002). *New tools for environmental protection: Education, information, and voluntary measures.* Washington, DC: National Academies Press, Division of Behavioral and Social Science and Education, National Research Council.

Dolnicar, S., & Leisch, F. (2008). An investigation of tourists' patterns of obligation to protect the environment. *Journal of Travel Research, 46*(4), 381–391.

Donohoe, H. M., & Needham, R. D. (2006). Ecotourism: The evolving contemporary definition. *Journal of Ecotourism, 5*(3), 192–210.

Dunlap, R. E., & McCright, A.M. (2008). A widening gap: Republican and Democratic views on climate change. *Environment: Science and Policy for Sustainable Development, 50*(5), 26–35.

Dunlap, R. E., & Van Liere, K. D. (1978). The "New Environmental Paradigm." *Journal of Environmental Education, 9*, 10–19.

Dunlap, R. E., Van Liere, K. D., Mertig, A. G., & Jones, R. E. (2000). Measuring endorsement of the new environmental paradigm: A revised NEP Scale. *Journal of Social Issues, 56*(3), 425–442.

Eagly, A. H., & Chaiken, S. (1993). *The psychology of attitudes.* New York, NY: Harcourt, Brace, & Jovanovich.

Edgell, M. C. R., & Nowell, D. E. (1989). The New Environmental Paradigm Scale: Wildlife and environmental beliefs in British Columbia. *Society and Natural Resources, 2*, 285–296.

Egan, A., & Jones, S. (1993). Do landowner practices reflect beliefs? *Journal of Forestry, 91*(10), 39–45.

El-Zein, A., Nasrallah, R., Nuwayhid, I., Kai, L., & Makhoul, J. (2006). Why do neighbors have different environmental priorities? Analysis of environmental risk perception in a Beirut neighborhood. *Risk Analysis, 26*(2):423–435.

Esty, D. C., Levy, M., Srebotnjak, T., & De Sherbinin, A. (2005). *Environmental sustainability index: Benchmarking national environmental stewardship*. New Haven, CT: Yale Center for Environmental Law and Policy.

Falk, J. H., & Heimlich, J. E. (2009). Who is the free-choice environmental education learner? In J. H. Falk, J. E. Heimlich, & S. Foutz (Eds). *Free-choice learning and the environment*. Lanham, MD: AltaMira Press.

Farmer, J., Knapp, D., & Benton, G. M. (2007). An elementary school environmental education field trip: Long-term effects on ecological and environmental knowledge and attitude development. *Journal of Environmental Education, 38*(3), 33–42.

Finger, M. (1994). From knowledge to action? Exploring the relationship between environmental experiences, learning, and behavior. *Journal of Social Issues, 50*(3), 141–160.

Fishbein, M. (2000). The role of theory in HIV prevention. *AIDS Care, 12*(3), 273–278.

Fishbein, M., Hennessy, M., Yzer, M., & Douglas, J. (2003). Can we explain why some people do and some people do not act on their intentions? *Psychology, Health & Medicine, 8*(1), 3–18.

Fletcher, R. (2009). Ecotourism discourse: Challenging the stakeholders theory. *Journal of Ecotourism, 8*(3), 269–285.

Fransson, N., & Gärling, T. (1999). Environmental concern: Conceptual definitions, measurement, methods, and research findings. *Journal of Experimental Psychology, 19*(4), 369–382.

Frick, J. Kaiser, F. G., & Wilson, M. (2004). Environmental knowledge and conservation behavior: Exploring prevalence and structure in a representative sample. *Personality and Individual Differences, 37*(8), 1597–1613.

Fritsch, A. J. (2009). Water and eco-spirituality. *Water Resources Impact, 11*(6), 9.

Geller, S. E. (1995). Actively caring for the environment: An integration of behaviorism and humanism. *Environment and Behavior, 27*(2), 184–195.

Grob, A. (1995). A structural model of environmental attitudes and behavior. *Journal of Environmental Psychology, 15*(3), 209–220.

Harvey, O. J. (1986). Belief systems and attitudes toward the death penalty and other punishments. *Journal of Personality, 54*(4), 659–675.

Harvey, O. J., Hunt, D.E., & Schroder, H. M. (1961). *Conceptual systems and personality organization.* New York, NY: John Wiley.

Heimlich, J. E., & Ardoin, N. M. (2008). Understanding behavior to understand behavior change: A literature review. *Environmental Education Research, 14*(3), 215–237.

Heimlich, J. E., & Harako, A. (1994). Teacher values in teacher recycling. *Environmental Education and Information, 13*(1), 21–30.

Hills, A.M. (1993). The motivational bases of attitudes towards animals. *Society & Animals*, *1*(2), 111–128.

Hines, J. M., Hungerford, H., & Tomera, A. N. (1987). Analysis and synthesis of research on responsible environmental behavior: A meta-analysis. *Journal of Environmental Education, 18*(2), 1–8.

Hodgkinson, S. P., & Innes, J. M. (2000). The prediction of ecological and environmental belief systems: The differential contributions of social conservatism and beliefs about money. *Journal of Environmental Psychology, 20*, 285–294.

Holt, J. (1988). *A study of the investigation of the effects of issue investigation and training on characteristics associated with environmental behavior on non-gifted eighth grade students* (Unpublished research paper). University of Southern Illinois, Carbondale, IL.

Hsu, S. J. (2004). The effects of an environmental education program on responsible environmental behavior and associated environmental literacy variables in Taiwanese college students. *Journal of Environmental Education, 35*(2), 37–48.

Hsu, S. J., & Roth, R. E. (1999). Predicting Taiwanese secondary teachers' responsible environmental behavior and associated environmental literacy variables. *Journal of Environmental Education, 30*(4), 11–18.

Hungerford, H. R., Tomera, A. N., & Sia, A. P. (1985–1986). Selected predictors of responsible environmental behavior: An analysis. *Journal of Environmental Education, 17*(2), 31.

Hungerford, H. R., & Volk, T. (1990). Changing learner behavior through environmental education. *Journal of Environmental Education, 21*(3), 8–21.

Hwang, Y. H., Kim, S. I., & Jeng, J. M. (2000). Examining the causal relationships among selected antecedents of responsible environmental behavior. *The Journal of Environmental Education, 31*(4), 19–25.

Ingelhart, R. (1995). Public support for environmental protection: Objective problems and subjective values in forty-three societies. *PS: Political Science & Politics, 28*(1), 57–72.

Jagers, S. C., & Matti, S. (2010). Ecological citizens: identifying values and beliefs that support individual environmental responsibility among Swedes. *Sustainability, 2*(4), 1055–1079.

Jensen, B.B. (2002). Knowledge, action, and pro-environmental behavior. *Environmental Education Research, 8*(3), 325–334.

Jurin, R. R. (1995). *College students' environmental belief and value structures, and relationship of these structures to reported environmental behavior* (Unpublished dissertation). The Ohio State University, Columbus, OH.

Jurin, R. R., & Fox-Parrish, L. (2008). Factors in helping educate about energy conservation. *Applied Environmental Education and Communication, 7*(3), 66–75.

Kaiser, F. G., Hubner, G., & Bogner, F. X. (2005). Contrasting the Theory of Planned Behavior with the Value-Belief-Norm model in explaining conservation behavior. *Journal of Applied Social Psychology, 35*(10), 2150–2170.

Kaiser, F. G., Wolfing, S., & Fuhrer, U. (1999). Environmental attitude and ecological behavior. *Journal of Environmental Psychology, 19*, 1–19.

Kaltenborn, B. P., Bjerke, T., Nyahongo, J. W., & Williams, D. R. (2006). Animal preferences and acceptability of wildlife management actions around Serengeti National Park, Tanzania. *Biodiversity and Conservation, 15*(14), 4633–4649.

Kollmuss, A., & J. Agyeman. (2002). Mind the gap: Why do people act environmentally and what are the barriers to pro-environmental behavior? *Environmental Education Research, 8*(3), 239–260.

Koslowsky, M., Kluger, A., & Yinon, Y. (1988). Predicting behavior: Combining intention with investment. *Journal of Applied Psychology, 73*(1), 102–106.

Kuhlemeier, H., Van den Bergh, H., & Lagerweij, N. (1999). Environmental knowledge, attitudes and behavior in Dutch secondary education. *The Journal of Environmental Education*, 30(2), 4–14.

Kusmawan, U., O'Toole, J. M., Reynolds, R., & Bourke, S. (2009). Beliefs, attitudes, intentions, and locality: The impact of different teaching approaches on the ecological affinity of Indonesian secondary school students. *International Research in Geographical and Environmental Education, 18*(3), 157–169.

Mainieri, T., Barnett, E. G., Valdero, T. R., Unipan, J. B., & Oskamp, S. (1997). Green buying: The influence of environmental concern on consumer behavior. *Journal of Social Psychology, 137*(2), 189–204.

Maloney, M. P., & Ward, M. P. (1973). Ecology: Let's hear from the people: An objective scale for the measurement of ecological attitudes and knowledge. *American Psychologist, 28*(7), 583–586.

Manoli, C. C., Johnson, B., & Dunlap, R. E. (2007). Assessing children's environmental worldviews: Modifying and validating the new ecological paradigm scale for use with children. *Journal of Environmental Education, 34*(4), 3–13.

Marcinkowski, T. (1989). An analysis of correlates and predictors of responsible environmental behavior. *Dissertation Abstracts International, 49*(12), 3677.

Mayes, G., Dyer, P., & Richins, H. (2004). Dolphin-human interaction: Changing pro-environmental attitudes, beliefs, behaviors and intended actions of participants through management and interpretation programs. *Annals of Leisure Research, 7*(1), 34–53.

Mayes, G., & Richins, H. (2009). Dolphin watch tourism: Two differing examples of sustainable practices and pro-environmental outcomes. *Tourism in Marine Environments, 5*(2), 201–214.

McCright, A. M., & Dunlap, R. E. (2010). The American conservative movement's success in undermining climate science and policy. *Theory, Culture & Society, 27*(2–3), 100–133.

McKenzie-Mohr, D., Nemiroff, L. S., Beers, L., & Desmarais, S. (1995). Determinants of responsible environmental behavior. *Journal of Social Issues, 51*(4), 193–156.

Menzel, S., & Bogeholz, S. (2010). Values, beliefs, and norms that foster Chilean and German pupils' commitment to protect biodiversity. *International Journal of Environmental and Science Education, 5*(1), 31–49.

Milfont, T. L., & Duckitt, J. (2010). The environmental attitudes inventory: A valid and reliable measure to assess the structure of environmental attitudes. *Journal of Environmental Psychology, 30*(1), 80–94.

Mobley, C. (2010). Exploring additional determinants of environmentally responsible behavior: The influence of environmental literature and environmental attitudes. *Environment and Behavior, 42*(4), 420–447.

Montano, D., Kasprzyk, D., von Haeften, I., & Fishbein, M. (2001). Toward an understanding of condom use behaviors: A theoretical and methodological overview of Project SAFER. *Psychology, Health & Medicine, 6*(2), 139–150.

Morgan-Brown, T., Jacobson, S. K., Wald, K., & Child, B. (2010). Quantitative assessment of a Tanzanian integrated conservation and development project involving butterfly farming. *Conservation Biology, 24*(2), 563–572.

Moseley, C., Huss, J., & Utley, J. (2010). Assessing K-12 teachers' personal environmental education teaching efficacy and outcome expectancy. *Applied Environmental Education and Communication, 9*(1), 5–17.

Muma, M., Martin, R., Shelley, M., & Holmes, Jr., L. (2010). Sustainable agriculture: Teacher beliefs and topics taught. *Journal of Sustainable Agriculture, 34*(4), 439–459.

Myers, O. E., Saunders, C. D., & Bexell, S. M. (2009). Fostering empathy with wildlife: Factors affecting free-choice learning for conservation concern and behavior. In J. H. Falk, J. E. Heimlich, & S. Foutz (Eds.), *Free-choice learning and the environment*. Lanham, MD: AltaMira Press.

National Research Council. (2009). Learning science in informal environments: People, places, and pursuits. Committee on Learning Science in Informal Environments. In P. Bell, B. Lewenstein, A. W. Shouse, & M. A. Feder (Eds), *Board on science education*. Washington, DC: The National Academies Press, Center for Education, Division of Behavioral and Social Sciences and Education.

Needham, M. D. (2010). Value orientations toward coral reefs in recreation and tourism settings: A conceptual and measurement approach. *Journal of Sustainable Tourism, 18*(6), 757–772.

Nixon, H., Saphores, J. M., Ogunseitan, O. A., & Shapiro, A. A. (2009). Understanding preferences for recycling electronic waste in California: The influence of environmental attitudes and beliefs on willingness to pay. *Environment and Behavior, 41*(1), 101–124.

Nolan, J. M. (2010). "An Inconvenient Truth" increases knowledge, concern, and willingness to reduce greenhouse gasses. *Environment and Behavior, 42*(5), 643–658.

Olofsson, A., Ohman, S. (2006). General beliefs and environmental concern: Transatlantic comparisons. *Environment and Behavior, 38*(6), 768–790.

Oreg, S., & Katz-Gerro, T. (2006). Predicting pro-environmental behavior cross-nationally: Values, the theory of planned behavior, and value-belief-norm theory. *Environment and Behavior, 38*(4), 462–483.

Parag, Y., & Eyre, N. (2010). Barriers to personal carbon trading in the policy arena. *Climate Policy, 10*(4), 353–368.

Paton, K., & Fairbairn-Dunlop, P. (2010). Listening to local voices: Tuvaluans respond to climate change. *Local Environment: The International Journal of Justice and Sustainability, 15*(7), 687–698.

Peyton, R., & Miller, B. (1980). Developing an internal locus of control as a prerequisite to environmental action taking. In A. Sacks (Ed.), *Current issues VI: The yearbook of environmental education and environmental studies* (pp. 173–192). Columbus, OH: ERIC/SMEAC.

Poortinga, W., Steg, L, & Vlek, C. (2004). Values, environmental concern, and environmental behavior: A study into household energy use. *Environment and Behavior, 36*(1), 70–93.

Prochaska, J. O. (1979). *Systems of psychotherapy: A transtheoretical analysis*. Homewood, IL: Dorsey Press.

Qablan, A. M., Al-Ruz, J. A., Khasawneh, S., & Al-Omari, A. (2009). Education for sustainable development: Liberation or indoctrination? An assessment of faculty members' attitudes and classroom practices. *International Journal of Environmental and Science Education, 4*(4), 401–417.

Ramsey, J. E. A. (1981). The effects of environmental action and environmental case study instruction on the overt environmental behavior of eighth grade students. *Journal of Environmental Education, 13*(1), 24–29.

Ramsey, J. M. (1989). *A study of the effects of issue investigation and action training on characteristics associated with environmental behavior in seventh grade students* (Unpublished doctoral dissertation). Southern Illinois University, Carbondale, IL.

Ramsey, J., & Rickson, R. (1977). Environmental knowledge and attitudes. *Journal of Environmental Education, 13*(1), 24–29.

Raudsepp, M. (2001). Environmental belief systems: Empirical structure and typology. *Trames, 5*(3), 234–254.

Reeder, G. D., & Brewer, M. B. (1979). A schematic model of dispositional attribution in interpersonal perception. *Psychological Review, 86*, 61–79.

Rickinson, M., Lundholm, C., & Hopwood, N. (2009). *Environmental learning. Insights for research into the student experience.* New York, NY: Springer.

Rise, J., Thompson, M., & Verplanken, B. (2003). Measuring implementation intentions in the context of the theory of planned behavior. *Scandinavian Journal of Psychology*, *44*, 87–95.

Roberts, J. A., & Bacon, D. R. (1998). Exploring the subtle relationship between environmental concern and ecologically conscious consumer behavior. *Journal of Business Research, 40*(1), 79–89.

Rokeach, M. (1968). *Beliefs, attitudes and values.* San Francisco, CA: Jossey-Bass Inc., Pub.

Rokeach, M. (2000). *Understanding human values: Individual and societal.* New York, NY: The Free Press.

Rosenstock, I. M. (1966). Why people use health services. *Milbank Memorial Fund Quarterly, 44*, 94–124.

Rundblad, G. (2008). The semantics and pragmatics of water notices and the impact on public health. *Journal of Water and Health, 6*, 77–86.

Schultz, P.W. (2000). Empathizing with nature: The effects of perspective taking on concern for environmental issues. *Journal of Social Issues*, *56*(3), 391–406.

Schultz, P.W. (2001). The structure of environmental concern: Concern for self, other people, and the biosphere. *Environmental Psychology, 21*(4) 327–339.

Schultz, P. W., Gouveia, V. V., Cameron, J. D., Tankha, G., Schmuck, P., & Franek, M. (2005). *Journal of Cross-Cultural Psychology, 36*(4), 457–475.

Schwartz, S. (1994). Are there universal aspects in the structure and content of human values? *Journal of Social Issues*, *50*(4), 19–45.

Severtson, D. J. (2009). The effect of graphics on environmental health risk beliefs, emotions, behavioral intentions, and recall. *Risk Analysis, 29*(11), 1549–1565.

Simpson, P. R. (1989). *The effects of an extended case study on citizenship behavior and associated variables in fifth and sixth grade students* (Unpublished doctoral dissertation). Southern Illinois University, Carbondale, IL.

Sivek, D. (1989). An analysis of selected predictors of environmental behavior of three conservation organizations. *Dissertation Abstracts International, 49*(11), 3322.

Smith, J. W., Burr, S. W., & Reiter, D. K. (2010). Specialization among off-highway vehicle owners and its relationship to environmental worldviews and motivations. *Journal of Park & Recreation Administration, 28*(2), 57–73.

Smith-Sebasto, N. J. (1995). The effects of an environmental studies course on selected variables related to environmentally responsible behavior. *Journal of Environmental Education, 26* (4), 30.

Spence, A., Poortinga, W., Pidgeon, N., & Lorenzoni, I. (2010). Public perceptions of energy choices: The influence of beliefs about climate change and the environment. *Energy & Environment, 21*(5), 385–407.

Stern, P. C. (2000). Toward a coherent theory of environmentally significant behavior. *Journal of Social Issues, 56*(3), 407–424.

Stern, P. C., & Dietz, T. (1994). The value basis of environmental concern. *Journal of Social Issues, 50*(3), 65–84.

Stern, P. C., Dietz, T., & Kalof, L. (1993). Value orientations, gender, and environmental concern. *Environment and Behavior, 25*(3), 322–348.

Stevenson, R. B. (2007). Schooling and environmental education: Contradictions in purpose and practice. *Environmental Education Research, 13*(2), 139–153.

Takacs-Santa, A. (2007). Barriers to environmental concern. *Human Ecology Review*, *14*(1), 26–38.

Tanner, C. (1999). Constraints on environmental behavior. *Journal of Environmental Psychology, 19*(2), 145–157.

Taylor, S., & Todd, P. (1995). Understanding household garbage reduction behavior: A test of an integrated model. *Journal of Public Policy and Marketing*, *14*(2), 192–204.

Teel, T.L., Manfredo, M.J., & Stinchfield, H.M. (2007). The need and theoretical basis for exploring wildlife value orientations cross-culturally. *Human Dimensions of Wildlife, 12*(5), 297–305.

Terry, D. J., & O'Leary, J. E. (1995). The theory of planned behavior. The effects of perceived behavioral control and self-efficacy. *British Journal of Social Psychology, 35,* 199–220.

Trafimow, D., Sheeran, P., Conner, M., & Finlay, K.A. (2002). Evidence that perceived behavioral control is a multidimensional construct. *British Journal of Social Psychology, 34,* 378–397.

Triandis, H. C. (1977). *Interpersonal behavior*. Monterey, CA: Brooks/Cole.

Tung, C., Huang, C., & Kawata, C. (2002). The effects of different environmental education programs on the environmental behavior of seventh-grade students and related factors. *Journal of Environmental Health, 64*(7), 24.

Van Birgelen, M., Semeijn, J., & Keicher, M. (2009). Packaging and pro-environmental consumption behavior: Investigating purchase and disposal decisions for beverages. *Environment and Behavior, 41*(1), 125–146.

Vaske, J. J., Donnelly, M. P., Williams, D. R., & Jonker, S. (2001). Demographic influences on environmental value orientations and normative beliefs about national forest management. *Society and Natural Resources, 14*(9), 761–776.

Vazquez Brust, D. A., & Liston-Heyes, C. (2010). Environmental management intentions: An empirical investigation of Argentina's polluting firms. *Journal of Environmental Management, 91*(5), 1111–1122.

Vignola, R., Koellner, T., Scholz, R. W., & McDaniels, T. L. (2010). Decision-making by farmers regarding ecosystem services: Factors affecting soil conservation efforts in Costa Rica. *Land Use Policy, 27*(4), 1132–1142.

Vining, J. (2003). The connection to other animals and caring for nature. *Human Ecology Review, 10*(2), 87–99.

von Haeften, I., Fishbein, M., Kasprzyk, D., & Montano, D. (2001). Analyzing data to obtain information to design targeted interventions. *Psychology, Health & Medicine, 6*(2), 151–164.

Wall, G. (1995). General versus specific environmental concern: A Western Canadian case. *Environment and Behavior, 27*(3), 294–316.

Weigel, R., & Weigel, J. (1978). Environmental concern: The development of a measure. *Environment and Behavior, 10*(1), 3–15.

Withers, L. E., & Wantz, R. A. (1993). The influence of belief systems on subjects' perceptions of empathy, warmth, and genuineness. *Psychotherapy, 30*(4), 608–614.

Wolsink, M. (2010). Near-shore wind power—Protected seascapes, environmentalists' attitudes, and the technocratic planning perspective. *Land Use Policy, 27*(2), 195–203.

Wright, J. M. (2008). The comparative effects of constructivist versus traditional teaching methods on the environmental literacy of post-secondary non-science majors. *Bulletin of Science, Technology, and Society, 28*(4), 324–337.

Xiao, C., & Dunlap, R. E. (2006-08-10). *Ecological worldview as the central component of environmental concern: Clarifying the role of the NEP.* Paper presented at the annual meeting of the American Sociological Association, Montreal Convention Center, Montreal, Quebec, Canada. Online <PDF>. 2008-12-12 from http://www.allacademic.com/meta/p93957_index.html.

Xiao, C., & Dunlap, R. E. (2007). Validating a comprehensive model of environmental concern cross-nationally: A US-Canadian comparison. *Social Science Quarterly, 88*(2), 471–493.

Xiao, C., & McCright, A. M. (2007). Environmental concern and sociodemographic variables: A study of statistical models. *The Journal of Environmental Education, 38*(2), 3–13.

Zelezny, L. C. (1999). Educational interventions that improve environmental behaviors: A meta-analysis. *Journal of Environmental Education, 31*(1), 5.

Zeppel, H. (2008). Education and conservation benefits of marine wildlife tours: Developing free-choice learning experiences. *Journal of Environmental Education, 39*(3), 3–18.

Zhang, H-X., & Wang, Y-M. (2009). Value orientation of tourist development and institutional innovation: Inspiration from the national park system in US. *Resources and Environment in the Yangtze Basin, 18*(8), 738–744.

Zint, M. (2002). Comparing three attitude-behavior theories for predicting science teachers' intentions. *Journal of Research in Science Teaching, 39*(9), 819–844.

28

Landscapes as Contexts for Learning

Carol B. Brandt
Temple University, USA

Introduction

As environmental educators, we find ourselves enmeshed with landscape in both our personal and working lives. Several years ago, I visited the Outer Hebrides in western Scotland, drawn to this landscape after reading *Soil and Soul*, a memoir that chronicled a Scottish grassroots environmental movement (McIntosh, 2004). As I walked the rugged coastline, I gazed upon breathtaking scenery that stirred my imagination. For residents of Isle of Lewis, this landscape is an indelible part of their collective identity—a fact that was revealed by the signs I saw posted in the window of homes and businesses: *No to the Turbines!* These signs protested the construction of a wind farm in nearby Stornaway. Locals saw the project as a threat to ecotourism; they feared that visitors like me might avoid Scotland if the towering wind generators marred the skyline. Similarly, ecologists speculated how the turbines would impact the nesting bird populations along the rocky shore. Others argued that it was yet another way for the energy hungry, industrialized south to colonize the land of the rural north.

As I spent the week walking, I came to appreciate the landscape of the western isles in a new way and to direct my gaze in a manner that I had not expected. I began to recognize where the terrain concealed evidence of the crushing poverty of the nineteenth century. I learned how the large estates of absentee landlords were a legacy of the brutal history of the enclosures. Similarly, the quiet villages with shuttered houses belied the recent migration that occurred during the present day exodus as residents moved to urban centers in search of employment. With each walk, landscape became more than my enhanced consciousness of the Scottish countryside; I began to deeply understand the complicated nature of the physical, natural, and human relations across scales of time and space. I came to grasp what I viewed as "landscape" was not only shaped through cultural narratives and international transactions, but also, it reflected larger macro-political and economic forces of global relations through time.

Moreover, walking the Scottish Hebrides helped me understand the ways that landscape is conceptualized in the southern Appalachian Mountains where I currently work. Home to the Scottish diaspora during the period of the highland clearances in the eighteenth and nineteenth centuries, the communities in my region view land as a symbol of ethnic and family pride. The forcible taking of common pastures and agricultural fields in Scotland resulted in a landless working class that immigrated to the mountains of Virginia, West Virginia, and North Carolina. Environmental education is particularly contentious in these Appalachian communities and viewed as "government meddling" or a way to usurp local control over shared resources. A landscape-based approach to environmental monitoring in the rivers of southern Virginia compelled me to reconsider the ways that land was conceptualized through time and space among the residents in these Appalachian communities (Brandt, 2009).

In this chapter, I examine the use of a landscape-based approach to environmental education as a context for learning that draws upon the fields of ecology, natural resource management, and human geography. I argue that a landscape-based approach takes into account the ways we use our gaze to interpret "place" and the meanings we ascribe to landscape. I begin by expanding a definition of landscape using examples across disciplinary fields from international contexts. I explicitly differentiate the ways in which a landscape-based approach differs other placed-based approaches to learning in environmental education. Next, I explore the ways that environmental education is incorporating transdisciplinary perspectives to use landscape as context for learning, and how fields such as human geography, natural resources management, and urban planning employ landscape-based strategies to transform how we, as environmental educators, approach environmental issues and the contentious politics of land worldwide. With rural and urban landscape as an object of, and site for learning, instructors can provide perspectives across scales of time and location, as well as offer space for critical reflection. I conclude by offering examples of how transdisciplinary approaches to studying landscape can advance new directions to our work in environmental education.

Defining Landscape

Recently, environmental educators have challenged the commitment of public schooling to eco-justice (Bowers, 2001; Gruenewald, 2003; McKenzie, 2008) by drawing attention to the situated nature of learning "in place" (see Greenwood & Smith, this volume). Certainly landscape is one element of defining place, but landscape implies *perspective*: an awareness of one's position and the ever-shifting gaze of the viewer (Certeau, 1984; Wylie, 2006, 2007). Among the literature examined for this chapter definitions of landscape vary widely, yet all of the definitions acknowledge the gaze extending across space and time, distinguishing it from a place-based approach. Some disciplines allied to environmental education (e.g., natural resources management) employ the concept of landscape to describe the visible features of natural open space in terms of aesthetic appeal, that emphasize the gaze and those emotions attached to viewing landscape. This connection of landscape to affect or human emotion is also caught up in the ways in which we reflect upon our individual and collective identity and the meaning one attaches to landscape (Nash, 1993; Walck, 2003). Similarly, positionality—an awareness of one's location and shifting perspective—underlies the disciplines of landscape ecology and cultural geography, which examine the spatial patterning of rural and urban development and ecological processes on various scales and organizational levels (Batterbury, 2001).

A landscape-based approach to environmental education differs from discussions of "place" in several ways. Place-based education has been critiqued for promulgating idealized notions of "place as a stable, bounded, and self-sufficient communal realm" (Nespor, 2008, p. 479). Certainly, the indiscriminant label of "place" has resulted in distracting educators from a closer analysis of the ways that environmental (in)justice has been culturally reproduced and sustained. While both place and landscape are concepts tied to geography, a landscape-based strategy for learning in environmental education emphasizes developmental spatial patterns across time. It also emphasizes reflexivity: the position of the observer and the means by which one recognizes these patterns. I argue that landscape-based education is as much about spatial perception as it is about comparative studies that tease apart the ways that particular environmental strategies have been culturally and social reproduced through time. While landscape-based education might be focused on the "local," these lessons are always situated within larger frames of movement, migration, economics, and politics.

In this section, I consider the various definitions surrounding landscape-based learning that emerged in my reading of this literature and discuss their implications for environmental education. Each approach to landscape implies a holistic view of the environment that is tied to ontology—being in the world, and epistemology—the ways in which we make knowledge about the environment. Furthermore, these definitions of landscape imply critical reflection concerning one's location and an understanding of the coconstitutive relationship between human practices and environment through time, at varying scales.

Visual perspective is essential to understanding landscape, yet landscape is more than a representational object imbued with an imposed cultural meaning (Wylie, 2006). Drawing upon the philosophical writing of Merleau-Ponty (1962), Wylie examines landscape as "being-in-the world," emphasizing embodied, visual subjectivity. Rejecting the notion of landscape as merely a representation of "space as external 'grid' or 'container' " (p. 521), Wylie pushes the reader to go beyond Cartesian or Euclidean conceptions of topology. He believes that an essential element to landscape and the gazing subject is *depth*—which he describes as engaging a subjective point of view, reflexivity, and detachment, all at the same moment. Wylie underscores the slippery nature of the concept of landscape, stressing those elements that are beyond our grasp: "Depth animates the world as a landscape open to exploration and ordering; it also hides the world, guarantees its ambiguity and complexity, and ensures that it remains beyond the reach of thought—whose goal is certainty and transparency" (p. 527).

Mackey (2002) takes up this notion of depth in the landscape when teasing apart the ways that landscape and her teaching are tied to ontology. Mackey recounts her experiences at an annual summer theatrical production with students at an outdoor theatre on the coast of Cornwall. She defines landscape as "that seamless combination of wilderness and human contrivance that gives a curious sense of safety-in-the-wild while we are usually caught up, however momentarily, in the resonant beauty of the topography" (p. 13). By turning one's gaze beyond oneself—to the land—it is possible to transform one's consciousness. This movement outward to the landscape simultaneously becomes a turn inward, whereby students found the landscape as a location whereby they could break out of a restrictive psychological path. In a sense, landscape is portrayed as transcendent and where being-in-landscape offered space for critical reflection.

Many environmental educators were first introduced to habitat conservation and landscape through *A Sand County Almanac* (Leopold, 1949/1970) in which the transcendent qualities of "the land" guided moral action. Using the concept of land in her teaching of natural resource management, Walck (2003) argues for creating a sense of responsibility for the common good, a land ethic, whereby the private ownership of resources within a circumscribed area is viewed as problematic. Walck develops activities for her students around a series of themes related to land and place that include moral, emotional, and aesthetic considerations, while still acknowledging the ways that land is connected to labor, the production of shared culture, and narratives of a particular place. Explicit in her assertions is a morally "right" way in which to make management decisions. Even though Leopold's land ethic was the cornerstone of an environmental movement in the United States (Knapp, 2005), it nevertheless speaks from an unquestioned (and uncritical) position of privilege and power.

For other educational researchers, landscape also has moral dimensions, including the basic human rights of access, health, safety, and self-governance. For Spirn (2005), landscape is both constituted by, and constitutive of, human activity, economic history, and environmental location. It is a "mutual shaping of the people and place" (p. 397) that are ordered through time and space. Spirn argues that in order to make informed, socially just decisions surrounding urban communities, one must be able to read the urban landscape. The vacant lots and deserted tenements in western Philadelphia (the United States) belie a long history of environmental and racial strife, a history that is not immediately discernable from the physical topography or built environment. She notes that portrayals of urban neighborhoods "are usually static snapshots of current conditions, narrowly framed" (p. 396) and ignore the economic and environmental history that shaped both people and location.

Few definitions of landscape in the literature integrate an awareness of economic history, production, and labor. Yet, Batterbury (2001) provides an uncommon and particularly insightful analysis through his examination of land rights and the economic history of African landscapes in Niger. Batterbury recommends that we take into account the "productive bricolage" (p. 438) of a location, in which households mix income from land and nonland based activities for subsistence in order to provide food and shelter for their families. Bricolage describes how livelihood is cobbled together from a range of resources, not always well chosen, but typically a result of one's local knowledge and experience. "Landscapes can be seen as the outcomes of the interplay of forces over time . . . but they also emerge as the result of scaled processes that interact in a world that is ecological complex: they have a nested political ecology" (pp. 439–440).

The economic histories of landscape are often couched within colonial histories of conquest, boundaries, borders, and nationalism as Nash (1993) recounts in her analysis of the modern Irish landscape. Nash notes how place names act as a "repository of community knowledge," yet the social memory of the landscape is marred through the loss of place names, only recently recovered through the research of historic maps and oral history. The Anglicization of place names and the remapping of Ireland in the nineteenth century along with the loss of the Gaelic language provide a backdrop for new forms of Irish nationalist discourse and the ways that the Irish landscape is currently represented in schooling. The British government claimed the power of representation, and with this renaming sought an erasure of Irish cultural identity. Similarly, Waterton (2005) chronicles the conflict between cultural resource managers of an archaeological preserve in Northumberland, UK, where the concept of landscape is also tangled up in issues of place, identity, and social history. Landscape is more than a physical backdrop to history and prehistory; landscape is the "site" upon which community groups vie for power over representation and resource management. "Landscapes cannot be objects simply understood, but instead exist as living, social processes with the ability to generate values through a community's knowledge of the past" (Waterton, 2005, p. 314).

Summary: Defining Landscape-Based Education Landscape is often thought in terms of "rootedness" and of being in a particular place. Yet, in modernity, the global flows of people, products, and information have certainly left their mark on local and regional landscapes (Appadurai, 1996). Rural and urban landscapes are shaped equally through the movement of people, as well as by plants and organisms deemed "invasive" or alien. Landscape is more than the merging of physical, biological, and cultural features; landscape is the result of the many meanings we give to locale and the particular ways we represent and think about an expanse of land, whether urban, rural, or wilderness. As noted above, definitions of landscape may vary, yet each approach to landscape positions the viewer both apart *and* within the expanse of environment. *Landscape is not static, but a concept that is shifting, and that affords a location for critical reflection. Landscape-based education is the study of developmental spatial patterns through time that allow us to assess the impact of particular human practices on the natural and physical environment.* Landscape as a context for learning in environmental education demands that we continually reposition our perspective as we come to understand our place in the world.

Learning in Landscapes

In reviewing the literature on landscapes as contexts for learning, this chapter draws from published studies in environmental education, science education, restoration ecology, environmental planning, landscape architecture, and natural resource management. Studies that focus on a holistic approach to the environment were particularly targeted for this overview. Published accounts of landscape-based learning around the world include research from North America, South America, Europe, and Asia. Unfortunately, environmental education research from Eastern Europe and Africa has a lower representation in international academic journals, and therefore are absent from this synthesis. Geospatial education (see Barnett) and indigenous knowledge (see Shava & Cavanaugh), which also take up landscape-based approaches to learning, are summarized elsewhere in this volume.

In this chapter, learning is viewed as sets of sociocognitive practices that are embodied, situated in time and space, and connected to roles, identities, and larger cultural practices (Calabrese, 2010; Lave, 1993). Most learning in environmental education occurs beyond the classroom in informal settings through multiple sensory modes. This context-bound, embodied construction of knowledge affords opportunities for learning that link to one's everyday life and sense of self. Brody (2005) contends that meaningful learning in environmental education is indelibly situated, tied both to experiences from the learner's past, as well as the physical and social context of the educational moment. In a sense, learning is a cognitive process in which one moves backward and forward in connecting experiences. Insight and new meaning occurs over time, and therefore, curricula must incorporate opportunities for critical reflection, dialog with others, and activity.

In reviewing this eclectic group of literature, six themes dominate the connection among situated learning, shifting perspectives of the environment, and cultural dimensions of landscapes. Central among these studies is the connection between childhood perceptions of landscape, affect, and emotional responses with one's lifelong identity and place. Landscapes as locations for critical reflection cross-cut much of this literature, as do research on learners' movement, mobility, and fieldtrips through landscapes. More unique are studies that examine the ways that scale, time, and space can be explored through landscape-based curricula. Several researchers describe a socioecological approach to learning as essential to sustaining biodiversity and particular bio-cultural landscapes. And finally, several studies point to the ways that citizenship and public engagement in environmental planning can be promoted through the use of landscape-based learning.

Childhood Perceptions and Lifelong Connections to Landscape A number of researchers contend that early experiences with nature shape individuals' subsequent involvement and advocacy of environmental issues. These studies focus on the ways that youth perceive landscape, their affectivity toward landscapes, and the ways youth use particular landscapes. These studies argue that when looking at the landscape through the eyes of a child, we may come to see how we might engage youth to be more involved with the out of doors, and thereby encourage greater environmental stewardship. In order to mobilize the public in conservation efforts, this research contends that environmental educators need to understand the diverse ways that youth self-identify and interact with rural and urban landscapes.

Arguing the saliency of childhood experiences in the environment, Tunstall, Tapsell, and House (2004) ask: What can be learned when placing cameras in the hands of children and asking them to photograph as they actively explore a landscape? In this study, Tunstall et al. utilized photography to engage a group of children (ages 9–11) to learn more about their perceptions of river landscapes in urban London. Drawing upon Gibson's (1979) notion of "affordances," the authors wanted to understand which features of the river landscape yielded interaction and engagement. Their analyses reveal the aesthetic appeal of particular elements of the environment and children's use of the landscape in imaginative play.

Similarly, when considering the meaning and role of landscape in public attitudes toward environmental issues, Roe (2007) argues that planners and educators should include children in their assessment. Youth have strong connections to neighborhood landscapes and offer a valuable contribution to the planning and management of local landscapes. One of the important findings of Roe's research in northeast England was to learn that children had a particularly strong attachment to their neighborhood landscapes. Children particularly wanted their opinions to be known when it came to the management of these spaces. Having "special places" with which they self-identified were a part of their imaginative play and were important to the children, especially those places not managed by adults.

Measham (2006) contends that in order to influence how the public interacts with environment, as researchers, we need to understand the ways that people rethink or relearn their interactions when confronted with new landscapes. In an interview study in north Queensland, Australia, Measham found that many participants cited past childhood experiences as being central to their orientation to new environments. These locations then, become touchstones as "primal landscapes" that are integrated into one's identity, and serve as the basis upon which one's experience in new environments is compared. From these data, Measham points to ways that environmental education can strategically intervene at key junctures in youth's lives to influence their public engagement with environmental issues. Likewise, Bizerril (2004) argues that understanding Brazilian youth's perceptions of the *cerrado*, the second largest biome after the Amazon rainforest, can assist schools in developing curricula that might influence positive attitudes toward conservation of biodiversity in this landscape.

Landscapes for Critical Reflection Several environmental educators point to the ways in which landscape is instrumental for encouraging critical reflection. Again, these kinds of curricula directly address shifting notions of time and space, and invite learners to expand the confines of their perspective.

In their understanding of landscapes, Payne and Wattchow (2009) argue for a slow pedagogy that can foster an appreciation of the embodied aspects of being and moving through particular places. Drawing upon feminist theory (e.g., Grosz, 2004) and phenomenology (e.g., Merleau-Ponty, 1962), an approach of slow pedagogy advocates methods in environmental education that reengage the sensuous, corporeal experience of being-in-location. Payne and Wattchow describe a third-year undergraduate curriculum using a slow ecopedagogy in the Australian landscape, in which students and instructors rearranged the curriculum so that they could undertake large blocks of time to become intimate with the Australian bush. Approaching landscape through a slow pedagogy encourages experiential meaning making, which Payne and Wattchow argue is an essential element of critical praxis. Moving beyond a place-based approach, Payne and Wattchow explore landscape and the imagination, as well as how time and space are organized through their teaching.

Experiencing landscape in an educational setting evokes both an individual and a collective response, as Brody (1997) found with a group of teachers studying a watershed in the northern Rocky Mountains of the United States. By walking the landscape, teachers were able to grasp the complexity and scale of the watershed beyond a simplistic understanding of water moving in a channel. Brody views the watershed in its figurative and literal meaning. As a metaphor, the watershed encompasses a space with borders where one can traverse psychological boundaries, and thus taking on a critical reflective stance, characteristic of a landscape-based approach. Alongside the interdisciplinary science of the watershed where the teachers examined water and nutrient cycling, biodiversity, and geologic processes, Brody also emphasized the diverse cultural meanings that the watershed held for each individual and the ones they forged as a group through their walk. With each new vista, the walking of this landscape challenged the learners and instructor to constantly shift their perspective in their appreciation of the watershed's complexity.

Using the landscape of the southwestern United States, Brandt (2004, 2006) guided university students in a capstone biology seminar to examine the cultural construction of scientific knowledge about plant communities. Each week, students spent several hours in conversation with ecologists and natural resource managers as well as local experts such as *curanderos* (traditional healers) and farmers as they explored particular landscapes: farm fields, orchards, mesa tops, and riparian zones. By walking with these experts across the terrain, students came to recognize the different cultural lens that each authority brought to view the landscape, thereby challenging dominant notions of Western science as an immutable truth.

Fieldtrips, Guides, and Mobility Movement through urban or rural landscapes is a common way most youth and adults engage in a learning experience. Field trips can involve long distances to completely new ecosystems or consist simply of walking outside the classroom door to experience local urban landscapes. In either case, landscapes as contexts for learning involve a heightened awareness of place through movement. More importantly, these school outings shape how learners engage with and think about the public spaces, environment, and landscape (Nespor, 2000).

In a week-long geography field course situated in the urban environment of Havana, the students (from the University of Manchester, UK) took on the role of guide, or an interpreter of city landscape (Coe & Smyth, 2010). More than just "being in the field," this course emphasized human movement through space, mobile technologies, and the urban environment. They sought to capture the ways that being in motion offered different experiences of place and thus, alternative understandings of location. More than just "walking and talking," students were encouraged to view the cityscape by examining patterns of movement in the urban environment. Through their research, students assumed the role of experts and became intimate with way finding through city streets, markets, and parks. Yet, the instructors did not critically reflect on how students became familiar with issues surrounding colonialism, global politics, and the problematic stance of being a guide in a foreign country.

Cultural geography emphasizes the ways in which landscapes are the product of human-environment interaction over time. In their work with undergraduates in southern Ohio in the central United States, Medley and Gramlich-Kaufman (2001) also used an interpretative approach, whereby students developed an environmental guide to a hiking route in a natural reserve. The process of building an interpretative guide required students to conduct both on-the-ground ecological inquiry and also demanded in classroom research with historic records of soils and vegetation maps. Medley and Gramlich-Kaufman (2001) assert that through the construction of the guide, students developed an appreciation of spatial patterns as they contrasted vertical structure against lateral relationships in the landscape. Moreover, the development of the guide required that students learn how to communicate their synthesis to a broad audience. Yet, this study too, offered little insight into the ways that students understand long-term economic impacts on landscape structure and land use history.

One way that some geographers have captured landscape is not through the visual, but through auditory experiences via sound walks. Butler (2006) discusses how sound is part of the semiotics of place, where "auditory communities" shape a sense of self and one's relationships with the natural world and built environment. Butler examines the way sounds from the interaction of human activity and the natural environment can evoke memory, and are also caught up in our notions of home and identity. Butler describes a sound walk he developed along the Thames in London, with maps, images, and sound recordings. Similarly, the Touring Exhibition of Sound Environments (2002) published recordings from the Isles of Lewis and Harris in the Outer Hebrides that documents the sounds of the *machair*, a rare coastal habitat, as a sound walk with local naturalists. This project also integrated local youth into their recordings through the use of "sound diaries" of their experiences and capturing children's indoor and outdoor lives on Lewis. By emphasizing sound and movement as means for coming "to know" a landscape, youth are positioned in a state of heightened awareness. Too, they recognize the ways that some sounds are indicative of past lifeways (e.g., the shuttle in a weaving loom, sheep herding on the moors) that are no longer economically viable.

Shifting Levels of Analysis: The Micro and the Macro Movement through a landscape is but one way to examine "depth" in knowing. Other environmental educators emphasize the importance of shifting scales of focus when studying environmental systems, landscapes, or watersheds. By moving between the micro and macro levels of analysis, learners can appreciate environmental change over time, as well as the inconspicuous—but crucial—components of an ecosystem.

Through their research on water science literacy with youth, Covitt, Gunckel, and Anderson (2009) found that few students could trace water flow and other materials through the watershed. While youth had some understanding of the visible elements of the water cycle, the invisible water movements (at micro *and* macro levels) were absent from their reasoning. The path of pollutants in solution and atmospheric evaporation at a microscopic scale were unrecognized by students, and similarly, they had difficulty conceptualizing water flowing through landscape-wide systems at the watershed level. Their research underscores the importance of scale in environmental education curricula for youth to grasp a complete understanding of the elements and physical processes involved in connected water systems. Covitt et al. argue that a comprehensive understanding of water through micro and macro processes is fundamental to an environmentally literate citizenry.

Brady and Brady (2009) use a landscape-based approach with students to examine the lasting effects of human land use history on local environments. Located in southern Ohio in the central United States, the students collected soil samples from sites that experienced coal strip mining between the early 1800s to 1950s and from sites that were unmined. "Every place has its history of human land use, and this history is imprinted in the biodiversity existing there" (p. 423). Their curriculum is intended assist students in understanding how land use practices alter biodiversity over time. Through a series of field experiences, students explore the legacy of land use by examining macrovertebrate diversity in soil samples across a number of sites.

Map making is another way to explore the dimensions of landscape scale and requires complex cognitive practices, as well as extensive social negotiation. In her ethnography of a community-based mapping project, Grasseni (2004) recounts how residents from a small valley in northern Italy conceptualized and coordinated their efforts to redraw a local map intended for the tourism industry. In the making of the new map, the recording and representation involved cognition, memory, and geographic imagination. She noted how meetings revolved around trying to verbalize and grasp one another's "cognitive maps," as speakers and listeners engaged in the "embodied imagination" (p. 702) of reliving the experience of walking upon a particular path. But moreover, they quickly came to understand that they had to "recalibrate," or rescale their map and movements along the paths for the imagined tourist. Their original efforts were too detailed and carried a layer of history that was meaningful to the residents, but irrelevant to newcomers. Their new map, completed at a new scale, captured general information whose details mask the complex deliberations in its making.

Socioecological Learning and Landscapes Several studies on learning in landscape explicitly approach education as a socioecological process. Not only is landscape viewed in terms of its physical, biological, and cultural features, learning and education are also considered to be fundamental to the continued existence of these particular landscapes and the resiliency of particular ecosystems. Not surprisingly, these examples of learning in landscape are derived from situations in community-based educational contexts beyond schools. Yet its importance for schooling is should not be minimized—youth play an important role in developing resiliency and continuity through community-based notions of care and stewardship.

Olsson, Folke, and Hahn (2004) view landscape an integrated social-ecological system requiring a deep historical awareness of the link between nature and culture. Their case study focused on the regional wetlands in southern Sweden used for grazing and haymaking, which require sustained management. Key in creating the comanagement system was the development of an ecomuseum that provided outdoor, onsite interpretation of the wetland's cultural history, geology, and ecology. An activity of the eco-museum was mapping land-use practices with the help of ecologists and local residents as a means to build local ecological knowledge of the wetlands management to inform policy. As a result of the mapping, specific types of agricultural practices used in cultivating the meadows were linked to biodiversity of the wetlands, trust was built among key actors, and knowledge necessary for a detailed policy plan was constructed. Collaborative learning and social networking while generating new knowledge provided the basis for developing a more resilient social-ecological landscape for the wetland, one that could more readily respond to increasing urbanization and governmental change.

The *satoyama*—a mosaic of fields, forest, villages, and wetlands—is a landscape that has been cultivated for thousands of years in Japan, and is integral to the Japanese national identity. The pressures of urbanization, industrial agriculture, and forest cutting have led to the loss of this cultural landscape and the biodiversity it supports. Kobori and Primack (2003) describe how one landscape conservation program recruited mixed age volunteers from the surrounding urban Yokohama to work with local farmers to restore the *satoyama* as a means of reconnecting with their Japanese agricultural heritage. Kobori & Primack argue that the *satoyama* can act as a model for restoring cultural ecosystems within high-density urban situations, especially if citizens are allowed volunteer opportunities to renew their sense of identification with traditional land use practices. The intergenerational education of the *satoyama* and its focus on historical reconstruction of landscapes compelled youth and adults to consider the impacts of rapid urbanization. As part of socioecological learning, participants engage in critical reflection of their role in maintaining this landscape for future generations.

Landscape, Citizenship, and Public Engagement In addition to engaging learners in scientific inquiry through environmental education, landscape as contexts for learning is also fundamental to engagement in public discourse and participation in conservation and environmental stewardship. Palmberg and Kuru (2000) describe an environmental education program in Finland, where youth (grades 1–6) learned basic camping and survival skills, information about the forest ecosystem, and the legal rights of public access to open space. These programs offered youth direct experience with a range of landscapes, and were intended to help them develop self-confidence in addition to learning responsible action in nature and environmental sensitivity. More than just being exposed to "primal landscapes" (as described by Measham, 2006), Palmberg and Kuru argued that these experiences laid the groundwork for future environmental decision making as well as "reactive empathy" (p. 35) and unconscious action toward being engaged in stewardship.

Democratic discourse and deliberative engagement allow participants to make sense of their world and to learn from others' perspectives, views, and knowledge. In her research with residents near Birmingham, UK, Petts (2007) observed how community members worked with land managers and ecologists to develop a land-use plan and water management in the urban floodplains. In their study of an urban river that was being regenerated for a nature reserve and riverside park, Petts focused on two types of learning: instrumental, whereby the learner acquires new skills or knowledge and thereby enhances individual understanding; and communicative, the ways in which a person works collaboratively to solve problems. Through a series of workshops, small group work, and daylong site visits, participants had ample time for informal discussion in the planning process. The onsite workshops enabled greater advocacy and a commitment to work through contentious issues instead of walking away from difficult conversations. The landscape provided various locations for perspective-taking that were essential to making the difficult deliberations possible.

Through the use of landscape, Walck (2003) works with managers and management students in higher education to make fiscal and environmental relationships more visible, concrete, and tangible – to assist them in recognizing the "common ground" on which economic and ecological decisions are made and how the common ground is the landscape. Walck points out that management decisions rapidly impact the integrity of ecosystems and open space habitat on the margins of urban and suburban environments. She argues that management decisions are often obscured by artificially separating economic and ecological conditions.

Similarly, Lackstrom and Stroup (2009) use greenways as educational sites for studying human-environment interactions, urban ecosystems, and the advancement of sustainable practices along urban river corridors. Riparian greenways offer a means to examine the built environment of a river, its historical uses, and alterations. In addition to onsite mapping, aerial photography provides learners a means to examine the relationships between the greenway's location, as well as political boundaries within the larger watershed. Moreover, an integration of these data allows an analysis of the ways that human activity has altered or degraded the environment through pollutions, channelization, or the introduction of nonnative species. Thus, greenways afford a means for learners to examine the magnitude of historic, economic, and political change on the landscape, as well as to understand the role in the sustainability of this urban environment.

Conclusions

> The landscape, in short, is not a totality that you or anyone else can look *at*, it is rather the world *in* which we stand in taking up a point of view on our surroundings. And it is within the context of this attentive involvement in the landscape that the human imagination gets to work in fashioning ideas about it. (Ingold, 2001, p. 207)

By engaging youth and adults with a landscape-based approach to environmental education, they discover locations for critical reflection that facilitate "depth" in their connections to particular site, or deeper understanding of the environment. Also, a landscape-based approach can offer safe spaces connected to memories of childhood—primal landscapes—that might promote imagination, freedom, and spontaneity through the multisensory experience of location. In addition, landscape-based approaches position learners in close proximity to pressing environmental issues in their communities, and reveal their impact across time and space. By examining landscapes through the scale of time and space, learners begin to be aware of their own position for understanding the environment as well as the ontologies and epistemologies of others.

Given the multidimensional nature of landscape-based learning, researchers have employed diverse methods to analyze the kinds of learning that occurs in these spaces. Many of the studies summarized for this chapter use a mixed methods approach to examine the types of learning. Surveys, interview studies, observations, written artifacts, video, photography, and ethnography describe and assess the affective and cognitive gains made when youth and adults engaged in landscape-based approaches to environmental education. Those studies that employ only the use of survey instruments were less fruitful. Missing from these published accounts, however, are longitudinal studies that demonstrate the long-term impact that learning in landscapes might have on subsequent participation in environmental conservation. Also needed are more studies of environmental curricula that incorporate an awareness of the political and economic processes shaping environments through history. Many published accounts of landscape-based learning stop short of exploring the ways that mapping, borders, boundaries, and territories are linked to historical conflicts, colonialism, and a struggle for control of natural resources. A landscape-based approach to learning incorporates the history and changes in the landscape over time, where the learner can begin to understand her or his own role with this and other landscapes.

This literature highlights the role of landscape-based environmental education (urban and rural) as contexts for learning with contributions from allied disciplines. In each study there exists a tension between viewing landscape and living in it—between "observation and inhabitation" (Wylie, 2007, p. 5). Landscape is connected to our sense of self (ontology) and knowledge construction (epistemology) and tied to notions of embodiment, dwelling, and community. Our gaze upon the landscape is shaped by our values, cultural norms, ideologies, and expectations. This approach to learning acknowledges the ways that natural processes *and* cultural practices shape the landscape, just as the landscape has simultaneously shaped its inhabitants. Landscape-based approaches to environmental education employ multimodal teaching strategies that invite transdisciplinary insights.

References

Appadurai, A. (1996). *Modernity at large: Cultural dimensions of globalization*. Minneapolis, MN: University of Minnesota Press.

Batterbury, S. (2001). Landscapes of diversity: A local political ecology of livelihood diversification in southwestern Niger. *Ecumene, 8*(4), 437–464.

Bizerril, M. X. A. (2004). Children's perceptions of Brazilian cerrado landscapes and biodiversity. *Journal of Environmental Education, 35*(4), 47–58.

Bower, C. (2001). *Educating for eco-justice and community*. Athens, GA: University of Georgia Press.

Brady, J. K., & Brady, J. C. (2009). Using field experiences to study the land-use legacy. *The American Biology Teacher, 71*(7), 419–423.

Brandt, C. B. (2004). A thirst for justice in the arid southwest: The role of epistemology and place in higher education. *Educational Studies, 36*(1), 93–107.

Brandt, C. B. (2006). Narratives of location: Epistemology and place in higher education. In G. Spindler & L. Hammond (Eds.), *Innovations in educational ethnography: Theory, methods and results* (pp. 321–344). Mahwah, NJ: Lawrence Erlbaum & Associates.

Brandt, C. B. (2009, September 22–23). *Water watchers: Coordinating the practices of environmental monitoring through protocols and social construction of citizenship*. Paper presented at the annual conference of Ethnography & Education, Oxford, UK.

Brody, M. (1997). Descending the watershed: Rethinking the "place" of curriculum. *Canadian Journal of Environmental Education, 2*, 114–131.

Brody, M. (2005). Learning in nature. *Environmental Education Research, 11*(5), 603–621.

Butler, T. (2006). A walk of art: The potential of the sound walk as practice in cultural geography. *Social & Cultural Geography, 7*(6), 889–908.

Calabrese, B. A., & Tan, E. (2010). We be burnin'! Agency, identity, and science learning. *Journal of the Learning Sciences, 19*, 187–229.

Certeau, M. D. (1984). *The practice of everyday life* (S. Rendall, Trans.). Berkeley, CA: University of California Press.

Coe, N. M., & Smyth, F. M. (2010). Students as tour guides: Innovation in fieldwork assessment. *Journal of Geography in Higher Education, 34*(1), 125–139.

Covitt, B. A., Gunckel, K. L., & Anderson, C. W. (2009). Students developing understanding of water in environmental systems. *Journal of Environmental Education, 40*(3), 37–51.

Gibson, J. (1979). *The ecological approach to visual perception*. London, UK: Lawrence Erlbaum.

Grasseni, C. (2004). Skilled landscapes: Mapping practices of locality. *Environment and Planning D: Society and Space, 22*, 699–717.

Grosz, E. (2004). *The nick of time: Politics, evolution, and the untimely*. Durham, NC: Duke University Press.

Gruenewald, D. (2003). Foundations of place: A multidisciplinary framework for place-conscious education. *American Educational Research Journal, 40*(3), 619–654.

Ingold, T. (2001). *The perception of the environment: Essays in livelihood, dwelling and skill*. London, UK: Routledge.

Knapp, C. E. (2005). The "I-Thou" relationship, place-based education and Aldo Leopold. *Journal of Experiential Education, 27*(3), 277–285.

Kobori, H., & Primack, R. B. (2003). Conservation for satoyama, the traditional landscape of Japan. *Arnoldia, 62*(4), 2–10.

Lackstrom, K., & Stroup, L. J. (2009). Using a local greenway to study the river environment and urban landscape. *Journal of Geography, 108*, 78–89.

Lave, J. (1993). The practice of learning. In S. Chaiklin & J. Lave (Eds.), *Understanding practice: Perspectives on activity and context* (pp. 3–31). Cambridge, UK: Cambridge University Press.

Leopold, A. (1970). *A Sand County almanac*. New York, NY: Ballantine. (Original work published 1949).

Mackey, S. (2002). Drama, landscape and memory: To be is to be in place. *Research in Drama Education, 7*(1), 9–25.

McIntosh, A. (2004). *Soil and soul: People versus corporate power*. London, UK: Aurum Press.

McKenzie, M. (2008). The places of pedagogy: Or, what we can do with culture through intersubjective experiences. *Environmental Education Research, 14*(3), 361–373.

Measham, T. G. (2006). Learning about environments: The significance of primal landscapes. *Environmental Management, 38*(3), 426–434.

Medley, K. E., & Gramlich-Kaufman, L. M. (2001). A landscape guide in environmental education. *Journal of Geography, 100*, 69–77.

Merleau-Ponty, M. (1962). *Phenomenology of perception*. London, UK: Routledge.

Nash, C. (1993). Remapping and renaming: New cartographies of identity, gender, and landscape in Ireland. *Feminist Review, 44*, 39–57.

Nespor, J. (2000). School field trips and the curriculum of public spaces. *Journal of Curriculum Studies, 32*(1), 25–43.

Nespor, J. (2008). Education and place: A review essay. *Educational Theory, 58*(4), 475–489.

Olsson, P., Folke, C., & Hahn, T. (2004). Social-ecological transformation for ecosystem management: The development of adaptive co-management of a wetland landscape in southern Sweden. *Ecology and Society, 9*(4). Retrieved from http://www.ecologyandsociety.org/vol9/iss4/art2

Palmberg, I. E., & Kuru, J. (2000). Outdoor activities as a basis for environmental responsibility. *Journal of Environmental Education, 31*(4), 32–36.

Payne, P. G., & Wattchow, B. (2009). Phenomenological deconstruction, slow pedagogy, and the corporeal turn in wild environmental/outdoor education. *Canadian Journal of Environmental Education, 14*, 15–32.

Petts, J. (2007). Learning about learning: Lessons from public engagement and deliberation on urban river restoration. *The Geographical Journal, 173*(4), 300–311.

Roe, M. (2007). Feeling "secrety": Children's view on involvement in landscape decisions. *Environmental Education Research, 13*(4), 467–485.

Spirn, A. W. (2005). Restoring Mill Creek: Landscape literacy, environmental justice, and city planning and design. *Landscape Research, 30*(3), 395–413.

Touring Exhibition of Sound Environments. (2002). *The sounds of Harris & Lewis: Machair Soundwalks. [CD]*. London, UK: Earminded.

Tunstall, S., Tapsell, S., & House, M. (2004). Children's perceptions of river landscapes and play: What children's photographs reveal. *Landscape Research, 29*(2), 181–204.

Walck, C. (2003). Using the concept of land to ground the teaching of management and the natural environment. *Journal of Management Education, 27*(2), 205–219.

Waterton, E. (2005). Whose sense of place? Reconciling archaeological perspectives with community values: Cultural landscapes in England. *International Journal of Heritage Studies, 11*(4), 309–325.

Wylie, J. (2006). Depths and folds: On landscape and the gazing subject. *Environment and Planning D: Society and Space, 24*, 519–535.

Wylie, J. (2007). *Landscape*. New York, NY: Routledge.

Section VI

Evaluation and Analysis of Environmental Education Programs, Materials, and Technologies and the Assessment of Learners and Learning

Michael Brody
Montana State University, USA

Martin Storksdieck
National Research Council, USA

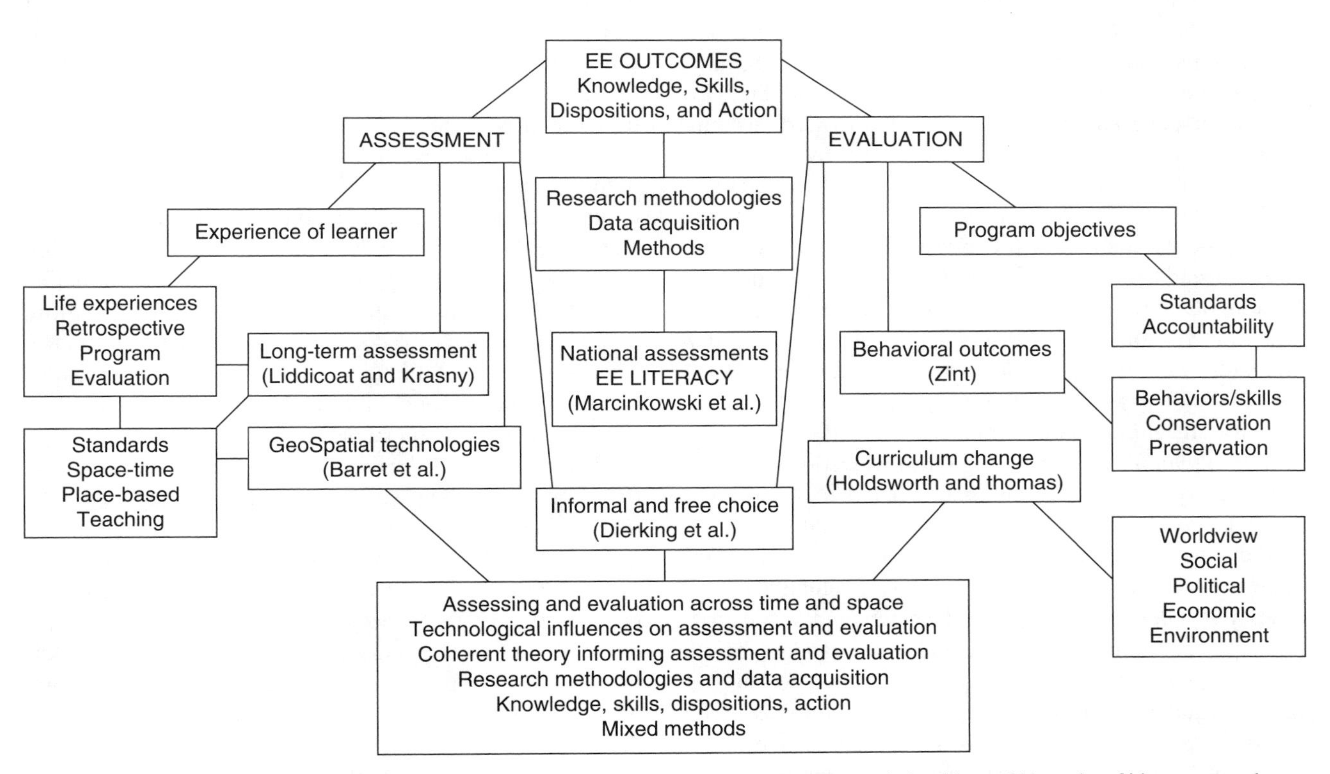

Figure VI.1 The above graphic organizer is the section six editors' interpretation of the concepts, relationships and hierarchy of ideas presented by the chapter authors. It attempts to highlight some of the main ideas but is not inclusive. It represents a part of a broader conception of EE as presented in the handbook and as such may not include every aspect of REE such as a social change agenda for research in EE. It does represent our interpretation of the main concepts found in section six.

Research related to assessment and evaluation in environmental education is diverse in its theoretical perspectives, methodologies, and methods, in part because of the broad landscape of goals and outcomes that are associated with environmental education or education for sustainability instructional programs and learning activities. These goals tend to go beyond the goals and outcomes defined for other areas of interest (or disciplines, if one would want to define environmental education (EE) as a discipline). One profound difference is the focus on behaviors associated with improving the state of affairs; this manifests itself when environmental educators aim at changing consumer behavior, all the way to perspectives that tie EE or UK department for education and skills (EFS) to wider societal reform efforts and transformational trends. In that sense, EE or EfS defines immediate to long-term outcomes at the level of the individual learner, groups, societies, and even the natural, build, and social environment overall. Outcomes for individual learners are defined in the cognitive, affective, and cognitive domain, and include broad conceptions of skills and behavior, ranging from thoughtful and reflective action to ritualized habits. Goals and outcomes that target behavior bridge individuals with a larger societal project that may be guided by desires for improved stewardship all the way to radical reorientations of our societies toward sustainability, thereby moving toward impacts (or even visions) that transcend far beyond a sum of individual outcomes.

The diverse perspectives on what EE or EfS are to achieve poses challenges for assessment, evaluation and research, and the chapters featured in this section represent these challenges at multiple levels. In EE, especially among practitioners, the terms "assessment," "evaluation," and "research" are often used somewhat interchangeably, even though they carry rather different (and somewhat complementary) meaning. Assessment concerns itself with the task of explaining what is going on in educational events. It helps explain how to outcomes related to "learning," "knowledge." "attitude," "disposition," "critical thinking," "reflexivity," or "behavior" are truly valid and reliable (or believable and trustworthy) constructs. Evaluation is an often misunderstood term. EE research and evaluation aims at constructing knowledge and understanding based on the values associated with learning outcomes of EE activities and then informs EE practice and future research. In some cases, REE related to assessment and evaluation strives to generalize, even if much research is actually conducted at the level of exploratory or initial stages and does not claim generalizability. We continue to see a myriad of single program evaluation studies submitted to EE research journals for publication.

There has been a significant emphasis on evaluation in EE especially in countries that favor educational policies tied to national, state, and local standards. In these standards based educational contexts, for example the United States, many of the assessment and evaluation efforts are tied to specific projects, and ask value-oriented questions about the nature, context, progress, or achievement of these projects, that is, what knowledge, skills, and dispositions did the learners exhibit in relation to the program goals and objectives. While evaluation may employ methods that are indistinguishable from those of research, by definition evaluation in EE programs aim at asking whether, and/or to what degree and why a particular project or program is achieving their mission, goals, outcomes, or impacts, or whether investments into these projects or programs were worthwhile (to name the most common guiding questions for EE evaluation studies).

Evaluation can occur at all stages of an EE project, beginning with challenging assumptions (front-end) to supporting the design and implementation of a project (formative, developmental) to determining degrees of success (outcome, summative). Each of these three phases of evaluation correspond to broader research areas, and, in fact, can benefit from findings that allow placing specific projects into a wider context of EE theories, principles, and conceptual constructs. Front-end evaluation in EE often corresponds to research that broadly characterizes groups of people, types of partnerships, effectiveness of instructional approaches, etc. but, based on existing literature, specifies what assumptions hold and what previous research may be relevant to a project. Formative evaluation similarly corresponds to research on effective practices, and summative evaluation differs from research only in the specific focus of the effort, and can, if done well and communicated to the field, itself add to the overall knowledge base. In addition to the typical summative evaluation studies in EE, front-end preassessments and formative assessments can be a valuable part of describing the entire picture of complex events which surround EE activities. The assessment of EE activities across entire programs can better inform, add depth and relevance, and increase the value of evaluation studies in EE.

Research and evaluation in EE or EfS is necessarily based on a wide range of methods and methodologies, in part because of the wide range of outcomes that stretch from raising awareness to behavior change, with much latitude for how each outcome of construct could or should be understood. Where does raising awareness bleed over to learning of facts or concepts? How do attitude, motivation, interest, disposition and identity link together, and what should be the focus of anyone study? What type of behavior is being investigated? Is it ritualized or reflective? Are we interested in the immediate or the long-term, or the immediate a predictor and best proxy for long-term? Do we focus on outcomes for the learner, the quality of the teaching/learning exchange, the broader societal context in which learning occurs, or the impacts on society or the environment? Furthermore, do we feel confident enough in our understanding of the research arena that we begin to seek definite answers or are we exploring a topic or issue for initial understanding or to provide deeper explanation on why things are they way research seems to portray results and conclusions? The authors of chapters in section six will further provide evidence that these questions in general remain unanswered and provide a framework for next steps in EE research related to assessment and evaluation.

Research and evaluation questions beget methodologies and methods. What may appear to outsiders as a hodgepodge of approaches used in EE and EfS is actually a reflection on rigor. Methodological frameworks range from ethnography to qualitative and quantitative social science and education research and corresponding methods because research questions differ from the descriptive that simply wants to understand the "what" to the inferential, through which we tend to generalize. Methods are then chosen accordingly, and given the complexity of the questions asked by evaluators or researchers, often include multiple methods, either to increase the validity of the findings or to cover range of logically linked questions (for instance, what outcomes where achieved to what degree and why?). In some approaches to evaluation, even that which couldn't be foreseen needs to be captured. When the question may focus less on the degree to which a project or program achieved its preset goals and outcomes, but instead wonders about the overall value that was created, so-called unintended outcomes that were not foreseen and not planned or programmed for can be taken into consideration for making evaluative statements (and those can be positive or negative).

The set of articles featured in this section covers much of the theoretical ground described above, and the following illustration is an attempt to visualize the connections between the papers and the broader context they (partially) represent.

A team of eleven environmental education researchers led by Tom Marcincowski provides us with an overview of four national assessments into the environmental literacy of school-aged children in different grades. Differences and similarities between the United States, Korea, Israel, and Turkey not only in the overall level of environmental literacy, but also in potential factors that may explain the differences has much utilitarian value: it can be used as a rough yardstick to compare oneself with others (useful in the policy arena), but more broadly, sensitizes us to the larger sociocultural context in which environmental education and learning occurs, thus challenging us in our attempt to find universal truth about ways in which various variables that compose environmental literacy (awareness, understanding, attitude, disposition, behavior) may be linked. Like any large-scale study on literacy of any type across populations and over time, the four national assessments challenge us in many ways: how do findings at this grain size translate down to the local and regional? What does it mean if we find significant but ultimately weak linkages among variables across entire populations? And what are the best indicators which, in measuring and tracking them, will provide value to the discussion of environmental literacy and its assessment?

Liddicoat and Krasny challenge our perspective on assessment by focusing on the need to take a long-term perspective. They introduce the concept of Significant Life Experiences, and ask retrospectively whether and how intense and outstanding experiences may shape the long-term perspective on environmental outcomes. The concept of SLE challenges the field of EE and EfS in many ways. It defines at least part of the value of EE and EfS as long-term and fundamental, and it poses a host of theoretical and methodological challenges, least of which is the difficulty to link experiences and outcomes when they are separated by long time frames and many intervening events, which are themselves seen as being part of reinforcing experiences. In such a situation, attribution is difficult and requires clear theories of action. In fact, as Dierking et al. argue in a later chapter, isolating the effects of specific events and experiences within a lifelong learning pathway may pose significant theoretical challenges. Some of the methodological issues have been addressed in other fields (for instance, research on world fairs, value of field trips and museum visits to interest in science), but more remain specific to this field, the most important one being: what would actually count as SLE. In fact, the challenge is that we have to simultaneously identify what they are (though their measured or perceived impact) and assess their impact in order to point to their value in EE and EfS. So far, longer-term outdoor or classroom experiences are often cited, and those challenge our very notion of what may count as significant.

Michaela Zint taps into the "gray" literature of evaluation studies in EE and EfS to ask fundamental questions about the cumulative value of summative evaluations and their potential role in contributing to the knowledge base in EE and EfS. This question is explored through ten exemplary evaluation studies that focused on behavioral outcomes. Main obstacles in aggregating across a variety of evaluations are not only the often very different nature of the evaluated experience, but also the lack of proper reporting of theoretical frameworks, methodologies, and methods that these evaluations are based on.

Holdsworth, Thomas, and Hegarty provide us with a very different angle on research and evaluation. Their article uses an assessment of EE and EfS curricula to focus on questions of Western worldview and education as perpetuating an existing insufficient system rather than to promote radical change in thinking in order to address the fundamental sustainability challenges. The chapter challenges our thinking about goals and outcomes by embedding EE and EfS into their larger social, political, economic, and environmental contexts. By asking fundamental questions about goals and objectives, Holdsworth, Thomas, and Hegarty also imply fundamental rethinking about the theoretical foundations in research and evaluation. Just like Liddicoat and Krasny ask questions about fundamental change in individuals, so do Holdsworth, Thomas, and Hegarty. And both cases challenge us in our thinking about immediate goals and outcomes, the level and significance of change we are seeking, and the appropriateness of educational experiences for the fundamental change being sought. Just like the examples given by Liddicoat and Krasny for SLE, the examples of transformative education experiences discussed in Holdsworth, Thomas, and Hegarty have to live up to high standards, the achievement of which is difficult to document.

In science, technological improvements often lead to leaps in understanding since they allow us to search, test, and discover in ways that were not previously possible. Michael Barnett and his colleagues ask whether geospatial technologies (i.e., ways to make the invisible visible in time and space) may play such a transformative role in environmental education. Barnett et al. provide an intriguing lens into possibilities for learner-centered explorations of linkages between natural, build, and social environmental layers. Based on a review of existing and future technology and evaluation and research findings in using these technologies in environmental education, Barnett et al. posit that these technologies do more than teach geography: they teach the very skills and habits of mind that create individual capacity to solve previously unknown problems. Technology can play a crucial role as an enabler of better learning and teaching strategies, and just as Holdsworth, Thomas, and Hegarty aim at "higher" goals, Barnett et al. provide a vision for developing competencies in individuals that goes beyond the usual cognitive, affective, contative, and behavioral, with profound consequences for assessment, evaluation, and research.

Finally, Dierking et al. discuss how the notion of informal or free-choice learning has implications for assessing evaluation outcomes. The concept of free-choice, self-directed or informal learning across the lifespan and across time and space poses similar challenges to assessment, evaluation, and research as previous chapters: what are goals and outcomes when people follow very individual pathways of learning, where each individual encounter changes the likelihood of the next encounter, where subsequent reinforcing experiences are outcomes and factors for learning, and where traditional approaches to assessment, by definition, are unsuitable? Where choice is paramount to understanding learners and their behavior, traditional methods to establish causation become inappropriate. In fact, it is the development of desires, motivations, interests, and identities that shapes the choices we make: choice is an outcome in and of itself. Dierking et al. base much of their argument on the rich literature on informal or free-choice science education and learning, where some dimensions of EE and EfS are less important or relevant. However, much can be learned by peeking over one's disciplinary borders.

This section makes no claim to cover the entire spectrum of research related to assessment and evaluation. Instead, the selection of chapters provides some insights (and examples) for challenges and accomplishments in selected areas of the authors' interest. The papers touch on the need for clear theoretical foundations and theories of action, for defining the grain size of one's goals and outcomes and for adjusting methodologies and methods accordingly. The chapters also discuss profound challenges when finding attribution across time and space and experiences. How do we know that a specific educational experience in EE of EfS had any particular effect when so much is going on around it? The section also illustrates that our knowledge base can grow significantly from evaluation studies, if those are conducted and reported in ways that makes them accessible to meta-analyses. Conversely, research and evaluation based on rigorous and theory-based assessment have the potential to support the development of future EE and EfS education efforts. Finally, technological innovations may have profound influences for teaching, learning, curriculum, or even assessment, but promise requires verification through thoughtful assessment and evaluation.

Much more could be said about REE related to assessment and evaluation and about appropriate methodologies and methods to tackle them. More could also be said about challenges to "traditional" research related to assessment and evaluation posed by holistic approaches, transdisciplinary research or the involvement of knowledge frameworks that differ markedly from current dominant academic ones (for instance indigenous ways of knowing). The same holds for strategies to embed assessment into the experience itself, or to develop nonintrusive, noninteruptive, and ecologically valid assessment systems that allow a deeper understanding of EE and EfS experiences; action research, observational methods, assessment through games and simulations or through embedded performance to mind. Ultimately researchers need to assess the receptivity of EE and EfS research and evaluation for diverse approaches and for advances in the field. How do we respond to our colleagues' work? How do we reconcile our varying definitions of rigor and appropriateness, become accepting for a multitude of approaches while remaining vigilant stewards of quality? Much remains to be done . . .

29

Research on the Long-Term Impacts of Environmental Education

Kendra Liddicoat
University of Wisconsin-Stevens Point, USA

Marianne E. Krasny
Cornell University, USA

Current widespread interest in the relationship of spending time in nature to the healthy development of children and adults throughout the life span has sparked reevaluation of environmental education (EE) approaches and research goals. One such reevaluation has focused on the importance of considering the lasting impacts of EE programs, as one approach for connecting children with nature. Among practitioners, there is renewed enthusiasm for goals that reach beyond influencing short-term behaviors or meeting science education standards, and include ideas such as "lighting a spark" or "inspiring lifelong environmental stewardship." In this chapter, we review the methods and findings from two research approaches that address this need to document long-term impacts of EE.

Our focus is on research conducted months and where possible years *after* an influential experience rather than at the conclusion of an EE program. The first approach, which we refer to as significant life events, focuses on life experiences that have led to later adoption of pro-environmental attitudes and lifestyles and involvement in environmental action. This literature includes both the narratives gathered by significant life experiences (SLEs) researchers and more quantitative studies linking childhood experiences with adult attitudes and behaviors. The second research approach is retrospective program evaluation, much of which is based on long-term memory theory. Most of these studies focus on outdoor education experiences. Taken together, these two bodies of literature offer insight into the types of experiences that promote environmental stewardship, whether environmental education programs provide such experiences, and how effective environmental education programs are in meeting their long-term stewardship behavior goals.

A discussion of long-term research must first consider how long is long-term. For research on significant life events, the time frame is necessarily quite fluid because study participants of different ages are being asked to reflect on experiences that occurred across their life span. In terms of program evaluation, time frames in the literature range from six months (e.g., Bogner, 1998; Smith-Sebasto & Oberchain, 2009) to over seventeen (e.g., Gass, Garvey, & Sugerman, 2003) and even forty-five years (Liddicoat, 2013), but there has been little discussion of how time frames were or should be selected. The most common determining factors likely have been logistical—availability of research participants and length of time allotted for the study.

For the purposes of this chapter, we have selected one year postprogram as our distinguishing line between short- and long-term studies. This fits the general distinction in the existing literature between studies gathering follow-up data for comparison to prepost test data and those focused specifically on memories and lasting impacts. As the body of literature grows, we would encourage researchers to choose the length of time between the experience and the research not just based on logistics, but also on theory and common sense. For example, developmental psychology suggests time frames over which participants are likely to have matured sufficiently to put environmental education learning into practice, and contextual factors that change over the life span (e.g., local regulations about recycling) also influence an individual's ability to act in an environmentally responsible manner. Other decisions about methodologies, for example the use of memory theory to provide insight into the stability and accuracy of memories and reflections over time, have implications for the ability to deal with the intervening variables that are unavoidably present in long-term research. Current literature on significant life events or lasting impacts provides limited explanation and justification for methodological decisions.

Significant Life Events Research

Significant life experiences (SLEs) research began with Tanner's (1980) qualitative study which asked environmental activists in the United States to reflect on what led them to their current role. Responses emphasized childhood experiences in nature. Additional studies were then conducted in multiple countries with environmental educators and activists using both qualitative and quantitative methods. A full review of the early studies (pre-1998) can be found in Chawla (1998a; 1998b) and Table 29.1 summarizes the more recent work. Overall, childhood outdoor experiences remained salient across many cultures; other significant influences on adult environmental attitudes and behaviors included work, other people, exposure to pollution, and education (in varying orders) (Chawla, 1999; Corcoran, 1999; Palmer & Suggate, 1996; Palmer et al., 1998; Palmer, Suggate, Bajd, & Tsaliki, 1998; Palmer, Suggate, Robottom, & Hart, 1999; Sward, 1999). Building on these studies, more recent work in Asia has explored similar questions using mixed methods with control groups. Furihata, Ishizaka, Hatakeyama, Hitsumoto, and Ito (2007) compared environmental educators (n=188) with general citizens (n=25) in terms of current responsible environmental behaviors and SLE. Few significant differences were found, perhaps due to the small size of the control group, but many of the SLE categories noted in other studies were evident in the narratives collected from a subgroup of the educators. Hsu (2009) also compared environmentally active citizens (n=277) with a control group of nonactive citizens (n=153) and found that seventeen SLEs, including time spent in nature, participation in environmental organizations (some of which provided opportunities for time in nature), friends involved in environmental organizations and action, and loss of beloved natural places, among others, could explain 55 percent of the variance in environmental activism. Hsu (2009) rightly claims that this figure is much higher than generally found in studies linking environmental literacy to environmental action, a finding with significant implications for environmental educators designing programs. However, this comparison suffers from the fact that SLE combines seventeen factors, which may be difficult to integrate into any one experience (supporting the need for cumulative experiences over a lifetime rather than one-time experiences). Interestingly, Hsu (2009) found that whereas contact with nature was the most often cited SLE among environmental activists, nonactivists also rated this factor as among the most significant of their life experiences. Thus in the comparison between the two groups, contact with environmental organizations and friends, and loss of a significant natural place, differed most between the activist and nonactivist group, although all seventeen factors differed significantly between the two groups. Such results point to the need for continued comparative research using controls.

Despite, or perhaps because of its breadth, the SLE literature has been criticized on a number of fronts. Critics have noted inconsistency in sample selection, data collection, and data analysis within and across studies and sites as well as a lack of attention to cultural differences among study participants (Chawla, 1998a; Dillon, Kelsey, & Duque-Aristizabal, 1999; Tanner, 1998). Considering the larger premise of SLE research, Gough (1999) also questions how much the childhood experiences of today's adults can tell us about the experiences of today's children. Sivek (2002) began to address this problem by interviewing and surveying high school students (rather than adults) about the significance of time outdoors, role models, and personality in determining current environmental sensitivity. His results mirrored other SLE research with time outdoors and people ranking most influential. Delving deeper with youth environmental leaders, Arnold, Cohen, and Warner (2009) again found that time outdoors and role models were influential while also emphasizing the importance of friends, peers, and youth conferences. However, a persistent issue for studies such as these that respond to Gough's (1999) critique is that while a shorter time frame between childhood and the evaluative research makes the findings more relevant to today's youth, it also limits the time during which such experiences could have had a significant influence on life.

SLE research has been primarily descriptive and interpretive, rather than strongly informed by existing theories. As Chawla (1998a) notes, a qualitative approach was beneficial because it added depth to our understanding of the individual and emotional side of environmentalism at a time when most research was quantitative. However, the self-referencing nature of the research has led to calls for greater consideration of theory, including autobiographical memory theory (Chawla, 1998a; N. Gough, 1999) and theory related to the role of identity and culture in research and practice (Dillon et al., 1999). Such theoretical work based on SLE data is beginning, with an emphasis on the process by which experiences with nature are influential. Chawla (2007) used attachment theory and ecological psychology in combination with her SLE data to propose a model illustrating the progressive benefits of free play in nature. In this model, mobility (freedom to explore), access to natural environments, and successful engagement with natural processes leads to a continuing spiral of learning and involvement, especially when supported by an interested adult. Considering experiences in both childhood and adulthood, James, Bixler, and Vadala (2010) have also proposed a model based on their own SLE interviews (n=61 including 10 control subjects). Their work suggests that there are four sequential stages of involvement with nature leading up to an adult natural history related vocation: (1) direct experience dominant, (2) emerging formalized skills, (3) role awareness, and (4) identity formation. Additional studies that refine, validate, and expand these models may offer a new direction for additional SLE research with a more direct application to environmental education.

Several recent studies have begun to use quantitative methods to gather data from the general public, thus looking beyond SLE's exclusive sampling of environmentalists. These studies mirror SLE results and, in some cases, begin to collect data specifically on environmental

TABLE 29.1
Research on Influential Life Experiences

Authors	Date	Methods	Site	Sample	Results
Palmer, Suggate, Bajd, and Tsaliki	1998	Autobiographical narratives	UK, Slovenia, Greece	Environmental educators, n=575	Most common influences in Slovenia were people, pollution, and childhood time in nature. Most common influences in Greece were pollution, childhood time in nature, and work.
Palmer et al.	1998	Autobiographical narratives	9 Countries	Environmentally active individuals, n=1259	Across sites (Australia, Canada, Greece, Hong Kong, Slovenia, South Africa, Sri Lanka, Uganda, and UK), the most influential factors in descending order were experiences of nature, people, education, witnessing negative situations, work, and media.
Palmer, Suggate, Robottom, and Hart	1999	Autobiographical narratives	UK, Australia, Canada	Environmental educators, n=363	Childhood time in nature, other people, work, and education were described most frequently as influences on current proenvironmental activities.
Sward	1999	Survey, structured interviews	El Salvador	Environmental professionals, n=17	In descending order, childhood time in nature with family or friends, witnessing environmental destruction, formal education, and organized outdoor experiences were described as significant influences.
Corcoran	1999	Autobiographical narratives	US	Environmental educators, n=510	Study explored individual narratives in greater depth to understand the variety of outdoor experiences and family/friends who were influential.
Chawla	1999	Open-ended interviews	Norway, US	Environmentalists, n=56	Most respondents described multiple influences with childhood experiences in nature, family, and membership in related organizations being mentioned most frequently.
Sivek	2002	Focus group, questionnaire	US	Youth environmentalists, n=20+64	Influences on current levels of environmental sensitivity were grouped by role models, environmental influences, and personality. Teachers, parents, time outdoors, and being outgoing were most frequently cited as influential.
Furihata, Ishizaka, Hatakeyama, Hitsumoto, and Ito	2007	Workshops, interviews	Japan	Environmental educators n=188, and other citizens n=25	SLEs occurring both in childhood and adulthood included (in descending order) experiences in nature, loss of nature, family, and books/media. Due to sample sizes, there were no significant differences between study groups.
Hsu	2009	Autobiographical narratives, surveys	Taiwan	Educators/civil servants n=40 narratives, n=81 and n=430 surveys	Narratives revealed 17 categories of influence which were shown to be significantly different among environmentally active and apathetic survey respondents.
Arnold, Cohen, and Warner	2009	Interviews	Canada	Young environmentalists (16–19 yrs), n=12	Influences on current environmental activism included people (parents, friends, role models, and teachers), outdoor experiences, school, and youth conferences.
Lohr and Pearson-Mims	2005	National telephone survey	US	Adults, n=2004	Active gardening in childhood, passive interaction with plants, environmental education, and a home with natural surroundings in childhood were correlated with positive attitudes toward trees and participation in gardening in adulthood.
Ewert, Place, and Sibthorp	2005	Scale (NEP)	US	College students, n=576	Appreciative outdoor experiences in childhood were related to ecocentric beliefs while consumptive outdoor experiences were related to anthropocentric beliefs with both variable combined explaining 8.4% of the variance.
Wells and Lekies	2006	National telephone survey	US	Adults, n=2004	Active participation with wild nature and to a lesser extent participation with domesticated nature in childhood predicted proenvironmental attitudes and behaviors in adulthood. Environmental education predicted neither.
Thompson, Aspinall, and Montarzino	2008	Focus groups, close-ended questionnaires	UK	Adults, n=798	Higher and lower frequency of visits to woodlands in childhood predict corresponding frequencies in adulthood. Demographic variables including gender and proximity to woodlands were also influential.

education. Survey research by Thompson, Aspinall, and Montarzino (2008) in Scotland (n=339) and England (n=459) found that frequency of childhood visits to green spaces was highly correlated to frequency of adult visits. Looking more closely at types of childhood outdoor experiences in relation to adult attitudes, Ewert, Place, and et al. (2005), found that for a sample of US college students (n=533), appreciative outdoor experiences in childhood were related to an ecocentric perspective while consumptive outdoor experiences were related to anthropocentric beliefs. Attitudes were measured using the New Environmental Paradigm scale (Dunlap, Van Liere, Mertig, & Jones, 2000). Interestingly, education and involvement in environmental/outdoor organizations were not significant predictors of ecocentric beliefs in Ewert et al. (2005), although involvement in environmental organizations and to a lesser extent education were significant factors differentiating environmental activists and nonactivists in a study in Taiwan (Hsu 2009).

Lohr and Pearson-Mims (2005) drew on results of a national survey (n=2004) and found that a variety of childhood experiences, including active participation with nature, EE and gardening, were correlated with positive views toward trees and interest in gardening in adulthood. Wells and Lekies (2006) used the same data set to examine the link between environmental experiences and adult attitudes and behaviors. Childhood participation with nature (wild and domestic) was significantly correlated with adult environmental attitudes and behaviors, but participation in childhood EE per se was not a significant predictor of either. Wells and Lekies (2006) suggest that EE may not have been properly operationalized to include hands-on, outdoor EE. Therefore, additional research is needed to understand the possible connection between specific types of EE and proenvironmental attitudes and behaviors.

Wells and Lekies (2006) and Ewert et al. (2005) begin to address concerns about the paucity of theory framing or emerging from significant life events research. Wells and Lekies (2006) draw on life-course perspective and Ewert et al. (2005) use the New Environmental Paradigm. Other theories brought to bear on SLE research include sense of place (Thompson et al., 2008). Each of these theories merits further investigation, especially in relation to their explanatory and predictive power. However, no consistent theoretical framework for SLE and related survey studies has emerged to date.

Taken together, SLE and quantitative large sample survey studies provide information on the types of childhood experiences that lead to adult environmental attitudes and behaviors. This big picture can help environmental educators appreciate the importance of time spent in nature and can encourage them to think broadly about the types of experiences their programs might incorporate. However, what the research summarized above does not do, is provide concrete information on whether specific types of environmental education programs, lessons, or activities have a long-term influence. For example, questions related to the long-term outcomes of engaging in issues-based lessons, or of the social component of an outdoor education experience, are not addressed. Gathering this information requires more focused research with individuals who are known to have participated in an environmental education experience and, therefore, have the potential to be influenced by it.

Retrospective Program Evaluation

The second major research tradition focusing on long-term impacts of EE is retrospective evaluation of specific educational programs, which is summarized in Table 29.2. In comparison to SLE research and surveys of the general public, retrospective evaluation research focuses on a clearly defined experience, and in cases where memory research is included in such studies, often goes deeper than the general EE program to focus on the specific activities. Where the EE programs are voluntary (e.g., participant-paid Outward Bound programs), information is gathered from participants with environmental interests, while in school-based programs such as those at outdoor education centers that serve entire grades, information is gathered from participants with a diversity of environmental perspectives.

In a major mixed methods study of Outdoor Adventure Education (OAE) programs, Kellert (1998) found that such programs had wide-ranging impacts as perceived and reported by participants. Impacts included increased interest in outdoor recreation activities, more positive environmental attitudes, commitment to conservation, desire to learn about nature, and personal well-being, self-confidence, and initiative. Few impacts on ecological knowledge or conservation behaviors were reported. Perhaps most strikingly, nearly all the survey respondents repeatedly referred to their OAE course as being a life changing experience and one of the most important experiences of their life (Kellert 1998). Other studies in similar settings have also documented lasting program impacts in terms of personal and professional growth but have not explored environmental attitude and behavior outcomes in depth (Ballard, Shellman, & Hayashi, 2006; Daniel, 2007; Gassner, Kahlid, & Russell, 2006).

To better understand how EE-related outcomes might be achieved through OAE, D'Amato, and Krasny (2011) asked past participants in Outward Bound and the National Outdoor Leadership School about the program attributes that contributed to their transformational experiences. Respondents indicated that living in pristine nature, experiencing a different lifestyle (and breaking with normal life), experiencing a community of individuals undergoing the same experience, and the intensity and challenge of the outdoor course, were the key factors contributing to their transformations. However, participants commonly experienced difficulties in transitioning back to their normal life in the absence of the oncourse community, factors which could influence the impact of outdoor experience

TABLE 29.2
Retrospective Program Evaluations

Authors	Date	Methods	Site	Sample	Results
Hanson	1993	Quantitative scale, up to 6 years postprogram	US	Youth (11–12 yrs), n=1349	Students who had participated in a greater number of energy conservation curriculum units reported greater energy conservation knowledge, attitudes, and behaviors.
Everson	2000	Telephone survey, 1–31 yrs postprogram	US	Adults, n=162	Participation in a month-long outdoor environmental science education program positively influenced subsequent outdoor recreation behaviors (ORB) and to a lesser extent environmentally responsible behaviors (ERB).
Gass	2003	Observation, telephone interviews, 17 yrs postprogram	US	Adults, n=16	Participants in a wilderness-based college orientation program indicated that the experience helped them challenge assumptions and develop a network of friends. The program continued to have a positive influence even postgraduation.
Peacock	2006	Focus groups	UK	Youth, n=108	Experience with school-based guardianship programs in wild areas positively impacted attitudes toward the environment, increased related knowledge, and may have influenced behavior.
Knapp and Benton	2005	Telephone interviews, 2 yrs postprogram	US	Adults, n=6	Adults who had participated in a one hour interpretive program recalled visual images, active and novel program components, and impressions of the interpreter.
Knapp and Benton	2006	Telephone interviews, 1 yr postprogram	US	Youth (10–11 yrs), n=10	Students who had participated in a week-long residential outdoor EE program recalled activities, specific program content, and emotional reactions to the experience.
Farmer, Knapp, and Benton	2007	Interviews, 1 yr postprogram	US	Youth (9 yrs), n=15	Students who had participated in a day-long EE program at a National Park recalled activities and knowledge learned. Data also revealed some impact on proenvironmental attitudes.
Liddicoat	2010	Interviews 13–45 yrs and 5–7 yrs postprogram	US	Adults n=45, youth n=52	Participants in a three-day residential EE program recalled the novel, active, social, and personally engaging components of the trip. Impacts included personal growth, new friendships, enthusiasm for natural settings, and science learning.

on longer-term environmental behaviors (D'Amato & Krasny, 2011). Looking at a program that incorporates both outdoor adventure and environmental science, Everson (2000) interviewed former participants in the Teton Science Schools month-long high school program up to thirty years after the experience. The data showed that the participants currently demonstrated positive outdoor recreation and environmentally responsible behaviors, and perceived that the program impacted their outdoor and, to a lesser extent, their environmental behaviors.

Turning to school-based environmental education, Hanson (1993) examined cumulative and enduring impacts on knowledge, attitudes, and behaviors related to energy conservation among sixth-grade students who had participated in multiple energy education programs throughout elementary school. Comparisons between participants and nonparticipants revealed consistently higher scores among participants in all three outcome areas (energy conservation knowledge, attitudes, and behaviors) and comparisons among participants showed that the greater the number of educational components experienced, the higher the scores. Peacock (2006) also looked at a program with repeated exposure, but rather than focusing on classroom activities, this study investigated an outdoor stewardship program in which classes repeatedly traveled to a land trust area. Follow-up interviews with high school students revealed a continued attachment to the natural area where they had conducted stewardship activities, continued visits (outside of school and with family and friends) to the stewardship site, concern about sustainability issues in their community, perceived increased self-esteem and other social skills due to the program, and knowledge of context specific information. Program participation did not seem to have led to greater concern or knowledge about larger, global environmental issues.

A major challenge inherent to retrospective evaluation studies is linking current attitudes and behaviors to specific experiences in the past, given months and often years of intervening experiences. As a result, some researchers have begun to draw on memory theory and gather episodic memories of EE experiences as evaluative data (e.g., Knapp & Benton, 2006; Liddicoat, 2013). Memories can be collected many years later and reveal what aspects of the program participants have retained over time (Knapp & Benton, 2005). Episodic memories are also, by definition, event specific (Baddeley, 2001), thus reducing some of the ambiguity regarding how closely linked the retrospective data are to the program itself. Through their specificity and in combination with existing theory, episodic memories also shed light on teaching practices that will promote long-term retention of knowledge, attitudes, and behaviors. Psychology research indicates that experiences that are novel, repetitive, active, and emotional are particularly memorable (Christianson & Safer, 1996; Herbert & Burt, 2004; Linton, 1982; Thompson, Skowronski, Larsen, & Betz, 1996; Zimmer et al., 2001). Retrospective research in EE can provide insight into how well our field is effectively incorporating these program characteristics to create lasting memories.

Research focused on subjects' remembering and reminiscing about an EE experience also provides an opportunity for past participants to reflect on why program elements were memorable to them and, with the benefit of hindsight, explain how the environmental education experiences under discussion have had a lasting impact. The potentially causal link between memories and continued impact is an area still open to exploration through retrospective studies of EE. Psychology theories suggest that episodic memories may serve a social function (fostering interactions through reminiscing), identity function (allowing for awareness of one's capabilities), or directive function (informing future actions) (Bluck, 2003). Episodic memories also can enable an experience to grow in impact over time as one is repeatedly reminded of what was learned through encounters that bring the experience to mind., a process called retrospective causality (Pillemer, 1998). Although theories can rarely be translated fully from one field to another, memory-based retrospective research in EE has the potential to make a valuable contribution to a body of literature in EE that has been criticized for being theory poor.

Memory studies in EE grew out of work in museum studies (e.g., Hudson & Fivush, 1991) and were initially conducted by Knapp and colleagues. Knapp and Benton (2006) observed a five-day Expedition Yellowstone! program for fifth graders, and then contacted participants by phone one year later. The former students (n=10) were able to relate specific information about games they played and to describe aspects of program content, as well as talk about the positive and negative emotional aspects of the trip. Interviews with past participants in single day EE programs (in Hoosier National Forest and the Great Smoky Mountains National Park) revealed a similar ability to recall specific program knowledge gained through activities (Farmer, Knapp, & Benton, 2007; Knapp & Benton, 2005). Memories of emotional experiences during these shorter trips did not emerge as a theme, perhaps because opportunities for this type of engagement are more limited in a day rather than overnight program (Liddicoat, 2013).

Extending the time frame between program experience and retrospective data collection, we (the authors of this chapter) have conducted additional memory-based research on residential outdoor EE programs for upper elementary school students ages ten to twelve (Liddicoat, 2013). An exploratory study conducted in 2005 with forty-five individuals who participated in a three-day natural history EE program between 1958 and 1992 revealed that not only do adults remember such short experiences, but they also recall specific lessons, people, and opportunities for personal growth. Impacts and memories were relatively consistent across the decades, including among current high school students (n=37) who were surveyed to validate the results from the older participants, in light of Gough's (1999) concern that significant experiences for today's youth may differ from those of adults. Most memorable were the active, social, personally challenging, and novel components of the trip. Stated impacts included becoming more knowledgeable about and comfortable in the outdoors, making and maintaining friends, and developing a sense of independence by spending the night away from home. Research at two additional sites with teens who had participated in similar residential outdoor EE programs five and seven years prior to the study revealed similar themes. These latter studies also explored in greater depth what function these memories served. For many participants, their memories served a social function enabling them to reminisce with friends and classmates, as well as a directive function by fostering an awareness of local natural history, increased interest in outdoor recreation, and more proenvironmental attitudes and behaviors. However, further research is needed to fully understand the link between the rich memories created by an EE experience and their usefulness to the individual who possesses them, as well as how individuals apply them toward proenvironmental behaviors despite potential constraints in their life or community.

The relative scarcity of retrospective evaluations designed to assess lasting impacts may reflect the difficulty of conducting such research. In our highly mobile society, locating participants years after an experience and convincing participants who disliked or have forgotten the experience to be surveyed or interviewed is difficult. This may introduce significant selection bias (Shadish, Cook, & Campbell, 2002), similar to that in SLE research. Internal validity is also compromised by issues of self-report, socially desirable responses, and possible intervening variables (Kerlinger & Lee, 2000). A focus on memory, which is inherently tied to a specific event, is personal, and is perhaps less seemingly value-laden, can help but does not fully remove these threats to validity. Memory also introduces new sources of error, related to the accuracy of memories and the varying speeds at which people forget positive, negative, and emotional experiences (Thompson et al., 1998). A focus on remembered events also emphasizes overt learning at the expense of learning that occurred without people realizing it.

Looking beyond methods and logistics, long-term research also faces the challenge of gathering data over a long period during which people, society, and education change. It may be that certain types of EE are more amenable to long-term evaluation. Understanding how nature study, which has been a cornerstone of our field for nearly a century, influenced participants forty years ago offers insight into the impacts of similar programs today. In contrast, decades-long studies of EE that focuses on environmental issues that change over time, such as point source and nonpoint source pollution, may be less feasible and relevant. EE evolves and thus long-term studies should carefully balance the need for extended duration and continued relevance.

Conclusion

Environmental educators speak of "planting a seed" in the hope that environmental behaviors will emerge later on. However, given the potentially long timeframe between when an individual participates in a program and demonstrates environmental behaviors, during which participants experience many intervening factors, documenting any changes in attitudes and behaviors through objective before and years after program measures presents numerous challenges. On the other hand, limiting ourselves to documenting immediate postprogram impacts risks the potential of missing out on behaviors that are only expressed years later, as a result of supportive social and political structures, social norms that differ with age cohort, cumulative experiences, and other factors. For this reason, the subfield of long-term EE research merits further examination.

Significant life events research focusing on environmentally active individuals and on the general public has revealed the importance of time spent in nature, within or outside of formal EE programs, in influencing later involvement in environmental recreation, activism, and choice of profession, as well as positive attitudes toward nature. Retrospective studies also have demonstrated the importance of repeated exposure to EE and other outdoor activities in influencing subsequent knowledge, attitudes, and behaviors, and memory studies have begun to suggest particular program elements that may be connected to content retention and lasting impacts. However, long-term research overall suffers from lack of a consistent theory, and in some empirical work, from inattention to any theoretical frameworks. This research also suffers threats to validity related to biased samples and intervening variables. Although not specifically addressed in this chapter, long-term research, similar to other types of EE research, also suffers from lack of agreed upon outcomes. For example, in the retrospective evaluation research conducted by the authors of this chapter, programs varied from a focus on teaching natural history and science knowledge to promoting future outdoor recreation. While the programmatic focus did influence what participants recalled, other aspects such as social interactions and opportunities for personal growth also emerged as memorable and influential (Liddicoat, 2013). The diversity of goals and outcomes found across our three research sites are but a small sample of the diversity of goals stated by EE programs around the world, which may be an impediment to moving forward in documenting EE's impacts beyond those specific to a particular program. This lack of ability to consistently document EE outcomes may in turn may prove a barrier to garnering political and public support for EE.

As the field of EE moves forward, researchers may want to work with practitioners to define program goals that reflect on significant life events and long-term impacts of EE programs. For example, if studies continue to suggest, as do the SLE studies and Ewert, Place, and Sibthorp (2005), that *time spent in nature*, especially with environmental organizations and with environmentally active friends, may be more important in influencing subsequent stewardship behaviors than *formal EE*, the design and location of EE programs may need to change. As the subfield of long-term studies evolves, researchers may want to consider combining the approaches reviewed in this chapter to further both practice and theory. When time and logistics allow, true longitudinal studies with pre-, post-, and follow-up data involving both objective measures and qualitative self-report may be particularly informative. One exemplary study of this sort was conducted by Gass et al. (2003) on the impacts of freshman orientation programs. Students were observed during the program, interviewed directly after it, and interviewed again seventeen years later, thus allowing the researchers to answer questions related to short and long-term impacts. A longitudinal study focused on memory with interviews at intervals over an extended period also could provide insight into how memories change over time and how they are linked to program impacts. Future studies may also combine SLE and memory approaches. In our own research attempting to link memory and outcomes of EE programs, we chose not to ask about influential experiences other than specific elements of the EE programs. However, adding information that would allow us to compare memories of an EE experience with other memories, and how respondents link various memories to behaviors, would create a more complete understanding of the impacts of EE programs and specific program elements.

References

Arnold, H. E., Cohen, F. G., & Warner, A. (2009). Youth and environmental action: Perspectives of young environmental leaders on their formative influences. *Journal of Environmental Education, 40*(3), 27–36.

Baddeley, A. (2001). The concept of episodic memory. *Philosophical Transactions: Biological Science, 356*(1413), 1345–1350.

Ballard, A., Shellman, A., & Hayashi, A. (2006). Collective meanings of an outdoor leadership program experience as lived by participants. *Research in Outdoor Education, 8*, 1–21.

Bluck, S. (2003). Autobiographical memory: Exploring its functions in everyday life. *Memory, 11*(2), 113–123. doi:10.1080/741938206

Bogner, F. X. (1998). The influence of outdoor ecology education on long-term variables of environmental perspective. *Journal of Environmental Education, 29*(4), 17–29.

Chawla, L. (1998a). Research methods to investigate significant life experiences: Review and recommendations. *Environmental Education Research, 4*(4), 383–397.

Chawla, L. (1998b). Significant life experiences revisited: A review of research on sources of environmental sensitivity. *Environmental Education Research, 4*(4), 369–382.

Chawla, L. (1999). Life paths into effective environmental action. *Journal of Environmental Education, 31*(1), 15–26.

Chawla, L. (2007). Childhood experiences associated with care of the natural world: A theoretical framework for empirical results. *Children, Youth and Environments, 17*(4), 144–170.

Christianson, S.-A., & Safer, M. A. (1996). Emotional events and emotions in autobiographical memories. In D. C. Rubin (Ed.), *Remembering our past: Studies in autobiographical memory* (pp. 218–243). Cambridge, UK: Cambridge University Press.

Corcoran, P. B. (1999). Formative influences in the lives of environmental educators in the United States. *Environmental Education Research, 5*(2), 207–220.

D'Amato, L. G., & Krasny, M. E. (2011). Outdoor adventure education: Applying transformative learning theory to understanding instrumental learning and personal growth in environmental education. *Journal of Environmental Education, 42*(4), 237–254.

Daniel, B. (2007). The life significance of a spiritually oriented, outward bound-type wilderness expedition. *Journal of Experiential Education, 29*(3), 386–389.

Dillon, J., Kelsey, E., & Duque-Aristizabal, A. M. (1999). Identity and culture: Theorizing emergent environmentalism. *Environmental Education Research, 5*(4), 395–405.

Dunlap, R. E., Van Liere, K. D., Mertig, A. G., & Jones, R. E. (2000). Measuring endorsement of the new ecological paradigm: A revised NEP scale. *Journal of Social Issues, 56*(3), 425–442.

Everson, M. (2000). *A long-term retrospective study of participants' perception about the influence of Teton Science School's flagship programs on environmental and recreational behaviors.* Unpublished master's thesis, Utah State University, Logan, UT.

Ewert, A., Place, G., & Sibthorp, J. (2005). Early-life outdoor experiences and an individual's environmental attitudes. *Leisure Sciences, 27*, 225–239.

Farmer, J., Knapp, D., & Benton, G. M. (2007). An elementary school environmental education field trip: Long-term effects on ecological and environmental knowledge and attitude development. *Journal of Environmental Education, 38*(3), 33–42.

Furihata, S., Ishizaka, T., Hatakeyama, M., Hitsumoto, M., & Ito, S. (2007). Potentials and challenges of research on "Significant Life Experiences" in Japan. *Children, Youth and Environments, 17*(4), 207–226.

Gass, M. A., Garvey, D. E., & Sugerman, D. A. (2003). The long-term effects of a first-year student wilderness orientation program. *The Journal of Experiential Education, 26*(1), 34–40.

Gassner, M., Kahlid, A., & Russell, K. (2006). Investigating the long-term impact of adventure education: A retrospective study of outward bound Singapore's classic 21-day challenge course. *Research in Outdoor Education, 8*, 75–93.

Gough, A. (1999). Kids don't like wearing the same jeans as their mums and dad: So whose "life" should be in significant life experiences research? *Environmental Education Research, 5*(4), 383–393.

Gough, N. (1999). Surpassing our own histories: Autobiographical methods for environmental education. *Environmental Education Research, 5*(4), 407–418.

Hanson, R. (1993). Long-term effects of the energy source education program. *Studies in Educational Evaluation, 19*, 363–381.

Herbert, D. M. B., & Burt, J. S. (2004). What do students remember? Episodic memory and the development of schematization. *Applied Cognitive Psychology, 18*, 77–88.

Hsu, S.-J. (2009). Significant life experiences affect environmental action: A confirmation study in eastern Taiwan. *Environmental Education Research, 15*(4), 497–517.

Hudson, J. A., & Fivush, R. (1991). As time goes by: Sixth graders remember a kindergarten experience. *Applied Cognitive Psychology, 5*, 347–360.

Kellert, S. R. (1998). *A national study of outdoor wilderness experience.* New Haven, CT: Yale School of Forestry and Environmental Studies and the Student Conservation Association.

Kerlinger, F. N., & Lee, H. B. (2000). *Foundations of behavioral research* (4th ed.). Orlando, FL: Harcourt College Publishers.

Knapp, D., & Benton, G. M. (2005). Long-term recollections of an environmental interpretive program. *Journal of Interpretation Research, 10*(1), 52–54.

Knapp, D., & Benton, G. M. (2006). Episodic and semantic memories of a residential environmental education program. *Environmental Education Research, 12*(2), 165–177.

Liddicoat, K. R. (2013). *Memories and lasting impacts of residential outdoor environmental education programs.* Unpublished doctoral dissertation, Cornell University, Ithaca, NY.

Linton, M. (1982). Transformations of memory in everyday life. In U. Neisser (Ed.), *Memory observed: Remembering in natural contexts* (pp. 77–91). New York, NY: W. H. Freeman.

Lohr, V. I., & Pearson-Mims, C. H. (2005). Children's active and passive interactions with plants influence their attitudes and actions toward trees and gardening as adults. *HortTechnology, 15*(3), 472–476.

Palmer, J. A., & Suggate, J. (1996). Influences and experiences affecting pro-environmental behavior of educators. *Environmental Education Research, 2*(1), 109–121.

Palmer, J. A., Suggate, J., Bajd, B., Hart, P., Ho, R. K. P., Ofwono-Orecho, J. K. W., . . . , Staden, C. V. (1998). An overview of significant life influences and formative experiences on the development of adults' environmental awareness in nine countries. *Environmental Education Research, 4*(4), 445–465.

Palmer, J. A., Suggate, J., Bajd, B., & Tsaliki, E. (1998). Significant influences on the development of adults' environmental awareness in the UK, Slovenia, and Greece. *Environmental Education Research, 4*(4), 429–444.

Palmer, J. A., Suggate, J., Robottom, I., & Hart, P. (1999). Significant life experiences and formative influences on the development of adults' environmental awareness in the UK, Australia, and Canada. *Environmental Education Research, 5*(2), 181–200.

Peacock, A. (2006). Changing minds: The lasting impact of school trips. Retrieved November 4, 2006, from http://www.nationaltrust.org.uk/main/w-schools-guardianships-changing_minds.pdf

Pillemer, D. B. (1998). *Momentous events, vivid memories.* Cambridge, MA: Harvard University Press.

Shadish, W. R., Cook, T. D., & Campbell, D. T. (2002). *Experimental and quasi-experimental designs for generalized causal inference.* Boston, MA: Houghton Mifflin.

Sivek, D. J. (2002). Environmental sensitivity among Wisconsin high school students. *Environmental Education Research, 8*(2), 155–170.

Smith-Sebasto, N. J., & Oberchain, V. L. (2009). Students' perceptions of the residential environmental education program at the New Jersey School of Conservation. *Journal of Environmental Education, 40*(2), 50–62.

Sward, L. L. (1999). Significant life experiences affecting the environmental sensitivity of El Salvadoran environmental professionals. *Environmental Education Research, 5*(2), 201–206.

Tanner, T. (1980). Significant life experiences: A new research area in environmental education. *Journal of Environmental Education, 11*(4), 20–24.

Tanner, T. (1998). Choosing the right subjects in significant life experiences research. *Environmental Education Research, 4*(4), 399–417.

Thompson, C. P., Herrmann, D. J., Bruce, D., Read, J. D., Payne, D. G., & Toglia, M. P. (1998). *Autobiographical memory: Theoretical and applied perspectives.* Mahwah, NJ: Lawrence Earlbaum Associates.

Thompson, C. P., Skowronski, J. J., Larsen, S. F., & Betz, A. L. (1996). *Autobiographical memory: Remembering what and remembering when.* Mahwah, NJ: Lawrence Earlbaum Associates.

Thompson, C. W., Aspinall, P., & Montarzino, A. (2008). The childhood factor: Adult visits to green places and the significance of childhood experience. *Environment and Behavior, 40*(1), 111–143.

Wells, N., & Lekies, K. (2006). Nature and the life course: Pathways from childhood nature experiences to adult environmentalism. *Children, Youth and Environments, 16*(1), 1–24.

Zimmer, H. D., Cohen, R. L., Guynn, M. J., Englecamp, J., Kormi-Nouri, R., & Foley, M. A. (Eds.). (2001). *Memory for action: A distinct form of episodic memory?* New York, NY: Oxford University Press.

30

Advancing Environmental Education Program Evaluation

Insights From a Review of Behavioral Outcome Evaluations

Michaela Zint
University of Michigan, USA

Introduction

This chapter provides an introduction to environmental education program evaluation, and based on a review of behavioral outcomes evaluations, illustrates contributions evaluation can make to the field. The chapter reflects the author's pragmatic desire (Creswell, 2009) to contribute to environmental education by advancing environmental education program evaluation research and practice.

Program evaluation involves systematically collecting information to judge a program's performance (Fournier, 2005) and should generate insights that are used to inform program decisions (Stufflebeam, Madaus, & Kellaghan, 2000). Historically, program evaluations have been driven by accountability requirements (Kellaghan, Stufflebeam, & Wingate, 2003) but it would be a mistake to think of program evaluation only in accountability terms. In fact, there are benefits to incorporating evaluation throughout environmental education programs' life cycles (Loomis, 2002; Wiltz, 2001). For example, front-end evaluations identify program participants' needs. Logic model and theory of change exercises clarify program outcomes and make programs' underlying assumptions explicit (Frechtling, 2007; Rogers, Hacsi, Petrosino, & Hubner, 2000). Process evaluations shed insight into how programs are actually implemented. And outcome and impact evaluations identify programs' short- and long-term benefits and should reveal what program characteristics these benefits can be attributed to (Patton, 2008). When incorporated in such ways, evaluations lead to improvements that help environmental education programs better meet their objectives (Birnbaum, 2010; Jacobson & McDuff, 1997; Jenks, Vaughan, & Butler, 2010; Norris & Jacobson, 1998).

Although research and evaluation use similar methods (Rossi, Lipsey, & Freeman, 2004), they differ in their purposes and vary in how their quality is assessed (Mathison, 2008; Patton, 2008). Research seeks to contribute to a particular body of knowledge and is judged on this basis. Evaluation meets the needs of a particular program and is judged in great part based on its utility (Joint Committee on Standards for Educational Evaluation, 1994). Thus, a quality research study may not be a quality program evaluation and vice versa. At the same time, however, research can inform program evaluations and findings from program evaluations, particularly when synthesized, can contribute to disciplinary bodies of knowledge (Chelimsky, 1997; Labin, 2008).

This chapter presents such a synthesis. More specifically, the chapter offers a systematic review of evaluations of environmental education programs published in peer reviewed journals that assessed participants' environmentally responsible behaviors; i.e., behaviors expected to contribute to the conservation or preservation of the environment. The synthesis focuses on behavioral outcome evaluations, because research on these particular environmental education outcomes is lacking (Leeming, Dwyer, Porter, & Cobern, 1993; Lucas, 1980), interest in behavioral outcomes is high among the conservation and preservation organizations that fund environmental education (Heimlich, 2010), and because researchers and evaluators are likely to continue to be called upon to assess changes in these difficult to measure outcomes (Fleming & Easton, 2010).

The articles referenced in, and reviewed for, this chapter were selected from among those identified by a manual examination of the table of contents of three prominent environmental education journals, *Applied Environmental Education and Communication, Environmental Education Research*, and *The Journal of Environmental Education,* by searching the databases Web of Science, Educational Abstracts, IngentaConnect, Educational Resources Information Center (ERIC), and Google Scholar for articles with evaluation or its derivatives in the title or abstract, and by searching "My Environmental Education Evaluation Resource Assistant,"

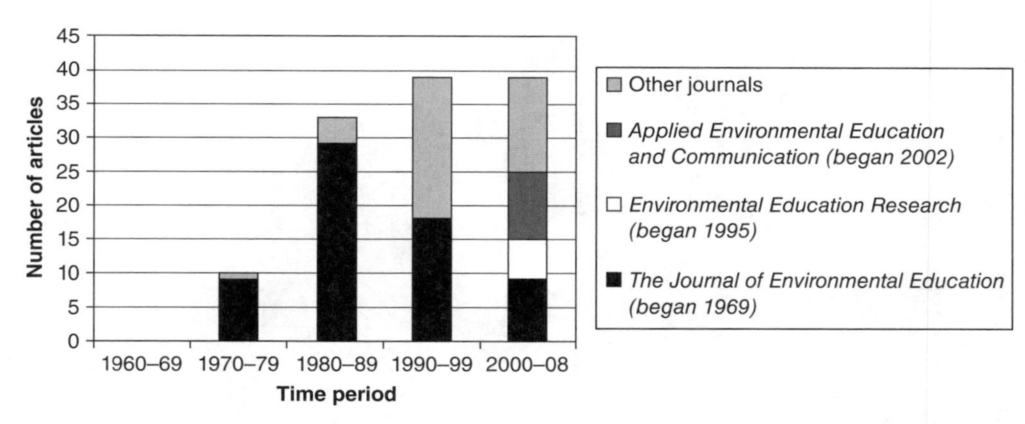

Figure 30.1 Total number of articles on environmental education program evaluation published in peer reviewed environmental education and other journals.

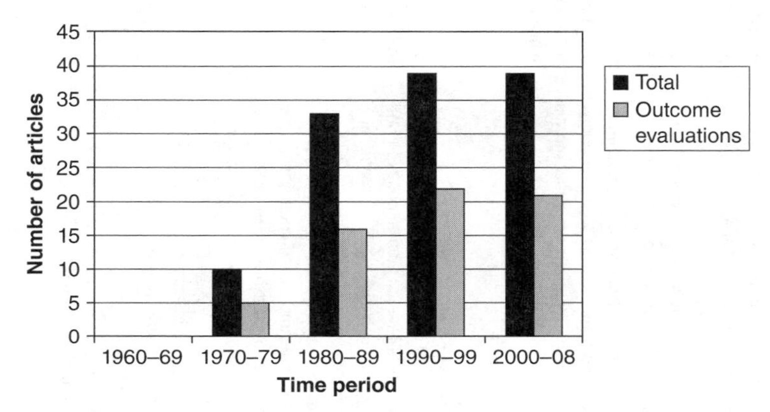

Figure 30.2 Number of outcome evaluations among the total number of articles on environmental education program evaluation published in peer reviewed environmental education and other journals.

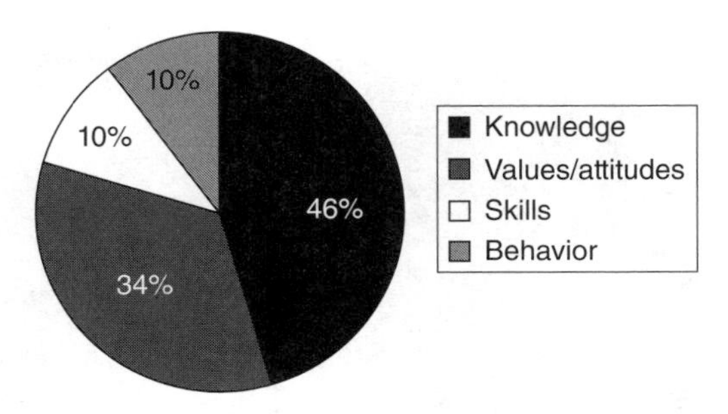

Figure 30.3 Different types of outcomes assessed by evaluations of environmental education programs published in peer reviewed journals (n=64).

www.meera.snre.umich.edu (Zint, 2010). Overall, the searches revealed a large number of articles on environmental education program evaluations (Figure 30.1), including three special issues on the topic in *The Journal of Environmental Education* (1982, 13(4)), *New Directions for Evaluation* (2005, 108), and the *Journal of Evaluation and Program Planning* (2010). The majority of the identified articles describe results from evaluations that assessed changes in environmental education program participants' outcomes, particularly on knowledge and attitude/value outcomes (Figures 30.2 and 30.3). The searches, however, also revealed many conceptual pieces that address a variety of evaluation topics and issues (e.g., Ebbutt, 1998; Loomis, 2002; Monroe, 2002) as well as a few that report on front-end (McDuff, 2002) and process evaluations (Ernst, 2005; Morgan & Soucy, 2006; Van Petegem, Blieck, & Pauw, 2007). Last, Carleton-Hug and Hug (2010) present a series of challenges and opportunities for evaluating environmental education programs based on a "cursory review" of environmental education evaluation literature published between 1994 and 2008 in the above three environmental education journals as well as based on their own experiences.

Synthesis of Behavioral Outcome Evaluations of Environmental Education Programs

Table 30.1 identifies the ten articles that report evaluation results on the behavioral outcomes (or lack thereof) of environmental education programs.[1] These ten articles were published between 1975 and 2010 in a diversity of journals, with five published in environmental education journals, two in environmental/conservation journals, and one each in an informal education, science education, and

TABLE 30.1
Overview of Reviewed Behavioral Outcome Evaluations of Environmental Education Programs (Listed by Year Published)

Author (Year) Type of Evaluator(s) Outcome(s) Evaluated	Program Evaluated and Program Implementation	Perspectives Informing Evaluation	Evaluation Methods	Behavioral Outcome Results	Stakeholder Involvement and Evaluation Use
Asch and Shore (1975) External and internal evaluators Behavior	Fifth grade boys from working class families in Montréal inner-city elementary school learned about environmental issues including how to address them in school and participated in service learning and field trips to a nature center as part of a two-year program. Program has explicit objective to change behavior.	EE evaluations and research on attitude outcomes.	Randomly selected treatment and comparison groups (with n=12 students each). The treatment group participated in the program, the comparison group only visited the nature center for 4 days. Observations of four different types of conservation and destructive behaviors in field setting by 6 independent male raters.	Treatment group more likely to engage in conservation behaviors and less likely to engage in destructive behaviors than comparison group, treatment group also more likely to engage in conservation behaviors than destructive behaviors.	
Irvine, Saunders, and Foster (1995) External and internal evaluators Behavior Knowledge	"Bog of Habits" game/exhibit at Brookfield Zoo, IL designed to change knowledge and behavior. Program has explicit objective for participants to consider alternative behavior. Observations revealed how visitors played the game.	Formative front-end and process evaluation informed by psychological research on preferred environment.	90 randomly selected visitors' interactions were observed and visitors were interviewed after the game. Intentions were measured as an indicator for behavior change.	60 percent of visitors learned about environmentally responsible behaviors and intended to try a new behavior.	Observations and other evaluation results were used to develop and improve the game.
Bogner (1999) External evaluator Behavior Knowledge Attitude	New, year-long endangered migrant bird species extracurricular program for secondary Swiss schools included classroom instruction, construction of nest-boxes for the migrant birds, sharing of observations with students in Senegal where birds spend the winter, and a field trip to observe the birds. Program has explicit objective to change behavior. Students' perceptions of the program are described.	EE research including Hungerford and Volk (1990) model and Leeming et al. (1993).	Quasi-experimental design (Cook & Campbell, 1979), including pre (4 weeks before)/post (4 weeks after to control for "post group euphoria") tests of treatment and comparison groups (10 classes of 226 students in the treatment and 75 students in the unexposed comparison group). Treatment groups were selected from 275 participating classes. Intentions were measured through previously tested, age appropriate instrument that included intention measures as indicators for behavior change.	Program increased students' intentions to engage in a variety of environmentally responsible behaviors.	Teachers' input was sought to validate measures.
Rovira (2000) External evaluator Behavior Attitude	Flexible four year old "Global Environmental Education Programme of Cornella" for primary and secondary students in Catalonia, Spain that sought to influence the behaviors of students and indirectly, that of their parents. Program does not have explicit objective to change behavior. Schools adapted the curriculum in ways that resulted in more and less "engaged" schools.	Sociological.	Quasi-experimental design tested treatment and comparison groups of students from primary and secondary as well as from working and middle class schools. Treatment and comparison groups were selected from "engaged" and "not engaged" schools, respectively. Seven focus groups with students and teachers (separately), again from the most and least engaged schools. The questions used to collect data for the evaluation measured self-reported behaviors.	Students' behaviors were determined by their families' working and middle class backgrounds rather than the program. Primary school students were more likely to engage in behaviors than secondary school students.	

Zint, Kraemer, Northway, and Lim (2002) External evaluators Behavior Knowledge Attitude Skill	Established Chesapeake Bay Foundation, MD youth (one-day, three-day, and two-week field trips, a curriculum, and pilot service learning project) and teacher (two-day curriculum workshop and five-day field in-service) programs. Programs are intended to influence variables in the Hungerford and Volk (1990) model and thus, lead to changes in participants' behaviors. Teachers' logs indicated that they implemented only a few of the intended activities and did not incorporate service learning.	EE research including Hungerford and Volk (1990) model and Leeming et al. (1993) as well as psychological behavior theory by Ajzen (1985).	Data were collected from current participants through pre, post and retention tests administered to program and comparison youth [i.e., quasi-experimental design based on Cook and Campbell (1979)] and their teachers as well as from past participants through mail surveys that included retrospective pretest questions. Simple random samples, censuses, and convenience samples ranged from 2 to 33 groups with 31–578 participants. Measures were adapted from EE and psychological behavior models. Reliabilities ranged from .44 to .91. Current participants' intentions and (changes in) self-reported behaviors were measured. Multilevel analyses were conducted to control for class or group effects.	Partly by drawing on the Hungerford and Volk (1990) model, the authors concluded that some youths' and many teachers' environmentally responsible behaviors increased as a result of the programs. Curriculum students' behaviors did not change but the curriculum was not enacted as intended.	Evaluation questions were determined by the program's manager and staff were involved in creating and administering data collection instruments, as well as in interpreting results. Results were used to improve programs, in part by providing students with longer, more in-depth experiences.
Covitt et al. (2005) External evaluators Behavior Knowledge Attitude	Environmental risk curriculum for US high school students. Program has explicit objective of changing decisions. Teacher logs revealed that only three treatment classes participated in all activities and the recommended action project. The remaining classes completed between two and seven out of eleven activities.	EE research including Hungerford and Volk (1990) model	Quasi-experimental design with treatment (532 students in 28 classes) and comparison groups (305 students in 20 classes). Students completed pre and posttests. Measures were adapted from previous EE research and evaluations. Intentions were measured as an indicator for behavior. Multilevel analyses were conducted to control for class effects.	Students intended to individually use what they learned to reduce risks to the environment posed by their actions but were less likely to intend to participate in collective environmental risk decisions.	
Morris, Jacobson, and Flamm (2007) External evaluators Behavior Knowledge Attitude	Three-year-old Florida "Manatee Watch" Program. Boaters received outreach kits along with a brief (less than one minute) information based talk on manatees. Program objective is to protect manatees. 48 percent reported that they used the outreach materials and 42 percent agreed that the materials helped them learn about manatees.	Accountability evaluation and EE research including Hungerford and Volk (1990) model and psychological behavior theory by Ajzen and Fishbein (1980).	Data were collected using a telephone survey. Individuals who received the kits (n=202) were compared to those who did not (n=297) (47 percent response rate). Participants were selected randomly from among boat registration records. Survey questions were adapted from a prior boater survey about manatees and included self-report behavior measures.	Participants' boating behaviors did not change.	

(Continued)

TABLE 30.1 (Continued)

Author (Year) Type of Evaluator(s) Outcome(s) Evaluated	Program Evaluated and Program Implementation	Perspectives Informing Evaluation	Evaluation Methods	Behavioral Outcome Results	Stakeholder Involvement and Evaluation Use
Negev, Sagy, Garb, Salzberg, and Tal (2008) External evaluators Behavior Knowledge Attitude	Israeli environmental education curriculum that focuses on increasing students' knowledge. Program does not have explicit objective to change behavior. Previous study revealed that schools only spent 1.5–3.5 hours/week on environmental science education. Students were also asked about the type and extent of nature outings.	EE research including Hungerford and Volk (1990) model	Data were collected through a stratified random sample with questionnaires administered to 1,591 sixth grade students in 39 schools and 1,530 twelfth grade students in 38 schools. Schools were representative of those in Israel (determined based on their socioeconomic characteristics). Prior environmental literary instruments developed by Wilke Hungerford, Volk, and Bluhm (1995), Nowak Wilke, Marcinkowski, Hungerford, and Mckeown-Ice. (1995), and Goldman, Yavetz, and Pe'er (2006) were adapted to measure environmental outcomes with reliabilities ranging from .63 to .82. Intentions and self-reported behaviors were measured.	Schools have a modest effect on environmentally responsible behaviors relative to sociodemographic factors and role models (i.e., children from middle class families were more likely to report engaging in behaviors than those from lower or higher socioeconomic groups, as were children with role models). sixth graders were also more likely to report engaging in behaviors than twelve graders.	The survey was developed in consultation with teachers, students, ecologists, survey research experts, and an advisory committee.
Schneller (2008) External evaluator Behavior	Six-year-old voluntary nonformal service learning program for secondary students in Pescadero, Mexico that included weekly after school meetings for two semesters and focused on a range of local issues complimenting existing community efforts. Program has explicit objective to change behavior. Students' reactions revealed insight into how program was implemented.	Environmental education, service learning research and evaluations.	This article focuses on reporting the evaluation's qualitative results (quantitative pre/post treatment/control results are indicated to be reported elsewhere). Data were collected through semistructured interviews based on a life history approach of 15 students in the 2004–2005 cohort and 23 students in the current 2006–2007 cohort as well as the principal, teachers, course developer, 7 parents of students, and several influential community members.	Participants engaged in "four increasing degrees of commitment" (i.e. types of behavior): making a lifestyle change, intergenerational learning, peer group tutoring, and voluntary participation in environmental community service. These behaviors were attributed to the program's experiential nature.	Program developer and teacher helped to write interview questions.
Flowers (2010) External evaluator Behavior Knowledge Attitude Skill	Over-ten-year-old fishing education program for primary and secondary students in Montana. Program has explicit objective to change behavior. On average participants experienced 1–2 hours of classroom activities and a 2 hour outdoor fishing trip.	Patton's (1997) Utilization-Focused Evaluation approach and EE research including Hungerford and Volk (1990) model	Data for this evaluation were collected from all of the students, their teachers and the instructors who participated or led the program during one year. A quasi-experimental design was used to collect posttests from 2,083 treatment students, retention-tests from 194 students, and 229 comparison students. 101 teachers (89 percent) completed online questionnaires and 16 program instructors were interviewed. The instrument was adapted from a previously developed one (Fedler, 2005). Intentions were measured as an indicator for behavior.	Program did not change students' intentions.	Stakeholders were involved in determining focus of "evaluation" and throughout the remaining "evaluation" process. Several examples of changes that were made by the teachers, instructors, and program coordinator based on the evaluation are provided.

Note: If information in this table is unclear or appears insufficient it is likely because the authors did not provide additional details or clarifications.

evaluation journal. One of the articles reported results from formative front-end and process evaluations of an interactive exhibit whereas the remaining articles focused on assessing summative outcomes of mainly established environmental education programs. Two of the articles were coauthored by collaborating internal and external evaluators and the remaining by external evaluators.

In addition to identifying the reviewed articles, Table 30.1 provides more in-depth information about each of the ten evaluations including whether or not the authors were internal or external evaluators, the outcomes they evaluated, the environmental education programs that were evaluated and their implementation, the perspectives that informed the evaluation, the evaluations' methods, behavioral outcome results, and information about the extent to which stakeholders were involved and evaluation results were used. Information about the internal or external nature of the evaluator is shared because each has its advantages and limitations that will affect the evaluation (Conley-Tyler, 2005). The programs that were evaluated and their implementation are described because this information is critical to interpreting evaluation results (Patton, 2008). The perspectives that informed the evaluation are identified because they influence which methods are selected and how results are interpreted (Creswell, 2009; Dillon & Wals, 2006). The evaluations' methods are described because without, for example, the use of comparison groups as part of experimental designs, of reliable and valid measures, or of analyses that take into account appropriate sampling units, outcome results can and should be questioned (Leeming et al., 1993). Behavioral outcome results are provided for the purpose of drawing conclusions about the potential of environmental education programs to change participants' behaviors. Information about stakeholder involvement is shared because such involvement tends to determine to what extent evaluation results are used (Patton, 2008). Last, utility is addressed because it is one of the primary ways that the quality of evaluations is judged (Joint Committee on Standards for Educational Evaluation, 1994). What follows is an overview of the reviewed behavioral outcomes evaluations based on the above characteristics.

What Outcomes Were Evaluated? All of the ten reviewed articles sought to evaluate changes in environmental education program participants' behaviors in addition to measuring changes in participants' knowledge, attitudes, and/or skills. The authors who addressed why they focused on these outcomes explained that they did so because the programs they evaluated tended to be based on the belief that changes in these variables lead to behavioral outcomes.

What Environmental Education Programs Were Evaluated and How Were They Implemented? The programs that were evaluated were conducted in several different countries including Canada, Israel, Mexico, Spain, Switzerland, and the United States. The majority targeted secondary students although some also targeted primary students. The remaining two programs were designed for families and adult boaters. The majority of the evaluated programs for students consisted of classroom instruction although several also included field trips and service learning. Families experienced an interactive exhibit and adult boaters received an outreach kit. The evaluated programs ranged from an extremely short one consisting of the dissemination of an information kit to two year experiences. One program was in its development, one in its first year, and the remaining programs existed for several years. Most but not all had explicit objectives to change participants' behaviors. The majority of the articles contained limited information about the programs' actual implementation or enactment and focused mostly on describing the intended program. However, there was one evaluation that used observation and two that asked teachers to keep logs of their enactment to obtain closer insight into how programs were actually implemented.

What Perspectives Informed the Evaluations? Similar to other environmental education researchers (Dillon & Wals, 2006) and with one exception, the authors were not explicit about their research methodologies, ideologies, or the evaluation approaches that informed their work. The exception consisted of an evaluation that was explicitly based on Patton's (1997) Utilization-Focused Evaluation approach. However, in light of the literature the authors drew on and their methods, it can be concluded that they mostly drew on a positivist or postpositivist worldview (Creswell, 2009). This conclusion is based on the fact that many of the authors used the Hungerford and Volk (1990) model and similar work to inform their understanding of the factors that influence individuals' behaviors, for tested instruments and scales, to inform data collection approaches, and to help make sense of, and draw conclusions about, the evaluations' results. One evaluation, for example, was designed around this model because the environmental education program had adopted this model, used previously tested measures of variables in the model, and drew on the model for judging the overall success of the program in achieving its behavioral outcome goal.

What Methods Did the Evaluations Use? Seven of the evaluations adopted primarily quantitative strategies, two primarily qualitative strategies, and one a mixed methods strategy (Creswell, 2009). Of the seven that adopted primarily quantitative strategies, four used a quasi-experiment design [two explicitly made reference to Cook and Campbell (1979)], one pursued an experimental design, and the remaining two conducted surveys. One of these evaluations also assessed changes in participants through a retrospective pretest approach (Pratt, McGuigan, & Katzev, 2000). The two evaluations that adopted primarily qualitative strategies collected their data through observations and interviews, with one employing a life history method.

The mixed methods evaluation collected data through a quasi-experimental design and focus groups. In addition, four evaluations triangulated data by collecting information from students as well as their instructors, parents, and community members. Two evaluations also incorporated a semilongitudinal strategy by collecting data from current and past program participants which has been rare among evaluations of environmental education programs (Engels & Jacobson, 2007).

Only one evaluation set up a context in which participants' actual behaviors were observed directly. Of the remaining evaluations, about half measured current participants' intentions to act and about half asked past participants to self-report about their behaviors. In most instances, the evaluations assessed changes in specific intentions or behaviors that participants could engage in and that were appropriate considering the programs' content (Heimlich & Ardoin, 2008). For example, the evaluation that sought to measure boaters' behaviors asked them about carrying nautical charts while boating, maintaining slower speeds while boating in shallow water, and watching out for manatees while boating in shallow water.

The articles were not always clear on how individuals were selected to participate in the evaluation but when it was, a census, random, purposeful sample, or a mix of these sampling methods was used. In one instance, a purposeful sample was selected based on enactment information. The evaluation with the smallest sample collected data from 24 individuals and the one with the largest sample from about 3,000 individuals. Six of the evaluations adapted previously tested instruments and two of these reported reliabilities for their intention or behavior measures. Only two of the evaluations conducted multilevel analyses, controlling for the fixed factor effects of students participating in classes or groups.

What Conclusions Were Drawn About the Program's Behavioral Outcomes? The majority of the authors concluded that the environmental education programs they evaluated changed participants' behaviors. The authors based their conclusions on changes in observed behaviors, increases in current participants' intentions, and changes in the behaviors reported by past participants. One program was also concluded to have had the unanticipated outcome of changing student participants' families' and community members' behaviors.

There were, however, also two evaluations whose authors concluded that the environmental education programs they evaluated did not change participants' behaviors. In one of these instances, the information-based nature of the program was identified as one of the reasons for why the program did not change participants' behaviors. In the other instance, the author provides evidence to suggest that students' behaviors were better explained by their sociodemographic background. Interestingly, another set of authors came to a similar conclusion, explaining that they felt the program's effects were modest relative to the influences of students' socioeconomic background and role models. These same two evaluations also found that younger students were more likely to report that they intended to, or changed their behaviors than older students.

Moreover, there were two evaluations that identified negative outcomes. In one evaluation, one of three intentions decreased as a result of the program and some students experienced increased feelings of fear and helplessness based on what they learned (Covitt, Gomez-Schmidt, & Zint, 2005). As part of the other evaluation, findings suggested that students who were more knowledgeable about environmental issues were slightly less likely to act (Negev et al., 2008).

In the few instances when authors collected enactment information, it helped them to explain their evaluations' findings. For example, as part of two evaluations, teachers' reported enactment indicated that they did not implement the intended number of activities and service learning experience, and this information explained why students' behaviors did not change or did not change more.

How Were Stakeholders Involved and Evaluation Results Used? The majority of the articles provided little information to suggest that stakeholders were involved in the evaluation other than to assist with creating or validating data collection instruments' measures. As would therefore be anticipated, the majority of the articles also did not share information on if, or how, evaluation results were used. However, there were some exceptions. Not surprisingly, for example, the authors of the formative front-end and process evaluation provided a variety of examples of how earlier tests were used to inform improvements. In addition, the evaluation based on Patton's (1997) Utilization-Focused Evaluation approach described how stakeholders were involved throughout the evaluation and gave examples of improvements that were implemented based on evaluation results. Last, one set of authors describe that they answered questions posed by the program's manager, involved staff in developing the instruments' questions as well as in interpreting results, and these authors also shared examples of how evaluation results were used.

On the Contribution of Evaluation to the Body of Knowledge on Environmental Education

As suggested by this review, environmental education program evaluations can provide valuable insights to the field. For example, the reviewed evaluations provide evidence that environmental education programs can change participants' behaviors and even participants' families and communities behaviors. As such these reviewed evaluations contribute to the limited body of research on the behavioral outcomes (Leeming et al., 1993) and intergenerational benefits (Duvall & Zint, 2007) of environmental education. In addition, the reviewed evaluations, especially when considered in combination, provide evidence in support of program characteristics likely to be unsuccessful and successful in fostering behavior change.

For example, programs are likely to be unsuccessful in changing behaviors if they lack clearly defined behavioral objectives, focus on general environmental knowledge and attitudes, and are top-down, passive, and short. In contrast, environmental education programs have the potential to change participants' behaviors if they have clearly defined behavioral objectives, are designed based on behavior theories and models, consider participants' needs, context, and background, incorporate experiential learning such as field trips and service learning, and include more activities and/or are longer in duration.

Consider, for example, the contrast between two programs conducted in Israel and Spain that had many unsuccessful program characteristics with one program in Mexico that had many of the characteristics attributed to successful programs. Neither of the former programs resulted in substantial changes in participants' behaviors whereas the latter did. Interestingly, the author of one of the former evaluations also noted that a school which incorporated a more participatory and extensive approach was able to overcome what she termed the "class effect" on behavior. Consistent with research (e.g., Johnson-Pynn & Johnson, 2005; Trewhella et al., 2005), these particular evaluations also support that environmental education programs designed to meet the needs of participants from low income or working class backgrounds have the potential to change these participants' as well as their families' behaviors, whereas programs that are designed to reflect higher income or middle class views are unlikely to.

In addition to contributing to the identification of environmental education program characteristics that have the potential to foster behavior change, the reviewed evaluations also offer other useful insights for the field. The former evaluations of environmental education programs in Israel and Spain, for example, suggest that participants and their families from middle class backgrounds are more likely to engage in environmentally responsible behaviors than those from working class backgrounds. Thus, the potential benefits of environmental education programs for middle class audiences may not be as great as of those for working class families, a conclusion also supported by an environmental education program evaluation that assessed knowledge outcomes (Powers, 2004b). These same two evaluations in Israel and Spain also found that younger students were more likely to act than older students, suggesting a greater need for environmental education programs that target secondary students.

Moreover, the review also illustrates how evaluations can be beneficial for uncovering, and thus providing the opportunity to rectify, undesirable outcomes. One of the evaluations, for example, revealed that a curriculum experience increased some students' feelings of fear and helplessness (Covitt et al., 2005), whereas the other evaluation identified a negative correlation between knowledge and behavior (Negev et al., 2008). The latter finding suggests that environmental education programs designed only to convey greater knowledge about environmental challenges may be less likely to change participants' behaviors. This could be the case, for example, if greater environmental knowledge leads individuals to feel more overwhelmed and helpless (Kaplan, 2000).

Last, there is the possibility for evaluations to contribute to environmental education theory (Chen, 1990). For example, among the reviewed evaluations, there was one which suggested the need for further testing of the Hungerford and Volk (1990) behavioral model. This particular model proposes that programs which target a series of entry-level, ownership, and empowerment variables will lead participants to form intentions which in turn, will change behaviors. As part of this evaluation, however, there was one program whose participants increased in all of the predicted variables but not in their intentions, whereas participants in another program who increased in only two of the predictors increased in their behavioral intentions (Zint, Kraemer, Northway, & Lim, 2002).

Advancing Environmental Education Program Evaluation Based on Insights From the Review of Behavioral Outcome Evaluations

Syntheses of past work should facilitate researchers' and evaluators' ability to benefit from past "lessons learned." Indeed among the reviewed evaluations, two authors were explicit about how a review of environmental education outcome research (Leeming et al., 1993) influenced their use of control groups, of valid and reliable measures, their attempts to collect follow-up data, and their use of multilevel analyses to control for participants' group experience. Based on the following discussion of the strengths and limitations of the reviewed behavioral outcome evaluations, it is hoped that researchers and evaluators will be able to similarly build on the experiences of those before them.

First, researchers and evaluators should be aware that past evaluations of environmental education programs have adopted a limited number of possible evaluation approaches (Stufflebeam, 2001). The majority of the reviewed evaluations were accountability evaluations based on quasi-experimental design strategies. Only one evaluation adopted a utilization-focused, and one evaluation a theory-based approach. There are a few examples of evaluations using alternative approaches within the larger environmental education program evaluation literature including participatory ones (McDuff & Jacobson, 2001; Somers, 2005). However, even within this broader literature, examples of evaluations using the many potential alternatives could not be identified. While this is not a problem per se, it is if approaches other than traditional ones are more appropriate given particular evaluation contexts. For example, when asked to evaluate behavioral outcomes of programs designed to enhance only participants' environmental knowledge and/or attitudes, the traditional accountability approach and its implied quantitative strategies may not be the most useful. Others have also voiced concern over this traditional approach to evaluating environmental education programs. Robottom (1985, p.32), for example, was among the first to critique what he termed "the hold of the dominant quantitative, scientific/analytic approach to evaluation." And more recently Fleming and Easton (2010) similarly voice concern that "early influences (Title III) persist with focus on behavioral objectives, experimental designs, standardized tests even if subsequent findings are not used and may be irrelevant to the field (p. 173)."

There are many reasons why these authors and many in the evaluation community question the traditional accountability approach to program evaluation. One of the primary reasons stems from the fact that these traditional evaluations tend not to involve stakeholders. Without such involvement, program evaluations are unlikely to meet stakeholders' needs and support the use of evaluations (Patton, 2008). Indeed the majority of the reviewed behavioral outcome evaluations that adopted a traditional approach tended not to involve stakeholders or involved them only to a limited extent. This appeared to be the case even with regard to determining what the evaluations of programs without clear objectives should focus on, a common challenge with evaluating environmental education programs (Carleton-Hug & Hug, 2010). In these instances, joint deliberations would likely have been particularly appropriate (Fien, Scott, & Tilbury, 2001; Monroe et al., 2005; Powers, 2004a). Not surprisingly, these evaluations also tended to be the ones that did not mention use of evaluation results. In contrast the two evaluations that reported involving stakeholders in a variety of ways throughout the evaluation process gave several examples of how findings and recommendations were used to improve the evaluated programs. As suggested by these results and again, as pointed out by Robottom (1985) some time ago, researchers and evaluators should give more consideration to whether evaluations should be about control and serve select self-interests or if they want environmental educators to have ownership over and be empowered by them. The latter not only has the potential to directly inform program decisions but can also influence how individuals think about programs and allow individuals to learn not just from the evaluation's findings but also its process (Clavijo, Fleming, Hoermann, Toal, & Johnson, 2005; Patton, 2008; Powell, Stern, & Ardoin, 2006; Somers, 2005).

Second, with regard to evaluation methods, the value of using a mixed method strategy is evident from this review as well as from evaluations of other environmental education program outcomes (Ernst, 2005; Powers, 2004b). The benefits of collecting both quantitative and qualitative data have been recognized by environmental education researchers some time ago (Hart & Nolan, 1999; Mrazek, 1993) and have also been widely accepted by the evaluation community (Coffman, 2002). Whereas almost all of the reviewed evaluations' authors collected at least some quantitative and qualitative data, the majority focused on reporting only one or the other. In addition, the majority of studies that used qualitative strategies relied on interviews, focus groups, and observations. There are, however, examples of alternative qualitative strategies such as case studies, concept mapping, and others in the broader environmental education program evaluation literature that researchers and evaluators could draw on (Christensen, Nielsen, Rogers, & Volkov, 2005; Storksdieck, Ellenbogen, & Heimlich, 2005; Thomas, 1990).

In addition, the review confirmed the importance of documenting the implementation of programs that are being evaluated or studied. As education researchers know, the enacted curriculum rarely reflects the intended curriculum. Moreover, in terms of ultimately helping to improve or replicate the program, it is essential to know what occurred as part of the program that can explain why it may have failed or succeeded in achieving its outcomes (Patton, 2008). As has also been the case in the evaluation community (King, Morris, & Fitz-Gibbon, 1987), this review revealed that the majority of past environmental education program evaluations did not sufficiently track the evaluated programs' implementation. In the two instances when program enactment was assessed, it provided critical insight into why particular environmental education programs did not meet their behavioral objectives. Moreover, as illustrated by another environmental education program evaluation (Van Petegem et al., 2007), if a program's implementation is not well understood, it is premature to attempt to evaluate its outcomes.

Researchers and evaluators can also learn from how past authors measured changes in behavioral outcomes. For one, the retrospective pretest approach (Pratt et al., 2000) can be useful for obtaining participants' perceptions of changes in their behaviors as a result of environmental education programs. In addition, the reviewed evaluations suggest that by measuring intentions and self-reported behaviors it is possible to obtain results that can lead to program improvements and inform the field. This observation is consistent with findings in environmental education and other research contexts that points to intentions as the single best predictor of behavior (Bamberg & Möser, 2007; Hungerford & Volk, 1990; Zint, 2002) and suggests that self-reported behaviors can be valid measures of actual behaviors (e.g., Fishbein & Pequegnat, 2000; Patrick et al., 1994). At the same time, however, there is also no question that these measures have their limitations (Camargo & Shavelson, 2009). Intentions, for example, may not lead to behaviors if there significant barriers to these behaviors and self-reported behaviors may not be highly correlated with actual behaviors (Bickman & Rog, 1998; Corral-Verdugo, 1997). One way to address the limitations of these measures is through triangulation (Cohen & Manion, 1986; Denzin, 1978). For example, several of the reviewed evaluations triangulated data by corroborating students' responses with responses from their teachers, families, and/or community members. Yet another way is draw on behavioral theories and models (Heimlich & Ardoin, 2008). One evaluation, for example, in addition to measuring changes in participants' intentions, also measured changes in the antecedents of a behavioral model and found this approach helpful in judging the extent to which environmental education program participants' behaviors may have changed. Importantly, the limitations of these survey measures can be avoided if it is possible to set up a means through which participants' behaviors can be directly observed unobtrusively (Camargo & Shavelson, 2009) as was the case in one of the reviewed evaluations. There may even be other creative ways to assess changes in behaviors such as by asking participants to document changes through the use of visual media (Janke, 1982).

As suggested by this discussion, researchers and evaluators could benefit from supporting each other. For example, efforts underway by researchers to share scales for assessing environmental outcomes (see www.conpsychmeasures.com) should reduce evaluators' measurement challenges and evaluators could introduce researchers to the approaches and standards that guide evaluations. Moreover, researchers and evaluators could benefit from collaborating with each other on particularly challenging research and evaluation questions such as ones that seek to assess the long-term outcomes and impacts of environmental education programs on the environment (Birnbaum & Mickwitz, 2009; Duffin, Murphy, & Johnson, 2008).

Systematic reviews and syntheses of evaluations can clearly benefit the field of environmental education and help to inform future evaluations as well as research. However, such reviews and syntheses are dependent on having access to evaluations through, for example, peer reviewed journals. It is therefore critical for more evaluations to be published (Carleton-Hug & Hug, 2010). Moreover, when evaluations are published, they should include information in addition to that provided by traditional research studies such as about the evaluation's approach, the implementation of the evaluated program, and about how results were used to guide program improvements.

Conclusion

The review of environmental education program evaluation literature conducted for this chapter, brought to mind the famous quote by George Santayana "*those who cannot learn from history are doomed to repeat it.*" As early as the 1970s, for example, this literature contains a reference expressing concern over environmental education programs lacking objectives (Putney & Wagar, 1973), an issue that continues to resurface among recently published evaluations (Carleton-Hug & Hug, 2010). Or consider the number of programs evaluated for behavioral outcomes that were based on the knowledge + attitude = behavior assumption despite its limitations (Heimlich & Ardoin, 2008; Hines, Hungerford, & Tomera, 1986/1987; Hungerford & Volk, 1990). And last, there is the continued emphasis on evaluations conducted for accountability purposes relying primarily on quantitative strategies over which O'Hearn (1982) was the first to raise concern in the early 1980s. Such evaluations are unlikely to meet evaluation standards (Joint Committee on Standards for Educational Evaluation, 1994). Meeting these standards is important to ensure that environmental education program evaluations are not discredited and do not result in broader questions about the quality of environmental education practice and research (Heimlich, 2010).

At the same time, the literature on environmental education program evaluation also contains very promising examples of evaluations that were used to improve programs. Particularly encouraging is the growing emphasis in the literature on exploring ways to increase environmental educators' evaluation competencies and their organizations' evaluation capacities and thus, the ability of environmental education programs to benefit from evaluation. For example, researchers and evaluators are beginning to explore how courses and resources are enhancing environmental educators' evaluation competencies (Fleming & Easton, 2010; Gayford, 2004; Monroe, Scollo, & Bowers, 2002; Zint, Dowd, & Covitt, 2011) and how organizations' evaluation capacities can be built based on needs assessments (McDuff, 2002) and through evaluation systems (Carleton-Hug & Hug, 2010; Powell et al., 2006). The reason this trend is encouraging that it in time for evaluation to become an integral part of environmental education programs. Evaluative thinking is essential for supporting the continuous improvement of the environmental education programs that are so desperately needed to help address environmental challenges.

Acknowledgments

I am extremely thankful for the extensive assistance and support provided by University of Michigan graduate students Brian T. Barch, John Franklin Cawood, Catherine Ruth Game, Jose Gonzalez, and Gillian Ream as well as the valuable feedback I received from the editors of this Handbook, particularly Dr. Michael Brody.

Note

1. It is likely that additional behavioral outcome evaluations of environmental education programs exist among the many evaluations that have not been published or shared publicly through other means.

References

Ajzen, I. (1985). From intentions to actions: A theory of planned behavior. In J. Kuhl & J. Beckmann (Eds.), *Action control: From cognition to behavior* (pp. 11–39). New York, NY: Springer-Verlag.

Ajzen, I., & Fishbein, M. (1980). *Understanding attitudes and predicting social behavior*. Englewood Cliffs, NJ: Prentice-Hall.

Asch, J., & Shore, B. (1975). Conservation behavior as the outcome of environmental education. *Journal of Environmental Education, 6*(4), 25–33.

Bamberg, S., & Möser, G. (2007). Twenty years after Hines, Hungerford, and Tomera: A new meta-analysis of psycho-social determinants of pro-environmental behaviour. *Journal of Environmental Psychology, 27*(1), 14–25.

Bickman, L., & Rog, D. J. (1998). *Handbook of applied social research methods.* Thousand Oaks, CA: Sage Publications.

Birnbaum, M. (2010). Interview with Brett Jenks, President and CEO of Rare. *Journal of Evaluation and Program Planning, 33*(2), 191–193.

Birnbaum, M., & Mickwitz, P. (2009). Environmental program and policy evaluation: Addressing methodological challenges. *New Directions for Evaluation, 122,* 105–112.

Bogner, F. X. (1999). Empirical evaluation of an educational conservation programme introduced in Swiss secondary schools. *International Journal of Science Education, 21*(11), 1169–1185.

Camargo, C., & Shavelson, R. (2009). Direct measures in environmental education evaluation: Behavioral intentions versus observable actions. *Applied Environmental Education and Communication, 8*(3–4), 165–173.

Carleton-Hug, A., & Hug, J. W. (2010). Challenges and opportunities for evaluating environmental education programs. *Journal of Evaluation and Program Planning, 33*(2), 159–164.

Chelimsky, E. (1997). The coming transformation in evaluation. In E. Chelimsky & W. Shadish (Eds.), *Evaluation for the twenty-first century* (pp. 1–26). Thousand Oaks, CA: Sage.

Chen, H. T. (1990). *Theory-driven evaluations.* Newbury Park, CA: Sage.

Christensen, L., Nielsen, J. E., Rogers, C. M., & Volkov, B. (2005). Creative data collection in nonformal settings. *New Directions for Evaluation, 108,* 73–79.

Clavijo, K., Fleming, M. L., Hoermann, E. F., Toal, S. A., & Johnson, K. (2005). Evaluation use in nonformal education settings. *New Directions for Evaluation, 108,* 47–55.

Coffman, J. (2002). *A conversation with Michael Quinn Patton.* Cambridge, MA: Harvard Family Research Project.

Cohen, L., & Manion, L. (1986). *Research methods in education.* London, UK: Croom Helm.

Conley–Tyler, M. (2005). A fundamental choice: Internal or external evaluation? *Evaluation Journal of Australasia, 4*(1–2), 3–11.

Cook, T. D., & Campbell, D. T. (1979). *Quasi-experimentation: Design analysis issues for field settings.* Chicago, IL: Rand McNally.

Corral-Verdugo, V. (1997). Dual 'realities' of conservation behavior: Self-reports vs. observations of re-use and recycling behavior. *Journal of Environmental Psychology, 17*(2), 135–145.

Covitt, B. A., Gomez–Schmidt, C., & Zint, M. T. (2005). An evaluation of the risk education module: Exploring environmental issues: Focus on risk. *Journal of Environmental Education, 36*(2), 3–13.

Creswell, J. W. (2009). *Research design: Qualitative, quantitative, and mixed methods approaches* (3rd ed.). Los Angeles, CA: Sage.

Denzin, N. (1978). *Sociological methods: A sourcebook* (2nd ed.). New York, NY: McGraw Hill.

Dillon, J., & Wals, A. E. J. (2006). On the danger of blurring methods, methodologies, and ideologies in environmental education research. *Environmental Education Research, 12*(3–4), 549–558.

Duffin, M., Murphy, M., & Johnson, B. (2008). *Quantifying a relationship between place-based learning and environmental quality: Final report.* Woodstock, VT: NPS Conservation Study Institute in cooperation with the Environmental Protection Agency and Shelburne Farms.

Duvall, J., & Zint, M. (2007). A review of research on the effectiveness of environmental education in promoting intergenerational learning. *Journal of Environmental Education, 38*(4), 14–24.

Ebbutt, D. (1998). Evaluation of projects in the developing world: Some cultural and methodological issues. *International Journal of Educational Development, 18*(5), 415–424.

Engels, C. A., & Jacobson, S. K. (2007). Evaluating long-term effects of the Golden Lion Tamarin environmental education program in Brazil. *Journal of Environmental Education, 38*(3), 3–14.

Ernst, J. (2005). A formative evaluation of the prairie science class. *Journal of Interpretation Research, 10*(1), 9–30.

Fedler, A. J. (2005). *An evaluation of youth participant outcomes from fully-implemented hooked on fishing not on drugs programs.* Gainesville, FL: Human Dimensions Consulting.

Fien, J., Scott, W., & Tilbury, D. (2001). Education and conservation: Lessons from an evaluation. *Environmental Education Research, 7*(4), 379–395.

Fishbein, M., & Pequegnat, W. (2000). Evaluating AIDS prevention interventions using behavioral and biological outcome measures. *Sexually Transmitted Diseases, 27*(2), 101–110.

Fleming, L., & Easton, J. (2010). Building environmental educators' evaluation capacity through distance education. *Journal of Evaluation and Program Planning, 33*(2), 172–177.

Flowers, A. B. (2010). Blazing an evaluation pathway: Lessons learned from applying utilization-focused evaluation to a conservation education program. *Journal of Evaluation and Program Planning, 33*(2), 165–171.

Fournier, D. M. (2005). Evaluation. In S. Mathison (Ed.), *Encyclopedia of evaluation* (pp. 139–140). Thousand Oaks, CA: Sage.

Frechtling, J. (2007). *Logic modeling methods in program evaluation.* San Francisco, CA: Jossey–Bass.

Gayford, C. (2004). A model for planning and evaluation of aspects of education for sustainability for students training to teach science in primary schools. *Environmental Education Research, 10*(2), 255–271.

Goldman, D., Yavetz, B., & Pe'er, S. (2006). Environmental literacy in teacher training in Israel: Environmental behavior of new students. *The Journal of Environmental Education, 38*(1), 3–20.

Hart, P., & Nolan, K. (1999). A critical analysis of research in environmental education. *Studies in Science Education, 34*(1), 1–69.

Heimlich, J. E. (2010). Environmental education evaluation: Reinterpreting education as a strategy for meeting mission. *Journal of Evaluation and Program Planning, 33*(2), 180–185.

Heimlich, J. E., & Ardoin, N. M. (2008). Understanding behavior to understand behavior change: A literature review. *Environmental Education Research, 14*(3), 215–237.

Hines, J. M., Hungerford, H. R., & Tomera, A. N. (1987). Analysis and synthesis of research on responsible environmental behavior: A meta-analysis. *The Journal of Environmental Education, 18,* 1–8. (Original work published 1986)

Hungerford, H. R., & Volk, T. L. (1990). Changing learner behavior through environmental education. *Journal of Environmental Education, 21*(3), 8–21.

Irvine, K., Saunders, C., & Foster, J. S. (1995). Using evaluation to guide the development of behavior change programs. *Visitor Studies: Theory, Research, and Practice, 8*(2), 47–56.

Jacobson, S. K., & McDuff, M. D. (1997). Success factors and evaluation in conservation education programmes. *International Research in Geographical and Environmental Education, 6*(3), 204–221.

Janke, D. (1982). The use and interpretation of visual evidence in evaluating environmental education programs. *Journal of Environmental Education, 13*(4), 41–43.

Jenks, B., Vaughan, P. W., & Butler, P. J. (2010). The evolution of Rare Pride: Using evaluation to drive adaptive management in a biodiversity conservation organization. *Journal of Evaluation and Program Planning, 33*(2), 186–190.

Johnson-Pynn, J. S., & Johnson, L. R. (2005). Successes and challenges in East African conservation education. *Journal of Environmental Education, 36*(2), 25–39.

Joint Committee on Standards for Educational Evaluation. (1994). *The program evaluation standards.* Thousand Oaks, CA: Sage.

Kaplan, S. (2000). New ways to promote proenvironmental behavior: Human nature and environmentally responsible behavior. *Journal of Social Issues, 56*(3), 491–508.

Kellaghan, T., Stufflebeam, D. L., & Wingate, L. A. (2003). Introduction. In T. Kellaghan, D. L. Stufflebeam, & L. A. Wingate (Eds.), *International handbook of educational evaluation* (pp. 1–8). Boston, MA: Kluwer Academic Publishers.

King, J. A., Morris, L. L., & Fitz-Gibbon, C. T. (1987). *How to assess program implementation*. Newbury Park, CA: Sage.

Labin, S. N. (2008). Research synthesis: Toward broad-based evidence. In L. Smith & P. R. Brandon (Eds.), *Fundamental issues in evaluation* (pp. 89–110). New York, NY: The Guildford Press.

Leeming, F. C., Dwyer, W. O., Porter, B. E., & Cobern, M. K. (1993). Outcome research in environmental education: A critical review. *The Journal of Environmental Education, 24*(4), 8–21.

Loomis, R. J. (2002). Visitor studies in a political world: Challenges to evaluation research. *Journal of Interpretation Research, 7*(1), 31–42.

Lucas, A. M. (1980). Science and environmental education: Pious hopes, self praise and disciplinary chauvinism. *Studies in Science Education, 7*(1), 1–26.

Mathison, S. (2008). What is the difference between evaluation and research—and why do we care? In L. Smith & P. R. Brandon (Eds.), *Fundamental issues in evaluation* (pp. 183–196). New York, NY: The Guildford Press.

McDuff, M. (2002). Needs assessment for participatory evaluation of environmental education programs. *Applied Environmental Education & Communication, 1*(1), 25–36.

McDuff, M. D., & Jacobson, S. K. (2001). Participatory evaluation of environmental education: Stakeholder assessment of the Wildlife Clubs of Kenya. *International Research in Geographical and Environmental Education, 10*(2), 121–148.

Monroe, M. C. (2002). Evaluation's friendly voice: The structured open-ended interview. *Applied Environmental Education and Communication, 1*(2), 101–106.

Monroe, M. C., Fleming, M. L., Bowman, R. A., Zimmer, J. F., Marcinkowski, T., & Washburn, J. (2005). Evaluators as educators: Articulating program theory and building evaluation capacity. *New Directions for Evaluation, 108*, 57–71.

Monroe, M. C., Scollo, G., & Bowers, A. W. (2002). Assessing teachers' needs for environmental education services. *Applied Environmental Education and Communication, 1*(1), 37–43.

Morgan, M., & Soucy, J. (2006). Usage and evaluation of nonformal environmental education services at a state park: Are anglers catching more than fish? *Environmental Education Research, 12*(5), 595–608.

Morris, J., Jacobson, S., & Flamm, R. (2007). Lessons from an evaluation of a boater outreach program for manatee protection. *Environmental Management, 40*(4), 596–602.

Mrazek, R. (Ed.). (1993). *Alternative paradigms in environmental education research*. Washington, DC: North American Association for Environmental Education.

Negev, M., Sagy, G., Garb, Y., Salzberg, A., & Tal, A. (2008). Evaluating the environmental literacy of Israeli elementary and high school students. *The Journal of Environmental Education, 39*(2), 3–20.

Norris, K. S., & Jacobson, S. K. (1998). Content analysis of tropical conservation education programs: Elements of success. *Journal of Environmental Education, 30*(1), 38–44.

Nowak, P., Wilke, R., Marcinkowski, T., Hungerford, H., & Mckeown-Ice, R. (1995). *The secondary school environmental literacy instrument*. Unpublished instrument.

O'Hearn, G. T. (1982). What is the purpose of evaluation? *Journal of Environmental Education, 13*(4), 1–3.

Patrick, D. L., Cheadle, A., Thompson, D. C., Diehr, P., Koepsell, T., & Kinne, S. (1994). The validity of self-reported smoking: A review and meta-analysis. *American Journal of Public Health, 84*(7), 1086–1093.

Patton, M. O. (2008). *Utilization-focused evaluation* (4th ed.). Thousand Oaks, CA: Sage.

Patton, M. Q. (1997). *Utilization-focused evaluation* (3rd ed.). Thousand Oaks, CA: Sage.

Powell, R. B., Stern, M. J., & Ardoin, N. (2006). A sustainable evaluation framework and its application. *Applied Environmental Education & Communication, 5*(4), 231–241.

Powers, A. L. (2004a). An evaluation of four place-based education programs. *Journal of Environmental Education, 35*(4), 17–32.

Powers, A. L. (2004b). Evaluation of one- and two-day forestry field programs for elementary school children. *Applied Environmental Education & Communication, 3*(1), 39–46.

Pratt, C. C., McGuigan, W. M., & Katzev, A. R. (2000). Measuring program outcomes: Using retrospective pretest methodology. *The American Journal of Evaluation, 21*(3), 341–349.

Putney, A. D., & Wagar, J. A. (1973). Objectives and evaluation in interpretive planning. *Journal of Environmental Education, 5*(1), 43–44.

Robottom, I. (1985). Evaluation in environmental education: Time for a change in perspective? *Journal of Environmental Education, 17*(1), 31–36.

Rogers, P., Hacsi, T., Petrosino, A., & Hubner, T. (Eds.). (2000). Program theory evaluation: Challenges and opportunities. *New Directions in Evaluation, 87*, 1–112.

Rossi, P. H., Lipsey, M. W., & Freeman, H. E. (2004). *Evaluation: A systematic approach* (7th ed.). Thousand Oaks, CA: Sage.

Rovira, M. (2000). Evaluating environmental education programmes: Some issues and problems. *Environmental Education Research, 6*(2), 143–155.

Schneller, A. J. (2008). Environmental service learning: Outcomes of innovative pedagogy in Baja California Sur, Mexico. *Environmental Education Research, 14*(3), 291–307.

Somers, C. (2005). Evaluation of the Wonders in Nature-Wonders in Neighborhoods conservation education program: Stakeholders gone wild! *New Directions for Evaluation, 108*, 29–46.

Storksdieck, M., Ellenbogen, K., & Heimlich, J. E. (2005). Changing minds? Reassessing outcomes in free-choice environmental education. *Environmental Education Research, 11*(3), 353–369.

Stufflebeam, D. L. (2001). Evaluation models. *New Directions for Evaluation, 89*, 7–98.

Stufflebeam, D. L., Madaus, G. F., & Kellaghan, T. (2000). *Evaluation models: Viewpoints on educational and human services evaluation*. Boston, MA: Kluwer Academic Publishers.

Thomas, I. G. (1990). Evaluating environmental education programs using case studies. *Journal of Environmental Education, 21*(2), 3–8.

Trewhella, W. J., Rodriguez–Clark, K. M., Corp, N., Entwistle, A., Garrett, S. R. T., Granek, E., . . . Sewall, B. J. (2005). Environmental education as a component of multidisciplinary conservation programs: Lessons from conservation initiatives for critically endangered fruit bats in the western Indian Ocean. *Conservation Biology, 19*(1), 75–85.

Van Petegem, P., Blieck, A., & Pauw, J. B. (2007). Evaluating the implementation process of environmental education in preservice teacher education: Two case studies. *The Journal of Environmental Education, 38*(2), 47–54.

Wilke, R., Hungerford, H., Volk, T., & Bluhm, W. (1995). *Middle school environmental literacy instrument*. Unpublished instrument.

Wiltz, L. K. (2001). *Proceedings of the Teton Summit for program evaluation in nonformal environmental education*. Kelly, Wyoming, May 19–22, 2000. (p. 47). Retrieved from ERIC database. ED453066.

Zint, M. (2002). Comparing three attitude-behavior theories for predicting science teachers' intentions. *Journal of Research in Science Teaching, 39*(9), 819–844.

Zint, M. (2010). An introduction to my environmental education evaluation resource assistant (MEERA), a web-based resource for self-directed learning about environmental education program evaluation. *Journal of Evaluation and Program Planning, 33*(2), 178–179.

Zint, M., Dowd, P., & Covitt, B. (2011). Enhancing environmental educators' evaluation competencies: Insights from an examination of the effectiveness of the My Environmental Education Evaluation Resource Assistant (MEERA) web site. *Environmental Education Research, 17*(4), 471–497.

Zint, M., Kraemer, A., Northway, H., & Lim, M. (2002). Evaluation of the Chesapeake Bay Foundation's conservation education programs. *Conservation Biology, 16*(3), 641–649.

31

National Assessments of Environmental Literacy

A Review, Comparison, and Analysis

Thomas Marcinkowski
Florida Institute of Technology, USA

Donghee Shin
Ewha Womans University, South Korea

Kyung-Im Noh
University of Connecticut, USA

Maya Negev
Tel Aviv University, Israel

Gonen Sagy
The Arava Institute for Environmental Studies, Israel

Yaakov Garb
Ben Gurion University, Israel

Bill McBeth
University of Wisconsin–Platteville, USA

Harold Hungerford
The Center for Instruction, Staff Development and Evaluation, USA

Trudi Volk
The Center for Instruction, Staff Development and Evaluation, USA

Ron Meyers
Ron Meyers and Associates, USA

Mehmet Erdogan
Akdeniz University, Turkey

Introduction

During the 1970s, several researchers around the world conducted large, national, or multistate assessments of the environmental knowledge and attitudes of K-12 students (e.g., Bohl, 1977; Eyers, 1976; Perkes, 1974; Richmond, 1977). These served as the first wave of national assessments in environmental education (EE). Of the national assessments that followed, some focused on similar learning outcomes (e.g., Makki, Abd-El-Khalick, & Boujaoude, 2003; Ndayitwayeko, 1995), while others began to expand this range of learning outcomes (e.g., Cortes, 1987; Kuhlmeier, Van Den Bergh, & Lagerweij, 2005; Nelson, 1997).

The purpose of this chapter is to review, compare, and analyze what may be viewed as the second wave of national assessments in EE. Since 2000, there have been at least four of these involving K-12 students: South Korea (2002–2003), Israel (2004–2006), the United States (2006–2008), and Turkey (2007–2009). What differentiates national assessments in this second wave from those in the first is the range of learning outcomes assessed; i.e., rather than assess only knowledge and attitudes, each assessed a wider range of environmental literacy (EL) components (i.e., knowledge, affect, skill, and participation learning outcomes).

To accomplish these purposes, this chapter is organized into three sections. The first contains a review of definitional features of EE and EL reflected in these studies. The second contains a summary of each of these four national assessments. The third section contains a brief discussion of these national assessments, and wider questions pertaining to national and international assessments, and to EL.

Definitional Features of Environmental Education and Environmental Literacy Relevant to These National Assessments

Each of the four national environmental literacy (EL) assessments included in this review was undertaken in the context of environmental education (EE), and was influenced by the evolution of thinking about EE and EL within each nation, as well as by how that thinking may have been influenced by related international developments. In a broad sense, three different sources have had a major influence on this thinking, notably (a) definitional statements about EE (the rhetoric of EE); (b) reviews of curriculum and program frameworks for EE (guides to practice in EE); and (c) reviews of published research and evaluation in and closely related to EE. There are apparent differences in the contribution of these sources to the design of each of these national assessments. However, there are noteworthy commonalities to the manner in which EL has been defined in these four national assessments, due, in part, to the influence of US publications pertaining to these sources. For this reason, this overview will begin with a brief review of these sources within the United States, and then review what is known about these sources in Korea, Israel, and Turkey.

There is a rich literature that details the historical evolution and definition of EE within the United States that, collectively, serves as the first of these influences on thinking about EE and EL. Within the United States and perhaps elsewhere, several educational movements preceded EE, notably nature study (ca. 1900), outdoor education (ca. 1920s), and conservation education (ca. 1930s). On the one hand, each movement shaped the purposes of, infrastructure for, and practices in EE (Stapp, 1974; Swan, 1984). On the other, some of the first efforts to define EE were attempts to differentiate it from these earlier movements (e.g., Hungerford, 1975; Schoenfeld, 1969; Tanner, 1974). During this period, there also were efforts to define EE in terms of what it was, primarily in terms of its purpose and scope. It was common for these early definitions to consist of one or two sentences (Disinger, 1983; Harvey, 1977a, 1977b; Schmeider, 1977). One of the earliest published definitions was developed by Stapp and his colleagues.

> Environmental education is aimed at producing a citizenry that is knowledgeable concerning the biophysical environment and its associated problems, aware of how to help solve these problems, and motivated to work toward their solution. (1969, p. 31)

This definition is relevant to these national assessments for several reasons. *First,* it states that the mission of EE is to help prepare a citizenry that can and will become involved in the prevention and resolution of environmental problems and issues (i.e., an environmentally literate citizenry). *Second,* it suggests that environmental literacy is comprised of cognitive, affective, and behavioral components. *Third,* it appears to have helped shape subsequent types of definitional statements about EE. These included sets of educational goals and objectives for the field (Harvey, 1977a, 1977b), notably those promulgated at the Belgrade Workshop and Tbilisi Conference (UNESCO, 1977, 1978), and sets of key characteristics (E. Hart, 1981; P. Hart, 1980) and guiding principles (UNESCO, 1977, 1978). Of these, the Tbilisi categories of objectives (i.e., awareness, knowledge, attitudes, skills, and participation) have become the most widely recognized and accepted definition of EE around the world. These Tbilisi goals and objectives were reaffirmed at numerous UN-sponsored meetings, including the United Nations Conference on Environment and Development (UNESCO, 1992), and guided UNESCO's International Environmental Education Programme through the 1990s. Delegates from Korea, Israel, and Turkey participated in the Tbilisi Conference, the UNCED Conference, and other UN-sponsored EE meetings around the world. By virtue of this, their exposure to US thinking about EE and, by inference, EL, can be traced back to the 1970s.

A second source of thinking about EE and EL reflects recommendations from the Tbilisi Conference, notably Recommendation No. 21: Research (UNESCO, 1978, p. 38). In accordance with this Tbilisi recommendation and as is apparent in this *Handbook*, the body of research and evaluation studies in and closely related to EE has grown over time. Both within and beyond the United States, efforts have been made to summarize the evidence from collections of these studies (e.g., Iozzi, 1984; Rickinson, 2001; R. Roth, 1976; R. Roth & Helgeson, 1972; Volk & McBeth, 1997; Zelezny, 1999), as well as from studies of active participation in environmental problem solving from related fields such as psychology and sociology (e.g., Bamberg & Moser, 2007; Hines, 1985; Hines, Hungerford, & Tomera, 1986/1987; Osbaldiston, 2004). Over time, two prominent messages emerged from these summaries of research: (1) some of the definitional features of EE and factors related to participation in environmental problem solving have been given substantial attention in EE research (e.g., knowledge and attitudes), while others have received very little attention (e.g., skills) (e.g., Hines, 1985; Iozzi, 1984; Rickinson, 2001; Volk & McBeth, 1997); and (2) there were more learning outcomes in EE and more factors related to environmental problem-solving than were apparent in the Tbilisi categories of objectives (e.g., beyond attitudes, affective outcomes included environmental sensitivity, locus of control, personal responsibility, and verbal commitment or intention).

The third source of thinking about EE and EL in the United States followed from Simmons' review and comparative analysis of conceptual and curriculum frameworks presented in widely disseminated materials and by state agencies (1995). The framework that Simmons synthesized from her review included seven major components that reflected and expanded upon the Tbilisi categories of objectives: affect, ecological knowledge, sociopolitical knowledge, knowledge of environmental issues, skills, additional determinants of environmentally responsible behavior, and behavior (pp. 54–55).

Within the United States, a number of efforts drew upon one or more of these sources to develop *environmental literacy frameworks* (e.g., Harvey, 1977a, 1977b; C. Roth, 1992; Simmons, 1995; Wilke, 1995). Implicitly, these efforts addressed recommendations from major EE conferences in the United States that the Tbilisi goal, objectives, and guiding principles should be further clarified for use by teachers and youth leaders (Gustafson, 1983, p. 112; Stapp, 1978, p. 71). Consequently, these EL frameworks included learning outcomes that closely reflect four of the five Tbilisi categories of objectives: knowledge, affect, skills, and behavior.

Several clarifications about the conception of EL embodied in these frameworks are relevant to these national assessments. *First*, behavior is construed in broad terms to encompass all familial and lifestyle decisions, and all forms of citizen participation, community service, and citizen action that, by intention or consequence, benefit the environment (i.e., what people do, individually and collectively). *Second*, by design, EL has three major thematic emphases apparent across the history of EE (Stapp, 1974; Swan, 1984) and in current EE practice (Scott & Gough, 2003), namely the natural world, environmental problems and issues, and sustainable solutions to them. *Third*, based on early work by Harvey (1977a, 1977b) and more recent work by C. Roth (1992), EL is commonly viewed as something that develops and can be developed over time, rather than as a result of a single program or a single year of schooling. *Fourth*, EL frameworks can be used to guide reviews of past research (e.g., Volk & McBeth, 1997) and new studies (e.g., Bogan, 1993; Hoffmann, 1998; Moseley, 1994; Todt, 1996; Willis, 1999), and therefore are subject to scrutiny and modification on the basis of new reviews of research and evaluation evidence. *Finally,* the developers of these EL frameworks recognize that the sources they drew from tend to reflect developed nation and, more specifically, US views on environmental problems and problem solving, and on educational theory and practice. To determine the relevance of these EL frameworks to other parts of the world, professionals must conduct their own analyses of documents pertaining to these sources. If and as appropriate, the results of these can be used to adapt EL frameworks for those nations and cultures.

When the researchers involved in the Korean, Israeli, and Turkish national assessment were asked about a national definition of EE or EL framework, each was very modest. Despite the absence of a nationally accepted definition of EE or a national EL framework in these nations, each of these national assessments was built upon prior educational research and development efforts (i.e., sources pertinent to two or three sources described above). However, rather than report on those here, the specific sources that each national research team consulted and used will be described in the next section as part of the summary for each of these national reports, thereby allowing those sources to be viewed "*in context*." Nonetheless, as will be apparent in these national reports, each of these research teams was familiar with: (a) the international and national definitions (rhetoric) of EE; (b) guides to and practices in EE within their nation; (c) reviews of individual studies or collections of studies in EE conducted within their nation; and (d) several of the EL frameworks developed from these sources within the United States.

A Summary of Four Recent National Assessments of Environmental Literacy

Summary of the Korean National Assessment Over 2002–2003 a national survey of the EL among third, seventh, and eleventh grade students in Korea was conducted, with support from the Korea Research Foundation (Chu et al., 2007; Chu, Shin, & Lee, 2005; Shin, Lee, Lee, Lee, & Lee, 2005). Faculty members and graduate students in the Department of Science Education, Dankook University, in Seoul, Korea, took a leading role in this study, with support from advisors at the Korea Institute of Curriculum and Evaluation and Florida Institute of Technology.

Origins and Rationales This study was initiated to understand what students know, how they feel, and how they act from the perspective of EL. It was a serious problem that there were no data about Korean students' status of EL prior to this. At the national level, this lack of understanding about the status of students' EL might lead EE policies in the wrong direction. Without an adequate diagnosis of educational status, any educational policy could cause unexpected educational results. Based on an adequate diagnosis of Korean students' EL, EE in Korea, including the national curriculum, teaching materials, and strategies, could be more effective.

Framework for Environmental Education and Environmental Literacy The research team was very familiar with the Tbilisi categories of objectives, and with other UNESCO-sponsored EE initiatives (Chu, et al., 2005; Shin, Lee, et al., 2005), as well as with the inclusion of EE in their national curriculum (Korean Ministry of Education, 1997, 2007). In addition, team members had conducted studies of and developed curricula in the area of earth science education, with specific attention to EE (Shin, 2001; Shin, Lee, et al., 2005). Further, with respect to relevant K-12 research, the team members were concerned that "few studies in Korea have been concerned for the relationship between environmental knowledge and attitudes, environmental attitude and behavior, and environmental skill and attitude, etc., [so] this study will analyze the relationship between [these] areas" (Shin, Chu, et al., 2005, p. 359). Finally, because there was no framework for EL in Korea at that time, the team accessed and conducted a comparative analysis of six EL frameworks published by US sources, including those by C. Roth (1992), Simmons (1995), and Wilke (1995) (Shin, Chu, et al., 2005, p. 359). On the basis of this analysis, they selected Simmons' (1995) framework as the guide for their national assessment (i.e., environmental knowledge, skills, affect, and behavior), which is consistent with the general goals of EE.

Methods This study investigated factors that may contribute to students' EL (Table 31.1). To understand the relationship between EE and science education, science-related predictors such as science content with preference, relation to science achievement, role of science and technology, and role of scientists were specifically analyzed. The role of science education in improving students' EL could be clarified through these results.

The development of the instrument was based on previous studies with similar purposes (Hines, et al., 1986/1987; Hsu, 1997; Hsu & Roth, 1998; Marcinkowski, 1998; Marcinkowski & Rehring, 1995; Sia, Hungerford, & Tomera, 1985/1986; Sivek & Hungerford, 1989/1990; Wisconsin Environmental Education Board, 1997). For each grade level, items used in previous studies were analyzed. The team members selected items for each of the five domains of EL, although they did not consider whether students learn them in school or whether items were tied to the seventh curriculum of Korea (Korean Ministry of Education, 1997, 2007). Selected items were translated into Korean, reviewed by two EE experts to establish content validity, and modified according to their recommendations.

Fifty or more students in the third, seventh, and tenth grade with similar socioeconomic and educational backgrounds to those in the main study participated in a pilot. Piloted items were analyzed using adjusted item-to-total correlations. By selecting items with reasonably strong item-to-total correlations ($r > 0.4$), researchers were more confident that selected items measured the same trait. The number of items in each instrument is summarized in Table 31.2. Environmental knowledge and skill items were multiple-choice, and environmental attitude and behavior items employed a four-point Likert-type scale.

Participating students were from the Seoul and Kyunggi-do areas. For sampling, students were divided into three categories on the basis of their living area: rural, urban, and metropolitan. Three schools at each grade level were sampled from each of these three areas. Samples were then drawn in proportion with the population of each area. A total of 2,993 students participated in this study (Table 31.3).

TABLE 31.1
Factors Affecting Environmental Literacy in This Study

Personal Background	Openness to the Environment	Relation to Science Subject
Gender	Experience of EE	Science content with preference
Students' socioeconomic status	Concern to the environment	Relation to science achievement
Parents' educational level	Source of environment information	Role of science and technology
Living area		Role of scientists

TABLE 31.2
Number of Assessment Items, by Domain

Grade \ Domain	Knowledge	Skills	Attitude	Behavior	Background Information	Total
3	24	7	22	16	12	71
7	27	9	27	25	14	102
11	24	7	27	28	14	100

TABLE 31.3
Sample Size, by Grade Level and by Area

Grade	Urban	Suburban	Rural	Total
3	475	400	94	969
7	464	400	123	987
11	559	335	143	1,037
Total	1,498	1,135	360	2,993

Data were collected from the third grade students to understand students' EL before they learn science at school, seventh grade students to understand students' EL after they learned science in elementary school, and eleventh grade students to understand students' EL after they learned science in middle school. SPSS 10.0 version was used to analyze these data. The results of descriptive analyses, in the form of standard scores, are summarized in Table 31.4, and indicate that there were no apparent differences between grades in any domain.

Results The correlation between domains of EL showed different trends (Table 31.5). Regardless of grade level, the highest correlations were found between environmental attitudes and behavior (.520–.661). The correlations between environmental knowledge and skills were relatively high (.337–.422), whereas low correlations were observed between environmental skills and behavior (.039–.082) and also between environmental knowledge and behavior (.095–.107). The correlations between environmental attitudes and knowledge and between environmental attitudes and skills were moderate.

Table 31.6 summarized the results of the analysis of factors that may affect students' EL. As indicated in the table, gender affected young students' EL. Female students in the third grade showed significantly higher scores on environmental knowledge, skills, attitudes, and behavior. In the seventh and eleventh grades, there were few significant differences in EL between genders, including environmental knowledge, in contrast to prior results showing male students' superiority in scientific knowledge (Shin & Noh, 2002).

The results indicated that parents' schooling plays an important role in students' EL, especially for younger students. Students who have college-graduate parents had significantly higher scores on environmental knowledge, skills, attitudes, and behavior. However, this parental influence was not apparent among seventh or eleventh graders.

Environmental education in schools also affected students' EL in all grades. Students who had experiences of EE in school had significantly higher EL literacy scores, particularly for environmental attitude and behavior.

On the other hand, students' science-related attributes had almost no affect on EL at any grade level. Neither students' cognitive nor affective perspectives in science had a relationship to their EL. However, third graders' attitudes and behavior were affected by their science achievement. This indicates that traditional science-related enterprises have not contributed to improving students' environmental literacy, which implies the need for changes in science education.

Discussion This study has significance as the first step to assess EL among Korean elementary and secondary school students. The results indicated that environmental behavior is highly correlated with attitude rather than knowledge.

TABLE 31.4
Standard Scores on Measures of Environmental Literacy, by Domain

Grade	Knowledge (*Mean±SD*)	Skills (*Mean±SD*)	Attitude (*Mean±SD*)	Behavior (*Mean±SD*)
3	49.83±7.48	50.00±7.38	50.12±9.03	49.94±8.69
7	50.02±8.77	49.57±7.41	50.18±9.38	49.72±9.14
11	49.97±8.26	50.02±8.03	50.16±9.12	49.10±9.38

TABLE 31.5
Correlations between Domains of Environmental Literacy (n=2,993)

	Grade	Knowledge	Skills	Attitude	Behavior
Knowledge	3				
	7				
	11				
Skills	3	.337			
	7	.422			
	11	.369			
Attitude	3	.238	.211		
	7	.252	.268		
	11	.290	.248		
Behavior	3	.107	.039	.562	
	7	.095	.082	.661	
	11	.097	.081	.520	

TABLE 31.6
Significant Factors Affecting Environmental Literacy*

	Knowledge			Skills			Attitudes			Behavior		
Factor	**3**	**7**	**11**	**3**	**7**	**11**	**3**	**7**	**11**	**3**	**7**	**11**
Gender	**F>M**	NS	NS	**F>M**	**F>M**	NS	**F>M**	NS	**F>M**	**F>M**	NS	NS
Living Area	NS	NS	NS	NS	NS	NS	NS	NS	NS	NS	NS	NS
Father's Schooling	**C>H**	**C>H**	NS	**C>H**	**C>H**	NS	**C>H**	**C>H**	NS	**C>H**	NS	NS
Mother's Schooling	**C>H**	**C>H**	NS	**C>H**	**C>H**	**C>H**	**C>H**	NS	NS	**C>H**	NS	NS
Experience of EE	NS	**yes>no**	NS	NS	**yes>no**	NS	**yes>no**	**yes>no**	**yes>no**	**yes>no**	**yes>no**	**yes>no**
Information Source	outdoor learning, books	newspaper/ magazine, books	family, field trip	outdoor learning, books	newspaper/ magazine, TV	newspaper/ magazine, family	outdoor learning, newspaper/ magazine	family, books	field trip, newspaper/ magazine	outdoor learning, newspaper/ magazine	newspaper/ magazine, books	field trip, newspaper/ magazine
Science Achievement	NS	NS	NS	NS	NS	NS	**G>NG**	NS	NS	**G>NG**	NS	NS
Science Subject of Preference	NS	NS	NS	NS	NS	NS	NS	NS	NS	NS	NS	NS
Opinion of Science	NS	NS	NS	NS	NS	NS	NS	NS	NS	NS	NS	NS

Notes: F= Female; M= Male; NS= Not Significant; C= College graduate; H= High school graduate; G= Good, NG: Not good; A > B = indicates significantly higher achievement than B at .05 significance level.

In Korea, although the *Environment* has been implemented as a separate subject at the secondary level since 1992, few schools have chosen it. Therefore it was recommended that science, in addition to EE, should play a major role in educating for environmental attitude and behavior. Consequently, the concept of Earth System was added to the national science curriculum (Korea Ministry of Education, 2007). This concept emphasizes the aesthetic aspects of the Earth, which enables students to have concerns for the beauty and value of the Earth. This study, which proposed the importance of EL, also affected EE curriculum reform in 2007. For instance, EL was included as one of the major goals in the National EE Curriculum.

Summary of the Israeli National Assessment A national assessment of the EL of sixth and twelfth grade school children in Israel was conducted over 2004–2006 by Tal (PI), Negev, Sagy, Garb, and Salzberg. A full report was presented to the Knesset—the Israeli parliament (Tal et al., 2007), and several subsequent articles have been published in 2010 (Negev, Sagy, Garb, Salzberg, & Tal, 2008; Sagy, Negev, Garb, Salzberg, & Tal, 2008a, 2008b).

Origins and Rationales This assessment was initiated in 2003 at the Arava Institute for Environmental Studies by Professor Tal to map for the first time the EL levels among school children in Israel. Furthermore, in the last decade EE programs had started in Israel, and there were no data regarding the EL levels of the students. The study was conducted in cooperation with Ben Gurion University, and supported by Brown University's Middle East Environmental Futures Project, The Lisa and Maury Friedman Foundation, and the GM foundation.

Framework for Environmental Education and Environmental Literacy Israel is a tiny country of 22,000 sq. kilometers (8,500 sq. miles) and seven million residents, located at the meeting point between three continents and characterized by diverse habitats. The combination of rapid population growth and European standards of living (Orenstein, 2004) has resulted in a broad range of environmental hazards (Tal, 2002).

Environmental education is not a separate subject in the Israeli education system. Instead, environmental knowledge is increasingly included in school subjects of science and technology and geography. Some 10 percent of schools choose to enhance their EE offerings, usually through external NGOs. For example, "Although social aspects of environmental issues are officially part of the primary school educational program, the emphasis appears to be heavily on scientific aspects of environmental issues" (Negev et al., 2008, p. 4). Further, "some Israeli high schools have an environmental science major . . . some 5 percent of Israel's 100,000 secondary students select this major" for their matriculation exams (Negev et al., 2008, pp. 4–5). Recently, the Ministry of Education declared 2009 a "green year in the education system," focusing on recycling. In terms of relevant research, the team drew upon studies by Blum (1986), Ben-Hur and Bar (1996), and Goldman, Yavetz, and Pe'er (2006) due to their relevance to the design of this national assessment. However, there was not a comprehensive EL framework in Israel at that time. Consequently, this team decided to use an operational definition based on Hungerford and Volk (1990), Marcinkowski (1998) and Simmons (1998), namely:

- *Knowledge*: general ecological principles, national and global issues;
- *Attitudes*: awareness, sensitivity, sense of responsibility, willingness to act; and
- *Behavior*: consumption, conservation, activism (Negev et al., 2008, pp. 5–6).

Methods Grade-specific assessment instruments (surveys) were prepared in consultation with an advisory committee comprising twenty experts from the Israel Ministries of Environment and Education, academia, NGOs, and K-12 schools. The first version of the survey drew heavily from prior EL surveys (Bluhm, Hungerford, McBeth, & Volk, 1995; Goldman et al., 2006; Marcinkowski & Rehrig, 1995). The knowledge questions were coordinated with the Ministry of Education standards for science and technology. All questions were phrased with a linguistic expert and with school children. The surveys were piloted in eight schools and modified accordingly. The surveys were translated from Hebrew to Arabic and then back to Hebrew twice by two translators for proper use among Arabic-speaking students.

In 2006, 1,591 sixth graders in thirty-nine schools and 1,530 twelfth graders in thirty-eight schools completed grade-specific surveys (i.e., at the end of primary and secondary education, respectively). These constituted a representative national sample of the formal education system, based on demographic data obtained from the Ministry of Education. These included size of town, performance level of the school, socioeconomic standing of the school and ethnic/religious identity of the school. Individual socioeconomic reporting was limited due to privacy restraints that are part of Israel's Ministry of Education external testing policies.

The surveys were distributed by a field team trained for this task. The children received a standardized explanation from the field team, emphasizing the importance of eliciting their views to improve EE, the need to take the survey seriously, the anonymity of the survey, and the difference between questions that required opinion and those involving objective knowledge. The vast majority of students completed the survey before the forty-five-minute period was over, and there were almost no disciplinary problems associated with survey administration.

Results The results (Table 31.7) suggest various levels of student knowledge on different environmental issues. For example, whereas approximately 80 percent of sixth graders answered correctly questions about bottle-deposit laws

TABLE 31.7
Achievement of 6th and 12th Grade Israeli Students on Measures of Knowledge, Attitudes, and Behavior

Parts of the Instrument	Grade	No. Items	Range	Sample Size		Mode	Median	M	SD
				n	Missing				
Knowledge	6	19 Items (50–68)	0–19.5	1797	121	10	9.5	9.34	3.13
	12	17 Items (69–86)	0–17	1530	0	12	11	10.43	3.32
Attitudes	6	18 Items (32–49)	18–108	1797	385	82	83	81.49	12.78
	12	17 Items (42–68)	17–102	1530	276	115	110	108.92	15.15
Behavior	6	8 Items (19–26)	8–40	1797	127	21	21	21.65	5.93
	12	9 Items (21–30)	9–45	1530	59	18	20	20.34	5.86

and recycling, only 25 percent answered correctly regarding the sources of global warming. As expected, knowledge scores were higher in older students for identical questions. For example, 55 percent of twelfth graders knew that most of the garbage in Israel is disposed in sanitary fills, while only 25 percent of sixth graders knew this, and the majority thought that most garbage is recycled. On a scale of 0–100, the overall knowledge score in sixth grade was 46.2 (SD17.35) and in twelfth grade, 62.1 (SD 19.38).

Attitudes regarding the environment were measured on a Likert-type scale (1- not true at all, to 6- very true) and were generally positive in both grades. Students indicated that they cared about open spaces (5.1 in 6th grade) and cleanliness (5.3 in 12th grade), and enjoyed nature (5.0 in 12th grade) and animals (5.5 in 6th grade). They were in favor of fines for polluters (5.0 in 6th), and feared the health consequences of pollution (5.0 in 6th grade, 4.7 in 12th grade). However, they were less willing to change their behavior in favor of public transport or to reduce consumerism.

Regarding behavior, most students saved water and electricity at least sometimes, and took part in other activities such as recycling, green consumerism, and activism even though recycling facilities and green products are not yet widely available in Israel.

While knowledge is the dimension of EL most emphasized in the Israeli curriculum, neither grade exhibited high scores. However, the environmental attitudes were generally positive. Behavior, the desired end point of EE, was generally low, especially when it required commitment.

Analysis of the correlations among the knowledge, attitudes and behavior dimensions revealed that knowledge is not correlated with attitudes and behavior (Table 31.8). This is consistent with other international findings (Courtenay-Hall & Rogers, 2002; Kollmuss & Ageyman, 2002; Olli, Grendstad, & Wollebaek, 2001). There were also significant positive correlations between knowledge and attitudes, and attitudes and behavior, but not between knowledge and behavior.

The research team also examined the association of informal spaces of learning to knowledge, attitudes, and behavior. These analyses revealed that informal spaces of learning have a major association with proenvironmental behavior (Figure 31.1), as well as with proenvironmental attitudes (Figure 31.2), but are barely associated with environmental knowledge. The extent of this influence on behavior is markedly large relative to the influence of demographic factors (Figure 31.3). In particular, children who reported participating in agriculture or having someone with whom they enjoyed studying nature have substantially higher scores on both proenvironmental behavior and attitude scales.

Finally, the team found that a mediating adult is highly correlated to EL, particularly attitudes and behavior, as shown in Table 31.9.

TABLE 31.8
Correlations Among Knowledge, Attitudes, and Behavior for 6th and 12th Grade Israeli Students

Correlated Variables	6th			12th		
	r	n	p	r	n	p
Behavior and knowledge	.04	1788	.0814	.04	1526	.0903
Attitudes and knowledge	.41	1788	<.0001	.23	1526	<.0001
Attitudes and behavior	.37	1783	<.0001	.56	1524	<.0001

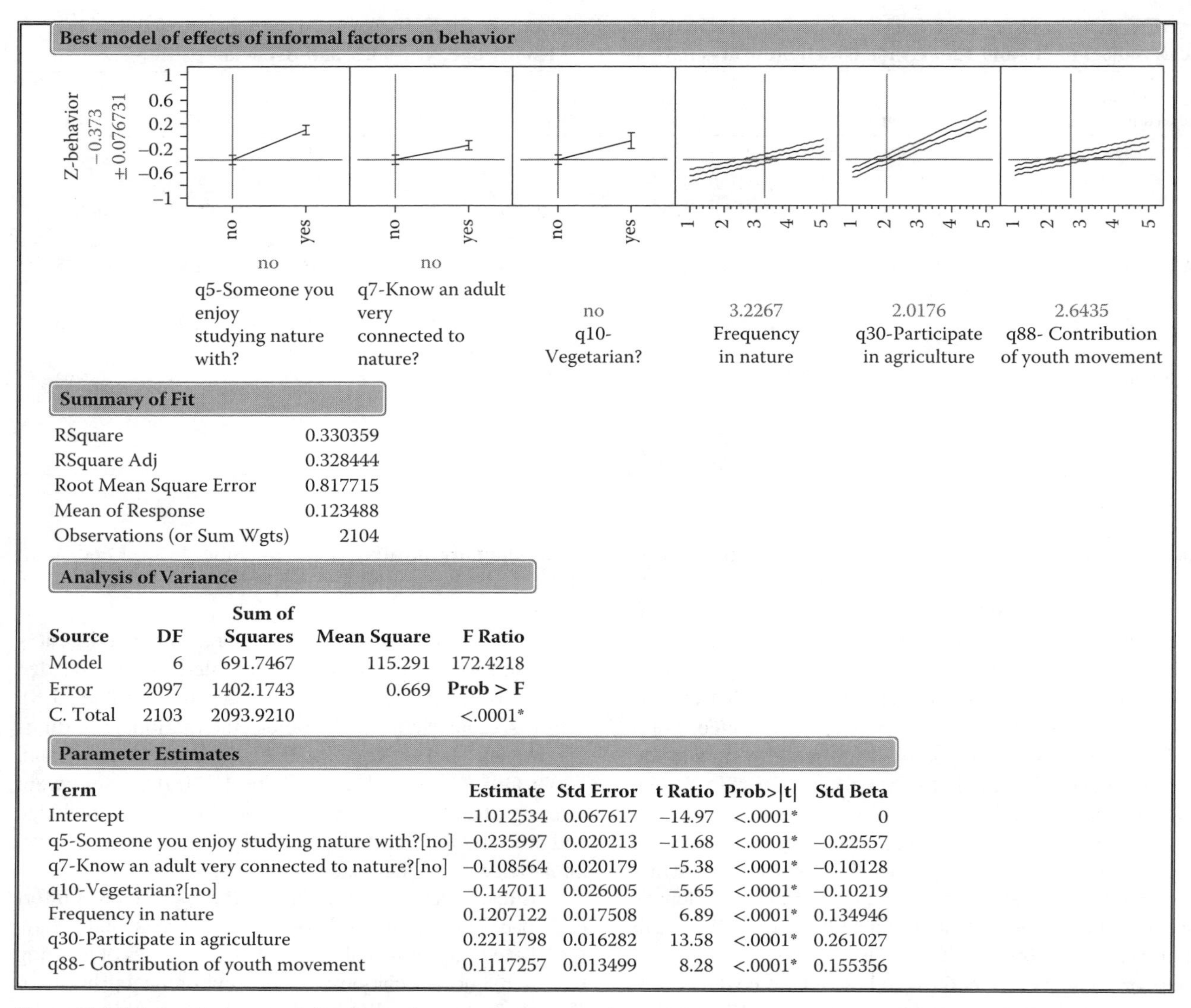

Summary of Fit

RSquare	0.330359
RSquare Adj	0.328444
Root Mean Square Error	0.817715
Mean of Response	0.123488
Observations (or Sum Wgts)	2104

Analysis of Variance

Source	DF	Sum of Squares	Mean Square	F Ratio
Model	6	691.7467	115.291	172.4218
Error	2097	1402.1743	0.669	**Prob > F**
C. Total	2103	2093.9210		<.0001*

Parameter Estimates

Term	Estimate	Std Error	t Ratio	Prob>|t|	Std Beta
Intercept	−1.012534	0.067617	−14.97	<.0001*	0
q5-Someone you enjoy studying nature with?[no]	−0.235997	0.020213	−11.68	<.0001*	−0.22557
q7-Know an adult very connected to nature?[no]	−0.108564	0.020179	−5.38	<.0001*	−0.10128
q10-Vegetarian?[no]	−0.147011	0.026005	−5.65	<.0001*	−0.10219
Frequency in nature	0.1207122	0.017508	6.89	<.0001*	0.134946
q30-Participate in agriculture	0.2211798	0.016282	13.58	<.0001*	0.261027
q88- Contribution of youth movement	0.1117257	0.013499	8.28	<.0001*	0.155356

Figure 31.1 Relationship between informal experiences and environmental behavior.

Discussion These findings revealed large gaps in environmental knowledge and a significant drop in environmental behavior among Israeli high school students. School background appeared to have only a modest effect on environmental attitudes and behavior among Israeli children, relative to other factors such as informal spaces and acquaintance with a person who is strongly attached to nature.

The time for the completing the survey by the children was limited to forty-five minutes per class; consequently, the surveys included a limited number of open-ended questions. The statistical analysis was limited due to the fact that the demographic data were obtained at the school level and not per student.

Future research using qualitative methods is needed to better understand these findings and the educational practices associated with them. One member of this research team (Sagy) has begun an in-depth study of schools with varied levels of EL in order to find school elements that influence EL among the pupils, while a second (Negev) is developing a multicultural approach to environmental education. In addition, Peled and Tal are involved in a study to characterize indicators for environmental literacy in Israel (Peled, 2010). These studies are considered a part of the national governmental and nongovernmental efforts in Israel to improve the national level of EL.

Recommendations from this national assessment are as follows: (1) implement a sustainability policy in the whole education system; (2) implement the existing recommendations of the Ministry of Education regarding environmental education; (3) prepare a pedagogical infrastructure for environmental education; (4) increase the exposure of students to environmental education; and (5), conduct further research including qualitative, case studies, and learning from successes.

Summary of the US National Assessment

In Spring 2007, a team of researchers conducted a national assessment of EL among sixth and eighth grade students

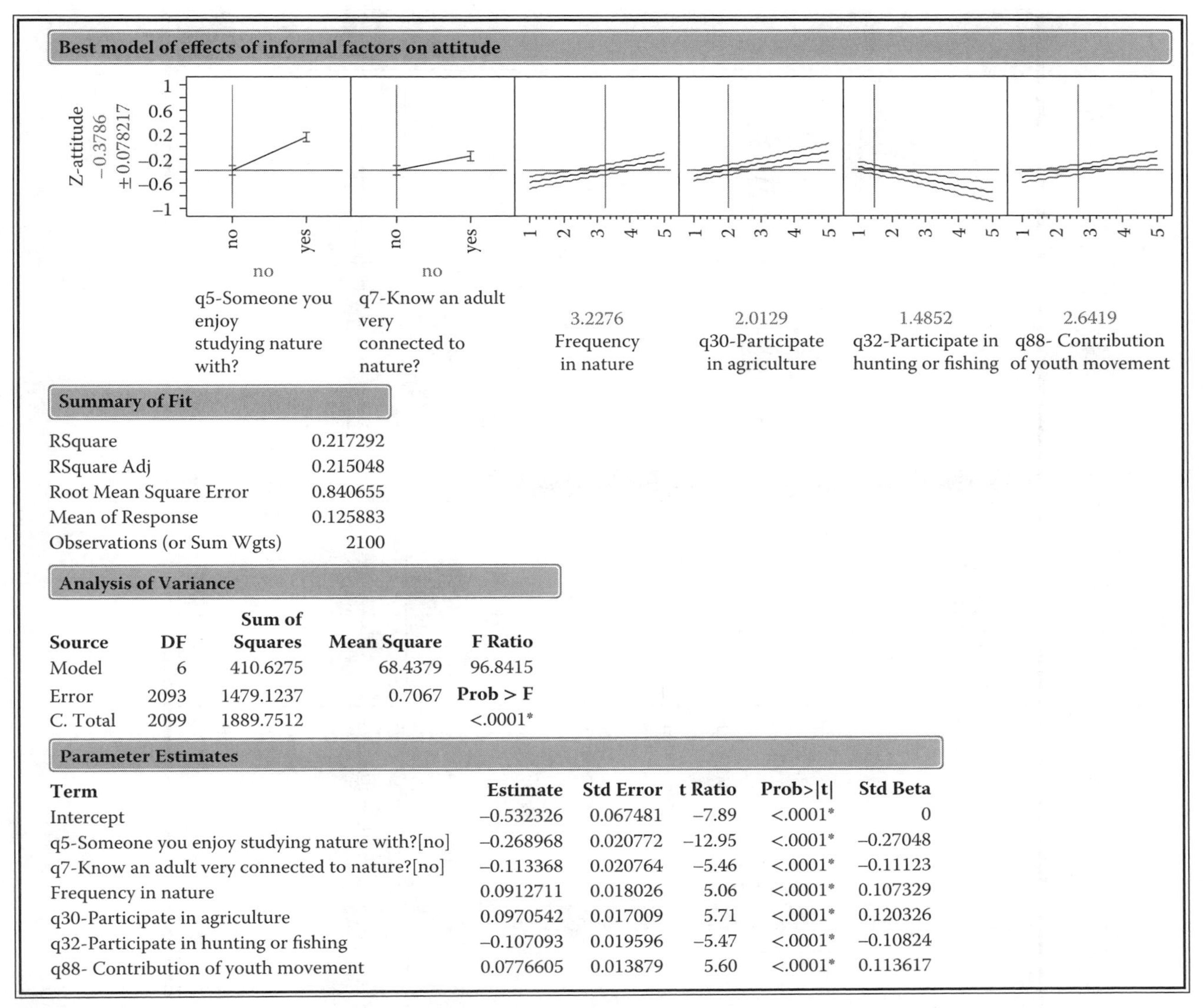

Summary of Fit

RSquare	0.217292
RSquare Adj	0.215048
Root Mean Square Error	0.840655
Mean of Response	0.125883
Observations (or Sum Wgts)	2100

Analysis of Variance

Source	DF	Sum of Squares	Mean Square	F Ratio
Model	6	410.6275	68.4379	96.8415
Error	2093	1479.1237	0.7067	Prob > F
C. Total	2099	1889.7512		<.0001*

Parameter Estimates

Term	Estimate	Std Error	t Ratio	Prob>|t|	Std Beta
Intercept	−0.532326	0.067481	−7.89	<.0001*	0
q5-Someone you enjoy studying nature with?[no]	−0.268968	0.020772	−12.95	<.0001*	−0.27048
q7-Know an adult very connected to nature?[no]	−0.113368	0.020764	−5.46	<.0001*	−0.11123
Frequency in nature	0.0912711	0.018026	5.06	<.0001*	0.107329
q30-Participate in agriculture	0.0970542	0.017009	5.71	<.0001*	0.120326
q32-Participate in hunting or fishing	−0.107093	0.019596	−5.47	<.0001*	−0.10824
q88- Contribution of youth movement	0.0776605	0.013879	5.60	<.0001*	0.113617

Figure 31.2 Relationship between informal experiences and environmental attitudes.

in the United States (McBeth, Hungerford, Marcinkowski, Volk, & Meyers, 2008). This study was funded by the United States Environmental Protection Agency's Office of Environmental Education and the National Oceanic and Atmospheric Administration, and was conducted with the cooperation of the North American Association for Environmental Education (NAAEE). This study yielded findings on a baseline level of EL among sixth and eighth graders, as well as data describing the schools, programs, teachers, and students involved in the study.

Origins and Rationales In 1990, a panel of professional environmental educators and researchers identified the need for national assessments of EL as part of a national research agenda for EE (Wilke, 1990). In a subsequent Delphi study to prioritize these research needs, the third highest need was for "a national cross-sectional study of the status of environmental literacy among K-12 students" (Saunders, Hungerford, & Volk, 1992, p. 3). Unfortunately, over the next decade, the only steps toward a national assessment of EL at any grade level in the United States were those reported by Wilke (1995), McBeth (1997), and Volk and McBeth (1997). Consequently, in its 2005 Report to Congress, the National Environmental Education Advisory Council recognized the ongoing need for, and therefore included a separate recommendation for, the conduct of a national measure of EL.

Framework for Environmental Education and Environmental Literacy Much of the literature that informed this research team's thinking about EE and environmental literacy was summarized in the second section of this chapter, *Definitional Features of Environmental Education and Environmental Literacy Relevant to These National Assessments*. Team members were very familiar with the evolution of thought about the goals, objectives, and definitional

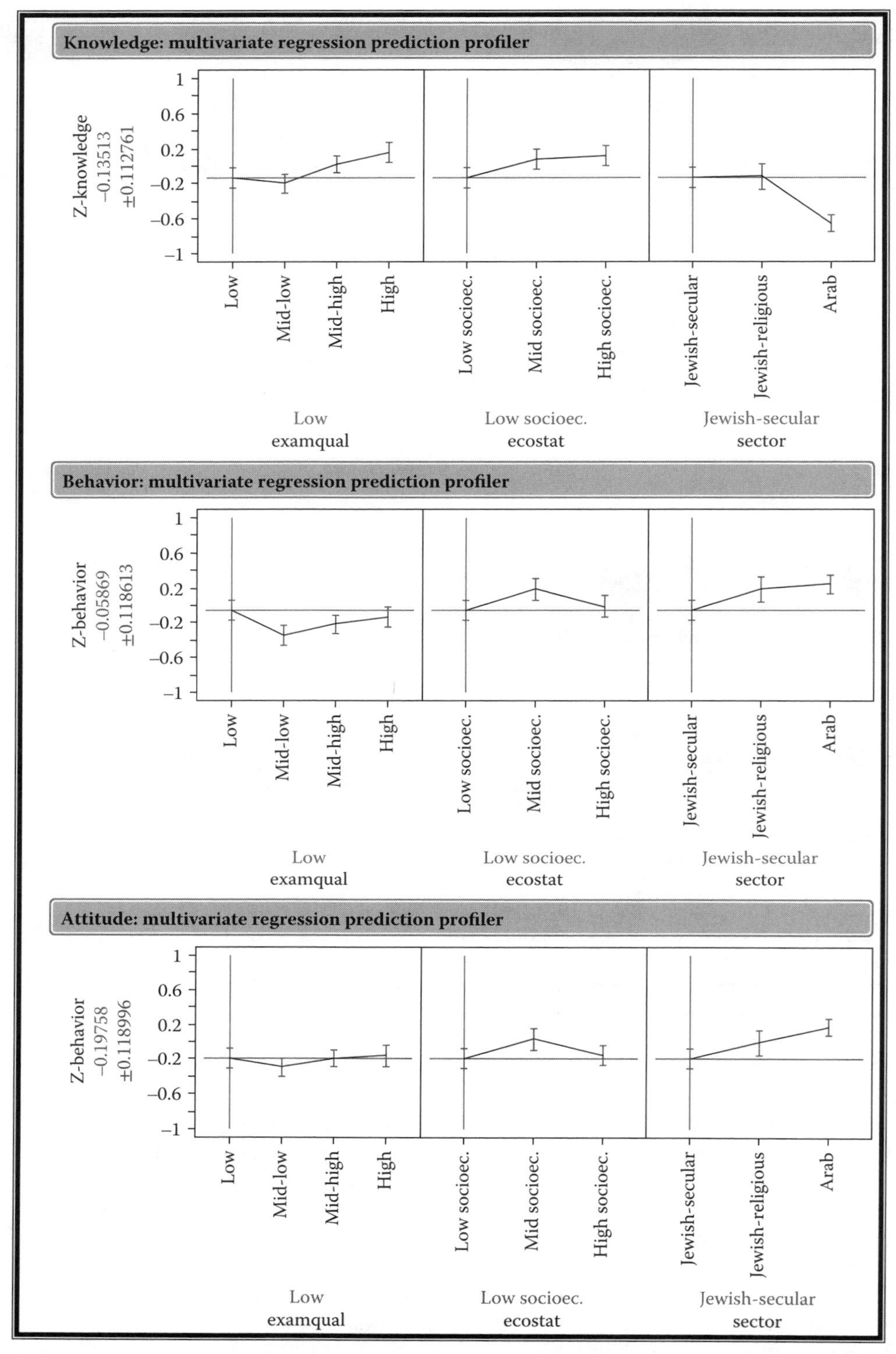

Figure 31.3 The relation between sociodemographic factors and environmental knowledge, attitudes, and behavior. (examqual = exam quality, or school achievements in national exams; ecostat = socioeconomic status; sector = ethnic sector).

features of EE. For example, at least two national conferences offered recommendations "that the Tbilisi goals and objectives be further clarified for use by teachers and youth leaders" (Gustafson, 1983, p. 112; Stapp, 1978, p. 71), and one of the few attempts to do so was undertaken by Hungerford, Peyton, and Wilke (1980). With respect to research, Hungerford served as dissertation advisor to Hines (Hines, 1985; Hines et al., 1986/1987), and other members of this team were involved in the preparation of other collections and reviews of research in EE (e.g., Iozzi, 1981,

TABLE 31.9
Correlations Between Environmental Literacy and Mediating Adults for Israeli 6th and 12th Graders, as Z Scores

	Knowledge		Attitudes		Behavior	
	6th	12th	6th	12th	6th	12th
There is someone I enjoy being in nature with	0.19	0.29	0.74	0.74	0.58	0.07*
There is someone I enjoy studying about nature with	0.02*	0.23	0.6	0.82	0.62	0.54
I know an adult who is very connected to nature	0.08*	0.07*	0.37	0.74	0.4	0.48

*$P > .05$.

1984; Marcinkowski & Mrazek, 1996; Volk & McBeth, 1997). Further, these researchers were very familiar with the results of Simmons' (1995) analysis of EE program and curriculum frameworks, having used that in some of their own work (e.g., Babulski, Gannett, Meyers, Peppel, & Williams 1999; Volk & McBeth, 1997). Finally, members of this researcher team were involved in the development of several of the aforementioned EL frameworks (e.g., Harvey, 1977a, 1977b; C. Roth, 1992; Wilke, 1995). The EL framework that guided this national assessment was developed and validated by Wilke (1995), although it was delimited for developmental reasons for use with middle grades students (e.g., Bluhm et al., 1995; McBeth, 1997).

Methods Two broad research questions guided the design and conduct of this study.

1. What is the level of EL of sixth and eighth grade students across the United States on each of the following variables:
   a. ecological knowledge?
   b. verbal commitment?
   c. actual commitment?
   d. environmental sensitivity?
   e. general environmental feelings ?
   f. selected environmental issue and action skills?
2. What is the general level of EL of sixth and eighth grade students across the United States?
   The research team agreed upon the following study delimitations.
   - This study was limited to students enrolled in public and private schools within the United States in the 2006–2007 school year.
   - For practical and financial reasons, the number of counties selected for school sampling purposes was limited to fifty.
   - Only those schools that had students in both sixth and eighth grade classes were eligible to be selected into this sample and participate in this study.
   - The population from which the study sample was drawn was limited to sixth and eighth grade students.
   - Only EL components identified in the research question above were surveyed.
   - The survey that gathered student data on these components of EL was designed to be administered in a fifty-minute time period (i.e., the common length of a school period).
   - Data were collected late in the school year to allow students as much time as possible to learn and mature developmentally.

The Middle School Environmental Literacy Survey (MSELS) utilized in this study measured four major components of environmental literacy: (a) ecological knowledge; (b) environmental sensitivity; (e) cognitive skills; and (f) behavior (Bluhm et al., 1995; McBeth, 1997; McBeth et al., 2008). As such, it included one or more measures in each of the four major domains that appear to be critical to EL: knowledge, affect, skill, and behavior (C. Roth, 1992; Simmons, 1995; Wilke, 1995).

A probability-proportional sample of fifty counties, and sixth and eighth grade classrooms within each, was identified by GfK-Roper. Onsite data collectors (researchers recruited through NAAEE's Research Commission) visited school sites, administered the assessment instrument, and collected demographic information related to the site. An assessment coordinator oversaw this effort, including the distribution of assessment materials, monitoring of data collection progress, and return of assessment materials to Florida Institute of Technology. There, data entry and analyses were completed, with major support from another Roper affiliate, the Center for Survey Research and Analysis.

Results Major findings pertaining to Research Question One are summarized in Table 31.10 and Table 31.11. Eighth graders (n = 962) evidenced higher means than sixth graders (n = 1,042) on Ecological Knowledge and on measures of two Environmental Issue and Action Skills, Issue Analysis and Action Planning. The sixth graders had higher means on Verbal Commitment, Actual Commitment, Environmental Sensitivity, General Environmental Feelings and on the Environmental Issues Skill of Issue Identification. Significant differences were observed between sixth and eighth graders on Ecological Knowledge, Verbal Commitment, Actual Commitment, Environmental Sensitivity, and Environmental Feeling. These differences favor the sixth graders on all variables except in the case of Ecological Knowledge.

Major findings for Research Question Two were reported in the form of Composite Scores. Environmental Literacy composite scores were derived by compiling scores for each of four major components of EL (i.e., possible composite score of 240, with a range from 24 to 240). The sixth grade mean composite score was 143.99 and the eighth grade mean composite score was 140.19,

TABLE 31.10
Summary of Descriptive Statistics, by Scale/Index, and by Grade, Using Weighted Data

Parts of the MSELI	No. Items	Range	Grade	Sample Size n	Sample Size Missing	Mode	Median	Mean	Std. Dev.
II. Ecological Foundations	17 Items (5–21)	0–17	6	934	108	13	12	11.24	3.26
			8	921	42	13	12	11.62	3.32
III. How You Think About the Environment	12 Items (22–33)	12–60	6	1000	42	44	44	43.89	8.88
			8	936	27	43	41	41.10	9.20
IV. What You Do About the Environment	12 Items (34–45)	12–60	6	974	68	40	39	38.44	9.15
			8	921	41	40	35	35.14	9.39
V. You and Environmental Sensitivity	11 Items (46–56)	11–55	6	978	63	31	33	32.54	7.47
			8	913	49	30	30	30.11	7.48
VI. How You Feel About the Environmental	2 Items (57–58)	2–10	6	987	55	10	9	8.14	2.00
			8	930	32	10	8	7.82	2.06
VII.A. Issue Identification	3 Items (59, 60, 67)	0–3	6	902	139	1	1	1.31	0.93
			8	885	77	1	1	1.29	0.95
VII.B. Issue Analysis	6 Items (61–66)	0–6	6	905	137	2	2	2.75	1.89
			8	869	93	1	3	2.86	2.00
VII.C. Action Planning	8 Items (68–75)	0–20	6	874	168	2	6.97	7.25	5.44
			8	820	142	2	7.00	7.86	5.64

TABLE 31.11
Results of T-test Comparisons of the 6th and 8th Grade Samples, Using Unweighted Data

Variables	Grade	N	Mean	SD	t	Prob *
I. Ecological Knowledge	6	1042	10.95	3.856	−2.836	0.0046**
	8	962	11.42	3.622		
II. Verbal Commitment	6	1033	43.83	8.942	6.987	0.000**
	8	953	41.01		9.067	
III. Actual Commitment	6	1023	38.23	9.037	8.045	0.000**
	8	942	34.89	9.350		
IV. Environmental Sensitivity	6	1021	32.47	7.384	6.937	0.000**
	8	942	30.14		7.449	
V. Environmental Feeling	6	993	8.26	1.940	4.527	0.000**
	8	930	7.85		2.033	
VI.A. Issue Identification	6	986	1.33	.938	0.802	0.4225
	8	928	1.29	.945		
VI.B. Issue Analysis	6	959	1.87	1.444	−0.489	0.6248
	8	902	1.91	1.470		
VI.C. Action Planning	6	878	7.55	5.401	−1.790	0.0736
	8	817	8.03	5.592		

**Note*: statistically significant at p = .00625

with a combined sixth and eighth mean composite score of 142.14. These scores all fall in the mid-range (97–168) of possible scores, reflecting a moderate level of EL. Of the four literacy components, the highest scores for the sixth and eighth grade were attained in Ecological Knowledge, with slightly lower scores obtained in Environmental Affect and lower scores, yet, obtained in Behavior. The lowest scores were observed in the component of Cognitive Skills.

Discussion A vital premise of this study was that the development of environmental literacy is complex, and attempts to develop it can take many forms. This study is important in that it provides a measure against which to compare current and future efforts in EE within the United States, such as has been undertaken by this research team in Phase 2 of this national assessment (2008–2010). In addition, further analyses of these data, in particular with respect to the classroom and teacher information, might shed light on the impacts of EE efforts, where it was present in these schools and classrooms.

Additional research is needed to identify factors that contribute to the disparities across variables measured by the *MSELS* (e.g., socioeconomic, educational, cultural, and developmental). Also, further research is needed to investigate schools that fell in either end of the EL spectrum. These schools, in particular, become "schools of interest" and a thorough study of them should reveal attributes that contributed to the rather obvious disparities regarding EL. This could help researchers and theorists identify promising educational practices as they relate to EL.

Summary of the Turkish National Assessment Over 2007–2009, Erdogan conducted a national assessment (survey) of the EL of fifth graders in Turkey. This study was approved by the Middle East Technical University (METU) Ethics Committee, and supported by the Turkish Educational Research and Development Directorate (EARGED). Subjects included 2,412 fifth graders drawn from twenty-six private schools in urban areas (21.6%) and fifty-two public schools (78.4%) from twenty-six provinces in Turkey. Students' age ranged from ten to eleven.

Origins and Rationale This national assessment was undertaken as a doctoral dissertation in the Department of Educational Sciences, Middle East Technical University (Erdogan, 2009). Prior to this, an analysis of research studies on EE and the existing literature in Turkey revealed the lack of a nationwide survey/study on EE and assessment of EL that could contribute to the development of a strong base for environmental policy and practice in the context of Turkish schools. Erdogan's ten-month visit with Dr. Marcinkowski (2006–2007), and his prior communications with Drs. Hungerford, Volk, and Marcinkowski encouraged him to pursue his idea of a national environmental literacy assessment in Turkey.

Framework for Environmental Education and Environmental Literacy In the Turkish Education System, there is no separate environmental education course and curriculum. Rather, in Ministry of Education curricula developed for elementary schools, it is readily apparent that environmental topics were integrated into different courses, taking advantage of the interdisciplinary nature of EE. However, these topics were not sufficiently incorporated until 1960s, consistent with trends and developments in EE in the United States and Europe. Consequently, primary school science curricula that had a greater emphasis on environmental topics were developed in 1992, 2000, and 2004 respectively.

Because research on EE and EL is in its infancy in Turkey, there has not been yet published any framework for either of them. In preparation for this national assessment, a content analysis of qualitative and quantitative K-8 EE research in Turkey published between 1997 and 2007 (53 studies) was conducted. For one part of this analysis, the EL framework proposed by Simmons (1995) and used later by Volk and McBeth (1997) was used to determine the extent to which these studies gathered evidence about students learning outcomes related to EL (Erdogan, Marcinkowski, & Ok, 2009)

The results of this review of research were used to revise and expand the subcomponents of EL developed by Babulski et al. (1999). Then, those subcomponents were used to analyze the extent to which objectives in five curriculum guidebooks used in obligatory science courses in elementary schools in Turkey included attention to the components of EL (Erdogan, Kostova, & Marcinkowski, 2009).

Finally, in Erdogan's (2009) dissertation study that served as this Turkish National EL assessment, the results of both of these analyses were used to modify Simmons' EL framework and the EL framework developed by Babulski et al. (1999) to guide this national EL assessment.

TABLE 31.12
Descriptions of the Sample (N=2412)

Variables	*f* (frequency)	% (percent)
Gender		
Female	1207	50
Male	1185	49.1
Missing	20	0.9
School Type		
Students in Public Schools	1891	78.4
Students in Private Schools	521	21.6
Participation in Preschool Education		
Students Who Took Preschool Ed.	1083	44.9
Students Who Did Not Take Preschool Ed.	1299	53.9
Missing	30	1.2
Residence		
Students in Urban Public Schools	1059	43.9
Students in Rural Public Schools	832	34.5
Students in Urban Private Schools	521	21.6
Income		
500 YTL and below	243	10.1
501–1000	321	13.3
1001–1500	117	4.9
1501–2000	63	2.6
2001 and above	158	6.6
I Do Not Know and Missing	1510	62.5

Method The Elementary School Environmental Literacy Instrument (ESELI) used in this study was developed in six steps, and piloted with 673 students (Erdogan & Ok, 2008). These steps were: (1) developing a conceptual framework for the instrument; (2) analysis of existing research in Turkey (Erdogan, Marcinkowski, et al., 2009); (3) analysis of national objectives for primary schools (Erdogan, Kostova, et al., 2009); (4) development of an item pool and constructing the instrument; (5) instrument review by experts in Turkey; and (6) piloting the instrument. The ESELI included five parts (81 items), each of which pertained to a component of EL, notably knowledge, affect, skills, and behavior. For the part that measured knowledge, items corresponded to national objectives. The name of the parts and subparts (factors), number and types of the items in each, and the reliability of the parts are presented in Table 31.13.

Results Each part of the ESELI included a different number of items and therefore had a different range of raw scores. The pros and cons of calculating EL composite scores were debated among members of the US research team (McBeth et al., 2008). They recognized that calculating composite scores would combine different types of metrics (measures) and mask differences among measures, but could be of interest to educational policy makers and administrators. Eventually, the procedures developed by McBeth et al. (2008) were used in this study (i.e., each of four sections contribute equally to EL composite scores through the use of multipliers), resulting in a maximum adjusted score of sixty for each of the four parts. These adjusted scores were then summed to yield a maximum composite score of 240 (with the range of 15–240). To aid in the interpretation of the students' composite EL score, this range was divided into three categories: low (15–90); moderate (91–165), and high (166–240). Similar procedures were applied to each part of ESELI. For the components Ecological Knowledge, Cognitive Skills, and Environmentally Responsible Behavior (ERB), the range of 0–60 was divided into three categories: low (0–20); moderate (21–40); and high (41–60). For the component Affect, the range of 15-60 was divided into three categories: low (15–30); moderate (31–45); and high (46–60).

In Table 31.14, the adjusted mean environmental knowledge score of the participants was 42.42 (SD 9.48), which fell in the high range (41–60), reflecting a high level of environmental knowledge. Of the 2,410 students, 1,607 students (66.7%) scored between 41 and 60 (i.e., in the high range). The knowledge level of 738 students (30.6%) was moderate, and only 65 students had low knowledge scores. The adjusted mean affective disposition score for participants was 52.27 (SD 9.07), which fell in the high range (46-60), reflecting a high level of affective disposition toward the environment. Of the 2,410 students, more than 2,000 (86%) had high affective disposition scores, and only 124 students had low affective disposition scores.

The adjusted mean cognitive skill score of participants was 24.67 (*SD* 14.69), which fell in mid-range (21–40), reflecting a moderate level of cognitive skills. Only 377 students' cognitive skill score fell into the high range (41–60), while the cognitive skill scores for 897 students' (37.2%) was moderate, and for 935 students' (39.5%) was low. However, 184 students did not complete this measure of cognitive skills, which were treated as missing data.

The adjusted mean ERB score of participants was 30.58 (*SD* 10.89), which fell in the mid-range (21–40), reflecting a moderate level of ERB. Scores for 1,581 students (65.6%) fell in this moderate range. Only 433 students (18%) were engaged in a high level of ERB, whereas 398 students (16.4%) were engaged in a low level of ERB.

The mean for environmental literacy composite scores was 149.66 (*SD* 26.19). This score fell in the mid-range (91–165), reflecting a moderate level of EL. Of the 2,226 students with composite EL scores, 1,545 (64.1%) had

TABLE 31.13
Names of, Items in, and Reliability for Parts of the ESELI

Part	Number and type of items	(α)
I. Demographic Information	11 (Varied)	—
II. The Test for Environmental Knowledge [TEK]	19 (Multiple Choice)	.69
	3 (T-F)	—
III. The Affective Disposition Scale [ADTES]		
Factor.1 Willingness to Take Environmental Action	5 (Likert Type)	.66
Factor.2 Environmental Attitudes	5 (Likert Type)	.63
Factor.3 Environmental Sensitivity	4 (Likert Type)	.58
IV. Children Responsible Environmental Behavior Scale [CREBS]		
Factor.1 Political Action	7 (Likert Type)	.91
Factor.2 Eco-management	6 (Likert Type)	.71
Factor.3 Consumer and Economic Action	5 (Likert Type)	.73
Factor.4 Individual and Public Persuasion	8 (Likert Type)	.81
V. The Issue Identification and Evaluation Skills Test (IIEST)	7 (Matching item)	.59
	1 (Open-ended)	—

TABLE 31.14
Students' Levels According to the Components of EL and Their Overall Level of EL

		Low	Moderate	High	Mean	Sd
Knowledge	Range	*(0–20)*	*(21–40)*	*(41–60)*		
	f	65	738	1607 (66.7%)	42.42	9.48
	%	(2.7%)	(30.6%)			
Affect	Range	*(15–30)*	*(31–45)*	*(46–60)*		
	f	124	213	2073	52.27	9.07
	%	(5.1%)	(8.8%)	(86%)		
Cognitive Skills*	Range	*(0–20)*	*(21–40)*	*(41–60)*		
	f	952	897	377	24.67	14.69
	%	(39.5%)	(37.2%)	(15.6%)		
ERB	Range	*(0–20)*	*(21–40)*	*(41–60)*		
	f	398	1581	433	30.58	10.89
	%	(16.4%)	(65.6%)	(18%)		
EL Composite Score	Range	*(15–90)*	*(91–165)*	*(166–240)*		
	f	22	1545	659	149.66	26.19
	%	(0.9%)	(64.1%)	(27.3%)		

*There are 184 missing items which were never replaced with mean.

scores that fell in the moderate range. On the other hand, more than a quarter of the participants (27.3%) scored in the high range, and only 22 students (0.9%) had scores in the low range of EL.

Discussions Interpretation of EL composite score which was calculated by combining all components of EL showed that the average EL score of students was 149 (SD = 26.19), suggesting moderate level of EL and many of the students (64.1%, n = 1545) fell in the mid-range (91–165) reflecting moderate level of EL.

This study highlighted the need for developing and/or contextualizing an EL framework for the Turkish context. In this regard, the findings of this study can provide a strong base for developing such a framework, and for furthering additional EL assessments with a wide range of participants. It is recommended that other Turkish researchers conduct nationwide EL assessments with middle school, high school and university students. The results of these studies would shed light on establishing and practicing a stronger EE policy in the Turkish Education system.

Closing Discussion

Each of these national assessments of EL was the first of its kind in that nation. All four were undertaken with substantial financial and institutional support, and three were undertaken by a research team (i.e., only Turkey's was not, although Erdogan did have support and assistance). The need for this support and teaming is apparent in the sample sizes (i.e., 2,000 to 3,250), and in the level of effort required to collect, edit, and analyze data on multiple components of EL from samples of this size.

The grade levels sampled and components of EL included in each national assessment are summarized in Table 31.15. While the Israeli, Turkish, and US assessment involved nationally representative samples, it is not evident that this was true for the Korean assessment. Further, while the samples in the Korean and Israeli studies may be considered cross-sectional, this is less true for the US study, and not true for the Turkish study. Consequently, differences in the number and level of grades from which subjects were drawn, as well as differences in EL components assessed and in EL measures used (Table 31.15, p. 223), make it virtually impossible to compare results, and difficult to draw anything other than very general and tentative conclusions, across these national assessments.

All four assessed one or more aspects of environmental knowledge. In the Israeli and US assessment, knowledge scores appeared to be moderate, while in the Turkish study they were relatively high. Further, in both the Israeli and US assessment, students in the higher grade outscored students in the lower grade.

All four studies assessed multiple dimensions of environmental affect, although in the Korean, Israeli, and Turkish assessments, scores on measures of each dimension were aggregated for reporting purposes. In the Israeli, US, and Turkish assessment, students at all grade levels tended to score high on measures of environmental affect.

Three of the four studies assessed students' cognitive skills. Scores in the US and Turkish assessments tended to be low. In the US assessment, scores on measures of specific skills favored sixth graders on one, and eighth graders on two of these.

With respect to behavior, again, construed and measured broadly, results were mixed. In the Israeli and

TABLE 31.15
Grade Levels and Components of Environmental Literacy in Each National Assessment*

	Korea	Israel	US	Turkey
Grade Levels	3, 7, 11	6, 12	6, 8	5
Domains and Components				
I. Knowledge				
a. Ecological Knowledge	X	X	X	X
b. Environmental Knowledge	X	X		X
II. Affect				
a. Environmental Sensitivity	X	X	X	X
b. Environmental Feelings			X	
c. Environmental Attitudes	X	X		X
d. Personal Responsibility	X	X		
e. Locus of Control/Efficacy	X	X		
f. Verbal Commitment/ Willingness	X	X	X	X
III. Cognitive Skills	X		X	X
IV. Behavior	X	X	X	X

Note: Additional sources on these national assessment were consulted for this table. These include: for Korea, Shin, Chu, et al. (2005); for Israel, Negev et al. (2008); for the United States, McBeth et al. (2008); and for Turkey, Erdogan (2009).

US assessments, students tended to score moderately high, with students in the lower grade outscoring students in the higher grade. In the Turkish assessment, scores appeared to be more moderate.

Finally, the Korean and Israeli assessments included analyses of correlations among domains or components of EL. Results from both assessments indicated that the relationship between attitude (affect) and behavior was reasonably high, while the relationship between knowledge and behavior was low. Results for the knowledge—attitude (affect) relationship were inconsistent (i.e., higher in the Israeli than in the Korean assessment). Thus, while each assessment was undertaken to provide unique insights into the status of EL in that nation, some comparative results such as these may be of some interest and relevance to those interested in these relationships.

These studies raise as many, if not more, questions than they answer. Certainly, these results are best understood in the multifaceted contexts in which these assessments were conducted. For example, differences in results pertaining to knowledge may well reflect differences in national curriculum guidelines, and correspondence between those guidelines and items in the measures used (i.e., content validity). In addition, because each assessment was viewed as a first step within each nation, the results of each will be most useful when they are compared to results of subsequent studies in each nation. It is noteworthy that follow-up studies are underway in Israel and the United States, as briefly described in the *Discussion* section for each of these national assessments. Beyond this, the results of each assessment, and of subsequent studies, will only be of value if they are communicated to and understood by those who shape educational policies in each nation. It is equally noteworthy that initial steps have been taken in each nation toward this end.

Beyond these national assessments, there have been few, if any, international assessments of EL to date. One of the studies that came closest to this was a subset of the 2006 Science Assessment conducted by the Program for International Studies Assessment for OECD (Programme for International Student Assessment [PISA], 2008). The 2006 science assessment included a subset of questions designed to assess knowledge, skill, and dispositions of more than 40,000 fifteen-year-olds from fifty-seven countries in the areas of environmental science and geoscience. A report of this assessment, entitled *Green at Fifteen?*, indicated that the conceptual framework and scales for the environmental science portion of the 2006 Science Assessment were developed post hoc (PISA, 2008, pp. 19–21), and that minimal attention was given to the body of published work in EE and on EL during the development or analysis process. The instrument used in this study focused on scientific knowledge, awareness and understanding of issues, sources of information, environmental concern, and optimism/pessimism (PISA, 2008, pp. 19–26). Thus, it appears that the national assessments summarized in this chapter included a wider range of EL components than the PISA 2006 Science Assessment (e.g., cognitive skills and behavior). At the same time, while the four national assessments summarized herein involved fairly extensive item and scale development procedures, that item development and data analysis process used in the PISA 2006 Science Assessment was more extensive and sophisticated. Only the Israeli and, perhaps, Korean national assessments involved as extensive an item translation process. Due to the relative strengths and limitations of these national assessments and of this PISA 2006 Science Assessment, initial steps have been taken through NAAEE to engage representatives of Canada, Mexico, the United States, and other entities in discussions of what the EE and science educational communities have to learn from and offer one another so as to overcome some of these limitations in future national and international assessments.

Due to limitations apparent in these comparisons, the chapter authors agreed to pose several recommendations. First, because these comparisons were broad and general, the authors thought it would be beneficial to conduct a more careful and detailed comparison of items used to assess common environmental literacy components. Specifically, if comparable items were used in two or more of these assessments, then members of these research teams should take the necessary steps to compare responses on those items across these national assessments. Second, if the results of such an item-by-item comparison warranted it (i.e., if there are comparable items, if estimates of validity and reliability exists either for the measures from which those items were drawn or for the items themselves, and if such comparisons of look promising), then members of these research teams would recommend selected items for inclusion in future national assessment of this kind so as to permit similar comparisons (i.e., such as those possible in *Green at FIfteen?*).

Third, the authors recognized that it is highly unlikely that there is only one EL framework appropriate to all nations, peoples, and times. Thus, while these assessments explored EL using a similar frame of reference (Simmons, 1995), from a broader perspective, each should be viewed as exploratory. In early discussions with researchers planning each of these assessments, Marcinkowski suggested that eventually, three vital steps needed to be taken to assess environmental literacy in a given nation in a meaningful or "grounded" manner:

- as has been attempted in the United States and to some extent these other nations, efforts must be taken to summarize available research and evaluation studies that pertain to learning outcomes in EE, as well as conduct a broad analysis of various EE goals, objectives, and frameworks-in-use so as to help determine which components should be included in an EL framework for that nation;
- once there is reasonable clarity about and agreement on what belongs in an EL framework, the results of these reviews and analyses, along with other sources of information, should be used to identify salient dimensions of each component for that nation and its peoples; and
- once there is reasonable clarity and agreement on how those EL components should be defined and understood, it is necessary to determine how each can best be assessed. This determination should consider: advances in traditional and alternative assessment methods appropriate for each component; which assessment methods are acceptable to peoples in that nation; and what is practical in light of conditions and constraints in a research setting.

Finally, it was apparent that each national assessment required a modest-sized research team. Across nations, the involvement of a team of researchers had numerous benefits, including securing needed support (e.g., governmental interest, institutional support, and funding), debating and critiquing proposed methods, and sharing the project workload. Since the completion of each assessment, it has been apparent that other researchers and graduate students have taken an active interest in methods used in and findings from their initial national assessment, and wished to extend this work (e.g., case studies of selected schools and classes in Israel, and planned for Phase 4 of the US national assessment). The authors have welcomed this interest. Consequently, they recommend that efforts be made to expand and strengthen research teams in each of these four nations, and that in nations in which such an assessment might be under consideration, that efforts be made to form such a research team.

References

Babulski, K., Gannett, C., Myers, K., Peppel, K., & Williams, R. (1999). *A white paper on the relationship between school reform and environmental education in Florida: Correlating Florida's sunshine state standards and an environmental literacy framework* (Unpublished research paper). Florida Institute of Technology, Melbourne, FL.

Bamberg, S., & Moser, G. (2007). Twenty years after hines, hungerford, and tomera: A new meta-analysis of psycho-social determinants of pro-environmental behavior. *Journal of Environmental Psychology, 27*(1), 14–25.

Ben-Hur, Y., & Bar, H. (1996). *Eichut hasviva be'beit hasefer* [Environmental in the school]. Jerusalem, Israel: Gutman Institute.

Bluhm, W., Hungerford, H., McBeth, W., & Volk, T. (1995). The middle school report: A final report on the development, pilot assessment of The Middle School Environmental Literacy Instrument. In R. Wilke (Ed.), *Environmental education literacy/needs assessment project: Assessing environmental literacy of students and environmental education needs of teachers; final report for 1993–1995* (pp. 8–29). (Report to NCEET/University of Michigan under US EPA Grant #NT901935-01-2). Stevens Point, WI: University of Wisconsin—Stevens Point.

Blum, A. (1986). Knowledge and approaches for ninth grade students toward environmental problems in Israel. In A. M. Meir & P. Tamir (Eds.), *Science teaching in Israel* (pp. 291–302). Jerusalem, Israel: Israeli Center for Science Teaching.

Bogan, M. (1993). Determining the environmental literacy of participating high school seniors from the hillsborough and pinellas county school districts in Florida: A curriculum study (Doctoral dissertation, University of South Florida). *Dissertation Abstracts International, 53*(8), 2609A. (UMI No. DA9235069)

Bohl, W. (1977). A survey of cognitive and affective components of selected environmentally related attitudes of tenth, and twelfth–grade students in six mideastern, four southwestern, and twelve plains and mountain states (Doctoral dissertation, Ohio State University, 1976). *Dissertation Abstracts International, 37*(8), 4717A. (UMI No. DBJ77–02352)

Chu, H., Lee, E., Ko, H., Shin, D., Lee, M., Min, B., & Kang, K. (2007). Korean Year 2 children's environmental literacy: A prerequisite for a Korean environmental education curriculum. *International Journal of Science Education, 29*(6), 731–746.

Chu, H., Shin, D., & Lee, M. (2005). Chapter 33: Korean students' environmental literacy and variables affecting environmental literacy. In S. Wooltorton & D. Marinova (Eds.), *Sharing wisdom for our future. Environmental education in action: Proceedings of the 2006 Conference of the Australian Association of Environmental Education*. Bellingen, Australia: Australian Association for Environmental Education.

Cortes, L. (1987). A survey of the environmental knowledge, comprehension, responsibility, and interest of the secondary level students and teachers in the Philippines (Doctoral dissertation, Michigan State University, 1986). *Dissertation Abstracts International, 47*(7), 2529A. (UMI No. DET87–13278)

Courtenay-Hall, P., & Rogers, L. (2002). Gaps in mind: Problems in environment knowledge-behavior modeling research. *Environmental Education Research, 8*(3), 283–297.

Disinger, J. (1983). *Environmental education's definitional problem* (ERIC Information Bulletin #2). Columbus, OH: ERIC Science, Mathematics, and Environmental Education Clearinghouse.

Erdogan, M. (2009). *Fifth grade students' environmental literacy and the factors affecting students' environmentally responsible behaviors* (Unpublished doctoral dissertation). Middle East Technical University, Turkey.

Erdogan, M., Kostova, Z., & Marcinkowski, T. (2009). Components of environmental literacy in elementary science education curriculum in Bulgaria and Turkey. *Eurasia Journal of Mathematics, Science, and Technology Education, 5*(1), 15–26.

Erdogan, M., Marcinkowski, T., & Ok, A. (2009). Content analysis of K–8 environmental education research studies in Turkey, 1997–2007. *Environmental Education Research, 15*(5), 525–548.

Erdogan, M., & Ok, A. (2008). Environmental literacy assessment of Turkish children: The effects of background variables. In I. H. Mirici, M. M. Arslan, B. A. Ataç, & I. Kovalcikova (Eds.), *Creating a global culture of peace: Strategies for curriculum development and implementation* (Vol. 1, pp. 214–227). Antalya, Turkey: Anıttepe Publishing.

Eyers, V. (1976). Environmental knowledge and beliefs among grade ten students in Australia (Doctoral dissertation, Oregon State University, 1976). *Dissertation Abstracts International, 36*(19), 6626A. (UMI No. 76–07747)

Goldman, D., Yavetz, B., & Pe'er, S. (2006). Environmental literacy in teacher training in Israel: Environmental behavior of new students. *The Journal of Environmental Education, 38*(1), 3–20.

Gustafson, J. (Ed.). (1983). *The first national congress for environmental education futures: policies and practices*. Columbus, OH: ERIC Science, Mathematics, and Environmental Education Clearinghouse.

Hart, E. (1981). Identification of key characteristics of environmental education. *The Journal of Environmental Education, 13* (1), 12–16.

Hart, P. (1980). Environmental education: Identification of key characteristics and a design for curriculum organization. (Doctoral dissertation, Simon Fraser University). *Dissertation Abstracts International, 40*(9), 4985–A.

Harvey, G. (1977a). A conceptualization of environmental education. In J. Aldrich, A. Blackburn, & G. Abel (Eds.), *A report on the north american regional seminar on environmental education* (pp. 66–72). Columbus, OH: ERIC Clearinghouse for Science, Mathematics, and Environmental Education.

Harvey, G. (1977b). Environmental education: A delineation of substantive structure (Doctoral dissertation, Southern Illinois University, 1976). *Dissertation Abstracts International, 38*(2), 611A. (UMI No. 77–16622)

Hines, J. (1985). An analysis and synthesis of research on responsible environmental behavior (Doctoral dissertation, Southern Illinois University at Carbondale, 1984). *Dissertation Abstracts International, 46*(3), 665A. (UMI No. DER85–10027)

Hines, J., Hungerford, H., & Tomera, A. (1987). An analysis and synthesis of research on responsible environmental behavior. *The Journal of Environmental Education, 18*(2), 1–8. (Original work published 1986).

Hoffmann, J. (1998). Identification of the critical elements of environmental literacy: A Delphi study (Doctoral dissertation, Texas A&M University–Commerce, 1997). *Dissertation Abstracts International, 58*(5), 4118B. (UMI No. DA9806300)

Hsu, S. (1997). An assessment of environmental literacy and analysis of predictors of responsible environmental behavior held by secondary teachers in Hualien county of Taiwan. (Doctoral dissertation, The Ohio State University, 1997). *Dissertation Abstracts International, 58*(5), 1646A. (UMI No. DA9731641)

Hsu, S., & Roth, R. (1998). An assessment of environmental literacy and analysis of predictors of responsible environmental behavior held by secondary teachers in the Hualien area of Taiwan. *Environmental Education Research, 4*(3), 229–249.

Hungerford, H. (1975). The myths of environmental education. *The Journal of Environmental Education, 7*(3), 21–26.

Hungerford, H., Peyton, R., & Wilke, R. (1980). Goals for curriculum development in environmental education. *The Journal of Environmental Education, 11*(3), 42–47.

Hungerford, H., & Volk, T. (1990). Changing learner behavior through environmental education. *The Journal of Environmental Education, 21*(3), 8–22.

Iozzi, L. (Ed.). (1981). *Research in environmental education, 1971–1980*. Columbus, OH: ERIC/SMEAC. (ERIC Document No. 214 762)

Iozzi, L. (Ed.). (1984). A summary of research in environmental education, 1971–1982. The second report of the National Commission on Environmental Education Research. *Monographs in Environmental Education and Environmental Studies, Volume 2*. Columbus, OH: ERIC Science, Mathematics, and Environmental Education Clearinghouse. (ERIC Document Reproduction Service No. ED259879)

Kollmuss, A., & Ageyman, J. (2002). Mind the gap: Why do people act environmentally and what are the barriers to pro–environmental behavior? *Environmental Education Research, 8*(3), 239–260.

Korean Ministry of Education. (1997). *The seventh school curriculum of the republic of Korea*. Seoul, Korea: Korean Ministry of Education Publication Service.

Korean Ministry of Education. (2007). *The revised seventh curriculum: Science*. Seoul, Korea: Korean Ministry of Education Publication Service.

Kuhlmeier, H., Van Den Bergh, H., & Lagerweij, N. (2005). Environmental knowledge, attitudes, and behavior in Dutch secondary education. *The Journal of Environmental Education, 30*(2), 4–14.

Makki, M., Abd–El–Khalick, E., & Boujaoude, S. (2003). Lebanese secondary school students' environmental knowledge and attitudes. *Environmental Education Research, 9*(1), 21–33.

Marcinkowski, T. (1998). Predictors of responsible environmental behavior: A review of three dissertation studies. In H. Hungerford, W. Bluhm, T. Volk, & J. Ramsey (Eds.), *Essential readings in environmental education* (pp. 247–276). Champaign, IL: Stipes Publishing.

Marcinkowski, T., & Mrazek, R. (Eds.). (1996). *Research in environmental education, 1981–1990*. Troy, OH: NAAEE.

Marcinkowski, T., & Rehrig, L. (1995). The secondary school report: A final report on the development, pilot testing, validation, and field testing of The Secondary School Environmental Literacy Assessment Instrument. In R. Wilke (Ed.), *Environmental education literacy/needs assessment project: Assessing environmental literacy of students and environmental education needs of teachers; Final Report for 1993–1995* (pp. 30–76). (Report to NCEET/University of Michigan under US EPA Grant #NT901935–01–2). Stevens Point, WI: University of Wisconsin—Stevens Point.

McBeth, W. (1997). A historical description of the development of an instrument to assess the environmental literacy of middle school students (Doctoral dissertation, So. Illinois University at Carbondale, 1997). *Dissertation Abstracts International, 58*(36), 2143–A. (UMI No. DA9738060)

McBeth, W., Hungerford, H., Marcinkowski, T., Volk, T., & Meyers, R. (2008). *National Environmental Literacy Assessment Project: Year 1, National baseline study of middle grades students. Final report.* (Report to the US Environmental Protection Agency, National Oceanic and Atmospheric Administration, and North American Association for Environmental Education under Grant #NA06SEC4690009). Retrieved from http://www.oesd.noaa.gov/NAEE_Report/Final_NELA%20minus%20MSELS_8–12–08.pdf

Moseley, C. (1994). Effects of a residential environmental science academy on the environmental literacy of eleventh and twelfth grade students. (Doctoral dissertation, Oklahoma State University, 1993). *Dissertation Abstracts International, 54*(10), 5081B. (UMI No. DA9407266)

National Environmental Education Advisory Council. (2005). *Setting the standard, measuring results, and celebrating successes. A report to congress on the status of environmental education in the United States* (EPA 240–R–05–001). Washington, DC: US Environmental Protection Agency.

Ndayitwayeko, A. (1995). Assessment and comparison of environmental knowledge and attitudes held by thirteenth grade general and technical education students in the republic of Burundi (Doctoral dissertation, The Ohio State University, 1994). *Dissertation Abstracts International, 55*(10), 3080A. (UMI No. DA9505265)

Negev, M., Garb, Y., Biller, R., Sagy, G., & Tal, A. (2010). Environmental problems, causes, and solutions: An open question. *The Journal of Environmental Education, 41(2), 101–115.*

Negev, M., Sagy, G., Garb, Y., Salzberg, A., & Tal, A. (2008). Evaluating the environmental literacy of Israeli elementary and high school students. *The Journal of Environmental Education, 39*(2), 3–20.

Nelson, W. (1997). Environmental literacy and residential outdoor education programs (Doctoral dissertation, University of La Verne, 1996). *Dissertation Abstracts International, 57*(10), 4314A. (UMI No. DA9708891)

Olli, E., Grendstad, G., & Wollebaek, D. (2001). Correlates of environmental behavior: Bringing back social context. *Environment and Behavior, 33*(2), 181–208.

Orenstein, D. (2004). Population growth and environmental impact: Ideology and academic discourse in Israel. *Population and Environment, 26*(1), 41–60.

Osbaldiston, R. (2004). Meta–analysis of the responsible environmental behavior literature (Doctoral dissertation, University of Missouri—Columbia, 2004). *Dissertation Abstracts International, 65*(8), 4340B. (UMI No. AAT 3144447)

Peled, E. (2010). *Developing indicators of environmental literacy among elementary and high school pupils is Israel* (Unpublished master's thesis). Technion–Israel institute of Technology, Haifa, Israel.

Perkes, A. (1974). A survey of environmental knowledge and attitudes of tenth–grade and twelfth–grade students from five Great Lakes and six far western states (Doctoral dissertation, Ohio State University, 1973). *Dissertation Abstracts International, 34*(8), 4914A. (UMI No. 74–3287)

Programme for International Student Assessment. (2008). *Green at Fifteen? How Fifteen–year–olds perform in environmental science and geoscience in PISA 2006.* Paris, France: Organisation for Economic Co–Operation and Development.

Richmond, J. (1977). A survey of the environmental knowledge and attitudes of fifth year students in England (Doctoral dissertation, The Ohio State University, 1976). *Dissertation Abstracts International, 37*(8), 5016A. (UMI No. 77–02484)

Rickinson, M. (2001). Special Issue: Learners and learning in environmental education: A critical review of the evidence. *Environmental Education Research, 7*(3), 208–320.

Roth, C. (1992). *Environmental literacy: Its roots, evolution, and directions in the 1990s.* Columbus, OH: ERIC Science, Mathematics, and Environmental Education Clearinghouse.

Roth, R. (1976). *A review of research related to environmental education, 1973–1976.* Columbus, OH: ERIC Science, Mathematics, and Environmental Education Clearinghouse. (ERIC Document Reproduction Service No. ED135647)

Roth, R., & Helgeson, S. (1972). *A review of research related to environmental education.* Columbus, OH: ERIC Science, Mathematics, and Environmental Education Clearinghouse. (ERIC Document Reproduction Service No. ED068359)

Sagy, G., Negev, M., Garb, Y., Salzberg A., Tal, A. (2008a). Environmental literacy: Results of a national survey in the Israeli education system (Hebrew). *Studies of the Management of Nature and Environmental Resources,* 6, 50–73.

Sagy, G., Negev, M., Garb, Y., Salzberg A., Tal, A. (2008). Trends of environmental education in Israel (Hebrew). *Studies of the Management of Nature and Environmental Resources, 6,* 35–49.

Saunders, G., Hungerford, H., & Volk, T. (1992). *Research needs in environmental education: A Delphi assessment. Summary Report.* Unpublished manuscript.

Schmeider, A. (1977). 1: The nature and philosophy of environmental education: Goals and objectives. In J. Aldrich & A. Blackburn (Eds.), *Trends in environmental education* (pp. 23–34). Paris, France: UNESCO.

Schoenfeld, C. (1969). What's new about environmental education? *Environmental Education, 1*(1), 1–4.

Scott, W., & Gough, S. (2003). Categorizing environmental learning. *NAAEE Communicator, 33*(1), 8.

Shin, D. (2001). Earth science in the perspectives of environmental education. *Journal of Korean Earth Science Society, 22*(2), 147–158.

Shin, D., Chu, H., Lee, E., Ko, H., Lee, M., Kang, K., . . . Park, J. (2005). An assessment of Korean students' environmental literacy. *Journal of the Korean Earth Science Society, 26*(4), 358–364.

Shin, D., Lee, Y., Lee, K., Lee, E., & Lee, K. (2005). Identification of the future–oriented earth science education placing on the "earth environment." *Journal of the Korean Association of Research in Science Education, 25*(1), 239–259.

Shin, D., & Noh, K. (2002). Korean students' scientific literacy. *Journal of the Association of Research in Science Education, 22*(1), 76–92.

Sia, A., Hungerford, H., & Tomera, A. (1986). Selected predictors of responsible environmental behavior: An analysis. *The Journal of Environmental Education, 17*(2), 31–40. (Original work published 1985).

Simmons, D. (1995). Working paper #2: Developing a framework for national environmental education standards. In *Papers on the development of environmental education standards* (pp. 10–58). Troy, OH: North American Association for Environmental Education.

Simmons, D. (1998). Education reform, setting standards, and environmental education. In H. Hungerford, W. Bluhm, T. Volk, & J. Ramsey (Eds.). *Essential readings in environmental education* (pp. 65–72). Champaign, IL: Stipes Publishing.

Sivek, D., & Hungerford, H. (1990). Predictors of responsible behavior in members of three Wisconsin conservation organizations. *The Journal of Environmental Education, 21*(2), 35–40. (Original work published 1989).

Stapp, W. (1974). Historical setting of environmental education. In J. Swan & W. Stapp (Eds.), *Environmental education* (pp. 42–49). New York, NY: J. Wiley and Sons.

Stapp, W. (Ed.). (1978). *From ought to action in environmental education: A report of the national leadership conference on environmental education.* Columbus, OH: ERIC Science, Mathematics, and Environmental Education Clearinghouse.

Stapp, W., et al. (1969). The concept of environmental education. *Environmental Education, 1*(1), 30–31.

Swan, M. (1984). Forerunners of environmental education. In N. McInniss & D. Albrecht (Eds.), *What makes education environmental?* (pp. 4–20). Medford, NJ: Plexus Publishing.

Tal, A. (2002). *Pollution in a promised land.* Berkeley, CA: University of California Press.

Tal, A., Garb, Y., Negev, M., Sagy, G., & Salzberg, A. (2007). *Environmental literacy in Israel's education system (Hebrew).* Retrieved from http://storage.cet. ac.il/CetForums/Storage/MessageFiles/7143/68938/Forum68938M89I0.pdf

Tanner, R. (1974). *Ecology, environment, and education.* Lincoln, NE: Professional Educator's Publications.

Todt, D. (1996). An investigation of the environmental literacy of teachers in south–central Ohio using the Wisconsin environmental literacy survey, concept mapping, and interviews (Doctoral dissertation, The Ohio State University, 1995). *Dissertation Abstracts International, 56*(9), 3526A. (UMI No. DA9544703)

UNESCO. (1977). *Trends in environmental education.* Paris, France: Author.

UNESCO. (1978). *Final Report: Intergovernmental Conference on Environmental Education.* Paris, France: Author

UNESCO. (1992). *Agenda 21: The United Nations programme of action from Rio.* New York, NY: United Nations Publications.

Volk, T., & McBeth, W. (1997). *Environmental literacy in the United States* (Report Funded by the US Environmental Protection Agency, and Submitted to the Environmental Education and Training Partnership, North American Association for Environmental Education). Washington, DC: North American Association for Environmental Education.

Wilke, R. (1990). Research in EE: Research and evaluation needs and priorities. *The [NAAEE] Environmental Communicator*, 6.

Wilke, R. (Ed.). (1995). *Environmental education literacy/needs assessment project: Assessing environmental literacy of students and environmental education needs of teachers; Final Report for 1993–1995.* (Report to NCEET/University of Michigan under US EPA Grant #NT901935–01–2). Stevens Point, WI: University of Wisconsin—Stevens Point.

Willis, A. (1999). A survey of the environmental literacy of high school junior and senior science students from a southeast Texas school district (Doctoral dissertation, University of Houston). *Dissertation Abstracts International, 60*(5), 1506A. (UMI No. DA9929299)

Wisconsin Environmental Education Board. (1997). *Are we walking the talk?* Stevens Point, WI: Wisconsin Center for Environmental Education.

Zelezny, L. (1999). Educational interventions that improve environmental behaviors: A meta–analysis. *The Journal of Environmental Education, 31*(1), 5–14.

32

Geospatial Technologies

The Present and Future Roles of Emerging Technologies in Environmental Education

Michael Barnett
Boston College, USA

James G. MaKinster
Hobart and William Smith Colleges, USA

Nancy M. Trautmann
Cornell University, USA

Meredith Houle Vaughn
San Diego State University, USA

Sheron Mark
Loyola Marymount University, USA

Introduction

The ability to use geospatial technologies to explore and analyze the world is no longer isolated to a few skilled scientists and researchers. Rather, such technologies are now available to nearly everyone. Over the past decade, consumer demand has skyrocketed for ways to manipulate and display geospatial information using global positioning systems (GPS) and geographic information systems (GIS) (Folger, 2008). For example, the integration of GPS data with digital maps has led to handheld and dashboard navigation devices used daily by millions of people worldwide. The release of Google Earth in 2005 made it possible for people from all walks of life to manipulate digital maps and geospatial data (Folger, 2008). The ability to swiftly and dynamically represent Earth's geography and scientific, social, political, economic, and environmental issues from a variety of perspectives creates powerful opportunities for teachers and students. Geospatial tools expand the scope of topics that students can explore, promote interdisciplinary learning, and change the way that students learn to reason about and interpret data (Audet & Abegg, 1996). In other words, anything that can be referenced to a specific geographic location becomes a candidate for investigation (Ramamurthy, 2006).

This chapter provides a snapshot of the current state of geospatial technologies and their use and impact in educational environments. The chapter is divided into several sections. First, given the rapid and innovative growth of this field, this chapter begins with a definition of terms and a description of the tools that comprise the field of geospatial technologies. This is followed by a discussion of why educators believe that geospatial technologies have so much potential for revolutionizing education, particularly environmental education. Next, a review and critique of the research is presented, focusing on the current status of research on teacher professional development and on student outcomes. The chapter closes with a discussion of future directions for both research and geospatial educational technology development.

What Is Geospatial Technology?

What Is Geospatial Technology? *Geospatial technology* includes geographic information systems (GIS), global positioning systems (GPS), virtual globes (Google Earth, ArcGIS Explorer, WorldWind, etc.), and web-based resources for the representation of geographic data. Geospatial technology is situated at the intersection between information technology and communication technology, using computer-generated maps to synthesize imagery data into compelling stories or evidence. Relatively recent advances in software and a significant push by software developers and vendors have dramatically increased the availability of these tools for use in K-12 settings. Geospatial technologies enable students to collect spatial data, create maps, and access much of the same data used by scientific and geographic professionals. Increasingly, such technology is becoming part of our daily lives for driving, public transportation, delivery services, weather, and an endless number of applications in science, business, and industry. Another significant contribution of geospatial technologies to science and environmental education is the availability of geo-referenced data that describe physical environments, natural resources, and socioeconomic activities associated with locations or regions. These geo-referenced data provide users with the opportunity for deep and interdisciplinary exploration. As a result, many educators believe that geospatial technologies have unique potential in science, environmental science, and environmental studies as they allow students to examine a particular location or place from a variety of perspectives with relative ease and to ask rather sophisticated questions regarding that place; something that was challenging or impossible just a few years ago.

Geographic Information Systems are one category of geospatial tools that can be used in teaching and learning across a range of subjects in the K-12 curriculum, but they are particularly well suited for studying environmental systems. A GIS is an integrated system of hardware and software designed to manage, manipulate, analyze, model, and display geospatial data. It can be used to solve important problems such as urban planning, development, and construction, determining the impact of a flood or a fire on humans, and identifying changes in land use. The power of a GIS lies in it its ability to combine geospatial information in unique ways—by layers or themes—and extract new ways of analyzing, synthesizing, or portraying data. For instance, a GIS analysis including the locations of highway intersections and the traffic flow they support would allow users to extract information enabling them to determine where to locate a business or for predicting the areas that are most likely to have poor air quality.

Most people consider *GIS mapping software* to be at the core of geospatial technology. ArcGIS from the Environmental Systems Research Institute (ESRI) is the industry leader in terms of desktop and server-based GIS software. Others include Idrisi, which integrates GIS and image processing more generally (Clark Labs, 2009) and Mapinfo, which is able to overlay maps and images from a variety of different sources and in different projections (Pitney Bowes, 2008). These GIS software enable users to overlay images and data that are "projected" in the same manner for proper alignment of geographic features or data points. At a basic level, GIS software enable users to turn data layers on and off, change the symbology within a data layer, and create customized maps. More advanced capabilities are offered by software packages such as ArcGIS, with an almost endless capacity to analyze data, represent changes over time, and model potential outcomes. A variety of books document use of this software to address issues in the social sciences, natural resources, and economics (Malone, Palmer, & Voigt, 2002; Milson & Alibrandi, 2008).

Global Positioning Systems can be used to accurately determine positions on the surface of the earth. They are most often thought about from the perspective of the handheld or portable units that we use to determine our location. These portable units rely on twenty-seven NAVSTAR satellites that orbit the earth (Steede-Terry, 2000). Using radio signals and a process known as trilateration, these GPS units determine the latitude and longitude of their current location. This technological system, including both hardware and software, enables professionals and amateurs to pinpoint geographic sites or features with considerable accuracy.

*Virtual Globes (*Google Earth, ArcGIS Explorer, Virtual 3D Earth, and WorldWind), have made a significant amount of geospatial data readily and freely available. These software packages enable users to manipulate and explore interactive representations of a three-dimensional Earth and the Earth's surface. Google Earth, the most popular of these visualization tools, features learning resources that are accessible through the software itself, within the Google Earth Community website (http://bbs.keyhole.com/ubb/), and through sharing of Google Earth files among users (.kml or .kmz files). The technological capacity of the virtual globe software continues to increase rapidly. For example, each of these programs are now able to easily integrate data layers created within GIS software. Something that was not possible two years ago.

In recent years, a number of exciting new software options have emerged that support widespread educational use of geospatial technologies. These include My World GIS (http://www.myworldgis.org/), Digital Worlds (http://www.esriuk.com/schools/) , CommunityViz (http://www.communityviz.com/), and National Geographic's Fieldscope (http://www.fieldscope.us/). Each of these customized tools combine the visualization and analytic capacities of the various geospatial software discussed above in ways that facilitate the use and analysis of relevant data (e.g., Edelson, Smith, & Brown, 2008)

Why Use Geospatial Technologies?

Berdarnz (2004) put forth three primary justifications for incorporating GIS into K-12 education. Even though her argument was targeted specifically at the integration of GIS, we feel that it is applicable to geospatial technologies more generally. Bednarz's (2004) argument included: (1) an educative justification based on GIS and the teaching and learning of geography and environmental education, (2) a workforce, and (3) a place-based learning justification. In our reading of the research we have added an additional justification, a scientific justification. We have folded this discussion into the workforce discussion because much of the discussion at the national and international level is on how to better prepare a scientifically literate workforce and as such the discussion of a scientific justification fit nicely in the workforce discussion. Using this framework as a conceptual lens, we examine each of these justifications for the use of geospatial tools in education.

Educative Justification Accompanying the rapid growth in development and use of geospatial technologies is an increase in knowledge regarding how to support and engage students in doing science (Bransford, Brown, & Cocking, 1999). In the recent document, *Taking Science to School* (National Research Council, 2008), strongly argued that in the doing of science the process of doing science and the content of science are deeply interconnected. This relationship between process and content is central to the educational research and application of geospatial technologies (T. Lee, Quinn, & Duke, 2006; National Council for Geographic Education, 2006; NSF Task Force on Cyberlearning, 2008; Wang, 2008). In fact, geospatial and environmental educators have been quick to recognize that the pedagogical power of geospatial technologies supports the arguments put forth in *Taking Science to School* regarding how to best engage students in scientific inquiry (McInerney, 2006). For example, using geospatial technologies in environmental education facilitates the process of scientific problem solving through the exploration of real world contexts and projects, providing students with opportunities to see the utility of such technologies in all phases of the scientific process. This is consistent with assertions by many scholars that if students are to develop rich, meaningful, and useable understandings of scientific phenomena, as well as the process of scientific knowledge construction, they need to be engaged in a scientific community of practice in which they are doing science in real world settings with tools similar to those being used by scientists and other professionals (S. A. Barab, Hay, Barnett, & Keating, 2000; Lave & Wenger, 1991).

Building on these perspectives, geospatial technologies have significant potential to engage students in locally relevant, interdisciplinary study of phenomena with direct impact in their daily lives (Barnett et al., 2006). For example, students can use geospatial technologies not only to map data and explore digital representations of Earth's surface, but also to visually explore relationships between various types of environmental and social data. By overlaying different data layers, students can identify areas of environmental concern in their own communities, such as steep slopes vulnerable to erosion, or areas of greatest benefit, such as habitat for a threatened or endangered species. Students also can examine relationships between biological and social variables. For example, in urban areas they might consider comparing the health of trees, shrubs, and other vegetation with neighborhood income levels. Such relationships can be explored not only visually, but by using the analytic capabilities of a GIS. "Queries" enable a user to filter a specific data set or create a new data layer based on the intersection of two or more layers. Such queries enable students to identify hidden patterns or difficult to see correlations in their data. In essence, geospatial technologies provide an ideal tool for students to use when visualizing data, posing their investigation, and addressing specific questions. These process skills are critically important as being able to visualize and manipulate data can increase students' ability to transfer new knowledge to novel situations (Bransford et al., 1999) and can help students understand complex science concepts and phenomena (Gordin & Pea, 1998).

Scientific and Workforce Justification In 2004, the US Secretary of Labor identified geospatial technologies as one of the three "most important emerging and evolving fields" for the future US workforce (with nanotechnology and biotechnology being the others). This recognition was due to geospatial technologies, particularly GIS, becoming indispensable tools for geoscientific exploration, commerce, and decision making in environmental and social sciences. For example, GIS is used by scientists to examine complex environmental problems including ground-level ozone, soil erosion, and sea-level rise due to global warming. It is also used in urban planning, in delineation of ecologically sensitive areas, and in monitoring of air, water, and land resources. A major reason for the growing use of GIS is the convenient framework that it provides for multidisciplinary analysis and synthesis, key aspects in exploring the frontiers of science. GIS has become a critical tool in environmental science because it involves a broad array of computer tools for mapping and managing geographically referenced data and for spatial analysis. As the demand for innovative GIS applications, services, and expertize grows, GIS will continue to play a pivotal role in shaping the cyber-infrastructure and data services in the environmental science industry.

Keeping pace with the demands of the US workforce requires integrating greater integration of these new and powerful tools into K-12 schools and other educational environments. At the same time, a similar transformation is needed in terms of the nature and type of knowledge and skills acquired by students in preparation for their future roles in the workplace. This need was documented nearly twenty years ago in a Department of Labor report (SCANS, 1991) and emphasized in the more recent Partnership for Twenty-first Century Skills (P2CS, n.d.). Both of these documents argue that a broad range of integrated skills in information technology, global business, teamwork, and creative entrepreneurship are needed in the workplace. In fact, Jenkins (2006) argues that youth should no longer be viewed as spectators or even consumers of information—rather, they are active participants and producers of knowledge, contributing to and shaping information through the use of information technology tools and resources. These types of knowledge-producing skills are supported by many of the educational projects that use geospatial technologies to engage students in the production and sharing of original questions with policy makers (Mitchell, Borden, & Schmidtlein, 2008).

Applying geospatial technology to environmental science projects enables students to explore the same issues and use the same tools as professional scientists, engineers, and resource managers (Ramamurthy, 2006). When students address real environmental problems, they take pride in doing something with value beyond their classroom and they have opportunities to observe, implement, and reflect on analytical approaches used in professional contexts (SERB, 2000). Middle and early high school students are ready to develop key knowledge, skills, attitudes, and awareness that provide the foundation for future careers. This does not necessitate adding new topics into over-packed curricula. Instead, a more effective approach is to integrate career awareness experiences into existing coursework through relevant projects in which students learn firsthand how science and technology apply to relevant, contemporary, real-world issues (Hurd, 2000; Kerka, 2000).

Place-Based Education Justification This final justification is particularly important for environmental educators, as place-based education has become a commonly used framework for thinking about teaching and learning in environmental education (e.g., Smith, 2007). As we have seen, the use of geospatial technologies affords teachers and students with an interdisciplinary approach to science that combines the power of science, *as a way of knowing*, with the direct impact of active learning, and being of service within the local community (Berkowitz, Nilon, & Hollweg, 2003). The ability to critically examine a specific place or location from a range of perspectives using geospatial technologies enables students to ask and answer many questions regarding their community such as: "How is land used in my area?," "How many supermarkets are within walking distance?," "Are there specific areas in my neighborhood where ozone concentrations are particularly high?," or "Where is air pollution the worst, and is there a relationship between these locations and the amount of vegetation?" In the past, a significant amount of time was required for students to explore and answer each of these questions. With today's technologies, however, each question can be quickly explored and/or answered, allowing students to investigate deeper and more challenging questions regarding the historical, economic, social, and scientific underpinnings of the nature and structure of the area in which they live.

Coupling geospatial technology with environmental science opens the door to local and regional investigations. Engaging in such studies can improve students' performance (Lieberman & Hoody, 1998) and help them connect new concepts to prior knowledge (Carlsen, 2001). Enabling students to understand the history, nature, and ecological and environmental value of a specific place is the stated goal of many environmental education programs and has been a focus of much of the research in the environmental education research field for over the past decade. Geospatial tools provide compelling new ways to reach this goal through analysis of the patterns, appearances, and behaviors of a particular place. In the United States, most classroom-based GIS projects involve studies of local communities (Audet & Ludwig, 2000). The emphasis on community-based studies is probably best explained by the deeply held beliefs of many teachers regarding the value and wisdom of grounding learning in the local environment and in students' experiences. Classroom use of GIS facilitates greater availability of local data and possibilities to link students with GIS users in local government and business.

Research Review

The methodology employed for this review entailed a search of peer-reviewed research and practitioner-based journals, along with peer-reviewed conference proceedings in the learning sciences, teacher education, environmental education, geography education, and educational technology. Additional sources were located through previous reviews focused on the use of geospatial technologies (e.g., Kerski, 2003). This search focused on identifying empirical studies that examined the use of geospatial technologies in K-16 settings. The studies selected included qualitative or quantitative empirical data collected from teachers or students. This selection process revealed 115 studies. These studies were then summarized with regards to their content, methods, results, and interpretations (see Table 32.1). In selecting studies for this review, each study was examined for the characteristics of "good empirical research" as described by Wideen, Mayer-Smith, and Moon (1998). That is, each study was examined for a clear statement of intentions, a coherent theoretical framework, detailed information about the study and the study context, methodology, sources of data, findings, whether those data justifies the findings, conclusions that integrated the original theoretical framework, interpretations of the findings, and implications regarding how the study's findings could impact either classroom practice, student learning, or teacher professional development. In adopting this stance, it is acknowledged that the review presented here may omit valuable and creative programs using geospatial technologies. In particular, we recognize that many recent projects, particularly those funded through the National Science Foundation's Innovative Technologies Experiences for Students and Teachers program (http://itestlrc.edc.org/itest-projects) and Advanced Technological Education program (http://www.atecenters.org/

TABLE 32.1
Empirical Research Studies Using GIS in Educational Settings

Authors	Year	Program/Research	Research Goals	Target Population	Methods	Findings
Akerson and Dickinson	2003	Master's level course for K-12 teachers on environmental issues and scientific inquiry	How does GIS improve teacher understanding of scientific inquiry and the nature of science?	K-12 Teachers	Mixed methods, focus on understanding nature of science	With time and support teachers can use GIS technology to enhance their inquiry science learning
Almquist et al.	2009	Summer research institutes to provide teachers with hands-on training using geospatial technologies, authentic research experiences with scientists, and to prepare teachers to bring new skills and knowledge into their own classrooms.	How does intensive professional development improve teacher use of GIS?	K-12 Teachers	Mixed methods, surveys and interviews	Almost all participating teachers designed and implemented inquiry-based learning activities in their own classrooms
Allen	Under review	Describes a place-based introductory GIS/GPS middle school curriculum unit in which students used measuring tools, GPS units, and My World GIS software to collect physical and spatial data of trees to create a Schoolyard Tree Inventory.	In what ways do the use of GIS tools enhance students' spatial awareness?	K-12 Students	Quantitative through the use of prepost exams	A statistically significant increase in students' spatial awareness was documented. A technology-based curriculum can significantly increase students' spatial awareness especially in a place and context relevant to each student.
Audet, Abbeg	1996	A research program to evaluate the differences between expert and novice GIS users.	In what ways can GIS learning experiences support problem solving skills?	K-12 Students, teachers, and GIS professionals	Qualitative methods including classroom observations, think-along transcripts and keystroke	GIS is a technology that supports the development of problem solving skills
Audet, Paris	1997	Report of in-depth case studies of programs recognized by professionals in the GIS community for their effective integration of GIs	What are the challenges and successes of implementation GIS in secondary education classrooms?	K-12 Students	Mixed methods using quantitative through the use of questionnaires	Successful GIS projects required time and significant use in teachers' classrooms. Developed the GIS readiness survey.
Baker, White	2003	Implementation of problem-based learning unit for eighth grade earth science.	In what ways does a GIS supported PBL unit improve student learning?	K-12 Students	Mixed methods through the use of a pretest, treatment, and posttest design. This affective component measured science and technology attitudes and self-efficacy.	GIS was shown to modestly improve integrated science process skills, especially data analysis (geographic and mathematical) activities. GIS was particularly suited to the exploration and analysis of large datasets collected by a great number of students.

(*Continued*)

TABLE 32.1 (Continued)

Authors	Year	Program/Research	Research Goals	Target Population	Methods	Findings
Barnett, Houle, Strauss	2008	Inner city high school program using GIS and computational modeling to help students understand urban ecology	In what was does a GIS enhanced environment support student understanding of content?	K-12 Students	Mixed methods through the use of prepost content exams and interviews	Computer-based modeling coupled with GIS improves students understanding of content.
Beckett, Shaffer	2005	A research study focused on developing students' understanding the ecology through participation in a technology-supported urban planning simulation	Did participants develop an understanding of the domain of ecology through participation in a simulation modeled on the real world tools and practices of urban planners?	K-12 Students	Qualitative in nature through observations and prepost test design	Students developed a deeper understanding of urban ecology and better understanding of how technology is used in urban planning
Bednarz, Audet	1999	A research project to assess the status of geographic information systems (GIS) training in preservice teacher education in the United States	What is the status of GIS in teacher preparation programs?	Preservice teachers, college	Qualitative using A semi-structured format with both open-ended and constructed interview questions and a survey of teacher educators	Teacher educators found it difficult to justify the inclusion of GIs in the preservice curriculum, No exemplary models for integrating GIS into preservice teacher preparation programs exist
Bodzin	2008	Using Google Earth as a tool in an afterschool science club to explore and learning about watersheds	What is the impact of using geospatial technologies when coupled with an outdoor learning environment on elementary student affective and learning outcomes?	K-12, focus on elementary	Mixed methods using surveys and observations	Analysis of the student artifacts revealed that participation in the long-term pond investigation promoted a sense of environmental stewardship and fostered responsible environmental behavior
Doering, Veletsianos	2007	Two specific geospatial technologies were integrated within an adventure learning curriculum-ArcExplorer Java Educationfor Educators (AEJEE) and Google Earth.	To better understand student experiences when learning geography when using geospatial technologies within their classroom while utilizing real-time authentic data from an adventure learning environment.	K-12 (middle school)	Qualitative data from seven student focus groups	Five major findings were reported: I. Assisted learners in developing a sense of place through the use of authentic preexisting data. II. Assisted learners in developing a sense of place through the use of authentic newly acquired data. III. Provided opportunities for the co-construction of knowledge through learner-created data. IV. Motivated students to explore geographic locations through ease of use. V. Assisted learners in understanding the geography of the region studied through the use of newly acquired data.

Authors	Year	Program/Research	Research Goals	Target Population	Methods	Findings
Hall-Wallace, McAullife	2002	Report on the development and implementation of investigations of student learning from a suite of GIS-based activities on plate tectonics and geologic hazards taught in a large introductory course for nonscience majors.	Can students' content knowledge be improved through engagement in GIS-based activities and is there a relationship between spatial thinking and GIS —based learning?	College	Mixed methods using classroom observations, student interviews and surveys, pre and posttests of knowledge, and measures of spatial skills	Improved content understanding and a positive correlations between students' spatial ability and performance on both the content exams.
Kerski	2003	Reports on the implementation and use of GIS in secondary classrooms as of 2003.	What is the status of GIS use in secondary classrooms?	K-12, high school focus	Quantitative, Survey	Teaching with GIS provides the opportunity for issues-based, student-centered, standards-based, inquiry oriented education, but its effectiveness is limited primarily by social and structural barriers such as lack of technology and teacher time.
Lam, Lai, Wong	2009	Research on the use of GIS in secondary geography curriculum in Hong Kong	What are teachers' view of including GIS in their curriculum?	K-12, high school	Qualitative, teacher interviews	Level of use of GIS in geography teaching in Hong Kong was low, despite the fact that teachers generally found GIS to be an important geographical skill. In addition, teachers' understanding of how GIS could be used in teaching to enhance student learning in geography was limited
J. Lee, Bednarz	2009	Describes the impact on spatial reasoning skills in a college level cartography course.	How and in what ways does GIS learning opportunities improve student spatial skills?	K-16, college	Quantitative, prepost exams	GIS learning helped students think spatially. These improvements were the result of the connection between students' GIS activities and experiences and the tested spatial thinking skill.
Giordano, Lu, Anderson, Fonstad	2007	Describes a capstone course for undergraduate students who were provided the opportunity to use GIS.	What are the geographical principles that are learned through application of a GIS in a cooperative environment?	Undergraduate	Qualitative with a reliance upon self-reports	Students generally found the course experience to be valuable.
MaKinster, Trautmann	In press	Implemented and researched a three-year professional development project focused solely on teaching science using geospatial technology.	How to improve and support teachers in implementing geospatial technologies in their classroom across disciplines?	In-Service Teachers	Qualitative, Case Studies	Teachers' perceived skills, knowledge, and self confidence with regard to integrating geospatial technology into their science teaching
Milson, Earle	2007	Explored the use of Internet-based geographic information systems (IGIS). Students used IGIS resources to explore different geographical regions.	What is the impact of using Internet based GIS tools to support student geographical learning?	K-12 Students	qualitative case study with additional data from focus groups	IGIS projects can lead to gains in students' cultural awareness and empathy for distant others.

(Continued)

TABLE 32.1 (Continued)

Authors	Year	Program/Research	Research Goals	Target Population	Methods	Findings
McClurg, Buss	2007	Describes a professional development experience of fifth to twelfth grade teachers in using geographic information systems (GIS) and global positioning systems (GPS) technologies to enhance classroom teaching and learning environments	What are the characteristics of effective professional development programs that focus on the use of GIS?	In-service teachers	Mixed methods with a focus on the use of surveys	Developed a set of guidelines for effective professional development for teachers who want to implement GIS projects in their classrooms.
Doering, Velestianos	2007	Reports on the use of adventure learning (AL)approach is a hybrid distance education approach that engaged students real-world issues through authentic learning experiences	What is the impact of using geospatial technologies such as Google Earth and ArcExplorer on improving students understanding an interpretation of data?	K-12	Qualitative data from the use of focus groups	Students found the coupling of real data collection and geospatial tools motivating and students sense of place improved.
Patterson, Reeve, Page	2003	Reports on the integration of GIS into an advanced placement classroom.	Whether exposure to GIS technologies would aid students in the learning of geographic principles?	K-12	Quantitative through the use of a prepost exam	Exposure to GIS does not improve geographic learning
Sasaki, et al.	2008	Examined the deployment of GIS education across the Japanese higher education system.	What are the challenges of successful implementation of GIS in higher education course?	Undergraduates	Quantitative through the use of surveys	Most GIS was used in lecture based settings with lab time devoted to GIS
Shinn	2006	Implementation of a learning project in a fourth grade classroom through the use of GIS technology	In what ways does GIS improve student geographic content knowledge and map skills?	K-5, elementary	Mixed methods using sketch maps and prepost content knowledge in geography and their map skills	Students showed gradual progress on gaining geographic content knowledge, the overall improvement on their map skills was rather rapid and sporadic
Tulloch, Graff	2007	describes a series of data-based Green Map learning exercises positioned within a problem based framework	What is the impact of using green mapping on students understanding of geography?	K-16	Qualitative through the use of case studies	Green Mapping supported students in learning about their surroundings and enhanced their appreciation regarding the natural resources around them.
West	2003	This paper discusses the role of GIS in the development of higher order thinking skills and in motivating student learning.	Can and in what ways does GIS enhanced classes improve student learning and motivation?	K-12	Quantitative through the use of surveys	Attitudinal improvements were found in three of these: perceived usefulness of computers, perceived control of computers and behavioral attitude to computers

Authors	Year	Program/Research	Research Goals	Target Population	Methods	Findings
Wilder, Brinkerhoff, Higgins	2003	Investigated the effects of using a long-duration, project-based science professional development model on the acquisition of declarative knowledge and basic terminology associated with the use of geographic information technologies	Does GIS PD improve the acquisition of basic terminology and improve teachers' confidence in using geospatial technologies?	In-service teachers	Quantitative through the use of prepost surveys	Participation improved significant difference in mean knowledge and terminology and teachers confidence about their use of geospatial technology in their classrooms.
Wiegand	2003	Students engaged in the creation of Choropleth maps.	In what ways GIS improve students' ability to create and interpret Choropleth maps?	K-12 Students	Qualitative through the use of discourse analysis as students worked	Students spent a high percentage of their time talking about cartographic strategy and a low percentage of time on technical aspects of GIS functionality. However, older students engaged in more reasoning and questioning.
Wiggles worth	2003	a research project conducted to investigate the strategies developed by middle school students to solve a route-finding problem using ArcView GIS software	In what ways does GIS help students improve their ability to read and navigate using maps?	K-12 Students	Qualitative using think along scripts and the use of a logical thinking assessment	Students use several different strategies to find the best route using maps and GIS technology supported the different strategies

centers.html) have yet to reach maturity for publication of their findings. However, focusing on empirical research within this context provides a stronger basis for discussing the design, use, and potential of geospatial technologies in environmental education. Below we review first on those studies focused on teacher professional development, then those studies focused on student learning outcomes, and finally summarize the successes and challenges faced by the scholars within this field of inquiry. While the majority of this research has not occurred within the field of environmental education, all are directly relevant and many projects included a focus on environmental education, environmental science, or specific environmental issues.

Teacher Professional Development

Designing and implementing effective professional development opportunities focused on using geospatial technologies to teach science is an emerging field. Science, geography, and environmental education communities have developed sophisticated understandings of the opportunities created by GIS (e.g., Alibrandi, 2003; English & Feaster, 2003; National Research Council, 2006) and the challenges that teachers face when trying to use GIS (Baker, Palmer, & Kerski, in press; Kerski, 2001; McClurg & Buss, 2007; Trautmann & MaKinster, in press). A more pressing need is to identify characteristics of effective professional development and to create theoretical and practical models through which educators can design, implement, and evaluate such programs.

Loucks-Horsley, Love, Stiles, Mundry, and Hewson, (2003) articulated an important framework for thinking about teacher professional development more generally and identified the primary characteristics of effective professional development programs. One of the many outcomes of this work has been a shift over the past ten years from an emphasis on one-time workshops or summer institutes to involving teachers in sustained professional development either individually (reviewed in MaKinster, Trautmann, & Barnett, in press) or as part of their entire school, such as NSF's Local Systemic Change initiatives (Banilower, Heck, & Weiss, 2007; Supovitz & Turner, 2000). During this same period, educators using GIS and other geospatial technologies also began to identify specific characteristics of professional development that effect change in classroom practice (e.g., Buss, McClurg, & Dambekalns, 2002; Coulter and Polman, 2004; Jeanpierre, Oberhauser, & Freeman, 2005; McClurg & Buss, 2007; Trautmann & MaKinster, in press).

Several studies provided insight into which teachers were interested in using GIS as a tool, in general, and specifically for teaching science. Kerski (2001) surveyed 422 teachers who owned mapping software (ArcView, Idrisi, or MapInfo) in an effort to discover why and how secondary teachers were using GIS in their curricula (conducted primarily in 1999). Although these teachers came from a variety of disciplines, the majority (36 percent) were science teachers. This study revealed that GIS technology is well suited for asking scientific and environmental questions, inquiry-based learning, and constructivist views of learning. Kerski also noted that reform-minded science teachers were most likely to have incorporated GIS technology into their classroom instruction or were most interested in doing so in the future. Findings from this survey illustrated the extent to which teaching with GIS creates challenges for teachers on multiple fronts (Kerski, 2001). Teachers identified training, time, hardware access, the availability of lessons, and access to data as some of the most common challenges to using GIS in their classrooms.

Baker, Palmer, and Kerski (in press) conducted a survey in 2004 with 186 teachers who attended GIS training workshops conducted by the authors between 1998 and 2004. Again, the largest proportion of these teachers taught science (32 percent), and they represented all of the most common science disciplines and courses. Sixty-two percent of the respondents had implemented the curriculum materials used in the training, and 66 percent developed their own GIS-based lessons. Common topics included global climate, weather events, water resources, ecology, conservation, agriculture, and pollution. The authors discussed how the availability of GIS lessons had increased dramatically over the previous five years, suggesting that implementation was easier for this group than for the teachers surveyed in 1999. These surveys are useful in assessing the extent to which short-term GIS training (1–5 day workshops) has lasting impacts on teachers. Other studies have revealed the opportunities and challenges of working with teachers over an extended period of time. The research reviewed below focuses on studies with specific emphasis on the design and implementation of long-term professional development experiences focused on teaching science using geospatial technologies.

Wilder, Brinkerhoff, and Higgins (2003) examined the extent to which a long-term professional development experience for science teachers resulted in an increase in self-confidence and declarative knowledge regarding geospatial technology skills. The authors provided professional development for twenty-seven teachers, spanning two years. The teachers attended a fifteen-day summer institute, two-day institutes in November and June, a second fifteen-day summer institute, and a final two-day institute in December of the second year. They were introduced to various geospatial applications, environmental and social issues, and many of the fundamental concepts underlying geospatial investigations and technology. Problem-based science (Blumenfeld et al., 1991) served as the central framework for the teacher professional development and for the experiences that the teachers provided their students.

Evaluation data for this project demonstrated growth in teachers' declarative knowledge and self confidence with regard to teaching with these technologies. Analysis of two eleven-item knowledge and terminology surveys demonstrated significant increases in declarative knowledge during both summers. Significant increases occurred during both summers, and these gains were retained from one year to the next. The authors found no significant positive or negative change in scores on the knowledge survey from the end of the first summer compared with the end of the second summer, suggesting that the teachers had at least retained their knowledge over time. The same patterns were evident in measures of teacher self-confidence, which were evaluated using two thirteen-item surveys administered on the same schedule.

The authors did not collect formal data on the extent to which teachers were using geospatial technology in their teaching, but anecdotal evidence suggested that this was occurring. Several teachers wrote grants for geospatial technology, purchased GPS units, and reported having their students conduct investigations. Two participants initiated school-wide professional development programs related to use of geospatial technologies, and a third began development of a departmental integrated science curriculum making use of these tools. While somewhat limited in scope, this study demonstrates the value of long-term professional development.

Coulter and Polman (2004) discussed their research within a professional development experience for teachers entitled *Mapping Our Environment*. Working from a design-based research perspective (S. Barab & Squire, 2004; Brown, 1992; Sandoval, 2004; Sandoval & Bell, 2004), the authors modified the institute from one year to the next in an effort to increase teacher implementation and outcomes. The project goal was to support teachers in conducting inquiry-based environmental lessons and units that incorporated GIS. Evaluation data from the 2002–2003 showed somewhat limited classroom implementation levels by the teachers who had participated in the 2002 summer institute. The authors discussed how this institute had focused on a wide variety of skills, science topics, and activities. As a result, the teachers did not have the time or support necessary to prepare curriculum plans and resources specific to their classroom curriculum. By incorporating the *Understanding By Design* curriculum development model (Wiggins & McTighe, 2005) and enabling teachers to practice with specific hypothetical examples, the teachers participating in the 2003 summer institute implemented GIS lessons and units to a greater extent and that reflected much higher levels of inquiry. Coulter and Polman (2004) found that the "operative variable appeared to be the extent to which the teachers were supported in their efforts to translate workshop learning into professional practice." In other words, professional development needs to build teachers' ability to use GIS software, model use of lessons that integrate GIS in teaching specific topics, and provide time for teachers to develop curricula specific to their own classroom, school, and state requirements.

Buss et al. (2002) conducted a five-year study focused on identifying the experiences and support necessary for teachers to effectively integrate GIS and related geospatial technologies into their teaching. The project team focused on developing collaborative relationships among University of Wyoming scientists and science educators with K-12 teachers throughout the state. Also based on a design experiment model (Brown, 1992), each set of workshops was conducted, evaluated, and revised in a cyclical or iterative process. Based on relevant literature, practical experience, and their own research, this five-year project reflected many of the recommendations made by Loucks-Horsley et al. (2003). For example, based on evaluative data and personal observations, the authors made numerous changes to specific professional development strategies and to the overall focus of the program. Furthermore, the authors implemented the following steps based on feedback from their teachers:

1. Created noncomputer-based introductions to GIS concepts,
2. Allowed students to create their own data layers,
3. Committed more time to discussing databases and queries,
4. Addressed and discussed state and national standards,
5. Learned how to customize existing lesson plans,
6. Used a web-based discussion board to engage in ongoing conversations and as a mechanism of support,
7. Involved teachers from other districts and content areas,
8. Conducted a field investigation that modeled data collection and mapping processes, and
9. Used a book that served as a resource throughout the workshop.

These and others changes resulted in a higher proportion of teacher cohort from the second year completing the series, developing lessons, and teaching those lessons in their classrooms compared to the teachers from the first cohort. The authors demonstrated that the participating teachers grew in their technological skills and the extent to which they saw geospatial technologies as useful classroom tools.

Building on this work, McClurg and Buss (2007) presented a professional development model focused on enabling fifth through twelfth grade teachers to use GIS and GPS. The authors created an experience that reflected lessons learned from the research by Buss et al. (2002) and also embodied professional development recommendations made by the National Staff Development Council (2001):

1. Participants joined a learning community that prioritized a nexus of technology-supported learning and real world experience.
2. Continuous support was available through the web site, e-mail, phone, and onsite availability of the professional development instructional team.
3. The professional development opportunities were spread out over the long term allowing time for reflection, practice, and integration.
4. The experience deepened participants' content knowledge and skill and introduced them to powerful tools to support learning in standard-based environments. (p. 85)

While designed for teachers from a variety of disciplines and grade levels (sixth-twelfth), over half of the 130 participants were science teachers.

McClurg and Buss (2007) clearly demonstrated the ways in which participating teachers developed knowledge, skills, and confidence within the realm of teaching with GIS. Teachers consistently rated their abilities much higher than at the beginning of the experience. The authors explained these outcomes in the context of "key components" of the professional development experience, many of which emerged as the result of ongoing formal and informal evaluative research:

1. Pacing—Adjusting the pace of instruction from an intensive week-long workshop to a series of three two-day sessions over a six-month period
2. Relevancy—Making explicit connections to state standards in terms of the examples and lessons used during the workshops
3. Relevant Data—Providing data sets that were easily accessible and ready to use in terms of their projection and symbology
4. Conceptual Introductions—Introducing the idea of GIS by using paper maps that displayed different themes/data as the basis for exploration and discussion
5. Support Structures—Providing a variety of personal and physical resources such as the website, e-mail access, onsite visits, manuals, and the loan of GPS units among others
6. Personal Technology—Requiring teachers to bring a school-owned laptop ensured that the teacher had the software, knew where data and related files were stored, and had practice on the computer that they would use for teaching
7. Motivators—All participating teachers were required to register for either a university graduate-level course or state sponsored continuing education credit

The design and implementation of this project and careful attention to the needs of the teachers resulted in an increase in teachers' knowledge and skills, as well as successful classroom implementation. The authors provide several examples of teachers who were able to create meaningful learning opportunities for their students.

Trautmann and MaKinster (in press) used similar strategies and many of the previously mentioned frameworks to craft and implement a three-year professional development project focused solely on teaching science using geospatial technology. Each year, a cohort of fifteen to twenty teachers participated in a five-day summer institute in July, a three-day institute in August, and four or five Saturday workshops throughout the academic year. Participating teachers taught general science, biology, earth science, environmental science, and various electives in sixth through twelfth grade. The project aimed to help teachers gain skill in using various geospatial technologies, model the use of project-based and inquiry-based lessons, and support the teachers in developing their own geospatial lessons and units that address concepts and topics in their curricula. The authors documented growth in the teachers' perceived skills, knowledge, and self confidence with regard to integrating geospatial technology into their science teaching. The teachers also were asked to estimate the impact on their students. All of the teachers reported that such lessons had moderately or greatly increased their students' science content knowledge, interest in the course, and awareness of the relevance of science. All but one felt that the geospatial lessons had increased their students' critical thinking skills and ability to think spatially. The authors present the cases of three teachers to illustrate the diversity of participants and the various challenges and successes they experienced in integrating geospatial technology into their science teaching.

Trautmann and MaKinster (in press) concluded their study by forwarding a model that they referred to as *flexibly adaptive professional development*. This is an approach to professional development that is designed to meet the individual needs of participants from multiple grade levels who teach a variety of subjects. Simultaneously supporting teachers from diverse backgrounds and with disparate needs is challenging and requires flexibility in terms of professional development strategies and implementation expectations.

Student Learning Outcomes

Like teacher professional development, the evidence of effectiveness of geospatial technologies on student learning is just starting to emerge. As recently as 2006, the National Research Council (2006) in their publication: "*Learning to Think Spatially: GIS as a Support System in the K-12 Curriculum*" noted that:

> Geographic information science *has significant but as yet unrealized potential* in the K-12 curriculum. . . . In principle, GIS reflects many of the ideals of exploration-driven, discovery-based, student-centered inquiry. Nonetheless, current GIS software is less well equipped for data exploration and hypothesis generation than for data analysis and presenting information. In addition, current GIS is too cumbersome and inaccessible for effective use in K–12 education (p 8).

These authors also noted, however, that existing GIS tools could (1) be supportive of the inquiry process; (2) be useful in solving problems in a wide range of real-world contexts; (3) facilitate learning transfer across a range of school subjects; and (4) provide rich, generative, inviting, and challenging problem-solving environments. Here, we evaluate the extent to which the use of geospatial technologies can achieve these educational goals, all of which are central to effective science and environmental education.

A wide range of projects have been implemented in classrooms. These range from relatively modest projects, such as use of GPS units for geocaching, to comprehensive, analytical neighborhood studies informed through use of GIS. What appears to be common across the projects is that by coupling geospatial technologies with sound pedagogical approaches (i.e., project-based learning), educators are finding these cutting-edge tools can provide the motivation and resources needed to go beyond the limits of a more traditional curriculum (Almquist et al., 2009; Barnett, Houle, & Strauss, 2008; Bodzin, 2008). In this section, we review the research that has been conducted on student learning outcomes.

Prevalence of Research on Student Outcomes Despite the fact that geospatial technology is critical to conducting so much of environmental science education, the use and implementation of the technology for supporting environmental education has largely been ignored by the mainstream environmental educational research community in the United States. For example, only one article appears in the *Journal of Environmental Education* (Bodzin, 2008) and none in the journal *Environmental Education Research.* Most of the research articles that explore classroom use of geospatial technologies have appeared in publications such as the *Journal of Geoscience Education, Journal of Geography Education,* and *International Research in Geographical and Environmental Education.* Most of this research has been conducted by scientists, evaluators on federally funded projects, geographers, or specialists in the field of geospatial technologies, most of whom lack a strong background in educational theory and research. Thus, it is not surprising that much of the research to date has suffered from a lack of rigorous methodological design and data analysis. The majority of the published articles reviewed for this chapter were not grounded in recent theories of learning, did not reference mainstream science and environmental education research literature, and did not advance our understanding of how to design and implement educational technologies in classroom settings. This situation has likely kept geospatial technologies from gaining greater acceptance as a potential learning technology in the science and environmental education fields. However, educational initiatives across the K-16 spectrum illustrate the utility of such technologies in projects ranging from the study of watersheds (Stylinski & Smith, 2006) to pollution analysis (Jenner, 2006) to using GIS for detailed spatial analyses (Oldakowski, 2001). Of the research that has been conducted, some studies show promising results regarding the impact on student learning outcomes. We divided these into two categories of student learning outcomes: (1) map and spatial skills, and (2) content and inquiry skills.

Supporting Map or Spatial Learning Skills A number of studies examined the impact of using geospatial technologies, particularly GIS, on students' map making and spatial analysis skills. Generally, much of this research found that students develop better geographic knowledge such as creating, reading, and analyzing maps (e.g., Shinn, 2006). In one of the more rigorous studies, Allen et al. (in press) developed a middle school curriculum in which students used GPS units and My World GIS software to collect physical and spatial data of trees to create a schoolyard tree inventory. Using a primarily quantitative research design, they found a statistically significant increase in students' spatial awareness as a result of geospatial technology use. However, one inference of particular importance to the environmental education community was that they believed the increase in spatial awareness likely was correlated to the relevance of the place of study to the student.

Shinn (2006) developed a GIS-enhanced instructional unit for fourth graders and evaluated the impact of this experience on students' geographic skills and knowledge. Using a mixed methods approach, she found that students demonstrated gradual progress in gaining geographic content knowledge and that their map skills improved; however, these outcomes were highly variable among students.

In a related study, Wigglesworth (2003) investigated the strategies developed by middle school students to solve route-finding problems using GIS software. He found that his students used one of the following: (a) a *visual strategy* of assessing which route seemed to be the shortest based only on a visual assessment, (b) a *transitional strategy* of using GIS tools to examine the route that had been selected visually, and (c) a *logical strategy* of developing hypotheses, testing a variety of routes, and conducting a variety of evidence-based comparisons before choosing a specific route. The most significant implication of this work was the extent to which it revealed how GIS technology supported the use and application of each of the aforementioned strategies, and thus supported multiple styles of geographic learning.

Many projects have been implemented at the university level (Lloyd, 2001; Mitchell et al., 2008; Stewart, Schneiderman, & Andrews, 2001; Tulloch & Graff, 2007); however, few have been evaluated critically. Much of the research conducted was performed by the course instructor and not grounded in an action-research framework, which is an essential approach to maintaining rigor within self-studies. Despite this problem, there have been some interesting results and in the following we review the studies that, based upon our analysis, maintained some degree of methodological rigor and focused on learning outcomes.

Hall-Wallace and McAullife (2002) conducted one of the few studies at the university level examining the relationship between spatial understanding and content knowledge. They reported on the development and implementation of a series of units in which college students used GIS technologies to investigate plate tectonics and geologic hazards. The researchers employed a mixed methods approach that included prepost content exams and interviews. They found that content understanding was positively correlated with improved spatial understanding. This finding could have significant implications regarding the use of geospatial technologies to support students' understanding of place in reference to the scientific or environmental concepts of focus within a lesson or unit.

In an environmental science-focused project, Tulloch and Graff (2007) developed a set of "Green Map" learning exercises for their university level courses that were based on a problem-based learning framework. The authors used a case-study approach to examine the impact of these exercises on students' understanding of geographic concepts. They found that students developed a deeper appreciation regarding the natural resources around them. In particular, they found that not only did using Google Earth seem less contrived than some other assignments, it provided students with a sense of place and facilitated "spatial cognition and critical inquiry" (p. 275).

In one of the more methodologically rigorous studies, J. Lee and Bednarz (2009) investigated the impact of using GIS in a college-level cartography course. The authors validated a spatial skills test through multiple testing iterations and found that use of GIS helped students to think spatially. The student improvements were directly related to the GIS learning experiences in these courses; however, there were several confounding variables and wide variability in student outcomes. As a result, the extent to which prior spatial reasoning experiences contributed to these findings could not be determined.

Supporting Content or Inquiry Skills Audet and Abbeg (1996) conducted one of the first studies on the impact of geospatial technologies. They focused how novices and experts using GIS solved problems differently. Using classroom observations, think-along protocols, and transcripts of interviews, the authors found that use of GIS supported the development of problem solving skills. These authors found that while experts relied almost solely on logical formulations during GIS investigations, novice GIS users used three discrete problem-solving styles: trial and error, spatial querying, and logical querying. Development of problem-solving expertize within novel users involved a gradual progression through several identifiable, intermediate stages of procedural knowledge.

Baker and White (2003) investigated the impact of using GIS technologies on science process skills and science attitudes of eighth grade students within a problem-based earth science unit. Using prepost assessments and comparison group data, they found significant improvements in these students' attitudes toward technology and in their self-efficacy toward science. For those students who used GIS software as part of this unit, the authors found modest but significant improvements in their ability to analyze geographic data.

Beckett and Shaffer (2005) developed an urban planning curriculum using a combination of GIS technologies and spreadsheets with the goal of engaging students as urban planners. Relying mostly on qualitative data, they found that enabling students to use the same tools as urban planners improved their understanding of the nature of the profession and the role of technologies such as GIS.

Similarly, Barnettet et al. (2008) used GIS technologies in combination with computer modeling to enable inner city high school students to evaluate the ecological value of trees in their neighborhoods. This was one of the only studies to report reliability and validity data of their research instruments. The authors used both a prepost test design and student interviews. They found that participation in a GIS-focused curriculum improved student interest, self-efficacy, and urban ecological content knowledge. Most importantly, the students reported becoming more interested in protecting and caring for their urban environment.

Doering and Velestianos (2007) investigated the impact of engaging students in the analysis of real data as a part of an adventure learning program. By enabling students to explore real data using GIS technology (Arc Explorer Java Edition for Education), they found that the use of preexisting and newly acquired data contributed to students developing a "sense of place." In addition, the ways in which the GIS enabled students to explore these data with relative ease motivated them to explore geographic locations. Furthermore, this analysis enabled the students to develop a more sophisticated understanding of the geography of the region they were exploring.

Bodzin (2008) developed an outdoor-based curriculum that focused on a long-term pond monitoring project, within an after-school club for urban fourth grade students. The students used web-based GIS maps and Google Earth to analyze a variety of local and regional data. The findings suggested that participation in this project enhanced environmental attitudes, improved students' sense of environmental stewardship, and fostered responsible environmental behavior.

West (2003) conducted surveys of students who had participated in a range of GIS-based activities taught by the himself or other teachers. In total, the author collected survey data from 109 students from two middle schools. He found that students' perceived usefulness of computers, perceived control of computers, and behavioral attitudes toward computers improved. Use of GIS encouraged the students to use higher level thinking skills. In conclusion, West speculated that improved student attitudes had resulted from using GIS for two interconnected reasons. First, it had increased the relevance of the geographical study and, second, it had allowed students to focus on specific problems and areas using GIS's ability to isolate and evaluate relationships within data.

Looking Across Teacher and Student Outcomes Despite the fact that a number of K-16 curriculum development and teacher professional development projects have resulted in a considerable number of teachers learning about geospatial technologies, the use of these tools have been relatively low, especially in environmental education classrooms. The complex and time consuming nature of the technology has presented major implementation challenges in classroom settings (Bednarz & Schee, 2006; J. Kerski, 2003; J. J. Kerski, 2001). The steep learning curve can be daunting, especially for teachers with limited computer experience. The current state of much of the geospatial software typically requires advanced skills to acquire and import suitable datasets and convert them into a format and size appropriate for classroom use (National Council for Geographic Education., 2006). Even with appropriate data in hand, many teachers lack the time to construct lessons and units that use real world data and are aligned with state science content and skills standards. These challenges have hindered widespread acceptance and classroom implementation, granting access to only a limited number of students (NRC, 2006).

In spite of these challenges, pioneering educators have successfully integrated geospatial technologies into K-12 teaching (Alibrandi, 2003; Barnett et al., 2008; Bodzin, 2008; DeMers & Vincent, 2007; Doering & Veletsianos, 2007). Much of the research on use of geospatial technologies has focused on student use of either Google Earth or ESRI's ArcView products. Research and development studies show that use of such technologies has potential to provide students in all grades with a rich, inviting, and challenging problem-solving environment (Akerson & Dickinson, 2003; Baker & White, 2003; Carlson, 2007; J. J. Kerski, 2007; National Research Council, 2006; Stubbs et al., 2007). In particular, promising new models have been advanced for the development and implementation of professional development programs for teachers (Trautmann & MaKinster, in press). If geospatial technologies are to become widely used in K-12 settings, the educational research community must draw on existing models of professional development and integrate the specific requirements and knowledge needed by teachers to implement geospatial activities in their classrooms.

The Future of Geospatial Technology in Environmental Education

In1994, the first United States conference on the educational application of GIS was held, sponsored by the National Science Foundation (NSF). During that conference, NSF Program Officer Gerhard Salinger asked a set of important and poignant questions to the then burgeoning geospatial education community:

> What is the learning that produces understanding of concepts and processes students should know and be able to apply? What insights does GIS allow that the other ways of learning do not? What is GIS going to allow in education that we cannot do in other ways? (Salinger, 1995, p. 24).

As documented above, significant progress has been made in answering these questions; however, in many ways the nature of the questions has shifted due to several factors. Current research and practice in this area has identified the following key contributions of geospatial technologies to science and environmental education:

a) Provides relatively easy access to large volumes of valuable scientific data,
b) Facilitates scientific visualization (graphic/cartographic display),

c) Enables visual and computational comparison of multiple data layers,
d) Facilitates spatial analysis and geo-processing of the scientific data,
e) Provides users with an understanding and sense of place, and
f) Facilitates the investigation of problems from interdisciplinary perspectives by the use of multiple data sources.

Perhaps a better overarching question in the contemporary context is: "How can one leverage and integrate a range of geospatial tools into science and environmental education in ways that support student learning, scientific inquiry, and spatial reasoning skills?" To address this question, rigorous research in at least two areas still needs to be conducted. One line of research should investigate how to design and implement learning environments that leverage geospatial technologies for students. The other line of research, which is somewhat better established, should focus on how best to design professional development to prepare teachers to implement these technologies in their classrooms.

Despite significant opportunities that these tools provide in terms of supporting science and environmental investigations, substantial limitations still exist. For example, most of the existing geospatial technology tools currently in use are not well equipped for the evaluation of hypotheses about issues that require understanding of temporal changes—a critical piece in understanding many scientific and geographic concepts. Furthermore, K-12 use of the tools often is limited to settings in which teachers or technology administrators have installed the software. This creates a significant impediment for many teachers (Kerski, 2001; NRC, 2006).

Some of these limitations and others are being addressed by the development of new technologies. For example, some software applications avoid the problem of requiring software to be installed by running within a web browser and requiring no additional software installation or support (see Fieldscope at http://www.fieldscope.us/). Furthermore, GPS and GIS technologies have become readily available on mobile devices such as the iPod Touch, iPhone and other cellular phones. Driving along a highway, it now is possible to use a mobile device to visualize your precise geologic location (i.e., how far to the nearest fault line), and access geographic information (see http://www.integrity-logic.com/GeologyCA/). Web-based and mobile geospatial technologies will increasingly provide access to a wealth of spatially referenced information to anyone, at any time, from anywhere. What is needed, however, are development efforts and research on how to create and assess educational applications of these increasingly powerful and ubiquitous tools.

While geospatial technologies clearly have much to offer in supporting the major goals of science and environmental educators, much of this potential has remained isolated to a relatively small group of educators, practitioners, and scholars. Their projects have resulted in promising results that suggest that geospatial technologies may have a significant role to play in the future of science and environmental education. To reach this potential, research is needed into how to design and implement user-friendly, pedagogically sound software for educational use. Curriculum development efforts need to incorporate geospatial technology into existing and future curricula, ensuring alignment with national standards and significant learning goals. Such learning goals include graphicacy, critical thinking, and citizenship skills (Bednarz, 2004). Continuing research also is needed into development of effective models for professional development related to teaching with these new and ever-emerging technologies.

Review of the research indicates significant gaps in our knowledge of how to leverage geospatial technologies in educational contexts. Exploration of the following questions will provide insight for science and environmental educators on the strengths, weaknesses, and mechanisms through which geospatial technologies can improve a typical learning environment.

Geospatial Technology and Student Learning: Unanswered Questions

1. How can we best create meaningful contexts for learning in science and environmental education using geospatial technology?
2. How do the unique capabilities of geospatial technologies intersect with research on student cognition?
3. In what way does learning environmental science through geospatial technology affect how students think about such concepts and processes?
4. How do geospatial technologies influence students' spatial thinking and ability to reason about a place?
5. What are the design principles to consider when creating curriculum materials that use geospatial technologies to teach environmental science?
6. How can we best use geospatial technology-based curricular materials and related resources to attract more students into academics and careers in environmental science fields?
7. In what ways does the use of geospatial technologies support students in doing environmental science research?
8. To what extent and in what ways do students who use geospatial technologies develop better understanding of environmental landscapes, conservation, and the complex factors that impact the environment?

Geospatial Technology and Teacher Practice Learning: Unanswered Questions

1. How does teaching with geospatial technologies impact teaching practices, such as pedagogical strategies, used or the manner in which classroom activities are carried out?

2. What are the unique aspects of teaching *science* using geospatial technology, compared to implementing this sort of technology in other topic areas?
3. How can we use geospatial technologies to effectively teach core environmental science topics?
4. How do geospatial technologies enable teachers to make connections to real-world contexts?
5. To what extent do teachers adapt curricular resources that use geospatial technology investigate environmental or scientific issues?

Teacher Professional Development: Unanswered Questions

1. What other design principles should be considered when designing teacher professional development experiences that focus on the use of geospatial technology?
2. What is unique about designing teacher professional development experiences that focus on the use of geospatial technology, compared with other disciplines?
3. How do impacts on teachers vary depending on the design and focus of professional development projects?
4. How do teachers integrate new knowledge of geospatial technologies into their existing content and pedagogical knowledge?
5. How can we best assess the impact of technology-focused professional development experiences on science teachers?
6. How can we best support teachers in adapting geospatial curriculum for their classroom contexts, students, and curriculum?

Research on the Design of the Next Generation of Geospatial Technologies

1. How can an integrative geospatial technology tool be designed to scaffold learners through the scientific inquiry process?
2. What aspects of existing geospatial technologies are needed in the next generation of tools, and which aspects are unnecessary?
3. What role is best played by mobile media devices such as iPhones and other rapidly developing handheld devices in a geospatial-focused learning environment?
4. How and in what ways can the emerging features and functions of Web 2.0 technology be coupled with geospatial technology to widen the access in environmental education to user-friendly geospatial tools?

In addressing these questions, it is crucial that rigorous research be conducted to establish the strengths and weaknesses of using geospatial technologies to support student learning. As noted earlier, much of the research to date has been conducted by noneducational researchers and has relied primarily on self-reports, small sample sizes, and use of the researcher's own students or curricular materials. This new body of research must be rigorous methodologically, grounded in learning theory, and based on a sound understanding of the science and environmental educational research literature.

The field of design-based research holds significant promise to move the burgeoning geospatial education field forward. Design-based research has evolved as a methodological framework that enables researchers to systematically investigate and adjust various aspects of an educational context to evaluate the impacts of those adjustments on outcomes (S. Barab & Squire, 2004; Brown, 1992). This type of research is valuable in investigating a novel context or technology because it focuses on understanding the challenges and nuances of a particular situation and on determining how to improve the design of an environment to support learning. Design-based research could provide a valuable framework for geospatial researchers because this type of technology is new and has just begun to get a foothold in classrooms. Current research suggests that classroom implementations of geospatial technology are fraught with a wide range of complex issues ranging from teacher efficacy to technological impediments to student learning styles. Research leading to understanding of how geospatial technologies can best leverage student learning will be far better equipped to answer Salinger's (1995) questions of what value is added when geospatial technologies are used in education.

References

Akerson, V. L., & Dickinson, L. E. (2003). Using GIS technology to support K–8 scientific inquiry teaching and learning. *Science Educator, 12*(1), 41–47.

Alibrandi, M. (2003). *GIS in the classroom: Using geographic information systems in social studies and environmental science [with CD–ROM]* New York, NY: Heinemann.

Almquist, H., Blank, L., Crews, J., Gummer, E., Hanfling, S., & Yeagley, P. (2009). *Embedding spatial technology in a field–based science education course for teachers.* Paper presented at the Society for Information Technology and Teacher Education International Conference 2009, Charleston, SC.

Audet, R. H., & Abegg, G. L. (1996). Geographic information systems: Implications for problem solving. *Journal of Research in Science Teaching, 33*(1), 21–45.

Audet, R., & Ludwig, G. (2000). *GIS in schools*. Redlands, CA: ESRI.

Baker, T. R., & Kerski, J. J. (in press). Lone Trailblazers: GIS in K-12 science education. In J. Makinster (Ed.), GIS in science inquiry. Springer.

Baker, T. R., & White, S. H. (2003). The effects of GIS on students' attitudes, self–efficacy, and achievement in middle school science classrooms. *Journal of Geography, 102*(6), 243–254.

Banilower, E. R., Heck, D. J., Weiss, I. R. (2007). Can professional development make the vision of the standards a reality? The impact of the National Science Foundation's local systemic change through teacher enhancement initiative. *Journal of Research in Science Teaching, 44*, 3, 375–395.

Barab, S. A., Hay, K. E., Barnett, M., & Keating, T. (2000). Virtual solar system project: Building understanding through model building. *Journal of Research in Science Teaching, 37*(7), 719–756.

Barab, S., & Squire, K. (2004). Design–based research: Putting a stake in the ground. *Journal of Learning Sciences, 13*(1), 1–14.

Barnett, M., Houle, M., & Strauss, E. (2008). *Using geographic information systems to support student learning through urban ecology.* Paper presented at the International Conference of the Learning Sciences, Utrecht, The Netherlands.

Barnett, M., Lord, C., Strauss, E., Rosca, C., Langford, H., Chavez, D., Deni, L. (2006). Using the urban environment to engage youth in urban ecology field studies. *The Journal of Environmental Education, 37*(2), 3–11.

Beckett, K. L., & Shaffer, D. W. (2005). Augmented by reality: The pedagogical praxis of urban planning as a parthway to ecological thinking. *Journal of Educational Computing Research, 33*(1), 31–52.

Bednarz, S. W. (2004). Geographic information systems: A tool to support geography and environmental education. *GeoJournal, 60*, 191–199.

Bednarz, S. W., & Audet, R. H. (1999). The status of GIS technology in teacher preparation programs. *Journal of Geography, 98*, 60–67.

Bednarz, S. W., & Schee, J. v. d. (2006). Europe and the United States: The implementation of geographic information systems in secondary education in two contexts. *Technology, Pedagogy and Education, 15*(2), 191–205.

Berkowitz, A., Nilon, C., & Hollweg, K. (Eds.). (2003). *Understanding urban ecosystems: A new frontier for science and education.* New York, NY: Springer.

Blumenfeld, P. C., Soloway, E., Marx, R. W., Krajcik, J. S., Guzdial, M., Palincsar, A. (1991). Motivating project-based learning: Sustaining the doing, supporting the learning. *Educational Psychologist, 26*(3–4), 369–398.

Bodzin, A. M. (2008). Integrating instructional technologies in a local watershed investigation with urban elementary learners. *Journal of Environmental Education, 39*(2), 47–57.

Bowes, P. 2008. *MapMarker Version 14 developer's guide.* Troy, NY: Pitney Bowes Software.

Bransford, J., Brown, A., & Cocking, R. (1999). *How people learn: Brain, mind, experience and school.* Washington, DC: National Academy Press.

Brown, A. L. (1992). Design experiments: Theoretical and methodological challenges in creating complex interventions in classroom settings. *The Journal of the Learning Sciences, 2*(2), 141–178.

Buss, A., McClurg, P., & Dambekalns, L. (2002). *An investigation of GIS workshop experiences for successful classroom implementation.* Retrieved from http://gis.esri.com/library/userconf/educ02

Carlsen, B. (2001). Mapinfo Professtional 6.5. *GeoWorld, 14*(10), 59.

Carlson, T. (2007). A field–based learning experience for introductory level GIS students. *Journal of Geography, 106*(5), 193–198.

Clark Labs. (2009). IDRISI. Worcester, MA: Author.

Coulter, B., & Polman, J. L. (2004). *Enacting technology–supported inquiry learning through mapping the environment.* Paper presented at the American Educational Research Association, San Diego, CA.

DeMers, M. N., & Vincent, J. S. (2007). Arcatlas in the classroom: Pattern identification, description, and explanation. *Journal of Geography, 106*(6), 277–284.

Doering, A., & Veletsianos, G. (2007). An investigation of the use of real-time, authentic geospatial data in the K–12 classroom. *Journal of Geography, 106*(6), 217–225.

Edelson, D. C., Smith, D. A., & Brown, M. (2008). Beyond interactive mapping: Bringing data analysis with GIS into the social studies classroom. In A. J. Milson & M. Alibrandi (Eds.), *Digital geography: Geospatial technologies in the social studies classroom* (pp. 77–98). Charlotte, NC: Information Age.

English, K., & Feaster, L. (2003). *Community geography: GIS in action.* Redlands, CA: ESRI.

Folger, P. (2008). *Geospatial information and geographic information systems (GIS): Current issues and future challenges.* Washington, DC: Congressional Report Service for Congress.

Giordano, A., Lu, Y., Anderson, S., & Fonstad, M. (2007). Wireless mapping, GIS, and learning about the digital divide: A classroom experience. *Journal of Geography, 106*(6), 285–295.

Gordin, D. N., & Pea, R. D. (1998). Prospects for scientific visualization as an educational technology. *The Journal of the Learning Sciences, 4*(3), 249–279.

Hall–Wallace, M. K., & McAuliffe, C. M. (2002). Design, implementation, and evaluation of GIS–based learning materials in an introductory geoscience course. *Journal of Geoscience Education, 50*(1), 5–14.

Hurd, P. D. (2000). Science education for the 21st century. *School Science and Mathematics, 100*(6), 282–288.

Jeanpierre, B., Oberhauser, K., & Freeman, C. (2005). Characteristics of professional development that effect change in secondary science teachers' classroom practices. *Journal of Research in Science Teaching, 42*(6), 668–690.

Jenkins, H. (2006). *Convergence culture: Where old and new media collide.* New York, NY: New York University.

Jenner, P. (2006). Engaging students through the use of GIS at Pimlico state high school. *International Journal of Environment and Pollution, 15*(3), 278–282.

Kerka, S. (2000). *Middle school career education and development.* Columbus, OH: Center on Education and Training for Employment.

Kerski, J. J. (2001). A national assessment of GIS in American high schools. *International Research in Geographical and Environmental Education, 10*(1), 72–84.

Kerski, J. J. (2003). The implementation and effectiveness of geographic information systems technology and methods in secondary education. *Journal of Geography, 102*(3), 128–137.

Kerski, J. J. (2007). *The implementation and effectiveness of geographic information systems technology and methods in secondary education.* Retrieved from http://gis.esri.com/library/userconf/proc01/professional/papers/pap191/p191.htm

Lam, C., Lai, E., & Wong, J. (2009). Implementation of geographic information system (GIS) in secondary geography curriculum in Hong Kong: Current situations and future directions. *International Research in Geographical and Environmental Education, 18*(1) 57–74.

Lave, J., & Wenger, E. (1991). *Situated learning: Legitimate peripheral participation.* New York, NY: Cambridge University Press.

Lee, J., & Bednarz, R. (2009). Effect of GIS learning on spatial thinking. *Journal of Geography in Higher Education, 33*(2), 183–198.

Lee, T., Quinn, M. S., & Duke, D. (2006). Citizen, science, highways, and wildlife: Using a web–based GIS to engage citizens in colleting wildlife information. *Ecology and Society, 11*(1), 11. Retrieved from http://www.ecologyandsociety.org/vol11/iss1/art11.

Lieberman, G. A., & Hoody, L. L. (1998). *Closing the achievement gap: Using the environment as an integrating context for learning.* San Diego, CA: State Education and Environment Roundtable.

Linehan, P. E. (2006). Learning geospatial analysis skills with consumer-grade GPS receivers and low cost spatial analysis software. *Journal of Natural Resources and Life Sciences Education, 35*, 95–100.

Lloyd, W. J. (2001). Integrating GIS into the undergraduate learning environment. *Journal of Geography, 100*(5), 158–163.

Loucks-Horsley, S., Love, N., Stiles, K. E., Mundry, S., & Hewson, P. W. (2003). *Designing professional development for teachers of mathematics and science.* Thousand Oaks, CA: Corwin.

Malone, L., Palmer, A., & Voigt, C. (2002). *Mapping our world: GIS lessons for educators.* Redlands, CA: ESRI Press.

McInerney, M. (2006). The implementation of spatial technologies in Australian schools: 1996–2005. *International Journal of Environment and Pollution, 15*(3), 259–264.

Milson, A. J., & Alibrandi, M. (Eds.). (2008). *Digital geography: Geo–spatial technologies in the social studies classroom.* Greenwich, CT: Information Age.

Mitchell, J. T., Borden, K. A., & Schmidtlein, M. C. (2008). Teaching hazards geography and geographic information systems: A middle school level experience. *International Research in Geographical and Environmental Education, 17*(2), 170–188.

National Council for Geographic Education. (2006, August 15–16). *integrating geographic information systems and remote sensing for technical workforce training at two–year colleges. National council for geographic education.* Paper presented at the Geospatial Workshop, Arlington, VA.

National Research Council. (2006). *Learning to think spatially: GIS as a support system in the K–12 curriculum.* Washington, DC: National Academies Press.

National Staff Development Council. (2001). *NSDC standards for staff development.* Retrieved from http://www.nsdc.org/standards/index.cfm

NSF Task Force on Cyberlearning. (2008). *Fostering learning in the networked world: Learning opportunity and challenge, a 21st century agenda for the national science foundation.* Arlington, VA: National Science Foundation.

Oldakowski, R. K. (2001). Activities to develop a spatial perspective among students in introductory geography courses. *Journal of Geography, 100*(6), 243–250.

Patterson, M. W., Reeve, K., & Page, D. (2003). Integrating geographic information systems into the secondary curricula. *Journal of Geography, 102*(6), 275–281.

Pitney Bowes MapInfo. 2008. *MapMarker Version 14 developer's guide.* Troy, NY: Pitney Bowes Software.

P2CS (n.d.). *Partnership for 21st century skills, 2007.* Retrieved from http://www.21stcenturyskills.org/

Ramamurthy, M. K. (2006). A new generation of cyberinfrastructure and data services for earth system science education and research. *Advances in the Geosciences, 8*, 69–78.

Salinger, G. (1995). *The charge from the National Science Foundation.* Paper presented at the first conference on the Educational Applications of Geographic Information Systems (EdGIS), Washington, DC.

Sandoval, W. (2004). Developing learning theory by refining conjectures embodied in educational designs. *Educational Psychologist, 39*(4), 213–223.

Sandoval, W., & Bell, P. (2004). Design–based research methods for studying learning in context: Introduction. *Educational Psychologist, 39*(4), 199–201.

SCANS. (1991). *What work requires of schools: A SCANS report for America 2000.* Washington, DC: US Department of Labor.

Shinn, E.–K. (2006). Using geographic information system (GIS) to improve fourth graders' geographic content knowledge and map skills. *Journal of Geography, 105*, 109–120.

Smith, G. A. (2007). Place-based education: Breaking through the constraining regularities of public school. *Environmental Education Research, 13*(2), 189–207.

Steede–Terry, K. (2000). *Integrating GIS and the global positioning system.* Redlands, CA: ESRI Press.

Stewart, M. E., Schneiderman, J. S., & Andrews, S. B. (2001). A GIS class exercise to study environmental risk. *Journal of Geoscience Education, 49*(3), 227–234.

Stubbs, H. S., Fowler, K., Ball, J., Singh, N., Whitaker, D., & Hawley, B. (2007). Educational environmental projects, using technology applications, for middle school students in formal and non-formal settings. *Meredian: A middle school computer technologies journal.* Retrieved from http://www.ncsu.edu/meridian/sum2003/gis/3.html

Stylinski, C., & Smith, D. (2006). *Connecting classrooms to real–world GIS–based watershed investigations.* Paper presented at the ESRI Educators User Conference, San Diego, CA.

Supovitz, J. A., & Turner, H. M. (2000). The effects of professional development on science teaching practices and classroom culture. *Journal of Research in Science Teaching, 37*(9): 963–980.

Trautmann, N. M., & MaKinster, J. G. (2010). Flexibly adaptive professional development in support of teaching science with geospatial technology. *Journal of Science Teacher Education, 21*(3), 351–370.

Tulloch, D., & Graff, E. (2007). Green map exercises as an avenue for problem–based learning in a data–rich environment. *Journal of Geography, 106*(6), 267–276.

Wang, S. (2008). *Formalizing computational intensity of spatial analysis.* Paper presented at the 5th International Conference on Geographic Information Science, Park City, UT.

West, B. A. (2003). Student attitudes and the impact of GIS on thinking skills and motivation. *Journal of Geography, 102*(6), 267–274.

Wideen, M., Mayer-Smith, J., and Moon, B. (1998). A critical analysis of the research on learning to teach: making the case for an ecological perspective on inquiry. *Review of Educational Research, 68*(2), 130–178.

Wiggins, G., & McTighe, J. (2005). *Understanding by design* (2nd ed.). Boston, MA: Prentice Hall.

Wigglesworth, J. C. (2003). What is the best route? Route–finding strategies of middle school students using GIS. *Journal of Geography, 102*(6), 282–291.

Wilder, A., Brinkerhoff, J. D., & Higgins, T. M. (2003). Geographic information technologies plus project-based science: A contextualized professional development approach. *Journal of Geography, 102*(6), 255–266.

33

Sustainability Education

Theory and Practice

Sarah Holdsworth, Ian Thomas, and Kathryn Hegarty
RMIT University, Australia

Introduction

Western society today is shaped by the social and political constructs of a neo-liberal capitalistic society where the wealth of nations and the optimal play of market forces dominate social and political agendas (Gough & Stables, 2008; UNEP 2002a). A dramatic change in our mindset and behaviors is required if this is to change, and education is one tool that shapes and informs how we think and act.

As a society we inherently believe in the value of education and the resultant capabilities that perpetuate social and technological improvement (Clinton, 2006). However, education and teaching is more than the forcing of accepted knowledge on students; it has more to do with an understanding of the world, the ability to reason, and the growth of character and personality. Universities have a key role to play in helping society move toward a more sustainable existence. While, many have changed their operational procedures, there has been less progress in the implementation of pedagogy, improved teaching methods and curricula relevant to the needs of sustainable development.

While the literature to date has focused on setting an international agenda for the integration of sustainability into public education, and offers principles that could form the foundations of education for sustainability, there is a lack of research on ways of implementing these principles. This chapter will explore the process required to successfully embed sustainability education, i.e., significant curriculum change, across universities. We argue that sustainability education is complex, poorly understood, and that mechanisms for a change in educational praxis requires understandings of organizational culture (Eckel & Kezar, 2003), the role of disciplinary dialogue and sense-making (Weick, 1995), staff politics (Arnold & Civian, 1997) and of external factors that affect change. This chapter identifies and discusses the following areas of sustainability educational praxis:

1. Curriculum: the development of course material to include capabilities so that graduates understand sustainability as a transformative concept, not simply one that reinforces the dominant world paradigm.
2. Learning and teaching: The development of teaching methods that allow for understandings of sustainability as a transformative concept i.e. reflective practice, self-evaluation (Schön, 1983).
3. Pedagogy: The development of educational approaches that shift recognized disciplinary boundaries.

Finally, the chapter discusses the potential application of sustainability education defined as third order transformative learning (Sterling, 2001). This has implications not only for universities but also within professional practice. If we understand how to create long-lasting change that promotes education for sustainability we will begin to build capacity for graduating professionals to better meet the documented demand, for sustainability/transferable skills, from a wide range of employers and professional organizations.

Education and Its Role in Society

Education and the Dominant Western Paradigm It is argued that the role of education in society is an ethical consideration, guiding the development of an understanding of the world, the ability to reason, and the growth of character and personality (Clinton, 2006; Huckle, 2005). However, the current approach to education in western society is founded on teaching facts; teaching learners what to think, founded primarily on what it is that their peers and the educators "think they know and know they believe (indoctrination)" (Henn & Andrews, 1997, para. 3). Subjects are taught separately by academic specialists in ever greater depth and detail (reductionism) (Henn & Andrews, 1997; Huckle, 2005; Murray & Murray, 2007; Murray, Brown, & Murray, 2007). Reflected in these traditional learning and teaching patterns are learning outcomes where individuals are trained to think and act with individual benefits as a focus for reinforcing the dominant social paradigm of western society (Kemmis & Smith, 2008; Sprigett, 2005).

Additionally, this has resulted in educators who are unable to make a distinction between "education," with individual and social purpose, and "schooling," which is the institutional formation of learners to attain approved learning outcomes; which may or may not be in the interests of the students themselves or the good of humankind (Kemmis, 2008b). Educators of this kind are unable to identify related distinctions between indoctrination and education (Kemmis, 2008b). Consequently, Kemmis and Smith (2008) that praxis in education today is "endangered" (p. 5) and is slowly amounting to educational *practice*, which is simply "following the rules" (p. 5). The authors define praxis as a particular kind of action that is morally committed and oriented and informed by traditions of a field. It is the kind of action people are engaged in when they think about what their actions will mean in the world. Praxis embodies the meaning of theoretical ideas of learning and teaching (pedagogy); the conscious reflection on practice (learning and teaching, and the development of curriculum that inform each other). The concept of practice, by contrast, has a broader application. Kemmis and Smith (2008a) define practice as more general and encompassing and apply it to a wide variety of actions and activities in social settings. Consequently, praxis is the reflection of an educator on their own pedagogy against their practice and subsequent development of curriculum.

Underpinning this lack of understanding of education is the loss of the traditions of educational studies and educational philosophy and theory, and the tradition of *Pedagogik* (Kemmis, 2008a). Historically, education (and *Pedagogik*) was once understood to have a double purpose: to educate for the development of a "good person" and an "educated person," and be in the interests of the individual and the interests of humankind (Kemmis, 2008b, p. 29). Understandings of these terms were not finite but reflective and continually questioned what "society" should look like, and this was open for consideration within every changing society. The loss of this educational philosophy has been attributed to the dominant scientific western worldview, which penetrates and shapes all areas of our society (Capra, 1975; Kemmis, 2008; Robottom & Hart, 1993; Sterling, 1996). The scientific worldview has a strong empiricist quality, which assumes that any knowledge obtained through the scientific method is objective, rational and true, and that this method is the only rational avenue to acquire knowledge (Capra, 1988).

Kemmis (2008b); Orr (2001); Robottom and Hart (1993); and Sterling (1996) believe this scientific worldview dominates the field of formal education, and is then reflected in society (Fricker, 1998). This positivist approach fundamentally seeks "to apply standards and methods of the natural sciences to the problems of education" (Robottom & Hart, 1993, p. 29). Kemmis (2009) argues that positivist approaches to educational thinking have legitimized the idea of the development of finite answers to tradition of *Pedagogik,* and therefore act to conserve existing institutional structures across all sectors of society including the field of education (Codd, 1982; Kemmis, 2008b). This kind of thinking is a direct product of the dominant motives and interests of a western worldview: efficiency, effectiveness and productivity. There is a bureaucratic appeal in positivistic approaches to education. Robottom and Hart argue that positivistic approaches to education have bureaucratic appeal guaranteeing "discipline in the workplace and contribut[ing] to a growing gulf between those who conceptualize tasks and those who execute them" (1993, p. 30). Education of this nature does not enable the development of skills or the ability for individuals to recognize the dominant ideology. Jucker (2002a) concludes that in its current conceptualization Western education has been successful in reinforcing an understanding of what is unsustainable rather than what is sustainable.

However, dominance of the scientific worldview in formal education is under increasing scrutiny, according to Capra (1982). Huckle (2005) suggest that new ways of thinking need to emerge so that we are able to explain and question our own worldviews and models of thinking, allowing deep reflection on personal and societal value positions. Lazlo (2006) and Robottom and Hart (1993) believe a new philosophical and conceptual framework (a new worldview) is emerging; one where new social, ethical, ecological, epistemological, and ontological problems are included. This has been reflected in alternative approaches to education, such as environmental education, global education, development education, peace education, citizenship education, human rights education, and multicultural education (Fien, Guevara, Lang, & Malone, 2004; Gough, 1997). The policy statements that emerged from the international conferences at Belgrade in 1975 and Tbilsi in 1977 established environmental education as a goal that should be pursued within all areas of education (Gough, 1997; Stevenson, 1987). The goals and objectives were to alleviate exploitation of the environment through social construction, to avoid the social injustices in the process of that reconstruction and to strengthen and encourage independent communities, locally and globally. Tension and conflict is often the result as these goals challenge the dominant worldview with a different values system (Gough, 1997).

However, some authors argue that environmental education, in its conceptualization, has failed in achieving its goals as it has been dominated by an underlying research paradigm, characterized by the scientific worldview (Robottom & Hart, 1993; Gough, 1997). Sustainability education (or education for sustainable development; Brundtland, 1987) has evolved to bridge the gap, aspiring to integrate social, economic, political and environmental issues (Tilbury & Cooke, 2005). However if sustainability education is to avoid the issues associated with environmental education, educators must recognize the different assumptions that inform their own and others' practice, and the lack of ability to identify and address inconsistencies and contradictions in their professional and personal lives (Robottom & Hart, 1993; Selby, 2006).

Sustainability Education and the Role of Universities For nearly twenty years through international conferences, publications and commitments, there has been increasing activity in moving toward the inclusion of environmental understanding and sustainability in universities; most recently with the declaration of a United Nations Decade of Education for Sustainable Development, starting in 2005. The United Nations Decade of Education for Sustainable Development (2005–2014): International Implementation Scheme (UNESCO, 2005, p. 7) states that Education for Sustainable Development (ESD):

> ESD is for everyone, at whatever stage of life they are. It takes place, therefore, within a perspective of lifelong learning, engaging all possible learning spaces, formal, non-formal, and informal, from early childhood to adult life. ESD calls for a reorientation of educational approaches—curriculum and content, pedagogy and examinations.

Education founded on a sustainability paradigm aims to develop skills and competencies that will allow students to seek out and examine their own frameworks for thinking (Cortese, 2003; Huckle, 2005; Holdsworth, Bekessy, Hayles, Mnguni, & Thomas, 2006a; Lang, 2004). The objective is for students to develop skills to critically enquire and systematically think about problems in ways that allow them to explore the complexity and implications of a more sustainable way of being (Sterling, 2001). Central to this approach is values and respect for us and our natural environment. The DESD breaks down the traditional educational scheme and promotes:

- Interdisciplinary and holistic: learning for sustainable development embedded in the whole curriculum, not as a separate subject;
- Values-driven: sharing the values and principles underpinning sustainable development;
- Critical thinking and problem solving: leading to confidence in addressing the dilemmas and challenges of sustainable development;
- Multimethod: word, art, drama, debate, experience, . . . different pedagogies which model the processes;
- Participatory decision making; learners participate in decisions about how they are to learn;
- Locally relevant; addressing local as well as global issues, and using the language(s) which learners most commonly use (UNESCO, 2005, p. 3).

Universities have a uniquely important role in the development of education for sustainability. Specifically:

> Universities produce those that will reproduce the existing power structures. Academics themselves create and run society's political and social institutions that in theory underpin and run our capitalist economy and technological direction, direct the world's media, and educate our students. (Jucker, 2002b, p. 242)

So universities, and the education they provide, offer the opportunity to lay the base of sustainability literacy that our society increasingly needs to move to more sustainable practices. Yet, while there is commitment to achieving institutional change for sustainability through educational praxis in universities (Thomas, 2004), few institutions have been able to achieve genuine educational change (Bekessy, Burgman, Wright, Filho, & Smith, 2003; Holdsworth, Wyborn, Bekessy, & Thomas, 2008).

Understandings of Sustainability and Education and Their Role in a Lack of Application in Universities In a broad sense, commitment to education underpinned by a sustainability paradigm reflects a moral responsibility to address a range of profound social and environmental challenges (Lozano-Garcia et al., 2008). However, there has been much discussion and debate about specific definitions of sustainable development and consequently the use of the term in an education context. Cotton, Warren, Maiboroda, and Bailey (2007), Filho (2000), and Jones Trier, and Richards (2008) argue that the introduction of sustainable development in higher education is constrained by ongoing confusion over terminology and controversy over whether sustainable development is a valid part of the curriculum. Gligo (1995); Huckle (2005); Huckle and Martin (2001); Jucker (2002a); Sterling (2001, 2003); and Robottom and Hart (1993) argue that the term sustainable development is contradictory and limiting, and that is often interchanged with sustainability. Cotton et al. (2007) recognize that the relationship between education for sustainable development and the broader concept of sustainable development is multifaceted and difficult to define, particularly in terms of operationalizing the many ways in which sustainable development concepts, principles and practice can be incorporated into learning and teaching.

This subjectivity and resultant diversity in understanding and application of sustainability in education is the central "problem" of ESD. At the heart of all definitions of sustainable development and sustainability, is the fact that our definitions and understanding of the terms are constructed from the way we understand and construct knowledge and then make meaning from the resultant language (Hegarty,

2008). Our understanding of the reality of the world and our actions within it are founded in and built around our own theoretical frameworks or worldview (Bawden, 1997).

Further, Fricker (2002) argues that the Brundtland definition of sustainable development is an oxymoron and is founded in the assumptions embedded in a scientific worldview, with notions of fiscal growth and advantage. The definition of sustainable development constructed from an ecocentric worldview would be starkly different with development perhaps being thought of as a pathway to a future where environmental, social and economic growth is synergistic, and is embedded with a completely different set of assumptions. Consequently, how we interpret and understand the concepts of sustainable development is a reflection of the ability of the individual to interpret and understand the way in which they and others construct and therefore understand these assumptions and resultant definitions (Hegarty, 2008). Given that the construction of knowledge and understanding is inherently linked to an individual's own identity and experience, discussions of definitions of sustainable development and sustainability are meaningless without context and examination of self (Hegarty, 2008; Sterling, 2001).

The implications for university curricular are significant since worldviews themselves must be consciously developed. For one, Bawden (2005) advocates that experiences (prior knowledge, beliefs, skills) have a major influence on what is learned, and how it is learned and contextualized. Consequently, education must recognize that there are multidimensional worldviews and those systemic approaches to education . . .

> . . . would then be grounded in the belief that epistemic foundations can be both challenged and changed through "movements to more advanced states" which themselves reflect complex evaluative positions involving epistemological, ontological and axiological features. (pp. 156–157)

Closely related, Reid (2002) suggests that sustainability and environmental education each:

i. encourage distinct notions of thinking, valuing and acting for teachers and learners;
ii. suggest specific priorities for thinking, valuing and acting in what is practiced as "education"; and
iii. invoke particular features of thinking, valuing and acting over others regarding what is fundamental to their distinctiveness from, and relationship to, each other and "education" more widely (Reid, 2002, p. 5).

These qualities are problematic as they translate into the development of pedagogy, practice, and curriculum that is directly underpinned by an individual's philosophical and epistemological foundation. In essence by adopting a particular paradigm, environmental education, and education for sustainable development are already ideologically predetermined (Robottom & Hart, 1993). Sterling (2001) recognizes three different approaches to education underpinned by a sustainability paradigm which embody different degrees of understanding, constructing and translating knowledge:

- Education *about* sustainability is "learning as maintenance" (p. 60), not challenging the current paradigm. This is first-order learning and has a context/knowledge basis, takes on some sustainability concepts easily inserted into "the existing educational paradigm" (p. 60).
- Education *for* sustainability is "an adaptive response that equates to second-order learning" (p. 60), based on values and capability. The existing paradigm reflects more thoroughly the ideas of sustainability, but is largely unchallenged. Education is founded on an identified set of values; knowledge and skills needed to achieve "learning for change" (p. 60), including the development of skills in critical and reflective thinking.
- Education *as* sustainability or sustainable education is third-order learning and change—a creative and paradigmatic response to sustainability. "This is a transformative, epistemic education paradigm, which is increasingly able to facilitate a transformative learning experience" (p. 61). This form of education is holistic, with learning approached as change requiring the engagement of the whole person and institution (Sterling, 2001).

If we are to evolve our education practice, which results in first order learning, to educational praxis, which results in third order learning, Sterling (2001) argues that we must understand and reflect on the way we construct knowledge, understand how this shapes our view of reality (ontology), and then how we translate this into the discovery of new knowledge. This builds on the work of Bawden (1997) who believes that it is through epistemic learning that we learn to appreciate the nature of the worldviews and paradigms which we hold as the contexts for what and how we know, and also that we learn how to both challenge and, if appropriate, change them. While Cotton et al. (2007) recognize this approach works well in assisting understanding of the different relations ESD may have with sustainable development, it does not diminish the extent of user interpretation, as they are all subjectively defined, and the way these are conceptualized and enacted is diverse. These issues have ramifications on the curriculum, pedagogy, and learning approaches in universities.

Discussion: The Application of Sustainability Education in Universities

Sustainability Education Pedagogy To achieve second and third order transformative learning we need to move beyond a reductionist pedagogy. Pedagogy for many educators is the set of principles and personal philosophies that guide their design and development of learning the art of teaching. For the development of a theory of practice and an effective pedagogy leading to quality of instruction, much more must be considered than mere development and design of learning activities. Essentially pedagogy includes:

- skills to impart knowledge, create quality learning environments for both a student's personal and academic success
- understanding and knowledge of the complexity of the learner and the learning environment and how people learn including the role of institutions regarding teaching and learning
- awareness of one's own philosophical beliefs and their impact on our teaching including the diversity of social, economic, cultural elements that an educator interacts with. Learning theories, educational theories, the underlying social environments and the interplay of context and content. (De Figueiredo Afonso, & da Cunha, 2002; Miriam-Webster, 2006; Siemens, 2004)

The literature on specific pedagogies for ESD reflects this and advocates for pedagogies that are interactive and student centered, typifying constructivist learning theories (Corneya & Ried, 2007; Tilbury & Cooke, 2005; Summers et al., 2005). Dialogical, critical, and active learning—deep learning—requires pedagogy in which teachers and students learn, reflect, and act together, and by doing so transform themselves and the world around them (Freire, 1972; Sterling, 2001; Warburton, 2003).

To facilitate this radical change Dawe, Jucker, and Martin (2005) have categorized three approaches to EfSD; the personal approach, connecting or reconnecting to reality, and holistic thinking. Piccinin (1997) argues for a learner-centered pedagogy. A learner-centered pedagogy is based on the needs of the learner rather than the needs of the teacher or the institution (Tam, 2000). The objectives of a learner-centered pedagogy are highly compatible with that described by Dawe et al. (2005) as sustainability education pedagogy, and begin to address the issues of praxis discussed by Kemmis and Smith (2008). An example of the adoption of such an approach can be founding the first undergraduate course "Sustainability: Society and Environment" (SSE) run in global studies, social science, and planning at RMIT University (Hegarty Thomas, Kriewaldt, Holdsworth, & Bekessy, 2011). A compulsory twelve-week course delivered through a lecture and tutorial/workshop format comprising three contact hours per week for seven professional degree programs. The course aims to develop in its students the content knowledge, and skills, which best enable the application of sustainability literacy with a focus on *applied professional* contexts. The course takes a "triple bottom line (TBL)," approach to the presentation of sustainability, and is currently embedded in a learner-centered philosophical framework to ensure the deep learning outcomes are achieved by all students. A number of pedagogical principles were prioritized:

- "The development of key academic and transferable skills for transition from first semester first year to the broader university learning environment
- Enabling the recognition of complex situated problems as significant in the application of sustainability knowledge
- Highlighting the role of a wide range of values and belief systems in measuring and evaluating decision making for sustainable futures
- Emphasizing the relationship of the sustainability studies course to the professional field of each student." (Hegarty et al., 2011 p. 8)

Sustainability Education: Learning and Teaching

Subject content for sustainability education must focus on the inter-relationships between environmental, economic and social factors, which Corneya and Reid (2007) argue are "complex and value-laden, and the terms used are open to differing interpretations" (p. 36). Associated methodologies should be grounded in: experimental and cooperative learning; systemic thinking, the clarification and judgment of values; ideology critique; critical reflection and creative thinking; the envisaging of sustainable futures, sensory, and empathic exercise; communication skills; learning as a continuous process for all (Elliot, 1991; Sterling, 1996). Additional characteristics that have been identified include: collaborative learning, problem solving skills to deal with complex real-life problems, creative thinking, personal and professional self-reflection, creative participation in interdisciplinary teams and learning from others (collaborative learning), holistic thinking, recognition and appreciation of environmental, social, political and economic contexts for each discipline and experiential learning by reconnecting to real-life situations (Dawe et al., 2005; Sterling, 2004; HEA, 2006).

Approaches to Learning and Teaching in sustainability education identified within the literature and reflecting the character of sustainability educational pedagogies have been described as allowing students to be able to experience a range of situation see Figure 33.1.

Consequently, for students to achieve these experiences educators are faced with a challenge in choosing and implementing teaching strategies given the value-laden nature of subject matter and the greater teaching expertise required to use "interactive" as compared with "didactic" teaching strategies (Corneya & Ried, 2007). The learning outcomes and associated skills related to these approaches are aligned with more than one pedagogy. However, if deep learning is to occur, then self-reflection and questioning of personal values and identity must begin with a learner-centered pedagogy, to which any pedagogy can be added to or merged with, assuming these concepts are discretely identifiable and definable.

Sustainability Education: Curriculum Development

Kemmis and Fitzclarence (1986) believe that central to the issues associated with curriculum is the problem of the relationship between theory and practice and the relationship between education and society. Curriculum can

Students should experience the following
- development of knowledge from the learning process;
- questioning of their assumptions;
- recognition that constructing knowledge involves critical analysis, dialogue, and reflection;
- development in complex reasoning;
- practice in demonstration of knowledge and skills;
- practice in transferable problem solving skills;
- practice in the recognition of values and how this relates to their own action;
- strategies for change;
- uncertainty in data, analysis, and decision making;
- critical analysis the theories, data, and values being presented to them;
- identify the connections between the principles of sustainable development and the disciplinary theory;
- ability to challenge injustice and inequalities;
- cooperation and conflict resolution;
- critical thinking;
- respect for people and things;
- ability to understand their own sense of identity and self-esteem;
- value and respect for diversity.
  (Holdsworth et al., 2006a, Parker, Wade, and Van Winsum, 2004, Parkin, Johnson, Buckland, and White, 2004)

Figure 33.1 Approach to learning and teaching in sustainability education.

be defined as "an attempt to communicate the essential principles of an educational proposal in such a form that it is open to critical scrutiny and capable of effective translation into practice" (Stenhouse, 1975, p. 4). Kemmis and Fitzclarence (1986) argue that this places emphasis on curriculum as a "bridge" between educational principles and educational practice and on the activities of consciously relating the two and reviewing the relationship between them in a notion of critical scrutiny which involves testing curriculum proposals and educational theories in practice. This is problematic as the nature of curriculum has moved from a focus on the practical to the theoretical (Schwab, 1969).

The current approaches to curriculum development as argued by Kemmis et al. (1996) and Schwab (1969) is on the development of curriculum that is founded on theory is incompetent "as a basis for wise educational practice" (Kemmis & Fitzclarence, 1986, p. 14). They argue for a return to the focus on "curriculum thinking and theorizing in which the 'arts of the practical' (the arts of moral and political argument) were more central to thinking about education" (p. 15). How sustainability is understood and practiced by the educator will influence the curriculum content that is considered relevant to disciplinary knowledge and practice. Yet irrespective of the discipline Fien (2001) argues that beneath the disciplinary knowledge must be an understanding of four interdependent systems that underpin the sustainability paradigm. The systems are:

1. Biophysical systems—which provide the life support systems for all life, human and nonhuman;
2. Economic systems—which provide a continuing means of livelihood (jobs and money)
3. Social and cultural systems—which provide ways for people to live together peacefully, equitably and with respect for human rights and dignity:
4. Political systems—through which power is exercised fairly and democratically to make decisions about the way social and economic systems use the biophysical environment (Fien, 2001, p. 4).

Some understanding of these systems is required so relevant content can be embedded into existing and new curriculum within all traditional disciplinary areas. Considering all the issues associated with curriculum it must be founded in several critical concepts, see Box Figure 33.2. As previously stated individual educators will use these concepts to shape their individual curriculum, the EfS literature reveals a number of examples of sustainability courses which illustrate these principles (Adomssent, Godeman, Leicht, & Busch, 2006; Alverez & Rogers, 2006; Barth, Godemann, Rieckmann, & Stoltenberg, 2007; Blewitt & Cullingford, 2004; Eisen & Barlett, 2006; Grant, 2009; Hayles & Holdsworth, 2008; Sterling & Thomas, 2006)

Sustainability Education and Universities: Where to From Here For sustainability education to become embedded in universities, a change in educational praxis is required, and the development of a new learning culture (Kemmis and Smith, 2008; Kemmis, 2009). This learning culture cannot be founded on academic tradition and principles of indoctrination, but needs to evolve out of an open-minded and participative process. Rather, Tilbury, Keogh, Leighton, and Kent (2005) argue that the development of sustainability curricula in universities needs to be accompanied by a process of "institutional strengthening and professional development in order for their principles to be translated into practice" (Tilbury et al., 2005, p. 40). Academic development is necessary to provide educators with the capacity for understanding sustainability as an overarching conceptual framework, which can be used to reconsider the way we think and act toward each other

- inter-disciplinary and intercultural practice
- based on discourse with much room for discussion, subject diversity, and cross-cutting topics
- holistic i.e. consisting of a mix of targeted activities, cognitive learning modules, and emotional and practical experiences
- general components and inter-connection of ecological/natural systems, social systems (including cultural and political), and economic systems and political; system to core curriculum
- holistic or systemic thinking (and analysis)
- key current and historical sustainability issues in their local, regional and international.
- curriculum should be made up of issues relevant to the discipline that explore society justice, diversity, and equity

(Baud, 2003; Fien, 2001; Holdsworth, Bekessy, Hayles, Mnguni, & Thomas, 2006b; Parker et al., 2004)

Figure 33.2 Key concepts of sustainability education curriculum.

and the planet. Academic development is also an important step in providing educators with the capacity to undertake sustainability educational praxis.

Despite recognition of the need for academic development, examples of programs are rare (Holdsworth et al., 2008; Thomas & Nicita, 2003). Barth et al. (2007) claim that essential to this is that the process itself is relevant and related to an academic's own sphere of influence and desires, but also related to individual and societal learning. An instance of an action learning research project undertaken at RMIT University, Australia during 2005 entitled "Beyond Leather Patches" (BELP) attempted to address these approaches. The research project was designed to embed sustainability principles in the curriculum of nontraditional disciplines. The project represented significant cross-campus collaboration between three Schools at RMIT: the School of Property Construction and Project Management; the School of Management; and the School of Social Science and Planning. Its focus was the creation of a holistic vision of sustainability, defined in the context of the disciplines involved, understood in relation to the limitations and opportunities presented in societal practice, and taught in a way that is progressive rather than reactive. The project provided assistance in the identification of systemic links across the schools where relevant sustainability capabilities and theory could be linked to course content. In turn there were opportunities for assisting academic development through creation of their own vision of sustainability, and shaping it in relation to their chosen discipline and its professional practice. The BELP project drew on the insights of previous attempts to address sustainability education at RMIT and sought to achieve lasting change in organizational structure/operations and curriculum content. The project resulted in several tangible outcomes including sixteen new and revised courses in a range of discipline areas, and the development of a flexible change framework to assist the establishment of sustainability content into curricula.

Research conducted by Holdsworth et al. (2006b, 2008, 2009) and Holdsworth (2010) to assess the success of the project approach identified that academic development should be founded on placing sustainability education praxis as the central goal of any academic development program; focusing on the development of an individual's own definition and understanding of sustainability/sustainable development. Allowing for the role of disciplinary and personal experience in their construction; as well as the exploration of the individual's worldview and how it informs their educational practice. The aim should be to achieve deep learning ("third order" learning) to achieve a shift of consciousness and changes in behavior and professional practice. Academic development must be instigated and supported from both the "top down" and "bottom up," and needs to be based on identified drivers of change and committed support structures (financially and collegially). Universities have a distinctive dominant culture because of their complex social organization (Hegarty, 2008). This must be understood as a means of minimizing the occurrence of cultural conflict, and to foster the development of shared goals (Hegarty, 2008).

Additionally further research into incentives institutional policies and action plans that will encourage sustainability education is required. Such as research into action research projects in curriculum change for sustainability, multidisciplinary and transdisciplinary practice. While, research and the development of models that link learning and teaching to successful and effective campus management initiatives will allow for experiential learning.

Conclusion: Now and Into the Future

Universities are beginning to develop and incorporate sustainability into education; however, to ensure we develop sustainability curriculum that is holistic, multidisciplinary and contextually relevant, we need our academic institutions to go beyond merely reflecting the priorities of today's society. Rather, graduates must be able to educate others to think innovatively and creatively with the ability to embed their own set of values in their professional practice.

Sustainability education requires students to develop meta-skills such as the ability to think critically about the nature of knowledge and about the ways in which knowledge is produced and validated. Skills and capabilities specific to each profession are also required. Educating for these new skills will require shifts in educational practice and the development of new curriculum. Educators must become aware of their disciplinary assumptions and traditions and recognize the priorities and values that are played out in the classroom (through the construction of curriculum, learning and teaching methods and the pedagogy that consciously informs this), which in turn influence their own and their students' personal and professional practice.

Improving learning and teaching practice requires educators to consider themselves as active learners who recognize how they construct their own understanding of knowledge. This is especially important for those engaging in sustainability education, given that the sustainability paradigm is contested and open to epistemological interpretations (McAlpine & Western, 2000).

A significant transformation is required for universities to meet the challenge of sustainability education. Yet, the theoretical and practical foundations required are still in their infancy and holistic models of sustainability education are yet to be developed. For sustainability education to become institutionalize into educational praxis within universities more research needs to be undertaken into the areas of appropriate disciplinary pedagogies, learning and teaching methods, and curriculum, along with academic development programs which work systemically in changing the university culture to support this new approach to education. The need for understanding change in universities, both curriculum change and organizational change, is paramount if sustainability education is to be successful.

References

Adomssent, M., Godeman, J., Leicht, A., & Busch, A. (Eds.). (2006). *Higher education for sustainability: New challenges from a global perspective.* Waldkirchen, Germany: UNESCO.

Alvarez, A., & Rogers, J. (2006). Going "out there": Learning about sustainability in place. *International Journal of Sustainability in Higher Education, 7,* (2), 176–188.

Arnold, G., & Civian, J. (1997). The ecology of general reform. *Change, 29*(4), *19–23.*

Barth, M., Godemann, J., Rieckmann, M., & Stoltenberg, U. (2007). Developing key competencies for sustainable development in higher education. *International Journal of Sustainability in Higher Education,* Bradford, UK, *8*(4), 416–441.

Bawden, R. (1997). "The community challenge: The learning response." *Invited Plenary Paper: Twenty-ninth Annual International Meeting of the Community Development Society.* Athens, Georgia, 27–30 July 1997.

Bawden, R. (2005). Systemic development at Hawkesbury: Some personal lessons from experience. *Systems Research and Behavioral Science, 22,* 151–164.

Baud, R. (2004). *Y.E.S.—Student Education in Sustainability* Public Education in a Knowledge Society: Creativity, Content, and Delivery Mechanisms Delhi Sustainable Development Summit, New Delhi, India, February 1/7/2004.

Bekessy, S., Burgman, M., Wright, T., Filho, W., & Smith, M. (2003). *Universities and sustainability,* Tela Series, 11. Australia: ACF.

Blewitt, J., & Cullingford, C. (2004). *The sustainability curriculum the challenge for higher education.* London, UK: Earthscan.

Brundtland, G. (Ed.). (1987). *Our common future: The world commission on environment and development.* Oxford, UK: Oxford University Press.

Capra, F. (1975). *The tao of physics.* London, UK: Wildwood House.

Capra, F. (1982). *The turning point—science, society and the rising culture.* London, UK: Wildwood House.

Clinton, A. (2006). *Why systems thinking is a critical skill.* Global Learning Inc. Retrieved April 10, 2006, from http://www.globallearningnj.org/glean07.htm.

Codd, J. (1982). Epistemology and the politics of educational evaluation, in Deakin University, *Curriculum Evaluation: Philosophical and Procedural Dilemmas,* ECS801 Curriculum Evaluation, Geelong, VIC: Deakin University.

Corneya, G., & Reid, A., (2007). Student teachers' learning about subject matter and pedagogy in education for sustainable development. *Environmental Education Research, 13*(1), 33–54.

Cortese, A. (2003). The critical role of higher education in creating a sustainable future. *Planning for Higher Education, 31(3),* 15–22.

Cotton, D., Warren, M., Maiboroda, O., & Bailey, I. (2007). Sustainable development, higher education, and pedagogy: A study of lecturers' beliefs and attitudes. *Environmental Education Research, 13*(5), 579–597.

Dawe, G., Jucker, R., & Martin, S. (2005). *Sustainability literacy in higher education: Current practice and future developments.* UK: The Higher Education Academy.

De Figueiredo, A. D., Afonso, A. P., & da Cunha, P. R. (2002). *Learning and education: Beyond the age of delivery.* Paper presented at International Conference on Engineering Education, Manchester, UK.

Eckel, P. D., & Kezar, A. (2003, March). Key strategies for making new institutional sense: Ingredients to higher education transformation. *Higher Education Policy, 16*(1), 39–53(15).

Eisen, A., & Barlett, P. (2006). The piedmont project: Fostering faculty development towards sustainability. *Journal of Environmental Education, 38*(1), 25–36.

Elliot, J. (1991). *Action research for educational change.* Buckingham: Open University Press.

Fien, J. (2001). *Education for sustainability: Reorientating Australian schools for a sustainable future.* Tela Series, 8, Australia: ACF.

Fien, J., Guevara, R., Lang, J., & Malone, J. (2004). Australian country report UNESCO–NIER Regional seminar on policy, research and capacity building for education innovation for sustainable development, ESCO *Australian National Commission. Education for sustainable development,* Tokyo, Japan.

Filho, L. W. (Ed.). (2000). Dealing with misconceptions on the concept of sustainability. *International Journal of Sustainability in Higher Education, 1*(1), 9–19.

Fricker, A. (1998). Measuring up to sustainability. *Futures*, 30, 367–375.

Fricker, A. (2002). The ethics of enough, *Futures, 34*(5), 427–433.

Freire, P. (1972). *Pedagogy of the oppressed.* UK: Herder & Herder.

Gligo, N. (1995). Monitoring for sustainability. In T. Trzyna (Ed.), *A sustainable world: Defining and measuring sustainability* (p. 17). Gland, Switzerland: IUCN.

Gough, A. (1997). *Education and the environmental policy, trends and the problems of marginalisation.* Australia: ACER.

Gough, S., & Stables, A. (Eds.). (2008). *Sustainability and security within liberal societies.* London, UK: Routledge.

Grant, M. (2009). Internationalising education for sustainability—the Youth Encounter on Sustainability (YES). In P. B. Corcoran & P. M. Osano (Eds.), *Young people, education and sustainable development: Exploring principles, perspectives, and praxis.* Wageningen, The Netherlands: Wageningen Academic Publishers.

Hayles, C. S., & Holdsworth, S. E. (2008). Curriculum change for sustainability. *The Journal for Education in the Built Environment, 3*(1), 25–48.

Hegarty, K. (2008). Shaping the self to sustain the other: Mapping impacts of academic identity in education for sustainability. *Environmental Education Research, 14*(6), 681–692.

Hegarty, K., Thomas, I., Kriewaldt, C., Holdsworth, S., Bekessy, S. (2011). Insights into the value of a 'stand-alone' course for sustainability education. *Environmental Education Research, 17*(4), 451–469.

Henn, C., & Andrews, C. (1997). *Why Systems Thinking is a Critical Skill,* Retrieved June 13, 2006, from http://www.globallearningnj.org/glean07.

Higher Education Academy (HEA). (2006). *Sustainable Development in Higher Education, Current Practice and Future Developments: A Progress Report for Senior Managers in Higher Educational,* Retrieved March 1, 2006, from http://www.heacademy.ac.uk/4074.htm.

Holdsworth, S., Bekessy, S., Hayles, C., Mnguni, P., & Thomas, I. (2006a). Beyond leather patches project for sustainability education at RMIT. In W. L. Filho & D. Carpenter (Eds.), *University sustainability in the Australasian university context* (pp. 107–28). Frankfurt: Peter Lang Scientific Publishers.

Holdsworth, S., Bekessy, S., Hayles, C., Mnguni, P., & Thomas, I. (2006b). Beyond leather patches: Sustainability education at RMIT University, Australia. In W. L. Filho (Eds.), *Innovation, education and communication for sustainable development* (pp. 153–176). Frankfurt: Peter Lang Scientific Publishers.

Holdsworth, S., Wyborn, C., Bekessy, S., & Thomas, I. (2008). Professional development for education for sustainability: How advanced are Australian universities? *International Journal of Sustainability in Higher Education, 9*(2), 131–146.

Holdsworth, S., Bekessy, S., & Thomas, I. (2009). Evaluation of curriculum change at RMIT: Experiences of the BELP project. *Reflecting Education, 5*(1), 51–72.

Holdsworth, S. (2010). Doctoral thesis: A critique of academic development in sustainability for tertiary educators. RMIT University, Australia.

Huckle, J. (2005). *Education for sustainable development: A briefing paper for the teacher training agency*. UK: The Teacher Training Agency.

Huckle, K., & Martin, A. (2001). *Environments in a changing world.* Harlow, UK: Prentice Hall.

Jones, P., Trier. C., & Richards, J. (2008). Embedding education for sustainable development in higher education: A case study examining common challenges and opportunities for undergraduate programs. *International Journal of Educational Research, 47*(6), 341–350.

Jucker, R. (2002a). "Sustainability? Never heard of it!" Some basics we shouldn't ignore when engaging in education for sustainability. *International Journal of Sustainability in Higher Education, 3*(1), 8–18.

Jucker, R. (2002b). *Our Common illiteracy: Education as if the earth and people mattered.* Germany: Peter Lang.

Kemmis, S. (2008a). Practice and practice architectures in mathematics education. Keynote address to the 31st annual Mathematics Education Research Group of Australasia (MERGA) Conference: Navigating Currents and Charting Directions, University of Queensland, St Lucia, 28 June–1 July 2008.

Kemmis, S. (2008b). *Research for practice: Knowing doing*, Paper presented at the *Pedagogy, Culture & Society* (journal) sponsored special conference on "researching practice," University of Gothenburg, September 13 2008.

Kemmis, S. (2009). *Sustaining practice: Towards a rich characterization of exemplary Education for Sustainability initiatives.* Australia: Charles Sturt University and the Australian Research Council.

Kemmis, S., & Fitzclarence, L. (1986). *Curriculum theorising beyond reproducing.* VIC: Theory Deakin University Press.

Kemmis, S., & Smith, T. (2008). *Enabling Praxis: Challenges for Education*. Rotterdam, The Netherlands: Sense Publishers.

Lang, J. (2004). Environmental education, education for sustainability and UN Decade of Education for Sustainable Development. Paper presented at the International Seminar/Workshop on Quality Environmental Education in Schools for a Sustainable Society, Cheongju, Korea, 24–28 August.

Lazlo, E. (2006). *The Chaos Point. The World at the Crossroads*. Charlottesville, VA: Hampton.

Lozano–Garcia, F., Gandara, G., Perrni, O., Manzano, M., Hernandez, D., Huisingh, D., (2008). Capacity building: A course on sustainable development to educate the educators. *International Journal of Sustainability in Higher Education, 9*(3), 257–281.

McAlpine, L., & Weston, C. (2000). Reflection: Issues related to improving professors' teaching and students' learning. *Instructional Science, 28*, 363–385.

Miriam–Webster. (2006). *Merriam–Webster Online Dictionary—pedagogy*. Retrieved October 8, 2006, from http://www.m–w.com/cgi–bin/netdict?pedagogy.

Murray, P., Brown, N., & Murray, S. (2007). Deconstructing sustainability literacy. *International Journal of Environmental, Cultural, Economic and Social Sustainability, 2*(7), 85–92.

Murray, P., & Murray, S. (2007). Promoting sustainability values within career–oriented degree programs: A case study analysis. *International Journal of Sustainability in Higher Education, 8*(3), 285–300.

Orr, D. (2001). Foreword. In S. Sterling (Ed.) *Sustainable education: Re-visioning learning and change. Schumacher Briefings,* no. 6. Bristol, UK: Green Books.

Parker, J., Wade, R., & Van Winsum, A. (2004). Citzenship, and community from local to global; Implications for higher education of a global citizenship approaches. In J. Blewitt & C. Cullingford (Eds.), *The sustainability curriculum; The challenge for higher education* (pp. 63–77). London, UK: Earthscan.

Parkin, S., Johnson, A., Buckland, H., & White, E. (2004). *Learning and skills for sustainable development: Developing a sustainability literate society*. London, UK: Higher Education Partnership for Sustainability (HEPS).

Piccinin, S. (1997). Making our Teaching More Student–Centred. OPTIONS. 1(5) Retrieved August 21, 2000, from http://www.uottawa.ca/academic/ cut/options/Dec_97/ Student_centred.htm.

Reid, A. (2002). On the possibility of education for sustainable development. *Environmental Education Research, 8*(1).

Robottom, I., & Hart, P. (1993). *Research in environmental education.* Australia: Deakin University Press.

Schön, D. (1983). *The reflective practitioner: How professionals think in action*. New York, NY: Basic Books.

Schwab, J. J. (1969). The practical a language for curriculum. *School Review, 78*, 1–24.

Selby, D. (2006). The firm and shaky ground of education for sustainable development. *Journal of Geography in Higher Education, 30*(2), 353–367.

Siemens, G. (2004). *Connectivism: A learning theory for the digital age*. Retrieved September 15, 2006, from http://www.elernspace.org/Articles/connectivism.htm–.

Sprigett, D. (2005). 'Education for Sustainability' in the business studies curriculum: A call for a critical agenda. *Business Strategy and the Environment, 14*, 146–159.

Stenhouse, L. (1975). *An introduction to curriculum research and development.* London, UK: Heinemann.

Sterling, S. (1996), Education in Change. Chapter 2. In J. Huckle & S. Sterling (Eds.). *Education for sustainability* (pp. 18–39), London, UK: Earthscan.

Sterling, S. (2001). *Sustainable Education Re–visioning Learning and Change*. Schumacher Briefings Number 6. UK: Green Books.

Sterling, S. (2003). Doctoral thesis: Whole systems thinking as a basis for paradigm change in education: Explorations in the context of sustainability, Retrieved June 1, 2007, from http://www.bath.ac.uk/cree/sterling.htm.

Sterling, S. (2004). An analysis of the development of sustainability education internationally: Evolution, interpretation and transformative potential. In J. Blewitt & C. Cullingford (Eds.), *The sustainability curriculum: The challenge for higher education* (pp. 43–62). London: Earthscan.

Sterling, S., & Thomas, I. (2006). Education for sustainability: The role of capabilities in guiding university curricula. *International Journal of Innovation and Sustainable Development (IJISD), 1*(4), 349–370.

Stevenson, R. (1987). Schooling and environmental education: Contradictions in purpose and practice. In I. Robottom (Ed.), *Environmental education: Practice and possibility* (pp. 69–82). Geelong, Australia: Deakin University Press.

Summers, M., Childs, A., & Corney, G. (2005). Education for sustainable development in initial teacher training: Issues for interdisciplinary collaboration. *Environmental Education Research, 11*(5), 623–647.

Tam, S. W. (2000). *Managing learner–centeredness: The role of effective student support in ODL.* A paper presented at ICDE Asian Regional Conference, New Delhi 3–5 November.

Tilbury, D., & Cooke, K. (2005). *A national review of environmental education and its contribution to sustainability in Australia: Frameworks for sustainability.* Canberra, ACT: Australian Government Department of the Environment and Heritage and Australian Research Institute in Education for Sustainability.

Tilbury, D., Keogh, A., Leighton, A., & Kent, J. (2005). *A national review of environmental education and its contribution to sustainability in Australia: Further and higher education.* Canberra, ACT: Australian Government Department of the Environment and Heritage and Australian Research Institute in Education for Sustainability (AIRES).

Thomas, I. G., & Nicita, J. (2003). Employers' expectations of graduates of environmental courses: An Australian experience. *Applied Environmental Education and Communication, 2*(1), 49–59.

Thomas, I. (2004). Sustainability in tertiary curricula: What is stopping it happening? *International Journal of Sustainability in Higher Education, 5*(1), 33–47.

UNEP. (2002a). *Global Environment Outlook 3: Past, present and future perspectives.* Nairobi, Kenya: United Nations Environment Programme.

UNESCO. (2005). *United Nations Decade for Education for Sustainable Development (2005–2014): Draft international implementation scheme,* Retrieved February 14, 2006, from http://unesdoc.unesco.org/images/0014/001403/140372e.pdf.

Warburton, K. (2003). Deep learning and education for sustainability. *International Journal of Sustainability in Higher Education, 4*(1), 44–56.

Weick, K. (1995). *Sensemaking in organizations*. Thousand Oaks, CA: Sage.

34

Learning From Neighboring Fields

Conceptualizing Outcomes of Environmental Education Within the Framework of Free-Choice Learning Experiences

Lynn D. Dierking
Oregon State University, USA

John H. Falk
Oregon State University, USA

Martin Storksdieck
National Research Council, USA

Overview

Increasing evidence suggests that the public constructs its understanding of science, technology, engineering, the environment, and mathematics (STEEM for the purpose of this chapter) over the course of their lives, gathering information from many sources and contexts, and for a diversity of reasons (Falk, Heimlich, & Foutz, 2009; Falk, Storksdieck, & Dierking, 2007). This perspective has profound implications for the conceptualization and assessment of environmental education (EE) or education for sustainability (EfS) impacts, the focus of this chapter. First we situate the discussion of environmental education outcomes within the framework of lifelong learning, with a particular focus on the contribution of free-choice environmental learning, learning that is driven by intrinsic rather than extrinsic motivations.[1] We then discuss implications for future evaluation practice, strategies, and tools, by exploring the following three questions: (1) Given the unique temporal and spatial nature of lifelong free-choice environmental learning, what evaluation strategies and tools incorporating this broad perspective have been used to assess STEEM efforts generally, and EE efforts in particular? (2) What does the research say about the efficacy, validity, and reliability of these strategies and tools? (3) What strategies and tools might be used more effectively in the future?

Lifelong Science/Environmental Learning

A growing body of evidence supports the assertion that science/environmental learning occurs in settings and situations outside of school, as well as across time and a variety of settings (Falk & Dierking, 2010; Heimlich & Falk, 2009). Thus to effectively assess, measure, and evaluate EE outcomes, one needs to situate EE learning within the context of lifelong learning—schooling contributes to this learning but does not account for all of it; out-of-school, free-choice learning is equally if not more important. Following from this perspective are a set of premises about the nature of learning and what counts and is considered valid evidence for learning. The most important aspect of this perspective is that STEEM learning in general, and environmental learning in particular, is viewed as a natural and fairly common result of living within a STEEM-rich world, an outcome of everyday life. Over the course of a lifetime, a person constructs his/her understanding of the world by connecting and building upon experiences they have in school, at work, and in free-choice learning settings and configurations. This cumulative process not only involves the acquisition of facts and concepts, though important, but also includes changes in interest, awareness, skills, behavior, attitudes, beliefs, habits of mind, in feelings, and emotions (Bell, Lewenstein, Shouse, & Feder, 2009; Falk & Dierking, 2000).

Lifelong, free-choice learning also has a sociocultural dimension, such as a person learning about and increasing his/her appreciation for one another and/or other cultures. This learning also has physical dimensions, critical when trying to understand EE outcomes which often are strongly connected to place and other physical characteristics of the environment or experience. There are also aesthetic/recreative outcomes such as renewing, refreshing, and/or restoring one self, often manifested by wonder, awe, joy, and pleasure; these are frequent outcomes in outdoor settings also.

Free-choice learning experiences may result in transformation/connection, significant changes in thinking, attitudes, beliefs, behaviors or habits of mind. Such learning may result because the learner was "primed and ready" for transformation through previous experiences or because unique experiences are so powerful that significant changes result. Of course, the impact of such "priming" experiences only are evident much later. Since most research and evaluation currently misses the contribution of individual EE experiences along a lifelong learning pathway they underestimate the value that even brief educational encounters may have.

For example, adults visit settings such as national parks, science centers, and botanical gardens to satisfy their intellectual curiosity, as well as to fulfill a need for relaxation, enjoyment, and even spiritual fulfillment (Ballantyne & Packer, 2009; Brody, Tomkiewicz, & Graves, 2002; Falk, 2006; Heimlich, Falk, Bronnenkant, & Barlage, 2005; Gammon, 1999; Kaplan, & Kaplan, 1982; Moussouri, 1997; Pekarik, Doering, & Karns, 1999). Adults take their children to these settings because they feel such experiences are worthwhile, educational, and fun, and that they and their children learn about STEM and the environment in the process (Borun, Chambers, & Cleghorn, 1996; Borun, Chamber, Dristas, & Johnson, 1997; Dierking, & Falk, 2003; Dierking, Luke, Foat, & Adelman, 2000; Falk, & Dierking, 1992; Rounds, 2004).

Adults also encourage their children to participate in a wide variety of after-school and extracurricular experiences including scouting and summer camp experiences, many of which also support environmental learning (e.g., Dierking & Falk, 2003; McCreedy & Zemsky, 2002; Ponzio & Marzolla, 2002; St. John et al., 2000). Adults and children also learn while engaged in personal investigations, through civic organizations and active leisure pursuits (Anderson, 1999; Anderson, Greeno, Reder, & Simon, 2000; Falk, 2002; Falk & Dierking, 1992, 2000, 2002). Similar motivations and findings can be ascribed to watching STEEM specials on television, using the Internet to access STEEM-related information, and engaging in STEEM-related hobbies and special interest groups such as birding, gardening, and hiking (Azevedo, 2005; Chadwick, 1998; Chadwick, Falk, & O'Ryan, 2000; Chan, 1996; Chien, 1996; Elder, Coffin, & Farrior, 1998; Eveland and Dunwoody, 1998; Gross, 1997).

A recent study by the National Research Council (NRC) on learning science in informal environments (Bell et al., 2009) reinforces the abundant evidence that free-choice EE efforts, even everyday experiences such as a walk in the park, contribute to people's knowledge and interest in STEM generally and the environment specifically. This chapter is based on a commissioned paper we authored to inform the NRC report on evaluation studies of free-choice/informal STEM programs, in particular those that tried to operationalize and measure STEM learning outcomes (Institute for Learning Innovation, 2007).[2] Another commissioned paper that informed the committee's work, entitled *Assessing Learning in Informal Science Contexts*, by Brody, Bangert, and Dillon (2007),[3] provides another approach and perspective.

Efficacy, Validity, and Reliability of Strategies and Tools Currently Being Used

Assessing the quality of current evaluation practice in informal science and environmental education was challenging for a number of reasons, not the least of which was getting access to studies themselves. We have outlined our approach and rationale for selecting evaluation studies in detail elsewhere (Institute for Learning Innovation, 2007). The final criteria for selecting studies were fairly basic: at a minimum they needed to have measured outcomes and followed rigorous standard in terms of methodologies chosen (for instance, adequate sample sizes). We analyzed studies that met these criteria and they provide a reasonable understanding of what is being accomplished in the ISE field, particularly given the current environment which typically includes inadequate funding for evaluation (this has been changing recently with the National Science Foundation's leadership), and for the most part, a lack of clear guidelines for what constitutes quality evaluation. NSF's Informal Science Education (ISE) Evaluation Framework was published in 2007 (Friedman, 2007) but it is too soon to determine whether it has improved the situation.

Emerging Trends The evaluation studies selected reflect the bare minimum expected but not necessarily more. Most studies were informal, lacked clearly delineated independent and dependent variables and sadly also lacked vision, measuring what *could* be measured rather than trying to develop new approaches to document what *should* be measured.

For the most part research and evaluation designs, methodologies, data collection, analysis and interpretation approaches have not evolved greatly over time although there has been some effort to move beyond measuring outputs and basic learner behavior and satisfaction. With the need to document not only whether, but how and why informal learning experiences are effective, the need for methodological explanatory power has increased somewhat in recent years, resulting in evaluators employing mixed methods and triangulation designs. There has also been a need to develop more rigorous quantitative scales specific to ISE, introducing item development and analysis to the field of ISE evaluation (this rigorous approach to instrument development has long been a part of more formal evaluation practice).

In addition, as qualitative approaches have been adopted, strategies for validating such data have also been integrated within designs. A variety of strategies can be used to ensure the validity of such approaches, including triangulation by method and analysis (content, cross-case, and recursive analysis), identification of patterns of contradiction, as well as coherence (Sewell, 2010), and periodic opportunities for study participants to review researchers' analysis and interpretive results to ensure validity from the perspective of the participants. Although a few evaluators still express preference for either quantitative or qualitative research/evaluation designs, mixed methods are far more common. Increasingly, at least the rhetoric of the field suggests that one must tailor the approach to the project and the types of questions being asked.

There is also a growing trend to embed evaluation (and research) more fully into the design and implementation of projects from the outset. In part this is because the whole approach to project design is becoming far more developmental and design-based, with the influence of ISE's new guidelines which necessitate a "backwards design" approach in which project staff first identify the intended impacts they want to accomplish and then move forward by targeting an appropriate audience, developing the project design, assembling the project team and any partners involved and delineate project deliverables. This is also a result of the growing field of learning sciences, whose researchers approach research from this perspective, and increasingly are working in the informal/free-choice STEEM learning arena.

Although for many years there been efforts to meaningfully integrate evaluation into the project development process, this trend seems more frequent, natural, and organic than earlier "team" efforts. Findings are being used to improve projects and there is increasing openness and acceptance that evaluation is helpful, that audience perspectives matter and that quality evaluation requires expertise and experience. This is resulting in new models of cooperation; for instance some projects are embedding rapid evaluation into the design and development process. The emphasis of rapid evaluation stresses capacity building and project staff involvement (even sometimes in data collection). Reports are delivered as bullets or PowerPoints, and followed up quickly in meetings or via phone by the evaluator and project team in order to discuss and integrate results quickly.

Research-based projects are also using iterative design-based approaches in an ongoing manner. The *Active Prolonged Engagement* (APE) project, a four-year NSF-funded effort at the Exploratorium, is an excellent example (Humphrey, Gutwell, & The Exploratorium APE Team, 2005). Both an exhibition development project and an audience research study, the primary aim of APE was to explore approaches that might shift the role of visitors from passive recipients of information to active participants in the learning experience. In projects like this, the documentation of the process, often including interviews and videotapes from focal participants, becomes an important finding. In the case of the APE project, thirty APE exhibits were developed and evaluated, a workshop was hosted by the Exploratorium, and a publication, conference presentations, and journal articles documenting findings resulted.

Our research also revealed a great deal of debate in the ISE field and the wider after school/family/community involvement and youth development arenas about both randomized control treatment (RCT) designs and standardized and tested outcome measures (a response to increasing accountability pressure). This debate is also alive and well in the social science research community as a whole (in fact there was an entire issue of *Educational Researcher* devoted to the topic a few years ago). Issues that have been raised in response to this pressure include the fact that learning is complex and multifaceted, individual, yet socially mediated, with in a rich physical setting. In addition, audiences are heterogeneous in terms of age, gender, background, and knowledge. All of these present tremendous challenges to creating valid studies that utilize RCT design.

There was also a great deal of discussion about what makes the informal/free-choice field of learning unique, particularly the effort of ISE of late to emphasize strategic impact, build on prior work and educational research, while also supporting projects at the frontiers which may be too exploratory to easily fit into a RCT design. There was concern that adherence to common metrics, whether intended or not, might ultimately limit the innovative and cutting edge nature of efforts in this arena. One other issue raised was the failure of RCT-designed experiments to acknowledge the reality first raised by Levins (1968) that it is impossible for any research design or model to simultaneously maximize precision, generalizability, and reality; at least one of these always has to be sacrificed. RCT designs typically sacrifice realism and/or generalizability at the expense of precision. This is an unfolding issue but another interesting point is that most people who say they are utilizing a RCT design are actually using a quasi-experimental design (Ferraro, 2005). Very few studies are able to truly use random selection because of the difficulties inherent in the process (Chamberlain, 2007). Needless to say, in free-choice learning self-selection is an inherent aspect of the learning process itself, requiring new approaches and new thinking about sampling designs.

Implications for Alternative Evaluation Approaches and Methods

In addition to analyzing current practice, our NRC paper addressed the contribution that evaluation has made to the field's understanding of the impacts of free-choice/informal STEEM learning experiences and suggests how evaluation and research practices can be enhanced in the future. Our conclusions and recommendation for evaluation and research practice, summarized in the following section, strongly called for the need for alternative approaches to evaluation in this arena. In part we felt alternative approaches were needed because the epistemological foundation of most traditional assessment approaches is positivist-behaviorist, raising concerns about their validity vis-à-vis free-choice/informal STEEM learning. Typically great effort is placed on choosing tools and strategies that are highly reliable (consistent over time of use and utilization by different evaluators/researchers). However, the approaches we suggest also highlight the importance of validity, that is, that one is using an approach or tool that actually "measures" a relevant outcome or set of outcomes. Unfortunately the coin of the realm in many evaluation studies is the highly reliable survey, which may have little validity, particularly when considering the unique personal and idiosyncratic nature of knowledge and knowing.

Traditional methods, either explicitly or implicitly, assume that everyone starts at the same place (e.g., "no or little knowledge") and ends at a similar place (e.g., "the correct answer"), as a result of a consistent and knowable learning intervention (e.g., the "educational stimulus"). By contrast, we suggest that alternative methods are needed that encapsulate and value the fact that lifelong STEEM learning is best described and operationalized from the perspective of the person or group as his/her or their self-defined understanding, what we have previously referred to as *working* knowledge (Falk et al., 2007). This asset-based approach to understanding evaluates what people know, understand and care about from their own perspective and assumes that everyone has some knowledge and understanding, even if that of the novice. Thus to meaningfully assess the impact of a STEEM-related environmental education effort, requires approaches and tools that analyze the what, where, when, why how and with whom of the effort. In addition to accounting for context, alternative evaluation strategies also are designed to recognize that learning is the result of contributions from a variety of experiences across a lifetime, that learners themselves vary considerably *and* that the nature of knowledge is uniquely personal and idiosyncratic. This is a tall order, but critical to effectively documenting impact.

Another subtle implication of this perspective is that although evaluation is often designed to assess the impact of a *specific* program or activity isolated in time and space, in order to demonstrate causal relationships between program activities and specific outcomes. This is not always the best approach (interestingly it often yields disappointing results, particularly when assessing short-term impacts of EE learning interventions that may be relatively brief in nature such as field trips, family vacations, or summer camp experiences). The cumulative and incremental nature of learning suggests that it is more beneficial to employ evaluation approaches and methods that attempt to document and "measure" the *contribution* that a *specific* program or activity makes to what a person *already* understands, knows, and feels about a particular topic, behavior, or attitude, as well as the connection that this effort has to other experiences in the learner's life, both before and after. The implication is that approaches should seek out and assess correlational relationships between and among learning experiences, rather than trying to make causal claims about the impact of any particular effort.

The white elephant in the room of course is what counts as impact, and given the diversity of EE or EfS learning experiences, and the depth of exposure learners may (or may not) have, what realistic impacts and outcomes might be. The following represent some criteria and a potential framework for defining outcomes within the social, cultural, and historical context of people's lives (Dierking, 2007; Dierking et al., 2002; Falk, 2009). Outcomes taking this understanding into account:

- align with prior experience, knowledge, and expectations of learners;
- are open-ended and flexible;
- are assessed at different points in time;
- require alignment with value judgments about effectiveness and need.

Ultimately, defining and prioritizing outcomes requires a value judgment; all outcomes are not created equal. In certain contexts, some outcomes will be more appropriate, valuable, or useful than others. After all, evaluation is a political process. In defining outcomes, the decision-maker is making a value judgment, based on their definition of the purpose and contribution this effort will make to participants and an EE professional is also communicating a strong message for why this effort or activity is important, in other words making a case for the unique contribution and niche of this experience. Thus an outcome should be relevant to the EE experience, and be something that is more optimally accomplished through that experience than another so that its contribution can be teased out. This last criterion is particularly important for the continued value of EE outcomes within society. EE professionals need to be able to point to their singular contribution within the menu of learning options from which people can choose. Ultimately, the discussion about outcomes is a discussion about the contribution of EE in society.

Two approaches we have championed are designed to try to account for the notion of broad self-defined outcomes and potential variability in outcomes between individuals, though each tackle the issue in a slightly different manner. The first, Personal Meaning Mapping (PMM), a modified concept mapping method developed specifically for use in informal education settings by Falk and colleagues at the Institute for Learning Innovation (Falk, Moussouri, & Coulson, 1998), is an all-encompassing approach to framing a study, collecting, analyzing, and interpreting data. The other strategy, data aggregation, is a technique for analyzing data in more sensitive ways that recognize the differing entering knowledge, understanding, attitudes, and behaviors of participants/learners.

Personal Meaning Mapping grew out of concerns raised about the validity of traditional assessment approaches. PMM is designed so that there is no specific "right" answer to demonstrate impact. Instead, this approach measures the unique conceptual, attitudinal, and emotional impacts of a self-defined learning experience on an individual, focusing both on *the degree of the change* but equally on *the nature of the change.* It enables investigators to probe meanings that cannot easily be gleaned from methods such as written questionnaires or surveys, structured or semistructured interviews, or observational methods (cf., Falk, 2003).

Personal Meaning Mapping has several potential advantages as an evaluation tool. It does not feel like a "test," and requires relatively little time on the part of the research participant. And since the methodology focuses on the person's own experiences and unique ways of articulating those experiences—the technique is enjoyed by most study participants. Its relativist-constructivist foundation also allows the researcher to negotiate more meaningful relationships with the people they are interviewing, and to develop a more personalized understanding of the learning of an individual. Importantly, PMM provides a valid way to understand the personal meaning people construct from learning experiences.

PMM data is analyzed along four independent dimensions: *Extent, Breadth, Depth and Mastery,* and for some topics such as those related to the environment, also along an emotional intensity dimension (Adelman et al., 2001; Dierking et al., 2004; Falk & Storksdieck, 2005). To date, the measure has been used to assess changes in individual's understanding of such broad topics as gems and minerals (Falk et al., 1998), First Nations topics and art (Adams, Falk, & Dierking, 1999, 2003); specific concepts such as basic biology knowledge (Falk & Storksdieck, 2005) conservation knowledge, behavior and attitudes (Adelman, Falk, & James, 2000; Dierking et al., 2004); as well as abstract ideas such as creativity (Falk, 2003) and art interpretation (Luke, Adams, Abrams, & Falk, 1998). It is also being used as a tool to document the long-term impact of free-choice science learning experiences on young women's interest, engagement, and participation in science communities, hobbies, and careers (McCreedy & Dierking, 2008; Dierking & McCreedy, 2010). Results show that not only is PMM a meaningful way to document impact, but it is highly valid *and* reliable. In only one instance has a validity issue arisen: Falk and Storksdieck (2005) found independence in only three out of four measures; the holistic measure, "mastery" was functionally identical to measure three, "depth." Researchers continue to use the approach, though mindful of this issue.

The other approach to dealing with potential variability in outcomes between and among individuals, aggregating data, is an approach to data analysis. As we have said throughout this chapter, the impact of any environmental education experience will vary widely because of the vast range of experiences, knowledge, attitudes, interests, and motivations of the learners involved. These differences directly affect how each guest perceives the experience, the sense he or she makes of the information and ultimately, the degree to which subsequent action is influenced. As suggested these outcomes will be highly personal and quite variable (Falk & Dierking, 2000).

To deal with this issue, researchers have had success aggregating data from groups of people with similar entering knowledge, behavior, and attitudes, demonstrating that changes are more discernible when people are grouped in meaningful ways in relationship to what they know. For example in a study at the National Aquarium in Baltimore, significant changes in knowledge, understanding, and attitudes emerged for similar groups of people that were not seen when the data were analyzed as a whole (Adelman et al., 2000). Based on this experience, researchers hypothesized that one of the challenges of measuring changes in conservation behavior might be diminished if it were possible to categorize people in meaningful ways related to their behavior. The Prochaska Model, adapted by Falk from the public health arena (1998), provides a theoretical framework which may be helpful. The continuum of behavior change proposed by the Prochaska Model allowed researchers to pool samples of various conservation behavior according to the stage a person was at, thus reducing the sample's variability. By gauging beginning self-reported behavior and change among more similar people discernable impact from the experience could be determined.

Future Practice

A variety of improvements in terms of strategies and tools that we felt could be made to better document the impact of ISE efforts also have relevance for the evaluation of EE impacts. The most important improvement that could be made is to recognize that lifelong free-choice learning is different from classroom learning and the field is well served to choose a different assessment path than that taken by schools (Ballantyne & Packer, 2009; Brody et al., 2007; Heimlich & Falk, 2009). The following are aspects to consider when evaluating learning that results from free-choice experiences:

- Understand the often complex agenda and the mixed motivations of free-choice learners. Free-choice learners who visit zoos, watch a documentary, or visit a national park often are trying to combine enjoyment and learning: for them, learning is fun and highly personal in nature (Packer, 2006).
- Help program developers understand their target audience and define what can be accomplished well. An activity that seeks to address everybody is most likely satisfying no one as it will likely fail to serve individual needs, wants or expectations sufficiently. Solid front-end evaluation or needs assessment is crucial and is best embedded in a development model that links audiences to expected outcomes through clearly defined experiences.

- Lifelong, free-choice learners are not empty vessels: they come with prior knowledge and understanding—sometimes with alternative conceptions and other times knowledge exceeding those designing the experiences. Learners bring with them awareness, attitudes, interest and intentions, and a lifetime of experiences leading to this moment. Again, solid front-end evaluation is needed to uncover these aspects of one's audience, particularly since we know that learners are not necessarily open to new ideas if not embedded into learners' pre-existing cognitive and emotive background.
- There is no such thing as an average person; assuming that learners in a nonschool setting are seeking the same outcomes from an experience is to set oneself up for failure. Rather than understanding the "audience," consider the many audiences—smaller groups with similar backgrounds can be addressed in similar ways, but it is important to remember that one size does not fit all. One potentially useful strategy for segmenting audiences has been Falk's (Falk, 2006, 2009; Falk, Heimlich, & Bronnenkant, 2008; Falk & Storksdieck, 2010) use of learners' identity-related motivations.
- Impact, or audience outcomes, in part should be defined by the target audience and carefully calibrated to what is possible to achieve for that particular audience: do not overestimate (or underestimate) the depth of impact you can achieve. Some audience members may be ready to change their behaviors, others may be skeptical and need more convincing, but both could be in the audience.
- The sum total of the experience is not just what happens at the time of the experience—much is determined by events that happen in the time before an individual participates, and equally importantly, in the time subsequent to the experience. Without this longer perspective one cannot hope to understand a learner's experience, thus studies are beginning to focus on science learning as part of lifelong learning; the perspective of "horizontal" learning.

There is also increasing understanding that the field needs to invest in longitudinal studies. Taking a more longitudinal approach to data collection allows researchers to get a more holistic view of the role of STEEM learning within peoples' lives. Most commonly, longitudinal designs involve conducting follow-up interviews with participants/users weeks, months, and even years after the experience and results suggest that long-term learning is a result. Researchers have repeatedly shown that many of the conversations that begin during a free-choice learning experience continue once people are back at home (Astor-Jack, Whaley, Dierking, Perry, & Garibay, 2007; Falk & Dierking, 1997; Leinhardt, Crowley, & Knutson, 2002). Ethnographic case studies that investigated a set of families who visited museums frequently, allowing for repeated observations and interviews before, during, and after museum visits (Ellenbogen, 2003), demonstrated that conversational connections between museum experiences and real-world contexts are frequent and yet must be examined carefully since the connections are not always obvious to those outside the family. This research was important because it also explored the life-wide notion of observing and talking to people in the various settings they experience in their lives. As a field, there is much to be learned by situating outcomes within larger frames of time, culture, and space.

Technology is influencing all aspects of our lives, so it should be no surprise that it is also having a tremendous impact on evaluation and research, both in terms of expanding data collection techniques, but also in terms of providing additional ways to represent and share research findings, ranging from embedded assessment in digital games and simulations, to online surveys, as well as GPS and RFID tags, digital audio and video recording for unobtrusive observations.

In an arena of increased accountability, many funders are requiring that projects demonstrate broad strategic impact. Thus, evaluators are increasingly being asked that the findings of their studies either contribute to the field (even some private funders are requiring that projects they fund demonstrate wider impacts and serve as proof of concept) or support the overall mission and goals of an organization, or in some cases show whether and to what degree, entire communities are affected. Thus the purpose of the evaluation becomes one of not merely improving an individual project as it is developed or assessing the degree to which an effort achieves its goals and objectives; rather, the evaluation assesses the overall usefulness of the project in reaching institutional or community goals. In this way, individual evaluation studies can become part of larger strategic assessments or ideally, the design for the evaluation is whole cloth within the strategic effort itself. In the United States, this approach is referred to as strategic or asset evaluation or institution-wide assessment and a recent *Curator* issue was devoted to the topic (Koster & Falk, 2007). Such an approach is referred to as "validation" in Europe, though the way validation is implemented in most European projects is that there is a summative evaluation team *and* a separate validation team.

The pressure to take such an approach is coming from a variety of directions, much of it external, including OMB requirements for federal agencies to demonstrate the impact of *program* areas rather than *individual* projects, or from private funders, and directors and boards at institutions seeking to know how their efforts address broader strategic needs in their local communities (rather than the traditional needs of PI's, project directors, or education directors, who have focused on whether and to what degree the project has accomplished its specific goals). Interestingly another driver for this kind of evaluation is increasingly the actual communities involved in the efforts. In particular, organizations supporting populations referred to as underserved are increasingly demanding that any efforts involving them be able to demonstrate real impact on the individuals and communities on which the effort is focused in order to continue participation (Reardon, Sorenson, & Clump, 2003). At the most extreme, some have argued that the ultimate measure of impact for an EE effort should be the resulting impact it has on the environment, for example, less negative climate change impacts. Of course, if we struggle to provide evidence for impacts on individuals, groups, and communities, are we in any position to assess impacts on something as complex as an ecosystem?

In Conclusion

Although historically in terms of its practice and efforts to measure its impact, STEEM education has focused on precollege and college level schooling, there is an opportunity to move beyond such narrow and potentially invalid approaches. With the recognition that only a part of the public's interest and knowledge about STEEM is shaped by compulsory schooling, one is freed and obliged to develop and apply methodologies, methods, and approaches that better reflect current understanding of learning and behavior change as part of a broad ecology of learning opportunities which are navigated daily by individual learners. By ignoring this opportunity, we are likely to both underestimate and undervalue what the public knows about the environment and the important role that lifelong learning plays in supporting this understanding. By situating environmental education outcomes within the framework of lifelong learning (no matter where it occurs), with a particular focus on the contribution of free-choice environmental learning, it is much more likely that the EE field will make progress in understanding how to support the learning and behavior change of citizens, critical as we face complex issues such as global climate change, loss of biodiversity, and an everincreasing demand on natural resources to fulfill human needs.

Notes

1. The authors use the term *free-choice* rather than *informal* learning because it better reflects the nature of the learning: nonlinear, open-ended, voluntary, self-directed, personal, ongoing, learner-centered, guided by the learner's needs and interests, contextually relevant, collaborative, with a high degree of choice as to what, when, how, why, where, and with whom to learn (Falk & Dierking, 2000). This definition also recognizes other disciplinary approaches such as anthropology and sociology so discourse, networks, communities of practice, apprenticeships, gaming, serious leisure, the study of hobbies, and so on are also considered.
2. http://www7.nationalacademies.org/bose/Institute_for_Learning_Innovation_Commissioned_Paper.pdf
3. http://www7.nationalacademies.org/bose/Brody_Commissioned_Paper.pdf

References

Adams, M., Falk, J. H., & Dierking, L. D. (2003). Things change: museums, learning, & research. In M. Xanthoudaki, L. Tickle, & V. Sekules (Eds.), *Researching visual arts education in museums and galleries: An international reader.* Amsterdam: Kluwer Academic Publishers.

Adelman, L. M., Dierking, L. D., Haley Goldman, K., Coulson, D., Falk, J. H., & Adams, M. (2001). *Baseline impact study.* Disney's Animal Kingdom: Conservation Station.

Adelman, L. M., Falk, J. H., & James, S. (2000). Assessing the National Aquarium in Baltimore's impact on visitor's conservation knowledge, attitudes and behaviors. *Curator: The Museum Journal, 43*(1), 33–62.

Anderson, D. (1999). *Understanding the impact of post–visit activities on students' knowledge construction of electricity and magnetism as a result of a visit to an interactive science centre* (Unpublished doctoral dissertation) Queensland University of Technology. Brisbane, Australia.

Anderson, D., Lucas, K., Ginns, I., & Dierking, L. (2000). Development of knowledge about electricity and magnetism during a visit to a science museum and related post–visit activities. *Science Education, 84*(5), 658–679.

Anderson, J. R., Greeno, J. G., Reder, L. M., & Simon, H. (2000). Perspectives on learning, thinking, and activity. *Educational Researcher, 29*(4), 11–13.

Astor–Jack, T., Whaley, K., Dierking, L., Perry, D., & Garibay, C. (2007). Investigating socially mediated learning. In J. Falk, L. Dierking, & S. Foutz (Eds.), *In principle, in practice.* Lanham, MD: AltaMira Press.

Azevedo, F. S. (2005). *Serious play: A comparative study of learning and engagement in hobby practices* (Unpublished doctoral dissertation). University of California, Berkeley.

Ballantyne, R., & Packer, J. (2009). Future directions for research on free–choice environmental learning. In J. H. Falk, J. E. Heimlich, & S. Foutz (Eds.), *Free–choice learning and the environment.* Lanham, MD: AltaMira Press.

Bell, P., Lewenstein, B., Shouse, A. W., & Feder, M. A. (Eds.). (2009). *Learning science in informal environments: People, places, and pursuits.* Washington, DC: Committee on Learning Science in Informal Environments, National Research Council.

Borun, M., Chambers, M., & Cleghorn, A. (1996). Families are learning in science museums. *Curator, 39*(2), 123–138.

Borun, M., Chamber, M. B., Dritsas, J., & Johnson, J. I. (1997). Enhancing family learning through exhibits. *Curator, 40*(4), 279–295.

Brody, M., Bangert, A., & Dillon, J. (2007). *Assessing science learning in informal settings.* Washington, DC: Commissioned paper for the National Research Council.

Brody, M., Tomkiewicz, W., & Graves, J. (2002). Park visitors' understandings, values and beliefs related to their experience at Midway Geyser Basin, Yellowstone National Park, USA. *International Journal of Science Education, 24*(11), 1119–1141.

Chadwick, J. (1998). *Public Utilization of Museum–based WorldWideWeb Sites* (Unpublished doctoral dissertation). University of New Mexico, New Mexico.

Chadwick, J., Falk, J. H., & O'Ryan, B. (2000). Assessing institutional websites: Summary of report. In Council on Library and Information Resources (Ed.), *Collections, content and the web.* Washington, DC: Council on Library and Information Resources.

Chamberlain, A. (2007). *Randomized control design issues.* Presentation by Anne Chamberlain, research scientist, Success for All Foundation, Baltimore, MD.

Chan, K. K. W. (1996). Environmental attitudes and behaviour of secondary school students in Hong Kong. *The Environmentalist, 16*, 297–306.

Chien, C. J. (1996). *Comparison of US science and environmental reporters' perceptual differences regarding factors affecting the quality of environmental stories.* Columbus: The Ohio State University: School of Natural Resources.

Dierking, L. D. (2007). Evidence & categories of informal science education impacts. In A. Friedman (Ed.), *A framework for evaluating impacts of informal science education projects.* Washington, DC: National Science Foundation.

Dierking, L. D., Adelman, L. M., Ogden, J., Lehnhardt, K., Miller, L., & Mellen, J. D. (2004). Using a behavior change model to document the impact of visits to Disney's Animal Kingdom: A study investigating intended conservation action. *Curator, 47*(3), 322–343.

Dierking, L. D., Cohen Jones, M., Wadman, M., Falk, J. H., Storksdieck, M., & Ellenbogen, K. (2002). Broadening our notions of the impact of free–choice learning experiences. *Informal Learning Review, 55*(1), 4–7.

Dierking, L. D., & Falk, J. H. (2003). Optimizing out–of–school time: The role of free–choice learning. *New Directions for Youth Development, 97*(Spring), 75–88.

Dierking, L. D., Luke, J., Foat, K., & Adelman, L. (2000). Families and free–choice learning. *Museum News, 80*(6), 38–43, 67.

Dierking, L. D., & McCreedy, D. (2008, March 30–April 2). *The impact of free-choice STEM experiences on girls' interest, engagement, and participation in science communities, hobbies and careers: Results of Phase 1*. Paper presented at the annual meeting of National Association for Research in Science Teaching, Baltimore, MD.

Doering, Z. D., & Pekakirk, A. J. (1996). Questioning the entrance narrative. *Journal of Museum Education, 21*(3), 20–22.

Elder, J., Coffin, C., & Farrior, M. (1998). *Engaging the public on biodiversity: A roadmap for education and communication strategies*. Madison, WI: The Biodiversity Project: 118. Retrieved from http://www.biodiversityproject.org/roadmap.pdf

Ellenbogen, K. M. (2002). Museums in family life: An ethnographic case study. In G. Leinhardt, K. Crowley, & K. Knutson (Eds.), *Learning conversations in museums*. Mahwah, NJ: Lawrence Erlbaum Associates.

Ellenbogen, K. M. (2003). From dioramas to the dinner table: An ethnographic case study of the role of science museums in family life. *Dissertation Abstracts International, 64*(03), 846A (University Microfilms No. AAT30–85758).

Eveland, W. P., & Dunwoody, S. (1998). Users and navigation patterns of a science World Wide Web site for the public. *Public Understanding of Science, 7*(4), 285–311.

Falk, J. H., & Dierking, L. D. (1992). *The museum experience*. Washington, DC: Whalesback Books.

Falk, J. H., & Dierking, L. D. (1997). School field trips: Assessing their long–term impact. *Curator, 40*(3), 211–218.

Falk, J. H. (1998). A framework for diversifying museum audiences: Putting heart and head in the right place. *Museum News, 77*(5), 35–43.

Falk, J. H. (Ed). (2001). *Free–choice science education: How we learn science outside of school*. New York, NY: Teacher's College Press, Columbia University.

Falk, J. H. (2002). The contribution of free–choice learning to public understanding of science. *Interciencia, 27*(2), 62–65.

Falk, J. H. (2003). Personal meaning mapping. In G. Caban, C. Scott, J. Falk, & L. Dierking (Eds.), *Museums and creativity: A study into the role of museums in design education* (pp. 10–18). Sydney, Australia: Powerhouse Publishing.

Falk, J. H. (2006). An identity–centered approach to understanding museum learning. *Curator, 49*(2), 151–166.

Falk, J. H. (2009). *Identity and the museum visitor experience*. Walnut Creek, CA: Left Coast Press.

Falk, J. H., & Dierking, L. D. (2000). *Learning from museums: Visitor experiences and the making of meaning*. Walnut Creek, CA: AltaMira Press.

Falk, J. H., & Dierking, L. D. (2002). *Lessons without limit: How free–choice learning is transforming education*. Lanham, MD: AltaMira Press.

Falk, J. H., & Dierking, L. D. (2010). The 95% solution: School is not where most Americans learn most of their science. *American Scientist, 98*, 486–493.

Falk, J. H., Heimlich, J. E., & Bronnenkant, K. (2008). Using identity-related visit motivations as a tool for understanding adult zoo and aquarium visitors' meaning-making. *Curator, 51*(1), 55–80.

Falk, J. H., & Storksdieck, M. (2005). Using the contextual model of learning to understand visitor learning from a science center exhibition. *Science Education, 89*, 744–778.

Falk, J. H., & Storksdieck, M. (2010). Science learning in a leisure setting. *Journal of Research in Science Teaching, 47*(2), 194–212.

Falk, J. H., Heimlich, J., & Bronnenkant, K. (2008). Using identity–related visit motivations as a tool for understanding adult zoo and aquarium visitor's meaning making. *Curator, 51*(1), 55–80.

Falk, J. H., Heimlich, J. E., & Foutz, S. (Eds). (2009.). *Free–choice learning and the environment*. Lanham, MD: AltaMira Press.

Falk, J. H., Moussouri, T., & Coulson, D. (1998). The effect of visitors' agendas on museum learning. *Curator, 41*(2), 106–120.

Falk, J. H., Storksdieck, M., & Dierking, L. D. (2007). Investigating public science interest and understanding: Evidence for the importance of free–choice learning. *Public Understanding of Science, 16*(4), 455–469.

Ferraro, P. J. (2005). *Are we getting what we paid for? The need for randomized environmental policy experiments in Georgia*. Atlanta, GA: H20 Policy Center, Georgia State University.

Friedman, A. (Ed.). (2007). *Framework for evaluating impacts of informal science education projects*. Washington, DC: National Science Foundation. Retrieved from http://insci.org/resources/Eval_Framework.pdf

Gammon, B. (1999). Everything we currently know about making visitor–friendly mechanical interactive exhibits. *Informal Learning Review, 39*, 1–13.

Gross, L. (1997). *The impact of television on modern life and attitudes*. Paper presented at 1997 International Conference on the Public Understanding of Science and Technology, Chicago, IL.

Heimlich, J. E., Falk, J. H., Bronnenkant, K., & Barlage, J. (2005). *Measuring the learning outcomes of adult visitors to zoos and aquariums* (Unpublished technical report). Annapolis, MD: Institute for Learning Innovation.

Heimlich, J. E., & Falk, J. H. (2009). Free–choice learning and the environment. In J. H. Falk, J. E. Heimlich, & S. Foutz (Eds.), *Free–choice learning and the environment*. Lanham, MD: AltaMira Press.

Humphrey, T., Gutwell, J., & The Exploratorium APE Team. (2005). *Fostering active prolonged engagement*. Walnut Creek, CA: Left Coast Press.

Institute for Learning Innovation. (2007). *Evaluation of learning in informal learning environments*. Invited paper for the Learning Science in Informal Environments National Committee. Washington, DC: National Academies of Sciences.

Jones, D., & Stein, J. K. (2005). *The Flandrau Science Center front–end evaluation* (Unpublished technical report). Institute for Learning Innovation, Annapolis, MD..

Kaplan, S., & Kaplan, R. (1982). *Cognition and environment*. New York, NY: Praeger.

Koster, E., & Falk, J. H. (2007). Maximizing the external value of museums. *Curator, 50*(2), 191–196.

Leinhardt, G., Crowley, K., & Knutson, K. (Eds.). (2002). *Learning conversations in museums*. Mahwa, NJ: Earlbaum.

Levins, R. (1968). *Evolution in changing environments*. Princeton, NJ: Princeton University Press.

Luke, J., Adams, M., Abrams, C., & Falk, J. (1998). *Art around the corner: Longitudinal evaluation report*. Annapolis, MD: Institute for Learning Innovation.

McCreedy, D., & Dierking, L. D. (2010, March). *Investigating informal science education on a large scale: Long-term impacts of free-choice Science, Technology, Engineering & Mathematics (STEM) experiences for girls*. Paper presented at the annual meeting of National Association for Research in Science Teaching, Philadelphia, PA.

McCreedy, D., & Zemsky, T. (2002). *Girls at the Center: Girls and adults learning science together*. Philadelphia, PA: The Franklin Institute/Girl Scouts of the USA.

Moussouri, T. (1997). *Family agendas and family learning in hands–on museums* (Unpublished doctoral dissertation). University of Leicester, Leicester, England.

Packer, J. (2006). Learning for fun: The unique contribution of educational leisure experiences. *Curator, 49*(3), 329–344.

Pekarik, A. J., Doering, Z. D., & Karns, D. A. (1999). Exploring satisfying experiences in museums. *Curator, 42*, 152–173.

Ponzio, R., & Marzolla, A. M. (2002). Snail trails and science tales: Inventing scientific knowledge. *Canadian Journal of Environmental Education, 7*(2), 37–43.

Reardon, K. M., Sorenson, J., & Clump, C. (2003). Partnerships with communities and neighborhoods. In B. Jacoby (Ed.), *Building partnerships for service– learning* (pp. 192–212). New York, NY: Jossey–Bass Press.

Rounds, J. (2004). Strategies for the curiosity–driven museum visitor. *Curator, 47*(4), 389–412.

Sewell, A. (2010). *Research methods and intelligibility studies*. New York, NY: Wiley.

St. John, M., Carroll, B., Hirabayashi, J., Huntwork, D., Ramage, K., & Shattuck, J. (2000). *The Community Science Workshops: A report on their progress*. Inverness, CA: Inverness Research Associates.

Part C

Issues of Framing, Doing, and Assessing in Environmental Education Research

Section VII

Moving Margins in Environmental Education Research

Constance L. Russell
Lakehead University, Canada

Leesa Fawcett
York University, Canada

Introduction

Whose voices have been heard in environmental education research? Whose stories have been told? By whom? How have these stories been gathered? How have they been represented? To which audiences? As we ponder the history and the future of environmental education research, it is vital to consider these questions, to examine what has been happening on the margins of the field and why these margins exist. As the title of this section introduction, "Moving Margins," implies, not only are we interested in "moving" as an adjective, that is, how the margins have shifted over the years, but also in "moving" as a verb, that is, how we might proactively work together to continue to move or remove those margins.

As other chapters in this *Handbook* illustrate, there are various histories that can be told about environmental education research. Histories are grounded in particular places—for example, a history of environmental education research in Canada (Hart, 1990; Russell, Bell, & Fawcett, 2000) has different touchstones than Australia (Gough, 1997), and environmental education research conducted and reported in English has a different flavor than that of other countries (Jickling, Sauvé, Brière, Niblett, & Root, 2010). Histories, therefore, depend very much on who has the privilege of telling the tale. As Donna Haraway (1989) so clearly demonstrated in her landmark genealogy of primatology, certain theories and research approaches gained prominence in that field, not necessarily because they were better than others, but because they were part of dominant culture; it should come as no surprise, for example, that Western primatologists reporting their results in English dominated the field in comparison to their Japanese counterparts even though Japanese research questions and relational methodology produced fascinating results.

As noted elsewhere in the *Handbook*, there have been significant methodological shifts in environmental education research. Qualitative research was once at the margins (Mrazek, 1993; Robottom & Hart, 1993, 1999) but certainly is no longer so; similarly, methodologies that were featured as "alternative" in a special issue of the *Canadian Journal of Environmental Education* (Russell & Hart, 2003) such as narrative inquiry, participatory research, and poststructuralist approaches have become less unusual. The methodological margins are constantly shifting. In this section, however, knowing that some of the methodological turns are addressed elsewhere in the book, our primary concern is particular voices that have been on the margins of environmental education research. We will begin by providing an overview of the chapters in this section, and then turn our attention to voices that are so far on the margins of environmental education research that there was insufficient literature to review to warrant an entire chapter on their own, but which nonetheless need to be heard.

The Chapters

Our section kicks off with a chapter by Annette Gough, who demonstrates that it was not until the late 1980s that gender was even on the map in environmental education and, even now, the literature is sparse especially when compared to other areas of educational research. Environmental education research focused on gender that has been done has occurred primarily in Australia and Canada; why is that? Gough makes clear that paying attention to gender and incorporating a feminist or ecofeminist lens to environmental education research is not merely about adding women's voices to the fray, although that in itself is important; the experiences of girls and women surely do matter and continuing to deny the importance of gender is decidedly not gender-neutral. Wary of essentializing women's ways of knowing, however, Gough calls for an analysis that makes space for the complexity of ways in which gender intersects with race, ethnicity, class, disability, and sexuality. She sees much promise in critical and poststructuralist feminist methodologies, but also asserts that a variety of methodologies and methods can be applied to the study of gender in environmental education research. Feminist research methods, she argues, are particularly helpful in allowing researchers to think and perceive differently and to ask questions that have received scant attention thus far in the field.

The next chapter by Soul Shava and the accompanying vignette by Greg Lowan-Trudeau turn the spotlight on indigenous environmental education research. Given the importance of place in indigenous environmental education, we asked each of them to pay particular attention to the contexts each knows well; in Shava's case that is southern Africa and in Lowan-Trudeau's case that is North America. In his chapter, Shava offers a historical analysis of the ways in which indigenous knowledges have been portrayed, including in various United Nations documents. Critiquing colonial anthropological and ethnographic research that positions indigenous peoples as either ignoble primitives or noble savages, he offers anticolonial and postcolonial responses and he highlights the work of indigenous scholars. Shava observes that there is a growing body of indigenous environmental education research being conducted in Africa, notably appearing in the *Southern African Journal of Environmental Education*. Nonetheless, there remains a relative dearth of indigenous environmental education research literature. Lowan-Trudeau, in reviewing literature from North America, comes to the same conclusion. In his vignette, Lowan-Trudeau begins with a discussion of foundational concepts in indigenous environmental education including the recognition of the colonial past and the hope for a decolonized future, the importance of holistic approaches, the role of elders, and the centrality of place. He then goes on to provide examples of contemporary indigenous environmental education practices, describes decolonizing indigenous research methodologies, and reviews the limited body of research literature. Both Shava and Lowan-Trudeau discuss pressing concerns and research needs. Shava remains wary of the consequences of indigenous knowledges being assimilated into dominant disciplinary and institutional discourses, and calls for environmental education research that explicitly addresses Indigenous content, epistemology, and research methodologies. Lowan-Trudeau identifies a number of research needs, including determining how Indigenous and nonindigenous peoples can respectfully engage with one another in environmental education research and the ways in which indigenous knowledge can become part of formal schooling for all students.

The next chapter, by Randy Haluza-DeLay, traces ideas about environmental racism, environmental justice, and ecojustice, and how these have, and have not, been taken up in environmental education research. Growing out of a social and academic movement that demonstrated that poor, racialized communities were disproportionately impacted by environmental degradation, it became clear that mainstream depictions of the environmental movement mostly reflected concerns of the white middle class in the so-called "developed" world. While environmental education practice has made significant efforts to address environmental injustice, most notably with the North American Association for Environmental Education (NAAEE) making "diversity and environmental justice" a priority, Haluza-DeLay notes that by NAAEE's own admission, environmental education has made only limited progress in addressing these inequities thus far. Further, he convincingly demonstrates that very little writing about environmental justice education is grounded in research. What research does exist is mostly situated in non-European and non-North American countries or in higher education sites. Haluza-DeLay concludes that environmental education in general and environmental education research in particular appears to remain afraid of being perceived as political and thus tends to shy away from explicitly naming things like racism, poverty, or capitalism as key contributors to environmental injustice.

Finally, turning to the voice of "nature" in environmental education research, Leesa Fawcett draws in particular on writing in environmental philosophy, environmental ethics, political ecology, and ecofeminism for conceptual and philosophical guidance. She makes what might appear, at first glance, to be a counterintuitive case: that environmental education has not sufficiently addressed the nature/culture divide. Tracing the ways in which anthropocentrism, biocentrism, and ecocentrism underlay different approaches to environmental education, she urges more attention be paid to fundamental epistemological and ontological assumptions about the relationships between humans and the rest of nature. Pointing to developments in humane education and environmental education that draws on human/animal studies, Fawcett finds particular inspiration in work that decenters anthropocentrism and disrupts false dualisms like nature/culture, and that works toward intersectional analyses that help build alliances between interdependent members of hybrid ecological communities.

Each of these chapters illustrates the exciting conceptual and methodological work that is happening on the edge of environmental education research. Since none of these areas have yet had sufficient research attention, it is no surprise that there exists a myriad of possibilities for future research. As but two examples, we personally would welcome research related to gender and environmental education research that critically interrogates from a feminist perspective the "domestic environmentalism" often touted in action approaches to environmental education (Lousley, 1999; MacGregor, 2006) and an analysis of the discourse of Mother Nature in environmental education (Roach, 2003; Russell, 2003). There remains so much more to research in environmental education regarding gender, environmental justice, Indigenous education, and human relationships to the rest of nature.

Who Is Missing Even at the Margins?

When we initially discussed what chapters we wanted to see in this section of the *Handbook*, a number of possibilities came to mind. Besides those that ended up being included in the section, we also thought chapters on environmental education research that featured class, disability, sexuality, and body size could be useful. However, based on the scanty or nonexistent attention paid to each in environmental education research, the resulting chapters would have been exceedingly short. The two of thus chose to instead briefly mention these in this section introduction in the hope that sincere and sustained attention is soon paid to each in future environmental education research.

To begin, we are disturbed by the continuing lack of attention to issues of social class in environmental education research. As Russell, Dillon, Fawcett, and Czank (2010) argued at the most recent American Educational Research Association conference, this is unconscionable given that one in two children live in poverty and there is increasing disparity between rich and poor (Shah, 2008), and when the environmental dimensions of poverty are coming to the fore in discussions of issues such as food security, access to clean water, waste management, and the predicted disproportionate impact of global warming on people who live in poverty. Research into school effectiveness shows that socioeconomic status alone is a massive predictor of school success (Thrupp & Lupton, 2006)—there is a growing body of literature in education asserting that class does indeed still matter (e.g., hooks, 2000; McLaren & Farahmandpur, 2001; Scott & Freeman-Moir, 2007; Thrupp, 2008).

To be fair, there has been a small amount of attention to class issues in environmental education. Examples include research that describes how important outdoor common spaces such as parks and school grounds are for poor children (Chawla, 2002; Malone, 2001; Thomson & Philo, 2004), that analyzes the role of class in schoolyard greening projects in Toronto (Dyment, 2005; Dyment & Bell, 2006), and that examines the "classed body" in outdoor education dependant on canoe tripping (Newbery, 2003). As well, the relatively new journal *Green Theory and Praxis: Journal of Ecopedagogy* has published a few articles that discuss the impacts of globalized capitalism on pedagogy (Fassbinder, 2008; Kahn, 2008a; Nocella & Walton, 2005). Further, as Haluza-DeLay noted in his chapter in this section of the *Handbook*, environmental justice clearly highlights the impacts of environmental degradation on poor, racialized communities; still, as he demonstrates, while there is some rhetoric about the importance of environmental justice education, there remains little research on the topic.

Turning our attention now to disability, environmental education research again does not compare favorably to the wider education field. In education, there has long been attention to "inclusive" practices that respond to the "special needs" of individual students; indeed, one would be hard pressed to find a teacher education program that did not address such concerns. The critical disability movement has interrogated these and other understandings of disability and offered intersectional analyses of the complex interplay of disability and race, class, gender, and sexuality (Davis, 2010; Pothier & Devlin, 2006); educational theory, practice, and research has responded in generative ways (Danforth & Gabel, 2006; Erevelles, 2000; Gabel, 2005). The many journals and books and teacher resources demonstrate the breadth and depth of the field.

The situation in environmental education research, however, only murkily reflects this general trend. Where one sees the most engagement with disability issues is in the outdoor experiential education field (Fox & Avramadis, 2003; Healy, Jenkins, Leach, & Roberts, 2001; McAvoy & Schleien, 2001; O'Connell & Breunig, 2003, 2005). In their review of the outdoor education research literature, Dillon et al. (2006) noted the continued barriers faced by students with disabilities in accessing field experiences; perhaps not surprisingly, then, outdoor education publisher Acorn Naturalists has offers for sale guidebooks to help practitioners involve students of a range of abilities in outdoor programs (e.g., Brannan, Fullerton, Arick, Robb, & Bender, 2003; Heath & Gilbert, 2001). On a more theoretical level, Newbery (2003) builds on insights from disability studies and feminist theories of the body to explore her own practice as a canoe guide, outdoor educator, and university professor, and to question the meanings of self, ability, gender, and class that are often produced during canoe trips. She also described how as a strong and proficient canoe tripper, she disrupted heterosexist norms of femininity.

Newbery's (2003) discussion of heterosexism in outdoor education is one of the few examples in the environmental education literature of attention to sexuality. Feminists working in and theorizing about outdoor experiential education were the first to engage seriously with this topic (see McClintock, 1996; Mitten, 1997; Warren & Rheingold, 1996). They critiqued the macho ethos of much outdoor education and the reification of traditional gender roles, noting how strong women in the field were often labeled "Amazons" (code word for lesbians) regardless of their sexuality (McClintock, 1996; Mitten, 1997). Russell, Sarick, and Kennelly (2002) and Gough and Gough (2003) decried the resounding heterosexualizaton of environmental education theory, practice, and research and advocated a "queering" of the field. Alas, there is as yet little evidence that this call has been heeded in a meaningful way. Sexuality thus remains firmly at the margins.

Finally, an area that has yet to receive any attention in environmental education theory, practice, or research is body size. "Fat studies" grew out of feminist examinations of the discourses of obesity (e.g., Bordo, 1993; Evans Braziel & LeBesco, 2001; Rothblum & Solovay, 2009) and calls for critical analyses of the politics of food and the interplay of body size and body image with gender, race, class, disability, sexuality, and age. Education has begun to take up these ideas (Guthman, 2009; Koppleman, 2009; Rawlins, 2008). We argue that environmental education research could benefit from engagement with fat studies for a number of reasons. First, attention to embodiment continues to be an area of interest in environmental education (Barrett, 2009; Payne, 1997). Second, "fit" bodies tend to be privileged in the practice and representations of outdoor education (Newbery, 2003). Third, the "obesity epidemic" is one of the dire consequences regularly mentioned in the "nature deficit disorder" literature (Louv, 2005). A critical analysis of fitness and obesity discourses in environmental education is thus long overdue.

Conclusion

The voices highlighted in this introductory chapter are marginalized to various degrees. While some, such as those working in fat studies, have not yet been heard at all in environmental education research, others such as feminists have been expressing their concerns for some time now. Gough (1994, 1999) made clear in her study of the "founding fathers" of environmental education how the discourse then was very much shaped by men. As she demonstrates in her chapter in this section, the problem persists. In that regard, one elephant in the *Handbook*, so to speak, is the fact that it has been edited by four white men of a certain vintage. We wish to make clear that we do not blame these four individual men for this situation since we know from personal experience that in the early days of the *Handbook*, efforts were make to redress this imbalance. Nonetheless, it *is* indicative of a persistent systemic problem in the field. Is it a surprise that, at the time of writing, seven of the ten editors/coeditors of the main English-language journals in the field (i.e., *Applied Environmental Education Research*, *Australian Journal of Environmental Education, Canadian Journal of Environmental Education, Environmental Education Research*, *Green Theory and Practice: Journal of Ecopedagogy, Journal of Environmental Education,* and *Southern African Journal of Environmental Education*) are men? How about the fact that all the current editors are white?

Environmental education needs to ask itself why particular problems of marginalization persist when there has been more movement in the wider education field, at least in some realms. Is the continuing marginalization of certain voices persisting because we are a relatively new, small field? Or are there specific hegemonic forces at play in the field? The two of us have no quick answers, but we maintain that this issue deserves serious and sustained thought and action. As contributors to this section have suggested, it is important to look at underlying root issues and the processes by which marginalization happens.

One of the areas where we think that environmental education research can lead educational research generally is in its attention to anthropocentrism and environmental justice. As well, since we are an inherently interdisciplinary field, intersectional analyses may make intuitive sense to many environmental education researchers (Kahn & Humes, 2009). Such intersubjective approaches (McKenzie, 2009) help us ensure we do not get mired in overly simplistic identity politics and also allows us to pay attention to larger forces at play like colonialism, imperialism, materialism, and capitalism (Gough, 2009; Greenwood, 2010; Kahn, 2008b, 2010; McKenzie, Hart, Bai, & Jickling, 2009; Peters & Araya, 2009).

In the conclusion to the recent book, *Fields of Green: Restorying Culture, Environment and Education*, Hart, McKenzie, Bai, and Jickling. (2009) discuss ways that environmental education might reposition itself for the tasks ahead: "The pedagogical process of education is so complex and so entrenched in mindsets that despite reimagining exercises, like stretching exercises, much of what happens remains unexamined, unintelligible, and unseeable. . . . Even where existing categories or frames no longer work, we have not learned how to think our way out of them" (p. 347). This statement reminds us that working to (re)move margins is a challenging process and it also keeps us humble. In this introduction, we highlighted a few voices that we assert deserve to be heard in environmental education research; had we written this introduction ten years ago, we likely would not thought to have discussed the potential of fat studies for the field. One implication of this observation is that, no doubt, there are other voices that we have not mentioned here that also deserve to be heard. Ten years from now we'll be wondering why we had not mentioned x or y.

Not so long ago, one of us wrote about the phrase, "race, class, gender and so forth" and the fact that queers were usually relegated to the "and so forth" category in environmental education research (Russell et al., 2002). A few years later, pondering the continuing absence of "nature" in environmental education research reports, one of us wrote, "I remain intrigued by who retains membership in the 'and so forth' category. Who does not quite make it into the lists of those silenced Others deserving to be heard?" (Russell, 2005, p. 434). This question, we argue, is one that needs to be continually asked; all of us must remain vigilant to who and how voices continue to be marginalized and work together to (re)move the margins.

References

Barrett, M. J. (2009). *Beyond human-nature-spirit boundaries: Researching with animate Earth* (Unpublished PhD dissertation) Faculty of Education, University of Regina, Saskatchewan, Canada. Available at: http://www.porosity.ca/

Bordo, S. (1993). *Unbearable weight: Feminism, western culture, and the body*. Berkeley, CA: University of California Press.

Brannan, S. A., Fullerton, A., Arick, J. R., Robb, G. M., & Bender, M. (2003). *Including youth with disabilities in outdoor programs.* Urbana, IL: Sagamore.

Chawla, L. (2002). *Growing up in an urbanized world.* London: UNESCO/Earthscan.

Danforth, S., & Gabel, S. (2006). *Vital questions facing disability studies in education.* New York, NY: Peter Lang.

Davis, L. (2010). *The disability studies reader* (3rd ed.). New York, NY: Routledge.

Dillon, J., Rickinson, M., Teamey, K., Morris, M., Choi, M. Y., Sanders, D., & Benefield, P. (2006). The value of outdoor learning: Evidence from research in the UK and elsewhere. *School Science Review, 87,* 107–111.

Dyment, J. E. (2005). "There's only so much money hot dog sales can bring in": The intersection of green school grounds and socio-economic status. *Children's Geographies, 3*(3), 307–323.

Dyment, J. E., & Bell, A. C. (2006). "Our garden is colour blind, inclusive, and warm": Reflections on green school grounds and social inclusion. *International Journal of Inclusive Education, 12*(2), 169–183.

Erevelles, N. (2000). Educating unruly bodies: Critical pedagogy, disability studies, and the politics of schooling. *Educational Theory, 50*(1), 25–47.

Evans Braziel, J., & LeBesco, K. (2001). *Bodies out of bounds: Fatness and transgression.* Los Angeles, CA: University of California Press.

Fassbinder, S. D. (2008). Capitalist discipline and ecological discipline. *Green Theory & Praxis: The Journal of Ecopedagogy, 4*(2), 87–101.

Fox, P., & Avramidis, E. (2003). An evaluation of an outdoor education program for students with emotional and behavioral difficulties. *Emotional and Behavioral Difficulties, 8*(4), 267–283.

Gabel, S. (2005). *Disability studies in education: Readings in theory and method.* New York, NY: Peter Lang.

Gough, A. (1994). *Fathoming the fathers in environmental education: A feminist poststructuralist analysis* (Unpublished doctoral dissertation). Deakin University, Geelong, Australia.

Gough, A. (1997). *Education and the environment: Policy, trends and the problems of marginalisation.* Melbourne, VIC: Australian Council for Educational Research.

Gough, A. (1999). Recognizing women in environmental education pedagogy and research: Toward an ecofeminist poststructuralist perspective. *Environmental Education Research, 5*(2), 143–161.

Gough, N. (2009). Becoming transational: Rhizhosemiosis, complicated conversation, and curriculum inquiry. In M. McKenzie, P. Hart, H. Bai, & B. Jickling (Eds.), *Fields of green: Restorying culture, environment, and education* (pp. 67–83). Cresskill, NJ: Hampton Press.

Gough, N., & Gough, A. (2003). Tales from Camp Wilde: Queer(y)ing environmental education research. *Canadian Journal of Environmental Education, 8,* 44–66.

Guthman, J. (2009). Teaching the politics of obesity: Insights into neoliberal embodiment and contemporary biopolitics. *Antipode: A Radical Journal of Geography, 41*(5), 1110–1133.

Greenwood, D. (2010). Nature, empire, and paradox in environmental education. *Canadian Journal of Environmental Education, 15,* 9–24.

Haraway, D. (1989). *Primate visions: Gender, race and nature in the world of modern science.* New York, NY: Routledge.

Hart, P. (1990). Environmental education in Canada: Contemporary issues and future possibilities. *Australian Journal of Environmental Education, 6,* 45–66.

Hart, P., McKenzie, M., Bai, H., & Jickling, B. (2009). Conclusion: Repositioning ourselves for the task ahead. In M. McKenzie, P. Hart, H. Bai, & B. Jickling (Eds.), *Fields of green: Restorying culture, environment, and education* (pp. 343–348). Cresskill, NJ: Hampton Press.

Healy, M., Jenkins, A., Leach, J., & Roberts, C. (2001). *Issues in providing learning support for disabled students undertaking fieldwork and related activities.* Gloucestershire, UK: University of Gloucestershire.

Heath, D., & Gilbert Almeras, B. (2001). *Access nature.* Washington, DC: National Wildlife Federation.

hooks, b. (2000). *Where we stand: Class matters.* New York, NY: Routledge.

Jickling, B., Sauvé, L., Brière, L., Niblett, B., & Root, E. (2010). The fifth World Environmental Education Congress, 2009: A research project. *Canadian Journal of Environmental Education, 15,* 47–67.

Kahn, R. (2008a). From education for sustainable development to ecopedagogy: Sustaining capitalism or sustaining life? *Green Theory and Praxis: The Journal of Ecopedagogy, 4*(1), 1–14.

Kahn, R. (2008b). Towards ecopedagogy: Weaving a broad-based pedagogy of liberation for animals, nature and the oppressed peoples of the earth. In A. Darder, R. Torres, & M. Baltodano (Eds.), *The critical pedagogy reader* (2nd ed.). New York, NY: Routledge.

Kahn, R. (2010). *Critical pedagogy, ecoliteracy and planetary crisis.* New York, NY: Peter Lang.

Kahn, R., & Humes, B. (2009). Marching out from Ultima Thule: Critical counterstories of emancipatory educators working at the intersection of human rights, animal rights, and planetary sustainability. *Canadian Journal of Environmental Education, 14,* 179–195.

Koppelman, S. (2009). Fat stories in the classroom: What and how are they teaching about us? In E. Rothblum & S. Solovay (Eds.), *The fat studies reader* (pp. 213–220). New York, NY: New York University Press.

Lousley, C. (1999). (De)politicizing the Environment Club: Environmental discourses and the culture of schooling. *Environmental Education Research, 5*(3), 293–304.

Louv, R. (2005). *Last child in the woods: Saving our children from nature-deficit-disorder.* Chapel Hill, NC: Algonquin Books of Chapel Hill.

MacGregor, S. (2006). *Beyond mothering earth: Ecological citizenship and the politics of care.* Vancouver, BC: University of British Columbia Press.

Malone, K. (2001). Children, youth, and sustainable cities. *Local Environments, 6*(1), 5–12.

McAvoy, L., & Schleien, S. (2001). The name assigned to the document by the author. This field may also contain subtitles, series names, and report numbers. Inclusive outdoor education and environmental interpretation. The entity from which ERIC acquires the content, including journal, organization, and conference names, or by means of online submission from the author. *Taproot, 13*(1), 11–16.

McClaren, P., & Farahmandpur, R. (2001). Teaching against globalization and the new imperialism: Toward a revolutionary pedagogy. *Journal of Teacher Education, 52*(2), 136–150.

McClintock, M. (1996). Lesbian baiting hurts all women. In K. Warren (Ed.), *Women's voices in experiential education* (pp. 241–250). Dubuque, IA: Kendall Hunt.

McKenzie, M. (2009). Pedagogical transgression: Toward intersubjective agency and action. In M. McKenzie, P. Hart, H. Bai, & B. Jickling (Eds.), *Fields of green: Restorying culture, environment, and education* (pp. 211–224). Cresskill, NJ: Hampton Press.

McKenzie, M., Hart, P., Bai, H., & Jickling, B. (2009). *Fields of green: Restorying culture, environment, and education.* Cresskill, NJ: Hampton Press.

Mitten, D. (1997). In the light: Sexual diversity and women's outdoor trips. *Journal of Leisurability, 24*(4), http://lin.ca/resource-details/2375

Mrazek, R. (1993). *Alternative paradigms in environmental education research.* Troy, OH: North American Association for Environmental Education.

Newbery, L. (2003). Will any/body carry that canoe? A geography of the body, ability, and gender. *Canadian Journal of Environmental Education, 8,* 204–216.

Nocella, A., & Walton, M. (2005). Standing up to corporate greed: The Earth Liberation Front as domestic terrorist target number one. *Green Theory & Praxis: Journal of Ecopedagogy, 1*(1), 1–18.

O'Connell, T., & Breunig, M. (2003). Accessible outdoor travel: "Batteries not included." *Pathways: Ontario Journal of Outdoor Education, 16*(4), 25–26.

O'Connell, T., & Breunig, M. (2005). Sense of community on integrated wilderness trips: A pilot study. *Research in Outdoor Education, 7*, 90–99.

Payne, P. (1997). Embodiment and environmental education. *Environmental Education Research, 7*(1), 68–88.

Peters, M., & Araya, D. (2009). Network logic: An ecological approach to knowledge and learning. In M. McKenzie, P. Hart, H. Bai, & B. Jickling (Eds.), *Fields of green: Restorying culture, environment, and education* (pp. 239–250). Cresskill, NJ: Hampton Press.

Pothier, D., & Devlin, R. (2006). *Critical disability theory: Essays in philosophy, politics and law*. Vancouver, BC: UBC Press.

Rawlins, E. (2008). Citizenship, health education, and the UK obesity "crisis." *ACME: An International E-Journal for Critical Geographies, 7*(2), 135–151.

Roach, C. (2003). *Mother/nature: Popular culture and environmental ethics*. Bloomington, IN: Indiana University Press.

Robottom, I., & Hart, P. (1993). *Research in environmental education: Engaging the debate*. Geelong, Australia: Deakin University Press.

Robottom, I., & Hart, P. (1999). Behaviorist environmental education research. *Journal of Environmental Education, 26*(2), 5–10.

Rothblum, E., & Solovay, S. (2009). *The fat studies reader*. New York, NY: New York University Press.

Russell, C. L. (2003). Book review: Mother/Nature: Popular culture and environmental ethics (C. Roach). *Journal of Environmental Education, 35*(1), 56–57.

Russell, C. L. (2005). "Whoever does not write is written": The role of "nature" in post-post approaches to environmental education research. *Environmental Education Research, 11*(5), 433–443.

Russell, C., Dillon, J., Fawcett, L., & Czank, J. (2010). *Social class and environmental education research: Missing voices, stilled lives*. Presentation at the American Educational Research Association (AERA) Annual Conference, Denver, April.

Russell, C. L., Bell, A. C., & Fawcett, L. (2000). Navigating the waters of Canadian environmental education. In T. Goldstein & D. Selby (Eds.), Weaving connections: Educating for peace, social, and environmental justice (pp. 196–217). Toronto, ON: Sumach Press.

Russell, C. L., & Hart, P. (2003). Exploring new genres of research in environmental education. *Canadian Journal of Environmental Education, 8*, 5–8.

Russell, C. L., Sarick, T., & Kennelly, J. (2002). Queering environmental education. *Canadian Journal of Environmental Education, 7*, 54–66.

Scott, A., & Freeman-Moir, J. (2007). *The lost dream of equality: Critical essays on education and social class*. Rotterdam: Sense.

Shah, A. (2008). Poverty facts and stats. [http://www.globalissues.org/article/26/poverty-facts-and-stats]

Thomson, J. L., & Philo, C. (2004). Playful spaces? A social geography of children's play in Livingston, Scotland. *Children's Geographies, 2*(1), 111–130.

Thrupp, M. (2008). Education's "Inconvenient Truth" part two: The middle classes have too many friends in education. *New Zealand Journal of Teachers' Work, 5*(1), 54–62.

Thrupp, M., & Lupton, R. (2006). Taking school contexts more seriously: The social justice challenge. *British Journal of Educational Studies, 54*(3), 308–328.

Warren, K., & Rheingold, A. (1996). Feminist pedagogy and experiential education: A critical look. In K. Warren (Ed.), *Women's voices in experiential education* (pp. 118–129). Dubuque, IA: Kendall Hunt.

35

Researching Differently

Generating a Gender Agenda for Research in Environmental Education

Annette Gough
RMIT University, Australia

Introduction

The need to change the perspectives from which we think and act is one of the foundations of environmental education. As Foucault (1984/1985/1990) wrote, "There are times in life when the question of knowing if one can think differently than one thinks, and perceive differently than one sees, is absolutely necessary if one is to go on looking and reflecting at all" (p. 8). In this instance, using feminist research strategies to generate a gender agenda provides the basis for different ways of thinking and perceiving in environmental education research.

The major contributions of feminist research, in all its many forms, have been to raise the question of epistemological claims, such as who can be an agent of knowledge, what counts as knowledge, what constitutes and validates knowledge, and what the relationship should be between knowing and being. Feminist questions put the social construction of gender at the center of research (Lather, 1991) for "what 'grounds' feminist standpoint theory is not women's experiences but the view from women's lives" (Harding, 1991, p. 269).

Feminism enables people to revision their world—"to know it differently than we have ever known it; not to pass on a tradition, but to break its hold on us" (Rich, 1972, p. 19). As Heilbrun (1999) noted, "Women began to portray the new possibilities that, as a result of feminism, they found themselves confronting. They began to question . . . all strictures about women and about the institutions in which women now, in even greater numbers, and in a state of awakening, found themselves" (p. 8).

Ecological feminists have embraced personal and political action to "fully engage in the interweaving of humour, irony, grace, resistance, struggle and transformation" (Sandilands, 1999, p. 210) to envision a more democratic future for all. And, occasionally, environmental education researchers whose work is informed by (at least aspects of) feminism are publishing their work (see, e.g., Barrett, 2005; Bodzin, Shiner Klein, & Weaver, 2010; Carrier, 2007; Di Chiro, 1987; Fawcett, 2000, 2002; Fontes, 2002; Gough, 1997, 1999a, 1999b, 1999c, 2004; Gough & Gough, 2003; Gough & Whitehouse, 2003; Hallen, 2000; Lotz-Sisitka & Burt, 2002; Lousley, 1999; Malone, 1999; McKenzie, 2004, 2005; Newbery, 2003; Russell, 2003; Russell & Bell, 1996; Russell, Sarick, & Kennelly, 2002; Wane & Chandler, 2002). But, despite these in-roads, the subject of gender remains marginal to much environmental education research.

Many researchers still consider a "human" subjectivity homogenous, ungendered, and unproblematic, whereas a vast edifice of sociological research reveals the opposite to be the case. Environmental education research remains bound with traditional epistemological frameworks of scientific research, which have, in Harding's (1987) words, "whether intentionally or unintentionally, systematically exclude(d) the possibility that women could be 'knowers' or *agents of knowledge*" (p. 3, emphasis in original). For example, writings on significant life experience research can be critiqued as remaining blind to gendered subjectivities (Gough, 1999b).

Addressing the balance is simply not a matter of "adding women" to traditional analyses. Rather, what is needed is a transformative process in which new empirical and theoretical resources are opened up to reveal new purposes and subjects for inquiry. And what needs to come under scrutiny is the implicit constitution of the assured, homogenous, and universalized human subject of much environmental research. "Human" identity as constituted through positivist research regimes is not inclusive of all the different ways of being in the world.

Much past (and present) environmental education research has analyzed only male experiences or has constructed universalized subjects, which are not distinguished as male or female. Yet, there is no universal "Man" who acts as a powerful agent on an equally symbolic "Environment"—except perhaps in the imaginations of writers who reproduce these discourses. "Man" is not a term that is logically inclusive of women. Early formulations of environmental education, such as the IUCN (1970) definition referred to as "the interrelatedness among man, his culture, and his biophysical surroundings" (as cited in Linke, 1980, pp. 26–27). Although more recent environmental education literature is gender neutral in its language, this too is a problem, as neutral voice is still interpreted as male by readers of both genders. As Cherryholmes (1993) argues, "texts that deny gender present themselves as generic. They pretend to speak the truth and truth is gender-neutral. Authoritative texts are distanced, objective, have a single voice (otherwise they would not be authoritative), are value-neutral, dispassionate and controlling" (p. 10). Perhaps the shackles of the past are proving difficult to shrug off, but the practice of creating gender-blind binaries is exclusive of lived experience.

In reality, we have culturally, racially, socioeconomically, and sexually (and so on) different people with fragmented identities whose experiences and understandings can only be constituted through the lenses of subjectivity. Given there is growing recognition that there is no one way of looking at the world, no "one true story," rather a multiplicity of stories, then we should look at a multiplicity of strategies for policies, pedagogies, and research in environmental education. These strategies should be strategies that are neither universal nor part of the dominant discourse, but strategies that are from the lives of the colonized and marginalized, including the lives of women.

However, to date, environmental education research has rarely addressed areas of different women's experiences and knowledges, which means many useful insights have not been adequately pursued. Environmental education research has ignored other aspects of human identities too, but these are beyond the scope of this chapter. Rather, the emphasis here is on women's experiences and knowledge and the perspective these bring to environmental education research.

Why Is a Feminist Perspective in Environmental Education Important?

Developing a feminist perspective in environmental education is important, because the vast majority of work in environmental education to date has been concerned with universalized subjects, rather than recognizing multiple subjectivities. It is time that we start generating different ways of knowing and seeing environments so we might understand human relationships with them better. As Brown and Switzer (1991) argued nearly two decades ago,

> Women and men contribute to maintaining environmental, economic and social sustainability in distinctive ways. For women these contributions are made through:
>
> - their public roles as the majority of the workforce in the health, education, welfare and service industries;
> - their private roles as care-givers, farm managers, educators of children, and the principal purchasers of food and consumer goods; and
> - the many public (paid) and private (unpaid) arenas where women have a major responsibility for the management of change and the transmission of social values. (p. iv)

Such a perspective is neither intended to essentialize women as caretakers of the earth's household, obsessed with green cleaners, nor intended to cast women as symbols of nature (for further discussion of these aspects, see, e.g., MacGregor, 2006; Warren, 1997). Rather, the intention in developing a feminist perspective in environmental education research is to recognize the complexity of human roles and relationships with respect to environments and that there are multiple subjectivities and multiple ways of knowing and interacting with environments that cannot be encapsulated within the notion of universalized subjects.

Most importantly, by pursuing a feminist perspective in environmental education research, we are able to construct "less partial, less distorted" (Harding, 1991) accounts of environments. Such a pursuit is also consistent with feminist praxis—a term that recognizes "a continuing feminist commitment to a political position in which 'knowledge' is not simply defined as 'knowledge *what*' but also as 'knowledge *for*'" (Stanley, 1990, p. 15, emphasis in original). It indicates rejection of the theory/research divide, "seeing these as united manual and intellectual activities which are symbiotically related (for all theorising requires 'research' of some form or another)" (p. 15). And it centers interest on methodological/epistemological concerns: "'how' and 'what' are indissolubly interconnected and . . . the shape and nature of the 'what' will be a product of the 'how' of its investigation" (p. 15). A central concern of feminist praxis is, thus, the reconstituting of knowledge. As Spender (1985) argues, "at the core of feminist ideas is the crucial insight that there is no one truth, no one authority, no one objective method which leads to the production of pure knowledge" (p. 5). I, therefore, argue that, through feminist research in environmental education, we will reconstitute knowledge about environments, and for environments.

Evidence that men and women do think differently about environmental issues is apparent in Brown's (1995) comparison of Australian women's priority concerns about the environment with a survey of issues prioritized in the scientific literature. Similarly, Zelezny, Chua, and Aldrich's (2000) review of thirty-two research studies on gender and environmental attitudes and behaviors between 1988 and 1998 found that, in most studies, females reported more proenvironmental behaviors than males and females expressed greater environmental concern.

That women have a different role with respect to the environment was recognized nearly two decades ago in *Agenda 21*, the report from the 1992 United Nations Conference on Environment and Development (UNCED), but there was a lack of reciprocity between the "Global Action for Women Towards Sustainable and Equitable Development" and the "Promoting Education, Public Awareness, and Training" chapters in *Agenda 21* (UNCED, 1992, Chapters 24 and 36, respectively). The "Women" chapter has as its overall goal achieving active involvement of women in economic and political decision making, with emphasis on women's participation in national and international ecosystem management and control of environmental degradation. One of its objectives for national governments was:

> To assess, review, revise and implement, where appropriate, curricula and other educational material, with a view to promoting the dissemination to both men and women of gender-relevant knowledge and valuation of women's roles through formal and non-formal education. (UNCED, 1992, para. 24.2(e))

Other objectives addressed topics such as increasing the proportion of women decision-makers, eliminating obstacles to women's full participation in sustainable development, achieving equality of access to opportunities for education, health, etc. for women, equal rights in family planning, and prohibiting violence against women. The activities for governments related to such objectives are broadly concerned with:

- achieving equality of opportunity for women, such as by eliminating illiteracy);
- increasing proportions of women as decision-makers in implementing policies and programs for sustainable development; and
- recognizing women as equal members of households, both with respect to workloads and finance.

Consumer awareness is particularly mentioned, as are "programmes to eliminate persistent negative images, stereotypes, attitudes and prejudices against women through changes in socialization patterns, the media, advertising, and formal and nonformal education" (UNCED, 1992, para. 24.3(i)). Here, women's contributions to society are recognized and valued as something different, rather than assuming that women will achieve equality simply through equal opportunity, although there are some elements of a liberal feminist view.[1]

Unfortunately, attention to gender was not matched in the "education" chapter of *Agenda 21*. Here, women were generally included with all sectors of society, although specific mention is made of the high illiteracy levels among women that need to be addressed (UNCED, 1992, para. 36.4(a)) in the objectives. In the activities, women are mentioned in the following terms: "Governments and educational authorities should foster opportunities for women in nontraditional fields and eliminate gender stereotyping in curricula" (UNCED, 1992, para. 36.5(m)). No mention is made of recognizing and valuing women's roles in promoting and achieving sustainable development, and the perspective seems, once more, that of liberal feminism, although indigenous peoples' experiences with and understanding of sustainable development is affirmed as playing a part in education and training (UNCED, 1992, para. 36.5(n)).

Gender equity is more closely related to education in the International Implementation Scheme for the United Nations Decade of Education for Sustainable Development (ESD) 2005–2014: "Environmental issues like water and waste affect every nation, as do social issues like employment, human rights, gender equity, peace and human security . . . Such issues are highly complex and will require broad and sophisticated educational strategies for this and the next generation of leaders and citizens to find solutions" (United Nations Educational, Scientific, and Cultural Organization [UNESCO], 2005b, Annex. I, p. 3). However, much of this higher profile comes because gender equity is integral to the World Declaration on Education for All and the Millennium Development Goals, which are part of the decade agenda: "Education for sustainable development is based on ideals and principles that underlie sustainability, such as intergenerational equity, gender equity, social tolerance, poverty alleviation, environmental preservation and restoration, natural resource conservation, and just and peaceable societies" (UNESCO, 2005b, Annex. II, p. 3). The earlier Draft International Implementation Scheme provided a more direct message: "In terms of ESD specifically, the full and equal engagement of women is crucial, first, to ensuring balanced and relevant ESD messages and, second, to give the best chance for changed behaviours for sustainable development in the next generation" (UNESCO, 2005a, p. 19). This statement definitely puts gender on the agenda—but how do we make it happen?

What Is Feminist Research?

According to Lather (1991), "to do feminist research is to put the social construction of gender at the centre of one's inquiry . . . feminist researchers see gender as a basic organising principle which profoundly shapes and/or mediates the concrete conditions of our lives" (p. 17). Feminist research is, thus, openly ideological, aiming to correct both the invisibility of female experience and its distortion.

Feminist educational research methods comprise many traditional forms, including statistical, interview, ethnographic, survey, cross-cultural, oral history, content analysis, case studies, and action research, and some feminist researchers have invented or are using newer forms as well (see Hesse-Biber, 2007). As Stanley (1990) argued, "there is no one set of methods or techniques . . . which should be seen as distinctly feminist. Feminists should use any and every means available" (p. 12).

According to Reinharz (1992), in her still useful text on feminist methods in social research, "feminist methodology is the sum of feminist research methods" (p. 240).

In making this statement, Reinharz epitomizes a problem identified by Harding (1987), "that social scientists tend to think about methodological issues primarily in terms of methods of inquiry" (p. 2). Indeed, "method" and "methodology" are terms frequently either intertwined, used interchangeably, or confused in feminist (and other genres of) research scholarship, and contestation abounds as to whether or not there is a feminist research method or methodology. For example, Harding (1987, 2007) argues against the idea of a distinctive feminist method of research, because it is a distraction from discussing the more interesting aspects of feminist research processes: the differences between method, methodology, and epistemology. As Harding (1987) argues,

> . . . it is new methodologies and new epistemologies that are requiring these new uses of familiar research techniques. If what is meant by a "method of research" is just this most concrete sense of the term, it would undervalue the transformations feminist analyses require to characterize these in terms only of the discovery of distinctive methods of research. (p. 2)

The confusion between method (techniques for gathering evidence), methodology (a theory and analysis of how research should proceed), and epistemology (issues about an adequate theory or justificatory strategy) is not the sole province of feminist research. Such confusions abound in nonfeminist research as well. In both feminist and nonfeminist research, "method" is often used to refer to all aspects of research, thus, making discussions about distinctiveness particularly difficult in regard to feminist research, because it undervalues the transformations happening in feminist research methods, methodologies, and epistemologies. According to Harding (1987), what makes feminist research distinctive is that it opens up:

- new empirical and theoretical resources (women's experiences),
- new purposes of social science research (for women), and
- new subject matter of inquiry (locating the researcher in the same critical plane as the overt subject matter).

Although early feminist research was largely positivistic, recent methodologies have been more concerned with "generating and refining . . . more interactive, contextualised methods in the search for pattern and meaning rather than for prediction and control . . . Hence feminist empirical work is multi-paradigmatic" (Lather, 1991, p. 18). Feminist research methodologies now include the whole range, from postpositivistic concerns with prediction through interpretive, constructivist, phenomenological, and ethnographic concerns to understand, to emanicipatory methodologies, such as critical, participatory, and action research, and postmodern concerns, such as poststructuralism and deconstruction (Hesse-Biber, 2007). Reinharz (1992, p. 240) proposed ten characteristics of feminist research (which she calls "methodology," but I prefer to call "approach"):

1. Feminism is a perspective, not a research method.
2. Feminists use a multiplicity of research methods.
3. Feminist research involves an ongoing criticism of nonfeminist scholarship [*To this I would add criticism of feminist scholarship too!*].
4. Feminist research is guided by feminist theory.
5. Feminist research may be transdisciplinary.
6. Feminist research aims to create social change.
7. Feminist research strives to represent human diversity.
8. Feminist research frequently includes the researcher as a person.
9. Feminist research frequently attempts to develop special relations with the people studied (in interactive research).
10. Feminist research frequently defines a special relation with the reader.

These characteristics share much in common with some environmental education research, particularly the use of a multiplicity of methods, adopting a transdisciplinary focus, and aiming to create social change.

What Feminist Research Has Already Been Undertaken in Environmental Education?

The English-language literature on feminist research in environmental education comes from the past two decades, but it is sparse and generally Australian or Canadian. One of the earliest articles is by Di Chiro (1987), which was written and published in Australia, even though she is American. Another is by Salleh (1989), an Australian ecofeminist and social theorist. Other early research that drew attention to gender differences in environmental knowledge, concerns, and behaviors includes Kremer, Mullins, and Roth (1990/1991, American), Fawcett, Marino, and Raglon (1991, Canadian), Hallam and Pepper (1991, British), Hampel, Holdsworth, and Boldero (1996, Australian), Pawlowski (1996, Polish), Russell and Bell (1996, Canadian), and Storey, da Cruz, and Camargo (1998, UK/Brazil). Other relevant literature from the 1990s, which specifically relates gender to environmental education, is Australian (including Barron, 1995; Brown & Switzer, 1991; Department of Prime Minister and Cabinet/Office of the Status of Women, 1992; Gough 1994, 1997, 1999a, 1999b, 1999c; Greenall Gough, 1993; National Women's Consultative Council [NWCC], 1992; Peck, 1992; Whitehouse & Taylor, 1996). During the past decade, Canadian writing about feminist environmental education research has continued with the work of Barrett (2005), Fawcett (2000, 2002), McKenzie (2004, 2005), and Russell (2003, 2005, 2006). Other relevant research has continued to be sparse (see, e.g., Carrier, 2007;

Gough & Whitehouse, 2003; Hallen, 2000; Newbery, 2003; Wane & Chandler, 2002).

The paucity of feminist research in environmental education could be considered surprising when related fields, such as science education, have a history of feminist research (see, e.g., Parker, Rennie, & Fraser, 1996). Feminist scholarship in science started about the same time as the ecofeminist movement[2] (in the early 1970s), but, perhaps because of the greater social status of science or because of some of the more extreme[3] writings of some ecofeminists, feminist research in science education has received a much higher profile and generated many more studies than in environmental education.

In her keystone paper, Di Chiro (1987) places a feminist perspective on environmental education within a socially critical framework. She grounds her ecofeminist perspective in radical and socialist feminism and asserts that:

> A feminist perspective [on] environmental education offers a more complete analysis of environmental problems and therefore a better understanding of those problems and their potential solutions. Such an analysis is political, in that it examines how power relations (in, for example, gender, class, race) shape the world in which we live; it asserts that the "polity" (human social world) determines and controls how this social world is and has been historically constructed and organised, and hence refutes the myth that the past and present state of the world is a "natural" and therefore justifiable progression. Moreover, environmental education's analysis of ocio-environmental problems is political in that it believes that if human social relations create the problems they can also change and improve them. (p. 40)

In particular, she argues that the environmental problem is socially constructed and should be viewed as a social problem, that environmental education should engage in a feminist critique of environmental problems and that it should engage in self-criticism "in order to understand how it is responsible as an educational enterprise for maintaining certain 'un-environmental' values and ideologies" (p. 41). Although others also argued the first two points to varying degrees, Di Chiro seems to be alone in asserting the need for environmental education to be self-critical, as well as socially critical.

An example of the type of feminist research that informs the discussions around the need for a gender agenda in ESD is Salleh's study from two decades ago. Salleh (1989) describes a group of upper working class women coming together "to see what might be done about household waste recycling in their local community" (p. 27). Few of the group had completed high school (only one had a university degree), all were older than thirty, and either were, or had been, married with children. All were already involved in some kind of ecologically sound practice at home. Opening up the issue with the technique of consciousness raising as a catalyst for moving from personal to political concerns, Salleh found that:

> . . . it was the tensions growing out of the consciousness-raising process itself that undermined the possibility of their participation in an environmental program . . . Their workforce and personal marginality were so severe that they lacked the necessary human support and self-assurance to transform their critical stance into a collaborative praxis. (p. 30)

Responses to the 1992 NWCC consultations on Australian women's priorities for environmental action support Salleh's findings. In Salleh's group, few women had completed high school and they continued to choose individual action, rather than social or political action, as their focus. In the NWCC (1992) study, with respect to women's priorities for action on environmental issues, there were noticeable differences in responses, depending on education level:

> The less education, the more likelihood that respondents would choose education or individual actions as the principal action for women, and the less likely they would suggest changing social frameworks or political action. More of the respondents who had not proceeded beyond school level gave individual or personal answers, than women from the other three levels of education. (p. 66)

Similar findings were reported by Wane and Chandler (2002) in their work with elderly rural Kenyan women who had no formal education but did have a deep understanding of their ecological situation. Wane and Chandler also raise questions about valuing indigenous women's knowledge: "What would transpire if indigenous women's ecological knowledges were included in environmental discourse and influenced curriculum, teaching, and learning?" (p. 92)

In posing this challenge, Wane and Chandler echo Hutcheon's (1991) question, "How do we construct a discourse, which displaces the effects of the colonizing gaze while we are still under its influence?" (p. 176). The task is to dismantle colonialism's system, expose how it has silenced and oppressed its subjects, and find ways for its subjects' voices to be heard, whether referring to the colonized or women anywhere. Many of these and related issues, such as the education levels of women and the importance of local knowledges, have been taken up in the rhetoric around the implementation of the Decade of ESD (see, e.g., UNESCO, 2005a, pp. 19–20); however, they have not had much influence on the environmental education research agenda to date.

The Potential of Feminist Research in Environmental Education

Although my personal disposition is toward critical and poststructuralist feminist research, all feminist research methodologies (and methods) can be applied to environmental education research contexts. We can participate in many types of research that put the social construction of gender at the center of the inquiry, whether seeking to predict, understand, emancipate, or deconstruct, and we need more stories from women relating to environments that we can use in environmental education.

Like McKenzie (2004, 2005) and Barrett (2005), I argue for poststructuralist research as the most promising approach for achieving the potential of feminist research in environmental education (Gough, 1997, 1999a, 1999b, 1999c). The power of feminist poststructuralist research in environmental education is that it calls into play a deconstructionist impulse that provokes consideration of the gendered positions made available to students and of understandings of gender identity. This approach is consistent with Davies (1994), who suggests a pedagogy informed by poststructuralist theory might begin:

> . . . with turning its deconstructive gaze on the fundamental binarisms of pedagogy itself: teacher/student, adult/child, internal/external, society/individual, reality/fiction, knower/known, nature/culture, objective/subjective [because] each of these underpin or hold together both what we understand as pedagogy and the discourses through which pedagogy is done. (p. 78)

Such an examination of the binaries of environmental education practices could also form the basis for a research agenda.

Another aspect of the appeal of feminist poststructuralist research as an approach for environmental education is summarized by Agger (1992), who argues that,

> Another primary aim of feminist cultural criticism is to decenter men from their dominance of various official canons and genres. Equally as troubling as the omission of women from the canon and from criticism is the installation of men as *those who speak for women*—universal subjects of world history. A good deal of poststructural feminist criticism has focused on the issue of the voices in which culture is expressed, the standpoints from which knowledge is claimed. (p. 118, emphasis in original)

Decentering the male perspectives that dominate environmental education discourses is a challenge for the future and will not be easy. As Felicity Grace (1994) argues, "Men's interests, women's interests and the common interest are all difficult contingent alliances, and the ideological claim to them forms part of continuing negotiations of power" (p. 19). Nevertheless, there is a need to tell "less partial, less distorted" stories in our research. As I previously noted, by studying women's experiences, we open up new empirical and theoretical resources and provide a new purpose for social science (for women rather than men), and a new subject matter of inquiry; although studying women is not new, it is new to study them "from the perspective of their own experiences so that women can understand themselves and the world" (Harding, 1987, p. 8).

Empowerment, and the search for more empowering ways to know, is a central concern of both critical and poststructuralist theorizing; it is also a goal shared by many environmental educators. Emancipatory (empowerment) research involves "analyzing ideas about the causes of powerlessness, recognizing systemic oppressive forces, and acting both individually and collectively to change the conditions of our lives" (Lather, 1991, p. 4), and Ellsworth (1989), for example, argues that "critical pedagogies employing this strategy prescribe various theoretical and practical means for sharing, giving or redistributing power to students" (p. 306). However, empowerment is not something done *to* or *for* someone; it is a process one undertakes for oneself in the development of a new relationship within his or her own particular contexts. We need research in environmental education that focuses on how women have been empowered, and are empowering themselves, rather than looking at universalized subjects; traditional research has focused on men's experiences and "asked only the questions about social life that appear problematic from within the social experiences that are characteristic for men" (Harding, 1987, p. 6).

Because of its political nature and concerns with empowerment, it is important to look at power and knowledge in the context of environmental education. This can be a focus of both critical and poststructuralist research. Power and knowledge, as they are exercized through discourses, are central aspects of poststructuralist theory: "discourses . . . are ways of constituting knowledge, together with the social practices, forms of subjectivity and power relations which inhere in such knowledges and the relations between them" (Weedon, 1987, p. 108).

The texts, myths, and meanings of our culture and our relationships with nature need deconstructed so we know the stories of which we are a part. Such deconstruction and critical analysis helps practitioners and students recognize whose interests are served at particular moments in environmental issues. It helps them understand that "it *does* make a difference who says what and when. When people speak from the opposite sides of power relations, the perspective from the lives of the less powerful can provide a more objective view than the perspective from the lives of the more powerful" (Harding, 1991, pp. 269–270, emphasis in original). The challenge is to encourage the development of alternative stories (discourses) by drawing attention to whose knowledge is legitimized and valorized in the power/knowledge structures of the dominant discourses. In particular, women's knowledge needs recognized and valued.

The dominant discourses in environmental education treat the subject of knowledge as homogeneous and unitary. Thus, in the behaviorist/individualist model that dominates much of these discourses, there is an emphasis on individuals having "the right behavior" and the knowledge of how to "get it right." This implies a power relationship in which some take it as their role to set out what those "right behaviors" are. However, it is no longer possible to find a universal subject: as subjects/agents of knowledge, we are all part of multiple, heterogeneous, and contradictory or incoherent positionings of race, class, gender, sexuality, disability, and ethnicity, and there is no one right way of knowing or behaving. The recent turn to "intersectional analyses" (e.g., Kahn & Humes, 2009) is relevant here, but it does raise the tension of how to do intersectional analyses without losing sight of particular concerns, such as gender.

Such multiple subjectivities are constantly achieved through relations with others (both real and imagined), which are themselves made possible through discourse. Accepting that the subjects of knowledge are multiple, rather than homogenous, unitary, and universal, has implications for curriculum, pedagogy, and research in environmental education. Exploring and developing such possibilities for environmental education is a challenge for the future, and it is one that must include the power and promise of feminist research.

Although poststructuralist research in environmental education is a challenge, I believe it offers much promise consistent with the stated goals of environmental education. The dominant discourses of environmental education recognize that the environment and environmental problems are complex, not simple. For example, the guiding principles of environmental education from Tbilisi (UNESCO, 1978, p. 27) refer to the multidisciplinary nature of the environment—"consider the environment in its totality—natural and built, technological and social (economic, political, technological, cultural-historical, moral, aesthetic)"—and "the complexity of environmental problems." Although not overt or necessarily intended and recognizing that they have also been interpreted as consistent with positivist and behaviorist perspectives, such a perspective could involve multiple readings or interpretations of the environment consistent with adopting poststructuralist pedagogy and research approaches. The multiple readings I suggest are not only of nature, but also of the individuals and groups concerned with the particular environment or environmental issue: there is a need to develop local or situated knowledges that disrupt oppressions. We should be listening to multiple stories in the spirit of a partnership ethic (and its precepts), rather than following, and an egocentric or homocentric ethic (Merchant, 1996).

Whatever the issue or environment, there are multilevel meanings of narratives and texts and multiple stories that can be told. There is not "one true story" about the environment. The knowledges involved in dealing with environments are multiple, involving both humans (in which each human is a multiple subject) and nonhuman nature (which also has a multiple subjectivity), and must be considered as such. Thus, poststructuralist pedagogy and research is also consistent with a partnership ethic[4] (and feminist research) in that it is concerned with listening to the voices of the marginalized, as well as those of the dominant discourses.

Poststructuralist research is also concerned with deconstructing power/knowledge relationships, which is also a goal of a partnership ethic in environmental education. As in critical research, it is important to analyze who has the power and what can be done to dismantle or subvert that power through developing counter-hegemonic and oppositional discourses. We need to know the stories of which we are a part and develop local knowledges.

Poststructuralist research consistent with a partnership ethic is also concerned with the liberation of nature and people. The goal is "to work toward a socially-just, environmentally sustainable world" (Merchant, 1996, p. 222). It is time to stop trying "from the outside, to dictate to others, to tell them where their truth is and how to find it" (Foucault, 1984/1985/1990, p. 9). By engaging in feminist and poststructuralist research in environmental education, we can come closer to achieving this goal, because we will have less partial and less distorted stories.

Notes

1. The term "liberal feminism" is often used to characterize the dominant form of feminism up to the 1960s. Its current form, inspired by the works of Simone de Beauvoir (*The Second Sex*, 1949/1953/1972) and Friedan (*The Feminine Mystique*, 1963), "emanates from the classical liberal tradition that idealizes a society in which autonomous individuals are provided maximal freedom to pursue their own interests . . . [and] endorses a highly individualistic conception of human nature" (Warren, 1987, p. 8). This conception locates our uniqueness as humans in our capacity for rationality and/or the use of language (Jaggar, 1983), and "when reason is defined as the ability to comprehend the rational principles of morality, then the value of individual autonomy is stressed" (Tong, 1989, p. 11). For liberal feminists, the attainment of knowledge is an individual project, and their epistemological goal is "to formulate value-neutral, intersubjectively verifiable, and universalizable rules that enable any rational agent to attain knowledge 'under a veil of ignorance'" (Warren, 1987, p. 9).

   Historically, liberal feminists have argued that women do not differ from men as rational agents and that it is only their exclusion from educational and economic opportunities that has prevented women from realizing their potential (Jaggar, 1983). Critiques of liberal feminism focus on its alleged tendencies to accept male values as human values, to overemphasize the importance of individual freedom over that of the common good, to adhere to normative dualism, and to valorize a gender-neutral humanism over a gender-specific feminism (Jaggar, 1983; Tong, 1989).
2. The term "ecofeminism" was coined in 1974 by Françoise d'Eaubonne "who called upon women to lead an ecological revolution to save the planet. Such an ecological revolution would entail new gender relations between women and men and between humans and nature" (Merchant, 1996, p. 5).
3. Here, I particularly refer to radical ecofeminism, because it overlooks "the historical and material features of women's oppression (including the relevance of race, class, ethnic, and national background), [and] it insufficiently articulates the extent to which women's oppression is grounded in concrete and diverse social structures" (Warren, 1987, p. 15).
4. I argue that it is time to move beyond utilitarian homocenterism and toward an environmental ethic for society along the lines of what Merchant (1992) calls a "partnership ethic" that "treats humans (including male partners and female partners) as equals in personal, household, and political relations and humans as equal partners with (rather than controlled-by or dominant-over) nonhuman nature" (p. 188).

References

Agger, B. (1992). *Cultural studies as critical theory*. London, UK: Falmer Press.

Barrett, M. J. (2005). Making (some) sense of feminist poststructuralism in environmental education research and practice. *Canadian Journal of Environmental Education, 10*, 79–93.

Barron, D. (1995). Gendering environmental education reform: Identifying the constitutive power of environmental discourses. *Australian Journal of Environmental Education, 11*, 107–120.

Bodzin, A. M., Shiner Klein, B., & Weaver, S. (2010). *The inclusion of environmental education in science teacher education*. Dordrecht: Springer.

Brown, V. A. (1995, 9–11 February). *Women who want the Earth: Managing the environment is a gender issue*. Paper presented at the Conference "Towards Beijing—Women, Environment and Development in the Asian and Pacific Regions," Victoria University of Technology, Melbourne, Victoria.

Brown, V. A., & Switzer, M. A. (1991). *Engendering the debate: Women and ecologically sustainable development*. Canberra, ACT: Office of the Status of Women, Department of the Prime Minister and Cabinet.

Carrier, S. J. (2007). Gender differences in attitudes toward environmental science. *School Science and Mathematics, 107*(7), 271–278.

Cherryholmes, C. H. (1993). Reading research. *Journal of Curriculum Studies, 25*(1), 1–32.

Davies, B. (1994). *Poststructuralist theory and classroom practice*. Geelong, Victoria: Deakin University Press.

de Beauvoir, S. (1972) *The second sex*. Harmondsworth: Penguin. (Original work published in French 1949 and English 1953)

Department of Prime Minister and Cabinet/Office of the Status of Women. (1992). *Women and the environment*. Canberra, Australian Capital Territory: Australian Government Publishing Service.

Di Chiro, G. (1987). Environmental education and the question of gender: A feminist critique. In I. Robottom (Ed.), *Environmental education: Practice and possibility* (pp. 23–48). Geelong, Victoria: Deakin University Press.

Ellsworth, E. (1989). Why doesn't this feel empowering? Working through the repressive myths of critical pedagogy. *Harvard Educational Review, 59*(3), 297–324.

Fawcett, L. (2000). Ethical imagining: Ecofeminist possibilities and environmental learning. *Canadian Journal of Environmental Education, 5*, 134–149.

Fawcett, L. (2002). Children's wild animal stories: Questioning interspecies bonds. *Canadian Journal of Environmental Education, 7*(2), 125–139.

Fawcett, L., Marino, D., & Raglon, R. (1991). Playfully critical: Reframing environmental education. In J. H. Baldwin (Ed.), *Confronting environmental challenges in a changing world*. Selected papers from the Twentieth Annual Conference of the North American Association for Environmental Education (pp. 250–254). Troy, OH: NAAEE.

Fontes, P. J. (2002). The stories (woman) teachers tell: Seven years of community-action-oriented environmental education in the north of Portugal. *Canadian Journal of Environmental Education, 7*(2), 256–268.

Foucault, M. (1990). *The use of pleasure (Volume 2 of the history of sexuality)*. New York, NY: Vintage. (Original work published in French 1984 and English 1985)

Friedan, B. (1963). *The feminine mystique*. New York, NY: W.W. Norton.

Gough, A. (1994). *Fathoming the fathers in environmental education: A feminist poststructuralist analysis* (Unpublished doctoral dissertation). Deakin University, Geelong, Australia.

Gough, A. (1997). *Education and the environment: Policy, trends and the problems of marginalisation*. Melbourne, Victoria: Australian Council for Educational Research.

Gough, A. (1999a). Recognizing women in environmental education pedagogy and research: Toward an ecofeminist poststructuralist perspective. *Environmental Education Research, 5*(2), 143–161.

Gough, A. (1999b). Kids don't like wearing the same jeans as their mums and dads: So whose 'life' should be in significant life experiences research? *Environmental Education Research, 5*(4), 383–394.

Gough, A. (1999c). The power and the promise of feminist research in environmental education. *Southern African Journal of Environmental Education, 19*, 28–39.

Gough, A. (2004). Blurring boundaries: Embodying cyborg subjectivity and methodology. In H. Piper & I. Stronach (Eds.), *Educational research: Difference and diversity* (pp. 113–127). Farnham: Ashgate.

Gough, N., & Gough, A. (2003). Tales from Camp Wilde: Queer(y)ing environmental education research. *Canadian Journal of Environmental Education, 8*, 44–66.

Gough, A., & Whitehouse, H. (2003). The "nature" of environmental education research from a feminist poststructuralist standpoint. *Canadian Journal of Environmental Education, 8*, 31–43.

Grace, F. (1994). Do theories of the state need feminism? *Social Alternatives, 12*(4), 17–20.

Greenall Gough, A. (1993). *Founders in environmental education*. Geelong, Victoria: Deakin University Press.

Hallam, N., & Pepper, D. (1991). Feminism, anarchy, and ecology: Some connections. *Contemporary Issues in Geography and Education, 3*(2), 151–167.

Hallen, P. (2000). Ecofeminism goes bush. *Canadian Journal of Environmental Education, 5*, 150–166.

Hampel, B., Holdsworth, R., & Boldero, J. (1996). The impact of parental work experience and education on environmental knowledge, concern and behaviour among adolescents. *Environmental Education Research, 2*(3), 287–300.

Harding, S. (1987). *Feminism and methodology*. Bloomington, IN: Indiana University Press.

Harding, S. (1991). *Whose science? Whose knowledge? Thinking for women's lives*. Ithaca, NY: Cornell University Press.

Harding, S. (2007). Feminist standpoints. In S. N. Hesse-Biber (Ed.), *Handbook of femininst resarch: Theory and praxis* (pp. 45–69). Thousand Oaks, CA: Sage.

Heilbrun, C. G. (1999). *Women's lives: The view from the threshold*. Toronto, ON: University of Toronto Press.

Hesse-Biber, S. N. (2007). Feminist research: Exploring the interconnections of epistemology, methodology, and method. In S. N. Hesse-Biber (Ed.), *Handbook of femininst research: Theory and praxis* (pp. 1–26). Thousand Oaks, CA: Sage.

Hutcheon, L. (1991). Circling the downspout of Empire. In I. Adam & H. Triffin (Eds.), *Past the last post: Theorizing post-colonialism and post-modernism* (pp.167–189). Hemel Hempstead: Harvester Wheatsheaf.

Jaggar, A. M. (1983). *Feminist politics and human nature*. Totowa, NJ: Rowman and Allanheld.

Kahn, R., & Humes, B. (2009). Marching out from Ultima Thule: Critical counterstories of emancipatory educators working at the intersection of human rights, animal rights, and planetary sustainability. *Canadian Journal of Environmental Education, 14*, 179–195.

Kremer, K. B., Mullins, G. W., & Roth, R. E. (1991). Women in science and environmental education: Need for an agenda. *Journal of Environmental Education, 22*(2), 4–6. (Original work published 1990)

Lather, P. (1991), *Getting smart: Feminist research and pedagogy with/in the postmodern*. New York, NY/London, UK: Routledge.

Linke, R. D. (1980). *Environmental education in Australia*. Sydney, New South Wales: Allen & Unwin.

Lotz-Sisitka, H., & Burt, J. (2002). Writing environmental education research texts. *Canadian Journal of Environmental Education, 7*(1), 132–151.

Lousley, C. (1999). (De)politicizing the Environment Club: Environmental discourses and the culture of schooling. *Environmental Education Research, 5*(3), 293–304.

MacGregor, S. (2006). *Beyond mothering earth: Ecological citizenship and the politics of care*. Vancouver, BC: University of British Columbia Press.

Malone, K. (1999). Environmental education researchers as environmental activists. *Environmental Education Research, 5*(2), 163–177.

McKenzie, M. (2004). The 'willful contradiction' of poststructural socio-ecological education. *Canadian Journal of Environmental Education, 9*, 177–190.
McKenzie, M. (2005). The 'post-post period' and environmental education research. *Environmental Education Research, 11*(4), 401–412.
Merchant, C. (1992). *Radical ecology: The search for a livable world.* New York, NY: Routledge.
Merchant, C. (1996). *Earthcare: Women and the environment.* New York, NY: Routledge.
National Women's Consultative Council. (1992). *A question of balance: Australian women's priorities for environmental action.* Canberra, Australian Capital Territory: National Women's Consultative Council.
Newbery, L. (2003). Will any/body carry that canoe? A geography of the body, ability, and gender. *Canadian Journal of Environmental Education, 8,* 204–216.
Parker, L. H., Rennie, L. J., & Fraser, B. J. (1996). *Gender, science and mathematics: Shortening the shadow*. Dordrecht: Kluwer.
Pawlowski, A. (1996). Perception of environmental problems by young people in Poland. *Environmental Education Research, 2*(3), 287–300.
Peck, D. (1992). Environmental education: Time to branch out. *The GEN, 1*(4).
Reinharz, S. (1992). *Feminist methods in social research.* New York, NY: Oxford University Press.
Rich, A. (1972). When we dead awaken: Writing as re-vision. *College English, 34*(1), 18–30.
Russell, C. (2003). Minding the gap between methodological desires and practices. In D. Hodson (Ed.), *OISE papers in STSE education* (Vol. 4, pp. 485–504). Toronto, ON: University of Toronto Press.
Russell, C. (2005). 'Whoever does not write is written': The role of 'nature' in post-post approaches to environmental education research. *Environmental Education Research, 11*(4), 433–443.
Russell, C. (2006). Working across and with methodological difference in environmental education research. *Environmental Education Research, 12*(3/4), 403–412.
Russell, C., & Bell, A. (1996). A politicized ethic of care: Environmental education from an ecofeminist perspective. In K. Warren (Ed.), *Women's voices in experiential education* (pp. 172–181). Dubuque, IA: Kendall Hunt.
Russell, C., Sarick, T., & Kennelly, J. (2002). Queering environmental education. *Canadian Journal of Environmental Education, 7*(1), 54–66.
Salleh, A. K. (1989). Environmental consciousness and action: An Australian perspective. *Journal of Environmental Education, 20*(2), 26–31.
Sandilands, C. (1999). *The good-natured feminist: Ecofeminism and the quest for democracy*. Minneapolis, MN: University of Minnesota Press.
Spender, D. (1985). *For the record: The making and meaning of feminist knowledge*. London, UK: Women's Press.
Stanley, L. (1990). *Feminist praxis: Research, theory and epistemology in feminist sociology*. New York, NY: Routledge.
Storey, C., da Cruz, J. G., & Camargo, R. F. (1998). Women in action: A community development project in Amazonas. *Environmental Education Research, 4*(2), 187–199.
Tong, R. (1989). *Feminist thought: A comprehensive introduction*. Boulder, CO: Westview Press.
United Nations Conference on Environment and Development. (1992). *Agenda 21*. Rio de Janeiro: UNCED. Final, advanced version as adopted by the Plenary on June 14.
United Nations Educational, Scientific, and Cultural Organization. (1978). Intergovernmental Conference on Environmental Education: Tbilisi (USSR) (pp. 14–26). October 1977. Final Report Paris, UNESCO.
United Nations Educational, Scientific, and Cultural Organization. (2005a). *United Nations Decade of Education for Sustainable Development 2005–2014. Draft Implementation Scheme*. Paris: Author.
United Nations Educational, Scientific, and Cultural Organization. (2005b). Report by the Director-General on the United Nations Decade of Education for Sustainable Development: International Implementation Scheme and UNESCO's Contribution to the Implementation of the Decade. Executive Board Meeting Paper 172 EX/11. Retrieved from portal.unesco.org/education/en/ev.php-URL_ID=36025&URL_DO=DO_TOPIC&URL_SECTION=201 html.
Wane, N., & Chandler, D. J. (2002). African women, cultural knowledge, and environmental education with a focus on Kenya's indigenous women. *Canadian Journal of Environmental Education, 7*(1), 86–98.
Warren, K. J. (1987). Feminism and ecology: Making connections. *Environmental Ethics, 9*(1), 3–20.
Warren, K. J. (1997). *Ecofeminism: Women, culture, nature*. Bloomington, IN: Indiana University Press.
Weedon, C. (1987). *Feminist practice and poststructuralist theory*. Oxford, UK: Blackwell.
Whitehouse, H., & Taylor, S. (1996). A gender inclusive curriculum model for environmental studies. *Australian Journal of Environmental Education, 12*, 77–83.
Zelezny, L., Chua, P., & Aldrich, C. (2000). Elaborating on gender differences in environmentalism. *Journal of Social Issues, 27*(3), 41–44.

36

The Representation of Indigenous Knowledges

Soul Shava
University of South Africa, South Africa

Introduction

One of the key concerns rising among indigenous peoples is the unequal and unjust **representation** of indigenous or local knowledges in relation to formalized, Western knowledge systems (Agrawal, 1996; Nader, 1996). This concern is directly related to the processes and impacts of colonization still evident even to this day. Concerns with representation include invalidation, devaluation, subjugation, appropriation, misappropriation, misrepresentation, marginalization, primitivization, decontextualization, exclusion, and rejection of indigenous knowledges that has been perpetuated primarily by modern, Western knowledge institutions and researchers (Dei, Hall, & Rosenberg, 2002; Hoppers, 2001, 2002; Hountondji, 1997; Masuku van Damme & Neluvhalani, 2004; Shava, 2000; Smith, 1999). In this chapter, I offer an analysis of some of the modes of representation of indigenous knowledges by modern institutions, with a particular focus on the southern Africa context (see the vignette by Lowan-Trudeau in this section for discussion specific to North American and other contexts).

Defining Indigenous Knowledge

The knowledges of indigenous peoples (that is indigenous ways of knowing and doing things) have been variably termed "indigenous knowledge" (IK), "indigenous knowledge systems" (IKS), "endogenous knowledge" (Crossman 2002; Hountondji, 1997), "local knowledge," "situated knowledge," "traditional knowledge," "traditional knowledge systems" (TKS), "traditional ecological knowledge" (TEK) (Inglis, 1993), and "traditional environmental knowledge" (Johnson, 1992). These terms are related analogues that basically attempt to define and give meaning to the knowledge of indigenous people. "Indigenous Knowledge" (IK) appears to be the most widely used internationally.

I have avoided the limitations of giving a concise and prescriptive definition of indigenous knowledge to fit all contexts, taking into consideration the contextual variability of indigenous knowledges. I instead focus on what I think are the key aspects that characterize indigenous knowledges and that I consider a definition of indigenous knowledge should incorporate. These are: **people**, **context** (place and time), **culture**, **language**, **knowledge** and **practices,** and **dynamism**.

Indigenous **people** (the knowers) are creators of indigenous knowledge; they give it discourse and meaning based on, and relating to, their experiences in interactions with their environment (the known) over time. The knowledge that indigenous people generate is embedded in their **culture** and embodied in their **practices**. This knowledge is transgenerational, transmitted from generation to generation orally (through narratives, stories/folklore, songs, and poetry), visually (through arts, such as "bushmen" paintings, writings, craft, cultural rituals, and dance), and practically (through doing and the artifacts associated with practice).

Language is the main medium for the representation and transmission of indigenous knowledge. Changing language usually results in modifications, accommodations, and loss of its fundamental features to fit the new language, resulting in distortion or loss through transmission and translation (Agrawal, 1995).

Because indigenous knowledge systems derive from different **locales/places** and different communities, they cannot be grouped as a collective, single entity under the commonly used unifying term "indigenous knowledge." Rather, they are **plurally definable as heterogeneous bodies of knowledge** or "indigenous knowledges."

Although rooted in history, indigenous knowledges are reflexive to changes over **time** in the lived environment and external influences, contacts, and interactions. They should not be rigidly held and perceived, rather analyzed to reveal the emergent processes of natural evolution of knowledge. This means indigenous knowledges are not static, stagnant, closed systems, but rather open, **dynamic systems** being transformed, created, and recreated in context (see Dei et al., 2002; Masuku, 1999; Masuku van Damme & Neluvhalani, 2004; Pottier, Bicker, & Sillitoe, 2003; Shava, 2000). In other words, there are both stable and transforming aspects within indigenous knowledges.

Historical Contextualizing of Indigenous Knowledge

The value of indigenous knowledge to local communities and beyond has been emphasized by various authors, including those discussed in this chapter. However, because of the diversity of indigenous peoples (and, hence, indigenous knowledges), I cannot comprehensively speak for all indigenous peoples. In this section, I attempt to review the development of representations of indigenous knowledges in both the global arena and in the southern African context with which I am most familiar.

International Conventions on Indigenous People and Indigenous Knowledge The emergence of indigenous issues in international political fora, as reflected in international policies and conventions, has made indigenous peoples increasingly aware that their localized struggles for representation are shared by other indigenous peoples throughout the world. This, in turn, has intensified the global proliferation of IK literature and the interchange of experiences among the world's indigenous communities that share similar struggles against colonization, marginalization, and misrepresentation.

Early international concerns about indigenous peoples have been exclusively devoted to human rights issues, rather than to knowledge. Indigenous people and their rights made their debut appearance in the international political arena in 1957 with the adoption of Convention 107 *Indigenous and Tribal Populations Convention* by the International Labour Organisation (ILO, 1957). This convention includes provisions on the protection of the rights of indigenous and tribal peoples. However, its main thrust was on the progressive integration of those groups into the life of their respective countries. In other words, it focused on the normalization of indigenous peoples through the nationalization and the "modernization" of indigenous people's ways of life, a process that ironically runs counter to protecting the identity of indigenous people, their knowledges, and their cultures.

In 1966, the United Nations (UN) General Assembly adopted the *International Covenant on Civil and Political Rights*. Although the covenant makes no specific reference to indigenous peoples, it does deal with ethnic, religious, and linguistic minorities in Article 27, which stipulates that, in states where minorities exist, they should have the right "to enjoy their own culture, to profess and practise their own religion, or to use their own language."

The 1980s marked the beginning of an era in which some UN agencies and international environmental organizations advocated for the recognition of the value of indigenous/traditional knowledge, particularly in biodiversity conservation management. This indicates a major shift in power/knowledge relations between scientific institutions and organizations and indigenous communities—from a view that, in the past, excluded and marginalized the knowledge of local communities, caricatured their livelihoods as destructive and unsustainable and their knowledge systems as unscientific, and promoted the hegemony of Western science to a view that represents indigenous communities and knowledges as valuable. This period also marked recognition in the international arena of the limitations and fallibility of Western science and the search for alternatives in other knowledges. An example of the shift or inversion in power/knowledge relationships is the focus on indigenous/traditional knowledge in Chapter 14 of the *World Conservation Strategy* (IUCN/UNEP/FAO/UNESCO, 1980). There is also a delicate terminological transition in international environmental organizations from an emphasis on "science" to the broader category of "knowledge" as an approach to addressing environment-development challenges (Martello, 2001).

In 1982, the UN Working Group on Indigenous Populations was established. Its main mandate was to prepare a draft Declaration on the Rights of Indigenous Peoples. Although this process started in 1985, this declaration was only recently approved by the UN General Assembly on September 7, 2007. The declaration recognizes that the "respect for indigenous knowledge, cultures and traditional practices contribute to sustainable and equitable development and proper management of the environment."

The 1992 Earth Summit (United Nations Conference on Environment and Development [UNCED], 1992) in Rio signified the entry of indigenous knowledge into mainstream international environmental and developmental discourse. Its key products included the Rio Declaration, the Earth Charter, Agenda 21, and the Convention on Biological Diversity.

The Rio Declaration on Environment and Development (UNCED, 1992a) acknowledges that:

> Indigenous peoples and their communities and other local communities have a vital role in environmental management and development because of their knowledge and traditional practices. States should recognize and duly support their identity, culture and interests and enable their effective participation in the achievement of sustainable development. (Principle 22)

The Earth Charter (UNCED, 1992b) affirms the right of indigenous people to their knowledge under provision 12b of the principles, which aims to: "Affirm the right of indigenous peoples to their spirituality, knowledge, lands and resources and to their related practice of sustainable livelihoods."

Agenda 21 has a specific focus on indigenous knowledge in Chapter 26, entitled, "Recognizing and Strengthening the

Role of Indigenous People and Their Communities" (UN, 1992a). The Convention on Biological Diversity (CBD) addresses indigenous/traditional knowledge in the following article:

(i) Each contracting party shall "subject to its national legislation, respect, preserve and maintain knowledge, innovations and practices of indigenous and local communities embodying traditional lifestyles relevant for the conservation and sustainable use of biological diversity and promote their wider application with the approval and involvement of holders of such knowledge, innovations and practices." (UN, 1992b, Article 8(j))

A key aspect of the CBD is equitable benefit sharing. This is designed to enable the distribution of the proceeds from commercial application of indigenous knowledge to directly benefit the local communities from which it derives (Article 19).

The alternative Nongovernmental Organizations (NGO) Treaty on Environmental Education for Sustainable Societies and Global Responsibility held during the Rio Summit of 1992 has as one of its principles that environmental education must "recover, recognize, respect, reflect and utilize indigenous history and local cultures as well as promote cultural, linguistic and ecological diversity."

The prominent idea of indigenous/traditional knowledges in most of the mentioned conventions is very instrumental and technicist in that they are perceived as mainly utilitarian in value and as a resource waiting to be extracted, documented, or data-based, codified, abstracted, decontextualized, institutionalized, comodified, universalized, and widely applied. This restructures and reorients indigenous knowledges into compartmentalized Western knowledge disciplines and somehow paradoxically runs counter to the holistic and process-oriented nature of indigenous knowledges recognized by the same international conventions (see Martello, 2001). This process also reveals the "scientization" of indigenous knowledges, which is how modern global institutions now assimilate indigenous knowledges into their own discourses.

(Colonial) Anthropological and Ethnographic Research

Early research on indigenous people and their knowledge was mainly by early Western travelers making the first accounts of "newly discovered" territories and by anthropologists and ethnographers. Perceptions of indigenous people and their knowledges by these early Western travelers and anthropologists have usually been read through the lenses of the Western contexts from which the authors originate (see Nader, 1996). These readings have taken two forms: i) a negative reading, depicting an undesired dark, primitive past, seen in the "ignoble primitive/savage" (see Levi-Strauss, 1966; and also Achebe, 1978; Tennant, 1994); and ii) a yearning for the rich, lost past perceived in the ideologically romanticized "other," commonly known as the "noble primitive" (Tennant, 1994).

Commenting on perceptions of the other as primitive, Kuklick (1991) says:

> When Daniel Gookin observed the Massachusetts Algoquins in 1674, he saw "as a mirror or looking glass, the woeful, miserable, and deplorable estate that sin reduced mankind [sic] unto naturally." His was a conventional European view: technologically unsophisticated, preliterate peoples were living in humankind's primeval condition, innocent of repressions necessary to civilisation. (p. 1)

It is from this European gaze of the "exotic other" that the "noble and ignoble savage" images tend to mirror either the "repulsive self" or the longing for the "lost self" of the author's European home context.

Writing on the perceptions of the African as the "primitive savage," Achebe (1978) posits that:

> . . . the West seems to suffer deep anxieties about the precariousness of its civilization and to have a need for constant reassurance by comparison with Africa. . . . Africa is to Europe as the picture is to Dorian Gray—a carrier onto whom the master unloads his physical and moral deformities so that he may go forward, erect and immaculate. (p. 13)

Paradoxically, it is also from anthropology and ethnography that early advocacy for indigenous people and their knowledges can be found. Some early colonial anthropological accounts that advocate for indigenous people include that of Malinowski (1948), who, in an ambivalent mix of denigration and praise, argued that the so-called primitive men [sic], or in his words "savages," were "endowed with an attitude of mind wholly akin to that of modern man of science!" (p. 26, exclamation mark in original).

Incidentally, some of the old, negative anthropological texts and terms are now used by indigenous communities in their efforts to reclaim their rights. Shaw (2003) contends that an interesting turn in the current era is that colonial anthropology is used to authenticate indigeneity as "much of the 'bad, old' anthropology remains crucial to indigenous peoples' legal and political struggles, or is helping to reinvent their societies in ways they desire" (p. 203). Masuku-Van Damme (personal communication, 2007), in her study of the Khomani San in the Kalahari, reports that they proudly identify themselves as the "Bushmen" [sic], a previously denigrating term, in a process of political self-definition. This situation posits anthropology in an ambivalent role in which it has both served the colonizer and the colonized.

However, advocacy for indigenous knowledge in its extreme form amounts to glorifying the lifestyles of indigenous people in comparison to the modern world, from which the romantic images of the "noble savage/primitive" derive. This includes idealized images of indigenous people as "minimal disturbers of nature" and "admirable scientist of the concrete" (Malkki, 1992, p. 29).

Anticolonial and Postcolonial Research The arena of indigenous knowledges (IK) has been the subject of renewed attention in environmental education, environmental management, and development disciplines during the past two decades (Brossius, 2004; Ellen, 2004; Nygren, 1999), and it has emerged as a new field unto itself. This renewed interest in IK appears to have two main drivers.

On the one hand, there is waning confidence in Western science as the prime source of solutions to the world's environmental problems. This is due to a gradual realization that Western science, although helpful in many realms, does not provide answers to all environmental problems (Hoppers, 2002; Le Grange, 2004). At times, Western science has given wrong answers, and some Western scientific solutions actually contribute to environmental crises (e.g., agrochemicals threatening species biodiversity and synthetic drugs having fatal side effects on the very same humans they are supposed to heal; see Shiva in Dei et al., 2002). The recognition of the fallibility of science has rejuvenated the search for alternative solutions to environmental problems in other previously marginalized knowledges.

Interest in IK, on the other hand, also stems from indigenous peoples themselves. This interest by indigenous peoples is both emancipatory and political. The focus on IK by indigenous people is a form of resistance, a search for identity and origin, a process of struggle for freedom from historical marginalization by dominating knowledge discourses (Dei, 2000) that is bound to, and that succeeds, liberation (Tambouku, 1999).

Modern Scientific Institutions Interest in IK by modern scientific institutions is, to a large extent, focused on the utilitarian aspects of these knowledges, including sustainable natural resource use and management, technical know-how (e.g., farmers' knowledge), and specialized knowledge, such as medicinal plant use. This has mainly been driven by interests from industry development, in particular, the search for "patentable" knowledge, and conservation of natural resources. Achebe (1978) anticipated this shift in the perception of other knowledges by Western scientific institutions in his suggestion that:

> Perhaps this is the time when it can begin, when the high optimism engendered by breakthrough achievements of Western science is giving doubt and even confusion. There is just a possibility that Western man [sic] may begin to look seriously at the achievements of other people. (p. 14)

The early emphasis on IK by Western scientific and development institutions was on documentation of this knowledge and making it accessible for use by scientific and development institutions. Warren, Slikkerveer, and Bronkesha (1995), in their book *The Cultural Dimension of Development: Indigenous Knowledge Systems*, call for the documentation of IK, warning that this knowledge is rapidly disappearing. However, that they view this knowledge as best used in the hands of Western institutions can be read from the statement: "By taking time and effort to document these systems, they become accessible to change agents and client groups" (p. xv).

Another danger of documentation is that it assumes indigenous knowledge can only come to be legitimized and validated if it conforms to experts' ideas of useful knowledge, which usually implies judging indigenous knowledges against the standard of Western science and discarding anything that does not fit these standards as unnecessary, excess baggage. There has been a more recent shift toward a focus on local community participation in development and conservation initiatives, and, with it, has emerged an associated plethora of participatory approaches and methodologies. This shift is characterized by such development texts as *Participating in Development: Approaches to Indigenous Knowledge* (Sillitoe, Bicker, & Pottier, 2002). The main aim of participatory approaches has been to empower local communities. However, problems still arise as to the level of participation, with concerns regarding participation becoming rhetorical—an appropriate terminology that does not match the way development workers perceive local communities or how they feel development should be done (see Chambers & Richards, 1995; Ellen, 2002).

The value of indigenous knowledge has also been acknowledged in environmental management, particularly with regards to biodiversity conservation. Speaking of traditional environmental knowledge and its possible role in environmental management, Johnson (1992) claims that:

> Today, a growing body of literature attests not only to the presence of a vast reservoir of information regarding plant and animal behaviour but also to the existence of effective indigenous strategies for ensuring sustainable use of local natural resources. (p. 3)

Relating to the economic potential of local knowledge of plant biodiversity, Balick and Cox (1996) claim the reasons ethnobotanists should focus attention on indigenous people are because indigenous cultures still retain much knowledge concerning plants and indigenous knowledge systems can guide the development of new crop varieties or medicines.

Commenting in a similar vein on the intertwined nature of knowledges, particularly the growth/development of science from the appropriation of indigenous knowledges, Nader (1996) states:

> . . . in spite of differences, there is the common theme of human societies doing science or accumulating knowledge by verifying observation, or by borrowing from

> others knowledge that works the same way. Thus one way of looking at modern science is as the ongoing result, though not cumulative result, of the discoveries, inventions, and collective sciences of others. We have been munching on each other for millennia. (p. 11)

Indigenous Scholars Interest in IK also stems from indigenous peoples themselves as "a response to the growing awareness that the world's subordinated peoples and their values have been marginalized – *that their past and present have been flooded out* by the rise in influence of Western industrial capital" (Dei et al., 2002, p. 6, emphasis in original). This interest in IK by indigenous scholars appears directly linked to liberation from Western colonial domination. For example, there was an increasing body of IK research in South Africa after the attainment of independence in 1994 (see an overview by Masuku van Damme & Neluvhalani, 2004), as there was in West Africa in the 1960s when most countries in that region attained independence (wa Thiongo, 1993). In their writing, indigenous scholars challenge the hegemony of Western scientific knowledge historically associated with Western colonial oppression. Similarly, the interactions of modern institutions with local communities in education, research, and development have been linked to a continued (postcolonial) marginalization of indigenous knowledge (Hountondji, 1997; McGovern, 1999; Rahnema & Bawtree, 1997) and the perpetuation of the dominant Western discourses.

In reference to the forms of representation by modern Western institutions (in Dei et al., 2002), reminiscing on his time teaching within the Canadian education system, states:

> I hear my students—especially though not exclusively those from minoritized groups—ask me why certain experiences and histories count more than others when 'valid' academic knowledge is being produced and validated. I hear students lament the effort it is taking for educators to recognize the powerful linkages between identity, schooling, and knowledge production. But more importantly, I hear my students worry how indigenous knowledges are being marginalised in the academy, and about the impact that the ranking of knowledges may well have on the prospects for educational transformation and social change. (pp. xi–xii)

Smith (1999), a New Zealand Aboriginal researcher, challenges Western dominance, knowledge appropriation, and knowing of the other in declaring:

> It galls us that Western researchers and intellectuals can assume to know all that it is possible to know of us, on the basis of their brief encounters with some of us. It appalls us that the West can desire, extract and claim ownership of our ways of knowing, our imagery, the things we create and produce, and then simultaneously reject the people who created and developed those ideas and seek to deny them opportunities to be creators of their own culture and own nations. (p. 1)

Reflecting on his educational experiences and how indigenous knowledges were viewed while growing up, Dei (2000) recalls:

> I come to a discussion of indigenous knowledges through an educational journey replete with experiences of colonial and colonizing encounters that left unproblematized what has conventionally been accepted as "in/valid knowledge." My early educational history was one that least emphasized the achievement of African peoples and their knowledges both in their own right, and also for their contributions to academic scholarship on world civilizations. Like many others, I engage the topic of "indigenous knowledges" with a deep concern about the historical and continuing deprivileging and marginalizing of subordinate voices in the conventional processes of knowledge production, particularly (but not exclusively), in Euro-American contexts. (pp. 112–113)

Dei (2000) argues that the interplay between different knowledges is one reason for integrating (recentering) indigenous knowledges in academic knowledge work. He suggests using indigenous knowledges for the political purposes of academic decolonization, that is as counterhegemonic knowledges to challenge imperial ideologies and colonial relations of knowledge production that continually characterize and shape academic practices.

Commenting on Western science research, Smith (1999) claims that, among indigenous peoples, the term "research" is inextricably linked to European imperialism and colonialism. She posits that:

> The word itself, "research," is probably one of the dirtiest words in the indigenous vocabulary. When mentioned in many indigenous contexts, it stirs up silence, it conjures up bad memories, it raises a smile that is knowing and distrustful. (p. 1)

Smith (1999) calls for creating indigenous research methodologies and approaches to research that are contextual to place and respectful of culture and for developing indigenous people as researchers.

Relaying the decontextualizing nature of modern education systems and their exclusion and invalidation of local contexts, Lizop (1997) points out that:

> As soon as the school is opened, it creates around itself a zone of cultural depression, as it were. Ask an African teacher what the cultural resources of his village are. He [sic] will answer: the school–and nothing else . . . Maybe the missionary, but often because he, too, is imported. But the market, the palaver tree, the dance, the song, the language of the tam-tam, the tales and proverbs, the historical legendary stories, the potter, the blacksmith, the weaver, are not for him sources of culture. The school acts as an instrument of humiliation. It establishes its empire upon destroying whatever it is not, whereas its mission should be to reveal to everyone all the riches and gifts they represent. (p. 157)

Evident here is the lack of contextual relevance of modern education systems and the subordination and exclusion of the educational role of the local community and its knowledge. Through formal education, people become alienated from their own culture and are absorbed into Western culture to the detriment of their own culture.

Reflecting on the marginalization of local knowledge in Westernized education systems in Africa and their possible role, Opoku (1999) laments:

> There is a tendency among many Africans who have formal schooling, as well as many foreigners, to think that those Africans who have not been to school and who usually live in the villages are ignorant and that those who are "educated," in the modern sense of the word, possess real and worthwhile knowledge. But such thinking is wrong, for there is knowledge which is not necessarily acquired in the classroom. Besides, our schools in Africa tend to make us ignorant of the knowledge which is the basis of the way of life of our respective societies, and the reason we go to school is to learn how others live, not how we live. (p. 43)

Opoku places the emphasis here on the need for education to be relevant to the immediate society of learners, the need to move beyond the enclosed classroom and the individual to engage the community context, and the need to realize the (possible) role of local community knowledge and expertise in education and development.

Commenting on colonial domination and oppression of self-representation, wa Thiongo (1993) speaks of postcolonial literature as:

> . . . that tradition of the struggle for the right to name the world for ourselves. The new tradition was challenging the more dominant one in which Asia, Africa and South America were always being defined from capitals of Europe by Europeans who often saw the world in colour-tinted glasses. (p. 3)

Wa Thiongo proposes "moving the centre from its location in Europe towards a pluralism of centres, themselves being legitimate location of the human imagination" (p. 8).

Speaking on processes of exclusion of indigenous knowledge by dominant Western knowledge, Shiva (1993) states that "local knowledge is made to disappear by simply not seeing it, by negating its very existence" (p. 9). She further contends that, when local knowledge does appear, it is denied the status of systematic knowledge and rendered "primitive" and "unscientific." Such rendering includes the emergence and proliferation of pseudo-scientific disciplines such as ethnobotany, ethnomathematics, and ethnomedicine.

Shiva (1993) claims that dominant scientific knowledge, therefore, breeds a "monoculture of the mind" by destroying the very conditions for alternatives to exist and calls for pluralism and democratization of knowledges against the disturbing tendency of Western institutions and the academy to divorce Western institutional discourse from the rest of the world to frame what counts as in/valid knowledge and to determine regimes of truth.

In analyzing the hegemony of Western science and how it has transformed power/knowledge relations, Shiva (2000) argues:

> The priorities of scientific development and R&D efforts, guided by a Western bias, transformed the plurality of knowledge systems into an hierarchy of knowledge systems. When knowledge plurality mutated into knowledge hierarchy, the horizontal ordering of diverse but equally valid systems of knowledge was converted into a vertical ordering of unequal systems, and the epistemological foundations of Western knowledge were imposed on non-Western knowledge systems with the result that the latter were invalidated. (p. vii)

Shiva argues that one knowledge system, the Western system and its epistemology, must not serve as the benchmark for all systems and that diverse knowledge systems should not be reduced to seeing the world through the logic of Western knowledge systems, taking into account the fallibility of Western knowledge systems and their grievous appropriation (through biopiracy and intellectual piracy) of other knowledges.

Analyzing the current and intended role of scientific research in Africa, Hountondji (1997) claims:

> . . . research is an activity oriented outwards, focused on the external world, ordered by and subordinate to external needs. Its focus is not inward. Its primary purpose is not, as it should be, to address issues raised, directly or indirectly, by African society itself. (p. 2)

Commenting on the prescriptive nature of Western driven development and its limitations in the Third World, Escobar (1997) holds:

> Development assumes a teleology to the extent that it proposes that the "natives" will sooner or later be reformed; at the same time, however, it reproduces endlessly the separation between reformers and those to be reformed by keeping alive the premises of the Third World as different and inferior, as having a limited humanity in relation to the accomplished European. (p. 93)

Indigenous Knowledge in Southern Africa

This section looks at indigenous knowledge in the pre- and postcolonial periods within the southern Africa context with which I am familiar. It should be noted that this periodic demarcation does not imply a disjuncture (lack of continuity) in indigenous knowledge literature, but is a socially framed heuristic, an attempt to historically contextualize it and reveal the transformative processes in the emerging IK discourse.

Colonial Literature The colonial era in southern Africa has been marked by two main forms of nonfiction writing that incorporate indigenous people and their knowledges. One form has been the appropriation of indigenous knowledge into Western scientific texts. The other form is that which describes the lives of indigenous peoples and their use of natural resources in their environment, often compared against the standard of European colonizers.

The appropriation of indigenous knowledge is exemplified by texts, such as Watt and Breyer-Brandwijk's (1962) *The Medicinal and Poisonous Plants of Southern and Eastern Africa*, which documents the use of plants by indigenous people as medicines, giving it their (the writers') own authorship and making it broadly available purportedly for the good of humankind (in this case, for the good of the settler communities).

Anthropological representations of indigenous peoples and their knowledge in Southern Africa include those that represent indigenous people as the primitive other. An example is Lee (1979) in his book, entitled *The !Kung San: Men, Women and Work in a Foraging Society*, in which he claims the purpose of his study was to look at a "contemporary hunting and gathering society from an evolutionary perspective" (p. xvii), revealing the study was aiming to look for a society that was "evolutionary primitive" compared to the society from which the author originates. The portrayal of the San as "foragers" in the title reduces them to the level of animals.

At the other extreme are texts that romanticize indigenous people and their relation to the environment. Describing the life of the Bushmen of the Walker and Richards (1975) claim:

> The Bushman's [sic] place in the ecology of the reserves is that of a predator on the ungulates, springhares, jackals, foxes, rodents, birds and insects which they hunt and eat. Bushmen are also the rivals of browsers and fruit-eaters for the edible plant foods. (p. 43)

This description posits "Bushmen" in the category of wildlife, rather than humans, relegating them to primitive beings in an imaginary human evolutionary line. However, in this case, "Bushmen" are not portrayed as "ignoble savages" but are romanticized as a natural part of the ecosystem, living in harmony with nature and the Bushman's [sic] environment, which, although harsh and restrictive to him, is "ecologically unaffected by his occupation over many years" (p. 43).

Postcolonial Literature Postcolonial writings on indigenous knowledge proliferate and come to bloom in the postindependence era of southern African states. Commenting on the impact of Western knowledge hegemony and the role of indigenous voices in education, Hoppers (2000) states:

> The African voice in education at the end of the twentieth century is the voice of the radical witness of the pain and inhumanity of history, the arrogance of modernization and the conspiracy of silence in academic disciplines towards what is organic and alive in Africa. It is the voice of "wounded healers" . . . struggling against many odds to remember the past, engage with the present and determine a future built on new foundations. . . . It exposes the established hegemony of Western thought, and beseeches it to feel a measure of shame and vulgarity at espousing modes of development that build on silencing of all other views and perceptions of reality . . . and dares educators to see the African child-learner not as a bundle of Pavlovian reflexes, but as a human being culturally and cosmologically located in authentic value systems. (p. 1)

Hoppers calls for making quality space available for the emergence and mainstreaming of the African voice in education.

Within environmental education, related views are shared. O'Donoghue (1994) notes how indigenous knowledge had previously been marginalized and how it is now a tool for liberation:

> Indigenous knowledge has historically been transformed to become both a tool of oppression and a voice within the struggle for liberation. These anomalies present the teacher with a challenge to use so-called indigenous/traditional knowledge as an enabling voice in a process of environmental education that is both transformative of and liberating for the cultural perspectives that are eroding the earth's capacity to sustain life. (p. 4)

Environmental education work on recontextualizing IK in the formal education curriculum and the reconceptualization of schools as existing within and linked to local communities is the subject of several theses at Rhodes University (Asafo-Adjei, 2004; Hanisi, 2006; Kota, 2006; Masuku, 1999; Shava, 2000), as well as a chapter by O'Donoghue, Lotz-Sisitka, Asafo-Adjei, Kota, and Hanisi (2007).

At the national level, efforts to bring in indigenous knowledge are evident in national policies and curricula that make specific reference to indigenous knowledge. In South Africa, through the IKS Policy for South Africa adopted by the cabinet in 2004, the government has registered its commitment to the recognition, promotion, development, protection, and affirmation of IKS. In Zimbabwe's National Environmental Education Policy and Strategies of 2004, one of the nine key National Objectives to be implemented across all sectors is "To protect and promote the use of indigenous knowledge systems."

There are also similar efforts at the regional level and beyond. Important points of reference here are the Environmental Education Association of Southern Africa's journal, the *Southern African Journal of Environmental Education* (SAJEE), and publications by the Southern African Development Committee Regional Environmental Education Programme (SADC-REEP). In the *Southern African Journal of Environmental Education*, evidence of the stronger indigenous voice is seen by the increase in the number of articles on indigenous knowledge (Impey, 2006; Jackson, 2007; Mokuku & Mokuku, 2004;

O'Donoghue, 2005; Price, 2005; Shava, 2005). Within the SADC-REEP, several projects emerged in the postcolonial era that foreground indigenous knowledge, including the indigenous knowledge series publications project that has seen the development of several IK learning support materials. The SADC-REEP, in collaboration with Rhodes University, also published the Environmental Education Association of Southern Africa (EEASA) Monograph: *Indigenous Knowledge in/as Environmental Education Processes* in 1999. This monograph is a diverse collection of short articles on IK practices, such as stories, narratives, and proverbs, and more conceptual papers.

Indigenous knowledge is an emerging issue in current deliberative discourses on education for sustainable development (ESD) in southern Africa. There is concern for the "'taken for granted' validity associated with scientific knowledge and information provided by scientific institutions on environmental issues and risks" (Lotz-Sisitka, Olvitt, Gumede, & Pesanaya, 2006a, p. 30). A key challenge identified by ESD practitioners is the "lack of capacity and research on mobilising IK in education in the context of ESD issues" (p. 32). The practitioners recognized local knowledges as being critical to contextual approaches to education and training of which they were in favor (Lotz-Sisitka, Olvitt, Gumede, & Pesanayi, 2006b). They pointed to the need to mobilize indigenous knowledges as a key feature of ESD. Also emerging within the postcolonial era in southern African is the indigenous knowledge journal, *Indilinga: African Journal of Indigenous Knowledge Systems* (http//www.indilinga.org.za), with a more embracing focus on indigenous knowledge issues in the region.

Beyond the region, numerous IK articles have been published (see the Lowan-Trudeau vignette in this section for mention of articles published outside of Africa). It should be noted that *Environmental Education Research*, arguably the journal with the greatest international readership, since its inception in 1995, has published only two articles that directly focus on indigenous knowledge. The first of these is an article by Reid, Tearney, and Dillon (2004), entitled "Valuing and utilizing traditional ecological knowledge: Tensions in the context of education and the environment." In it, Reid and colleagues point out that it is outsiders, rather than indigenous people, who usually conceptualize what traditional ecological knowledge (TEK) is and they portend that the key ingredient of environmental educators' interests in TEK is its "value through utility." They also argue that TEK, like other knowledges, should be critiqued and can be studied and explained. Masuku van Damme and Neluvhalani published an article in the same journal in 2004, entitled, "Indigenous knowledge in environmental education processes: Perspectives on a growing arena," which provides an overview of research activities and debates that have characterized the emergence and growth of indigenous knowledge in/as environmental education processes within southern Africa.

Conclusion

The arena of Indigenous Knowledges within environmental education and development is quite complex and related to the intertwined nature of indigenous people and their local context, on one hand, and the colonizer and their empires, on the other hand. It includes issues of **politics of power**, such as democracy, social justice, ethics and values, inclusivity, diversity, and plurality. It also includes **knowledge representation issues** around educational quality and contextual relevance, validity, and acknowledging different ways of knowing (plural epistemologies). Among indigenous peoples, the thrust is that indigenous knowledges are a counterhegemonic challenge to Western scientific knowledges and should be given space and voice in local and global discourses, after long periods of subjugation, in mainstream knowledge discourses.

Linked to this resuscitation of indigenous knowledges is a notable growing concern for **contextual and epistemological relevance** of local educational processes. There are apparent shifts in the perception, representation, and application of indigenous knowledges by Western scientific institutions, among the increasing frequency and amplitude of counter-hegemonic indigenous voices, coming from its validity and possible applications in environment management, development, and education disciplines. However, there is still the looming danger of assimilation of indigenous knowledges into dominant disciplinary and institutional discourses.

There have also been shifts in the use of the term "indigenous," from its negative connotations associated with colonialism in which the colonizer defines the other as primitive, ignorant, and inferior to its reappropriations by indigenous peoples and application as a signifier of identity, self-determination, autonomy, and difference. Another notable change is the shift in discourse from "indigenous knowledge" to "indigenous knowledges" and from "indigenous people" to "indigenous peoples," indicating the recognition of the existence of plural forms of indigenous knowledges from the diverse communities of indigenous peoples that exist in real life contexts, rather than the unified representation normalized by Western knowledge institutions.

It should be noted that the coverage of IK in most environmental education research journals and publications, including *Environmental Education Research*, remains scanty (exceptions are the *Southern African Journal of Environmental Education* and, to some extent, the *Canadian Journal of Environmental Education*). This situation needs to be redressed. Far more research and writing covering content, epistemology, and research methodologies directly addressing IK as it relates to environmental education research is necessary.

References

Achebe, C. (1978). An image of Africa. *Research in African Literatures, 9*(1), 1–15.

Agrawal, A. (1995). Indigenous and scientific knowledge: Some critical comments. *IK Monitor, 3*(3), 3–6.

Agrawal, A. (1996). A sequel to the debate (2): A response to certain comments. *IK Monitor, 4*(2), 17–18.

Asafo-Adjei, R. (2004). *From imifino to umfuno. A case study foregrounding indigenous agricultural knowledge in school-based curriculum development* (Unpublished master's in education thesis). Rhodes University, Grahamstown, South Africa.

Balick, M. J., & Cox, P. A. (1996). *Plants, people and culture: The science of ethnobotany*. New York, NY: Scientific American Library.

Brossius, J. P. (2004, March 17–20). *What counts as indigenous knowledge in global environmental assessments and conventions?* Paper presented to the Plenary Session on "Integrating Local and Indigenous Perspectives into Assessments and Conventions," at conference Bridging Scales and Epistemologies: Linking Local Knowledge and Global Science in Multi-Scale Assessments, Biblioteca Alexandrina, Alexandria, Egypt.

Chambers, R., & Richards. P. (1995). Preface. In D. M. Warren, L. J. Slikkerveer, & D. Bronkesha (Eds.), *The cultural dimension of development: Indigenous knowledge systems* (xiii-xiv). London, UK: Intermediate Technology Publications.

Crossman, P. (2002). *Teaching endogenous knowledge in South Africa: Issues, approaches and aids*. Pretoria, South Africa: Centre for Indigenous Knowledge, Department of Anthropology, University of Pretoria.

Dei, G. J. S. (2000). Rethinking the role of indigenous knowledge in the academy. *International Journal of Inclusive Education, 4*(2), 111–132.

Dei, G. J. S., Hall, B. L., & Rosenberg, D. G. (2002). *Indigenous knowledges in global contexts: Multiple readings of our world*. Toronto, Ontario, Canada: University of Toronto Press.

Ellen, R. (2002). Déjà vu, all over again, again: Reinvention and progress in applying local knowledge to development. In P. Sillitoe, A. Bicker, & J. Pottier (Eds.), *Participating in development: Approaches to indigenous knowledge*. London, UK: Routledge.

Ellen, R. (2004). From ethnoscience to science, or "what indigenous knowledge debate tells us about how scientists define their project." *Journal of Cognition and Culture, 4*(3), 409–450.

Escobar, A. (1997). The making and unmaking of the third world through development. In M. Rahnema, & V. Bawtree (Eds.), *The post-development reader*. London, UK: Zed Books.

Environmental Education Association of Southern Africa. (1999). *EEASA Monograph No.3: Indigenous knowledge in/as environmental education processes*. Howick, South Africa: EEASA, SADC REEP and WESSA.

Hanisi, N. (2006). *Nguni fermented foods: Mobilising indigenous knowledge in the life sciences* (Unpublished master's in education thesis). Rhodes University, Grahamstown, South Africa.

Hoppers, C. A. O. (2000). African voices in education: Retrieving the past, engaging the present, and shaping the future. In P. Higgs, N. C. G. Vkalisa, T. V Mda, & N. T. Assie-Lumumba (Eds.), *African voices in education*. Lansdowne, India: Juta.

Hoppers, C. A. O. (2001). Indigenous knowledge systems and academic institutions in South Africa. *Perspectives in Education, 19*(3): 73–85.

Hoppers, C. A. O. (2002). *Indigenous knowledge and the integration of knowledge systems: Towards a philosophy of articulation*. Claremont, South Africa: New Africa Books.

Hountondji, P. (Ed.). (1997). *Endogenous knowledge: Research trails*. Dakar, Senegal: CODESRIA.

Impey, A. (2006). Musical construction of place: Linking music to environmental action in the St. Lucia wetlands. *Southern African Journal of Environmental Education, 23*, 92–106.

Inglis, J. T. (1993). *Traditional ecological knowledge: Concepts and cases*. Ottawa, Ontario, Canada: International Development Research Centre.

International Labour Organisation. (1957). *Indigenous and tribal populations convention (Convention 107)*. Geneva, Switzerland: General Conference of the ILO.

Jackson, M. G. (2007). Learning to think differently. *Southern African Journal of Environmental Education, 24*, 82–89.

Johnson, M. (1992). *Lore: Capturing traditional environmental knowledge*. Ottawa, Ontario, Canada: IDRC.

Kota, L. (2006). *Local food choices and nutrition: A case study of amaRewu in the consumer studies curriculum* (Unpublished master's in education thesis). Rhodes University, Grahamstown, South Africa.

Kuklick, H. (1991). *The savage within: The social history of British anthropology, 1855–1945*. Cambridge, UK: Cambridge University Press.

Lee, R. B. (1979). *The !Kung San: Man, women and work in a foraging society*. Cambridge, UK: Cambridge University Press.

Le Grange, L. (2004). Western science and indigenous knowledge: Competing perspectives or complementary frameworks? Perspectives on higher education. *South African Journal of Higher Education, 18*(3), 82–91.

Levi-Strauss, C. (1966). *The savage mind*. London, UK: Weidenfeld & Nicolson.

Lizop, E. (1997). Schools as instruments of humiliation. In M. Rahnema & V. Bawtree (Eds.), *The post-development reader*. London, UK & New Jersey, US: Zed Books.

Lotz-Sisitka, H., Olvitt, L., Gumede, M., & Pesanayi, T. (2006a). *ESD practice in southern Africa: Supporting participation in the UN Decade of Education for Sustainable Development*. Howick, South Africa: SADC REEP.

Lotz-Sisitka, H., Olvitt, L., Gumede, M., & Pesanayi, T. (2006b). *Policy support for ESD in southern Africa: Supporting participation in the UN Decade of Education for Sustainable Development*. Howick, South Africa: SADC REEP.

Malinowski, B. (1948). *Magic, science and religion and other essays*. Glencoe. IL: Free Press.

Malkki, L. (1992). National Geographic: The rooting of peoples and the territorialization of national identity among scholars and refugees. *Cultural Anthropology, 7*(1), 24–44.

Martello, M. L. (2001). A paradox of virtue?: "Other" knowledges and environment-development politics. *Global Environmental Politics, 1*(3), 114–141.

Masuku, L. S. (1999). *The role of indigenous knowledge in/for environmental education: the case of a Nguni story in the schools water action project* (Unpublished master's in education thesis). Rhodes University, Grahamstown.

Masuku Van Damme, L. S., & Neluvhalani, E. F. (2004). Indigenous knowledge in environmental educational processes: Perspectives on a growing research arena. *Environmental Education Research, 10*(3), 353–370.

McGovern. (1999). *Education, modern development, and indigenous knowledge: An analysis of academic knowledge production*. New York, NY: Garland.

Mokuku, T., & Mokuku, C. (2004). The role of indigenous knowledge in biodiversity conservation in the Lesotho Highlands: Exploring indigenous epistemology. *Southern African Journal of Environmental Education, 21*, 37–49.

Nader, L. (1996). *Naked science: Anthropological inquiry into boundaries, power and knowledge*. New York, NY: Routledge.

Nygren, A. (1999). Local knowledge in the environment-development discourse. *Critique of Anthropology, 19*(3), 267–288.

O'Donoghue, R. (1994). *Story, myths, competing perspectives and ways of thinking about environmental education*. An unpublished working document, Natal Parks Board, Pietermaritzburg.

O'Donoghue, R. (2005). Cholera in KwaZulu-Natal: Probing institutional governmentality and indigenous hand-washing practices. *Southern African Journal of Environmental Education, 22*, 59–72.

O'Donoghue, R., & Janse van Rensburg, E. (1999). Indigenous myth, story, and knowledge in/as environmental education processes. In *EEASA Monograph No.3: Indigenous knowledge in/as environmental education processes*. Howick, South Africa: Environmental Education Association of Southern Africa (EEASA), South African Development Community—Regional Environmental Education Programme (SADC REEP), and Wildlife and Envioronment Society (WESSA).

O'Donoghue, R., Lotz-Sisitka, H., Asafo-Adjei, R., Kota, L., & Hanisi, N. (2007). Exploring learning interactions arising in school-in-community contexts of socio-ecological risks. In A. E. J. Wals, (Ed.), *Social learning: Towards a sustainable world*. Wageningen, The Netherlands: Wageningen Academic Publishers.

Opoku, K. A. (1999). Hearing the crab's cough: Indigenous knowledge and the future of Africa. In J. Z. Z. Matowanyika (Ed.), *Hearing the crab's cough: Perspectives and emerging institutions for indigenous knowledge systems in land resources management in Southern Africa*. Harare, Zimbabwe: IUCN Regional Office for Southern Africa.

Pottier, J., Bicker, A., & Sillitoe, P. (2003). *Negotiating local knowledge: Power and identity in development*. London, UK: Pluto Press.

Price, L. (2005). Playing musement games: Retroduction in social research, with particular reference to indigenous knowledge in environmental and health education. *Southern African Journal of Environmental Education, 22*, 87–96.

Rahnema, M., & Bawtree, B. (Eds.). (1997). *The post-development reader*. London, UK: Zed Books.

Ramose, M. B. (2004). In search of an African philosophy of education. *South African Journal of Higher Education, 18*(3),138–160.

Reid, A., Tearney, K., & Dillon, J. (2004). Valuing and utilizing traditional ecological knowledge: Tensions in the context of education and the environment. *Environmental Education Research, 10*(3), 237–254.

Shava, S. (2000). *The use of indigenous plants as food by a rural community in the Eastern Cape: An educational exploration* (Unpublished master's in education thesis). Rhodes University, Grahmstown, South Africa.

Shava, S. (2005). Research on indigenous knowledge and its application: A case of wild food plants of Zimbabwe. *Southern African Journal of Environmental Education, 22*, 73–86.

Shaw. (2003). Whose knowledge for what politics. *Review of International Studies, 29*, 199–221.

Shiva, V. (1993). *Monocultures of the mind: Perspectives on biodiversity and biotechnology*. London, UK, & New York, NY: Zed Books.

Shiva, V. (2002). Introduction. In G. J. S. Dei, B. L. Hall, & D. G. Rosenberg (Eds.), *Indigenous knowledges in global contexts: Multiple readings of our world*. Toronto, Ontario, Canada: University of Toronto Press.

Sillitoe, P., Bicker, A., & Pottier, J. (2002). *Participating in development: Approaches to indigenous knowledge*. London, UK: Routledge.

Smith, L. T. (1999). *Decolonizing methodologies. Research and indigenous peoples*. London, UK: Zed Books.

Tambouku. (1999). Writing genealogies: an exploration of Foucault's strategies for doing research. *Discourse: Studies in the Cultural Politics of Education. 20*(2), 201–217.

Tennant, C. (1994). Indigenous peoples, international institutions, and the international legal literature. *Human Rights Quarterly, 16*(1), 1–57.

United Nations. (1992a). *Agenda 21: The United Nations programme of action from Rio*. New York, NY: United Nations Department of Public Information.

United Nations. (1992b). *Convention on biological diversity*. Geneva, Switzerland: Interim Secretariat for the Convention on Biological Diversity.

United Nations. (2007). *United Nations declaration on rights of indigenous peoples*. New York, NY: UN General Assembly.

United Nations Conference on Environment and Development. (1992a). *Earth Charter*. Rio de Janeiro, Brazil: Author.

United Nations Conference on Environment and Development NGO Forum. (1992b). Treaty on environmental education for sustainable societies and global responsibility. The NGO Forum Alternative Treaties. Rio de Janeiro, Brazil: UNCED.

Walker, C., & Richards, D. (1975). *Walk through the wilderness*. Cape Town, Johannesburg, & London, UK: Purnell.

Warren, D. M., Slikkerveer, L. J., & Bronkesha, D. (1995) *The cultural dimension of development: Indigenous knowledge systems*. London, UK: Intermediate Technology Publications.

wa Thiongo, N. (1993). *Moving the centre: The struggle for cultural freedoms*. London, UK: James Curry.

Watt, J. M., & Breyer-Brandwijk, M. G. (1962). *The medicinal and poisonous plants of southern and eastern Africa: Being an account of their medicinal and other uses, chemical composition, pharmacological effects and toxicology in man and animal*. Edinburgh, UK, & London, UK: E.S. Livingstone.

World Conservation Union, United Nations Environmental Programme, World Wildlife Fund, Food and Agricultural Organisation of the United Nations, and United Nations Educational, Scientific and Cultural Organisation. (1980). *World conservation strategy: Living resource conservation for sustainable development*. Gland, Switzerland: Author.

37

Educating for Environmental Justice

Randolph Haluza-DeLay
The King's University College, Edmonton, Canada

Mark G. ("Grover") was constantly in trouble at the environmental education program of a major environmental organization at which we worked. He was a great teacher, especially with the younger kids. Maybe the kids liked him because of his rather elfin manner. Maybe it was because he was so full of delight, plus his encyclopedic knowledge of the local ecosystem, plus excellent pedagogy, plus being able to whistle in birds.

What got Grover in trouble was his tendency to discuss the destruction of Amazonian rainforests during those bird classes, and mention of the forced extraction of Native Americans when the class was discussing local forests. He pushed the program's director to buy local foods and argued the necessity to get kids to think about what their families bought and owned. He almost got fired when he criticized "capitalism." He believed these issues were a fundamental part of environmental education, because they were linked to environmental degradation; however, his director disagreed. He almost got fired again when a parent complained his child came home wanting to "live simply so others could simply live." Grover's end came when he let it be known that he thought poverty was an environmental issue.

Grover's problem in the mid-1980s is the same today: environmental education has primarily ignored what has come to be known as environmental justice. Environmental education now encompasses more than nature study or wildlands; to some extent, it includes the ways society is organized, with topics such as lifestyle choices, land sprawl, urban development, pollution, toxic wastes, and climate change. Environmental justice points out that environmental problems disproportionately affect some groups more than others and asks why? Because environmental education has not embraced environmental justice, it remains an ineffectual band-aid on the wounds of the earth and its inhabitants.

Extensive research and practical experience conclusively identifies that environmental hazards and amenities are distributed unevenly:

- Early research in the United States by the United Church of Christ (UCC, 1987) showed race was the most significant variable associated with the location of uncontrolled hazardous waste sites. Twenty years later, a follow-up indicated little had changed (Bullard, Mohai, Saha, & Wright, 2007).
- Extensive research has now shown that exposure to environmental "bads," such as toxics, poor air quality, hazardous waste sites, and workplace safety, are disproportionately borne by poor and racialized people and communities (Agyeman, 2005; Agyeman, Cole, Haluza-DeLay, & O'Riley, 2009; Brulle & Pellow, 2006; Bullard, 1990; Mohai, Pellow, & Roberts, 2009).
- A 1991 presidential order in the United States required the Environmental Protection Agency (EPA) to include environmental justice concerns in its environmental impact assessments. A similar order is on the books in the United Kingdom.
- Global climate change affects some nations far more than others (Roberts & Parks, 2007). That most of these nations, such as the Maldives, Bangaldesh, Sub-Saharan Africa, or Arctic Inuit, have had little share in the benefits of industrial causes of the anthropogenic portion of climate change illustrates the unfairness or injustice of climate change. "Climate justice" has been a rallying cry at recent global climate change conventions.

These are some environmental issues raised by what is known as the environmental justice movement. For the most part, environmental educators and environmental education researchers have ignored these sorts of issues. In doing so, environmental educators have reduced the scope of environmental sustainability and missed opportunities to connect with more people and potential allies among a broader reach of civil society organizations and other educators.

This chapter substantiates some of these claims and, eventually, demonstrates that environmental education must include environmental justice as central to environmental problem solving. First, I describe the concept of environmental justice and related terms and the social science research literature that supports them. A fuller understanding of environment and environmental justice requires attention to the political-economic systems that negatively impact both natural environments and marginalized social groups, as well as attention to some of the theory of what constitutes justice. As Brulle and Pellow (2006) stated in a recent review of the literature, "[the] politics of the distribution of the fruits of economic production is overlaid with the politics of the distribution of environmental pollution, producing environmental injustice" (pp. 108–109). Second, I examine the education and environmental education research literature regarding educating for environmental justice. A review shows that, although there is often mention of equity and justice, rarely does the environmental education research literature go beyond mere mention.

What Is Environmental Justice?

A number of terms have been used in the literature and by social movement actors, including environmental justice, environmental racism, and environmental inequality. *Environmental racism* represented the first, and still salient, way of understanding certain relationships between environmental exposure and social factors. Beginning in the 1980s, research began to show that environmental hazards were disproportionately located in, or disproportionately affected, racialized communities—primarily black, hispanic, or Native American (Cole & Foster, 2001). The research began to extend the conclusions to other forms of social inequality, such as socioeconomic status. Gender causes other variations, with women often more impacted by some forms of environmental risk, including chemicals during pregnancy, but also relative to their position as caregivers (Salleh, 2009), whereas working class men face occupational safety environmental hazards. These patterns were evident in other nations, such as the United Kingdom (Bulkeley & Walker, 2005), Canada (Agyeman et al., 2009), and globally (Hossay, 2006) as well.

The term *environmental justice* quickly developed into the predominant way of expressing environment-related societal inequities. As Pellow (2000) defines it, "An environmental injustice occurs when a particular group . . . is burdened with environmental hazards" (p. 582). For some time, the US research focus was on delineating whether race or class issues were the primary causes of environmental burdens. In a recent summary, Mohai, Pellow, and Roberts (2009) conclude, "If we want to understand the causes of environmental inequality, we need to know what role both race and class play because disparities have been found along both dimensions" (p. 411). Regardless of the precise term used, the benefit of the environmental justice lense is to link environmental concerns with other societal processes that produce social inequality and recognize these as structural questions about the distribution of power and resources in society (Pellow, 2000).

Another benefit of the environmental justice lense is to expand the scope of what counts as "the environment." Initially, the commonly understood meaning of an "environmental issue" has been associated with "nature." It is, however, this very construction of "the environment" that environmental justice activists have challenged. For example, urban activists may lump natural space conservation with overall quality of life issues that include adequate housing and safe neighborhoods or may advocate lead remediation or pollution issues as health rather than environmental matters. Safety and security were key "environmental" issues for Detroit children (Wals, 1996). Some children's experience of "nature" is limited to ground-level pollution, lead exposure, cockroaches, and mold (Strife & Downey, 2009). To some degree, the mainstream environmental movement has acceded to this trend, but it still faces criticism for lack of partnership with new constituencies and on non-"nature" issues (Pellow & Brulle, 2005; Sandler & Pezzullo, 2007). Beyond the particulars of what constitutes an "environmental issue" and beyond the injustices of distribution of hazards, environmental justice activists and scholars have also expanded their attention to the social and political processes in operation.

It takes very little effort to show that all is not equal, even in a democratic society. Most theories of justice emphasize that social goods should be equitably distributed and that inequitable distribution is unfair. This is called *distributive* or *substantive justice*. Income, safe housing, effective police, safety from toxic exposure, and adequate food are examples of social goods. They may be considered fairly distributed on the basis of several different types of criteria, for example, *merit* (e.g., hard work gets a person a higher income), *need* (e.g., all should have enough food), *equality* (e.g., everyone gets a vote), or a variety of other measures. Much of the environmental justice research has focused on cases of distributive injustice, primarily because it is easiest to empirically correlate with demographic variables. Explanations of these distributional disparities also overlap (Mohai, Pellow, & Roberts, 2009). Some explanations have focused on systematic discrimination, say of racialized communities or poorer communities and neighborhoods as "sacrifice areas" (Hooks & Smith, 2004). Economically oriented explanations focus on market dynamics alone, such as the lower costs associated with lower environmental standards in marginalized communities or reduced public resistance to undesirable land uses. Sociopolitical explanations focus on power differentials or political resources that result in uneven distribution.

The last explanation initiates consideration of a second aspect of environmental justice. *Procedural* or *participatory justice* remedies inequities in engagement and participation in social or political processes (Hillman,

2002; Schlosberg, 2005; Taylor, 2000). Those with limited access to information, democratic participatory opportunities, or power to shape discourse or government decisions are less able to defend themselves or their communities from negative environmental effects. Documents written in technical language require higher levels of education, again causing participatory injustice. Technical hearings require knowledge of specific procedures. Social capital differences disable access to politicians or business leaders. Media are more favorable to particular manners of advocacy. In these ways, participatory injustice reproduces additional forms of inequality.

This latter dimension of justice emphasizes the need to go beyond the *what* of environmental injustice. Injustice may be done not just in the things that happen, but in *how* they happen. This awareness leads toward an extended analysis of the processes of democratic participation in contemporary late modern societies and the effect of social hierarchies in the construction of environmental problems. Unfortunately, procedural justice is much more difficult to explain, measure, and ensure than material inequality.

Environmental Justice or Eco-Justice?

Scholars have begun to emphasize a third dimension of justice, which is central to an important debate in environmental education. Bowers has argued strenuously, in a great variety of books and essays, that environmental education should be educating for eco-justice (e.g., Bowers, 2001). Much of the work in environmental justice is a fairly traditional political-economic analysis of disparities regarding environmental goods and services. But these disparities rest on cultural assumptions and culturally specific social institutions. The existing social and legal systems, as well as predominant concepts of "justice" and "environment," are neither universally held nor ontologically normative. Thus, another aspect of environmental justice is *recognitional justice*, that is, injustices related to the misrecognition or devaluing of particular social forms and cultural worldviews. For example, extending concerns for justice to nonhuman portions of creation is a facet of a holistic and relational worldview, such as often expressed in Aboriginal worldviews (e.g., Simpson, 2002; see introduction in Agyeman et al., 2009). How can nonhuman others be treated with fairness or be given political standing? Can (or should) we see the trees, sun, moon, or other creatures as fellow citizens or even brothers and sisters?

The legitimation processes by which meaning is attributed to these questions are as fundamental to the quest for justice, equality, and environmental protection as distributive and participatory dimensions. Although distributional equality is crucial, exclusive focus on it narrows the scope of justice. Similarly, inclusion of procedural justice dimensions—although an improvement—still limits participation to those who are allowed, by the system itself, to participate in the manner of the system. It does not ask the more fundamental questions of "Who is the system?" and "How did the system come to be this way?"

Schlosberg (2005), following from political theorists like Young and Fraser, argues:

> If differences are constituted in part by social, cultural, economic, and political processes, any examination of justice needs to include discussions of the structures, practices, rules, norms, language, and symbols that mediate social relations. (p. 99)

Bowers' argument about environmental education is that the Western, European-derived worldview is founded on particular root metaphors that ground that way of thinking. Education, imbued with the cultural values that are part of the problem, has exacerbated the environmental crisis. Educating for eco-justice, Bowers argues, extends justice to the environment, not just humans, and requires examination of the root metaphors, symbols, norms, and practices of society. Even emancipatory and critical pedagogy have come under Bowers' attack, being described as having, at best, an orientation to the earth of benign neglect and an anthropocentric focus inadequate for environmental restoration. According to him, even Freirian pedagogy is founded in Western imperial rationality and represents a form of domination that surfaces as colonialism and domination of the earth (Bowers & Apffel-Marglin, 2005). It is this latter critique that once led McLaren (1991) to dismiss Bowers as a "patrician critic."

The debate between these positions has grown increasingly vituperative. Trying to moderate it, Gruenewald invited both Bowers and McLaren to contribute essays to a journal special issue. Gruenewald (2003) had earlier developed an interesting synthesis of a "critical pedagogy of place" that emphasized the bioregionalist focus on coming to know deeply one's local place (reinhabitation) with social justice (decolonization) and drew heavily on both McLaren and Bowers. Bowers (2005) directly challenged the "eco-socialism" argument of McLaren and Houston (2004), who then attacked back (Houston & McLaren, 2005). Gruenewald (2005) tried to demonstrate common ground and value in both scholarly approaches:

> McLaren, Houston, and other critical social theorists view environment mainly in relation to large-scale political economy, and Bowers and other critical ecological theorists view environment mainly in relation to local bioregion and culture. Both views, it seems to me, are necessary to grasp the interrelationship among environment, culture, and economics. (p. 210)

Martusewicz (2005) disagreed. She argued that Houston and McLaren's "analysis begins from the assumption that the exploitation of the natural world rests on the exploitation of humans," which she argued reinscribed a difference between environmental justice—"unequal distribution of harmful environments among people"—and eco-justice—"justice toward nature" (p. 218). In her view, too often, a critical approach ignores the deep-seated cultural frames that produce political-economic structures, as well as symbolic but no less materially consequential dominations of the natural world. In some important ways, Bowers' arguments parallel the recognitional dimension of justice described, in that misrecognition of the roots of the system occlude answers to, "How did the system get this way?" Eventually, Gruenewald also found himself on the receiving end of an attack from Bowers (2008). The debate between Bowers and McLaren continued in the journal *Capitalism Nature Socialism* (see Mauro, 2008). Mueller (2008) jumped into the fray and argued that Bowers' version of eco-justice is an improvement on environmental justice.

Recognizing the continuing unsettled character of this debate, in the remainder of this essay, the acronym EJE will be used for educating for eco-/environmental justice. Other terms have also been used in an attempt to be more inclusive. Coining the term "ecopedagogy," Kahn (2008) presented a Freiran educational philosophy of social justice in the "age of ecocatastrophe" in a chapter that was included in the second edition of *The Critical Pedagogy Reader*; the first edition had no environmental articles whatsoever. Both the chapter and subsequent book (Kahn, 2010) are tremendous works of educational philosophy with a core theme of social justice. For the most part, however, all of these discussions represent exhortations to get on with the task of including justice dimensions in environmental education and there remains little *research* literature on such practices.

The patterned character of environmental injustices points out that the issues are structural. As Agyeman (2005) comments, "The purveyors of environmental bads, such as large multinationals, are favored in pluralistic decision-making processes because of their disproportionate influence, economic muscle, and knowledge" (p. 3). Consequently, solutions to both environmental and justice problems require addressing social and political-economic factors, as well as ecological knowledge. These factors—congregating in a system of extraction, production, and consumption oriented around the maximization of capital accumulation—degrade environmental conditions at the same time that they contribute to social inequities. From this perspective, it could be argued that the most pressing and basic environmental issue is social inequality (Pellow, 2009). For the efficacy of environmental education, the morality of justice, and the need for sociocultural and political-economic analyses of environmental problems, EJE should be central to all environmental education. Ecological degradation is first and foremost a social violence.

Further, it is possible that EJE could learn much from other forms of education grounded in social justice movements. Nordstrom (2008) indicates there are similarities between environmental education and multicultural education. Wenden's (2004) contributors explicitly link social and ecological issues as part of a culture of peace. A specifically Canadian collection spans a breadth of educational forms that often link justice and environment (Goldstein & Selby, 2000). Such contributions variously report on practices, policy regimes, and context, showing the grounds for educational innovation are always shifting. Again, however, very little of this writing is grounded in *research*.

How Does Environmental Education Fare?

The North American Association for Environmental Education (NAAEE) has made "diversity and environmental justice" one of its priorities. Nevertheless, by its own admission, environmental education has made only small steps in this regard (Lozar Glenn, 2009). Environmental education is not alone in this, as many environmental nongovernmental organizations (ENGO) exhibit similar superficiality (Sandler & Pezzullo, 2007). Barriers include a weak understanding of the social and cultural positions of minority groups or lack of attention to racialization. The mainstream environmental movement has largely been white and middle class, which is not to deny that nonwhites or nonmiddle classes are also concerned about the environment (Lee, 2008). Jafri's (2009) research with Toronto ENGOs showed that, even in that very multicultural city, ENGOs sought to "hire diversity." Diversity often meant simplistic recruitment of visible minorities, rather than investigating organizational practices for possible barriers or undergoing organizational changes for more genuine inclusion. A study of diversity factors for employment with ENGOs found differences betweens whites and minorities, especially pronounced on factors that would improve diversity in both recruitment and retention (Taylor, 2007). Race, power, and culture remain largely unproblematized in both environmental education and the general environmental movement.

The field of environmental education often sees "state of the field" reports or calls to pay attention to trends, challenges, or opportunities (e.g., Gonzalez-Gaudiano, 2006; Marcinkowski, 2010; Potter, 2010). Few mention environmental justice or do so only superficially. For example, although Cole (2007) refers to environmental justice as one of the frameworks with which to expand environmental education, it is not described beyond the article's abstract, nor do the central concerns of environmental justice figure into the article. When McKeown-Ice and Dendinger (2000) identified concepts from the social sciences that were requisite for understanding environmental problems, nothing about differential distribution of environmental amenities or hazards, or inequities of any sort, were mentioned.

Another research team provides one of the few assessments of EJE in common environmental education curricula (Kushmerick, Young, & Stein, 2007). In the 224 lesson plans from such curricula as Project Wild and World Wildlife Fund, the research team found there was little explicit environmental justice framing. They called this a "missed opportunity" and suggested that many of the lessons could be easily adapted for EJE inclusion. Lewis and James (1995) pointed out this same weakness in environmental education some time ago.

Agyeman (2002, 2003a) asked what would be needed to orient environmental education toward environmental justice: a shift in policy/practices/curriculum or a shift in paradigm? He argued the need for "culturing environmental education," by which he meant paying attention to the differences produced by varying life experiences, social location, or ethnic affiliation. Alternate voices are needed, because the voices that presently dominate are unlikely to move the field of environmental education in new directions. According to many analysts, because environmental education often lacks a cultural awareness, it often lacks a critical social analysis, leading to a techno-fix approach. As Smith and Williams (1999) state, "regulating toxins" is different than "healing creation." The *Canadian Journal of Environmental Education* (CJEE) appears to have taken up the challenge of culturing environmental education more than other environmental education research journals, perhaps owing to the explicitly multicultural policies of the Canadian nation-state, as well as increasing social recognition of Aboriginal peoples and indigenous education, as well as a more expansive social safety net that reduces social distance, such as poverty, and opens more possibilities for other voices to gain a position in the field.

The 2002 and 2003 issues of CJEE were deliberately devoted to descriptions of alternate approaches to environmental education and the ways such voices have been accorded status in environmental education. Although not explicitly about environmental justice, inclusion of Aboriginal (e.g., Cole, 2002; Simpson, 2002), gay and lesbian (Gough et al., 2003; Russell, Sarick, & Kennelly, 2002), and diverse cultural approaches (Kato, 2002; Wane & Chandler, 2002) are forms of participatory and recognitional justice-work, and each of these types of justice figure in papers in both issues. Other essays expressed inclusion of disability and gender (e.g., Newbery, 2003). Of particular note were the essays on cultural diversity. James (2003) described means to design research to include marginalized peoples with a lengthy (and rare for environmental education research) discussion of environmental racism. Among James' conclusions is the existence of a lacuna in environmental education regarding social power; that those with existing status in the field make research decisions may mean that topics central to marginalized people are underrepresented. This could be among the reasons for the weak representation of research on EJE in the environmental education literature. Marouli (2002) insisted environmental education must take account of cultural diversity. Placing her assessment in the framework of "education for sustainability," Marouli described multicultural environmental education. Her research identified three primary strategies among programs she identified as multicultural environmental education: cultural pluralism, making global connections, and environmental and social justice.

Multicultural environmental education has sometimes been touted as a way to ensure environmental education includes environmental justice concerns. However, Agyeman (2003a) and others insist it is important to avoid "adjective adjective education." Adding adjectival modifiers makes the concerns of the subfield peripheral to the mainstream of the field. Acknowledging the centrality of cultural pluralism and environmental justice to environmental education, Kaza (1999, 2002) showed how it is possible to do EJE with primarily white, middle-class university students. Kaza presents a model for moving into the recognitional domain of environmental justice and assessing the cultural base and manifested power in existing social structures. Still, such models need a stronger research basis.

Another form of environmental education that addresses environmental justice concerns might be urban environmental education, but it raises the same cautions (Agyeman, 2003b). Access to nature in urban areas is unequally distributed. Schoolyard greening in Toronto was affected by socioeconomic status (Dyment, 2005); this is one of the astonishingly few studies to bring such an important social characteristic into environmental education research. Access to funds, volunteers, and the skills necessary to green the schoolyard were a function of socioeconomic status. Even more significant may be the perception of the importance of green schoolyards—nature is still considered a "frill" by many. In many cities, natural spaces are lacking in poorer areas. Given the benefits of access to nature and the negative ways lower income already degrades health, education, and life chances, researchers state, "unfortunately, it appears that the children who might benefit the most from nature are the ones who live where it is glaringly missing" (Castonguay & Jutras, 2009, p. 107). Other research has sought to elicit urban youth's perceptions of nature and draw forth implications for environmental education (Wals, 1996; Wilhelm & Schneider, 2005). A close reading of such research shows it rarely attends to the social forces that have eliminated urban nature.

Most references to environmental justice in the education research literature attended to environmental issues in non-European and non-North American countries. Examples include a description of environmental education for democracy and social justice in Costa Rica (Locke, 2009) and combining indigenous ecological science and intercultural education within a framework that protects local communities from excess intrusion by powerful and globalized capital in Ecuador (Schroder, 2006). Another study showed how educating for social justice in the context of local ecologies can proceed in Peru and then made comparison to a North American project training preservice teachers for global education (Alsop, Dippo, & Zandvliet, 2007). A topic of another chapter in this book, research has shown that environmental education with Aboriginal peoples has to be contextualized in ways that allow recognitional justice and facilitate inclusion and participation (Lowan, 2009; Swayze, 2009).

It is clear that EJE must be more central to environmental education. Although environmental justice has often been an urban movement, EJE should not merely be for urban citizens. If only "diverse youth" or urban youth are getting a form of EJE, then broader analysis of environmental problems and solutions are weakened. Furthermore, environmental education will remain marginalized, as it will be seen as secondary and a "frill" to the "real" needs of these constituencies. Children lack the resources to address problems arising on the sociopolitical level. The same holds true of adults when participatory injustices are factored into analyses of urban environmental and social problems.

Beyond the Environmental Education Literature

Beyond the environmental education literature, some attention to EJE is found in other fields of education. For example, *science education* is often perceived as closely associated with environmental education. Science education may be resistant to the inclusion of social concerns, but EJE may hold promise, as it clearly demonstrates the interpenetration of environmental risk (discernible through scientific techniques) and social demographics. This is the argument made by Mueller (2009) and others. Another study demonstrated how this worked with a geographic information systems lesson correlating toxic releases with available demographic data (Stewart, Schneiderman, & Andrews, 2001). Similarly, science education can accomplish learning objectives and serve as local data collection for environmental justice (Connors & Perkins, 2009). The subsequent list includes other school subjects and some examples in which there is EJE practice:

- *social studies*, particularly in the realm of environmental education for democratic practice;
- *arts education* (e.g., Hicks & King, 2007) which has been used extensively for environmental education, to the extent that Wilcox (2009) even describes dance education as benefiting environmental justice and Sullivan et al. (2008) utilize community theatre;
- *language arts* (e.g., Cutter-MacKenzie, 2009; Gaard, 2008; Howard, 2007);
- *religious education*, as the term "eco-justice" was used by church groups as early as 1973 (Gibson, 2004) and such groups may already be used to the moral discourse of justice (Haluza-DeLay, 2008); and
- *early childhood education,* including an influential editor who argues strenuously that environmental education must be critical of social power (Siraj-Blatchford, 2009), showing the field has come a long ways since Wilson (1996) argued there were indeed distributional inequities in environmental risks.

Schooling is only one site for learning. Most of the environmental justice literature cites informal learning, social learning, or community-based education strategies as crucial to mobilization around issues. Public health documents in the United States consistently emphasize environmental justice in their prescriptions, although it is unclear how this is occurring on the ground (Brulle & Pellow, 2006). Charley, Dawson, Madsen, and Spykerman (2004) present a Navajo educational program on uranium risks as a model for other EJE programs. Unfortunately, such purposive education is not likely to help environmental justice become more central in environmental education for nonaffected communities or learners. To remedy this problem, numerous scholars have suggested or developed programs to facilitate experiential learning and place-based education while addressing local issues. Such programs fall under such rubrics as community-based education, participatory action, service learning, and action research and community problem-solving (ARCPS; Stapp, Wals, & Stankorb, 1996). The ARCPS model was often represented in Marouli's (2002) study on multicultural environmental education. Community projects with students are beneficial, but challenging, and must be carefully organized to provide benefit to the students and community alike (Rao, Acury, & Quandt, 2004). Part of the impetus of environmental justice activism is the need to collect data and information on local concerns that are often not acknowledged by other authorities. Therefore, community-based participatory learning methods are aligned with organizing principles of the environmental justice movement.

Community members must also learn as they navigate the effort to protect themselves and their environments, making community education and social movement learning domains of research relevant for EJE. Considerable research has been done in community education (Tilbury & Wortman, 2008). "Learning in action" during social movement organizing and activism has been extensively studied in adult education. Both are genres of the broad field of social learning, which may be defined as "collective action and reflection that occurs among different individuals and groups as they work" (Keen, Brown, & Dyball, 2005, p.4; Wals, 2007). Social learning acknowledges the situated contexts of learning and the interactional effects of shared knowledge production. Assessments of the environmental justice movement show it to be both a source and site of learning (Agyeman, 2005; Hill, 2003; Hillman, 2002). Environmental education may not be able to "be untangled from the ideology of liberal culture where, for instance, expanding the curriculum is positioned as a substitute for the unpleasantness of political struggle" (Hill, 2003, p. 31). This discussion argued that EJE needs to go beyond information-driven approaches and educate for participatory justice and cultural recognition and that doing so may help environmental education develop in more effective ways to address the deeper roots of environmental and social problems.

Finally, more research and reflection appears to have been done on educating for environmental justice in higher education circles than in other forms of education.

Kaza's (1999, 2002) assessments of her pedagogical practice are insightful. Adamson, Evans, and Stein (2002) include four chapters on university-level pedagogy in their environmental justice reader. So too do several of the chapters in other edited collections (e.g., Smith & Williams, 1999; Wenden, 2004). Gordon (2007) used environmental racism as a means to address racialization on the university campus. Given the levels of environmental impact in poorer southern communities, other teacher educators assessed the benefit of an eco-justice perspective for recruitment of science teachers (Hodges & Tippins, 2009). Again, these are reflections on practice, and more research is needed.

Conclusion

The environmental education literature, in general, and the environmental education research literature, in particular, have only laterally considered the systematic ways environmental degradation unfairly impacts particular social groups. Conversely, there has been a vociferous discussion of "eco-justice" as extending considerations of fair treatment beyond humans and to nature. For the most part, environmental education literature points out the need for social equity, but researchers have not extensively examined the extent to which environmental educators are including justice or equity concerns, or educational outcomes related to environmental justice in mainstream environmental education. The weakness in research on educating for eco-justice/environmental justice is an area for scholars to correct.

The reasons for this weakness are difficult to discern. In the environmental education literature, the word "equity" appears to be used more often than "justice." The discursive consequences of this language should be unpacked by research in the field. Could it be that the former is a more polite word? Democracy is founded on an ethical norm of equity; perhaps to demonstrate forms of injustice may undermine assumptions of democratic equality. Such observations are speculative, because, as this review has shown, research on environmental justice in the context of environmental education is not extensive. However, it does allude to the inherently political character of the contents and processes of education. A valuable research program examines current environmental educators' conceptions of "justice" and "equity," as well as understanding of sociopolitical issues, such as racism, multiculturalism, the historical development of urban areas, the economic foundations of poverty and social class, and a host of other societal facets associated with, as well as reinforce both social inequality and environmental degradation.

Another reason often suggested for the weak inclusion of environmental justice concerns in environmental education, and by extension for the minimal research on educating for environmental justice, is the relatively minor involvement of visible minorities in the environmental education field. Numerous observers have argued that the field tends to reproduce itself, its curricula, and its research foci and that nonmajority researchers would bring new questions and new topics to the research (Agyeman, 2003a; Cole, 2002; James, 2003; Lee, 2008; Taylor, 2007). This is a process of "culturing" and "decolonizing" research (Agyeman, 2002; Cole 2004; Smith, 1999) and includes both changes in research topics and methodologies. As the North American Association for Environmental Education recognizes, increasing the diversity of environmental educators is central to the vitality of education to effectively address environmental degradation. Not only will solutions to environmental problems not succeed without equity at their core, but the resources from people of different social backgrounds or cultural practices may be beneficial to new and more effective ways of dealing with environmental issues. Such resources are a key area for environmental education research to examine. Similarly, researchers should examine the diversity assumptions of current environmental educators to discern potential barriers to genuine inclusion of a broader spectrum of people in the field.

Drawing on the scholarship reviewed in this chapter, there are several key conclusions if environmental education is to educate for environmental justice. First, knowledge about society, its history, its structures, and the interplay of social and economic processes with ecological and environmental matters must be an important part of environmental education. Much of the content of environmental education still relies on natural sciences and a transmission/instrumental method of instruction. Second, environmental education must include education for citizenship. In other words, environmental education content should include acquisition of abilities to speak and write about what is being learned, to do analyses of the discourses, and to engage with institutions that appear to be causing social and environmental damage. Third, to teach and learn about social or environmental justice implies a moral duty to act on what becomes known as unjust. Fourth, by the choice to include environmental justice-related content, environmental education will face even more criticism that it is politicized. EJE is a process to help people construct, critique, and transform their worlds in more fair and more sustainable ways.

Education for environmental justice should be central to environmental education theory, practice, and research. A broad vision of sustainability can expand the role of environmental education, encompassing the traditional content of environmental education and giving deliberate attention to forms of social injustice and their causes. More research on environmental justice education will likely show that weak inclusion of environmental justice education reduces environmental education's effectiveness at genuine sustainability and reduces its appeal with stakeholders who do not immediately see environmental concerns as part of their pressing issues. Educating for environmental justice is a positive, proactive vision for moving forward to create just and sustainable communities and an area ripe for research growth.

References

Adamson, J., Evans, M. M., & Stein, R. (Eds.). (2002). *The environmental justice reader: Politics, poetics and pedagogy*. Tucson, AZ: University of Arizona Press.

Agyeman, J. (2002). Culturing environmental education: From first nation to frustration. *Canadian Journal of Environmental Education, 7*(1), 5–12.

Agyeman, J. (2003a). "Under-participation" and ethnocentrism in environmental education research: Developing "culturally sensitive research approaches." *Canadian Journal of Environmental Education, 8*, 81–95.

Agyeman, J. (2003b). The contribution of urban ecosystem education to the development of sustainable communities and cities. In A. R. Berkowitz, C. H. Nilon, & K. S. Hollweg (Eds.), *Understanding urban ecosystems: A new frontier for science and education* (pp. 450–464). New York, NY: Springer.

Agyeman, J. (2005). *Sustainable communities and the challenge of environmental justice*. New York, NY: New York University Press.

Agyeman, J., Cole, P., Haluza-DeLay, R., & O'Riley, P. (Eds.). (2009). *Speaking for ourselves: Environmental justice in Canada*. Vancouver, BC: University of British Columbia Press.

Alsop, S., Dippo, D., & Zandvliet, D. B. (2007). Teacher education as or for social and ecological transformation: Place-based reflections on local and global participatory methods and collaborative practices. *Journal of Education for Teaching, 33*(2), 207–223.

Bowers, C. A. (2001). *Educating for eco-justice and community*. Athens, GA: University of Georgia Press.

Bowers, C. A. (2005). How Peter McLaren and Donna Houston, and other "green" Marxists contribute to the globalization of the West's industrial culture. *Educational Studies, 37*(2), 185–195.

Bowers, C. A. (2008). Why a critical pedagogy of place is an oxymoron. *Environmental Education Research, 14*(3), 325–335.

Bowers, C. A., & Apffel-Marglin, F. (Eds.). (2005). *Rethinking Freire: Globalization and the environmental crisis*. Mahwah, NJ: Lawrence Erlbaum Associates.

Brulle, R. J., & Pellow, D. N. (2006). Environmental justice: Human health and environmental inequalities. *Annual Review of Public Health, 27*, 3.1–3.22.

Bulkeley, H., & Walker, G. (2005). Environmental justice: A new agenda for the UK. *Local Environment, 10*(4), 329–332.

Bullard, R. (1990). *Dumping in Dixie*. Boulder, CO: Westview Press.

Bullard, R. D., Mohai, P., Saha, R., & Wright, B. (2007). *Toxic wastes and race at twenty 1987–2007: Grassroots struggles to dismantle environmental racism in the United States*. Cleveland, OH: United Church of Christ Justice and Witness Ministry.

Castonguay, G., & Jutras, S. (2009). Children's appreciation of outdoor places in a poor neighborhood. *Journal of Environmental Psychology, 29*(1), 101–109.

Charley, P. H., Dawson, S. E., Madsen, G. E., & Spykerman, B. R. (2004). Navajo uranium education programs: The search for environmental justice. *Applied Environmental Education & Communication, 3*(2), 101–108.

Cole, A. G. (2007). Expanding the field: Revisiting environmental education principles through multidisciplinary frameworks. *Journal of Environmental Education, 38*(2), 35–44.

Cole, P. (2002). Land and language: Translating Aboriginal cultures. *Canadian Journal of Environmental Education, 7*(1), 67–84.

Cole, P. (2004). Trick(ster)s of Aboriginal research: Or how to use ethical review strategies to perpetuate cultural genocide. *Native Studies Review, 15*(2), 7–30.

Cole, L., & Foster, S. (2001). *From the ground up: Environmental racism and the rise of the environmental justice movement*. New York, NY: New York University Press.

Connors, M. M., & Perkins, B. (2009). The nature of science education. *Democracy & Science, 18*(3), 56–60.

Cutter-Mackenzie, A. (2009). Multicultural school gardens: Creating engaging garden spaces in learning about language, culture, and environment. *Canadian Journal of Environmental Education, 14*, 122–135.

Dyment, J. E. (2005). "There's only so much money hot dog sales can bring in": The intersection of green school grounds and socio-economic status. *Children's Geographies, 3*(3), 307–323.

Gaard, G. (2008). Toward an ecopedagogy of children's environmental literature. *Green Theory & Praxis, 4*(2), 11–24.

Gibson, W. E. (Ed.). (2004). *Eco-justice: The unfinished journey*. New York, NY: State University of New York Press.

Goldstein, T., & Selby, D. (Ed.). (2000). *Weaving connections: Educating for peace, social and environmental justice*. Toronto, ON: Sumach Press.

Gonzalez-Gaudiano, E. J. (2006). Environmental education: A field in tension or in transition? *Environmental Education Research, 12*(3), 291–300.

Gordon, J. (2007). What can white faculty do? *Teaching in Higher Education, 12*(3), 337–347.

Gough, N., Gough, A., Appelbaum, P., Appelbaum, S., Doll, M. A., & Sellers, W. (2003). Tales from Camp Wilde: Queer(y)ing environmental education research. *Canadian Journal of Environmental Education, 8*, 44–66.

Gruenewald, D. A. (2003). The best of both worlds: A critical pedagogy of place. *Educational Researcher, 32*(4), 3–12.

Gruenewald, D. A. (2005). More than one profound truth: Making sense of divergent criticalities. *Educational Studies, 37*(2), 206–215.

Haluza-DeLay, R. B. (2008). Churches engaging the environment: An autoethnography of obstacles and opportunities. *Human Ecology Review, 15*(1), 71–81.

Hicks, L. E., & King, R. J. H. (2007). Confronting environmental collapse: Visual culture, art education, and environmental responsibility. *Studies in Art Education, 48*(4), 332–335.

Hill, R. J. (2003). Environmental justice: Environmental adult education at the confluence of oppressions. *New Directions for Adult and Continuing Education, 99*, 27–38.

Hillman, M. (2002). Environmental justice: A crucial link between environmentalism and community development? *Community Development Journal, 37*(4), 349–360.

Hodges, G., & Tippins, D. J. (2009). Using an eco-justice perspective to inform science teacher recruitment and retention in the rural black belt region of Georgia. *Rural Educator, 30*(3), 1–3.

Hooks, G., & Smith, C. L. (2004). The treadmill of destruction: National sacrifice areas and Native Americans. *American Sociological Review, 69*(4), 558–575.

Hossay, P. (2006). *Unsustainable: A primer for global environmental and social justice*. London, UK: Zed Books.

Houston, D., & McLaren, P. (2005). Response to Bowers: The "nature" of political amnesia: A response to C.A. "Chet" Bowers. *Educational Studies, 37*(2), 196–206.

Howard, P. (2007). The pedagogy of place: Reinterpreting ecological education through the language arts. *Diaspora, Indigenous, and Minority Education, 1*(2), 109–126.

Jafri, B. (2009). Rethinking "green" multicultural strategies. In J. Agyeman, P. Cole, R. Haluza-DeLay, & P. O'Riley (Eds.), *Speaking for ourselves: Environmental justice in Canada* (pp. 219–232). Vancouver, BC: University of British Columbia Press.

James, K. (2003). Designing research to include racial/ethnic diversity and marginalized voices. *Canadian Journal of Environmental Education, 8*, 67–79.

Kahn, R. (2008). Towards ecopedagogy: Weaving a broad-based pedagogy of liberation for animals, nature and the oppressed peoples of the earth. In A. Darder, R. Torres, & M. Baltodano (Eds.), *The critical pedagogy reader* (2nd ed.). New York, NY: Routledge.

Kahn, R. (2010). *Critical pedagogy, ecoliteracy and planetary crisis*. New York, NY: Peter Lang.

Kato, K. (2002). Environment and culture: Developing alternative perspectives in environmental discourse. *Canadian Journal of Environmental Education, 7*(1), 110–116

Kaza, S. (1999). Liberation and compassion in environmental studies. In G. A. Smith & D. R. Williams (Eds.), *Ecological education in action: On weaving education, culture and the environment* (pp. 143–160). Albany, NY: State University of New York.

Kaza, S. (2002). Teaching ethics through environmental justice. *Canadian Journal of Environmental Education, 7*(1), 99–109.

Keen, M., Brown, V. A., & Dyball, R. (Eds.). (2005). *Social learning in environmental management: Towards a sustainable future.* London, UK: Earthscan.

Kushmerick, A., Young, L., & Stein, S. E. (2007). Environmental justice content in mainstream US, six to twelve environmental education guides. *Environmental Education Research, 13*(3), 385–408.

Lee, E. B. (2008). Environmental attitudes and information sources among African American college students. *Journal of Environmental Education, 40*(1), 29–42.

Lewis, S., & James, K. (1995). Whose voice sets the agenda for environmental education? Misconceptions inhibiting racial and cultural diversity. *Journal of Environmental Education, 26*(3), 5–12.

Locke, S. (2009). Environmental education for democracy and social justice in Costa Rica. *International Research in Geographical and Environmental Education, 18*(2), 97–110.

Lowan, G. (2009). Exploring place from an Aboriginal perspective: Considerations for outdoor and environmental education. *Canadian Journal of Environmental Education, 14*, 42–58.

Lozar Glenn, J. M. (2009). *Still developing the toolbox: Making EE relevant for culturally diverse groups.* Washington, DC: North American Association for Environmental Education.

Marcinkowski, T. J. (2010). Contemporary challenges and opportunities in environmental education: Where are we headed and what deserves our attention? *Journal of Environmental Education, 41*(1), 34–54.

Marouli, C. (2002). Multicultural environmental education: Theory and practice. *Canadian Journal of Environmental Education, 7*(1), 26–42.

Martusewicz, R. A. (2005). On acknowledging differences that make a difference: My two cents. *Educational Studies: A Journal of the American Educational Studies Association, 37*(2), 215–224.

Mauro, S. E.-D. (2008). Beyond the Bowers-McLaren debate: The importance of studying the rest of nature in forming alternative curricula. *Capitalism Nature Socialism, 19*(2), 88–95.

McKeown-Ice, R., & Dendinger, R. (2000). Socio-political-cultural foundations of environmental education. *Journal of Environmental Education, 31*(4), 37–45.

McLaren, P. (1991). Critical pedagogy: Constructing an arch of social dreaming and a doorway to hope. *Journal of Education, 173*(1), 9–34.

McLaren, P., & Houston, D. (2004). Revolutionary ecologies: Ecosocialism and critical pedagogy. *Educational Studies: A Journal of the American Educational Studies Association, 36*(1), 27–45.

Mohai, P., Pellow, D., & Roberts, J. T. (2009). Environmental justice. *Annual Review of Environment and Resources, 34*(1), 405–430.

Mueller, M. (2008). Eco-justice as ecological literacy is much more than being "green!" *Educational Studies, 44*(2), 155–166.

Mueller, M. (2009). Educational reflections on the "ecological crisis": Eco-justice, environmentalism, and sustainability. *Science & Education, 18*(8), 1031–1056.

Newbery, L. (2003). Will any/body carry that canoe? A geography of the body, ability, and gender. *Canadian Journal of Environmental Education, 8*, 204–216.

Nordstrom, H. K. (2008). Environmental education and multicultural education: Too close to be separate? *International Research in Geographical and Environmental Education, 17*(2), 131–145.

Pellow, D. N. (2000). Environmental inequality formation: Toward a theory of environmental justice. *American Behavioural Scientist, 43*(4), 581–601.

Pellow, D. N. (2009). "We didn't get the first 500 years right, so let's work on the next 500 years": A call for transformative analysis and action. *Environmental Justice, 2*(1), 3–6.

Pellow, D. N., & Brulle, R. (Eds.). (2005). *Power, justice, and the environment: A critical appraisal of the environmental justice movement.* Cambridge, MA: MIT Press.

Potter, G. (2010). Environmental education for the twenty-first century: Where do we go now? *Journal of Environmental Education, 41*(1), 22–33.

Rao, P., Arcury, T. A., & Quandt, S. A. (2004). Student participation in community-based participatory research to improve migrant and seasonal farmworker environmental health: Issues for success. *Journal of Environmental Education, 35*(2), 3–15.

Roberts, J. T., & Parks, B. C. (2007). *A climate of injustice: Global inequality, North-South politics, and climate policy.* Cambridge, MA: The MIT Press.

Russell, C., Sarick, T., & Kennelly, J. (2002). Queering environmental education. *Canadian Journal of Environmental Education, 7*(1), 54–66.

Salleh, A. (2009). *Eco-sufficiency and global justice: Women write political ecology.* London, UK: Pluto Press.

Sandler, R., & Pezzullo, P. (2007). *Environmental justice and environmentalism: The social justice challenge to the environmental movement.* Cambridge, MA: The MIT Press.

Schlosberg, D. (2005). Environmental and ecological justice: Theory and practice in the United States. In J. Barry & R. Eckersley (Eds.), *The state and the global ecological crisis* (pp. 97–116). Cambridge, MA: MIT Press.

Schroder, B. (2006). Native science, intercultural education and place-conscious education: An Ecuadorian example. *Educational Studies, 32*(3), 307–317.

Simpson, L. (2002). Indigenous environmental education for cultural survival. *Canadian Journal of Environmental Education, 7*(1), 13–25.

Siraj-Blatchford, J. (2009). Editorial: Education for sustainable development in early childhood. *International Journal of Early Childhood, 41*(2), 9–22.

Smith, L. T. (1999). *Decolonizing methodologies: Research and indigenous peoples.* London, UK: Zed Books.

Smith, G. A., & Williams, D. R. (Eds.). (1999). *Ecological education in action: On weaving education, culture and the environment.* Albany, NY: State University of New York.

Stapp, E. B., Wals, A. E. J., & Stankorb, S. L. (1996). *Environmental education for empowerment: Action research and community problem solving.* Ann Arbor, MI: Global Rivers Environmental Education Network.

Stewart, M. E., Schneiderman, J. S., & Andrews, S. B. (2001). A GIS class exercise to study environmental risk. *Journal of Geoscience Education, 49*(3), 227–234.

Strife, S., & Downey, L. (2009). Childhood development and access to nature: A new direction for environmental inequality research. *Organization Environment, 22*(1), 99–122.

Sullivan, J., Petronella, S., Brooks, E., Murillo, M., Primeau, L., & Ward, J. (2008). Theater of the oppressed and environmental justice communities: A transformational therapy for the body politic. *Journal of Health Psychology, 13*(2), 166–179.

Swayze, N. (2009). Engaging indigenous urban youth in environmental learning: The importance of place revisited. *Canadian Journal of Environmental Education, 14*, 59–73.

Taylor, D. (2000). The rise of the environmental justice paradigm: Injustice framing and the social construction of environmental discourses. *American Behavioral Scientist, 43*(4), 508–580.

Taylor, D. (2007). Diversity and equity in environmental organizations: The salience of these factors to students. *Journal of Environmental Education, 39*(1), 19–43.

Tilbury, D., & Wortman, D. (2008). How is community education contributing to sustainability in practice? *Applied Environmental Education & Communication, 7*(3), 83–93.

United Church of Christ Commission for Racial Justice (UCC). (1987). *Toxic wastes and race in the United States: A national report on the racial and socio-economic characteristics of communities with hazardous waste sites*. New York, NY: United Church of Christ Commission for Racial Justice.

Wals, A. E. J. (1996). Back-alley sustainability and the role of environmental education. *Local Environment: The International Journal of Justice and Sustainability, 1*(3), 299–316.

Wals, A. E. J. (Ed.). (2007). *Social learning: Towards a sustainable world*. Wageningen: Wageningen Academic Publishers.

Wane, N., & Chandler, D. J. (2002). African women, cultural knowledge, and environmental education with a focus on Kenya's indigenous women. *Canadian Journal of Environmental Education, 7*(1), 86–98.

Wenden, A. (Ed.). (2004). *Educating for a culture of social and ecological peace*. Albany, NY: State University of New York Press.

Wilcox, H. N. (2009). Embodied ways of knowing, pedagogies, and social justice: Inclusive science and beyond. *NWSA Journal, 21*(2), 104–120.

Wilhelm, S. A., & Schneider, I. E. (2005). Diverse urban youth's nature: Implications for environmental education. *Applied Environmental Education & Communication, 4*(2), 103–113.

Wilson, R. (1996). Healthy habitats for children. *Early Childhood Education Journal, 23*(4), 235–238.

38

Indigenous Environmental Education Research in North America

A Brief Review

GREG LOWAN-TRUDEAU
University of Northern British Columbia, Canada

Indigenous environmental education is a rapidly expanding area of practice in North America (Turtle Island) and other areas of the world. Programs teaching indigenous knowledge and philosophies for the benefit of both indigenous and nonindigenous students have recently flourished. Descriptions of these programs abound in the literature, along with publications that discuss indigenous ecological philosophies and wisdom. Well-known indigenous scholars, such as Gregory Cajete and Angayuqaq Oscar Kawagley have firmly established the field of indigenous environmental education. Other indigenous scholars like Cole (1998, 2002a, 2002b; O'Riley & Cole, 2009) have also challenged conventional academic standards through creative publications inspired by the oral history tradition.

However, reports on *research* that formally examine indigenous environmental educational practices and philosophies are limited. In this chapter, I briefly highlight key concepts in indigenous environmental education, provide examples of contemporary programs, discuss indigenous research methodologies, and review the limited body of research literature currently available.

Foundational Concepts

Mi'kmaq scholar Battiste (1998) states, "Aboriginal peoples throughout the world have survived five centuries of the horrors and harsh lessons of colonization. They are emerging with new consciousness and vision" (p. 16). The revitalization of Aboriginal languages, epistemologies, and pedagogies, recognizing the importance of the land, privileging indigenous voices, the involvement of elders in education, and indigenous control of indigenous education, are key factors in the decolonization process (Battiste, 1998, 2005; Goulet, 2001; Hampton, 1999; Kirkness, 1998; Simpson, 2002). Anishnaabe scholar Cavanagh (2005) suggests:

> Creating relevant educational programs and curricula requires that, at the outset, Aboriginal peoples take a step away from the current system . . . We must recreate an Aboriginal education system founded on traditional values ensuring cultural continuity throughout. The option then exists to expand on and/ or include other existing methodologies. (p. 236)

Okanagan scholar Armstrong (1987) states that, prior to colonial intervention, indigenous education was intended to equip children with a holistic set of skills necessary to live healthy and successful lives. Anishnaabe environmental educator Simpson (2002) also highlights the holistic nature of indigenous education:

> Employing Indigenous ways of teaching and learning, including ceremonies, dreams, visions and visioning, fasting, storytelling, learning-by-doing, observation, reflecting, and creating, not only allows students to share and learn in a culturally inherent manner, but also reinforces the concept that Indigenous knowledge is not only content but also process. (p. 18)

Cavanagh (2005) expands on the indigenous concept of holistic education in his work on developing an Aboriginal education framework:

> The traditional approach to education was premised on the value of developing creative, responsible, and balanced individuals equipped to handle the challenges they would face within Creation. It required the commitment of teachers, family, community and Nation, and was responsive to the student and her educational desires and needs . . . Traditionally, new teachings were imparted to the student when she was ready, that is, when the student had achieved certain levels of understanding. (p. 234)

Simpson (2002) and others (e.g., Battiste, 1998, 2005; Graveline, 1998) also emphasize the importance of including elders in all aspects of indigenous education programs.

She outlines their importance as keepers of traditional knowledge and culture and highlights the support and guidance elders provide not only for students, but also for young educators. Further, Simpson describes the importance of structuring programs to include elders on a regular basis, not just as occasional guest speakers. Battiste (2005) discusses the importance of a generational transfer of knowledge: "By building relationships with the land and its inhabitants, they [Elders] come to understand the forces around them. Each generation then passes their knowledge and experience of the social and cultural contexts of their ecological origins to succeeding generations" (p. 122).

Alaskan scholars Barnhardt and Kawagley (2005) also emphasize the ancient relationships of indigenous peoples with specific geographical areas:

> Indigenous peoples throughout the world have sustained their unique worldview and associated knowledge systems for millennia, even while undergoing major social upheavals as a result of transformative forces beyond their control . . . The depth of Indigenous knowledge rooted in the long inhabitation of a particular place offers lessons that can benefit everyone, from educator to scientist, as we search for a more satisfying and sustainable way to live on this planet. (p. 9)

As Barnhardt and Kawagley (2005) suggest, indigenous people developed highly sophisticated and intimate understandings, over thousands of years, of the lands we in North America presently inhabit. Indigenous North American scholars, such as Kawagley, Cajete, and LaDuke have added a strong voice to contemporary environmental scholarship. Their theories and activism provided a firm foundation that has fostered the growth of Indigenous environmental scholarship not only in North America, but also internationally. The revival and sharing of indigenous epistemologies has provided deep insight for indigenous and nonindigenous peoples alike into the ancient symbiotic relationship of indigenous peoples with the landscape of Turtle Island. Eminent Tewa scholar and educator Cajete (1999) explains:

> The Americas are an ensouled and enchanted geography, and the relationship of Indian people to this geography embodies a "theology of place," reflecting the very essence of what may be called spiritual ecology . . . Through generations of living in America, Indian people have formed and been formed by the land . . . The land has become an extension of Indian thought and being because, in the words of a Pueblo elder, "It is this place that holds our memories and the bones of our people . . . This is the place that made us." (p. 3)

The result of this connection is similar to what many contemporary Western scholars might call a "sense of place," a strong feeling of connection to a particular geographical area (e.g., Barnhardt, 2008; Curthoys, 2007; Watchow, 2006). According to Cajete (1999), this deep feeling of relationship results in a feeling of shared responsibility for the nonhuman world: "Indigenous people felt responsibility not only for themselves, but also for the entire world around them" (p. 11).

Well-known Anishnaabe activist, educator, and politician LaDuke (2002) emphasizes the ancient roots of indigenous knowledge and argues for its inclusion in contemporary ecological discourse:

> Traditional ecological knowledge is the culturally and spiritually based way in which indigenous people relate to their ecosystems. This knowledge is founded on spiritual-cultural instructions from "time immemorial" and on generations of careful observation within an ecosystem of continuous residence. I believe that this knowledge represents the clearest empirically based system for resource management and ecosystem protection in North America . . . Frankly, these native societies have existed as the only example of sustainable living in North America for more than 300 years. (p. 78)

An increasing number of indigenous and nonindigenous scholars support the perspective presented by LaDuke. For example, the late Nakoda Chief, Snow (1977/2005) suggests that the future success of our society requires the combined wisdom of Aboriginal and non-Aboriginal cultures.

Indigenous Environmental Education in Practice

Indigenous environmental education is a growing field of practice in North America with a limited, but growing, body of literature. Many organizations currently deliver programs designed to share indigenous knowledge. Although some programs aim to share indigenous knowledge with indigenous students only, others are open to both indigenous and nonindigenous students. A number of programs also attempt to blend indigenous knowledge with Western approaches.

Simpson's (2002) description of her experiences as an environmental educator in Canada is one of the most comprehensive descriptions of indigenous environmental education available. Based on extensive experience with several programs, Simpson relates key features she believes should be part of any indigenous environmental education program: supporting decolonization, grounding programs in indigenous philosophies of education, allowing space for the discussion and comparison of indigenous and Western epistemologies, emphasizing indigenous ways of teaching and learning, creating opportunities to connect with the land, employing indigenous instructors as role models, involving elders as experts, and using indigenous languages when possible.

Barnhardt and Kawagley (2005; see also Barnhardt, 2008) describe the Alaska Rural Systemic Initiative (AKRSI), a program developed in 1995 in consultation with indigenous Alaskan elders. The intent of the program was to enhance indigenous students' experiences in the formal school system by integrating local Indigenous values and knowledge into school curricula across Alaska. Barnhardt and Kawagley (2005) report significantly improved educational experiences and academic success for participating students. The AKRSI now serves as a catalytic model for other programs across North America.

"Two-Eyed Seeing," viewing the world simultaneously through both Western scientific and Aboriginal lenses to form a focused and unified vision, is a theory developed by Mi'kmaq Elder Albert Marshall (Lefort & Marshall, 2009). A recent issue of *Green Teacher* (Grant & Littlejohn, 2009) also highlighted the concept of Two-Eyed Seeing; several educational programs that strive to embody Two-Eyed Seeing were profiled (e.g., Bartlett, 2009; Hatcher & Bartlett, 2009a, 2009b; Kazina & Swayze, 2009; Myers, 2009; Snively, 2009). For example, Hatcher and Bartlett (2009a, 2009b) describe "integrative science units" they have developed through Cape Breton University's Integrative Science program, an innovative educational research initiative. In their courses, they attempt to integrate Western and indigenous Mi'kmaq science approaches with students ranging in age from those in elementary school to those attending a university.

Indigenous educators Kazina and Swayze (2009) also relate their experiences with "Bridging the Gap," an inner-city program in Winnipeg that works with both Aboriginal and non-Aboriginal youth. Bridging the Gap strives to integrate Western and Aboriginal approaches to learning about the natural world. Based on their description and another article by Swayze (2009), it appears that Kazina and Swayze are experiencing success. Kazina and Swayze (2009) demonstrate the cultural awareness they instill in their students through lessons, such as the offering of tobacco and how to respectfully approach the elders, who are a strong part of their program.

Snively (2009; see also Snively & Corsiglia, 2001), originally from the United States, teaches at the University of Victoria in British Columbia. While relating her experiences as a teacher/educator interested in blending Western and indigenous approaches to science and environmental education, she observes:

> Cross-cultural science education is not merely throwing in an Aboriginal story, putting together a diorama of Aboriginal fishing methods, or even acknowledging the contributions Aboriginal peoples have made to medicine. Most importantly, cross-cultural science education is not anti-Western science. Its purpose is not to silence voices, but to give voice to cultures not usually heard and to recognize and celebrate all ideas and contributions. It is as concerned with how we teach as with what we teach. (p. 38)

The programs and approaches described by Snively, Bartlett, Swayze, and others are inspiring and exciting. However, although there is a growing body of literature on indigenous environmental education in Canada and elsewhere, few in-depth studies have been completed on indigenous environmental educators, their programs, and the deeper societal implications of their work.

Developing Indigenous Research Methodologies

The historic misuse and abuse of research conducted with indigenous peoples is well documented (Deloria, Jr., 1997; Lassiter, 2000; Tuhiwai Smith, 1999). A history of positivist anthropological and ethnographic approaches has left indigenous peoples worldwide wary of researchers in general, but especially nonindigenous researchers. A growing number of researchers are attempting to address these kinds of concerns by embodying Indigenous values in their research practices.

Indigenous researchers, such as Maori scholar Tuhiwai Smith (1999) of Aotearoa/New Zealand, are guiding the development of research approaches by and with indigenous peoples. Tuhiwai Smith developed a methodology called Kaupapa Maori, a form of Maori-centered research that is an example of an empowering approach to Indigenous research. Kaupapa Maori is research conducted by Maori people, for Maori people, to explore topics of concern to Maori people. Central to conducting respectful indigenous research are key concepts, such as reciprocity, researcher positioning, and embodying cultural traditions in research methods.

Creswell (2002) describes reciprocity as a mutually beneficial, reciprocal relationship between a researcher and the people with whom they are working. This is an especially important characteristic in contemporary indigenous research because of the historically uneasy relationship between researchers and indigenous peoples (Lassiter, 2000; Tuhiwai Smith, 1999; Wilson, 2008). Reciprocity extends to postdata collection when it is important for the researcher to share findings with participants and their communities to ensure its accuracy and to gain approval for its use, especially if sensitive cultural information is shared. Graveline (1998) would describe this as enacting "First Voice," making space for historically oppressed peoples to speak for themselves.

Absolon and Willet (2005) also suggest that positioning yourself culturally is an integral part of relationship-building in many indigenous societies and is, therefore, especially important in research projects involving Indigenous peoples. Positioning means introducing yourself to your research participants and, later, your audience (Bolak, 1997; Tuchman, 1972/1996). For example, I explicitly inform participants that I am a Canadian of Métis, Norwegian, and German descent. This helps them understand my perspective and background and allows them to position themselves accordingly. I did this in my master's research (Lowan, 2009) and found it very helpful; several participants commented that they felt more at ease discussing culturally related topics knowing my background.

Some indigenous researchers are also seeking ways to embody cultural traditions in their research methods. For example, in my master's research, I provided all the participants with institutional consent forms, as well as an offering of tobacco, a traditional method in many North American indigenous cultures for establishing reciprocity when important personal or cultural knowledge is being shared (Lickers, 2006).

Indigenous research is a dynamic field. Innovative theories and reports of new studies are steadily emerging in the literature. However, one area that remains under-represented is Indigenous environmental education. The following is a brief overview of some studies conducted in the area.

Examples of Current Research

One study that investigated Indigenous environmental education is Takano's (2005) examination of a community-developed, land-based cultural education program based in Igloolik, Nunavut. Takano, a researcher of Japanese descent, participated in Paariaqtuqtut, a 400 kilometer journey through the community's ancestral territory in May 2002. Paariaqtuqtut aims to connect young people with cultural skills and teachings in a land-based context. Community leaders and elders developed the program motivated by a concern that many youth were losing their traditional connections to land and culture. Elders observed that this leads to youth feeling lost between two worlds—disconnected from their own community and culture, yet unprepared to live in the Western world. Takano reported several youth participants felt that Paariaqtuqtut had helped them reconnect with their land and culture and contributed to stronger feelings of self-esteem.

Another study that examined indigenous environmental education is Mohawk scholar Longboat's (2008) doctoral research. Longboat studied the interconnection of indigenous languages and cultures with their surrounding ecological systems. He argues that the preservation of indigenous cultures and languages is intimately linked to preserving ecological systems and vice-versa. Longboat is the director of Trent University's indigenous environmental studies program and a leading voice in indigenous environmental education in Canada.

Root's (2010) recent master's study also provides an interesting perspective. Root, a Euro-Canadian, explored the experiences of white outdoor and environmental educators attempting to decolonize their own teaching practices. She highlights the importance of relationship building as a key component of successful intercultural partnerships and emphasizes the importance of recognizing traditional indigenous connections to specific geographical territories.

A multicultural team of nonindigenous researchers from Carleton University is currently working with the community of Cape Dorset in the Canadian Arctic (Ip, Grimwood, Kushwawa, Doubleday, & Donaldson, 2008). They have developed a program with local Inuit youth and elders that involves learning about and monitoring local plants, using a combination of modern photography and traditional Inuit knowledge.

My own work has also contributed to the field of Indigenous environmental education in North America. My master's study (Lowan, 2009) explored the development of Indigenous outdoor and environmental education programs through a lense of decolonization. One of the key findings from that study was a strong interest from both indigenous and nonindigenous participants for increased intercultural learning opportunities. This theme provided the impetus for my doctoral study, which examined the ecological identities, philosophies, and teaching practices of intercultural environmental educators in Canada who blend Western and indigenous approaches (Lowan, 2012).

The studies described are all very recent. Indigenous environmental education is an increasingly, widely practiced, but relatively unexplored, area of research. Examples of potential future research areas include exploring the role of nonindigenous people in indigenous environmental education, how to respectfully engage with and share indigenous knowledge, exploring protocols for engaging with different types of indigenous knowledge (e.g., ceremonial, historical, ecological), the role of indigenous knowledge in formal schooling for both indigenous and nonindigenous students, and the experiences of recent immigrants to North America who engage with indigenous knowledge. As the theories of leading indigenous theorists and educators are increasingly embraced and applied in practice, countless more research opportunities will surely emerge.

References

Absolon, D., & Willett, C. (2005). Putting ourselves forward: Location in Aboriginal research. In L. Brown & S. Strega (Eds.), *Research as resistance: Critical, indigenous, and anti-oppressive approaches* (pp. 97–126). Toronto, ON: Canadian Scholars' Press.

Armstrong, J. C. (1987). Traditional indigenous education: A natural process. *Canadian Journal of Native Education, 14*(3), 14–19.

Barnhardt, R. (2008). Creating a place for indigenous knowledge in education: The Alaska native knowledge network. In D. A. Gruenewald & G. A. Smith (Eds.), *Place-based education in the global age: Local diversity.* New York, NY: Taylor & Francis.

Barnhardt, R., & Kawagley, A. O. (2005). Indigenous knowledge systems and Alaska native ways of knowing. *Anthropology and Education Quarterly, 36*(1), 8–23.

Bartlett, C. (2009). Mother earth, grandfather sun. *Green Teacher, 86*, 29–32.

Battiste, M. (1998). Enabling the autumn seed: Toward a decolonized approach to Aboriginal knowledge, language, and education. *Canadian Journal of Native Education, 22*(1), 16–27.

Battiste, M. (2005). You can't be the global doctor if you're the colonial disease. In L. Muzzin & P. Tripp (Eds.), *Teaching as activism, equity meets environmentalism* (pp. 121–133). Montreal and Kingston, ON: McGill-Queen's University Press.

Bolak, H. C. (1997). Studying one's own in the Middle East: Negotiating gender and self-other dynamics in the field. In R. Hertz (Ed.), *Reflexivity and voice* (pp. 95–118). Thousand Oaks, CA: Sage Publications.

Cajete, G. (1999). "Look to the mountain": Reflections on indigenous ecology. In G. Cajete (Ed.), *A people's ecology: Explorations in sustainable living* (pp. 2–20). Santa Fe, NM: Clearlight Publishers.

Cavanagh, R. (2005). The Anishanaabe teaching wand and holistic education. In L. Muzzin & P. Tripp (Eds.), *Teaching as activism, equity meets environmentalism* (pp. 232–253). Montreal and Kingston, ON: McGill-Queen's University Press.

Cole, P. (1998). An academic take on "indigenous traditions and ecology." *Canadian Journal of Environmental Education, 3*, 100–115.

Cole, P. (2002a). Aboriginalizing methodology: Considering the canoe. *International Journal of Qualitative Studies in Education, 15*(4), 447–459.

Cole, P. (2002b). Land and language: Translating aboriginal cultures. *Canadian Journal of Environmental Education, 7*(1), 67–85.

Creswell, J. W. (2002). *Educational research: Planning conducting and evaluating quantitative and qualitative Research*. Upper Saddle River, NJ: Pearson Education Inc.

Curthoys, L. P. (2007). Finding a place of one's own. *Canadian Journal of Environmental Education, 12*(1), 68–79.

Deloria, V., Jr., (1997). Conclusion: Anthros, Indians, and planetary reality. In T. Biolsi & L. Zimmerman (Eds.), *Indians and anthropologists: Vine Deloria Jr. and the critique of anthropology* (pp. 211–221). Tucson, AZ: University of Arizona Press.

Goulet, L. (2001). Two teachers of Aboriginal students: Effective practice in sociohistorical realities. *Canadian Journal of Native Education, 25*(1), 68–82.

Grant, T., & Littlejohn, G. (Eds.). (2009). Two eyed seeing. *Integrative Science. Green Teacher*, 86.

Graveline, F. J. (1998). *Circle works: Transforming Eurocentric consciousness*. Halifax, UK: Fernwood Press.

Hampton, E. (1999). Towards a redefinition of Indian education. In M. Battiste & Y. Barman (Eds.), *First Nations education in Canada: The circle unfolds* (pp. 5–46). Vancouver, BC: UBC Press.

Hatcher, A., & Bartlett, C. (2009a). MSIT: Transdisciplinary, cross-cultural science. *Green Teacher, 86*, 7–10.

Hatcher, A., & Bartlett, C. (2009b). Traditional medicines: How much is enough? *Green Teacher, 86*, 11–13.

Ip, M., Grimwood, B., Kushwawa, A., Doubleday, N., & Donaldson, S. G. (2008). Photos and plants through time: Monitoring environmental change in the Canadian Arctic with implications for outdoor education. *Pathways: The Ontario Journal of Outdoor Education, 21*(1), 14–18.

Kazina, D., & Swayze, N. (2009). Bridging the gap: Integrating Indigenous knowledge and science in a non-formal environmental learning program. *Green Teacher, 86*, 25–28.

Kirkness, V. (1998). Our peoples' education: Cut the shackles; Cut the crap; Cut the mustard. *Canadian Journal of Native Education, 22*(1), 10–16.

LaDuke, W. (2002). *The Winona LaDuke reader: A collection of essential writings*. Penticton, BC: Theytus Books.

Lassiter, L. E. (2000). Authoritative texts, collaborative ethnography, and Native American studies. *American Indian Quarterly, 24*(4), 601–614.

Lefort, N., & Marshall, A. (2009). *Learning with the world around us: Practicing two-eyed seeing.* Paper presented at the fifth World Environmental Education Congress, Montreal, PQ.

Lickers, M. (2006). *Urban Aboriginal leadership* (Unpublished master's thesis). Royal Roads University, Victoria, BC.

Longboat, D. R. (2008). *Owehna'shon: a (The islands) the Haudenosaunee archipelago: The nature and necessity of bio-cultural restoration and revitalization* (Unpublished PhD dissertation). York University, Canada.

Lowan, G. (2009). Exploring place from an Aboriginal perspective: Considerations for outdoor and environmental education. *Canadian Journal of Environmental Education, 14*, 42–58.

Lowan, G. (2012). Expanding the conversation: Further explorations into Indigenous environmental science education theory, research, and practice. *Cultural Studies in Science Education, 7*, 71–81.

Myers, D. (2009). Two-eyed seeing in a school district. *Green Teacher, 86*, 39–40.

O'Riley, P., & Cole, P. (2009). Coyote and raven talk about the land/scapes. In M. McKenzie, P. Hart, H. Bai, & B. Jickling (Eds.), *Fields of green: Restorying culture, environment, and education* (pp. 125–134). Cresswell, NJ: Hampton.

Root, E. (2010). This land is our land: This land is your land? The decolonizing journeys of white outdoor environmental educators. *Canadian Journal of Environmental Education, 15*, 103–119.

Simpson, L. (2002). Indigenous environmental education for cultural survival. *Canadian Journal of Environmental Education, 7*(1), 13–35.

Snively, G. (2009). Money from the sea: A cross-cultural indigenous science activity. *Green Teacher, 86*, 33–38.

Snively, G., & Corsiglia, J. (2001). Discovering indigenous science: Implications for science education. *Science Education, 85*(1), 6–34.

Snow, J. (2005). *These mountains are our sacred places*. Calgary, AB: Fifth House (Original work published 1977).

Swayze, N. (2009). Engaging indigenous urban youth in environmental learning: The importance of place revisited. *Canadian Journal of Environmental Education, 14*, 59–73.

Takano, T. (2005). Connections with the land: Land-skills courses in Igloolik, Nunavut. *Ethnography, 6*(4), 463–486.

Tuchman, B. (1996). Distinguishing the significant from the insignificant. In D. K. Dunaway & W. K. Baum (Eds.), *Oral history: An interdisciplinary anthology* (2nd ed., pp. 94–98). Lanham, MD: Altamira Press (Original work published 1972).

Tuhiwai Smith, L. (1999). *Decolonizing methodologies: Research and indigenous peoples*. London, UK, and New York, NY: Zed Books; Dunedin, New Zealand: University of Otago Press.

Watchow, B. J. (2006). *The experience of river places in outdoor education: A phenomenological study* (Unpublished PhD dissertation), Monash University, Australia.

Wilson, S. (2008). *Research is ceremony: Indigenous research methods*. Winnipeg, MB: Fernwood.

39

Three Degrees of Separation

Accounting for Naturecultures in Environmental Education Research

Leesa Fawcett
York University, Canada

Introduction

The overall theme of this philosophically grounded chapter revolves around the intense rupture of human relationships from various "natures"[1] and the different ways in which this shows up in environmental education research. The premise follows Williams' (1980) idea that, "If we alienate the living processes of which we are a part, we end, though unequally, by alienating ourselves" (p. 84). Starting with what must be, inevitably, a partial and incomplete historical position on research in environmental education, I trace the philosophical context, methodological outcomes, and ontological repercussions of how past research in the field has grappled with human-nature relationships. Reflecting on current research trends, I attend to their productive and disruptive endeavors. Finally, this chapter ends with an exploration of possible futures and transformative moments, at a potentially new beginning.

The guiding questions are: "How are particular relationships between humans and their natural environments epistemologically framed and why does that matter for research in the field of environmental education?" Ultimately, this chapter is about ethical inquiries and how those inquiries are politically and socially enacted in research. Jickling's (2009) work has shown how "ethics inquiry in an educational context can be conceived as a means to explore controversy, dissonance, unconventional ideas, and to imagine new possibilities" (p. 209).

To disrupt the teaching/research dualism, I have not differentiated between the two, but review scholarly teaching in both domains seeking work that offers theoretical and methodological insights into the debates about anthropocentrism in research. Anthropocentrism refers to the consideration of humans as the center of all valuing. Like a fish in water, the environmental education field rarely discusses anthropocentrism explicitly—it is the taken-for-granted water in which we all swim. In this chapter, I endeavor to experiment with theoretical explorations about and across the nature/culture divide to aspire to better research. Using narrative experiments, Gough (2009) argues that, "In theoretical inquiry an essay can serve similar purposes to an experiment in empirical research—a methodical way of investigating a question, problem or issue" (p. 69). He "understands environmental education research to be a species of curriculum inquiry" (p. 69), and warns against the dangers inherent in the globalization of knowledge production. With globalized knowledge and comodified curriculums, there is the danger of reproducing anthropocentric positions without acknowledging specific teaching contexts. This chapter, searches for frameworks that nourish the field of environmental education in diverse ways—it is not about finding one value center for all research to radiate from.

Anthropocentric—Biocentric—Ecocentric

It is easy to say that nature/culture dualisms have been rampant in past environmental education research, but that fact has changed throughout the young history of contemporary Western environmental education. Specific authors' voices surfaced in the past twenty years, and their ideas have traversed the spectrum from anthropocentric to biocentric to ecocentric, with differing results. For example, the environmental education literature on sense of place has surged within the last decade and exemplifies the changing focus from seeing land and ecosystem as principally a resource to a bioregional ecological literacy approach (Curthoys & Cuthbertson, 2002). Expanding the conversation, Greenwood (2009) says "the literature around place-based education is self-consciously nonanthropocentric" (p. 275). This chapter focuses primarily on crossing the human/animal and culture/nature boundaries, which impinge on the habitat and "placefulness" debates. Almost thirty years ago, when environmental education was still a fledgling academic field, renowned Canadian naturalist Livingston (1981) wrote: "The direction seems to lie in the compliant acceptance by individual human beings of membership—which is to say, 'place'—in the beauty that is life process" (p. 117). If humans are not the center, where do we belong?

Reflecting on the dilemmas caused by ethnocentrism, androcentrism, phallocentrism, and Eurocentrism, feminist philosopher Plumwood (1997) proclaimed that "concepts of centrism have been at the heart of modern liberation politics and theory" (p. 328). To begin, we need to deconstruct the key terms in the anthropocentric debates. *Anthropocentrism* assumes human interests matter the most and environments are important only in terms of how they instrumentally fulfill human needs, desires, and goals.[2] As Callicott (2002) says, "A kind of unapologetic, uncritical anthropocentrism—sometimes characterized as strong anthropocentrism—still dominates environmental economics and most of public environmental policy" (p. 232). I would venture that, in many places in the world, it also dominates educational policy, in general, and environmental education policy, in particular, if such policies exist. Callicott (2002) goes on to describe weak anthropocentrism as a resourcist valuing of nature for its raw materials and psychological and spiritual benefits. I also suggest that who gets to count as the human center is also very specifically raced, gendered, and classed—not all humans are considered, and included, as fully *human* in strong or weak anthropocentrism's orbit. For example, in an article on queering environmental education, the authors talk about women, people of color, and queers not being considered human (Russell, Sarick, & Kennelly, 2002).

Consequently, a person can ask what might a liberatory form of anthropocentrism looks like? Plumwood (1997) advocated for a kind of liberation anthropocentrism in contrast to cosmic anthropocentrism. She argued that cosmic models of anthropocentrism were detached from place, that they were everywhere at once, like the globalization of knowledge production. Plumwood's main point is that we need to take responsibility for solely human perspectives on nature to enact a liberating form of anthropocentrism. She writes, "If human-centredness similarly structures our beliefs and perceptions about the other which is nature, it is a framework for generating ecological denial and ecological blindness in just the same way that ethnocentrism is a framework for generating moral blindness" (p. 344).

At this point, it is important to note that anthropocentrism is not the same as *anthropomorphism*, which entails the attribution of human characteristics to the more-than-human (Abrams, 1996) world. One can be anthropomorphic without ascribing to anthropocentrism. As Livingston (1994) asserted:

> In defense of anthropomorphism, however, it must be said that we have no real alternative. Being no more than human, we apprehend the world in human terms. My dogs apprehend the world in dog terms. They "canimorphize" *me*, as a member of their social organization. Neither they nor I discredit the other in so doing, however, and neither wants or needs to have it both ways. (p. 82)

Clearly, not all acts of anthropomorphizing are equal in terms of the differential power relations enacted between humans and other life forms. (I have discussed this topic at length elsewhere; see Fawcett, 1989, 2002).

Going beyond anthropocentrism to value all life and the act of being alive is a tenet of *biocentrism*. Callicott (2002) believes biocentrism as "the proposition that all living beings possess intrinsic value and merit respect—builds upon the theory of animal rights" (p. 241). Following Aldo Leopold's notion of a land ethic, Callicott (2002) differs dramatically from many ecofeminists and humane educators, because he argues that biocentrism's focus on individual animals does not adequately address environmental concerns. Conversely, humane educators point out that their concerns, such as factory farming, international trade in animals, fur farming, and trapping, all harm individuals and have significant holistic environmental impacts (Selby, 1995).[3]

Ecocentrism, according to Eckersley (1992), recognizes the value and moral standing of all ecosystems, regardless of their value to humans. As with biocentrism, Callicott (2002) updates the idea to take into account Leopold's land ethic by arguing that "a more dynamic ecocentrism would morally limit anthropogenic change and disturbance to normal temporal and spatial scales, thus to preserve the land's health and beauty" (p. 241). In environmental philosophy and applied ethics, the debates about anthropocentrism, biocentrism, and ecocentrism storm across the literature terrain and show up in environmental education research. Often, however, the trouble is that these debates appear as unstated, implicit, or hidden assumptions that underpin the research process without reflexive dialogue.

Historical Roots: Objects, Selves, Miracles

As one way to investigate the historical roots of environmental education research, I employ Evernden's (1988, 1992) differentiation between *nature-as-object*, *nature-as-self*, and *nature-as-miracle*. *Nature-as-object* is the dominant and familiar belief that nature is a storehouse of resources, a "bare-bones nature with no subjectivity and no personal variables at all: just stuff" (1988, p. 11). The legacies of positivist science education and resource management framed nature as a storehouse of objects to be used and managed, which, in turn, effected research questions, designs, and methods. This, too, was an era of mostly empirical/quantitative environmental education research often grounded in positivist methodologies (Hart, 1996) and anthropocentric underpinnings. There were exceptions to this trend, such as Rejeski's (1982) exploratory Piagetian developmental study of children's ideas about nature.

Nonetheless, there was resistance to the idea of nature being just a series of resources and growing curiosity about people's attitudes toward nature. Swan (1975) argued that, during Leopold's conservation career in the 1920s, "a variety of educational movements emerged. Three of these—nature education, conservation education, and outdoor education—are related and partially overlap, and to an important degree are forerunners of environmental education in their contribution to both its content and methods" (p. 6). To varying degrees, each of these movements has been concerned with attitude change and skill development, relying on individualistic behavioristic models of human-nature relationships. For example, in an overview to an early publication, *What Makes Education Environmental*, Stapp and VandeVisse (1975) wrote, "If an important 'root cause' of our environmental crisis is the lifestyle of our people, then education should be concerned with the development of values, belief, and attitudes that reflect the necessity of our living in harmony with our environment" (p. xiv). The arrival of socially critical environmental education in the 1990s (Robottom & Hart, 1995) helped lead us out of the behavioristic conundrum toward an opening up of qualitative research questions and methods in environmental education.

Historically, the homogenizing tendencies embedded in referring to "our environment" often go largely unquestioned. This has been well taken up in recent decades by the environmental justice movement (Taylor, 2000) wherein issues of class, ethnicity, ability, and sexuality are accounted for, and these have slowly trickled into environmental education research. Critical questioning includes "whose environment, whose nature, who benefits from particular answers, and why?" As well, indigenous environmental knowledges have a tremendously long continuous praxis of honoring relations between humans, other living beings, and natural communities (Haudenosaunee Environmental Task Force, 1999; LaDuke, 2002; O'Riley & Cole, 2009). Despite the decidedly unique community-nature relationships philosophically and spiritually embedded in indigenous knowledge, unfortunately, there is a dearth of indigenous environmental education research.

Moving away from the limited view of *nature-as-object* may guide us toward a larger circle of care and compassion. *Nature-as-self* arises when humans lose the idea "that we are merely skin-encapsulated egos, and realize that we actually have a 'field of care' in which we dwell, which makes us literal participants in the existence of all beings, then we will realize that to harm nature is to harm ourselves" (Evernden, 1988, p. 12). People can see traces of *nature-as-self* in the ways in which humane education has explicitly stated its obligation to Schweitzer's (1923) initial formulation of a reverence for life and the importance of a circle of compassion (Schweitzer [as cited in Selby, 1995] 1995; Weil, 1998, 2004). Broadly defined, humane education commits to a life-affirming ethic, interconnectedness, values clarification, and democratic principles (Selby, 1995, p. 49).

Research in humane education is still emergent. There is Oakley's (2009) excellent research literature review on the educational and ethical dilemmas associated with school dissection practices. Concurrently, exemplary scholarship remains mostly at the conceptual level, including the work by Kahn and Humes on ecopedagogy and total liberation pedagogy (Humes, 2008; Kahn, 2008, 2010; Kahn & Humes, 2009). At a broader level than humane education, there is a lively body of research that exists at the intersection of environmental education and human-animal studies (e.g., Fawcett, 2002; Russell, 2005; Warkentin, 2002; Watson, 2006) that shares many of the same commitments as humane education and moves beyond the notion of *nature-as-self* into wider circles.

In another vein, with their imaginative forays into ecological learning and teaching, van Matre's early *Acclimatization* books were original and engaged across nature-human-animal boundaries. Later, van Matre's Earth Education became an American educational voice in the broad, deep ecology movement.[4] Often, deep ecology is associated with *nature-as-self* and is usually interpreted as an ecocentric approach, favoring ecological integrity over individual interdependence. Van Matre (1990), illustrating the rift between two types of interdependence, wrote:

> Talk about confusion. I remember one professional who said he was doing environmental education because he was working with the concept of interdependence. When we asked him what he was doing to get this concept across, he replied that he had everyone in his group try to stand at the same time on top of an old tree stump, and to do so they had to "interdepend" in holding everyone together. . . . he didn't seem to grasp at all that he was talking about sociological interdependence, not ecological interdependence. (p. 10)

Here, we witness a breech in which sociological and ecological interdependence are ripped asunder. Livingston (1994) believed deeply in the sociality of all living beings and argued against constructed divisions like this most of his life. He insisted that all life is socially organized through prevailing processes of cooperation and compliance. Nevertheless, although the aforementioned stump stance could be construed as a weak, disrespectful excursion into socioecological knowledge-making, perhaps in a highly individualistic society the actual experience of a group holding itself together on top of an old life form has some salient learning implications.

Nature-as-extended-self has been the platform of the deep ecology movement, but ecological feminists have critiqued this form of identification for its emphasis on personal transformation at the expense of political and social structures and inequities (Kheel, 1990; Plumwood, 1993). Plumwood (1993) warned that the resolution of dualism does not require merger or the simple erasure of the boundary between the colonizer and the colonized. Similarly, Evernden (1992) questioned how different these two positions really are: *Nature-as-object* begs the question "What's in it for me?" whereas *nature-as-self* asks, "What is this to me?" As Clark, Brody, Dillon, Hart, and Heimlich (2007) have claimed, the questions we ask in environmental education profoundly influence the research process; the questions themselves need to grow and change so research design and methods can complement the tone and direction of evolving research. Easier said than done, and, as Clark et al. point out, the research process contains messy things, such as "method surprises, problematic answers and publication dilemmas" (p. 110).

Nature-as-miracle refers to the wondrous, the inexplicable, and the unpredictable and asks the metaphysical question, "What is this?" (Evernden, 1988). It is not meant to be miraculous in an exclusively religious, hand-of-God sense. As a scientist thinking about miracles, Eisley (1978) wrote:

> Since . . . the laws of nature have a way of being altered from one generation of scientists to the next, a little taste for the miraculous in this broad sense will do us no harm. We forget that nature itself is one vast miracle transcending the reality of night and nothingness. We forget that each one of us in his [sic] personal life repeats that miracle. (p. 291)

One of the emergent possibilities in human relationships is to be in awe, to wonder, to be enchanted by encounters and experiences with other life. Certainly, indigenous cultures have had respectful and integral relationships with the surrounding plant and animal life, as the *Haudenosaunee Creation Story and Thanksgiving Address* illustrate.[5] Beyond awe and empathy lies the terrain of participatory consciousness—forms of sensory, somatic knowing, embodied knowing that does not dissolve into nature just like-me, or nature as transcendent force.

Bai's (2009) research cultivates the enchantment of the universe and people's "animated sensuous perceptions of the world" (p. 141). She traces the history of Plato's logocentrism:

> In short, Plato was heading a major epistemological revolution that changed the very texture, tone and colour of human consciousness: from the sensuous, emotive, empathic participatory mind or consciousness to the conceptual, abstract, and analytic rational mind or discursive consciousness. Ever since Plato we have been under a spell, not of the sensuous, but of the discursive. (p. 139)

Bai's excavation of classical Greek texts tells us that this was a drastic change for "Athenians who were still quite steeped in the tradition of Homeric participatory consciousness. This kind of consciousness is prone to merge with anything presented to them through poetic rendition" (p. 139). Bai and Scutt's (2009) conceptual research further explores ideas about Buddhism, mindfulness, and education.

Also in environmental education, there is a comparable philosophical line of playfulness and responsibility in Payne's creative gnome stories (Payne & Wattchow, 2009) in terms of miraculous, magical, and unknowable aspects of environmental education and the interplay between imagination and phenomenology. Cornell's (1989) flow learning is another method of learning and teaching environmental and outdoor education that combines a sense of the miraculous in nature along with ecological knowledge.

Multi-Centered Debates: Current Research Trends

Feminism and, particularly, ecological feminism have labored well in the service of environmental and social justice. Haraway (2000) talks about inseparable *nature-cultures* that she described as:

> . . . act(s) of faith in worldliness where the fleshy body and the human histories are always and everywhere enmeshed in the tissues of interrelationships where all the relators aren't human . . . where the categorical separation of nature and culture is already a kind violence, an inherited violence anyway. (p. 106)

Bowers (1993, 1997, 2003) has written extensively on the anthropocentric tendencies in critical pedagogy and the harm they cause. Bell and Russell (2000) productively examine some of the anthropocentric traps associated with poststructural approaches to critical pedagogy, while maintaining that Bowers "understates the extent to which these assumptions are being questioned within critical pedagogy" (p. 193). Nonetheless, they ask, "How is educating for freedom predicated on the exploitation of the nonhuman?" (p. 193) and "What meanings and voices have been pre-empted by the virtually exclusive focus on humans and human language in a human-centred epistemological framework?" (p. 189). They propose a decentering of anthropocentric assumptions about language, agency, and meaning in future environmental education research. In a similar vein, I asked how we "acknowledge other animals/beings as subjects of lives we share, lives that parallel and are interdependent in profound ways?" (Fawcett, 2000, p. 140).

These questions remain barely explored in environmental education research. A decade after Bell and Russell (2000) and Fawcett (2000) asked these questions, a 2009 international environmental education research symposium hosted one session on human/animal research and pedagogy that clarified the continuing marginalization of animal others in environmental education research (see Oakley et al., 2010).

Methodological Excursions

Fortunately, early on, the field of environmental education was reminded by researchers Robottom and Hart (1995) of the importance of ontological questions in research, even if the place of humans in relation to the vast majority of nature has not yet received sufficient attention. Payne (2003) wisely warns that "the *question of being-a-researcher* needs answering before methodological choices are made in *doing* research about the most appropriate starting points needed for entering into and enquiring about 'other' subject's experiences" (p. 187). The research methods employed by those who actually question the validity and meaning of the nature-culture divide have been intriguing, playful, and underrepresented in the literature.

Pivnick's (2003) paper optimistically equates an ecological worldview[6] with a decentered human position as part of the web of life (p. 145). Alas, she fails to recognize that human beings could also act as though they are the most important part of the web, even if they are not at the center of it. Otherwise, Pivnick's (2003) thoughtful work evokes the characteristics of vulnerability, humility, uncertainty, compassion, and respect as some of the challenges of living an ecological worldview, while trying to do environmental education research. She points to the following research methods as resonating with an ecological worldview: hermeneutics, phenomenology, participatory action research, feminist methods, ethnographies, and narrative inquiries. That being said, although these methods have the potential to defy anthropocentric boundaries, they do not inherently do so.

Environmental philosopher Weston opens his 1994 book, *Back to Earth*, with reference to Martin Buber's ideas about teaching and learning through universal reciprocity. Weston (2004) argues for ways of teaching and learning *wild* environmental education to counter cultures disconnected from the larger living systems. He claims environmental ethics has the potential to remake all of ethics, so perhaps it is not so surprising that the same could be true of the relationship between environmental education, specifically, and education, in general. The voice of Russell (2005) has been the clearest call to interrogate the place of, and space for, "'nature' in multivocal representations of research" (p. 433) to "trouble the 'nature'/culture divide" (p. 439). Russell (2005) calls for a bold imagining and enacting of "nature" as co-constructor in environmental education research.

The environmental education research on nature and children continues to offer some of the most intriguing methods and findings. One of the earliest studies that actually asked children what they thought about nature was by Rejeski (1982); he used children's drawings to interpret their relationships to nature through a Piagetian framework of child development. Fawcett (2002) explored children's relationships to common and familiar animals through interviews, problem-solving, and children's drawings and stories; this research culminated in a description of children's concepts about kinship imaginaries. More recently, Watson (2006) investigated children's ideas about nature, while in a camp setting using a combination of phenomenography and ethnography. Relying on a theoretical framework from Deleuze and Guattari's (2002) work, Watson explores ideas about being human, "where deleuzeoguattarian thought models becoming as the radically nonsubjective view of the alliances that people may form with women, animals, vegetables, molecules, ad infinitum" (p. 137). Watson (2006) discusses how "it is through the act of becoming-camper, becoming-animal, becoming-place that campers have the opportunity to redefine their relationships with others they encounter" (p. 138). Watson's findings are significant, because they disrupt the fixity of anthropocentric models of relationships to nature spawned by adult cultures and exemplify how they are partially resisted by children's culture. Likewise, Cutter-Mackenzie's (2009) research on multicultural school gardening innovatively employed children as researchers and collected data through journaling, photography, and peer interviews. She observed that "the children were seeking a nature or environment connection . . . particularly one that made them feel empowered in their environmental behaviours and actions" (p. 133).

Warkentin (2002, 2007, 2009) researched informal learning and ethical affordances in a myriad of spaces and places in which humans pay to interact with whales. Her work stands out not only because she consistently attended to the agency and resistance of the whales, but also because she directly questioned why the whales were never considered as subjects in any institutional ethical review process. The applicability of Warkentin's research to environmental education is evident in her purposeful entwining of teaching, ethics, language, and imagination. She defines "a particular biocentric ethics as a process of intersubjective empathy through embodied experiences with other beings, and suggests both an emphasis on spoken/written metaphor and a practice of 'metaphorical imaginative embodiment' as potential ways of nurturing such an ethics" (2002, p. 241). Clearly, this research goes beyond biocentrism to a more nuanced place of intersubjective agency and action.

McKenzie, Hart, Bai, and Jickling's (2009) edited book, *Fields of Green: Restorying Culture, Environment, and Education*, is a refreshing addition to environmental education, in part, because it illustrates and celebrates a multiplicity of research methods and practices without positing one true way. McKenzie et al. sought "more complicated conversations, more embodied and sensuous subjectivities, more adventurous wanderings and connective intelligence, more geophilosophical notions of place . . . to provoke and disrupt desires and dreams as specific social identities are constituted and negotiated" (p. 343). They were successful in their quest.

In *Fields of Green*, Gonzalez-Gaudiano and Buenfil-Burgos (2009) aptly demonstrate the benefits of a contested state of environmental education whereby "the proliferation, ambiguity, and the very open-ended character of environmental education entails a productive political, ethical, and epistemic possibility for the permanent construction of its identity" (p. 97). However, it is not clear to me that we need a permanent identity for environmental education research.

Still, Gonzales-Gaudiano and Buenfil-Burgos (2009) are right to value a contested definition of environmental education. Drawing on Foucault's work on governmentality, Ferreira (2009) asks, "How positions compete for dominance in a field, and most importantly, how do these positions have effects that govern what is possible to think and what is possible to do in environmental education" (p. 608). In the myriad ways in which power flows through a discipline like environmental education, Ferrieira's work leads me to wonder how we discipline students, teachers, and researchers to acknowledge their own experiences and to think and act across the nature/culture divide. If the divide is permeable, how does our research change and reflect this porosity? (Barrett, 2009). For example, Wilson's (1984) theory of biophilia as "an innate tendency to focus on life and lifelike processes" (p. 1) certainly crosses the anthropocentric divide and points to an evolutionary continuum. There are postsecondary educators who champion biophilia (e.g., Kellert, 1993), whereas others are concerned about biophilia's "innate-ness" and associated notions of biological determinism and sociobiology.[7] For the critics, the dangers of genetic inscription lie in standardizing and normalizing certain Western relationships to all environments and peoples. Ghosts of eugenics past and Social Darwinism float by.

The So What: Hybrid Ecological Communities of Environmental Justice

There is a need to go beyond the debates and posturing to undertake more innovative and empirical research in naturecultures of environmental education. Kahn and Hume's (2009) call for intersectional antispeciesist pedagogy illuminates paths for further research in the area of humane education. To broaden environmental education research efforts in this area beyond the formal school system (also beyond school board worries about insurance and liability claims) and into the vast arena of informal learning is crucial. As long-time host of the radio show "Animal Voices," Lauren Corman's ground-breaking public pedagogy "demonstrates to other social justice and environmental movements that many animal activists and scholars are not single-issued in their approaches" (Kahn & Hume, 2009, p. 189).

In his year review of thirty years of environmental education research, one of the challenges Scott (2009) outlines is for "more understanding across cultures about who we are and what we know" (p. 155). Berger's (2008) ground-breaking research on Inuit schooling demonstrates Scott's point, as Berger shows that teachers must make a clear choice between working for or against Eurocentric assimilation, and one of the determining factors is the support of the cultural importance of land skills and the teachings of elders, since intergenerational learning and traditional ecological knowledge matter profoundly to the sustainability of those communities. Berger's research also epitomizes the kind of liberatory and coalitional anthropocentric approach Plumwood (1997) was talking about as she argued that "the speech of the liberation supporter, such as the white antiracist or the male feminist, has always been an essential part of effective liberation politics" (p. 350).

In Whiteside's (2002) examination of the influential differences between English and French ecological theorists, he comprehensively argues that a diverse range of French theorists resisted entangled debates about anthropocentrism. Instead, historically, they favored noncentered ecological arguments and the entwining of nature and humanity. He elaborates:

> French green theorists tend to study how conceptions of nature and human identity intertwine. They elaborate green thought more often by *reciprocally problematizing* "nature" and "humanity" than by refining the distinction between them. . . . They maintain that what "nature" is shifts in relation to epistemological, social and political-ethical changes. Noncentred ecologists see "nature" as multiform and as inextricably confounded with humanity's projects and self-understandings. (p. 3)

Given the anthropocentric debates do not seem to have propelled us significantly further along sustainable or inspiring paths, I suggest the French stream with a socially inclusive, multiform noncentered approach to "nature" is echoed in Haraway's (2000) radical call for attention to naturecultures—boding promising similar and synergistic approaches across cultures.[8]

What is often left out of political ecological accounts is the sheer interdependencies that all living, decomposing, and dying entails. Just as many intricate ecological processes are becoming endangered, so is our collective time-linked knowledge of them. Qualities of relationality are at the heart of any good research and do not necessarily require complex methods. Weston's (1996) radical call to *deschool environmental education* was also a critique of how "schools remain profoundly conservative social institutions, and so remain profoundly human-centered as well. We might expect a significant degree of anthropocentrism" (p. 39). Although I agree we should still expect the dominance of anthropocentrism, we can desire and throw aspects of biocentrism and ecocentrism in the mix. Why do we not expect at least three degrees of separation in our environmental education research, with methodologies that enact at least two more ethical stances than anthropocentrism, such as biocentrism and ecocentrism? Bell's (1997) essay on learning through natural history offers an antidote to an anthropocentric diet, persuasively stating: "When taken up as an opportunity for fully embodied participation in the more-than-human world, natural history can offer an alternative to the fragmented, rationalistic, decontextualized experiences which characterize modern schooling" (p. 132). She illustrates the power of natural history to help us "strive to learn, in our hearts, heads and every limb that we belong and participate in a more-than-human world" (p. 141).

In the juncture in which environmental education meets natural history and ethics, the depth of questioning through naturecultures surrenders unexpected insights, such as Jardine's (2009) birding and cicada lessons:

> A birding lesson: I *become* someone through what I know. . . . Another birding lesson: If this place is fouled by the (seeming) inevitabilities of "progress" the cost of that progress is always going to be part of my life that is lost. . . . But it betrays another little birding lesson: that we are their relations as much as they are ours that we are thus caught in whatever regard this place places on us. (pp. 156–157)

Jardine's (2009) ethical linking of place-based learning to social relations can be tied to democratic intentions as it was earlier with the definition of humane education. In a persuasive argument for the importance of lifelong learning in environmental education, Gough, Walker, and Scott (2001) point out that the primary focus for much educational work in the environmental field has been on young people in the formal school setting as "citizens-in-the-making," rather than "employees-in-the-making" (p. 180). Given learning happens everywhere, in checkout lines at the supermarket, on public transit, in washrooms, and so on, then environmental education research needs to go beyond the formal school system to more deeply explore lived examples of the sociality of nature and our complex ecological embeddedness. Learning urban natural history, knowing when the flowers bloom, when the pollinating insects begin working in neighborhoods, when and where the spring migrations peak, and when birds are in danger of hitting brightly lit downtown buildings—these are all learning scenarios currently happening and in need of research.

Britzman (2009) talks about the emotional vulnerability and uncertainty inherent in teaching and learning, and she places education between criticism and disillusion, and hope and illusion. She writes about education as one of the impossible professions: "The central theme of the "Very Thought of Education" begins with the idea that, like the dream, education requires association, interpretation, and a narrative capable of bringing to awareness, for further construction, things that are farthest from the mind" (p. viii). Our research in environmental education needs to continue to *reflect* this vulnerability and to *refract* the disillusion that makes human relationships with natures a-relational, disembodied matters of unjust resource allocation, so we can keep dreaming narratives of naturecultures.

Realizing the assumptions underlying naturecultures multiplies inclusions. Multiplying who is included in environmental education research and whose lives matter may result in unexpected forms of rhizomatic social change, ideas I have discussed elsewhere (Fawcett, 2009). The branching roots from one center of interest to another can pop up in fertile places. Like community centers, naturecultures offer a collaborative multifaceted notion of what a center is and how it functions for the flourishing of a particular interdependent group in a specific location with a distinct history. Including social and environmental justice issues in environmental education research shifts the historical center of anthropocentrism and opens it up to the vistas offered by biocentric, ecocentric, and other centered and noncentered perspectives still imperceptible to our collective imaginations. Human bodies, neighborhoods, and societies are hybrid ecological communities largely cooperating, sometimes in parasitic relationships, often in symbiotic relationships. Holding on to our ecological embeddedness and advocating for environmentally just conditions is a prime directive for environmental education research. The idea of multiple centers of interest that can form alliances will surely enliven environmental education research in the decades to come.

Acknowledgments

Many thanks to Traci Warkentin and Sue Ruddick and anonymous reviewers for invaluable and challenging editorial suggestions.

Notes on Contributor

Leesa Fawcett is an Associate Dean and Associate Professor, as well as the Coordinator of the Graduate Diploma in Environmental and Sustainability Education, in the Faculty of Environmental Studies at York University in Toronto, Canada.

Notes

1. Williams (1980) described "Nature" as one of the most complicated words in the English language. For an extended discussion of the social constructions of nature see Evernden's (1992) *The Social Creation of Nature*.
2. Eckersley (1992) divided anthropocentrism into various forms, such as resource conservation, human welfare ecology, and preservationism.
3. The debates between land ethicists and animal liberationists get played out, for example, when the killing of one species is allowed to "protect" another, such as an endangered species.
4. The United States versions of deep ecology and Scandinavian versions are quite different. For a discussion of deep ecology in outdoor education philosophy and research, see Hallen (2000).
5. Haudenosaunee Environmental Task Force (1999).
6. The notion of an ecological worldview leaves me uneasy: whose ecology and whose view from where? The changing particularities of interdependent relationships between life, place, culture, and sociopolitics render an ecological worldview too generic.
7. For example, at the North American Association of Environmental Education's (NAAEE) research symposium in 2009, there was a discussion of biological determinism and homophobia in relation to E.O. Wilson's theories.
8. This is not to say that the French have figured it all out; as Whiteside (2002) points out, there is a distinct lack of concern for wilderness in French thinking.

References

Abrams, D. (1996). *The spell of the sensuous: Language in a more-than-human world.* New York, NY: Pantheon Books.

Bai, H. (2009). Re-animating the universe: Environmental education and philosophical animism. In M. McKenzie, P. Hart, H. Bai, &

B. Jickling (Eds.), *Fields of green: Re-storying culture, environment and education* (pp. 135–149). Cresskill, NJ: Hampton Press.

Bai, H., & Scutt, G. (2009). Touching the earth with the heart of enlightened mind: The Buddhist practice of mindfulness for environmental education. *Canadian Journal of Environmental Education, 14,* 92–106.

Barrett, M. J. (2009). Beyond human-nature-spirit boundaries: Researching with animate Earth (Unpublished PhD dissertation). Faculty of Education, University of Regina, Saskatchewan, Canada.

Bell, A. (1997). Natural history from a learner's perspective. *Canadian Journal of Education, 2,* 132–144.

Bell, A., & Russell, C. (2000). Beyond human, beyond words: Anthropocentrism, critical pedagogy, and the poststructuralist turn. *Canadian Journal of Education, 25*(3), 188–203.

Berger, P. (2008). *Inuit visions for schooling in one Nunavut community* (Unpublished PhD dissertation). Faculty of Education, Lakehead University, Thunder Bay, Ontario, Canada.

Bowers, C. A. (1993). *Education, cultural myths and the ecological crisis.* Albany, NY: State University of New York Press.

Bowers, C. A. (1997). *The culture of denial.* Albany, NY: State University of New York Press.

Bowers, C. A. (2003). Can critical pedagogy be greened? *Educational Studies, 34,* 11–21.

Britzman, D. (2009). The very thought of education: Psychoanalysis and the impossible professions. New York, NY: State University of New York Press.

Callicott, J. B. (2002). Environmental ethics. In P. Timmerman (Vol Ed.), *Encylopedia of global environmental change* (pp. 231–242). Toronto, ON: John Wiley & Sons.

Clark, C., Brody, M., Dillon, J., Hart, P., & Heimlich, J. (2007). The messy process of research: Dilemmas, process and critique. *Canadian Journal of Environmental Education, 12,* 110–126.

Cornell, J. (1989). *Sharing the joy of nature: Nature activities for all ages.* Nevada City, CA: Dawn Publications.

Curthoys, L., & Cuthbertson, B. (2002). Listening to the landscape: Interpretive planning for ecological literacy. *Canadian Journal of Environmental Education, 7*(2), 224–240.

Cutter-Mackenzie, A. (2009). Multicultural school gardens: Creating engaging garden spaces in learning about language, culture, and environment. *Canadian Journal of Environmental Education, 14,* 122–135.

Deleuze, G., & Guattari, F. (2002). *A thousand plateaus: Capitalism and schizophrenia* (B. Massumi, Trans.). Minneapolis, MN: University of Minnesota Press.

Eckersley, R. (1992). *Environmentalism and political theory.* Albany, NY: State University of New York Press.

Eisley, L. (1978). *The star thrower.* San Diego & London: Harcourt Brace Jovanovich.

Evernden, N. (1985). *The natural alien.* Toronto, ON: University of Toronto Press.

Evernden, N. (1988). Nature in industrial society. In S. Jhally & I. Angus (Eds.), *Cultural politics in contemporary America.* New York, NY: Routledge, Chapman, & Hall.

Fawcett, L. (1989). Anthropomorphism: In the web of culture. *Undercurrents: A Journal of Critical Environmental Studies, 1*(1), 14–20.

Fawcett, L. (2000). Ethical imagining: Ecofeminist possibilities and environmental learning. *Canadian Journal of Environmental Education, 5,* 134–149.

Fawcett, L. (2002). Children's wild animal stories: Questioning interspecies bonds. *Canadian Journal of Environmental Education, 7*(2), 125–139.

Fawcett, L. (2009). Feral sociality and (un)natural histories: On nomadic ethics and embodied learning. In M. McKenzie, P. Hart, H. Bai, & B. Jickling (Eds.), *Fields of green: Re-storying culture, environment and education* (pp. 227–237). Cresskill, NJ: Hampton Press.

Ferreira, J. (2009). Unsettling orthodoxies: Education for the environment/for sustainability. *Environmental Education Research, 15*(5), 607–620.

Gonzalez-Gaudiano, E., & Buenfil-Burgos Nidia, R. (2009). The impossible identity of environmental education: Dissemination and emptiness. In M. McKenzie, P. Hart, H. Bai, & B. Jickling (Eds.), *Fields of green: Restorying culture, environment and education* (pp. 97–108). Cresskill, NJ: Hampton Press.

Gough, N. (2009). Becoming transnational: Rhizosemiosis, complicated conversation, and curriculum inquiry. In M. McKenzie, P. Hart, H. Bai, & B. Jickling (Eds.), *Fields of green: Restorying culture, environment and education* (pp. 67–81). Cresskill, NJ: Hampton Press.

Gough, S., Walker, K., & Scott, W. (2001). Lifelong learning: Towards a theory of practice for formal and non-formal environmental education and training. In B. Jensen, K. Schnack, & V. Simovska (Eds.), *Action competence revisited.* Copenhagen: Royal Danish School of Educational Studies.

Greenwood, D. (2009). Place: The nexus of geography and culture. In M. McKenzie, P. Hart, H. Bai, & B. Jickling (Eds.), *Fields of green: Restorying culture, environment and education* (pp. 271–281). Cresskill, NJ: Hampton Press.

Hallen, P. (2000). Ecofeminsim goes bush. *Canadian Journal of Environmental Education, 5,* 150–166.

Haraway, D. (2000). *How like a leaf.* New York, NY: Routledge.

Hart, P. (1996). Problematizing inquiry in environmental education. *Canadian Journal of Environmental Education, 1,* 56–88.

Haudenosaunee Environmental Task Force. (1999). *Words that come before all else: Environmental philosophies of the Haudenosaunee.* Cornwall Island: North American Traveling College.

Humes, B. (2008). Moving toward a liberatory pedagogy for all species: Mapping the need for dialogue between humane and anti-oppressive education. *Green Theory & Praxis: The Journal of Ecopedagogy, 4*(1), 65–85.

Jardine, D. (2009). Birding lessons and the teaching of cicadas. In M. McKenzie, P. Hart, H. Bai, & B. Jickling (Eds.), *Fields of green: Restorying culture, environment and education* (pp. 155–159). Cresskill, NJ: Hampton Press.

Jickling, B. (2009). Environmental education research: To what ends? *Environmental Education Research, 15*(2), 209–216.

Kahn, R. (2008). Towards ecopedagogy: Weaving a broad-based pedagogy of the liberation for animals, nature and the oppressed peoples of the Earth. In A. Darder, M. Baltodano, & R. Torres (Eds.), *The critical pedagogy reader* (2nd ed., pp. 522–540). New York, NY: Routledge.

Kahn, R. (2010). *Critical pedagogy, ecoliteracy and planetary crisis: The ecopedagogy movement.* New York, NY: Peter Lang.

Kahn, R., & Humes, B. (2009). Marching out from Ultima Thule: Critical counterstories of emancipatory educators working at the intersection of human rights, animal rights, and planetary sustainability. *Canadian Journal of Environmental Education, 14,* 179–195.

Kellert, S. (1993). The biological basis for human values of nature. In S. Kellert & E. O. Wilson (Eds.), *The biophilia hypothesis* (pp. 42–69). Washington, DC: Island Press.

Kheel, M. (1990). Ecofeminism and deep ecology: Reflections on identity and difference. In I. Diamond & G. Orenstein (Eds.), *Reweaving the world* (pp. 128–137). San Francisco, CA: Sierra Club Books.

LaDuke, W. (2002). *The Winona LaDuke reader: A collection of essential writings.* Penticton, BC: Theytus Books.

Livingston, J. (1981). *The fallacy of wildlife conservation.* Toronto, ON: McClelland & Stewart.

Livingston, J. (1994). *Rogue primate: An exploration of human domestication.* Torotno, ON: Key Porter Books.

McKenzie, M., Hart, P., Bai, H., & Jickling, B. (2009). *Fields of green: Re-storying culture, environment and education.* Cresskill, NJ: Hampton Press.

Oakley, J. (2009). Under the knife: Animal dissection as a contested school science activity. *Journal for Activist Science & Technology Education, 1*(2), 59–67.

Oakley, J., Watson, G., Russell, C., Cutter-McKenzie, A., Fawcett, L., Kuhl, G., . . . Warkentin, T. (2010). Animal encounters in environmental education research: Responding to the "question of the animal." *Canadian Journal of Environmental Education, 15,* 86–102.

O'Riley, P., & Cole, P. (2009). Coyote and raven talk about environmental justice. In J. Agyeman, P. Cole, R. Haluza-DeLay, & P. O'Riley (Eds.), *Speaking for ourselves: Environmental justice in Canada* (pp. 233–251). Vancouver, BC: UBC Press.

Payne, P. (2003). Postphenomenological enquiry and living the environmental condition. *Canadian Journal of Environmental Education, 8,* 169–190.

Payne, P., & Wattchow, B. (2009). Phenomenological deconstruction, slow pedagogy, and the corporeal turn in wild/outdoor. *Canadian Journal of Environmental Education, 14,* 15–32.

Pivnick, J. (2003). In search of an ecological approach to research: A meditation on Topos. *Canadian Journal of Environmental Education, 8,* 143–154.

Plumwood, V. (1993). *Feminism and the mastery of nature*. London, UK: Routledge.

Plumwood, V. (1997). Androcentrism and anthropocentrism: Parallels and politics. In K. Warren (Ed.), *Ecofeminism: Women, culture, nature* (pp. 327–355). Bloomington, IN: Indiana University Press.

Rejeski, D. (1982). Children look at nature: Environmental perception and education. *Journal of Environmental Education, 13*(4), 27–40.

Robottom, I., & Hart, P. (1995). Behaviourist EE research: Environmentalism as individualism. *Journal of Environmental Education, 26*(2), 5–9.

Russell, C. (2005). Whoever does not write is written: The role of "nature" in post-post approaches to environmental education research. *Environmental Education Research, 11*(4), 433–443.

Russell, C., Sarick, T., & Kennelly J. (2002). Queering environmental education. *Canadian Journal of Environmental Education,7*(1), 54–66.

Scott, W. (2009). Environmental education research: 30 years on from Tbilisi. *Environmental Education Research, 15*(2), 155–164.

Selby, D. (1995). *Earthkind: A teacher's handbook on humane education*. Staffordshire, UK: Trentham Books Ltd.

Stapp, W., & VandeVisse, E. (1975). Overview. In N. McInnis & D. Albrecht (Eds.), *What makes education environmental* (pp. xii–xiv). Washington, DC: Environmental Educators.

Swan, M. (1975). Forerunners of environmental education. In N. McInnis & D. Albrecht (Eds.), *What makes education environmental* (pp. 4–20).Washington, DC: Environmental Educators.

Taylor, D. (2000). The rise of the environmental justice paradigm: Injustice framing and the social construction of environmental discourses. *American Behavioural Scientist, 43*(4), 508–580.

Van Matre, S. (1990). *Earth education: A new beginning*. Warrenville, IL: The Institute for Earth Education.

Warkentin, T. (2002). It's not just what you say, but how you say it: An exploration of the moral dimensions of metaphor and the phenomenology of narrative. *Canadian Journal of Environmental Education, 7*(2), 241–255.

Warkentin, T. (2007). Captive imaginations: Affordances for ethics, agency and knowledge-making in whale-human encounters (Unpublished PhD dissertation). Faculty of Environmental Studies, York University, Toronto, Ontario, Canada.

Warkentin, T. (2009). Whale agency: Affordances and acts of resistance in captive environments. In S. McFarland & R. Hediger (Eds.), *Animals and agency: An interdisciplinary exploration* (pp. 23–43). Leiden: Brill.

Watson, G. (2006). Wild becomings: How the everyday experience of common wild animals at summer camp acts as an entrance to the more-than-human world. *Canadian Journal of Environmental Education, 11,* 127–142.

Weil, Z. (1998). Humane education: Charting a new course. *The Animals Agenda,* September/October 19–21.

Weil, Z. (2004). *The power and promise of humane education*. Gabriola, BC: New Society Publishers.

Weston, A. (1994). *Back to Earth: Tomorrow's environmentalism.* Philadelphia, PA: Temple University Press.

Weston, A. (1996). Deschooling environmental education. *Canadian Journal of Environmental Education, 1,* 35–46.

Weston, A. (2004). What if teaching went wild? *Canadian Journal of Environmental Education, 9,* 31–46.

Whiteside, K. (2002). *Divided natures: French contributions to political ecology.* Cambridge, MA & London: The MIT Press.

Williams, R. (1980). Ideas of nature. In *Problems in materialism and culture* (pp. 68–85). London and New York: Verso.

Wilson, E. O. (1984). *Biophilia: The human bond with other species.* Cambridge, UK: Harvard University Press.

Section VIII

Philosophical and Methodological Perspectives

PAUL HART
University of Regina, Canada

Introduction

Bridging the Self Across Relational Inquiry to Action with Culture in Mind This section of the *Handbook* brings together a wide range of philosophical perspectives and methodological approaches that play a part in characterizing the environmental education research field. We recognize this field within the broad fields of educational and social science research that, over several decades, has quite rapidly diversified as it engaged the complexities of human social (and environmental) relations. In these days of paradigm evolution and proliferation, scholars may be challenged in tracking developments across, for example, feminist, poststructural, or cultural theory as they impact methodologies that range from ethnography and phenomenology through various narrative and action-oriented approaches, as well as increasingly sophisticated applications of applied science techniques of analysis. Given such diversity, it should not be surprising to find novice as well as experienced researchers learning how to operate in spaces where ideas about certainty and stability of method are at once and always in question.

A challenge in bringing this section together, given the impossibility of providing a comprehensive representation of genres of inquiry, was to display an array of approaches that could provide openings to the diverse nature of inquiry as it relates to issues of environment and sustainability. Beyond the obvious problem of representation, there was also the problem of abstraction, faced by synoptic texts. The only reasonable way to avoid the potential paralysis posed by these difficulties was to select authors who would challenge readers to look beyond their own perspectives, that is, to direct attention away from solidification of individual positioning and toward openings for thinking about imaginative possibilities. So, in introducing ideas constructed within these few chapters, we anticipate that readers will make up their own minds about the broadening bandwidths of environmental education research and use these papers to find their own ways not only to question the orthodoxy of extant methods but also to move toward those emergent approaches that can better meet their needs.

Environmental education research is deeply embedded within the discussions and debates of educational and social inquiry. Multiple philosophical positions, theoretical frames, and research methodologies and methods work, along with personal predispositions and interests, to govern an array of conceptual and practical choices. Each of the chapters in this section was selected to represent a part of a spectrum of thinking about these choices. In recognizing the ways in which these grounding perspectives were conceptualized, contextualized, and represented, we can see how they have exerted a profound influence on the ways that knowledge in fieldwork was conceived and constructed. We can see how really understanding any environmental education inquiry requires understanding of its onto-epistemological grounding. In McKenzie's (2004) view, environmental educators, once pulled only by modern metanarratives to "explain" the social-environmental, are now subject to the necessary uncertainties and ambiguities of the willful contradiction of a more poststructurally informed socioecological education. Thus how we proceed, as environmental education researchers, in the face of less certain, but more theoretically and practically active and often qualitative forms of inquiry, demands our full engagement not only in our inquiries but also in the issues that inevitably surround them.

One way to proceed is to assume responsibility for making our positions on research perspectives more visible for ourselves and more explicit for others, as authors have illustrated in this section. Getting clear about our onto-epistemological assumptions can work to deepen our engagement, to be more conscious of multiple layers of meaning, to critique our own interpretations and representations. This grounding work seems crucial in environmental education research that is designed intentionally to stretch the boundaries of current thinking. If environmental education research can be conceptualized in terms of multiple ways of knowing, just as environmental education as a field contradicts the dominant cultural discourse (Fien, 1993; McKenzie, 2004; Stevenson, 1987, 2007), then it may be perceived capable of questioning forms of knowledge construction itself. In fact, readers may find that reading the subtexts of the papers assembled here reveals considerable questioning of taken-for-granted cultural, as well as inquiry, discourses. In Phillip Payne's chapter, for example, we see a certain proclivity to base environmental education historical inquiries within socioecological frames of human experience. This paper seems intent on re-engaging debate, within the environmental education research community, based in a kind of postcritical phenomenological praxis that accounts as much for the value of experiential-relational knowing-seeing-being (with environment) as for knowledge of the propositional kind.

What is interesting about Payne's paper is the challenge to redirect our environmental education inquiries toward an aesthetic of embodied experience. In questioning how we as humans have been abstracted through cultural discourses, Payne finds ecophenomenological entry points for methodologies that can help us get inside our socially constituted, subjective selves. He urges us to consider postcritical ecophenomenological frames that direct researchers to attend to inquiry that can remain prediscursive through its reflexivity. As Payne says, following Grosz (2004), we need to understand not only how culture inscribes bodies . . . but what these bodies are that inscription is possible. And, although he somewhat eschews poststructural work that seems all too preoccupied with language (and mind frames over embodied ones), it seems to me that he is in many respects not too distant from questions posed by Davies' poststructural-feminist-informed questions concerning how we can come to construct ourselves as environmentally oriented people, educators, and researchers.

Davies' paper gently introduces researchers to tough questions concerning certain conceptual and methodological spaces, the thinking about inquiry, opened by feminist-poststructural perspectives. These questions challenge environmental education researchers to question our taken-for-granted assumptions that work to structure existing business-as-usual values as normal. They direct attention to normative cultural meanings that, when rendered visible, reveal how power can act to open, as it can to close, the possibility of overcoming "common sense." And these feminist-poststructural perspectives can offer conceptual strategies for finding ways to act differently, much the same as Payne encourages researchers to exert agency once one recognizes what is "really" going on in Western "democracies." In different ways, each of these researchers draws attention to how relational ways of knowing can mobilize our creative capacities to rethink ourselves in the world—spatially, materially, and conceptually—that is, as a concrete doable process of "becoming different." While Davies views such relational inquiry in terms of the methodological praxis involved in collective biography, the importance of relational epistemology as a basis for participatory action research, as discussed by both Bradbury-Huang and Long and by Stevenson and Robottom, is remarkably similar. Both genres are grounded historically in critical approaches to research and pedagogy where individualized inquiry work is no longer sufficient for addressing complex, interdisciplinary problems of people-society-environment. Payne, Davies, Bradbury-Huang and Long, Stevenson and Robottom, Barratt Hacking, Cutter-Mackenzie and Barratt, and McKenzie et al., each in their own way, advocate for opening up the ontological (re)positioning work required for such relational work in environmental education. This is the deep work that seriously considers "knowledges of being" (i.e., ontologies) that previously were assumed to belong to the other—as the other's marks of identity. The idea that this work now becomes collectively imagined, as coming to know ourselves in relation to people and the planet, works toward curriculum and pedagogy as building communities of "action."

Arguing that educators, like those involved in other organizations, need enlightened educational practices such as those participatory processes that can generate "actionable knowing," Bradbury-Huang and Long frame action research as a form of collaborative ecological inquiry for sustainability. They argue that precisely because conventional education or educational inquiry seems unequal to the task of sustainability challenges, locked within industrial educational models of propositional knowledge transmission, forms of participation through action inquiry can apply experiential knowing in new and necessary ways. Notwithstanding the range of perspectives found within participatory action research, their concrete examples position action research exactly where both Stevenson and Robottom have for many years, from a base in socially-critical theory. Stevenson and Robottom articulate this position in terms of historical and theoretical characteristics that no longer treat people as the objects of inquiry but as responsible agents capable of participating in changing/improving their own circumstances.

This position is well developed in respect of children as researchers by Barratt Hacking, Cutter-Mackenzie, and Barratt who develop a case for applications of environmental education research methodology that actively involve children in all phases of inquiry. Given the emphasis over the years within environmental education research on the importance of formative experience in the early years in establishing lifelong perspectives (exemplified in the special edition of *Environmental Education Research, 13*[4]), this chapter lets readers into challenges of inquiry involving children-as-researchers, as a recent turn of perspective. Barratt Hacking et al. raise issues of researcher positioning within the intersubjectivity of power relations in all qualitative forms of inquiry. They raise issues of identity-subjectivity and agency in young people that impact not only their active participation in social/environmental issues but also in participatory forms of inquiry. They provide examples of fieldwork in England and in Australia that reflect sensitivities and invoke serious thought about researcher responsibility in creating conditions for ethically appropriate ways to make children's involvement meaningful to the children themselves. But they also provide insights into processes of intergenerational learning that raise fundamental questions about inquiry processes that may be useful for those planning for fieldwork but also for those interested in thinking differently about the philosophical commitment required to conduct qualitative inquiries that demand critical reflection.

What the paper by McKenzie et al. adds to this emphasis on "relational inquiry" within research based in experiential learning, it seems to me, is found in their concern about creating conditions for self exploration of the meaning of experiences, that is, in identity work in personal-social-environmental relations that can lead to action. This concern, similar to that of Davies and the two chapters on action research, is then illustrated in interactive dialogue about how common research interests are interpreted by various levels of participants, including high school students, undergraduate and graduate students, and faculty. The paper represents a discussion about how intellectual-emotional engagement is achieved as an evolving dialogue that implicated general ideas of socioecological learning. In this way, rather than simply talking about inquiry in the abstract, the paper performed the inquiry without extolling the virtues of theory. The idea of using metaphor can either be lost or found by readers who have a mind to extend their ideas about self and discourse. And, there are strong arguments for the value of relational learning exemplified by doing the methodology while discussing why and how it was being performed. Obvious connections to inquiry-based relational learning within an action research frame, collective storied (almost biographical) work, directly connect methodology to intergenerational critical pedagogy, thus demonstrating the value of critically reflecting on methodology as you do it. The McKenzie et al. paper pushes the boundaries of educational inquiry across relationships in ways that use potential power relations to raise participant consciousness about how it is being performed within a learning environment that works actively to demonstrate how collegiality can empower individual action.

Michael Peters' paper takes each of these papers concerned with methodological grounding to the level of philosophical inquiry and the broad conceptual issues that must be addressed at the intersections of philosophy (i.e., ecosophy), ecology and globalization (see also, Jickling & Wals, 2008). The paper explores some difficult territory where conceptual work involves large (scale) systems. It is at this level of thought that issues like sustainability will be transposed within wider political frames. This final chapter of the section portends radical new approaches for environmental education research, approaches that, while they acknowledge the need to draw from a wide range of multidisciplinary methodologies, also need to be capable of responding to learning needs that cross cultural boundaries where pedagogical difference is more than educational in nature.

Raising issues of educational inquiry framed in creative imagining (which is approached in McKenzie, Bai, Hart, & Jicking, 2009), Peters' main concern is applying these explorations to learning in open knowledge economies. His argument is based on connections that must be made between spaces of public knowledge and public action if environmental education is to be valued within new social practices of production. Not unlike Payne, Peters raises concerns about issues of slow-fast, public-private continua that will require creative thinking about the kinds of inquiry needed to manage change both temporally and spatially. Research cultures, particularly those of environmental education research, had better be prepared to focus on questions of information in relation to what can count as knowledge in the discursive production of ourselves within political rationalities and knowledge cultures. Knowledge cultures (like environmental education) risk colonization, he argues, if researchers within these fields are incapable of capturing the complexities of these exchange systems. In light of the need for more awareness of global economic systems, Peters thus proposes new relevance for forms of inquiry that can expose such systems at levels beyond the practical working context.

While it seems important to recognize the importance of some discursive threads that may be seen to connect ideas of people writing here, it also seems important to allow recognition of blank spots and difference in inquiry that need to be part of a larger picture of environmental education research. Clearly, one of the strengths of environmental education research has been its tendency to embrace different stories built on unique and important histories that, somewhat akin to environmental advocacy, have contributed to diversity in form and method of inquiry. Looking beyond this introductory *Handbook*, discussion more explicitly directed toward feminist, GLBTQ(O), race-based, postcolonial and indigenous cultural studies is necessary. Arguments familiar to researchers in the social sciences and to educational research more generally need more airtime in environmental education research. In particular, we encourage arguments that resonate across onto-epistemological perspectives, such as those that engage relations between power and the knowledge produced by research, critical and disruptive texts, and studies that work across familiar methodological categories and signal oft-repeated tensions and contradictions, for example, among insider/outsider or researcher/researched. There is space now for collaborative work that resists "hardening of the categories," embraces new forms of representation, authenticity of multiple voices and perspectives, and calls for more activist, community-based approaches.

While the chapters in this section explicitly seek to raise broad questions, embrace diversity, build coalitions across methodological categories, they must also engage these at different levels of complexity. In a call for intelligibility within educational inquiry, St. Pierre (2000) asks how one can become available to intelligibility across discursive formations that may not be available across philosophical perspectives. As Lather (1996) says, focus on issues of methodology often diverts attention from more fundamental issues of epistemology (and, as Payne argues, ontology). The problem for *Handbook* sections like this one, that cannot accommodate existing methodological categories comprehensively, must raise awareness of the inquirer's responsibilities to get beyond the rhetoric and read the philosophy underlying methodology so that critiques are informed by the original literature (see Peters, 1996, 1999).

At a practical level, questions of actual field methods and processes remain somewhat underrepresented, undoubtedly a problem for those concerned about fieldwork and the disposition of findings. The focus here on matters of conceptualizing, contextualizing, designing and representing research as worthy of attention was a conscious choice based on the assumption that the complex tasks of grounding environmental education inquiry in philosophical and methodological work must precede more practical tasks of fieldwork. As Agee (2009) argues, poorly conceived inquiry is likely to create problems that affect all subsequent stages of inquiry. So while no single volume can address all of the methodological and practical options, part of the strengthening process involves researchers in questioning how particular theoretical positionings, however tacit, predispose researchers to see and do things in particular ways and position themselves toward certain aims. It is hoped that in presenting several of these positionings, environmental education researchers will be more predisposed to thoughtfully explore their own positionings.

This section was constructed on the assumption that there are different legitimate ways of understanding reality (e.g., the new environmental paradigm versus the dominant social paradigm), different truths and different ways of knowing. Whatever your beliefs about what counts as legitimate inquiry in education, the literature now includes multiple discourses about knowledge and how we can come to know. And because this literature raises such onto-epistemological questions, the *Handbook* section includes philosophical and methodological perspectives that range across interpretive, critical and postcritical approaches in recognition that different kinds of research questions demand different approaches. So, for those readers contemplating dissertation/thesis work or formal project research, it may be useful to consider educational inquiry as a critically reflexive process where the researchers' credibility rests on the specifics of place and the people inhabiting a place. Given that the traditions from which qualitative inquiry sprang placed social processes/interaction at the center, our research questions are conceptualized and contextualized to perform different functions and it is to these functions that chapters in this section speak. For example, the problem of the utility of particular methods is implicitly a function of power in the political construction (i.e., conceptualization) of research questions. These basic problems govern questions of positioning, voice, difference, empowerment, and hence constructions of what can count as inquiry—a question to which all discussions of inquiry must ultimately return.

Thus, the focus of this section of the *Handbook* on larger questions of theory and methodology should be read as a signification of the problems of educational research as philosophical in origin and substance. Attention to the effects of one's own arguments and explicit discussion of those effects is prerequisite to all subsequent research activity including problems of ethics, relations between researchers and researched, problems of analysis and representation of findings, and problems of the production of discourse. Recognition of one's ability to sort through the complex layering and diverse perspectives that now characterize environmental education research is crucial to improving the thinking and hence the practice of inquiry in this field. This section signifies the importance of directing attention to questions of academic credibility, competition/cooperation and critique/reflexivity. It concerns the need to develop the ability to articulate one's position relative to the historical and traditional perspectives that impact the field of environmental education research to the degree that one can challenge and move beyond existing frames as well as to recognize the political landscape (the discursive structures) within which our work is embedded as a means to changing what counts (sacred stories) in the interest of those who live there.

References

Agee, J. (2009). Developing qualitative research questions: A reflective process. *International Journal of Qualitative Studies in Education, 22*(4), 431–447.

Fien, J. (1993). *Education for the environment: Critical curriculum theorizing and environmental education*. Geelong, VIC, Australia: Deakin University Press.

Grosz, E. (2004). *The nick of time: Politics, evolution, and the untimely*. Durham, NC: Duke University Press.

Jickling, B., & Wals, A. (2008). Globalization and environmental education: Looking beyond sustainable development. *Journal of Curriculum Studies, 40*(1), 1–22.

Lather, P. (1996). *Methodology as subversive repetition: Practices toward a feminist double science*. Paper presented at the annual meeting of the American Educational Research Association, New York, NY.

McKenzie, M. (2004). The "willful contradiction" of poststructural socio-ecological education. *Canadian Journal of Environmental Education, 9*, 177–190.

McKenzie, M., Bai, H., Hart, P., & Jickling, B. (Eds.). (2009). *Fields of green: Restorying culture, environment and education*. Cresskill, NJ: Hampton Press.

Peters, M. (1996). *Poststructuralism, politics and education.* Westport, CT: Bergin & Garvey.

Peters, M. (1999). (Posts-) modernism and structuralism: Affinities and theoretical innovations. *Sociological Research Online*, *4*(3), http://www.socresonline.org.uk/4/3/peters.html

St. Pierre, E. (2000). The call for intelligibility in postmodern educational research. *Educational Researcher, 29*(5), 25–28.

Stevenson, R. (1987). Schooling and environmental education: Contradictions in purpose and practice. In I. Robottom (Ed.), *Environmental education: Practice and possibility* (pp. 69–82). Geelong, VIC, Australia: Deakin University Press.

Stevenson, R. (2007). Schooling and environmental/sustainability education: From discourses of policy and practice to discourses of professional learning. *Environmental Education Research, 13*(2), 265–285.

40

(Un)timely Ecophenomenological Framings of Environmental Education Research

Phillip G. Payne
Monash University, Australia

One-Ear

One-ear is a furry female grey kangaroo. A passing buck must have nibbled off part of her left ear in the frenzy of mating, well before she hopped into the place we now share. She hopped into my world four blisteringly hot summers ago when only parched, cracking clayish dirt surrounded our home—a modern but rustic dwelling built from one hundred year old handmade "reds" adjacent to a recently declared national park in central Victoria. Her hopping was earnest but hesitating, scanning, and foraging for any remnant weeds to nourish her depleted body. I later realized she was also feeding her utterly vulnerable joey, suckling in her pouch, probably unsuccessfully, given the absence of any grass. We'd endured a lengthy decade-long drought (sic)—its dusty dryness fills the nose and empties the mouth. Or, the arid consequences of anthropogenic climate change where the fire-threatening number of days in a year predicted to exceed 40 degrees Celsius will rise over the next two decades from about 10 to nearly 30 . . . but kangaroos do not count, in more ways than one. The trees and kangaroos know the bush much better than we do—their space is diminishing, as are the sources of replenishment, hence One-ear's summer bold entry to our place. Once she even came inside the house when our then eighteen-year-old daughter left the front door wide open!

I watched her closely that first year of what proved to be an ongoing cross species encounter over the four years now. The initial glances we exchanged were from afar—through our front window which she often peered into, or from outside in the wilting garden where, strangely, she didn't seem to mind my presence as those innocent or bewildered glances turned to face-to-face, body to body gazing. Perhaps her patience tapped my curiosity. I'd seen her on different occasions, not noticing her pouched joey. But when I did see the bulge move, I recall that somewhere, or was it time, an anxiety welled from my stomach. I'd heard elsewhere that mother kangaroos can't reproduce for a number of seasons if an in-pouch joey dies. But that is probably a myth, or story. I don't care as her presence seduces my interest and worry. One-ear is desperate for food, as I guess her joey is and who, occasionally has raised her little grey face from the precarious security of the pouch. Her dark, innocent eyes meet my gaze, time and time again. I had also noticed how frequently in the heat of the evening One-ear licked the short grey-brown bristly fur on her front quarters and constantly dribbled in what I presume were bodied attempts to cool off. She doesn't move much even when I gently, silently venture towards her. Kangaroos have poor eyesight so their looking is a prolonged and intense one. So how is it they know how and when to jump high over the fencing wire that surrounds our place and whose thin, jagged strands separate the national park from my property—a legacy of the previous owners. I've cut down those nasty strands after finding other snared kangaroos' mangled bodies, sometimes beheaded by introduced foxes or pet dogs on the nightly roam for a feed.

I eventually throw out some old bread, soaked in water. Our dam, emptied of any water for three years, once served as a watering hole for the mob of kangaroos who in the calmer heat of the hot, summer evenings would drink from and wet-down their lean, blood-rich hind and front quarters. The bread encourages One-ear to revisit the next evening. Bolder. I'm surprised, intrigued, and a bit annoyed at myself. Feeding a wild animal doesn't sit well with my thoughts, a disembodied type of thinking trained by my much study of environmental ethics. I later learn that with urbanization and its encroachment on kangaroo (and other species') habitat that feeding bread can lead to gum disease and jaw problems, given their customary diet of grass, or whatever they can find when the combination of prolonged drought and overheated dryness forces them into gardens, flower and vegetable plots. Some people shoot the intruders. I live in this bush place retreat only on the weekends when I return from my academic work in the city—so I rationalize I'm not damaging her, but occasionally helping the joey get through the stifling heat—at least until some late Autumn tufts of grass reappear.

A few weeks later, with the intense summer heat about to give way to its autumn cooling, a clearly hungry One-ear eats the soggy cool bread out of my hand. We creep up on each other, still exchanging glances—me stooping to keep my body lower than hers so as to not scare her—she (and joey) desperately hungry. Have I trained part of the wildness and otherness out of her? I'm delighted and annoyed, still.

Three summers later, joey, now "Grey-face," and One-ear return annually to my/our place's desert. It's even hotter now—a 48-degree Celsius day with 70-kilometer per hour Northerlies fuelling a blitzkrieg of fires around Victoria. Their devastating consequences spread instantaneously as global news. One hundred and seventy three human lives were lost, as were uncounted animal lives, and with massive "collateral" property and "nature" damage. One-ear is still too bold, and hungry, and shriveling in the heat—something I suspect a somewhat nervous Grey-face is yet to fully experience, in the same way, as three years ago when she was being nourished in that first intense summer for her by One-ear's share of a meager handout of bread. Over the years, I have learned to communicate with One-ear and Grey-face. My body crouched, as if to hop, with arms angled upwards to my head, palms intermittently twitching outwards to face her—occasionally mimicking a magpie's warble or kookaburra's laugh, or perhaps a threatening human sound like a car's engine or chainsaw. Each kangaroo ear can independently sweep around a 150-degree arc to maximize protection from possible predators. The glances and gazes remain—the eyes mean a lot! We also share a limited range of oral communications. The "dsttdd-dsttdd-dsttdd" to say "hello." Or the "uerrrgggggghhhh-uerrrggggghhhh" that I sometimes "grunt" (like an unwanted buck) to discourage One-ear's "overly aggressive" feeding from my hand. Over the years, I've taken a few heart-felt scratches on my wrists as she paws for the wet bread. Her nose twitches in my presence; why I don't know and when she is close I detect smells I've never encountered mixed with the unmistakable traces of urine. Again, I don't care; nor does she.

A now almost independent Grey-face occasionally follows One-ear on to the more-parched clay of our place. But I'm forgetting, as we are prone to do. It's not really our place, is it? It's an ongoing negotiation over time-space of our (currently) lived senses of time, space, and bodies represented as in-between nature and culture. Our place slowly started a long, long recylical time ago with/in the original Jaara Jaara's Country and the later Dja Dja Wurrung language group, even well before them in nature's own time-space and, most recently in the accelerated linear time and "place" of the first whites who colonized the area/land and called it "Mandurang" (the indigenous term meaning black cicada) in 1852 for agricultural purposes so as to provide food for the gold diggers and that riches "rush" and crash into country and nature. So "our place" might only be as aesthetically, ethically, and politically good as the sensitive and sensible custodianship offered by I/we to that time-space.

Before the ghastly, tragic, deathly fires we had experienced a relentless "record-breaking" three consecutive days measured by scientific means above 43 degrees Celsius. Those normally cool, locally handmade reds my house is built from eventually uncomfortably soak up such a prolonged heating; but we will not buy an air conditioner. My body is a better measure of the combination of ambient air temperature, hot northerly breeze, changes in humidity, ground conditions, cloud cover, and solar intensity. The dam is dry, cracked and wrinkled as each minute clay grain is sucked even drier and smaller of any remnant moisture. As it does with my lungs when I venture for more-than-a-few-minutes in reexploring the national park place surrounding "my" home. Yet, the climate change skeptics prevail politically while Copenhagen is a global and moral failure and betrayal. I still throw out some wet bread. And One-ear is still panting and dribbling a great deal. Almost finally, I never see One-ear and Grey-face together. Another mother with joey in-pouch has taken their place, there is no rain or grass, but will I feed them?

Ecophenomenology as Normative Reflexivity

A specific aim of the above *story* is to tap into the reader's sensitivities and sensibilities about our individual and collective *being "of"* nature as well as in, with, about, or for particular aspects of the "environment," a term that just doesn't do justice to what I'm trying to express and reveal. In general, I describe a deeply felt and somewhat historical relationship of a number of *beings*, animate and inanimate, but always "living." The felt and lived episodic interactions between One-ear and I started spontaneously in the present on the urban/rural and culture/nature "edge" with a repertoire of face-to-face glances, as they occurred in a particular site called home.[2] Over more time, the *ofness* of our human and animal natures described above has ultimately anthropomorphized into one of those derivative forms of a once "nature" and "country" now fragmented into many "natures" by the dehumanizing and disenchanting excesses of modernity—the possessively "our place" that surrounds the house and home in which I live and fleetingly *belong*, as it is shared temporarily with One-ear and Grey-face.

The first narrative, and three others following, are an attempt to recapture and sequentially represent this ofness of nature as a form of body-time-space *unfolding* of experience and manner of *ecobecoming*. Pedagogically, *ecobecoming*, potentially, is an end-in-view for environmental education and, if so, sets in motion the intellectualized need for revitalized *framings* of inquiry so as to better appreciate and understand that now (un)timely *imaginary* of becoming other/more-than-what we-are now (e.g., Grosz, 1999). As far as texts go, the four narratives included in this chapter mirror a deeper meaning, value and usefulness of the ways inquiry into the *nature of experience* and *experience of nature* and their *relations* advances a vitalistic dimension of environmental education research and, indeed, critical approaches to curriculum theory and pedagogical development.

In particular, the four stories used in this chapter recall and retell the somaesthetic and ecopolitical sensibility about how living, breathing, feeling and meaning/sense-making *beings* are an "inalienable presence" of phenomenal "things themselves" that are "already there" (Merleau-Ponty, 1962, pp. vii–ix). That is, we tend to forget what we are; this chapter partially serves as an (un)timely remembrance of our "ofness" of natures and, therefore, posthuman worthiness of reinclusion in any framing in environmental education research about the researcher and the researched. I locate this vital presence *of* bodied beings and things of nature somewhat pragmatically in the recent social construction of place (Tuan, 1977). In doing so, I stress the *intercorporeality* or intersomacity of animate beings existing of and within proximal aspects of nature, our own and others, be they human and/or more/other/wilder-than-human. My aim, therefore, is to reveal the mutually constitutive "nature," to play on that term, of an ecocentrically disposed and re/soma/moralized imaginary of such still elusive natures via an ecologically attuned interpretation, or *hermeneutic,* of such a soma/intercorporeality and its/their time-space movements, richness, and complexity (e.g., Abram, 1996; Sheets-Johnstone, 2009). The narrative disclosing of the unfolding phenomena of an "ecological" self embodied in and enframed by nature's past, present, and imagined ecological affordances is, for the purposes here, an ontologically *and* epistemologically revealing interpretation of many natures. This phenomenological, epistemological, and ontological, and methodologically inclusive entrée to our *ofness of nature* is barely visible in most current accounts of, and future imperatives for the framing of and methodological development of environmental education research, curriculum theorizing and pedagogical innovation.

An *(eco)*[4] *phenomenological* approach to the (un)timely reframing of environmental education and its research addresses this lack because of phenomenology's historical interest in retrieving the *lived* nature of human experience, the role of perception in sense and meaning-making, and conscience, consciousness and understanding of them, and the responsibilities for such relations such insights might invoke.[5] Mention here of body(ies) and their intercorporeality must be qualified immediately due to the availability of the term "somaesthetics" and the strong contribution it makes to the stronger democracy of fully bodied selves, and others, entailed in its fleshing out in the everyday (e.g., Jay, 2002; Shusterman, 2008). The notion of somaesthetics strongly incorporates the intrinsically meaningful aspects of a renewed human understanding (Johnson, 2008), the reclaiming of the sensuousness of nature experiences (Abram, 1996) and, inevitably, the conceptualizing and contextualizing of a somaesthetics and ecopolitics pertinent to the critical and justice aspirations of environmental education and its research.

Somaesthetics, as a conceptual driver of the four stories used in this chapter underpins the complementary theoretical exegesis here of the role and rationale for ecophenomenology but is responsive to what has become known in some quarters of theory as the "corporeal turn" (Sheets-Johnstone, 2009). And, as I have attempted in a somewhat transdisciplinary manner in each and all of the four narratives, the turn to corporeality invokes other turns now prominent in philosophy and theory, such as the *spatial, wild, and animal.* Each of those turns in theory is gaining some attention in environmental education research but they are not well linked to interpretations or explanations of the felt/lived embodied body and its primacy of movement (Johnson, 2008) in the transdisciplinary narrative form and ecophenomenological framing of inquiry and research pursued here.

Some cautions. The narrative writing, or textualized form of production of knowledge, of my animated or animal-like relationship, in and over time-space and in place with One-ear and Grey-face, is descriptive, expressive, and, unavoidably, reflexive in a normative sense. It is critical of the orthodox and invariably anthropocentric and allegedly neutral meanings attached to phenomenologically descriptive and methodologically bracketed accounts of particular, ordinary lived experience.[6] Normatively, the ecological/ecocentric spin my text introduces for critical purposes blurs the customary phenomenological distinction of the *is* of description and the critical *ought*, the latter unfolding sensibly beyond the pretension of neutrality in each and all of the narrative productions. To those theoretical turns I employ more broadly in amplifying the somaesthetic role of ecophenomenology, methodological innovation can also be added to the framing envisaged here. In phenomenological circles sometimes used in education research, we are indebted to Max van Manen's (1990) formulations of the science of lived, human experience and its associated pedagogical tactfulness. Beyond the still anthropocentric applications of van Manen's work, somatically driven phenomenological innovation consistent with the purposes of this chapter can be found in, for example, Sarah Pink's (2009) approach to the doing of *sensory* ethnography. Pink advises the researcher to use all of his/her body senses in gathering and interpreting data about the phenomena/on under investigation. Pink asks the researcher to do the same with the researched. Moreover, better known to environmental education researchers, David Abram's (1996) notion of *synaesthesia* is a more ecological and *ecocentric* account of the fusion of the senses where that vital, feeling body(ies) of/in nature outlined above in the narrative introduction and can be seen to act as a source and means of datum about self, nature, species, and others, and their relationality. This fusion, seen within the various turns used in this exegesis, stands against the tendencies of modern positivist science to reduce things—be it bodies, the world, nature and so on to, for example, five disaggregated and individualized sensory inputs. *Synamnesia,* Abram reminds us, is the forgetting of the senses and invoking of the loss of meaning and value in what it is to be human. One-ear's story, therefore, is a somaesthetic account of a sensuous and intercorporeal unfolding of our relational and collective being. Or, if you like, a foray into the human condition and its ofness of nature and, for pedagogical tact here, an expression and illustration of one form of eco*becoming.*

The isness of the ofness of experience sensitized by the various turns outlined above gestures to the ecocentric qualification of that somaesthetic experience.[7] The obvious limitation is that such a narrative account of the unfolding of experience and a meaningful ecobecoming can only ever be partially represented in and as (anthropocentric/anthropormorphizing) text because many of the somaesthetic and synaesthetic qualities, moral or somaesth etic intuitions and conundrums, and ecopolitical imperatives of such ontologically attuned inquiry defy that fullest expression in their conversion to the language of environmental education research (e.g., Payne, 2005a). Inevitably, the textual production of that nature/experience with One-ear and its theorization slides inexorably into the *embodied mind* or *mind embodied* tensions that have occupied other researchers grappling with a nondualist and ecological ontology (e.g., Gallagher, 2005; Johnson, 2008; Lakoff & Johnson, 1999; Weiss & Haber, 1999).

This chapter, therefore, offers a starting point only in the quest for an ecocentric framing of ecologically oriented research, its conceptualization, contextualization, representation and legitimation and associated quest for value and usefulness. In summary, the mood, manner, form, and type of researcher reflexivity entailed by the ecophenomenological framing advanced above is neatly introduced by Merleau-Ponty (1962, p. vii). He recommended that we place in "abeyance the numerous assertions arising out of the natural attitude, the better to understand them . . ." so as to reachieve "a direct and primitive contact with the world." Gaston Bachelard's (1958/1964) views are also helpful. He asserted the phenomenological hermeneutic constituted a major critique of conventional forms of knowledge production.

> A philosopher who has evolved his entire thinking from the fundamental themes of the philosophy of science, and following the main line of the active, growing rationalism of contemporary science as closely as he could, must forget his learning and break with all his habits of philosophical research, if he wants to study the problems posed by the poetic imagination (p. xv).

One-ear's story, hopefully, is descriptively and evocatively *critical* in two senses if, indeed, a renewed imaginary in environmental education research is now needed (McKenzie, Hart, Bai, & Jickling, 2009). Or there is a need to reconcile the means and ends of research development in environmental and sustainability education (Reid, 2009). The main warrant of ecophenomenology and the turns suggested above supporting such a frame is, emphatically, a need to better understand the ecocentric nature of experience. This need becomes clearer if there is a willingness or concession to suspending various assumptions in reachieving or retrieving, or reimagining and reclaiming our relationships (or connections, or place) with/in/of nature. Such a concession enables us to pursue the ontologically and epistemologically *revealing* research reframings and methodologies that help transfer inquiry into narrative productions about such *unfolding* and *ecobecoming*.[8]

Why critical? First, in moving slightly beyond the normal researcher bracketing of a mere description only of the intercorporeal essence or nature of the interactions between me and kangaroos the normatively reflexive eco-textualization and historicizing of that experiencing hopes to tactfully (van Manen, 1990) arouse "other" wise in the reader a range of intuitions, sensitivities, memories, and, inevitably, sensibilities. The story is pedagogical and political, even to the extent that it highlights the problematic of the consequences of anthropogenic climate change on the manner in which we live in "place" given the complexity of the social representations of the phenomena of climate change (e.g., Gonzalez-Gaudiano & Meira-Cartea, 2010). The revealing and reclaiming I undertake is not so much about One-ear and prevailing drought conditions in most of Australia,[9] both unfamiliar to many readers, but more so about the unfolding *relational,* and *ecological* dimensions of the intercorporeal dimensions of our individual *being of nature* and collective *ecobecoming in "natures"* (e.g., Embree, 2003). Bodied, synaesthetic accounts of our individual, collective, and historical being and becoming in time-space are, invariably, missing from the discourse of education but are, indeed, belatedly finding a more *eco*aestheticized presence in the discourse of environmental education research (e.g., Dunlop, 2009).

Second, an ecocentric approach flagged by the inclusion in environmental education research of ecophenomenology's interest in the body in time-space, through the corporeal, spatial wild and animal turns, supported by sensory-methodological innovation hinted at above is highly suggestive of a significant move in the framing of research that goes well beyond the emphasized orthodoxy of heavily rationalized approaches to research, treatments and analyses of data as well as mind-focused teaching, learning and knowing, and their corresponding research designs in (environmental) education. If so, ecophenomenology, as suggested here in the form of an ecopedagogical narration of human-environment *unfolding* and *ecobecoming*, via four limited stories, signals for inclusion a broader and critical role for a richer, less instrumentalized, more somaestheticized meaning-making purpose of education.

Third, to be sure, the injection of a critical purpose in an ecophenomenology and intercorporeal manner is unabashedly soma-political for us academics/researchers. Deconstruction via the linguistic and textual turn in theory best represented by poststructuralism is a valuable intellectual strategy in decentering imagining differently and incorporating the "other" whose presence in such texts is too often absenced, and will be addressed later. This line of thought has a primary emphasis on texts but may, however, have little appeal to the researched, or learners and the public beyond those occupying the academic role (Archer, 2000). Its hopeful praxis remains unclear in environmental education research. To be clear, ecophenomenology, as a bodied, experiential, and methodologically lived negation of those dualisms that occupy much textually focused deconstructive attention can be located sensibly and, eventually, conceptually in an amalgam and reassemblage of the corporeal, spatio-temporal, animal and linguistic turns in theory (e.g., Grosz, 2004; Law, 2004). The stories told then, by both researchers and participant inquirers, may be far more accessible to a much wider audience. Stories have effect as well as value. The intersection of these turns in an ecophenomenological de/reconstruction promises a more ecological, democratic, and everyday form of intercorporeal responsiveness, relationality, and praxis. It troubles our currently problematic of *being* a researcher of the researched (e.g., Neilson, 2010), as flagged in the approach to and narrative representation of the lived nature of an unfolding relationship of *becoming* more/other with, for example, One-ear.

Leucy as Human-Induced Climate Change

Sandi Kogtevs is a wildlife rehabilitator who lives near fire-affected areas. Leucy—a twelve-year-old female kangaroo, was held at the animal shelter for some unexplained reason. Leucy followed Sandi everywhere. If (human) visitors arrived, Leucy would stand with them in a circle "waiting to have her photo taken—she even developed that[10] 'smile' for the occasion!" On that terrible bush fire day in which so many lives were lost and damage to nature occurred, the screaming wind drowned out Sandi's calls to her. Two days later, Sandi found Leucy—terribly weak, with badly burned paws and tail, in the scorched tree forest, her young joey nearby, dead but not burnt. A grieving Sandi somehow managed to carry the large and heavy Leucy back to the home. Despite constant care and hand feeding of grapes and other special foods, Leucy stopped eating due to a caustic smoke inhalation affected throat. Sandi sat with her till she died (Decortis, 2010).

Ted Toadvine's (2009, p. 6) *Merleau-Ponty and the Philosophy of Nature* addresses three interrelated questions. They are, "what is the nature of experience?" and "what is the experience of nature?" and "what is their relation?" Their nexus around the *relational ofness of the nature of experience* reminds us of Dewey's (1938/1988, p. 31) prescient call for an "intelligent theory" or "philosophy of experience" in education. Those early formulations in the 1970s of environmental education emphasized notions like experiential learning, education, and interdisciplinarity. The still unfulfilled challenge of developing a theory of experience in education for different contexts and circumstances remains critically relevant to researching the policy, curriculum, and pedagogical assumptions and imperatives of environmental education for what they say, or don't, about the meanings of "experience" and the "learning" role and value of it (e.g., Rickinson, 2001; Rickinson, Lundholm, & Hopwood, 2010). The interdisciplinary task of theorizing experience remains elusive, or is avoided, even by specialized advocates of experiential education who now concede this lack or gap (e.g., Fox, 2008; Itin, 2008), let alone by those in environmental education who might promote a needed ecocentric vision of educative experience, as pursued here that brings the ecophenomenology and anthropology of experience (e.g., Abram, 1996) into dialogue with Dewey's call and the various turns recommended here in, for example, descriptively revealing the soma-time-space *unfolding* of experience with One-ear. In that textualized representation of knowledge, I have tried to represent an (my) episodic and serial somaesthetic experience of nature as a nexus of the *nature of experience* and the *experience of nature*, following Toadvine, in the specific living of that relational experience over time-space but in the now socially/textually constructed context of place.

Notwithstanding the limitation of such knowledge representations, all of the narratives in this chapter highlight the ontology-epistemology interfaces Toadvine directs us to, but in the ecological sense of those interfaces. They, it must be said, resist the dominant (Western) ontological presumptions rife in education research that split mind-body and I–we–world into separate entities. This enduring Cartesian mindset hampers our best efforts to practically foster or conceptually imagine ecocentricity along the somaesthetic and inter/transdisciplinary lines described above while also negating efforts to reconcile theory-practice gaps in curriculum and pedagogy. Hence, the narratives in this chapter are part of a Deweyan-like experience theory concerning the nondualist intersections of experience and nature. They are indicative of a phenomenological ecology (Brown & Toadvine, 2003; Toadvine, 2009). The same sort of logic can be found elsewhere in curriculum theory and pedagogical development in what is a "critical ecological ontology for inquiry" (e.g., Payne, 1997, 2006). The mutually constitutive somaesthetic ecologies, ontologies, and epistemologies are discernible everywhere in the urban everyday where traces and memories of nature live resiliently (e.g., Lefebvre, 2004; Lingis, 2007) and are available for critical investigation and revealing. Working continually toward these deeper layers of ontological unfolding, or corporeal phenomenology as a phenomenological ecology underscores the broader ecocentric purposes of this chapter about the pressing need in environmental education research, curriculum theory, and ecopedagogical innovation to incorporate the lived somaesthetics and ecopolitics of ecologically attuned and afforded experiences.

Finally, by way of lengthy introduction to the role, potential and rationale for ecophenomenological framings, it is important to acknowledge that some undertheorized aspects of it can be found in the historical conceptions and constructions of environmental education and its research.[11] One key example is the action research and participatory processes, principles, and praxis emphasized in the socially critical perspective (e.g., Fien, 1993; Robottom, 1987). Later, Ian Robottom and Paul Hart (1993) outlined the initial case for incorporating ontological considerations into the nascent scholarly concerns the field had about the role of epistemology and methodology in knowledge production efforts. Traces of the ecophenomenological appreciation of the deeply felt/lived nature of (environmental) experience can be found in the "significant life experience" literature (Chawla, 2002; Palmer, 1998; Tanner, 1980). "Currents" or undercurrents of it can also be found in Lucie Sauvé's (2005)[12] mapping of pedagogical movements in the field. Methodologically emergent indicators are displayed in autoethnographic "memory work" (e.g., Kaufman, Ewing, Hyle, Montgomery, & Self, 2001), ethnography (Blum, 2008), narrative inquiry (Hart, 2003), and autobiography (Doerr, 2004). A handful of publications employ ecopoetical, ecocomposition, ecoart, and/or ecocriticism-like methods and representations. Poetic/illustrative examples of nature as lived sensuously and, therefore relationally are occasionally published (Berryman, in Sauvé, 2005; Cole, 1998; Dunlop, 2009; Jardine, 1998) but, self-consciously, avoid the theoretical exegesis included here. These more aesthetic ends-in-view for research in environmental education remain a challenge (Reid, Payne, & Cutter-Mackenzie, 2010).

Notwithstanding critiques of some of the above *positionings* in environmental education research, the hidden or nascent value of those studies that have focused broadly on experience(s) but perhaps not the nature of it/them, lies in conceding the place of perceptual, embodied, and relational experiences in settings, places, and versions of nature of importance to the aspirations of the field. The problem is not so much the critique of these positions but more so the need to get at the relations between the nature of experience and experience of nature, following Toadvine, given that it has not properly been examined, articulated or represented and theorized in environmental education. This comment clearly presumes the want of a still absent ecocentric theory of experience, following Dewey in dialogue with Toadvine and Abram, from the *inside* of environmental education by the community of environmental education scholars, researchers, pedagogues, and practitioners (Reid & Scott, 2006). More generally, it also suggests the difficulty we have in shedding or suspending the exclusive anthropocentric assumptions and intentions that dog the field, carefully noting Merleau-Ponty and Bachelard.

The methodological, interpretive, and representational creation of a lived duality of ontology and epistemology called for above and illustrated empirically (and somewhat autoethnographically and ethnographically) via the narratives told about One-ear, Grey-face, and Leucy go beyond the sources suggested immediately above. The empirical quest of "soma-time-space" theory and philosophy of experience/nature, crucial to the way curriculum and pedagogy might be reconceived, constructed, and critiqued, is more-than-challenging but possible. Thus, the value and probable metausefulness for the purposes of this chapter of conceiving the ecophenemenological narratives and exegesis as an "assemblage" (Law, 2004) of an imaginary framing device that is located heuristically within the various soma/eco-oriented turns outlined above as they, in turn, might critically elaborate and resensitize conventional interpretive methods like [sensory] ethnography, narrative inquiry, collective memory work, autoethnography, ethnography, and literary approaches like environmental criticism and ecocomposition in anticipation of a very somaesthetically revitalized environmental ethics and ecopolitics, or "ecoliteracy."

Mindful, therefore, of Robottom and Hart's (1993) prescient call for including epistemological and ontological considerations in the socially critical approach to research (see also, Lotz-Sisitka, 2009; Price, 2009) it is important to note, in summary, where environmental education research as a metanarrative/story currently sits ecophenomenologically, following Merleau-Ponty and others. Martin Heidegger (1927/1962, p. 60) preferred, "Only as phenomenology, is ontology possible." Heidegger (p. 62) went on to argue that the way of accessing and revealing, or making known, an ontology of being was that its phenomenology is a *hermeneutic* of "working out the conditions on which the possibility of any ontological investigation depends." Paul Ricouer's (1992) *Oneself as Another* is illustrative.[13] Thus, in this chapter, the (un)timely but overdue employment of an ecophenomenological framing for research responds in a partial but focused way to those difficult challenges listed above via the complementarities of the four narrative and theoretical exegesis. Outside the limited presence of ecophenomenology in environmental education research noted above, excellent examples of this type of inquiry and representation of experience lived can be found, for example, in Alphonso Lingis' (2007) recent work on the "first person singular." He writes poetically about being here, time, about space, about the everyday and its urbanity or cosmopolitanism, about voice and visions in ways that crystallize the intentions of this chapter but in disparate ways.

In all, Toadvine's (2009, p. 6) ecophenomenology calls for a richer "philosophy of nature that includes ontological, epistemological, aesthetic and theological dimensions." There are many interesting challenges for environmental education research. An ecophenomenological orientation to framing of research (and critical interrogations of curricula, pedagogy, texts, etc.) does indicate a "methodological" approach and means of ontologically and epistemologically accessing the underlying ecocentric nature of our experiences of natures. Additional pointers are still needed to return this text to those normative grounds of a relational experience as they were lived ecologically in my everyday encounters with One-ear and Grey-face. Here, I am drawing some threads of the exegesis back into the illustrative narratives as part of the anticipation of another remembrance of Merleau-Ponty's (2003) notions of the chiasm and flesh.

One pointer most worth highlighting is the *feelingness* of our vital, animate, intentional being (e.g., Bai, 2009) intertwined with the "other" of various natures, afforded through their times, according to the spaces presenced and absenced and the becomings opened "wildly" (Griffiths, 2006) or "ferally" (Fawcett, 2009) in the experience of such things. This sensuously wild/other ofness of nature and the world, in the contexts along the spectrum of the natural and cultural represented here, is the deeply felt, raw primordiality of the intercorporeal and cross species relations of human self and kangaroo placed within the proximal temporal and spatial conditions, as theorized narratively through Merleau-Ponty, Heidegger, Dewey, Toadvine, Lingis, Johnson, Shusterman, and Sheets-Johnstone, to name a few. This pointer to a resensitized primordiality of our *unfolding* condition of our *being* highlights the significance of the ecophenomenology framing as a necessary condition of an *ecobecoming* that is eventually supported by the textually constituted act of offering a normatively reflexive description of my/our experienced relations of/in/with/for nature. That is, for environmental education research, the incorporation of the ecophenomenological framing is a candidate for stronger "commensurability and coherence" of the purposes, means and ends (in view) of such research, and is responsive to some longer term concerns expressed about the field's development (e.g., Hart, 2005; Reid, 2009).

Absence

Winters, as they are so named, at our place (sic) partially disrupt the persistence and immediate consequences of the dry summer heat. Some rain temporarily disrupts the anthropogenic drought. One-ear and Grey-face's summer presencing turns to absence. In the gullies, on the spurs and amongst the trees around this cultural space, there is now plenty of grass for them to graze upon, to replenish, renourish, and to renew. I don't see that but I sense it. My presence isn't needed. Absence and silence about such things like growth and death reminds us of the importance of such things. I don't miss One-ear and Grey-face because I know they are there, somewhere, living out their lives in ways that I don't need to comprehend but, if I try, can imagine. The harshness of sound from the overflying sulphur-crested cockatoos, and echoed laughter of kookaburra pairs up and down the valley, in the dimmed winter light contrasts sharply with my socially constructed tuning of what counts as music. Occasionally kangaroo mobs bound hastily through the ground-level of our place, or is it sp(l)ace and I keep an anticipatory eye out for the momentary prospect of a reunion with their being, or with One-ear and Grey-face. But, if not, that promise can wait to next summer's constancy of their droughted presence. The moistness of the now damp surrounds to our sp(l)ace has a peculiar aromatic that contrasts with summer's delicious lemon-scented gumness—but is best not to consider, too much, because their seasonal time will come.

We feel things unfold in absence that presence doesn't always reveal. We imagine that which isn't there—fondly; we become other for time and memory are always present. I chat with my daughter, and some colleagues, about these things. Its inadequate because the ecophenomenological/ontological construction ecocentrically found in these things, and their intercorporeal relationality, eludes or evades this text. But, at least, language helps make the partial case for that which eludes it.

Editorially and interpretively, the above narrative pushes beyond the animality and cross-species focus privileged in the two earlier narratives. One-ear and Grey-face are still (textually) positioned among other things. The "land" and its scape, and history, or my/our sp(l)ace, and there presencing and absencing of things in themselves are equally deserving of attention, but it is hard. The first narrative established the significance of the moral proximity of the "face-to-face"—the glancing, gazing, and rehearsing of inter/soma/corporeal nature of experience with One-ear. It was alert to Levinas's prioritizing of a preontological *being for* (Llewelyn, 2003; Payne, 2009). The *ofness* of that *being for* is experientially crucial to the "horizon" of how that initial ecophenomenological framing can be stretched imaginatively, temporally, and spatially, into the *unfolding and ecobecoming* emphasized now. And depicted above in reintroducing, for example, how presence, nonpresence, absence, and silence might be incorporated "other" wise as things too in the textualizing of an ecocentric somaesthetics of experience.

Flagged earlier, the later Merleau Ponty's (1968, 2003) ontologically central but elusive notion of the *chiasm* is conceptually important. It helps make, again, the case to consider the normativity of reflexivity in inquiry, research, and critique (and their representations) appropriate to exploring the pedagogically tactful question of incorporating presence and absence in the stories we might narrate about *how* we might become more-than-what-we-are-now in relation to various natures, such as sp(l)ace. Merleau-Ponty's chiasm is an "intertwining" or "overlapping and crisscrossing" (Carman, 2008) of the preconscious perceptual and intentional ofness of our bodies and world, as those terms illuminate notions like the "body schema" (cf. image) and *flesh* in orienting our bodied selves (with)in various ecological affordances "out there," both presenced and absenced. The opening narrative is as suggestive as words/texts can be of this presenced ontological ofness of beings where the lives of my "self" and One-ear twine or intersect into each other and episodically overlap and crisscross according to the presenced affordances in which we "found" our selves, primarily in sp(l)ace. The narrative above stretches this chiastic and fleshy presence into their temporal-spatial absences, silences, and other potential "others/wilds" that phenomenologically and ontologically provide keener insights into our soma/ethical comportment, provided we are enabled and open to it. Nikolas Rose's (1996) call for a reassembling of the ontological self in research, alongside John Law's (2004) method assemblage mentioned earlier become useful additions to the ecophenomenological framing.

Law (2004, p. 161) is unequivocal: method assemblage is, "the process of crafting and enacting the necessary boundaries between presence, manifest absence and Otherness." Method assemblage is generative or performative, producing absence and presence. More specifically, it is the crafting or bundling of relations in three parts: (a) whatever is in-here or present . . .; (b) whatever is absent but also manifest (that is, it can be seen, is described, is manifestly relevant to presence); and (c) whatever is absent but is other because, while necessary to presence, it is also hidden, repressed or uninteresting.

The opening narrative dwells on Law's (a). The latter story on (b) and (c). It focuses more enigmatically on, for example, the presencing and absencing, seeing/not seeing and hearing/silencing of One-ear in a more spatially and temporally diffuse sense of the horizons of nature(s), as does the yet-to-be-revealed mystery of a concluding narrative in this chapter. Here, in the *unfolding* chiastic of the narrative representations there is a more open/wild/other ecophenomenological approach and ecocentric inclination, given the unfolding of circumstance and context. This opening to the unfolding addresses Merleau-Ponty and Bachelard's (methodological) invitation to reflect upon and "suspend" as best we can the rational assumptions we make about the phenomena under study. There were a number of intuitive, spontaneous, embodied experiences highlighted in the introductory narrative about the presencing of the relation of the experiencing of nature. The above story emphasizes some of these enigmatic "others" found in the phenomenological/existentially experienced feelingness, absencing, invisibility, hiddenness and stillness of things that still reoccur over time-space in the episodic nature of the encounters in sp(l)ace with One-ear, Greyface in nature, or anticipations thereof. Perhaps the gaps and "in-betweens" explored all too briefly here (in the third narrative) and their silences are worthy of much greater attention. If so, the textual representation of absence provided an additional glimpse of the ontology of my *ecobecoming* and its time-space-place *unfolding* as a form of somaesthetic (transformative!) *being* (Maitland, 1995). The nature of those experiences and experiences of nature were, basically, where our human and other than human bodies were rawly presenced and absenced to each other, as intercorporeal, where the politics of silence prevailed (Sim, 2006).

In addition to the stretched ontological politic of reassembling oneself (Grosz, 2004; Rose, 1996) undertaken, normatively and reflexively, about the absences, silences and (un)timeliness signaled above, Brown and Toadvine (2003, p. xi) view ecophenomenology as a retrieval of our forgetting reminds us of another of Dewey's many concerns about the need for an experience theory that contests the stripping of the qualities of experience. Abram's (1996) *Spell of the Sensuous* is a rare treat because his anthropology of experience delves into the sensory basis of our individual and collective *being-in-the-world.* Abram sheds needed light on those Deweyan qualities of experience that we have absenced from our perceptual selves, and how we might go about reembodying and reinhabiting aspects of nature most responsive to the inevitable call for how a philosophy of nature and experience of it might reframe environmental education and its research. Similarly, within that ontological politic, Grosz's (2004, p. 2) account of the *untimely* is a lengthy excursus into matters political. Grosz questioned prevailing notions of agency, structure, geography, and culture. In recanting some of her earlier work in *Volatile Bodies,* Grosz argued that the reintroduction of time and bodies into the politics of ontology and, therefore, phenomenology of experience and nature ". . . serves as a reminder . . . to those interested in feminism, antiracism, and questions of the politics of globalization." To which I would add environmentalism and the fully bodied democracy offered up by somaesthetics (Jay, 2002). Grosz maintained that theorists of the social, the political, and the cultural, notwithstanding the absence of the ecological, ". . . have forgotten a crucial dimension of research" of concepts that inform such politics and serve as a "remembrance of what we have forgotten." With a particular interest in time oriented concepts like *unfolding* and *ecobecoming*, embedded in the scaffolding effect of the narratives presented in this chapter, Grosz's notion of remembrance seems to echo Heidegger's and Ricouer's concern about the importance to phenomenology (of the body, self) for revealing the "conditions" of our being. Grosz's contemporary focus included, "the nature, the ontology of the body, the conditions under which bodies are enculturated, psychologized, given identity, historical location, and agency."

To be sure, the raw experiences of One-ear, the constructed narratives, her absences and silences are emphatic parts of any remoralized and reimagined narrative concept of selfhood that, for example, Alasdair MacIntyre (1984) and Ricoeur (1992), among others, makes the compelling ontologically hermeneutic case for. Thus, to return, ecophenomenological framing method assemblages and constellations of turns in policy, research, curriculum, and pedagogy might well dwell on the presence and absence of such *conditions* and constructions of presences and absences in any unfolding narrative they might seek to tell about the *ecobecoming* of the researcher and the researched.

Ecophenomenology's (un)timely Intervention in Environmental Education Research

While Merleau-Ponty didn't openly address the ethical and political implications of his work that might inform an ecophenomenological framing, others who are engaging ecophenomenology do see the importance of meaning and values existing intrinsically or inherently and, therefore, ontologically in nature (Langer, 2003; Wood, 2003). The momentary "glance," another dimension of the "face" (Casey, 2003) in-between One-ear and myself was repeated on numerous occasions; later as gazes. Grey-face is so named. This chiasmic ofness of our intercorporeal relations might be most what (pedagogically) is needed to flesh out how inquiry might proceed in a normatively reflexive manner—more viscerally, intuitively and perceptually in revealing the moral ontology of a somaesthetics and ecopolitics of, in and for or "facing" nature. Acamporo's (2006) ontology deals with interspecies ethics, part of the "animal turn" in theory and through which I "felt" such a relational ethic (and politic) with One-ear, Grey-face and empathy for Leucy and her "keeper." Toadvine's (2009) effort to reclaim a philosophy of nature is also a startling critique of Western environmental ethics. That field, he proposes, is primarily founded on a combination of (environmental) problem solving pragmatism and logical positivism. As such, he argues, alternatives exist.

Indeed, the ecocentric need for a humanly constructive, socioecological theory of education and research development is being foreshadowed. Outlined above, using an ecophenomenological framing to invoke a method assemblage in lieu of a constellation of various turns in theory, the combination of narratives and exegesis employed in this chapter strenuously incorporates a somaesthetics and ecopolitics in sp(l)ace but also beyond. The historical remembering of our sensuousness and vitality expressed throughout this chapter inform Dewey's ignored call for an intelligent theory of experience. Their recall and reclaiming are highly suggestive of how a resensitized and heightened normative reflexivity might valuably and usefully be approached in critically retheorizing environmental education and its research.

Reid and Scott (2006, p. 332) asked that we reflect on the maturity, dynamics, and balance of the field's research endeavors, identity, value, and usefulness (see also, e.g., Jickling, 2005; Russell & Hart, 2003). They suggest much research is not yet capable of driving needed self and social transformation and environmental change. Reid and Scott identified a number of potential sources of the field's current lacks and gaps. These included the conservative push for policy-driven evidence rather than research that can inform policy; the orthodox planning of research that is safe, convenient, and conventional and is reflected in the modesty of the research questions asked, methodologies employed, and politics of inquiry. Reid and Scott concluded the field's effect on policy and educational reform is not self evident, nor is it used or appreciated in related fields of inquiry and endeavor. They probe the need for "harder-to-reach" varieties of research and theory, and how research impact might follow. This probe draws upon a fundamental tension they discern in the field—its explanatory and interpretive powers (or lack of) being derived from theory drawn from the *outside in* as opposed to the reverse—from the field's more indigenous *inside out*. Here, from somewhere *in-between*, I do both in commencing the work of drawing upon (eco)phenomenology from the *outside* and, following Dewey's *inside* challenge for an intelligent theory in education incorporates basic insights worthy of initial attention in ecocentrically developing a comprehensive theory, or ecology of a somaesthetics and ecopolitics of environmental education and its research, including a reassembling of the researcher and researched.

We might ask, therefore, if there is something distinctive, or *original*, in environmental education and its research that can encourage researchers to generate adequate, valuable and useful ecopedagogies, curriculum theory and views of knowledge production, representation and transfer. This question is posed and probed against the corrosive rise of "postintellectual" *conditions* of our academic and scholarly work (S. Cooper, 2002; James & McQueen-Thomson, 2002) whose effects or consequences in environmental education research have been remarked upon by, for example, Reid and Scott (2006). The question of originality, or quest for it *in* environmental education research, demands a more fully or ecologically responsive accounting for the ecocentric-driven triad of ontology-epistemology-methodology in research and is worthy of both internal attention in the field and broader concern to the development of educational research (Reid, 2009). How might we in environmental education, for example, examine and explain the demise of the concept of (educative) experience in education? What are the qualities of educative experience when we consider the "place" and role of "meaning-making" and its values as part of learning, knowing, thinking (e.g., D. Cooper, 2006)? Or, what are the consequences of a lack of progress in articulating an ecocentric notion of inter/transdisciplinarity and, by implication, addressing it head on in environmental education and its research, noting the glimpses of it in Sauvé's currents?[14]

Most of all given the concerns outlined above about the value and usefulness of environmental education research, it is now (un)timely to "(re)engage the debate" (Robottom & Hart, 1993) about that triad, a meaningful and democratic theory of ecocentric experience and notions of inter/transdisciplinarity in research, policy, curricula and pedagogy. Various constellations of theoretical turns and assemblages of methods might assist. A normatively reflexive ecophenomenological framing will help start up these deliberations because phenomenology's great contribution is that it starts with experience as it is lived primordially–or naturally, perhaps authentically. It tries to go to the core of our being and becoming. If so, such a debate will interrogate the research questions we pose, the tensions between conceptualizations and contextualizations of such questions, the renewed need for curriculum theory and pedagogical development as they might normatively and reflexively connect with bodies, learners, teachers, educators and, more broadly, the public. Reengaging debate is politically important and should most likely be informed by a transdisciplinary notion of ecocentrism, as it is ontologically and epistemologically revealing in experience, perhaps via the framing of inquiry outlined above. Here, I have indicated how such a framing shapes the epistemological and, inevitably, methodological orientations of environmental education research. Inevitably, that debate about the intersections of the nature of experience and experience of nature and their relations will return to examining how the ontology-epistemology-methodology triad might be reworked into our understandings of the experiences and positionings of the researched and the researcher in the research.

A primary purpose of this chapter about the place and potential of ecophenomenology in environmental education research is working the *in-between* spaces between the binary frames of *inside out* and *outside in.* On one hand, we are now grasping normatively and reflexively for an ecocentric inspired theorization of *experience, education*, and, even, *the body* in time-space, in all its soma, moral, social and political comportments. On the other, I argue that the intercorporeal other and wild of *ecophenomenology*, the *sensuous* and *perceptual*, and relationally *more-than-human and animate nature* provides a warrant for an original and potentially indigenous frame for environmental education inquiry, as well as for conceiving and constructing pedagogy, curriculum, and policy. Thus, with the notion of ecophenomenology mostly *in* view, and Dewey's challenge inspiring us, this partial and selective conversation of those two starting points for debate does stake a claim for the importance of an ecophenomenological framing for environmental education research.

Ecophenomenological Framings of Environmental Education Research

Most of all, ecophenomenology provides us with a storytelling frame that reminds us that the somaestheticized and memoried body is a sentient, sensuous, and sensible source, or genesis, of preconceptual meaning-making of, in and about environmental relations. It has a great deal to say ontologically and, therefore, epistemologically about a philosophy of nature. My aim in this chapter is to highlight how this vital setting serves as an active experiential and existential site *of* and for inquiry in and with various natures and environments in which the researcher and researched are positioned, live and feel and *unfold* and, therefore, can interpret the time-space sensitivities and sensibilities of their individual and collective *ecobecoming*. The interpretive and explanatory purpose of combining story as narrative and exegesis is to provide a conceptual resource that helps us (a) appreciate our *ofness* of natures while (b) reversing our alienation and abstraction *from* constructed natures, understood here anthropogenically as the *ecologically problematic human condition.*

As an orientation to, ontologically as well as epistemologically, what we are *of*, our natures, and who we might recover, or reclaim from the bewildering often contradictory barrage of sociocultural constructions and representations, (eco)phenomenology provides an "other" vantage point about the nature of experience of nature. Various epistemologically or/and methodologically prioritized constructions we take as normal or natural might be reflexively and normatively assessed—in the keenest, praxical sense of the commitment of the term critical to various justices, based upon a more liberated, transformative ecological self indicated here. Ecophenomenology promises little more than that start, essentially it nonnaively can act in a soma/intercorporeal manner as a memory of sensuous, animated natures—the human, other-than, and more-than that not only recover the ecology of our *being* in the world and its chronic abstraction (James, 2006) in the modern/postmodern but also and therefore the ecocentric possibility of our *unfolding* and *becoming.* The *slow* notion of an *unfolding ecobecoming* might inform an ecopolitics of ontological resistance to the excesses in (environmental) education research and pedagogical practices of "fast" learning solutions to socioenvironmental problems and issues (e.g., Payne, 1997, 2005b; Payne & Wattchow, 2009; see also Huebner, 1967/1987). They effectively reduce education to the hegemony/authority of rational learning and knowing, and related pedagogies, ordained in the heavily abstracted and disembodied world whose normalization and naturalization we have come to accept. And, if so, hampers the normatively reflexive posing of valuable and useful questions and, inevitably, the commensurability of research purposes, means and ends now quested for in parts of our own discourse. To those ends-in-view for inquiry into and critique of pedagogical, curricula, and policy development, the ecophenomenological framing demands complementary support from a range of critical insights and theories committed to the various justices education might encourage.

It would have been disingenuous to not have acknowledged the limitations of ecophenomenology, and its role and position in environmental education research, as we should expect in positing any theoretical, philosophical, methodological, and pedagogical vantage point. To reiterate, the ecopheneomenological framing is a normatively reflexive starting pointed needed in debating research, how it is framed and conducted, and for what means and ends. Here, the exegesis draws upon a range of turns and assemblages that also are partially embedded in each and over the four narratives. Indeed, ecophenomenology is hardly a method or methodology, as indicated by Merleau-Ponty and Bachelard, among others. It is an approach, a sensitizer, that helps the researcher and researched live the inquiry. And, I have only hinted it through Leucy's testimonial and some existing examples in environmental education research, some of the broader everyday, social, and historical problems and possibilities, potentials, and imaginaries of this ecophenomenological starting point. Otherwise, this chapter might be interpreted as privileging yet another hyper-intensified form of individualism, and methodological individualism. Even methodological nationalism, given the framing here deals with the basis of felt, lived experience irrespective of the various cultural/historical overlays that shape such "experience" and demand description, normative reflexivity and critique. The development of method assemblages and constellations in support of the ecophenomenenological framing outlined here is strongly encouraged, as is the pluralistic perspective of returning ontology and its politics, and relations with epistemological considerations to the center stage of research (Lotz-Sisitka, 2009; Payne, 1997) and debate (Robottom & Hart, 1993) in ways that tackle the means-end questions (Reid, 2009). Indeed, the fuller elaboration of a socioecophenomenology beckons. What this chapter does recommend is the ecocentric need for incorporating far more assertively a transdisciplinary somaesthetics and ecopolitics of educative experience in our scholarly framings, conceptualizations, contextualizations, representations, and legitimations of the purposes, means, and ends-in-view of environmental education and its research.

Grey-Face

Many months later in my time but summer again in seasonal time, time alive might have finished for One-ear. I haven't seen her for some time. I'm saddened. The memory gnaws. Another fully grown kangaroo regularly comes to the front of the encultured time-space I still call a home and a place, an unthinking habit. It too seems bold, just like One-ear. I watch carefully through the window over consecutive hot, dusty evenings. She keeps returning. Only until I go outside, stand still and she doesn't back off, do I think this might be a fully grown Grey-face. Her face and gaze are steady, as are mine. I watch her closely—she peers at me in much the same way One-ear's joey did three summers ago. The peer, not a glance, combines the quizzical and the familiar. She shakes her head and ears . . . again I think the same as some time ago, mimicking her mother, or in response to my headshakes. I approach slowly, once again with body crouched, as before and will next Summer when the dusty clay returns one more time, forever. She is skittish, as I remember her being as a nervous just out-of-the-pouch joey, but far more so than One-ear, notwithstanding her obvious differences to the mob. I hold out some wet bread to see what will happen. She leans forward, slowly and, I suppose, a little more nervously then pushes off from that third leg, her thick tail resting on the dirt. And circles back, sideways, in slow withdrawal then, head and ears turned, inches forward, back towards me—face-to-face, ever so tentatively with those dark little eyes in the face-to-face glancing and gazing—me at her and her at me. This is Grey-face. She eats and wants more, just like One-ear. I'm silent, lost in it all—my body, hers, the memory, One-ear's presence/absence, the time-space. Each evening and day this is repeated and increases as the chronic dry heat bites, sucks, swallows, and exhales.

The unfolding of time-space and bodies of One-Ear, Grey-face and myself "of" this world of natures demand that the first narrative be shifted to the start of this "(un) timely" chapter and, back on the plastic keyboard, replace it with this to bring a temporary closure to an account of the role and justification for ecophenomenology to be included in our thinking, framings, conceptualizations, contextualizations, and representations of research that is, at the one simultaneous time but for and many spaces and scapes, an intersection of the aesthetic, ethical, and political—again not very somaesthetically well for stories, words, and texts are indicators only of that individual, collective, and relational experience of our natures' being and ecobecoming.

Notes

1. Thanks to Paul Hart for his indefatigable support of the writing of this chapter in his role as colleague and section editor; Stephen Smith and Alan Reid for helping me refine or clarify various ideas herein; and Lindsay Fitzclarence for comments on an earlier draft.
2. Casey (2003) makes the preliminary ecophenomenological case for the ethical and political importance of the "glance" in that it provides access to the feeling of the other, senses the less manifest aspects of the other, is an instant sign of what is present and presenced by the other, and allows for exchange and, therefore, gives witness to that other in a way of being for the other best linked with Levinas (Calmarco, 2008). The characteristics of the glance are evidenced in the narrative about One-ear, it being written before this reading and incorporation of Casey's sketch. From a different vantage point, Keltner's (2009) science of a meaningful life points to the "goodness" of the face and gesture.
3. From the Greek, *soma,* meaning of the body. Somaesthetics will be used throughout this chapter to represent the intersection of the bodied aesthetic/feelingness and its moral bearing or *being for* as an antecedent for the taking of ethical responsibility for the other (see also Payne, 2009).
4. Eco, as a prefix used in various expressions like "ecological consciousness" signals the ecocentric qualification of terms commonly used in education, like "literacy" and "pedagogy" Pedagogy typically represents the transaction of teaching/learning in an orthodox anthropocentric manner of human to human interactions and relations (i.e., teacher and child). If used without qualification in environmental education, the term continues to anthropomorphize the relational nature of humans and their other-than-human beings. In this chapter, the constant use of eco (and soma) flags the need for greater textual alertness, at the very least, as well as interpretation by the reader so as to move us ecocentrically and "posthumanly" to, possibly, *become* more-than-what-we-currently-are.
5. See Husserl, Heidegger, Sartre, Merleau-Ponty, and Levinas as the key proponents of phenomenology in the 20C.

6. Numerous other narratives surround those included here. I could tell the story of children ("alien" to Australia) visiting our place and spending many hours playing, exploring, and canoeing around an island in the dam, from which the kangaroos drink. They found poisonous redback spiders, marveled at the bird-life, waded in muddy clay, all the time smiling. Or my ailing neighbor whose old dog I take for a walk around her property. "Wags" can't chase the kangaroos but still manages an occasional bark.
7. Acampora's (2006) phenomenological philosophy of the body, or corporeal phenomenology, is one example of how the ontological and epistemological intersections of our animality with the animality of others can give "common" rise to an animal and interspecies ethics. More broadly for the "other" of natures see, for example, Lingis (2007).
8. Complementary normatively reflexive and ideological/textual/praxical critiques in environmental education research—about the "everyday," incorporating the sociocultural-global-ecological, intergenerational, education policy, curriculum, and pedagogical—are desperately needed for educationally ecocentric ends-in-view and to which the incorporation of ecophenomenological framings are crucially important, if "assemblages" (Law, 2004, see later) inclusive of "turns" in theory is a precondition).
9. Notwithstanding anthropogenic climate change, see above!
10. A "full face" photo of Leucy accompanies Decortis's (2010) article.
11. Progress on ontology, ecocentrism, and somaesthetics has been relatively slow in the general field of education, despite a history of phenomenology in education than is longer than what we think (e.g., Huebner, 1967/1987; Kneller, 1984). Pinar and Reynold's (1992) edited collection about the importance of phenomenology to curriculum as "text" is one of the more significant contributions of this genre to inquiry. Norman Denzin and Yvonna Lincoln's (2005) otherwise magisterial scoping of qualitative research in the social sciences included in its fourty-four chapters only one that is dedicated to the environment. Brady's (2005) chapter champions a "poetics" view of planetary being, where the poetical provides a hermeneutic style of writing for meaning-making, less literal, and sympathetic to the phenomenological tradition and its interest in alternative forms of textual representation. Lincoln's own contribution to the case for qualitative research addresses the "challenge to and from the phenomenological paradigms," where her use of the term phenomenology is used in a manner so generalized that the myriad meanings of phenomenology are obscured. Nonetheless, Lincoln's main argument supports the contention of this chapter about the importance of phenomenology and its politics and, by implication, the necessity and importance of ecophenomenology to environmental education and its research.
12. Sauvé's (2005) mapping of fifteen currents in environmental education is a useful indicator of the relative status of different versions of environmental education. In her mapping of those currents, three are identified as being dominated by "sensorial" and "experiential" means. One other is classified as "holistic, organic, intuitive, creative" which, presumably, includes the sensorial and experiential while two others are identified as "praxic" which is probably experiential and, possibly, sensorial. Some others include "affective," and "spiritual" but are not necessarily experiential or sensorial. Sauvé's mapping is both useful and ambiguous, and further development is encouraged. Overall, it seems as if at least six of the currents emphasize the "direct" sensorial and (temporarily) "lived experience" of an environmental education whose "roots," arguably, lie in a phenomenological tradition, approach or framing. Yet, the number of published works identifying or reflecting an (eco)phenomenological framing is very small. Nine currents are identified by Sauvé as stressing "cognitive" development. A large number of published works substantively and methodologically address this presumption of rational "cognitive growth" and "knowledge" outcomes for environmental education and from environmental education research.
13. See, Ricoeur's (1992, pp. 297–356) account of an ontology of self whose phenomenological investigation intersects three problematics—reflective analysis, a determination about selfhood by way of its contrast with sameness and dialectic with otherness, where self-interpretation is an unfolding and a phenomenology of the body provides the recourse for otherness.
14. This tokenism, and reductionism of the body in education, is ironic. Embodied, experiential meaning-making is, undoubtedly, a precursor of cognitive "learning" and knowing (e.g., Gallagher, 2005). While cognitive learning is emphasized in a number of Sauvé's (2005) currents, meaning-making, meaningfulness and "meaning" sourced in the bodily relation with time-space is rarely included in the way we are (rationally) led to nonsensuously think about pedagogy, educating and researching.

References

Abram, D. (1996). *The spell of the sensuous: Perception and language in a more-than-human world.* New York, NY: Pantheon.

Acamporo, R. (2006). *Corporal compassion: Animal ethics and philosophy of body.* Pittsburgh, PA: University of Pittsburgh Press.

Archer, M. (2000). *Being human: The problem of agency.* Cambridge, UK: Cambridge University Press.

Bachelard, G. (1964). *The poetics of space: The classic look at how we experience intimate places.* Boston, MA: Beacon Press. (Original work published 1958).

Bai, H. (2009). Reanimating the universe: Environmental education and philosophical animism. In M. McKenzie, P. Hart, H. Bai, & B. Jickling (Eds.), *Fields of green: Restorying culture, environment, and education* (pp. 135–152). Cresskill, NJ: Hampton Press.

Blum, N. (2008). Ethnography and environmental education: Understanding the relationships between schools and communities in Costa Rica. *Ethnography and Education, 3*(1), 31–48.

Brady, I. (2005). Poetics for a planet: Discourse on some problems of being-in-place. In N. Denzin & Y. Lincoln (Eds.), *The sage handbook of qualitative research* (3rd ed., pp. 979–1026). Thousand Oaks, CA: Sage.

Brown, C., & Toadvine, T. (Eds.). (2003). *Eco-phenomenology: Back to the earth itself.* Albany, NY: SUNY Press.

Calmarco, M. (2008). *Zoographies: The question of the animal from Heidegger to Derrida.* New York, NY: Columbia University Press.

Carman, T. (2008). *Merleau-ponty.* London, UK: Routledge.

Casey, E. (2003). Taking a glance at the environment: Preliminary thoughts on a promising topic. In C. Brown & T. Toadvine (Eds.). *Eco-phenomenology: Back to the earth itself* (pp. 187–210). Albany, NY: SUNY Press.

Chawla, L. (Ed.). (2002). *Growing up in an urbanizing world.* London, UK: Earthscan.

Cole, P. (1998). An academic take on "indigenous traditions and ecology." *Canadian Journal of Environmental Education, 3,* 100–115.

Cooper, D. (2006). *A philosophy of gardens.* Oxford, UK: Clarendon Press.

Cooper, S. (2002). Post-intellectuality? Universities and the knowledge industry. In S. Cooper, J. Hinkson, & G. Sharp (Eds.). *Scholars and entrepreneurs: The university in crisis* (pp. 207–233). Carlton, Melbourne, VIC: Arena Publications.

Decortis, M. (2010, Autumn). *Wild matters* (p. 19). *Greens Victoria News.* Melbourne, VIC: The Greens.

Denzin, N., & Lincoln, Y. (Eds.). (2005). *The Sage handbook of qualitative research* (3rd ed.). Newbury Park, CA: Sage.

Dewey, J. (1988). Experience and education. In J. A. Boydston (Ed.). *The later works, 1925–1953: John Dewey* (Vol. 13, 1–62). Carbondale, IL: Southern Illinois University Press. (Original work published 1938.)

Doerr, M. (2004). *Currere and the environmental autobiography: A phenomenological approach to the teaching of ecology.* New York, NY: Peter Lang.

Dunlop, R. (2009). Primer: Alphabet for the new republic. In M. McKenzie, P. Hart, H. Bai, & B. Jickling (Eds.), *Fields of green: Restorying culture, environment, and education* (pp. 11–63). Cresskill, NJ: Hampton Press.

Embree, L. (2003). The possibility of a constitutive phenomenology of the environment. In C. Brown & T. Toadvine (Eds.). *Eco-phenomenology: Back to the earth itself* (pp. 37–50). Albany, NY: SUNY Press.

Fawcett, L. (2009). Feral sociality and (un)natural histories. In M. McKenzie, P. Hart, H. Bai, & B. Jickling (Eds.), *Fields of green: Restorying culture, environment, and education* (pp. 227–236). Cresskill, NJ: Hampton Press.

Fien, J. (1993). *Education for the environment: Critical curriculum theorizing and environmental education.* Geelong, VIC: Deakin University Press.

Fox, K. (2008). Rethinking experience: What do we mean by this word "experience"? *Journal of Experiential Education, 31*(1), 36–54.

Gallagher, S. (2005). *How the body shapes the mind.* Oxford, UK: Clarendon Press.

Gonzalez-Gaudiano, E., & Meira-Cartea, P. (2010). Climate change education and communication: A critical perspective on obstacles and resistances. In F. Kagawa & D. Selby (Eds.), *Education and climate change: Living and learning in interesting times* (pp. 13–34). London, UK: Routledge.

Griffiths, J. (2006). *Wild: An elemental journey.* London, UK: Penguin.

Grosz, E. (1999). (Ed.). *Becomings: Explorations in time, memory, and futures.* New York, NY: Cornell University Press.

Grosz, E. (2004). *The nick of time: Politics, evolution, and the untimely.* Durham, NC: Duke University Press.

Hart, P. (2003). *Teachers' thinking in environmental education, consciousness and responsibility.* New York, NY: Peter Lang.

Hart, P. (2005). Transitions in thought and practice: Links, divergences, and contradictions in post-critical inquiry. *Environmental Education Research, 11*(4), 391–400.

Heidegger, M. (1962). *Being and time.* New York, NY: HarperSanFrancisco. (Original work published 1927).

Huebner, D. (1967/1987). Curriculum as concern for man's temporality. *Theory into Practice, 26*, 324–331.

Itin, C. (2008). Reasserting the philosophy of experiential education as a vehicle for change in the twenty-first century. In K. Warren, D. Mitten, & T. Loeffler (Eds.), *Theory and practice of experiential education* (pp. 135–148). Boulder, CO: Association for Experiential Education.

James, P. (2006). *Globalism, nationalism, tribalism: Bringing theory back in.* London, UK: Sage.

James, P., & McQueen-Thomson, D. (2002). Abstracting knowledge formation: A report on academia and publishing. In S. Cooper, J. Hinkson, & G. Sharp (Eds.), *Scholars and entrepreneurs: The university in crisis* (pp. 183–206). Carlton, Melbourne, VIC: Arena Publications.

Jardine, D. (1998). Birding lessons and the teachings of cicadas. *Canadian Journal of Environmental Education, 3,* 92–99.

Jay, M. (2002). Somaesthetics and democracy: Dewey and contemporary body art. *Journal of Aesthetic Education, 36*(4), 55–69.

Jickling, B. (2005). A decade past and a decade to come. *Canadian Journal of Environmental Education, 10,* 5–7.

Johnson, M. (2008). *The meaning of the body: Aesthetics of human understanding.* Chicago, IL: The University of Chicago.

Kaufman, J., Ewing, M., Hyle, A., Montgomery, D., & Self, P. (2001). Women and nature: Using memory work to rethink our relationship to the natural world. *Environmental Education Research, 7*(4), 359–377.

Keltner, D. (2009). *Born to be good: The science of a meaningful life.* New York, NY: W. W. Norton.

Kneller, G. (1984). *Movements of thought in modern education.* New York, NY: Macmillan.

Lakoff, G., & Johnson, M. (1999). *Philosophy in the flesh: The embodied mind and its challenge to Western thought.* New York, NY: Basic Books.

Langer, M. (2003). Nietzsche, Heidegger, and Merleau-Ponty: Some of their contributions and limitations for environmentalism. In C. Brown & T. Toadvine (Eds.). *Eco-phenomenology: Back to the earth itself* (pp. 103–120). Albany, NY: SUNY Press.

Law, J. (2004). *After method: Mess in social science research.* London, UK: Routledge.

Lefebvre, H. (2004). *Rhythmanalysis: Space, time and everyday life.* London, UK: Continuum.

Lingis, A. (2007). *The first person singular.* Evanston, IL: Northwestern University Press.

Llewelyn, J. (2003). Prolegomena to any future phenomenological ecology. In C. Brown & T. Toadvine (Eds.), *Eco-Phenomenology: Back to the earth itself* (pp. 51–72). Albany, NY: SUNY Press.

Lotz-Sisitka, H. (2009). Why ontology matters to reviewing environmental education research. *Environmental Education Research, 15*(2), 165–176.

MacIntyre, A. (1984). *After virtue: A study in moral theory.* Notre Dame, IN: University of Notre Dame.

Maitland, J. (1995). *Spacious body: Explorations in somatic ontology.* Berkeley, CA: North Atlantic Books.

McKenzie, M., Hart, P., Bai, H., & Jickling, B. (Eds.). (2009). *Fields of green: Restorying culture, environment, and education.* Cresskill, NJ: Hampton Press.

Merleau-Ponty, M. (1962). *The phenomenology of perception.* London, UK: Routledge.

Merleau-Ponty, M. (1968). *The visible and the invisible.* Evanston, IL: Northwestern University Press.

Merleau-Ponty, M. (2003). *Nature.* Evanston, IL: Northwestern University Press.

Neilson, A. (2010). *Disruptive privilege, identity, and meaning: A reflexive dance of environmental education.* Rotterdam, The Netherlands: Sense.

Palmer, J. (1998). *Environmental education in the twenty-first century: Theory, practice, progress, and promise.* London, UK: Routledge.

Payne, P. (1997). Embodiment and environmental education. *Environmental Education Research, 3*(2), 5–34.

Payne, P. (2005a). Lifeworld and textualism reassembling the researcher/ed and "others." *Environmental Education Research, 11*(4), 413–431.

Payne, P. (2005b). "Ways of doing" learning, teaching, and researching. *Canadian Journal of Environmental Education, 10,* 108–124.

Payne, P. (2006). Environmental education and curriculum theory. *Journal of Environmental Education, 37*(2), 25–35.

Payne, P. (2009). Postmodern oikos. In M. McKenzie, P. Hart, H. Bai, & B. Jickling (Eds.), *Fields of green: Restorying culture, environment, and education* (pp. 309–322). Creskill, NJ: Hampton Press.

Payne, P., & Wattchow, B. (2009). Phenomenological deconstruction, slow pedagogy, and the corporeal turn in wild environmental/outdoor education. *Canadian Journal of Environmental Education, 14,* 15–32.

Pinar, W., & Reynolds, W. (Eds.). (1992). *Understanding curriculum as phenomenological and deconstructed text.* New York, NY: Teachers College Press.

Pink, S. (2009). *Doing sensory ethnography.* London, UK: Sage.

Price, L. (2009). Playful musement. In M. McKenzie, P. Hart, H. Bai, & B. Jickling (Eds.), *Fields of green: Restorying culture, environment, and education* (pp. 111–120). Creskill, NJ: Hampton Press.

Reid, A. (2009). Environmental education research: Will the ends outstrip the means? *Environmental Education Research, 15*(2), 129–154.

Reid, A., Payne, P., & Cutter-Mackenzie, A. (2010). Openings for researching environment and place in children's literature: Ecologies, potentials, realities and challenges. *Environmental Education Research, 16*(3–4), 429–461.

Reid, A., & Scott, W. (2006). Researching education and the environment: Retrospect and prospect. *Environmental Education Research, 12*(3–4), 571–587.

Rickinson, M. (2001). Special issue. Learners and learning in environmental education: A critical review of the evidence. *Environmental Education Research, 7*(3), 208–320.

Rickinson, M., Lundholm, C., & Hopwood, N. (2010). *Environmental learning: Insights from research into the student experience*. Dordrecht, The Netherlands: Springer.

Ricoeur, P. (1992). *Oneself as another.* Chicago, IL: The University of Chicago Press.

Robottom, I. (Ed.). (1987). *Environmental education: Practice and possibility.* Geelong, VIC: Deakin University Press.

Robottom, I., & Hart, P. (1993). *Research in environmental education: Engaging the debate.* Geelong, VIC: Deakin University Press.

Rose, N. (1996). *Inventing our selves: Psychology, power, and personhood.* Cambridge, UK: Cambridge University Press.

Russell, C., & Hart, P. (2003). Exploring new genres of inquiry in environmental education research. *Canadian Journal of Environmental Education, 8,* 5–8.

Sauvé, L. (2005). Currents in environmental education: Mapping a complex and evolving pedagogical field. *Canadian Journal of Environmental Education, 10,* 11–37.

Sheets-Johnstone, M. (2009). *The corporeal turn: An interdisciplinary reader.* Exeter, UK: Imprint Press.

Shusterman, R. (2008). *Body consciousness: A philosophy of mindfulness and somaesthetics*. Cambridge, UK: Cambridge University Press.

Sim, S. (2006). *Manifesto for silence: Confronting the politics and culture of noise.* Edinburgh, UK: Edinburgh University Press.

Tanner, T. (1980). Significant life experiences: A new research area in environmental education. *Journal of Environmental Education, 11*(4), 20–24.

Toadvine, T. (2009). *Merleau-Ponty's philosophy of nature.* Evanston, IL: Northwestern University Press.

Tuan, Y.-F. (1977). *Space and place: The perspective of experience.* Minneapolis, MN: University of Minnesota Press.

van Manen, M. (1990). *Researching lived experience: Human science for an action sensitive pedagogy.* Albany, NY: SUNY Press.

Weiss, G., & Haber, H.-F. (1999). *Perspectives on embodiment: The intersections of nature and culture.* London, UK: Routledge.

Wood, D. (2003). What is ecophenomenology? In C. Brown & T. Toadvine (Eds.), *Eco-phenomenology: Back to the earth itself* (pp. 211–234). Albany, NY: SUNY Press.

41

Children as Active Researchers

The Potential of Environmental Education Research Involving Children

ELISABETH BARRATT HACKING
University of Bath, UK

AMY CUTTER-MACKENZIE
Southern Cross University, Australia

ROBERT BARRATT
Bath Spa University, UK

Introduction

The following chapter explores the potential of undertaking environmental education research which involves children as active researchers. Throughout the chapter the term "research involving children" is used to describe any research that includes children whether the child be object, subject, participant, or independent researcher. Although a range of approaches to research involving children are considered including the philosophies that underpin them and the issues arising from their implementation, the authors have elected to focus on discussing, justifying, and presenting research in which children are actively involved in research and not just objects of research[1]. Environmental education research involving children tends to be dominated by children as objects of research and some have argued for alternative approaches to be considered (Barratt & Barratt Hacking, 2008; Barratt Hacking & Barratt, 2009; Cutter-Mackenzie, 2009; Fleer & Quinones, 2009; Rickinson, 2001). Two key and interrelated arguments for actively involving children in environmental education research are presented, firstly, those to do with the general arguments for involving children in research including rights of children to participate in matters of relevance to them and secondly, those to do with the unique perspectives that children can bring to bear as experts, actors, and stakeholders in their own and other environments.

The chapter draws upon traditions in social and educational research that have informed the authors' approaches to research and the development of research in environmental education more broadly. It begins by presenting a continuum of methodological approaches used in research involving children in order to illustrate the range of approaches adopted by researchers. The chapter is also informed by the place and role of children in social science research; the discussion includes the notion of the child as researcher or as researched and the emerging child-framed or oriented research approaches. This is followed by a critique of ecological and cultural perspectives on child development and the insights these perspectives can offer to environmental education research involving children.

Two examples of environmental education research projects are presented in order to illuminate this methodological discussion. Both projects involved children as active researchers and each was undertaken in a different cultural context (country). The first example, *Listening to Children—Environmental Perspectives and the School Curriculum*, was a project involving ten- to twelve-year-old children in England. The research focused on children's experience of the local community and environment, and how they make sense of this in relation to both their lives and the school curriculum. The second example presents research involving primary and secondary aged children investigating sustainability practices in their households and schools as part of a broader project, the *Sustainables Challenge*, in Australia. An analysis of these projects is presented in order to support readers in understanding how and why methodological decisions were made in relation to the context and purpose of each project.

The chapter concludes by reflecting on the strengths of each project's methodological approaches and proposing four fundamental positions that influence the integrity of research involving children as active researchers. These include: (1) adopting a cyclical research process; (2) developing adults' understandings of researching with children; (3) dedicating time for children to develop as researchers; and (4) assuring children's welfare through professional and ethical practices. Finally, the chapter summarizes the spectrum of approaches to research involving children, proposes a set of considerations for readers interested in pursuing research involving children as active researchers, and reviews the case for involving children as active researchers in environmental education research.

Perspectives on Research Involving Children from Educational and Social Science Research

Methodological Approaches in Research Involving Children Figure 41.1 reflects the ways in which children are involved and/or "treated" in educational and social science research. This continuum reflects a timeline of developments in research involving children starting with research *on* children then research *with* children and, most recently, research *by* children. Involvement ranges from children as objects of research (research *on* children) through to partners in research (research *with* children) and finally, leaders of research (research *by* children). Some describe these three approaches as "paradigms" and the developments in their use as "paradigm shifts" (Christensen & James, 2000, p. 3). In educational research a shift away from research *on* children to research *with* and *by* children originated in the child rights movement and is reflected in a growth of interest in pupil voice research (Cook-Sather, 2006; Fielding, 2004b; Flutter & Rudduck, 2004; Rudduck, 2007; Waller, 2006; Whitty & Wisby, 2007). The focus on pupil voice was "in reaction against the traditional exclusion of young people from dialogue and decision making about issues of schooling" (Cook-Sather, 2007, p. 391) and can also be applied to other settings for children and youth. In its most extended forms such research is participatory in that it is undertaken *with* and/or *by* pupils, representing a shift from simply listening to children (consultation) to enabling their role in establishing or contributing to research foci, data gathering, interpretation and recommendation/action. Research *with* and *by* children has also focused attention on the possibilities of empowering and raising the social consciousness of children and young people in order to transform their own lives (Alderson, 2001; Cook-Sather, 2007; Fielding, 2001; Kellett, 2005a, 2005b).

Here, the range of approaches is presented along a "continuum" of changing practice to support the reader in understanding the origins, nature, and evolution of research involving children. Although this is presented as a timeline, research involving children currently employs each of the three main approaches and remains largely positioned toward the left of the continuum.

The Child as Researcher or Researched

> Respect for children's participation [in research] recognises them as subjects rather than objects of research, who speak in their own right and report valid views and experiences. (Williamson, 2005, p. 1)

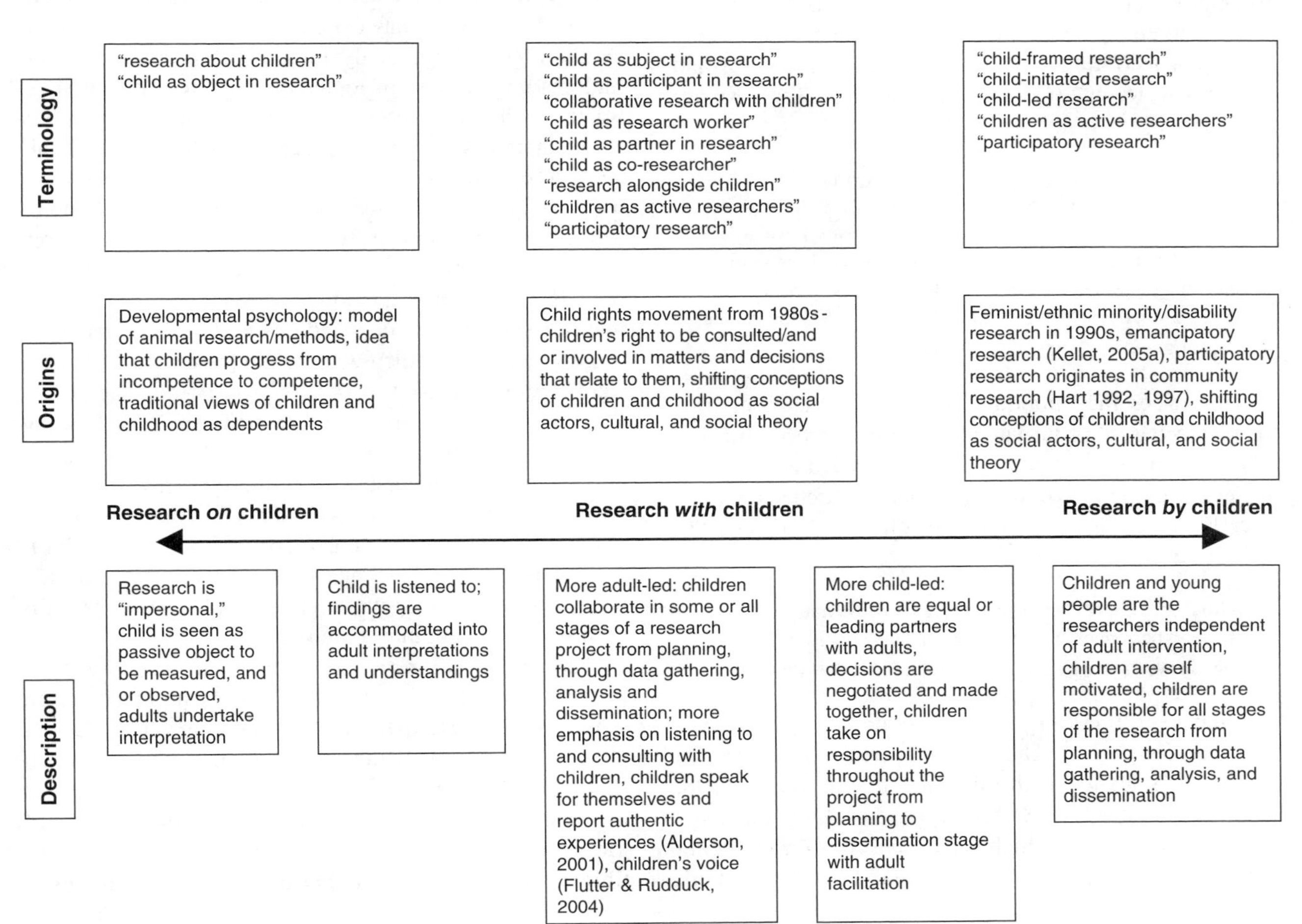

Figure 41.1 A continuum of research involving children (adapted from Barratt Hacking & Barratt 2007).

Children and young people are one of the most heavily researched groups in society; however, there is very little research in which children play an active and meaningful role (European Network of Ombudspersons for Children, 2007). Although there have been attempts to get beyond observation to extended dialogue and to report findings in children's own words, the process still tends to be controlled by professional adults, that is, teachers and researchers (Clark, Dyson, Meagher, Robson & Wootten, 2001; Kellett, 2005a). Kellett (2005a, p. 2) argues that children should be "acknowledged as experts on their own lives and if adults genuinely want to understand children and childhood, better ways to seek out child perspective and unlock child voice must be sought."

In recent years, educational researchers have begun to involve children as participants and/or coresearchers (Bell, 2008; Morrow, 2008; Skelton, 2008). This reflects a growing interest in children's participation in schools and schooling, or "pupil voice"; "there is clear evidence that the political and social climate has begun to warm to the principle of involving children and young people (in schools)" (Flutter & Rudduck, 2004, p. 139). However, research agendas involving children are still often conceived of and led by adults. Criticism of pupil voice research argues that in its limited forms, for example, as consultation, pupils are acting only as informants. It is argued that opportunities for pupils' to act as interpreters and coconstructors should be provided, "if we wish actually to engage with the unfamiliar rather than simply redefine it according to the givens of our own outlook and in our own terms" (Cook-Sather, 2007 p. 397). Kellett (2005a, p. 3) also cautions:

> Indeed, in many instances adult-led research about children is undertaken in power-laden settings with a captive child audience—such as schools. Questions have to be asked about the validity of child research (and the knowledge it generates) in context where adults control children's time, occupation of space, choice of clothing, times of eating—even their mode of social interaction.

This raises the important issue of the positioning of the researcher (the adult) and the researched (the child) as othered from each other; what some have called the intersubjective dynamics of the researcher-researched relationship (Finlay & Gough, 2003). Lincoln and Denzin (2000, p. 1050) ask:

> Who is the other? Can we ever hope to speak automatically of the experiences of the Other, or an Other? And if not, how do we create a social science that includes the Other?

Lahman (2008) argues that children are "always Othered." as they are the least powerful participants in research. In dealing with this challenge Mandell (1998; cited in Lahman, 2008) recommends taking the *least adult role* in research which some have criticized as being naïve and wishing away the complexity between adult and child perspectives (Christensen & James, 2000). Lahman (2008) suggests that as an attempt in addressing intersubjectivity and remaining reflexive, researchers must relate to children as fellow human beings. She further describes this:

> I feel much may be gained by approaching a research relationship with this stance equally in hand with the idea of child as always Othered. Knowing the child will remain Othered is not the same as actively treating them as Other. In fact the acknowledgement of child as Other is a step closer to understanding and engaging with children intersubjectively. (Lahman, 2008, p. 293)

That said, there is a growing body of literature calling for more child framed or oriented research approaches where children are positioned as active participants and/or coresearchers. This reflects paradigm shifts in educational and social science research (Figure 41.1) that have involved repositioning children and that recognize the potential of research *with* and *by* children (Alderson, 2000; Christensen & James, 2000; Kellett, 2005a, 2005b; Kirby, 2004). "The idea that children are social actors with a part to play in their own representation . . . is at last beginning to be absorbed into mainstream social science thinking" (Christensen & James, 2000, p. xi).

Child Framed or Oriented Methodologies and Designs

Research is an everyday experience in most schools, with children collecting, analyzing, and reporting research findings on a variety of topics ranging from pet surveys, waste audits to interviews with community members about their environmental practices in the home. Similarly, in environmental education programs children can be participants and researchers in environmental monitoring and improvement, for example, of habitats (Stapp, Wals, & Stankorb, 1996). However, such research is often seen as practice or as a medium for enabling the development of skills and understandings within the curriculum or program, "rather than worthwhile [research] in its own right" (Alderson, 2001, p.143).

A number of approaches are advocated for framing and orienting research with children. Figure 41.2 presents a model for understanding children's involvement in research, highlighting two dominant approaches for framing and orienting research, namely adult-led and/or child-framed research. These approaches incorporate varying levels and degrees of participation with respect to children's involvement in research.

The variety of research methods utilized by children is comprehensive, ranging from basic observation, interviewing participants, photographing and videoing phenomena to reporting and disseminating research findings. There have been some criticisms of the use of images and photographs to represent the "voice" of the child (photovoice[2]); "photographs seem susceptible to naïve and realist interpretations superimposed

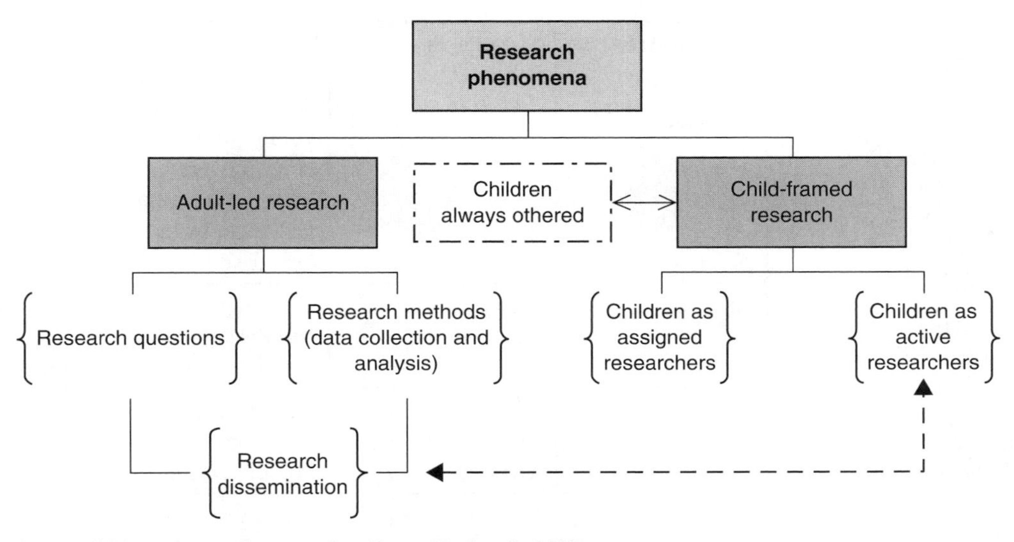

Figure 41.2 Adult and child framed research approaches (Cutter-Mackenzie, 2009).

by those who are 'reading' them" (Piper & Frankam, 2007, p. 385). However, where children take, explain, and interpret their own photographs the risk of the image being used to speak *for* the child may, to some extent, be obviated. The level of complexity of research methods utilized by children also varies with some child-framed research approaches considered nonparticipatory and participatory. Using Hart's (1997) ladder of participation (see also Arnstein, 1969), manipulation (e.g., pretending research is done by children), decoration (e.g., children used to help strengthen a research agenda), and tokenism (e.g., children given no or little choice in research) would be considered nonparticipatory child framed research models (Figure 41.3).

At the higher rungs of the ladder (levels 4 to 8) there are a number of preconditions for the successful involvement of children and young people. These are that the children and young people have:

- access to those in power and to relevant information;
- a genuine choice between distinctive options;
- a trusted, independent person to provide support; and
- a means of redress for an appeal or complaint (Hart, 1997).

More recently, Hart (2008) points to the need for alternate models (e.g., Shier, 2001; Treseder, 1997) that explore the relationship between different aspects of children's participation and have relevance in different cultural contexts. This acknowledges the problems of applying Western notions and democratic models of participation to other cultural and political contexts.

Participatory child framed research models, using Hart's ladder of participation and Tresedar's degrees of participation (Treseder, 1997), would include five layers of participation as shown in Figure 41.4.

In practice Alderson's (2001) research provides some examples of research active children. Detailed examples of methodologies relating to research active children in the context of environmental education research are presented later in this chapter.

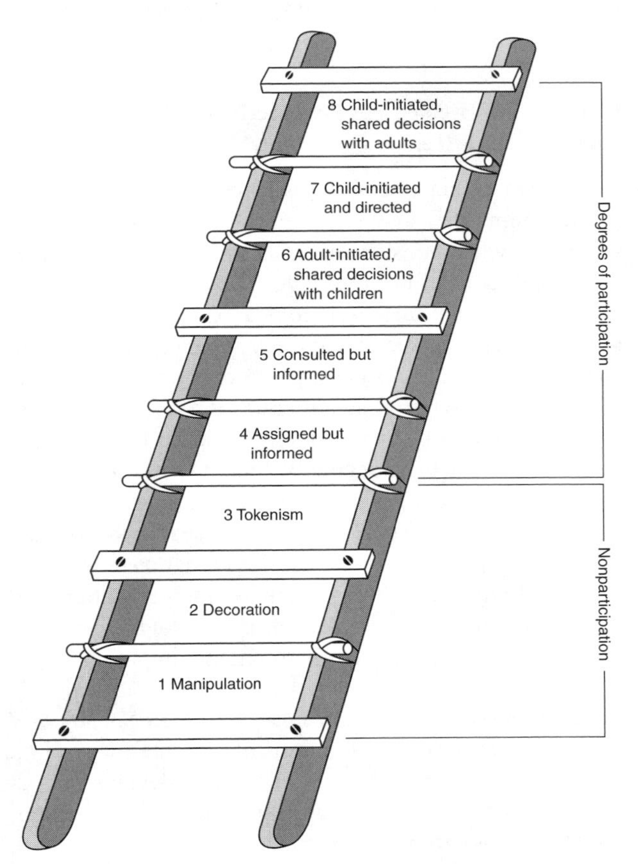

Figure 41.3 Hart's ladder of participation (Hart, 1992, p. 8). Image reproduced with permission from the UNCIEF Innocenti Research Centre.

	Children as Active Researchers	*Young people-initiated, shared decisions with adults*	This happens when projects or programs are initiated by young people and decision making is shared between young people and adults. These projects empower young people while at the same time enabling them to access and learn from the life experience and expertise of adults.
		Young people-initiated and directed	This step is when young people initiate and direct a project or program. Adults are involved only in a supportive role.
		Adult-initiated, shared decisions with young people	Occurs when projects or programs are initiated by adults but the decision-making is shared with the young people.
		Consulted and informed	Happens when young people give advice on projects or programs designed and run by adults. The young people are informed about how their input will be used and the outcomes of the decisions made by adults.
	Children as Assigned Researchers	*Assigned but informed*	This is where young people are assigned a specific role and informed about how and why they are being involved.

Figure 41.4 Child framed participatory research (adapted from The Freechild Project, 2008).

Cultural and Ecological Perspectives on Research Involving Children

An understanding of children's place in society and role in the community and environment are important starting points when considering children's position in (environmental education) research. The following section explores how cultural and ecological perspectives can illuminate the child's social and environmental involvement and experience. This discussion offers two sets of insights for developing environmental education research involving children. Firstly, by considering notions of the child's individual agency and secondly, by exploring the significance of the child's environment and everyday context in their development. These ideas raise questions about the extent to which children should be recognized as actors and stakeholders in relation to environmental matters and from here how their involvement in environment-related research can contribute to environmental improvement.

Cultural Perspectives on Children's Role in Society and Environment Traditional notions of childhood and child development are rooted in the idea that "development is a staged process whether with respect to physical, moral, social, emotional or intellectual capacity" (Lansdown, 2005, p. 9) and that adults support the child from incompetence to competence in this process. The influence of these ideas is far reaching in terms of a typical view of children as "welfare dependents" in many [Western] cultures, that is, vulnerable, incapable, and in need of guidance and protection by adults (Willow, Marchant, Kirby, & Neale, 2004, p. 7). Equally such ideas have influenced how children are represented in research as objects so "infantilizing them, perceiving and treating them as immature, and in so doing producing evidence to reinforce notions of their incompetence" (Alderson, 2000, p. 243).[3]

Such views have been under scrutiny in recent years, for example, educational philosophers have suggested that observers should not predetermine a child's development and identity (Borgon, 2007). In the past twenty years cultural theory has offered its own critiques on such views arguing, for example, that: children (1) grow up in specific cultural environments that influence their development; (2) have strengths and competencies which they use to make contributions to their social world; and (3) possess individual agency and thus are capable of interpreting and influencing their own lives (James & Prout, 1997; see also Hedegaard and Fleer's critique of notions of child development using cultural-historical perspectives, 2008). Following from this is the idea that children should be viewed as "young citizens who are entitled to respect and participation" (Willow et al., 2004, p. 7). Such ideas suggest that children have a significant role to play in their own representation and in research that matters to them and indeed have a right to do so.

While supporting research that actively involves children and in adopting a sociohistorical approach Fleer and Quinones (2009) question the assumptions of the youth voice and associated children as researchers movement. They argue that "the concept of 'hearing children's voices' must be understood within its own cultural communities" (p. 88) that is, largely European and European heritage communities where children have not traditionally held responsibilities and/or have been separated out, for example, through childcare or sleeping arrangements. In such communities youth voice has proliferated as 'a way of reintegrating children into communities when undertaking research' (p. 88). This raises questions about the universal applicability of the youth voice and children as active researchers movement across cultures and societies. There is also the question of how youth voice and therefore children as active researchers approaches can flourish in less democratic and more authoritative, hierarchical contexts where less privileged groups and individuals are marginalized in decision making (Hart, 2008).

Ecological Perspectives on Children's Role in Society and Environment Within an ecological perspective, context and experience are also viewed as significant influences on child behavior, children's relationship to the world and thus children's development. This perspective views the child within the family, social, environmental and cultural context; it emphasizes the significance of the child's everyday experience in physical and social settings in relation to their learning and development (Baacke, 1985; Bronfenbrenner, 1979). The ecological model of child development originally proposed by Bronfenbrenner was later developed by Baacke to illustrate the socioecological zones of community and space which a child inhabits (Figure 41.5).

This model has drawbacks in respect of its universal applicability; although the model can be applied to many children in industrialized and affluent societies it has less relevance to children living in marginal communities, for example, street children, children living through conflict and war and those living in intractable poverty (see, e.g., Bartlett & Minujin, 2009). Nevertheless, the idea that child development is affected by the child's everyday environment experience has wider relevance. Of particular interest are Baacke's ideas on the way in which the child goes beyond simply experiencing these zones to acting as an agent within them, developing their own meanings through experience and appropriating zones for their own purposes. It is these zones or spaces that contribute to children's learning and development and that children have influence over (Baacke, 1985; Barratt & Barratt Hacking, 2008; Bronfenbrenner, 1979; Jans, 2004). In this two way process children both learn from and contribute to community and social spaces. As "experts" in these spaces children are agents of change and are therefore able to make important contributions to knowledge advancement through research.

Child–Environment Relationships The emphasis of such cultural and ecological theories on child development and children's role in society and environment have focused more on social and community environments than on natural environments. Sauvé's model of personal and social development extends such ideas by emphasizing the importance of experience in the natural, biophysical environment (2001). Sauvé's model stresses the role of interaction proposing,

> . . . three closely linked spheres of interaction: interaction with oneself (for construction of one's own identity); interaction with others (for construction of relations with

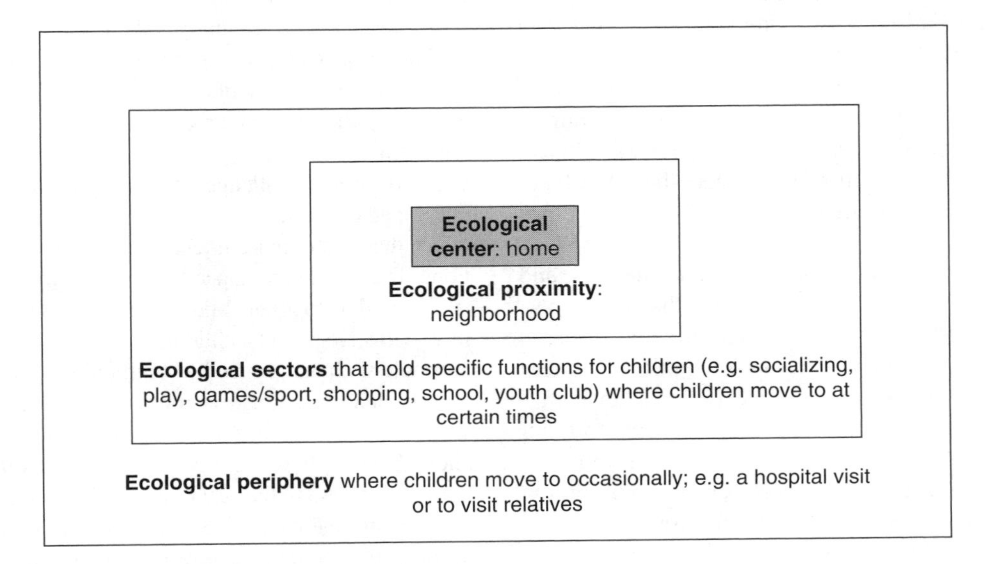

Figure 41.5 A representation of the ecological zones that children occupy (adapted from Baacke, 1985).

> other human beings); and interaction with the shared "home of life", *Oïkos*, the setting for both ecological and economic education, where the sense of "being-in-the-world" is enriched by the person's relations with the "non-human world." (Sauvé, 2002, p. 1)

Sauvé further argues that it is through experience in the local environment that children start to develop environmental responsibility.

> The local context is the first crucible for the development of environmental responsibility in which we learn to become guardians, responsible users and builders of Oikos, our common "home of life." (Sauvé, 2002, p. 2)

Further, Chawla's "significant life experience" research suggests that positive experiences of nature in childhood contribute to an interest in the environment and environmental care as adults (see, e.g., Chawla, 2006, 2007).

The philosophy of phenomenology sheds further light on child development and the environment-child relationship, in this the child's environment, community and space is viewed as an emotionally bounded area, to which the child (or individual) has a strong emotional and physical relationship and *sense of place* (Tuan, 1977; see also Seamon, 1979; Somerville, Power, & de Carteret, 2009).

> People are not separate from their worlds; rather, they are immersed through an invisible net of bodily, emotional, and environmental ties. (Seamon, 1984, p. 135)

Those interested in place studies, geographers and phenomenologists have long argued that we need to understand locality and region in terms of the insider, that is, the resident or the user of the place or region (Buttimer, 1980); journals such as *Children's Geographies* and *Children, Youth and Environments* have made significant contributions to an understanding of place in respect of children's views and experiences. The place-based education movement builds on the importance of local places for children by using the local community and natural surroundings as a context for learning (see, e.g., Gruenewald & Smith, 2008; place-based Education Evaluation Collaborative website; Promise of Place website).

This discussion of child development theories and child-environment relationships suggests that the opportunity for children to engage in environmental research would seem to have genuine relevance for children. These theoretical perspectives further suggest that as "insiders," "experts" and therefore "stakeholders" of places, communities and environments children should have opportunities to learn more about them and work toward environmental and community improvement through research. Such opportunities offer interesting possibilities for children to participate in local development projects.

Two Examples of Research Projects Involving Children as Active Researchers

In the following section two research projects are introduced. In both cases children were involved as active researchers in partnership with adult researchers. Although there were instances of independent "research *by* children," essentially both projects should be classified as "research *with* children" (Figure 41.1) due to the ongoing involvement of adult researchers. A critical commentary on the methodological approaches adopted in each research project is presented in tabular form. Aspects of each project are analyzed using a series of criteria (Tables 41.1 and 41.2). The tabular analysis focuses on the role of children (and adults) in the research and how this illustrates the preceding discussion on research involving children. The section concludes with a reflective discussion of what has been learned about methodological design in this area. For readers who wish to pursue details of each project in more depth key supporting references are provided below.

Example 1 Summary of "Listening to Children; Environmental Perspectives and the School Curriculum." (Table 41.1)

"Listening to Children: Environmental Perspectives and the School Curriculum" (L2C) was a UK research council project based in schools in a socially and economically deprived urban area in England. This one-year, participatory research project was undertaken by a research team comprising of eleven- to twelve-year-old children and their seventeen- to eighteen-year-old mentors, teachers, a parent, university researchers, and community representatives. It focused on ten- to twelve-year-old children's experience of their local community and environment, how they perceive, and act within it and how they make sense of this in relation both to their lives and the school curriculum. In developing this research project with children there was an attempt to engage children's untapped potential as local community stakeholders.

The project was set in the context of an 11–19 secondary school that serves an urban community on the northeastern edge of a large conurbation in southwest England. This area exhibits the sorts of social, economic, environmental and educational challenges that many urban communities in developed economies face, and associated concerns about the effect that such environments have on their young populations. The school recognized the need to involve local people in school development, and the project provided an opportunity to involve parents, children and local voluntary bodies in curriculum change related to both citizenship and education for sustainability. The research was a twelve month study of children's local environmental and community perspectives and behaviors, how they make sense of this in relation both to their lives and to the school curriculum, and how children's local environmental perspectives might become a part of their curriculum experience. The research highlighted the desire that young people have for schools to address community issues

TABLE 41.1
Application of Research Involving Children. "Listening to Children; Environmental Perspectives and the School Curriculum," England

Aspect of Research Project	Application in Practice	Commentary
Overall research program	Project within a UK research council program (economic and social research council/ESRC) Environment and Human Behavior. (RES-221-25-0036, ESRC Environment and Human Behavior Program).	Nationally funded project; all proposals reviewed by experts in the field. Contracted and time limited (twelve months).
Who was involved	Team of child and adult researchers including 16, eleven- to twelve-year-olds (8 girls and 8 boys), their seventeen-year-old mentors (2 boys, 2 girls), 4 university academics, 2 school teachers, and 1 parent. One secondary school, one university.	Several schools were approached and the project was negotiated in relation to the school calendar. One secondary school felt the project matched their mission statement.
Research Context	The area surrounding the school faces a number of social, economic, and environmental challenges, for example, air pollution, road safety and crime. There is local concern for both educational achievement and the quality of the urban environment in which children are growing up. The study school is a mixed comprehensive school with approximately 1,100 students aged 11–18 and seventy teaching staff. Pupils come from a range of socioeconomic backgrounds including educationally and economically disadvantaged. The school is working hard to address a number of educational challenges, for example, high absenteeism, poor progression to post-16 and beyond, and underachievement of pupils against national/regional standards. One of the school's great strengths is its ethos of support for pupils.	Educational achievement and the quality of the urban environment are not seen to be related by the community. The project set out to reconcile children's understanding of this relationship. The school recognized the value of engaging children in projects beyond the formal curriculum. Senior management support is key to the success of any school-based project.
Methodological frame	"Collaborative research with children" "participatory research" "community-based research" "children as active researchers"	This was new practice for the school children and school staff. Time was taken to introduce children and staff to the ideas underpinning collaboration, participation and community-based research.
Project aims	The project set out to consider how children's environmental experiences can be incorporated into the curriculum, exploring how participation in community-based research can help children to become environmentally conscious and active citizens, and so contribute toward more sustainable urban environments.	The aims of the project were refined by children and adopted by the project. This created tensions between the original project proposal and the need to respond to the wishes of the children. This led the project in a slightly different direction challenging the project preconceptions of the university and school staff.
Research questions	1. How do eleven- to twelve-year-old children experience and think about their local environment? 2. How can schools make use of this new body of data in order to increase the relevance of the curriculum to children, their families and the local community? 3. How can schools make use of these emerging data in innovation related to citizenship and sustainable development? 4. How can schools provide for children, parents, community workers and teachers to participate in the curriculum development process in order to develop a curriculum that relates to children's local environment perspectives and help shape behavior?	The research questions were remodeled by children using children's own language.
Identification of research focus and questions	The research focus and questions were identified prior to children's involvement in order to obtain funding. However, the development of L2C's research design was emergent; the project started with research aims and questions but not with a fully determined research design. The approach developed through a process of discussion and negotiation with children. Children were involved in reframing and adapting aims and research questions.	Ideally children would have been involved in framing the research focus and questions; the research funder was a barrier in this case. The project has established that children are able to frame research ideas and future projects will take this into consideration.

(*Continued*)

TABLE 41.1 (Continued)

Aspect of Research Project	Application in Practice	Commentary
Ethical issues and consent	The idea to open up the project to volunteers from the whole year group through an assembly attempted to avoid coercion and provide relevant information as a basis for volunteering. The issue of consent was explored openly with the volunteers; discussion included negotiating how to address each other. (The outcome was that all children and adults, with the exception of the teacher researchers, would be known by their first names). Consent to participate in the research was a negotiated process. This began at the start of the project and continued in an ongoing way as the nature of children's involvement evolved in this emergent design Children's right to withdraw from the project was made clear from the outset, repeated in writing to parents/ carers and revisited at different stages of the research. Children were offered counseling/debriefing if they decided to withdraw. (One child chose to withdraw and requested debriefing.)	Prior to the involvement of children the teacher and university researchers agreed that obtaining reliable informed consent in a school setting was an issue. We explored how this might be approached including how to inform children about the research project in a way that is meaningful and relevant to the needs of eleven- to twelve-year-olds so they understand the risks involved and expectations. Children were clear that prior to consent they needed to believe the project would make a difference to the school ethos and curriculum and, if possible, their family and community life.
Phases of the research	The two main phases of the research *Phase 1*: establishing and planning the project then gathering and analyzing evidence about the nature of ten- to twelve-year old pupils' local environment experience (May 2004–December 2004); *Phase 2*: using the evidence gathered and analysed in Phase 1 to develop, implement, and evaluate a curriculum project (January 2004–May 2005).	Each phase of the research was defined by children and fitted naturally with the emergent design of the project.
Research design and methods	The research design was a mixed method approach, generating both qualitative and quantitative data. The children devised and developed each of the data collection methods to gather data from their peers and younger children from a feeder primary school as follows. • Children's photographic and video diaries of their local experience (using mobile phone technology/digital cameras) • Children's personal maps (plus questionnaire) which the children termed the "little map" for all the year group at secondary school (ninety children) and twenty primary school children • Children's group map drawn on a very large sheet of paper by children who live in the same neighborhood ("big map") • Group interviews to discuss the little and big maps • Parents' carers' (and grandparents) group interview • Curriculum audit (for teachers)	There was a concern to represent the authentic voices and experiences of children and their families equally, alongside educationalists and local authority officials. Therefore children in the study school were involved in devising research instruments and leading meetings. . . Children's health and safety was a key area of concern; some data collection methods were rejected or adapted to minimize risks. For example, children's independent use of video cameras in the local parks could put children at risk from theft, instead they used their own mobile telephone cameras.
Research preparation with children	Children and adults were supported in developing research skills throughout the life of the project with dedicated "research preparation" events. For example, a day's conference at the university focused on data collection. With advice from experienced researchers, the children were able to consider the research methods and tools that they would employ; children chose to adopt and modify methods and devise their own methods (e.g., using mobile phone technology to record environment experience diaries). Children were also able to develop skills in using equipment such as dictaphones and digital cameras. In a further example a two-day data analysis conference was held later in the project. The children were supported in their analysis by adult researchers and four seventeen-year-old mentors. These older students were able to offer a range of expertise to the data analysis including their greater experience in the local community and their own research and ICT skills.	The use of older students as "mentors" in the research preparation process seems to have been an important step in promoting children's confidence and independence as researchers and reducing reliance on adults. The adult researchers became more distant from the research process as the older students took on a key support role. The adults also developed new skills and understandings in relation to research involving children. This included new "child friendly" ways of collecting data, for example, the "big" map "little" map approach, and the reliability of children's research instruments.

Aspect of Research Project	Application in Practice	Commentary
Analysis	A data analysis framework was introduced over a two day conference by one of the university researchers as follows. This framework was refined and developed by children. • You will need to read, look at maps, listen to the audiotape or look at the video in order to familiarize yourself with the data. Relook at the data, start to think about what is important. • Make points about what strikes you as important on sticky "post it" notes • Talk to your partner add your notes to his/her notes • Make one set of notes for the whole group using the lap top • Organize your notes into different groups of ideas. The children worked mostly independently of adults for the two days. Part way through the two days each small group met with a sample of children from whom they had gathered data to corroborate the findings. The children presented their interim findings at the end of the conference to an audience of child researchers, teachers, university researchers, senior managers and parents.	The adult members of the research team were surprised by the way in which the children approached this new and complex task. Children were both creative in their analysis and challenging in their approach to the data.
Findings	Following the two day analysis conference and further analysis the children's findings included: • children have an intricate knowledge of their local community and attach great importance to local environment quality. • children have difficulty taking action to achieve what they want for their local environment; they do not know how to go about it. • children's local knowledge/ideas for improvement are not included in the school curriculum. However, children feel strongly that school should support them in achieving their goals. *"We should try to get even more involved with our community . . . like kids maybe building something inside the school like composting and show it to the adults . . . so we can bring change into the community."* (Gemma, 11 years) (For further details of the findings see references for Barratt, Barratt Hacking)	The children's findings revealed that children can operate safely in the local environment. While some of the children's local concerns reflect other research findings, e.g. traffic dangers, it is clear that there are barriers preventing children from pursuing their local aspirations. Children's local knowledge is mostly unknown by the school—and hence not used. The children have a strong desire to be involved in local improvement. The opportunity for parents, community partners, children and teachers to be involved in community and curriculum discussion was pioneering for this school.
Dissemination	Planning how and where dissemination should take place was an ongoing task for the children and adults involved. The importance of children's role in this was agreed alongside the impact children could have on both adult and child audiences. Consideration was afforded to different forms of dissemination for the range of audiences, not least the children who had been researched; those examples included: • PowerPoint presentations to the school senior management team • sending an e-mail to the local Member of Parliament (MP) and later meeting with him • giving interviews reported in local newspapers and the school newsletter • planning and implementing a children's conference to which all eleven- to twelve-year-old children, teachers, and local community officials and academics were invited • making a DVD film including a future scenario acted and filmed in the local park • making an oral presentation to a local authority conference • academic/professional publications (dissemination by adult researchers).	Three important conditions for children's successful participation in dissemination emerged through the project experience. (i) children's authentic views and words are represented (ii) children consider forms of dissemination that are concomitant with children's own interests and skills (iii) children play a key role in deciding how and when ideas emerging from the project should be disseminated. A number academics, policy and professional publications and conference papers were produced. Where children were not able to be involved children's authentic views and words were represented and acknowledged.

(Continued)

TABLE 41.1 (Continued)

Aspect of Research Project	Application in Practice	Commentary
Outcomes	*By the latter stages of the project the children had developed:* • knowledge and understanding about the local area and its community and about local democratic processes • research skills including using technology to gather and analyze data and disseminate findings about the local environment • a sense of responsibility for the future of the local community/ environment and had started to take action to achieve their aspirations • confidence and skills to participate in a democratic change process (in a school context and to some extent in the local community) • *and the teachers/school had developed:* • new knowledge about the locality and children's local experience • commitment to using local perspectives within the school curriculum including through children's community research • a desire to maintain a forum for children to discuss local environment/community issues together with matters that relate to learning and the school curriculum.	At the end of the project semi-structured interviews (group and individual) were carried out with children and adults that had been involved seeking to critically appraise (1) the project experience of those involved, and (2) the impact of the project on the school, its curriculum and those involved. The children welcomed the opportunity to reflect on their project experience and future possibilities.

within the curriculum and for schools to play a much more significant role in community development (see also Barratt & Barratt Hacking, 2008; Barratt, Barratt Hacking, & Scott, 2005; Barratt Hacking & Barratt, 2007; Barratt Hacking & Barratt, 2009; Barratt Hacking, Scott, & Barratt, 2007; Barratt Hacking et al., 2005; Barratt Hacking et al., 2007; Scott, Barratt Hacking, & Barratt, 2007).

Example 2 Sustainables Research Challenge, Australia (Table 41.2)

The "Sustainables Challenge" project was a statewide (Victorian) government initiative funded in 2006–2008 (three years) linked to Victoria's Sustainability frameworks, namely: *Our Environment, Our Future* (State of Victoria Department of Environment and Sustainability, 2006) and *Learning to Live Sustainably* (Department of Environment and Sustainability & Department of Primary Industries, 2005). The central ethos of the project was to advocate ten simple sustainability actions "that schools, community and environmental educators can use . . . to teach students about the importance of living sustainably" (Victorian Government, 2007). They were:

1. Take a four minute power shower.
2. Turn off lights and appliances at the switch when not in use.
3. Buy renewable energy, by signing up to Green Power with your electricity provider.
4. Buy the most energy and water efficient appliances you can afford.
5. Take reusable bags with you when you go shopping
6. Put your food or plant scraps in the compost or a worm farm.
7. Look for products without unnecessary packaging.
8. Walk, cycle, or use public transport when you can - and leave the car at home.
9. Grow plants native to your area in your garden.
10. Go chemical free when you clean (Gould League, 2009).

From 2006 to 2008, 120 schools participated in the project which was delivered by the Gould League involving teacher professional development, two-term curriculum program (six months), theater performance (by the Sustainables family) and ongoing support from the Gould League. Cutter-Mackenzie was engaged to research the Sustainables Challenge program from 2006 to 2008; in effect to gauge the impact of the program against its stated objectives. The research design included three distinct phases (Figure 41.6).

In order to gain a deeper understanding of the implementation of the Sustainables Challenge program, students were invited to be researchers. Over the course of the project, fifteen schools (nine primary schools and six secondary schools) volunteered to be part of the Sustainables Challenge Think Tanks. Five schools participated per year for one school year which coincided with the year that these schools focused on the Sustainables Challenge project. Fifteen to twenty students (from different year levels) participated in each school research group

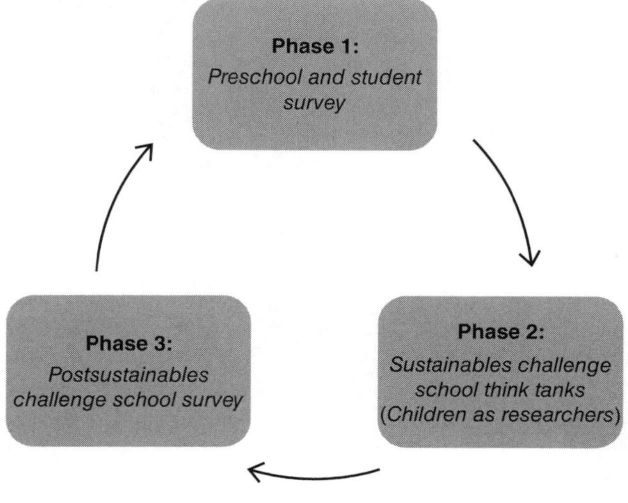

Figure 41.6 Research phases.

(think tank), approximately 250–300 student researchers altogether over the full duration of the project.

As shown in Figure 41.7, the Sustainables Challenge Think Tank (Phase 2) consisted of three stages. Stage 1 involved initial school sessions where the children (student researchers) carried out "mock" research assignments alongside University researchers. These sessions served as training exercises and were essential preparation for the second stage, *children as researchers*. For this particular stage the children were invited to undertake a research challenge. As part of this challenge children were asked to research their family's sustainability actions at school and home. This data formed the basis of discussion at the final think tank session (stage 3) which also involved the children's parents/guardians. The children's research was analyzed using qualitative thematic analysis techniques (Lofland & Lofland, 1995) which they commenced at the final think tank session.

Reflections on the Two Projects Reflecting on the experience of the two projects has identified a number of practical ways in which involving children as active researchers overcomes some of the limitations of conventional research *on* children including the more limited types of pupil voice research. It should be noted that both projects were set in the context of wider programs and therefore the research focus was not determined by children entirely but in both cases children influenced and shaped the direction of the research throughout the research process; this avoided the adult determined foci prevalent in research involving children. Thus, the approaches adopted in both projects represent a combination of decision sharing (between children and adults) together with children framing and playing an active leadership role in the research; this represents rungs 7 and 8 of Hart's ladder (1997) at different stages in the research process with an element of rung 6 (Figure 41.2 and see also Figure 41.3). The research foci of the two projects overlapped in respect of personal, family and local community experience, one in relation to sustainability, the other, children's perspectives and behavior in the local environment. Children's involvement in establishing and refining the foci demonstrated the significance and relevance of the research process for children.

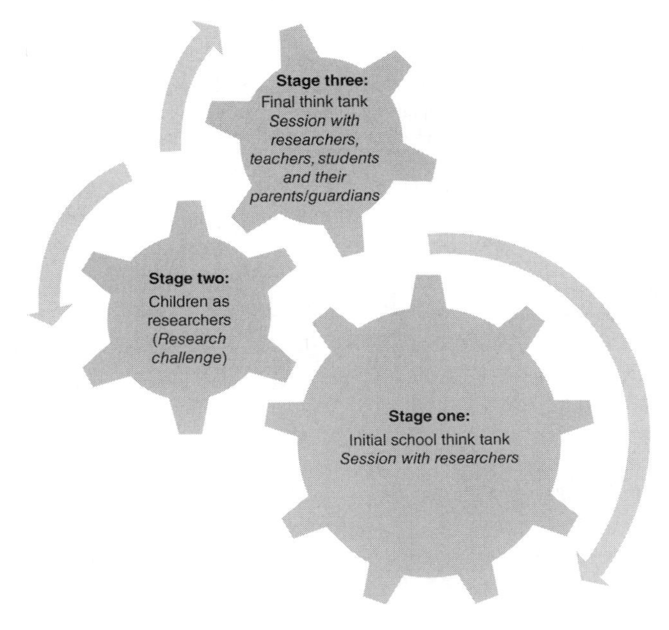

Figure 41.7 Think tank phases.

Both of the projects were located in schools involving partnerships between university researchers, children and teachers; the school context necessitated careful planning within each project to enable children's genuine consent and participation. Schools represent a "consenting environment" (Barratt Hacking & Barratt, 2009) and many researchers have argued for a more rigorous approach to obtaining reliable consent in these settings (see, e.g., David, Edwards, & Alldred, 2001; Heath, Charles, Crow, & Wiles, 2007; Wiles, Heath, Crow, & Charles, 2005). The experience of each project also demonstrated that time is required to enable children to participate genuinely in research. In each case the research evolved through different phases in which children took increasing responsibility as their confidence, understandings, and research skills developed and adult responsibility changed to that of research facilitator. Important research preparation activities developed children's understandings, skills and confidence as researchers, for example, skills in using research equipment, interviewing peers and adults and analyzing qualitative data (see Tables 41.1 and 41.2: Research Preparation Activities).

The child researchers in each project developed and employed mostly qualitative methods to document experience and phenomena including video diaries and journals, photographs, map making, and interviews. This reflects children's ability to design "appropriate and innovative research tools which help to engage young respondents in research" (Kirby, 2004, p. 276). The balance of visual and verbal methods also ensured that research tools were "child friendly," that is, accessible to children of different abilities. The adult researchers provided children with a framework for analyzing the range of qualitative data; within this the children had the freedom to develop their own analytical tools and interpretations. The children's role in analysis enabled the children to develop new insights and avoid adult interpretations that draw on predetermined ideas (Borgon, 2007; Cook-Sather, 2007; Save the Children, 2004).

Through the research experience the child researchers developed new knowledge and skills. It became evident that the children's knowledge, and how they use it, is different to that of adults' and not necessarily known by adults. For example, the Listening to Children project found that children's local knowledge "is generated through exploration and play, passed to the children from their peers and families through stories, and is renewed through contact with each other, with older children, with adults, and by being in the locality" (Barratt Hacking et al., 2007, p. 134). In each project the

TABLE 41.2
Application of Research Involving Children. "Sustainables Research Challenge," Australia

Aspect of Research Project	Application in Practice	Commentary
Overall research program	The "Sustainables Challenge" project was a statewide (Victorian) government initiative funded in 2006–2008 (three years). Cutter-Mackenzie was engaged to independently research the Sustainables Challenge program during this period; effectively to measure the impact of the program against its stated aims.	The broader project was funded by the State Government of Victoria (Department of Sustainability and Environment and Sustainability Victoria). The Gould League delivered the program.
Who was involved	120 primary *(years 3–6: ages 9–12)* and secondary schools *(years 8–10: ages 13–15)* altogether participated in the Sustainables Challenge project. All schools participated in the broader research project, although only fifteen participated in the Think Tank Phase (children as researchers) of the research.	While only fifteen schools participated in the Think Tank phase over a three-year period, seventy-five schools volunteered which was unanticipated given the difficulties often encountered when recruiting schools to participate in research.
Research Context	The "Sustainables Challenge" project was an initiative linked to Victoria's Sustainability frameworks, namely: *Our Environment, Our Future* (State of Victoria Department of Environment and Sustainability, 2006) and *Learning to Live Sustainably* (Department of Environment and Sustainability & Department of Primary Industries, 2005).	The fifteen schools were selected via random selection among various regions of Victoria including: inner city; greater metropolitan; provincial; and rural/regional.
Methodological frame	Children as Active Researchers	None of the schools had previously participated in research where the children were researchers. For many of them this was their motivation for being involved. One school principal commented that research is often *"done to children, rather than by them."*
Project aims	The primary purpose of the research component of the broader project was to measure the impact of the program against its stated aims (for specific aims see Gould League, 2009).	The broader project aims and research aims were synonymous. The project aims were rephrased into appropriate research questions.
Research questions	The key research questions for the Think Tank (children as researchers) phase of the research project were: How sustainable am I? How sustainable is my family? How sustainable is my school? How sustainable is my community?	The research questions were written by the children in the first year of the project, making the language of the questions child-friendly.
Identification of research focus and questions	Prior to the commencement of the research project one research question was devised (for the Think Tank phase), namely how sustainable is my school? In the first year of the project the children themselves devised the remaining questions which were utilized in all proceeding years of the project. A basic three- staged research design was applied as mentioned earlier, but the children elected methods from the qualitative methodological framework presented.	Presenting a broad question to the children appeared to be effective with respect to devising more specific and contextual questions.

Aspect of Research Project	Application in Practice	Commentary
Ethical issues and consent	Children as researcher methodologies immediately raise ethical issues. As Masson (2005, p. 232) identifies "children and young people are rarely free to decide entirely for themselves whether to participate in research." Typically a parent or guardian (also referred to as "gatekeepers" by Masson) provides consent on behalf of their child. Given these concerns a consent process was created where both the child and their parent/guardian were required to give consent in order for the child to participate in the research. While a standard child friendly consent form was used children's verbal and nonverbal body language was also observed that may have indicated that they did not want to participate. This process enabled the child to decide whether or not to participate in the project.	At one research site two parents provided consent for their child to participate in the project; however the child himself declined to participate (verbally). In this instance the wishes of the child were respected.
Phases of the research	The Think Tank component of the project was phase 2 of the larger research project. There were 3 phases altogether (see Figure 41.6). Phase 2 though incorporated 3 stages as depicted in Figure 41.7.	Stages 1 and 3 were set in place as stage 1 involved essential research methodology training and stage 3 (referred to as the final think tank session) was structured as a space and opportunity for children to present, discuss and analyze their findings alongside their parents and teachers. Stage 2 was designed by the children drawing on various qualitative methodologies.
Research design and methods	The overall research design of the project incorporated combined qualitative and quantitative approaches. However, the Think Tank phase (Phase 2) was primarily qualitative. Given the large quantitative aspect of the broader research project occurring before and after Phase 2, qualitative methodologies were employed.	While a quantitative dimension may have added depth to the research (phase 2) it was not feasible to incorporate given the complex task of research methodology training.
Research preparation with children	Stage 1 involved initial school sessions where the children (student researchers) carried out "mock" research assignments alongside university researchers. These sessions were essentially research methodology workshops where children learned: qualitative research methodologies; interviewing skills (including interviewer-participant relationships, consent, asking and framing questions and writing/recording notes); using digital recorders; operating digital cameras and video cameras as a way of documenting phenomena; and journal writing as a medium for narrative inquiry. These sessions served as training exercises and were essential preparation for the second stage, *children as researchers*. For this particular stage the children were invited to undertake a research challenge. As part of this challenge children were invited to investigate the research questions as presented earlier. This data formed the basis of discussion at the final think tank session (stage 3) which also involved the children's parents/guardians serving as a means of triangulation.	The workshops typically involved two to three university researchers working with 10–15 children. This ratio allowed for intensive group and individual work where children developed their research skills. Given the mixture of ages in the group younger and older children peer mentored. As the workshops progressed the university researchers increasingly played a facilitator role.

(*Continued*)

TABLE 41.2 (Continued)

Aspect of Research Project	Application in Practice	Commentary
Analysis	The children's research was analyzed using qualitative thematic analysis techniques. The children did the preliminary analysis of their findings at the final think tank session (whole school day). In groups they reviewed all data initially (categorizing it initially) and then utilized Lofland and Lofland's (1995) methods of "scissors, circling, and filing/coding" in determining themes and anomalies in the data. This is a similar logic to that of contemporary software like NVIVO.	University researchers led the children through this process of data analysis with respect to the techniques described by Lofland and Lofland (1995). In hindsight the children should have been given more flexibility in considering data analysis strategies. However, although a framework was created the children devised innovative ways of sorting, categorizing, and coding data (e.g., categorizing the data by sustainable action and then by research question).
Findings	While the children's research revealed that participation in the Sustainables Challenge led to direct behavior change to implement new sustainable actions in the home, there was also significant evidence that many students lived in households where many of the sustainable actions were well established prior to participating in the Sustainables Challenge program. Students often claimed ownership of "simpler" energy and water efficiency actions (turn off switches, use half-flush, take four-minute showers and put buckets under taps). However they also frequently reported a link between these sustainable actions and reduced utility bills, suggesting that parents may be motivated by potential financial savings as a key reason for implementing sustainable actions at home. Conversely, very few students reported the sustainable action relating to "green energy," which has a direct household impact by increasing utility costs. This area requires further investigation, specifically focusing on parents' motivation for environmental behavior change. Students also suggested that parents have generally been the catalysts in relation to sustainable actions requiring significant financial outlays or technical set-up (buying energy/water efficient appliances, installing water tanks and installing hoses to reuse washing machine water). This is contrary to past research identifying children as key catalysts for sustainable change in their households as Ballantyne, Connell, and Fien, (2006, p. 414) explain: The process of intergenerational influence whereby school students act as catalysts of environmental change among their parents and other community members could be a powerful but, as yet, untapped means of addressing current environmental problems. While there is most certainly intergenerational influences (between child-parent and parent-child) concerning environmental change, the significant lifestyle changes that Ballantyne et al. (2006) refer to are more likely to be instigated by parents (adults) suggesting that community environmental education programs are having success. This trend has been identified through the Sustainables Challenge since 2006 (Cutter-Mackenzie, 2006, 2007, 2008).	

Aspect of Research Project	Application in Practice	Commentary
Dissemination	An important consideration of research involving children is dissemination of research findings. In consultation with the children we elected to produce a research DVD (for each year of the project) which documented the Sustainables at the Think Tanks schools, in addition to providing the formal academic report and related publications making it accessible to both children and adults (teachers). The children also presented findings via local newspaper articles, radio interviews, school assemblies and performances, school newsletters and school websites.	All schools (120) received a copy of the research DVD. All children who participated also received a DVD and a certificate of recognition. The DVD was also provided to all key stakeholders, including the Victorian Government, Sustainability Victoria, Department of Sustainability and Environment and the Gould League. Many children commented when receiving their DVDs that they felt that they had made a difference. One student said *"it makes me feel like I have done something; something for the world and it makes you feel good about yourself that you have done it" (year 5 student).*
Outcomes	Many teachers, children and parents reported the profound impact that the research had on the wider Sustainables Challenge project and their consequent sustainable actions: *"It sure has motivated them and their families too. I only had three researchers in my grade and it made a difference. They loved it, being researchers. I think that the children that did it got much more out of it." Teacher* *"Being a researcher definitely made my family more aware." (Student)* *"I think the project has most profoundly affected me (the mother). I think that I was pretty unaware of many things and have taken steps to change our household habits and will continue to do so." (Parent)*	At the end of the final think tank session all children, their parents/guardians and teachers participated in an informal interview about the research process. This was an important evaluation aspect of the research process allowing the researchers to hear directly from all participants (as noted adjacent).

children were able to make an impact on environmental decision making in their school, home or community either through personal action or through dissemination. In both projects the children took an active role in decision making about project dissemination and in dissemination activities. The active involvement of children in disseminating research findings ensured that children's authentic views and words were represented and further, that forms of dissemination drew upon children's own interests and skills and were relevant to both child and adult audiences. For example, children performed in school assemblies and in making DVDs; they also gave interviews and wrote articles for newspapers and school newsletters (See Tables 41.1 and 41.2: Dissemination). This contrasts with conventional research involving children which is typically disseminated by adults for adults, for example, in academic papers.

The adult researchers also developed new knowledge and skills both in relation to the research foci and in terms of learning about how to research with children; this latter aspect of researcher learning, particularly how to avoid imposing the adult viewpoint, is framed by Cook-Sather (2007) as "translation" which, she claims,

> raises questions of how we as researchers interpret and render ourselves and how we can work with students to interpret and render themselves in educational contexts and in analyses of those (p. 396).

Reflecting on the successes and challenges of the two projects has highlighted four fundamental positions to ensure the integrity of research involving children as active researchers. These positions can also be supported by wider experience of research involving children.

1. The research process for each project is presented in a linear format (Tables 41.1 and 41.2); however, research involving children as active researchers necessitates a cyclical process of iteration. It is important to rationalize the direction and purpose of the research engagement and to continuously reference research questions, methodological design and the emerging data. In the research examples this was achieved through referential corroboration with all parties.
2. It is important for adult researchers to evaluate their own experience of the research process throughout the life of a project which involves children as active researchers. This can be achieved through a dialogue which focuses on the relationship between the adults, the children and the research project. In practice this would involve children in chairing meetings which focus on monitoring and reviewing the interests and engagement of all parties in the research.
3. Time is an important factor influencing the success of research that actively involves children. In the early stages of a project time is needed for children

to develop trust both in the adult researchers and in the research process, for example, to be reassured that children's aspirations will be addressed and that children will influence the research direction. Children also need time to develop an understanding of the research process, develop research and participation skills, discuss and make reasoned research decisions and learn how to research alongside adults.

4. Adult researchers have the responsibility to adopt a professional and ethical code of practice in shaping and understanding the emerging relationships between themselves, children and the research project. Children's welfare, health, and safety must be afforded the highest priority. Of particular concern in environmental education research is the opportunity for children to collect data outdoors; this must be discussed with all parties and carefully managed in order to mitigate risks to children's psychological and physical welfare.

Conclusion

While this chapter has focused on an emerging methodological approach in environmental education research, that is, research that involves children as active researchers, it has also introduced a range of paradigms and methodological approaches available to researchers undertaking research involving children (Figure 41.1). Environmental education research involving children has largely treated children as objects of research (Barratt Hacking et al., 2007; Cutter-Mackenzie, 2009) reflecting the dominant research *on* children approach. It has been argued that adult observations and interpretations of childhood and children's experiences are likely to be accommodated in predetermined ideas about childhood and child development (Borgon, 2007; Christensen & James, 2000). Adult interpretations may be better avoided through more child-focused approaches, for example, children's involvement in interpreting their own experiences.

In educational research the "new wave" (Fielding, 2004b) of pupil voice research has shifted the position of the child from informant to interpreter (Cook-Sather, 2007) and led to the more active involvement of children in the analysis and improvement of classroom and school conditions (Cook-Sather, 2006; Fielding, 2001, 2004a; Thiessen & Cook-Sather, 2007; Rudduck, 2007). A recent example of such work in environmental education research gives new insights into pupil perspectives and experiences of learning for sustainability in English schools (Gayford, 2009). Nevertheless, when undertaking pupil voice research in environmental education it is important to bear in mind the associated ethical and methodological issues (Cairns, 2009; Cook-Sather, 2007; Currie, Kellie, & Pomerantz, 2007; Fielding, 2007; James, 2007). Examples would include: (1) how to deal with contradictory positions or partial stories in circumventing the researcher's search for coherence (Currie et al., 2007; Cairns, 2009); (2) how to avoid totalizing children's voices when children are not a homogenous group (Cook-Sather, 2007; James, 2007); and (3) how to embrace less privileged voices (Cook-Sather, 2007; James, 2007; Malewski, 2005).

Readers wishing to adopt this methodological approach may like to consider a series of questions when developing their research plans with children (Table 41.3), adapting this to reflect the norms of child-adult relationships in their own cultural and social settings and the funding conditions (where appropriate). The table illustrates key considerations when planning this type of research; these are presented in relation to a typical, if not extended, research process "through paying great attention to the initial and follow up stages" (Alderson, 2001, p. 144). A set of questions are offered as prompts for readers' to evaluate and validate their research plans in relation to an "ideal" process for planning research involving children as active researchers.

Those interested in this methodological approach continue to explore how to engage children in an authentic research experience; there is growing interest in developing this approach through researching "how to research with children" (Barratt Hacking & Barratt, 2009; Barratt Hacking et al., 2007; Cutter-Mackenzie, 2009; National Foundation for Educational Research, undated; Punch, 2002). Areas of interest include: (1) developing ethical frameworks (Bell, 2008; Halman, 2008; Morrow, 2008; Skelton, 2008); (2) how to promote genuine participation in school settings (Barratt Hacking & Barratt, 2009; Kirby, 2001); (3) developing children's research methods (Barratt Hacking & Barratt, 2009; Kellett, 2005a); and (4) adopting a cultural-historical framework in developing the "children as researchers" methodological approach (Hedegaard & Fleer, 2008). Researchers are also considering and implementing methodologies that are appropriate for researching children as researchers, for example action enquiry, participatory research and participatory action research (Ataöv & Haider, 2006; Barratt Hacking & Barratt, 2009) and discourse analysis and participant interviews. Participant interviews were used in both project examples discussed in this chapter in order to evaluate the research experience, approach and impact.

This chapter and the research project examples have shown that research involving children as active researchers is characterized by children partnering adults or adopting a leading role in all stages of a research process. This chapter has made a case for actively involving children in terms of addressing children's rights (Alderson, 2000; Hart, 1997) and also recognizing that children are experts in their everyday experience, have a particular knowledge set in relation to the (local) environment and act as agents within it (Baacke, 1985; Barratt & Barratt Hacking, 2008; Cutter-Mackenzie, 2009; Jans, 2004). The insights and perspectives that children can therefore bring to bear in environmental education research are far reaching in terms of children's local expertise and agency and also their capacity to contribute to the development

TABLE 41.3
Considerations in Developing Environmental Education Research Involving Children as Active Researchers

Research Process	Validation
1. Involving children	Is there an opportunity for children to volunteer? Has there been a satisfactory consultation process with children, their carers and the organization (if relevant)? Have children and their careers been properly informed prior to their consent? Are children assured of their anonymity/confidentiality? Is advice sought about children's psychological and physical welfare in considering their decision to be involved? Is account taken of the personal nature of data/experience for children; are children offered support in relation to this? Are children's aspirations for the research shared? Is care taken not to raise children's expectations in relation to their aspirations for change when this might not be possible? Does the research offer the possibility of a positive and meaningful outcome in relation to children's aspirations? Does the child have the opportunity to shape decisions about their rights and responsibilities in the research and those of the adults (i.e. is a "local democracy" established, Barratt and Barratt Hacking, 2009)? Is there support on offer if the child chooses to withdraw; does the child have someone they can talk to?
2. Exploring what research is	Has time been given for children to properly consider their thinking about what research is and why we do it? (Are children's prior experiences of research activity, e.g. enquiries at school, drawn upon)? Are the principles of research understood by all participants? For example, that: (i) there is a defined and systematic process to seek new knowledge (ii) there is a question that you try to answer (iii) there are appropriate methods to answer your question (iv) the importance of corroborating data is agreed (v) findings are based on data (vi) there is an ethical framework to govern your research approach including in terms of the ethics of researching people and environments.
3. Field of study, research design and context	Is the field of study negotiated and/or adapted with children? Does it address children's interests and aspirations? Are children's ideas and concerns taken seriously and incorporated into the research design? Do children have a say in bounding the scope of the research? Is the research context of interest and relevance to children, for example, a local place or space or an issue of concern? Will the research allow children to explore matters of importance to them? Is the research context accessible and safe for children?
4. Research questions	Are the research questions devised or modified by children? Are the research questions framed in children's own words/language? Are the research questions of interest and importance to children?
5. Methodology and methods	Do children have the opportunity to innovate, devise and/or select methodology and methods? Are tools accessible to all children regardless of their ability/age/literacy skills? Are verbal and nonverbal methods considered? In the case of verbal methods are children's own words used?
6. Data collection and storage	Are children involved in data collection? Are children supported in how to use new equipment and technology for collecting data? Has there been a full discussion of health and safety issues when collecting data, for example, outdoors? Are risks planned for and managed safely? Have the children agreed an ethical framework for gathering data from people/ children (including taking images) and for gathering data in the environment? Do children have a say in where and how data is organized and stored? Are children informed about data protection issues to support them in making data collection and storage decisions? Do the children have access to the data they have collected?
7. Data analysis and findings	Has time been given to talk to children about what data analysis is? Do children have the opportunity to talk about and devise their own ways of analyzing data? Are children offered support in learning about procedures and methods of data analysis? Do children have the opportunity to corroborate and check their findings? Are the children content that the findings reflect the data and its analysis?

(*Continued*)

TABLE 41.3 (Continued)

Research Process	Validation
8. Reporting and dissemination	Do children talk about how they would like to report their findings? Do reports incorporate children's voices (verbal and nonverbal)? Do children have opportunities to report verbally, nonverbally and through both formal and creative presentations? Do children have opportunities to identify and decide on the methods and for as for dissemination? Is there a child-receptive audience? Can children talk with other children about the process and findings? Are children's findings disseminated using children's voices? Has care been taken to discuss children's expectations for the outcomes of dissemination?
9. Outcomes	Is there the opportunity for the children and adults involved in the research project to reflect on the outcomes of the research (with other children and adults involved) including (i) new knowledge arising from the research (ii) their personal growth/learning (iii) the positive impact of the research, for example, on a community or wildlife or environment (iv) how the research might have been improved (v) future research directions (vi) how children/adults might use/build on their new skills in the future?

(Adapted from Barratt Hacking and Barratt, 2007).

of innovative and inclusive research methods. The analysis of the authors' projects demonstrates that children can develop and use innovative child friendly research methods; findings suggest that children have different experiences of the environment to adults and value different aspects of it. Children's active involvement in the research projects made this knowledge more visible, developed children's transferable skills and motivated and engaged children. The projects also demonstrated that children's research can enhance their contribution to school and community decision making albeit this necessitates efforts to rebalance the typical power relations between adults and children in school and community settings.

In concluding this chapter, we challenge environmental education researchers to further consider, discuss and critique children's roles in research. We propose that this debate focuses on the potential of children as collaborators in research rather than objects of investigation or discussion. The research methodology "children as active researchers" has received limited attention in environmental education research. Researchers, teachers, politicians, policy makers and curriculum developers demonstrate concern about the sustainability of children's futures and identify children as primary participants in environmental education. Nevertheless, children are not often positioned as researchers who can bring valid and new views or voices to educational practice and policy. As Williamson (2005, p. 1) points out "to involve children more directly in research can rescue them from the silence and exclusion, and from being represented, by default, as passive objects."

Notes

1. The scope of the chapter does not extend to children's participation in environmental education programs, for example, monitoring and improving habitats; while important and worthwhile the authors would characterize this type of activity as environmental education practice, rather than children as active researchers as defined in this chapter.
2. The term "photovoice" has been used to describe the use of photographs to represent "voice."
3. It is important to acknowledge here that children's opportunity to participate as researchers naturally would vary depending on the social, cultural, religious, and historical context where the research is actually taking place (Fleer & Quinones, 2009).

References

Alderson, P. (2000). Children as researchers: The effects of participation rights on research methodology. In P. Christensen & A. James (Eds.), *Research with children, perspectives and practices* (pp. 241–257). New York, NY: Falmer Press.

Alderson, P. (2001). Research by children. *International Journal of Social Research Methodology, 4*(2), 139–153.

Arnstein, S. R. (1969). A ladder of citizen participation. *Journal of the American Planning Association, 35*(4), 216–224.

Baacke, D. (1985). *Die 13-bis 18-jährigen: Einführung in probleme des jugendalters [13 to 16 year olds: Introduction to problems of adolescence].* Weinheim: Beltz.

Ballantyne, R., Connell, S., & Fien, J. (2006). Students as catalysts of environment change: A framework for researching intergenerational influence through environmental education. *Environmental Education Research, 12*(3–4), 413–427.

Barratt, R., Barratt Hacking, E., & Scott, W. A. H. (2005). Listening to children (L2C): A collaborative school-based research project. *Teaching Geography, 30*(3), 137–141.

Barratt, R., & Barratt Hacking, E. (2008). A clash of worlds: Children talking about their community experience in relation to the school curriculum. In A. D. Reid, B. B. Jensen, J. Nikel, & V. Simovska (Eds.), *Participation and learning. Perspectives on education and the environment, health and sustainability* (pp. 285–298). Dordrecht: Springer.

Barratt Hacking, E., Scott, W. A. H., Barratt, R., Talbot, W., Nicholls, D., & Davies, K. (2005). Listening to children: Using students' local environment experience in curriculum planning. *Environmental Education, 79*(July), 8–10.

Barratt Hacking, E., Scott, W. A. H., Barratt, R., Talbot, W., Nicholls, D., & Davies, K. (2007). Education for sustainability: Schools and their communities. In J. Chi-Lee & M. Williams (Eds.), *Environmental and geographical education for sustainability: Cultural contexts* (pp. 123–137). New York, NY: Nova Science Publishers, Inc.

Barratt Hacking, E., Scott, W. A. H., & Barratt, R. (2007). Children's research into their local environment: Stevenson's gap and possibilities for the curriculum. *Environmental Education Research, 13*(2), 225–244.

Barratt Hacking, E. & Barratt, R. (2007). *Children researching their urban environment: Developing a methodology*. Paper presented at ASPE/ BERA Seminar (Special Interest Group on Primary Teachers' Work), Educational Research with Children, Methodological Issues. July 12, 2007, University of Bath.

Barratt Hacking, E. & Barratt, R. (2009). Children researching their urban environment: Developing a methodology. *Education 3–13, 37*(4), 371–384.

Bartlett, S., & Minujin, A. (2009). The everyday environments of children's poverty. *Children, Youth and Environments, 19*(2), 1–11. Retrieved February 23, 2010, from http://www.colorado.edu/journals/cye

Bell, N. (2008). Ethics in child research: Rights, reason and responsibilities. *Children's Geographies, 6*(1), 7–20.

Borgon, L. (2007). Conceptions of the self in early childhood: Territorializing identities. *Educational Philosophy and Theory, 39*(3), 264–274.

Bronfenbrenner, U. (1979). *The ecology of human development*. Cambridge, MA: Harvard University Press.

Buttimer, A. (1980). Home, reach, and a sense of place. In A. Buttimer & D. Seamon (Eds.), *The human experience of space and place* (pp. 166–187). London, UK: Croom Helm.

Cairns, K. (2009). A future to voice? Continuing debates in feminist research with youth. *Gender and Education, 21*(3), 321–335.

Chawla, L. (2006). Learning to love the natural world enough to protect it. *Barn, 2*, 57–78.

Chawla, L. (2007). Childhood experiences associated with care for the natural world: A theoretical framework for empirical results. *Children, Youth and Environments, 17*(4), 144–170.

Christensen, P., & James, A. (2000). *Research with children, perspectives and practices*. New York, NY: Falmer Press.

Clark, J., Dyson, A., Meagher, N., Robson E., & Wootten, M. (2001). *Young people as researchers: Possibilities, problems and politics*. York, UK: Youth Work Press.

Cook-Sather, A. (2006). Sound, presence, and power: "Student voice" in educational research and reform. *Curriculum Inquiry, 36*, 359–390.

Cook-Sather, A. (2007). Resisting the impositional potential of student voice work: Lessons for liberatory educational research from poststructuralist feminist critiques of critical pedagogy. *Discourse: Studies in the Cultural Politics of Education, 28*(3), 389–403.

Currie, D. H., Kelly, D. M., & Pomerantz, S. (2007). Listening to girls: Discursive positioning and the construction of self. *International Journal of Qualitative Studies in Education, 20*(4), 377–400.

Cutter-Mackenzie, A. (2006). *2006 Sustainables Challenge Research Report*. Melbourne, VIC: Gould Group.

Cutter-Mackenzie, A. (2007). *2007 Sustainables Challenge Research Report*. Melbourne, VIC: Gould Group.

Cutter-Mackenzie, A. (2008). *2008 Sustainables Challenge Research Report*. Melbourne, VIC: Sustainability Victoria.

Cutter-Mackenzie, A. (2009). *Children as researchers: Exploring the possibilities and challenges in environmental education*. Paper presented at the Fifth World Environmental Education Congress: Earth, Our Common Home.

David, M., Edwards, R., & Alldred, P. (2001). Children and school-based research: 'Informed consent' or 'educated consent'? *British Educational Research Journal, 27*(3), 347–365.

Department of Environment and Sustainability & Department of Primary Industries. (2005). *Learning to live sustainably*. Melbourne, VIC: Victorian Government.

European Network of Ombudspersons for Children. (2007). Researching Children Conference 2007. Retrieved December 1, 2008 from http://www.ombudsnet.org/enoc/resources/infodetail.asp?id=14996

Fielding, M. (2001). Students as radical agents of change. *Journal of Educational Change, 2*, 123–141.

Fielding, M. (2004a). Transformative approaches to student voice: Theoretical underpinnings, recalcitrant realities. *British Educational Research Journal, 30*, 295–311.

Fielding, M. (2004b). "New wave" student voice and the renewal of civic society. *London Review of Education, 2*, 19–217.

Fielding, M. (2007). Beyond "voice": New roles, relations, and contexts in researching with young people. *Discourse: Studies in the Cultural Politics of Education, 28*(3), 301–310.

Finlay, L., & Gough, B. (2003). *Reflexivity: A practical guide for researchers in health and social sciences*. Oxford, UK: Blackwell.

Fleer, M., & Quinones, G. (2009). A cultural-historical reading of 'children as researchers'. In M. Fleer, M. M. Hedegaard, & J. Tudge, (Eds.), *Constructing childhood: Global-local policies and practices. World Year Book Series* (pp. 86–107). New York, NY: Routledge. Retrieved July 22, 2010, from http://books.google.co.uk/books?id=6fNNzTqEhmkC&pg=PA86&lpg=PA86&dq=Fleer+and+Quinones&source=bl&ots=8Xr_bnzKZh&sig=PWOvlUNtWgDOBdaFLXopsqDpVcI&hl=en&ei=H1yIS4LZKsa04Qb62ZWvDw&sa=X&oi=book_result&ct=result&resnum=3&ved=0CAsQ6AEwAg#v=onepage&q=&f=false

Flutter, J., & Rudduck, J. (2004). *Consulting pupils: What's in it for schools?* London, UK: Routledge Falmer.

Gayford, C. (2009). *Learning for sustainability from the pupils' perspective*. Godalming, Surrey: Worldwide Fund for Nature.

Gould League. (2009). The Sustainables Challenge. Retrieved April 1, 2009, from http://www.gould.org.au/html/JoinTheSustainablesChallenge.asp

Gruenewald, D. A., & Smith, G. A. (Eds.). (2008). *Place-based education in the global age: Local diversity*. New York, NY: Lawrence Erlbaum.

Hart, R. (1992). Children's Participation: From tokenism to citizenship. Innocenti Essays, Num. 4. Florence: UNCIEF, International Child Development Centre.

Hart, R. A. (1997). *Children's participation: The theory and practice of involving young citizens in community development and environmental care*. London, UK: Earthscan.

Hart, R. A. (2008). Stepping back from "the ladder": Reflections on a model of participatory work with children. In A. Reid, B. B. Jensen, J. Nikel, & V. Simovska (Eds.), *Participation and learning: Perspectives on education and the environment, health and sustainability* (pp. 19–31). Dordrecht: Springer.

Heath, S., Charles, V., Crow, G., & Wiles, R. (2007). Informed consent, gatekeepers and go-betweens: Negotiating consent in child- and youth-orientated institutions. *British Educational Research Journal, 33*(3), 403–417.

Hedegaard, M., & Fleer, M. (2008). *Studying children: A cultural-historical approach*. London: Open University Press.

James, A., & Prout, A. (1997). A new paradigm for the sociology of childhood? Provenance, promise and problems. In A. James & A. Prout (Eds.), *Constructing and reconstructing childhood* (pp. 7–34). London, UK: RoutledgeFalmer.

James, A. (2007). Giving voice to children's voices: Practices and problems, pitfalls, and potentials. *American Anthropologist, 109*(2), 261–272.

Jans, M. (2004). Children as citizens: Towards a contemporary notion of child participation. *Childhood, 11*(1), 27–44.

Kellett, K. (2005a). *How to develop children as researchers*. California, US: Paul Chapman Publishing.

Kellett, M. (2005b). Children as active researchers: A new research paradigm for the twenty-first century? ESRC National Centre for Research Methods, NCRM/003. Retrieved June 22, 2008 from http://oro.open.ac.uk/7539/1/MethodsReviewPaperNCRM-003.pdf

Kirby, P. (2001). Participatory research in schools. *Forum, 43*(2), 74–77.

Kirby, P. (2004). Involving young people in research. In B. Franklin (Ed.), *The new handbook of children's rights, comparative policy and practices* (pp. 268–284). London, UK: Routledge.

Lahman, M. (2008). Always othered. Ethical research with children. *Journal of Early Childhood Research, 6*(3), 281–300.

Lansdown, G (2005). *The evolving capacities of the child.* Florence, Italy: UNICEF.

Lincoln, Y. S., & Denzin, N. K. (2000). The seventh moment: Out of the past. In N. K. Denzin & Y. S. Lincoln (Eds.), *Handbook of qualitative research* (2nd ed., pp. 1047–1065). Thousand Oaks, CA: Sage.

Lofland, J., & Lofland, L. H. (1995). *Analyzing social settings: A guide to qualitative observation and analysis* (3rd ed.). Belmont, CA: Wadsworth.

Malewski, E. (2005). Epilogue: When children and youth talk back: Precocious research practices and the cleverest voices. In L. D. Soto & B. B. Swadener (Eds.), *Power and voice in research with children* (pp. 177–190). New York, NY: Peter Lang Publishing.

Mandell, N. (1988). The least adult role in studying children. *Journal of Contemporary Ethnography*, 16: 433–437.

Masson. (2005). Researching children's perspective: Legal issues. In K. Sheehy, M. Nind, J. Rix, & K. Simmons (Eds.), *Ethics and research in inclusive education: Values into practice* (pp. 23–241). New York: RoutledgeFalmer.

Morrow, V. (2008). Ethical dilemmas in research with children and young people about their social environments. *Children's Geographies, 6*(1), 49–61.

National Foundation for Educational Research. (undated). Listening to children and young people reading list. Retrieved July 31, 2009 from http://www.nfer.ac.uk/nfer/index.cfm?2BAF4157-A3E9-9DD9-C2F0–EBE3F0628611

Piper, H., & Frankham, J. (2007). Seeing voices and hearing pictures: Image as discourse and the framing of image-based research. *Discourse: Studies in the Cultural Politics of Education, 28*(3), 373–387.

Place-based Education Evaluation Collaborative website http://www.peecworks.org/index

Promise of Place website http://www.promiseofplace.org/

Punch, S. (2002). Research with children: The same or different from research with adults? *Childhood, 9*(3), 321–341.

Rickinson, M. (2001). Learners and learning in environmental education: A critical review of the evidence. *Environmental Education Research, 7*(3), 207–320.

Rudduck, J. (2007). Student voice, student engagement, and school reform. In D. Thiessen & A. Cook-Sather (Eds.), *International handbook of student experience in elementary and secondary schools* (pp. 587–610). Dordrecht, The Netherlands: Springer.

Sauvé, L. (2001). L'éducation relaitve à l'environment. Une dimension esentielle de l'éducation fondementale. In C. Goyer & S. Laurin (Eds.), *Entre culture, compétence et contenu: La formation fondamentale, unespace é redéàfinir* (pp. 293–318). Montreal, Canada: Logique.

Sauvé, L. (2002). Environmental education: Possibilities and constraints. *Connect, 27*(1–2), 1–4. Retrieved June 22, 2008 from http://unesdoc.unesco.org/images/0014/001462/146295e.pdf

Save the Children. (2004). *So you want to involve children in research?* Sweden: Save the Children.

Scott, W. A. H., Barratt Hacking, E., & Barratt, R. (2006). *Listening to children: Environmental perspectives and the school curriculum.* End of award report. Retrieved July 27, 2009 from http://www.esrcsocietytoday.ac.uk/ESRCInfoCentre/ViewOutputPage.aspx?data=v9XrjLJ6xhGpro5KhccM%2fxusZ8ZrylGfjhU7bITjc3tHtG5Po3WfHz3rbaa2VTqoRgsXWD%2bRk2YUQppkD1KRYQzxgDcyUo-CSSErr–nRTtH5Yc7Vpj8%3d&xu=0&isAwardHolder=&isProfiled=&AwardHolderID=&Sector=

Scott, W. A. H., Barratt Hacking, E., & Barratt, R. (2007). ESD: Bringing students' community experience into schools. In R. McKeown (Ed.), *Good practices in education for sustainable development: Teacher education institutions.* Good Practices Paper No. 1. Paris: UNESCO.

Seamon, D. (1979). *A geography of the lifeworld.* London, UK: Croom Helm.

Seamon, D. (1984). Phenomenologies of environment and place. *Phenomenology and Pedagogy, 2*(2), 130–135. Retrieved March 16, 2008 from http://www.phenomenologyonline.com/articles/seamon.html

Shier, H. (2001). Pathways to participation. *Children and Society, 15*, 107–117.

Skelton, T. (2008). Research with children and young people: Exploring the tensions between ethics, competence and participation. *Children's Geographies, 6*(1), 21–36.

Somerville, M., Power, K., & de Carteret, P. (Eds.). (2009). *Landscapes and learning: Place studies for a global world.* Rotterdam: Sense Publishers.

Stapp, W. B., Wals, A., & Stankorb, S. (1996). *Environmental education for empowerment: Action research and community problem solving.* Dubuque, IA: Kendall/Hunt Publishing.

State of Victoria Department of Environment and Sustainability. (2006). *Our environment, our future: Sustainability action statement.* Melbourne, VIC: Victorian Government.

The Freechild Project. (2008). Ladder of participation. Retrieved February 1, 2008 from http://www.freechild.org/ladder.htm

Thiessen, D., & Cook-Sather, A. (2007). *International handbook of student experience in elementary and secondary school.* Netherlands: Springer.

Treseder, P. (Ed.). (1997). *Empowering children and young people: Training manual promoting involvement in decision-making.* London, UK: Save the Children in association with Children's Rights Office.

Tuan, Yi-Fu. (1977). *Space and place: The perspective of experience.* Minneapolis, MN: University of Minnesota Press.

Victorian Government. (2007). The sustainables household challenge. Retrieved August 14, 2008 from http://www.dse.vic.gov.au/thesustainables/teachers.htm

Waller, T. (2006). 'Don't come too close to my octopus tree': Recording and evaluating young children's perspectives of outdoor learning. *Children, Youth and Environments, 16*(2), 75–104. Retrieved February 23, 2010 from http://www.colorado.edu/journals/cye

Whitty, G., & Wisby, E. (2007). Whose voice? An exploration of the current policy interest in pupil involvement in school decision making. *International Studies in Sociology of Education, 17*(3), 303–319.

Wiles, R., Heath, S., Crow, G., & Charles, V. (2005). *Informed consent in social research: a literature review.* Southampton, UK: ESRC National Centre for Research Methods. Retrieved April 3, 2008 from http://eprints.ncrm.ac.uk/85/1/MethodsReviewPaperNCRM-001.pdf

Williamson, B. (2005). Young people as researchers. Futurelab. Retrieved June 24, 2009 from http://www.futurelab.org.uk/resources/publications-reports-articles/web-articles/Web-Article530

Willow, C., Marchant, R., Kirby, P., & Neale, B. (2004). *Young children's citizenship: Ideas into practice.* York, UK: Joseph Rowntree Foundation. Retrieved March 9, 2009 from http://www.jrf.org.uk/sites/files/jrf/1859352243.pdf

42

Collaborative Ecological Inquiry

Where Action Research Meets Sustainable Development

Hilary Bradbury-Huang
Oregon Health & Science University, USA

Ken Long[1]
US Army Command and General Staff College, USA

> *We cannot regard truth as a goal of inquiry. The purpose of inquiry is to achieve agreement among human beings about what to do, to bring about consensus on the ends to be achieved and the means used to achieve those ends. Inquiry that does not achieve coordination is not inquiry, but simply wordplay.*
>
> —Richard Rorty, 1999.

We have inherited an educational model that generally separates thinking and doing. It has been suggested that conventional education is a product of the machine logic of the industrial era, with its concern for predictability, reliability, and external validity (Senge et al., 2000). The action research paradigm, on the other hand, offers a complementary system of knowledge creation that addresses complex sustainability challenges in an integrative way. As recently articulated by the international community of action researchers:

> At this time we are called to engage with unprecedented challenges that are interrelated and compounding; challenges such as poverty and injustice, climate change, globalization, the regulation of science and technology, the information and communication technology revolution, inequalities and fundamentalisms of all types. Conventional science and its conduct are part of these problems. Action researchers, therefore, are concerned with the conduct and application of research. We acknowledge the complexity of social phenomena and the nonlinearity of cause and effect and see that the best response to such complexity is to abandon the notion of understanding as a product of the enterprise of a lone researcher, and to engage local stakeholders, particularly those traditionally excluded from being part of the research process, in problem definition, research processes, interpretation of results, design for action, and evaluation of outcomes. In this way, we step beyond what has been labeled "applied research," into the democratization of research processes and program design, implementation strategies, and evaluation. (Bradbury-Huang and advisory editorial board of *Action Research* journal, 2009)

Conventional research and education has certainly produced many material advantages for the inhabitants of the Western world, but it appears unequal, on its own, to the task of providing practical and emancipatory pathways around the systemic sustainability challenges we face. In fact, all major life systems are now declining (Hawken, 2009; World Watch Institute [WWI], 2008).

We may therefore see conventional education at the college level as both a culprit in our current crisis and, more usefully, as a potent vector for positive change. Stringer (2004) underscores the importance of using action research in guiding school improvement because, he argues, to change culture, we must change schools. His work in transforming the nascent education systems of East Timor and Aboriginal areas of Australia, demonstrates how action research can effectively engage the struggle over dominant values by pragmatic methods, such as improving teacher efficacy and through that, the capacity of a community of practice in producing their desired educational outcomes.

Of course, attempts at education reform are hardly new. However, given the extent of our sustainability challenge, rarely have the stakes been higher. We need enlightened practice with regard to the kinds of processes that will incorporate multiple ways of knowing and points of view. We need to reform the industrial age processes of mass education, which otherwise will continue to produce consumerism, generate attention to immediate goals, and a competitive drive that divides the world into winners and losers. It's not as if we can afford to have any losers in the challenge to craft sustainable processes. However we do need safe spaces to develop, practice and master new ways of grappling with longer term, systemic challenges. The action research model can serve as a critical component in education toward sustainability.

What Is Action Research: A Note on Historical Background

Action research is an orientation to knowledge creation that arises in a context of practice and requires researchers to work *with* practitioners. Unlike conventional social science, its purpose is not primarily or solely to understand social arrangements, but also to effect desired change as a path to generating knowledge and empowering stakeholders. We may therefore say that action research represents a *transformative orientation to knowledge creation in that action researchers seek to take knowledge production beyond the gate-keeping of professional knowledge makers.*

Action research is frequently defined as "a participatory process concerned with developing practical knowing in the pursuit of worthwhile human purposes. It seeks to bring together action and reflection, theory and practice, in participation with others in the pursuit of practical solutions to issues of pressing concern to people and more generally the flourishing of individual persons and their communities, (Reason & Bradbury, 2008, p. 4).

The roots of action research, especially in the Southern Hemisphere, firmly embrace the liberationist influences of Karl Marx and Paolo Freire. In the social science departments of the Northern Hemisphere, action research is more often said to have originated in the 1950s with the social-psychology work of Kurt Lewin. It must be noted that there are questions about whether it was John Collier or Kurt Lewin who originally coined the term.

Today action research is receiving resurgent interest especially in the "helping" professions, i.e., in the fields of education, social work, international development, healthcare. Action research also goes by different names, e.g., in healthcare it is known better as "community-based participatory research for health" (Minkler & Wallerstein, 2008), and the overall approach is associated with increasing popular and large National Institute of Health grants in the United States to make healthcare research more impactful.

Why Action Research Is Appropriate for Moving Toward Sustainability

Open systems theory (Ison, 2008) offers important insights into achieving sustainable environments, processes, organizations, communities, and people. It emphasizes the importance of robustness, adaptability, coordination, cooperation, and goal-setting, and the ability to learn from feedback signals. The dynamic and nonlinear quality of our social and physical environments reward people and organizations that cooperate, to act locally in support of global values. The systems movement shares a long history of success with action research, and suggests that quality should be pursued in many places simultaneously: as individual actors with the capacity to make a difference through our own action; in the quality of our relationships with others and in our concerted, cooperative actions; in the nature and quality of the groups we form to leverage our common interests for mutual, sustainable benefit; in the goals and values we promote to improve the environment we live in for the long term; in the ways we collaborate collectively to set and seek goals with morally-informed processes. The literature reveals action research to be especially well suited within education for improving our knowledge and practice in all of these dimensions.

At its core, action research encourages collaborative learning and stakeholder oriented approaches to creating systems wide desired futures. It privileges the importance of experiential learning (Kolb, 1984) for generating knowledge in the context of action. It proceeds not from novel questions, but from stakeholders' practical needs and involves multiple ways of generating and expressing knowledge through cycles of reflection on action. Its quality is assessed by a spectrum of concerns, from meeting practical needs in a partnership mode, to producing knowledge for stakeholders beyond the immediate effort.

Figure 42.1 offers some key concepts from a systems approach to educating around sustainability issues. Noted along the arrow from left to right are the powerful determiners of educational experience, essentially defining the norms of the educational experience. In that context come the educator and her/his students who work together to bring both conceptual and practical knowledge to the fore. As a consequence the organizational and larger environment is impacted. The figure reminds us of the stakeholders that play an important role in any action research learning or research efforts, and how project design considerations must include scope, stakeholders, resources and values.

In the past decade we see an increase in the types of studies and literature appearing at the intersection of environmental/sustainability education and action research. The common themes of these studies are the coalitions of community members creating action-oriented groups that transcend conventional boundaries and channels of power. For example, Henderson, Hawthorne, and Stollenwerk (2000), found that action research can bring diverse groups together creatively to support cross-cultural collaborations in environments where conventional problem solving methods fail. This follows an explicit emphasis on developing cross cutting networks of practice as outlined in McCaleb (1997) who found action research particularly effective in bringing together families, teachers, students, and members within the local community to shape their learning experiences in ways that reinforced important local values, particularly when the local community was richly diverse with competing values. Action research was therefore effective in promoting the role of teachers as change agents.

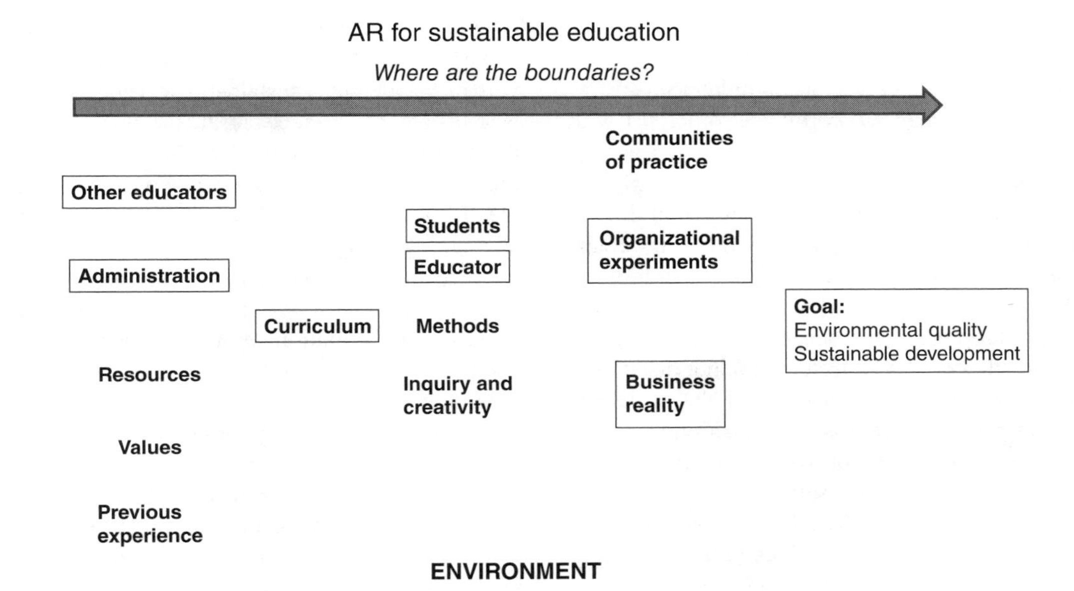

Figure 42.1

Additionally we see in this effort that Sagor (2000) describes the propensity of action research to transform the curriculum and leadership processes of schools, and praises its power to create networks of shared values from among schools, families, and local communities, removing barriers between these disparate groups. He highlights the action research process' nonmechanistic and locally adaptive properties, and how it is well-suited to incorporate and leverage dynamic human and environmental potentials, while contributing to the enduring infrastructure changes that are needed to lock-in progress.

Ainscow, Booth, and Dyson (2004) emphasize the importance of making lasting changes driven from the bottom-up in collaborative action research. They show how collaborative action research is adept at provoking and surfacing contentious social and moral beliefs about which we can explicitly engage to promote personal growth and organizational change. Craig and Ross (2008) focus on action research as an ethical means of improving teacher efficacy as curriculum shapers and knowledge makers. They find a wide variety of ways in which action research empowers teachers to create, find and adaptively apply quantitative findings to local circumstances. Schubert (2008) addresses how action research can surface issues of co-opting curriculum as an instrumentality of control, and how action research can set the conditions for a principled discourse about power, authority, validity, and standards. Action research provides a robust framework for rapidly and effectively organizing a variety of change agents around shared local values, in an iterative process that will vary based on local circumstances, yet sharing a central core of criteria concerned with results, methods, quality and principles.

The paper proceeds with examples of the application of action research based education, which are then used to underscore key principles and methodological characteristics.

Core Characteristics and Quality Criteria of Action Research

Leaders in the global community of action research (Bradbury-Huang and advisory editorial board of *Action Research* journal, 2009) describe the work of action research as providing models for increasing the relevance of conventional social research to wider society:

> ". . . what makes our work fundamental to the revitalization of social research more generally, we argue, lies in its orientation towards taking action, its reflexivity, the significance of its impacts and that it evolves from partnership and participation."

The core characteristics of action research inform the emerging dialogue centering on the quality criteria whereby action research projects in a variety of research settings may be evaluated.

By *partnership and participation* we are referring to the quality of the relationships we form with primary stakeholders and the extent to which all stakeholders are appropriately involved in the design and assessment of inquiry and change. By *actionable* we refer to the extent to which work provides new ideas that guide action in response to need as well as our concern with developing action research crafts of practice in their own terms. By *reflexive* we mean the extent to which the self is acknowledged as an instrument of change among change agents and our partner stakeholders. By *significant* we mean having meaning and relevance beyond an immediate context in support of the flourishing of persons, communities, and the wider ecology.

When we think of actionability in the context of sustainability, we see how *practical needs* of stakeholders comprise financial (economic), human-social, and environmental concerns. While it is easy to give up trying to focus on all three in preference for one or two of these concerns, it is therefore important that the leaders of the project maintain the larger focus on "significance."

But what does it mean to work *in partnership* with stakeholders in sustainability efforts? There are a variety of possibilities. Some partnerships exist simply to complete a single loop project, e.g., students mobilize to bring organic food to their campus, in which they focus mostly on technical, tactical learning rather than looking into the values that shape the context (Argyris & Schoen, 1974). In single loop work, the deeper values driving the status quo are neither deeply reflected upon nor discussed; they are simply replaced by a new set, if the project is successful. Generally speaking, however, action researchers now agree that some attempt to integrate both single and double loop reflection are useful for the work over time (Reason & Bradbury, 2001). Double loop emphases imply grappling with deeper values and offer a way in which the emancipator aims of action research can be added to a focus on practical outcomes. Double loop projects therefore require more systematic development of cycles of action and reflection so that the values that are driving actions may be recognized more clearly. Part of the complex decision making around research design and focus therefore concerns creating balanced partnerships among those with differing perspectives.

Power sharing is typically easier to talk about than to do well. Some partnerships are characterized by a seamless balance of power among stakeholders. More common, however, is to have partnerships that are less balanced, e.g., a project may rely heavily on the skills of expert researchers to shape and carry it. The latter type of partnership does not automatically weaken research, since the expert or academic can choose to remain democratic within a project's scope. (Breu & Hemingway, 2005). On the other hand, one sees partnerships in which the action researcher is dominated by the practitioner interests forcing a subordination of an inquiry perspective. Facilitators of action research will usually offer repeated gentle reminders of the power dynamics of which partners are consciously and often unconsciously a part. A more balanced partnership, as Phillips suggests, "involves the building of industry clusters and social capital, and action-oriented communities that can embrace change," (Phillips, 2005, p. 548)."

We now turn to offer two illustrations of executive level education in which action research methods are exemplified. Both suggest the possibility of a college/academic institution itself operating as an agent of change by offering a relatively neutral site in which stakeholders across government, civil society, and business can convene and in which the rich mix of researchers and practitioners can work to generate cycles of reflection on action.

Illustration I: Educating for Sustainable Systems

In November 2008, twenty-five people gathered in Southern California for the annual strategic retreat of PLANET—the US association of landscape professionals. Mostly business owners themselves, these leaders would grapple together with the challenge of sustainability. Their goal was to define best sustainable practices for their industry and promulgate them to industry stakeholders. The work was timely and pressing. In the area in which they were meeting, for example, 70 percent of water was used for landscaping despite an ongoing drought and growing population.

In reality the meeting had begun many months earlier with a phone call to the first author, requesting help with design and facilitation of the meeting using an "action research" approach (which they admitted had come highly recommended but was entirely unfamiliar to them). The first phone call quickly required a second teleconference, because at the heart of the action research approach is a need to involve the stakeholders. Stakeholders, i.e., anyone with a stake in their work, meant expanding the usual retreat population to include voices for environmental impact, employee health and welfare, soil scientists, etc. PLANET was not used to conducting its annual retreat with "outsiders" and so for the first time since inception in the 1970s, water regulators were to be invited into the meetings, customer surveys were to be done beforehand, experts from different disciples were also invited to share about specific best practices with urban ecology. It was agreed that a research agenda, to support the association's move to sustainability, would be part of the strategic outcome. Some of that research would be conventional, e.g., research done by professionals that would lead to peer-reviewed measures for a set of predefined technical questions. Some of that research, however, would involve collaborative experiments by the businesses themselves, thus continuing the action research process of reflecting on action in the context of practice. Some of that research would also be a mix of the two types.

The three day executive seminar followed the cycle of experiential learning (Kolb, 1984) and much attention went to creating a strong relational space to start (Senge, Lichtenstein, Kaeufer, Bradbury, & Carroll, 2007) with many opportunities for peer interaction.

After a brief review of goals for the days ahead, a lengthy "check in" of all present began in which participants shared about their business and their experience of sustainability issues to date. This had the effect of allowing participants become fully present to what they hoped to accomplish (Isaacs, 1999) while highlighting the complementary skills and wisdom in the room. Interactive reflection was then invited on six interdependent "actionable questions" concerning sustainability. These questions were developed (drawing much on the work of *The Natural Step*, Robert, 2002) through reflection on the fact that Earth is a closed system and that in natural systems there is no waste (e.g., trees "inhale" CO_2 and "exhale" O_2; rain evaporates to become a cloud that in turn rains). In contrast to nature, conventionally designed enterprise produces a lot of waste (landfill, gas combustion) that burdens us with unnecessary environmental pollution, illness and clean up costs. The questions forced attention on how to change current practices. The questions prompted design of a more sustainable system for PLANET practitioners and also suggest generic categories that any system may engage when undertaking action research for sustainability.

In keeping with a desire to move people from reflective observation to abstract conceptualization, the participants were asked to gain familiarity with the six sustainability questions by first applying them to a lifecycle analysis of a cotton T-shirt. The shirt is both tangible (many participants admit to owning too many) and a manifestation of ecological, technical, social, and organizational resources (Rivoli, 2005). In teams, participants discussed how the beginnings of the shirt emerges from the cotton fields of Texas (made possible by tax subsidies and herbicides), gets transformed in the factories of Shanghai (filled with low paid women, mostly erstwhile rural dwellers), and is transported back to destination stores, via a long, fossil- fueled cargo transshipment and distribution network. The ensuing team learning—the participants rendered the complexity of social, financial, and environmental inputs to a simple T-shirt on flip charts—opened eyes both to the challenge and opportunity that a conversation about sustainability implied. Thus a shared foundation was been created for moving away from the relatively simpler task of solving others sustainability problems (namely the T-shirt purveyors!), to conceptualizing their own system better, and in turn, moving into "execution" mode by considering how each question would help them devise a path forward for their own businesses. The second day started therefore, with the formation of execution teams that worked in increasing depth until the close of the first phase of the action research to articulate, e.g., a "2020 Sustainable Vision for PLANET," a list of best practices for current landscaping work, and defining a research agenda and process in support of the sustainability agenda.

In the final day, teams clarified opportunities, new business models were identified to meet short and longer term profit targets, customers could be educated to demand alternatives, and employees could become part of the green collar economy (Jones, 2008).

Illustration II: Sustainable Enterprise Executive Roundtable ("SEER")[2]

SEER originated among research faculty at University of Southern California (USC) whose initial collaboration at the Center for Sustainable Cities resulted, first, in a proposal to the U.S. Environmental Protection Agency (EPA) to fund work aimed at developing a model that described how value was generated or destroyed around use, reuse, and recycling of aluminum based products originating in the Pacific Rim. As part of this proposal, the EPA required letters from business partners expressing their willingness to allow access to data. Initial conversations with the business partners sparked an interest among researchers in going beyond primarily or solely describing the aluminum products networks to actively engaging these potential champions in efforts to align their overlapping systems around sustainable outcomes.

The nexus of the business network was readily identified as the Port of Los Angeles (referred to as "POLA") because it is a hub of importing aluminum and exporting recyclables. Thus the researchers' design focused increasingly on drawing together businesses whose interests and activities overlapped, even as their jurisdictional authorities did not. The port itself was very willing to support the work, not least for its own self-interest, among other motives. As a landlord, however, it has limited direct control over the activities of its users (e.g., the ocean carriers who rent space at dock).

The founding corporate cohort included: Port of LA, Disney (entertainment company), Toyota (automotives), Mattel (toy makers), CDM (an engineering company), YTI/NYK (Ocean Carriers/logistics), Volvo (truck makers), and Waste Management (trash haulers). To have the work last, both regional sustainability needs and individual participants' business needs would have to be met.

The companies' participants were introduced to a widely accepted framework developed by the Natural Step, an international think tank whose work on operationalizing sustainability had proven useful to many international companies. In this framework, an *integrated* approach balances concern for a healthy system, access to sustainable feedstocks of all kinds, and a profitable marketplace with the concept of not passing along environmental cleanup costs to future generations. Therefore going beyond the general definition of sustainability offered by the Brundtland commission, the researchers introduced the Natural Step's "system conditions" for achieving sustainability. A more operational definition was also introduced: *a sustainable system is one in which neither materials from the Earth's crust (e.g., oil) nor man-made toxins systematically increase, and in which the health of green spaces and basic human needs are secured.*

From this more operational definition, six concrete categories were derived, within which the participants could articulate "design questions" whose function was to get at what it might mean for the shared cargo transshipment system to move toward sustainability.

SEER Research Results

From a shared "mental model" based on the six systems redesign questions (Table 42.1), participants moved into productive conversation about what a sustainable end state might look like. It was agreed that future SEER projects would take steps toward reducing reliance on fossil fuel and toxins and move toward increasing green spaces and healthier communities. Working off the "same page" allowed participants to identify leverage points for sustainable change in the places in which they had accountability. Importantly, the question of how mutual benefit can

TABLE 42.1
Systems Redesign Questions for Sustainability

1. Carbon reduction How does the project result in systematic decrease of dependence on fossil fuels?
2. Toxicity reduction How does the project result in systematic decrease of production and use of bio-accumulating materials? e.g., how does it reduçe reliance on man-made toxins and plastics?
3. Increase in global and native ecosystem health How does the project result in systematic increase of bio-productive spaces? e.g., how does it reinvigorate unproductive land, set aside more habitat, increase forests, and coral reef?
4. Ethical engagement of stakeholders How does the project result in increase of social inclusion and access to resources? For example, how does it support environmental justice (EJ)?
5. Partnership in organizational networks How does the project increase collaboration within value networks so that organizations can do together what no one can do alone?
6. Leadership How does the project go about systematically increasing our ability to "walk the talk"—demonstrating and cultivating our own and others' leadership so that our actions are consistent with the demands of a sustainable world?

accrue from working collaboratively around some of the challenges identified moved the conversation from merely understanding toward *action that might make a difference*.

Along with the one focused collective project—developing a system's carbon calculator—, much "haphazard" learning was reported along the way, including numerous bilateral projects developed to address the interests of two companies. For example, Waste Management and Port of LA saw that natural gas captured from Waste Management's landfills could be used to fuel port vehicles and support the move away oil dependence. Additionally, guests invited from participant company's supply chains also brought new and different conversations to meetings, e.g., a representative from PepsiCo brought an idea for true recycling of plastic bottles (different from the typical down-cycling that occurs when bottles are made into something less economically valuable). Sparking a larger conversation about waste reduction also spearheaded a new project inside Mattel to rethink packaging design with end-of-life in mind. Like so many learning experiences, projects were emergent and their eventual value is hard to know immediately. Having patience appears to have been pragmatically useful for those involved.

Insights About Methodology: The Principles of Action Research in Action as Quality Criteria

Let's see how the principles of Action Research unfold in our two illustrative examples:

Partnership and participation: In the SEER project, we see a broad coalition of stakeholders spanning, academia, private and public organizations, ranging from multinational corporations to individual concerned citizens participated in meaningful ways to set agendas, scope problems, design interventions, engage in dialogue, define measures of effectiveness and performance, establish timelines, commit resources, engage in dialogue and inquiry and collaboratively produce results in a set of ongoing change projects. In the PLANET project, a voluntary coalition of concerned individuals and organizations expanded their network of concern to find partners across tradition boundaries to generate new insights and commitments to action.

Actionable: Both projects focused on actions and solutions that were suitable (appropriate for the problems identified), feasible (within current resource constraints) and acceptable (agreeable to the stakeholder community). The SEER project examined working business processes and standards to find improvements, while the PLANET project took action on industry-wide standards that could have immediate payoffs.

Reflexive: Both projects explicitly made changes in not only how the people and organizations viewed the situation, but in how they framed entire problem and opportunity sets, and expanded the range of behaviors that could be undertaken now and in the future. The SEER project envisioned new definitions for partnership and values for a community, while the PLANET project reinvented the terms for partnership and engagement with organizations that hadn't been at the table before. The projects included work on self, processes and cognition in addition to the direct work on the external problem.

Significant: Both projects made commitments and took action to make a meaningful difference in the present. The SEER project not only examined environmental quality that included meaningful change in product engineering and supply chain waste management but reached beyond to consider life cycle costs, life space impacts, and the health and welfare of the community they lived and worked in. The PLANET project considered the professional standards of an entire industry, how to educate their members professionally and how to reframe the boundaries and scope of their work.

In both we see effort to go beyond *propositional knowledge*. Heron and Reason (2008) extrapolate from the paradigm of experiential learning to define four general types of knowing. They argue for less emphasis on propositional knowing that currently dominates conventional education (and strongly aligns with the assimilating tendencies of most professional academics) to make room for experiential, practical and presentational knowing. Experiential knowing is the essence of experience in everyday life. It is our interpersonal relationships with others and how we interpret interactions with them. Our interpretations are based upon our collective experiences, but we must reflect on these experiences to learn more about our understanding of them. With every inclusive meeting, whether with stakeholders or coresearchers, each person will bring their own experiences, which shape their paradigms, tactics, or openness to communications and learning. Experiential knowing is the basis of all knowing, in that, without the experiences that have shaped our lives, we could not assimilate or construct meaning for interpersonal relationships nor our relationship to the world.

A group exercise, such as T-shirt in the PLANET illustration, or a port cruise to allows the huge scale of current production mechanisms to become real to participants (who otherwise know only their small portion of the larger system), aids a group to approach issues as systems, since it enables people to name the many variables of the systems problem, what systems thinker Russ Ackoff refers to as "the mess." However the "messy map" that results can seem overwhelming. Therefore the concept of "leverage point" is also crucial. These are places for intervention in the system that can lead to resolution. Differentiating between quick fix points (which can bolster confidence) and long term solution points is critical.

This, in turn, speaks to the importance of multiple perspectives and creating an atmosphere where all can contribute.

Dynamics of Action Research Across Time and Scale

Reason and Bradbury (2008) suggest that the most compelling and enduring action research will engage three modes of research/practice in one project, namely first, second and third person research/practice (Chandler & Torbert, 2003).

First-person action research/practice is the researcher's inquiring approach to his or her own life, the demand to make choices and act with awareness and understanding and to assess effects in the outside world while acting. First person research practice brings inquiry into more and more of our moments of action in the whole range of everyday activities. This inquiry is a foundational practice to monitor the impact of our behavior (Marshall & Mead, 2005).

Second-person action research/practice addresses our ability to inquire face-to-face with others into issues of mutual concern—for example in the service of improving our personal and professional practice both individually and separately. Second person inquiry starts with interpersonal dialogue and includes the development of communities of inquiry and learning organizations.

Third-person research/practice aims to extend these otherwise small scale projects to create a wider impact. Third person strategies aim to create a wider community of inquiry involving persons who, because they cannot be known to each other face-to-face (say, in a large, geographically dispersed corporation), have an impersonal quality. As Gustavsen points out, action research will be of limited influence if we think only in terms of single cases. We need to create a series of events interconnected in a broader stream of social movements or social capital (Gustavsen, 2003a, 2003b). Writing and other reporting of the process and outcomes of inquiries are important forms of third person inquiry.

Adding these components, perhaps especially first person work, is unusual in conventional education which largely bypasses engaging the learner in understanding and managing the primary learning instrument, namely her/his own mind. Many practices can be useful here, from journaling to self analysis of communication styles. Learning to meditate, while perhaps too radical a suggestion for mainstream executive education, may also provide the most robust transformations over time, as seen in studies at Fairfield University in Iowa, where meditation is a mandatory part of the educational endeavor.

TABLE 42.2
Contrasting Actionable and Conventional Education

Actionable Education	Conventional Education
Oriented around what is needed	Oriented around predefined, intellectual puzzle
Experiential	Conceptual
Emergent	Predefined by curriculum
Inquiry focused	Answers focused
Questions to stakeholders generate insight and action.	Questions of stakeholders are aimed at understanding and description only.
Interacting teams	Individualistic
Implications for practice interwoven	Implications for practice may be omitted
Invites in stakeholders	Stakeholders are kept beyond the classroom
First, second, third person research/practice	Third person research

First person research practice is best conducted in the company of friends and colleagues who can provide support and challenge. As a practice, first person work may indeed evolve into a second-person collaborative inquiry process. Third person research, which is not based in rigorous first person inquiry into one's purposes and practices, is open to distortion through unregulated bias. Finally, third person work draws attention to how to inform/impact a group of people not originally part of the project. A classic way to do that is to engage in the peer review process. To take just one example from the cases mentioned above, the carbon modeling work from SEER is moving through peer review (cf. Madachy et al., 2008). This third person work makes public the System Dynamics Modeling developed for the executive workshops. The workshops themselves constitute second person work that facilitated inquiry and initiatives in others, which, in turn, rested on the learning facilitator's attention to self-inquiry. That first-person work was critical for understanding how to best facilitate the group's willingness to seek benefit by joining with others in collective inquiry for support and challenge in developing their experiences and skills. This is but one illustration of the action research principle that cultivating personal insight and skills (first person work) is as important as learning how to facilitate the best group or team outcomes (second person work) which together allow for knowledge to be shared across the community of inquiry.

Why Critics of Action Research Should Rethink Their Concerns.[3]

There tend to be two common misconceptions of action research. Let's call the first the *"I intend to show my results to the folks in charge—that's action research,"* misconception. Keeping "at-a-distance" from practitioners and limiting interaction to those with formal power is not action research. Action research emerges from working with practitioners, hence the core emphasis on "partnership and participation."

Simply offering one's insights to practitioners about how change might best happen is not action research either. Consider for a moment when and how practitioners do engage with actionable knowledge—isn't it more often through personal experiential learning? Experiential learning can be usefully shaped by vicarious or substitute experience in association with (in partnership with!) a researcher who is close to the practitioner in their context of practice. It is naïve to believe there will be genuine interest in your work among practitioners if they have been treated as an afterthought.

But the second is more problematic—it's the "Action Research Work Is Just Sloppy Social Science" misconception. It helps to admit that there is indeed sloppy conventional *and* sloppy action research. However referring to sloppy conventional work as "action research" is mendacious. Based on the definitions above, and in much more detail elsewhere (see Reason & Bradbury, 2001, 2008), we hope such a mistaken identity can be unmasked. Beyond that however we may indeed imagine that action research is well designed and executed but draws confusion and disdain from conventional social scientists simply because it is different. When the quality of action research is to be evaluated using the criteria of conventional research only, aspects of actionability and significance the work can be overlooked and more narrow definitions of quality (qua validity) might lead to a negative evaluation. In simple terms we cannot compare apples and oranges, or, more properly, as we are reflecting on paradigmatic difference, we cannot compare apples and blue. Action researchers may have to seek approbation from conventional scholars but ideally we may use our time and talent to do excellent action

TABLE 42.3
Steps to Collaborative Ecological Inquiry

Phase 1: Engaging Stakeholders in codiagnosis and collaborative learning
- What need is pressing for the stakeholders?
- Define the sustainability objectives.
- Crystallizing the need to be addressed.
- Create messy map—Mapping all variables associated with the challenge
- Stakeholder diagnosis—
- Define Goals—short term (small wins) and longer term infrastructure.
- Extending epistemology—including multiple ways of knowing.
- Listening/learning two ways (interviews and focus groups).
- Designing for full system engagement.

Phase 2: Harmonizing the system
- Reconnecting the system parts so they can learn and innovate.
- Designing new structures.
- Using the action research quality criteria to consciously exercise a midpoint correction.
- Quality of partnership; quality of actionability; quality of reflexivity; quality of significant outcomes.

Phase 3: Ensuring learning over time
- Developing documentary (paper/video) records for stakeholders.
- Developing papers for a global community.

research whose very success for stakeholders involved can become its own measure.

We must also consider that seeking to increase the generalizability of action research jeopardizes the partnership with practitioners. However we can also imagine the work extending through time. Action researchers can do more to develop post intervention insights by articulating propositions based on the partnership phase. In this way the complementarity of the paradigms may be developed in active partnerships between action and conventional researchers.

Rather than solving merely intellectual problems, the central concern of conventional research, action research invites motivated participants to tackle status quo "realities"—made amenable to tackling because they are exposed as human constructions. In turn, this means engaging the self (first person work), getting connected and engaging others (second person work). Action research is therefore critiqued for placing dangerous stress on the pursuit of conventional objectivity, a claim which simply ignores the many problems that exist for researchers' claims to objectivity for work done in any social system. At the same time, action research shows both respect for and continuity with the long tradition of scientific method and maintains rigor in pursuit of partial objectivity (Haraway, 1991), reconceived as "quality choicepoints" (Bradbury & Reason, 2001). Concern for objectivity must always be part of enlightened education, otherwise it risks devolving to reify whatever fundamentalisms hold sway in the dominant community. Education has always been a political act.

As action research becomes more mainstream, a new critique must also be noted, namely the dangers of drifting from pragmatic concern toward philosophical discussions about differences with conventional social science. With this comes a consequent need to assess the extent to which materiality has become overlooked in the development of nuanced methods of inquiry and sense-making (Boje, 2010). In the same way that a vigorous and contentious public debate signifies the health of the society, it is important for constructive and timely criticism such as this engage our attention and passion.

Yet, the will to practical results will likely never fade. Therefore even conventional social science will keep open a place for dialogue with the actionable, holistic, dialogical orientation to science. A richer understanding of quality (beyond validity) along with a richer practice that leads to research of consequence in which actionability is considered central, can be accomplished (Bradbury-Huang, 2010).

Conclusion

Social challenges that are systemic and interdependent, continue to grow inexorably; sustainability issues are also deeply interdependent, comprising many vectors of unsustainability (poverty, species decline, social discord, etc). Quite simply stated, most living systems are now in decline, e.g., ocean health, greenhouse gases, forests, fish population (Hawken, 2009; WWI, 2008). The decline is largely the result of affluence that is predicated on the throughput of environmentally toxic materials, technologies driven by consumer oriented economies. As Max Weber predicted, technical rationality holds increasing sway in modern society (Habermas, 1984), and yet technical rationality lacks answers to some of our most pressing problems.

At least some of the fault for our predicament is due to the paradigm of conventional education. Preferring to study things as "objectively" as possible, conventional education has eschewed the messy and emergent arena of how to engage change. Knowledge, in the conventional paradigm, is limited to *conceptual* knowledge about something external to the self. This knowledge prepares students to respond to intellectual puzzles rather than building their capacity to take needed actions. Action research is human centered inquiry. It can complement the conventional paradigm by insisting that we take the impact of humans into consideration, not as a given but as a system that is open to change. Results driven, action research is "knowledge-in-action."

The emphasis of action research on democratic engagement, communication, the primacy of local knowledge, and the commitment to action inevitably will surface issues and friction between members of the polity. It releases and in fact encourages discourse in the struggle over ideas and goals. It reveals areas where agreement can be quickly converted into action, and provides a structure for assessing and evaluating results in a variety of locally appropriate and satisfying methods. We would expect that it also surfaces areas where good hearted people can disagree on principle and process as part of the ongoing community discussion about ends, ways and means. Important topics on the table for the action research in sustainable education community to engage include:

Our transformation from industrial era society to a sustainable postindustrial era, requires a transformation of education. The action research paradigm is suggested as an important component of this transformation to increased adaptability between people and our larger ecology. This transition is likely the work of generation.

Notes

1. The authors are grateful to our colleague Jeff Couch for work on an earlier version of this manuscript.
2. This project is described in detail in Bradbury Huang, H., 2010. "Sustainability as Collaboration: The SEER Case." *Organization Dynamics*. 39(4).
3. This section draws from an essay: Bradbury Huang, H. 2010. What is Good Action Research? Why the Resurgent Interest? *Action Research Journal*. 8(1) 1–11.

References

Ainscow, M., Booth, T., & Dyson, A. (2004). Understanding and developing inclusive practices in schools: A collaborative action research network. *International Journal of Inclusive Education, 8(2)*, 125–140.

Argyris, C., & Schön, D. A. (1974). *Theory in practice: Increasing professional effectiveness*. San Francisco, CA: Jossey Bass.

Boje, D. (2010). A call for inquiry: Materiality in action research. Colorado Technical University presentation. Colorado Springs, CO.

Bradbury, H. (2004). Doing work that matters despite the obstacles: An interview with Riane Eisler. *Action Research Journal, 2*(2), 209–227.

Bradbury, H. (2008). Quality and "Actionability": What action researchers offer from the tradition of pragmatism. In A. B. Shani, S. A. Mohrman, W. A. Pasmore, B. Stymne, & N. Adler (Eds.), *Handbook of collaborative management research*. London and Los Angeles: Sage Publications.

Bradbury, H., Mirvis, P., Neilson, E., & Pasmore, W. (2008). Action research at work: Creating the future following the path from Lewin. *The SAGE handbook of action research participative inquiry and practice* (pp. 77–92). Los Angeles, CA: Sage Publications.

Bradbury, H., & Reason, P. (Eds.). (2001) *Handbook of action research: Participative inquiry and practice*. London: Sage Publications.

Bradbury-Huang, H., & All the Advisory Journal Editors of Action Research. (2009). Transforming the generation and application of knowledge: A manifesto on quality in action research. Retrieved from http://arj.sagepub.com

Bradbury-Huang, H. (2010). Sustainability by collaboration: The SEER case. *Organizational Dynamics,* 39(4), 335-344.

Breu, K., & Hemingway, C. (2005). Researcher–practitioner partnering in industry-funded participatory action research. *Systemic Practice and Action Research, 18*(5), 437–455.

Chandler, D. E., & Torbert, W. R. (2003). Transforming inquiry and action: Interweaving twenty-seven flavors of action research. *Action Research, 1*(2), 133–152.

Craig, C., & Ross, V. (2008). Cultivating the image of teachers as curriculum makers. In F. Connelly, F. Ming, & J. Fillion (Eds.), *The SAGE handbook of curriculum and instruction* (pp. 282–305). Los Angeles, CA: Sage Publications.

Eikeland, O. (2001). Action research as the hidden curriculum of the western tradition. *The handbook of action research*. US/UK.: Sage Publications.

Friedmann, J. (1987). *Planning in the public domain: From knowledge to action*. Princeton, NJ: Princeton University Press.

Gustavsen, B. (2001). Theory and practice: The mediating discourse. *The handbook of action research*. US/UK.: Sage Publications.

Gustavsen, B. (2003a). Action research and the problem of the single case. *Concepts and Transformation,* 8(1), 93-99.

Gustavsen, B. (2003b). New forms of knowledge production and the role of action research. *Action Research,* 1(2), 153-64.

Habermas, J. (1984). *The theory of communicative action* (T. McCarthy, Trans.). Boston. MA: Beacon Press.

Haraway, D. J. (1991). *Simians, cyborgs, and women: The reinvention of nature*. New York, NY: Routledge.

Hawken, P. (2009). Graduation Address to the Class of 2009 Students at University of Portland.

Henderson, J., Hawthorne, R., & Stollenwerk, D. (2000). *Transformative curriculum leadership* (2nd ed.). Upper Saddle River, NJ: Merrill Prentice Hall.

Heron, J. (1996). *Cooperative inquiry: Research into the human condition*. New York, NY: Sage.

Heron, J., & Reason, P. (2008). Extended epistemology with cooperative inquiry. In P. Reason & H. Bradbury (Eds.), *The SAGE handbook of action research participative inquiry and practice* (pp. 366–380). Los Angeles, CA: Sage Publications.

Isaacs, W. (1999). *Dialogue and the art of thinking together*. New York and London: Doubleday.

Ison, R. (2008). Systems thinking and practice for action research. *The handbook of action research participative inquiry and practice* (pp. 366–380). Los Angeles, CA: Sage Publications.

Jones, V. (2008). *Green collar economy: How one solution can fix America's two biggest problems*. New York, NY: HarperCollins Publishers.

Kolb, D. (1984). *Experiential learning: Experience as the source of learning and development*. Englewood Cliffs, NJ: Prentice Hall.

Kolb, D. (2005). *The Kolb learning style inventory*. Boston, MA: Hay Group, Inc.

Madachy, R., Haas, B., Bradbury, H., Newell, J., Rahimi, M., Vos, R., . . . Wolch, J. (2008). *Achieving sustainable development in Southern California: Collaborative learning through system dynamics modeling*. Paper presentation at INCOSE conference: "Systems Engineering for the Planet." Retrieved from http://arsecc.net

Marshall, J., & Mead, G. (2005). Editorial: Self-reflective practice and first-person action research. *Action Research, 3*(3), 235–244.

McCaleb, S. (1997). *Building communities of learners: A collaboration among teachers, students, families and community*. Mahwah, NJ: Lawrence Erlbaum.

Minkler, M., & Wallerstein, N. (2008). *Community-based participatory research for health: From process to outcomes*. San Francisco, CA: Jossey Bass.

Mogensen, F., & Schnack, K. (2010). The action competence approach and the "new" discourses of education for sustainable development, competence, and quality criteria. *Environmental Education Research, 16*(1), 59–74.

Phillips, F. (2005). Toward an intellectual and theoretical foundation for shared prosperity. *Systemic Practice and Action Research, 18*(6), 547-568.

Reason, P., & Bradbury, H. (Eds.). (2001, 2008). *The handbook of action research*. US/UK.: Sage Publications.

Rivoli, P. (2005). *The travels of a T-shirt in the global economy: An economist examines the markets, power, and politics of world trade*. Hoboken, NJ: John Wiley & Sons.

Robert, K.-H. (2002). *The natural step story: Seeding a quiet revolution*. Gabriola Island, BC: New Society Publishers.

Rorty, R. (1999). *Philosophy and social hope*. London, UK: Penguin Books.

Sagor, R. (2000). *Guiding school improvement with action research*. Alexandria, VA: Association for Supervision and Curriculum Development.

Schubert, W. (2008). Curriculum theory. In F. Connelly, F. Ming, & J. Fillion (Eds.), *The SAGE handbook of curriculum and instruction*. (pp. 391–414). Los Angeles, CA: Sage Publications.

Scientific Research on Maharishi Transcendental Meditation and TM-SIDHI Program, Volumes 1–5, Retrieved from http://www.maharishi.org/tm/research/508_studies.html

Senge, P., Cambron-McCabe, N., Lucas, T., Smith, B., Dutton, J., & Kleiner, A. (2000). *Schools that learn. A fifth discipline fieldbook for educators, parents, and everyone who cares about education*. New York, NY: Doubleday.

Senge, P., Lichtenstein, B., Kaeufer, K., Bradbury, H., & Carroll, J. (2007). Collaborating for systemic change: Conceptual, relational, and action domains for meeting the sustainability challenge. *Sloan Management Review, 48*(2), 44–53.

Sterling, S. (2004). Higher education, sustainability, and the role of systemic learning. In P. B. Corcoran & A. E. J. Wals (Eds.), *Higher education and the challenge of sustainability: Problematics, promise and practice* (pp. 49–70). The Netherlands: Kluwer Academic Publishers.

Stringer, E. (2004). *Action research in education*. Columbus, OH: Merrill Prentice Hall.

World Watch Institute. (2008). State of the World: Vital signs. Norton and Company: New York. See also State of the World: Vital Signs Trends on Line. Retrieved from http://www.worldwatch.org/vsonline.

43

Critical Action Research and Environmental Education

Conceptual Congruencies and Imperatives in Practice

Robert B. Stevenson
James Cook University, Australia

Ian Robottom
Deakin University, Australia

In this chapter, we examine the use of action research or participatory (action) research approaches to environmental education. We begin by offering a conceptualization of our critical view of action research by identifying what we consider its key characteristics. We then use these characteristics to analyze the ways in which it can be viewed as different from other research genres and to argue the conceptual congruency between critical action research and a critical orientation to environmental education. Three case studies then follow of the use of action research in environmental education projects in Australia, Europe, and an Australia–South Africa partnership. Finally, drawing from these case studies, we identify four imperatives for action research in environmental education: those of authentic active participation (beginning with agenda setting), contextual connections, relational practice, and individual, interpersonal and institutional capacity building.

Conceptualizing Action Research

The now extensive literature on action research, or participatory (action) research, makes clear that it is not a discrete term for one particular form of inquiry. There are a diverse range of meanings and interpretations of action research, but also some commonalities and recurrent themes. Action research developed in different places and circumstances from different traditions and with different emphases. It not only has a long history in education, being relevant to all educators (e.g., teachers, administrators, nature interpreters, museum guides) and other educational stakeholders and partners, but also has an important history in many other fields, such as social work, health care, management and, more recently, natural resource management.

A significant part of this history in education is the work of the late Lawrence Stenhouse, a noted British curriculum scholar, who advocated the role of teachers as researchers in Britain. He believed that "curriculum research and development ought to belong to the teacher" (p. 142) because

> . . . the uniqueness of each classroom setting implies that any proposal—even at school level—needs to be tested and verified and adapted by each teacher in his own classroom. The ideal is that the curricular specification should feed a teacher's personal research and development programme through which he is progressively increasing his understanding of his own work and hence bettering his teaching. . . . (Stenhouse, 1975, p. 143)

Stenhouse's work was influential in shaping action research work in Australia and the teacher researcher movement in the UK and the United States. Educational action research became defined as simply as "research into practice, by practitioners, for practitioners" (Grundy & Kemmis, 1981), and more specifically as "a systematic and intentional inquiry by teachers in order to make sense of their practices and improve them" (Lomax, 1994, p. 115). These definitions make clear that there is a central positioning of practitioners or anyone working as an "insider" in an educational setting that is a common characteristic of most versions. External stakeholders or partners, such as university researchers, may be involved as collaborators or co-participants in the inquiry, because they either have an interest in the problem or bring relevant skills or resources (Anderson, Herr, & Nihlen, 1994), but action research should involve educators (and other inside stakeholders) as the primary actors, with any outsiders participating as collaborators at the invitation of the inside researcher(s). In other words, irrespective of the involvement of outsiders, an essential characteristic for us is that the research agenda and process is controlled by those individuals or groups who are engaging an issue of their practice. Put simply, action research involves participants in *researching their own practice*, not researching other people's practice.

Second, as the name implies, action research is *action-oriented*. Unlike other research genres which frown on intervention in the research setting and where the researcher is conducting an inquiry in order for other people to later decide what action(s)—if any—should be taken, action research involves taking immediate action to address a particular concern or situation, usually in cycles that integrate action and reflection. Neither action nor reflection, however, should be isolated or haphazard, but a systematic and deliberative form of inquiry should be followed.

Methodologically as a form of inquiry, action research is characterized by a *systematic process of recursive cycles of action and reflection that is responsive to evolving understandings and circumstances*. Kurt Lewin, who has been credited by many as the founder of action research, developed a theory of action research based on a spiral of planning, fact-finding, and execution which was later interpreted by Kemmis and McTaggart (1988) as recursive cycles of planning, acting, observing, and reflecting (McNiff, 2002). This process, which has become a common methodology for action research, is iterative in that plans, actions, and goals are continually changed and refined as new learning occurs as a result of completed cycles. The cycles, however, are rarely as straightforward, linear, or tidy in practice as the simple description may make them seem. Cycles frequently become intermingled and multiple cycles often take place in parallel. Yet, the benefit of the (seemingly straightforward) four phase cycles as a heuristic is that it is compatible with the rhythms of thoughtful teaching and what reflective practitioners do intuitively. At the same time, it pushes the researcher toward a more systematic process, while allowing for a questioning of the taken-for-granted and a continual revisiting of visions, issues, and practices. Thus, the methodology of action research recognizes that our understanding of educational practices and situations is always tentative or "partially correct and partially in need of revision" (Noffke, 1995, p. 5).

A fourth characteristic is that action research is *concerned with the practical*—"the everyday practical problems experienced by teachers, rather than the 'theoretical problems' defined by pure researchers within a discipline of knowledge" (Elliott, 1978, p. 356). The focus on practical problems is not intended to suggest that action research is only concerned with techniques, strategies, or "nuts and bolts," or to exclude an intellectual question about a problematic situation that may provide the starting point. Nevertheless, for most the concern is with practical issues associated with human actions and social situations. The purposes or intent of action researchers, as reflected in the scope of their endeavors, range widely and differ in relation to the kinds of issues addressed and transformations sought.

In keeping with Stenhouse's (1981) view that research is "systematic self-critical inquiry made public," another important characteristic is that the *intentions, process, and outcomes of action research are publicly shared*. First, this is necessary so that one's knowledge claims and theories can be scrutinized by others. Issues of trust and willingness to subject one's thoughts and practices to critical scrutiny are obviously major concerns that must be addressed for a critical dialogue to take place. Second, making inquiry public for other people concerned by and interested in the concern and practices being studied can be viewed as a professional responsibility and obviously a necessary one if the work is to contribute to a professional knowledge base.

Critical Action Research

Having described five important general features of action research, it is important to now examine some distinguishing characteristics of our preferred version which has been labeled "critical action research." Some versions of action research focus on solving local problems by following the usual procedures of traditional positivistic or, more often, interpretive naturalistic research, but on a scale that is sufficiently small for practitioners to use. For example, a common approach in education, often termed classroom action research, typically involves the use of interpretive modes of inquiry and qualitative data collection by teachers in order to make decisions about how to improve their own classroom practices (Kemmis & McTaggart, 2000). This approach parallels traditional research in that data are collected in order to make decisions about what actions to take to try to be more effective. Unfortunately, many of these versions we have observed also focus exclusively on the technical aspect of teaching strategies without examining one's purposes—or the broader contextual factors that shape teaching and schooling.

Such versions do not challenge the epistemological assumptions of traditional research. Stenhouse viewed curriculum as a process, rather than a product, in which teachers engage in curriculum theorizing by generating and testing their theories from their efforts to change curriculum practice. Thus, theory is constructed from practice rather than generated externally (in universities and policy bodies) and then applied by teachers in their classrooms. Teacher action researchers construct new understandings of their practice from self-critically reflecting on their experience of their contextually situated practice in relation to their espoused intentions and values, which in turn leads to new actions and further critical reflections on those actions. This process is based on the premise that knowledge arises from and for action. Such knowledge has been referred to as coming to know through experiential (Heron & Reason, 1997) or personal ways of knowing that are derived from tacit knowledge (Polanyi, 1962). In summary, in critical action research there is a dialectical relationship between knowledge and action such that understandings and actions emerge in a continuous process or cycle of action and reflection (McNiff, 2002; Noffke, 1995). Action research, therefore, we believe involves *developing understanding of both one's theory and practice from critical reflection on action*.

There is also an assumption in the kinds of classroom action research just described that significant improvements in teaching can be made without the need for broader institutional and community support and social change—contradicting the literature on educational change (Kemmis & McTaggart, 2000). Stephen Kemmis added an important dimension to Stenhouse's work by drawing attention to "the socially and politically constructed nature of educational practice" (McNiff with Whitehead, 2002, p. 45). This involves recognizing the contextual factors that influence curriculum, teaching, and educational change and therefore the need to address the broader structures and conditions in which educational practices take place. Organizational structures and cultures, power relations within both institutions and the larger society, and political and educational discourses are all treated as important influences on educational practices and situations. Besides reflecting inwardly on intentions and actions, reflection and analysis are directed outward toward the factors and situations that circumscribe practice. Given that an improved situation may be necessary for improved practice to be possible, reflection should be aimed at ways of removing, reducing, or working within and around influences that result in "unjust, irrational, unproductive or unsatisfying" ways of thinking and working (Kemmis & McTaggart, 2005, p. 567). This kind of critical reflection demands the incorporation of a concern for broader social analyses for understanding and attempting improvements. This means not only trying to be more effective, but also more just and rational. Thus, the focus is on efforts to remake and reframe one's situation by addressing distortions, contradictions, incoherencies, and injustices that have been socially or historically constructed (Kemmis & McTaggart, 2000). Critical (action and) reflection therefore can be characterized as including *the examination of the justice and rationality of one's practices and the ways in which they are shaped by structures in the broader society* (Kemmis & McTaggart, 2005).

Other lists of characteristics can be found in the critical action research literature (see Kemmis & McTaggart, 2005; Anderson, Herr, & Nihlen, 2007). However, to summarize, in our view action research involves:

- practitioner-researchers in studying their own practice;
- an action orientation;
- a systematic process of recursive cycles of action and reflection that is responsive to evolving understandings and circumstances;
- a concern for practical issues;
- developing understanding of both one's theory and practice from critical reflection on action;
- critical reflection which includes examining the justice and rationality of practices and the ways in which they are shaped by broader social structures; and
- public sharing of the intentions, processes, and outcomes of the research.

Although, as we've pointed out, there are issues of contention within each of these characteristics, they help to distinguish action research from other genres of research.

Critical Action Research and Other Research Genres

Drawing from the seven characteristics that have been identified, action research can be summarized as differing from other research genres in at least four significant ways: (1) the role of the researcher(s) as also the subject(s) of study; (2) the inclusion of interventions or actions as part of the research process; (3) the epistemological position that understanding can emerge from action, rather than only inform action; and (4) the recognition that practitioners can generate knowledge and theory, not just specialist external researchers, usually based in universities.

The conceptualization of knowledge use and the place of actions or interventions in action research are quite different from other research genres. Academic research, particularly within an empirical-analytic tradition, is usually conducted within a knowledge creation, dissemination, and utilization model in which educators are assumed to be the end users, consumers, or implementers of the products of research. The task of inquiring into and theorizing about educational practice is viewed as the responsibility of the academic researcher and not the practitioner. The latter is expected to be involved only in the utilization phase and to instrumentally use particular propositional knowledge generated by research on the assumption that such knowledge generated in other educational settings should be "used to direct specific decisions and/or interventions" (Estabrooks, 2001, pp. 283–284).

This assumption can be questioned on at least three grounds. First, educators work in complex situations where most practices and circumstances are filled with rich sets of particulars (including conflicting information) that researchers are unable to take into account but must inform educators' (often dilemma-ridden) decisions about their educational practices. In other words, research-generated knowledge is only one of a number of factors that educators have to consider in making pedagogical or leadership decisions. Second, research on teacher thinking indicates that "teachers develop and hold implicit theories about their students, about the subject matter they teach, and about their roles and responsibilities and how they should act" (Clark, 1986 cited in Eraut, 1994, p. 72). In other words, educators are constantly involved in interpreting their world and "theorizing," albeit usually implicitly, about their intentions and actions. Simply stated, educational practitioners are engaged, like social scientists, in drawing inferences and making judgments based upon their interpretations of social reality (Codd, 1989). Third, as constructivist learning theory makes clear, individuals' existing knowledge influences their understanding and interpretation of new knowledge. In this view, knowledge is treated not as static but as constantly being constructed and reconstructed by the user.

Action research builds on the propensity of many educators to continually reflect on and improve their work, but provides a more systematic, more rigorous, and more collaborative means of doing so. The process of recursive cycles of action and reflection enables educators to inquire into and be constantly constructing and reconstructing knowledge of their own practice as they try out different ways of improving their practice consistent with the unique context in which they each work.

Unlike traditional positivist research in which the goal is to prove something (or more accurately, disprove or disconfirm a hypothesis) as representing a universal law or generalization, action research seeks to directly improve practice, or some aspect of it, in a particular situation. And unlike interpretive research where the purpose is limited to understanding and interpreting a situation, the intent of action research is to produce change as well as understanding. In other words, action researchers are not trying to generate laws or just understand a situation better, but are working in a situation to improve it. They are trying to make their practice and their situation better—both more effective and, for critical action researchers, more just.

Yet, the epistemological belief underlying action research that understanding can both emerge from and inform action can challenge the privileged position of university researchers as the sole producers of knowledge, as well as that of centralized educational policymakers as the authorities on curriculum, teaching, and educational reform. Recognizing that practitioners can be engaged in knowledge generation can threaten to alter existing relations of power.

Only a relatively short time ago, the scientific method was generally regarded as the only legitimate approach to systematic inquiry or research. Since then, beginning in the 1960s, a "linguistic and cognitive turn has swept the social sciences and humanities" (Reason & Bradbury, 2001). The cognitive turn "focused on the cognitive structures (schema or mental models) which allow us to make sense of the world" while the linguistic turn "looked at the hitherto underestimated role of language in our construction of our world in which we are always seeking to make (or give) sense" (p. 5). This evolution and acceptance of other research genres has accelerated over the past decade.

New research paradigms usually emerge in response to perceived limitations of and challenges to existing ones. The postmodern/poststructuralist emphasis on the metaphor of text is seen as limiting by some because there is little concern for the relationship of discourse, narrative, or the crisis of representation to knowledge in action (Reason & Bradbury, 2001). And as Lather adds: "The question of action . . . remains largely under-addressed within postmodern discourse" (1991, p. 12). Now, argue Reason and Bradbury, it is time for "the action turn." Apparently, the editors of the above mentioned handbook agree as they have added a participatory (action research) paradigm to the three paradigms that were identified in the first edition of the handbook—nearly twenty years after Lewin and Stenhouse's work was shaped into a claim for action research as a distinct form of educational research (Carr & Kemmis, 1983). Clearly, the acceptance of new research genres does not happen overnight.

The Coherence of Action Research and Environmental Education

As already argued, action research is about learning for and from inquiry into the intentional transformation of practice—as well as the transformation of educational situations that circumscribe and constrain practice (Kemmis & McTaggart, 1988). Put simply, it is about improving a situation and/or practice of concern.

Environmental education is also concerned with transforming situations and practices—with transforming ecologically unsustainable situations and practices and the values that underlie individual and public decision making, from those which aid and abet ecological (and human) degradation to those which support a sustainable planet in which all people live with human dignity. Such transformations of environmental situations and practices are of course political in nature, but the politically (and socially) constructed nature of educational practice is also made explicit in the critical or emancipator conceptualization of action research developed by the influential Deakin University group of action research scholars in Australia. They argued that the purpose of transforming practices and situations is to create not only a better or more effective educational system, but a more just and compassionate society—to which we would add, a more ecologically sustainable society.

The correspondence in the goals of critical orientations to action research and environmental education (EE) is matched by a shared participatory or action taking dimension as well as a methodological approach grounded in critical inquiry. Action research offers a systematic process of change through critical inquiry on explicit interventions or actions. EE has been similarly conceptualized in EE/Education for Sustainability (EfS) discourse, at least by socially critical scholars, as a process of critical inquiry into environmental issues and concerns and the taking of actions to address those issues or concerns. For example, besides developing knowledge of and sensitivity to environmental concerns, EE is intended to offer opportunities to thoughtfully and critically appraise environmental situations, to make informed decisions about such situations and to develop the capacity and commitment to act in ways that sustain and enhance the environment (Stevenson with Stirling, 2010).

Finally, there is a shared democratic intent in both critical action research and socially critical EE of enabling people to be their own agents of change and to be responsive to changing conditions and problems rather than having change imposed on them. In other words, action research offers a methodology to assist individuals, groups, and communities to develop the capacity to bring about change in their own situations and practices (Ferreira, Ryan, Davis,

Cavanagh, & Thomas, 2009) such that they become more ecologically sustainable and morally just.

Case Studies of Action Research Approaches in Environmental Education

Action research has been used in many environmental education projects as an integrated approach to systematic inquiry (i.e., research) and educational change. We describe three cases to illuminate the kinds of educational situations in which action research can be productively employed. Our intention in this section is not to include full case studies as instances of action research in environmental education, but rather to draw on a number of already-published accounts of such cases to both exemplify characteristics of action research and to extend the conversation about the problematic nature of these characteristics. In doing this we are adopting an approach of reflection upon case study practice, in which "findings from case studies . . . can serve as a heuristic in the form of analytical constructs or categories that readers can use to reflect on their practice, particularly to 'help them grasp in descriptive and explanatory ways certain aspects of their work that were previously inaccessible (Zeichner & Liston, 1996, p, 30)'" (Stevenson, 2004, p. 46).

The cases we draw on in this section involve respectively preservice teacher education, curriculum, and professional development of primary and secondary teachers, and professional development of environmental education staff in tertiary institutions. They are:

- ARIES mainstreaming sustainability into preservice teacher education project;
- the European Environment and School Initiatives (ENSI) Project (REFS); and
- the Australia/South Africa Links (AusLinks) Project (REFS)

Mainstreaming Sustainability into Teacher Education Programs[1] Action research was selected as the methodology for a federally funded project sponsored by the former Australia Research Institute on Education for Sustainability (ARIES) to pilot a participatory system-wide model for embedding or mainstreaming sustainability into teacher education programs. The model, developed from an earlier literature and document review study, was premised on the need for broad engagement with key change agents *across* the wider teacher education system as well as the active participation of stakeholders *within* the system. An action research process was chosen because in the first instance the study was not only trying to understand how change is effected, but was also seeking to create change by intervening in and transforming a situation of concern, namely the lack of adequate attention to the preparation of future teachers in environmental sustainability education. The second stated reason for using action research was its perceived conceptual congruence with education for sustainability, as well as with systems thinking. This congruence included the shared characteristics of "critical reflection; and systemic enquiry with a focus on improving a situation of concern" (p. 6) through "an iterative approach to learning-based change" (p. 17) driven by the participants who are viewed as having "the capacity to bring about change within their own situations" (Ferreira et al., 2009, p. 17).

The pilot study was carried out from March to October 2008 (the period of funding support which meant a limitation on the number of action cycles that could be carried out) and focused on a practical and two-fold concern: how can we better connect and engage relevant stakeholders? And how can a combined action research and whole-of-system approach create organizational and systemic change for mainstreaming sustainability in preservice teacher education? In the first phase of the project, a mapping of the teacher education system at the national and state levels enabled the identification, and determination of relationships of influence, of key government (e.g., federal and state departments of education and environment, the state board of teacher registration) and nongovernment organizations (e.g., professional teacher associations and community organizations with interests and expertise in sustainability) as well as key stakeholders (principally teacher education academics and students). This resulted in key change agents and stakeholders being invited to assist in working for change within one or more of three layers of action research that were designated at national, state, and institutional levels. The intent was to engage as many relevant and influential groups as possible in an ambitious effort to bring about system-wide change.

A state level group of project leaders from each of five universities in the state of Queensland, which shared the same teacher registration and policy context for their programs, formed one community of inquiry and reflective practice. These leaders established institutional groups of participants and stakeholders with whom they shared the outcomes of mapping the broader teacher education system in order to then map their institutional subsystem. At the national level, federal government department representatives declined to participate in the action research because they "did not see themselves as 'directly' involved in preservice teacher education" (Ferreira et al., 2009, p. 60). However, as important stakeholders, teacher education students from across the five institutions participated as one group in the research and change process. They used social media to facilitate their conversations and deliberations and had the opportunity to present and debate, at a state Student Forum, their vision of sustainability, the role of education in creating a sustainable future and their ideas about the place of sustainability in teacher preparation. This led to the development of a charter expressing students' concerns about the lack of education for sustainability in their teacher education programs which was presented to the Minister of Education who attended the meeting (Brevitt, 2009).

A range of strategic actions and interventions occurred at both the state and institutional levels. Besides the mapping exercise described, project leaders worked with participants to locate their institutional situation within the roles and relationships of the larger system, to identify what parts they could directly affect, and to develop and compile visions of sustainability in teacher education from which shared visions were created. The questions that guided this research group's reflections on and analysis of the outcomes of their action cycles addressed the three goals of improving practice, improving understanding of practice, and improving the situation in which the practice occurs (Kemmis & McTaggart, 1988, 2000). These questions, which were addressed in the final report of the project, included:

- Has the situation of concern improved?
- Has the understanding (learning) by the practitioners in the situation improved?
- Has the practice of the action researchers improved?
- Has the understanding of the (action) research practice by the practitioners improved?

The major conclusions drawn in responding to these four questions are summarized here. First, the reported outcomes of the action cycles enacted during the year included improving the situation of concern with respect to engaging and connecting stakeholders, developing relationships and networks, and improving and creating new lines of communication across academic, government, and nongovernment stakeholders. Thus, the project leader group's concern for connecting and engaging was largely addressed. The other concern of creating systemic change for mainstreaming sustainability in preservice teacher education was more complex and challenging and not surprisingly generated more limited and mixed results. One limitation was the way many stakeholders from the broader system viewed their role as either not in need of change or not directly involved in teacher education and hence more attention was needed "in establishing a shared vision and clarifying their respective role in the preservice teacher education system" (p. 60). Unfortunately, funding ended before additional participatory action research cycles could be enacted to respond more fully to this concern, highlighting the importance of ongoing support for sustaining this kind of inquiry and change process.

The Environment and School Initiatives (ENSI) Project The ENSI project was founded in 1986 under the auspices of the Organization for Economic Cooperation and Development's Center for Educational Research and Innovation (OECD-CERI). The project has evolved over time and is now viewed as an international network which has supported educational developments, environmental understanding, and active approaches to teaching and learning, through research and the exchange of experiences internationally. The association aims at supporting educational and pedagogical developments that, via research and international exchange of experiences, promote insight into learning for sustainable development, environmental studies, active forms of learning and teaching, as well as education for citizenship (http://www.ensi.org/About_ENSI/). While the history of ENSI has seen several changes in focus, its defining characteristics were perhaps most clearly evident in the first two phases, the first of which is described as follows at the ENSI website:

> **Phase 1: 1986–1988**
> **ENSI as OECD/CERI Innovation exchange project**
>
> ENSI was established in 1986 . . . with eleven participating countries focusing on the promotion of environmental awareness and such "dynamic qualities" as initiative, independence, and the readiness to accept responsibility. At that time, these relatively new school requirements were not normally related to each other. ENSI centered on the concept of environmentally oriented project teaching, which offered potential for the development of human creativity, intelligence, and organizational skills.
>
> It was self-evident that the experiences of the fundamental partnership (i.e. of teachers and pupils), were of decisive importance in this respect. This led to the development of a number of case studies in some countries, which were written by teachers using an action research method.
>
> Main publication/conference report: Environment, Schools and Active Learning, An OECD/CERI-Report. OECD/CERI, Paris, 1991, ISBN 92-64-13569-3

The second phase of ENSI is perhaps its most relevant to a chapter on action research in environmental education. In this phase, the project overtly adopted action research as the preferred methodology for participants active in EE "on the ground" to develop reflective accounts of their work for publication in a range of case studies for the ENSI project.

The ENSI project has been highly productive in terms of reports and publications. The following account draws on an article published by one of the present authors (Robottom) who had five years' experience in this project as a country representative and consultant; the other author (Kyburz-Graber) has over two decades of experience, formerly as a country representative and later as project director.

> The ENSI project is based on the assumption that there is a strong case for a form of environmental education premised upon active learning rather than upon the transmission of knowledge, supported by a form of professional development similarly premised upon participatory action research rather than upon instrumentalist, centrally orchestrated teacher in-services (Kyburz-Graber & Robottom, 1999).

> Thus the ENSI project is seen as a marked alternative both to conventional environmental education as "education *about* and *in* the environment," and to conventional approaches to professional development (which ultimately see the problem of improving environmental education as one of central development of policy and materials, followed by their delivery to schools and teachers for implementation).
>
> In terms of its subject matter emphasis, the ENSI project favors the promotion of activities related to the development of "dynamic qualities in environmental education" such as initiative, independence, commitment, and readiness to accept responsibility. Of more relevance to the focus of this article, the process of the ENSI project, as originally conceived and described in publications since its origination (Elliott, 1995), explores the role of participatory research-based curriculum and professional development strategies, with a specific focus on the establishment of action research as a basis for linking curriculum and professional development in the field of environmental education.
>
> Since its origin thirteen years ago, the project has sought to support teachers to adopt a research perspective on their environmental education work and to help them prepare written accounts of this work . . . (Robottom & Kyburz-Graber, 2000).

In terms of outcomes, the ENSI project has produced reviews of policy developments in several participating countries, as well as country reports that present case studies of actual environmental education practice. One of the features of these country reports, which are essentially case studies of professional development based on action research in environmental education, is that they indicate a wide range of interpretations of action research and an equally wide range of environmental education practices in schools and other educational settings (see, e.g., Kyburz-Graber et al, 1995; Robottom, 1993). Uniformity of curriculum materials was neither an aspiration nor an achievement of the ENSI project (Kyburz-Graber & Robottom, 1999).

Australia/South Africa Institutional Links (AusLinks) Project The project entitled "Educating for Socio-Ecological Change: Capacity-building in Environmental Education, focusing on South Africa's tertiary educators," was funded by AusAID and administered by IDP Education Australia as one of its Australia/South Africa Institutional Links projects.

The overall focus of the AusLinks project is the professional development of new and existing environmental education staff in participating tertiary institutions. The project was organized into four activities which aimed to develop curricula and materials (Activity 1), enhance existing programs by reviewing courses and planning new courses (Activity 3), and enhance research capacity through reviewing, and developing research supervision strategies and resources (Activity 4). Activity 2, which we will focus on here, sought to enhance research and professional capacity by working with colleagues in a process framed by participatory action research principles aimed at the development of original case studies of changing environmental education practice.

The process by which these principles were enacted is explained in a paper presented by Activity 2 participants at the 1999 annual conference of their national professional association:

> In small groups participants shared with others a number of relevant environmental and environmental educational issues. It was decided by consensus that the activity would involve the development of case studies related to the professional contexts of participants and located in the geographical context of their respective workplaces. Guidelines and frameworks were provided by activity coordinators which provided initial structure and ideas. This process was to include development of photographic records of people, places, contexts, and activities to be shared at the next meeting. The project provided participants with cameras for this purpose.
>
> At the second meeting held in the Northern Province, each participant tabled for discussion photographs of aspects of the issues they felt were important in conveying the meaning and significance of the case under study in their own context. Other Activity 2 participants provided feedback on these illustrated reports.
>
> At the third meeting captions were written for the photographs by individual participants and these were shared and discussed with other participants who then made input into the further development and improvement of the text. At this meeting participants also began to develop case study commentaries for presentation at the next meeting.
>
> At the fourth meeting the draft case studies included the commentaries (five to ten pages) and photographs with captions (five to ten lines). These draft case studies were circulated among at least two other participants who provided critical feedback verbally and in the form of annotations on the text (LeGrange et al., 1999).

Further descriptions of the AusLinks project, and an account of emerging issues associated with this approach, are presented in Lotz and Robottom (1998) and in LeGrange, Makou, Neluvhalani, Reddy, and Robottom (1999).

Characteristics of Action Research in Environmental Education As stated at the beginning of this chapter, action research has diverse interpretations. Some of these differences in the way participants interpret and enact action research are demonstrated in the just described case studies. Even within the ENSI project, there is a range of purposes to which action research has been put. For example, a recent account of ENSI-related action research activities (Kyburz-Graber, Hart, Posch, & Robottom, 2006) reveals a range of expressions of action research reported by participants in the project including:

- Action research as quality assurance;
- Action research as curriculum evaluation;
- Action research as an approach to innovation; and
- Action research as program improvement.

Broad applications of a concept like action research may of course lead to dilution of its potency. There may be a tendency for an educational activity to be reported as an instance of action research on the basis of its dependence on practical activity of some kind, without due regard to the "political theory" (an interest in enhancement of social justice in the social settings of educational work) that drove original conceptions of action research. It is therefore important to look beyond the surface layers of purported action research projects to consider what may be learned about more specific operational aspects of projects that may preserve this political theory. We argued earlier for certain essential characteristics of action research. Our intention now is to draw on the case studies presented above to indicate some of the ways in which these characteristics may be operationalized in practice and some of the issues associated with this operationalization.

The Imperative of Active Authentic Participation (Beginning with Agenda Setting) Participatory research (usually in the form of action research) is central to the ARIES, ENSI, and AusLinks projects. In the ENSI project action research was used as a means for teachers to develop curriculum strategies that involve students in a dynamic learning process focused on concrete environmental problems and issues encountered in their own communities at the local level. Unsurprisingly, a fundamental characteristic of action research projects is that it is action-oriented—it entails activity of some kind. It cannot be conducted solely from the armchair of the self-satisfied meta-theorist.

The key point here is that action research entails authentic involvement by participants in all phases of the research, including conceptualization of the research problem, data collection, data analysis, report writing, and dissemination of results. In the common situation where external university-based researchers work with classroom-based practitioners in environmental education there may be a tendency for the research to involve classroom participants at the level of data collection (either actively collecting data themselves or passively as the subjects of researcher observation) in project work driven by the preexisting research questions of the university-based researcher. If action research is to have any meaning, participants ought to have an enshrined role in the setting of research questions as well. In the AusLinks project, the participatory tone of the project was set early as the university-based researchers came to the project at the invitation of participants themselves, who already had articulated a research problem in the need to reconstruct higher education in the aftermath of the dismantling of the previous apartheid dispensation of education. A key operational aspect of the ensuing research was the provision by the researchers of cameras for use by the participants in capturing images depicting specific social/educational issues that they wished to address—a means of ensuring that research agenda-setting was "owned" by participants. A period of three months was available for this issue-identification phase—a period in which the researchers were absent—for groups of participants to themselves determine and represent topics and issues to be explored in collaboration with the researchers on their return. This created the conditions for internal ownership of the project from the outset, and this prefigured all subsequent phases of the research.

In contrast, in the ARIES project, the funding agency demanded the prespecification of project outcomes prior to the formation of the groups of participants. This precluded the opportunity for all participants to be engaged in and assume ownership of developing a shared research agenda or focus of concerns. As a result levels of commitment to the action research process, and to the project's aims, varied as people had different reasons and agendas for participating in the project. This problem highlights the importance in action research of engaging participants from the very beginning of formulating a project while revealing the dilemma of such an approach usually not fitting with the requirements of most funding agencies which violate this fundamental principle of action research.

The Imperative of Relational Practice As we've explained, learning and constructing meaning from actions and reflections is central to action research. Contemporary learning theories of social constructivism and sociocultural or situative perspectives emphasize that learning is social or collaborative in nature. In other words, communication and dialogue that occur through social interactions provide opportunities for meaningful learning. Such opportunities occur in everyday communities of practice where learning is valued (Greeno, Collins, & Resnick, 1996) and relationships of trust, openness, and transparency are established.

In the ARIES project, workshops, online discussion groups, and shared databases were employed to build shared understandings of sustainability and education for sustainability within and across each institution. The state level project leader group also shared their institutional level experiences and reflections via monthly phone conferences and several face-to-face meetings. Although some members of this group knew each other quite well prior to this project, a couple did not know and were not known by these members and so the first in-person meeting of the group early in the project was important for establishing trust and a mutual understanding of conceptions of action research and EfS. Efforts were made to maintain and build on these relationships through the regular two hour phone conferences. As a result of this relationship building, an open and supportive network of researchers working collaboratively was created in which ideas were shared and challenged and resources were garnered from each other. Similarly, the high level of trust that was built among participants in the AusLinks project enabled collaborative critiques to be provided of the draft case studies that were central to producing rich and thoughtful final texts.

In contrast, in some institutional settings in the ARIES project, this level of relationships within the action research group was unable to be created which resulted in tensions that are not uncommon in collaborative action research efforts. For example, some of the teacher educators apparently believed they were already reflective practitioners and therefore did not feel the need to engage systematically in the action research cycles. For this and other reasons data collection was inconsistent and variable, even when participants were prompted to respond to questions that were distributed to facilitate their reflections on actions being taken. In other words, the relationships were not sufficiently strong to create a common feeling of obligation or commitment to the collective good of the group. Or, it may suggest that the lack of an authentic, self-identified and motivating substantive issue (as an imperative for collaborative research) resulted in differential commitment once participants realized that they had some familiarity with the process.

These experiences and resulting insights from systematic and coordinated but emergent and flexible research approaches generally enabled the participants to improve their practice by encouraging them to collaborate more extensively and intensively. These examples illustrate that learning and changes in practice are dependent on close collegial relationships and meaningful social interactions within communities of practice.

The Imperative of Contextual Connections Environmental educational issues cannot be handled simply in terms of planning by objectives but rather by a (re)searching process promoted by action research. In all three projects, action research is the medium by which participants research their own geographical, social, political, and educational contexts in developing approaches or case studies for use in their own curricula in environmental education. In each project, this process has yielded valuable instances of home-grown contextual environmental education. Each project demonstrates the positive role that action research can play in the original development of environmental education programs in particular school or university settings—programs that are highly contextual in nature.

These projects adopted action research as a medium or process for critically reflecting on the meaning and significance of practice and practical settings and sought to recognize identity, biography, and context in their method. We stated earlier that "praxis" is an important element of action research—indeed, that praxis is the key operational driver of action research. In the AusLinks project's Activity 2 "praxis" was defined as "a reflective interaction between personal professional theory, personal professional practice and the professional settings within which these are intelligible." The matching of a contextualized methodology with what was perceived as a contextualized subject matter was an important feature in the rationale of this project and the ENSI project, which both emphasized the praxis-based process of curriculum development of particular case studies above an interest in widely disseminating these case studies for adoption and adaptation elsewhere.

In the case of the ARIES project, the systems approach that was adopted enabled participants to problematize the issue of context and to uncover the relationships between different contexts as they identified multiple layers of contexts with their immediate setting embedded within larger contexts. Personal professional theory and practice was embedded within a program setting of collective professional theories and practices that in turn was circumscribed by both institutional and state policy and accountability contexts. Not only did different groups of participants critically reflect on the meaning of embedding sustainability in preservice teacher education in their particular state or institutional context, but the state group of institutional project leaders was able to reflexively use their action research experiences at both the state and institutional levels to envision possibilities while improving understanding and even working around constraints. Dissemination of ideas and resources among this group were not for the purpose of unreflective adoption or adaptation in their different local contexts but for stimulating different understandings and interpretations of what is being done and might be done in their own professional setting. In this way action research was building capacity for contextualized change.

The issue of the tension between universalism and contextuality turns on what happens from the point of original development. In all three projects, the main purpose of participatory or action research concludes in the development of localized programs. In other projects, the purpose of the research extends to serving as a mechanism for wider dissemination of such programs beyond the context of their development. Sometimes, action research is expanded to include the process of disseminating, for example, a curriculum package for more universal adoption or adaptation.

Consideration of the experiences of the AusLinks and ENSI projects shows some of the tensions between an interest in preserving the contextuality of environmental education through the adoption of participatory action research as a means of curriculum development, and an interest in universal dissemination that tends to be exhibited by international development agencies. When the status of the curriculum package exceeds that of the process of its development, contextuality can be compromised. In our view, ENSI and AusLinks are two projects that succeeded in avoiding being compromised in this way, while the ARIES project was comprised in terms of ongoing action cycles when the funding agency failed to support the next stage of the project because of its interest in expanding the participation to other sites in order to create a wider or more universal implementation of sustainability in preservice teacher education. This suggests a third imperative—the imperative of continuing support and capacity—if action research approaches to environmental education program development are to be sustained.

The Imperative of Individual, Interpersonal and Institutional Capacity Building It became evident in the ARIES project from a mapping exercise in an initial cycle that understandings of sustainability, systems theory, and action research differed quite substantially among the participants in the project. This illuminated the need to build capacity for change within participants by developing conceptual understandings of EfS, systems thinking, organizational change, and action research. As a result, workshops were organized on these topics. Critical reflection on data collected on subsequent actions at both institutional and state levels revealed the importance of key agents for change being able to leverage for change across the different strata of the teacher education system. These revelations included the need for individuals with the willingness, confidence, and conceptual capacity to initiate activities and galvanize others to bring about change. The extent of such capacity contributed to the degree to which action research led to more effective and extensive embedding of sustainability in teacher education. The importance of capacity building was recognized from the beginning of the AusLinks project as a key goal was to enhance (action) research and professional capacity in postapartheid South Africa, which was addressed in part by considering the principles of action research in developing the project.

Institutional support and capacity was found to both contribute to facilitating action research and to be facilitated by its practice. Participation in the ARIES project contributed to enhancing the status and support for EfS within at least one of the participating universities, while the existing and tentatively emerging institutional recognition of the importance of education for sustainability as a cross-cutting theme in teacher education contributed to creating conditions in which action research to further investigate and strategize for its mainstreaming could be supported.

The capacity for engaging in critical action research requires: individuals with an understanding of its purpose and process, and a language, analytical framework and disposition to be self-critically reflective; collaborative groups or communities of practice in which personal support is balanced with constructive critique; and institutional structures and norms for collaboration (Stevenson, 1995). Specific attention to developing each of these individual, interpersonal and institutional capacities is necessary if action research is to be a productive and sustained approach to learning and change in environmental education.

These four imperatives that emerged across the three case studies extend and deepen our understanding of the characteristics and conditions that are vitally important for sustaining action research as a viable approach to environmental/sustainability education. Reflexivity on future action research approaches to environmental/sustainability education should add to this understanding.

Notes

1. This case study has been written from a full report of the project, which can be found at: http://www.aries.mq.edu.au/projects/preservice2/index.php, and from conversations with one of the authors and one of the institutional project leaders.

References

Anderson, G., Herr, K., & Nihlen, A. (1994). *Studying your own school: An educators' guide to qualitative practitioner research.* Thousand Oaks, CA: Corwin Press.

Anderson, G., Herr, K., & Nihlen, A. (2007). *Studying your own school: An educators' guide to practitioner action research.* Thousand Oaks, CA: Sage.

Brevitt, J. (2009). Evanescent attention: Auto(eco)biography of my education for sustainability. (Unpublished honors thesis). James Cook University.

Carr, W., & Kemmis, S. (1983). *Becoming critical: Knowing through action research.* Geelong, Victoria: Deakin University Press.

Codd, J. (1989). Educational leadership as reflective action. In J. Smyth (Ed.), *Critical perspectives on educational leadership* (pp. 157–178). London, UK: Falmer.

Elliott, J. (1995). Reconstructing the environmental education curriculum: Teachers' perspectives. In OECD, *Environmental Learning for the 21st Century* (pp. 13–29). Paris: OECD.

Elliott, J. (1978). What is action research? *Journal of Curriculum Studies, 10*(4), 355–357.

Eraut, M. (1994). The acquisition and use of educational theory by beginning teachers. In G. Harvard & P. Hodkinson (Eds.), *Action and reflection in teacher education.* Norwood, NJ: Ablex.

Estabrooks, C. (2001). Research utilization and qualitative research. In J. Morse, J. Swanson, & A. Kuzel (Eds.), *The nature of qualitative evidence* (pp. 275–298). Thousand Oaks, CA: Sage.

Ferreira, J., Ryan, L., Davis, J., Cavanagh, M., & Thomas, J. (2009). *Mainstreaming sustainability into pre-service teacher education in Australia.* North Ryde, NSW: Australian Research Institute in Education for Sustainability, Macquarie University.

Greeno, J., Collins, A., & Resnick, L. (1996). Cognition and learning. In D. Berliner & R. Calfree (Eds.), *Handbook of Educational Psychology* (pp. 15–46). New York, NY: Macmillan.

Grundy, S., & Kemmis, S. (1981). Educational action research in Australia: The state of the art (an overview). In S. Kemmis et al. (Eds.), *The action research reader*. Geelong, Victoria: Deakin University Press.

Heron, J., & Reason, P. (1997). A participatory inquiry paradigm. *Qualitative Inquiry, 3*, 274–294.

Kemmis, S., & McTaggart, R. (2005). Participatory action research? In N. Denzin & Y. Lincoln (Eds.), *The Sage handbook of qualitative research* (3rd ed.). Thousand Oaks, CA: Sage Publications.

Kemmis, S., & McTaggart, R. (2000). Participatory action research? In N. Denzin & Y. Lincoln (Eds.), *Handbook of qualitative research* (2nd ed. pp. 138–157). Thousand Oaks, CA: Sage Publications.

Kemmis, S., & McTaggart, R. (1988). *The action research planner*. Geelong, Victoria: Deakin University Press.

Kyburz-Graber, R. & Robottom, I. (1999) The OECD-ENSI Project and its Relevance for Teacher Training Concepts in Environmental Education. *Environmental Education Research, 5*(3), 273–291.

Kyburz-Graber, R., Hart, P., Posch, P., & Robottom, I. (Eds.). (2006). *Reflective practice in teacher education: Learning from case studies of environmental education*. Bern, NY: Peter Lang.

Lather, P. (1991). *Getting smart: Feminist research and pedagogy with/in the postmodern*. New York, NY: Routledge.

LeGrange, L., Makou, T., Neluvhalani, E., Reddy, C., & Robottom, I. (1999). Professional Self-Development in Environmental Education: the case of the 'Educating for Socio-Ecological Change Project'. Paper presented at the Annual conference of the Environmental Education Association of Southern Africa, Grahamstown, South Africa, September 7–9.

Lomax, P. (1994). Standards, criteria and the problematic of action research within an award bearing course. *Educational Action Research, 2*(1), 113–126.

Lotz, H., & Robottom, I. (1998). "Environment as Text": Initial insights into some implications for professional development in environmental education. *Southern African Journal of Environmental Education. 18*, 19–28.

McNiff, J. with Whitehead, J. (2002). *Action research: Principles and practice* (2nd ed.). London, UK: RoutledgeFalmer.

Noffke, S. (1995). Action research and democratic schooling: Problematics and potentials. In S. Noffke & R. Stevenson (Eds.), *Educational action research: Becoming practically critical*. New York, NY: Teachers College Press.

Polanyi, M. (1962). *Personal knowledge: Towards a post-critical philosophy*. Chicago, IL: University of Chicago Press.

Reason, P., & Bradbury, H. (2001). *Handbook of action research: Participative inquiry and practice*. London, UK: Sage Publications.

Robottom, I. (1993). *Policy, practice, professional development and participatory research: Supporting environmental initiatives in Australian schools: An Australian report to the Environment and School Initiatives (ENSI) Project*. Geelong, Victoria: Centre for Studies in Mathematics, Science and Environmental Education, Deakin University.

Robottom, I., & Kyburz-Graber, R. (2000). Recent International Developments in Professional Development in Environmental Education: Reflections and Issues. *Canadian Journal of Environmental Education, 5*, 1–19.

Stenhouse, L. (1975). *An introduction to curriculum research and development*. London, UK: Heinemann.

Stenhouse, L. (1981). What counts as research? *British Journal of Educational Studies, 24*(2), 103–113.

Stevenson, R. B., & Stirling, C. (2010). Environmental learning and agency in diverse educational and cultural contexts. In R. Stevenson & J. Dillon, J. (Eds.), *Engaging environmental education: Learning, culture, and agency* (pp. 219–238). Rotterdam: Sense Publishers.

Stevenson, R. (2004). Constructing knowledge of educational practice from case studies. *Environmental Education Research, 10*(1), 39–52.

Stevenson, R. (1995). Action research and supportive school contexts: Exploring the possibilities for transformation. In S. Noffke & R. Stevenson (Eds.), *Educational action research: Becoming practically critical*. New York, NY: Teachers College Press.

Zeichner, K., & Liston, D. (1996). Historical roots of reflective teaching. *Reflective teaching: An introduction*. Mahwah, NJ: Lawrence Erlbaum Associates (Chapter 2, pp. 8–22).

44

A Feminist Poststructural Approach to Environmental Education Research

Bronwyn Davies
University of Melbourne, Australia

When we approach environmental education from a feminist poststructural perspective, what conceptual and methodological spaces are opened up for research? The first conceptual and methodological space might be described as one in which the binary separation between the one who researches and the object of research is deconstructed. Environmental education is not understood, then, to exist prior to and independent of the research, but to be itself opened up by the gaze of the researcher. The second space is one which opens the researcher to being changed through the research. In this second space, Helene Cixous, a feminist poststructuralist writer, reflecting in conversation with Mireille Calle-Gruber, asks *why do we live?* Her answer is: *I think: to become more human: more capable of reading the world, more capable of playing it in all ways. This does not mean nicer or more humanistic. I would say: more faithful to what we are made from and to what we can create* (Cixous & Calle-Gruber, 1997, p. 30). What I hope to show in this chapter is that this position, that we live in order to be able to create, and that we can do so by developing our skills in reading the world, in approaching the world from multiple points of view, and by paying greater attention to the specificity of our material bodies, is of central importance in feminist poststructural research into environmental education.

Framing

What is feminist poststructural theory? Historically it appeared after, and in response to liberal feminism, which mobilized a humanist discourse of individual rights, and to radical feminism, which celebrated womanhood and mounted a major critique of patriarchal power (Davies, 2000a). Each of these earlier feminisms opposed exclusionary practices based on gender and the prevailing negative constructions of women and girls. In contrast, feminist poststructural theory turns its attention to the constitutive power of discourse with its capacity to make social structures and practices appear normal and desirable (Davies, 2003a, 2003b, 2004, Gannon & Davies, 2007). It turns its attention to discourse, to the constitutive effect of discourse, that is, the way in which subjects and objects are shaped, and the means by which they are made intelligible:

> Discourse is not merely spoken words, but a notion of signification which concerns not merely how it is that certain signifiers come to mean what they mean, but how certain discursive forms articulate objects and subjects in their intelligibility. In this sense "discourse" is not used in the ordinary sense . . . Discourse does not merely represent or report on pregiven practices and relations, but it enters into their articulation and is, in that sense, productive. (Butler, 1995)

Feminist poststructural theory challenges the normative power of social/discursive structures and practices. It calls into question the individuality of the liberal humanist subject by deconstructing binary categories such as self/other, male/female, inside/outside, making visible the way they are constituted, and questioning the discursive practices through which they are made to appear natural and inevitable. Linked to the deconstruction of the male/female binary is the broader challenge to binary thought, to mind/body binaries, for example (Wilson, 2004), and to human/nonhuman binaries (Davies, 2000b). It thus opens a space of critique in which the effects of discourse are made both visible and revisable.

That deconstructive work of feminist poststructural theory shows, among other things, how relations of power are constructed and maintained by granting rationality to the dominant half of any binary (men, mind, human), and in contrast, how the subordinate term (women, body, nonhuman), is marked as other, as lacking, and as needing to be managed and controlled. Through examining the ways discourse inscribes itself in our minds/bodies, feminist poststructural theory shows how it is that power works, not by shaping us against our will, but by governing the soul. Power works at the level of desire, persuading us to take up dominant discourses as our own. As Butler (1997, p.14) explains:

> Power acts on the subject in at least two ways: first, as what makes the subject possible, the condition of its possibility and its formative occasion, and second, as what is taken up and reiterated in the subject's own acting. As a subject *of* power (where "of" connotes both "belonging to" and "wielding"), the subject eclipses the conditions of its own emergence; it eclipses power with power. . . . the subject emerges both as the *effect* of a prior power and as the *condition of possibility* for a radically conditioned form of agency.

Feminist poststructural theory thus makes visible the everyday, normative cultural meanings given to gendered bodies, to the environment (including human and nonhuman, and animate and inanimate beings), and to the relations between them (Davies & Gannon, 2009). But more, as Butler points out, it offers conceptual strategies for transgression, for finding other ways to speak and write against the grain of normative, naturalized constructions. Language is thus understood, on the one hand, as a powerful constitutive force, shaping what we understand as possible, and shaping what we desire within those possibilities. And on the other, in Cixous's words, the "intoxicating" power of language is able to be mobilized, creatively, to open up the not-yet-known (Cixous & Calle-Gruber, 1997, p. 22).

The meaning given to agency in this perspective is of critical importance. Agency lies in the power of the individual or the collective to, in a sense, catch discourse in the act of shaping or limiting them, and to find ways of thinking or acting differently. In this understanding, agency becomes recognition of the power of discourse, a recognition of one's love of, immersion in and indebtedness to that discourse, and also a fascination with the capacity to create new life-forms, life-forms capable of disrupting old constructions of gender, of power, and of exploitation (Davies, 2000b). This is what Cixous means by "more capable of reading the world, more capable of playing it in all ways" (1997, p. 30).

One way of reading and playing in the context of environmental education that feminist poststructural theory opens up, is to mobilize a creative capacity to rethink ourselves spatially and materially. How can we move beyond the separation of the human from other life forms? Wilson, for example, draws attention to the way the category "human" has been constructed, making it utterly distinct from other life forms. She points out that, "By discounting biology from its constitutive relations with other ontological systems, biology becomes isolated and destitute" (2004, p.70). She deconstructs the human/nonhuman binary, pointing out that we limit our evolutionary capacity if we accept *human* as the dominant term, separated from and superior to other ontological systems:

> Darwin's system of evolution specifies the ontological coimplication of animals, man, plants, rocks, and emotions. Each mode of materiality is built through its complicitous relations to others, and heredity is governed by a heterogeneous set of forces . . . By accentuating the structural intimacy of biology and psychology in *The Expression of the Emotions in Man and Animals*, Lorenz hints at one of the most underexamined aspects of Darwin's work: that evolution . . . is radically heterogeneous; certainly it is biological, but it is also psychological, cultural, geological, oceanic, and meteorological. (Wilson, 2004, p. 69)

Human existence in this understanding is not an existence that is separate from other forms of existence. Human, animal, earth—all exist, and exist in networks of relationality and dependence (Davies & Gannon, 2009). Wilson gives examples of ways humans are not actually separate: the mouth to anus tube that runs through us, for example, allows (depends on) the world to run through us, becoming part of us; and brains hold fish and reptilian ancestry. Such examples throw into vivid disarray the sacralization and separation of "humanity." Wilson also opens up our thinking about the mind/body binary, emphasizing the importance of openness and movement between the two: "Neurological obligation, then, is one way of understanding a relation between psyche and soma in which there is a mutuality of influence, a mutuality that is interminable and constitutive" (Wilson, 2004, p. 22).

In order to rethink those habituated modes of thought through which we constitute ourselves as human, and as separate to others, including nonhuman others, we need to unsettle the sedimented practices of categorical thought. Difference has, for a long time, been understood as categorical difference; the other is discrete and distinct from the self, with difference lying in the other; their identity constructed through a string of binaries in which their sameness as, or difference from, oneself is made real. Deleuze (2004) offers another approach to difference in which difference comes about through a continuous process of becoming different, of differentiation. Massey (2005, p. 21) describes these two approaches as:

1. "discrete difference/multiplicity (which refers to extended magnitudes and distinct entities, the realm of diversity)," and
2. "continuous difference/multiplicity (which refers to intensities, and to evolution rather than succession)."

In the first approach difference is being "divided up, a dimension of "separation," while in the second, Deleuzian approach, difference is "a continuum, a multiplicity of fusion." Deleuze wishes "to instate the significance, indeed the philosophical primacy, of the second (continuous) form of difference over the first (the discrete) form. What is at issue is an insistence on the genuine openness of history, of the future" (Massey, 2005, p. 21). As Williams (2003, p. 60) points out, for Deleuze "real difference is a matter of how things become different, how they evolve and continue to evolve beyond the boundaries of the sets they have been distributed into." And in this openness to evolution there is important work to be done in generating a new understanding of relationality, of understanding what it means to be in relation to, and known through the other—the human other and earth other (Davies, 2000b; Davies & Gannon, 2009). Cixous speaks of a complex balance between knowing and not knowing oneself in relation to the other. She develops the terms *positive incomprehension* to capture something of the movement and openness of differentiation, once differentiation is no longer individualized and separated off from the other, where we are no longer fixated and capturing the essence of self and other in their difference but on being open to the unknown, to the impossibility of knowing, and simultaneously to the beauty of moments of insight into the being of the other:

> [*Positive incomprehension*] is perhaps what we discover in love; or in friendship-love: the fact that the other is so very much other. Is so very much not-me. The fact that we can say to each other all the time: here, I am not like you. And this always takes place in the exchange, in the system of reflection where it is the other we look at—we never see ourselves; we are always blind; we see of ourselves what comes back to us through (the difference of) the other. And this is not much. We see much more of the other. Or rather, on the one hand, we see an enormous amount of the other; and on the other hand, at a certain point we do not see. There is a point where the unknown begins. The secret other, the other secret, the other itself. The other that the other does not know. What is beautiful in the relation to the other, what moves us, what overwhelms us the most—that is love—is when we glimpse a part of what is secret to him or her, what is hidden, that the other does not see; as if there were a window by which we see a certain heart beating. (Cixous & Calle-Gruber, 1997, pp. 16–17)

In order to open up the creative forces that enable us to evolve, continuously, beyond the fixities and limitations of the present moment we need to both turn our attention to what we are made of, our material continuity and "ontological coimplication" with others, including nonhuman others, and to open ourselves to multiple points of view, while deconstructing the sacralization and ascendance of humanity.

Doing (Pedagogy)

The primary implication of this approach is the valuing of the capacity to know oneself as being-in-relation, and to listen, both to oneself and to the other (Nancy, 2007a). In Rinaldi's (2006, p.114) words: "Competent listening creates a deep opening and predisposition toward change." Such listening is not categorical and self-referencing (how is this separate being categorically the same as or different from me?) but involves listening to thought with an openness of mind, with an openness to differentiation:

> *Listening to Thought* is not the spending of time in the production of the autonomous subject (even an oppositional one) or of an autonomous body of knowledge. Rather, to listen to Thought, think beside each other and beside ourselves, is to explore an open network of obligations that keeps the question of meaning open as a locus of debate. Doing justice to Thought, listening to our interlocutors, means trying to hear that which cannot be said but that which tries to make itself heard. (Readings, 1996, p. 165)

Environmental education, in this framework, would not be "lessons about the environment" where the teacher passes on the already-known to receptive, listening students, where the already-known, the already-thought, makes invisible the "ontological coimplication of animals, man, plants, rocks, and emotions" (Wilson, 2004, p. 69). It would be about openness to multiple ways of being, to evolving, of being engaged in a process of differentiation that is never complete. It would entail developing an understanding of oneself as coimplicated in the other with a mutuality of influence that is both constitutive and interminable. The initial question becomes, between teachers and their students, what is this place that we are constituting with, and in relation to, each other? How are our practices constitutive of each other, and how do they close down or open up the possibilities of differentiation and evolution? How are discourses at work on us and on our desire, that limit or potentiate our capacity for making viable lives in a viable world?

Nancy's (2007b) concept of "world-making," in contrast to "globe-making" takes up some of these questions. Education in a globalized world has been taken over by neoliberal technologies of government (Davies & Bansel, 2007). Teachers have typically been encouraged to be "compliant labourers, delivering curriculum using best practice strategies and having their work checked by quality control testing tied to objective standards" (Coulter & Wiens, 2002, p. 23). Dahlberg and Moss (2005, p. 9) see the quality practices of neoliberal education systems as attempts not to educate, but to reproduce and to control the already known:

> Measuring quality is just one of a variety of technologies deployed to regulate practice in [schools]. [Schools] are increasingly bounded by other normalizing frameworks—either required by government or offered by experts: standards, curricula, accreditation, guidelines on best practice, inspection, audits, the list rolls on . . . What these normalizing technologies have in common is an administrative logic, an intention and capacity to govern more effectively by ensuring that correct outcomes are delivered.

Globalization, or globe-making, produces over-regulated pedagogical practices focused on the production of generic, predictable individuals, responsive to the forces of government and geared to forging docile bodies, bodies "that may be subjected, used, transformed and improved" (Foucault, 1977, p. 136). Globe-making dismantles the social in favor of the economic. "World-making" as Nancy (2007b) defines it, in marked contrast, begins with self-in-relation to the other. It requires openness to new directions and possibilities, not mandated by governmental imperatives, but emergent from the specificity of particular places in the world. It focuses on engagement with the other, where other includes nonhuman, earth others, and on anticipating the eruption of the new. It has an unpredictable appearance, maintaining a crucial reference to the world as a space of relationality and as a space for the construction and negotiation of meaning. Against the disembodied and dislocated "everywhere and anywhere" of the global, Nancy (2007b) emphasizes the specificity of subjects in their particular time and space–of "this place, here." World-making is not focused on controlling the future. Its question is, rather, how can we give ourselves, open ourselves, in order to look ahead of ourselves, where nothing is yet visible.

What might pedagogy look like if it was not dominated by the forms of thought generated within globe-making discourses that construct land and its inhabitants as exploitable others?

Advocates of early childhood education have taken a significant lead in reconceptualizing pedagogy in feminist poststructuralist terms. Over the last decade they have shifted from a focus on cognition, and school readiness to the context of ethical and political debates in which the nature of childhood is to be challenged and rethought (Lenz-Taguchi, 2009). Rather than children being acted upon in the sphere of early childhood education with the goal of producing predefined outcomes, the child is positioned in this new thinking, as citizen, and as coconstructor of identity and culture (Dahlberg & Moss, 2005; Dahlberg, Moss, & Pence, 1999; Moss & Petrie, 2002). This shift requires very different work from teachers. "Implicit in this turn to active ethical practice is trust in the ethical capacities of individuals, their ability to make judgments rather than simply apply rules" (Dahlberg & Moss, 2005, p. 2). Their approach requires "listening, reflection, interpretation, confrontation, discussion, judgments open to question: there is no escaping the provisionality of this practice or its messiness" (2005, p. 13). In my work with primary school children (2003b) I explore ways of working with primary school children that enable them to develop different ways of being, through inventing, inverting and breaking open those structures and practices through which gender difference (as discrete difference) is created and maintained.

Such pedagogy, at all levels, is about movement, about folding and unfolding, about openness to difference and to change (Springgay, 2008). As such, there must be openness and a strategy of attention to the many small moments of deindividualization, in which both teachers and students escape "the limits of the individual" (Roffe, 2007, p. 43). These small escapes, these lines of flight, these slides toward the not-yet-known, are moments of becoming, in which there is also a "constitution of new ways of being in the world, new ways of thinking and feeling, new ways of being a subject" (Roffe, 2007, p. 43). In such pedagogical spaces, creative energies are mobilized through the unfolding relations between self and other, where "other" includes other selves, as well as the political, geographical, architectural and aesthetic spaces that they are constituting and, simultaneously, being constituted by. It is a relational space with a focus on listening, and on strategies of attention; it is defined by openness and softness, and also by the management of conflicts and contradictions lying at its heart (Ceppi & Zini, 1998). Fidyk observes that it is within such spaces that "we are enabled to 'return' to ourselves, to be true to ourselves, to the belonging-together of things that go on without us, without our doing. Thus, we know ourselves in terms of our relations rather than substance so that personal identity appears as emergent and contingent, defining and defined by interactions with the surrounding space" (Fidyk, 2003, n.p.).

Doing (Research)

Feminist poststructural research rejects the "evidence-based" discourse, recently mandated in the United States, and made popular in globe-making contexts (Davies, 2003c). It moves beyond the realism of evidence-based practice, with its construction of the researcher as separate from the field of research yet at the same time having a god's eye capacity to see and know all. Its orientation is not to the positivist power to predict the future, but to opening up the future—to differentiation and to the evolution of new ways of being. Feminist poststructural research recognizes the coimplication of the researcher in the field of study, and requires both a commitment to a position of *positive incomprehension* and to *reflexive, ethical responsibility*. As St Pierre writes, "we are ethically bound to pay attention to how we word the world . . . Poststructuralism does not allow us to lay the blame elsewhere, outside our own daily activities, but demands that we examine our own complicity in the maintenance of social injustice," and in this context, in the maintenance of exploitation (St Pierre, 2000, p. 484). Feminist poststructural research seeks to make visible the way the world is made real, not just by research participants, but by other researchers working in the particular field of inquiry (Ellwood & Davies, 2009; Stronach & MacLure, 1997), and to open up thought both in the research context and in the broader context of research reporting.

The particular research methodology that I will discuss here, that meets some of these criteria, is collective biography. Collective biography (Davies, 2000b, 2008a; Davies & Gannon, 2006, 2009) works with oral accounts of memories—memories that are made relevant in the participants' discussion of and reading about the topic they have collectively chosen to take up for their study. The participants, usually a small group of eight, or fewer, become coresearchers with the research leader. The research strategy focuses on the memory stories that are told by the participants in a workshop setting where the group, including the research leader, pays particular attention to the detail of each story, questioning each storyteller in turn, in order to be able to envisage "what happened," and the affect and emotions in each memory. The particular listening strategies enable each participant to see their own stories as continuous with others rather than as discrete and separate. Each story is then written, avoiding clichés and explanations, then read out loud to the group, again with an emphasis on listening. The listeners are asked to note any words or phrases that prevent them from being in the moment that the story evokes. The story is then discussed in detail, with participants working together to find the words that capture for writer and listeners the precise, embodied *haecceity* of the remembered moment (Halsey, 2008). Each story is then rewritten to rid it of those words and phrases that hold the readers/listeners out.

Each participant's stories are an attempt to express a moment-of-being in such a way that it holds the precise *haecceity* or thisness of the remembered moment. The technique of listening that participants adopt requires a particular close attention to each other's moments-of-being, so that each person is vividly present as a singularity. Each member of the group shares the "inner resonances" (Nancy, 2007a 42) of the world as it is encapsulated in the moments-of-being of each of the others. At the same time, listening and documentation, as they are practiced in collective biography, enable a taking-up of the being of the other in such a way that each dismantles in some part the normative, habituated repetitions through which identity-as-discrete-other takes place (Deleuze, 2004; Davies, 2008b). Through the particular mode of telling, listening, questioning, writing, and reading, a shift takes place. The memories are no longer told and heard as just autobiographical (that is, an assemblage of already known stories that mark one individualized person off from the next), but as opening up for, and in, each other, knowledges of being that previously belonged only to the other, as that other's marks of identity. In working collectively with memories, we live intimately within our own bodies, and our bodies take on the intimate knowledge of each other's being. Each subject's specificity in its very particularity, in its sensory detail, becomes, through this process, the collectively imagined detail through which we know ourselves as human, even as more human—as humans-in-relation.

The practices of collective biography build a community in the sense that Ceppi and Zini use that term, through "the willingness to listen and be open to others . . . [and through] respect for differences, however they may be expressed . . . [through a] sense of empathy, a closeness that creates bonds, that enables each group member to recognize the other and to recognize him/herself in the other" (Ceppi & Zini, 1998, n.p.). When the other is different, and has made different, initially unimaginable choices, the processes of collective biography enable each participant to know—through attending to affect, to emotion, to voice, to images, to the specificity of the other—the rich and surprising multiplicity of their own and others' being. It is this specific move that opens up a new mode of ethical reflexivity (Davies, 2006, 2008c).

In the collective biography story that I include here the preschool child listens to the sound of the piano and the triangle, the teacher and the other children. In this moment listening involves, in Nancy's words, stretching the ear "—an expression that evokes a singular mobility, among the sensory apparatuses, of the pinna of the ear—it is an intensification and a concern, a curiosity or an anxiety" (Nancy, 2007a, p. 5).

The teacher opened up a box of musical instruments. She asked who would like the tambourines, who would like the drums, and last of all, who would like the triangles. The small girl had never played with any of these instruments, so she did not raise her hand. The triangles were given to the last ones left who had chosen no instruments. The triangles seemed inferior, the small girl thought, when compared to the drums. The teacher demonstrated how each instrument was to be played. The triangle must be held so, by the string, and struck just so with the small metal stick. Then the teacher sat down at the piano and played them the tune they were to accompany. Then again, with the children this time, and the noise was terrible, the children seeming to ignore completely the sound coming from the piano. The small girl carefully hit her triangle, but the sound was ugly and flat. The other triangle children ran their stick around the triangle hitting all the sides, laughing, making the triangle fly off in all directions with a jangling sound muddled up with the whack and thump of the drums and the terrible jingling of the tambourines. She anxiously watched the other children's wild experimentation with their instruments until, suddenly, she could see that she must loosen her grasp on the stick before the triangle would sing. When the piano started again she noticed the sound of her triangle came after the note she was supposed to accompany. She listened hard, focused only on the piano, the triangle and the stick. The teacher repeated the tune. The small girl found she had to begin to strike not when the piano note came, but the moment before it came. Her body discovered exactly the moment the stick must begin its descent in order for the two sounds to come together. The sound of the piano and the triangle exactly together made a warm feeling in the small of her back that ran down the back of her legs and into her shoes.

In this moment the child listens for meaning, but much more than this, she listens for sound. Her listening generates something new, not located in the teacher or in her, but in-between her, the other children, the teacher, the piano and the triangle. Her attentive listening is not just with ears for the teacher's meaning or intention. She listens with her whole body to the vibration of the piano and the timing of the beat. Her body discovers a new way of moving in relation to the other, where the other includes both humans and nonhumans. Together they produce a new sound—a new capacity for engagement. The child has listened with all her being, and an emergent self finds itself cocreating a new event that she could not have imagined beforehand. In Nancy's words:

> To listen is to enter that spatiality by which, at the same time, I am penetrated, for it opens up in me as well as around me, and from me as well as toward me: it opens me inside me as well as outside, and it is through such a double, quadruple, or sextuple opening that a "self" can take place. To be listening is to be at the same time outside and inside, to be open from without and from within, hence from one to the other and from one in the other. (Nancy, 2007a, p. 14)

In environmental education classes both students and teachers could explore giving voice to the human, nonhuman and earth others coimplicated in their own being, their own community (Davies, 2008b). Because the stories are documented through this process of writing and rewriting, it can also be an extraordinary resource for researchers, through which they can explore, in relation with their participants, their coimplication in the being of others in their specific place. Learning to listen and moving beyond individualized, hierarchical, globe-making modes of identity is central both to feminist poststructural theory, and to the specific demands of environmental education.

Assessing

I have discussed in detail elsewhere the question of what makes research legitimate (Davies, 2003c, 2009). The assessment of research that adopts the kind of thinking and practices elaborated in this chapter would not be based on the standard evidence-based logic that requires the researcher to set up a project with prespecified, measurable outcomes. Rather the assessment would look for commensurability between the conceptual framework and the research practices. It would look for research that is geared toward listening to the not-yet-heard, even the not-yet-hearable, and that opens up spaces for thought that were previously closed (that closure being due to language, to habits of research/writing/practice, to invisible practices of normalization and categorization). It would look for research that moves beyond description and repetition, and that recognizes the ways subjects are caught up in discourse, in social relations, in history. Assessment of collective biographical research would look for the ways in which it has succeeded in making visible relations among individual subjects (both human and not), and the ways in which it has unmoored unreflected, habituated beliefs and practices. It would look at the ethics of the research practices asking in what ways the researcher was open to the participants, able to work with them to capture the *haecceity* of their moments of being, and to what extent the participants were able to make visible and meaningful "their ontological coimplication with others" (Wilson, 2004, p. 69). It would as in what ways creating a viable world with others has been enhanced through the research process. Assessment would not look for the extent to which the researcher has successfully separated her- or himself off from the research context, but would look for evidence of relationality, of listening in order to hear the as-yet-unhearable. It would look for respect for the other's unknowable difference, and for openness to seeing the specificity of the other in new ways. Those new ways would, ideally, involve not objectification but positive incomprehension (Cixous & Calle-Gruber, 1997), not regulation and control but openness to the not-yet-known, and finally, not prediction but generation.

References

Butler, J. (1995). For a careful reading. In S. Benhabib, J. Butler, D. Cornell, & N. Fraser (Eds.), *Feminist contentions: A philosophical exchange* (pp. 127–143). New York, NY: Routledge.

Butler, J. (1997). *The psychic life of power: Theories in subjection*. Stanford, CA: Stanford University Press.

Ceppi, G., & Zini, M. (Eds.). (1998). *Children, spaces, relations: Metaproject for an environment for young children*. Milan: Domus Academy Research Center.

Cixous, H., & Calle-Gruber, M. (1997). *Rootprints: Memory and life writing*. London, UK: Routledge.

Coulter, D., & Wiens, J. R. (2002). Educational judgment: Linking the actor and the spectator. *Educational Researcher, 31*(4), 15–25.

Dahlberg, G., & Moss, P. (2005). *Ethics and politics in early childhood education*. London, UK: Routledge/Falmer.

Dahlberg, G., Moss, P., & Pence, A. (1999). *Beyond quality in early childhood education and care*. London, UK: Routledge/Falmer.

Davies, B. (2000a). *A body of writing 1989–1999*. Walnut Creek, CA: Alta Mira Press.

Davies, B. (2000b). *(In)scribing body/landscape relations*. Walnut Creek, CA: Alta Mira Press.

Davies, B. (2003a). *Frogs and snails and feminist tales: Preschool children and gender* (2nd ed.). Cresskill, NJ: Hampton Press.

Davies, B. (2003b). *Shards of glass: Children reading and writing beyond gendered identities* (2nd ed.). Cresskill, NJ: Hampton Press.

Davies, B. (2003c). Death to critique and dissent? The policies and practices of new managerialism and of "evidence-based practice." *Gender and Education, 15*(1), 89–101.

Davies, B. (2004). Introduction: Poststructuralist lines of flight in Australia. *International Journal of Qualitative Studies in Education, 17*(1), 3–9.

Davies, B. (2006). Collective biography as ethically reflexive practice. In B. Davies & S. Gannon (Eds.), *Doing collective biography*. Maidenhead, Berkshire: Open University Press.

Davies, B. (2008a). Practicing collective biography. In A. Hyle & J. Kauffman (Eds.), *Dissecting the mundane: International perspectives on memory-work* (pp. 58–74). Lanham, MD: University Press of America.

Davies, B. (2008b). Writing on an imminent plane of composition: Opening oneself to difference. In A. Jackson & L. Mazzei (Eds.), *Voice in qualitative inquiry: Challenging conventional, interpretive and critical conceptions in qualitative research*. New York, NY: Routledge.

Davies, B. (2008c). Rethinking "behaviour" in terms of positioning and the ethics of responsibility. In A. M. Phelan & J. Sumsion (Eds.), *Critical readings in teacher education: Provoking absences* (pp. 173–186). The Netherlands: Sense Publishers.

Davies, B. (2009). Legitimation in post-critical, post-realist times, or whether legitimation? In Paul Hart (Ed.), *Sage companion to research.*

Davies, B., & Bansel, P. (Eds.). (2007). Neoliberalism and education [Special issue]. *International Journal of Qualitative Studies in Education, 20*(3), 247–260.

Davies, B., & Gannon, S. (Eds.). (2006). *Doing collective biography.* Maidenhead, Berkshire: Open University Press.

Davies, B., & Gannon, S. (2009). *Pedagogical encounters.* New York, NY: Peter Lang.

Deleuze, G. (2004). *Difference and repetition.* (P. Patton, Trans.). London, UK: Continuum.

Ellwood, C., & Davies, B. (forthcoming). What counts as bullying? Questions of intentionality, relations of power and repetition. *Qualitative Research in Psychology.*

Foucault, M. (1977). *Discipline and punish: The birth of the prison.* Harmondsworth: Penguin Books.

Fidyk, A. (2003). Attunement to landscape: Dis/Composure of self. *Educational Insights, 8*(2). Retrieved January 25, 2009 from http://www.ccfi.educ.ubc.ca/publication/insights/v08n02/contextualexplorations/curriculum/fidyk.html

Gannon, S., & Davies, B. (2007). Postmodern, poststructural, and critical perspectives. In S. Nagy Hesse-Biber (Ed.), *Handbook of feminist research: Theory and praxis* (pp. 71–106). Thousand Oaks, CA: Sage.

Halsey, M. (2008). Molar ecology: What can the (full) body of an ecotourist do? In A. Hickey-Moody & P. Malins (Eds.), *Deleuzian encounters: Studies in contemporary social issues* (pp. 135–150). Houndmills, UK: Palgrave Macmillan.

Lenz-Taguchi, H. (2009). *Going beyond the theory/practice divide in early childhood education: Introducing an intra-active pedagogy.* London, UK: Routledge/Falmer Press.

Massey, D. (2005). *For space.* London, UK: Sage Publications.

Moss, P., & Petrie, P. (2002). *From children's services to children's spaces: Public policy, children and childhood.* London, UK: Routledge/Falmer.

Nancy, J.-L. (2007a). *Listening.* (C. Mandell, Trans.). New York, NY: Fordham University Press.

Nancy, J.-L. (2007b). *The creation of the world or globalization.* In F. Raffoul & D. Pettigrew (Trans.). Albany, NY: State University of New York Press.

Readings, B. (1996). *The university in ruins.* Cambridge, MA: Harvard University Press.

Rinaldi, C. (2006). *In dialogue with Reggio Emilia: Listening, researching and learning.* London, UK: Routledge.

Roffe, J. (2007). Politics beyond identity. In A. Hickey-Moody & P. Malins (Eds.), *Deleuzian encounters: Studies in contemporary social issues* (pp. 40–49). Houndmills, UK: Palgrave Macmillan.

Springgay, S. (2008). *Body knowledge and curriculum: Pedagogies of touch in youth and visual culture.* New York, NY: Peter Lang Publishing.

St Pierre, E. A. (2000). Poststructural feminism in education: An overview. *International Journal of Qualitative Studies in Education, 13*(5), 477–415.

Stronach, I., & Maclure, M. (1997). *Educational research undone.* Milton Keynes, UK: Open University Press.

Williams, J. (2003). *Gilles Deleuze's difference and repetition: A critical introduction and guide.* Edinburgh: Edinburgh University Press.

Wilson, E. A. (2004). *Psychosomatic: Feminism and the neurological body.* Durham, NC: Duke University Press.

45

Suited

Relational Learning and Socioecological Pedagogies

Marcia McKenzie, Kim Butcher, Dustin Fruson, Michelle Knorr, Joshua Stone, Scott Allen, Teresa Hill, Jeremy Murphy, Sheelah McLean, Jean Kayira, and Vince Anderson
University of Saskatchewan, Canada

It smells sweet yet refreshing, almost like sweetgrass and peppermint. Each branch has hundreds of tiny, delicate leaves pointing upward. The entire plant is so soft it feels like velvet. As I observe it, I cannot decide on the colour. It is a pale green, that is definite; but it has tints of red coming through—on the stem, on the tips of the leaves . . . Closer down to the base, a yellow tint reveals itself on a few of the leaves. The delicate leaves are astonishing. (field journal entry, Bailey, "Place, Experience, and the Socio-Ecological" course, August 2009)

This chapter builds on a simple premise that learning occurs relationally, and that this has important implications for teaching and research in relation to social and ecological justice. We, as beginning teachers, experienced teachers/graduate students, and teacher educators, have been investigating this focus together over the past two years, and offer our collective research stories here as a way to further examine relational experiences of learning about socioecological pedagogies (see also McKenzie, 2008, 2009). Rather than telling you about participatory research methodology, we seek to show you a working example highlighting the processes and outcomes of one such endeavor.

In total, eight of us began our teacher education program at the University of Saskatchewan in Saskatoon, Canada in September, 2008; and became part of an inaugural "Social and Ecological Justice Teacher Education Cohort," initiated and in part taught by two of us as faculty and seconded faculty (Marcia & Sheelah). This cohort experience involved taking several classes together through the first of two years of the program with a particular emphasis on both social and ecological issues and their relationship to each other (e.g., intersectional issues of race, gender, sexuality, class, ability, globalization, anthropocentrism, consumption, community, and place). During this first year we also together began a participatory research project with an aim of shifting our own and others' practices of teaching and learning for social and ecological justice. This research group, which additionally included two graduate students and prior teachers (Vince & Jean), involved monthly meetings through 2009, undertaking interviews with teacher candidate peers, administering surveys to teacher candidates and secondary students, transcribing and analyzing data, and participating in multiple days of collective data analysis.[1] In this paper we draw on this analysis of survey, interview, and focus group data collected from sixty teacher candidates and secondary school students; subsequently collected data from the field journals of teacher candidates participating in a experiential socioecological pedagogy course; and use that data to support a central focus on a collaborative inquiry project undertaken by four of us entitled "Suited."

The "Suited" inquiry project was undertaken by Kim, Dustin, Michelle, and Josh as an assignment for Marcia's EFDT 101: Introduction to Education class in September–December 2008. The assignment itself was based on a series of interviews with and observation of a secondary student named Diego who was a student in Sheelah's "Citizen's Inspiring Change" grade twelve course, which also ran from September–December 2008. All teacher candidate authors were observers of that course and in the school in which it was housed once a week during the 2008–2009 academic year; and thus were able to observe parallels between the aims and activities of that grade twelve course and their own teacher education "Social and Ecological Justice Cohort." The "Suited" project was also developed in relation to a selection of articles read in the EFDT 101 class on topics such as binaries (Davies, 1989), discourse and agency (McKenzie, 2006), the truth about stories (King, 2003), communities of practice (Gustavson, 2007), and the arts and imagination (Greene, 2000).

We understand that power operates in this collaborative research through the initial "Suited" project undertaken as coursework, in the obvious ways that theory informs our empirical questions and data, in the writing and practice of the ideas below, in who has more or less purchase in different domains of academia and teacher practice, in the fact that nine out of eleven of us are white and that all of us experience relative privilege as educated learners/teachers living in the west and global north. Like others that have written about participatory forms of research, we realize that power is inevitable in any collaborative inquiry as it operates across and within research participants and projects (e.g., Kesby, 2005). This paper is about our collaborative learning about how learning works. We do not offer a thin seamless account, but rather seek to share a thicker patchwork of stories of learning and action. Through this, we hope to show the possibilities of participatory research as relational socioecological pedagogy.

Our Questions and Answers: How Does Socioecological Learning Work?

> *It was the first time in university that I really felt, really engaged, and that it was worth something. That project made us feel like we were part of something bigger, and more important than just a grade, a mark, or getting it done in time. It was something deeper. And that's where my interests lie. (Michelle on "Suited" inquiry project, January 2009)*

> *What are the strategies that you use to create these experiences? Like, these are the ideas that you come across that make you socially and ecologically just. So how do you create that in the classroom? That you can't spoon feed it into kids, but how do you create the environment so that it happens? (Josh, January 2009)*

Our research group meetings began in January 2009 during the start of the teacher candidates second term of their program, and following the completion of their inquiry projects for Marcia's class in the first term. We began by discussing our reasons for wanting to participate in an outside-of-class collaborative research group investigating teaching and learning for social and ecological justice. The explanations that Michelle and Josh provide above were echoed across the group: to be more engaged ourselves and to better engage our students in learning for social and ecological justice.[2] Through further discussion, we developed the following as our shared research focus:

> *What experiences allow teacher candidates and students to explore understandings of, and make connections among, self, community/local, global, and ecological issues? More specifically, we're interested in encouraging through our pedagogy/research*:
>
> (i) *self-exploration and meaning, and connecting that to broader contexts, and*
> (ii) *empowering socially and ecologically conscious and active individuals and communities. (January 16, 2009)*

Several days later, Josh circulated an additional question by e-mail, which we then discussed in a later meeting as included below:

> *We are concerned with the journey, the small slice of time in which students become socially aware and can negotiate in and out of discourse. Is this aha or awakening experience one of the mind or the heart? Why?*

Marcia: So that's a great question that kind of helps start to unpack that idea of "experience." What kinds of experiences? Are they cognitive, are they emotional? And how do they affect how we understand the world? . . .

Kim: I don't even know if it is, one sort of slice of time. Or if it is continual, it's never ending, it's not going to finish when you are done this course. It might be sort of one small event or big event or one instance that kind of twigs something either in your heart or in your mind that makes you aware of something, and then you go from there and keep building on it . . .

Marcia: So would you say the small slice of time you are thinking of a particular experience? Like seeing a film or—

Josh: Or seeing an injustice or something, and then—

Marcia: Going camping or whatever.

Josh: Going traveling or whatever . . . You know, is it the thought that makes you change, or is it the feeling that makes you change? Yeah, I would say both, it's a combination . . . So I think if we understand those experiences better, it would be easier for us to create those sorts of experiences for our students . . .

Michelle: I think it has to trigger their heart. And I mean like, not passion, but just yearning or sadness or longing or something like that. I think it needs to be involved more so at the heart than in the intellect . . . Like of all the stuff we've learned, all the experiences we have had, I can remember emotions, or people, or feelings, way more than I can the actual dialogue of the intellect, right? (January, 2009)

This discussion begins to flesh out an understanding of socioecological learning as occurring "more so at the heart than in the intellect," or perhaps as Riceour writes, in "feeling as thought made ours" (Fawcett, 2009). Yet, as the below sections will elaborate further, a critical intellectual engagement is also necessary:

Scott: Would you not need to have some sort of cognitive ability to handle some of the issues you're going to be dealing with?—the discourse requires specific movement within. And, if you're not cognitively able to do that, you have to learn it . . . Without the cognitive training you can only go so far on emotion.

Josh: I also think there is a deliberate conscious choice to be vulnerable and listen and weigh some sort of evidence or assess the situation and say there is some sort of inequality here, something's not right. It's a choice. We can be confronted with all of these inequalities, but it is a choice to be vulnerable and put out of comfort. (January 2009)

These comments by Scott and Josh synthesize the core of much work on antiracist and other forms of anti-oppressive education, including some socioecological pedagogies, in which the focus (or "discourse" as Scott puts it) is on learning skills of critique in order to be able to deconstruct trajectories of cause and responsibility (Kahn, 2010; St. Denis & Schick, 2003). And as this work also highlights, as Josh points out, this is difficult work that necessitates vulnerability and discomfort in examining one's own complicity in various cultural patterns and social norms (Boler, 1999; Britzman, 2007; Felman, 1992).

In addition to discussing aspects of how socioecological "learning" occurs through embodied and emotional experience as well as critical engagement (McKenzie, 2008), we were interested in what makes subsequent socioecological "action" possible as part and outflow of that learning (see part two of the research question). As a range of previous research in socioecological education has suggested, the social context of a collective group enables action in ways that are not possible or are less likely by an individual on their own (Reid, Jensen, Nikel, & Simovska, 2007; Wals, 2007). Our research data and our own experiences as a group also suggest that "knowing that others are out there" can provide powerful support for enabling action and continued learning in one's own life and teaching practices. But, we are getting ahead of our story . . .

The remainder of the paper centers on the "Suited" inquiry project undertaken by Kim, Dustin, Michelle, and Josh based on their discussions and observations with secondary student, Diego. The project took the form of a hand-stitched, leather-bound book which tells the story of Diego's social and ecological justice learning and agency through narrative, photos, and footnotes which reference observation and interview data. Also figuring centrally in the rest of the paper is a co-interview conducted with Kim, Michelle, Josh, and Marcia following completion of the "Suited" project, in which the process and outcomes of the work are discussed in relation to the larger participatory research project on socioecological learning. To elaborate components of the discussion, some additional research data is included.

Wardrobe as metaphor for relational learning: suited for action

Prologue

> *This is Diego's story. It starts one morning at 6:45. The alarm rings and Diego turns over, slams his hand down on the clock, and covers his head with his blanket. He dreads mornings . . . A good cold shower will get him ready for the day. He stretches and hears the phone ring . . . He can't find it. Piles of clothes. Piles of papers. Ring. Magazines. Books. Ring. No phone. He knows exactly who it is and what they want. There it is. He doesn't want to answer but knows he must. He flicks it open . . . "Yeah. What do I need to do?" Diego asks, putting on his favorite hat.*

Marcia: Why the clothes?

Michelle: It was all a metaphor.

Josh: They are not just clothes, they are, you are becoming an agent and once you're an agent you have the power to use agency. And so these are the tools that you need to use agency. So they are very much his training . . . So once he has that, then he has one of the tools to become an agent and use his agency to negotiate the world . . .

Josh: So everything in the story has some meaning . . . And, it's a secret agent story and what meanings that has, and the clothes, and the rooms that he goes to, and the people he talks to, and the things he gets. And you know, we went over everything as a group, everything that everyone said, and made sure that this is what we wanted to convey.

Kim: There wasn't anything that we put in the book that doesn't matter, or that is extra. Everything has a place. (January 2009)

Shirt and Pants: Stories to Live In

> *After getting an initial call from Whitehead, Diego heads to school: Diego walks up the path to BRC. Otherwise*

Figure 45.1 The structures of schooling.

known as the Brain Reconstruction Center. He swiftly opens the door and walks in. Everything at the Institute is familiar, but something feels different today. He walks down the corridor. He passes by the familiar posters on the wall. He's seen them a million times, read their messages, but today something is off. He walks past the office, past the library, past the Commons. As he rounds the corner he comes face to face with Agent Whitehead.

"Where have you been hiding?" Whitehead asks . . . "You know me. I've been here all along. You've been in hiding all these years, since you were a young boy. You just haven't realized it. Don't be afraid, you know what you're doing. It will prepare you for the rest of your life. Are you ready for your mission?"

"What mission? What am I supposed to do?"

"Look at the bigger picture Diego."

"But where do I start?"

"Deconstruct your experience, Diego. And follow the music."

Josh: Right, we call it an institute, and it is the school. And it is kind of like an institute, like we do the building to make it look imposing and, you know, it's almost like a jail . . . That's the first thing that he learns is to deconstruct the messages that are around the school . . . We have really mixed feelings about schools, you can have good learning there, but there is a lot of—

Kim: Institutionalization.

Josh: Yeah, built-in discourse, you know. The fact that the building is built like that, and the posters around, and the community . . .

Marcia: How are you thinking about discourse, maybe how do you define discourse?

Josh: Discourse is just the language that surrounds a culture or an ideology or—. It's the parameters of, or it's the way you talk about something, and the way it talks you. (January 2009)

Various discourses, understandings, or orientations to the world are learned literally through language—the words we use and the relationships they imply and embody. But discourses, or the culturally and place-specific stories we live by, are also passed on to us and by us through material practices, such as the ways in which we build our schools and communities, the everyday practices we negotiate the world through, the ways in which we engage with one another and with places. In a collaborative interview with Teresa, Scott explains how such cultural practices form the ways in which we understand and interact with the "environment," and also how change requires the ability to see alternatives:

Scott: Culture is a big factor in what you are capable of, and how you make a difference in social and ecological terms. Like if you see your parents throwing garbage away, or letting the car run for half an hour in the summer when no one is in it, or stupid things that are wasting, like watering at the heat of the day. And what are you going to do? You're probably going to do the same thing, because that's what you know . . . You're going to carry on those practices, until you are shown something different. (In interview with Teresa, December 2008)

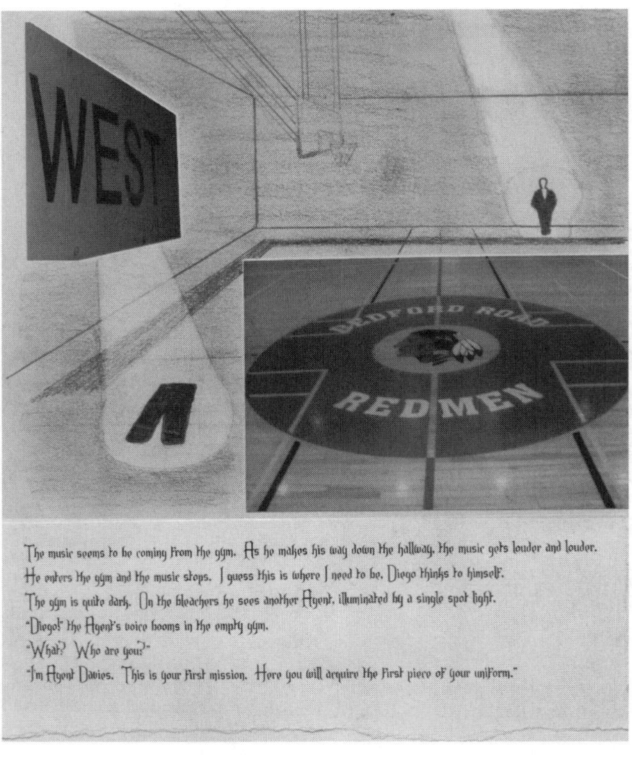

Figure 45.2 Learning about ambiguity.

The music leads Diego to the gym, where he learns via an encounter with Agent Davies (1989), that he does not need to see the world in black and white.[3] Selecting what turn out to be the "perfect" gray pants, Diego moves on to choose a shirt in the nurse's office.

Josh: There is an illusion of choice in discourse, and if you take the antithetical position you are still talking about the same thing even though it seems like you are not. And so, yeah we wanted to show that the discourse is saying you have choice, but really you don't. For example, you have a religious person that says I believe in God, but you have someone who says I am an atheist I don't believe in God, but they are still talking about it.

Marcia: Defining themselves in relation to God.

Josh: Right. And so you're never outside of discourse.

Kim: I think it was Josh that brought up the idea of an injection, sort of, the idea is injected into you.

Josh: It becomes part of you, and you can't, it's part of you who you are . . .

Kim: I mean, he didn't get discourse injected into him, and he chose not to see when Agent McKenzie said, "Go ahead and pick one, there are lots of colors [of shirts]." And Diego recognizes that they are *not* colors, that they are white. And so I think he chooses *not* to get that injection of discourse and go along with that sort of dominant idea of whatever situation . . . And one of our footnotes [at the bottom of this page]: "Field Notes from October—Teacher makes students repeat their presentation if it wasn't what he was looking for." And, Diego was on the honor roll . . . And, I think he was aware that maybe grades aren't sort of a motivating factor for him; but, it was still, he was still excited at winning.

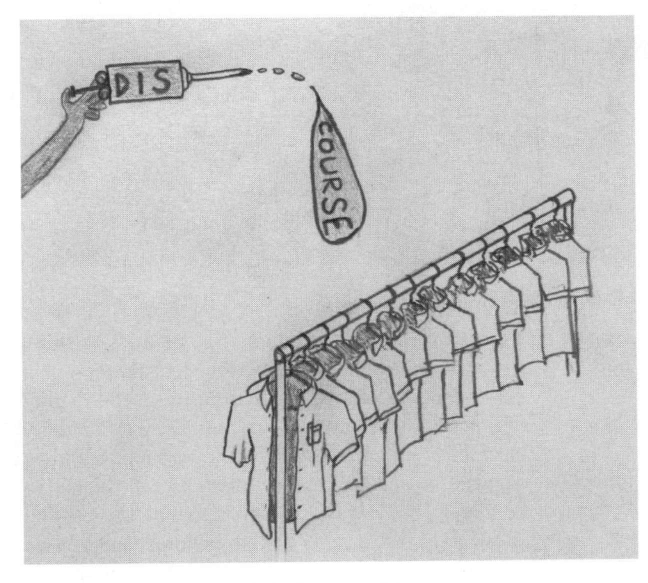

Figure 45.3 Discourse and the illusion of choice.

Marcia: Did you feel that he was struggling between discourses?

Josh: Yeah, he totally was. Sheelah was opening things up for him in his classes, but still he was excited about, I don't know [previous ways of understanding the world via grades, etc.]—

Kim: And I think he was so close with his parents that I think, his parents sounded really great. I never met them, but I think he was sort of struggling between things that his parents would say to him and things that he would learn at school from Sheelah or other teachers.

Shoes, and Toes: The Places We Are

Michelle: We had a disagreement over the shoes. Dustin didn't think we should use the Nike brand. And I believe that we argued that we actually should use something that is so common and so pop culture orientated.

Figure 45.4 Stories heard and told.

Kim: Well it goes back to the previous theories, that he is trying to fight against being sort of lumped in with youth culture, he is trying to get away from that. But, he is still buying the brand name clothes.

Josh: Playing video games.

Michelle: Agent King, of course, *The Truth About Stories* (King, 2003) is all about "your story". . .

Kim: And the shoes were just what we choose because stories—

Michelle: They are rooted in something . . .

Kim: And [shoes] can take you to different places, they can transport you to other worlds and experience how other people live, and how other people see the world. So his shoes enable him to sort of walk in those different—

Michelle: Well, it's like the classic thing about, you don't judge somebody until you have walked a mile in their moccasins, or something. So that kind of fit in with Agent King as well . . .

We read Thomas King's (2003) *The Truth About Stories* in our "Introduction to Education" teacher education class, as did Diego in his grade twelve "Citizens Inspiring Change" class. Through the telling of stories about his own life, about the politics and representations of what it means to be Aboriginal in Canada, King poignantly shows how we *are* the stories we know, hear, and tell. And while we are constrained by those stories and the misunderstandings and harms they bear and perpetuate, there is always the hope of learning or relearning different stories: "Take this story, for example," he says, "Do with it what you will, but don't say you never heard it." As Kim explains, "I think that stories are a form of discourse, so they can be transformational and they can take you out of what you know. And they can place you into a different discourse." In October of 2008, the two classes of teacher candidates and grade twelve students spent a morning together discussing the meanings and possible implications of Thomas King's book, *The Truth About Stories.*

Overall one of the biggest things I have learned in this class is to talk less and to,

Listen More

and to remember,

"Wherever you go, there you are"

(field journal entry, Kara, "Place, Experience, and the Socio-ecological" course, August 2009)

Michelle: I think this was your idea Kim to come to the camp, to have the shape shifting into a classroom above the campfire. That was a great idea . . .

Figure 45.5 Learning in place.

Josh: This is representative of how, of what it feels like, the feelings that he had about Sheelah's class.

Michelle: There's no power. They are all in a half moon formation, deliberately.

Josh: Right, like they are all equals sitting in a circle, and the sort of sharing of feeling.

Kim: And getting out of the familiar. Getting out of the classroom, and going on a field trip somewhere.

Michelle: It was his favorite thing to do. He loved that, he learned more than anything that way.

Josh: And all the people are wearing different colored shirts. And so they all come with their discourses, that it's all on an equal plane I guess. There is romance and discipline here too, it's both. And they are both fully colored in [heart and square of "romance" and "discipline/precision"][4] . . . He is fully into both learning intellectually and emotionally (multiple speakers).

Michelle: And that page represented that more than any other time throughout the whole book.

The elder looks to the heavens and tells the story not of the past but of space. Diego is content to listen for a while, and when the story is finished the campfire burns out slowly and Diego is back in the classroom. He looks around and realizes is that everything is the same as it was, except for his shoes.

Diego: I get it, the shoes both take me to any space that I will experience and to whatever story I hear. (Interview with Kim and Michelle, January, 2009)

Changing Jackets: Context and Community

Kim: Yes, here he is in his community of organized occupational practice: The petroleum transfer engineer.

Figure 45.6 Communities of practice.

Josh: This is one of his communities of practice.[5] I think it was Dustin didn't like the fact that the co-op [a provincial cooperative gas station] was a community of practice. And I think it was because a community of practice is something that you are supposed to be engaged in, and he was only there because he had to be there. That it was you were there by choice, and so we argued that he *is* there by choice, and you *do* interact with people. And it is a defined community.

Michelle: And it really fit into the jacket motif of the story, so we kind of needed. The story had to propel itself further, so . . .

Marcia: So, "This is not my only jacket": Did you see him taking on different roles, various responsibilities?

Kim: Just that a jacket is easy to change, and it was representative of how he is in each different situation.

So, it's like you would speak with your friends in one way, but when you're speaking to your professor, you use a different language, a different kind of language. So, that's what his jacket represented, that he could enter different spaces and, sort of, put on the jacket and be who he needs to be in each community of practice, and then change that if he needs to.

Josh: It's who you are on the outside. Your jacket is who you are on the outside.

Marcia: And how you can be a different person in a different context? . . .

Michelle: And that every experience he goes into, he is going to wear some of that on his body. That was our idea with that. But he can't shake any experience he has had totally. It will be evident on him or in him.

Josh: Once he understood discourse, his shirt could change color according to which story he's telling or—I don't know if we mentioned that in the story.

Michelle: It's kind of in the next page with the jacket. He takes off the co-op jacket and there is a stain. It's kind of—

Marcia: You can take the boy out of the co-op, but you can't take the co-op out of the boy.

Building on earlier sections of the book that introduce the concept of discourse and the possibility of working through and across understandings and practices of the world, as well as the role of place and travel through space as part of developing or solidifying new or old stories; here the role of community is introduced as another means through which we are made or remade. As teacher candidates, we explored the language of "communities of practice" through our reading of Leif Gustavson's (2007) book, *Youth Learning on their Own Terms: Creative Classroom Practice,* an ethnography of the creative learning of three youth through the various communities of practice they are involved in. The book suggests how collective contexts enable learning and support the rehearsal of, and feedback on, forms of practice specific to that community, including language, social norms, historical knowledge, artistic skills, and so on. In the above dialogue about the Suited book, there is a sense of the complexity with which communities operate through and on us, in that while "a jacket [or community] is easy to change," they also leave "stains." In socioecological pedagogy contexts, this requires considering both how our students are affected by the communities they are or are not a part of beyond "school," as well as how their educational contexts themselves may better operate as productive and enabling communities of practice.

In many ways, the Suited project group, our participatory research group, and the broader Social and Ecological Justice Cohort have functioned as communities of practice supporting learning and action, including toward social and ecological justice. As Michelle explains, "It's fairly astonishing to me how much of an extended family we have become . . . When you have that connection, and you have heart in a room already, then it just makes your learning sort of effortless" (January 2009). We can see this too in the dialogues above, in the ways that Suited group members finish each others' sentences, and have developed a shared consciousness of sorts in relation to the project. In another conversation Kim says,

Kim: I don't even feel like I've really had that experience yet, of meeting somebody that really inspired me, other than in the cohort, but—

Michelle: Maybe *this* is your experience.

Kim: Yeah, exactly. (January 2009)

The Magic Hanky: Imagination and the Not Yet

Kim: I think this came out at like midnight in the art room at the university.

Michelle: One of those moments . . .

Kim: I think because we just wanted to, we all kind of found [Maxine] Greene, that video especially[6], really interesting, and kind of out there, and a little bit crazy . . .

Michelle: Of course the handkerchief was pretty apparent, he had it all along but he didn't know it.

Josh: It represents his imagination, yeah, and his person.

Kim: And he finds it under his hat, right? And then she [Greene] disappears in another puff of smoke (laughing).

Michelle: We had fun with this one.

Kim: Yeah, this was all just about imagination and that he doesn't need to get his imagination extrinsically—

Michelle: From somebody else.

Josh: And this is fully romantic, like your imagination is . . .

Michelle: And he is still creating his own life, his own pathway, his own choices. So it's—

Kim: And I think along with the "I am what I am not yet" (Sartre in Greene, 2000), I think he knows, [Diego]

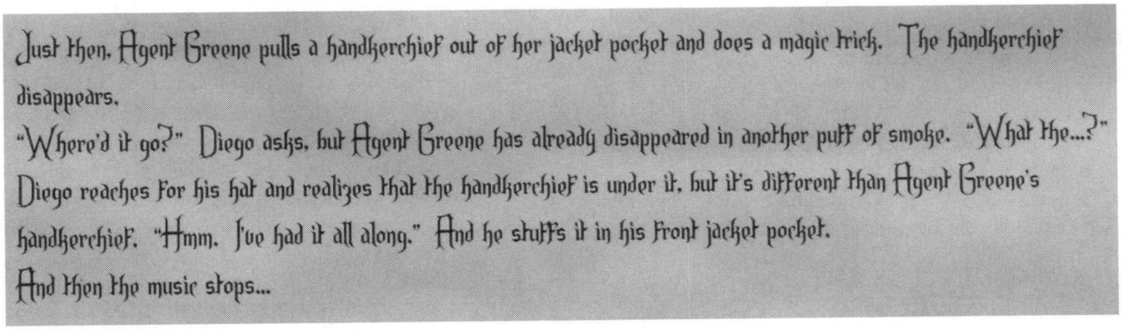

Just then, Agent Greene pulls a handkerchief out of her jacket pocket and does a magic trick. The handkerchief disappears.

"Where'd it go?" Diego asks, but Agent Greene has already disappeared in another puff of smoke. "What the...?"

Diego reaches for his hat and realizes that the handkerchief is under it, but it's different than Agent Greene's handkerchief. "Hmm. I've had it all along." And he stuffs it in his front jacket pocket.

And then the music stops...

Figure 45.7 Imagination and the arts.

has a really solid sense of who he is, but he is still discovering it and he is still trying to figure out where to go with that . . .

Josh: It is important for us to have imagination as one of the tools to negotiate the world.

In her work and life Maxine Greene illuminates the role of the arts in eliciting imagination for what could be. Through the defamilization processes of art, Greene (2000) writes, "I find myself revising, and now and then renewing, the terms of my life" (pp. 4–5). The educative task, for Greene, is in part the creation of situations that enable the release of "this kind of reshaping imagination," in which students "are moved to *begin* to ask, in all the tones of voice there are, 'Why?'" (pp. 5–6).

The "Suited" book itself combines material from observations and interviews with Diego, photos of him and the places he spends time, drawings, symbolism, script, all in the hand-stitched leather book:

Michelle: And I think it was the method that we chose to show it in too. I think that book was something different about it . . . It was a creation. Like we created that. That was nothing. That was a piece of leather in a hide shop. Do you know what I mean? It's a total creation, it's not like a baby, but, sort of. It's our baby, right? I am really proud of it . . .

Josh: This is an example of the kind of assignment that embodies *heart* and mind. You have to understand the articles, but it's also not a regular writing assignment, because it's sort of self-directed, right? It's an example of inquiry. And so, we had a stake in it, you have some pride. And when it takes on meaning of its own, then we try harder.

Shades of Agency: Making Change

Kim: And then the music stops at the end of—

Michelle: Imagination, and there is a reason for that.

Kim: Well, that was sort of the cue that he was done, he had obtained all of his articles of clothing and all of the tools that he needs to go out into the world . . .

Kim: Whitehead says, there is just one more thing you need, a device that you can use any way you want. Sometimes the world can be confusing for agents, the sunglasses will help you see the world in your own way.

Marcia: So sort of his agency—

Michelle: And then, "Welcome to the agency."

Josh: The ability to see the world with all the tools he has gained.

Kim: And to use those tools to break it down and understand it in his own way.

Michelle: So it starts with "Deconstruct your experience Diego" [at the beginning of the book], back to the full circle.

Kim: To sort of having the tools to be able to do that.

Figure 45.8 Shades of agency.

When the book is finished, Kim and Michelle go back to the school to show it to Diego. After Diego has had an opportunity to read through it, we ask:

Michelle: So how do you feel about being an agent?

Diego: Pretty cool. The whole thing kind of reminded me of the "Matrix" just because of that whole agent thing.

Michelle: Cool. We didn't plan for that.

Kim: So what do you think it means, in the context of this story, what do you think it means to be an agent?

Diego: Someone who can teach others and, sort of train people to do, not exactly what to do in life because I don't want to tell people what to do, but to sort of guide them.

Kim: Okay, and that maybe you can do that for yourself, you can act for yourself?

Diego: Yeah. Yeah.

Michelle: The last device that you have to get from agent Whitehead, why do you think that we chose the sunglasses for the last thing?

Diego: Maybe it's just kind of like, story-wise, without the sunglasses on I see what other people want me to see, but when I put the sunglasses on I see what I want to see.

Michelle: Exactly.

Diego: That's what I connected with.

> *"I'll never know everything I need to know. I'll never be done learning. I don't think my training will ever end, but I have the tools to navigate through the world," Diego responds.*

Josh: When he puts his sunglasses on he is socially and ecologically just . . .

Michelle: In hindsight, we could have done something, one more page.

Kim: One more page, draw some more trees.

Figure 45.9 No exit.

Participatory Inquiry and Relational Learning

Marcia: Thinking about these different pieces, you know, trying to go gray instead of black or white, the importance of the imagination, and working through some of these ideas, do you feel like *you* have gained different tools than you had before? . . .

Kim: Well, and I don't think in my head, we didn't intend to make the story of Diego a parallel of what we have learned (multiple speakers), but I think it happened. Yeah, we certainly didn't intend to, we never said, we learned these things, let's make Diego learn these things too. But looking back yeah I think (multiple speakers) . . .

Josh: We definitely wanted him to go from not being socially aware to being a socially, ecologically justice-oriented person. And what is the process of that, and the articles were the way to do that . . . Yeah we wanted him to do that, I guess I didn't really notice that we did too . . .

Michelle: I've said it before, and I'll stand by it, collectively, this was a bigger and better project than we ever could have done independently, and that is without a doubt in my mind . . . It was bigger than just an assignment, it just became something bigger. And then the title, we came to it at 11:25 at night or something. We were throwing a lot of stuff around, then Dustin just threw out "Suited."

Josh: And we all just said, it fit.

Michelle: It just fit everything. "Suited." He got "Suited." It was just perfect. So I can't really explain the experience because it was, I don't want to be all in the sky, but it was beyond us independently . . . This is going to stay with me, probably indefinitely.

We have tried to show through the stories of Diego and those of ourselves, that socioecological learning happens through relation as we engage in opportunities to better understand the discourses, stories, and practices we live by and perpetuate; when we encounter astonishing places, people, and experiences that teach us other possible ways of being; as we intentionally participate in communities that support the practice of things we have learned or are unlearning; as we creatively engage the imagination in ways that enable us to see different questions, new and old possibilities. Experiences of hearing and telling stories, being in places with their delicate leaves, participating in communities and friendship, and viewing and making art, can all be considered relational socioecological pedagogies. This is not a recipe for learning, but rather an overlapping patchwork that is already what life is about (Stewart, 2007). In realizing that more clearly, we gain new eyes, ones that can then enable us to engage ourselves and our students more intentionally and more passionately with these possibilities, and which can only lead to more learning on their parts and ours . . .

Marcia: Now that you have identified those themes for me, I mean you kind of structured the book around that. And they are great.

Michelle: That was our starting point, that was our outline. It was huge.

Marcia: And they really tie into Josh's question, and the question that we formulated the other week [as a research group] . . .

Josh: It is an example of what we are talking about in the larger research question: The experience of becoming socially and ecologically just.

Marcia: Yeah, you guys have basically, through the readings and Diego's experience, you basically came up with an answer to that question in a sense: What are the experiences that cause people to engage meaningfully? You have told a story about that in a way, right?

Michelle: Well, our work here is done. Look at that research team, our work is over. (laughter) (January, 2009)

By sharing stories from our participatory research and learning, we have sought to show the relational quality and possibilities of socioecological education. Unlike research that analyzes and summarizes from a birds-eye view, this is a ground-up approach that seeks to explore the textures and bumps of relational learning, including as research methodology. We define our research as participatory because it has involved a range of its "subjects" as co-researchers in developing and undertaking research on a topic we are all committed to, and because we seek to together make change through our research toward more engaging and ethical teaching—our own and that of others. As our research data on the challenges of socioecological pedagogy have also shown us, there are many barriers to such teaching. Particularly a lack of collective support for enacting change can cause new learning and passions to be overwhelmed by bureaucracy, fear, or the power of the status quo everyday. And if this research story seems overly optimistic in the face of those realities, it is this hope and collective power we want to share and build.

> *The research that the group did concerning learning and teaching for the social and ecological most definitely affected my teaching during internship. The knowledge that I learned through interviewing and analyzing data allowed me to be aware of how social and ecological justice is understood by others . . . it became evident that every individual understood social and ecological justice on different levels, in different contexts, and in relation to altering cognitive and/or emotional states. Therefore, during my internship, I approached social and ecological issues using interactive methods and self-directed inquiry . . . I have also changed as a person after taking part in this project . . . I have definitely "gone green" and influenced my friends and family to do the same. I am more sensitive to social injustice and never let things slide. It is not just issues of blatant injustice, such as a racist comment, that I address immediately in a school setting. I go out of my way to do small but powerful things . . . Everything you do adds up to make change. When you are in a leadership role such as teachers are, what you do has a huge impact. (Teresa, Research Notes, May 2010)*

> *Watching others, listening to their comments, I came to understand issues like ecological justice and sustainability better than I did before. I will then return to my own circle of relationships and pass on this influence. That is what I valued most . . . It wasn't just about experiencing different places and observing. It was about building relationships with each place, person, object, environment, and state of being . . . I felt powerful relationship to people and place and I know that by continuing to hold onto and build off of these relationships I will grow even more . . . (field journal entry, Josh, "Place, Experience, and the Socio-ecological course, August 2009)*

Notes

1. All data were collected by teacher candidates; with a graduate student and one teacher candidate undertaking interview transcription; and teacher candidates, a graduate student, and faculty undertaking data analysis.
2. Of note is that one of the "Suited" inquiry project students left the program after the first semester, and that one of the other original eight research group members also left the program at the end of term two. The cohort drew its students from secondary humanities teacher candidates, most of who had previous degrees in areas such as anthropology, commerce, and English. For some, participating in the research group seemed to offer a place to seek a level of engagement and learning that was missing in much of their program more broadly.
3. The "Suited" group situated this part of the book in the gym with the "Redman" symbol emblazed on the floor, a choice of mascot that had elicited much discussion and controversy at the school.
4. Reading Alfred North Whitehead (1929) in another class, many of us resonated with his ideas that education requires "romance" before and in combination with the "precision" and "generalization" processes of the intellect. Indeed, these various forms of engagement are marked throughout the "Suited" project in small heart, square, and triangle representations on each corner of the pages of the 20 × 10 centimeter leather-bound hand-stitched book following the life and learning of Diego.
5. Diego's other "communities of practice" included his music friends/band, as well as the members of Sheelah's "Citizens Inspiring Change" integrated grade twelve class with whom he took a lead in organizing a city-wide sustainability conference as well as other events.
6. Hancock, M. (2001). *The life of Maxine Greene: Exclusions and Awakenings.*

References

Boler, M. (1999). *Feeling power: Emotions and education.* New York, NY: Routledge.

Britzman, D. (2007). Teacher education as uneven development: Towards a psychology of uncertainty. *Leadership in Education, 10*(1), 1–12.

Davies, B. (1989). *Frogs and snails and feminist tales: Preschool children and gender* (pp. 1–152). Sydney, NSW: Allen and Unwin. *2nd Edition* (2003). Cresskill, NJ: Hampton Press.

Fawcett, L. (2009). Feral sociality and (un)natural histories: On nomadic ethics and embodied learning. In M. McKenzie, P. Hart, H. Bai, & B. Jickling (Eds.), *Fields of green: Restorying culture, environment, and education* (pp. 227–236). Cresskill, NJ: Hampton Press.

Felman, S. (1992). Education and crisis, or the vicissitudes of teaching. In S. Felman & D. Laub, *Testimony: Crises of witnessing in literature, psychoanalysis, and history* (pp. 1–56). New York, NY: Routledge.

Greene, M. (2000). *Releasing the imagination: Essays and education, the arts, and social change.* San Francisco, CA: Jossey-Bass.

Gustavson, L. (2007). *Youth learning on their own terms: Creative practices and classroom teaching.* London, UK: Routledge.

Hancock, M. (2001). *The life of Maxine Greene: Exclusions and awakenings.* Hancock Productions

Kahn, R. (2010). *Critical pedagogy, ecoliteracy, and the planetary crisis: The ecopedagogy movement.* New York, NY: Peter Lang.

Kesby, M. (2005). Retheorizing empowerment-through-participation as a performance in space: Beyond tyranny to transformation. *Signs: Journal of Women in Culture and Society, 30*(4), 2039–2065.

King, T. (2003). *The truth about stories: A native narrative.* Toronto, ON: Anansi Press.

McKenzie, M. (2006). Three portraits of resistance: The (un)making of Canadian students. *Canadian Journal of Education, 29*(1), 199–222.

McKenzie, M. (2008). The places of pedagogy: Or, what we can do with culture through intersubjective experiences. *Environmental Education Research, 14*(3), 361–373.

McKenzie, M. (2009). Pedagogical transgression: Toward intersubjective agency and action. In M. McKenzie, P. Hart, H. Bai, &

B. Jickling (Eds.), *Fields of green: Restorying culture, environment, and education* (pp. 211–224). Cresskill, NJ: Hampton Press.

Reid, A., Jensen, B., Nikel, J., & Simovska, V. (2008). *Participation and learning: Perspectives on education and the environment, health and sustainability.* Berlin: Springer.

St. Denis, V., & Schick, C. (2003). What makes antiracist pedagogy in teacher education difficult? Three popular ideological assumptions. *Alberta Journal of Educational Research, XLIX*(1), 55–69.

Stewart, K. (2007). *Ordinary affects.* Durham, NC: Duke University Press.

Wals, A. (Ed.). (2007). *Social learning toward a more sustainable world: Principles, perspectives, and praxis.* Wageningen, The Netherlands: Wageningen Academic Publishers.

Whitehead, A. N. (1929). *The aims of education, and other essays.* New York, NY: Mentor.

46

Greening the Knowledge Economy

Ecosophy, Ecology, and Economy

Michael A. Peters
University of Waikato, New Zealand; University of Illinois, Urbana-Champaign, USA

Introduction

This chapter discusses methodology in environmental educational research by employing an approach from green philosophy (ecosophy) and green political economy to examine some wider conceptual issues concerning learning processes within the "knowledge economy" (Peters, 2010).[1] This constitutes "wide-canvas research" with a visionary element that is designed to demonstrate the importance of philosophical research in relation to broad conceptual questions that attempt to look for the connections and integrations among ecosophy, ecology, and economy in an approach that highlights education and learning as the central human activities that can enhance sustainability in its ecological, economic, and educational forms. It is also an example of *linked-up policy* analysis (cf. linked-up government) based upon the understanding and integration of large systems. What this age demands more than ever is an understanding not simply of systems in natural, social, and geopolitical environments and their interrelations but also the logic of large-scale system-events and their impacts for humanity. In the economic and political realm as social scientists we need to know more about the logic of large-scale events governing system failures such as the collapse of the Soviet system in 1989 and the collapse of neoliberal global financial system in 2009. The social sciences have not been good at predicting or analyzing these kinds of events which demand a better interface between social and natural sciences and their mediation and understanding through new mathematical and computational theories of complex systems, of complexity and chaos, and of the difficulties with formal mathematical modeling and simulation.[2]

The approach in this chapter is an example of doing environmental education research with complex issues of conceptualization, contextualization, representation, and legitimization. The approach is a combination of different methodologies and perspectives drawing on both philosophical scholarship and argument, and the tradition of radical political economy, applied to the contemporary policy problems of the knowledge economy and the role of education and learning within it. The chapter attempts to demonstrate the need for understanding the importance of philosophical argumentation to strategic research interests in environmental education and to building a case for understanding the concept of environment as a suitable perspective in order to trace the complex ecologies comprising knowledge societies and economies. The argument is made that the most sustainable and "productive" interface in advanced postindustrial societies in the twenty-first will be that between the knowledge and the "green economy"—what I refer to as the "greening of the knowledge economy." This chapter also demonstrates the need for an eclectic, synthesizing, synoptic, vision-based, and radically multidisciplinary approach that draws upon a range of methodologies such as network analysis and systems thinking especially in relation to distributed cognition, media, knowledge, pedagogy, and energy systems.

President Barack Obama in his address on the US economy at Georgetown University (April 14, 2009) laid out five pillars of the new foundations for recovering the American dream: new rules for Wall Street and greater regulation of finance capitalism with less emphasis of manipulation of numbers and more emphasis on making; investment in education at all levels and the preparation of students for the twenty-first century; the promotion and investment in clean-green energy technologies designed to utilize renewable resources and promote energy efficiencies while reducing the dependency on Middle-East oil, reforming the health care system (Medicare, Medicaid), reducing inflated costs and providing a system of universal provision, and reducing the deficit and creating a sustainable economic future for America. For Obama's administration these five pillars are the basis of long-term economic sustainability signaling a deliberate move away from the speculative bubble of an unregulated neoliberal finance capitalism that led to the worst global recession since the end of World War II and historic number of foreclosures and job losses.

The so-called "financialization of capitalism" led to the rise of speculative finance culture that benefited hedge fund and Wall Street financiers at the expense of the rest of the population causing credit and finance imbalances and system crises. The spectacular growth of finance capital based itself upon the selling of financial derivatives, credit-default swaps, and securitized risk products. Managed hedge funds that were packaged and sold on resulted in overvalued assets and a labyrinthine maze where it was no longer possible to fathom who owned the risk any longer. This new speculative finance culture collapsed distinctions between commercial and investment banking (beginning with the rescinding of the 1933 Glass-Speigel Act in 1994) and gave way to excessive profits, massive fraud, and a crisis of markets and financial institutions. It also imperiled the architecture and ecology of the whole global economy leading George Soros to call it the "era of the destruction of capital." Some thirty trillion was wiped off equity assets; a further thirty trillion was wiped off the books through lost production, the subprime mortgage market and bail-out attempts to ring-fence other toxic assets. Together the financial crisis and global climate-change and broader ecological challenges demand a new model of how America and the world pursues economic prosperity with a greater emphasis on long-term sustainability, state-centric policies and greater regulation aimed at reinvestment in public infrastructure as well as education, health and renewal energy forms.

The neoliberal era had encouraged a form of socio-economic evolution of wage-laboring "man" of industrial capitalism into global postindustrial "smart investor" with a balanced portfolio. The Obama ecological era promises to place the emphasis once again on an economic identity based of "making" rather than "speculating," on diligence, hard work, and community rebuilding rather than becoming a landlord with a portfolio of investment properties, able to retire early and live off investment returns. For Middle America and for the UK middle class—so-called Anglo-American model of capitalism—the dream of easy returns from smart investment and continuous monitoring of stock markets has evaporated. By contrast the Obama era emphasizes an age of renewed collective responsibility based on ecological, market and social sustainability. The efficient market thesis has been replaced by an acknowledgment of market failure essential to both ecological economics and to the "sign," symbolic, or knowledge economy.

This chapter is structured into the three sections. First, it discusses the significance of network analysis as a broad methodology that provides the basis in terms of policy for yoking large systems together—ecosophy, ecology and economics and social, ecological, and economic sustainability. Second, it outlines the green critique of neoliberalism to focus on conceptions of the green economy. Third, the chapter provides a "systems view" of the environment and environmental ethics that must guide the green economy, and finally it discusses aspects of ecopolitics, green capitalism, and environmental education.

The Network Logic of the Knowledge Economy

Network logic is increasingly the basis of economic transactions and social life.[3] The fact is that neoclassical economics does not understand this new logic. Network logic embodies a set of rules for the ordering, distribution and dissemination of knowledge and information and help to structure new forms of organization, decision making, coordination, and collective action. In communication networks the dynamic of information is one of openness and with the right kinds of transparency it would be possible to rebuild trust in public institutions. New communication networks also permit increased capacity for increased coordination such as that evidenced in the "open access" and "creative commons" movements, which also links with innovation considered as a networked endeavor. There are important implications here for regulation, accountability and ownership, and dwell on "network citizens" who will be able "to participate in the creation of new decision-making capabilities as well as understanding their informal power and responsibilities" (McCarthy et al., 2002). Power structures and shapes the contours of networks determining the entry points and conditions that define structural advantage (Castells, 2002).

The network perspective entails viewing natural and social systems as networks—molecules as networks of atoms, brains as neural networks, organisms as networks of cells, organizations as networks of jobs, economics as networks of organizations, and ecologies as networks of organisms. Thus, it is not just the composition of elements of the system but rather how they are configured and what kind of relations exist. Network analysis is, therefore, nonreductionistic and holistic with a emergent-properties orientation. In this perspective the structure of the system largely determines the outcomes or performance of the system and the individual position in the system determines both the opportunities and constraints encountered. Against the mainstream then the network perspective is nonatomistic and nonindependent, where individual are studied as they are embedded in the web of social relations and have direct influence on one another.[4]

In this context it is important to note that the concept of information considered in terms of today's scholarship in biology still is open to interpretation and checking even though its role is central. This is also a technical issue that requires careful scrutiny. Queiroz, Emmeche, and El-Hani (2005) adopting a semiotic approach reassesses and reconsiders the role of information in living systems, stating:

> "Information" is a concept which is very important but problematic in biology (see Jablonka, 2002; Griffiths, 2001; Oyama, 2000; Sarkar, 1996; Stuart, 1985). The

concept of information in biology has been recently a topic of substantial discussion (see, e.g., Adami, 2004; Godfrey-Smith, 2000; Jablonka, 2002; Sarkar, 2000; Smith, 2000; Sterelny, 2000; Wynnie, 2000). Furthermore, the evolution of new kinds of information and information interpretation systems in living beings has received a great deal of attention recently (See, e.g., Jablonka, 1994; Jablonka, Lamb, & Avital, 1998; Jablonka & Szathmáry, 1995; Smith & Szathmáry, 1995, 1999). It is even the case that the evolution of different ways of storing, transmitting, and interpreting "information" can be treated as a major theme in the history of life (Jablonka, 2002; Smith & Szathmáry, 1995, 1999).[5]

The network approach lends itself to interesting applications in political economy. For example, Granovetter (1985) argues that the concept of *homo economicus* in economics is extremely undersocialized because it ignores the importance of personal contacts and social networks, that is, the embeddedness of economic transactions in social relations. By doing so, economics ignores the incentives to mutual cooperation and its individualism based on rational choice theory is unable to provide an analysis of flows of information between actors which are used to make decisions of mutual gain and are endogenous to the social network. He argues that "the behavior and institutions to be analyzed are so constrained by ongoing relationships that to construe them as independent is a grievous misunderstanding" (p. 481). He goes on to argue, "Despite the apparent contrast between under- and oversocialized views, we should note an irony of great theoretical importance: both have in common a conception of action and decision carried out by atomized actors. In the undersocialized account, atomization results from narrow pursuit of self-interest; in the oversocialized one, from the fact that behavioral patterns have an internalized and ongoing social relations thus have only peripheral effects on behavior" (p. 485). The assumption of fully rational agents pursuing their own self-interest must be embedded in the social networks in which they are involved and make decisions. Granovetter's views reinforce what has now become widely known that many beneficial economic transactions are constituted by informal means involving trust, reputation, cooperation, and obligation (see also Granovetter, 1973, 1983). Social scientists like Coleman (1988) and Putman (1993, 2000) have argued that social interactions and network closure—dense connections between network participants—are key determinants in fostering trust and cooperative relationships.

The concept of the network was developed in the 1920s to describe communities of organisms linked through food webs and its use became extended to all systems levels: cells as networks of molecules; organisms as networks of cells; ecosystems as networks of individual organisms (Barabasi, 2002; Capra, 1996). The network pattern is one of the very basic patterns of organization of all living systems whose key characteristics is self-generation—the continual production, reproduction, repair, and regeneration of the network. This is where an ecological economics must be properly based. The notion of networks also has recently been used to describe society and to analyze a new social structure based on networking as a new form of organization (Castells, 1996, 2002). On the strong view social networks are self-generating networks of communication that unlike biological networks operate in the nonmaterial realm of meaning rather than matter yet like biological networks they form multiple feedback loops, which become self-generating, producing a shared or common context of meaning that we call culture. It is through this networked culture that individuals acquire their identities as members of the social network (Bateson, 1972; Capra, 2002; 2004). Castells argues that the proper identification of our society is in terms of its specific networked social structure, which provides the structural basis for globalization, the form of new organization (including political institutions), and the reconstruction of civil society.

The new science of networks offers strong methodological and epistemological promise across the social sciences with an apparently easy applications to economics and education with a focus on learning and knowledge networks, especially where these connect with issues of "innovation" and come into play within a "knowledge economy." Network science also has gathered a new fillip with the application of statistical modeling and developments in discrete mathematics to "small-world" analysis of complex systems—a form of analysis that is described as "new" and taken to depart in terms of its scope and power from traditional social network analysis. In short, network theory is pictured as attaining the status of a mega-paradigm in the social sciences as a form of social theory and analyses that in part gains its epistemological status from the influence of gestalt psychology and European structuralism promising a kind of empiricism which is both holistic and relational, and, thus, poses a challenge to all forms of epistemological atomism based on the individual as the basic unit of analysis, including rational choice theory. Both Gestalt psychology and structuralism that spawned relational systems and genetic epistemologies offered not only a relational account of structures (of the whole and its parts) but also seemed to offer the possibility of accounting for the genesis and transformation of structures. Yet the tangled genealogies of the emergence of the field are difficult to describe and there is doubt over to what extent we might talk of the different strands of network theory as comprising a coherent program or even sharing similar epistemological assumptions. Network theory has also been referred as a "new science" (Watts, 2004) characterized in terms of the mathematicization of method especially in relation to "small world" analysis of complex networks, yet it is not clear where formalization of methods, led by mathematicians and physicists, actually constitute a "new science" in the same way that any formalization of a discipline, say, for example economics, constitutes a "new science."

There are interesting similarities and a core set of shared concepts between policy discourses of the knowledge economy and the green economy that provide useful, new, and constructive economic imaginaries. These new imaginaries at the social scientific utilize an interdisciplinary set of concepts, theories, and approaches that has the potential to map a new reality and evolutionary stage in cultural and economic development of networked postindustrial knowledge societies.

The Green Critique of Neoliberalism

The green critique is not merely a negative account of neoclassical assumptions or simply an updating of economics according to the debates of the 1980s and after. It also constitutes a positive moment that provides important directions for the future. I have called these directions the "greening the knowledge economy" by which I mean a constellation after the "second industrial divide" of a synergistic relation between two mega-trends, imperatives and forces that acting upon one another become a significant trajectory for postindustrial knowledge economies. The tradition of economics of information and knowledge now is a well-documented field that coalesces with other disciplines to define the discourse of the knowledge economy (see Peters, 2008; Peters & Besley, 2006). This discourse both predates and postdates neoliberalism although it has also been given a neoliberal reading by world policy agencies like the World Bank based on a version of human capital theory with investment in key competencies and neoliberal restructuring of education based on principles of deregulation, privatization, and the introduction of student loans.

The neoliberal reading is also sometimes associated with the growth of sign economies and financialization of the global economy (Forster, 2007). Yet the neoliberal reading is only one reading and it does not analyze or identify the notion of knowledge as a global public good that demands government intervention designed to protect the public domain. The neoliberal reading does not take into account or try to explain the fundamental differences between the traditional industrial economy and the knowledge economy except by reference to pure rationality assumptions that do not sit well or apply within networked environments or merging distributive knowledge ecologies. In these "ecological" environments none of the elements of *homo economicus* focusing on individuality, rationality, and self-interest apply. The neoliberal reading does not understand how knowledge as a commodity behaves differently from other commodities. Neither does it recognize the parallel discourse of the "knowledge society" that begins in the sociological literature on postindustrialism in the early 1960s which is often directed at concerns about new forms of stratification, universal access to knowledge, and the role and significance of knowledge workers and institutions (Peters & Besley, 2006). Finally, the neoliberal reading is stuck temporally in the 1990s and does not take account of the movement toward various forms of the open economy signified in the creative economy, the learning economy, the open science economy (see Peters, 2009, 2010).

Conceptions of the Green Economy

The ideology of neoliberalism is in tatters and the free market belief system based on privatization and deregulation is in disrepute. The global economic and financial crisis sparked by the US subprime collapse in the housing market and the associated credit squeeze have destabilized the world economy and vindicated those who criticized neoliberal policies as unsustainable. As a consequence the world is facing the worst global economic, financial and social crisis since the Great Depression. It is estimated that some $35 trillion in assets has been lost or wiped off national accounts. Globally, we face a *new age of poverty* with rising employment in both the developing and developed worlds and a huge increase in numbers living on less than two dollars a day. At the same time the planet Earth faces multiple environmental crises including those of climate change, dwindling carbon-based energy resources, food security, polluted water, and air. In this chaotic environment policymakers and politicians are launching programs that emphasize sustainability and attempt to build on the promise and prospect of the "green economy."

President Barack Obama has made it clear that he sees a direct link between America's strategic and long-term economic interests, climate change, and renewable and clean energy technologies. In his first budget he allocated over $100 billion in green investments in energy efficient buildings, alternative energy technologies, and better, more effective forms of public transport. He and his team have also devoted resources to a retooled, reimagined auto industry and is calling on Congress to cut carbon emissions. Part of Obama election plan was to create five million "green" jobs in a decade. It is also clear that there has been a "greening" of the massive stimulus package.

Brussels and the EU also have been touting the "green economy" building it into the Lisbon agenda and targeting Europe to become a leader in green-innovation. Major are also caught up in this new shift: oil companies define themselves as "energy companies of tomorrow," and every industry from automakers to fast-food to fashion are painting themselves as green and sustainable. Climate change industries and alternative energy technologies combined now match software and biotech industries and some argue that the world is on the edge of a new "Kondratiev" wave of economic development—the age of sustainability and renewal resources, the economics of abundance.

The United Nations Environmental Program launched its Green Economy Initiative (GEI) in October, 2008 "aimed at" growing the green shots of tomorrow's economy. *A Global Green New Deal* (United Nations Environment Programme [UNEP], 2009) prepared by Edward B. Barbier from the

Department of Economics & Finance at the University of Wyoming, suggest that out of crisis we face a global opportunity to launch a "Global Green New Deal" (GGND) as the right mix of policies that can encourage recovery, growth, and sustainability. The report argues we cannot simply expect to resume business-as-usual. We need to embrace the three objectives of a Global Green New Deal (GGND): Revive the world economy, create employment opportunities, and protect vulnerable groups; Reduce carbon dependency, ecosystem degradation, and water scarcity; and Further the Millennium Development Goal of ending extreme world poverty by 2015. The report highlights the economic and employment implications of greening the energy sector including the alleged fact that, "Green energy initiatives have the potential to save the US economy an average of US $450 million per year for every US$1 billion invested" and that "the renewable energy sector of China has a value of nearly UD $17 billion and already employs close to one million workers" (p. 10). Low-carbon transport strategies can also stimulate growth and create jobs and, "There is a link between reducing ecological scarcity and improving the livelihoods of the poor." (p. 11).

Despite efforts over the past decades to demonstrate the interdependency between the environment and human well-being, the environment continues to receive marginal attention in economic policymaking. Worldwide, billions of dollars are spent annually to subsidize carbon-emitting fossil fuels. Meanwhile, investment in renewable energy remains inadequate, posing a threat to affordable and secure energy supply. Investment in the agricultural sector including water and soil conservation has actually declined in the last ten years in the developing world, threatening food security when the world's major food producers are subsidized to turn food into biofuels.

The fundamental problem with the neoliberal reading of the knowledge economy is that it does not recognize the way in which conceptions of the green economy now offer both new strategic and policy directions in ways that reinforce and interact dynamically with the knowledge economy.[6] The new ecologies of knowledge networks, distributive knowledge and learning systems that emphasize a social mode of production and focus on the flows of knowledge in a relational analysis stands opposed to the methodological individualism that motivates forms of neoclassical economics and neoliberalism based upon controlling assumptions of homo economicus: individuality, rationality, and self-interest.

Milani (2000), for instance, in his *Designing the Green Economy: The Postindustrial Alternative to Corporate Globalization* argues that the ecological economy is an authentic postindustrialism based on principles of regeneration and sustainability aimed at quality of life, community rebuilding, and environmental renewal. The green economy is based on the recognition of ecological principles of self-organization, protection of diversity, and the enhancement of network flows.[7] Neoclassical economics based on rationalistic and reductionist assumptions does not have the conceptual or philosophical resources to recognize the significance of natural assets, their relational contexts and their renewable and dynamic environments that presupposed elements of the ecosystem: throughput, distributive development, feedback, and scale (see Daly, 2003). Founded on the work of Boulding (1978), Georgescu-Roegen's bioeconomics (1971; see also Mayumi, 2001), and Daly (1999) ecological economics addresses the interdependence of human economies and natural ecosystems and has strong connections with both green economics and ecology with the focus on networks.[8]

Environmental Ethics: From Anthropocentrism to Systems[9]

As the renowned theoretical physicist, Stephen Hawking indicates in a lecture "On the Beginning of Time," "All the evidence seems to indicate, that the universe has not existed forever, but that it had a beginning, about fifteen billion years ago. This is probably the most remarkable discovery of modern cosmology. Yet it is now taken for granted."[10] He outlines how the discussion whether or not the universe had a beginning persisted through the nineteenth and twentieth centuries and was conducted on the basis of theology and philosophy on the basis of anthropocentric assumptions with little consideration of observational evidence partly because of the poor unreliability of cosmological evidence up until very recently. "Big Bang," the name for a cosmological model of the universe coined by Fred Hoyle for a theory he did not believe, began with observations by Edwin Hubble and his discovery of evidence for the continuous expansion of the universe. In essence, the theory is based notably on observations of the Cosmic Microwave Background Radiation, large-scale structures, and the redshifts of distant supernovae (see Ross, 2008). The technical details need not detain us here as there are many good accounts of the standard model. What is important for our purposes is to note the shift from a set of anthropocentric assumptions to a theory based on observation and its importance for providing an observational and empirical basis for an environmental ethics based on the existence, life, scale, and longevity of the sun at the center of our solar system. This feature requires some comment because it is an unusual claim to consider the way in which empirical matters to some extent determine the philosophical nature of environmental ethics even where the notion of ethics in relation to the environment is also unclear. Yet it seems clear that environmental ethics as the theory of environmental right conduct or the environmental good life (where the notion of life itself is, definitionally, at stake) rests fundamentally upon the notion of "environment" and how we understand it.

Environmental ethics has been slow to develop and has suffered from anthropocentrism or "human-centeredness" embedded in traditional western ethical thinking that has assigned intrinsic value only to human beings considered as separate moral entities from their supporting environment.

The difficulty is whether such anthropocentric accounts can reconceive the relations between human beings and their environment and if so, whether the concept of environment might be taken in an extraterrestrial sense as applying to our solar system with the sun at the center. This seems more like the environmental package that has a kind of systemic wholeness and integrity as a system with the energy source at its center without which life would not be possible.

If we are to accept this more inclusive notion of environment that decenters Earth within the solar system, then the notion of environment has to be renegotiated as one that dynamically also includes the life span of the solar system. One of the advantages of this definitional move is to resituate human beings in relation to the "environment" out of which they emerged in a number of evolutionary steps toward complex intelligent life forms and systems, and into which they will finally remerged. When environmental ethics emerged in the 1970s it began to call for a change of values based on ecological understandings that emphasized the interconnectivity of all life and thereby issues a challenge to theological, philosophical, and scientific accounts that posited individual moral agents as separate from and logically prior to their environment. This challenge drew on early environmental studies, and prompted the emergence of ecology as a formal discipline and deep ecology, as well as feminist, new animism, and later social ecology and bioregional accounts, sought to dislodge anthropocentric accounts that gave intrinsic value to human beings at the expense of the moral value of living systems (Brenan & Lo, 2008). While this insight does not establish what kind of environmental ethical theory one should adopt it does establish the prima facie case that traditional theories of ethics have been unable to talk about the environment in ethical terms. This is largely because they have been bolstered by deep anthropocentric assumptions that are embedded in earlier modern, scientific accounts of "nature," and also in the nature of industrial capitalism (Merchant, 1990; White, 1967).

Ecopolitics, Green Capitalism, and Environmental Education

Hardt and Negri (2001) writing in *The New York Times* over a decade ago recognized that the rainbow protests at the Genoa G8 "world" summit were united in the belief "that a fundamentally new global system is being formed" and that "(t)he world can no longer be understood in terms of British, French, Russian, or even American imperialism." They maintained that no longer can national power control or order the present global system and that, while the protests often appear anti-American, they are really directed at the larger power structures. Hardt and Negri while understanding the geopolitical decline of America—what Fareed Zarakia calls "the post-American age"—they do not fully account for the rise of China's industrial power which, while only a third the size of the US economy, is growing at three to four times the rate and industrializing at such a rapid pace that it poses a direct threat to resource exponential depletion and huge problems of pollution.

Yet it seems as if the environmental protests are to be successful they must win the same kind of battles for democracy at the global level that ordinary people as citizens won at the level of the nation-state, over three hundred years ago. And since those first democratic revolutions, movements of various kinds—civil rights, antiracism, antiwar, women's rights, children's rights, animal rights, environmental protests—have progressively enfranchised ever larger groups of the world's populations, although not inevitably or without struggle or reversals. Hardt and Negri point out the salient fact that "this new order has no democratic institutional mechanisms for representation, as nation-states do: no elections, no public forum for debate." And they go on to describe the antiglobalization protestors as a coalition united against the neoliberal capitalist globalization, but not against the forces or currents of globalization *per se*. The environmental protest movement is not isolationalist, separatist, or nationalist. Rather, as Hardt and Negri claim, the environmental movement want to *democratize globalization*—to eliminate the growing inequalities between nations and to begin to expand the possibilities for self-determination. Thus, "antiglobalization" really is a false description of this movement and better descriptions are "global justice movement," "anticorporate globalization," or "movement against neoliberal globalization." The movement is against financial deregulation of markets that led directly to the speculative subprime prime bubble disaster and the financial crisis of Western capitalism that quickly followed. It also finds additional theoretical support in the works of a variety of environmental philosophers and eco-feminists like Shiva (1997) whose *Biopiracy: the Plunder of Nature and Knowledge* documents the plunder of natural resources belonging to indigenous peoples.

It is interesting to reflect on the movement before the current crisis: against all odds, against the power of supranational forces, people in the street at Genoa, for example—and earlier in a series of locations at Gothenburg, Quebec, Prague, and Seattle—still believed in a form of resistance in the name of a better future. They believed, against all propagandizing and media control, in the story of democracy and in the seeds that were sown for emancipation and self-determination over three centuries ago. As Hardt and Negri remarked a new species of political activism has been born, reminiscent of the "paradoxical idealism of the 1960s." Such protest movements are part of democratic society even though they are unlikely to provide the practical blueprint for the future. Yet they create political desires for a better future and, remarkably, unify disparate interests and groups—unionists, ecologists together with priests and Communists—in openness toward defining the future anew in democratic terms.

Gitlin (2001) also clearly considers the present-day movements evident at Genoa as a successor movement to the student movements of the 1960s and 1970s—one that he claims has already engaged more activists over a longer period of time and one he predicts will be longer-lived. Gitlin, similarly, pictures the protestors as "creating a way of life," although he profiles the protestors as engaging in the debate about the meaning of Europe, seemingly truncating its obvious more global aspects outside Europe. He also questions the antiglobalization label, drawing attention to anticapitalist revolutionaries, reformists who demand to "Drop the Debt," and anarchists bent upon violence. The new face of protests is a *composite* of different types: anarchist and Marxists, globalists, health-issue advocates, environmentalists, and consumer advocates. The protest groups can be analyzed in terms from violent to nonviolent: Black Blocs (anarchist and Marxists who wear black masks); those who claim to be nonviolent but often provoke retaliation such as Globalize Resistance, Reclaim the Streets, Tute Bianche (Luca Casarini), and Ya Basta!; decidedly nonviolent groups ranging from celebrities to religious leaders, including AIDS activists, ATTAC (Bernard Cassen and Susan George), CAFOD, Christian Aid, Cobas, Confédération Paysanne (José Bové), various consumer groups, Drop the Debt (Bono), Greenpeace, La Via Campesina, Oxfam, Rainforest Action Network, Roman Catholic Church, War on Want, WWF.

The Peoples' Global Action (PGA) network is the attempted worldwide coordination of radical social movements especially around mass decentralized rallies against globalization summits of the G-8 and most recently aimed at the London G-20 summit and the World Bank/IMF meeting in Washington in 2009. An essential aspect of emerging ecopolitics is its networked activism (Juris, 2008) and at the broadest level some have interpreted "The Struggle for the World" as the political mission to defend their distinctive identities against modernity's homogenizing processes. As Lindholm and Zúquete (2010) argue activist groups across the political spectrum—Zapatistas, the French National Front, Slow Food, rave subculture, and al-Qaeda—share many fundamental characteristics, goals, and attitudes. Against this antimodernist philosophy I would argue for a postpositivist scientific ecology that is "capable of questioning the foundations of modernity and contesting its logic in the very name of science" (Sachs, 1991, p. 254) to overcome the deep ambivalence between ecology as a scientific discipline, on the one hand, and as a metaphysical world-view and popular form of political activism, on the other.

The opposition in question might be better seen in terms of a broader philosophical position that lines up science on one side with a mainstream, "no-limits-to-growth" economics of development (read "modernization"), reflecting Enlightenment (and Eurocentric) assumptions about "change" and "progress" against a Romanticist antimodernism that, by contrast, attempts to hold onto organicist metaphors, resists the instrumental rationality that characterizes the perceived positivism of the sciences, and courts "deep ecology" principles, "local knowledge," and the naturalism of other cultures. This deep philosophical ambivalence which originates within Enlightenment culture hints at a conceptual and epistemological tug-of-war that has its genealogy, at least in the modern *episteme* (to use a Foucauldian term), from the days well before the disciplinary formation of scientific ecology in the early twentieth century. Understanding this opposition—the whole intersecting matrix of grand narratives of modernism and its oppositional antimodernist counter-narratives—which, incidentally is still very much part of the ongoing "culture wars" of the early twenty-first century, is fundamental to understanding how we might break free of this controlling dualism and move beyond modernity. Ecology, in its very conception and in its current disciplinary understandings, harbors and mirrors the wider contestation and struggle between the cultural forces of modernity and antimodernity. This broader philosophical dualism is also reflected in a series of oppositions between "environment" and "development," "global knowledge," and "local knowledge"; First World (modern) state governance of the relations between humanity and nature, on the one hand, and Third World (premodern) subsistence, on the other; and between scientific and political models for environmental education and sustainable development education.

At any event, contemporary ecopolitics must come to terms with the scramble for resources that increasingly dominate the competitive motivations and long-range resource planning of the major industrial world powers. There are a myriad of new threats to the environment that have been successfully spelled out by ecophilosophers that have already begun to impact upon the world in all their facets. First, there is the depletion of nonrenewable resources and, in particular, oil, gas, timber, and minerals. Second, and in related-fashion, is the energy crisis itself upon which the rapidly industrializing countries and the developed world depend. Third, is the rise of China and India with their prodigious appetites that will match the United States within a few decades in a rapacious demand for more of everything that triggers resource scrambles and the heavy investment in resource-rich regions such as Africa. Fourth, global climate change will have the greatest impact upon the world's poorest countries, multiplying the risk of conflict and resource wars. With these trends and possible scenarios only a better understanding of the environment can save us and the planet. A better understanding of the earth's environmental system is essential if scientists in concert with politicians, policy makers and business leaders are to promote green exchange and to ascertain whether green capitalism strategies that aim at long-term sustainability are possible.

The energy crisis may be a blessing in disguise for the US Rifkin (2003) envisions a new economy powered by hydrogen that will fundamentally change the nature of our market, political, and social institutions as we approach the end of the fossil-fuel era, with inescapable consequences for industrial society. New hydrogen fuel-cells are now being pioneered which together with the design principles of smart information technologies can provide new distributed forms of energy use. Friedman (2008) also argues the crisis can lead to reinvestment in infrastructure and alternative energy sources in the cause of nation-building. Education has an important role to play in the new energy economy both in terms of changing worldview and the promotion of a green economy but also in terms of R&D's contribution to energy efficiency, battery storage and new forms of renewable energy.

At this stage of the world's development with space travel, planetary exploration, satellite communications systems in space, and scientific probing of the beginnings of the universe, concept of environment itself needs radical extension to the solar system and universe. Increasingly, although it is still early days, the earth needs to be thought not just as Gaia, as an organic living system but also as part of a larger, more broadly embracing environmental system. The notion that the environment is a dynamic concept, of which we are a part, is the central understanding of a greening of capitalism. Sustainable prosperity becomes possible with a shift to knowledge and creative economies based on services and clean, efficient technologies, although the ecological society depends on a broad consensus over the nature of the market and the economic system: What are the conflicts between the market and ecological economics? (Daly & Farley, 2004). Does sustainability imply "limits" and to what extent? (Greenwood, 2007). Can Green Capitalism 2.0 solve the looming biocrisis within the constraints of a green mixed economy? "Natural capital," the self-renewing ecosystem on which all wealth depends, is the basis of green capitalism and we need to develop democratic and participatory means by which to encourage and pursue it. This is one of the great tasks facing education at all levels in the twenty-first century.

Notes

1. The derivation of the English meaning of the prefix "eco" is based on the French *eco-*, Latin - *oeco* from the Greek *οικος (oikos)* meaning *"house," "household,"* or *"dwelling place."* Ernst Haeckel used the term "ecology" (*oikos-logos*) in the 1870s to describe the relationship of living organisms to their environment. Economy is also derived from the Greek *oikos* together with *nomos* (law; regulate) and *nomia* (stewardship, managing). *A Dictionary of Prefixes, Suffixes, and Combining Forms* based on *Webster's Third New International Dictionary Unabridged*, 2002 (p. 16), gives the following entry: "**ec-** *or* **eco-** *also* **oec-** *or* **oeco-** *or* **oiko-** *combining form* earlier also *yco-*, fr. MF? LL@ MF *yco-*, fr. LL *oeco-, oiko-*, fr. Gk *oik-, oiko-*, fr. *oikos* house, habitation **1a:** household *eco*nomy: **1b:** economic and *eco*-cultural: **2:** habitat or environment esp. as a factor significantly influencing the mode of life or the course of development *eco*species: *eco*system: *ec*ad: **3:** *ec-* or eco- : ecological or environmental *eco*catastrophe," at http://www.spellingbee.com/pre_suf_comb.pdf. There are good reasons both etymological and conceptual for examining the root prefix constructions of "ecosophy," "ecological," and "economy." This paper draws on Peters (2009), Peters and Araya (2009), and Peters and Hung (2009). See also Peters and Besley (2006), Peters (2007), Kapitzke and Peters (2007), and Peters, Marginson, and Murphy (2009).
2. Complexity theory is a broad term used for a research approach to problems in diverse disciplines (physics, chemistry, molecular biology, meteorology, economics, sociology, psychology and neuroscience) based on nonlinear, nondeterministic systems evolution. Cybernetic, catastrophe, chaos, and complexity are forms of thinking that historically have attempted to theorize these phenomena (see, e.g., Amaral & Ottino, 2004; Cilliers, 1998; Prigogine, 1997). In particular, see the special issue and monograph *Complexity and the Philosophy of Education* in Mason, 2008; and my essay "Complexity and Knowledge Systems" (Peters, 2008).
3. See the huge and growing field of network economics at http://www2.sims.berkeley.edu/resources/infoecon/Networks.html.
4. This basic description is taken from http://www.analytictech.com/networks/topics.htm.
5. Please refer to the original articles for references.
6. Obama's green capitalism based on green energy policies is in part a response to the problem of global climate change but also, I would argue, also an ecological understanding of the global financial crisis and the undesirable network effects of financialization of the global economy. Obama's policies offer the possibility of for a new wave of growth based on clean-green technologies for a low-carbon economy and forms of economic sustainability based on renewal resources.
7. See the website on green economics at http://www.greeneconomics.net/ and the site on ecological economics at http://www.greeneconomics.net/.
8. See http://en.wikipedia.org/wiki/Ecological_economics.
9. This section and the draws on Peters and Hung (2009) and my entries in *New Learning: A Charter for Change in Education* at http://education.illinois.edu/newlearning/.
10. See http://www.hawking.org.uk/lectures/bot.html.

References

Adami, C. (2004). Information theory in molecular biology. *Physics of Life Reviews, 1*, 3–22.

Amaral, L. A. N., & Ottino, J. M. (2004). Complex networks: Augmenting the framework for the study of complex networks. *The European Physical Journal B, 38*, 147–162. doi:10.1140/epjb/e2004-00110-5

Barabasi, A. –L. (2002). Linked: *The new science of networks*. New York, NY: Perseus Publishing.

Bateson, G. (1972). *Steps to an ecology of mind: Collected essays in anthropology, psychiatry, evolution and epistemology*. New York, NY: Chandler.

Boulding, K. E. (1978). *Ecodynamics: A new theory of societal evolution*. Beverly Hills, CA: Sage.

Brenan, A., & Lo, Y.-S. (2008). Environmental ethics, Stanford encyclopaedia of philosophy. Retrieved from http://plato.stanford.edu/entries/ethics-environmental/

Capra, F. (1996). *The web of life. A new synthesis of mind and matter*. London: Harper Collins.

Capra, F. (2002). Living networks. In H. McCarthy, P. Miller, & P. Skidmore (Eds.), *Network logic who governs in an interconnected world?* Retrieved from http://www.demos.co.uk/catalogue/networks/

Capra, F. (2004). *Hidden connections*. London: Harper Collins.

Castells, M. (1996). *The rise of the network society*. Oxford: Blackwell.

Castells, M. (1997). *The power of identity: Economy, society and culture*. Oxford: Blackwell.

Castells, M. (2002). Afterword: Why networks matter. In H. McCarthy, P. Miller, & P. Skidmore (Eds.), *Network logic who governs in an interconnected world?* Retrieved from at http://www.demos.co.uk/catalogue/networks/

Cilliers, P. (1998). *Complexity and postmodernism: Understanding complex systems*. London, UK: Routledge.

Coleman, J. (1988). Social capital in the creation of human capital. *American Journal of Sociology*, *94*(Suppl.), S95–S120.

Daly, H. (1999). *Ecological economics and the ecology of economics*. London, Edward Elgar.

Daly, H. E., & Farley, J. (2003). *Ecological economics: Principles and applications*. Washington, D. C: Island Press.

Forster, J. B. (2007). The financialization of capitalism. *Monthly Review, 58*, 11.

Friedman, T. (2008). *Hot, flat and crowded: Why we need a green revolution—and how it can renew America*. London: Farrar, Straus and Giroux.

Georgescu-Roegen, N. (1971). *The entropy law and the economic process*. Cambridge, MA: Harvard University Press.

Gitlin, T. (2001). Having a riot. *Newsweek,* 48–49.

Godfrey-Smith, P. (2000). Information, arbitrariness, and selection: Comments on Maynard Smith. *Philosophy of Science, 67*(2), 202–207.

Granovetter, M. (1973). The strength of weak ties. *American Journal of Sociology, 78*(6), 1360–1380.

Granovetter, M. S. (1983). The strength of weak ties: A network theory revisited. *Sociological Theory, 1*, 201–233. Reprinted in Marsden, Peter, V., & Lin, N. (Eds.). (1982). *Social structure and network analysis*. Sage.

Granovetter, M. (1985). Economic action and social structure: The problem of embeddedness. *American Journal of Sociology, 91*(3), 481–510.

Greenwood, D. (2007). The halfway house: Democracy, complexity, and the limits to markets in green political economy. *Environmental Politics, 16*(1), 73–91.

Griffiths, P. (2001). Genetic information: A metaphor in search of a theory. *Philosophy of Science, 68*(3), 394–403.

Jablonka, E. (1994). Inheritance systems and the evolution of new levels of individuality. *Journal of Theoretical Biology, 170*, 301–309.

Jablonka, E. (2002). Information: Its interpretation, its inheritance, and its sharing. *Philosophy of Science, 69*, 578–605.

Jablonka, E., Lamb, M. J., & Avital, E. (1998). Lamarckian' mechanisms in Darwinian evolution. *Trends in Ecology and Evolution, 13*, 206–210.

Jablonka, E., & Szathmáry, E. (1995). The evolution of information storage and heredity. *Trends in Ecology and Evolution, 10*, 206–211.

Juris, J. S. (2008). *Networking futures: The movements against corporate globalization*. Durham, NC: Duke University Press.

Kapitzke, C., & Peters, M. A. (Eds.). (2007). *Global knowledge cultures*. Rotterdam: Sense Publishers.

Lindholm, C., & Zúquete, J. P. (2010). *The struggle for the world: Liberation movements for the 21st century*. Stanford: Stanford University Press.

Mason, M. (Ed.). (2008). *Complexity theory and the philosophy of education, educational philosophy and theory*. Special issues (Vol. 40, No. 1). Oxford, UK: Blackwell.

Maynard Smith, J. (2000). The concept of information in biology. *Philosophy of Science, 67*(2), 177–194.

Maynard Smith, J., & Szathmáry, E. (1995). *The major transitions in evolution*. Oxford: W. H. Freeman.

Mayumi, K. (2001). *The origins of ecological economics: The bioeconomics of Georgescu-Roegen*. Routledge: Taylor & Francis.

McCarthy, H., Miller, P., & Paul Skidmore, P. (Eds.). (2002). *Network logic who governs in an interconnected world?* Retrieved from http://www.demos.co.uk/catalogue/networks/

Merchant, C. (1990). *The death of nature: Women, ecology, and the scientific revolution*. New York, NY: HarperOne.

Michael, H., & Antonio, N. (2001). *What the protesters in Genoa want. New York Times*. Retrieved from http://www.nytimes.com/2001/07/20/opinion/what-the-protesters-in-genoa-want.html

Milani, B. (2000). *Designing the green economy: The postindustrial alternative to corproate globalization*. Lanham, MD: Rowman & Littlefield Publishers.

Oyama, S. (2000). *The ontogeny of information: Developmental systems and evolution* (2nd ed.). Cambridge: Cambridge University Press.

Peters, M. A. (2007). *Knowledge economy, development and the future of higher education*. Rotterdam: Sense Publishers.

Peters, M. A. (2008). Complexity and knowledge systems. In M. Mason (Ed.), *Complexity theory and the philosophy of education, educational philosophy and theory*. Special issues (Vol. 40, No. 1). Oxford, UK: Blackwell.

Peters, M. A. (2009). Knowledge economy and scientific communication: Emerging paradigms of "Open Knowledge production" and "Open Education." In M. Simons, M. Olssen, & M. A. Peters (Eds.), *Re-reading education policies; A handbook studying the policy agenda of the 21st century* (pp. 311–336). Rotterdam: Sense.

Peters, M. A. (2010). Three forms of knowledge economy: Learning, creativity, openness. *British Journal of Educational Studies, 58*(1), 67–88.

Peters, M. A., & Besley, T. (A.C.). (2006). *Building knowledge cultures: Education and development in the age of knowledge capitalism*. Lanham, Boulder, NY, Oxford: Rowman & Littlefield.

Peters, M. A., & Araya, D. (2009). Network logic: An ecological approach to knowledge and learning. In M. McKenzie, H. Bai, B. Jickling, & P. Hart (Eds.), *Fields of green: Re-imagining education*. New Jersey: Hampton Press.

Peters, M. A., & Britez, R. (2008). *Open education and education for openness*. Rotterdam: Sense.

Peters, M. A., & Hung, R. (2009). Solar ethics: A new paradigm for environmental ethics? In E. J. Gonzalez-Gaudiano & M. A. Peters (Eds.), *Environmental education today: Identity, politics and citizenship*. Rotterdam & Taipei: Sense Publishers.

Peters, M. A., Marginson, S., & Murphy, P. (2009). *Creativity and the global knowledge economy*. New York, NY: Peter Lang.

Peters, M. A., & Roberts, P. (2009). *The virtues of openness*. Boulder, CO: Paradigm Publishers.

Prigogine, I. (1997). *The end of certainty*. New York, NY: The Free Press.

Putnam, R. (2000). *Bowling alone: The collapse and revival of American community*. New York, NY: Simon and Schuster.

Putnam, R., Leonardi, R., & Nanetti, R. (1993). *Making democracy work: Civic traditions in modern Italy*. Princeton: Princeton University Press.

Queiroz, J., Emmeche, C., & El-Hani, C. N. (2005). Information and semiosis in living systems: A semiotic approach. *Semiotics, Energy, Evolution, 5*(1), 60–90.

Rifkin, J. (2002). *Hydrogen economy. The creation of the world-wide energy web and the redistribution of power on earth*. New York, NY: Putnam.

Ross, M. (2008). *Expansion of the Universe—Standard Big Bang Model*. Retrieved from http://arxiv.org/PS_cache/arxiv/pdf/0802/0802.2005v1.pdf

Sachs, W. (1991). Environment and development: The story of a dangerous liaison. *The Ecologist, 21*(6), 252–257.

Sarkar, S. (1996). Biological information: A skeptical look at some central dogmas of molecular biology. In S. Sarkar (Ed.), *The philosophy and history of molecular biology: New perspectives*. Dordrecht: Kluwer.

Sarkar, S. (2000). Information in genetics and developmental biology: Comments on Maynard Smith. *Philosophy of Science, 67*(2), 208–213.

Sterelny, K. (2000). The 'genetic program' program: A commentary on Maynard Smith on information in biology. *Philosophy of Science, 67*(2), 195–201.

Stuart, C. I. J. M. (1985). Bio-informational equivalence. *Journal of Theoretical Biology, 113*, 611–636.

United Nations Environment Programme. (2009). *A global green new deal*; prepared by Edward B. Barbier. Retrieved from www.unep.ch/.../Green%20Economy/UNEP%20Policy%20Brief%20Eng.pdf

Watts, D. (2003). *Six degrees: The science of a connected age*. New York, NY: W.W. Norton & Company.

White, L. (1967). The historical roots of our ecological crisis. *Science, 55*, 1203–1207.

Wynnie, J. A. (2000). Information and structure in molecular biology: Comments on Maynard Smith. *Philosophy of Science, 67*(3), 517–526.

47

Preconceptions and Positionings

Can We See Ourselves Within Our Own Terrain?

PAUL HART
University of Regina, Canada

In the introduction to this section, environmental education research was acknowledged as an interdisciplinary landscape characterized by multiple philosophical positions, theoretical frames, and research methodologies and methods. And, as the papers reveal, it is a landscape that does not intentionally privilege one philosophical perspective or methodological strategy over another. Each research program takes its lead from the nature and interests of the research questions posed. If such is the case, then the chapters in this section may be viewed as examples that range across diverse research interests so that readers can learn from the experiences of those already invested within particular genres or who tend to find themselves attracted to certain philosophical tendencies and associated methodological approaches. While it was impossible to be comprehensive across categories, it was possible to represent a reasonable array of perspectives intentionally focused on the mutually participative character of the anthropological and sociological, that is, across interpretive and postcritical perspectives that may have been somewhat underrepresented in the past.

However, it seems worth considering that, if the field of environmental education research can be conceptualized in terms of multiple epistemological positionings, then just as environmental education as a field is conceived from a paradigmatic position that challenges the dominant social paradigm (Fien, 1993; McKenzie, 2004; Stevenson, 1987, 2007), environmental education research in fact may be seen to consciously privilege those forms of inquiry that question what counts as knowledge as well as how we might come to construct it. In reading the subtexts of the contributed papers, can we see in the inquiry processes a high degree of researcher consciousness of epistemological/ontological positionings as they impact design choices and subsequent fieldwork as well as critical reflection on the entire research process? For example, if the researcher's interest is in understanding the meaning of environmental education-related experiences "with" children or the ethical position or worldview of young adults, then various interpretive forms of inquiry may arguably be most appropriate (e.g., using ethnographic, phenomenological, or narrative approaches). If the interest is in critically engaging research participants in inquiring into their own social/educational conditions with a view to improve them, then participatory or relational forms (e.g., action research) may be appropriate. And, if the interest is in examining how subjectivities are construed in terms of certain cultural discourses, or in how major ideas such as "education for sustainability" are enacted through power structures of government or international declarations with contexts of globalization and knowledge economics, then forms of document or discourse analysis may obtain.

In each case we recognize different epistemological positions, each based on different understandings of social reality, as part of the inquiry process. And it is precisely these differences that have spawned what has been characterized as paradigm wars (Gage, 1989) and subsequently has generated an ongoing paradigm dialogue (Denzin, 2008). And, it is because these differences go beyond the level of methods debates or methodological positioning, that is, because they represent fundamental epistemological positions, that this section contains arguments beyond older discussions of the relative merits of quantitative and qualitative approaches. If knowledge of human social affairs is socially construed and therefore somewhat relative to people and context, then there are philosophical grounds for diverse methodological approaches and their application in particular methods.

So, beyond the more obvious debates concerning methods, mixing of methods or the value of particular methods for studying human phenomena, there are fundamental philosophical arguments that must be engaged by environmental education researchers (St. Pierre, 2000). This grounding work of qualitatively informed educational research is evident in the array of papers selected to engage readers in theory-into-practice framings that range across interpretive, critical, and postcritical perspectives, including, for example, those focused in feminist, poststructural, and cultural studies. It is evident in Payne's exploration of a socioecologically based phenomenology of experience, intent on locating methodological means of (re)engaging our collective consciousness of our responsibilities for the planet. He articulates how the theoretical ground was constructed in support of an ecophenomenological orientation to environmental education inquiry, a distinctively ecocentric focus which he believes is required to approach the deep meaning of people's experience with nature. Payne's challenge to environmental education researchers is to redirect their inquiries toward the aesthetics of embodied experience as spatial and temporal phenomena that are so often abstracted away in the business of social life governed by the reification of absolutes and the (false) promise of certainty within speed-time cultures and globalized educational spaces.

This topic of globalized cultures is central to Peters' argument that methodological and methods work be contextualized and conceptualized within forms of philosophical inquiry as underliers of the expanding territory characterized as the knowledge economy. In addressing environmental education research in terms of large-scale systems issues necessitating interdisciplinary learning within social collaboratives capable of engaging real-world issues, Peters, somewhat like Payne, argues for forms of inquiry capable of getting inside personal (i.e., phenomenological) embodied ways of learning/knowing, but in ways that can also attend to learning in open knowledge economies based on connections between public knowledge and public action. Thus Peters and Payne, each in their own way, raise issues about slow-fast and private-public continua that require creative thinking about what kinds of research methodologies are needed to capture deeply personal, and at the same time, large time-space scales. In other words, both stress the need to interact methodologically with environmental change, in ways that, rather than simply quantifying outcomes, can account for the holistic, relational nature of personal change within socially networked and globalized systems of communication.

Barratt-Hacking et al. and McKenzie et al.'s papers resonate with Payne's focus on experiential dimensions of learning where a very practical purpose necessitates a kind of inquiry that can engage ways of thinking about learning/knowing that involve participatory-action work aimed at building interpersonal relations. Although seemingly somewhat less focused on the need to deepen intersubjective, precognitive, and embodied personal experiences, the value of their work lies in what is called "actionable knowledge" for educational and environmental change. Much like Stevenson and Robottom and Bradbury-Huang and Long, who focus on the pragmatics of action, they argue that conventional education seems unequal to the crucial task of systemic sustainability challenges, most likely resulting from educational models inadequate in their epistemological origin. Thus environmental education researchers are becoming more conscious of constructing their inquiries in ways governed by what Barratt-Hacking et al. describe as challenges for education at the most basic levels of early childhood research. The idea that early learning, no less than adolescent or adult learning, involves questions of the positioning of the researcher within the intersubjectivity of researcher-participant relations that work to replace deficit models of developmental educational psychology as well as deficit models of educational research is implied in each of the chapters of this section of the *Handbook*.

Given critiques of both the research practice and the pedagogical models of current educational programs across each of the chapters, it seems important to consider what McKenzie et al. refer to as methodologies directed toward more situated critical ecopedagogical experiences. Their work seems to engage the complex interrelational discourse required in certain educational experiences that enable those kinds of learning/knowing envisioned by Peters, Barratt-Hacking et al., and Bradbury-Huang and Long in relation to social-ecological issues. The paper provides a concrete example of inquiry-based learning connected to relational methodology that has forms of community-based socioecological activities as outcomes. What we appear to have here, and in Barratt-Hacking et al. and Bradbury-Huang and Long, are illustrations of what Peters, Payne, Stevenson and Robottom, and Davies desire in terms of educational change. What begin as personal narratives of experience, on the ground, can become shared, relational experiences—aware of power operating yet using it to inform learning embedded in systems/discourses. Whereas traditional research practices tend to work to control thought and practice in participatory relational forms, such power relations are rendered less powerful by collegial, action-oriented processes that can empower individual and collective action for change. The work is participatory but also involves levels of self-exploration around personal-social relations, embedded more consciously within the educational and communicative systems that are, by their nature, rendered more susceptible to exposure by critical pedagogic strategies. Participation, in the postcritical frame, becomes much more complex as a subjectification process.

Davies' paper picks up on this process of subjectification, extending discussions about "actionable knowledge" by directing attention to taken-for-granted assumptions that work almost invisibly to structure the dominant social and educational paradigms as normal, those normative cultural meanings that operate to rationalize our wants and desires. Her paper is generative of re-engagement with "conceptual strategies" for transgression, for finding ways of "becoming" and acting different in both the practices of environmental education and environmental education research. Taken as more generic processes and strategies of agency—those collective listening, interacting, critiquing, reflexive techniques of relational engagement—the chapter captures and recapitulates many of the theories and practices of other papers in this collection. For example, how can we find ways of "going inside ourselves" (following Payne) in order to rethink ourselves more aware of what it means to know ourselves in relation to others (human and beyond human others)? How can we engage in personal or social action strategies as adapted to meet the needs of messy sociopolitico-cultural situations? This kind of epistemological positioning "inside work," so necessary to the "outside action work" needed to reframe educational and larger social systems, makes the point of this entire section on philosophical and methodological positioning that aims to serve environmental education inquiry as well as to critically (re)evaluate its purposes and perspectives.

So framed, environmental education may now be characterized as a field of inquiry more complex and more open to multiple ways of knowing/being. We can see how Davies, Payne, McKenzie et al. and others in this section do the groundwork as ontological work of positioning. They encourage research that is at once open to thinking about being yet understanding oneself as coimplicated relationally in other discourses that we are more conscious of Davies invokes authors that, like Payne, focus inward, like Peters, outward, and like McKenzie et al., Barratt-Hacking et al., Bradbury-Huang and Long, and Stevenson and Robottom engage the practical, each aware of the methodological warrant of strategies (telling, listening, questioning, enacting memories, critiquing) as processes more capable of opening up more creative and imaginative spaces for (re)constituting our environmental education research in personal, yet collective, ways, imagined as we come to rethink what really matters in building communities of practical action.

In this evolving field of environmental education research, we can see an array of complex forces in tension through internal dynamics (i.e., paradigm proliferation) and external interventions (i.e., across the disciplines). Change within environmental education research has occurred through both structural forces within environmental education research itself (at least since the 1990s [see Mrazek, 1993]) as well as through forces acting on it, imported from the wider disciplines of education and the social sciences. We have, as a field of inquiry, engaged in onto-epistemic questions as well as questions of what works, and pragmatic discussions of what use? or toward what ends? We have talked about the kinds of direct educational use that can be made of our research by teachers, how systems of control have been established over what counts as research, what research is funded and what is published. And in various ways we have engaged in processing how this research is held accountable within the complexities of people and politics against a background of larger educational frames of accountability and direction.

In so doing we have learned to recognize that in the very act of raising these notions of multiple perspectives of complexity and uncertainties within recent proliferation of perspectives and methodologies, that this will be taken by some to mean a lack of clarity or rigor, thus finding reason to discredit it. Much has been written about the conservative backlash against the widening discourses of research in education (Lincoln, 2010; Lincoln & Cannella, 2004) and the politics of evidence (Denzin, 2009). Much has been made about the value of inquiry where the field of education and, by implication, environmental education research, is legitimated on the autonomy of knowledge as valid, reliable and generalizable. The academic view in this frame seems prone to economic and political influences, what Grenfell and James (2004) have labeled as a kind of colonizing of the academic arena by controlling the means of legitimation. These developments can be understood as attempts to control the nature of inquiry because knowledge itself is controlled by what Popkewitz (2007) calls Napoleon's move, that is, by simply discrediting some forms of inquiry while promoting others.

Discussions within the chapters of this section of the *Handbook* recognize the need for practical applicability (i.e., usefulness) but in terms defined by issues surrounding the politics and ethics of evidence and the value of environmental education research in addressing matters of equity and social and environmental justice (see Lather, 2004, 2006). As Denzin (2009) indicates, the controversies surrounding the standards for assessing quality are forms of interpretive practice (using moral and ethical criteria) that enact a politics of evidence and truth. Environmental education researchers looking at this picture are obliged to recognize the importance of more consciously (i.e., philosophically) grounding their particular research process.

Whether their project adopts interpretive (e.g., ethnographic, phenomenological or narrative), critical (e.g., relational, activist) or postinformed approaches to educational questions, environmental education researchers will be confronted with contrasting arguments (e.g., relational, activist) or contesting perspectives to their positionings. Environmental education researchers need to be aware of major theoretical standpoints (e.g., critical realist, pragmatist) and to challenge themselves, as authors do here, to think across boundaries, traditional definitions and familiar spaces, to critically reflect on where they think they have positioned themselves, and to illustrate their facility with the propositional literature of the field. They will be called upon to articulate the processes by which they came to their research questions—how they think about their project process in particular ways—in relation to major ideas of the field. They must be prepared to discuss and debate philosophical issues more fundamental than the study (field) itself as well as methodological issues as they relate their ideas to those of the larger educational and social science fields. In particular, they must learn how to critically reflect on ways they may be looking at different things and looking at them differently. They must look at the language used as it relates to the methodological and theoretical discourses and set out this language in terms of major concepts, theories and methodological approaches, recognizing that different perspectives require critical reflexivity. They must accept critique on the grounds that it is inevitable and necessary for improvement, in thought and practice, recognizing that their communities of practice will ultimately decide on what counts as meaningful and thus how to serve legitimate ways forward within the area/field.

As a field of inquiry, environmental education research may be viewed as a continuum that includes those who choose to work within and for the status quo, using applied scientific approaches to generate evidence-based outcomes and generalizable results that may be seen to provide direct measurable benefits to the schools and less formal programs with high educational value. The continuum also includes those who choose to work toward educational change from more critical, participatory and postinterpretive social research perspectives despite the backlash from the established controlling bodies within American and British educational research systems. Learning to recognize where one wants to position themselves within the discursive structures and political influence (i.e., ideological power and perspectives) that underpin issues of legitimation is part of educational and environmental education research in these days of change.

Part of the challenge lies in recognizing why researchers are required to discuss the usefulness and meaning of their inquiries, at different levels of discourse and in terms of multiple ways of knowing. Realizing that it is always possible to generate different positionings within fields of inquiry and that the field itself requires more engagement with those processes of critical reflexivity that are capable of turning the field back on itself, we learn how to embrace communities of practice beyond individual programs of inquiry. Keeping in mind that we are always operating within discourses that work to orient our thought (and hence our agency), we can learn how to engage in the wider community processes that facilitate wider understandings of our own field of environmental education research, to adjust to multiple perspectives and to work to legitimate our own activities across diverse boundaries and less certain terrain.

As illustrated by the papers presented in this volume, reshaping the field of environmental education research has proceeded by expanding the range of what counts as inquiry and reconstructing what can count as quality. In this process there remains little doubt about the highly charged politics of such shifts in discursive structure and practices. Researchers are obliged to find their way through shifting sets of expectations, conceptualizations and representations and to think carefully and thoughtfully, perhaps differently, about their own positioning and their own intentions, and about how they relate to others including research participants as well as other researchers who are rethinking the field itself as they rethink their own research programs. Awareness of the deeper philosophical groundings and implications of their conceptual and methodological choices is now part of the inquiry process. As Grenfell and James (2004) put it, it is not a question of whether user and practitioner engagement is a "good" thing, but rather what epistemological/ontological expectations are placed upon it. So, we try to be more "savvy" of our preconceptions of the field as well as our positioning within which we work to legitimate our works in context. Without that, well, we can no longer go there.

References

Denzin, N. (2008). The new paradigm dialogs and qualitative inquiry. *International Journal of Qualitative Studies in Education, 21*(4), 315–325.

Denzin, N. (2009). The elephant in the livingroom: Or extending the conversation about the politics of evidence. *Qualitative Research, 9*(2), 139–160.

Fien, J. (1993). *Education for the environment: Critical curriculum theorizing and environmental education.* Geelong, Victoria, Australia: Deakin University Press.

Gage, N. (1989). The paradigm wars and their aftermath: A "historical" sketch of research and teaching since 1989. *Educational Researcher, 18*(7), 4–10.

Grenfell, M., & James, D. (2004). Change in the field—changing the field: Bourdieu and the methodological practice of educational research. *British Journal of Sociology of Education, 25*(4), 507–523.

Lather, P. (2004). Scientific research in education: A critical perspective. *British Educational Research Journal, 30*(6), 759–772.

Lather, P. (2006). Foucauldian scientificity: Rethinking the nexus of qualitative research and educational policy analysis. *International Journal of Qualitative Studies in Education, 19*(6), 783–792.

Lincoln, Y. (2010). "What a long, strange trip it's been …": Twenty-five years of qualitative and new paradigm research. *Qualitative Inquiry, 16*(1), 3–9.

Lincoln, Y., & Cannella, G. (2004). Dangerous discourses: Methodological conservatism and governmental regimes of truth. *Qualitative Inquiry, 10*, 5–14.

McKenzie, M. (2004). The "willful contradiction" of poststructural socio-ecological education. *Canadian Journal of Environmental Education, 9*, 177–190.

Mrazek, R. (Ed.). (1993). *Pathways to partnerships: Coalitions for environmental education.* Selected conference proceedings of the twenty-second annual conference of the North American Association for Environmental Education, Washington, DC.

Popkewitz, T. (2007). Alchemies and governing: Or, questions about the questions we ask. *Educational Philosophy and Theory, 39*(1), 64–83.

Stevenson, R. (1987). Schooling and environmental education: Contradictions in purpose and practice. In I. Robottom (Ed.), *Environmental education: Practice and possibility* (pp. 69–82). Geelong, Victoria, Australia: Deakin University Press.

Stevenson, R. (2007). Schooling and environmental/sustainability education: From discourses of policy and practice to discourses of professional learning. *Environmental Education Research, 13*(2), 265–285.

St. Pierre, E. (2000). The call for intelligibility in postmodern educational research. *Educational Researcher, 29*(5), 25–28.

Section IX

Insights, Gaps, and Future Directions in Environmental Education Research

48

The Evolving Characteristics of Environmental Education Research

Robert B. Stevenson
James Cook University, Australia

Justin Dillon
King's College London, UK

Arjen E.J. Wals
Wageningen University, The Netherlands; Cornell University, USA

Michael Brody
Montana State University, USA

Over the last forty years of environmental education research, which this handbook can only partially capture, there have been significant changes and shifts within this specialized and still emerging field. These changes need to be examined within the multiple and interconnected layers of contexts that shape the field, including the broader fields of education and education (and social science) research and within the broader global society. Several authors, as Annette Gough notes in the introduction to the first section, have emphasized the importance of understanding different histories of environmental education research, as well as how a history influences their own scholarship. These histories are grounded in particular places and the particular people privileged to tell a historical story. In this introductory chapter to the concluding section, we exercise, from the privileged position of editors of this volume, the opportunity to (re)present our voices on what characterizes environmental education research, the contextual influences, and current trends shaping the field. In the final chapter in this section we examine emerging technological, cultural, and educational landscapes that we see as shaping the field into the future.

In the chapter 50, Reid and Payne note that "teasing out the adequacies and logic of a particular vantage point of inquiry or framing of research is no small task." Researchers, particularly those new to the field, are faced with a multitude of philosophical stances, methodological perspectives, and ontological foci. The point is, as Reid and Payne suggest: "A dynamic and critical diversity in environmental education research implies choice [. . .] because of a historicized sense of the field's workings, achievements and shortcomings." The question, though, is how do you make a choice? And having made a choice, how does one go about taking a position on "how one reflexively views the history, ethics and politics of the field's knowledge and various 'impacts,'" as Reid and Payne put it, when you are new to the field, or, indeed if you have spent many years being encultured into a particular way of seeing, reading, and doing research?

Given that different perspectives on and approaches or entry points to identifying the characteristics of a field are possible, what choices did we make in framing the research selected to be reported in this *handbook* and what choices do we make now in offering our analysis of the results of that selection? Examining the substantive areas of focus is one approach which is largely reflected in the structure and many different sections of this handbook (e.g., environmental ethics, discourses and policies, curriculum and pedagogy, learning processes, assessment). The authors of the next chapter identify in a table a set of (nine) other categories of substantive areas of foci of EE research which are modeled after Tsai and Wen (2005, cited in Reid 2011) who mapped science education research in international journals. Reid and Scott point out that "such tables do little to engage reflexive debate about research aims and knowledge interests, and by extension, which research makes a difference."

Most of the issues that are the subject of debate in environmental education research are also represented in the handbook, both in the sectional structure (e.g., theoretical orientations, philosophical and methodological perspectives, and marginalization of particular groups) and within the analyses of many previous chapters. The field of environmental education has been a ferment of debate on many of these issues since the 1970s when the term came into vogue, some of which around methodological issues correspond to the so-called "paradigm wars" in the broader arena of educational research. While acknowledging that the field has evolved (and on occasion, metamorphized) discursively, conceptually, and pragmatically, Reid and Scott draw on Gough's (2004) argument that "a generative strategy is to consider the 'blind spots' and 'blank spots'" and "to invite reflection on the trajectories (intellectually, historically, geographically . . .) of the substantive and methodological aspects of inquiry." Blind spots are described as those areas in which we know enough to pose questions but not answer, while blank spots are those we don't know well enough to even ask.

Reid and Scott argue that it "is really up to the reader to critically appraise needs and identify priorities." Although not disagreeing with that statement, we take the view that those new to, and some outside, the field are interested in hearing perspectives on "blind and blank spots" while arguments identifying worn down or "bald" spots of inquiry should challenge the thinking of experienced researchers. Therefore, we take a somewhat different approach in this chapter and offer an analysis of how the environmental education research presented in this handbook might inform, illuminate, and provide insights for both environmental education theory and practice, as well as for future research.

A categorizing of characteristics, or in Hart's (2003) words "naming and framing," is intended as a synthetic search for a coherent meaning of environmental education research in terms of its evolving characteristics at this point in time. However, we also hope to open the representations of environmental education research presented in this chapter to critical debate and for interrogating future research in the field (both locally and globally), as well as contributing to new research possibilities (Hart, 2003). Therefore, we invite the reader to look "for new spaces that are not represented by [our] existing" representations (Hart, 2003, p. 246). As Reid and Scott point out in the third chapter in this section, the challenge is to sustain conversations about research ideas, interests, and values in order to identify future directions for the field. And like Reid and Payne, we "want to open up, rather than close down a conversation on what is demarcated, sketched and/or fantasized as the historic and contemporary environmental education research field."

Evolution of the Substantive Focus of Environmental Education Research

The chapters in this handbook illustrate how far environmental education research has evolved from an applied science positivistic orientation dominated by efforts to identify relationships among environmental knowledge, attitudes, and behaviors (Hart & Nolan, 1999; Rickinson, 2001) to a range of more diverse approaches. The long dominant focus of environmental education research on individual behavior change has contributed, at least in part, to the persistent but ill-founded assumption of many environmental education and interpretation programs that an increase in environmental awareness will produce a change in attitude which will lead directly to more environmentally responsible behavior. Research within this paradigm has revealed that this is far too simplistic an explanation of how people's behavior changes (Heimlich, Chapter 27; Heimlich & Ardoin, 2008). Empirically, researchers have concluded that "[e]nvironmentally significant behavior is dauntingly complex, both in its variety and in its causal influences" (Stern, 2000, p. 421).

Prior to the emergence of this broader understanding of behavior change from those working within a logical positivism framework, there emerged a new direction or focus in environmental education research, beginning in Australia in the late 1980s and closely followed in Canada and South Africa, to what was considered a more holistic theoretical approach. Critics of the individualistic behavioral research focus argued, generally from a critical theory perspective, for the necessity of recognizing the influence of wider social structures and institutional arrangements on individuals' behavioral choices and the need to frame environmental education more broadly to also consider the role of education in addressing changes to such structures and arrangements (Gough, 1992; Robottom, 1987). Questions were also raised about the appropriateness of casting education in the position of social engineering in which certain decision makers decide how others should behave. More recently, the additional concern has been raised about the lack of certainty about what counts as the "best" behavior from an environmental or sustainability point of view.

Subsequently, for many the goals of environmental education were seen as less instrumental and deterministic and more democratic and transformative in terms of "engaging people in existential questions about the way human beings and other species live on this Earth" (Jickling & Wals, 2008, p. 18) and empowering them to work individually and collectively toward their visions of more sustainable communities and societies. Typically, as described in numerous scholarly articles, international reports, and policies, the role of education is now conceptualized to provide opportunities for young and old to develop the capacity to think critically, ethically, and creatively in appraising environmental situations; to make informed decisions on those situations; and to develop the capacity and commitment to act individually and collectively in ways that sustain and enhance the environment (Stevenson, 2010).

This reorientation in theorizing the purposes of environmental education has shifted more research agendas toward the role and processes of learning, a noticeable absence from the field before 2000 (Rickinson, 2001). Approaches to environmental learning processes now recognize that worldviews and belief systems shape individuals' understanding and interpretation of environmental issues and mediate their environmental behaviors. Therefore recent environmental education research is examining how learning shapes beliefs and how those beliefs are connected to individuals' decisions and behaviors.

Instead of focusing on what others have predetermined as environmentally desirable behavior, some authors emphasize the central role of engagement and what contributes to individuals' intellectual and emotional engagement in socioecological issues. This increased attention to engagement in environmental learning has resulted in a greater focus on the agency of children, including issues of their identity-subjectivity and active participation in all phases of inquiry. Positioning children in such roles is consistent with calls for treating all people in inquiry as responsible agents capable of participating in changing and improving their circumstances.

Part of understanding environmental learning also involves understanding students' personal and emotional responses to learning and environmental issues. Such issues, as these authors reveal, can evoke strong emotional reactions. The current scientific and public attention to climate change has also resurfaced criticisms in an earlier era that environmental education projects a fear and despair scenario and that instead feelings of hope and optimism need to be developed. The transition in discourse from environmental education to education for sustainable development was seen by some as predicated in part on this concern (Smyth, 1995). As climate change education, and even education for climate change adaptation, is now becoming part of the discourse, similar concerns are being voiced that such a focus is leading to increasing states of depression and action paralysis. Several contributors emphasize not only the intellectual engagement of people in socioecological issues, but also their emotional engagement.

Methodological Evolution of the Field

The field of environmental education research was somewhat slow to embrace the contestation of logical positivism as the dominant approach to educational inquiry and the emergence of postpositivist methodologies that began taking place in the 1970s and '80s in the broader field of educational research. However, after noting the "need to explore multiple frames of reference to more fully understand human-environment relationships," an extensive review of environmental education research at the end of the last century (Hart & Nolan, 1999) concluded that "environmental education research can now be characterized as engaging positivist, interpretivist/constructivist, critical and feminist/postmodernist perspectives." This methodological diversity is consistent with qualitative research no longer being at the margins as these other genres "such as narrative inquiry, participatory research and poststructuralist approaches have become less unusual." Methodological margins, as noted by the Section 7 editors, are constantly shifting. This trend is exemplified in Section 7 where relational epistemologies, for example, are framing such approaches as feminist poststructuralism and participatory action research. Davies, for example, poses feminist poststructuralist inspired questions about how we can construct ourselves as environmentally oriented educators and researchers in terms of "coming to know ourselves in relation to people and the planet."

The acceptance and use of genres such as case studies, narrative inquiries, autobiographies, and participatory action research enabled a focus on illuminating the rich contextual nature of various student and community-based environmental education experiences. In addition, the complex web of factors involved in the relationships among knowledge, experience, beliefs, values, and actions also became evident from multiple methodological approaches. These revelations have contributed to the shift in focus in focus in environmental education inquiry described above from knowledge, attitudes and behaviors (and more recently, specific beliefs) to examining how people make sense of their surroundings. Drawing particularly on interpretive and critical studies, traditional school practices have been challenged by scholars in attempting to "articulate unclear relationships of educational experiences and environmental sensibilities" (Hart & Nolan, 1999, p. 33). This shift might also be attributed, at least in part, to recognition of the complexity of human-nonhuman relationships and the diversity of positions about approaches to ameliorating environmental issues and to the argument that worldviews shape our approaches to environmental education research (Payne, 1995) and practice (Stevenson, 1987, 2007). It can perhaps also be attributed in part to the interpretive turn to viewing knowledge as socially constructed and tentative and problematic and to the emergence of constructivist theories of learning.

An Emerging Interest in Worldviews

Worldviews and belief systems in environmental education extend to assumptions about environment, education, and (their intersection in) environmental education, as well as about research itself. Specifically, worldviews of interest to environmental education researchers encompass environment-related beliefs and ideologies (e.g., about human-environment relationships, philosophical, and political positions on environmental and sustainability issues) and educational beliefs (e.g., about how people learn) and ideologies (e.g., the role of education). Debates drawing on the field of education have been identified as involving (alternative and emerging) positions on: (a) purposes and histories of education; (b) how and where people learn; (c) role of local and wider contexts in shaping education and schooling; (d) links between education and the wider world; and (e) how education is researched and developed and how insights from research and practice contribute to the debate about education reform (Reid, 2011). Debates drawing on the environment field focus on both particular issues (e.g., biodiversity, climate change, (mis)management of natural resources, and quality of air, land, and water) and broader worldviews and philosophical positions concerning human-nonhuman relationships. The challenges of representing and reporting research about the environment/

education and theory/practice relationship at conceptual and pedagogical levels are explored in this handbook.

Environmental worldviews appear to be linked in some way to individual identity: to personal, historical/biographical, cultural, geographical, spiritual, existential, political, and professional identities. The complexity and uncertainty wrought by globalization and the rapid pace of technological and social change is also producing enormous cultural shifts which include a search for meaning and affiliation in locally defined identities (Hargreaves, 1994). A search that could be argued has played out in environmental education with the (re)emergence of place-based approaches (see chapters by Smith and Greenwood). Research is now exploring how identity, such as sense of place, shapes our relations with the planet and building communities of action, including through the lens and context of landscape. The many different dimensions of identity, as listed above, we believe offer fertile ground for future research on the nature of these environmental worldview-identity relationships.

Other Emerging Interests: Culture, Language, and Discourse

Emerging from the chapters in this handbook is a sense that the increased recognition that the cultural and biophysical contexts in which people live and work shape their thinking and behavior has extended more recently to cultural research within the field. The editors sought contributions to this handbook from a wide diversity of cultural contexts (and were only partially successful) and from a diversity of authors with respect to their worldviews. The result is chapters from fifteen different countries and cultural settings which indicate that environmental education research has continued to expand the diversity of its worldviews and methodological approaches. The authors have raised such questions as how culture mediates the development of a relationship with the land and sense of place and inscribes our embodied experiences. At the same time, the observations made over ten years ago not surprisingly still hold that the field is at different stages of an evolutionary process in different parts of the world and that a multitude of cultural factors face researchers in "working toward 'global sustainability' or 'best practice' ideals" as well as ensuring "that 'western' goals and evaluative criteria for environmental knowledge and actions are not applied globally" (Hart & Nolan, 1999, p. 22). Creating transcultural spaces for environmental education research in this era of globalization are specifically proposed as one way of responding to these concerns, while new or hybridized spaces for transboundary learning, which we discuss in the final chapter, are another.

The extensive attention in the handbook to interrogating the relationship of language and discourse to the framing of environmental education theory, policy, indigenous perspectives, curriculum practice, and professional development, highlights the extent to which critical discourse analysis has become a characteristic of the field. Again, this attention to discourse represents a response to understandings of how language shapes our thinking and frames and constrains the kinds of decisions and actions we consider possible. In other words, our worldviews are (often inadequately) represented in our language and discourse but language and (especially dominant) discourses also shape our worldviews. An example of this phenomenon emerges from consideration of the language of "nature" and "culture" which can shape our view of a separation between the nonhuman and human world. Yet there is no absolute nature "out there" and so the language of "naturecultures" connotes a different relationship, one in which humans and nonhumans are always acting on each other.

Finally, a continuing and important blind spot warrants mention. It can also be argued that a historical focus on individual behavior change that had largely neglected the role of social structures in influencing and constraining behavior had depoliticized environmental education and environmental education research in the past. The emergence, and in several countries dominance (e.g., Australia), of critical approaches to EE has fostered increased scholarly attention to the nature and role of political interests and unequal relationships of power in environmental and educational decision making as well as analyzes of the discourse and actions in those contexts, although it has tended to remain apolitical in many countries, most notably in the United States. Equity and social justice issues, their intersection with environmental issues, and their interplay with race, class, and gender, and more recently, sexuality, have received considerable attention in education research in recent decades, but as Russell and Fawcett point out, little work has been directed in environmental education to environmental justice issues. Political rationalities are now identified as in need of examination and more civically engaged scholarship provides opportunities for research to have greater influence on critical thinking about environmental issues. Yet to date the urgency of such issues as climate change which is having a profound impact not only on the research practices of conservation biologists and ecologists but also on their political identities and activism does not seem to be reflected in the work of environmental education researchers (Kelsey, personal communication).

Researchers' Ontological Positioning

A distinguishing feature of environmental education inquiry that has previously been claimed is that scholars in this field tend to hold a more eco-philosophical worldview than a traditional rationalist worldview of science (Hart & Nolan, 1999), an observation confirmed by many (but not all) chapters in this handbook. It has also been argued by the same authors that there is an alignment of this eco-philosophical worldview with new paradigm research, particularly socially critical and participatory approaches. However, in the case of both ontologies and methodologies, there is not a universal narrative or orientation. On the contrary, as will be apparent to readers of large portions of this text, there are a diversity of worldviews and methodological approaches within these nonrationalist and nonpositivist positions as well as, especially in the case of evaluation research, a traditional scientific paradigm approach to some important questions, such as to what extent and in what contexts are particular programs and approaches effective and for whom? This grounding in multiple philosophies, theoretical frames, and research methodologies is leading to a diverse array of theoretical and practical choices for environmental education research (Hart, 2003).

Reid and Payne focus their gaze on "opening up questions of how we (as authors/readers) train our attention on what researchers are prepared to talk about and prosecute as their political and theoretical projects in this field." In constructing their reflections on the *handbook*, they draw extensively on the work of Paul Hart who, as they point out, has written that, "All we are doing, as reflexive researchers, is to write in ways that reveal the limits of our knowledge, our political orientation and other dimensions of self, in ways that reveal the discourses that shape our work and open possibilities for thinking about our work as we get on with it" (2005, p. 399).

Reid, Payne, and Hart all posit openness and reflexivity as crucial values for environmental education researchers and we would echo that sentiment (indeed, why would we invite commentaries on the *handbook* otherwise?). These two characteristics, openness and reflexivity, suggest that researchers cannot be neutral in their work and that their background and biases affect what, who, and how they research. It is probably fair to say that the authors in this *Handbook* might vary in the degree to which they agree with that statement. A critical questioning of ontological positioning is evident in many chapters, especially in Sections 5 and 7, but absent in others. Even if we agree with the statement completely, returning to Hart's justification of being openly reflexive for a moment, one might ask how do we know for certain what are the determinants of "the discourses that shape our work"? How, then, do we know what shapes our work consciously and unconsciously? And how sure can readers be in their interpretation of the limits of our knowledge and the impact of our life-histories on our choice of topics and authors in this *handbook*?

Conclusion

We hope the *handbook*, as Paul Hart (2003) stated in reflexively examining his reviewing of environmental education research at the end of the last century, "may serve to open rather than close in terms of a plurality of articulations and differences, acknowledging that there is so much more to say about environmental education research." We concur with his intent to "open up the research for critical debate so that new and different questions can be asked by other 'new' voices" (Hart, 2003, p. 246). It is also our wish, as we argue in the final chapter in this final section, to see environmental education research break away from the classic separations of disciplines, generations, cultures and formal, informal, and nonformal learning, and reveal more compelling possibilities for addressing sustainability issues and the challenges of learning to live more sustainably.

The *handbook* offers several histories, each representing different understandings of what happened when and why. Suggesting another opportunity that needs to be explored carefully, Reid and Payne present a "key challenge" in asking "whether this research field will make strategic and good use of corpus and bibliometric approaches." Such approaches to inquiry, they argue

> can help a research community trace and map concepts, sources and theoretical framings, including the methodological orientations and interpretive approaches/representations most often used in and across this field, be that within, say, a genealogical or geophilosophical framing, or those that relate to and incorporate strongly those traditional concerns of critical perspectives—class, gender, race.

The point about the importance of not forgetting the histories of the field is well taken. It is a point which has validity for existing senior researchers to do something about as their careers begin to fade away. It also has implications for new researchers who need to be aware of where the field came from and how it has evolved. The relationship between more experienced and less-experienced researchers in terms of their responsibilities towards the history of the field is critical because, as Reid and Payne acknowledge, "a key mechanism for crafting, framing and constructing [. . .] knowledge" is that "Doctoral researchers and their supervisors or panels variously reproduce, extend, or break with assumptions, traditions and practices of environmental education research." Without a sound grasp of the history of the field, its future may slip quietly away.

References

Gough, A. (1992). Sustaining development of environmental education in national political and curriculum priorities. *Australian Journal of Environmental Education, 8*, 115–131.

Hargreaves, A. (1994). *Changing teachers, changing times: Teachers' work and culture in the postmodern age*. New York, NY: Teachers College Press.

Hart, P. (2003). Reflections on reviewing educational research: (Re) searching for value in environmental education research. *Environmental Education Research, 9*(2), 241–256.

Hart, P. (2005). Transitions in thought and practice: Links, divergences and contradictions in post–critical inquiry. *Environmental Education Research, 11*(4), 391–400.

Hart, P., & Nolan, K. (1999). A critical analysis of research in environmental education. *Studies in Science Education, 34*, 1–69.

Heimlich, J., & Ardoin, N. (2008). Understanding behavior to understand behavior change: A literature review. *Environmental Education Research, 14*(3), 215–237.

Jickling, B. (1992). Why I don't want my children to be educated for sustainable development. *Journal of Environmental Education, 23*(4), 5–8.

Jickling, B., & Wals, A. (2008). Globalization and environmental education: Looking beyond sustainable development. *Journal of Curriculum Studies, 40*(1), 1–21.

Payne, P. (1995). Ontology and the critical discourse of environmental education. *Australian Journal of Environmental Education, 11*, 83–105.

Reid, A. (2011). Environmental education debate. In J. Newman (Ed.), *Green Education: An A-Z Guide* (pp. 149–155). Thousand Oaks, CA: Sage Publications.

Robottom, I. (Ed.). (1987). *Environmental education: Practice and possibility*. Geelong, Victoria: Deakin University Press.

Rickinson, M. (2001). Learners and learning in environmental education: A critical review of the evidence. *Environmental Education Research, 7*(3), 207–320.

Smyth, J. (1995). Environment and education: A view of a changing scene. *Environmental Education Research, 1*(1), 3–20.

Stern, P. (2000). Toward a coherent theory of environmentally significant behavior. *Journal of Social Issues, 50*(3), 65–84.

Stevenson, R. (1987). Schooling and environmental education: Contradictions in purpose and practice. In I. Robottom (Ed.), *Environmental education: Practice and possibility* (pp. 69–82). Geelong, Victoria: Deakin University Press. (Reprinted in *Environmental Education Research, 13*(2), 139–153.)

Stevenson, R. (2007). Schooling and environmental/sustainability education: From discourses of policy and practice to discourses of professional learning. *Environmental Education Research, 13*(2), 265–285.

Stevenson, R., & Stirling, C. (2010). Environmental learning agency in diverse cultural contexts. In R. Stevenson & J. Dillon (Eds.), *Engaging environmental education: Learning, culture and agency*. Rotterdam: Sense Publishers.

49
Identifying Needs in Environmental Education Research

Alan Reid
Monash University, Australia

William Scott
University of Bath, UK

Introduction

This chapter revisits a retrospective analysis and prospective critique of work published in the journal, *Environmental Education Research* (Reid & Scott, 2009). It starts by exploring the role of metaphors as tools for reasoning about the development of a research field, with a particular focus on the notions of blind spots, blank spots, and bald spots in research. The chapter illustrates how the needs identified in our earlier analysis compare with those articulated in this handbook. We also use Bourdieu's notions of *doxa* and *illusio* to engage wider questions about purposes and ways of mapping and framing this research field, with a view to identifying trends, directions, futures, and challenges.

Metaphors as Tools for Reasoning

A handbook is as much a manual as a cookbook the sole proviso in preparing a meal.

We deliberately start with a book-based analogy because nonliteral thinking is part of our stock in trade as social science researchers. Indeed, as a key component of the conceptual traffic employed in doing and critiquing environmental education, it has become a feature of many discussions of research orientations and directions. Whether it is on the value of natural capital to ecological economics, the earth as a garden to be cultivated, the individual as the basic social unit, or science as the most powerful and legitimate form of knowledge, such (root) metaphors can exert a powerful influence on our thinking and practice in environmental education and its research (see Åkerman, 2005; Bell, 2005; Bowers, 2001; Raven, 2006; Wals, 2010).

This is because metaphors allow us to play with and test our conceptions and discourses: they promise to make the expected strange as much as the queer perhaps uncannily familiar. If they don't become set in stone, so to speak, such nonliteral and indirect approaches to reasoning about a field of inquiry might also foster critical and creative awareness: of what has come to dominate the horizons of our understandings as well as the vessels for thinking about research practices and priorities for this particular field.

Thus our opening truism serves as a prompt to consider the everyday lived experiences of researchers and how they/we might selectively use and critically interpret a collection such as this. As the social scientist Michel de Certeau argues, the "art and craft" of cooking can display an inventiveness that doesn't conform to the steps laid out in a recipe. In fact, Certeau's (1984) studies of everyday practices highlight a range of fluid and ephemeral activities that are not always bidden let alone captured by theory-driven accounts of how the everyday is framed and experienced. Whether it be in the context of cooking or reading, in fact creativity often takes place at the margins of our "books"—as we learn, swap, revise, and improvise recipes—or even, perhaps, research.

The notion of trajectories is important to us too, given the burden of this final section and the need to consider directions, challenges, and futures for environmental education research. Today's academic work, including that represented within the handbook's various sections, is usually prepared for publication with the intention that it becomes an authoritative referent for future generations and programs of research. This is just as those of the distant and more recent past (given its various moments and configurations) serve to anchor (or perhaps disrupt) the logics of today's research practices, self-evidences and habits of thought, as well as yesteryear's. Thus rather than viewing a field of research through a spatial lens and hence solely in terms of terrains, configurations, and boundaries, the idea of "futures" (as in those probable, possible, plausible, and preferred) invites consideration of the resources, dynamics, and parameters impinging on the "present" and thus, how we might identify and explicate research needs and priorities. In other words, readers of this section are invited to contribute to a broader debate about how we govern, discipline, and inspire what has counted as compelling and worthwhile research projects and programs, and what could count for today and tomorrow (Reid & Nikel, 2003).

In what follows, we mold and stretch such analogical thinking further by unweaving and reweaving a range of analytical strands about themes and needs in environmental education research, including its directions and challenges. Our goal remains the same throughout: first, to explore how texts such as this might offer ways to conceive, design, conduct, legitimate, and reflect on the directions, challenges, and futures of environmental education research; but also, to invite reflection on what this might entail for identifying needs and priorities for the field at this particular juncture. We start by considering what might appear to be bracketed in and out of discussions about substance and directions for research in this field.

Topic and Content as Typical Categories for Classifying and Mapping Environmental Education Research

Critical reading and writing requires working not just with but also against the grain of material presented in this handbook. Our inspiration for this approach is Jon Wagner's (1993) commentary on the goals and limits of the knowledge constructed and presented as educational research. As Noel Gough (2004) notes, Wagner champions a modest framing: that educational research is more about the reduction of ignorance than the unwitting assertion of truth-producing practices. It is a trope that resonates with a wider contemporary concern summarized by the slogan of "against methodolatry" [a portmanteau of method and idolatry]: in effect, the trumping in a research publication of right or rigorous method as principal guarantor of truth for a substantive area of inquiry ("orthopraxy?").

This tendency to trump or claim a triumphant method/ology (the elision can be deliberate) usually comes at the expense of critical and creative deliberations about other aspects of the research, researcher, and researched. Namely, in the framing of an account of research, reflexivity on the part of the researcher demands clarification be offered about the underlying assumptions and presuppositions of dominant (and/or preferred) research ontologies, epistemologies, methodologies, and axiologies.

Take Table 49.1, a typical mapping of the substantive foci of environmental education research. While the table illustrates a content analysis approach to categorizing environmental education research reports, it offers little clarification of the "ologies" associated with these foci. Modeled after Tsai and Wen (2005) who set out to "map" science education research in international research journals, it should also be recognized that such a classification proceeds from an inductive logic. In this case, it is derived from recording presence rather than absence in the publications record, and is constructed to express distinct thematics rather articulate relational or blurred categories.

Authors of such a table usually eschew incorporating aspects which could address which themes are over or under-represented in a field (either analytically or based

TABLE 49.1
Topic and Content as Typical Categories for Classifying and Mapping Environmental Education Research Reports

Category	Research Topic	Content
1	Teacher education	Preservice and continuing professional development of teacher and teaching material; teacher education programs; teacher education reform, action research
2	Teaching and teachers	Teacher cognition; pedagogical knowledge; forms of knowledge representation; teacher thinking; teachers' knowledge; teaching strategies [teachers' values, interests, motivation, attitudes, reasoning]
3	Learning about concepts—students' conceptions and conceptual change	Methods for investigating students' knowledge and understanding; students' alternative conceptions, instructional approaches for conceptual change; conceptual development
4	Learning about learners and learning—classroom contexts and learner characteristics	Student interest and motivation; learning environment; individual differences; learning approaches; teacher–student interaction; field-based experiences; affective dimensions of environment/sustainability learning (inc. attitude development, values,); social, political, and economic factors; argumentation [inc. reasoning, decision making]
5	Goals and policy, curriculum, evaluation, and assessment	Curriculum development, implementation, dissemination, and evaluation; social analysis of curriculum; assessment; teacher evaluation; educational measurements; curriculum policy and reform [links to other curriculum areas, e.g., health, . . .]
6	Cultural, social, and gender issues	Multicultural and bilingual issues; ethnic issues; gender issues; comparative studies; issues of diversity related to environment/sustainability teaching and learning
7	History, philosophy, epistemology, and the nature of inquiry	Historical issues; philosophical issues; epistemological issues; ethical and moral issues; nature of inquiry; research methods, and theoretical issues concerning research
8	Educational technology	Computers; interactive multimedia; video; integration of technology into teaching
9	Informal learning	Environment/sustainability learning in informal contexts (e.g., museums, outdoor settings); "practical work and fieldwork"; public awareness of environment/sustainability

Note: after Tsai and Wen (2005)

on the metrics of published researches and their trends), or how else they might be represented (e.g., in webs or nodes), and even, which themes are consistently attractive and amendable to research and professional development (cf. Robottom, 1992). Indeed, while these may be commented on elsewhere in such work, the critical point remains: it is often the table that forms the focus of attention and likely reproduction by other scholars rather than the framing and elaboration that surround it.

To concede to this tabular logic, for environmental education research there will be some, including ourselves, who may want to set aside these concerns for the time being. There will be some who see such a table as requiring tallies of how often or from where or by whom the categories are populated, or using χ^2 tests to examine differences between different category variables. It is equally possible to respond that more research in Category 7 is required (on history, philosophy, epistemology and the nature of inquiry). While some might also recognize the consistently high profile of work in Category 3 in this field (on learning about concepts—students' conceptions and conceptual change), particularly by those trained or interested in psychological research. And this may lead to arguments as to whether research in Category 3 has priority in informing, updating and improving environmental education practice over other category possibilities.

Yet it remains that despite their seductions and apparent amenity, such tables do little to engage reflexive debate about research aims and knowledge interests, and by extension, which "research makes a difference" in and of itself, or in conjunction with other studies or aspects (e.g., on participatory and collaborative research, see McKenzie, 2009). Nor does such a table invite consideration of whether a plurality of research must continually be championed (probably in the name of academic freedom) even if questions of normativity, urgency and impact are pressed on or within the field (see, e.g., Jickling, 2009, revisiting matters of "means-ends" congruence in environmental education research).

Nevertheless we note that this handbook includes helpful commentary on the empirical and theoretical aspects of research in Categories 5 (goals and policy, curriculum, evaluation and assessment) and 6 (cultural, social and gender issues) and that this particularly welcome when so little is published from these categories in research journals. But we must remain cautious here. We should ignore neither the publishing strategies, abilities, goals, and tribulations of authors nor the distinctive modes of knowledge production and outlets for such inquiries (e.g., given the peer review process, training and expertise of scholars, and perceived value of particular journals), all of which, and more, suggest critical attention be given to the robustness of any "data" that provide the raw material for categorizations.

Blind Spots, Blank Spots, Bald Spots

How else might we identify research priorities then? While we can recognize that the field has evolved (and on occasion, metamorphized) discursively, conceptually, and pragmatically, for Noel Gough (2004), a generative strategy is to consider the "blind spots" and "blank spots" some researchers may harbor about their foci and practices of inquiry. Here, Gough draws on Wagner's work to invite reflection on the trajectories (intellectually, historically, geographically . . .) of the substantive and methodological aspects of inquiry. Thus, alongside a direct critique of selected studies argued to be less or more useful to the environmental education research community, Gough (2004, p. 5) notes:

> In Wagner's schema, "materials" relevant to questions already posed can be seen as filling in blank spots in emerging social theories and conceptions of knowledge;" in other words, what we "know enough to question but not answer" are our blank spots. Materials that provoke researchers "to ask new questions illuminate blind spots, areas in which existing theories, methods, and perceptions actually keep us from seeing phenomena as clearly as we might." What we "don't know well enough to even ask about or care about" are our blind spots (p. 16).

In order to trace the logics and commitments expressed in research further, Wagner's visual analogy could be extended: for example, might we also be harboring "bald spots" in this research community? That is, might we need to carefully consider those aspects of inquiry worn down if not literally depilated by having the same questions or approaches unremittingly pursued (as these relate, e.g., to studies in Categories 1–3 in Table 49.1)? As in the case of "methodolatry," the charge to be reckoned with is that some researchers appear to put a preferred method or approach unrelentingly first, or unremittingly hold to a particular theory or framing of research, or write or present with undeviating focus. The critical question, noting we all start from somewhere and can only do so much, is whether this is done wittingly or not, and to what effect.

In Reid and Scott (2009), we began to explore these issues by raising the distinction between research that represents "low hanging fruit" in this field in contrast to its "harder-to-reach varieties." Thus, the argument goes, as an appreciation blossoms of phenomena, theory, empirics, and other researchable themes, we might also wear any "research hat" lightly, being prepared to move on to other scenarios and challenges in and for this field as the situation demands. Of course, the rub is this might require unweaving various commitments and investments in theory, practice, networks, and projects (see also Dillon & Wals, 2006, on the risks of blurring methods, methodologies, and ideologies in the field's research as the field changes), the "terrains" of which we now consider.

A Spotter's Guide on Needs and Directions

In an introduction and endpiece to an edited collection on what was published in 1995–2004 of the journal, *Environmental Education Research*, we created what amounts to our own "spotter's guide" on needs and directions (Reid & Scott, 2009). As the outgoing and incoming editors of the journal, working in consultation with board members we organized the collection around six themes in its articles as follows:

1. Environmental education and ESD: *tension* or *transition*?
2. Locating the *environmental* in environmental education research
3. *Doing* environmental education research
4. Environmental learning as *process* and *outcome*
5. Environmental education *for* . . .
6. *Developing* environmental education research

Alongside republishing two exemplar papers on each theme, we also invited a range of international commentators of diverse background, interest, and status to offer vignettes about the themes and in light of that also offer a series of recommendations about research foci and research approaches for the future. From their commentaries, we identified three key organizing ideas to the recommendations, in terms of:

1. Environmental education research as connected across interests, preferences, approaches, time, and distance.
2. Environmental education research foci that needed attention.
3. Sustainable development as inherently a learning process that needs researching by/with those involved in the dynamics of such learning.

We reframed each recommendation in the form of an expression of need for environmental education research. The full set is set out in the appendix, now labeled rather than bulleted for ease of reference. Our hope was that these might be of value to the field more broadly, not just *Environmental Education Research*, including to the field's meetings and networks, research programs and its many other journals (e.g., notwithstanding issues associated with the sampling, time frame, editorial policies, funding, intended audience, and submissions to each one).

Having read through the other sections of this handbook as it was being finalized, we can appreciate that some of these "needs" are being addressed well, some have yet to be, even as new ones are also advanced, for example, in relation to the second area, the value of fatness studies, animality studies, and ecotheological inquiries to environmental education researches, and thus these represent gaps in the earlier analysis. More specifically, given our experience as editors and contributors to a range of research events and publications in this field, in this chapter, we want to provide an update on the analysis and identify particular needs that appear to us most apparent from comparing the snapshot from 1995 to 2004 with the work published since then across the field, and as represented in this handbook.

From the first set of needs, we would highlight a need for the research community to continue to: *m) study and learn from research in wider education, learning and environmental contexts*. We single out this one because it serves to identify a "bald spot," that environmental education research continues to be primarily focused on educational institutions, usually schools and increasingly higher education institutions. In relation to this handbook, we recognize that addressing this might require giving greater attention to the themes covered by Arenas and Londoño, on vocational and technical education; Hoeffel et al., on environmental education, participation, and autonomy in rural areas; and Dierking and Storksdieck, on free-choice learning. It might also require further longitudinal and interdisciplinary work on learners' lives across the livespan.

From the second, we would want to encourage further work that: d) *stimulates environmental education research inquiries, theory-driven applications, and reflexive scholarship on key educational, environmental, and socioeconomic issues, and between researchers and others in complementary fields*. We would also welcome the field continuing to: e) *encourage and support research in environmental education that can shed light on the intersection and relationship between race/power/culture and action/experience/technology, while not adopting uncritical, remedial or under-representation approaches*. In relation to this handbook, for the first, this could include attending to some of the "blind spots" identified in the themes of Berryman and Sauvé on languages and discourses of education, environment, and sustainable development; Greenwood, on place-conscious education; and Davies, on the values and prospects of feminist poststructuralist approaches. For the second, it could include addressing those themes touched on by Shava among others on indigenous knowledge; Fawcett on nature-culture; and Haluza-DeLay on environmental justice, community, race, and class.

From the third, we would underscore a need in this field to: *c) achieve a deeper understanding of the relationship between learning, society, and sustainability*. Again, in relation to this handbook, this might include attending to the "blank spots" suggested by Barratt Hacking et al. on the challenges of undertaking environmental learning research involving children; Lotz-Sisitka et al. on regional and international curriculum and learning research foci; and Stevenson and Robottom on the status and demands of critical action research in environmental education.

But we must call a halt there, on three grounds. First, Chapter 47 discusses these and related matters in further detail. Second, these selections represent an insider's view, not just to the many papers submitted to the field's research journals, but also to the preparation and finalization of this handbook. While third, it is really up to the reader to critically appraise needs, and identify priorities, within and without the notion of a research field. Yet given the distance traveled thus far, which tools might be used to achieve this?

The "Field" of Environmental Education Research

To return to our opening reflections, we revisit the notion of the "field" of educational research and its many and diverse laborers (e.g., Grenfell & James, 2004). When these constructs are invoked in academic contexts, it is our view that we should acknowledge not only their academic credentials but also their metaphorical qualities. Thus we regard acknowledging the slippery, amorphous aspects of a "field of inquiry" a helpful step in a series of possible analytical moves and abstractions for understanding the field. Whether it is from a sociological perspective on knowledge production, with Pierre Bourdieu's work as a key point of reference on the conditions and dispositions of those working with/in academia, a field is always taken to operate across a dynamic range of conditions and circumstances and of intellectual systems, orthodoxies, and priorities—yet it remains, at its core, a metaphor. On the one hand, in analytical mode, it can create a space for some degree of traction and sense of transcendence from the immediate and local in our reflections and analysis of broader patterns and directions in environmental education research (e.g., Reid, 2009; Scott, 2009). Yet on the other, if we are to heed Certeau's wider concerns about overstating the affordances and insights of a metaphor, the notion of field might also simplify and possibly reify conditions on the ground, leading, for example, to "impositions" and "colonizations" by powerful ideas and others of the local scene (see, e.g., O'Donoghue & Russo, 2004).

To illustrate what is at stake, consider Chapter 22. Here, Shallcross and Robinson set out to question postmodernist and constructivist approaches to environmental education and its research and their apparent dominance in current work. Their intention is to challenge the "doxas" these approaches have created (e.g., the relativizing of knowledge claims) as well as their effects on what they would want to see contested in the field of practice (e.g., to challenge behaviorist or positivist approaches to environmental education, but yet not necessarily presenting a compelling alternative; see also Wals, 2010).

Using that chapter to exemplify our approach to critically interpreting an "overt recipe," calculative rationality's or even hidden agenda of a handbook, we focus on the notion of *doxa* in this chapter rather than Shallcross and Robinson's arguments to force reflection on our larger point. Bourdieu calls researchers to critically consider those "sets of inseparable cognitive and evaluative presuppositions whose acceptance is implied in membership itself" (of the field) (Bourdieu, 2000, p. 100). The scope of this call is significant because, as Ringer (1997) comments, the prereflexive aspects of doxa are formed and often internalized during the induction of researchers into a field. This is typically through doctoral studies, reading lists, supervisors and writing partners, conference presentations and first publications, and thus how they/we understand and conform with (or not) the game that is played in a particular academic field (see, e.g., Nikel et al., 2010). And what appears to be a key game in this field, as addressed by this handbook? That via research, environmental educators should be able to harness insights into how teaching and learning create and direct the necessary flows between knowledge, attitudes, and behaviors, which in this case, lead to the widespread betterment of human and ecological conditions and justice on this planet.

For Bourdieu, such a doxa is part of what gives a field its reality, purpose, and boundaries: of what is assumed, pursued, questioned, and questionable within and outwith the field. Also, and crucially, it invites us to consider the tacit if not explicit criteria for recognition of its members and relatedly, establishing and governing its priorities (Ringer, 1997, p. 5):

> the "intellectual field" (is) a constellation of positions that are meaningful only in relation to one another, a constellation further characterized by differences of power and authority, by the opposition between orthodoxy and heterodoxy, and by the role of the cultural preconscious, of tacit "doxa" that are transmitted by inherited practices, institutions, and social relations . . .

Thus if we continue to think along the grain of Bourdieu's analysis of researchers and their academic research fields—particularly in terms of researcher relations to their "research objects," their research community and to the larger society that they might wish to understand, inform, or transform—the *control* of the (evolving) doxa of one's field is paramount.

By control though, Bourdieu is not suggesting domination or censorship (legitimate concerns though these are), particularly in relation to the goals and configurations of the field's so-called "gatekeepers" (its professors, editors, conference convenors, supervisors, funding agencies, publishers, etc.). Rather, control is essentially viewed as an aspect of research community reflexivity. For the health and productivity of a research community, it must create and sustain those conditions that invite permanent and critical investigation of its doxa by its members (among themselves, as well as with interested outsiders, see, e.g., Reid & Payne, 2011). Thus they—we are invited to debate that which is the cognitive and evaluative presuppositions that constitute the doxa of this field, including the inertias, tensions, and momentum of said doxa and field.

To illustrate, for Stephen Gough (2006) this will involve continuing to locate, appraise, and problematize the *environmental* in environmental education research in ways that do not automatically default to discourses of environmental education, education research, or environmental research, nor allow matters of sustainability to drive such reflection—even while each has its own importance. For Paul Hart (2005), the notion of post-post inquiry (after postmodernism, poststructuralism, postcolonialism, and posthumanism, for instance) can serve to invite consideration of whether researchers in this field must reconcile their epistemologies and methodologies with shifts in ontologies, or perhaps (better?) "work the hyphens" and intersections associated with the contradictions and divergences that are emerging. While for Jasmin Godemann (2008), it will require addressing how we divide and integrate diverse forms of knowledge (e.g., practical, everyday and academic) in the face of disciplinary bases and structures and repeated calls for transdisciplinarity in academia, including in environmental education research.

A second abstracting step we can note stems from these possibilities for discussion. It concerns research communities and how they develop and shift in their configurations and constituencies over time and space, perhaps in step but not necessarily converging or diverging with immediate and wider research "conditions," "circumstances," "structures," and "systems." These are discussed in part in Chapter 47, in the full knowledge that these features and their relations remain too vast to ever be grasped or manipulated solely by an individual educational researcher alone (Bridges, 2006).

In effect, the lesson is against hubris—or more positively, the value of humility, as Fasal Rizvi (2008) argues, in proposing a suite of epistemic virtues in a learning and learned academic community. Thus, noting the globalizing and internationalized conditions penetrating many local and national research communities, for Rivzi the epistemic virtues associated with cosmopolitanism and de-parochialization serve to draw critical attention to a researcher's social imaginaries, relationalities, dispositions, and habits of practice. For example:

- How do/might you/we handle knowledge (others and our own: do we work with the same theory year on year, import exotic or comprehensible theory, let theory do the thinking for us, develop new theorizations, . . . e.g.)?
- How and when can you/we demonstrate criticality about the fallible nature of our knowledge claims (e.g., is this expressed as a methodological issue, or ontological, epistemological, axiological . . . in a research paper)? And,
- Is considering the historical aspects of knowledge as well as the possibility of imagining different and more desirable futures a compelling and salutary regulative ideal in our accounts of research (cf. Cooper, 2002, on the role of humility and mystery in framing and proceeding with inquiry)?

Thus, although the entry points, affordances, histories, and challenges of environmental education research (and other "fields of green"—McKenzie et al., 2009) remain the "stuff" of private as well as public conversations that also serve to shape researchers' professional orientations and intellectual projects, to forego reflection and deliberation on these aspects risks imprudent philosophies and practices of said research. The challenges issuing from Bourdieu and Rizvi are to ensure those dispositions and virtues are developed and sustained collectively and individually in contributing to the fashioning of the future of research (its research questions, methodologies, networks, resistances, etc.). This is so that a research community can continue to grasp the import of its relationalities, systems and contexts, and not just its findings and research foci, constraints, or its barriers. Whether the emphasis is on key ideas, affiliations, values, interests, or the positions of agents within the field of inquiry, the broader and usually harder work is to sustain awareness and conversations about these in identifying and adjudicating challenges, directions, and futures for the field.

A third step, which again follows Bourdieu's lead, is something of the flipside to the above. It requires a reflexivity that notes the ingredients to the collective self-deception or *illusio* of a field. While Rizvi omits this from his disquisition on virtues in times of increasing connectivity and interdependence, for Bourdieu the concept runs to the heart of a research community as it highlights where and how the stakes and rules of a field impel conformity to an established order. In other words, an unchallenged doxa leads to statis and paralysis in the intellectual development of a field, for example, about the priority of developing and/or critiquing knowledge, attitude and behavior models to change environmental actions (see Heimlich & Ardoin, 2008; Kollmuss & Agyeman, 2002).

The ingredients to such an "illusio" can be recognized in a range of ways; for Bourdieu, it is important to identify that which seems to favor those who already occupy well-established positions or have the dispositions and thus wherewithal to trade with the intellectual and social capital the doxa represents for any particular pocket of academia. Equally, a technique that can help highlight the illusio is to deconstruct the "tabulations" created, reproduced, or circulated in environmental education research (and by extension perhaps, some of the epistemic "vices" of the field?).

As initially illustrated above, if we now consider multiple column-organized tables, we can ask to what degree the interests, processes, traditions, and representations are provisional (or not), open-ended (or not), and self-critical (or not), and hence what they imply about our own and others' research (cf. Biesta, 2007). More concretely, we can consider the archetypal example of (perhaps inadvertently) privileging the right-hand column or last entry in a table (by not discussing or critiquing it!) when the cell or column summarizes a range of ideas, examples, possibilities, etc. Another is to consider the dichotomies or weightings of a table, if these occur (if at all)—does the table format suggest all items are equal or successive, for example?

Such critical clarifications also speak to the broader questions of awareness of doxa and illusio in a field, particularly in accounting for the explicit as well as the perhaps nondeclared limits for what has been conceived, practiced and then constructed as knowledge to then produce a credible representation of inquiry in environmental education research. Thus, if we are to continue to deconstruct such rhetorical devices as part of our wider deliberations on the challenges, futures, and direction of the research field, we could consider the following points for discussion:

a. Should we expect any categorization and hence differentiation and appreciation of the research programs in this field to attend to the core logics available to environmental education researchers in relation to their research goals? *For example, pursuing generalization, causal attribution, prediction, illustration, interpretation, intervention, evaluation, participation, action, emancipation, argument . . .*
b. When the most cited articles in a field are usually meta-research types (literature reviews, research synthesis, conceptual analysis, typology/model/theory construction, and philosophical/logical/normative argumentation), is a focus of such categorizations on the substantive misplaced? *For example, might they (also) engage how researchers test or generate theory as explanations in answering their research questions*?
c. What do we really know, and want to know, about how the categories are populated? *For example, in terms of: which context/s the research relates to, at which level/s it takes place, and with or by whom* (cf. children-as-researchers, practitioner research, design-based research, doctoral research, research programs, evaluation, academic knowledge workers (tenured, contract . . .)?

Repetition and Difference?

Our underlying point is that it is not just what one reads, but how one reads a handbook, that is crucial to any question of needs and priorities. It is impracticable to proceed as if readers would ever agree about the paradigms, purposes, and practices of environmental education research, as suggested by this handbook and its contributions, let alone needs or priorities. Whether that is because of the contributions included, or the reading habits and practices undertaken, is a moot point. Some do intentionally point (if not allude) to an intractable range of different and incompatible directions for the field, but what else might be made of this, given the three steps noted above?

Let us return to Certeau's (1984) critical interest in the strategies and tactics operating in the "invention of the quotidian." Whether it be the "everyday experiences" of talking, walking, cooking, dwelling, or reading, a critical theorization and analysis of our "everydays"—which for us, includes researching—must invite questions of why the taken-for-granted and our assumptions ("doxa") can end up be(com)ing "just so" (see Certeau et al., 1998).

This critical work includes acknowledging those active resistances to what might often be regarded as essentially passive cultural activities (such as reading) (so far so Bourdieu). But it also involves considering their creative appropriations and extraneous "social" uses (and thus beyond, say, Foucault) to consider how they might be subverted or diverted away from what is intended or expected of a text (or indeed us as scholars). For instance: reading and writing about research to ends unanticipated by the author of that which we read (ibid., xi-xxiv).

A supposition here, which we celebrate, is that authors (still) have an element of choice in what to say and how they will say in languaging their accounts of research. Equally, so do readers in searching for and deriving (and improvising and transforming) meanings from these pages, particularly in terms of challenges, futures, and directions. Thus readers come to author their own understandings and critical engagements with what is written or tabulated in a handbook, and thus never simply "download" an intended meaning. And this will include—hopefully—embracing, challenging, and surpassing the readings of the field and its needs and priorities offered herein.

This is an interactional, nondeterministic and messy prospect, particularly for a chapter that may be supposed to offer something incisive about futures, directions, and challenges. Indeed we must not ignore that some of the discourses at work in these pages anticipate some degree of normativity or discipline to the environmental education research field and its laborers. Yet it remains that there simply is no escape from an academic field to some overarching "outside" from which we can survey the scene disinterestedly, let alone achieve a freedom from discipline within the disciplines so to speak (see Bourdieu, 2000; Bridges, 2006). Instead, we must work from the "inside outwards," as in this chapter.

Doing so we have been faced with a typically troublesome Foucauldian-style challenge: to work through what it might mean to respond to the fundamental call in this handbook, *to act responsibly (and probably differently) in our environmental education research*. Put otherwise, and in terms already asked in this handbook, can we be more "ethical" in constituting and being constituted by the field, from the other to the self, and vice versa, given that we invariably remain someone else's Other (cf. Ricoeur, 1992)?

Chapter 47 in this section explores what this might mean given diverse starting assumptions and logics within and across the handbook's sections. As a critical reading it is important to show this reflexivity if we are to attest that this field produces knowledge worth knowing in ways worth discussing beyond its own borders, notwithstanding the postintellectual and sociolinguistic troubles of an "international" field of research in environmental education.

In this chapter, we have continued our dual-level explorations by sketching how we might be(come) critical of the knowledge represented in a research handbook, not just to be critical for its own sake, but with the aforementioned ethical horizon in mind. As such, this involves continuing to take collective stock of the discourses and research projects that have come to constitute this field. And hence, in line with our opening metaphor, reflect on what we prepare, ingest, and digest as researchers under the assumption that we still want a critically revised menu to "reimagine" the field and ourselves as researchers and a field into the future (see also Gough, 1999; Hart, 2003; Hart & Nolan, 1999; McKenzie et al., 2009; Mrazek, 2003).

Thus if we return again to the presences and absences in how Table 49.1 is populated, on the one hand we might consider which conferences and gray literatures (e.g., of doctoral theses, working papers, and policy reports) demonstrate similar nodes and concentrations of themes as journals and handbooks, and which don't, in this as well as proximal (and more distant) fields of inquiry. On the other, as members of the field shift in terms of their membership and interests, will such categories be necessarily reinvented or for that matter, wither? Again, all can be grist to the mill of reflexivity. Since the 1990s, for example, Category 9, on *Informal Learning*, has been argued in some quarters to be better served by the notion of "free-choice learning" (e.g., Kola-Olusanya, 2005). Equally, philosophies and practices of the Nordic notions of *Friluftsliv* (translated loosely as a particular form of "outdoor recreation") or even "action competence" would suggest distinctive, perhaps nontransferable cultural conditions for some of the pedagogical priorities ascribed to informal and formal environmental education (see Læssøe & Öhman, 2010). While to return to Stephen Gough's profile of interests, matched also in his recent published output (e.g., 2009), it is not just locating or problematizing the environmental in our research that may be at issue in mapping and framing the field, but creating spaces for other ways of conceiving and researching this notion in environmental education research, for example, in relation to philosophical and pragmatic interests in liberty, security, investment, risk, and coevolution.

With food and recipes, tastes may change, new ingredients and techniques emerge, while distinctive traditions of diet become relatively scarce, passé or hybridized. Readers and writers of research handbooks, like cookbooks, might do well to bear these analogies in mind, as well as their limitations. It remains though that neither book demands continuous, uniform, or progressive forms of engagement; they often sit on a shelf until need or curiosity requires them. This is even as authors and editors hope to inform, structure, and challenge what is commonplace and innovative as the "art and craft" of research. Equally, a cookbook, like a handbook, is shaped and disrupted by wider resources and forces of habit, fashion, innovation, and expediency—no matter whether cooking together or alone is seen as primarily for one or another's nourishment and good health or an expression of common human endeavor (cf. Gough & Reid, 2000, on environmental education research as profession, science, art, and/or craft).

To return to our opening comments, there really is no guarantee or requirement that any selections will prove more nutritious or satisfying for the research community (its participants or audiences). However, we do discern some common threads among most of these areas of study. Their corresponding "recipes," perhaps better, improvisations, are not quickly prepared, they may invoke "ingredients" and "tastes" not usually on the "menu" in this field, and will often require time and persistence to develop themselves as substantial components to any new diet or regimen of environmental education research. Perhaps now this first international handbook of research is complete, we do well to hope that its readers use it as a springboard to further innovation and better practice in the research field.

References

Åkerman, M. (2005). What does "natural capital" do? The role of metaphor in economic understanding of the environment. *Environmental Education Research, 11*(1), 37–52.

Bell, D. (2005). Environmental learning, metaphors, and natural capital. *Environmental Education Research, 11*(1), 53–69.

Biesta, G. (2007). Bridging the gap between educational research and educational practice: The need for critical distance. *Educational Research and Evaluation, 13*(3), 295–301.

Bourdieu, P. (2000). *Pascalian mediations*. Cambridge, UK: Polity Press.

Bowers, C. A. (2001). How language limits our understanding of environmental education. *Environmental Education Research, 7*(2), 141–151.

Bridges, D. (2006). The disciplines and discipline of educational research. *Journal of Philosophy of Education, 40*(2), 259–272.

Certeau, M. de. (1984). *The practice of everyday life* (S. Rendall, Trans.). Berkeley, CA: University of California Press.

Certeau, M. de., Giard, L., & Mayol, P. (1998). *The practice of everyday life, Volume 2: Living & cooking* (T. J. Tomasik, Trans.). Minneapolis, MN: University of Minnesota Press.

Cooper, D. (2002). *The measure of things: Humanism, humility, and mystery*. Oxford, UK: Oxford University Press.

Dillon, J., & Wals, A. E. J. (2006). On the danger of blurring methods, methodologies, and ideologies in environmental education research. *Environmental Education Research, 12*(3–4), 549–558.

Gage, J. (1986). Why write? In: A. R. Petrosky & D. Bartholomae (Eds.), *The teaching of writing* (pp. 8–29). Eighty-fifth Yearbook of the National Society for the Study of Education. Chicago, IL: University of Chicago Press.

Godemann, J. (2008). Knowledge integration: A key challenge for transdisciplinary cooperation. *Environmental Education Research, 14*(6), 625–641.

Gough, A. (1999). Recognizing women in environmental education pedagogy and research: Towards an ecofeminist poststructuralist perspective. *Environmental Education Research, 5*(2), 143–161.

Gough, N. (2004). Environmental education research: Producing "truth" or reducing ignorance? Paper presented at *Effective Sustainability Education: What Works? Why? Where Next? Linking Research and Practice* 18–20 February, Sydney.

Gough, S. (2006). Locating the environmental in environmental education research: What research, and why? *Environmental Education Research, 12*(3–4), 335–343.

Gough, S. (2009). Co-evolution, knowledge and education: Adding value to learners' options. *Studies in Philosophy and Education, 28*(1), 27–38.

Gough, S., & Reid, A. (2000). Environmental education research as profession, as science, as art, and as craft: Implications for guidelines in qualitative research. *Environmental Education Research, 6*(1), 47–57.
Grenfell, M., & James, D. (2004). Change in the field—changing the field: Bourdieu and the methodological practice of educational research. *British Journal of Sociology of Education, 25*(4), 507–523.
Hart, P. (2003). Reflections on reviewing educational research: (re) searching for value in environmental education. *Environmental Education Research, 9*(2), 241–256.
Hart, P. (2005). Transitions in thought and practice: Links, divergences, and contradictions in post–critical inquiry. *Environmental Education Research, 11*(4), 391–400.
Hart, P., & Nolan, K. (1999). A critical analysis of research in environmental education. *Studies in Science Education, 34*, 1–69.
Heimlich, J., & Ardoin, N. (2008). Understanding behavior to understand behavior change: A literature review. *Environmental Education Research, 14*(3), 215–237.
Jickling, B. (2009). Environmental education research: To what ends? *Environmental Education Research, 15*(2), 209–216.
Kola-Olusanya, A. (2005). Free-choice environmental education: Understanding where children learn outside of school. *Environmental Education Research, 11*(3), 297–307.
Kollmuss, A., & Agyeman, J. (2002). Mind the gap: Why do people act environmentally and what are the barriers to pro-environmental behavior? *Environmental Education Research, 8*(3), 239–260.
Læssøe, J., & Öhman, J. (2010) Learning as democratic action and communication: Framing Danish and Swedish environmental and sustainability education. *Environmental Education Research, 16*(1), 1–7.
McKenzie, M. (2009). Scholarship as intervention: Critique, collaboration, and the research imagination. *Environmental Education Research, 15*(2), 217–226.
McKenzie, M., Hart, P., Bai, H., & Jickling, B. (Eds.). (2009). *Fields of green: Restorying culture, environment, and education*. Cresskill, NJ: Hampton Press.
Mrazek, R. (Ed.). (2003). The decade following alternative paradigms in environmental education research. *Children, Youth and Environments, 13*(1). Retrieved from http://www.colorado.edu/journals/cye/13_1/retrospectives/CYE_AuthorRetrospective_Mrazek.htm
Nikel, J., Teamey, K., Hwang, S., Pozos–Hernandez, B., Reid, A., & Hart, P. (2010). Understanding Others, understanding ourselves: Engaging in constructive dialogue about process in doctoral study in (environmental) education. In B. Stevenson & J. Dillon, (Eds.), *Environmental education: Learning, culture and agency* (pp. 167–198). Dordrecht: Sense.
O'Donoghue, R., & Russo, V. (2004). Emerging patterns of abstraction in environmental education: A review of materials, methods, and professional development perspectives. *Environmental Education Research, 10*(3), 331–351.
Raven, G. (2006). Methodological reflexivity: Towards evolving methodological frameworks through critical and reflexive deliberations. *Environmental Education Research, 12*(3), 559–569.
Reid, A. (2009). Environmental education research: Will the ends outstrip the means? *Environmental Education Research, 15*(2), 129–153.
Reid, A., Jensen, B. B., Nikel, J., & Simovska, V. (Eds.). (2008). *Participation and learning: Perspectives on education and the environment, health and sustainability*. New York, NY: Springer.
Reid, A., & Nikel, J. (2003). Reading a critical review of evidence: Notes and queries on research programs in environmental education. *Environmental Education Research, 9*(2), 149–165.
Reid, A., & Payne, P. (2011). Producing knowledge and (de)constructing identities: A critical commentary on environmental education and its research. *British Journal of Sociology of Education, 32*(1), 155–165.
Reid, A., & Scott, W. (Eds.). (2009). *Researching education and the environment: Retrospect and prospect*. London, UK: Routledge.
Ricoeur, P. (1992). *Oneself as another*. Chicago, IL: University of Chicago Press.
Riggins, S. (Ed.). (1997). *The language and politics of exclusion: Others in discourse*. London, UK: Sage.
Ringer, F. (1997). *Max Weber's methodology*. Cambridge, MA: Harvard University Press.
Rizvi, F. (2008). Epistemic virtues and cosmopolitan learning. *The Australian Educational Researcher, 35*(1), 17–35.
Robottom, I. (1992). Matching the purposes of environmental education with consistent approaches to research and professional development. *Australian Journal of Environmental Education, 8*, 133–146.
Scott, W. (2009). Environmental education research: thirty years on from Tbilisi. *Environmental Education Research, 15*(2), 155–164.
Tsai, C.-C., & Wen, M. L. (2005). Research and trends in science education from 1998 to 2002: A content analysis of publication in selected journals. *International Journal of Science Education, 27*(1), 3–14.
Wagner, J. (1993). Ignorance in educational research: Or, how can you not know that? *Educational Researcher, 22*(5), 15–23.
Wals, A. E. J. (2010). Between knowing what is right and knowing that is it wrong to tell others what is right: On relativism, uncertainty, and democracy in environmental and sustainability education. *Environmental Education Research, 16*(1), 143–151.

Appendix

1. Environmental education research as connected across interests, preferences, approaches, time, and distance.

   There is a need to:

   (a) be open to new or unfamiliar ways of doing environmental education research, while constructively engaging with work already archived.
   (b) acknowledge the value of the range of knowledge traditions (positivist, postpositivist, interpretivist, critical . . .) and the drawing on of a range of perspectives (feminist, ethnic, cultural . . .) to environmental education research in ways that are consistent with both uniquely and collectively constituted standpoints, and a commitment to respecting conflicting perspectives under conditions of uncertainty in environmental education research.
   (c) build bridges between different kinds of research approach and design and see how they might be usefully combined or integrated, working with methodological, epistemological, and ontological difference to create space where diversity can flourish.
   (d) employ and research multiple perspectives to the point that members of different communities of research practice can understand and engage with one another in spite of or through their differences.
   (e) be more inclusive and pragmatic in our approaches to research, with research driven by questions rather than by preferred methods or methodologies.
   (f) avoid imposing an accommodation or reconciliation of multiple paradigms of thought, but to recognize them as unique, historically situated forms of insight, where if touchstone is required, it might be sought at the level of the "meta-paradigmatic."

(g) apply tests of originality, validity, rigor, and quality that are appropriate to context and approach.
(h) acknowledge that understandings can grow when emerging theories and a background set of theories interact, and when theories are problematized and revised and background knowledge is reviewed and developed.
(i) see ourselves as members of a pluralistic community of reflective scholars, and not as aggregates of individuals, competing camps, or a messy field of multiple unconnected research paradigms without common interests.
(j) be open-minded, respectful, and generous, acknowledging the goodwill and intelligence of each other, and the value of other views and their underlying claims, while being critical and exercising judgment.
(k) move beyond debates about research guidelines and turn our attention to ways of supporting and developing research by critically and reflexively reviewing what research we do, and how and why we do it.
(l) have the courage to discuss the "holes" in research, and research that doesn't go to plan, and to have openness on the part of reviewers and publishers to a range of reporting outcomes.
(m) study and learn from research in wider education, learning, and environmental contexts.

2. Environmental education research foci needing attention

There is a need to:

(a) research what we know and how we have come to know this, and how such knowing (linguistic, embodied, . . .) serves to contribute to who we are, and might yet become.
(b) understand what distinguishes different theoretical approaches to teaching and research and their implications for practice, and develop research reports that will help advance this understanding.
(c) focus attention on approaches to research and pedagogy that interrogate how dominant educational (or environmental) discourses make it difficult or (im)possible for teachers and students to engage in environmental education, and with transitions in the field toward ESD.
(d) stimulate environmental education research inquiries, theory-driven applications and reflexive scholarship on key educational, environmental and socioeconomic issues, and between researchers and others in complementary fields.
(e) encourage and support research in environmental education that can shed light on the intersection and relationship between race/power/culture and action/experience/technology, while not adopting uncritical, remedial, or under-representation approaches.
(f) circumvent some of the illusions of existing cultural ideals through empirical and reflexive co-engagement in and with the world, the social politics of the day, and the complexities of the histories shaping our surroundings.
(g) acknowledge that processes of environmental education research should provide opportunities for learning for all those involved: researchers, practitioners, and other users.
(h) work toward environmental educational and research discourses being informed by practitioners and practice in diverse cultural contexts, starting with real issues and experiences in local communities, and what (young) people particularly think, feel, and hold to be the most appropriate courses of action.
(i) develop open forums that provide opportunities for novice researchers to present and publish their work.

3. Sustainable development as inherently a learning process that needs researching by/with those involved in the dynamics of such learning.

There is a need to:

(a) acknowledge that we currently know little about the nature and dynamics of learning in relation to sustainable development/ESD.
(b) recognize that we engage in an incremental and somewhat erratic process of learning our way into an uncertain future, where a key aspect is the development of learners' abilities to make sound choices, and where the imperative is to learn our way forward.
(c) achieve a deeper understanding of the relationship between learning, society, and sustainability. In order for our understanding of learning as process and outcome to develop and deepen, we need to become more:
   - reflexive about what we mean by learning, and wide-ranging in where, when, and how we seek to research such learning;
   - sophisticated in our use of theory (learning, social, cultural, environmental . . .) and existing traditional and nontraditional forms of knowledge;
   - creative in how we seek to integrate knowledge generation with knowledge transformation/utilization.
(d) identify leading epistemological theories, and exactly what is different and similar about each, in order to help make better sense of an evidential base (qualitative and quantitative) that can be used to identify valuable curriculum and learning foci.

(e) recognize the tendency for the theorizing and reporting of environmental concerns and issues to be stripped of their historical complexities, and for a global agenda of concerns to be imposed on local, culturally situated, issues.
(f) shift from making policy pronouncements on what ESD should be, as abstractions decontextualized from contexts of practice, that are to be enacted by educators who have not participated in the formulation of goals and concepts, to co-constructing and recontextualizing policy discourses that engage educators in the discourse so it is constructed with them rather than for them, and is contextualized historically, ecologically, pedagogically, and politically.

50

Handbooks of Environmental Education Research

For Further Reading and Writing

Alan Reid
Monash University, Australia

Phillip G. Payne
Monash University, Australia

Introduction

The best international research handbooks gather an array of scholarly work sourced from a diversity of geo-cultural settings and perspectives. They also offer compelling analysis of a field's past and prospects particularly in terms of the substance and sufficiency of its researches. Indeed, an authoritative handbook is expected to make—as well as break—particular frameworks of interpretation and explanation of research in the field, and thus engage discussion of not just its possible but also its preferred futures.

Leafing through this first international research handbook dedicated to environmental education we see a wide range of chapters and sections authored by leading and emerging scholars who seek to make a contribution to the field that is "international" in its implications and perhaps "global" in what its publication in this handbook legitimizes. Undoubtedly, the question of the nature, scope, and reach of a large volume such as this is fraught with challenges and difficulties. For example, from where authors sit, write and research, some contributors bear down on or advance particular perceptions and expectations of programs of inquiry while others seek to deconstruct or broaden projects carried out in the name of environmental education research, usually as critical friends to such endeavors.

The chapters in Section 9 continue in this vein, by inviting readers to dig deeper and longer into the research cited and discussed in this handbook, and of course, expected because of it. As noted in an accompanying chapter, the handbook's sections provide a range of expositions and commentaries on various possibilities and limitations for inquiry, the diverse foci and framings of which serve to foreground some forms and features of inquiry as much as they might seek to sideline others.

Most usually (it is assumed) this dynamic and critical diversity in the handbook is offered in light of a historicized sense of the field and not just because of its "requisite variety" (Hart, 2000). However in this chapter, we must also recognize the contingencies and constraints in producing a coherent research handbook (in terms of its size, audience, coverage, etc.). For example, sections and chapters won't automatically coalesce theoretically and/or empirically in the minds of readers to present what its editors set out to achieve: namely, to provide a valuable and coherent account and creative/critical framing of the research field. Indeed, contributions can only ever partially represent not just the diverse sources, contexts and histories of what passes "locally" or might count "globally" as this particular field of inquiry, but also its trans-geographical and trans-historical purposes, means and ends-in-view (Reid, 2009).

As noted in Chapter 49, this is partly because of the sheer plurality and dynamism of a relatively young interdisciplinary field like environmental education research. But it might also require recognizing that amid the diversity and wealth of its research objects, subjects, and relations that its researchers pursue, various inertias and dearths remain. A recent account illustrates how these primarily relate to capacities, conceptual and procedural apparatus, contexts and priorities of those working in this field. It also raises the nagging question of whether the ends envisaged for this field of inquiry risk "outstripping the means" (Reid, 2009).

Another approach, as we pursue in this chapter, is to engage how we might understand environmental education research as to how arguments are argued, and evidence used and interpreted, when we reflect on how contributors derive recommendations for the field. Indeed, Gage (1986, p. 24) writes that the "road to a clearer understanding is travelled on paper. It is through an attempt to find words for ourselves in which to express related ideas that we often discover what we think." So in this chapter, we focus on critical questions that illuminate where we might go in our further reading and writing of environmental education research, and how demonstrating reflexivity will be a core feature of such activity.

Digging Over Environmental Education Research

Reflexivity as a state of mind as well as a set of practices involves addressing those ways of thinking and theorizing that are available (and valuable) to those participating in this research field. This can be pursued taxonomically and programmatically, as well as critically and creatively for understanding and developing an active and engaging field of inquiry (e.g., Connell, 1997; Dillon & Wals, 2006; Reid & Nikel, 2003).

Whichever route is traveled, in the first instance it requires recognition of a commonplace assumption: that research is about making "the world intelligible" to others and ourselves. In the second, in being in some sense "answerable" to the world we inhabit, it may even mean attempting to "make a difference" to that researched circumstance, context and, indeed, the "world."

Typically the first has entailed preparing and publishing reports, commentaries, and reviews of our findings as well as considering our successes and failures at sense-making as part of furthering the open, public account of inquiry. Yet is often the second aspect that has come to eclipse the first, fuelling debate about the adequacies and challenges of doing environmental education research, rooted in, for example, questions about how one views the ethics and politics of the field's knowledge.

Thus in this handbook, for Leigh Price, both assumptions relate to the claim that "we need research to be useful to us" but in different ways. Price reminds us of the notion of researcher integrity: for example, are our research accounts truthful and faithful to theory and data, ourselves and audiences (public and specialist), but also, if so/as such, from within and across which cultural constellations and framings?

That the particular cultures, discourses, and priorities of environmental education research(ers) might possibly be alien(ating) to others (researchers, participants, practitioners, funders, policy makers, partners . . .) should not be passed over hastily though in developing this point. The broader frame of social research has been thought of as *about, of, upon* and *into* a subject, as well as research *for* and research *through*. In this volume, we can recognize aspects of this in the contributions by Gough, Heimlich et al., Brandt, Le Grange, Bonnett, and Jickling and Wals, respectively. Such distinctions highlight that an emergent and dynamic environmental education research field remains relatively un-normalized even as it risks being destabilized by an overemphasis on definitional matters and practice. In other words, convergence is possible in research methods and framings even as it is not necessary in terms of philosophical groundings, projects or outputs, locally to internationally, be these now or then as well as into the future.

Why does this matter? Because as Lotz-Sisitka et al. and Shallcross and Robinson show—and the potential tensions between the two aspects demonstrate—environmental education researchers often hold to a distinction that sees them/us caught between sometimes competing aspirations. First, to change local lifestyles and lifeworlds by, for example, resolving pressing environmental problems (a cultural role for research, or, research for the "public good"). Second, seeking to understand the appeal, processes and outcomes of environmental education approaches and "interventions" within a broader sweep of sociocultural activity (i.e., a technical role for research, evidenced by "reasoning together" in critical, supportive and evaluative accounts of practice and inquiries into educational practice—see Biesta, 2007; de Vries, 1990).

Pursuing one or both roles is not exclusive to environmental education research, nor is it as simple as it seems particularly if we want research "to make a difference." Reflexivity about this requires another set of recognitions about individual projects and the field more generally, this time in relation to a potential conceit intimated in our nod that denies exclusivity.

The conceit is to pretend that a text about research, whether authored at some distance from an abstracted object of inquiry or closely and intimately involving those researched (including via the possibility of self-study), is ever "ours" alone. Of course it is never so (just like language). Neither are the uses and ends to which research is put even if the authors believe these to be compelling, useful and exact(ing). In each instance we have to recognize that researchers and their research programs, practices and outcomes, are historically, geographically, and culturally interwoven into a dynamic web of (sometimes dense, demanding, cooperative, transient) articulations and negotiations of a field of inquiry—including among the research community, its paymasters and publics.

As we discuss this further in Chapter 49, we focus here on a few examples. Consider Liddicoat and Krasny, who state "nature study . . . has been a cornerstone of our field for nearly a century" (p. 293) Some readers will press the question of which other cornerstones can be identified or imagined as present and absent in the interdisciplinary and shifting landscapes (and histories) of environmental education. Related questions include who decides and determines the cornerstones? And, are all such "cornerstones"—and by extension, "border markers"—permanently emplaced or moveable, literally if not metaphorically, when it comes to delimiting the field? While not to forgo that this is a research handbook, what is the role of research in this: is all research assumed to be part and parcel of these decisions and determinations—leading, following or mirroring contemporary trends, for example—or might some of it (which?) remain somehow separate?

Such arguments are not new and have been rehearsed before in other ways. Some employ philosophical inquiry, others action research, and of course there are yet other ways to address these matters. Their crux is about furthering our awareness of continuity and discontinuity, and hence the flux and coevolution of a research field, particularly when framed by a logic of research objects, subjects and relations. In other words, what counts—and from/for whom—is neither static nor fixed: as to the field's interests, resources, narrations, self-understandings, repressions and relations (e.g., Reid & Nikel, 2003; Scott, 2009).

Acknowledging flux and coevolution invites a questioning of what we have become accustomed to as environmental education research, including within the pages of this handbook. The argument goes, if we are to self-consciously continue and perhaps think and act otherwise in light of such recognitions and shifts, we might also have to confront or suspend the "comfort of the familiar" in this field (its range of research purposes, questions, languages, discourses, and framings). This is because, as noted in Chapter 49, reflexivity entails both critically and creatively engaging the field and its *doxa* and *illusio* to make deeper and richer sense of how environmental education researchers can seek critical guidance, challenge and inspiration for their inquiries and programs of research particularly through recourse to a handbook.

Barratt-Hacking et al.'s commentary on researching with children, for example, might stimulate curiosity about views and practices as to the cogeneration of knowledge in this field, including attending to the intersubjectivity of researchers relating to their fellow human beings (i.e., "self-in-relation to the other"). They regard these as lacunae in this field particularly as these relate to matters of the principles, theories and practices associated with the role and positioning of children-as-researchers, exploring whether this is because of the prevailing assumption that "children are 'always Othered' as they are the least powerful participants in research." They also raise the question of the potential disparities in research accounts if these are always presented in linear formats, contrasting this with the experience of research processes that are layered, iterative and reciprocal between researcher/s and researched. Finally, they discern another prospective gap concerning researcher ethics and codes of practice. In this field, these remain largely predicated on adults-as-researchers using "adult" understandings of research methods and methodologies, including as these relate to ethical practices.

They are not alone in advocating further deliberation and development of our thinking and practice in designing and conducting "better" environmental education research. Along with Stevenson and Robottom on action research and environmental education, Bradbury-Huang and Long engage similar lines of critique to Barratt-Hacking et al., although there "the Other" is argued to be the practitioner. Bradbury-Huang and Long press that research objects in environmental education research must not be solely understood in terms of conceptual development and congruence but rather in terms of liberatory action. And instead of suggesting divergence in the two aspects noted above, participatory action research affords the possibility of revitalizing social research and stakeholder partnerships, particularly toward sustainability.

Stepping back then, noting early childhood education considerations (e.g., Davis, 2009) are also making inroads into environmental education research, perhaps these chapters will suggest where genuine innovations are to be had in how inquiries in the field are conceived, designed and conducted, particularly in light of such critiques? Maybe some researchers will need to reimagine what it is to be childlike in the way their inquiries are framed and conducted, particularly if an intergenerational warrant presses down on the field's future (Payne, 2010).

Offering "Bright Line Certainties From the Fog of Life?" Research as Inevitably Complicit in Abstraction

As illustrated above, one of our major intentions for this chapter is to begin a process of annotating the handbook's text in relation to some of its research intertexts and metatexts. Our direct request is that with this chapter, readers reflect on the conceptualization and language of the research discussed throughout this handbook, particularly as to its aims and claims, key concepts and theorizations, reasoning and registers, analytical and expressive moves, if not the organization and patterning of ideas and examples in constructing scholarly argument. Actively doing so, we believe, puts the field in a stronger position for critically engaging the debates (e.g., Robottom & Hart, 1993) and claims made about the achievements and shortcomings of this field of inquiry over time, and as these relate to local, national and international levels.

Thus to critically engage implications for the field, rather than, say, simply reading these off from various chapters, some readers will note that a range of contributions are written as if the *audience* is not that of researchers but environmental educators or the field of environmental education alone. See, for example: Kyburz-Graber on "socioecological approaches"; N. Gough on "thinking globally"; Le Grange on "the politics of needs and sustainability educational"; Arenas and Londoño on connecting vocational and technical education with sustainability; and Barnett et al. on the teacher professional development and student outcomes in relation to the use and potential of geospatial educational technology development.

Equally there are contributions that are written as if research and practitioner communities should embody a particular *relation* to each other. Typically, it may be to be indifferent, critical, or championing toward each other in terms of a specific principle, theory, perspective or practice of environmental education. See, for example: Greenwood, on place-conscious education; Bai and Romanycia when applying Buddhist principles to education, ethics, and ecology; Olvitt on environmental ethics as processes of open-ended, pluralistic, deliberative inquiry; and Stevenson and Robottom on critical action research by practitioners to develop environmental education practice and theory.

There are also contributors who reflect on the form, sufficiency, quality, and extent of knowledge and evidence for an *argument*. For example, some note unargued assumptions in environmental education and its research by offering evaluative comments about their strengths and weaknesses; others discuss the plausibilities, support, gaps, leaps, (in)consistencies, and other alternative or equally compelling arguments in practice and scholarship. Examples to note here include: Berryman and Sauvé, on the language and discourses of education, environment and sustainable development; Stevenson on "tensions and pretensions" in critically and civically engaged policy scholarship; and, in distinguishing between propositional and experiential knowing, Bradbury-Huang and Long, and Payne. Lee, Wang, and Yang illustrate this concern too in writing on the relative lack of empirical studies of environmental education in China when compared with practice publications and policy statements; as do Gonzalez Gaudiano & Lorenzetti, on revitalizing the counterhegemonic challenges to the empirical-analytical tradition in environmental education research internationally and in Latin America. We also note Gonzalez, Gaudiano, & Lorenzetti argue for being more inclusive and pragmatic in the choice of research topics, to better respond to the priorities of the field "instead of clinging to theoretical approaches."

Reading and Writing Research Within Unsettled, Dynamic Constellations

Taking another step back in our reading and writing, we also acknowledge that this contribution sits alongside wider calls within these pages for critical attention to the processes and infrastructures, and the resources and products, of research in this field.

Paul Hart's introduction to and summary of Section 8, p. 418, on *Philosophical and Methodological Perspectives* are lucid exemplars of such work, by highlighting what is variously habituated and deconstructed as research practice in this field. As Hart shows, to carry this call forward in the research community requires careful consideration of the availability of and constraints on various modes, logics, and outcomes of research in the various instances of this field. This serves to open up discussion as to how these are embedded within particular constellations of the field's resources and capacities, not just its research subjects, objects, and relations, as noted in the examples above. Elsewhere contributors and section editors variously argue for similar foci of attention: to what is (being) lost or found, silenced or articulated, privileged or renounced, binarized or intersected, actioned or desired.

For instance, Shallcross and Robinson argue for multimodal methods to environmental education research because these should embrace and integrate different modes of representation. They argue these prove more congruent with reality than only a textualized representation of research and reality. In effect, for Shallcross and Robinson, it demands research that includes visual methods not only as a supplement to written text accounts but increasingly on its own merits (see, e.g., www.audiovisualthinking.org). Thus Shallcross and Robinson make the case for greater and more sophisticated use of stories, vignettes, units of engagement and analysis, pictures, diagrams, graphs, visual tapestries, performances, scripts, fictions and "fact"ions.

The possibilities that visual methods in environmental education research afford to inquiry raise wider questions of how readers and viewers of research approach an array of research challenges associated with the "crisis of representation" and nontextualized accounts of research. For example, visual methods can (re)open questions about traditions, options, and decisions about research design and representation that "word and/or number" forms of research may appear to suggest are relatively settled. These include questions of: verisimilitude, censorship and copyright, appropriation of reality, positioning of data and research/researched/researcher, and loss and gains to be had from the transformations and representations of data and theory in conventional (or less so, perhaps including ironic and playful) modes.

Russell and Fawcett, in introducing Section 8, relocate and reconfigure such concerns to questions of the "moving of margins" in environmental education research. They position them beyond familiar questions of marginalization, inclusion, and identity politics to note issues of intersectional analyses and intersubjective approaches. But, they caution, researchers must never lose sight of the particular and transectional in all these latter modes, such as issues concerning various waves of theorizing and experiencing gender and its analysis. Thus Russell and Fawcett also invite critical reflection on how age, employment status, income, family, body form (such as the health-at-every-size discourse and fatness studies) are visible in environmental education research accounts, and whether these aspects are also fertile ground for inspiring innovative research in this field.

Finally, we note some contributors and section editors invite readers to reflect on the ways modes and logics of inquiry are variously identified, selected, positioned, invested in, developed, and applied. As Hart puts it in his introduction to Section 7, "attention to the effects of one's own arguments and explicit discussion of those effects is prerequisite to all subsequent research activity including problems of ethics, relations between researchers and researched, problems of analysis and representation of findings and problems of the production of discourse." Thus, as part of the wider call of this handbook, Hart argues there is a need to get "clear about our onto-epistemological assumptions . . . to deepen our engagement, to be more conscious of multiple layers of meaning, to critique our own interpretations and representations." Which Berryman and Sauvé echo, quoting from Price's (2005) paper on "social epistemology and its politically correct words," to state:

> Different cultures will provide different language resources, different histories, different geographical potentials, constraints and evocative imagery; thus the same phenomenon, mobilized by people from different cultural, geographical, and historical heritages may have significantly different characteristics. (Price, 2005, p. 95)

Taking these examples together then, in our view one of the principal achievements of this handbook has been the attempt to revisit challenges presented as the "harder to reach varieties" of research thinking and practice (Reid & Scott, 2009), and related questions about the field's development as surfaced through reflexivity about the "knowledge" that a truly international collection can more assertively flag. As the quote and illustrations above show, this often includes in relation to the "ecological ontologies" and geo-cultural/historical epistemologies of the field, and its intergenerational prospects, in what is unavoidably now, following the insights of Lyotard (1984), a globalizing discourse of environmental education research and its "knowledge conditions."

Reading Between the Lines

Readers of a thoroughly critical handbook then should be able to identify tensions in what is uttered, refuted, affirmed, supplemented, relied upon, presupposed, and sought as worthwhile research. They should also be able to identify how these tensions are often consequential to certain (a priori) assumptions or judgments expressed in the initial framing and conceptualization of that which is being represented locally, culturally, regionally, internationally or globally as valuable, worthwhile, or useable knowledge in, of, with and for the field.

Critical questions emerging from this declaration of tensions to be dealt with include the sociolinguistic aspects at play in the field of inquiry as raised by Price above, even as we might also note those chapters that are positioned within perhaps familiar and Northern "psy"-focused frames. Others set out to challenge these presumed frames (given for example, ecophenomenological, socioecological, and ecopolitical concerns) or work from other geo-cultural, onto-epistemologically ethico-political and geographical places or normatively reflexive configurations within the empirical and theoretical landscapes to which environmental education is mapped (culturally, historically, politically, philosophically), as illustrated below.

This insistence on reflexivity, about framing, methodology and their normativities, throughout the handbook's pages invites us to critically engage the ways we speak of and to the various networks of research and researchers that contribute to and challenge environmental education inquiry. Be that in terms of understandings, scaffoldings, negotiations, proprieties, characterizations, narrations, positionings, attestations, intentions, stories and expectations (interests and desires), . . . the overarching point is, be it in practice and/or theory, there are many ways of engaging and doing research in this field, and we do a disservice to the field to maintain otherwise. As Hart argues (ibid.), acknowledging—even celebrating this critically—is key for the further development of the field given pressing questions of "academic credibility, competition/cooperation and critique/reflexivity." It also directly translates into the gravity of historicizing tasks such as literature review (original not just secondary texts), particularly if one is to demonstrate criticality and creativity in relation to the scholarly field.

Thus, as any brief consideration of chapters will show, despite some of the strengths of the handbook noted above, on occasion, some contributors do seem to resort to a descriptive and informational tone about environmental education and its research[1]. At the same time, others articulate a plethora of established and emerging perspectives on research from a range of backgrounds and settings. These are now most typically marked by the qualifier of a "critical" this or "post" that. Whatever the configuration or pattern though, these too also invite considerations of what is underrepresented—and perhaps by extension overrepresented—in the methods, methodologies, epistemologies, ontologies, research questions, strategies, and programs of the fields.

In other words, accompanying all perspectives—qualified or not, as subtext or not—are biases, assumptions, and challenges about the goals, traditions, and practices of inquiry taking place, or are possible, in this field. For instance, as noted in Chapter 49, Lotz-Sisitka et al. concede that tabularizing traditions of research, thematic areas, and theoretical influences (their Tables 1 and 2) may speak little to the "thicker" interpretive aspects of context or the standing of exemplar studies identified therein.

Globalizing the Research Imagination?

The challenge here (again, often expressed through reference to reflexivity) is how writers in environmental education research might better account for the "heft" of the thematics, niches, or provenance to their inquiries, be that within or beyond the interests (and even lives) of those working in the local research community from which the research arises, or of which it is intended to illuminate. Reflecting on such matters opens up (once again) a space for critically discussing the interests and positionings of environmental education researchers in the past, now and for the future. Indeed, we deliberately flag our interest in how this handbook, at a meta level of representing and legitimizing the field in an "internationalized" and globalized way, also portrays a wider "critical theorization of knowledge" and its value(s) and integrities (and usefulness to different interests) in a rapidly changing and fluid/mobile postmodernity. In other words, the broader politics, economy, and globalization of knowledge production interpenetrate this field, even as we create and maintain our own (Kenway & Fahey, 2009). If another research imaginary is pursued then, will it be one that avoids an enduring anthropocentrism and social/cultural centeredness, by extending into the socioecological or ecocentric.

More immediately and circumstantially, following Hart, this might follow an epistemological line of debate, such as in relation to the partialities and particularities of our "given" norms, narratives, constructions, models, and the "rhythms" and "regimes" of truth. But also, it could be in terms of the roles and forms of the knower/s and the (un)known/s, and the orders and facets of knowledge we are drawn (or perhaps aspire) to in and for this field.

For example, when we consider the meta level requirement for critical theorizations of knowledge generated and/or produced in, through and by environmental education research, an open question raised by Price is whether the vantage point of critical realism is the natural successor to poststructuralism in environmental education research. Given how other knowledge theories like postpositivism as applied to empirical-analytical approaches have been critiqued and reworked through the short history of environmental education research (e.g., Connell, 1997), does each theorization or knowledge vantage point, as the way many framings occur, only make sense in reaction to its "logical predecessor?" (As with postmodernism, by grasping the roots and limits of modernist theory and analysis?) Put differently, do/must we contrarian-like only define ourselves (in/through research and representations in handbooks) simply by what we are opposed to? If so, does this automatically lead to the much-maligned "dialogue of the deaf," even as there is some felt need for the field to have a degree of narrative continuity, notwithstanding internal critique and reflexivity?

Equally, at a "micro" level concerned more directly with agential ways of knowing, seeing and doing, what might we make of the lack of attention to research on the role of emotion, control beliefs, attributions, willingness and self-esteem, as identified by Zeyer and Kelsey in this field? Zeyer and Kelsey raise whether teachers and learners are being charged with addressing "environmental depression and ecological passivity" as well as feelings of being overwhelmed or hopeless, angered, shamed or bored during "education," not just environmental education. For Lundholm et al., contributing to a research agenda in this area requires expert and collaborative inquiries that attend to the value-ladenness of environmental issues, students' conceptions of values in subject matter and in relation to themselves, as well as students' emotions and values as part of the learning process. Both chapters underline the importance of clarity about working assumptions in scoping such research: for instance, when the voice of the learner is severely neglected in curriculum development and research; when learning is understood to be multimodal and interactional (i.e., leaky and messy rather than mono-modal or transmissive); and, when experience and emotion are understood to be as important to the mediation of students' learning as questions of curricular context or program planning.

Beyond some of the meta, meso, and micro layerings of knowledge generated internally and "within" the field that we have briefly introduced above, we might also note that by manifesting a diversity of views, perspectives and "consciousness" about environmental education research, some contributors have offered their contributions to the research field largely from an "outside in"[2] position, while others attempt to do so from the "inside out" (Marcinkowski et al., Chapter 31 on environmental literacy assessments; Lowan, Chapter 44, on indigenous environmental education research, . . .). There are also those who see their work as contributing to the "in-betweens" of "intersubjectivities," "border crossings," and "transdisciplinarities."[3] As such we are witness to a broad range of cultural and critical commitments and meta, meso, and micro standpoints as to how objectivity and subjectivist points of view are expected, practiced, opened up, or challenged for this field and via this handbook. These include and implicate those advocating research that gives life and space to perspectives variously self-defined by theoretical labels such as the phenomenological, postpositivist, liberal, critical, cultural and poststructural,[4] particularly when these are championed as able to address asymmetrical power gradients in research, such as between researchers and researched, and addressed through participatory approaches. Equally, it may also challenge the derivations of ideas from the literature predominantly rooted in one cultural context to the exclusion of many others legitimate and possible.[5]

Research That "Marks a Difference"?

This chapter then has become increasingly marked by opening up questions of how we (as authors/readers) train our attention on what researchers are prepared to talk about and prosecute as their political and theoretical projects, and even what is demarcated, sketched, and/or fantasized as the historic and contemporary environmental education research field. Central to this work is the question of the status of knowledge in environmental education research as represented by the purposes, section thematics and sequencing, and specific chapter contents of this handbook.

Thus in both taking stock and looking ahead in this final section, it is timely to ask: will a twin track of traditional and innovative approaches to research allow each to redress the other in this field, such that a plural approach offers alternative and dialectic modes of thinking to our understandings of the issues? What sorts of knowledge reconciliations—in framings, approaches, perspectives, and posing of research questions against enduring and intractable issues/problems in environmental education research—will help clarify some ends-in-view and the commensurabilities of the field's purposes, means, and such ends that might be more accessible as ushered in by the publication of this first international handbook? Thus, when some talk of theory-into-practice or practice-into-theory framings, without dialogue these are likely to remain incommensurable, particularly if viewed as taking place across postpositivist, interpretive, critical and postcritical forms of inquiry. The hard but necessary work is to keep talking to each other across such divides (see, e.g., Peters, Chapter 40, on attempts to de-anthropocentrize theory and observation in cosmological studies).

A key issue in Section 8 on margins and marginalization speaks to this challenge too: on the status of indigenous ways of knowing in environmental education and its research. As researchers we have to continue to ask, on what grounds are these ways treated as equally valid (or not) in inquiry to Western scientific approaches? Is it that they are in some sense irrevocably different as the term "paradigmatic" articulates, even when dialogue is possible? Might a monist epistemology simply refuse such difference? Or a relational epistemology reinscribe them and thus suggest dependencies not just interdependencies? The point being, the logics of the very terms of our epistemological and philosophical principles have effects not just on research design and objects but also on positions and conversations about research.

Thus, while Zeyer and Kelsey can point to a "rhetoric of othering" (Riggins, 1997) in student discourses to legitimize and position their actions and inaction, we note they also invoke postecologism (e.g., Blühdorn, 2002, 2011) as a concept since it offers them a critical interpretive frame for understanding life in postindustrial societies (cf. Certeau, 1986 on "discourse on the Other," and Russell & Fawcett, on the marginalization of other-than-human voices in research). Postecologism focuses on the normalization (and naturalization) of the paradoxical politics of unsustainability. Blühdorn's thesis diagnoses structures and expressions of apathy, denial, and skepticism of unsustainability as taking place within a series of simulative aspects of lifeworlds lived out within a postecological matrix, e.g., a mantra of reduce, reuse, and recycle that no longer refuses or refutes a culture of consumerism. According to Blühdorn, this matrix impels a continued commitment to consumerism and hedonistic lifestyles, a persistent belief in technological fixes, and increasingly a dislike—and in some instances a disdain—for nongovernmental actors (NGOs, social movements, and local green politics: e.g., as "idealistic troublemakers"). While somewhat inevitably, Blühdorn (2011) also asks serious questions of and about the role of democracy (and indeed social science inquiry and empirical insight/qualification) in terms of its and their normalization of the paradoxical and globally paralyzing politics of unsustainability.

In other words, if there is some kernel or skerrick of interpretive global "truth" in the postecologism thesis, we might well ask, is it now time for environmental education research to inquire not only into questions of foundations and the effectiveness of environmental education but also into their shadows or counterparts and mirrors in a "postintellectual" intellectual "climate change" in universities governed by neoliberal means of knowledge production and surveillanced/policed by its audit/managerial cultures (Apple, 2005)? Put simply, what is the "match" or "fit" of the current conditions in which research, as both the generation and now "production line" of knowledge (and accompanying normative reflexivities, addressed above) and the changing nature of the ecologically problematic human condition, as suggested by Bluedhorn's view that the (global) politics of unsustainability have normatively been (epsitemologicaly) naturalized? Indeed, do we (mis)recognize ourselves (Cooper, 2002) as environmental education researchers with a keen if not passionate interest in the field's development within such a changing knowledge climate: for example, what we can learn from "fast" practice or "quick" theory in research, other adjectival educations, and wider sociopolitical analysis of the nature and cultures we live with(in)?

The Duties of Research

Taking conceptual and discursive complexities and contradictions in a handbook at face value then (perhaps by somewhat blithely honoring instead of critically and creatively listening to and "interrogating" diversity, alternatives and difference), has its risks. Even to its critical friends, it can appear one step away from intuiting only irrevocable contradiction and conflict, or disconnection and incoherence, in the collection, and by extension, the research field to which the handbook refers. To its detractors, it may be little more than evidence of the field losing sight of its bearings and foundations (e.g., why not prioritize engagement with and relevance to educators, practice and professional knowledge in a research handbook?). As in the case of the emergence of education for sustainable development—not the subject of this handbook, although clearly of interest to some contributions—these pages could be considering the politics, priorities, and pragmatics of what gets thought and funded as research about this aspect of the "adjectival educations" and broadly, the "socioecological"—environmental literacy, place-based education, environmental education, ESD, . . .? While to its pragmatically inclined, talk of complexity, diversity, and ambiguity might suggest a dereliction of duty.

As Chawla (2003, para. 1) commented on the *Alternative Paradigms for Environmental Education Research* publication from NAAEE (Mrazek, 1993):

> In the decade since 1993, the urgency of achieving William Stapp's goals for environmental education- of producing a knowledgeable, aware citizenry who is motivated to work toward the solution of environmental problems- has intensified, while by all evidence, the opportunities for children to know and care for the environment through community-based projects, field trips, and informal exploration of their neighborhoods appears to be steadily eroding under the pressure of tightening school budgets, the tyranny of standards-driven testing and teaching, sprawl, and parents' fears of letting their children range freely. In this context, debates about one research paradigm or the other appear academic in a way that now appears to me out of touch with the contemporary reality of our need to know as much as we can about how to achieve Stapp's goal by every research means and method possible. To the extent that the collection clarified the definition and use of different research approaches, it remains useful. To the extent that some contributions advocated one approach to the exclusion of others, they appear unrealistic.

Two decades down the line, will this charge apply to this handbook too? We now offer a closer reading of the first section of the handbook, "Part A. Conceptualizing Environmental Education as a Field of Inquiry," to consider the issues, before concluding the chapter.

When Is History, History?

The first section of Part A of the handbook concerns: "Historical, contextual and theoretical orientations of environmental education research." We note the introduction to the section does not discuss to what these qualifiers to the orientations refer, nor how research spaces are created or how we conceptualize gaps and silences in the orientations of the field even as these are the themes of the section (such as by focusing on the "location" and distribution of research topics, researcher, and researched voices). However Annette Gough in Section I does draw on Hart and Nolan's account (1999, p. 41) of needs for environmental education research as well as her own conspectus (A. Gough, 1999, p. 153); to wit, it needs to:

- begin to address critical and feminist and postmodern challenges
- strive for more in-depth qualitative analyses, and
- move outside the academy and develop partnerships with schools and communities. (Hart & Nolan, 1999, p. 41)

While as "guiding principles for research in environmental education," Gough reiterates earlier points (Gough, 1999, p. 153), namely, to:

- recognize that knowledge is partial, multiple, and contradictory;
- draw attention to the racism and gender blindness in environmental education;
- develop a willingness to listen to silenced voices and to provide opportunities for them to be heard; and
- develop understandings of the stories of which we are a part and our abilities to deconstruct them.

She also surmises that the chapters in the first section display:

> a move away from the "longing for 'one true story' that has been the psychic motor for Western science" (Harding, 1986, p. 193), and a recognition of the need to be aware of the historical, contextual and theoretical orientations to environmental education that inform and influence our work as researchers.

The first chapter of that section, on "The Emergence of Environmental Education Research: a 'History' of the Field," offers a similar point of interest to this section, given its focus on how we write about research in and for this field. Again, given our interest above in probing questions of the narrative continuity of the field, notwithstanding the importance of an "internal" and "external" reflexivity at meta, meso, and micro levels, intergenerationally, and geo-cultural/historically epistemologically, we also note Gough hasn't discussed what a history is or might be, although she does chart changes in the orientation of environmental [sic. education?] research, or "EE research shifts," via cataloguing key documents and events about environmental education emanating largely from international meetings, e.g., UNESCO, UNEP.

As noted above by Chawla, some environmental educators and researchers expect research to follow the lead of these documents and their definitions of the field of environmental education such that research focuses on the effectiveness of environmental education to shape human behaviors, actions and values, albeit largely within an individualized frame for response and action (e.g., Hungerford & Volk, 1990). Yet this focus and frame, Gough argues, has proven moribund. She charges it positions legitimate research largely within a positivist and/or postpositivistic fold, and once there, the key goal of environmental education has been the acquisition of responsible environmental behaviors. The effect is that the fields of behavioral and social psychology become authoritative in framing the logics of pedagogical reasoning, practice and research. Yet following its critique by Robottom (e.g., 1992, 1993) and Hart (e.g., Robottom & Hart, 1993, 1995), among others, she states with approval:

> The argument advanced by Connell (1997, p. 130) opens up space for legitimizing all types of research methodologies where researchers "do what they do well and where methodologies are selected to meet clearly identified research needs, balanced with a clear understanding of the social, political and philosophical contexts in which they are located."

For us then, this opening chapter illustrates some of the key challenges and roles of a handbook in stimulating and demonstrating an historical consciousness of the field. For example, in terms of doctoral preparation and "learning the ropes," which methods and methodologies are selected from, reproduced, criticized, or extended, and on what grounds in the "histories of the field" presented to its initiates? To whom will it matter if Gough's "history" falls short "of the present," trailing out in the early 2000s, or if both Gough and Ferreira are bound (by handbook publication schedules, by critical interest, by availability?) to omit reference to recent collections such as David Zandvliet's (2009) *Diversity in Environmental Education Research*?

Equally, if we take the "post" demands of historical and material detail seriously, will it be important to note that a key driver in the development of the critical approaches identified by Gough and Ferreira was the gradual international access to and acceptance of ideas associated with the classbook and distance learning materials that arose from the Deakin-Griffith project for Masters environmental education students. Added to that might be the travel and exchanges among staff associated with that program (by John Fien, Annette Gough, Noel Gough, Ian Robottom, Bob Stevenson . . .) from Australia, and those heading there too, or finding solace in that material? While to note the devil in the detail, must we continue to ask which studies are actually reported and selected in constructing any particular "history" or reader of the field, and thus whether those citations that make it to the pages of a handbook are not only intellectually but statistically representative of, say, the long tail of doctoral projects and gray literature associated with the field?

A key challenge then is whether this research field will make strategic use of corpus and bibliometric studies. These approaches can help a research community trace and map concepts and sources and theoretical framings, methodological orientations, and interpretive approaches/representations most often used in and across this field, be that within, say, a genealogical or geophilosophical framing, or those that relate to and incorporate strongly those traditional concerns of critical perspectives—class, gender, race.

Arguably another gap emerges not in the handbook's "toolbox" this time but in the uses to which tools are put. If we consider "post–post" research in environmental education, as represented by the special issue of *Environmental Education Research* on postcritical inquiry, there is a range of philosophical and pragmatic issues to consider. The collection, edited by Paul Hart (2005), was initiated by an essay from Marcia McKenzie, with responses from a range of researchers, and a coda by the essay author. Hart's editorial (p. 396/7) notes that:

> In recognizing both the constructed and real aspects of the phenomena we study, post–post researchers are coming to acknowledge limitations of those forms of the posts that insist everything is a construct, an illusion that linguistifies "being" completely. Such a stance, however, also eschews forms of naive realism where the reality of social structure is unquestioned.

We note here that while McKenzie's (2005) call for better ends-means congruence in environmental education research echoes Robottom's and Hart's and Nolan's and A. Gough's earlier calls (among others), Hart's editorial to the collection provides a delicate rider to McKenzie's argument. Hart relays how Russell's (2005) response to the essay draws attention to how theoretical commitments write and shape not just our histories but also our experiences and sense of agency in being in the natural world via acts of environmental and outdoor education. Hart (p. 396) also discerns something significant in Payne's (2005) response piece to McKenzie (2005), to the effect that:

> researchers must be clear about their theoretical stance which allows simultaneously for both the reality of our experiences/relationships with others and environments, as well as the socially produced character of our researchers' judgments about them. (p. 396)

He adds:

> What this means for environmental education research, from the perspective of those researchers situated in the middle of "ambivalent post/neo colonial terrain," when social and environmental issues are in high relief, is perhaps not so much "working the ruins" (poststructurally) as "working the theories" (postcritically). "What matters" in environmental education research are deliberations that involve the ethics of everyday activity as coupled with those social perspectives that work to reinstate agency beyond narrow anthropomorphism through more reflexive actions in education. Environmental education research, as a more critically reflexive process, could have many variants, a certain requisite variety perhaps, but characterized by promising orientations for reinstating agency, as a social construction embedded in geopolitical contexts and sociocultural historic spaces and places as sites for "situated learning." (p. 397)

Moreover, Hart's editorial sets out a careful line to steer between (beyond?) absolutist and relativistic philosophies for environmental education and its research, including:

> . . . in reauthorizing ourselves through practice rather than confession, researchers can demonstrate understanding of means of production, that is, be aware of the possibilities for appropriation and know the constraints of methodological techniques as well as the power relations of location and position. Although there is risk in assuming that epistemological authority must necessarily entail social/moral inequality of worth between research and participants, researchers know some things about participants that they don't know (as they know things we don't). We must not, therefore, confuse positioning with morality and telling the self with decentering the self. (p. 399)

While in closing, Hart avers:

> All we are doing, as reflexive researchers, is to write in ways that reveal the limits of our knowledge, our political orientation and other dimensions of self, in ways that reveal the discourses that shape our work and open possibilities for thinking about our work as we get on with it. (ibid.)

We draw on Hart extensively here because his work foreshadows many of the themes we have wanted to discuss in the concluding part of this chapter. For example, Chapter 4, p. 38, by Noel Gough, on "Thinking Globally in Environmental Education: A Critical History" like Kyburz-Graber's before him on "Socioecological Approaches to Environmental Education," enact this reflexive sensibility even as their chapters are primarily about environmental education rather than research *per se*. Indeed, Gough draws out implications for "the blind spots that might still remain in the vision of even the most culturally sensitive scholars." Focusing his account around reflections on the sociology, culture, and anthropology of scientific knowledge, Gough's conclusion resonates with the tone and intent of Hart's commentary:

> . . ."thinking globally" in environmental education research might best be understood as a process of constructing transcultural "spaces" in which scholars from different localities collaborate in reframing and decentering their own knowledge traditions and negotiate trust in each other's contributions to their collective work. For those of us who work in Western knowledge traditions, a first step must be to represent and perform our distinctive approaches to knowledge production in ways that authentically demonstrate their localness. We might not be able to speak—or think—from outside our own Eurocentrism, but we can continue to ask questions about how our specifically Western ways of "acting locally" (in the production of knowledge) might be performed *with* other local knowledge traditions. By coproducing global knowledge in transcultural spaces, we can, I believe, help to make both the limits *and strengths* of the local knowledge tradition we call Western science increasingly visible.

In direct contrast to the style of the other chapters in Section 1, Marcinkowski et al. write on, "Selected Trends in Thirty Years of Doctoral Research in Environmental Education in the U.S." They use content analysis to report a series of clusters and structures of interest in the dissertations the researchers have audited. As we comment on Table 1 in Chapter 49, here readers may want to ask themselves whether the patterns and gaps Marcinkowski et al. discern extend to other geographical and linguistic regions, as well as whether the findings are best represented in tables given the availability and power of graphical visualizations? For this chapter though, we invite readers to consider whether a key question for further research raised by their results is how the trends relate to the actual drivers and structures of doctoral studies? For example, the graduate schools and traditions from which theses emerge, the expertize accessed and supervision experienced by the doctoral students, the demographics of the researchers (gender, age, career histories, . . .), and the networks sought and within which they participate(d) then and as postdocs. These can all have subsequent or structural implications for the future of research and professional development in this field (cf. Nikel et al., 2010).

We raise this because as broached above, a key mechanism for producing knowledge in this field concerns doctoral training and education. Doctoral researchers and their supervisors or panels variously reproduce, extend, or break with assumptions and practices of environmental education research. Again, corpus analysis and citation analysis can be helpful here in tracing the trajectories, clusters and ruptures in traditions, ideas and foci, as well as revealing the extant literature bases (≠ reading lists) available to doctoral researchers, their doctoral programs and mentors (e.g., the national and international research journals, the within and across subdisciplinary reading taking place—i.e., the sociolinguistics of research reading and writing).

As Nikel et al. (2010) note in relation to doctoral studies they undertook in the UK as overseas students, a culturally reflexive turn invites us to move beyond asking what we study to also engaging how we understand and experience the layerings and intersections in doctoral work, often personally and not just professionally. These vary at different stages in the progress of doctoral work, and in their cases, often as "outsiders becoming insiders" professionally as well as culturally during, and perhaps beyond, a doctoral stage. Again, noting the reflexivity trope above, this includes opening up discussions about the complexities of gender, identity, agency, power relations and career progression. Nikel et al. (2010) write on this in relation to those mundane, transformative or painful learning experiences and encounters that took place during their doctoral research processes as a vehicle for their own knowledge production and identity formation. And like Ferreira, they note this can lead to a productive, if not troublesome attempt to make visible the relations of power and the rationalities through which we are governed and govern ourselves as doctoral, emerging and more established scholars and researchers.

Conclusion

As we have shown in this chapter, environmental education scholarship and research are quite properly viewed as socially constructed and therefore open to discussion, contestation and change. Thus, for an international research community, a normatively reflexive account of its researches should be able to further our collective understanding of:

(a) how and in what ways a research-driven handbook with a warrant and claim on being international represents and, therefore, partially (re)constitutes and legitimizes a "report" on the status of knowledge in the field
(b) the geo-cultural/historical contexts and conditions and ontological-epistemological formations and social relations of individual and group intellectual/knowledge work and their "identities" that non-Anglophone and Anglophone researchers work with/in, are ineluctably shaped by and inform/embody certain horizons/imaginaries;
(c) the almost invisible emergence of normalized and naturalized postecological paradoxes and globalizing realities against and for with such normative reflexivity of the field would do well to reengage its debates and terms of reference as ends-in-view
(d) the drivers, infrastructures, discourses, technics, and experiences of research and research community around the globe
(e) the challenges and tensions of increasingly abstracted and socially extended "knowledge/research" networking, collaboration, internationalization, homogenization and prescription in a globalized *and* localized academy, and

(f) the narrative continuity(ies) (and discontinuity(ies) of the field of environmental education research in retrospectively and prospectively working through the changing intergenerational and globalizing conditions of knowledge generation, value, integrity, usefulness within postmodernity's productive/performative imperatives.

In other words, when we work through the issues and trajectories of research, researchers and researched as represented in this first *International Handbook of Research in Environmental Education*, we can understand the field is presented with direct challenges to its present and its future from the realms of sociolinguistics and geo-epistemology. Namely, can the ontology-epistemology-methodological triad also be normatively and reflexively framed in ways that more clearly, generously and effectively represent: i) the "simultaneous" demands/imperatives or hopes/aspirations for local, national, regional and global interpretations, descriptions, representations and legitimizations in the next international handbook; ii) the historical and intergenerational nature of the way research occurs, is presented and legitimized as a status of knowledge within the field's narrative continuities and discontinuities; and iii) research leading to that, as played out in journals/conferences, edited books, along the way?

Given this critical interest, it is perhaps of little surprise that some of the contributions to the handbook focus more on props and targets for readymade theories about experiences, motivations, dispositions, educations, learnings, ethics, identity, knowledge, affect, skills, behaviors, and actions. This is particularly clear when we consider what appears to count as their "determinants" over the lifes pan (e.g., reference to, or omission in such accounts of, youth, familial or kin factors such as: parents' level of schooling, childhood significant life experiences, cultural traditions or values, informal learning, mediating adults).[6] Equally, there are others represented here, writing out of dissatisfaction and/or hope, who seek more compelling patterns or intersections of constructs, theorizations, and projects in this field. Their accounts seem preoccupied with the need to clarify the normative and axiological "force" and potentially positive (constructive) power of the field. These (critical) "values" of the (reflexive) field are often argued by invoking a broader and richer frame that desires or anticipates environmental education and its research "making a difference" (usually in terms of longed for outcomes defined ontologically, conceptually, pedagogically, behaviorally, culturally, politically, or socioecologically . . .) to the environmental (and) educational challenges the peoples of this world do, and might, face.[7]

To summarize, as might have been expected, contributors to the various sections of this handbook offer diverse ways of thinking, understanding and evaluating the state of the field—discursively, philosophically, and normatively. In some senses discrete, in others continuous, only some of these ways can be reconciled with each other (e.g., methodological conservatism with radicalism? naïve relativist and critical realist conceptions of truth?).[8] And this is even as the "bandwidth" and "needs" of the field broaden and "intensify" for environmental education research.

Notes

1. Zint on program evaluation starts by outlining the features of ten behavioral outcome evaluations, and then concludes by stressing the importance of meeting evaluation systems and standards, or tying evaluation goals to models of competence, behavior, capacity, etc. In contrast, Leigh's response to Shallcross and Robinson highlights the reductionism inherent in much social world research, e.g., in preparing and selecting an interpretation of data, reducing explanations to that of textual explanations alone, or offering an interpretation as a description when the truthfulness of either is unequivocal.
2. For example, Davies, on how feminist poststructural perspectives can direct attention to normative cultural and pedagogical structures and meanings, acts and channels of power, and the "common sense" of "subjectification" and "transgression" of existing research purposes and perspectives through, for example, simply, deeply listening and attending to the voices and sounds of "others"; Peters, on strategic research interests emerging from network analysis, and the emergent properties, intersections and tensions of "ecosophy," ecology, economics, and sustainability, multidisciplinary pedagogies in environmental education given a green critique of neoliberalism and philosophical inquiries about emerging forms of learning in "open knowledge economies" for public knowledge and public action; Bradbury-Huang and Long, from management education and action research, the importance of relational epistemologies as a basis for participation and partnership that pursues collaborative ecological inquiry for sustainability; Barnett et al. on access to geospatial technologies in and outwith the classroom; and Dierking et al. on aggregating data and iterative and recursive analysis of informal science and environmental education programs, experiences and evaluations.
3. For example, Barratt Hacking et al. in focusing on children-as-researchers enjoying meaningful, relevant research involvement, integrity and independence, positive power relations and intergenerational forms of perspectives sharing and learning (i.e., different to research on and with children to by children, and adult-as-researchers of environmental education only); McKenzie et al. on methodologies for inquiry-based, relational, community-based intergenerational learning; Payne, on the value of (normatively reflexive) ecophenomenological framings to our research and method assemblages, including for ecocentric theories of experience, education, and scholarly inquiry.
4. For example, Bonnett on normative arisings given a Heideggerian frame for understanding our being in the world and hence the "presence of nature" in environmental education; Jickling and Wals on liberal ethical perspectives in curriculum; Liddicoat and Krasny given a focus on significant life experiences; Russell and Fawcett, on nonanthropocentric research, fat studies, queer theory.
5. For example, Marcinkowski et al. on doctoral studies; or noting the first wave of national assessments in environmental education in the United States, and then international assessments of environmental literacy in the United States, Korea, Israel, and Turkey; and Hoeffel et al. from Brazilian sources on participation and participatory forms of research in developing country contexts when discussing

research on disruptions to livelihoods, knowledge, participation in environmental decision-making, and autonomy in postcolonized regions of Brazil.

6. See Storksdieck and Brody's helpful comments in the introduction to Section 6 on the risks of confusing goals for evaluation (e.g., given demands for accountability) with research (e.g., to test hypotheses or build theory), seeking generalizability on inadequate grounds, and overconfidence in the methodological rigor and explanatory power of an inquiry and/or research design (e.g., in triangulation of findings and mixed methods approaches to data analysis, the units of analysis and attributions of effects, effect sizes and longevity, interactions and "intervening variables").
7. See, for example, Holdsworth et al., on the social and political constructs of neo-liberal capitalistic society, the limitations of a positivistic scientific worldview when linked to conserving existing institutional structures, and how their vision of sustainability education praxis requires attention to curriculum, organizational culture, staff politics, and disciplinary dialogue and sense-making, among others.
8. Bradbury-Huang and Ken Long open Chapter 36 with a Rorty (1999) epigraph that roundly rejects realism: "We cannot regard truth as a goal of inquiry. The purpose of inquiry is to achieve agreement among human beings about what to do, to bring about consensus on the ends to be achieved and the means used to achieve those ends. Inquiry that does not achieve coordination is not inquiry, but simply wordplay." (p. xxv).

References

Apple, M. (2005). Education, markets, and an audit culture. *Critical Quarterly, 47*(1–2), 11–29.

Biesta, G. (2007). Bridging the gap between educational research and educational practice: The need for critical distance. *Educational Research and Evaluation, 13*(3), 295–301.

Blühdorn, I. (2002). Unsustainability as a frame of mind and how we disguise it: The silent counter–revolution and the politics of simulation. *The Trumpeter, 18*, 1–11.

Bludhorn, I. (2011). The politics of unsustainability: COP15, post-ecologism, and the ecological paradox. *Organization & Environment, 24*(1), 34–53.

Certeau, M. de (1986). *Heterologies: Discourse on the other.* (B. Massumi, Trans.) Minneapolis, MN: University of Minnesota Press.

Chawla, L. (2003). Environmental education research: A decade later—an update of my review of alternative paradigms in environmental education research. *Children, Youth and Environments, 13*(1). Retrieved from http://www.colorado.edu/journals/cye/13_1/retrospectives/CYE_ReviewerRetrospective_Chawla.htm

Connell, S. (1997). Empirical–analytical methodological research in environmental education: Response to a negative trend in methodological and ideological discussions. *Environmental Education Research, 3*(2), 117–132.

Cooper, S. (2002). Post intellectuality?: Universities and the knowledge industry. In S. Cooper, J. Hinkson, & G. Sharp (Eds.), *Scholars and entrepreneurs: The university in crisis* (pp. 207–232). Fitzroy, VIC: Arena Publications.

Davis, J. (2009). Revealing the research "hole" of early childhood education for sustainability: A preliminary survey of the literature. *Environmental Education Research, 15*(2), 227–241.

Dillon, J., & Wals, A. E. J. (2006). On the danger of blurring methods, methodologies, and ideologies in environmental education research. *Environmental Education Research, 12*(3–4), 549–558.

Gage, J. (1986). Why write? In A. R. Petrosky & D. Bartholomae (Eds.), *The teaching of writing (pp. 8–29). 85th Yearbook of the National Society for the Study of Education*. Chicago, IL: University of Chicago Press.

Gough, A. (1999). Recognizing women in environmental education pedagogy and research: Towards an ecofeminist poststructuralist perspective. *Environmental Education Research, 5*(2), 143–161.

Harding, S. (1986). *The science question in feminism.* Ithaca, NY: Cornell University Press.

Hart, P., & Nolan, K. (1999). A critical analysis of research in environmental education. *Studies in Science Education, 34*, 1–69.

Hart, P. (2000). Requisite variety: The problem with generic guidelines for diverse genres of inquiry. *Environmental Education Research, 6*(1), 37–46.

Hart, P. (2003). Reflections on reviewing educational research: (Re) searching for value in environmental education. *Environmental Education Research, 9*(2), 241–256.

Hart, P. (2005). Transitions in thought and practice: Links, divergences, and contradictions in post–critical inquiry. *Environmental Education Research, 11*(4), 391–400.

Hungerford, H., & Volk, T. (1990). Changing learner behavior through environmental education. *Journal of Environmental Education, 21*(3), 8–21.

Kenway, J., & Fahey, J. (Eds.) 2009. *Globalizing the research imagination.* London, UK: Routledge.

Lyotard, J–F. (1984). *The postmodern condition: A report on knowledge.* Minneapolis, MN: University of Minnesota Press.

McKenzie, M. (2005). The "post–post period" and environmental education research. *Environmental Education Research, 11*(4), 401–412.

Mrazek, R. (Ed.). (1993). *Alternative paradigms in environmental education research.* Troy, OH: NAAEE Monograph Series.

Nikel, J., Teamey, K., Hwang, S., & Pozos–Hernandez, B., Reid, A., & Hart, P. (2010). Understanding others, understanding ourselves: Engaging in constructive dialogue about process in doctoral study in (environmental) education. In B. Stevenson & J. Dillon, (Eds.), *Environmental education: Learning, culture and agency* (pp. 167–198). Dordrecht, Netherlands: Sense.

Payne, P. (2005). Lifeworld and textualism: Reassembling the researcher/ed and "others." *Environmental Education Research, 11*(4), 413–431.

Payne, P. (2010). Remarkable–tracking, experiential education of the ecological imagination. *Environmental Education Research, 16*(3), 295–310.

Price, L. (2005). Social epistemology and its politically correct words: Avoiding absolutism, relativism, consensualism, and vulgar pragmatism. *Canadian Journal of Environmental Education, 10*, 94–107.

Reid, A. (2009). Environmental education research: Will the ends outstrip the means? *Environmental Education Research, 15*(2), 129–153.

Reid, A., & Nikel, J. (2003). Reading a critical review of evidence: Notes and queries on research programs in environmental education. *Environmental Education Research, 9*(2), 149–165.

Reid, A., & Scott, W. (Eds.) (2009). *Researching education and the environment: Retrospect and prospect.* London, UK: Routledge.

Riggins, S. (Ed.). (1997). *The language and politics of exclusion: Others in discourse.* London, UK: Sage.

Robottom, I. (1992). Matching the purposes of environmental education with consistent approaches to research and professional development. *Australian Journal of Environmental Education, 8*, 80–90.

Robottom, I. (1993). Beyond behaviorism: Making EE research educational. In R. Mrazek (Ed.), *Alternative paradigms in environmental education research* (pp.133–143). Troy, OH: North American Association for Environmental Education.

Robottom, I., & Hart, P. (1993). *Research in environmental education: Engaging the debate.* Geelong, VIC: Deakin University.

Robottom, I. & Hart, P. (1995). Behaviorist EE research: Environmentalism as individualism. *Journal of Environmental Education, 26*(2), 5–9.

Rorty, R. (1999). *Philosophy and social hope*. New York, NY: Penguin Books.

Russell, C. (2005). "Whoever does not write is written": The role of "nature" in post–post approaches to environmental education research. *Environmental Education Research, 11*(4), 433–443.

Scott, W. (2009). Environmental education research: 30 years on from Tbilisi. *Environmental Education Research, 15*(2), 155–164.

Vries, G. H. de (1990). *De ontwikkeling van wetenschap* [The development of science]. Groningen: Wolters–Noordhoff.

Zandvliet, D. B. (Ed.). (2009). *Diversity in environmental education research*. Rotterdam, Netherlands: Sense.

51

Tentative Directions for Environmental Education Research in Uncertain Times

Arjen E.J. Wals
Wageningen University, The Netherlands; Cornell University, USA

Robert B. Stevenson
James Cook University, Australia

Michael Brody
Montana State University, USA

Justin Dillon
King's College London, UK

Shifting Contextual Landscapes and Influences

In this concluding chapter to the *Handbook*, we identify current directions in, and future challenges and opportunities for environmental education research. We begin by tracing some of the broader and critical changes in society, particularly the influence of information and communication technologies. We then examine the implications for future directions for research on learning and living sustainably in times of accelerating and continuous change, and increasing complexity, risk, and uncertainty. We suggest three tentative directions for consideration by researchers: (i) connecting biophilia and videophilia, (ii) creating spaces for hybrid learning, and (iii) strengthening engaged scholarship with a planetary conscience.

As was stated in the introductory chapter to this closing section, many authors in this *Handbook* have identified how the increased recognition that the cultural and biophysical contexts in which people live and shape our thinking and behavior. That recognition, they suggest, has resulted in an increase in cultural research within the field of environmental education.

Many significant contextual influences have framed or constituted the issues examined in many of the preceding chapters. The increasing presence of information and communication technologies (ICTs) has become an almost taken for granted influence that has shaped the field of environmental education research. However, although some scholars have written about this influence (e.g., Freier & Kahn, 2009) it is quite striking that in the fifty chapters preceding this final one, the phenomenon is hardly mentioned. In this closing chapter, then, we will allocate some overdue space to the rise of the digital age and with that an increase of boundary-crossing and the emergence of hybrid learning configurations which we suspect will affect the way we do research and influence future environmental education research agendas.

It would be difficult to argue against an observation that we are living in times of accelerating environmental, technological, and cultural change, resulting in increased levels of complexity, uncertainty and contestation of claims (Wals & Corcoran, 2012). Many environmental changes are connected to major anthropogenic impacts on the planet's ecosystems and associated life-support systems, thereby increasing levels of risk to all life-forms (Beck, 1992, 2008) and necessitating mitigation and adaptive responses. Our focus here, however, is on changes in the way we interact, communicate and learn (which is not to say that the latter changes do not have an impact on the former).

We begin with a few examples of these critical changes. Although the times have always been changing, the speed and reach of change now is unique in history as we have entered the digital age, a time of hyperconnectivity, expansive mobility and the condensation of time and space. Tredinnick (2008) provides a rich and thorough analysis of the complexities and the impact of digital technology on the way we live. These complexities and impacts are not confined to the Western world (United Nations Global Alliance for ICT and Development, 2009). Kenyan farmers who, due to shifts in climate and changing weather patterns, can no longer rely on the return of certain migratory birds to determine when to plant their seeds, now use cell phones to find localized weather forecasts to aid their decision making. The Doppler radar-based weather report is rapidly replacing the farmers' indigenous understanding of the weather and the planting seasons. In Ghana, *Esoko*, a widely used digital agricultural market information platform has been created as a response to the explosive growth of cellular services in the country. *Esoko* is managed on the web and delivered via cell phones. Individuals, agribusiness, government, and projects use *Esoko* to collect and send out market data using text messaging. The *Esoko* platform provides automatic and personalized price alerts, buying and selling offers, bulk SMS messaging, stock counts and SMS polling. [For more examples of ICTs impacting the lives of the rural poor living in non-Western increasingly globalized contexts, see, for instance, *The Coming Revolution in Africa* (Zachary, 2008) and Kituyi-Kwake and Adigun (2008)].

Another example of how society is changing is that many of us are losing our spatial orientation ability. We no longer need to develop a sense of place as we increasingly use GPS to work out how to get from one place to another (see, e.g., Girardin & Blatt, 2010). A final, related, example is that in the Netherlands, the vast majority of school children still cycle to school but increasingly wear "ears" and headsets and many text or phone as they cycle. These young people seem to pay less and less attention to the physical world that glides by. Given this hyperconnectivity, it is no surprise that people are becoming more and more disconnected from nature—in part, because nature seems to be left behind in our high-paced digital age.

To function in the digital age requires multi-tasking, whereas connecting with the natural world requires focused attention and engagement of all our senses and capacities. There is a paradox here that needs to be addressed. On the one hand there are claims of "psychic numbing" or overstimulation that diminishes our capacity to feel (van Boeckel, 2009). As Richard Louv, author of *Last Child in the Woods* argues: "Nature is increasingly an abstraction you watch on a nature channel" (Louv, 2005). So one might conclude that "biophilia" is being displaced by "videophilia," a word that refers to the new human tendency to focus on sedentary activities involving electronic media, such as watching TV/movies, surfing on the Internet or playing video games (Zaradic & Pergams, 2007). Yet, on the other hand, electronic media provide access, not previously available to millions, to a wide range of natural phenomena and environmental issues that stimulate young people's interests in the nonhuman world and its conservation. In addition, new forms of communication and social networking contribute to the creation of learning communities and political movements for taking action on environmental issues although one could argue that the digital age, so far, has served the economy better than the environment as it has also led to easier access to consumer goods and consumption through, for instance *e*-based advertising and marketing (McAllister, 2011).

From our own professional world we know that academics rarely visit libraries or use interlibrary loan systems when they want to read literature in their field. Instead, they conduct a search using Google or Google Scholar to get an instant result of reasonable quality that usually provides them with the information and references they need for their next paper. Information appears to be everywhere and available from almost anywhere. This increased accessibility may lead to improved efficiency and productivity but does not necessarily provide more meaningful results or deeper understanding. In the words of biologist E.O. Wilson: "We are drowning in information while starving for wisdom" (Wilson, 2001, p. 269).

One undesirable consequence of hyperconnectivity is that misinformation and unfounded claims rapidly spread, sometimes virally. Time, careful thought, and reflexivity—characteristics not often associated with the rapidity of the digital age—are required to sort through vast and often conflicting sources of information and knowledge claims. As a result, the public can become confused about environmental issues, a situation which some argue leads to inaction justified by the question: "Why should I change as long as scientists do not agree about what is actually happening to the planet?"

This contestation of knowledge occurs daily in all fields of science but it is particularly common in areas dealing with health, biodiversity, climate change, poverty alleviation, natural resource management, environmental management, sustainable development and so on. This uncertainty leaves the public, at best, more critical and understanding of the dynamic and political nature of knowledge generation and science but, at worst and more likely, confused, cynical and disengaged. Again this situation poses a challenge to environmental educators and environmental education researchers in terms of how they might handle controversy and contestation (Oulton, Dillon, & Grace, 2004).

Another consequence of the hyperconnectivity of the digital age is a reconfiguring of the interpersonal "structure" of society as connections between people multiply and change in nature. In the digital age, people's perceived competence and social capital are a product of the connections they have formed (Siemens, 2005). Siemens cites Karen Stephenson who argues that: "Experience has long been considered the best teacher of knowledge. Since we cannot experience everything, other people's experiences, and hence other people, become the surrogate for knowledge. 'I store my knowledge in my friends' is an axiom for collecting knowledge through collecting people (Stephenson, undated)." The process of "collecting people" occurs daily through social networks such as Facebook and LinkedIn. This process also represents a form of boundary-crossing in that through a network of nodes and relationships people gain access to a wider world of increased possibilities and expansive choices. Again, these restructurings of learning, communication, and social relations provide both opportunities and challenges to environmental education researchers. This *Handbook*, for example, would not have been possible without the editors drawing on their networks of colleagues and collaborators.

Implications for Environmental Education Research

As we have already suggested, environmental education researchers will need to consider these shifts in knowledge creation, human interaction, and human-environment interactions if their research is to be relevant to our changing times. An obvious question that arises is: What might be some of the implications of living and learning in times of accelerating change, uncertainty, risk, complexity and contestation of knowledge for environmental education research? We offer three topics for consideration: (i) connecting biophilia and videophila, (ii) creating spaces for hybrid learning, and (iii) strengthening engaged scholarship with a planetary conscience. The first two topics follow from our brief analysis of how the digital age is shifting the landscape of environmental educators and environmental education researchers, the third topic follows from the need for transformative research with societal relevance that many authors in this *Handbook* explicitly or implicitly call for.

Connecting Biophilia and Videophilia As the use of ICTs and the worldwide web have become woven into the daily fabric of our lives, we may reach a stage, if we have not already, where we cannot fathom our life without them. The danger herein lies in that, as much as we are a technologically literate species, we are also animate human beings who have come of age through deep and intimate daily contact with each other and with an embodied, physical, natural and, often, wild world. This is a point that echoes Phillip Payne's argument for redirecting our environmental education inquiries toward the aesthetics of embodied experience, including turning our gaze toward how culture inscribes bodies.

One approach that environmental education researchers might consider would be to study ways in which these ever-present technologies and cyberspaces can be used to help people (re)gain a deeper and more empathetic contact with each other and with the world. For instance, when looking at the hundreds of thousands of apps that have been developed in the past few years one might conclude that most of them have been designed to make life easier (convenience), more enjoyable (entertainment) or to tempt people to buy something new or extra (consumption). Yet different kinds of apps have also being developed with a natural and environmental focus: to help people save fuel and reduce their carbon footprints (e.g., carpool apps, home energy apps), to aid young and old to get to know their physical surroundings better (e.g., plant and animal recognition apps, Monument apps, which provide rich information about phenomena encountered); to help children to learn about the natural world (e.g., biodiversity apps); and to encourage people to eat healthier and more responsibly (e.g., FishPhone apps, Foodprint apps).

To what extent, though, are app designers utilizing the knowledge generated by environmental education researchers about how people learn about these issues? And conversely, to what extent are environmental education scholars investigating the potential of these apps for fostering deep learning about environmental issues? What are the kinds of collaborative partnerships that could be created to contribute to the design of high quality, engaging and authentic apps that could enable citizens to investigate environmental issues, help monitor changes in the environment and provide data that can benefit environmental and environmental education research?

Another possibility to consider is exploring the implications and potential of new technologies such as the use of geospatial technologies, haptic technology, web-based applications, such as Google Earth, and social networking. Here, too, environmental educators and researchers alike need to ask what opportunities and, indeed, as discussed earlier in this chapter, what threats do these phenomena offer them? The opportunities come with some questions that have already been raised elsewhere in this book such as: How does culture mediate the influence of technological environments on (children's) development of sense of place or in Thomashow's (1995) terms, ecological identity? What kinds of further inequities and environmental injustices might arise when people familiar with the digital world benefit more from access to new technology-driven learning environments?

Creating Spaces for Hybrid Learning Learning, in what Beck calls a "risk society" (Beck, 1992, 2008) requires "hybridity" and synergy between multiple actors as well as the blurring of outdated notions such as formal and informal education. Opportunities for new types of learning expand with an increased permeability between disciplines, generations, cultures, institutions and sectors. Through this hybridity and synergy, new spaces might open up that would allow transformative learning including intergenerational and intercultural learning to take place.

Such spaces might open up a range of opportunities: alternative paths of development; new ways of thinking, valuing and doing; participation minimally distorted by power relations; pluralism, diversity and minority perspectives; deep consensus (but also for respectful disagreement and differences); autonomous and deviant thinking; self-determination; and contextual differences. "Transformative" here refers to a shift or a switch to a new way of being and seeing O'Sullivan, 2001). Environmental educators have an important role to play in conceptualizing these "hybrid learning configurations" (Wals, Kupper, & Lans, 2011) and in monitoring and evaluating the kind and quality of learning taking place within those configurations, and developing a better understanding of the way in which they can be designed and supported.

The increasing complexity of societies, the interdisciplinary nature of people-society-environment relationships, the problems faced at local and global scale, and the uncertainty of their solutions or resolutions also call for new spaces for collaborative approaches to research. Consequently, the classic distinctions and boundaries between different educational sectors, especially the formal and informal, are increasingly breaking down as the former, at its worst, is recognized as being too structured and predetermined and therefore constraining and insufficient for addressing sustainability issues.

Technological innovations and the resulting hyperconnectivity have expanded the sites at which education and learning occur with a concomitant decrease in the significance of the role of formal education and schooling and the emergence of popular culture, in the form of the Internet, music, film/video, television, as educational text or "places" of information acquisition and knowledge production for young people (Weis & Dimitriadis, 2008). Physical-virtual-community-school hybridizations of learning spaces are emerging and are likely to accelerate in terms of their diversity of form and the extent of their educational significance. Similar hybridized environments and new spaces are also needed for environmental education research that embraces the authenticity of multiple voices and cultural and theoretical perspectives, new forms of representation, and more activist, community-based approaches. Such research also assumes different boundaries between science and society, researcher and researched, theory and practice, and other, what might come to be seen as false, dichotomies. Boundary-crossing and the blurring of boundaries appear essential to allow for new forms of understanding and knowledge creation to emerge (Davidson-Hunt & Berkes, 2003; Wals, 2010).

Strengthening Engaged Scholarship with a Planetary Conscience A question for academics working in a globalizing university system, is whether our institutions provide sufficient space to develop and strengthen these transformative niches in both research and education that allow for hybrid learning between multiple actors to emerge. Today's universities are, to some extent, caught in between two trends—a rather hegemonic one and a somewhat marginal but emerging one. The hegemonic trend very much builds upon a model of fragmentation, management, control, and accountability. Elsewhere Peters and Wals (2012) refer to this as "science as commodity." The emerging but still marginal trend they refer to as "science as community." The latter conceptualization of science is based on integration, autonomy, learning, and reflexivity. The "science as community" perspective is born out of a recognition that the nature of the types of issues needing to be addressed is such that more of the same will not do the trick (Jones, Selby, & Sterling, 2010).

The emphasis on academic output, as expressed by the number of peer-reviewed articles published in high impact ISI journals, as is often the case within a "science as commodity" perspective, arguably leads to a disconnect between universities and the communities they are supposed to serve (Unterhalter & Carpentier, 2010). This phenomenon is propelling a countermovement or, perhaps we might say a "rebirth" as such scholarship is not new, of what is called "civically engaged scholarship." Engaged scholarship tends to focus on existentially relevant issues, the cocreation of knowledge with different groups in society in which scientific knowledge is one form of knowledge among several others (e.g., local and indigenous knowledge), in which the quality of the scholarly work is not just assessed by scientific peers but by an extended group that includes other societal groups, particularly those affected the most by the research. Finally, such scholarship recognizes that research in itself has a pedagogical or emancipatory end in that it seeks to affect change in the situations or issues at stake and the people involved.

Wals and Dillon point out that this countermovement is illustrated by the revival of university science shops (www.livingknowledge.org) and by the emergence of a number of new networks of community-engaged universities (e.g. the Talloires Network) and centers of expertise focusing on sustainability issues such as the Regional Centres of Expertise (RCEs) in which universities are partners in a network of NGOs, civil society organizations, community groups, schools, etc. (Mochizuki and Fadeeva, 2008). Wals and Dillon (*ibid.*) mention that within these networks and centers, a range of participatory forms of research can be found, such as: action research (Reason & Bradbury, 2001), citizen science (Irwin, 1995), and transdisciplinary research (Hirsch Hadorn et al., 2008).

Several contributors to this *Handbook* have emphasized ideas underpinning engaged scholarship, without, however, using the term. They have also referred to the generation of trust and transformation using different but related concepts and methodologies such as feminist poststructuralism, critical eco-pedagogies, collaborative ecological inquiry, and critical action research or by seeking to overcome the marginalization of certain groups in society. Stevenson explicitly speaks of a "civically engaged scholarship" which is not only concerned with the production of knowledge but also with reflecting on its use and impact and expanding the (public) dialogue or scope of conversation beyond the walls of the academy to all the many publics in any society. These contributors to the *Handbook*, and others, join a much larger community of engaged (educational) researchers seeking justice, transformation, democracy, equity, and so on. What is important to note though, is that they still occupy a crucial niche within that critical and engaged community in that they do their work with, to paraphrase David Orr (2004), the Earth in mind. To do that also requires a "planetary conscience" and the ability to develop an empathic understanding of the nonhuman and the more-than-human world, as several contributors in this *Handbook* emphasize.

Where Next for Environmental Education Research?

A concern about the environment and unsustainability is no longer the exclusive domain of an activist movement in society or of a well-demarcated niche in science. There appears to be a movement from "margin to mainstream." In the world of business, for instance, and in vocational education "sustainability," "corporate social responsibility" and "the triple bottom line of People, Planet and Profit/Prosperity (Triple P)" are becoming part of the conversation, sometimes for strategic reasons, to help solidify old principles and routines, and sometimes out of a genuine concern for the well-being of the planet and the realization that old principles and routines are no longer tenable.

In the world of science, many journals which traditionally paid little attention to environmental and sustainability issues are now publishing special issues on those topics. High status educational journals, such as the *Journal of Curriculum Studies*, regularly publish research on curriculum development, teaching, and learning in the context of environmental education and education for sustainable development (ESD). In the years 2009 and 2010, seven research journals outside the field of environmental education or ESD published a special issue on sustainability and/or sustainable development (e.g., *International Review of Education*). More recently, in 2011, the *American Psychologist*, the flagship journal of the American Psychological Association, published a special issue on "Psychology and Global Climate Change" (Volume 66, number 4). Closely related to the field of environmental education new journals have emerged such as the *Journal of Sustainability in Higher Education*.

Partly as a result of the widening interest in and concern about issues of nature, environment, climate change, biodiversity, sustainability, etc., scholars from within the environmental education research are stepping beyond the field to reach new audiences via journals beyond the traditional set of environmental education research journals. These shifting publishing trends raise the question: where does this leave environmental education research as a separate entity in the social sciences? Has it played its role in history and have we now arrived in a new phase where researchers with quite different backgrounds but with a common interest in understanding these existentially relevant issues and exploring ways to address them, are all engaged in what we might call environmental or sustainability education research, but what they might call something else? Will all education become environmental (Bowers et al., 2011; Orr, 2004). Or is there still that special niche that was created over forty years ago when the *Journal of Environmental Education* was first launched in North America?

With the publication of this, the first *International Handbook of Environmental Education Research* supported by the American Education Research Association, it seems a niche still exists. Environmental education research continues to adapt to the rapidly changing external circumstances and to grow in strength as the years go by. Its evolution is also driven and guided by critically reflective and reflexive practices which ensure that the field remains open and responsive to change.

The scale of the changes that have taken place emerges from the pages of the volume. The research described and analyzed in the preceding fifty chapters covers a wide range of methodologies and involves a number of traditional and innovative methods. Its quality varies substantially and that variation is one of the major challenges facing the field. It is a challenge that has been picked up by the editors of the major environmental education research journals and it will no doubt continue to exercise their minds in the years to come.

We have offered some thoughts about future directions. Our belief is that researchers would benefit from turning their gaze to critically examine the ways in which new and evolving technologies and cyberspaces can be used to help people (re)gain a deeper and more empathetic contact with each other and with the world. We believe that hybridized environments and new spaces are needed for environmental education research that would embrace the authenticity of multiple voices and cultural and theoretical perspectives, new forms of representation, and more activist, community-based approaches. Finally, we believe that there is a need for the field to adopt a critically engaged scholarship that would focus on existentially relevant issues and the cocreation of knowledge with different groups in society.

As we have argued above, such scholarship recognizes that environmental education research has a pedagogical or emancipatory end in that it seeks to affect positive change in the situations or issues at stake and the people involved. Although it is true that nothing dates faster than yesterday's visions of the future, we suggest that the agenda that we have identified offers new directions for the field and fresh challenges for all who work in it. The stakes could not be higher.

Acknowledgment

We would like to thank Ellen Field for her helpful comments on an earlier version of this chapter.

References

Beck, U. (1992). *Risk society: Towards a new modernity*. London, UK: Sage.

Beck, U. (2008). *World at risk*. Cambridge, UK: Polity Press.

Bowers, C. A., Jucker, R. Ishizawa, J., & Rengifo, G. (2011). *Perspectives on the ideas of Gregory Bateson, ecological intelligence, and educational reforms*. Eugene, OR: Eco-Justice Press.

Davidson-Hunt, I., & Berkes, F. (2003). Learning as you journey: Anishinaabe perception of social-ecological environments and adaptive learning, *Ecology and Society, 8*(1), 5. Retrieved August 19, 2011, from http://www.ecologyandsociety.org/vol8/iss1/art5/

Freier, N. G., & Kahn, P. H., Jr. (2009). The fast-paced change of children's technological environments. *Children, Youth and Environments, 19*(1), 1–11.

Girardin, F., & Blat, J. (2010) The co-evolution of taxi drivers and their in-car navigation systems. *Pervasive and Mobile Computing, 6*(4), 424–434.

Hirsch Hadorn, G., Hoffmann-Riem, H., Biber-Klemm, S., Grossenbacher, W., Joye, D., Pohl, C., . . . Zemp, E. (Eds.). (2008). The emergence of transdisciplinarity as a form of research. *Handbook of Transdisciplinary Research* (pp. 19–39). Dordrecht: Springer.

Irwin, A. (1995). *Citizen science: A study of people, expertise, and sustainable development.* London: Routledge.

Jones, P., Selby, D., & Sterling, S. (Eds.). (2010). *Sustainability education: Perspectives and practice across higher education.* London: Earthscan.

Kituyi-Kwake, A., & Adigun, M. (2008). Analyzing ICT use and access amongst rural women in Kenya. *International Journal of Education and Development using Information and Communication Technology, 4*(4), 127–147. Retrieved August 17, 2011, from ProQuest Education Journals. (Document ID: 2412007081).

Louv, R. (2005). *Last child in the woods: Saving our children from nature-deficit disorder*. Chapel Hill, NC: Algonquin Books.

McAllister, M. (2011). Consumer culture and new media: Commodity fetishism in the digital era. In S. Papathanassopoulos (Ed.), *Media perspectives for the twenty-first century*. London, UK: Taylor & Francis.

Mochizuki, Y., & Fadeeva, Z. (2008). Regional centres of expertise on education for sustainable development (RCEs): An overview. *International Journal of Sustainability in Higher Education, 9*(4), 369–381.

Orr, D. (2004). *Earth in mind*. Washington, DC: Island Press.

O'Sullivan (1999). *Transformative learning: Educational vision for the twenty-first century.* Toronto, ON: OISE/UT Press.

O'Sullivan (2001). *Expanding the boundaries of transformative learning*. New York, NY: Palgrave.

Oulton, C., Dillon, J., & Grace, M. (2004). Reconceptualizing the teaching of controversial issues. *International Journal of Science Education, 26*(4), 411–423.

Peters, S., & Wals, A. E. J. (2012). Learning and knowing in pursuit of sustainability: Concepts and tools for trans-disciplinary environmental research. In M. Krasny & J. Dillon (Eds.), *Trans-disciplinary environmental education research.* London: Taylor & Francis.

Reason, P., & Bradbury, H. (2001). *Handbook of action research: Participative inquiry and practice*. London, UK: Sage Publications.

Siemens, G. (2005). *Connectivism: Learning as network-creation.* Printed online by The American Society for Professional Development (ASPD). Retrievable February 11, 2011 from www.astd.org/LC/2005/1105_seimens.htm

Thomashow, M. (1995). *Ecological identity: Becoming a reflective environmentalist*. Cambridge, MA: MIT Press.

Tredinnick, L. (2008). *Digital information culture: The individual and society in the digital age*. Witney, UK: Chandos Publishing.

United Nations Global Alliance for ICT and Development (2009). *White Paper Information & Communication Technologies (ICT) in Education for Development*. (Publication No. 09–48710). New York, NY: UN Printing Office.

Unterhalter, E., & Carpentier, V. (Eds.). (2010). *Global Inequalities and Higher Education: Whose interests are we serving?* London, UK: Palgrave.

Van Boeckel, J. (2009). Arts-based environmental education and the ecological crisis: Between opening the senses and coping with psychic numbing. In B. Drillsma-Milgrom & L. Kirstinä (Eds.), *Metamorphoses in children's literature and culture* (pp. 145–164). Turku, Finland: Enostone.

Wals, A. E. J. (2010). "Message in a bottle: Learning our way out of unsustainability," Inaugural address held upon accepting a Professorship and UNESCO Chair in Social Learning and Sustainable Development, May 27, 2009, Wageningen, the Netherlands, Wageningen University Publications, Wageningen, accessible via:http://groundswellinternational.org/2010/12/14/message-in-a-bottle-learning-our-way-out-of-unsustainability/, last accessed August 3, 2011.

Wals, A. E. J., Kupper, H., & Lans, T. (2011). Blurring the boundaries between vocational education, business, and research in the agri-food domain. *Journal of Vocational Education and Training, 64*(1), 3–23.

Wals, A. E. J. & Corcoran, P. B. (2012). *Learning for sustainability in times of accelerating change*. Wageningen: Wageningen Academic Publishers.

Weis, L., & Dimitriadis, G. (2008). Dueling banjos: Shifting economic and cultural contexts in the lives of youth. *Teachers' College Record, 110*(10), 2290–2316.

Wilson, E. O. (2001). *Consilience, the Unity of Knowledge*. New York, NY: Alfred Knopf.

Zachary, G. P. (2008). The coming revolution in Africa. *The Wilson Quarterly, 32*(1), 50–66.

Zaradic, P. A., & Pergams, O. R. W. (2007). Videophilia: Implications for childhood development and conservation. *Journal of Developmental Processes, 2*, 130–144.

Author Index

Note: Page numbers followed by "n" refer to footnotes.

A

Abd-El-Khalick, E., 310
Abdi, A., 195
Abegg, G.L., 331
Abram, D., 95, 98, 117, 118, 135, 426, 428, 431
Abrams, C., 363
Abrams, D., 410
Absolon, D., 406
Acampora, R., 432, 435n7
Achebe, C., 386, 387
Ackett, W.A., 247
Adami, C., 500
Adams, M., 363
Adamson, J., 400
Adelman, L.M., 360, 363
Adey, P., 254
Adigun, M., 543
Adomssent, M., 354
Afonso, A.P., 353
Agee, J., 422
Ageyman, J., 317
Agger, B., 380
Agnostopoulous, D., 153
Agrawal, A., 384
Agyeman, J., 17, 24, 55, 151, 201, 266, 394–400, 523
Aikenhead, G., 206, 207
Aikenhead, G.S., 29
Ainscow, M., 461
Airasian, P., 49
Aivazidis, C., 196
Ajzen, I., 264, 268–9, 301
Åkerman, M., 135, 518
Akerson, V.L., 335, 344
Albas, A.H., 19
Alblas, A., 199
Alderson, P., 439–42, 454
Aldrich, C., 376
Alexander, J.C., 231
Alexander, P.A., 247
Alibrandi, M., 332, 339, 344
Allan, S., 487
Alldred, P., 449
Allen, J.B., 264
Almquist, H., 335, 342
Al-Ruz, J.A., 264
Alsop, S., 246, 398
Álvarez, A., 175
Alvesson, M., 225, 226
Amaral, L.A.N., 505n2
Ames, R.T., 34
Anderson, C.W., 280
Anderson, D., 163–6, 360
Anderson, G., 469, 471
Anderson, J.R., 360
Anderson, S., 337
Anderson, V., 487
Anderson, W., 35
Andrade, B., 172
Andree, M., 206
Andrews, C., 350
Andrews, K., 201
Andrews, S.B., 343, 399
Ángel, M.A., 174
Anscombe, G.E.M., 227
Anthwal, A., 263
Anthwal, S., 263
Antikainen, A., 138
Appadurai, A., 277
Apple, M., 102, 535
Araya, D., 372, 505n1
Archer, M., 428
Arden, M.A., 269
Ardoin, N., 306, 513, 523
Ardoin, N.M., 24, 262, 268, 304, 307
Arenas, A., 163, 166, 167
Arendt, H., 108, 111
Argyris, C., 258, 462
Arick, J.R., 371
Armitage, C.J., 268, 269
Armon-Jones, C., 206
Armstrong, J.C., 404
Armstrong, V., 223, 224
Arnold, G., 349
Arnold, H.E., 290, 291
Arnstein, S.R., 441
Arruda, R.V., 232
Ary, D., 46
Asafo-Adjei, R., 390
Asch, J., 300
Aspinall, P., 291, 292
Astor-Jack, T., 364
Audet, R., 334
Audet, R.H., 331, 335, 336, 343
Avanzi, M.R., 173
Averill, J., 206
Avital, E., 500
Avramidis, E., 371
Ayair, U., 115, 116
Azevedo, F.S., 360

B

Baacke, D., 443, 454
Babulski, K., 321, 323
Bachelard, G., 427
Bacon, D.R., 264
Baddeley, A., 294
Bai, H., 101–5, 412, 413, 421, 430
Bailey, L., 52
Bajd, B., 290, 291
Baker, T.R., 335, 339, 340, 343, 344
Balakrishnan, U., 50
Balick, M.J., 387
Ballantyne, R., 199, 244, 268, 360, 363, 452
Ball, S., 109, 147, 148, 150
Ball, S.J., 222
Ball, T., 115, 119
Bamberg, S., 47, 57, 266, 269, 306, 311
Bandura, A., 268, 269
Bangert A., 360
Banilower, E.R., 339
Banks, M., 225
Bansel, P., 482
Barabasi, A.-L., 500
Barab, S., 340, 346
Barab, S.A., 333
Bar, H., 316
Barlage, J., 360
Barlett, P., 354
Barnes, D., 27
Barnett, E.G., 262
Barnett, M., 331, 333, 336, 342–4
Barnhardt, R., 405, 406
Barratt Hacking, E., 247, 438, 439, 443, 448–9, 454–6
Barratt, R., 247, 438, 439, 443, 448, 449, 454–6
Barrett, M.J., 198, 372, 375, 378, 380, 414
Barron, D., 378
Barroso, J.C., 176n2
Barrow, R., 69
Barth, M., 354, 355
Bartlett, C., 406
Bartlett, S., 443
Bartosh, O., 216
Bassey, M., 46
Basso, K., 93
Bateson, G., 89, 500
Batterbury, S., 276, 277
Battersby, J., 247
Battiste, M., 404, 405
Baud, R., 355
Baudrillard, J., 227
Bauman, Z., 18, 83, 116, 117, 119
Bawden, R., 352
Bawtree, V., 388
Beane, J.A., 247
Beatty, A., 222

Bechtel, R.B., 264
Beckett, K.L., 336, 343
Beck, U., 112, 256, 542, 544
Bednarz, R., 337, 343
Bednarz, S.W., 332, 336, 344, 345
Beeman, C., 80, 82, 83
Beers, L., 267
Beers, M., 226
Bekessy, S., 351
Bell, A., 375, 378, 412, 414
Bell, A.C., 141, 369, 371
Bellah, R., 213
Bell, D., 135, 518
Bell, N., 440, 454
Bell, P., 340, 360
Bem, D.J., 263, 268–70
Benasayag, M., 138
Benayas del Álamo, J., 176n2
Benayas, J., 175, 176n2
Bender, M., 371
Ben-Hur, Y., 316
Bennell, P., 164
Bennett, D., 52
Bennett, S., 198
Benton, G.M., 264, 293, 294
Berenguer, J., 263
Berg, A., 94
Berg, B.L., 208
Berger, P., 414
Berger, P.L., 142
Beringer, A., 63
Berkes, F., 253, 258, 545
Berkowitz, A., 334
Berk, R., 27–9
Berlin, B., 37
Bernard, H., 46
Bernstein, B., 194, 195
Berryman, T., 79, 139
Berry, T., 98
Berry, W., 93, 94
Bertely Busquets, M., 172
Besley, T., 501, 505n1
Best, H., 264
Betz, A.L., 294
Bexell, S.M., 264
Bhabha, H.K., 42n25, 194
Bicker, A., 385, 387
Bickman, L., 306
Biersta, G., 195, 198, 199
Biesta, G., 90, 109–11, 523, 530
Billig, S., 218
Birnbaum, M., 298, 307
Bishop, K., 152
Bizerril, M.X.A., 278
Bjerke, T., 264
Blackmore, S., 227
Blanchard, K., 264
Blankenship, G., 50
Blank, M., 94
Blat, J., 543
Blaug, M., 164
Blenkinsop, S., 80, 82, 83
Blevins, J.E.K., 50
Blewitt, J., 354
Blieck, A., 299
Blühdorn, I., 207, 208, 210, 535
Bluhm, W., 302, 316, 321
Blum, A., 316
Blumenfeld, P.C., 340
Blum, N., 429
Bodzin, A.M., 336, 342, 344, 375
Bogan, M., 312
Bögeholz, S., 24, 264
Bogner, F.X., 244, 269, 289, 300
Bohl, W., 310
Boje, D., 467
Bolak, H.C., 406
Boldero, J., 378
Boler, M., 489
Bolscho, D., 24
Bond, N., 268
Bonnes, M., 264
Bonnett, M., 87–90, 112, 119, 126
Bookchin, M., 135
Booth, A.L., 50
Booth, T., 461
Borda, O.F., 234
Borden, K.A., 334
Borden, R., 264, 267
Bordo, S., 104, 372
Borgmann, A., 101
Borgon, L., 442, 449, 454
Borradori, G., 227
Borun, M., 360
Boujaoude, S., 310
Boulding, K.E., 502
Bourdieu, P., 224, 522, 524
Bourke, S., 264
Bousquet, W., 55
Boutin, F., 166
Bowe, R., 147, 148, 150
Bower, C., 276
Bowers, A.W., 307
Bowers, C., 136
Bowers, C.A., 80, 96, 99n3, 135, 167, 168, 396, 412, 518, 546
Boyden, S., 15
Boyden, S.V., 14
Boyle, R.A., 246
Bradbury, H., 259, 460, 462, 465, 466, 467, 472, 545
Bradbury-Huang, H., 459, 461, 467
Brady, I., 435n11
Brady, J.C., 280
Brady, J.K., 280
Brandão, C.R., 234
Brandt, C.B., 275, 279
Brannan, S.A., 371
Bransford, J., 333
Bremner, A., 264
Brenan, A., 503
Brennan, M.J., 133–4, 139
Breu, K., 462
Breunig, M., 371
Brevitt, J., 473
Brewer, M.B., 263
Breyer-Brandwijk, M.G., 390
Briceno Lobo, C., 82
Bridges, D., 523, 524
Briggs, H., 247
Bright, A., 201
Brinkerhoff, J.D., 339, 340
Britzman, D., 415, 489
Brody, M., 1, 137, 244, 278, 279, 285, 360, 363, 412, 512, 542
Brody, S.D., 63
Bronfenbrenner, U., 443
Bronkesha, D., 387
Bronnenkant, K., 360, 364
Brook, A., 247
Brookfield, S., 27
Brossius, J.P., 387
Brown, A., 333
Brown, A.L., 340, 346
Brown, C., 428, 431
Browne-Nuñez, C., 264
Brown, M., 332
Brown, M.L., 51
Brown, T.J., 264
Brown, V.A., 376, 378, 399
Brulle, R.J., 135, 395, 399
Brunelle, R., 139
Bucheit, J., 45, 46, 48–50
Buehl, M.M., 247
Buenfil-Burgos Nidia, R., 413
Buhrs, T., 50
Bulkeley, H., 395
Bullard, R.D., 394
Burbules, N.C., 27–9
Burr, S.W., 264
Burt, J., 375
Burt, J.S., 294
Buss, A., 339–41
Butcher, J., 71, 78, 79
Butcher, K., 487
Butera, F., 210
Butler, J., 480, 481
Butler, P.J., 298
Butler, T., 279
Buttimer, A., 444
Buxton, C., 219
Bybee, R., 24

C
Cabral, N.R.A.J., 232
Cairns, K., 454
Cajete, G., 404, 405
Calabrese, B.A., 278
Calabrese Barton, A., 219
Calixto, R., 172
Calle-Gruber, M., 480–2, 485
Callicott, J.B., 33, 410
Calmarco, M., 434n2
Calvino, I., 135
Camargo, C., 306, 307
Camargo, R.F., 378
Campbell, C., 123, 126, 134, 156
Campbell, D.T., 294, 300, 303
Canhedo, S.G. Jr., 232
Cannella, G., 509
Cano, L., 176n2
Cano Muñoz, L., 176n2
Cans, R., 135
Cantrell, D.C., 18
Capra, F., 137, 258, 350, 500
Cardoso, R.C.L., 232
Carew, A.L., 197
Caride Gómez, J.A., 138
Carleton-Hug, A., 299, 306, 307
Carlsen, B., 334
Carlson, D., 152
Carlson, T., 344
Carman, T., 430
Carpentier, V., 258, 545
Carr, A., 264
Carrier, S.J., 375, 379
Carrus, G., 264

Carr, W., 161, 472
Carson, R., 14, 80
Carvalho, L.M., 173
Carvalho, P.F., 233
Casey, E., 93, 94, 432, 434n2
Castells, M., 499, 500
Castillo, A., 176n4
Castonguay, G., 398
Castro, M.L., 232
Cavalari, R.M.F., 173
Cavanagh, M., 472–3
Cavanagh, R., 404
Cavern, L., 196
Ceppi, G., 483, 484
Chadwick, J., 360
Chaiken, S., 263
Chamberlain, A., 361
Chamber, M.B., 360
Chambers, M., 360
Chambers, R., 387
Chandler, D.E., 465
Chandler, D.J., 375, 379, 398
Chang, C-Y., 180
Chang, M-H., 183, 186
Chang, N-C., 182
Chang, T., 180, 181, 183
Chan, K.K.W., 360
Chan, S.L., 182
Charles, V., 449
Charley, P.H., 399
Chatelain, L., 50
Chawla, L., 24, 47, 55, 59, 215, 290, 291, 371, 429, 444, 535
Chelimsky, E., 298
Chenery, M., 52
Cheney, J., 42n24, 71, 80, 83, 117–19
Chen, H.T., 305
Cherem, G., 47, 48
Cherryholmes, C., 33
Cherryholmes, C.H., 376
Chien, C.J., 360
Child, B., 264
Childress, R., 46, 55
Childs, A., 182, 185
Chinien, C., 166
Chinnery, A., 104
Chou, J., 179
Cho, Y., 257
Christensen, L., 306
Christensen, P., 439, 440, 454
Christianson, S.A., 294
Chua, P., 376
Chu, H., 312
Chu, H.-E., 210
Cilliers, P., 505n2
Civian, J., 349
Cixous, H., 480–2, 485
Clark, C., 412
Clark, J., 440
Clavijo, K., 306
Cleghorn, A., 360
Clinton, A., 349
Cloete, E., 19
Clover, D., 263
Clump, C., 364
Coates, P., 135
Cobern, M., 47
Cobern, W., 211
Cocking, R., 333
Cock, J., 119
Codd, J., 350, 471
Code, L., 39, 40
Coe, N.M., 279
Coffin, C., 360
Coffman, J., 306
Cohen, F.G., 290, 291
Cohen, L., 306
Cole, L., 395
Coleman, J., 216, 500
Cole, P., 135, 394, 397, 398, 400, 404, 411, 429
Collins, A., 476
Collins, D., 16
Colucci-Gray, L., 19
Conley-Tyler, M., 303
Connell, S., 17, 18, 57, 247, 530, 534, 536
Conner, M., 268, 269
Connors, M.M., 399
Cooke, K., 351
Cook-Sather, A., 439, 440, 449, 453–4
Cook, T.D., 294, 300, 303
Cooper, D., 523
Cooper, P., 245
Cooper, S., 432, 535
Coquet, J.-C., 135
Corcoran, P.B., 82, 290, 291, 542
Cordano, M., 264, 269
Corey, D., 48
Cornell, C., 234
Cornell, J., 412
Corneya, G., 353
Corney, G., 182, 185, 197
Corraliza, J.A., 263
Corral-Verdugo, V., 264, 306
Corsiglia, J., 406
Cortese, A., 351
Cortes, L., 310
Costa, V.B., 206, 207
Cotton, D., 351, 352
Cotton, D.R.E., 249
Cottrell, S.P., 269
Coulson, D., 362
Coulter, B., 339, 340
Coulter, D., 482
Courtenay-Hall, P., 267, 317
Covitt, B.A., 280, 301, 304, 305
Cox, P.K., 387
Coyle, K., 207
Craig, C., 461
Cranton, P., 257
Creswell, J.W., 298, 303, 406
Cronon, W., 135
Crossman, P., 384
Cross, R.T., 24
Crow, G., 449
Crowley, K., 364
Culen, G., 199, 267
Culen, G.R., 17
Cullingford, C., 354
Cunha, P., 233
Cunningham, D.J., 26
Curran, S., 223, 224
Currie, D.H., 454
Curthoys, L., 409
Curthoys, L.P., 405
Cushman, J., 71
Cuthbertson, B., 409
Cutter-Mackenzie, A., 399, 413, 438, 441, 452, 454
Czank, J., 371

D

Da Cruz, J.G., 378
da Cunha, P.R., 353
Dahlberg, G., 482–3
Daly, H., 502
Daly, H.E., 502, 505
D'Amato, L.G., 292
Dambekalns, L., 339
Danermark, B., 200
Danforth, S., 371
Daniel, B., 292
Danter, E.H., 263, 269
Darier, É., 76
Datnow, A., 148
David, M., 449
Davidson, A.L., 207
Davidson-Hunt, I., 253, 258, 545
Davies, B., 380, 480–5, 487, 491
Davies, P., 250
Davis, J., 472, 531
Davis, L., 371
Dawe, G., 353
Dawson, S.E., 399
Day, J., 245
de Alba, A., 172
Dean, M., 64, 66, 67
de Beauvoir, S., 381n1
de Carteret, P., 444
de Certeau, M., 518, 524, 535
Decortis, M., 428, 435n10
De Figueiredo, A.D., 353
de Freitas, D., 173
Dei, G.J.S., 384, 385, 387–8
Delaney, R., 223, 227
Delanty, G., 5, 69
Deleuze, G., 41n3, 75, 76, 227, 413, 481, 484
Delisio, E.R., 166
Delizoicov, D., 173
Deloria, V. Jr., 406
Delory-Momberger, C., 138
Del Pilar Jimenes Silva, M., 24
del Rey, A., 138
Delshad, A.B., 264
DeMers, M.N., 344
Demo, P., 232
Dendinger, R., 397
Denis, V., 488
Denzin, N., 57, 58, 306, 435n11, 440, 507, 509
de Oliveira, H.T., 173
Derrida, J., 226, 227
Dervisoglu, S., 264
Descola, P., 135
De Sherbinin, A., 264
Des Jardins, J.R., 119, 221
Desmarais, S., 267
DeVault, M., 143
Devlin, R., 371
de Vries, G.H., 530
Dewey, J., 27, 76, 87, 213, 428
Di Chiro, G., 375, 378, 379
Dickinson, L.E., 335, 344
Diegues, A.C., 232
Dierking, L.D., 359–60, 362–4, 365n1
Dietrich, K.A., 257

Dietz, T., 264–5
Dillon, J., 1, 13, 57, 74, 109, 185, 239, 243, 253, 259, 290, 303, 360, 371, 391, 412, 512, 520, 530, 542
Dimitriadis, G., 545
Dionne, J., 234
Dippo, D., 167, 398
Disinger, J., 46, 49, 52, 54, 55, 311
Disinger, J.F., 16, 17
Doering, A., 336, 338, 344
Doering, Z.D., 360
Doerr, M., 429
Dolnicar, S., 264
Donaldson, S.G., 407
Donnelly, M.P., 264
Donohoe, H.M., 264
Dorn, T., 46
Doubleday, N., 407
Douglas, J., 269
Downey, L., 395
Doyle, W., 245
Dritsas, J., 360
Driver, R., 247
Dubs, R., 27
Duckitt, J., 264
Duffin, M., 215, 218, 307
Duffy, T.M., 26
Duke, D., 333
Duke, D.F., 51
Dunlap, R.E., 263–5, 292
Dunlop, R., 427, 429
Dunwoody, S., 360
Duque-Aristizabal, A.M., 290
Durkheim, E., 213
Duvall, J., 304
Dwyer, W., 47
Dyball, R., 399
Dyer, P., 268
Dyment, J., 201
Dyment, J.E., 371, 398
Dyson, A., 440, 461

E
Eagly, A.H., 263
Easton, J., 298, 306, 307
Ebbutt, D., 299
Eblen, R.A., 33
Eblen, W.R., 33
Eckel, P.D., 349
Eckersley, R., 410, 415n2
Edelson, D.C., 332
Edgell, M.C.R., 264
Edwards, A., 166
Edwards, R., 449
Efklides, A., 246, 250
Egan, A., 264
Egan, K., 71
Ehrlich, P.R., 14
Eisen, A., 354
Eisley, L., 412
Eisner, E.W., 106, 138
Ekström, M., 200
Elder, J., 360
El-Hani, C.N., 499
Ellenbogen, K., 306
Ellenbogen, K.M., 364
Ellen, R., 387
Elliot, J., 198, 353, 470, 475
Ellsworth, E., 380
Ellwood, C., 483
Elmore, R., 148
El-Zein, A., 264
Embree, L., 427
Emekauwa, E., 217
Emmeche, C., 499
Encalada, M., 175
Engelhardt, J., 45
Engels, C.A., 304
English, K., 339
Ennis, R.H., 27
Entwistle, N., 245
Eppert, C., 195
Eraut, M., 471
Erdogan, M., 310, 323, 324, 326
Erduran, S., 29
Erevelles, N., 371
Erickson, F., 223, 245
Ernst, J., 299, 306
Escobar, A., 389
Estabrooks, C., 471
Esteva, G., 97
Esty, D.C., 264
Eulefeld, G., 24
Evans Braziel, J., 372
Evans, J., 15, 222
Evans, M.M., 400
Eveland, W.P., 360
Evernden, N., 102, 118, 135, 140, 410–12, 415n1
Everson, M., 293
Ewert, A., 291, 292, 295
Eyers, V., 310
Eyre, N., 264

F
Fadeeva, Z., 253, 545
Fadini, A.A.B., 231, 233
Fahey, J., 533
Fairbairn-Dunlop, P., 264
Fairclough, N., 141
Falk, J.H., 268, 359–60, 362–4, 365n1
Falomir, J.M., 210
Farahmandpur, R., 371
Farley, J., 505
Farmer, J., 264, 293, 294
Farrior, M., 360
Fassbinder, S.D., 371
Fawcett, L., 369, 371, 375, 378, 409–13, 415, 430, 488
Feaster, L., 339
Feder, M.A., 359
Fedler, A.J., 302
Felman, S., 489
Fensham, P., 23, 149
Ferguson, T., 135
Fernández, C.A., 172
Ferng, J-Y., 186
Ferrand, J.L., 264
Ferraro, P.J., 361
Ferreira, J., 63–5, 69, 74, 75, 198, 200, 414, 472, 473
Ferreira, J.G., 182
Ferreira, L.C., 232
Fidyk, A., 483
Fielding, M., 439, 454
Fien, J., 13, 18, 34, 36, 75, 129, 165, 195, 199, 247, 306, 350, 354, 355, 420, 429, 507
Filho, L.W., 351
Finger, M., 267
Finlay, K.A., 268
Finlay, L., 440
Fiore, Q., 41n7
Fischman, G.E., 222, 223, 228
Fishbein, M., 255, 264, 268–9, 301, 306
Fiske, J., 223
Fitzclarence, L., 353–4
Fitz-Gibbon, C.T., 306
Fivush, R., 294
Flammer, A., 210
Flamm, R., 301
Fleck, L., 173
Fleer, M., 438, 442, 443, 454, 456
Fleming, L., 298, 306, 307
Fleming, M.L., 306
Fletcher, R., 264
Flowers, A.B., 302
Flutter, J., 439, 440
Foat, K., 360
Folger, P., 331
Folke, C., 280
Fonstad, M., 337
Fontes, P.J., 375
Forster, J.B., 501
Foster, J., 201
Foster, J.S., 300
Foster, P., 20
Foster, P.J., 164
Foster, S., 395
Foucault, M., 13, 63–6, 94, 130, 375, 381, 483
Fournier, D.M., 298
Foutz, S., 359
Fox, H., 119
Fox, K., 428
Fox, P., 371
Fox-Parrish, L., 264
Frankam, J., 441
Frankel, J., 46
Fransson, N., 264
Fraser, B.J., 379
Fraser, N., 128–30
Frechtling, J., 298
Freeman, C., 339
Freeman, H.E., 298
Freeman-Moir, J., 371
Freier, N.G., 542
Freire, P., 236, 353
Fricker, A., 50, 352
Frick, J., 267
Friedan, B., 381n1
Friedman, A., 360
Friedman, T., 505
Frieze, I.H., 269
Fritsch, A.J., 264
Fromm, E., 101
Fruson, D., 487
Fuentes, A.S., 172
Fuhrer, U., 267
Fullerton, A., 371
Funderburk, R., 52
Furihata, S., 290, 291

G
Gaard, G., 399
Gabel, S., 371
Gage, J., 529
Gage, N., 507
Gallagher, S., 427, 435n14

Gallimore, R., 224
Gammon, B., 360
Gamoran, A., 215
Gannon, S., 480–3
Garb, Y., 310
Garibay, C., 364
Gärling, T., 264
Garvey, D.E., 289
Gaskell, G., 235
Gass, M.A., 289, 293, 295
Gassner, M., 292
Gauntlett, D., 225, 226
Gau, T.S., 179
Gayford, C., 231, 234, 307, 454
Gay, L., 49
Geddis, A.N., 29
Geertz, C., 93, 207, 222, 225, 226
Geller, S.E., 264
Gelobter, M., 51
Georgescu-Roegen, N., 502
Gerstenmaier, J., 26
Gessler, R., 25
Gibson, J., 278
Gibson, W.E., 399
Gil, A.C., 234
Gilbert Almeras, B., 371
Gilligan, C., 15
Giordan, A., 24
Giordano, A., 337
Girardin, F., 543
Gitlin, T., 504
Givvin, K.B., 223
Glasser, H., 256, 258
Gligo, N., 351
Glock, H-J., 74, 75
Godemann, J., 523
Godfrey-Smith, P., 500
Goetz, T., 207
Goldman, D., 302, 316
Goldman-Segall, R., 225–6
Goldsmith, E., 14
Goldstein, .T., 397
Gomez-Schmidt, C., 304
González-Gaudiano, E., 138, 139, 142, 152, 171, 172, 176, 176n4, 413, 427
Gonzalez-Gaudiano, E.J., 397
Goodson, I.F., 14, 15
Goodwin, R., 126
Goonatilake, S., 37
Gordin, D.N., 333
Gordon, C., 63, 65
Gordon, J., 400
Gough, A., 2, 13–16, 18, 19, 34, 63, 135, 147–9, 151, 152, 184, 185, 105, 198, 200, 290, 294, 350, 351, 369, 371, 372, 375, 378–80, 513, 525, 536, 537
Gough, B., 440
Gough, N., 14, 18, 19, 24, 33, 39, 42n22, 151, 197, 198, 200, 290, 371, 372, 375, 398, 409, 513, 519, 520, 537
Gough, S., 13, 78, 79, 83, 108, 135, 147, 150, 153, 154, 197, 201, 202, 243, 312, 349, 415, 522, 525
Goulet, L., 404
Grace, F., 380
Graff, E., 338, 343
Gramlich-Kaufman, L.M., 279
Grande, S., 97
Granfield, R., 153
Grange, J., 135
Granovetter, M., 500
Grant, M., 354
Grasseni, C., 280
Graveline, F.J., 404, 406
Graves, J., 360
Gray, C., 16
Greenall, A., 14, 18, 20n1
Greenall Gough, A., 18, 19, 198, 378
Greene, M., 487, 493
Greeno, J., 476
Greeno, J.G., 360
Greenwood, D., 97, 99n3, 372, 409, 505
Greenwood, D.A., 24, 93
Greig, S., 34
Grendstad, G., 317
Grenfell, M., 509, 510, 522
Griffiths, J., 430
Griffiths, P., 499
Grimwood, B., 407
Grob, A., 267
Grossenbacher-Mansuy, W., 25
Gross, L., 360
Gross, M., 48
Grosz, E., 278, 420, 426, 428, 431
Gruenewald, D., 80, 82, 83, 93, 95, 98, 219, 276
Gruenewald, D.A., 195, 201, 396, 444
Gruenwald, D., 153
Grundy, S., 469
Guajardo, F., 214
Guattari, F., 41n3, 75, 76, 131, 413
Guenther, L., 201
Gumede, M., 391
Gunckel, K.L., 280
Gupta, N., 263
Gur-Ze'ev, I., 136
Gustafson, J., 312, 320
Gustavsen, B., 465
Gustavson, L., 487, 493
Guthman, J., 372
Gutiérrez, J., 176n2
Gutwell, J., 361
Guzmon, K., 45

H

Haber, H.-F., 427
Habermas, J., 196, 467
Haber, S., 135
Hacsi, T., 298
Haeberli, R., 25, 26
Haglund, L., 244
Hahn, T., 280
Hallam, N., 378
Hall, B.L., 384
Halldén, O., 244, 245
Hallen, P., 375, 379, 415n4
Hall, R., 223
Hall, S., 222, 226
Hall-Wallace, M.K., 337, 343
Halsey, M., 484
Haluza-DeLay, R., 394, 399
Hamilton, D., 13
Hamilton, L., 128
Hamilton, M.C., 15, 20n1
Hammerman, D., 52
Hammerman, W., 52
Hammersley, M., 20
Hammond, W.F., 135
Hampel, B., 378
Hampton, E., 404
Ham, S., 52
Hancock, M., 496n6
Hanh, T.N., 105
Hanisi, N., 390
Hanke, U., 245
Hannigan, J.A., 256
Hanson, R., 293
Harako, A., 265
Haraway, D., 369, 412, 414
Haraway, D.J., 467
Hardin, G., 14
Harding, S., 36, 39, 375, 376, 378, 380, 536
Hardt, M., 503
Hardy, J., 198, 200
Hare, M., 257
Hargreaves, A., 515
Hargrove, E., 24, 75
Harré, R., 206
Harrison, P., 101
Harrison, R., 135
Hart, E., 311
Hart, E.P., 45–7, 49, 52, 54, 56–8
Hart, P., 2, 13, 16–19, 24–7, 29, 57, 69, 74, 134, 136, 151, 176, 196–7, 200, 243, 290, 291, 306, 311, 350–2, 369, 372, 410–13, 419, 421, 429, 430, 432, 434, 434n1, 476, 507, 513–16, 523, 525, 529, 531, 536, 537
Hart, R., 94
Hart, R.A., 439, 441, 443, 449, 454
Harvey, G., 311, 312, 321
Harvey, O.J., 263
Hatakeyama, M., 291
Hatcher, A., 406
Hattingh, J., 79, 116, 117
Hautecoeur, J.P., 167
Hawken, P., 459, 467
Hawthorne, R., 460
Hawthorne, S., 34
Hay, K.E., 333
Hayles, C.S., 354
Hayles, N.K., 39
Hazlett, J.S., 20
Heal, F.A., 47
Healy, M., 371
Heath, D., 371
Heath, S., 449
Heck, D.J., 339
Hedegaard, M., 442, 454
Hegarty, K., 349, 352, 353, 355
Heidegger, M., 111, 429
Heid, H., 24
Heilbrun, C.G., 375
Heimlich, J., 241, 412, 513, 523
Heimlich, J.E., 24, 239, 262, 265, 268, 298, 304, 306, 307, 359, 360, 363, 364
Held, V., 104
Helfenbein, R., 94
Helgeson, S., 47, 311
Hellden, G.F., 196
Heller, C., 135
Hemingway, C., 462
Henderson, J., 460
Henderson, K., 184, 185
Henn, C., 350
Hennessy, M., 269
Hennessy, S., 223, 227
Henry, J., 161

Herbert, D.M.B., 294
Hernández, N., 176n2
Hernandez Rojas, L., 82
Heron, J., 465, 470
Herr, K., 469, 471
Heshusius, L., 221
Hess, D.J., 39
Hesse-Biber, S.N., 377, 378
Hesselink, F., 52–3, 78
Hester, J.B., 51
Hewson, P.W., 339, 341
Heymann, F., 117, 119
Heyneman, S.P., 164
Hickory, S., 71
Hicks, D., 34
Hicks, L.E., 399
Hiebert, J., 224, 225, 228
Higgins, T.M., 339, 340
Higgs, A.L., 197
Hill, R.J., 399
Hill, T., 487
Hillman, M., 395–6, 399
Hills, A.M., 264
Hines, J., 311, 313, 320
Hines, J.M., 23–4, 47, 57, 266, 269, 307
Hirsch, G., 25
Hirsch Hadorn, G., 24, 259, 545
Hirt, P.W., 50
Hitsumoto, M., 291
Hobson, M., 226, 227
Hodges, G., 400
Hodgkinson, S.P., 264
Hoeffel, J.L., 231, 233, 235
Hoermann, E.F., 306
Hofer, K., 26, 28
Hoffer, T., 216
Hoffmann, J., 312
Hoffmann, P.A., 108
Hogan, R., 201
Hohenstein, J., 254
Holdsworth, R., 378
Holdsworth, S., 349, 351, 354, 355
Holdsworth, S.E., 354
Hole, R., 223
Hollingworth, H., 223
Hollweg, K., 334
Holman, S., 227
Holmes, L. Jr., 264
Holt, J., 267
Hoody, L., 217
Hoody, L.L., 334
Hooks, B., 371
Hooks, G., 395
Hopkins, C., 79, 139
Hopkins, D., 225
Hoppers, C.A.O., 384, 387, 390
Hoppin, R., 215
Hopwood, N., 239, 243–5, 262
Horvat, R.E., 47
Hossay, P., 395
Houle, M., 331, 336, 342
Hountondji, P., 384, 388, 389
House, M., 278
Houston, D., 396
Houtsonen, J., 138
Howard, P., 399
Howe, R.W., 16, 17
Hsu, S., 313
Hsu, S-J., 182, 185, 267, 290–2
Huang, C., 267
Huang, C-C., 186
Huang, H-P., 184, 185
Huang, Y., 184, 185
Huberman, A.M., 225
Hubner, G., 269
Hubner, T., 298
Huckle, J., 17, 18, 34, 75, 126, 198, 350, 351
Huckle, K., 351
Hudson, J.A., 294
Huebner, D., 433, 435n11
Hug, J.W., 299, 306, 307
Hui, D., 138
Humes, B., 372, 381, 411
Humm, M., 20n1
Humphrey, T., 361
Hungerford, H., 302
Hungerford, H.R., 17, 24, 25, 63, 151, 181, 255, 266–7, 300–3, 305–7, 310, 311, 313, 316, 319, 536
Hung, R., 505
Hunt, D.E., 263
Huotelin, H., 138
Hurd, P.D., 334
Huss, J., 264
Husting, C., 166
Hutcheon, L., 379
Hutchinson, S.M., 182
Hwang, J.J., 180
Hwang, Y.H., 267

I
Iacofano, D.S., 50
Iervolino, A.S., 235
Impey, A., 390
Inglehart, R., 269
Inglis, J.T., 384
Ingold, T., 281
Innes, J.M., 264
Iozzi, L., 46–50, 55, 59, 311, 320
Iozzi, L.A., 13, 17
Ip, M., 407
Irvine, K., 300
Irwin, A., 259, 545
Irwin, R., 130
Isaacs, W., 462
Ishizaka, T., 291
Ison, R., 460
Itin, C., 428
Ito, S., 291

J
Jablonka, E., 500, 501
Jackman, W., 52
Jackson, M.G., 390
Jacobs, J.K., 223, 228
Jacobs, L., 46
Jacobson, S., 301
Jacobson, S.K., 264, 298, 304, 305
Jafri, B., 397
Jagers, S.C., 263
Jaggar, A.M., 383n1
Jakobsen, L., 200
James, A., 244, 439, 440, 442, 454
James, B., 400
James, D., 509, 522
James, P., 432
James, S., 363
Jami, C., 36
Jamison, A., 35
Janke, D., 307
Janousek, S., 168
Janse van Rensburg, E., 198
Jans, M., 443, 454
Jardine, D., 415, 429
Jay, M., 426, 431
Jay, T., 41n13
Jeanpierre, B., 339
Jeng, J.M., 267
Jenkins, E.W., 210
Jenkins, H., 333
Jenks, B., 298
Jenner, P., 342
Jenq, C.S., 185
Jensen, B., 195, 198, 201, 489
Jensen, B.B., 17, 24, 108, 222, 243, 265
Jensen, D., 103
Jickling, B., 24, 63, 69, 71, 72, 74–6, 78, 79, 83, 113, 115, 117, 119, 126, 131, 136, 150, 151, 185, 197, 201, 255–6, 369, 409, 413, 421, 432, 513, 520
Jickling, R., 57
John, M., 360
Johnson, B., 264, 307
Johnson, J.I., 360
Johnson, K., 305
Johnson, L.R., 306
Johnson, M., 135, 384, 387, 426, 427
Johnson-Pynn, J.S., 305
Johnson, R., 57
Johnston, J., 16
Jonassen, D.H., 26
Jones, P., 13, 351, 545
Jones, R.E., 264, 292
Jones, S., 264
Jones, V., 463
Jonker, S.A., 264
Jucker, R., 350, 351, 353
Judson, G., 71
Jurin, R.R., 263, 264
Juris, J.S., 504
Jutras, S., 398

K
Kadji-Beltran, C., 186
Kahlid, A., 292
Kahn, P.H. Jr., 542
Kahn, R., 371, 372, 381, 397, 411, 414, 488
Kai, L., 264
Kaiser, F.G., 267, 269
Kalof, L., 265
Kaltenborn, B.P., 264
Kapitzke, C., 505
Kaplan, R., 360
Kaplan, S., 305, 360
Kapoor, D., 194
Karlsson, J., 200
Karns, D.A., 360
Karr, W., 27
Kasprzyk, D., 268–9
Kato, K., 398
Katzev, A.R., 303
Katz-Gerro, T., 269
Kaufman, J., 429
Kauppila, J., 138
Kawagley, A.O., 404–6
Kawata, C., 186, 267
Kayira, J., 487

Kaza, S., 398, 400
Keating, T., 333
Keeley, B., 137
Keen, M., 257, 399
Keicher, M., 265
Kellaghan, T., 298
Kellert, S., 414
Kellert, S.R., 292
Kellett, K., 439, 440, 454
Kellett, M., 439–40
Kelsey, E., 142, 206, 207, 239, 290
Keltner, D., 434n2
Kemmis, S., 27, 161, 350, 353, 354, 469–72, 474
Kendall, G., 66
Kennedy, K.J., 185
Kennelly, J., 371, 375, 398, 410
Kenway, J., 533
Kerka, S., 334
Kerlinger, F.N., 295
Kerski, J., 334, 337, 344
Kerski, J.J., 339, 340, 344, 345
Kervorkian, K., 207
Kesby, M., 488
Ketlhoilwe, M.J., 198, 200
Kezar, A., 349
Khasawneh, S., 264
Kheel, M., 411
Khun, T., 134
Kierkegaard, Soren., 227
Kim, K., 263
Kim, S.I., 267
Kincheloe, J.L., 163, 167
King, H., 254
King, J.A., 306
King, R.J.H., 399
King, T., 487, 491
Kirby, P., 440, 442, 449, 454
Kirk, J., 52
Kirkness, V., 404
Kituyi-Kwake, A., 543
Klein, J.T., 257
Klein, P.A., 57
Kluger, A., 267
Knapp, C.E., 277
Knapp, D., 13, 264, 293, 294
Kneller, G., 435n11
Knorr, M., 487
Knudtson, P., 36
Knuth, R.A., 26
Knutson, K., 364
Kobori, H., 280
Koellner, T., 264
Kohák, E.V., 103
Kola-Olusanya, A., 525
Kolb, D., 460, 462
Kollmuss, A., 17, 55, 151, 201, 262, 266, 317, 523
Kool, R., 69, 207
Koppelman, S., 372
Korten, D., 101
Koslowsky, M., 267
Koster, E., 364
Kostova, Z., 323
Kota, L., 390
Kowalewski, D., 201
Kraemer, A., 305
Krasny, K., 243
Krasny, M., 201
Krasny, M.E., 19, 289, 292
Kraus, M., 53
Kremer, K.B., 378
Kuhlemeier, H., 244, 267
Kuhlmeier, H., 310
Kuklick, H., 386
Kundera, M., 223, 226
Kunstler, J., 93
Kuru, J., 281
Kushmerick, A., 397
Kushwawa, A., 407
Kusmawan, U., 264
Kyburz-Graber, R., 23, 24, 26–9, 198, 211, 474–6

L
Labin, S.N., 298
Lackstrom, K., 281
LaDuke, W., 405, 411
Laesse, J., 201
Lagerweij, N., 244, 267, 310
Lagos, D.A., 175
Lahman, M., 440
Lai, E., 337
Lai, K.C., 180, 244
Lake, D., 135
Lakoff, G., 135, 427
Lamarche, T., 137
Lamb, M.J., 500
Lam, C., 337
Lam, C.C., 186
Langer, M., 432
Lang, J., 351
Lansdown, G., 442
Larsen, S.F., 294
Larson, J.O., 208
Lassiter, L.E., 406
Læssøe, J., 525
Lather, P., 19, 198, 375, 377, 378, 380, 422, 472, 509
Latouche, S., 139
Latour, B., 36, 41n12, 42n23, 42n26
Laugksch, R.C., 24
Laval, C., 137, 139
Lave, J., 278, 333
Laville, C., 234
Law, J., 42n27, 428, 429, 431, 435n8
Lazaridou, M., 196
Lazlo, E., 350
Leach, J., 247
LeBesco, K., 372
Lederman, N.G., 29
Lee, C., 46, 47
Lee, E., 52
Lee, E.B., 397, 400
Lee, H.B., 295
Lee, J., 247, 337, 343
Lee, J.C.K., 179–80, 184–6
Lee, M., 312
Leeming, F., 47
Leeming, F.C., 298, 303–5
Lee, R.B., 390
Lee, T., 333
Lefebvre, H., 428
Lefort, N., 406
Le Goff, J.-P., 137
Le Grange, L., 71, 108, 110, 112, 126, 137, 198, 387, 475
Lehmann, C., 166
Lehmann, J., 24
Lehmann Pollheimer, D., 25
Leinhardt, G., 364
Leisch, F., 264
Leiss, W., 135
Lekies, K., 291, 292
Lemke, J., 27, 29
Lenz-Taguchi, H., 483
Leopold, A., 69, 76, 82, 93, 117, 276
Leunes, B.L., 51
Levinas, E., 111, 112
Levin, M., 257
Levin, R., 361
Levinson, B., 147–9
Levi-Strauss, C., 386
Levitt, T., 41n7
Levy, M., 264
Lewenstein, B., 359
Liang, M-H., 178, 186
Lickers, M., 407
Liddicoat, K., 289, 293–5
Lidstone, J., 244, 247
Lieberman, G., 217
Lieberman, G.A., 334
Lima, F.B., 231, 234, 235
Lim, M., 219, 305
Limon Luque, M., 250
Lincoln, Y., 57, 58, 435n11, 509
Lincoln, Y.S., 440
Lindemann-Matthies, P., 25
Lindholm, C., 504
Lingard, B., 147, 148, 150, 152, 153
Lingis, A., 428, 429, 435n7
Linke, R.D., 15, 20n2, 372
Lin, M.R., 183
Linsenbardt, C., 45
Linton, M., 294
Lipsey, M.W., 298
Lister, R., 195
Liston, D., 473
Liston-Heyes, C., 264
Liu, C-C., 178, 185
Liu, C-H., 186
Lively, C., 52
Livingston, J., 83, 409–11
Livingston, J.A., 135, 139
Lizop, E., 388
Llewelyn, J., 430
Lloyd, W.J., 343
Locke, S., 398
Lofland, J., 449, 452
Lofland, L.H., 449, 452
Lohr, V.I., 291, 292
Lomax, P., 469
Londoño, F., 163
Longboat, D.R., 407
Long, K., 459
Loomis, R.J., 298, 299
Lorenzetti, L., 171, 173
Lorenzoni, I., 264
Lotz, H., 475
Lotz-Sisitka, H., 71, 117, 119, 165, 191, 199, 200, 375, 390, 391, 429, 434
Loucks-Horsley, S., 339, 341
Louden, W., 160
Lousley, C., 135, 370, 375
Louv, R., 213, 372, 543
Lovelock, J., 221
Love, N., 339, 341

Loving, C.C., 211
Lowan, G., 398, 404, 407
Lo, Y.-S., 503
Lozano-Garcia, F., 351
Lozar Glenn, J.M., 397
Lucas, A.M., 14, 16, 74, 298
Lucas, K.B., 27
Luckmann, T., 142
Ludwig, G., 334
Lui, K.C.W., 182
Luke, J., 360, 363
Lundegard, I., 197
Lundholm, C., 239, 243–7, 250, 262
Lupele, J., 198–200
Lupton, R., 371
Lutz, C., 206
Lu, Y., 337
Lyotard, J.-F., 533

M

MacGregor, S., 370, 376
Machado, M.K., 231, 233
MacIntyre, A., 431
Mackey, B., 82
Mackey, S., 276
Maclean, R., 165
Maclure, M., 483
Macy, J., 101, 104, 207
Madachy, R., 466
Madaus, G.F., 298
Madden, T.J., 268
Madsen, G.E., 399
Mainieri, T., 262, 267
Makhoul, J., 264
MaKinster, J.G., 331, 337, 339, 341–2, 344
Makki, M., 310
Makou, T., 198, 475
Malancharuvil-Berkes, E., 138
Malandrakis, G., 201
Malewski, E., 454
Malinowski, B., 386
Malkki, L., 387
Malone, K., 200, 371, 375
Malone, L., 332
Maloney, M.P., 264
Mandell, N., 440
Mandl, H., 26
Manfredo, M.J., 264
Manion, L., 306
Mann, S., 38
Manoli, C.C., 264
Manteaw, B., 153
Mante, J.O.Y., 50
Marchant, R., 442
Marcinkowski, T., 17, 45–50, 57, 267, 310, 313, 316, 320–1, 323
Marcinkowski, T.J., 397
Marginson, S., 505
Marino, D., 378
Marker, M., 97
Mark, S., 331
Marks, H., 215
Marouli, C., 398, 399
Marquardt, M., 257
Marshall, A., 406
Marshall Egan, T., 257
Marshall, J., 465
Martello, M.L., 385, 386
Martin, A., 351
Martin, G.C., 1
Martin, J.R., 102, 104
Martin, R., 264
Martin, S., 353
Martusewicz, R.A., 396
Marx, R.W., 246
Marzolla, A.M., 360
Mason, L., 250
Mason, P., 224, 226
Massey, D., 481
Masuku, L.S., 385, 390
Masuku Van Damme, L.S., 384–6, 388, 391
Matheny, K., 46
Mathews, F., 88
Mathison, S., 298
Matthäus, H., 236
Matti, S., 263
Mauro, S.E.D., 397
Mau, T., 24
Maxwell, J.W., 225
Mayer, M., 256
Mayer-Smith, J., 334
Mayes, G., 263, 268
Maynard Smith, J., 500
Mayo, E., 226
Mayumi, K., 502
McAllister, M., 543
McAlpine, L., 356
McAuliffe, C.M., 337, 343
McAvoy, L., 371
McBeth, B., 310
McBeth, W., 24, 46, 47, 55, 58, 311, 312, 316, 319, 321, 323, 326
McCaleb, S., 460
McCarthy, H., 499
McClaren, P., 371
McClintock, M., 371
McClurg, P., 339
McCoy, L., 143
McCreedy, D., 360, 363
McCright, A.M., 263–5
McDaniels, T.L., 264
McDonald, C., 166
McDonnell, P., 224
McDuff, M., 299, 307
McDuff, M.D., 298, 305
McGuigan, W.M., 303
McInerney, M., 333
McIntosh, A., 275
McIntyre, D., 245
McKenzie, M., 19, 24, 97, 99n3, 244, 276, 372, 375, 378, 380, 413, 419–21, 427, 487, 489, 507, 520, 523, 525, 537
McKenzie-Mohr, D., 267
McKeown-Ice, R., 397
McKeown, R., 139
McLaren, P., 396
McLaughlin, M., 148
McLean, S., 487
McLuhan, M., 41n7
McMillan, V.M., 197
McMurtry, J., 101
McNaught, C., 24, 200
McNerney, P., 166
McNiff, J., 470, 471
McPeck, J., 27
McQueen-Thomson, D., 432
McTaggart, R., 161, 470–2, 474
McTighe, J., 340
Mead, G., 465
Meagher, N., 440
Measham, T.G., 278, 281
Medley, K.E., 279
Megid, J., 173
Meira-Cartea, P., 427
Meira, P., 172
Meisner, M., 135
Melaville, A., 94
Menzel, S., 264
Merchant, C., 104, 381, 381n2, 381n4, 503
Merleau-Ponty, M., 135, 276, 278, 426, 427, 430
Mertig, A.G., 264, 292
Meyer, J.H.F., 246
Meyer, M.A., 106
Meyers, R., 310
Mezirow, J., 257
Michelsen, G., 24
Mickwitz, P., 307
Milani, B., 502
Milburn, G., 69
Miles, M.B., 225
Milfont, T.L., 264
Millar, R., 247
Miller, B., 266
Miller, D., 128, 129
Milson, A.J., 332
Minkler, M., 460
Minner, D.D., 57
Minton, T.G., 53
Mintzes, J., 201
Minujin, A., 443
Mitchell, C.A., 197
Mitchell, J.T., 334, 343
Mitten, D., 371
Mobley, C., 263
Mochizuki, Y., 253, 545
Mogensen, F., 198
Mohai, P., 394, 395
Mokuku, C., 390
Mokuku, T., 201, 390
Monroe, M.C., 299, 306, 307
Montano, D., 268–9
Montarzino, A., 291–2
Mony, P., 262
Moon, B., 334
Moore, D., 48
Moorefield, D.L., 50
Moran, A., 102
Morehouse, B.J., 19
Morgan-Brown, T., 264
Morgan, M., 299
Morgan, P.A., 51
Morris, J., 301
Morris, L.L., 306
Morrow, V., 440, 454
Mortari, L., 197
Mortimer, E., 27
Moscovici, S., 135
Moseley, C., 264, 312
Moser, G., 47, 57, 264, 269, 311
Möser, G., 306
Mosley, M., 138
Moss, P., 482–3
Moulin, A.M., 36
Moussouri, T., 360, 362
Moyano, E., 175
Mrazek, R., 13, 16, 24, 27, 46–50, 57, 74, 176, 200, 306, 321, 369, 509, 525, 535

Mueller, M., 397, 399
Mugerauer, R., 135
Mugny, G., 210
Mukute, M., 199–201, 231, 237
Mullins, G.W., 378
Muma, M., 264
Mundry, S., 339, 341
Murphy, J., 487
Murphy, M., 307
Murphy, P., 505
Murray, P., 350
Murray, S., 350
Murthy, D., 222, 227
Myers, D., 406
Myers, O.E., 264

N

Nader, L., 384, 386, 387
Naess, A., 88, 117, 118, 197
Nagel, M., 244, 247
Nancy, J.-L., 482–5
Nash, C., 276, 277
Nasrallah, R., 264
Næss, A., 83, 84, 111
Ndayitwayeko, A., 310
Neale, B., 442
Needham, M.D., 264
Needham, R.D., 264
Negev, M., 302, 304, 305, 310, 316, 326
Negri, A., 503
Neilson, A., 428
Nelson, W., 310
Neluvhalani, E., 19, 198, 201, 475
Neluvhalani, E.F., 384, 385, 388, 391
Nemiroff, L.S., 267
Nespor, J., 97, 245, 276, 279
Neugebauer, S.R., 99n9
Newbery, L., 371, 372, 375, 379, 398
Newmann, F., 215
Nicita, J., 355
Nickel, J., 24
Nielsen, J.E., 306
Nihlen, A., 469, 471
Nikel, J., 243, 489, 519, 522, 530, 531, 538
Nikkel, J., 108
Nilon, C., 334
Nixon, H., 265
Nocella, A., 371
Noddings, N., 71, 75, 104, 138
Noffke, S., 470
Noguera, A.P., 174
Noh, K.-I., 310, 314
Nolan, J.M., 264
Nolan, K., 2, 13, 18, 19, 45–7, 49, 57–8, 243, 306, 513–15, 525
Nordstrom, H.K., 397
Norland, E., 52
Norris, K.S., 298
Northway, H., 305
Novick, P., 42n19
Novitz, D., 227
Nowak, P., 302
Nowell, D.E., 264
Nozaki, Y., 153
Nuwayhid, I., 264
Nyahongo, J.W., 264
Nyberg, D., 131
Nygren, A., 387

O

Oakley, J., 411, 412
Oberchain, V.L., 289
Oberhauser, K., 339
Ochs, K., 228
O'Connell, T., 371
O'Donoghue, R., 19, 24, 117, 120, 134, 197, 200, 201, 390, 522
Odum, E., 35
O'Farrell, C., 65, 65
O'Garro Joseph, G., 138
Ogbuigwe, A., 117
Ogunseitan, O.A., 265
O'Hearn, G.T., 307
Öhman, J., 198, 201, 525
Ohman, S., 264
Ok, A., 323
Olbrechts-Tyteca, L., 143
Oldakowski, R.K., 342
O'Leary, J.E., 268
Olli, E., 317
Olofsson, A., 264
Olsen, N., 153
Olsson, P., 280
Olvitt, L., 115, 120, 165, 391
Onwuegbuzie, A., 57
Onyenekwu, C.M., 51
Opoku, K.A., 389
Orams, M., 199
Oreg, S., 269
Orenstein, D., 316
O'Riley, P., 394, 404, 411
Orr, D., 350, 545, 546
Orr, D.W., 101, 102
Ortiz, B., 172
Ortiz, J.A., 50
Orwell, G., 78
O'Ryan, B., 3
Osbaldiston, R., 47, 58, 311
Osborne, J., 29
Oskamp, S., 262
Osterlind, K., 244
Östman, L., 201
O'Sullivan, E., 76, 257
O'Toole, J.M., 264
Ottino, J.M., 505
Oulton, C., 543
Oyama, S., 49

P

Pace, P., 224
Packer, J., 199, 244, 268, 360, 363
Page, D., 338
Page, S., 257
Pahl-Wostl, C., 257
Palmberg, I.E., 281
Palmer, A., 332, 339, 340, 429
Palmer, J., 18
Palmer, J.A., 185, 290, 291
Pan, H-P., 182
Parada, V., 264
Parag, Y., 264
Parker, J., 354, 355
Parker, L.H., 379
Parkin, S., 354
Park, M., 165
Parks, B.C., 394
Park, V., 148
Paton, K., 264
Patrick, D.L., 306
Patterson, M.W., 338
Patton, M.Q., 45, 46, 298, 302–4, 306
Paul, R., 27
Pauw, J.B., 299
Pavlova, M., 167–8
Pawlowski, A., 378
Pawson, R., 259
Payne, P., 141, 195, 197, 198, 201, 243, 244, 372, 412, 427, 428, 430, 433, 434, 434n3, 514, 522, 529, 531, 537
Payne, P.G., 278, 424
Peacock, A., 293
Pea, R., 223
Pea, R.D., 333
Pearson-Mims, C.H., 291, 292
Peat, F.D., 37, 42n15
Peck, D., 378
Pe'er, S., 302, 316
Peirce, C.S., 33
Pekarik, A.J., 360
Pekrun, R., 207
Peled, E., 318
Pelicioni, M.C.F., 235, 236
Pellow, D.N., 394–5, 397, 399
Pell, R.G., 210
Pence, A., 483
Pepper, D., 135, 378
Pequegnat, W., 306
Perelman, C., 143
Pergams, O.R.W., 543
Perkes, A., 310
Perkins, B., 399
Perry, D., 364
Perry, R.P., 207
Peruzzo, C.M.K., 232
Pesanayi, T., 199, 231, 391
Peters-Grant, V.M., 50
Peters, M., 41n3, 176n4, 372, 422
Peters, M.A., 498, 501, 505
Peters, R.S., 69, 75, 87
Peters, S., 535
Petitjean, P., 36
Petrella, R., 137
Petrie, P., 483
Petrosino, A., 298
Petts, J., 281
Peyton, B., 151
Peyton, R., 266, 320
Peyton, R.B., 181
Phelan, P., 207
Philippi, A. Jr., 236
Phillips, D., 228
Phillips, F., 462
Phillips, M., 218
Philo, C., 371
Piaget, J., 142
Piccinin, S., 353
Pickett, S.T.A., 35
Pidgeon, N., 264
Pierre, E., 422, 507
Pierre, E.A., 483
Pike, G., 34
Pillemer, D.B., 294
Pilon, A.F., 236
Pimbert, M.P., 232
Pimm, D., 222
Pinar, W., 432n11
Pinar, W.F., 39

Pinheiro, J.Q., 264
Pink, S., 223, 226, 426
Pintrich, P., 246
Pintrich, P.R., 246, 247
Piper, H., 441
Pivnick, J., 413
Place, G., 292, 295
Plant, M., 153
Plowman, L., 223, 224
Plummer, R., 243
Plumwood, V., 76, 83, 118, 119, 410, 411, 414
Polanyi, K., 139
Polanyi, M., 470
Polman, J.L., 339, 340
Pomerantz, S., 454
Ponzio, R., 360
Poortinga, W., 264, 265
Popham, W.J., 194
Popkewitz, T., 131, 134, 136, 138, 161, 509
Popkewitz, T.S., 195, 197, 198
Porter, B., 47
Posch, P., 19, 24, 26, 198, 256, 476
Postman, N., 136, 247
Pothier, D., 371
Potter, G., 397
Pottier, J., 385, 387
Powell, P., 264, 267
Powell, R.B., 306, 307
Power, A., 215
Power, K., 444
Powers, A., 218
Powers, A.L., 305–6
Pradenas, L., 264
Prakash, M., 97
Pratt, C.C., 306
Pretty, J.N., 232
Price, L., 141, 197–200, 391, 429, 532
Price, R.F., 24
Priess, J., 52
Prigogine, I., 505
Primack, R.B., 280
Prochaska, J.O., 269
Profeit-LeBlanc, L., 71
Prosser, J., 225
Prout, A., 244, 442
Punch, S., 454
Putnam, R., 213, 500
Putney, A.D., 307
Pyle, R.M., 98

Q
Qablan, A.M., 264
Quandt, S.A., 399
Queiroz, J., 499
Quiamzalde, A., 210
Quinn, M.S., 333
Quinones, G., 438, 443, 456

R
Raglon, R., 378
Rahnema, M., 388
Ramage, M., 257
Ramamurthy, M.K., 331, 334
Ramirez Ramirez, E., 82
Ramsey, J., 25, 266, 267
Ramsey, J.E.A., 267
Ramsey, J.M., 17
Rao, P., 399
Rapport, D., 257
Ratcliffe, M., 24
Rauch, F., 198
Raudsepp, M., 264
Raven, G., 518
Rawling, R., 199
Rawlins, E., 372
Raymbnd, L., 264
Razevieh, A., 46
Reardon, K.M., 364
Reason, P., 259, 460, 462, 465, 466, 467, 470, 472, 545
Reboul, O., 135
Reddy, C., 198, 475
Reder, L.M., 360
Reeder, G.D., 263
Reeve, K., 338
Rehrig, L., 313, 316
Reid, A., 24, 108, 176, 197, 201, 243, 352, 353, 391, 427, 429, 430, 432, 434, 434n1, 489, 514, 518–22, 525, 529–31, 533
Reid, W.A., 41n1
Reigota, M., 173
Reinharz, S., 377, 378
Reis, J.C., 231
Reiter, D.K., 264
Rejeski, D., 410, 413
Relph, E., 93
Renner, M., 166
Rennie, L.J., 379
Resnick, L., 476
Resnick, L.B., 26
Reynold, W., 435n11
Reynolds, R., 264
Reynolds, W.M., 39
Rgozzi, M.J., 79
Rheingold, A., 371
Rich, A., 375
Richards, D., 390
Richards, P., 387
Richardson, R.J., 234
Richins, H., 263, 268
Richmond, J., 310
Rickinson, M., 4, 13, 19, 46, 47, 55, 58, 195, 201, 202, 239, 243–5, 249, 253, 262, 311, 428, 438, 513
Rickson, R., 266
Ricoeur, P., 142, 429, 431, 435n13, 524
Rideout, B.E., 196, 199
Rifkin, J., 504
Rigendinger, L., 24
Riggins, S., 535
Riggins, S.H., 208
Rinaldi, C., 482
Ringer, F., 522
Rise, J., 269
Rist, G., 139
Rivoli, P., 463
Rizvi, F., 147, 148, 150, 152, 153, 523
Roach, C., 370
Robb, G.M., 371
Robert, K.-H., 462
Roberts, J.A., 264
Roberts, J.T., 394, 395
Robinson, J., 222, 224
Robinson, V., 154
Robottom, I., 16–19, 24–7, 29, 57, 74, 123, 134, 136, 151, 156, 158, 176, 181, 186, 195, 196–8, 200, 234, 290, 291, 306, 350–2, 369, 411, 412, 429, 432, 434, 469, 474–6, 513, 520, 531, 536
Robottom, R., 126
Robson, E., 440
Rocchio, R., 52
Rode, H., 24
Roe, M., 278
Roffe, J., 483
Rog, D.J., 306
Rogers, C.M., 306
Rogers, J., 354
Rogers, L., 267, 317
Rogers, P., 298
Rogers, R., 138
Rogoff, B., 255
Rokeach, M., 262, 263, 268, 269
Rolston, H., 135
Romanycia, S., 101
Rorty, R., 102, 459, 540
Rose, G., 223, 225
Rose, N., 65, 67, 431
Rosenberg, D.G., 384
Rosenstock, I.M., 268, 269
Rosiek, J., 250
Ross, M., 502
Ross, V., 461
Rossi, P.H., 298
Rost, J., 24
Roth, C., 312, 313, 321
Roth, R., 47, 311, 313
Roth, R.E., 182, 267, 378
Roth, W.-M., 27, 206, 210
Rounds, J., 360
Rousseau, J.-J., 87
Roux, C.L., 182
Rovira, M., 300
Rudduck, J., 245, 439, 440, 454
Rudsberg, K., 198, 201
Ruíz, D., 175
Rundblad, G., 264
Russell, C.L., 72, 75, 141, 369–72, 375, 378, 398, 410–13, 432, 537
Russell, K., 292
Russo, V., 522
Ruthenberg, I.M., 175
Ryan, G., 46
Ryan, L., 472
Ryu, H.-C., 63

S
Sachs, W., 139, 232, 504
Sadler, T.D., 24, 29, 211
Safer, M.A., 294
Sagor, R., 461
Sagy, G., 310, 316
Saha, R., 394
Said, E.W., 194
Salinger, G., 344, 346
Salleh, A., 395
Salleh, A.K., 378, 379
Salmon, S.D., 166
Sandell, K., 201
Sandilands, C., 375
Sandler, R., 395, 397
Sandoval, W., 340
Sanera, M., 71
Santana, L.C., 173
Santangelo, R., 45
São Paulo., 233

Saphores, J.M., 265
Sardar, Z., 36
Sarick, J., 398
Sarick, T., 371, 372, 375, 410
Sarkar, S., 499, 500
Sato, M., 52, 148–51
Saul, J.R., 83
Saunders, C., 300
Saunders, C.D., 264
Saunders, G., 319
Sauvage, J.L., 50
Sauve, L., 429, 435n12, 435n14
Sauvé, L., 23, 25, 27, 126, 135, 139, 140, 142, 179, 234, 444
Sawicki, V., 264
Sayer, A., 70, 120
Scheffler, I., 135
Scheja, M., 244
Schenke, A., 167
Scherer, R., 264
Schick, C., 488
Schleien, S., 371
Schlosberg, D., 395–6
Schmeider, A., 311
Schmidtlein, M.C., 334
Schmidt, P., 266
Schnack, K., 24, 195, 198, 201
Schneider, I.E., 398
Schneiderman, J.S., 343, 399
Schneller, A.J., 302
Schoenfeld, C., 52–3, 311
Schoenfeld, C.A., 14
Scholz, R.W., 264
Schön, D., 349
Schön, D.A., 462
Schor, J.B., 101
Schreiner, C., 210
Schreuders, P.D., 166
Schroder, B., 398
Schroder, H.M., 263
Schubert, W., 461
Schubert, W.H., 18
Schudel, I., 119
Schultz, P.W., 264, 265
Schutz, P.A., 207
Schwab, J.J., 16, 354
Schwartz, S., 264, 269
Scollo, G., 307
Scott, A., 371
Scott, B., 250
Scott, D., 194
Scott, P., 27, 247
Scott, W., 13, 24, 78, 79, 83, 108, 126, 135, 147, 150, 153, 154, 176, 197, 202, 243, 306, 312, 414, 415, 429, 432, 518, 520–2, 531, 533
Scott, W.A.H., 448
Scutt, G., 412
Seamon, D., 444
Searcy, C., 166
Searle, J., 227
Segerstrom, J., 164
Sehlola, M.S., 108
Selby, D., 13, 34, 156, 195, 351, 397, 410, 411, 545
Semeijn, J., 265
Senechal, E., 219
Senge, P., 459, 462
Severtson, D.J., 264
Sewell, A., 361
Seybold, H., 24
Sfard, A., 195
Shadish, W.R., 294
Shaffer, D.W., 336, 343
Shah, A., 371
Shalem, Y., 111
Shallcross, T., 222, 224
Shanahan, M., 246
Shapiro, A.A., 265
Sharma, A., 263
Sharpe, G., 52
Shava, S., 384, 385, 386, 390, 391
Shavelson, R., 306, 307
Shaw, J., 71
Sheeran, P., 268
Sheets-Johnstone, M., 426
Shelley, M., 264
Shepard, L.A., 76
Shepardson, D., 201
Sherran, P., 269
Shi, C., 182
Shier, H., 441
Shin, D., 310, 312–14, 326
Shiner Klein, B., 375
Shinn, E.-K., 338, 342, 343
Shipp, K., 257
Shiva, V., 387, 389, 503
Shore, B., 300
Shor, I., 167
Shouse, A.W., 359
Shulman, L., 245
Shultz, J., 245
Shusterman, R., 426
Sia, A., 313
Sia, A.P., 23–4, 267
Sicard, M., 142
Siegel, H., 27
Siemens, G., 353, 543
Sikes, P., 159
Sillitoe, P., 385, 387
Silo, N., 198
Silva, R.L.F., 173
Simmons, D., 46, 55, 59, 312, 313, 316, 321, 323, 327
Simmons, I.G., 135
Simmons, M.L., 24, 247
Simon, H., 360
Simon, R.I., 167
Simon, S., 29
Simovska, V., 108, 243, 489
Simpson, L., 396, 398, 404, 405
Simpson, P.R., 267
Sim, S., 431
Sinatra, G., 246
Singh, M., 142
Sinha, J.B.P., 264
Siraj-Blatchford, J., 399
Sivek, D., 267, 313
Sivek, D.J., 290, 291
Sjøberg, S., 210
Skelton, T., 440, 454
Skinner, B.F., 254
Sklar, H., 214
Skoldberg, K., 225, 226
Skolimowski, H., 103, 105, 106
Skopelti, I., 250
Skowronski, J.J., 294
Slikkerveer, L.J., 387
Slote, M.A., 104
Smith, C., 245
Smith, C.L., 395
Smith, D., 342
Smith, D.A., 332
Smith, G., 80, 95, 99n3, 195, 201, 213, 215
Smith, G.A., 334, 398, 400, 444
Smith, J., 52
Smith, J.W., 264
Smith, L.T., 98, 384, 388, 400
Smith, M., 500
Smith, S., 434n1
Smith-Sebasto, N.J., 196, 247, 267, 289
Smith, T., 350, 353, 354
Smyth, F.M., 279
Smyth, J., 152, 514
Snively, G., 406
Snow, J., 405
Sobel, D., 94, 95, 207, 215
Soffar, A.J., 51
Soja, E., 42n25, 94
Soledi, S.W., 166
Sollart, K., 168
Somers, C., 52, 305, 306
Somerville, M., 94, 257, 444
Sorensen, C., 46
Sorenson, J., 364
Sorial, G., 166
Souchon, C., 24
Soucy, J., 299
Souza, M.P., 232
Sparrman, A., 226
Spence, A., 264
Spender, D., 376
Spero, V., 45, 46, 48–50
Spirn, A.W., 277
Spork, H., 74, 75, 78, 126
Sprigett, D., 350
Springgay, S., 483
Spykerman, B.R., 399
Squella, M.P., 175
Squire, K., 340, 346
Srebotnjak, T., 264
Stables, A., 126, 135, 150, 152, 153, 349
Stadel, M., 45
Stake, R., 159
Stankorb, S., 440
Stankorb, S.L., 399
Stanley, L., 376, 377
Stapp, E.B., 399
Stapp, W., 24, 46, 49, 52, 149, 311, 312, 320, 411
Stapp, W.B., 15, 17, 440
Steede-Terry, K., 332
Steg, L., 265
Steiner, M., 34
Stein, R., 400
Stein, S.E., 397
Stenhouse, L., 354, 469, 470
Stephen, C.R., 53
Stephen, P., 223, 224
Sterelny, K., 500
Sterling, S., 182, 349–54, 545
Sternäng, L., 244
Stern, M.J., 306
Stern, P., 513
Stern, P.C., 264–5, 269

Stevenson, R.B., 1, 4, 13, 18, 28, 95, 99n3, 123, 137, 139, 140, 142, 147, 150, 152–4, 186, 194, 196, 243, 262, 350, 420, 469, 472, 473, 478, 507, 512–14, 542
Steverson, B.K., 50
Stewardson, G.A., 166
Stewart, K., 495
Stewart, M.E., 343, 399
Stigler, J.W., 224
Stiles, K.E., 339, 341
Stilson, J., 47, 48
Stimpson, P., 186
Stinchfield, H.M., 264
Stirling, C., 472
Stollenwerk, D., 460
Stone, J., 487
Storey, C., 378
Storksdieck, M., 285, 306, 359, 363
Strauss, E., 336, 342
Strife, S., 395
Stringer, E., 459
Stronach, I., 483
Stroup, L.J., 281
Stuart, C.I.J.M., 499
Stubbs, H.S., 344
Stufflebeam, D.L., 298, 305
Stylinski, C., 342
Sugerman, D.A., 289
Suggate, J., 290, 291
Sullivan, J., 399
Summers, M., 182, 185, 353
Supovitz, J.A., 339
Sureda, J., 176n2
Sutton, M., 147–9
Suzuki, D., 36
Swan, M., 46–9, 52, 311, 312, 411
Sward, L., 47
Sward, L.L., 290, 291
Swayze, N., 398, 406
Switzer, M.A., 376, 378
Sykes, H., 34, 36, 247
Szathmáry, E., 500

T
Takacs-Santa, A., 264
Takano, T., 407
Tal, A., 316
Tambiah, S.J., 42n25
Tamoutseli, K., 224
Tam, S.W., 353
Tanner, C., 263, 266
Tanner, R., 53, 311
Tanner, T., 289, 290, 429
Tapsell, S., 278
Tarrant, M., 201
Taschner, S.P., 233
Tate, W.F., 95
Taylor, C., 101
Taylor, D., 396, 397, 400, 411
Taylor, E.W., 257
Taylor, L.H., 94
Taylor, M., 200
Taylor, S., 269, 378
Tearney, K., 391
Teel, T.L., 264
Teng, M., 178
Tennant, C., 386
Terrón, E., 172
Terry, D.J., 268
Thayer, H.S., 41n4
Thiessen, D., 454
Thiollent, M., 234
Thomashow, M., 544
Thomas, I., 349, 351, 354
Thomas, I.G., 306, 355
Thomas, J., 472–3
Thompson, C.P., 294, 295
Thompson, C.W., 290–2
Thompson, M., 269
Thompson, S.J., 51
Thomson, J.L., 371
Thrupp, M., 371
Tidball, K., 201
Tilbury, D., 139, 156, 158, 168, 169, 179, 184, 185, 306, 351, 354, 399
Tilden, F., 52
Tilley, N., 259
Tippins, D.J., 400
Titz, W., 207
Toadvine, T., 428, 429, 431, 432
Toal, S.A., 306
Tobias, S., 207
Todd, F., 27
Todd, P., 269
Todt, D., 312
Tomera, A., 311, 313
Tomera, A.N., 24, 266, 267, 307
Tomkiewicz, W., 244, 360
Tong, R., 381n1, 381n2
Torbert, W.R., 465
Torres Carrasco, M., 124
Towers, J., 222
Trafimow, D., 268
Trainer, T., 151
Traoré, A., 79
Trautmann, N.M., 331, 337, 339, 341, 342, 344
Tredinnick, L., 542
Tremblay, G., 218
Treseder, P., 441
Tressler, K., 201
Trewhella, W.J., 305
Triandis, H.C., 268, 269
Tsai, C.C., 512, 519
Tsaliki, E., 290, 291
Tsang, E.P.K., 182
Tschakert, P., 257
Tschapka, J., 256
Tuan, Yi-Fu., 444
Tuchman, B., 406
Tuhiwai Smith, L., 406
Tulloch, D., 338, 343
Tung, C., 267
Tung, C-Y., 186
Tunstall, S., 278
Turnbull, D., 37, 39
Turner, H.M., 339
Tyler, R., 194

U
Ulanowicz, R.E., 41n11
Ulewicz, M., 222
Unipan, J.B., 262
Unterhalter, E., 258
Utley, J., 264

V
Valdero, T.R., 262
Valentin, L., 173
Vamvakoussi, X., 250
van Birgelen, M., 265
van Boeckel, J., 543
Van Den Bergh, H., 244, 267, 310
van der Leij, T., 137
van der Schee, J., 344
Van Harmelen, U., 81, 82
van Kampen, P., 52–3
van Kempen, P.P., 78
Van Liere, K.D., 264, 292
van Manen, M., 426, 427
Van Matre, S., 411
van Mierlo, B.C., 259
Van Petegem, P., 299, 306
VandeVisse, E., 411
Vare, P., 250
Vaske, J.J., 264
Vaughan, P.W., 298
Vazquez Brust, D.A., 264
Veletsianos, G., 336, 338, 344
Venegas, K.R., 99n9
Verdugo, V.C., 264
Verplanken, B., 269
Verran, H., 42n21
Vignola, R., 264
Vilela de Araujo, M., 82
Villegas, M., 99n9
Vincent, A., 127, 130
Vincent, J.S., 344
Vining, J., 264
Vlek, C., 265
Voelker, A.M., 47
Vogel, S., 135
Voigt, C., 332
Volet, S., 246, 250
Volkov, B., 306
Volk, T., 24, 25, 46, 47, 55, 58, 199, 255, 266–7, 302, 310–12, 316, 319, 323, 536
Volk, T.L., 17, 24, 63, 300–3, 305–7
von Haeften, I., 268–9
Von Secker, C., 219
Vosniadou, S., 240, 250, 253
Vygotsky, L., 255

W
Wagar, J.A., 307
Wagner, J., 38, 519
Walberg, H.J., 253
Walck, C., 276, 281
Wald, K., 264
Walker, B., 24
Walker, C., 390
Walker, G., 395
Walker, K., 19, 27, 198, 415
Walker, K.A., 247
Walker, L.M., 247
Wall, G., 265
Wallace, C., 245
Wallen, N., 46
Waller, T., 439
Wallerstein, N., 460
Wals, A.E.J., 1, 13, 19, 24, 25, 27, 53, 57, 69, 72, 74, 75, 78, 108, 109, 116–19, 131, 137, 184, 185, 195, 198, 199, 201, 202, 243, 253, 255–9, 303, 395, 398–9, 421, 440, 489, 512, 513, 518, 520, 530, 542, 544, 545
Walsh, P., 69, 76
Walton, M., 371

Waltz, M.E., 47, 48
Wane, N., 375, 379, 398
Wang, H., 195
Wang, S., 333
Wang, S.H., 183
Wang, S.M., 179, 185
Wang, Y.-M., 264
Wantz, R.A., 263
Warburton, K., 353
Ward, M., 18
Warkentin, T., 411, 413
Warner, A., 290, 291
Warren, D.M., 387
Warren, K., 76, 83, 371
Warren, K.J., 376, 381n1, 381n3
Warriner, D., 138
Wasmer, C., 186
Watchow, B.J., 405
Waterton, E., 277
wa Thiongo, N., 388, 389
Watson, G., 411, 413
Wattchow, B., 278, 412, 433
Watt, J.M., 390
Watts, D., 500
Watts, M., 246
Weakland, J.P., 82
Weaver, S., 375
Weber, M., 67
Weedon, C., 380
Wegener, D.T., 264
Weick, K., 349
Weigel, J., 264
Weigel, R., 264
Weil, Z., 411
Weingartner, C., 247
Weis, L., 153, 545
Weiss, G., 427
Weiss, I.R., 339
Welcomer, S., 264
Wells, N., 291, 292
Wen, M.L., 512, 519
Wenden, A., 397, 400
Wenger, E., 243, 333
Wensveen, L., 126
Werner Zentner, K., 24
Wesley, P., 199
West, B.A., 338, 344
Weston, A., 69, 71, 76, 80, 83, 103, 105, 116–19, 413, 414
Weston, C., 356
Whaley, K., 364
Whately, M., 233
Wheeler, K., 14
Whitehead, A.N., 87, 496n4
Whitehead, J., 471
Whitehouse, H., 19, 198, 375, 378, 379
White, L., 34, 503
White, P.S., 35
White, S.H., 335, 343–4
Whiteside, K., 414, 415n8
Whiteside, K.H., 135
Whitty, G., 439
Wickham, G., 65
Wickman, P., 197
Wideen, M., 334
Wiens, J.R., 482
Wiggins, G., 340
Wigglesworth, J.C., 339, 343
Wilcox, H.N., 399
Wilder, A., 339, 340
Wiles, R., 449
Wilhelm, S.A., 398
Wilke, R., 151, 302, 312, 313, 319
Willett, C., 406
Williams, D.C., 51
Williams, D.R., 264, 398, 400
Williams, J., 481
Williams, M., 186
Williamson, B., 439, 456
Williams, R., 69, 409, 415n1
Willis, A., 312
Willow, C., 442
Wilson, D., 165
Wilson, E.A., 480–2, 485
Wilson, E.O., 414, 543
Wilson, M., 267
Wilson, R., 399
Wilson, S., 406
Wiltz, K., 52
Wiltz, L.K., 298
Wingate, L.A., 298
Winter, C., 78–80
Winzler, E., 47, 48
Wisby, E., 439
Withers, L.E., 263
Wolfensberger, B., 26–9
Wolfing, S., 267
Wolf, J.E., 166
Wollebaek, D., 317
Wolsink, M., 264
Womersley, J., 14
Wong, J., 337
Wood, D., 432
Woods, R., 69
Wootten, M., 440
Worster, D., 35
Wortman, D., 399
Wren, L., 42n15
Wren, Y., 228
Wright, B., 394
Wright, J.M., 264
Wyatt, R.C., 51
Wylie, J., 276, 282
Wynne, B., 37
Wynnie, J.A., 500

X

Xiao, C., 264, 265

Y

Yang, G., 184–6
Yavetz, B., 302, 316
Yeh, J.-C., 186
Yeh, M.S., 183
Yencken, D., 34–40, 247
Yen, H.-W., 186
Yinon, Y., 267
Yocco, V., 262
Yore, L.D., 185
York, K.J., 57
Young, J.S., 52
Young, L., 397
Young, M., 101
Young, O.R., 24
Yu, C.-C., 180
Yu, L., 182
Yzer, M., 269

Z

Zachariou, A., 186
Zachary, G.P., 543
Zandvliet, D.B., 13, 398, 536
Zaradic, P.A., 543
Zeichner, K., 473
Zeidler, D., 24
Zeidler, D.L., 29
Zelezny, L., 24, 47, 311, 376
Zelezny, L.C., 267
Zemsky, T., 360
Zeng, H.Y., 184
Zeppel, H., 268
Zeyer, A., 206, 208, 210, 211
Zhang, H.X., 264
Zhu, H.-X., 182, 185
Ziedler, D.L., 247
Zimmer, H.D., 294
Zimmerman, M.E., 221
Zini, M., 483
Zint, M., 263, 299, 301, 304–6
Zint, M.T., 304
Zúquete, J.P., 504
Zwicky, J., 75

Subject Index

Boldface page numbers refer to figures and tables. Page numbers followed by "n" refer to footnotes.

A
academic development program, 355
academic engagement/achievement, place-based education on, 215–18
accommodation process of learning, 254–5
actionable education, 462, **465**
"actionable knowledge," 508
action research, 459; approaches in EE, 473–8; characteristics and quality criteria of, 461–2; characteristics of, 471, 475–6; coherence of, 472–3; conceptualization, 469–70; critics of, 466–7; defined, 460; dynamics of, 465–6; features of, 470; principles of, 464–5; and research genres, 471–2; and sustainability, 460–1
action research and community problem-solving (ARCPS) model, 399
Active Prolonged Engagement (APE) project, 361
adult-derived literature on environmental education, 245
AERA *see* American Educational Research Association
Africa, Westernized education systems in, 389
Agenda, 13, 21
Alaska Rural Systemic Initiative (AKRSI), 405–6
alternative evaluation approaches/methods, implications for, 361–3
ambiguity, learning about, **490**
American Educational Research Association (AERA), 95
American Educational Research Association (AERA) Books Editorial Board, 2, 3
androcentrism, 36
Anglo-American model of capitalism, 499
anthropocentrism, 127, 372, 409, 410; multi-centered debates, 412; to systems, environmental ethics from, 502–3
anthropogenic impacts on environment, 195
anthropological research, 386–7
anti-colonial research, 387–9
antiglobalization protestors, 503
APE project *see Active Prolonged Engagement* project
Appalachian communities, 275
archaeology, 65
ARCPS model *see* action research and community problem-solving model
ARIES *see* Australia Research Institute on Education for Sustainability
arts education, 399
assimilation processes of learning, 254–5
attitude toward behavior, 268
AusLinks project *see* Australia/South Africa Institutional Links project
Australian landscape, ecopedagogy in, 279
Australia Research Institute on Education for Sustainability (ARIES), 473, 476, 477; participation in, 478
Australia/South Africa Institutional Links (AusLinks) project, 475, 476; and ENSI projects, 477–8
authentic postindustrialism, 502
autobiographical memory theory, 290
autonomous acculturation, 210

B
Back to Earth (Weston), 413
baseline indicators, 168–9
behavior: beliefs and *see* beliefs; flowchart, **267**; oversimplified model of, **266**; prediction models, 263; pro-environmental, 265–6
behaviorist theories of learning, 254
beliefs: and behavior, 263–4; to environmental behavior, 264; environmental concerns and, 264–5; foundation of, 262–3; to intention and behavior, 266–9
Big brother Earth Charter, 81–2
Big brother sustainable development, 78, 79
Bildung, 111
biocentrism, 410
biodiversity, **129**; loss of, 25–6
biological classification systems, 37
biophilia, 544
biophysical environment, 15, 20n2
Blackfoot knowledge traditions, 37
Board on International Comparative Studies in Education, 222
Bonn Declaration, 13, 19
Both Gestalt psychology/structuralism, 500
Bourdieu's analysis of researchers and research fields, 522
Brazil, environmental education in, 171, 173
Brundtland Commission report, 126, 128
brute fact, 227

C
Canadian education system, 388
Canadian Journal of Environmental Education (CJEE), 398
CANS *see* Conservation Areas National System
Cantareira Reservoir Water System, 233
capitalism: Anglo-American model of, 499; consumeristic, 101; green, 503–5, 505n6
carbon sink concept, 38
CBD *see* Convention on Biological Diversity
central core beliefs, 262
child–environment relationships, 443–4
child-framed research approaches, 440–1, **441, 442**
children: digital ethnographies of, 225; education, 531; in environmental education research, 438; learning, 253–4; as researcher, 438, 439–40, 449, 454, **455–6**, 539n3; role in society and environment, 442–3
China: EE objectives in, 181; EE policies in, 179–80; green schools in, 184–5
Chinese communities, environmental education in, 178
Chinese Society for Environment Education (CSEE), 181–2
citation analysis, 538
citizen participation/stewardship, 218–19
civically engaged scholarship, 153, 545
classical content analysis, 46
classic theories of learning, 253–4
classroom: curriculum, conceptualization of, 245; performance, self-evaluation of, 222
climate change, 38; anthropogenic, 424, 427; human-induced, 428–30
climate justice, 394
Closing the Achievement Gap: Using the Environment as an Integrating Context for Learning (Lieberman and Hoody), 216–17
cognitive skill, 322, 324
cognitivist theories of learning, 254–5
collaborative action research, 461
collaborative ecological inquiry, **466**
collecting people process, 543
collections of dissertation research: in environmental education, 46, **47**; geographic features of, **50**; by sector, **53**; by study category, **51**
collective biography, 483–4; assessment of, 485
Colombia, environmental education in, 174–5
colonialism, 388
colonial literature, 389–90
color code dissertations, 46
COMIE *see Consejo Mexicano de Investig ación Educativa*

communication networks, 499
communities in decision making, 232
"communities of practice," **492**
community autonomy in Moinho D'Água project, 231, 235–6
community-based education, 399
Community-based School Environmental Education (CO-SEED) program, 214, 218
competence based education, 137
complexity theory, 505n2
concurrent educational trend movements, 54
Consejo Mexicano de Investigación Educativa (COMIE), 171, 172
Conservation Areas, creation of, 232–3
Conservation Areas National System (CANS), 232
conservation education, 53
constrained constructivism, 39
constructivist: epistemology, 27; learning theory, 26, 109, 110, 471
constructivist approaches, 197; to environmental education, 522
consumerism, 209
consumeristic capitalism, 101
contemporary concepts, Yencken's formulation of, 35
contemporary environmentalism, 35
content analysis of dissertation studies: concurrent educational trend movements, 54; by field and type, 49–52; framework for, 46; goal orientations and learning outcomes, 55–7, **56**; historical educational trend movements, 52–4, **54**; implications and recommendations, 58–9; K-12 sector, 54–5; research methodology, 57–8, **58**; trends in dissertations by sector, 52, **53**
content skills, 343–4
conventional education, 459, **465**
conventional instrumental theories of learning, 258
Convention on Biological Diversity (CBD), 386
corporeal phenomenology, 428, 435n7
corpus analysis, 538
CO-SEED program *see* Community-based School Environmental Education program
critical action research *see* action research
critical discourse analysis in education, 138
critical pedagogy, 27–8, 138
critical reflection in socioecological approaches, 29
critical sociocultural/sociological approaches, 148
critical theory, 74, 75, 94
critical thinking, 75; concepts of, 57
critical-transformative, environmental thinking style, 174
cross-boundary learning, 240
cross-cultural science education, 406
cross-cutting analysis, 244
CSEE *see* Chinese Society for Environment Education
cultural approach, 210, 211
cultural border crossing on science teaching, 207–8, 210
cultural clash, concept of, 210
cultural context of learning, 244
Cultural Dimension of Development: Indigenous Knowledge Systems, The (Warren), 387
cultural diversity, 160
cultural fingerprints, 36
cultural frames, implications of, 206
cultural perspectives on children's role, 442–3
cultural relativism, 39
cultural translation, 18
curricular context, 248
curriculum: defined, 349, 353–4; guidelines for EE, 180–1; processes, conceptualization of, 245; research in EE, 191–3
curriculum development: key strategies for, 167; in sustainability education, 353–4

D

debate week—planting citizenship strategy, 235
Decade of Education for Sustainable Development (DESD), 78
decision-making: communities in, 232–3; public, process of, 219
Declaration on the Rights of Indigenous Peoples, 385
decolonization, 99; as educational aim, 96–7
Decolonizing Methodologies: Research and Indigenous People (Smith), 98
deep ecology, 411
degradation, environmental, 207
deliberation, 118
democratization policy processes, 153–4
derived beliefs, 263
desert theory, 127
Designing the Green Economy: The Postindustrial Alternative to Corporate Globalization (Milani), 502
digital ethnography: advantages of, 227; of children, 225
dismantling, systemic, 102
Dissertation Abstracts International (DAI), 46; geographic scope of, 49; sections of, 48
dissertation research: categories, 59; collections in environmental education, 48–9; geographic scope of, 49; purpose of, 45–6; review limitations, 49–52
dissertation studies, content analysis of *see* content analysis of dissertation studies
distributive justice, 395
diversity: cultural, 160; and environmental justice, 397
domestic environmentalism, 370
dominant discourse, components of, 133
"double loop," learning, 258
dynamic constellation, reading and writing research, 532–3
dynamic "Wailing Wall" strategy, 235

E

early childhood education, 399
Earth Charter, 77, 80–1, 385; Big brother, 81–2; cautions, 83–4; enabling thought and action—beyond, 82–3; in environmental ethics, **81**; limited freedom—freedom bounded by, 82
Eastern culture, environmental philosophies of, 35
Eastern environmental thoughts, merits and reciprocal effects of, 34
ecocentric approach, 427
ecocentric nature of experience, 427
ecocentrism, 127, 410
ecocriticism, 140
ecofeminism, 381n2, 381n3
eco-justice, 396–7
Ecological & Environmental Education SIG, 3
ecological ethics, 104
ecological feminists, 375
ecological learning, 247–8
ecological model of child development, 443, **443**
ecological problems and natural world, 207
ecological sustainability, 129, **129**; context of, 159
ecological thinking style, 173
ecologism, 164
ecology education, 53
ecomuseum, 280
economic sustainability, 163
economic transaction, 109
ecopedagogy, 397
ecophenomenological orientation, 508
ecophenomenology, 429, 430; of environmental education research, 432–4; as normative reflexivity, 425–8
ecophobia, 207
ecopolitics and environmental education, 503–5
eco-scientism, 210; Swiss school culture of, 208
eco-scientistic culture, 211
ecosystem, 35–6
educating for eco-/environmental justice (EJE), 397, 398; environmental education literature, 399
education: African voice in, 390; developments in, 255–6; in environmental education research, 74–5; environmental ethics and, 75–6; idea of holistic education, 90–2; implications for, 130–1; language of, 71; learning, instrumental forms of, 255; place-based *see* place-based education; research journals in, 133; in society, 28; for sustainable development, 126, 130
educational action research: definition, 469
educational discourses: for sustainable development, 140; systematic analysis of, 136
educational ideologies, 151
educational languages, systematic analysis of, 136
educational movements, historical, 52–4
educational policy, scholarship, 147–8
education for environment "and" sustainable development (EFE and SD), 126
education for sustainability (EfS), 2, 147–8, 287, 288, 472, 477, 478, 507;

discourse of, 123, 124; national policies in, 151; research and evaluation in, 286
education for sustainable development (ESD), 5, 13, 19, 77, 78, 108, 147–8, 182, 186, 351, 377, 391; challenges of, 158; defined, 156; definitional features of, 54; discourse of, 123–5; example of, 159; national policies in, 152; Panel's seven domains of, 180; professional development in, 160; research, 221
education policy processes, environmental and sustainability, 148–9
educative justification for environmental education, 333
educators, professional development for, 158
EE *see* environmental education
EEAG *see* Environmental Education Activities Group
EEASA *see* European Early American Studies Association
EE/ESD *see* environmental education and education for sustainable development
EIC *see* environment as an integrating context
EJE *see* educating for eco-/environmental justice
EL *see* environmental literacy
Elementary School Environmental Literacy Instrument (ESELI), 324, **324**
emancipatory forms of learning, 109
emergence of environmental education, 13, 14
emerging trends, 360–1
emission trading, 38
emotion: environment and, 206, 211; of students, 246–7
empathy tasks, 249
empirical-analytical approaches, 534
empirical research, 334, 339
empiricism, 500
empowerment: of environmental conduct, 63, 67; feminist perspective, 380; methodological precaution, 65–6
energy education, 54
ENSI *see* Environment and School Initiatives Project
environment: developments in, 255–6; and emotions, 206
environmental behavior, 63, 255–6, 265; personal and situational variables, 263
environmental beliefs, 263
environmental concerns: and beliefs, 264–5; ethical basis of, 87
environmental conduct, governing of, 63–64
environmental degradation, 207
environmental despair, awareness of, 207
environmental discourses, 134
environmental education (EE): action research approaches in, 473–8; analytics of government, 66–7; approaches, 197; Apps, 544; challenges of, 158; characteristics of, 1–2; in Chinese communities, 178; coherence of, 472–3; critical theory, 69; curriculum research in, 191–3; definition of, 1; dimensions of, 133; discourse of, 10, 124–4; emergence of, 13, 14; ethics and, 70–1; evaluation and analysis of, 285–8; fare, 397–9; feminist perspective in, 376–7; feminist research in, 377–81; future of geospatial technology in, 344–6; geospatial technologies in, 333–4; goals of, 23, 63, 536; guidelines and curricula, 180–1; institutional and professional, 181–2; lack of, spatial orientation ability, 543; language and discourse, 141–3; language of, 71, 112–13, **137**, 139–41; learning processes in, 239–41; literature, 399–400; national assessments of, 310, 311–12; objectives of, 15; policies and practices, 125; policies in China, 179–80; policies in Hong Kong, 180; policies in Taiwan, 178–9 professional development in, 161; reevaluation of, 289; research, 16–19, 171, 195, 199; strengthening of, 545; styles in, 173; substantive structure for, 136; tentative directions for, 542; theoretical direction for, 95–6; theory, 305; worldviews in, 514–15
Environmental Education Activities Group (EEAG), 234
environmental education and education for sustainable development (EE×ESD), 222, 226–8; multi-modal research in, 223–4; process and principles, 224–6
environmental education curriculum, 198; thematic areas in, 201; traditions of, 196, 199
environmental education "for" sustainable development (EEFSD), 126
Environmental Education: Improving Student Achievement (Bartosh), 216
environmental education program evaluation, 298; behavioral outcome evaluations of, 299–304, **300–302**, 305–7; former evaluation of, 305; knowledge on, 304–5; in peer reviewed journals, 298–9, **299**
environmental education programs, 269; behavioral outcome evaluations of, 299–304, **300–302**, 305–7; former evaluation of, 305
environmental education research, 16–18, 91, 537; blind spots, blank spots and bald spots, 520; challenges of, 523; children as active researcher, 438, **455–6**; classifying and mapping, 5**19**, 519–20; contextual orientations, 10; culture, language and discourse, 515; digging over, 530–31; diversity in, 512; duty of, 535–6; ecophenomenology of, 429, 432–4; empirical and theoretical aspects of, 520; evolution of substantive focus of, 513–14; "field," of, 522–4; globalization of, 533–34; historical orientations, 9–10; historical roots of, 410–12; implications for, 11, 544–5; journals, 546; moving margins in, 369–72; nature and children, 413; normative dimensions of, 69–73; notions of, 18–19; patterns and directions in, 522; philosophical and methodological perspectives, 419–422; recommendations for, 521; simplified representations of, 228; status of knowledge in, 534–5; theoretical orientations, 10–11, 12; thinking and practice in, 518; visual methods in, 532
environmental education researchers, 95, 96, 99n5; Payne's challenge to, 508
environmental education research projects: *Listening to Children: Environmental Perspectives and the School Curriculum* (L2C), 438, 444, **445–8**; sustainables research challenge, Australia, 448, **450–3**
environmental educators, 120, 256
Environmental Educators' Initiatives (EEI), 181, 182
environmental ethics, 115; from anthropocentrism to systems, 502–3; as attentive and pluralistic, 118; as deliberative enquiry, 118–21; Earth Charter in, **81**; and education, 75–6; as open-ended processes, 117
environmental inequality, 395
environmentalism, domestic, 370
environmental issues, 24–5, 112, 113; education and, 89; and environmental education, 26–8; students' emotions and values, 246; views on, 249
environmental justice, 394–7; hybrid ecological communities of, 414–15
environmental knowledge, 113
environmental languages, 134
environmental learning, 70, 72, 113, 514; approaches to, 243–4; aspects of students in, 246–9; in formal settings, 249; forms of, 244; implications for research and practice, 250–1; models of, 243; power of, 120; research, 244–5; rise of, 109–10; in South Africa, 110
environmental literacy (EL), 310; correlations between domains of, **314**; factors affecting, **315**; grade levels and components of, **325**; and mediating adults, **317**; national assessments of, 310, 311–12, 325–7
Environmentally Responsible Behavior (ERB), 324
environmentally responsible citizen, defined, 17
environmental nongovernmental organizations (ENGO), 397
environmental philosophies of Eastern culture, 35
Environmental Protected Areas (EPA), 232
Environmental Protection Administration (EPA), 178, 179
environmental racism, 395, 398, 400
environmental sustainability, 163; VTE programs, 166
environmental theory, need and, 127, **128**
Environment and School Initiatives (ENSI) Project, 474–5, 476; AusLinks and, 477–8

environment as an integrating context (EIC), 217, 219
Environment, Education and Society in the Asia-Pacific: Local Traditions and Global Discourses (Yencken), 34
Environment, Population and Development (EPD), 182
environment-related educational work: ambiguities in, 158; discourse trends in, 156–7
EPA *see* Environmental Protected Areas; Environmental Protection Administration
EPD *see* Environment, Population and Development
epistemological beliefs, 240, 250
epistemology, 173–4, 197, 199; based ethics, 117; constructivist, 27; environmental problems, 25; of socioecological approaches, 28
ERB *see* Environmentally Responsible Behavior
ESD *see* education for sustainable development
Esoko, 543
ethical concerns, environmental basis of, 87
ethical dimensions, 65; of local environmental issues, 120
ethics: based epistemology, 117; defined, 103, 117; environmental education and, 70–1
ethics of care, 104
ethnographic research, 386–7
ethnography, digital *see* digital ethnography
Euclidean conceptions of topology, 276
Eurocentrism, 36, 41
European Early American Studies Association (EEASA), 391
European imperialism, 37, 388
evaluation of environmental education, 285–8, 360
evidence-based practice, realism of, 483
expansive care consciousness, generating, 104
expectancy theory, 268
experiential learning, 465
extrinsic motivation, 255

F
feminism, 375
feminist educational research methods, 377
feminist methods in social research (Reinharz), 377
feminist perspective in environmental education, 376–7
feminist poststructuralist: approaches, 521; research in environmental education, 380, 381
feminist poststructural theory, 480, 481; pedagogy, 482–3; research, 483–5
feminist research, 377–8; in environmental education, 378–81; strategy to environmental education research, 375
feminist scholarship in science, 380
field framing, environmental education, 15–16
"financialization of capitalism," 499
first order beliefs, 263
flexibly adaptive professional development, 342
focal groups method, 235
formal education, 350; sector, 52
free-choice learners, 364
free-choice learning, 359, 525; experiences, 360, 363; settings, 268
future evaluation practice, implications for, 359

G
gender equity, 377
genealogy, 65
General Education and Training (GET), 113
geographic information systems (GIS), 331, 332
geospatial education, 277
geospatial technology, 331, 344, 346; research review, 334–9; and student learning, 345; and teacher practice learning, 345–6; usage of, 332–4
geospatial tools, education, 331, 332, 333, 334
GGND *see* Global Green New Deal
GIS *see* geographic information systems
Glass-Speigel Act, 499
global environmental discourses, 34
Global Green New Deal (GGND), 502
global positioning systems (GPS), 331, 332
global warming, 38
GPPT *see* Green School Partnership Project in Taiwan
GPS *see* global positioning systems
green capitalism, 503–5, 505n6
green critique of neoliberalism, 501
green economy, conceptions of, 501–2
Green Economy Initiative (GEI), 501
green school, 184
Green School Partnership Project in Taiwan (GPPT), 183, 185

H
haecceity, 484, 485
help program developers, 363
heterosexism, 371
heuristic, 76–7; Earth Charter *see* Earth Charter; education for sustainable development *see* education for sustainable development; at work, 77–83
higher education programs, 54
historical educational movements, 52–4
"history of the present": characteristic concerns of, 65; for environmental education, 13, 19–20, 63, 64
homo economicus concept, 500
Hong Kong: EE policies in, 180; green schools in, 185
human capital theory, 501
humane education, 411
human-induced climate change, 428–30
humanism, 111; crisis of, 112
humanity, success for, 105
hybrid ecological community of environmental justice, 414–15
hybrid learning, spaces creation for, 544–5
hydrogen fuel-cells, 505
hyper-connectivity: consequence of, 543; technological innovations and, 545

I
IEEP *see* International Environmental Education Programme
IK *see* indigenous knowledge
IKS *see* indigenous knowledge systems
immigration problems, 209
India, VTE in, 164
indicators, types of, 168–9
indigenous environmental education, North America, 405–6; current research examples in, 407; foundational concepts in, 404–5
indigenous environmental knowledge, 411
indigenous inhabitation, 98, 99n9
indigenous knowledge (IK), 277; anthropological and ethnographic research, 386–7; anti-colonial and post-colonial research, 387–9; definition of, 384; international conventions on, 385–6; in Southern Africa, 389–91
indigenous knowledge systems (IKS), 384
indigenous people: anthropological and ethnographic research, 386–7; anthropological representations of, 390; diversity of, 385; international conventions on, 385–6
indigenous research, methodology, 406–7
indigenous scholars, 388
individual learner, 247–8
influential approaches in environmental education curriculum research, 196
influential life experiences research, **291**
informal conversation method, 235
informal education sector, 52
Informal Science Education (ISE): evaluation, 360; impact of, 363
informal science, evaluation practice in, 360
informal sectors, 52
inquiry skills, 343–4
instrumental beliefs, 263
instrumental theories of learning, 259
Integrally Protected Conservation Areas, 232
Integrative Model of Behavioral Predication, 269
intellectual colonization, 34, 38
intention, determinants of, 268, 269
intent-oriented behaviors, 265
interdisciplinary approaches, 54
Intergovernmental Panel on Climate Change (IPCC), 37
intermediate-authoritative beliefs, 262
intermediate axiology, environmental theory, 127
International Environmental Education Programme (IEEP), 33–4
International Labour Organisation (ILO), on purposes of VTE, 164–5
international research community, 538
International Union for the Conservation of Nature (IUCN), 149–50
interpretive walks method, 235
interviews, standard practice of transcribing, 223
intrinsic motivation, 255
Iowa Test of Basic Skills (ITBS), 216
IPCC *see* Intergovernmental Panel on Climate Change
Israeli national assessment, 316–18
IUCN *see* International Union for the Conservation of Nature

J
Judeo-Christian traditions, environmentally damaging behaviors, 35

K
Kaupapa Maori, 406
K-12 education, 332
knowledge economy, 137; network logic of, 499–501
knowledge system, pluralism and democratization of, 389
knowledge, value of, 92
Korean national assessment, 312–16
K-12 sector, trends in, 54–5
!Kung San: Men, Women and Work in a Foraging Society, The (Lee), 390
Kyoto Climate Change Summit, 38

L
landscape-based approach, 275; childhood perceptions and lifelong connections to, 278; citizenship and public engagement, 281; for critical reflection, 278–9; definition of, 276–7; to environmental education, 276; field trips, guides, and mobility, 279–80; learning in, 278–81; micro- and macro- analysis, 280; socio-ecological, 280–81
language: arts, 399; of environmental education, 71; transmission of indigenous knowledge, 384
language of environmental education, 112–13, 139–41; rise of learning in, **137**
language of environmental learning, rise of, 109–10
languages of education, 135–8; rise in, **140**
languages of environment, 134–5
Latin-American, environmental education research, 171, 175
learner-centered pedagogy, 353
learning: about ambiguity, **490**; accommodation process of, 254–5; assimilation processes of, 254–5; associated forms of, 257; behaviorist theories of, 254; classic theories of, 253–4; cognitivist theories of, 254–5; conceptualization of, 245; constructivist theories of, 514; "double loop," 258; emancipatory forms of, 109; environmental *see* environmental learning; instrumental theories of, 259; in landscapes, 278–81; and motivation, 255; outcomes, 55–7, **56**; programs, 110; questioning relevance, 247–8; role of feelings and emotions in, 250; settings, free choice, 268; socioecological, 280–81; students' emotions and values, 246; in sustainability education, 353; theory of, 109, 110
learning process, 116; in EE, 239–41; socio-ecological approaches, 26
learning research: thematic areas in, 201; traditions of, 196, 199
liberal-arts approach, 196–7
liberal feminism, 377, 381n1, 480
lifelong science/environmental learning, 359–60
life-world culture: of post-ecologism, 208–9; *vs.* school culture, 210
Likert-type scale, 313, 317
linked-up policy analysis, 498
Linnaean taxonomy, 37
Listening to Children: Environmental Perspectives and the School Curriculum (L2C), 438, 444, **444–48**, 448
low-carbon transport strategies, 502

M
male perspectives in environmental education, 375–6, 380
map/spatial learning skills, 342–3
Marine Education, 54
Massachusetts Comprehensive Assessment System (MCAS) tests, 218
Medicinal and Poisonous Plants of Southern and Eastern Africa, The (Watt and Breyer-Brandwijk), 390
memory theory, long-term, 289, 294
metaphors: for reasoning, 518–19; wardrobe as, 489–94
Mexico, environmental education research in, 171, 172
Middle School Environmental Literacy Survey (MSELS), 321, 322
MMA *see* multimodal analysis
modern education systems, 388–9
modernity, defined, 101
MoinhoD'Água project, 234–6; social environmental transformation in, 233–4
moral extensionism, 127
moral responsibility, concepts of, 119
motivation, learning and, 255
MSELS *see* Middle School Environmental Literacy Survey
multicentered debates, anthropocentrism, 412
multicultural environmental education, 398
multimodal analysis (MMA) in EE/ESD, 223, 225, 227, 228
multi modal methodology, 222
multimodal research in EE/ESD, 223–4
multi-perspectival approach, 223
mutual anticipation, 88

N
NAAEE *see* North American Association for Environmental Education
National Action Plan for Education for Sustainable Development (NAPESD), 156
national assessments: of EE and EL, 310, 311–12; Israeli, 316–18; Korean, 312–16; Turkish, 323–5; U.S., 318–22
National Curriculum Handbook, 79, 80
National Environmental Education Guidelines (NEEG), 181
national policies in ESD/EfS, 152
National Qualifications Framework (NQF), 110, 165
National Research Council (NRC), 361; on learning science, 360
National Science Foundation (NSF), 344
naturalistic generalization, 159, 160
natural science research, 29
nature: childhood participation with, 292; and environmental education reevaluation, 289
nature-as-extended-self, 411
nature-as-miracle, 412
nature-as-object, 410, 411
nature-as-self, 411
needs talk: emergence of, 128–9; and sustainable development, 129–30
NEEG *see* National Environmental Education Guidelines
neighborhood and self, 88
neoclassical economics, 502
neoliberalism, 109; expansion of, 110; green critique of, 499; ideology of, 501
neoliberal reading, 501, 502
network logic of knowledge economy, 499–501
network theory, 500
New Environmental Paradigm (NEP) Scale, 264
New Hampshire Educational and Improvement and Assessment Program (NHEIAP) tests, 218
New Social Studies of Childhood (NSSC), 244
NGOs *see* non-governmental organizations
niche areas of environmental education, 199–202
No Child Left Behind legislation, 215
non-desert theory, 127
non-formal education sector, 52
non-formal learning, 109
non-governmental organizations (NGOs), 52, 209
normative conception of education, 87, 91, 92
normatization, 102
North America, indigenous environmental education research in, 404–7
North American Association for Environmental Education (NAAEE), 110, 370, 397
NQF *see* National Qualifications Framework
NRC *see* National Research Council
NSSC *see* New Social Studies of Childhood

O
OAE programs *see* Outdoor Adventure Education programs
objectives of environmental education, 15
ontology, 428, 429, 432; environmental problem, 25; perspective, 197, 199; positioning in research, 515–16; of socio-ecological approaches, 28
oppositional discourses, 128, 130
organic agronomy, 209
organizational change for professional development, 161
outcome indicators, 169
Outdoor Adventure Education (OAE) programs, 292
outdoor education, 53, 289, 290, 292–5

P
participatory action research, 472, 474–6; literature on, 469
participatory inquiry and relational learning, 495–6
participatory justice, 395–6
participatory social environmental diagnosis in MoinhoD'Água project, 234–5
pedagogical insights, development of, 222
Pedagogik, 350

pedagogy, 95–6, 349, 352–3, 434n4, 482–3; critical, 27–8
PEEC *see* Place-based Education Evaluation Collaborative
peer-reviewed journals, environmental education program evaluation in, 298–9, **299**
peer-review system, 258
People's Global Action (PGA) network, 504
People's Republic of China (PRC), 178
perceived behavioral control, 268
performance indicators, 169
peripheral-inconsequential beliefs, 263
Personal Meaning Mapping (PMM), 362, 363
personal professional theories, 159–60, 477
perspectivity theory technologies, 225
phenomenological approach, 426
phenomenology, philosophy of, 444
philosophical inquiry, 530
photovoice, 440, 456n2
Piagetian model, 254
place-based education, 94, 95, 98, 191–2, 213, 276; on academic engagement and achievement, impact of, 215–18; justification for environmental education, 334; in practice, 214–15; preparing students to, 219–20; reform effort, 217
Place-based Education Evaluation Collaborative (PEEC), 217–18
place-based educators, 215
place-based learning, impact of, 217
place-conscious education, 93, 94, 95, 96
place-conscious learning, critical questions for, 97–9
placed-based approaches, 275
PLANET project, 464, 465
pluralism, 117–18
PMM *see* Personal Meaning Mapping
policy interpretations and responses, 153
political discourse, emergence of needs talk as, 128–9
political theory, 476
Population Education, 54
Port of Los Angeles (POLA), 464
post apartheid curriculum framework, 110
postcolonial literature, 390–1
postcolonial research, 387–9
post-ecologism, 210, 535; life-world culture of, 208–9
postindustrialism, authentic, 502
postmodernist approaches to environmental education, 522
postnormal environmental education, 257–8
postnormal transformative research, 258–9
Power over Power (Nyberg), 131
practice: place-based education in, 214–15; of transcribing interviews, 223; visual recordings of, 228
Pre-K-12 schools, 52
procedural justice, 395–6
productivism, 164
pro-environmental behavior, 17, 265–6
professional development, 159; context and culture in, 160; for educators, 158; in EE/ESD, 161; form of, 162; programs for teachers, 344; recommendations for, 341
Program for International Student Assessment (PISA), 137
protestors, antiglobalization, 503
Public and Private High Schools: The Impact of Communities (Coleman and Hoffer), 216
public decision making, process of, 219
Putonghua, 178

Q

qualitative content analysis, 208
qualitative methodologies, 57, 290
quantitative large sample survey studies, 292
quantitative methodologies, 57, 290
quasi-experiment design, evaluations using, 303, 304, 305
quasi-methodology, 66

R

racism, 209; environmental, 395, 398, 400
randomized control treatment (RCT) design, 361
rational autonomous, 111
rationality of rule, 64
RCT design *see* randomized control treatment design
reading, neoliberal, 501, 502
realist evaluation, 259
reality, 102
reasoned action, theory of, 268–9, **268**
reasoning, metaphors for, 518–19
rebirth phenomenon, 545
reciprocity, researcher and people, 406
recognitional justice, 396
red-green environmentalism, 74
reevaluation of environmental education, 289
reflective teacher, 222
Reflexive Monitoring and Evaluation (RMA), 259
Regional Environmental Education (REE), **285**
reinhabitation as educational aim, 96–7
relational ethics *see* ethics of care
relational learning: participatory inquiry and, 495–6; wardrobe as metaphor for, 489–95
"Relevance of Science Education (ROSE), The" project, 210
religious education, 399
reluctant holism, 127
renormatization, 102
reprivatization discourses, 128, 130
Research Commission, 48
research in environmental education, 16–18; notions of, 18–19; reviews of, 46, **47**
researching beliefs, 263
researching learners, 244
research in science education, 16
responsible environmental behavior model, 266, **266**, 269
retrospective pretest approach, 303, 306
retrospective program evaluation, 289; of educational programs, 292–5, **293**
Rio Atibainha Reservoir, 233
Rio Declaration on Environment and Development, 385
riparian greenways as educational site, 281
risk society, 112
RMA *see* Reflexive Monitoring and Evaluation
Rokeach centrality model, 262–3
rural areas, VTE schools in, 166

S

SADC-Regional Environmental Education Programme (SADC-REEP), 390–1
SAJEE *see* Southern African Journal of Environmental Education
salvation discourse, 134, 136, 139
Sand County Almanac, A (Leopold), 276
satoyama, 280–1
Sauvé's model, 443
scholarships, civically engaged, 545
school-based environmental learning, 245
school culture *vs.* life-world culture, 210
schooling, structures of, **489**
school subjects, trends in K-12 sector by, 54–5
Science Assessment, 2006, 326
science education, 399
science learning, problems of, 206
science teachers, 211
science teaching, cultural border crossing on, 207–8, 210
science, technology, engineering, the environment and mathematics (STEEM), 365; learning, 359, 362, 364
scientific justification for environmental education, 333–4
scientism, 211
SEAMEO Search for Young Scientists (SSYS) Congress, 157
SEER *see* sustainable enterprise executive roundtable
self: and environment, 88; integrity of, 90; as part of "greater whole," 89–90
self-evaluation of classroom performance, 222
self-regulation theory, 269
Significant Life Experiences (SLE): challenges, 287; research, 289–92
sink metaphor, 40
skepticism, 209
SLE *see* Significant Life Experiences
slogan system, role of, 161
social constructivism, 39
social context of learning, 244
social educational intervention in Moinho D'Água project, 231, 234
social environmental context, participation in, 231–3
social environmental diagnosis, 231; participatory, 234–5
social environmental pre-diagnosis in Moinho D'Água project, 231, 234
social environmental transformation, historical process of, 233–4
social justice principle, **129**
social justice theory, 127, **127**
social learning, 118–19, 137; theory, 257
social politics in professional development, 160–1
social pressure, behavior, 268
social representations of environment, 135
social science research, 439–42
social studies, 399
social sustainability, 163; VTE programs, 166
society, education role in, 349–51

socioconstructivist: education, 76; theory of learning, 110
sociocultural approach, 150; to educational policy, 153–4
sociocultural theories of learning, 239, 255
socioecological approaches, 24–5; to environmental education, 26–8; in learning and research, 28; learning processes and, 26; nature of, 24; potential of, 29–30
socioecological issues, critical reflection on, 28
socioecological learning: aspects of, 489; work, 488–9
socioecological pedagogy, 488; challenges of, 496; contexts, 493
somaesthetics, 426, 427, 434n3
South Africa: environmental learning in, 108, 10; indigenous knowledge in, 389–91
South East Asian Ministers of Education Organization (SEAMEO), 157
Southern African Journal of Environmental Education (SAJEE), 390
spatial orientation ability, 543
spatial visualisation, 39
Spell of the Sensuous, The (Abram), 98
spiritual ecology, 405
Spotter's guide on needs and directions, 521–2
SSYS Congress *see* SEAMEO Search for Young Scientists Congress
STEEM *see* science, technology, engineering, the environment and mathematics
STEM programs, 360
Stockholm declaration, demand in principle, 19, 23
student learning: geospatial technology and, 345; outcomes, 342–4
students' environmental learning: dealing with emotions and values, 246–7; negotiating viewpoints with teachers, 248–9; questioning relevance, 247–8
substantive foci of environmental education research, 519
substantive justice, 395
Suited inquiry project, 487, 489, 496n2
sustainability, 256; action research and, 460–1; assessment strategies for, 168–9; education for, 126, 255; expressions of, 166–7; issues, 253; systems redesign questions for, **464**; into teacher education programs, 473–4; VTE for, 163
sustainability education, 349; application of, 352–3; concepts of, **355**; curriculum development, 353–4; learning and teaching in, 353, **350**; politics of needs, 128–9; role in, 351–2; and role of universities, 351; in vocational and technical education, 124
sustainability education pedagogy, 352–3
sustainability values (VSD), 185
sustainable development: Big brother, 78, 79; cautions, 83–4; concept of, 149; definition of, 352; education for, 77–9, **78**, 126; enabling thought and action—beyond, 80; issues in, 13; languages of, 139–41; limited freedom—freedom bounded by, 79–80; needs talk and, 129–30; values in, **129**
sustainable enterprise executive roundtable (SEER), 463; research results, 463–4
Sustainables Challenge project, Australia, 448–9, **450–3**, 453–4
sustainable system, educating for, 462–3
Swiss school culture of eco-scientism, 208
symbolic beliefs, 263
systemic dismantling, 102
systemic sustainability challenges, crucial task of, 508
systems ecology, 36

T
Taiwan: EE objectives in, 181; EE policies in, 178–9
Taiwan Sustainable Campus Program (TSCP), 182–3
Tbilisi Declaration, 149, 151; environmental education, 2
teacher: education, 167–8; practice learning, geospatial technology and, 345–6; professional development, 339–42, 346; science, 211; students negotiating viewpoints with, 243, 248–9
teacher education programs: sustainability into, 473–4
teaching beliefs of ESD (TESD), 185
teaching in sustainability education, 353
TEK *see* traditional ecological knowledge
tensions: over educational ideologies, 151; in policy processes, 152, 154
TESD *see* teaching beliefs of ESD
Thai fish feeding project, environmental education, 159, 161
thinking globally, 38–41; in environmental education, 33–6; Turnbull's analysis for, 40; Western science, 36–8
Three Ecologies, The (Guattari), 131
Time to Act (1989), 180
TKS *see* traditional knowledge systems
TL *see* transformative learning
traditional ecological knowledge (TEK), 384, 391, 405
traditional knowledge, 384, 385, 386
traditional knowledge systems (TKS), 384
traditions of environmental education curriculum, 196, 199
transcribing interviews, standard practice of, 223
transformation of environmental conduct, 63
transformative education, 76
transformative learning (TL), 240, 257
Transtheoretical Model of Behavior Change, 269
trilateration, 332
triple bottom line (TBL) approach, 353
Truth About Stories, The (King), 491, 492
TSCP *see* Taiwan Sustainable Campus Program
Turkish Education System, 324
Turkish national assessment, 323–5
Turnbull's analysis for "thinking globally," 40
Turnbull's approach, 39

U
UNCED *see* United Nations Conference on Environment and Development
UNESCO, 19, 147, 148; conferences, 151; on purposes of VTE, 164–5
United Nations Climate Change Conference in Copenhagen, 38
United Nations Conference on Environment and Development (UNCED), 13
United Nations Environmental Program (UNEP), 501
United Nations Environment Program, 1
University Microfilms International (UMI), 47, 48
urbanization, 233
urban natural history, 415
U.S. national assessment, 318–22
Utilization-Focused Evaluation approach, 303, 304

V
values education, 82
videophilia, 544
virtual globes, 331, 332
visual data, technical skills of analyzing, 228
visual methods in environmental education research, 224, 532
visual recordings for research purpose, 222–9
vocational and technical education (VTE), 124, 163; debates about, 164–5; expressions of, 166–7
VSD *see* sustainability values
VTE *see* vocational and technical education

W
Wagner's schema, 38
Wagner's visual analogy, 520
wardrobe as metaphor for relational learning, 489–95
Washington Assessment of Student Learning (WASL), 216
Western environmentalism, 34, 35
Westernized education systems in Africa, 389
Western Knowledge Systems, 384
Western science, cultural determinants of, 36
Western scientific solutions to environmental crisis, 387
wide-canvas research, 498
workforce justification for environmental education, 333–4
World Conservation Strategy (WCS), 139
World Summit on Sustainable Development in 2002, 13
World Wildlife Fund (WWF), 209

Y
Yencken's formulation of contemporary concepts, 35
Youth Learning on their Own Terms: Creative Classroom Practice (Gustavson), 493

Z
Zeppel's model, 268
zero-order beliefs, 263
Zimbabwe's National Environmental Education Policy and Strategies, 390

The Editors

Robert B. Stevenson is Professor and Tropical Research Leader (Education for Environmental Sustainability) in The Cairns Institute and School of Education at James Cook University in Australia, where he also is Director of the Centre for Research and Innovation in Sustainability Education. He is an executive editor of the *Journal of Environmental Education* and has served on the editorial boards of all the major English language journals in environmental education. He was Chair of AERA's Ecological and Environmental Education Special Interest Group in 2006–2007 and received the North American Association for Environmental Education Outstanding Contributions to Research Award in 2010. Stevenson's research has focused on the relationships among theory, policy and practice in environmental/sustainability education and its history and marginalized status as an educational reform in K-12 schools. His current research interests focus on the current and potential spaces and approaches for learning about issues of sustainability.

Michael Brody is a faculty member in the College of Education, Health and Human Development at Montana State University, where he teaches courses in science, education and research at the graduate and undergraduate levels. He received his Ph.D. from Cornell University in science and environmental education, Master of Science in Biology from the University of New Hampshire and Bachelor of Science in Biology and Secondary Education from Boston College. Brody is a research associate of the Museum of the Rockies in Bozeman, Montana, and received the North American Association for Environmental Education Outstanding Contributions to Research Award in 2006.

Justin Dillon is Professor of Science and Environmental education at King's College London. He is head of the Science and Technology Education Group and co-leader of the Centre for Research in Education in Science, Technology, Engineering and Mathematics. His research interests include science engagement in museums, science centers and botanic gardens, learning outside the classroom, students' interests and aspirations, and teacher identity and development. Justin was Chair of AERA's Ecological and Environmental Education Special Interest Group in 1998–1999. He was elected President of the European Science Education Research Association in 2007–2011 and was elected a Fellow of the Linnean Society of London in 2005.

Arjen E.J. Wals is Professor of Social Learning and Sustainable Development at Wageningen University in The Netherlands. He also holds an adjunct professorship at Cornell University and is a UNESCO Chair in the same field. His research focuses on learning processes that contribute to a more sustainable world. Wals obtained his Ph.D. in 1991 with a Fulbright fellowship at the University of Michigan in Ann Arbor under the guidance of the late Bill Stapp, one of the founding fathers of the field of environmental education. Arjen maintains a blog at www.transformativelearning.nl.

The Contributors

Scott Allen
University of Saskatchewan
Canada

Vince Anderson
University of Saskatchewan
Canada

Alberto Arenas
University of Arizona
USA

Heesoon Bai
Simon Fraser University
Canada

Michael Barnett
Boston College
USA

Robert Barratt
Bath Spa University
UK

Elisabeth Barratt Hacking
University of Bath
UK

Tom Berryman
Université du Québec à Montréal
Canada

Michael Bonnett
Cambridge University and University of Bath
UK

Hilary Bradbury-Huang
Oregon Health & Science University
USA

Carol B. Brandt
Temple University
USA

Michael Brody
Montana State University
USA

Jennifer Bucheit
Florida Institute of Technology
USA

Kim Butcher
University of Saskatchewan
Canada

Amy Cutter-Mackenzie
Southern Cross University
Australia

Bronwyn Davies
University of Melbourne
Australia

Lynn D. Dierking
Oregon State University
USA

Justin Dillon
King's College London
UK

Jennifer Engelhardt
Florida Institute of Technology
USA

Mehmet Erdogan
Akdeniz University
Turkey

Almerinda B. Fadini
São Francisco University
Brazil

John H. Falk
Oregon State University
USA

Leesa Fawcett
York University
Canada

Jo-Anne Ferreira
Griffith University
Australia

John Fien
RMIT University
Australia

Dustin Fruson
University of Saskatchewan
Canada

Yaakov Garb
Ben Gurion University
Israel

Edgar González Gaudiano
Universidad Veracruzana
Mexico

Annette Gough
RMIT University
Australia

Noel Gough
La Trobe University
Australia

David A. Greenwood
Lakehead University
Canada

Katherine Guzmon
Florida Institute of Technology
USA

Randolph Haluza-DeLay
The King's University College
Edmonton, Canada

Paul Hart
University of Regina
Canada

Kathryn Hegarty
RMIT University
Australia

Joe E. Heimlich
Ohio State University
USA

Teresa Hill
University of Saskatchewan
Canada

João Luiz de Moraes Hoeffel
FAAT College (NES/FAAT)
São Francisco University
Brazil

Sarah Holdsworth
RMIT University
Australia

Nick Hopwood
University of Technology, Sydney
Australia
Oxford University
UK

Meredith Houle Vaughn
San Diego State University
USA

Harold Hungerford
The Center for Instruction, Staff Development and Evaluation
USA

Bob Jickling
Lakehead University
Canada

Jean Kayira
University of Saskatchewan
Canada

Elin Kelsey
Royal Roads University
Canada
James Cook University
Australia

Mphemelang Ketlhoilwe
University of Botswana
Botswana

Michelle Knorr
University of Saskatchewan
Canada

Marianne E. Krasny
Cornell University
USA

Regula Kyburz-Graber
University of Zurich
Switzerland

Lesley Le Grange
Stellenbosch University
South Africa

Lee Chi Kin John
Hong Kong Institute of Education
China

Kendra Liddicoat
University of Wisconsin-Stevens Point
USA

F.B. Lima
São Francisco University
Brazil

Christine Linsenbardt
Florida Institute of Technology
USA

Fernando Londoño
University of Arizona
USA

Ken Long
US Army Command and General Staff College
USA

Leonir Lorenzetti
Universidade do Vale do Rio do Peixe
Brazil

Heila Lotz-Sisitka
Rhodes University
South Africa

Greg Lowan-Trudeau
University of Northern British Columbia
Canada

Cecilia Lundholm
Stockholm University
Sweden

M.K. Machado
FAAT College (NES/FAAT)
Brazil

James G. MaKinster
Hobart and William Smith Colleges
USA

Thomas Marcinkowski
Florida Institute of Technology
USA

Sheron Mark
Loyola Marymount University
USA

Bill McBeth
University of Wisconsin–Platteville
USA

Marcia McKenzie
University of Saskatchewan
Canada

Sheelah McLean
University of Saskatchewan
Canada

Ron Meyers
Ron Meyers and Associates
USA

Preethi Mony
Columbus
USA

Jeremy Murphy
University of Saskatchewan
Canada

Maya Negev
Tel Aviv University
Israel

Kyung-Im Noh
University of Connecticut
USA

Lausanne Olvitt
Rhodes University
South Africa

Phillip G. Payne
Monash University
Australia

Michael A. Peters
University of Waikato
New Zealand
University of Illinois, Urbana-Champaign
USA

Alan Reid
Monash University
Australia

J.C. Reis
São Francisco University
Brazil

Mark Rickinson
Oxford University
UK

John Robinson
Robinson Consulting
UK

Ian Robottom
Deakin University
Australia

Serenna Romanycia
Simon Fraser University
Canada

Constance L. Russell
Lakehead University
Canada

Gonen Sagy
The Arava Institute for Environmental Studies
Israel

Richard Santangelo
Florida Institute of Technology
USA

Lucie Sauvé
Université du Québec à Montréal
Canada

William Scott
University of Bath
UK

Tony Shallcross
University of Hull
UK

Soul Shava
University of South Africa
South Africa

Donghee Shin
Ewha Womans University
South Korea

Gregory A. Smith
Lewis & Clark College
USA

Vanessa Spero-Swingle
University of Florida
USA

Marianne Stadel
Florida Institute of Technology
USA

Robert B. Stevenson
James Cook University
Australia

Joshua Stone
University of Saskatchewan
Canada

Martin Storksdieck
National Research Council
USA

Ian Thomas
RMIT University
Australia

Nancy M. Trautmann
Cornell University
USA

Trudi Volk
The Center for Instruction,
Staff Development and Evaluation
USA

Arjen E.J. Wals
Wageningen University
The Netherlands
Cornell University
USA

Wang Shun Mei
National Taiwan Normal University
Taiwan

Yang Guang
Capital Normal University
China

Victor Yocco
Ohio Department of Public Safety
USA

Albert Zeyer
University of Zurich
Switzerland

Michaela Zint
University of Michigan
USA